Electronic Communication

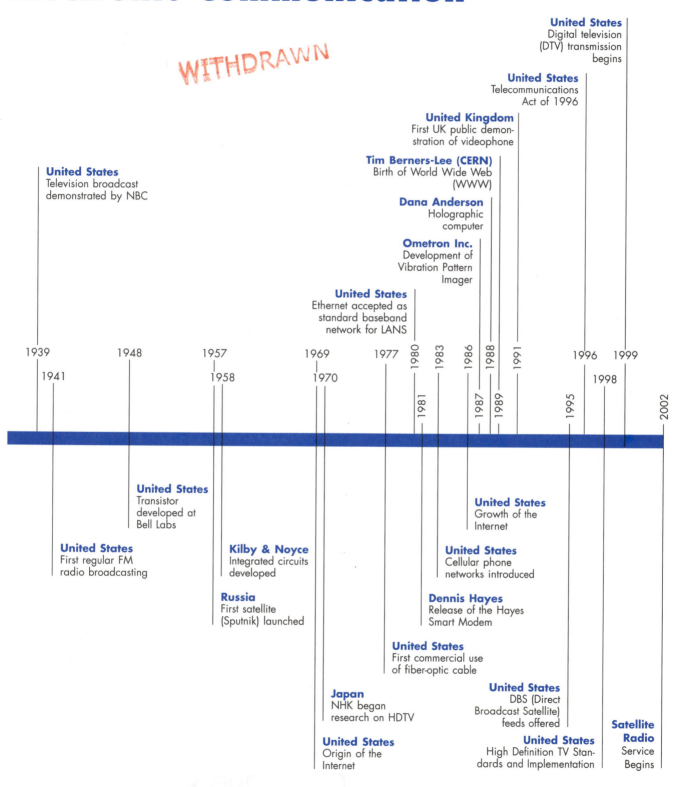

- **1939 — United States:** Television broadcast demonstrated by NBC
- **1941 — United States:** First regular FM radio broadcasting
- **1948 — United States:** Transistor developed at Bell Labs
- **1957 — Russia:** First satellite (Sputnik) launched
- **1958 — Kilby & Noyce:** Integrated circuits developed
- **1969 — United States:** Origin of the Internet
- **1970 — Japan:** NHK began research on HDTV
- **1977 — United States:** First commercial use of fiber-optic cable
- **1980 — United States:** Ethernet accepted as standard baseband network for LANS
- **1981 — Dennis Hayes:** Release of the Hayes Smart Modem
- **1983 — United States:** Cellular phone networks introduced
- **1986 — United States:** Growth of the Internet
- **1987 — Ometron Inc.:** Development of Vibration Pattern Imager
- **1988 — Dana Anderson:** Holographic computer
- **1989 — Tim Berners-Lee (CERN):** Birth of World Wide Web (WWW)
- **1991 — United Kingdom:** First UK public demonstration of videophone
- **1995 — United States:** DBS (Direct Broadcast Satellite) feeds offered
- **1996 — United States:** Telecommunications Act of 1996
- **1998 — United States:** High Definition TV Standards and Implementation
- **1999 — United States:** Digital television (DTV) transmission begins
- **2002 — Satellite Radio:** Service Begins

Modern Electronic Communication

Eighth Edition

Jeffrey S. Beasley
New Mexico State University
Gary M. Miller

Upper Saddle River, New Jersey
Columbus, Ohio

Library of Congress Cataloging-in-Publication Data

Beasley, Jeffrey S., 1955–
Modern electronic communication / Jeffrey S. Beasley, Gary M. Miller.--8th ed.
p. cm.
Miller's name appears first on earlier ed.
Includes index.
ISBN 0-13-113037-4
1. Telecommunication 2. Telecommunication. I. Miller, Gary M., 1941- II. Title.

TK5101.B327 2005
621.382--dc22 2004040080

Editor in Chief: Stephen Helba
Assistant Vice President and Publisher: Charles E. Stewart, Jr.
Assistant Editor: Mayda Bosco
Production Editor: Alexandrina Benedicto Wolf
Design Coordinator: Diane Ernsberger
Cover Designer: Kritina Holmes
Cover art: Digital Vision
Production Manager: Matt Ottenweller
Marketing Manager: Ben Leonard

This book was set in Times Roman by The GTS Companies/York, PA Campus. It was printed and bound by R.R. Donnelley & Sons Company. The cover was printed by Phoenix Color Corp.

Pearson Education Ltd.
Pearson Education Singapore Pte. Ltd.
Pearson Education Canada, Ltd.
Pearson Education—Japan
Pearson Education Australia Pty. Limited
Pearson Education North Asia Ltd.
Pearson Educación de Mexico, S.A. de C.V.
Pearson Education Malaysia Pte. Ltd.

10 9 8 7 6 5 4 3 2 1
ISBN 0-13-113037-4

Dedicated to my family,
Kim, Damon, and Dana
Jeffrey S. Beasley

Dedicated to the youth of the world,
Especially my favorites,
Evan, Maia, Willo, Kevin, Richard, and Luca
Gary M. Miller

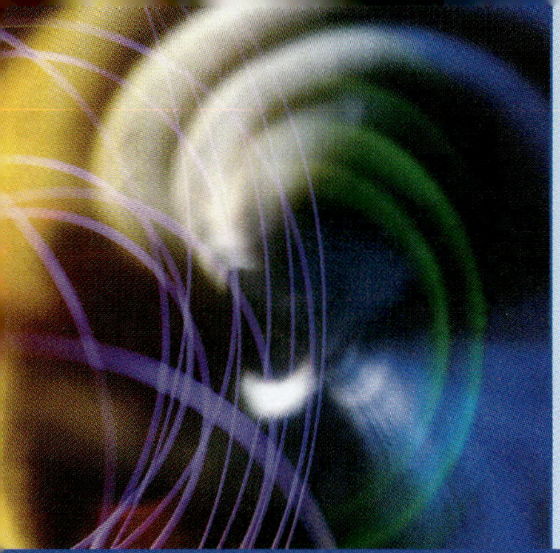

Preface

We are excited about the many improvements to this edition of *Modern Electronic Communication,* and we trust you will share in our enthusiasm as they are briefly described. The eighth edition maintains the tradition of the seventh, including up-to-date coverage of the latest in electronic communications, readable text, and many features that aid student comprehension.

This edition has greatly expanded the discussion on digital communications. In fact, a new chapter on wireless digital communications has been added. Chapter 10 focuses on the spread-spectrum techniques used to transport today's wireless digital data. This chapter includes Electronics Workbench™ Multisim simulations of the key components of a spread-spectrum communication system. A section on Orthogonal Frequency Division Multiplexing (OFDM) system—a wireless digital communication technique commonly used in wireless networking—is also included.

Features

- The most up-to-date treatment of digital and data communications
- Expanded use of Electronics Workbench™ Multisim in spread-spectrum communications
- Extensive troubleshooting sections
- Numerous questions and problems for each chapter, including Questions for Critical Thinking, designed to sharpen the users' analytical skills
- Circuits in the text are simulated on a full-function Electronics Workbench (EWB) Multisim CD. Additional circuits provide interactive, hands-on troubleshooting exercises.
- Key terms and definitions highlighted in the margins as they are introduced
- Extensive problem sets
- Updated photos of typical industrial equipment
- Chapter outlines, objectives, and key terms identified at the beginning of each chapter
- Summary of key points at the end of each chapter
- Complete directory of acronyms and abbreviations
- Comprehensive glossary

Partial Listing of New Material in the Eighth Edition

- Expanded coverage of wireless digital communications
- Expanded discussion of pseudonoise (PN) codes
- A detailed examination of Direct Sequence Spread Spectrum (DSSS)
- Extensive discussion on spreading the digital signal
- Electronics Workbench Multisim simulation of key spread-spectrum communication techniques
- Coverage of Orthogonal Frequency Division Multiplexing (OFDM)
- Numerous examples of OFDM transmitted data
- Updated data sheets, circuit examples, and discussion of the following:
  TDA1572T AM receiver
  AD630 Balanced Modulator/Demodulator
  MAX 2606 Single-Chip FM Transmitter
  Squelch techniques
  AD8369 Digitally-Controlled Variable Gain Amplifier
  MT8964 Codec
  MAX3451 USB Transceiver
  TDA8961 ATSC Digital Terrestrial TV demodulator/decoder
- Expanded coverage of fiber optics

Illustration of Features

CHAPTER OPENER—Each chapter begins with a color photo related to content, a chapter outline, a list of objectives, and key terms being introduced. An example is shown below.

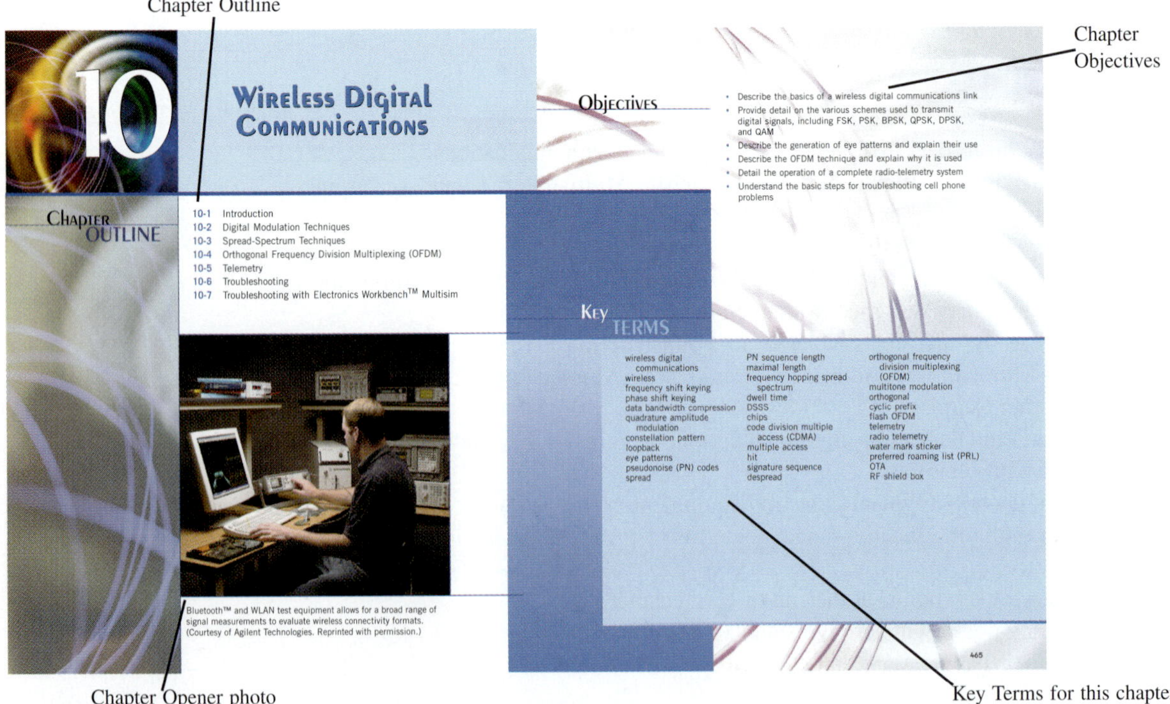

WORKED EXAMPLES—Numerous worked-out examples are included in every chapter, as shown below. These examples reinforce key concepts and aid in subject mastery.

TROUBLESHOOTING—Every chapter contains an extensive troubleshooting section. An illustration is provided below. Notice that areas of expected student mastery are highlighted. Students are very interested in applying knowledge gained by "fixing" real-world systems. Their comprehension is improved in this process. Equally important, employers and accrediting agencies strongly encourage emphasis on troubleshooting skills.

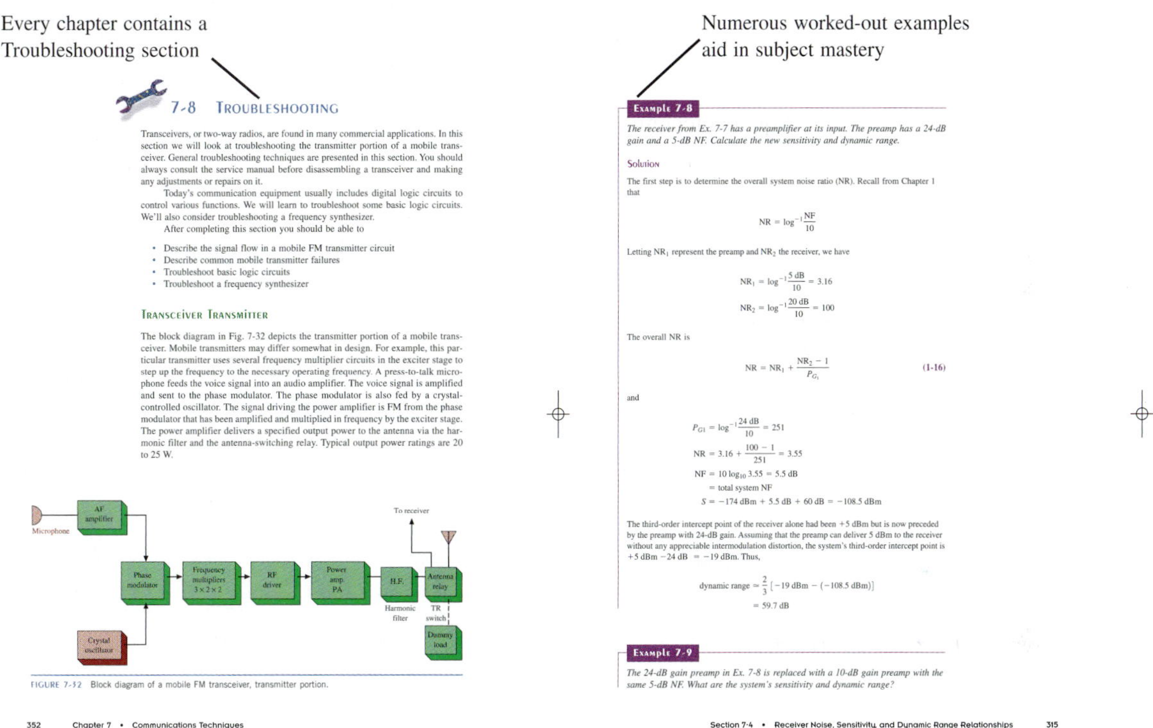

Every chapter contains a Troubleshooting section

7-8 TROUBLESHOOTING

Transceivers, or two-way radios, are found in many commercial applications. In this section we will look at troubleshooting the transmitter portion of a mobile transceiver. General troubleshooting techniques are presented in this section. You should always consult the service manual before disassembling a transceiver and making any adjustments or repairs on it.

Today's communication equipment usually includes digital logic circuits to control various functions. We will learn to troubleshoot some basic logic circuits. We'll also consider troubleshooting a frequency synthesizer.

After completing this section you should be able to

- Describe the signal flow in a mobile FM transmitter circuit
- Describe common mobile transmitter failures
- Troubleshoot basic logic circuits
- Troubleshoot a frequency synthesizer

TRANSCEIVER TRANSMITTER

The block diagram in Fig. 7-32 depicts the transmitter portion of a mobile transceiver. Mobile transmitters may differ somewhat in design. For example, this particular transmitter uses several frequency multiplier circuits in the exciter stage to step up the frequency to the necessary operating frequency. A press-to-talk microphone feeds the voice signal into an audio amplifier. The voice signal is amplified and sent to the phase modulator. The phase modulator is also fed by a crystal-controlled oscillator. The signal driving the power amplifier is FM from the phase modulator that has been amplified and multiplied in frequency by the exciter stage. The power amplifier delivers a specified output power to the antenna via the harmonic filter and the antenna-switching relay. Typical output power ratings are 20 to 25 W.

FIGURE 7-32 Block diagram of a mobile FM transceiver, transmitter portion.

352 Chapter 7 • Communications Techniques

Numerous worked-out examples aid in subject mastery

EXAMPLE 7-8

The receiver from Ex. 7-7 has a preamplifier at its input. The preamp has a 24-dB gain and a 5-dB NF. Calculate the new sensitivity and dynamic range.

Solution

The first step is to determine the overall system noise ratio (NR). Recall from Chapter 1 that

$$\mathrm{NR} = \log^{-1}\frac{\mathrm{NF}}{10}$$

Letting NR_1 represent the preamp and NR_2 the receiver, we have

$$\mathrm{NR}_1 = \log^{-1}\frac{5\ \mathrm{dB}}{10} = 3.16$$

$$\mathrm{NR}_2 = \log^{-1}\frac{20\ \mathrm{dB}}{10} = 100$$

The overall NR is

$$\mathrm{NR} = \mathrm{NR}_1 + \frac{\mathrm{NR}_2 - 1}{P_{G_1}} \qquad \text{(1-16)}$$

and

$$P_{G1} = \log^{-1}\frac{24\ \mathrm{dB}}{10} = 251$$

$$\mathrm{NR} = 3.16 + \frac{100-1}{251} = 3.55$$

$$\mathrm{NF} = 10\log_{10}3.55 = 5.5\ \mathrm{dB}$$

$$= \text{total system NF}$$

$$S = -174\ \mathrm{dBm} + 5.5\ \mathrm{dB} + 60\ \mathrm{dB} = -108.5\ \mathrm{dBm}$$

The third-order intercept point of the receiver alone had been +5 dBm but is now preceded by the preamp with 24-dB gain. Assuming that the preamp can deliver 5 dBm to the receiver without any appreciable intermodulation distortion, the system's third-order intercept point is +5 dBm − 24 dB = −19 dBm. Thus,

$$\text{dynamic range} \simeq \frac{2}{3}[-19\ \mathrm{dBm} - (-108.5\ \mathrm{dBm})]$$

$$= 59.7\ \mathrm{dB}$$

EXAMPLE 7-9

The 24-dB gain preamp in Ex. 7-8 is replaced with a 10-dB gain preamp with the same 5-dB NF. What are the system's sensitivity and dynamic range?

Section 7-4 • Receiver Noise, Sensitivity, and Dynamic Range Relationships 315

TROUBLESHOOTING—WITH ELECTRONICS WORKBENCH™ MULTISIM

Every chapter ends with an EWB circuit simulation and troubleshooting exercise as well as end-of-chapter exercises incorporating Electronics Workbench Multisim.

Troubleshooting with Electronics Workbench™ Multisim is featured in this edition

18-12 TROUBLESHOOTING WITH ELECTRONICS WORKBENCH™ MULTISIM

The concept of preparing a system design for a fiber installation was presented in this chapter. This section presents a simulation exercise of a system design. Open the file **Fig18-31.ms7 (.msm)** on your EWB Multisim CD. This exercise provides you with the opportunity to study a fiber-optic system design in more depth. The circuit for the light-budget simulation is shown in Fig. 18-31.

Electronics Workbench™ Multisim does not contain simulation models or instruments for lightwave communications, but with a little creativity, a system design for a fiber installation can be modeled. This example is patterned after Fig. 18-23. The function generator models the output of a fiber-optic transmitter. The generator is outputting a square wave to model the pulsing of light. The settings for the function generator for three possible operating levels have been provided.

1. The maximum received signal level (RSL): −27 dBm
2. The designed operating level: −31.6 dBm
3. The minimum received signal level (RSL) for a BER of 10^{-9}: −40 dBm

A 16-dB T-type attenuator has been provided to simulate the fiber cable and splice loss. The system is terminated with a 600-Ω resistor for consistency with the analog model, but this resistor does not exist in a real optical system. A voltage-controlled sine-wave oscillator has been provided to simulate the optical receiver. The settings for the voltage-controlled sine-wave oscillator are shown in Fig. 18-32.

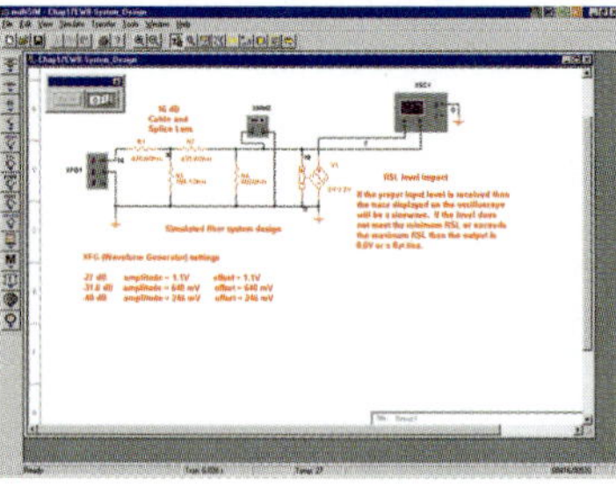

FIGURE 18-31 The Multisim circuit for the light-budget simulation.

872 Chapter 18 • Fiber Optics

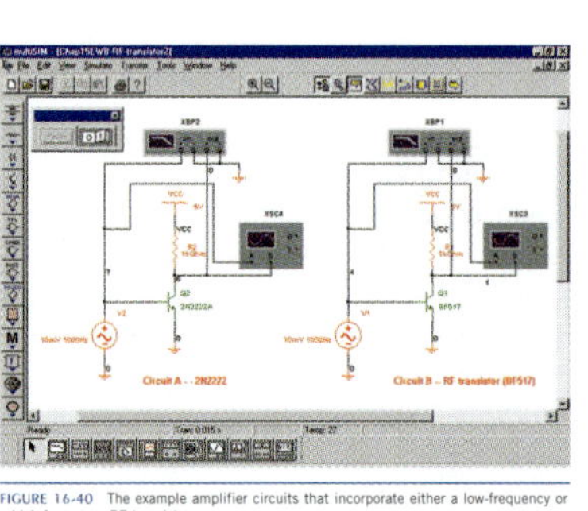

FIGURE 16-40 The example amplifier circuits that incorporate either a low-frequency or a high-frequency RF transistor.

35.5 MHz, whereas circuit B, which is using the BF517 RF transistor, has a 3-dB upper cutoff frequency of about 240 MHz. This demonstrates the vast improvement in the frequency response of an amplifier with the use of an RF circuit.

The following exercises provide you with an opportunity to explore the characteristics of an RF inductor and troubleshoot an RF amplifier.

ELECTRONICS WORKBENCH™ EXERCISES

1. Open the file **FigE16-1.msm** in your EWB CD. This circuit provides a comparison of an ideal and an RF inductor. Determine the upper 3-dB cutoff frequencies for the inductors. (194 kHz, approx. 1.5 GHz)
2. Open the file **FigE16-2.msm** in your EWB CD. Determine the resonant frequency of this dipole antenna. (f = 1.071 GHz).
3. Open the file **FigE16-3.msm** in your EWB CD. Determine if the RF amplifier is working properly. If it isn't, locate and correct the fault and retry the simulation. Report on your findings.

SUMMARY

In Chapter 16 we studied microwaves and lasers. We learned that microwaves share many properties with light waves. The major topics you should now understand include:

Summary 771

Each chapter contains Electronics Workbench™ exercises

FULL-COLOR FORMAT—Color is used throughout as an aid to comprehension and to make the material more visually stimulating. A representative use of color is shown below.

KEY TERMS DEFINED—The important new terms and concepts are defined in the margins near where they are introduced in the text. An illustration is shown below. Having the key terms presented in this way allows the student to quickly access, review, and understand new concepts and terminology.

Full-color photos enhance the text

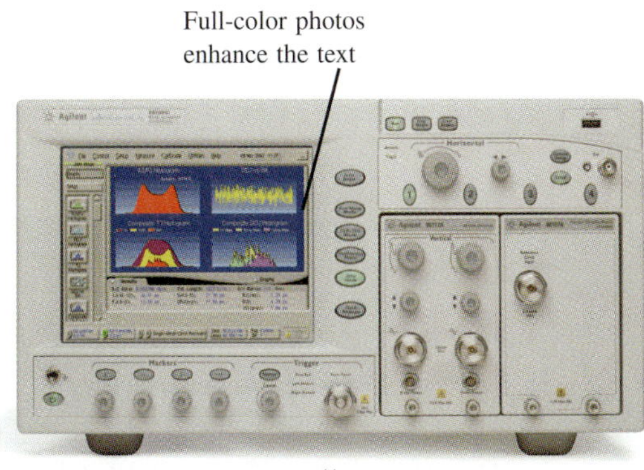

(a)

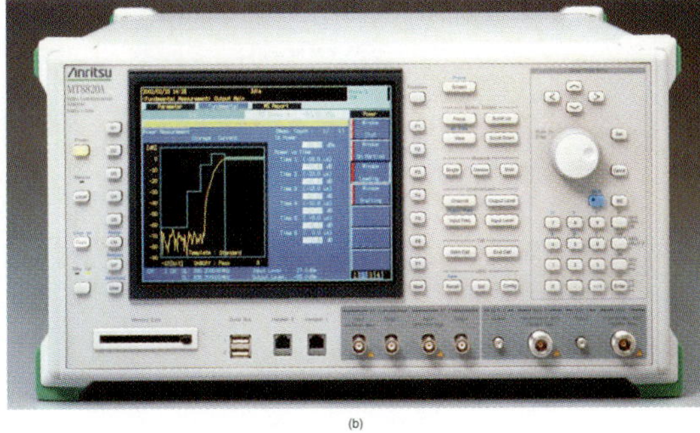

(b)

(a) The 86100C digital communications analyzer with jitter analysis offers breakthrough speed, accuracy, and affordability. (Courtesy of Agilent Technologies. Reprinted with permission.) (b) The MT8802A radio communications analyzer was designed to support the test needs of the manufacturing, R&D, and maintenance markets. (Courtesy of Anritsu Company. Reprinted with permission.)

242 Chapter 5 • Frequency Modulation: Transmission

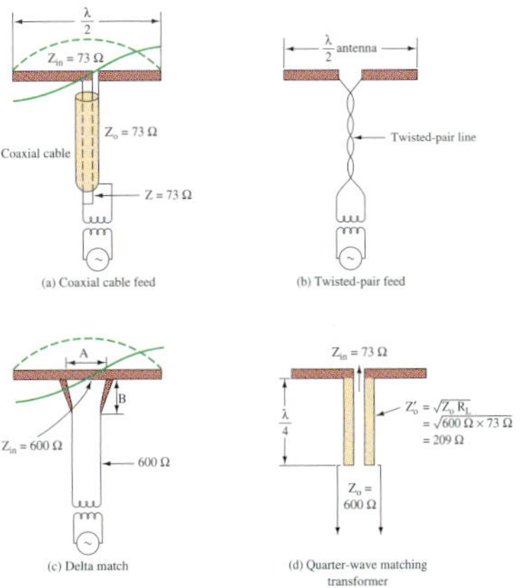

FIGURE 14-10 Feeding antennas with nonresonant lines.

antenna. This method of connection produces no standing waves on the line when the line is matched to a generator. Coupling to a generator is often made through a simple untuned transformer secondary.

Another method of transferring energy to the antenna is through the use of a twisted-pair line, as shown in Fig. 14-10(b). It is used as an untuned line for low frequencies. Due to excessive losses occurring in the insulation, the twisted pair is not used at higher frequencies. The characteristic impedance of such lines is about 70 Ω.

Delta Match

When a line does not match the impedance of the antenna, it is necessary to use special impedance matching techniques such as those discussed with Smith chart applications in Chapter 12. An example of an additional type of impedance matching device is the **delta match,** shown in Fig. 14-10(c). Due to inherent characteristics, the open, two-wire transmission line does not have a characteristic impedance

Delta Match
an impedance matching device that spreads the transmission line as it approaches the antenna

Section 14-4 • Antenna Feed Lines 659

Full-color format is used throughout, enhancing illustrations and highlighting key terms

END-OF-CHAPTER MATERIAL—Each chapter concludes with a summary of key concepts, an extensive problem set, a section entitled "Questions for Critical Thinking," and chapter exercises incorporating Electronics Workbench™ Multisim. See below for an illustration of how this material is presented. The questions and problems are very comprehensive and are keyed to the appropriate chapter section. An asterisk next to the question number indicates that a particular question has been provided by the FCC as a study aid for licensing examinations. In addition, the answer to quantitative problems is provided in parentheses following the question. Worked-out solutions to selected problems are available in the Instructor's Manual.

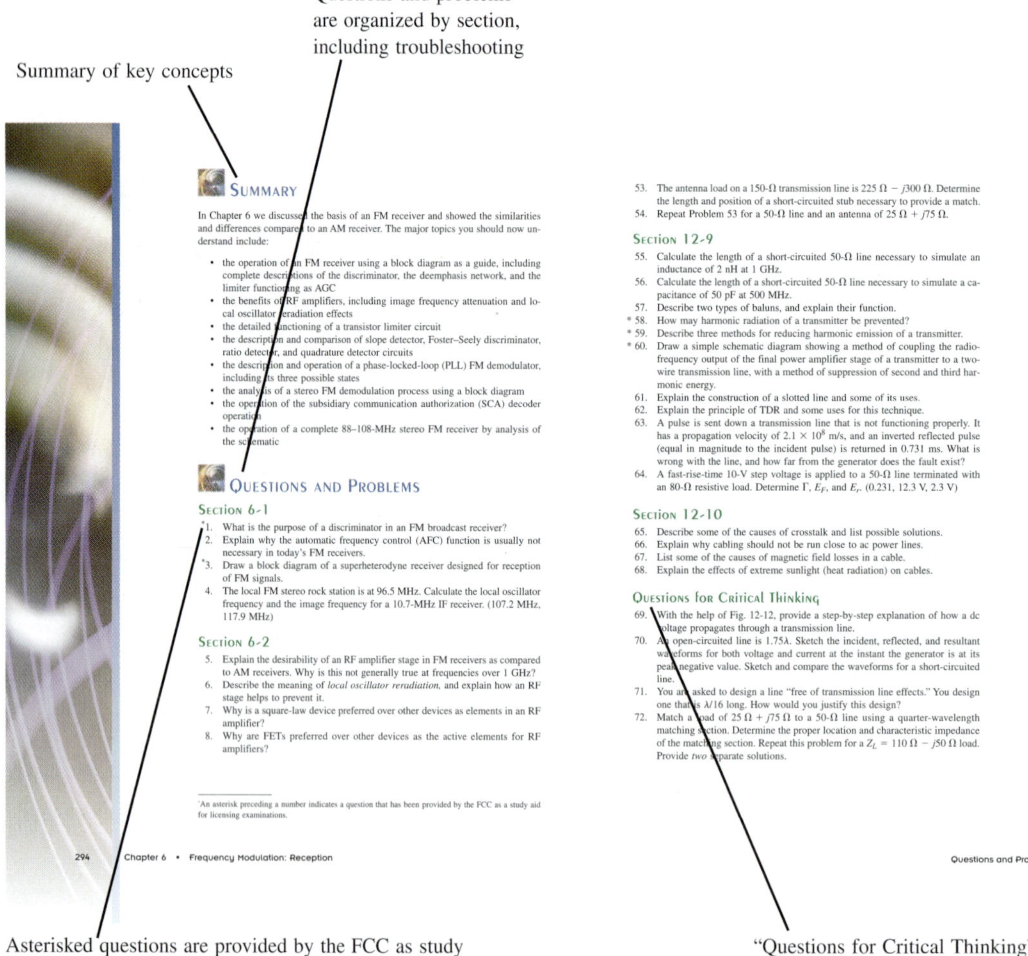

Questions and problems are organized by section, including troubleshooting
Summary of key concepts
Asterisked questions are provided by the FCC as study aids for licensing exams
"Questions for Critical Thinking" further develop the student's analytical skills

SUMMARY

In Chapter 6 we discussed the basis of an FM receiver and showed the similarities and differences compared to an AM receiver. The major topics you should now understand include:

- the operation of an FM receiver using a block diagram as a guide, including complete descriptions of the discriminator, the deemphasis network, and the limiter functioning as AGC
- the benefits of RF amplifiers, including image frequency attenuation and local oscillator reradiation effects
- the detailed functioning of a transistor limiter circuit
- the description and comparison of slope detector, Foster–Seely discriminator, ratio detector, and quadrature detector circuits
- the description and operation of a phase-locked-loop (PLL) FM demodulator, including its three possible states
- the analysis of a stereo FM demodulation process using a block diagram
- the operation of the subsidiary communication authorization (SCA) decoder operation
- the operation of a complete 88–108-MHz stereo FM receiver by analysis of the schematic

QUESTIONS AND PROBLEMS

SECTION 6-1

*1. What is the purpose of a discriminator in an FM broadcast receiver?
2. Explain why the automatic frequency control (AFC) function is usually not necessary in today's FM receivers.
*3. Draw a block diagram of a superheterodyne receiver designed for reception of FM signals.
4. The local FM stereo rock station is at 96.5 MHz. Calculate the local oscillator frequency and the image frequency for a 10.7-MHz IF receiver. (107.2 MHz, 117.9 MHz)

SECTION 6-2

5. Explain the desirability of an RF amplifier stage in FM receivers as compared to AM receivers. Why is this not generally true at frequencies over 1 GHz?
6. Describe the meaning of *local oscillator reradiation,* and explain how an RF stage helps to prevent it.
7. Why is a square-law device preferred over other devices as elements in an RF amplifier?
8. Why are FETs preferred over other devices as the active elements for RF amplifiers?

*An asterisk preceding a number indicates a question that has been provided by the FCC as a study aid for licensing examinations.

294 Chapter 6 • Frequency Modulation: Reception

53. The antenna load on a 150-Ω transmission line is $225\ \Omega - j300\ \Omega$. Determine the length and position of a short-circuited stub necessary to provide a match.
54. Repeat Problem 53 for a 50-Ω line and an antenna of $25\ \Omega + j75\ \Omega$.

SECTION 12-9

55. Calculate the length of a short-circuited 50-Ω line necessary to simulate an inductance of 2 nH at 1 GHz.
56. Calculate the length of a short-circuited 50-Ω line necessary to simulate a capacitance of 50 pF at 500 MHz.
57. Describe two types of baluns, and explain their function.
*58. How may harmonic radiation of a transmitter be prevented?
*59. Describe three methods for reducing harmonic emission of a transmitter.
*60. Draw a simple schematic diagram showing a method of coupling the radio-frequency output of the final power amplifier stage of a transmitter to a two-wire transmission line, with a method of suppression of second and third harmonic energy.
61. Explain the construction of a slotted line and some of its uses.
62. Explain the principle of TDR and some uses for this technique.
63. A pulse is sent down a transmission line that is not functioning properly. It has a propagation velocity of 2.1×10^8 m/s, and an inverted reflected pulse (equal in magnitude to the incident pulse) is returned in 0.731 ms. What is wrong with the line, and how far from the generator does the fault exist?
64. A fast-rise-time 10-V step voltage is applied to a 50-Ω line terminated with an 80-Ω resistive load. Determine Γ, E_F, and E_r. (0.231, 12.3 V, 2.3 V)

SECTION 12-10

65. Describe some of the causes of crosstalk and list possible solutions.
66. Explain why cabling should not be run close to ac power lines.
67. List some of the causes of magnetic field losses in a cable.
68. Explain the effects of extreme sunlight (heat radiation) on cables.

QUESTIONS FOR CRITICAL THINKING

69. With the help of Fig. 12-12, provide a step-by-step explanation of how a dc voltage propagates through a transmission line.
70. An open-circuited line is 1.75λ. Sketch the incident, reflected, and resultant waveforms for both voltage and current at the instant the generator is at its peak negative value. Sketch and compare the waveforms for a short-circuited line.
71. You are asked to design a line "free of transmission line effects." You design one that is $\lambda/16$ long. How would you justify this design?
72. Match a load of $25\ \Omega + j75\ \Omega$ to a 50-Ω line using a quarter-wavelength matching section. Determine the proper location and characteristic impedance of the matching section. Repeat this problem for a $Z_L = 110\ \Omega - j50\ \Omega$ load. Provide *two* separate solutions.

Questions and Problems 609

GLOSSARY AND ACRONYMS—The end-of-book material includes an extensive glossary and list of acronyms. These important tools are illustrated below. Acronyms are widely used in electronic communications and are often a source of confusion for students. This listing solves the problem by offering a quickly accessible description.

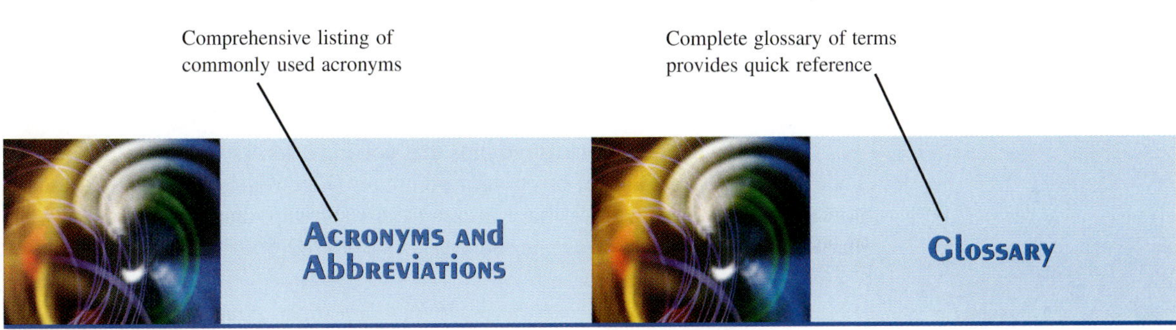
Comprehensive listing of commonly used acronyms
Complete glossary of terms provides quick reference
Acronyms and Abbreviations
Glossary

A

AAL	ATM adaptation layer
AC	alternating current
ACA	adaptive channel allocation
ACIL	trade association (formerly the American Council of Independent Laboratories)
ACK	acknowledgment
ACL	advanced CMOS logic
ACR	attenuation and crosstalk measurement
A/D	analog-to-digital
ADC	analog-to-digital converter
ADCCP	advanced digital communications control protocol
ADSL	asymmetric digital subscriber line
AF	audio frequency
AFC	automatic frequency control
AFSK	audio-frequency shift keying
AGC	automatic gain control
AIAA	American Institute of Aeronautics and Astronautics
AlGaAs	aluminum gallium arsenide
ALC	automatic level control
ALU	arithmetic logic unit
AM	amplitude modulation
AMI	alternate mark inversion
AML	automatic-modulation-limiting
AMPS	Advanced Mobile Phone Service
ANSI	American National Standards Institute
APC	angle-polished connectors
APD	avalanche photodiode
AP-S	Antennas and Propagation Society
ARPA	Advanced Research Projects Agency (now DARPA)
ARQ	automatic repeat request
ARRL	American Radio Relay League
ASCII	American Standard Code for Information Interchange
ASIC	application-specific integrated circuit
ASK	amplitude-shift keying
ASSP	application-specific standard products
ATC	adaptive transform coding
ATE	automatic test equipment
ATG	automatic test generation
ATM	asynchronous transfer mode
ATSC	Advanced Television Systems Committee
ATV	advanced television
AWGN	additive white Gaussian noise

B

B	byte
BAW	bulk acoustic wave
BBNS	broadband network services
BCC	block check character
BCCH	broadcast control channel
BCD	binary-coded decimal
B-CDMA	broadband CDMA
BCI	broadcast interference
BeCu	beryllium copper
B8ZS	bipolar 8 zero substitution
BER	bit-error rate
BERT	bit-error-rate tester
BFO	beat-frequency oscillator
BiCMOS	bipolar-CMOS
BIOS	basic input/output system
BIS	buffer information specification
B-ISDN	broadband integrated-services digital network (an ATM protocol model)
BJT	bipolar junction transistor
BPSK	binary phase-shift keying
BRI	basic-rate interface
BS	base station
BSC	base-station controller
BSS	Broadcasting Satellite Service

879

acoustic coupler supports a telephone handpiece and uses sound transducers to send and receive audio tones

acquisition time amount of time it takes for the hold circuit to reach its final value

ACR manufacturer combined measurement of attenuation and crosstalk. A large ACR indicates greater bandwidth

AC3 the Dolby Laboratory's audio compression technique for digital television

ADSL provision of up to 1.544 Mbps from the user to the service provider and up to 8 Mbps back to the user from the service provider

advanced mobile phone service (AMPS) cellular mobile radio that uses 12-kHz peak deviation channels, which are spaced 30-kHz apart in the 800–900-MHz band

Advanced Television Systems Committee (ATSC) developed to make recommendations for advanced television in the United States

air interface used by PCS systems to manage the transfer of information

algorithms a plan or set of instructions to achieve a specific goal

alias frequency an undesired frequency produced when the Nyquist sampling rate is not attained

aliasing errors that occur when the input frequency exceeds one-half the sample rate

aliasing distortion the distortion that results if Nyquist criteria are not met in a digital communications system using sampling of the information signal; the resulting alias frequency equals the difference between the input intelligence frequency and the sampling frequency

AMI alternate mark inversion

amplitude companding process of volume compression before transmission and volume expansion after detection

amplitude compandored single sideband (ACSSB) sideband transmission with speech compression in the transmitter and speech expansion in the receiver

amplitude modulation (AM) the process of impressing low-frequency intelligence onto a high-frequency carrier so that the instantaneous changes in the amplitude of the intelligence produce corresponding changes in the amplitude of the high-frequency carrier

anechoic chamber a large enclosed room that prevents reflected electromagnetic waves and shields out interfering waves from the outside world; used for radiation measurements

angle modulation superimposing the intelligence signal on a high-frequency carrier so that its phase angle or frequency is altered as a function of the intelligence amplitude

antenna a device that generates and/or collects electromagnetic energy

antenna array group of antennas or antenna elements arranged to provide the desired directional characteristics

antenna coupler an impedance matching network in the output stage of an RF amplifier or transmitter that ensures maximum power is transferred to the antenna by matching the input impedance of the antenna to the output impedance of the transmitter

antenna gain a measure of how much more power in dB an antenna will radiate in a certain direction with respect to that which would be radiated by a reference antenna, i.e., an isotropic point source or dipole

antialiasing filter a sharp-cutoff low-pass filter used to make sure no frequencies above one-half the sampling rate reach the ADC converter

aperture time the time that the S/H circuit must hold the sampled voltage

apogee farthest distance of a satellite's orbit to earth

Armstrong transmitter FM transmitter that uses a phase modulator to feed the intelligence signal through a low-pass filter integrator network to convert PM to FM

array a group of antennas or antenna elements arranged to provide the desired directional characteristics

887

CD-ROM INCLUDED—Over 90 percent of the circuits from the text plus additional circuits for troubleshooting are provided on the Electronics Workbench® Textbook Edition for Multisim 7 packaged with the text.

Multisim® is a schematic capture, simulation, and programmable logic tool used by college and university students in their course of study of Electronics and Electrical Engineering. The circuits on the CD in this text were created for use with Multisim software.

Multisim is widely regarded as an excellent tool for classroom and laboratory learning. However, no part of this textbook is dependent upon the Multisim software or provided files. These files are provided at no extra cost to the consumer and are for use by anyone who chooses to utilize Multisim software.

The first 25% of the circuits on the CD included with this text are already rendered "live" for you by Electronics Workbench in the *Textbook Edition of Multisim 7,* enabling you to do the following:

- Manipulate the interactive components and adjust the value of any virtual components.
- Run interactive simulation on the active circuits and use any pre-placed virtual instruments.
- Run analyses.
- Run/print/save simulation results for the pre-defined viewable circuits.
- Create your own circuits up to a maximum of 15 components.

The balance of the circuits requires that you have access to Multisim 7 in your school lab (the Lab Edition) or on your computer (Electronics Workbench Student Suite). If you do not currently have access to this software and wish to purchase it, *please call Prentice Hall Customer Service at 1-800-282-0693 or send a fax request to 1-800-835-5327.*

If you need *technical assistance or have questions concerning the Multisim software,* contact Electronics Workbench directly for support at (416) 977-5550 or via the EWB website located at www.electronicsworkbench.com.

Ancillary Package

- *Laboratory Manual* with accompanying CD-ROM, by Mark E. Oliver and Jeffrey S. Beasley, ISBN 0-13-170265-3
- Online Instructor's Manual featuring:
  Chapter Overviews
  Worked-out solutions to quantitative problems in the text
  PowerPoint slides of all figures in the text
- Prentice Hall TestGen, a computerized test bank

Acknowledgments

Many people have provided constructive criticism for the earlier seven editions of *Modern Electronic Communication* and we truly appreciate the input that everyone has made. A special thank you to Jim Andress, Dr. Russ Jedliuka, Dr. Ray Lyman, and Shannon Gunaji for their significant contributions to the eighth edition. We would like to thank the reviewers of this edition for their valuable feedback: Pradeep Bhattacharya, Southern University, LA; David Mayo, South Georgia Technical College; Randall Moser, Sr., Pennsylvania College of Technology; Michael Smith, Perry Technical Institute, WA; and Nick Smith, Education America, TX.

We would like to thank our publisher Charles Stewart, assistant editor Mayda Bosco, and production editor Alex Wolf for their editorial support and production coordination.

Finally, we'd like to thank our families for their continuing support and patience.

Jeffrey S. Beasley and Gary M. Miller

Brief Contents

Contents

Modern Electronic Communication

1 Introductory Topics

Chapter Outline

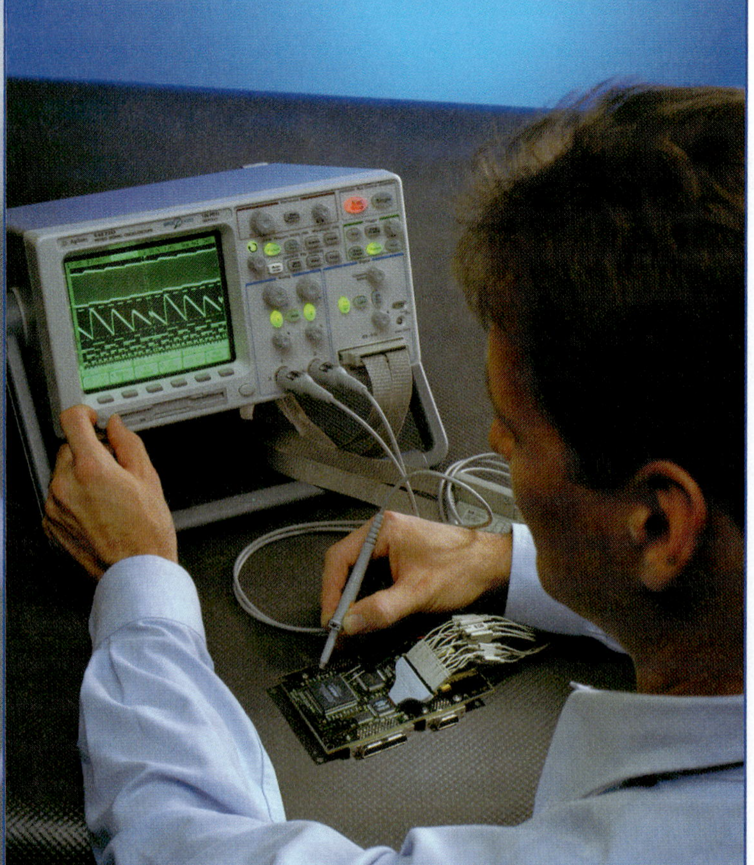

Agilent's digital os cilloscopes includ easy-to-use featur high bandwidth, MegaZoom deep memory, high sam pling rates, and in tegrated logic timi channels. (Courtes of Agilent Technol gies. Reprinted wi permission.)

Objectives

- Describe a basic communication system and explain the concept of modulation
- Develop an understanding of the use of the decibel (dB) in communications systems
- Define *electrical noise* and explain its effect at the first stages of a receiver
- Calculate the thermal noise generated by a resistor
- Calculate the signal-to-noise ratio and noise figure for an amplifier
- Describe several techniques for making noise measurements
- Explain the relationship among information, bandwidth, and time of transmission
- Analyze nonsinusoidal repetitive waveforms via Fourier analysis
- Analyze the operation of various *RLC* circuits
- Describe the operation of common *LC* and crystal oscillators

Key Terms

modulation
intelligence signal
intelligence
demodulation
transducer
dB
dBm
0 dBm
dBm(600)
dBm(75)
dBm(50)
dBW
dBμV
electrical noise
static
external noise
internal noise
wave propagation
atmospheric noise
space noise
solar noise
cosmic noise
Johnson noise
thermal noise
white noise
low-noise resistor
shot noise
excess noise
transit-time noise
signal-to-noise ratio
noise figure
noise ratio
octave
Friiss's formula
device under test
tangential method
information theory
channel
Hartley's law
Fourier analysis
FFT
frequency domain record
aliasing
quality
leakage
dissipation
resonance
tank circuit
poles
constant-*k* filter
m-derived filter
roll-off
stray capacitance
oscillator
flywheel effect
damped
continuous wave
Barkhausen criteria
frequency synthesizer

1-1 Introduction

This book provides an introduction to all relevant aspects of communications systems. These systems had their beginning with the discovery of various electrical, magnetic, and electrostatic phenomena prior to the twentieth century. Starting with Samuel Morse's invention of the telegraph in 1837, a truly remarkable rate of progress has occurred. The telephone, thanks to Alexander Graham Bell, came along in 1876. The first complete system of wireless communication was provided by Guglielmo Marconi in 1894. Lee DeForest's invention of the triode vacuum tube in 1908 allowed the first form of practical electronic amplification and really opened the door to wireless communication. In 1948 another major discovery in the history of electronics occurred with the development of the transistor by Shockley, Brattain, and Bardeen. The more recent developments, such as integrated circuits, very large-scale integration, and computers on a single silicon chip, are probably familiar to you.

The rapid transfer of these developments into practical communications systems linking the entire globe (and now into outer space) has stimulated a bursting growth of complex social and economic activities. This growth has subsequently had a snowballing effect on the growth of the communication industry with no end in sight for the foreseeable future. Some people refer to this as the age of communications.

The function of a communication system is to transfer information from one point to another via some communication link. The very first form of "information" electrically transferred was the human voice in the form of a code (i.e., the Morse code), which was then converted back to words at the receiving site. People had a natural desire and need to communicate rapidly between distant points on the earth, and that was the major concern of these developments. As that goal became a reality, and with the evolution of new technology following the invention of the triode vacuum tube, new and less basic applications were also realized, such as entertainment (radio and television), radar, and telemetry. The field of communications is still a highly dynamic one, with advancing technology constantly making new equipment possible or allowing improvement of the old systems. Communications was the basic origin of the electronics field, and no other major branch of electronics developed until the transistor made modern digital computers a reality.

Modulation

Modulation
process of putting information onto a high-frequency carrier for transmission

Intelligence Signal
the low frequency information that modulates the carrier

Intelligence
low-frequency information modulated onto a high-frequency carrier in a transmitter

Demodulation
process of removing intelligence from the high-frequency carrier in a receiver

Basic to the field of communications is the concept of modulation. **Modulation** is the process of putting information onto a high-frequency carrier for transmission. In essence, then, the transmission takes place at the high frequency (the carrier) which has been modified to "carry" the lower-frequency information. The low-frequency information is often called the **intelligence signal** or, simply, the **intelligence.** It follows that once this information is received, the intelligence must be removed from the high-frequency carrier—a process known as **demodulation.** At this point you may be thinking, why bother to go through this modulation/demodulation process? Why not just transmit the information directly? The problem is that the frequency of the human voice ranges from about 20 to 3000 Hz. If everyone transmitted those frequencies directly as radio waves, interference would cause them all to be ineffective. Another limitation of equal importance is the virtual impossibility of transmitting such low frequencies since the required antennas for efficient propagation would be miles in length.

The solution is modulation, which allows propagation of the low-frequency intelligence with a high-frequency carrier. The high-frequency carriers are chosen such that only one transmitter in an area operates at the same frequency to minimize interference, and that frequency is high enough so that efficient antenna sizes are manageable. There are three basic methods of putting low-frequency information onto a higher frequency. Equation (1-1) is the mathematical representation of a sine wave, which we shall assume to be the high-frequency carrier.

$$v = V_P \sin(\omega t + \Phi) \tag{1-1}$$

where v = instantaneous value
V_P = peak value
ω = angular velocity = $2\pi f$
Φ = phase angle

Any one of the last three terms could be varied in accordance with the low-frequency information signal to produce a modulated signal that contains the intelligence. If the amplitude term, V_P, is the parameter varied, it is called amplitude modulation (AM). If the frequency is varied, it is frequency modulation (FM). Varying the phase angle, Φ, results in phase modulation (PM). In subsequent chapters we shall study these systems in detail.

Communications Systems

Communications systems are often categorized by the frequency of the carrier. Table 1-1 provides the names for various frequency ranges in the radio spectrum. The extra-high-frequency range begins at the starting point of infrared frequencies, but the infrareds extend considerably beyond 300 GHz (300×10^9 Hz). After the infrareds in the electromagnetic spectrum (of which the radio waves are a very small portion) come light waves, ultraviolet rays, X rays, gamma rays, and cosmic rays.

Table 1-1 Radio-Frequency Spectrum

Frequency	Designation	Abbreviation
30–300 Hz	Extremely low frequency	ELF
300–3000 Hz	Voice frequency	VF
3–30 kHz	Very low frequency	VLF
30–300 kHz	Low frequency	LF
300 kHz–3 MHz	Medium frequency	MF
3–30 MHz	High frequency	HF
30–300 MHz	Very high frequency	VHF
300 MHz–3 GHz	Ultra high frequency	UHF
3–30 GHz	Super high frequency	SHF
30–300 GHz	Extra high frequency	EHF

Figure 1-1 represents a simple communication system in block diagram form. Notice that the modulated stage accepts two inputs, the carrier and the information (intelligence) signal. It produces the modulated signal, which is subsequently amplified before transmission. Transmission of the modulated signal can take place by any one of four means: antennas, waveguides, optical fibers, or transmission lines. These four modes of propagation will be studied in subsequent chapters. The receiving unit of the system picks up the transmitted signal but must reamplify it to compensate for attenuation that occurred during transmission. Once suitably amplified, it is fed to the demodulator (often referred to as the detector), where the information signal is extracted from the high-frequency carrier. The demodulated signal (intelligence) is then fed to the amplifier and raised to a level enabling it to drive a speaker or any other output transducer. A **transducer** is a device that converts energy from one form to another.

Transducer
device that converts energy from one form to another

Many of the performance measurements in communication systems are specified in dB (decibels). Section 1-2 introduces the use of this very important concept in communication systems. This is followed by two basic limitations on the performance of a communications systems: (1) electrical noise and (2) the bandwidth of frequencies allocated for the transmitted signal. Sections 1-3 to 1-6 are devoted to these topics because of their extreme importance.

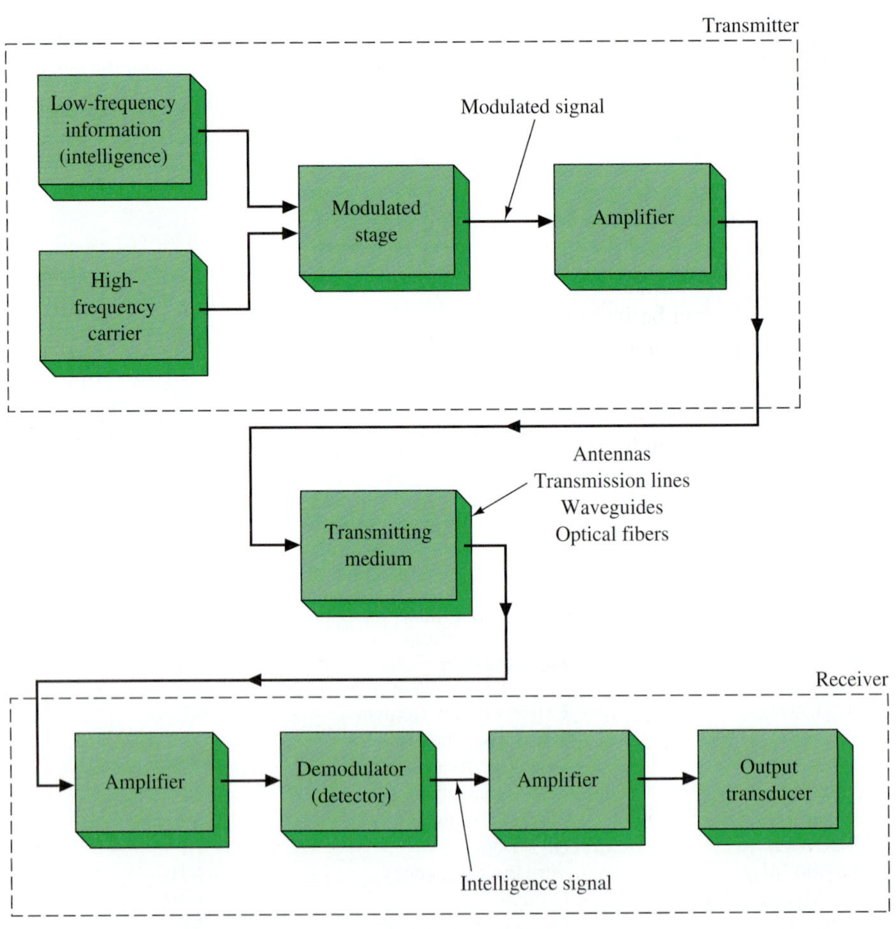

FIGURE 1-1 A communication system block diagram.

1-2 The dB in Communications

Decibels (dBs) are used to specify measured and calculated values in noise analysis, audio systems, microwave system gain calculations, satellite system link-budget analysis, antenna power gain, light-budget calculations, and many other communications system measurements. In each case, the dB value is calculated with respect to a standard or specified reference.

dB (decibel)
relative unit of measurement used frequently in electronic communications to describe power gain or loss

The dB value is calculated by taking the log of the ratio of the measured or calculated power (P_2) with respect to a reference power (P_1) level. This result is then multiplied by 10 to obtain the value in dB. The formula for calculating the dB value of two ratios is shown in Eq. (1-2). Equation (1-2) is commonly referred to as the *power ratio form* for dB.

$$\text{dB} = 10 \log_{10} \frac{P_2}{P_1} \tag{1-2}$$

By using the power relationship $P = V^2/R$, the relationship shown in Eq. (1-3) is obtained:

$$\text{dB} = 10 \log_{10} \left(\frac{V_2^2/R_2}{V_1^2/R_1} \right)$$

Let $R_1 = R_2$:

$$\text{dB} = 10 \log_{10} \frac{V_2^2}{V_1^2} \tag{1-3}$$

Note that we have assumed that the resistances (R_1 and R_2) are equivalent; therefore, these terms can be ignored in the dB power equation. This is a reasonable assumption in most communication systems since maximum power transfer (a desirable characteristic) is obtained when the input and output impedances are matched. Equation (1-3) can be modified (using a property of logarithms) to provide a relationship for decibels in terms of the voltage ratios instead of power ratios. This is called the *voltage gain equation* and is shown in Eq. (1-4).

$$\text{dB} = 20 \log_{10} \left(\frac{V_2}{V_1} \right) \tag{1-4}$$

Applying the dB Value

The dB unit is often used in specifying input- and output-signal-level requirements for many communication systems. When making dB measurements, a reference level is specified or implied for that particular application. An example is found in audio consoles in broadcast systems, where a **0-dBm** input level is usually specified as the required input- and output-audio level for 100% modulation. Notice that a lowercase *m* has been attached to the dB unit. This indicates that the specified dB level is relative to a 1-mW reference.

dBm
dB level using a 1-mW reference

0 dBm
1 mW measured relative to a 1-mW reference

In standard audio systems **0 dBm** is defined as 0.001 W measured with respect to a load termination of 600 Ω. A 600-Ω balanced audio line is the standard for professional audio, broadcast, and telecommunications systems. However, 0 dBm is not exclusive to a 600-Ω impedance.

Example 1-1

Show that when making a dBm measurement, a measured value of 1 mW will result in a 0 dBm power level.

Solution

$$\text{dB} = 10\log_{10}\frac{P_2}{P_1} = 10\log_{10}\frac{1\text{ mW}}{1\text{ mW}} = 0\text{ dB} \quad \text{or} \quad 0\text{ dBm} \qquad \textbf{(1-2)}$$

This result, 0 dB, is expressed as 0 dBm to indicate that the result was obtained relative to a 1-mW reference.

It can be shown that the voltage measured across a 600-Ω load for a 0-dBm level is 0.775 V. This value can be obtained by first modifying Eq. (1-2) by inserting the 1 mW value for P_1, as shown.

$$\text{dB} = 10\log_{10}\left(\frac{P_2}{P_1}\right)$$

where $P_2 = \frac{V_2^2}{600}$

$P_1 = 0.001\text{ W}$

Since 1 mW is the specified reference for dBm, the voltage reference for 0 dBm can be developed as follows:

$$0\text{ dBm} = 10\log\frac{V_2^2/600}{0.001}$$

$$0\text{ dBm} = \log\frac{V_2^2/600}{0.001}$$

$$\log^{-1}(0\text{ dBm}) = \frac{V_2^2/600}{0.001}$$

$$1 = \frac{V_2^2/600}{0.001}$$

$$0.6 = V_2^2$$

$$V_2 = 0.77459$$

dBm(600)
decibel measurement using a 1-mW reference with respect to a 600-Ω load

The voltage value 0.77459 (0.775 V) is the reference for 0 dB with respect to a 600-Ω load when a voltage measurement is used to calculate the **dBm(600)** value. The dBm(600) term indicates that this measurement or calculation is made using a 1-mW reference with respect to a 600-Ω load.

$$\text{dBm}(600) = 20\log_{10}\left(\frac{V_2}{0.775}\right) \qquad \textbf{(1-5)}$$

Example 1-2 demonstrates how to solve for the voltage value (V_2) if a +8-dBm level is specified.

Example 1-2

A microwave system requires a +8-dBm audio level to provide 100% modulation. Determine the voltage level required to produce a +8-dBm level. Assume a 600-Ω audio system.

Solution

Since this is a 600-Ω system, use the 0.775-V reference shown in Eq. (1-5).

$$\text{dBm}(600) = 20 \log_{10}\left(\frac{V_2}{0.775}\right) \qquad \textbf{(1-5)}$$

$$+8 \text{ dBm} = 20 \log \frac{V_2}{0.775}$$

$$0.4 = \log \frac{V_2}{0.775}$$

$$\log^{-1}(0.4) = \frac{V_2}{0.775}$$

$$V_2 = 1.947 \text{ V}$$

Thus, to verify that a +8-dBm level is being provided to the input of the microwave transmitter, approximately 1.95 V must be measured across the 600-Ω input.

The term *dBm* also applies to communication systems that have a standard termination impedance other than 600 Ω. For example, many communication systems are terminated with 75 Ω. The 0-dBm value is still defined as 1 mW, but it is measured with respect to a 75-Ω termination instead of 600 Ω. Therefore, the voltage reference for a 0-dBm system with respect to 75 Ω is obtained by solving for V in the expression $P = V^2/R$ as shown:

$$V = \sqrt{PR} = \sqrt{(0.001)(75)} = 0.274 \text{ V}$$

To calculate the voltage gain or loss with respect to a 75-Ω load, use Eq. (1-6). This value is specified as **dBm(75)** to indicate that this measure was made or calculated using a 1-mW reference relative to a 75-Ω load.

dBm(75)
a measurement made using a 1-mW reference with respect to a 75-Ω load

$$\text{dBm}(75) = 20 \log_{10} \frac{V}{0.274} \qquad \textbf{(1-6)}$$

Fifty-ohm systems are usually used in radio communications. The dBm voltage reference for a 50-Ω system is

$$V = \sqrt{PR} = \sqrt{(0.001)(50)} = 0.2236 \text{ V}$$

dBm(50)
a measurement made using a 1-mW reference with respect to a 50-Ω load

To calculate the voltage gain or loss expressed in dB for a 50-Ω system [**dBm(50)**], use Eq. (1-4) with $V_1 = 0.2236$. This relationship is shown in Eq. (1-7).

$$\text{dBm}(50) = 20 \log_{10}\frac{V}{0.2236} \qquad \textbf{(1-7)}$$

dBW
a measurement made using a 1-W reference

It is common for power to be expressed in watts instead of milliwatts. In this case the dB unit is obtained with respect to 1 W and the dB values are expressed as **dBW.**

0 dBW is defined as 1 W measured with respect to a 1-W reference.

Remember, dB is a relative measurement. As shown by Eqs. (1-1) and (1-3), both power and voltage gains can be expressed in dB relative to a reference value. In the case of dBW, the reference is 1 W; therefore, Eq. (1-1) is written with 1 W replacing the reference P_1. This gives Eq. (1-8).

$$\text{dBW} = 10 \log_{10}\frac{P_2}{1\ W} \qquad \textbf{(1-8)}$$

In some applications, it may be necessary to convert from one reference dB to another. Example 1-3 demonstrates how to convert from dBm to dBW.

Example 1-3

A laser diode outputs +10 dBm. Convert this value to
(a) watts.
(b) dBW.

Solution

(a) Convert +10 dBm to watts. Substitute and solve for P_2:

$$+10\ \text{dBm} = 10 \log\frac{P_2}{0.001}$$

$$\log^{-1}(1) = \frac{P_2}{0.001} \Rightarrow 10 = \frac{P_2}{0.001} \qquad \textbf{(1-2)}$$

$$P_2 = 0.01\ \text{W}$$

(b) Convert +10 dBm to dBW.

$$\text{dBW} = 10 \log\frac{0.01\ \text{W}}{1\ \text{W}} = -20\ \text{dBW} \qquad \textbf{(1-8)}$$

dBμV
a measurement made using a 1-μV reference

It is common with communication receivers to express voltage measurements in terms of **dBμV,** dB-microvolts. For voltage gain calculations involving dBμV, use Eq. (1-4) and specify 1 μV as the reference (V_1) in the calculations, as shown in Eq. (1-9).

$$\text{dB}\mu\text{V} = 20 \log_{10}\frac{V_2}{1\ \mu\text{V}} \qquad \textbf{(1-9)}$$

Table 1-2 Conversion Table for Common dBm Values

Common dBm Values	Equivalent Voltage Level (600 Ω)	Equivalent Voltage 75 Ω	Equivalent Voltage 50 Ω	Watts	dBW
38	61.560 V	21.765 V	17.761 V	6.3	8
30	24.508 V	8.665 V	7.071 V	1.0	0
20	7.750 V	2.740 V	2.236 V	1.00×10^{-1}	−10
15	4.358 V	1.541 V	1.257 V	3.16×10^{-2}	−15
10	2.451 V	0.866 V	0.7071 V	1.00×10^{-2}	−20
8	1.947 V	0.688 V	0.5617 V	6.31×10^{-3}	−22
6	1.546 V	0.547 V	0.4461 V	3.98×10^{-3}	−24
2	0.976 V	0.345 V	0.2815 V	1.58×10^{-3}	−28
1	0.870 V	0.307 V	0.2509 V	1.26×10^{-3}	−29
0	0.775 V	0.274 V	0.2236 V	1.00×10^{-3}	−30
−1	0.691 V	0.244 V	0.1993 V	7.94×10^{-4}	−31
−2	0.616 V	0.218 V	0.1776 V	6.31×10^{-4}	−32
−6	0.388 V	0.137 V	0.1121 V	2.51×10^{-4}	−36
−10	0.245 V	86.65 mV	70.7 mV	1.00×10^{-4}	−40
−15	0.138 V	48.72 mV	39.8 mV	3.16×10^{-5}	−45
−20	77.5 mV	27.40 mV	22.4 mV	1.00×10^{-5}	−50
−35	13.78 mV	4.872 mV	3.98 mV	3.16×10^{-7}	−65
−50	2.45 mV	866.5 μV	0.707 mV	1.00×10^{-8}	−80
−70	0.2451 mV	86.65 μV	70.7 μV	1.00×10^{-10}	−100

There are many applications using decibels in calculations involving relative values. The important thing to remember is that a relative reference is typically specified or understood when calculating or measuring a decibel value. Table 1-2 is a conversion table for many common dBm values. A conversion table is provided for dBm, voltage, and watts for 600-Ω, 75-Ω, and 50-Ω systems. Additionally, a list of common decibel terms is provided in Table 1-3 (page 12).

1-3 Noise

Electrical noise may be defined as any undesired voltages or currents that ultimately end up appearing in the receiver output. To the listener this electrical noise often manifests itself as **static.** It may only be annoying, such as an occasional burst of static, or continuous and of such amplitude that the desired information is obliterated.

Electrical Noise
any undesired voltages or currents that end up appearing in a circuit

Static
electrical noise that may occur in the output of a receiver

Noise signals at their point of origin are generally very small, for example, at the microvolt level. You may be wondering, therefore, why they create so much trouble. Well, a communications receiver is a very sensitive instrument that is given a very small signal at its input that must be greatly amplified before it can possibly drive a speaker. Consider the receiver block diagram shown in Fig. 1-1 to be representative of a standard FM radio (receiver). The first amplifier block, which forms the "front end" of the radio, is required to amplify a received signal from the radio's antenna that is often less than 10 μV. It does not take a very large dose of undesired signal (noise) to ruin reception. This is true even though the transmitter output may be many thousands of watts because, when received, it is severely attenuated. Therefore, if the desired signal received is of the same order of magnitude as

Table 1-3 dB Reference Table

Term	Description
dBm	The dB using a 1-mW (0.001-W) reference, which is the typical measurement for audio input/output specifications. This measurement is also used in low-power optical transmitter specifications.
dBm(600)	The standard audio reference power level defined by 1 mW measured with respect to a 600-Ω load. This measurement is commonly used in broadcasting and professional audio applications and is a common telephone communications standard.
dBm(50)	The standard defined by 1 mW measured with respect to a 50-Ω load. This measurement is commonly used in radio-frequency transmission/receiving systems.
dBm(75)	The standard defined by 1 mW measured with respect to a 75-Ω load.
dBmW	The generic form for a 1-mW reference, also written as dBm. This term usually has an inferred load reference, depending on the application.
dBW	A common form for power amplification relative to a 1-W reference (usually 50 Ω). Typical applications are found in specifications for radio-frequency power amplifiers and high-power audio amplifiers.
dBμV	A common form for specifying input radio-frequency levels to a communications receiver. This is called a decibel-microvolt, where $1\ \mu V = 1 \times 10^{-6}$ V.
dBV	The decibel value is obtained with respect to 1 V.
dBV_{RMS}	A dB value measured relative to 1 V_{RMS}, where 0 dB = 1 V_{RMS}. This value is sometimes used to define measurements in FFT frequency analysis, as described later in this chapter.
dB/bit	A common term used for specifying the dynamic range or resolution for a pulse-code modulation (PCM) system such as a CD player. This reference is defined by 20/log(2)/bit = 6.02 dB/bit.
dBi	Decibel isotropic, or gain relative to an isotropic radiator, as described in Chapter 14. It is used as the reference when defining antenna gain.
dB/Hz	Relative noise power in a 1-Hz bandwidth. This term is used often in digital communications and in defining a laser's relative intensity noise (RIN). For a laser system, this is an electrical, not an optical, measurement. A typical RIN for a semiconductor laser is −150 dB/Hz.
dBmV	A cable TV standard that uses a reference of 1 mV across 75 Ω.

the undesired noise signal, it will probably be unintelligible. This situation is made even worse because the receiver itself introduces additional noise.

The noise present in a received radio signal that has been introduced in the transmitting medium is termed **external noise.** The noise introduced by the receiver is termed **internal noise.** The important implications of noise considerations in the study of communications systems cannot be overemphasized.

External Noise
noise in a received radio signal that has been introduced by the transmitting medium

Internal Noise
noise in a radio signal that has been introduced by the receiver

External Noise

Human-Made Noise The most troublesome form of external noise is usually the human-made variety. It is often produced by spark-producing mechanisms such as engine ignition systems, fluorescent lights, and commutators in electric motors. This noise is actually "radiated" or transmitted from its generating sources through the atmosphere in the same fashion that a transmitting antenna radiates desirable electrical signals to a receiving antenna. This process is called **wave propagation** and is the subject of Chapter 13. If the human-made noise exists in the vicinity of the transmitted radio signal and is within its frequency range, these two signals will

Wave Propagation
movement of radio signals through the atmosphere from transmitter to receiver

"add" together. This is obviously an undesirable phenomenon. Human-made noise occurs randomly at frequencies up to around 500 MHz.

Another common source of human-made noise is contained in the power lines that supply the energy for most electronic systems. In this context the ac ripple in the dc power supply output of a receiver can be classified as noise (an unwanted electrical signal) and must be minimized in receivers that are accepting extremely small intelligence signals. Additionally, ac power lines contain surges of voltage caused by the switching on and off of highly inductive loads such as electrical motors. It is certainly ill-advised to operate sensitive electrical equipment in close proximity to an elevator! Human-made noise is weakest in sparsely populated areas, which explains the location of extremely sensitive communications equipment, such as satellite tracking stations, in desert-type locations.

Atmospheric Noise **Atmospheric noise** is caused by naturally occurring disturbances in the earth's atmosphere, with lightning discharges being the most prominent contributors. The frequency content is spread over the entire radio spectrum, but its intensity is inversely related to frequency. It is therefore most troublesome at the lower frequencies. It manifests itself in the static noise that you hear on standard AM radio receivers. Its amplitude is greatest from a storm near the receiver, but the additive effect of distant disturbances is also a factor. This is often apparent when listening to a distant station at night on an AM receiver. It is not a significant factor for frequencies exceeding about 20 MHz.

Atmospheric Noise
external noise caused by naturally occurring disturbances in the earth's atmosphere

Space Noise The other form of external noise arrives from outer space and is called **space noise.** It is pretty evenly divided in origin between the sun and all the other stars. That originating from our star (the sun) is termed **solar noise.** Solar noise is cyclical and reaches very annoying peaks about every 11 years.

All the other stars also generate this space noise, and their contribution is termed **cosmic noise.** Since they are much farther away than the sun, their individual effects are small, but they make up for this by their countless numbers and their additive effects. Space noise occurs at frequencies from about 8 MHz up to 1.5 GHz (1.5×10^9 Hz). While it contains energy at less than 8 MHz, these components are absorbed by the earth's ionosphere before they can reach the atmosphere. The ionosphere is a region above the atmosphere where free ions and electrons exist in sufficient quantity to have an appreciable effect on wave travel. It includes the area from about 60 to several hundred miles above the earth (see Chapter 13 for further details).

Space Noise
external noise produced outside the earth's atmosphere

Solar Noise
space noise originating from the sun

Cosmic Noise
space noise originating from stars other than the sun

Internal Noise

As stated previously, internal noise is introduced by the receiver itself. Thus, the noise already present at the receiving antenna (external noise) has another component added to it before it reaches the output. The receiver's major noise contribution occurs in its very first stage of amplification, where the desired signal is at its lowest level, and noise injected at that point will be at its largest value in proportion to the intelligence signal. A glance at Fig. 1-2 should help clarify this point. Even though all following stages also introduce noise, their effect is usually negligible with respect to the very first stage because of their much higher signal level. Note that the noise injected between amplifiers 1 and 2 has not appreciably increased the noise on

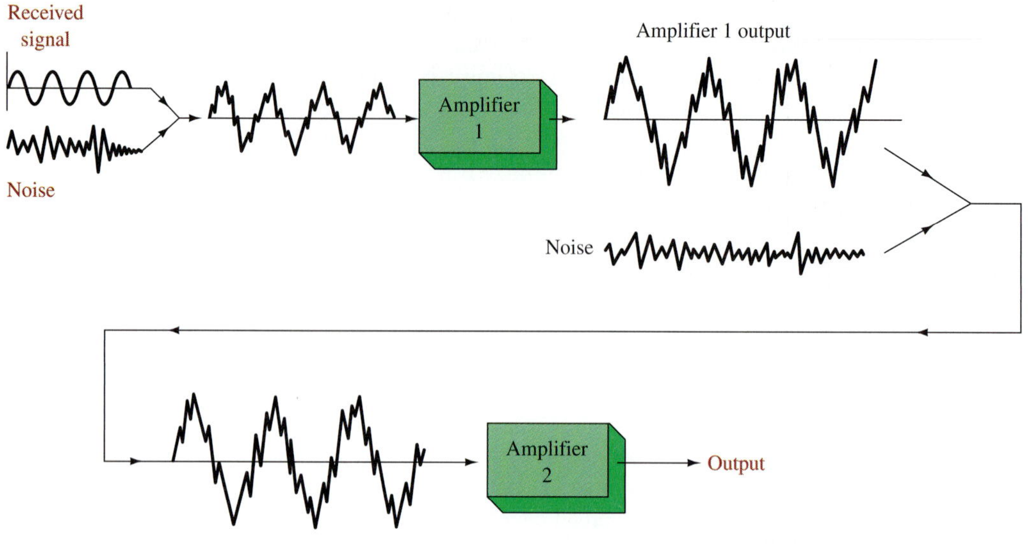

FIGURE 1-2 Noise effect on a receiver's first and second amplifier stages.

the desired signal, even though it is of the same magnitude as the noise injected into amplifier 1. For this reason, the first receiver stage must be carefully designed to have low noise characteristics, with the following stages being decreasingly important as the desired signal gets larger and larger.

Johnson Noise
another name for thermal noise, first studied by J. B. Johnson

Thermal Noise
internal noise caused by thermal interaction between free electrons and vibrating ions in a conductor

White Noise
another name for thermal noise because its frequency content is uniform across the spectrum

Thermal Noise There are two basic types of noise generated by electronic circuits. The first one to consider is due to thermal interaction between the free electrons and vibrating ions in a conductor. It causes the rate of arrival of electrons at either end of a resistor to vary randomly, and thereby varies the resistor's potential difference. Resistors and the resistance within all electronic devices are constantly producing a noise voltage. This form of noise was first thoroughly studied by J. B. Johnson in 1928 and is often termed **Johnson noise.** Since it is dependent on temperature, it is also referred to as **thermal noise.** Its frequency content is spread equally throughout the usable spectrum, which leads to a third designator: **white noise** (from optics, where white light contains all frequencies or colors). The terms *Johnson, thermal,* and *white noise* may be used interchangeably. Johnson was able to show that the power of this generated noise is given by

$$P_n = kT\,\Delta f \tag{1-10}$$

where k = Boltzmann's constant (1.38×10^{-23} J/K)
T = resistor temperature in kelvin (K)
Δf = frequency bandwidth of the system being considered

Since this noise power is directly proportional to the bandwidth involved, it is advisable to limit a receiver to the smallest bandwidth possible. You may be wondering how the bandwidth figures into this. The noise is an ac voltage that has random instantaneous amplitude but a predictable rms value. The frequency of this noise voltage is just as random as the voltage peaks. The more frequencies allowed

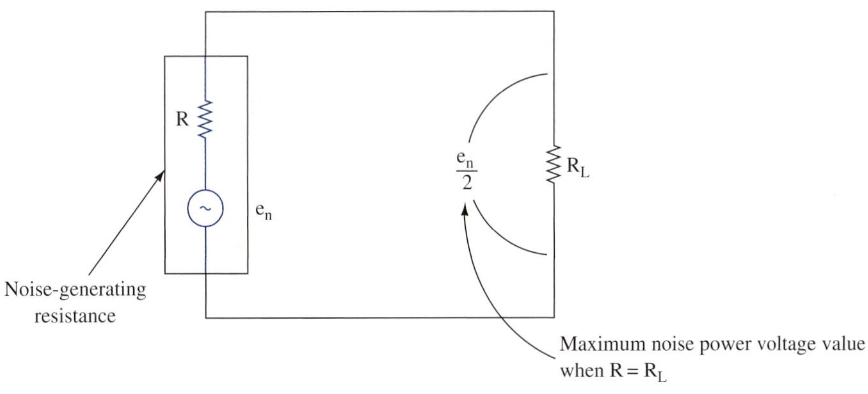

FIGURE 1-3 Resistance noise generator.

into the measurement (i.e., greater bandwidth), the greater the noise voltage. This means that the rms noise voltage measured across a resistor is a function of the bandwidth of frequencies included.

Since $P = E^2/R$, it is possible to rewrite Eq. (1-10) to determine the noise voltage (e_n) generated by a resistor. Assuming maximum power transfer of the noise source, the noise voltage is split between the load and itself, as shown in Fig. 1-3.

$$P_n = \frac{(e_n/2)^2}{R} = kT\,\Delta f$$

Therefore,

$$\frac{e_n^2}{4} = kT\,\Delta f R$$

$$e_n = \sqrt{4kT\Delta f R} \tag{1-11}$$

where e_n is the rms noise voltage and R is the resistance generating the noise. The instantaneous value of thermal noise is not predictable but has peak values generally less than 10 times the rms value from Eq. (1-11). The thermal noise associated with all nonresistor devices is a direct result of their inherent resistance and, to a much lesser extent, their composition. This applies to capacitors, inductors, and all electronic devices. Equation (1-11) applies to copper wire-wound resistors, with all other types exhibiting slightly greater noise voltages. Thus, dissimilar resistors of equal value exhibit different noise levels, which gives rise to the term **low-noise resistor;** you may have heard this term before but not understood it. Standard carbon resistors are the least expensive variety, but unfortunately they also tend to be the noisiest. Metal film resistors offer a good compromise in the cost/performance comparison and can be used in all but the most demanding low-noise designs. The ultimate noise performance (lowest noise generated, that is) is obtained with the most expensive and bulkiest variety: the wire-wound resistor. We use Eq. (1-11) as a reasonable approximation for all calculations in spite of these variations.

Low-Noise Resistor
a resistor that exhibits low levels of thermal noise

Example 1-4

Determine the noise voltage produced by a 1-MΩ resistor at room temperature (17°C) over a 1-MHz bandwidth.

Solution

It is helpful to know that $4kT$ at room temperature (17°C) is 1.60×10^{-20} Joules.

$$\begin{aligned} e_n &= \sqrt{4kT\,\Delta f R} \\ &= [(1.6 \times 10^{-20})(1 \times 10^{6})(1 \times 10^{6})]^{\frac{1}{2}} \\ &= (1.6 \times 10^{-8})^{\frac{1}{2}} \\ &= 126\ \mu\text{V rms} \end{aligned} \tag{1-11}$$

From the preceding example we can deduce that an ac voltmeter with an input resistance of 1 MΩ and a 1-MHz bandwidth generates 126 μV of noise (rms). Signals of about 500 μV or less would certainly not be measured with any accuracy. A 50-Ω resistor under the same conditions would generate only about 0.9 μV of noise. This explains why low impedances are desirable in low-noise circuits.

Example 1-5

An amplifier operating over a 4-MHz bandwidth has a 100-Ω source resistance. It is operating at 27°C, has a voltage gain of 200, and has an input signal of 5 μV rms. Determine the rms output signals (desired and noise), assuming external noise can be disregarded.

Solution

To convert °C to kelvin, simply add 273°, so that K = 27°C + 273° = 300 K. Therefore

$$\begin{aligned} e_n &= \sqrt{4kT\,\Delta f R} \\ &= \sqrt{4 \times 1.38 \times 10^{-23}\ \text{J/K} \times 300\ \text{K} \times 4\ \text{MHz} \times 100\ \Omega} \\ &= 2.57\ \mu\text{V rms} \end{aligned} \tag{1-11}$$

After multiplying the input signal e_s (5 μV) and noise signal by the voltage gain of 200, the output signal consists of a 1-mV rms signal and 0.514-mV rms noise. This is not normally an acceptable situation. The intelligence would probably be unintelligible!

Shot Noise
noise introduced by carriers in the *pn* junctions of semiconductors

Transistor Noise In Ex. 1-5, the noise introduced by the transistor, other than its thermal noise, was not considered. The major contributor of transistor noise is called **shot noise.** It is due to the discrete-particle nature of the current carriers in all forms of semiconductors. These current carriers, even under dc conditions, are not moving in an exactly steady continuous flow since the distance they travel varies due to random paths of motion. The name *shot noise* is derived from the fact that when amplified into a speaker, it sounds like a shower of lead shot falling on a metallic surface. Shot noise and thermal noise are additive. Unfortunately, there is no valid formula to calculate its value for a complete transistor where the sources of shot noise are the currents within the emitter–base and collector–base diodes. Hence, the device user must refer to the manufacturer's data sheet for an indication of shot noise characteristics. The methods of dealing with these data are covered in Sec. 1-5. Shot noise generally increases proportionally with dc bias currents except in MOSFETs, where shot noise seems to be relatively independent of dc current levels.

Frequency Noise Effects Two little-understood forms of device noise occur at the opposite extremes of frequency. The low-frequency effect is called **excess noise** and occurs at frequencies below about 1 kHz. It is inversely proportional to frequency and directly proportional to temperature and dc current levels. It is thought to be caused by crystal surface defects in semiconductors that vary at an inverse rate with frequency. Excess noise is often referred to as *flicker noise, pink noise,* or $1/f$ noise. It is present in both bipolar junction transistors (BJTs) and field-effect transistors (FETs).

Excess Noise
noise occurring at frequencies below 1 kHz, varying in amplitude inversely proportional to frequency

At high frequencies, device noise starts to increase rapidly in the vicinity of the device's high-frequency cutoff. When the transit time of carriers crossing a junction is comparable to the signal's period (i.e., high frequencies), some of the carriers may diffuse back to the source or emitter. This effect is termed **transit-time noise.** These high- and low-frequency effects are relatively unimportant in the design of receivers since the critical stages (the front end) will usually be working well above 1 kHz and hopefully below the device's high-frequency cutoff area. The low-frequency effects are important, however, to the design of low-level, low-frequency amplifiers encountered in certain instrument and biomedical applications.

Transit-Time Noise
noise produced in semiconductors when the transit time of the carriers crossing a junction is close to the signal's period and some of the carriers diffuse back to the source or emitter of the semiconductor

The overall noise intensity versus frequency curves for semiconductor devices (and tubes) have a bathtub shape, as represented in Fig. 1-4. At low frequencies the excess noise is dominant, while in the midrange shot noise and thermal noise predominate, and above that the high-frequency effects take over. Of course, tubes are now seldom used and fortunately their semiconductor replacements offer better noise characteristics. Since semiconductors possess inherent resistances, they generate thermal noise in addition to shot noise, as indicated in Fig. 1-4. The noise characteristics provided in manufacturers' data sheets take into account both the shot and thermal effects. At the device's high-frequency cutoff, f_{hc}, the high-frequency effects take over, and the noise increases rapidly.

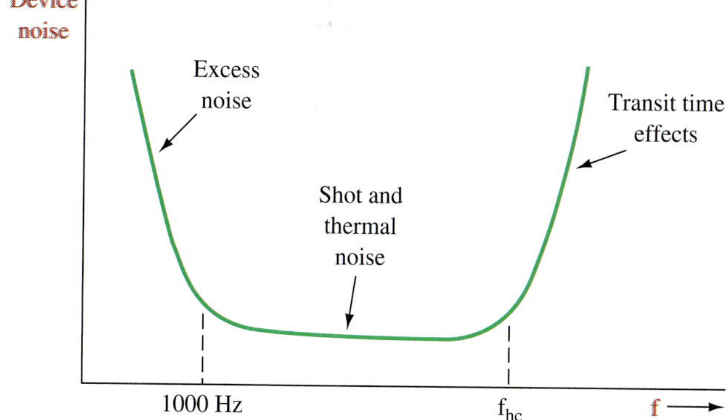

FIGURE 1-4 Device noise versus frequency.

1-4 NOISE DESIGNATION AND CALCULATION

SIGNAL-TO-NOISE RATIO

Signal-to-Noise Ratio
relative measure of desired signal power to noise power

We have thus far dealt with different types of noise without showing how to deal with noise in a practical way. The most fundamental relationship used is known as the **signal-to-noise ratio** (*S/N* ratio), which is a relative measure of the desired signal power to the noise power. The *S/N* ratio is often designated simply as *S/N* and can be expressed mathematically as

$$\frac{S}{N} = \frac{\text{signal power}}{\text{noise power}} = \frac{P_S}{P_N} \tag{1-12}$$

at any particular point in an amplifier. It is often expressed in decibel form as

$$\frac{S}{N} = 10 \log_{10} \frac{P_S}{P_N} \tag{1-13}$$

For example, the output of the amplifier in Ex. 1-5 was 1 mV rms and the noise was 0.514 mV rms, and thus (remembering that $P = E^2/R$)

$$\frac{S}{N} = \frac{1^2/R}{0.514^2/R} = 3.79 \quad \text{or} \quad 10 \log_{10} 3.79 = 5.78 \text{ dB}$$

NOISE FIGURE

Noise Figure
a figure describing how noisy a device is in decibels

S/N successfully identifies the noise content at a specific point but is not useful in relating how much additional noise a particular transistor has injected into a signal going from input to output. The term **noise figure** (NF) is usually used to specify exactly how noisy a device is. It is defined as follows:

$$\text{NF} = 10 \log_{10} \frac{S_i/N_i}{S_o/N_o} = 10 \log_{10} \text{NR} \tag{1-14}$$

Noise Ratio
a figure describing how noisy a device is as a ratio having no units

where S_i/N_i is the signal-to-noise power ratio at the device's input and S_o/N_o is the signal-to-noise power ratio at its output. The term $(S_i/N_i)/(S_o/N_o)$ is called the **noise ratio** (NR). If the device under consideration were ideal (injected no additional noise), then S_i/N_i and S_o/N_o would be equal, the NR would equal 1, and NF = 10 log 1 = 10 × 0 = 0 dB. Of course, this result cannot be obtained in practice.

EXAMPLE 1-6

A transistor amplifier has a measured S/N power of 10 at its input and 5 at its output.
(a) Calculate the NR.
(b) Calculate the NF.
(c) Using the results of part (a), verify that Eq. (1-14) can be rewritten mathematically as

$$\text{NF} = 10 \log_{10} \frac{S_i}{N_i} - 10 \log_{10} \frac{S_o}{N_o}$$

Solution

(a)
$$NR = \frac{S_i/N_i}{S_o/N_o} = \frac{10}{5} = 2$$

(b)
$$NF = 10 \log_{10} \frac{S_i/N_i}{S_o/N_o} = 10 \log_{10} NR \quad \textbf{(1-14)}$$

$$= 10 \log_{10} \frac{10}{5} = 10 \log_{10} 2$$

$$= 3 \text{ dB}$$

(c)
$$10 \log \frac{S_i}{N_i} = 10 \log_{10} 10 = 10 \times 1 = 10 \text{ dB}$$

$$10 \log \frac{S_o}{N_o} = 10 \log_{10} 5 = 10 \times 0.7 = 7 \text{ dB}$$

Their difference (10 dB − 7 dB) is equal to the result of 3 dB determined in part (b).

The result of Ex. 1-6 is a typical transistor NF. However, for low-noise requirements, devices with NFs down to less than 1 dB are available at a price premium. The graph in Fig. 1-5 shows the manufacturer's NF versus frequency characteristics for the 2N4957 transistor. As you can see, the curve is flat in the midfrequency range (NF ≈ 2.2 dB) and has a slope of −3 dB/octave at low frequencies (excess noise) and 6 dB/octave in the high-frequency area (transit-time noise). An **octave** is a range of frequency in which the upper frequency is double the lower frequency.

Octave
range of frequency in which the upper frequency is double the lower frequency

Manufacturers of low-noise devices usually supply a whole host of curves to exhibit their noise characteristics under as many varied conditions as possible. One of the more interesting curves provided for the 2N4957 transistor is shown in Fig. 1-6. It provides a visualization of the contours of NF versus source resistance and dc collector current for a 2N4957 transistor at 105 MHz. It indicates that noise operation at 105 MHz will be optimum when a dc (bias) collector current of about 0.7 mA

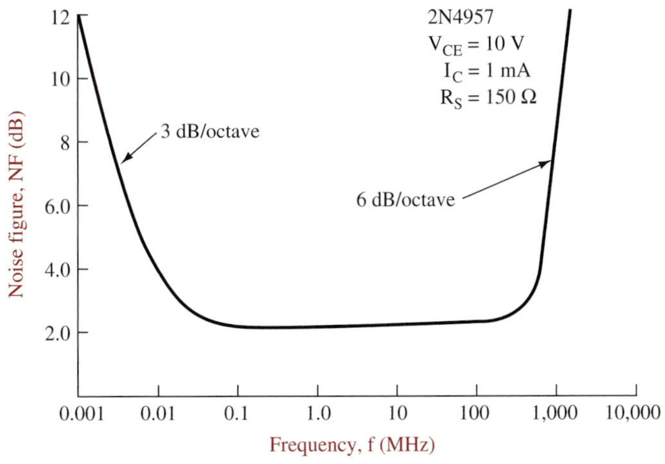

FIGURE 1-5 NF versus frequency for a 2N4957 transistor. (Courtesy of Motorola Semiconductor Products, Inc.)

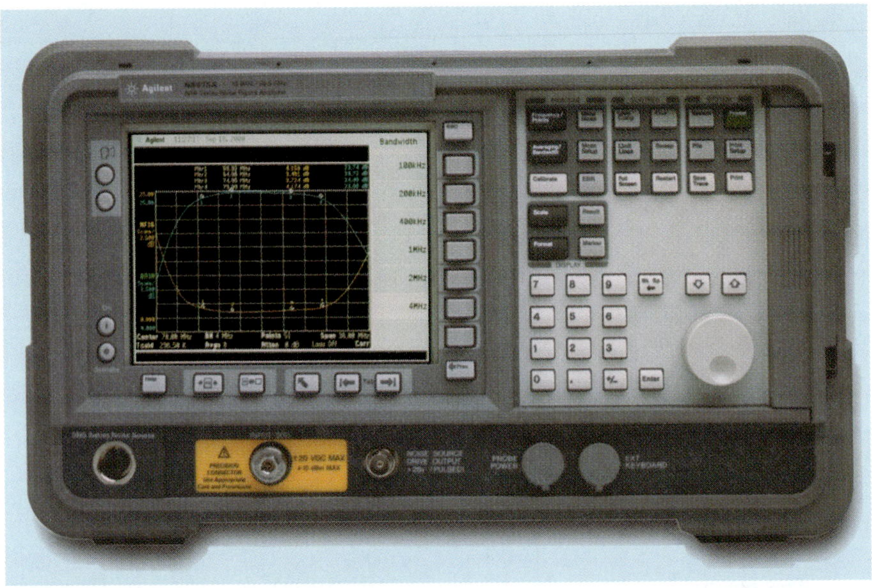

The Agilent N89752 is a high-performance noise figure analyzer designed to make fast, accurate, and repeatable noise figure measurements. (Courtesy of Agilent Technologies. Reprinted with permission.)

and source resistance of 350 Ω is utilized because the lowest NF of 1.8 dB occurs under these conditions.

The current state of the art for low-noise transistors offers some surprisingly low numbers. The leading edge for room temperature designs at 4 GHz is an NF of about 0.5 dB using gallium arsenide (GaAs) FETs. At 144 MHz, amplifiers with NFs down to 0.3 dB are being employed. The ultimate in low-noise-amplifier (LNA) design utilizes cryogenically cooled circuits (using liquid helium). Noise figures down to about 0.2 dB at microwave frequencies up to about 10 GHz are thereby made possible.

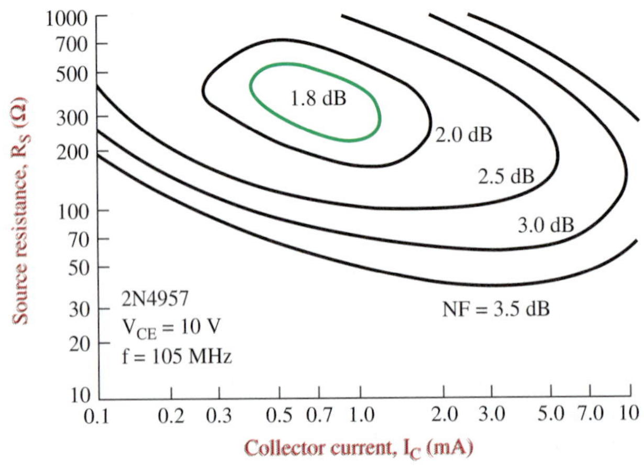

FIGURE 1-6 Noise contours for a 2N4957 transistor. (Courtesy of Motorola Semiconductor Products, Inc.)

Reactance Noise Effects

In theory a reactance does not introduce noise to a system. This is true for ideal capacitors and inductors that contain no resistive component. The ideal cannot be attained, but fortunately resistive elements in capacitors and inductors usually have a negligible effect on system noise considerations compared to semiconductors and other resistances.

The significant effect of reactive circuits on noise is its limitation on frequency response. Our previous discussions on noise have assumed an ideal bandwidth that is rectangular in response. Thus, the 10-kHz bandwidth of Ex. 1-7 implied a total passage within the 10-kHz range and zero effect outside. In practice, *RC*-, *LC*-, and *RLC*-generated passbands are not rectangular but slope off gradually, with the bandwidth defined as a function of half-power frequencies. This is detailed in Sec. 1-6. The equivalent bandwidth (Δf_{eq}) to be used in noise calculations with reactive circuits is given by

$$\Delta f_{eq} = \frac{\pi}{2} \text{BW} \qquad \textbf{(1-15)}$$

where BW is the 3-dB bandwidth as shown in Sec. 1-7 for *RC*, *LC*, or *RLC* circuits. The fact that the "noise" bandwidth is greater than the "3-dB" bandwidth is not surprising. Significant noise is still being passed through a system beyond the 3-dB cutoff frequency.

Noise Due to Amplifiers in Cascade

We previously specified that the first stage of a system is dominant with regard to noise effect. We are now going to show that effect numerically. **Friiss's formula** is used to provide the overall noise effect of a multistage system.

Friiss's Formula
method of determining the total noise produced by amplifier stages in cascade

$$\text{NR} = \text{NR}_1 + \frac{\text{NR}_2 - 1}{P_{G_1}} + \cdots + \frac{\text{NR}_n - 1}{P_{G_1} \times P_{G_2} \times \cdots \times P_{G(n-1)}} \qquad \textbf{(1-16)}$$

where NR = overall noise ratio of *n* stages
P_G = power gain ratio

Example 1-7

A three-stage amplifier system has a 3-dB bandwidth of 200 kHz determined by an LC-tuned circuit at its input, and operates at 22°C. The first stage has a power gain of 14 dB and an NF of 3 dB. The second and third stages are identical, with power gains of 20 dB and NF = 8 dB. The output load is 300 Ω. The input noise is generated by a 10-kΩ resistor. Calculate

(a) the noise voltage and power at the input and the output of this system assuming ideal noiseless amplifiers.

(b) the overall noise figure for the system.

(c) the actual output noise voltage and power.

Solution

(a) The effective noise bandwidth is

$$\Delta f_{eq} = \frac{\pi}{2}\,\text{BW} \qquad \textbf{(1-15)}$$
$$= \frac{\pi}{2} \times 200 \text{ kHz}$$
$$= 3.14 \times 10^5 \text{ Hz}$$

Thus, at the input,

$$P_n = kT\,\Delta f$$
$$= 1.38 \times 10^{-23} \text{ J/K} \times (273 + 22) \text{ K} \times 3.14 \times 10^5 \text{ Hz} \qquad \textbf{(1-10)}$$
$$= 1.28 \times 10^{-15} \text{ W}$$

and

$$e_n = \sqrt{4kT\ \Delta f R} \qquad \textbf{(1-11)}$$
$$= \sqrt{4 \times 1.28 \times 10^{-15} \times 10 \times 10^3}$$
$$= 7.15\ \mu\text{V}$$

The total power gain is 14 dB + 20 dB + 20 dB = 54 dB.

$$54 \text{ dB} = 10 \log P_G$$

Therefore,

$$P_G = 2.51 \times 10^5$$

Assuming perfect noiseless amplifiers,

$$P_{n\text{ out}} = P_{n\text{ in}} \times P_G$$
$$= 1.28 \times 10^{-15} \text{ W} \times 2.51 \times 10^5$$
$$= 3.22 \times 10^{-10} \text{ W}$$

Remembering that the output is driven into a 300-Ω load and $P = V^2/R$, we have

$$3.22 \times 10^{-10} \text{ W} = \frac{(e_{n\text{ out}})^2}{300\ \Omega}$$
$$e_n = 0.311 \text{ mV}$$

Notice that the noise has gone from microvolts to millivolts without considering the noise injected by each amplifier stage.

(b) Recall that to use Friiss's formula, ratios and not decibels must be used. Thus,

$$P_{G_1} = 14 \text{ dB} = 25.1$$
$$P_{G_2} = P_{G_3} = 20 \text{ dB} = 100$$
$$\text{NF}_1 = 3 \text{ dB} \qquad \text{NR}_1 = 2$$
$$\text{NF}_2 = \text{NF}_3 = 8 \text{ dB} \qquad \text{NR}_2 = \text{NR}_3 = 6.31$$

$$\text{NR} = \text{NR}_1 + \frac{\text{NR}_2 - 1}{P_{G_1}} + \cdots + \frac{\text{NR}_n - 1}{P_{G_1} P_{G_2} \cdots P_{G(n-1)}} \qquad \textbf{(1-16)}$$
$$= 2 + \frac{6.31 - 1}{25.1} + \frac{6.31 - 1}{25.1 \times 100}$$
$$= 2 + 0.21 + 0.002 = 2.212$$

Thus, the overall noise ratio (2.212) converts into an overall noise figure of 10 $\log_{10}$ 2.212 = 3.45 dB:

$$\text{NF} = 3.45 \text{ dB}$$

(c)

$$\text{NR} = \frac{S_i/N_i}{S_o/N_o}$$

$$P_G = \frac{S_o}{S_i} = 2.51 \times 10^5$$

Therefore,

$$\text{NR} = \frac{N_o}{N_i \times 2.51 \times 10^5}$$

$$2.212 = \frac{N_o}{1.28 \times 10^{-15}\,\text{W} \times 2.51 \times 10^5}$$

$$N_o = 7.11 \times 10^{-10}\,\text{W}$$

To get the output noise voltage, since $P = V^2/R$,

$$7.11 \times 10^{-10}\,\text{W} = \frac{e_n^2}{300\,\Omega}$$

$$e_n = 0.462 \text{ mV}$$

Notice that the actual noise voltage (0.462 mV) is about 50% greater than the noise voltage when we did not consider the noise effects of the amplifier stages (0.311 mV).

Equivalent Noise Temperature

Another way of representing noise is by equivalent noise temperature. It is a convenient means of handling noise calculations involved with microwave receivers (1 GHz and above) and their associated antenna system, especially space communication systems. It allows easy calculation of noise power at the receiver using Eq. (1-2) since the equivalent noise temperature (T_{eq}) of microwave antennas and their coupling networks are then simply additive.

The T_{eq} of a receiver is related to its noise ratio, NR, by

$$T_{\text{eq}} = T_0(\text{NR} - 1) \tag{1-17}$$

where T_0 = 290 K, a reference temperature in kelvin. The use of noise temperature is convenient since microwave antenna and receiver manufacturers usually provide T_{eq} information for their equipment. Additionally, for low noise levels, noise temperature shows greater variation of noise changes than does NF, making the difference easier to comprehend. For example, an NF of 1 dB corresponds to a T_{eq} of 75 K, while 1.6 dB corresponds to 129 K. Verify these comparisons using Eq. (1-17), remembering first to convert NF to NR. Keep in mind that noise temperature is not an actual temperature but is employed because of its convenience.

Example 1-8

A satellite receiving system includes a dish antenna (T_{eq} = 35 K) connected via a coupling network (T_{eq} = 40 K) to a microwave receiver (T_{eq} = 52 K referred to its input). What is the noise power to the receiver's input over a 1-MHz frequency range? Determine the receiver's NF.

Solution

$$P_n = kT\,\Delta f \qquad \textbf{(1-10)}$$
$$= 1.38 \times 10^{-23}\,\text{J/K} \times (35 + 40 + 52)\,\text{K} \times 1\,\text{MHz}$$
$$= 1.75 \times 10^{-15}\,\text{W}$$

$$T_{eq} = T_0(\text{NR} - 1) \qquad \textbf{(1-17)}$$
$$52\,\text{K} = 290\,\text{K}(\text{NR} - 1)$$
$$\text{NR} = \frac{52}{290} + 1$$
$$= 1.18$$

Therefore, NF = 10 $\log_{10}$ (1.18) = 0.716 dB.

Equivalent Noise Resistance

Manufacturers sometimes represent the noise generated by a device with a fictitious resistance termed the equivalent noise resistance (R_{eq}). It is the resistance that generates the same amount of noise predicted by $\sqrt{4kT\Delta fR}$ as the device does. The device (or complete amplifier) is then assumed to be noiseless in making subsequent noise calculations. The latest trends in noise analysis have shifted away from the use of equivalent noise resistance in favor of using the noise figure or noise temperatures.

SINAD

When the effects of noise and distortion on an amplifier or receiver are of interest, a specification called SINAD is used. Distortion introduced by a receiver is not random like noise but its effect on the intelligibility in the output is similar. For this reason, many radio receivers are rated using SINAD. This is especially true for FM receivers.

$$\text{SINAD} = 10 \log \frac{S + N + D}{N + D} \qquad \textbf{(1-18)}$$

where S = signal power out
N = noise power out
D = distortion power out

When measuring SINAD, an RF signal modulated by a 400-Hz or 1-kHz audio signal is usually applied to the receiver. The receiver output power is measured to give $S + N + D$. Then a highly selective filter is used to eliminate the 400-Hz or 1-kHz audio output. This leaves just the $N + D$ output, which is measured. SINAD can then be calculated using Eq. (1-18).

Example 1-9

A receiver is being tested to determine SINAD. A 400-Hz audio signal modulates a carrier that is applied to the receiver. Under these conditions, the output power is 7 mW. Next a filter is used to cancel the 400-Hz portion of the output, and then an output power of 0.18 mW is measured. Calculate SINAD.

Solution

$$S + N + D = 7\text{ mW}$$

$$N + D = 0.18\text{ mW}$$

$$\text{SINAD} = 10\log\frac{S + N + D}{N + D} \qquad \textbf{(1-18)}$$

$$= 10\log\frac{7\text{ mW}}{0.18\text{ mW}}$$

$$= 15.9\text{ dB}$$

1-5 Noise Measurement

Noise measurement has become a very sophisticated process. Specialty noise-measuring instruments that offer many computer-controlled functions are available for thousands of dollars. If you become involved with a large number of measurements, you will become familiar with some of these instruments. In this section we look at some general methods of noise measurement that can be accomplished with relatively standard laboratory instrumentation. A simple and reliable method of noise measurement is the case where the signal is equal to the noise. At some convenient point in the system, a power meter is connected and a reading taken of the noise with no signal input. Then an input signal is raised in power level until the monitored power rises by 3 dB (i.e., doubled). At this point the power level of the signal source is noted. This is equal to the effective input noise level of the system.

Noise Diode Generator

Another noise measurement technique involves using a diode to generate a known amount of noise. In this technique the output impedance of the diode noise generator circuit is matched into the amplifier under test. In these types of measurements, the amplifier is commonly called the **device under test,** or simply *DUT*. The procedure is first to measure the noise power output of the DUT when the dc current to the noise diode is zero. The dc current is then increased until the DUT noise power output is exactly doubled from the original value. The diode dc current is then used in the following equation to determine the noise ratio of the DUT:

Device Under Test
an electronic part or system that is being tested

$$\text{NR} = 20I_{\text{dc}}R \qquad \textbf{(1-19)}$$

where R is the input impedance of the DUT and the temperature is 290 K (approximately room temperature). The reader is referred to "Semiconductor Noise Figure Considerations," Application Note AN-421 from Motorola Semiconductor Products, Inc. for a derivation of this surprisingly simple and useful relationship.

Example 1-10

An amplifier has an impedance of 50 Ω. Using a matched-impedance diode noise generator, it is found that the DUT has doubled noise output power when the diode has a dc current of 14 mA. Determine the NR and NF for the DUT.

Solution

$$\begin{aligned} \text{NR} &= 20 I_{dc} R \qquad \textbf{(1-19)} \\ &= 20 \times 14 \text{ mA} \times 50 \ \Omega \\ &= 14 \end{aligned}$$

$$\begin{aligned} \text{NF} &= 10 \log_{10} \text{NR} \qquad \textbf{(1-14)} \\ &= 10 \log_{10} 14 \\ &= 11.46 \text{ dB} \end{aligned}$$

Notice that in Ex. 1-10, the NR is numerically equal to the diode's current in mA. This occurs when the DUT has an impedance of 50 Ω—a most convenient situation since many RF amplifier systems are designed with a 50-Ω impedance. Keep in mind that NR is a dimensionless ratio, however, and not measured in mA.

Tangential Noise Measurement Technique

Meters capable of accurately measuring the very low levels involved with noise measurements tend to be expensive and of limited use with regard to other applications. A dual-trace oscilloscope with high sensitivity is an exception to this limitation. Unfortunately, a direct noise reading from the scope results in errors for two reasons:

1. Noise is of a highly random nature and is not sinusoidal. Since rms values are required for noise calculations, the conversion from scope peak-to-peak values by dividing by $2\sqrt{2}$ is not accurate.
2. Since the noise peaks are random, their visibility on the scope is influenced by factors such as the scope's intensity setting, the persistence of the CRT's phosphor, and the length of the observation.

The two displays shown in Fig. 1-7 show exactly the same noise signal at two different intensity settings. The measurement can be erroneous by as much as 6 dB. A specially developed technique, known as the **tangential method,** reduces the possible error to less than 1 dB. The noise signal is connected to both channels of a dual-

Tangential Method
method of measuring the amplitude of noise on a signal using an oscilloscope display

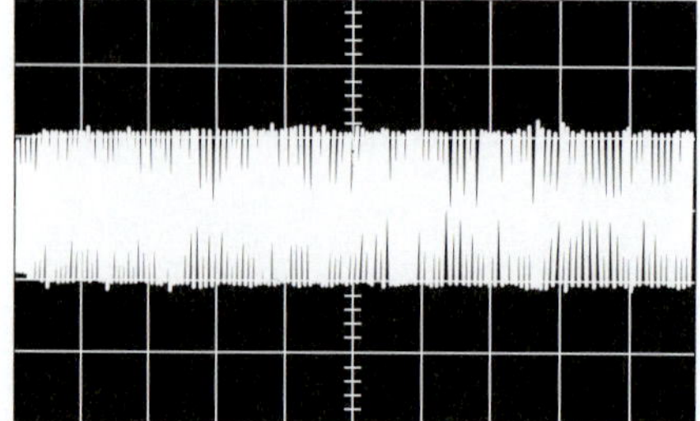
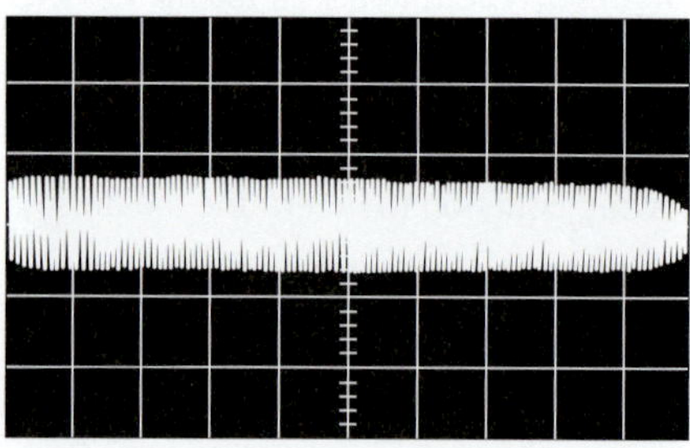

FIGURE 1-7 Scope display of the same noise signal at two different intensity settings. (Courtesy of Electronic Design.)

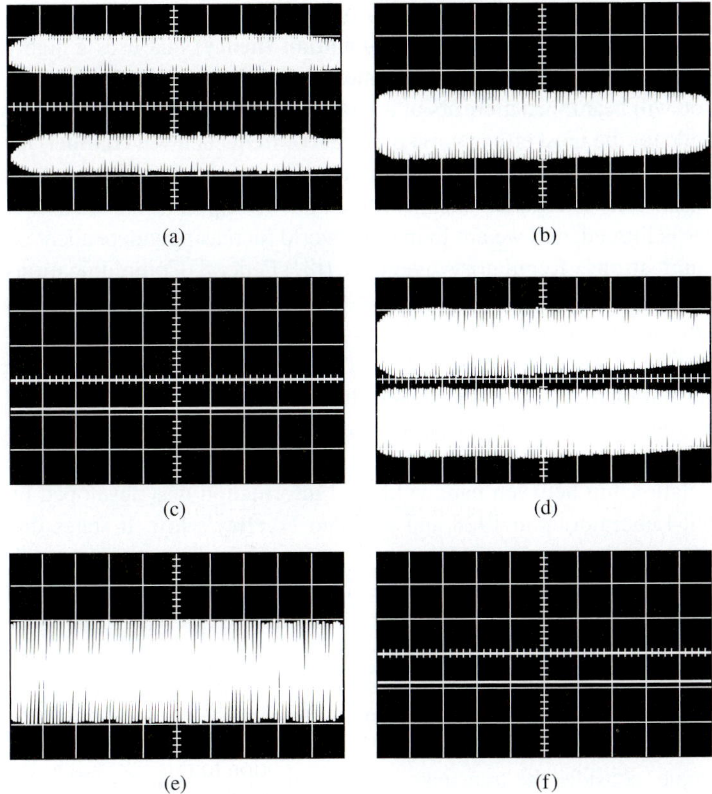

FIGURE 1-8 (a) With the tangential method, the noise signal is connected to both channels of a dual-channel scope used in the alternate-sweep mode. (b) The offset voltage is adjusted until the traces just merge. (c) The noise signal is then removed. The difference in the noise-free traces is twice the rms noise voltage. (d, e, f) This is repeated at a different intensity to show that the method is independent of intensity. Scope settings are: horizontal = 500 ms/cm, vertical = 20 mV/cm. (Courtesy of Electronic Design.)

trace scope with alternate sweep capability. As shown in Fig. 1-8(a), the two displayed signals are set up with both channels identically calibrated. Then their vertical position is adjusted until the dark band between them just disappears [Fig. 1-8(b)]. Now the noise signal input to both channels is removed, and the resulting separation represents twice the rms noise. In this case (with a vertical sensitivity of 20 mV/cm), the rms noise is 0.8 cm × 20 mV/cm ÷ 2, or 8 mV rms. Repeating this process with a different scope intensity setting [Figs. 1-8(d), (e), and (f)] yields the same result.

1-6 INFORMATION AND BANDWIDTH

In Sec. 1-1 it was mentioned that there are two basic limitations on the performance of a communications system. By now you should have a good grasp on the noise limitation. Quite simply, if the noise level becomes too high, the information is lost. The other limitation is the bandwidth utilized by the communications system. Stated simply once again, the greater the bandwidth, the greater the information that can be

Information Theory
concerned with optimization of transmitted information

Channel
a band of frequencies

transferred from source to destination. The study of information in communications systems is a science in itself (given the title **information theory**) that uses a highly theoretical method of analysis. It is beyond our intentions here, but if you pursue advanced studies, you will hear much more about it. Information theory is the study of information to provide for the most efficient use of a band of frequencies (a **channel**) for electrical communications. Additional information theory is provided in Chapter 9.

You might ask: Why is efficient channel utilization so important? The band of usable frequencies is limited, and we are living in a world increasingly dependent on electrical communications. Regulatory agencies (the Federal Communications Commission [FCC] in the United States) allocate the channel that may be used for a given application in a given area. This is done to minimize interference possibilities that will exist with two different signals working at the same frequency. The information explosion of recent years has taxed the total available frequency spectrum to the point where getting the most information from the smallest range of frequencies is in fact quite important.

Hartley's Law
information that can be transmitted is proportional to the product of the bandwidth times the time of transmission

A formal relationship between bandwidth and information was developed by R. Hartley of Bell Laboratories in 1928 and is called **Hartley's law.** It states that the information that can be transmitted is proportional to the product of the bandwidth utilized times the time of transmission. In simpler terms it means the greater the bandwidth, the more information that can be transmitted. Expressed as an equation, Hartley's law is

$$\text{information} \propto \text{bandwidth} \times \text{time of transmission} \quad \textbf{(1-20)}$$

As an example, consider the transmission of a musical performance. The full amount of information available to the human ear is contained in the range of frequencies from just above 0 Hz up to about 15 kHz. The allocated bandwidth of standard AM stations is about 30 kHz. On the other hand, FM stations are allocated a larger bandwidth (200 kHz), which allows the full amount of information (up to 15 kHz) to be reproduced at the receiver. This helps explain the better fidelity available with FM as compared to AM in our two basic commercial radio bands.* This is an example of greater bandwidth allowing a greater information capability and substantiates Hartley's law.

Understanding the Frequency Spectra

As stated by Hartley's law, the bandwidths of communication systems impose limitations on their information capacity. For example, the AM band has inherent limitations on its information capacity due to limited bandwidth. While the AM band may be suitable for audio transmission, transmission of a television system over the bandwidth allocated for AM transmission would hardly be acceptable. The United States allocates a 6-MHz bandwidth per channel for analog television transmission. Obviously, TV must require a great deal more information capacity than AM radio (AM radio bandwidth = 30 kHz). Television transmission uses a bandwidth 200 times that used for an AM radio-band transmission. This significant increase in bandwidth requirement is primarily due to the complexity of the video signal. The video signal contains many high-frequency components, including the color subcarrier (a sinusoid) and the luminance (black/white information), which contains many pulse-type wave-

* This example has been oversimplified for reasons that will become obvious as you study AM and FM in subsequent chapters. Its conclusion remains valid, however.

forms. It will be shown that a pulse-type waveform requires a much larger bandwidth for transmission than a sinusoid at the same frequency.

A method of analyzing complex repetitive waveforms is known as **Fourier analysis.** It permits any complex repetitive waveform to be resolved into a series of sine or cosine waves (possibly infinite in number for an ideal system with infinite bandwidth) and possibly a dc component (when necessary). The mathematical tool provided in Fourier analysis helps one to understand the meaning of harmonics and the complex waves of which they are a part and also to obtain insight into factors relating to distortion effects.

Fourier Analysis
method of representing complex repetitive waveforms by sinusoidal components

The expressions for selected periodic waveforms are provided in Table 1-4 on page 30. The Fourier series for a square wave, shown in Table 1-4(c), is made up of a summation of sinusoids multiplied by a constant $4V/\pi$. Note that each consecutive sinusoid is increasing in frequency.

$$\sin \omega t + \frac{1}{3}\sin 3\omega t + \frac{1}{5}\sin 5\omega t + \cdots$$

The frequency sin ωt is called the fundamental frequency of the waveform. The component $\frac{1}{3}$ sin $3\omega t$ is called the third harmonic. Sin $5\omega t$ is considered the fifth harmonic, and so on, until the bandwidth of the system is reached. The $\frac{1}{3}$ and $\frac{1}{5}$ values simply indicate that the amplitude of each harmonic is decreasing as the frequency increases.

In can be shown with a math software package that the series expressions for complex repetitive waveforms are indeed constructed of a series of sinusoids consisting of its fundamental frequency and many harmonic frequencies. Figure 1-9 shows the construction of a complex repetitive waveform: a square wave. It consists of its fundamental frequency (or first harmonic) as well as a multiple of harmonic frequencies. Figure 1-9(a) shows the fundamental frequency. Figure 1-9(b) shows the addition of the first and third harmonics, and Figure 1-9(c) shows the addition of the first, third, and fifth harmonics.

Although somewhat distorted, Figure 1-9(c) is beginning to resemble a square wave. This required the addition of third and fifth harmonics to the fundamental frequency. With the addition of more harmonics, the wave rapidly approaches an ideal square wave. This is demonstrated in Figs. 1-10(a) and (b), where 13 and 51 harmonics are included. These figures show that the square wave is better defined as the bandwidth is increased, at the expense of additional bandwidth. Of course no

Voltage

Time

(a)

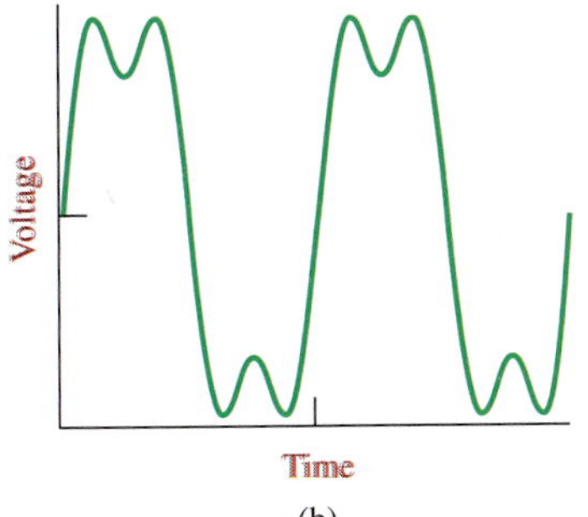

(b)

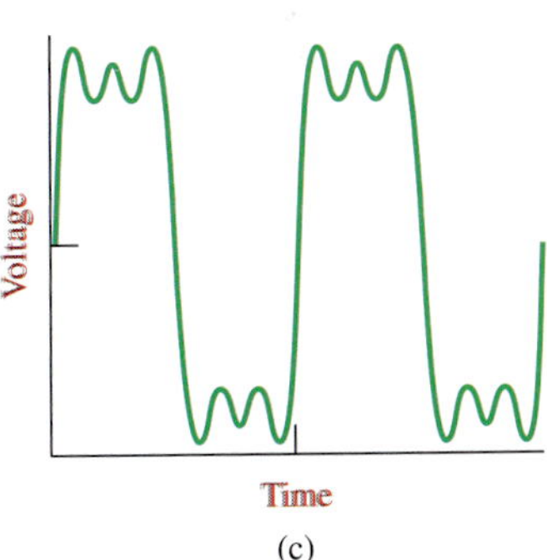

(c)

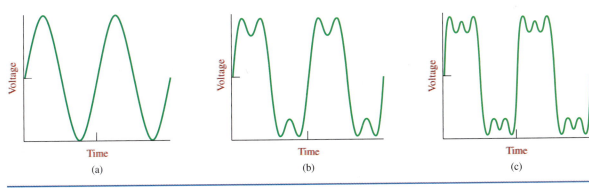

FIGURE 1-9 (a) Fundamental frequency (sin ωt); (b) the addition of the first and third harmonics (sinωt + $\frac{1}{3}$ sin $3\omega t$); (c) the addition of the first, third, and fifth harmonics (sin ωt + $\frac{1}{3}$ sin $3\omega t$ + $\frac{1}{5}$ sin $5\omega t$).

Table 1-4 Fourier Expressions for Selected Periodic Waveforms, $f = 1/T$, $2\pi f = \omega$

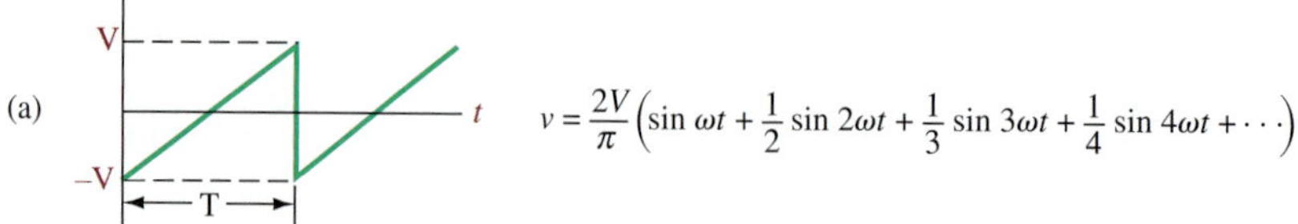

(a) $$v = \frac{2V}{\pi}\left(\sin \omega t + \frac{1}{2}\sin 2\omega t + \frac{1}{3}\sin 3\omega t + \frac{1}{4}\sin 4\omega t + \cdots\right)$$

(b) $$v = \frac{2V}{\pi}\left(\sin \omega t - \frac{1}{2}\sin 2\omega t + \frac{1}{3}\sin 3\omega t - \frac{1}{4}\sin 4\omega t + \cdots\right)$$

(c) $$v = \frac{4V}{\pi}\left(\sin \omega t + \frac{1}{3}\sin 3\omega t + \frac{1}{5}\sin 5\omega t + \cdots\right)$$

(d) $$v = V\frac{\tau}{T} + 2V\frac{\tau}{T}\left[\frac{\sin \pi(\tau/T)}{\pi\tau T}\cos \omega t + \frac{\sin 2\pi(\tau/T)}{2\pi(\tau/T)}\cos 2\omega t + \frac{\sin 3\pi(\tau/T)}{3\pi(\tau/T)}\cos 3\omega t + \cdots\right]$$

(e) $$v = \frac{8V}{\pi^2}\left[\cos \omega t + \frac{1}{(3)^2}\cos 3\omega t + \frac{1}{(5)^2}\cos 5\omega t + \cdots\right]$$

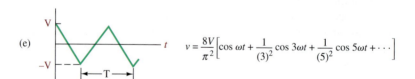

(f) $$v = \frac{2V}{\pi}\left[1 + \frac{2\cos 2\omega t}{3} - \frac{2\cos 4\omega t}{15} + \cdots (-1)^{n/2}\frac{2\cos n\omega t}{n^2-1}\cdots\right] (n \text{ even})$$

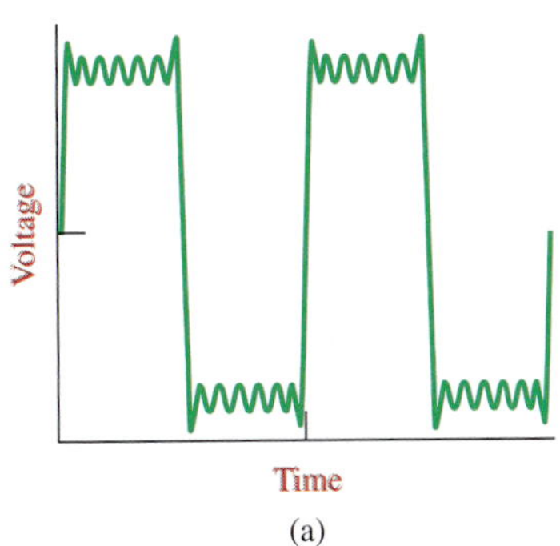

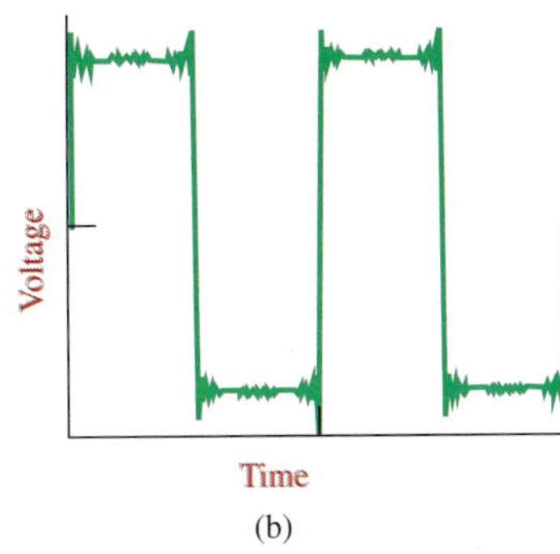

FIGURE 1-10 Square waves containing: (a) 13 harmonics; (b) 51 harmonics.

transmission media is ideal; therefore, it should be expected that some loss of signal will occur. This loss in information results in a square wave with edges that are not as sharp as the ideal.

The previous discussion demonstrates that the square wave consists of many harmonic frequencies, which underscores the importance of providing a communications system with sufficient bandwidth to pass the minimal required information.

The application of Fourier analysis when using oscilloscopes and spectrum analyzers is provided through the use of the fast Fourier transform **(FFT).** The FFT is a commonly used signal-processing technique that converts (transforms) time-varying signals to their frequency components. The FFT uses sampled (discrete) values to generate the frequency information. (See Chapter 8 for a discussion of PCM, converting an analog signal to its digital value.) The FFT algorithm (mathematical routine) then converts the sampled information into its frequency components.

FFT
a technique for converting time-varying information to its frequency component

Examples of obtaining the FFT for a 1-kHz sinusoid and a 1-kHz square wave are given next. It has already been shown that the square wave requires a significant number of harmonics (bandwidth) for it to be generated, whereas the sine wave contains only one frequency component.

A 1-kHz sinusoid was input into a Tektronix TDS 340 Digital Sampling Oscilloscope, which has the FFT math option. Figure 1-11(a) shows the 1-kHz sinusoid and the resulting FFT of the sinusoid. As shown in Fig. 1-11(a), the horizontal display for the FFT of the 1-kHz waveform has been set to 500 Hz per division, which is indicated as the *frequency step* (500 Hz/division). The spike of the FFT waveform represents the input sine-wave frequency and is two divisions from the start of the FFT **frequency domain record,** or at a frequency of 2×500 Hz, or 1 kHz, which is the frequency of the input sine wave (1 kHz). The start of the frequency domain record always begins at DC, or 0 Hz. The amplitude of the spike is expressed in dBV rms when a measurement is being made. The term dBV rms expresses the measured value relative to 1 V rms. In this case, no vertical value or scale is specified. The noisy information below the 1-kHz frequency spike is just that, noise.

Frequency Domain Record
data points generated by the time to frequency conversion using the FFT

The input signal is being sampled at a rate of 20 kS/s, or 20,000 samples per second. The minimum sample frequency must be at least twice the frequency being analyzed, or in this case 2×1 kHz, or 2 kS/s (2000 samples per second). The minimum sample frequency is called the Nyquist sampling rate. These concepts

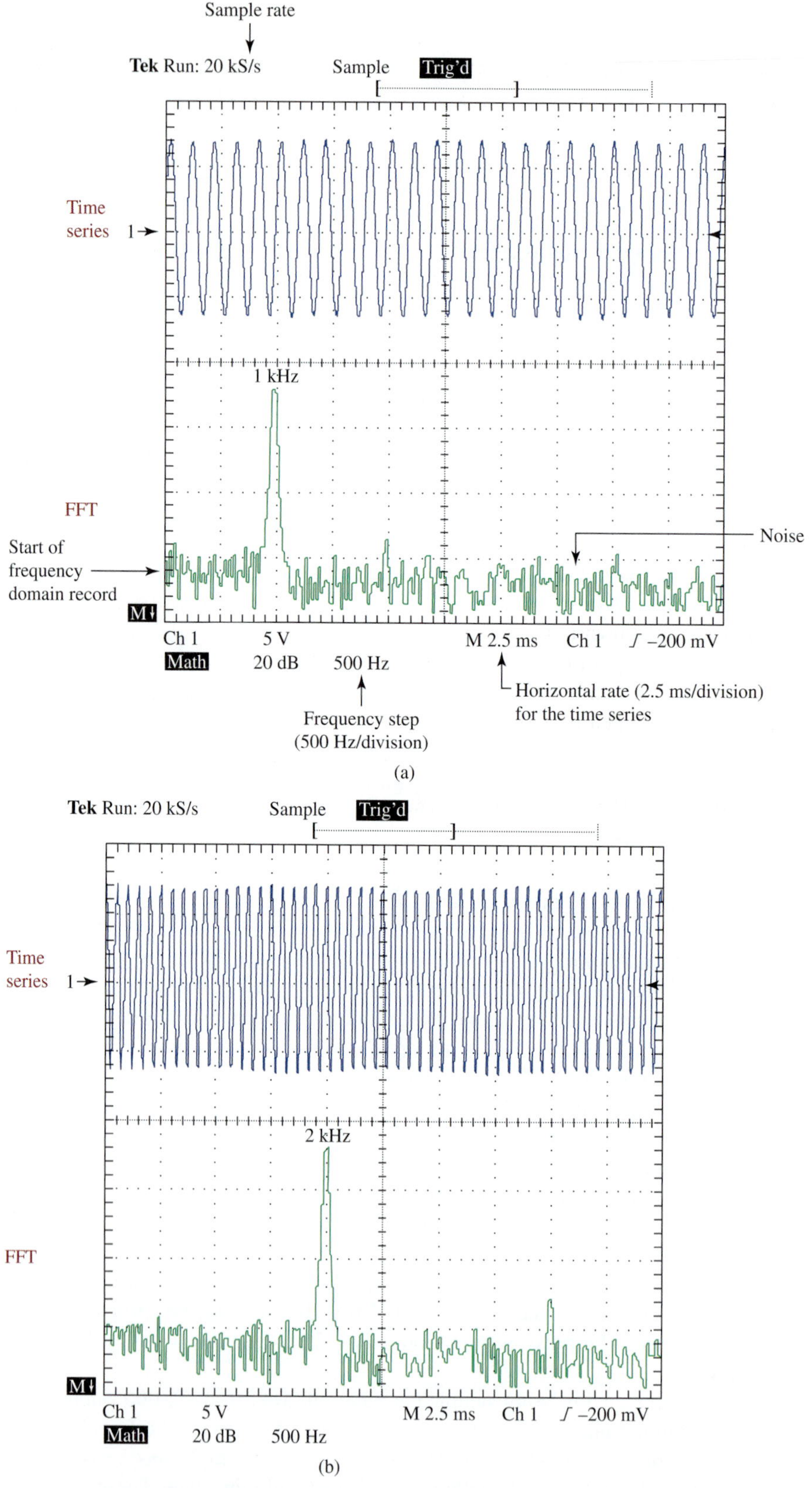

FIGURE 1-11 (a) A 1-kHz sinusoid and its FFT representation; (b) a 2-kHz sinusoid and its FFT representation.

are explained in detail in the pulse-code modulation section in Chapter 8. In this example, a sample rate of 20 kS/s will not introduce any errors. If an incorrect sample rate is selected, then **aliasing,** or undersampling, is created; the resulting signal waveform will be distorted and incorrect frequencies will be displayed. An example of an improperly selected sample rate and a distorted waveform is shown in Sec. 1-9.

Aliasing
errors that occur when the input frequency exceeds one-half the sample rate

Another example of reading the FFT information is shown in Fig. 1-11(b). In this case, a 2-kHz sinusoid was input into the oscilloscope. The horizontal display for the FFT is still set to 500 Hz per division. The frequency component displayed is 4 × 500 Hz, or 2 kHz, once again the frequency of the input sinusoid.

Next, a 1-kHz square has been input into the oscilloscope and the FFT option selected. The repetitive square wave is shown at the top of Fig. 1-12. The FFT of the square wave is shown at the bottom. The horizontal scale is 2.5 kHz per division. Notice that the waveform contains several frequency components. In fact, the first frequency component is at 1 kHz, the second is at 3 kHz, the third is at 5 kHz, and so on. This picture shows that the square wave contains the multiple harmonics (first, third, fifth, ...). This is the result predicted by Eq. (c) in Table 1-4, where it states that a square wave is constructed by a series of sine waves of increasing frequency and decreasing amplitude.

The square wave shown in Fig. 1-12 has nice sharp edges. The high-frequency components, as shown in the FFT, indicate the contribution of the higher harmonic values to generating a well-shaped square wave. What if the square wave were transmitted through a bandwidth-limited channel such as a telephone voice channel, which is band-limited to about 3 kHz? A 1-kHz square wave was fed through the

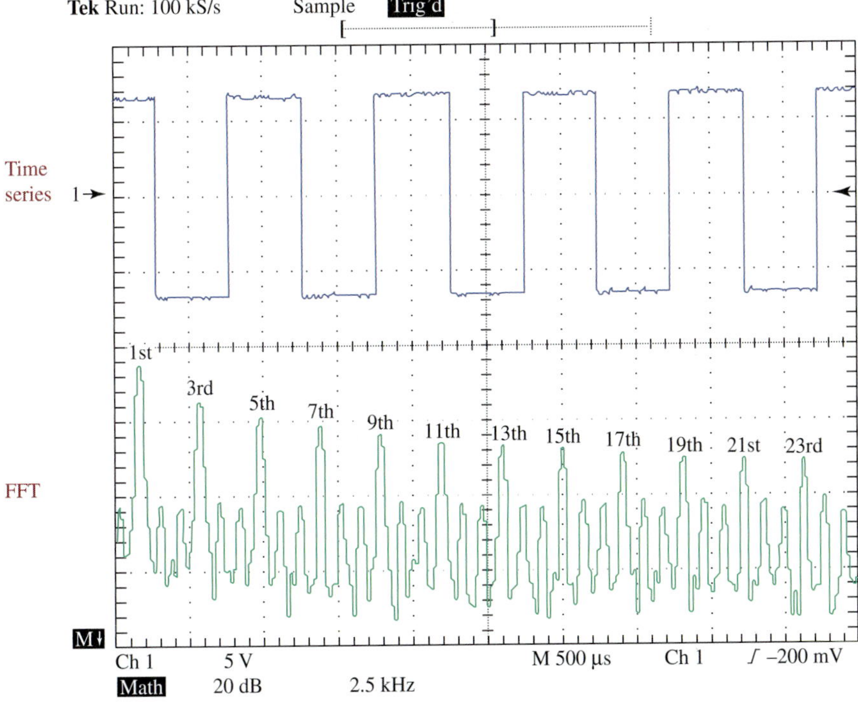

FIGURE 1-12 A 1-kHz square wave and its FFT representation.

low-pass filter shown in Fig. 1-13(a) to simulate transmission through a bandwidth-limited channel. Note the poor quality of the square wave, shown in Fig. 1-13(b). The FFT of the waveform shows that the higher harmonic values are severely attenuated, meaning that there is a loss of detail (information) through this bandwidth-limited system. These data can still be used in communications, but if the frequency information lost in the channel is needed for proper signal or data representation, then the loss of information may not be recoverable. For an analog system, this loss of information could result in noise and distortion. For a digital system, the loss of information could result in an increased bit error rate (BER). For the square wave, some of the features of the original signal (i.e., sharpness of the square wave) are lost. Signal processing may be required to regenerate the square wave.

This discussion introduced the fundamentals of frequency analysis in communication systems. It should be understood that Fourier analysis is more than just a useful mathematical equation representing a wave. The sine or cosine waves are physically real and measurable, as demonstrated by the FFT examples.

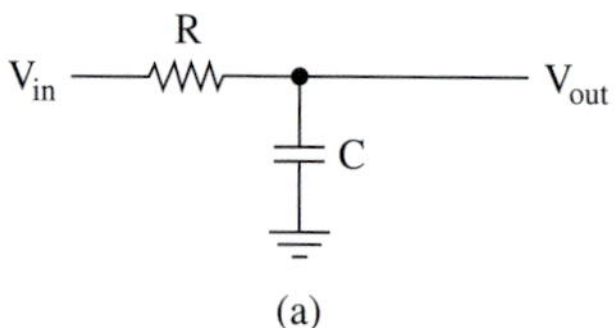

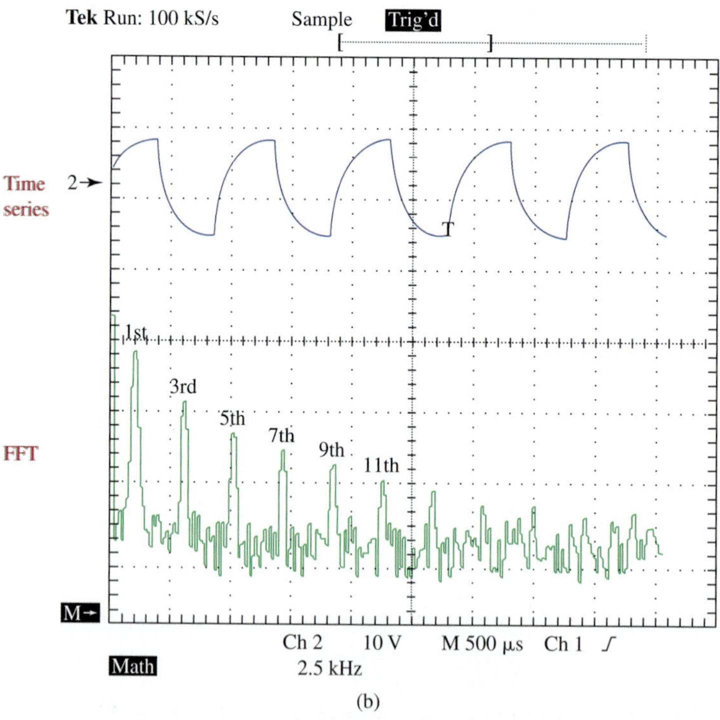

FIGURE 1-13 (a) A low-pass filter simulating a bandwidth-limited communications channel; (b) the resulting time series and FFT waveforms after passing through the low-pass filter.

1-7 *LC* Circuits

The remaining sections of this chapter cover some basic characteristics of *LC* circuits and oscillators. This material may be a review for you, but its importance to subsequent communication circuit study merits inclusion at this time.

Practical Inductors and Capacitors

Practical inductors (also referred to as *chokes* or *coils*) used at RF frequencies and above have an inductance rating in henries and a maximum current rating. Similarly, capacitors have a capacitance rating in farads and a maximum voltage rating. When selecting coils and capacitors for use at radio frequencies and above, an additional characteristic must be considered—the **quality** (Q) of the component. The Q is a ratio of the energy stored to that which is lost in the component.

Quality
ratio of energy stored to energy lost in a component

Inductors with axial leads or surface mount device (smd) capability. (Courtesy of API Delevan, Inc.)

Leakage
loss of electrical energy between the plates of a capacitor

Inductors store energy in the surrounding magnetic field and lose (dissipate) energy in their winding resistance. A capacitor stores energy in the electric field between its plates and primarily loses energy due to **leakage** between the plates.

For an inductor,

$$Q = \frac{\text{reactance}}{\text{resistance}} = \frac{\omega L}{R} \tag{1-21}$$

where R is the series resistance distributed along the coil winding. The required Q for a coil varies with circuit application. Values up to about 500 are generally available.

For a capacitor,

$$Q = \frac{\text{susceptance}}{\text{conductance}} = \frac{\omega C}{G} \tag{1-22}$$

where G is the value of conductance through the dielectric between the capacitor plates. Good-quality capacitors used in radio circuits have typical Q factors of 1000.

Dissipation
inverse of quality factor

At higher radio frequencies (VHF and above—see Table 1-1) the Q for inductors and capacitors is generally reduced by factors such as radiation, absorption, lead inductance, and package/mounting capacitance. Occasionally, an inverse term is used rather than Q. It is called the component **dissipation** (D) and is equal to $1/Q$. Thus $D = 1/Q$, a term used more often in reference to a capacitor.

Resonance

Resonance
balanced condition between the inductive and capacitive reactance of a circuit

Resonance can be defined as a circuit condition whereby the inductive and capacitive reactance have been balanced ($X_L = X_C$). Consider the series *RLC* circuit shown in Fig. 1-14. In this case, the total impedance, Z, is provided by the formula

$$Z = \sqrt{R^2 + (X_L - X_C)^2}$$

An interesting effect occurs at the frequency where X_L is equal to X_C. That frequency is termed the resonant frequency, f_r. At f_r the circuit impedance is equal to the resistor value (which might only be the series winding resistance of the inductor). That result can be shown from the equation above because when $X_L = X_C$, $X_L - X_C$ equals zero, so that $Z = \sqrt{R^2 + 0^2} = \sqrt{R^2} = R$. The resonant frequency can be determined by finding the frequency where $X_L = X_C$.

$$X_L = X_C$$

$$2\pi f_r L = \frac{1}{2\pi f_r C}$$

$$f_r^2 = \frac{1}{4\pi^2 LC}$$

$$f_r = \frac{1}{2\pi\sqrt{LC}} \quad \text{(Hz)} \tag{1-23}$$

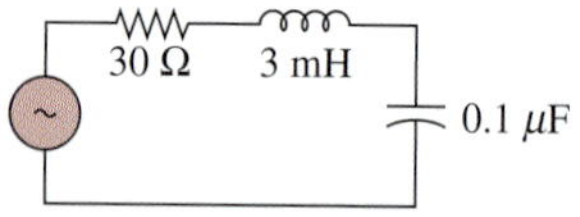

FIGURE 1-14 Series *RLC* circuit.

Example 1-11

Determine the resonant frequency for the circuit shown in Fig. 1-14. Calculate its impedance when $f = 12$ kHz.

Solution

$$f_r = \frac{1}{2\pi\sqrt{LC}} \tag{1-23}$$

$$= \frac{1}{2\pi\sqrt{3\text{ mH} \times 0.1\ \mu\text{F}}}$$

$$= 9.19\text{ kHz}$$

At 12 kHz,

$$X_L = 2\pi fL$$
$$= 2\pi \times 12\text{ kHz} \times 3\text{ mH}$$
$$= 226\ \Omega$$

$$X_C = \frac{1}{2\pi fC}$$
$$= \frac{1}{2\pi \times 12\text{ kHz} \times 0.1\ \mu\text{F}}$$
$$= 133\ \Omega$$

$$Z = \sqrt{R^2 + (X_L - X_C)^2}$$
$$= \sqrt{30^2 + (226 - 133)^2}$$
$$= 97.7\ \Omega$$

This circuit contains more inductive than capacitive reactance at 12 kHz and is therefore said to look inductive.

The impedance of the series *RLC* circuit is minimum at its resonant frequency and equal to the value of *R*. A graph of its impedance, *Z*, versus frequency has the shape of the curve shown in Fig. 1-15(a). At low frequencies the circuit's impedance is very high because X_C is high. At high frequencies X_L is very high and thus *Z* is high. At resonance, when $f = f_r$, the circuit's $Z = R$ and is at its minimum value. This impedance characteristic can provide a filter effect, as shown in Fig. 1-15(b). At f_r, $X_L = X_C$ and thus

$$e_{out} = e_{in} \times \frac{R_2}{R_1 + R_2}$$

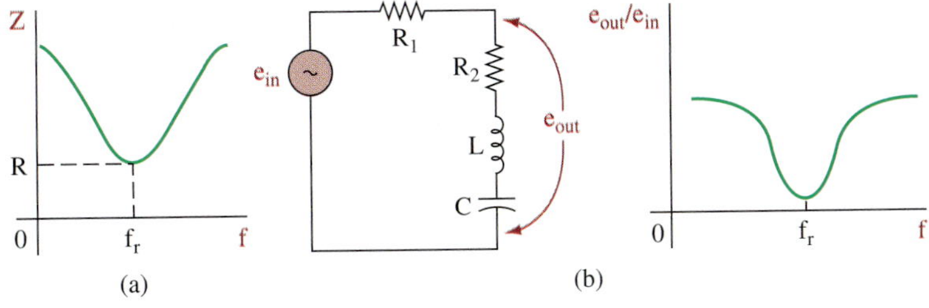

FIGURE 1-15 Series *RLC* circuit effects.

by the voltage-divider effect. At all other frequencies, the impedance of the *LC* combination goes up (from 0 at resonance) and thus e_{out} goes up. The response for the circuit in Fig. 1-15(b) is termed a band-reject, or notch, filter. A "band" of frequencies is being "rejected" and a "notch" is cut into the output at the resonant frequency, f_r.

Example 1-12 shows that the filter's output increases as the frequency is increased. Calculation of the circuit's output for frequencies below resonance would show a similar increase and is left as an exercise at the end of the chapter. The band-reject, or notch, filter is sometimes called a trap because it can "trap" or get rid of a specific range of frequencies near f_r. A trap is commonly used in a television receiver, where rejection of some specific frequencies is necessary for good picture quality.

Example 1-12

Determine f_r for the circuit shown in Fig. 1-15(b) when $R_1 = 20\ \Omega$, $R_2 = 1\ \Omega$, $L = 1$ mH, $C = 0.4\ \mu$F, and $e_{in} = 0$ mV. Calculate e_{out} at f_r and 12 kHz.

Solution

The resonant frequency is

$$f_r = \frac{1}{2\pi\sqrt{LC}} \tag{1-23}$$
$$= 7.96 \text{ kHz}$$

At resonance,

$$e_{\text{out}} = e_{\text{in}} \times \frac{R_2}{R_1 + R_2}$$
$$= 50 \text{ mV} \times \frac{1\ \Omega}{1\ \Omega + 20\ \Omega}$$
$$= 2.38 \text{ mV}$$

At $f = 12$ kHz,

$$X_L = 2\pi fL$$
$$= 2\pi \times 12 \text{ kHz} \times 1 \text{ mH}$$
$$= 75.4\ \Omega$$

and

$$X_C = \frac{1}{2\pi fC}$$
$$= \frac{1}{2\pi \times 12 \text{ kHz} \times 0.4\ \mu\text{F}}$$
$$= 33.2\ \Omega$$

Thus,

$$Z_{\text{total}} = \sqrt{(R_1 + R_2)^2 + (X_L - X_C)^2}$$
$$= \sqrt{(20\ \Omega + 1\ \Omega)^2 + (75.4\ \Omega - 33.2\ \Omega)^2}$$
$$= 47.1\ \Omega$$

and

$$Z_{out} = \sqrt{R_2^2 + (X_L - X_C)^2} = 42.2\ \Omega$$

$$e_{out} = 50\text{ mV} \times \frac{42.2\ \Omega}{47.1\ \Omega}$$

$$= 44.8\text{ mV}$$

LC Bandpass Filter

If the filter's configuration is changed to that shown in Fig. 1-16(a), it is called a bandpass filter and has a response as shown at Fig. 1-16(b). The term f_{lc} is the low-frequency cutoff where the output voltage has fallen to 0.707 times its maximum value and f_{hc} is the high-frequency cutoff. The frequency range between f_{lc} and f_{hc} is called the filter's bandwidth, usually abbreviated BW. The BW is equal to $f_{hc} - f_{lc}$, and it can be shown mathematically that

$$\text{BW} = \frac{R}{2\pi L} \tag{1-24}$$

where BW = bandwidth (Hz)
R = total circuit resistance
L = circuit inductance

The filter's quality factor, Q, provides a measure of how selective (narrow) its passband is compared to its center frequency, f_r. Thus,

$$Q = \frac{f_r}{\text{BW}} \tag{1-25}$$

As stated earlier, the quality factor, Q, can also be determined as

$$Q = \frac{\omega L}{R} \tag{1-26}$$

where ωL = inductive reactance at resonance
R = total circuit resistance

As Q increases, the filter becomes more selective; that is, a smaller passband (narrower bandwidth) is allowed. A major limiting factor in the highest attainable Q is

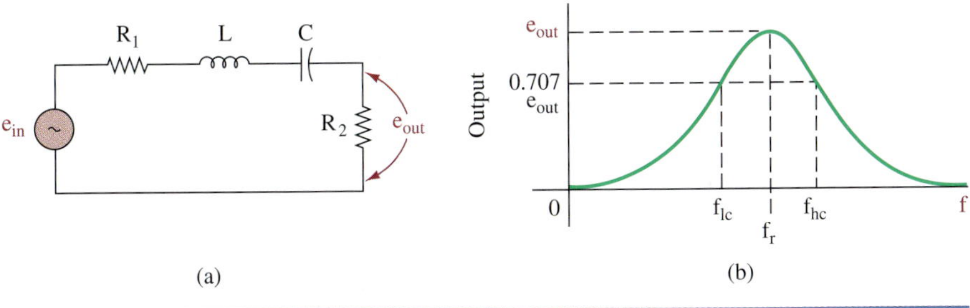

FIGURE 1-16 (a) *LC* bandpass filter and (b) response.

the resistance factor shown in Eq. (1-21). To obtain a high Q, the circuit resistance must be low. Quite often, the limiting factor becomes the winding resistance of the inductor itself. The turns of wire (and associated resistance) used to make an inductor provide this limiting factor. To obtain the highest Q possible, larger wire (with less resistance) could be used, but then greater cost and physical size to obtain the same amount of inductance is required. Quality factors (Q) approaching 1000 are possible with very high quality inductors.

Example 1-13

A filter circuit of the form shown in Fig. 1-16(a) has a response as shown in Fig. 1-17. Determine the
(a) bandwidth.
(b) Q.
(c) value of inductance if $C = 0.001\ \mu F$.
(d) total circuit resistance.

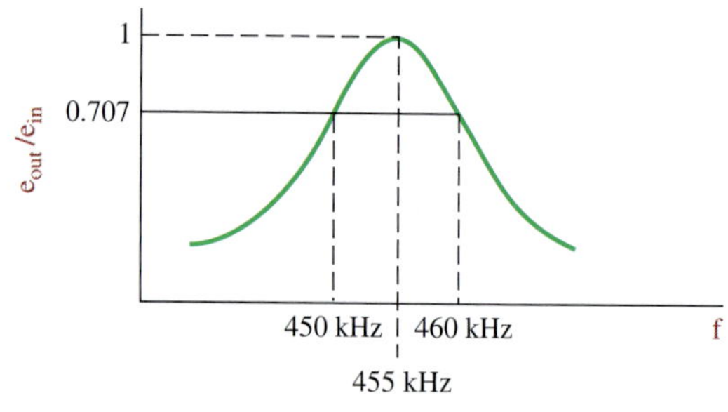

FIGURE 1-17 Response curve for Ex. 1-13.

Solution

(a) From Fig. 1-17, the BW is simply the frequency range between f_{hc} and f_{lc} or 460 kHz − 450 kHz = 10 kHz.
(b) The filter's peak output occurs at 455 kHz.

$$Q = \frac{f_r}{\text{BW}} \tag{1-25}$$
$$= \frac{455\ \text{kHz}}{10\ \text{kHz}}$$
$$= 45.5\ \text{kHz}$$

(c) Equation (1-23) can be used to solve for L because f_r and C are known.

$$f_r = \frac{1}{2\pi\sqrt{LC}} \tag{1-23}$$
$$455\ \text{kHz} = \frac{1}{2\pi\sqrt{L \times 0.001\ \mu\text{F}}}$$
$$L = 0.12\ \text{mH}$$

(d) Equation (1-24) can be used to solve for total circuit resistance because the BW and L are known.

$$\text{BW} = \frac{R}{2\pi L} \tag{1-24}$$

$$10 \text{ kHz} = \frac{R}{2\pi \times 0.12 \text{ mH}}$$

$$R = 10 \times 10^3 \text{ Hz} \times 2\pi \times 0.12 \times 10^{-3} \text{ H}$$
$$= 7.52\ \Omega$$

The frequency-response characteristics of LC circuits are affected by the ratio of L and C. Different values of L and C can be used to exhibit resonance at a specific frequency. A high L/C ratio yields a more narrowband response, while lower L/C ratios provide a wider frequency response. This effect can be verified by examining the effect of changing L in Eq. (1-24).

Parallel *LC* Circuits

A parallel LC circuit and its impedance versus frequency characteristic is shown in Fig. 1-18. The only resistance shown for this circuit is the inductor's winding resistance and is effectively in series with the inductor as shown. Notice that the impedance of the parallel LC circuit reaches a maximum value at the resonant frequency, f_r, and falls to a low value on either side of resonance. As shown in Fig. 1-18, the maximum impedance is

$$Z_{\text{max}} = Q^2 \times R \tag{1-27}$$

Equations (1-23) to (1-25) and (1-21) for series LC circuits also apply to parallel LC circuits when Q is greater than 10 ($Q > 10$), the usual condition.

The parallel LC circuit is sometimes called a **tank circuit.** Energy is stored in each reactive element (L and C), first in one and then released to the other. The transfer of energy between the two elements will occur at a natural rate equal to the resonant frequency and is sinusoidal in form.

Tank Circuit
parallel *LC* circuit

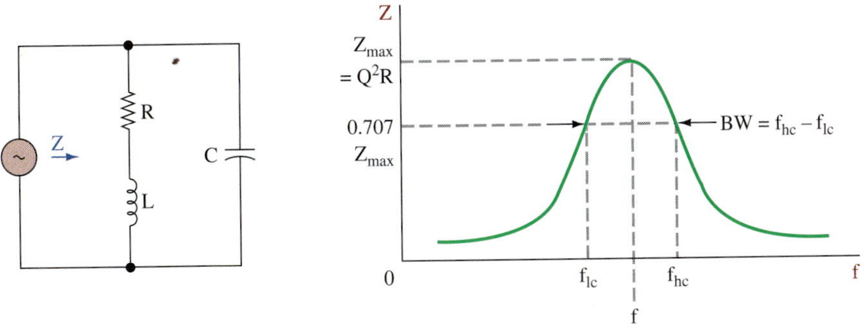

FIGURE 1-18 Parallel *LC* circuit and response.

Example 1-14

A parallel LC tank circuit is made up of an inductor of 3 mH and a winding resistance of 2 Ω. The capacitor is 0.47 μF. Determine

(a) f_r.
(b) Q.
(c) Z_{max}.
(d) BW.

Solution

(a)

$$f_r = \frac{1}{2\pi\sqrt{LC}} \qquad \textbf{(1-23)}$$

$$= \frac{1}{2\pi\sqrt{3\text{ mH} \times 0.47\ \mu\text{F}}}$$

$$= 4.24\text{ kHz}$$

(b)

$$Q = \frac{X_L}{R} \qquad \textbf{(1-21)}$$

$$\text{where } X_L = 2\pi fL = 2\pi \times 4.24\text{ kHz} \times 3\text{ mH} = 79.9\ \Omega$$

$$Q = \frac{79.9\ \Omega}{2\ \Omega} = 39.9$$

(c)

$$Z_{\text{max}} = Q^2 \times R \qquad \textbf{(1-27)}$$

$$= (39.9)^2 \times 2\ \Omega$$

$$= 3.19\text{ k}\Omega$$

(d)

$$\text{BW} = \frac{R}{2\pi L} \qquad \textbf{(1-24)}$$

$$= \frac{2\ \Omega}{2\pi \times 3\text{ mH}}$$

$$= 106\ \text{Hz}$$

Poles
number of *RC* or *LC* sections in a filter

Constant-*k* Filter
filter whose capacitive and inductive reactances are equal to a constant value *k*

***m*-Derived Filter**
filter that uses a tuned circuit to provide nearly infinite attenuation at a specific frequency

Roll-off
the rate of attenuation in a filter

Types of *LC* Filters

There is an endless variety of filters in use. At frequencies below 100 kHz, *RC* circuit configurations are used. The bulk and expense of the inductors needed at low frequencies limit their use. As we get above 100 kHz, the size of the required inductors becomes small enough so that *LC* combinations are used. Many filters use more than one *RC* or *LC* section to achieve the desired filtering. The number of *RC* or *LC* sections in a filter is referred to as the number of **poles** in the filter.

The two basic types of *LC* filters are the constant-*k* and the *m*-derived filters. The **constant-*k* filters** have the capacitive and inductive reactances made equal to a constant value *k*. The ***m*-derived filters** use a tuned circuit in the filter to provide nearly infinite attenuation at a specific frequency. The rate of attenuation is the steepness of the filter's response curve and is sometimes referred to as the **roll-off.**

It depends on the ratio of the filter's cutoff frequency to the frequency of near infinite attenuation—the *m* of the *m*-derived filter.

LC filters of the constant-*k* or *m*-derived types are further defined by the names of four persons who developed and first analyzed various *LC* configurations:

1. Butterworth
2. Chebyshev
3. Cauer (often referred to as elliptical)
4. Bessel (also called Thomson)

The study and design of *LC* filters is a large body of knowledge, the subject of many textbooks dedicated to filter design. Numerous software packages are also available to aid in their design and analysis.

High-Frequency Effects

At the very high frequencies encountered in communications, the small capacitance and inductance created by wire leads is a problem. Even the capacitance of the wire windings of an inductor can cause problems. Consider the inductor shown in Fig. 1-19. Notice the capacitance shown between the windings. This is termed **stray capacitance.** At low frequencies it has a negligible effect, but at high frequencies capacitance no longer appears as an open circuit and starts affecting circuit performance. The inductor is now functioning like a complex *RLC* circuit.

Stray Capacitance undesired capacitance between two points in a circuit or device

A simple wire exhibits a small amount of inductance. The longer the wire, the greater the inductance. At low frequencies this small inductance (usually a few nanohenries) looks like a short circuit and has no effect. At radio frequencies, however, this can be a problem because the unwanted inductive reactance goes up in value directly as the frequency goes up. Similarly, the stray capacitance between two wires becomes a problem as frequency goes up because it no longer looks like an open circuit. For these reasons it is important to minimize all lead lengths in RF circuits. The use of surface-mount components that have almost no leads except metallic end pieces to solder to the printed circuit board are very effective in minimizing high-frequency problems.

The high-frequency effects just discussed for inductors and capacitors also cause problems with resistors. In fact, at high frequencies the equivalent circuit for a resistor is the same as for an inductor. This is shown in Fig. 1-20.

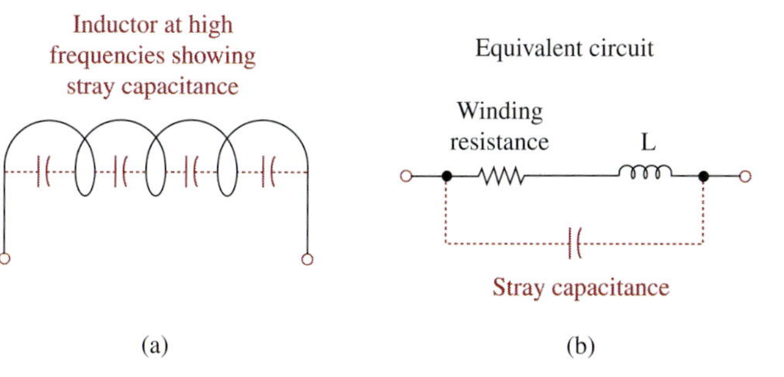

FIGURE 1-19 Inductor at high frequencies.

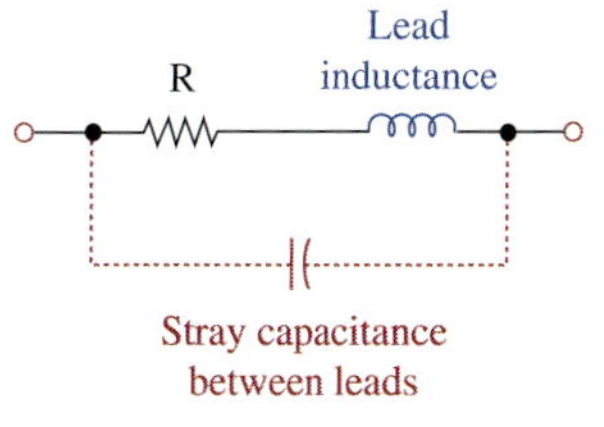

FIGURE 1-20 Resistor at high frequencies.

1-8 OSCILLATORS

Oscillator
circuit capable of converting electrical energy from dc to ac

The most basic building block in a communication system is an **oscillator.** An oscillator is a circuit capable of converting energy from a dc form to ac. In other words, an oscillator generates a waveform. The waveform can be of any type but occurs at some repetitive frequency.

A number of different forms of sine-wave oscillators are available for use in electronic circuits. The choice of an oscillator type is based on the following criteria:

1. Output frequency required.
2. Frequency stability required.
3. Is the frequency to be variable, and if so, over what range?
4. Allowable waveform distortion.
5. Power output required.

These performance considerations, combined with economic factors, will dictate the form of oscillator to be used in a given application.

LC OSCILLATOR

Flywheel Effect
repetitive exchange of energy in an *LC* circuit from the inductor to the capacitor and back

Damped
the gradual reduction of a repetitive signal due to resistive losses

The effect of charging the capacitor in Fig. 1-21(a) to some voltage potential and then closing the switch results in the waveform shown in Fig. 1-21(b). The switch closure starts a current flow as the capacitor begins to discharge through the inductor. The inductor, which resists a change in current flow, causes a gradual sinusoidal current buildup that reaches maximum when the capacitor is fully discharged. At this point the potential energy is zero, but since current flow is maximum, the magnetic field energy around the inductor is maximum. The magnetic field no longer maintained by capacitor voltage then starts to collapse, and its counter EMF will keep current flowing in the same direction, thus charging the capacitor to the opposite polarity of its original charge. This repetitive exchange of energy is known as the **flywheel effect.** The circuit losses (mainly the dc winding resistance of the coil) cause the output to become gradually smaller as this process repeats itself after the complete collapse of the magnetic field. The resulting waveform, shown in Fig. 1-21(b), is termed a **damped** sine wave. The energy of the magnetic field has been converted into the energy of the capacitor's

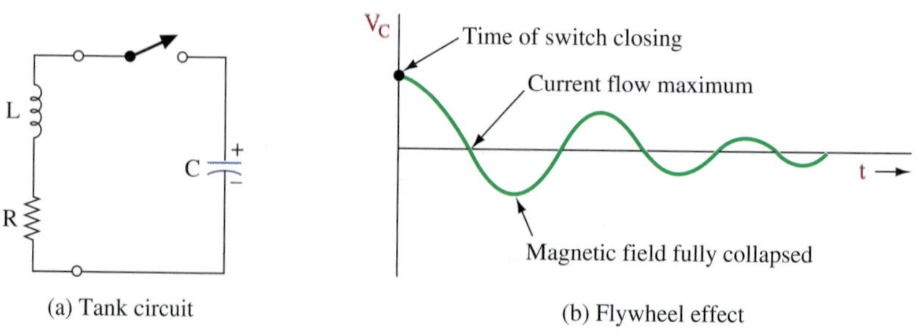

FIGURE 1-21 Tank circuit flywheel effect.

electric field, and vice versa. The process repeats itself at the natural or resonant frequency, f_r, as predicted by Eq.(1-23):

$$f_r = \frac{1}{2\pi\sqrt{LC}} \tag{1-23}$$

For an LC tank circuit to function as an oscillator, an amplifier is utilized to restore the lost energy to provide a constant-amplitude sine-wave output. The resulting "undamped" waveform is known as a **continuous wave** (CW) in radio work. The most straightforward method of restoring this lost energy is now examined, and the general conditions required for oscillation are introduced.

Continuous Wave undamped sinusoidal waveform produced by an oscillator in a radio transmitter

The LC oscillators are basically feedback amplifiers, with the feedback serving to increase or sustain the self-generated output. This is called positive feedback, and it occurs when the fed-back signal is in phase with (reinforces) the input signal. It would seem, then, that the regenerative effects of this positive feedback would cause the output to increase continually with each cycle of fed-back signal. However, in practice, component nonlinearity and power supply constraints limit the theoretically infinite gain.

The criteria for oscillation are formally stated by the **Barkhausen criteria** as follows:

Barkhausen Criteria two requirements for oscillations: loop gain must be at least unity and loop phase shift must be zero degrees

1. The loop gain must equal 1.
2. The loop phase shift must be $n \times 360°$, where $n = 1, 2, 3, \ldots$.

An oscillating amplifier adjusts itself to meet both of these criteria. The initial surge of dc power or noise in the circuit creates a sinusoidal voltage in the tank circuit at its resonant frequency, and it is fed back to the input and amplified repeatedly until the amplifier works into the saturation and cutoff regions. At this time, the flywheel effect of the tank is effective in maintaining a sinusoidal output. This process shows us that too much gain would cause excessive distortion and therefore the gain should be limited to a level that is just greater than or equal to 1.

Hartley Oscillator

Figure 1-22 shows the basic Hartley oscillator in simplified form. The inductors L_1 and L_2 are a single tapped inductor. Positive feedback is obtained by mutual inductance effects between L_1 and L_2 with L_1 in the transistor output circuit and L_2 across the base–emitter circuit. A portion of the amplifier signal in the collector circuit (L_1) is returned to the base circuit by means of inductive coupling from L_1 to L_2. As always in a common-emitter (CE) circuit, the collector and base voltages are 180° out of phase. Another 180° phase reversal between these two voltages occurs because they are taken from opposite ends of an inductor tap that is tied to the common transistor terminal—the emitter. Thus the in-phase feedback requirement is fulfilled and loop gain is of course provided by Q_1. The frequency of oscillation is approximately given by

$$f \simeq \frac{1}{2\pi\sqrt{(L_1 + L_2)C}} \tag{1-28}$$

and is influenced slightly by the transistor parameters and amount of coupling between L_1 and L_2.

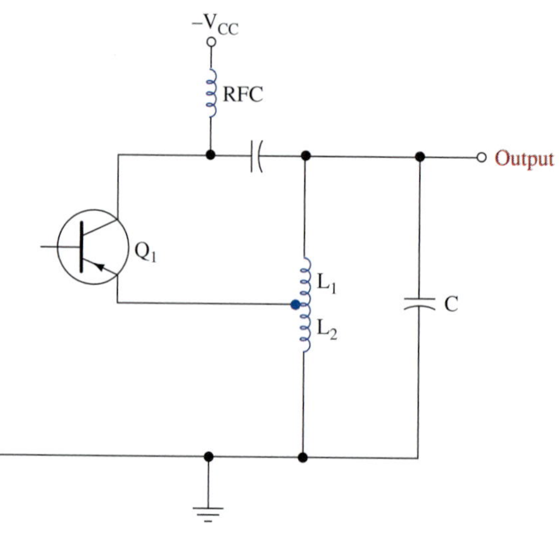

FIGURE 1-22 Simplified Hartley oscillator.

Figure 1-23 shows a practical Hartley oscillator. A number of additional circuit elements are necessary to make a workable oscillator over the simplified one used for explanatory purposes in Fig. 1-22. Naturally, the resistors R_A and R_B are for biasing purposes. The radio-frequency choke (RFC) is effectively an open circuit to the resonant frequency and thus allows a path for the bias (dc) current but does not allow the power supply to short out the ac signal. The coupling capacitor C_3 prevents dc current from flowing in the tank, and C_2 provides dc isolation between the base and the tank circuit. Both C_2 and C_3 can be considered as short circuits to the oscillator's frequency.

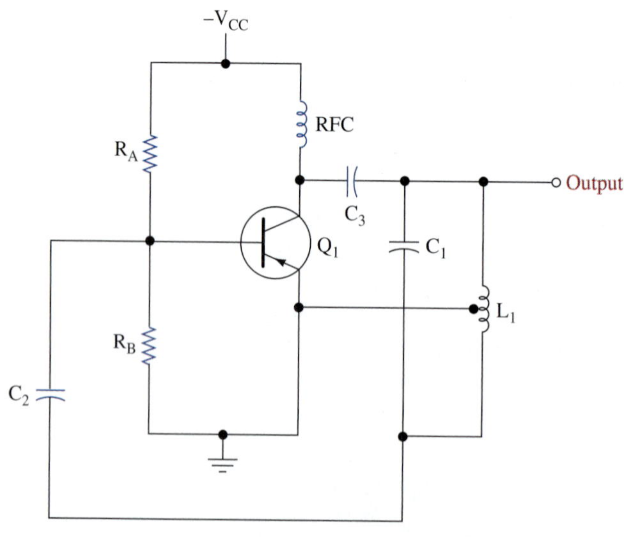

FIGURE 1-23 Practical Hartley oscillator.

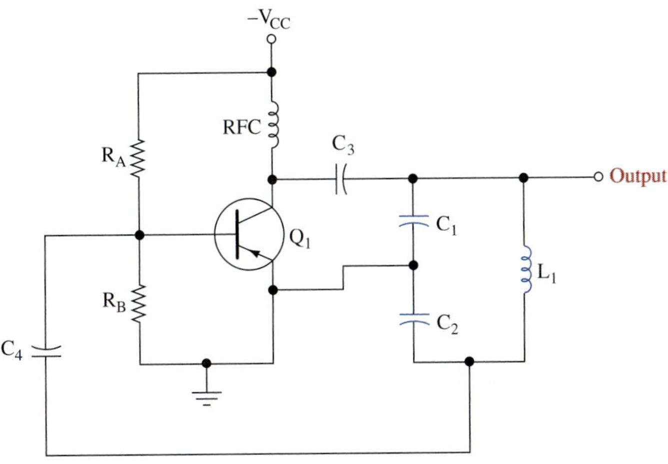

FIGURE 1-24 Colpitts oscillator.

Colpitts Oscillator

Figure 1-24 shows a Colpitts oscillator. It is similar to the Hartley oscillator except that the tank circuit elements have interchanged their roles. The capacitor is now split, so to speak, and the inductor is single-valued with no tap. The details of circuit operation are identical with the Hartley oscillator and therefore will not be explained further. The frequency of oscillation is given approximately by the resonant frequency of L_1 and C_1 in series with the C_2 tank circuit:

$$f \simeq \frac{1}{2\pi\sqrt{[C_1C_2/(C_1 + C_2)]L_1}} \tag{1-29}$$

The performance differences between these two oscillators' forms are minor, and the choice between them is usually made on the basis of convenience or economics. They may both provide variable oscillator output frequencies by making one of the tank circuit elements variable.

Clapp Oscillator

A variation of the Colpitts oscillator is shown in Fig. 1-25. The Clapp oscillator has a capacitor C_3 in series with the tank circuit inductor. If C_1 and C_2 are made large enough, they will "swamp" out the transistor's inherent junction capacitances, thereby negating transistor variations and junction capacitance changes with temperature. The frequency of oscillation is

$$f = \frac{1}{2\pi\sqrt{L_1C_3}} \tag{1-30}$$

and an oscillator with better frequency stability than the Hartley or Colpitts versions results. The Clapp oscillator does not have as much frequency adjustment range, however.

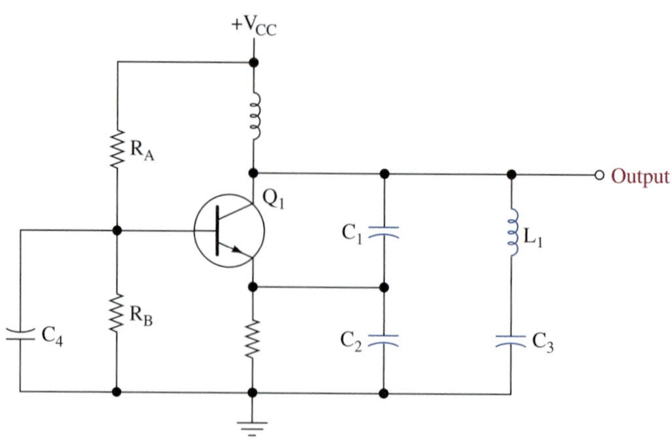

FIGURE 1-25 Clapp oscillator.

The *LC* oscillators presented in this section are the ones most commonly used. However, many different forms and variations exist and are used for special applications.

Crystal Oscillator

When greater frequency stability than that provided by *LC* oscillators is required, a crystal-controlled oscillator is often utilized. A crystal oscillator is one that uses a piezoelectric crystal as the inductive element of an *LC* circuit. The crystal, usually quartz, also has a resonant frequency of its own, but optimum performance is obtained when it is coupled with an external capacitance.

The electrical equivalent circuit of a crystal is shown in Fig. 1-26. It represents the crystal by a series resonant circuit (with resistive losses) in parallel with a capacitance C_p. The resonant frequencies of these two resonant circuits (series and parallel) are quite close together (within 1%) and hence the impedance of the crystal varies sharply within a narrow frequency range. This is equivalent to a very high *Q* circuit, and in fact crystals with a *Q*-factor of 20,000 are common; a *Q* of up to 10^6 is possible. This compares to a maximum *Q* of about 1000 with high-quality *LC* resonant circuits. For this reason, and because of the good time and temperature stability characteristics of quartz, crystals are capable of maintaining a frequency to $\pm 0.001\%$ over a fairly wide temperature range. The $\pm 0.001\%$ term is equivalent to saying ± 10 parts per million (ppm), and this is a preferred way of expressing such very small percentages. Note that $0.001\% = 0.00001 = 1/100{,}000 = 10/1{,}000{,}000 = 10$ ppm. Over very narrow temperature ranges or by maintaining the crystal in a small temperature-controlled oven, stabilities of ± 0.01 ppm are possible.

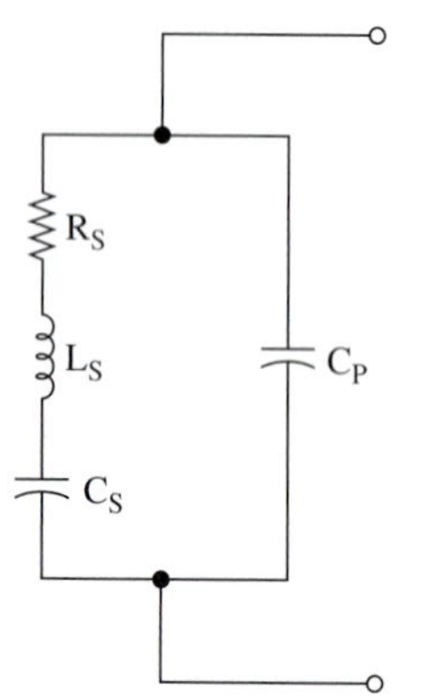

FIGURE 1-26 Electrical equivalent circuit of a crystal.

Crystals are fabricated by "cutting" the crude quartz in a very exacting fashion. The method of "cut" is a science in itself and determines the crystal's natural resonant frequency as well as its temperature characteristics. Crystals are available at frequencies of about 15 kHz and up, with higher frequencies providing the best frequency stability. However, at frequencies above 100 MHz, they become so small that handling is a problem.

Crystal oscillator. (Courtesy of Oscillatek/John Mutrix.)

Crystals may be used in place of the inductors in any of the *LC* oscillators discussed previously. A circuit especially adapted for crystal oscillators is the Pierce oscillator, shown in Fig. 1-27. The use of an FET is desirable because its high impedance results in light loading of the crystal, provides for good stability, and does not lower the *Q*. This circuit is essentially a Colpitts oscillator with the crystal replacing the inductor and the inherent FET junction capacitances functioning as the split capacitor. Because these junction capacitances are generally low, this oscillator is effective only at high frequencies.

The HA7210 IC can easily be used to build a crystal oscillator. It uses the Pierce oscillator circuitry just discussed with the crystal connected between pins 2 and 3. This is shown in Fig. 1-28, where a 1-MHz crystal oscillator with both

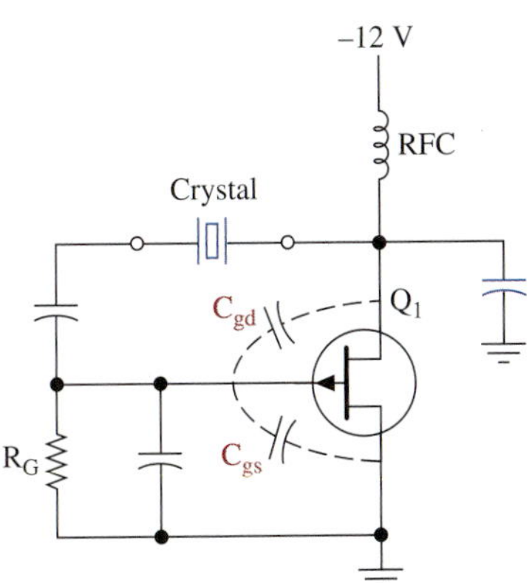

FIGURE 1-27 Pierce oscillator.

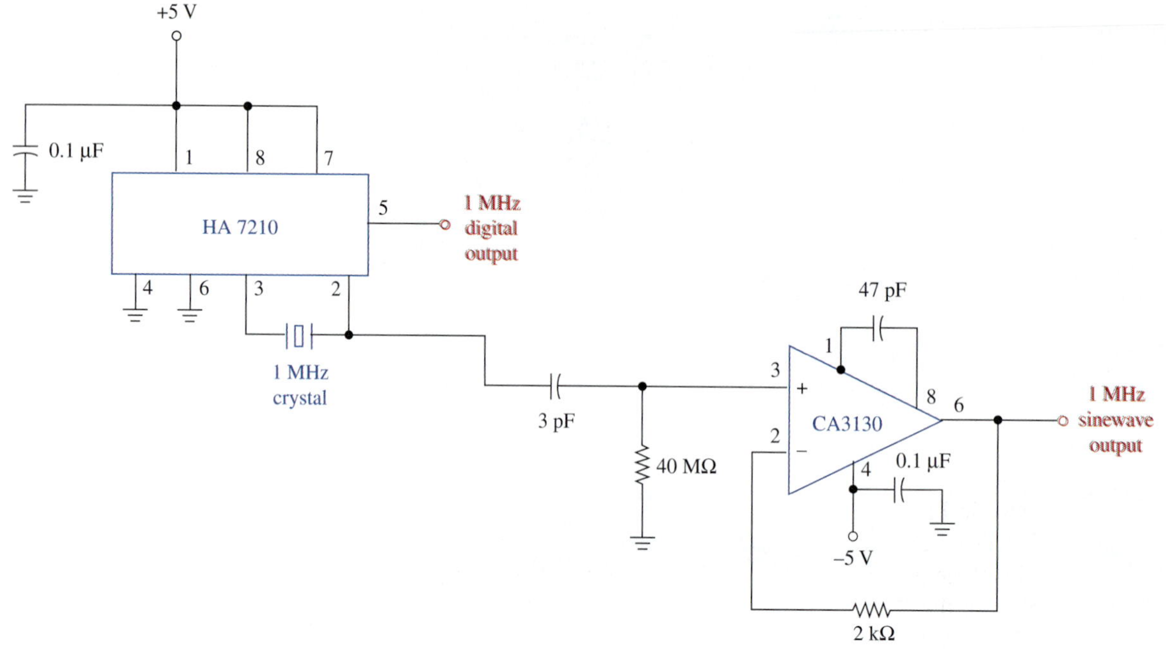

FIGURE 1-28 IC crystal oscillator.

digital and sine-wave outputs are provided. The CA3130 IC is a high-input impedance amplifier that prevents loading down the signal at pin 2 of the HA7210. If just a digital clock signal is required, the CA3130 and its related circuitry are not necessary.

Frequency Synthesizer oscillator that generates a wide range of output frequencies using one reference crystal oscillator

In Chapter 7, the use of a basic crystal oscillator in a **frequency synthesizer** is explained. The synthesizer generates a wide range of frequencies using a single-crystal oscillator as a basic reference. The various output frequencies have the same accuracy and stability as the crystal oscillator.

Crystal oscillators are available in various forms depending on the frequency stability required. The basic oscillator shown in Fig. 1-27 (often referred to as CXO) may be adequate as a simple clock for a digital system. Increased performance can be attained by adding temperature compensation circuitry (TCXO). Further improvement is afforded by including microprocessor (digital) control in the crystal oscillator package (DTCXO). The ultimate performance is attained with oven control of the crystal's temperature and sometimes also includes the microprocessor control (OCXO). These obviously require significant power to maintain the oven at some constant elevated temperature. A comparison of the four types of commonly available crystal oscillators is provided in Table 1-5.

Crystal Test

Crystal oscillators may fail to operate because of faulty design or failed crystals. The circuit shown in Fig. 1-29 works well as a tester for a wide variety of crystals and ceramic resonators over the 40-kHz to 20-MHz range. See Sec. 4-3 for a discussion of ceramic resonators.

Table 1-5 Typical Cost/Performance Comparison for Crystal Oscillators

	Basic Crystal Oscillator (CXO)	Temperature Compensated (TCXO)	Digital TCXO (DTCXO)	Oven-Controlled CXO (OCXO)
Frequency stability from 0 to 70°C	100 ppm	1 ppm	0.5 ppm	0.05 ppm
Frequency stability for one year at constant temperature	1 ppm	1 ppm	1 ppm	1 ppm
Price	\$2–\$10	\$10–\$50	\$25–\$100	\$100 and up

The oscillator in Fig. 1-29 is a Pierce type that operates at the crystal's parallel resonant frequency and presents about 30 pF capacitance to the crystal. The CD4007A contains three pairs of complementary MOSFETs with the first (input at pin 6) functioning as the Pierce oscillator. The second (input at pin 3) drives a 200–500-μA meter movement. The resistor *R* is selected to provide about 90% deflection with an active (good) crystal. The "tuning" meter from a discarded stereo is usually ideal for this application. The other complementary pair (input at pin 10) provides a low-impedance output that can drive a frequency counter or provide a connection for an oscilloscope.

The crystal being tested can be inserted in the crystal holder or connected with alligator clips. The input MOSFETs are well protected from electrostatic and leakage damage.

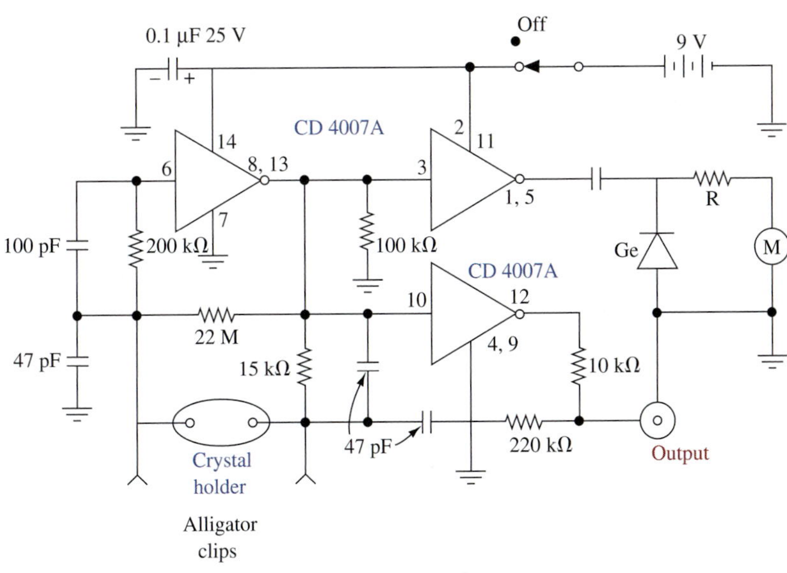

FIGURE 1-29 Crystal test circuit.

1-9 TROUBLESHOOTING

Because of the increasing complexity of electronic communications equipment, you must have a good understanding of communication circuits and concepts. To be an effective troubleshooter, you must also be able to isolate faulty components quickly and repair the defective circuit. Recognizing the way a circuit may malfunction is a key factor in speedy repair procedures.

After completing this section you should be able to

- Explain general troubleshooting techniques
- Recognize major types of circuit failures
- List the four troubleshooting techniques
- Test for a defective crystal
- Test for defective capacitors and inductors
- Understand digital sampling oscilloscope waveforms

GENERAL TROUBLESHOOTING TECHNIQUES

Troubleshooting requires asking questions such as: What could cause this to happen? Why is this voltage so low/high? Or, if this resistance were open/shorted, what effect would it have on the operation of the circuit I'm working on? Each question calls for measurements to be made and tests to be performed. The defective component(s) is isolated when measurements give results far different than they would be in a properly operating unit, or when a test fails. The ability to ask the right questions makes a good troubleshooter. Obviously, the more you know about the circuit or system being worked on, the quicker the problem will be corrected.

Always start troubleshooting by doing the easy things first:

- Be sure the unit is plugged in and turned on.
- Check fuses.
- Check if all connections are made.
- Ask yourself, "Am I forgetting something?"

Basic troubleshooting test equipment includes:

- a digital multimeter (DMM) capable of reading at the frequencies you intend to work at.
- a broadband oscilloscope, preferably dual trace.
- signal generators, both audio and RF. The RF generator should have internal modulation capabilities.
- a collection of probes and clip leads.

Advanced test equipment would include a spectrum analyzer to observe frequency spectra and a logic analyzer for digital work. Time spent learning your test equipment, its capabilities and limitations, and how to use it will pay off with faster troubleshooting.

Always be aware of any possible effects the test equipment you connect to a circuit may have on the operation of that circuit. Don't let the measuring equipment change what you are measuring. For example, a scope's test lead may have a capacitance of several hundred picofarads. Should that lead be connected across the output of an oscillator, the oscillator's frequency could be changed to the point that any measurements are worthless.

If the equipment you're troubleshooting employs dangerous voltages, do not work alone. Turn off all power switches before entering equipment. See additional comments on safe procedures in Sec. 2-8.

Keep all manuals that came with the equipment. Such manuals usually include troubleshooting procedures. Check them before trying any other approaches.

Maintain clear, up-to-date records of all changes made to equipment.

Replace a suspicious unit with a known good one—this is one of the best, most commonly used troubleshooting techniques.

Test points are often built into electronic equipment. They provide convenient connections to the circuitry for adjustment and/or testing. There are various types, from jacks or sockets to short, stubby wires sticking up from PC boards. Equipment manuals will diagram the location of each test point and describe and sometimes illustrate the condition and/or signal that should be found there. The better manuals indicate the proper test equipment to use.

Plot a game plan or strategy with which you will troubleshoot a problem (just as you might with a car problem).

Use all your senses when troubleshooting:

Look—discolored or charred components might indicate overheating.

Smell—some components, especially transformers, emit characteristic odors when overheated.

Feel—for hot components. Wiggle components to find broken connections.

Listen—for "frying" noises that indicate a component is about to fail.

Reasons Electronic Circuits Fail

Electronic circuits fail in many ways. Let's look at some major types of failures that you will encounter.

1. Complete Failures Complete failures cause the piece of equipment to go totally dead. Equipment with some circuits still operating has not completely failed. Normally this type of failure is the result of a major circuit path becoming open. Blown (open) fuses, open power resistors, defective power supply rails, and bad regulator transistors in the power supply can cause complete failures. Complete failures are often the easiest problems to repair.

2. Intermittent Faults Intermittent faults are characterized by sporadic circuit operation. The circuit works for awhile and then quits working. It works one moment and doesn't work the next. Keeping the circuit in a failed condition can be quite difficult. Loose wires and components, poor soldering, and effects of temperature on sensitive components can all contribute to intermittent operation in a piece of communications equipment. Intermittent faults are usually the most difficult to repair since troubleshooting can be done only when the equipment is malfunctioning.

3. Poor System Performance Equipment that is functioning below specified operational standards is said to have poor system performance characteristics. For example, a transmitter is showing poor performance if the specifications call for 4 W of output power but it is putting out only 2 W. Degradation of equipment performance takes place over a period of time due to deteriorating components (components change in value), poor alignments, and weakening power components.

Regular performance checks are necessary for critical communications systems. Commercial radio transmitters require performance checks to be done on a regular basis.

4. **Induced Failures** Induced failures often come from equipment abuse. Unauthorized modifications may have been performed on the equipment. An inexperienced technician without supervision may have attempted repairs and damaged the equipment. Induced failures can be eliminated by exercising proper equipment care. Repairs should be done or supervised by experienced technicians.

Troubleshooting Plan

Experienced technicians have developed a method for troubleshooting. They follow certain logical steps when looking for a defect in a piece of equipment. The following four troubleshooting techniques are popular and widely used to find defects in communications equipment.

1. **Symptoms as Clues to Faulty Stages** This technique relates a particular fault to a circuit function in the electronic equipment. For example, if a white horizontal line were displayed on the screen of a TV brought in for repair, the service technician would associate this symptom with the vertical output section. Troubleshooting would begin in that section of the TV. As you gain experience in troubleshooting you will start associating symptoms with specific circuit functions.

2. **Signal Tracing and Signal Injection** Signal injection is supplying a test signal at the input of a circuit and looking for the test signal at the circuit's output or listening for an audible tone at the speaker (Fig. 1-30). This test signal is usually composed of an RF signal modulated with an audible frequency. If the signal is good at the circuit's output, then move to the next stage down the line and repeat the test. Signal tracing, as illustrated in Fig. 1-31, is actually checking for the normal output signal from a stage. An oscilloscope is used to check for these signals. However, other test equipment is available that can be used to detect the presence of output signals. Signal tracing is monitoring the output of a stage for the presence of the expected signal. If the signal is there, then the next stage in line is checked. The malfunctioning stage precedes the point where the output signal is missing.

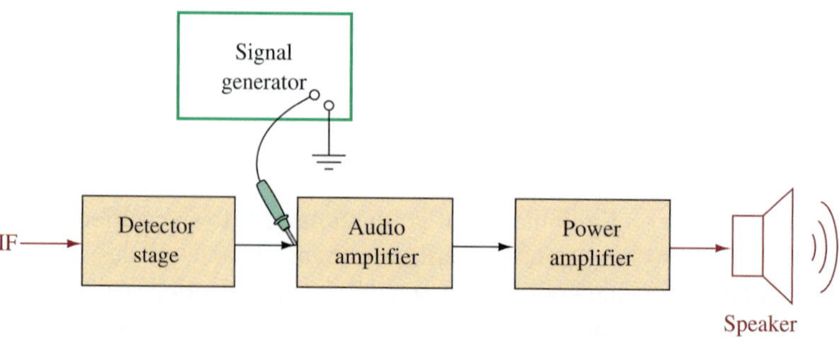

FIGURE 1-30 Signal injection.

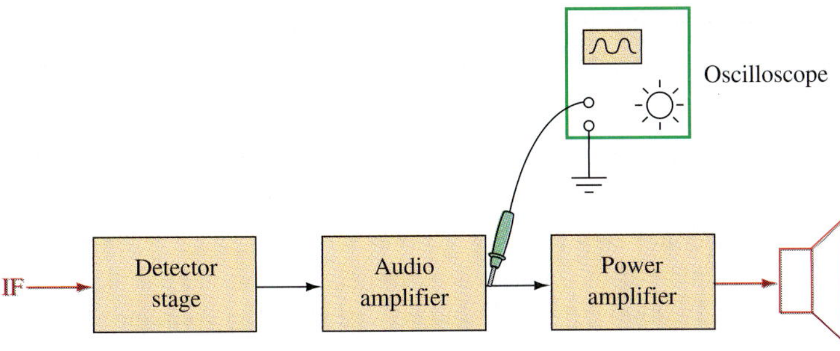

FIGURE 1-31 Signal tracing.

3. Voltage and Resistance Measurements Voltage and resistance measurements are made with respect to chassis ground. Using the DMM (digital multimeter), measurements at specific points in the circuit are compared to those found in the equipment's service manual. Service manuals furnish equipment voltage and resistance charts or print the values right on the schematic diagram. Voltage and resistance checks are done to isolate defective components once the trouble has been pinpointed to a specific stage of the equipment. Remember that resistance measurements are done on circuits with the power turned off.

4. Substitution Another method often used to troubleshoot electronic circuits is to swap a known good component for the suspected bad component. A warning is in order here: The good component could get damaged in the substitution process. Don't get into the habit of indiscriminately substituting parts. This method works best when you have narrowed the failure down to a specific component.

Testing a Crystal

An oscillator with a bad crystal may not oscillate at all, may be erratic, or may not oscillate at the correct frequency. One common crystal failure mode is a broken or corroded internal connection. Or if the crystal has been dropped, it may be cracked.

Figure 1-32 shows how to make a simple test to determine quickly the condition of the crystal. Normally, a crystal oscillator will oscillate at a slightly higher frequency than the crystal's series resonant point. If you can find the series resonant point of the crystal, you know the crystal is good.

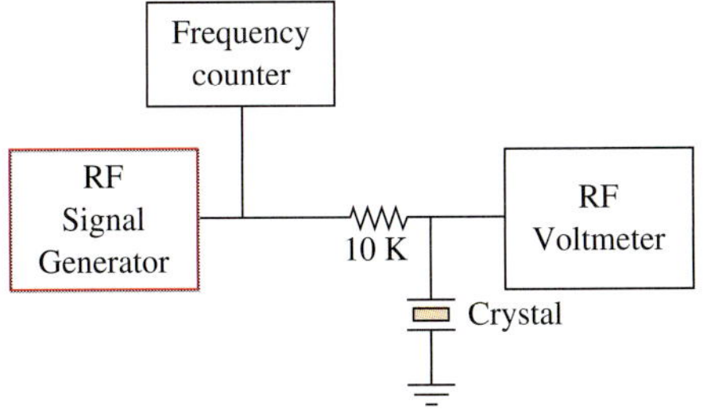

FIGURE 1-32 Crystal test.

Recall that at the series resonant point, the crystal should have a very low resistance, on the order of 100 Ω. At other frequencies, the crystal impedance should be quite high.

The generator should be very carefully tuned across the specified frequency of the crystal. If the crystal is operating properly, the voltmeter will show a dramatic dip at the series resonant point. Remember that the crystal is a very high Q device, and tuning the signal generator will have to be done carefully.

Because the impedance of the crystal is extremely high at the parallel or antiresonant point, perhaps 50,000 Ω, there should be a peak on the voltmeter at a frequency just slightly above the series resonant point. You should look for the series resonant point first because it is easier to find.

The voltage across a broken crystal will not change much as the generator frequency is varied. Internal connection problems could cause erratic operation. Corrosion problems will cause the resonant frequency to shift from the specified value.

Testing Oscillator Capacitors

The capacitors associated with the crystal or inductor together with the inductor determine the exact frequency of oscillation. This type of capacitor will seldom show a short, but it can become sensitive to temperature and shock or change value with age.

In the Clapp circuit shown in Fig. 1-33, C_3 is primarily responsible for setting the frequency. While observing the frequency with a counter, cool the capacitor with an aerosol spray sold for cooling electronic equipment. Defective capacitors will generally change value suddenly and shift the frequency a good bit when cooled. If C_3 is open, the circuit probably will not oscillate at all.

In the Clapp circuit, C_1 and C_2 are primarily responsible for providing the proper amount of feedback to allow oscillation. If either of these capacitors fails, the oscillator will not work. An oscilloscope connected to the collector of Q_1 should show a high-quality sine wave. C_1 and C_2 do have some effect on the frequency and should not be excluded from suspicion if the frequency is not correct.

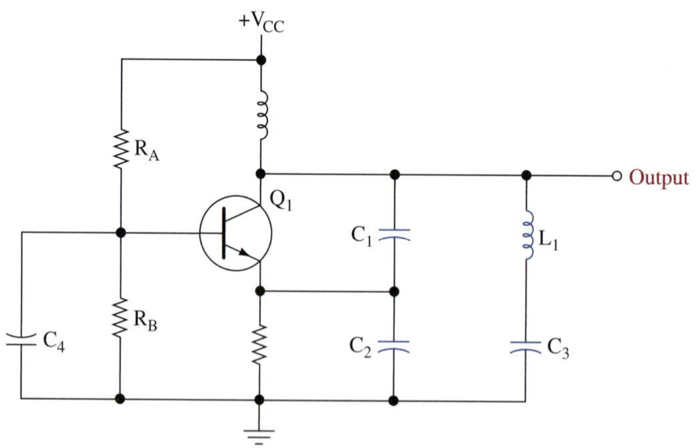

FIGURE 1-33 Clapp oscillator.

Testing Oscillator Inductors

A shorted or open inductor will completely kill an oscillator. Inductors can be easily checked for an open circuit with an ohmmeter, though the ohmmeter will not detect a shorted turn. A short in the inductor is best detected with a Q-meter or impedance bridge.

Understanding Digital Sampling Oscilloscope Waveforms

The waveforms created by the improper setup of the sampling frequency of a digital sampling oscilloscope (DSO) can lead to strange waveforms, confusion, and errors. The minimum sample frequency of the DSO must be set to at least twice the maximum input frequency. This is called the Nyquist sampling frequency. If the sample frequency is too low, then the resulting waveform will be greatly distorted and will not truly reflect the waveform being measured. Additionally, the frequency information generated by the FFT of the input signal will not be accurate. In fact, the FFT will indicate a frequency that does not occur in the measured signal.

For example, a 12.375-kHz sinusoid was input into channel 1 of a DSO. The sample frequency of the DSO was set to 10 kS/s. The minimum sample frequency should have been at least 24.75 kS/s to meet the Nyquist sample frequency criteria. Figure 1-34 shows the resulting time series and its FFT. Notice the extreme distortion of the time series. The input signal is a sinusoid, but the picture appears to contain amplitude variations and possibly more than one frequency. The FFT indicates that a 2.375-kHz signal is being sampled, which is not correct. The 2.375-kHz signal, generated by the selection of an improper sampling frequency, results from the 10-kHz

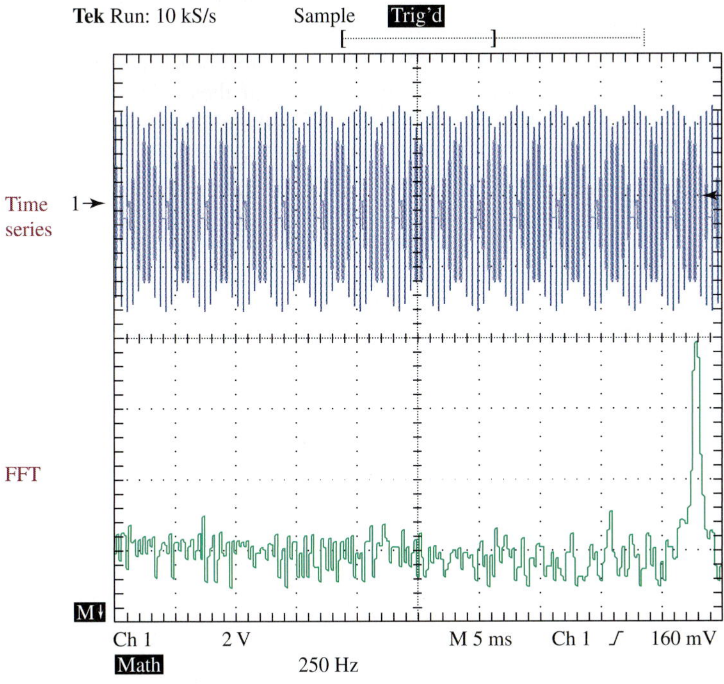

FIGURE 1-34 The time series (top) and the FFT (bottom) for a 12.375-kHz sinusoid with the sample rate set to 10 kS/s.

sampling frequency "mixing" with the 12.375-kHz sinusoid and the frequency difference of 2.375 kHz (12.375 kHz − 10 kHz) being generated. The concept of sampling is explained in greater detail in Sec. 8-3, and the mathematical relationship defining the mixing of two frequencies is introduced in Chapter 2. Basically, when two frequencies are "mixing" together, as in the case of sampling a 12.375-kHz sinusoid with a 10-kS/s sample frequency, two frequencies are generated, the sum of 10 kHz + 12.375 kHz, or 22.375 kHz, and a difference frequency of 12.375 kHz − 10 kHz, or 2.375 kHz. The FFT shows that a 2.375-kHz frequency (difference) was generated, which is the difference frequency. A 22.375-kHz frequency was also generated but is not shown on this display. Knowing that both a 2.375-kHz and a 22.375-kHz signal were generated helps to explain the complex-looking waveform and why the time series waveform appears to contain two frequency components.

1-10 TROUBLESHOOTING WITH ELECTRONICS WORKBENCH™ MULTISIM

This text presents computer simulation examples of troubleshooting and analyzing electronic communications circuits and concepts using Electronics Workbench Multisim. Examples are presented for each chapter on an important topic covered in that chapter. Electronics Workbench provides a unique opportunity for you to examine electronic circuits and concepts in a way that reflects techniques used for analyzing and troubleshooting circuits and systems in practice. The use of Electronics Workbench provides you with additional hands-on insight into many of the fundamental communication circuits, concepts, and test equipment while improving your ability to perform logical thinking when troubleshooting circuits and systems. The test equipment tools available in Electronics Workbench reflect the type of tools that are commonly available on well-equipped test benches.

An introduction to many fundamental concepts in communications was presented in Chapter 1. The topics included the dB, noise, oscillators, *LC* circuits, and frequency spectra. The first Electronics Workbench example in this text reinforces the concepts presented in the section on understanding the frequency spectra. This particular example demonstrates that a complex waveform such as a square wave generates multifrequency components called harmonics. A spectrum analyzer is used in Electronics Workbench to observe and analyze the spectral content of a square wave.

To begin this exercise, start Electronics Workbench Multisim and open the file called **Fig1-35.ms7 (.msm)** that is found on the Electronics Workbench (EWB) Multisim CD-ROM packaged with the text. It is a simple circuit containing a 1 kHz square-wave generator connected to a 1 kΩ resistive load. The circuit is shown in Fig. 1-35.

Begin the simulation by clicking on the **start simulation** button. Verify that the function generator is outputting a 5-V square wave at 1 kHz by viewing the trace with the oscilloscope. The oscilloscope display can be opened by double-clicking on the oscilloscope icon. The oscilloscope display is shown in Fig. 1-36. Measurement features for the oscilloscope are introduced in Section 2-9.

Next, double-click on the spectrum analyzer. In a few seconds, the spectrum analyzer will sample and build the image shown in Fig. 1-37. Each spike in the waveform shows a frequency component or harmonic of the square wave. The concept of a square wave containing multiple frequency components was presented in Section 1-6. An oscilloscope image of a 1-kHz square wave and its corresponding FFT spectrum were presented in Fig. 1-12.

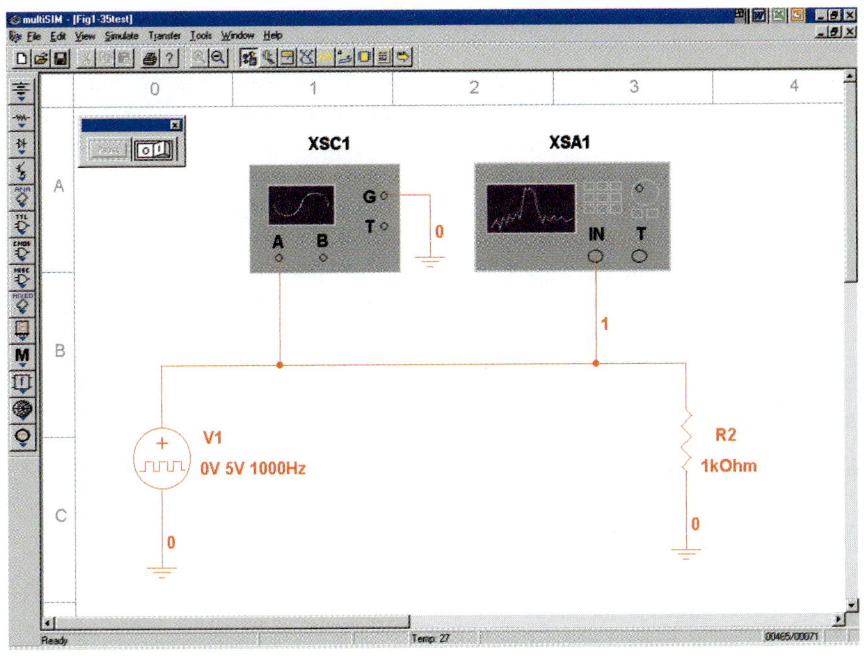

FIGURE 1-35 The Multisim component view of the test circuit used to demonstrate the frequency spectra for a square wave.

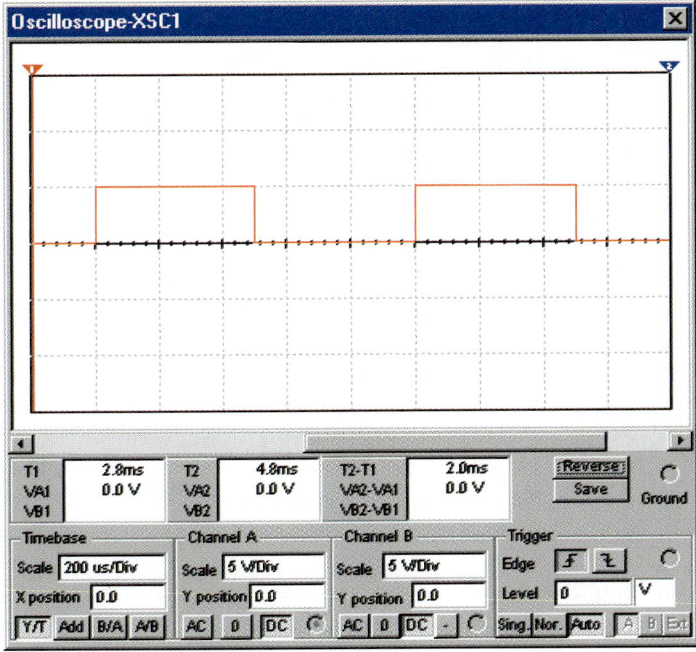

FIGURE 1-36 The Multisim oscilloscope image of the square wave from the function generator.

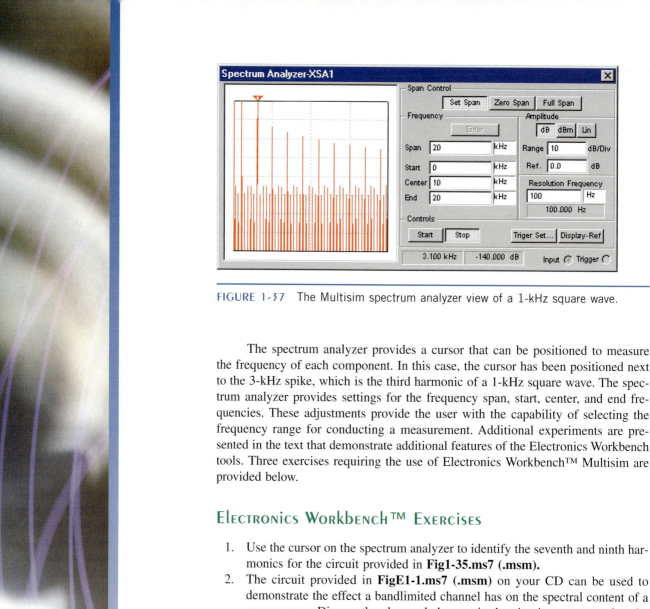

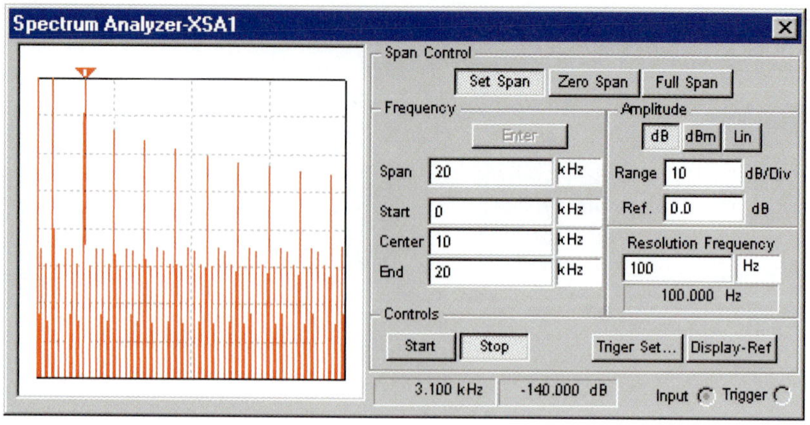

FIGURE 1-37 The Multisim spectrum analyzer view of a 1-kHz square wave.

The spectrum analyzer provides a cursor that can be positioned to measure the frequency of each component. In this case, the cursor has been positioned next to the 3-kHz spike, which is the third harmonic of a 1-kHz square wave. The spectrum analyzer provides settings for the frequency span, start, center, and end frequencies. These adjustments provide the user with the capability of selecting the frequency range for conducting a measurement. Additional experiments are presented in the text that demonstrate additional features of the Electronics Workbench tools. Three exercises requiring the use of Electronics Workbench™ Multisim are provided below.

Electronics Workbench™ Exercises

1. Use the cursor on the spectrum analyzer to identify the seventh and ninth harmonics for the circuit provided in **Fig1-35.ms7 (.msm).**
2. The circuit provided in **FigE1-1.ms7 (.msm)** on your CD can be used to demonstrate the effect a bandlimited channel has on the spectral content of a square wave. Discuss the observed changes in the circuit as compared to the circuit provided in **Fig1-35.ms7 (.msm).**
3. The circuit provided in **FigE1-2.ms7 (.msm)** on your CD contains a 100-kHz square wave. Change the settings on the spectrum analyzer so that the first, third, fifth, seventh, and ninth harmonics are displayed on the screen. The solution is provided on your EWB **FigE1-2-solution.ms7 (.msm).**

Summary

In Chapter 1 the concept of a communication system was introduced. Decibels and the effects of electrical noise were explained, and *LC* circuits and oscillators were discussed. The major topics you should now understand include:

- the use of the decibel in communications
- the function and basic building blocks of a communication system

- the need for modulation/demodulation in a communications system
- the difference between the carrier wave and intelligence wave and their importance
- the effects and analysis of electrical noise in a communications system
- the performance of signal-to-noise ratio and noise figure calculations
- the performance of electrical noise measurements on a communications system
- the makeup of nonsinusoidal waveforms
- the mathematical analysis of waveforms using Fourier analysis
- the analysis of *LC* filters
- the understanding of common oscillator types, including Hartley, Colpitts, Clapp, and crystal varieties

Questions and Problems

Section 1–1

1. Define *modulation.*

*2. What is *carrier frequency*?

3. Describe the two reasons that modulation is used for communications transmissions.
4. List the three parameters of a high-frequency carrier that may be varied by a low-frequency intelligence signal.
5. What are the frequency ranges included in the following frequency subdivisions: MF (medium frequency), HF (high frequency), VHF (very high frequency), UHF (ultra high frequency), and SHF (super high frequency)?

Section 1–2

6. A signal level of 0.4 μV is measured on the input to a satellite receiver. Express this voltage in terms of dBμV. Assume a 50-Ω system. (−7.95 dB$_\mu$V)
7. A microwave transmitter typically requires a + 8-dBm audio level to drive the input fully. If a +10-dBm level is measured, what is the actual voltage level measured? Assume a 600-Ω system. (2.45 V)
8. If an impedance matched amplifier has a power gain (P_{out}/P_{in}) of 15, what is the value for the voltage gain (V_{out}/V_{in})? (3.87)
9. Convert the following powers to their dBm equivalents:
   (a) p = 1 W (30 dBm)
   (b) p = 0.001 W (0 dBm)
   (c) p = 0.0001 W (−10 dBm)
   (d) p = 25 μW (−16 dBm)
10. The output power for an audio amplifier is specified to be 38 dBm. Convert this value to (a) watts and (b) dBW. (6.3 W, 8 dBW)
11. A 600-Ω microphone outputs a −70-dBm level. Calculate the equivalent output voltage for the −70-dBm level. (0.245 mV)
12. Convert 50-μV to a dBμV equivalent. (34 dBμV)

---

* An asterisk preceding a number indicates a question that has been provided by the FCC as a study aid for licensing examinations.

13. A 2.15-V rms signal is measured across a 600-Ω load. Convert this measured value to its dBm equivalent. (8.86 dBm (600))
14. A 2.15-V rms signal is measured across a 50-Ω load. Convert this measured value to its dBm(50) equivalent. (19.66 dBm(50))

Section 1–3

15. Define *electrical noise*, and explain why it is so troublesome to a communications receiver.
16. Explain the difference between external and internal noise.
17. List and briefly explain the various types of external noise.
18. Provide two other names for Johnson noise and calculate the noise voltage output of a 1-MΩ resistor at 27°C over a 1-MHz frequency range. (128.7 μV)
19. The noise produced by a resistor is to be amplified by a noiseless amplifier having a voltage gain of 75 and a bandwidth of 100 kHz. A sensitive meter at the output reads 240 μV rms. Assuming operation at 37°C, calculate the resistor's resistance. If the bandwidth were cut to 25 kHz, determine the expected output meter reading. (5.985 kΩ, 120 μV)
20. Explain the term *low-noise resistor.*
21. Determine the noise current for the resistor in Problem 18. What happens to this noise current when the temperature increases? (129 pA)
22. The noise spectral density is given by $e_n^2/\Delta f = 4kTR$. Determine the bandwidth Δf of a system in which the noise voltage generated by a 20-kΩ resistor is 20 μV rms at room temperature. (1.25 MHz)

Section 1–4

23. Calculate the *S/N* ratio for a receiver output of 4 V signal and 0.48 V noise both as a ratio and in decibel form. (69.44, 18.42 dB)
24. The receiver in Problem 23 has an *S/N* ratio of 110 at its input. Calculate the receiver's noise figure (NF) and noise ratio (NR). (1.998 dB, 1.584)
25. An amplifier with NF = 6 dB has S_i/N_i of 25 dB. Calculate the S_o/N_o in dB and as a ratio. (19 dB, 79.4)
26. A three-stage amplifier has an input stage with noise ratio (NR) = 5 and power gain (P_G) = 50. Stages 2 and 3 have NR = 10 and P_G = 1000. Calculate the NF for the overall system. (7.143 dB)
27. A two-stage amplifier has a 3-dB bandwidth of 150 kHz determined by an *LC* circuit at its input and operates at 27°C. The first stage has P_G = 8 dB and NF = 2.4 dB. The second stage has P_G = 40 dB and NF = 6.5 dB. The output is driving a load of 300 Ω. In testing this system, the noise of a 100-kΩ resistor is applied to its input. Calculate the input and output noise voltage and power and the system noise figure. (19.8 μV, 0.206 mV, 9.75×10^{-16} W, 1.4×10^{-10} W, 3.6 dB)
28. A microwave antenna (T_{eq} = 25 K) is coupled through a network (T_{eq} = 30 K) to a microwave receiver with T_{eq} = 60 K referred to its output. Calculate the noise power at its input for a 2-MHz bandwidth. Determine the receiver's NF. (3.17×10^{-15} W, 0.817 dB)
29. A high-quality FM receiver is to be tested for SINAD. When its output contains just the noise and distortion components, 0.015 mW is measured. When the desired signal and noise and distortion components are measured together, the output is 15.7 mW. Calculate SINAD. (30.2 dB)
30. Explain SINAD.

Section 1–5

31. Calculate the noise power at the input of a microwave receiver with an equivalent noise temperature of 45 K. It is fed from an antenna with a 35 K equivalent noise temperature and operates over a 5-MHz bandwidth. (5.52×10^{-15} W)
32. Calculate the minimum signal power needed for good reception for the receiver described in Problem 31 if the signal-to-noise ratio must be not less than 100 : 1. (5.52×10^{-13} W)
33. Calculate the NF and T_{eq} for an amplifier that has $Z_{in} = 300\ \Omega$. It is found that when driven from a matched-impedance diode noise generator, its output noise is doubled (as compared to no input noise) when the diode is forward biased with 0.3 mA. (2.55 dB, 232 K)
34. Describe the procedure used for noise measurement using the noise diode generator.
35. Describe what is known as a DUT.
36. Describe the procedure for noise measurement using the tangential technique.

Section 1–6

37. Define *information theory.*
38. What is Hartley's law? Explain its significance.
39. What is a *harmonic*?
40. What is the seventh harmonic of 360 kHz? (2520 kHz)
41. Why does transmission of a 2-kHz square wave require greater bandwidth than a 2-kHz sine wave?
42. Draw time- *and* frequency-domain sketches for a 2-kHz square wave. The time-domain sketch is a standard oscilloscope display while the frequency domain is provided by a spectrum analyzer.
43. Explain the function of Fourier analysis.
44. A 2-kHz square wave is passed through a filter with a 0- to 10-kHz frequency response. Sketch the resulting signal, and explain why the distortion occurs.
45. A triangle wave of the type shown in Table 1-4(e) has a peak-to-peak amplitude of 2 V and $f = 1$ kHz. Write the expression $v(t)$, including the first five harmonics. Graphically add the harmonics to show the effects of passing the wave through a low-pass filter with cutoff frequency equal to 6 kHz.
46. The FFT shown in Fig. 1-38 was obtained from a DSO.
    (a) What is the sample frequency?
    (b) What frequency is shown by the FFT?
47. Figure 1-39 was obtained from a DSO.
    (a) What are the frequencies of the third and fifth harmonics?
    (b) This FFT was created by inputting a 12.5-kHz square wave into a DSO. Explain where 12.5 kHz is located within the FFT spectrum.

Section 1–7

48. Explain the makeup of a practical inductor and capacitor. Include the quality and dissipation in your discussion.
49. Define *resonance* and describe its use.
50. Calculate an inductor's Q at 100 MHz. It has an inductance of 6 mH and a series resistance of 1.2 k. Determine its dissipation. (3.14×10, 0.318×10^{-3})

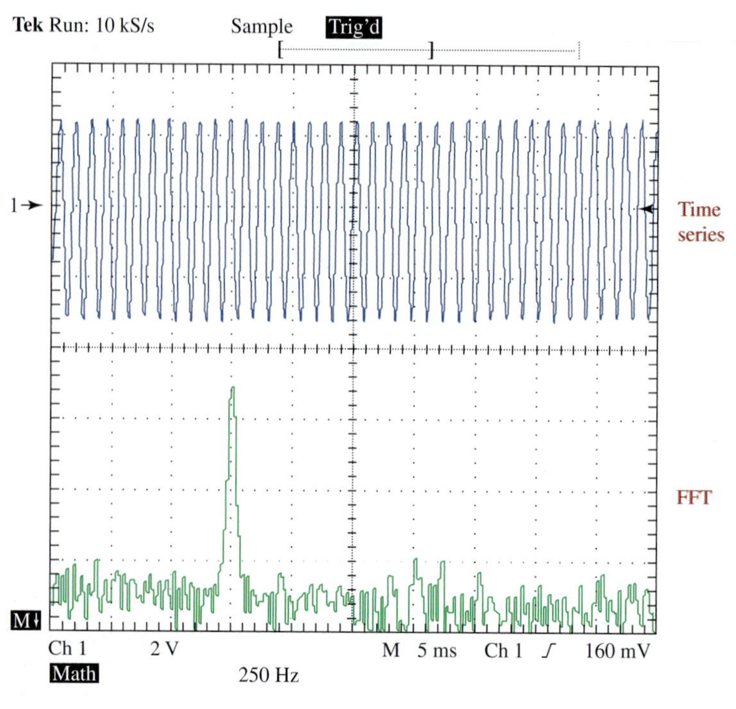

FIGURE 1-38 FFT for Problem 46.

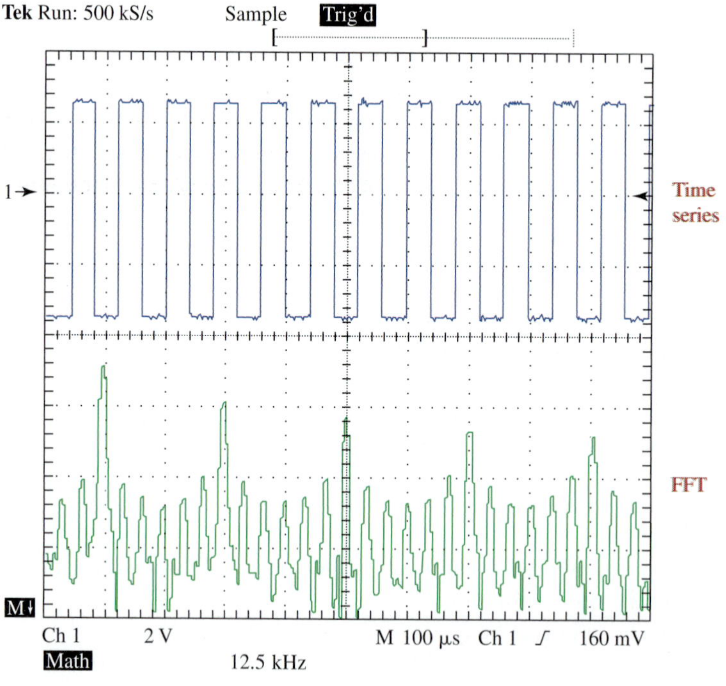

FIGURE 1-39 FFT for Problem 47.

51. Calculate a capacitor's Q at 100 MHz given 0.001 μF and a leakage resistance of 0.7 MΩ. Calculate D for the same capacitor. (4.39×10^5, 2.27×10^{-6})
52. The inductor and capacitor for Problems 50 and 51 are put in series. Calculate the impedance at 100 MHz. Calculate the frequency of resonance (f_r) and the impedance at that frequency. (37.5 kΩ, 65 kHz, 1200 Ω)
53. Calculate the output voltage for the circuit shown in Fig. 1-15 at 6 kHz and 4 kHz. Graph these results together with those of Ex. 1-12 versus frequency. Use the circuit values given in Ex. 1-12.
54. Sketch the e_{out}/e_{in} versus frequency characteristic for an LC bandpass filter. Show f_{lc} and f_{hc} on the sketch and explain how they are defined. On this sketch, show the bandwidth (BW) of the filter and explain how it is defined.
55. Define the quality factor (Q) of an LC bandpass filter. Explain how it relates to the "selectivity" of the filter. Describe the major limiting value on the Q of a filter.
56. An FM radio receiver uses an LC bandpass filter with f_r = 10.7 MHz and requires a BW of 200 kHz. Calculate the Q for this filter. (53.5)
57. The circuit described in Problem 56 is shown in Fig. 1-18. If C = 0.1 nF (0.1×10^{-9} F), calculate the required inductor value and the value of R. (2.21 μH, 2.78 Ω)
58. A parallel LC tank circuit has a Q of 60 and coil winding resistance of 5 Ω. Determine the circuit's impedance at resonance. (18 kΩ)
59. A parallel LC tank circuit has L = 27 mH, C = 0.68 μF, and a coil winding resistance of 4 Ω. Calculate f_r, Q, Z_{max}, the BW, f_{lc}, and f_{hc}. (1175 Hz, 49.8, 9.93 kΩ, 23.6 Hz, 1163 Hz, 1187 Hz)
60. Explain the significance of the k and m in constant-k and m-derived filters.
61. Describe the criteria used in choosing either an RC or LC filter.
62. Explain why keeping lead lengths to a minimum is important in RF circuits.
63. Describe a pole.
64. Explain why Butterworth and Chebyshev filters are called constant-k filters.

Section 1–8

65. Draw schematics for Hartley and Colpitts oscillators. Briefly explain their operation and differences.
66. Describe the reason that a Clapp oscillator has better frequency stability than the Hartley or Colpitts oscillators.
67. List the major advantages of crystal oscillators over the LC varieties. Draw a schematic for a Pierce oscillator.
68. The crystal oscillator time base for a digital wristwatch yields an accuracy of $\pm$15 s/month. Express this accuracy in parts per million (ppm). ($\pm$5.787 ppm)

Section 1–9

69. List and briefly describe the four basic troubleshooting techniques.
70. Describe the disadvantages of using substitution at the early stages of the troubleshooting plan.
71. Explain why resistance measurements are done with power off.
72. Describe the major types of circuit failures.

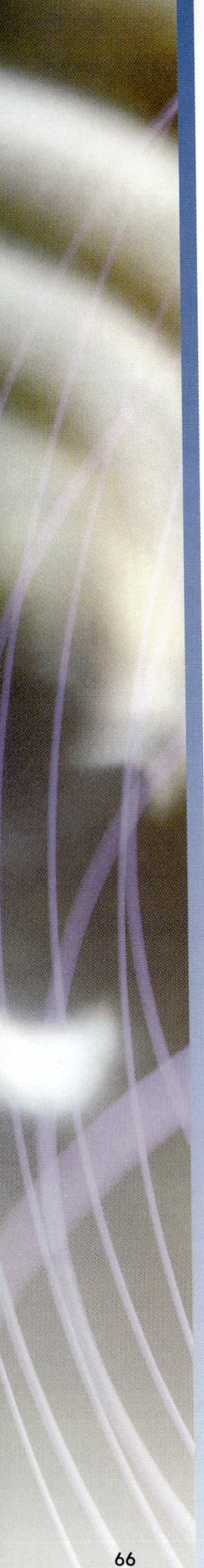

73. Describe when it is more appropriate to use the signal injection method.
74. What would the output of the Clapp oscillator in Fig. 1-33 look like if C2 was open?
75. In the crystal test setup shown in Fig. 1-32, explain the difference in output at the series and parallel resonant frequencies.

Questions for Critical Thinking

76. You cannot guarantee perfect performance in a communications system. What two basic limitations explain this?
77. You are working on a single-stage amplifier that has a 200-kHz bandwidth and a voltage gain of 100 at room temperature. The external noise is negligible. A 1-mV signal is applied to the amplifier's input. If the amplifier has a 5-dB NF and the input noise is generated by a 2-kΩ resistor, what output noise voltage would you predict? (458 μV)
78. How does equivalent noise resistance relate to equivalent noise temperature? Explain similarities and/or differences.
79. Describe a situation in which you would use the Barkhausen criteria for oscillation. How would positive feedback be involved in your use of these criteria?

2 Amplitude Modulation: Transmission

Chapter Outline

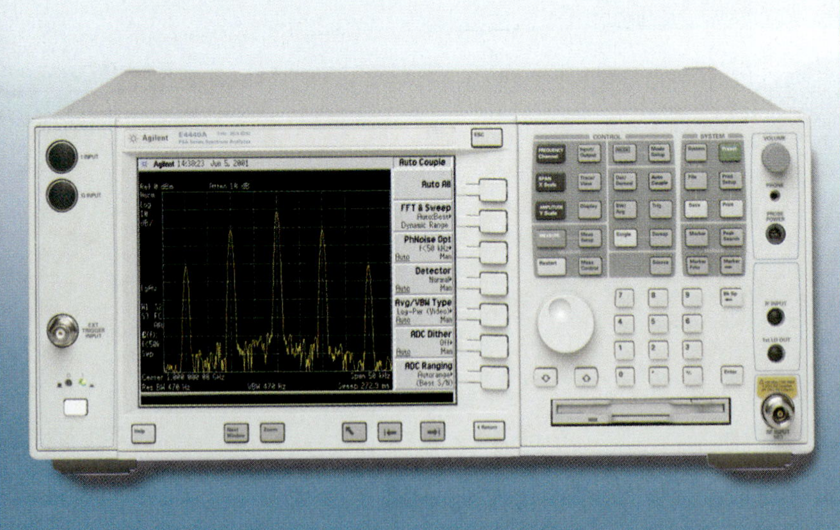

A captured RF spectrum using the Agilent 4440A PSA series spectrum analyzer. (Courtesy of Agilent Technologies. Reprinted with permission.)

Objectives

- Describe the process of modulation
- Sketch an AM waveform with various modulation indexes
- Explain the difference between a sideband and side frequency
- Analyze various power, voltage, and current calculations in AM systems
- Understand circuits used to generate AM
- Determine high- and low-level modulation systems from schematics and block diagrams
- Perform AM transmitter measurements using meters, oscilloscopes, and spectrum analyzers

Key Terms

modulation
nonlinear device
upper sideband
lower sideband
percentage modulation
modulation index
modulation factor
overmodulation
sideband splatter
base modulation
high-level modulation
low-level modulation
neutralizing capacitor
parasitic oscillations
modulated amplifier
driver amplifier
keying
low excitation
downward modulation
spectrum analyzer
spurious frequencies
spurs
noise floor
relative harmonic distortion
total harmonic distortion
dummy antenna

2-1 INTRODUCTION

The reasons that modulation is used in electronic communications have previously been explained as:

1. Direct transmission of intelligible signals would result in catastrophic interference problems because the resulting radio waves would be at approximately the same frequency.
2. Most intelligible signals occur at relatively low frequencies. Efficient transmission and reception of radio waves at low frequencies is not practical due to the large antennas required.

Modulation
impressing a low-frequency intelligence signal onto a higher-frequency carrier signal

The process of impressing a low-frequency intelligence signal onto a higher-frequency "carrier" signal may be defined as **modulation.** The higher-frequency "carrier" signal will hereafter be referred to as simply the carrier. It is also termed the radio-frequency (RF) signal because it is at a high-enough frequency to be transmitted through free space as a radio wave. The low-frequency intelligence signal will subsequently be termed the "intelligence." It may also be identified by terms such as modulating signal, information signal, audio signal, or modulating wave.

Three different characteristics of a carrier can be modified to allow it to "carry" intelligence. Either the amplitude, frequency, or phase of a carrier are altered by the intelligence signal. Varying the carrier's amplitude to accomplish this goal is the subject of this chapter.

2-2 AMPLITUDE MODULATION FUNDAMENTALS

Combining two widely different sine-wave frequencies such as a carrier and intelligence in a linear fashion results in their simple algebraic addition, as shown in Fig. 2-1. A circuit that would perform this function is shown in Fig. 2-1(a)—the

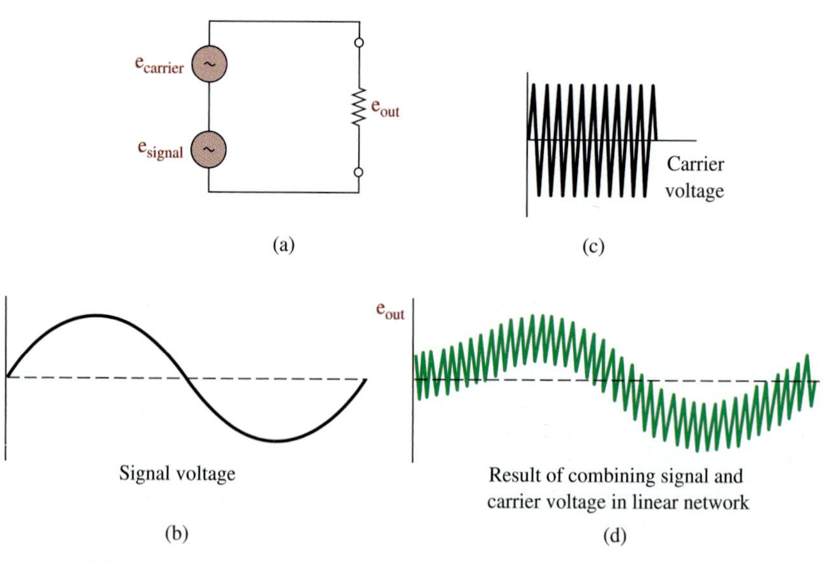

FIGURE 2-1 Linear addition of two sine waves.

two signals combined in a linear device such as a resistor. Unfortunately, the result [Fig. 2-1(d)] is *not* suitable for transmission as an AM waveform. If it were transmitted, the receiving antenna would be detecting just the carrier signal because the low-frequency intelligence component cannot be propagated efficiently as a radio wave.

The method utilized to produce a usable AM signal is to combine the carrier and intelligence through a **nonlinear device.** It can be mathematically proven that the combination of any two sine waves through a nonlinear device produces the following frequency components:

Nonlinear Device characterized by a nonlinear output versus input signal relationship

1. A dc level
2. Components at each of the two original frequencies
3. Components at the sum and difference frequencies of the two original frequencies
4. Harmonics of the two original frequencies

Figure 2-2 shows this process pictorially with the two sine waves, labeled f_c and f_i, to represent the carrier and intelligence. If all but the $f_c - f_i$, f_c, and $f_c + f_i$ components are removed (perhaps with a bandpass filter), the three components left form an AM waveform. They are referred to as:

1. The *lower-side frequency* $(f_c - f_i)$
2. The *carrier frequency* (f_c)
3. The *upper-side frequency* $(f_c + f_i)$

Mathematical analysis of this process is provided in Sec. 2-4.

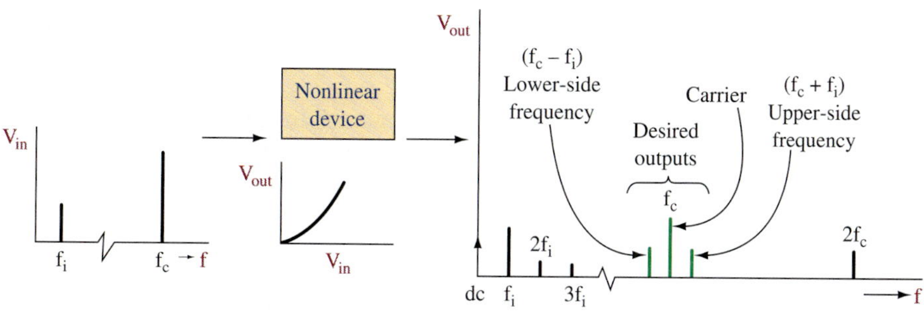

FIGURE 2-2 Nonlinear mixing.

AM Waveforms

Figure 2-3 shows the actual AM waveform under varying conditions of the intelligence signal. Note in Fig. 2-3(a) that the resultant AM waveform is basically a signal at the carrier frequency whose amplitude is changing at the same rate as the intelligence frequency. As the intelligence amplitude reaches a maximum positive value, the AM waveform has a maximum amplitude. The AM waveform reaches a minimum value when the intelligence amplitude is at a maximum negative value. In Fig. 2-3(b), the intelligence frequency remains the same, but its amplitude has been increased. The resulting AM waveform reacts by reaching a larger maximum value and smaller minimum value. In Fig. 2-3(c), the intelligence amplitude is reduced and its frequency has gone up. The resulting AM waveform, therefore, has reduced

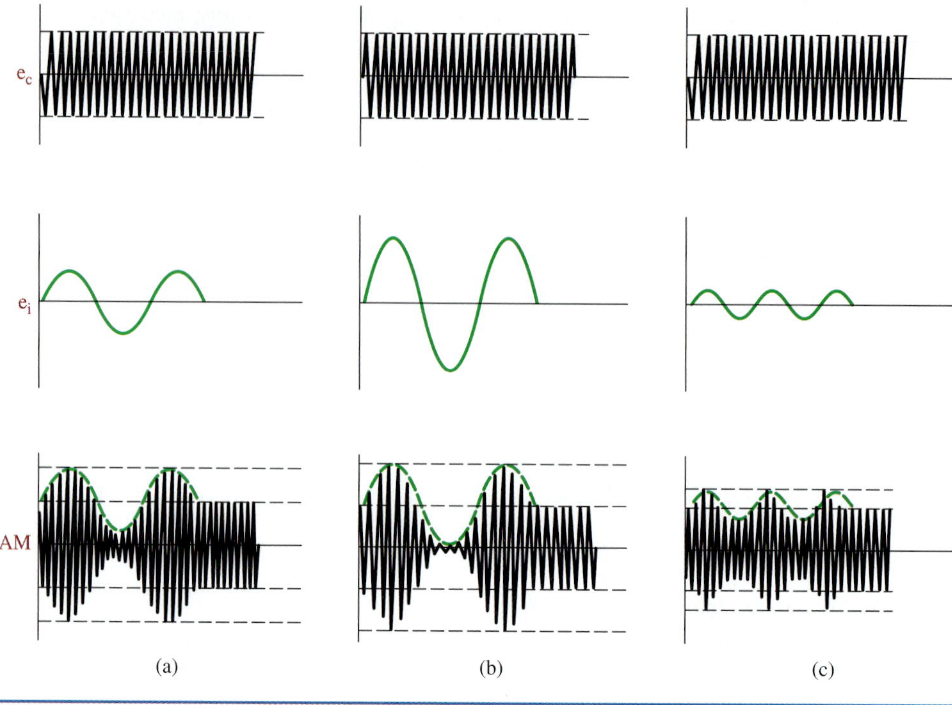

FIGURE 2-3 AM waveform under varying intelligence signal (e_i) conditions.

maximums and minimums, and the rate at which it swings between these extremes has increased to the same frequency as the intelligence signal.

It may now be correctly concluded that both the top and bottom envelopes of an AM waveform are replicas of the frequency and amplitude of the intelligence (notice the 180° phase shift). However, the AM waveform does *not* include any component at the intelligence frequency. The equation for the AM waveform (envelope) is provided in Eq. (2-1).

$$e = (E_c + E_i \sin \omega_i t)\sin \omega_c t \tag{2-1}$$

where E_c = the peak amplitude of the carrier signal
E_i = the peak amplitude of the intelligence signal
$\omega_i t$ = the radian frequency of the intelligence signal
$\omega_c t$ = the radian frequency of the carrier signal
$\omega = 2\pi f$

This equation indicates that an AM waveform will contain the carrier frequency plus the products of the sine waves defining the carrier and intelligence signals. Based on the trigonometric identity,

$$(\sin x)(\sin y) = 0.5 \cos(x - y) - 0.5 \cos(x + y) \tag{2-2}$$

where x is the carrier frequency and y is the intelligence frequency. The product of the carrier and intelligence sine waves will produce the sum and differences of the two frequencies. If a 1-MHz carrier were modulated by a 5-kHz intelligence signal, the AM waveform would include the following components:

$$1\text{ MHz} + 5\text{ kHz} = 1{,}005{,}000\text{ Hz (upper-side frequency)}$$
$$1\text{ MHz} = 1{,}000{,}000\text{ Hz (carrier frequency)}$$
$$1\text{ MHz} - 5\text{ kHz} = 995{,}000\text{ Hz (lower-side frequency)}$$

This process is shown in Fig. 2-4. Thus, even though the AM waveform has envelopes that are replicas of the intelligence signal, it does *not* contain a frequency component at the intelligence frequency.

The intelligence envelope is shown in the resultant waveform and results from connecting a line from each RF peak value to the next one for both the top and bottom halves of the AM waveform. The drawn-in envelope is not really a component of the waveform and would not be seen on an oscilloscope display. In addition, the top and bottom envelopes are *not* the upper- and lower-side frequencies, respectively. The envelopes result from the nonlinear combination of a carrier with two lower-amplitude signals spaced in frequency equal amounts above and below the carrier frequency. The increase and decrease in the AM waveform's amplitude is caused by the frequency difference in the side frequencies, which allows them alternately to add to and subtract from the carrier amplitude, depending on their instantaneous phase relationships.

The AM waveform in Fig. 2-4(d) does not show the relative frequencies to scale. The ratio of f_c to the envelope frequency (which is also f_i) is 1 MHz to 5 kHz, or

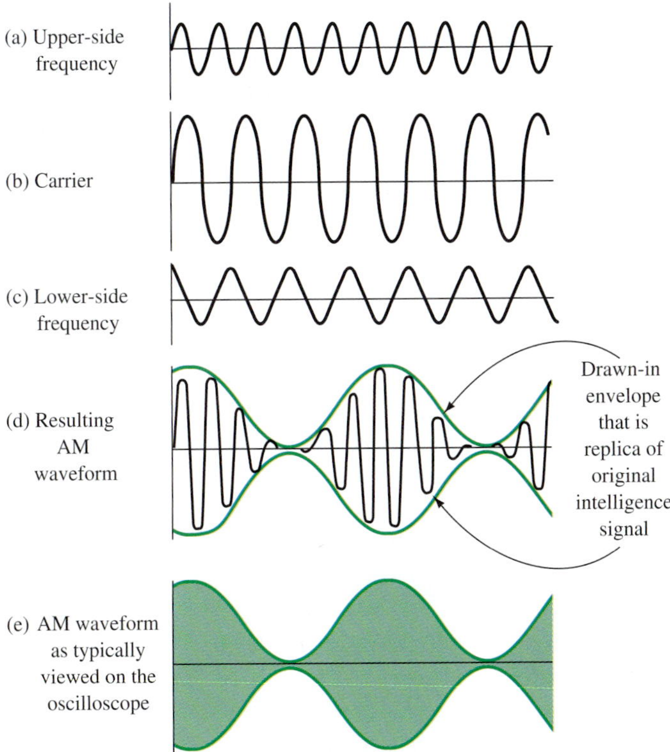

FIGURE 2-4 Carrier and side-frequency components result in AM waveform.

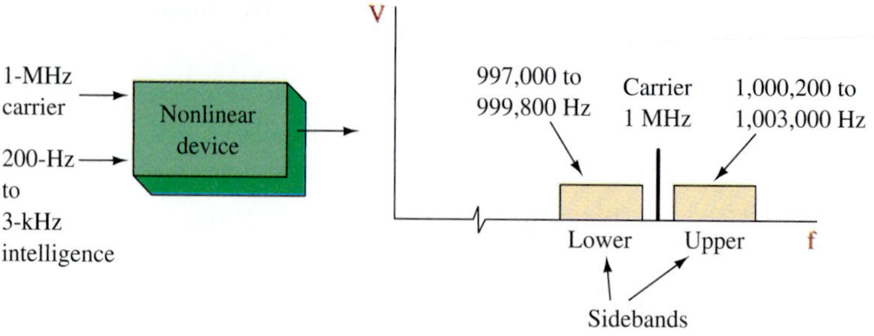

FIGURE 2-5 Modulation by a band of intelligence frequencies.

200 : 1. Thus, the fluctuating RF should show 200 cycles for every cycle of envelope variation. To do that in a sketch is not possible, and an oscilloscope display of this example, and most practical AM waveforms, results in a well-defined envelope but with so many RF variations that they appear as a blur, as shown in Fig. 2-4(e).

Modulation of a carrier with a pure sine-wave intelligence signal has thus far been shown. However, in most systems the intelligence is a rather complex waveform that contains many frequency components. For example, the human voice contains components from roughly 200 Hz to 3 kHz and has a very erratic shape. If it were used to modulate the carrier, a whole *band* of side frequencies would be generated. The band of frequencies thus generated above the carrier is termed the **upper sideband,** while those below the carrier are called the **lower sideband.** This situation is illustrated in Fig. 2-5 for a 1-MHz carrier modulated by a whole band of frequencies, which range from 200 Hz up to 3 kHz. The upper sideband is from 1,000,200 to 1,003,000 Hz, and the lower sideband ranges from 997,000 to 999,800 Hz.

Upper Sideband
band of frequencies produced in a modulator from the creation of sum-frequencies between the carrier and information signals

Lower Sideband
band of frequencies produced in a modulator from the creation of difference frequencies between the carrier and information signals

EXAMPLE 2-1

A 1.4-MHz carrier is modulated by a music signal that has frequency components from 20 Hz to 10 kHz. Determine the range of frequencies generated for the upper and lower sidebands.

SOLUTION

The upper sideband is equal to the sum of carrier and intelligence frequencies. Therefore, the upper sideband (usb) will include the frequencies from

$$1{,}400{,}000 \text{ Hz} + 20 \text{ Hz} = 1{,}400{,}020 \text{ Hz}$$

to

$$1{,}400{,}000 \text{ Hz} + 10{,}000 \text{ Hz} = 1{,}410{,}000 \text{ Hz}$$

The lower sideband (lsb) will include the frequencies from

$$1{,}400{,}000 \text{ Hz} - 10{,}000 \text{ Hz} = 1{,}390{,}000 \text{ Hz}$$

to

$$1{,}400{,}000 \text{ Hz} - 20 \text{ Hz} = 1{,}399{,}980 \text{ Hz}$$

This result is shown in Fig. 2-6 with a frequency spectrum of the AM modulator's output.

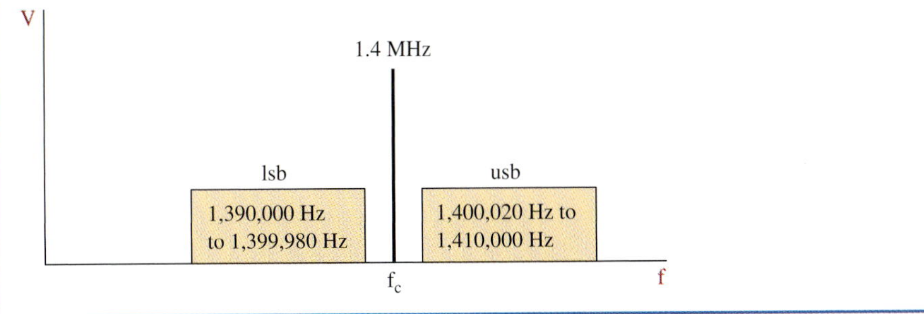

FIGURE 2-6 Solution for Ex. 2-1.

Phasor Representation of AM

It is often helpful to use a phasor representation to help understand generation of an AM signal. For simplicity, let's consider a carrier modulated by a single sine wave with a 100 percent modulation index ($m = 1$). Remember that the AM signal will therefore be composed of the carrier, the usb at one-half the carrier amplitude with frequency equal to the carrier frequency plus the modulating signal frequency, and the lsb at one-half the carrier amplitude at the carrier frequency minus the modulation frequency. With the aid of Fig. 2-7 we will now show how these three sine waves combine to form the AM signal.

1. The carrier phasor represents the peak value of its sine wave. The upper and lower sidebands are one-half the carrier amplitude at 100 percent modulation.

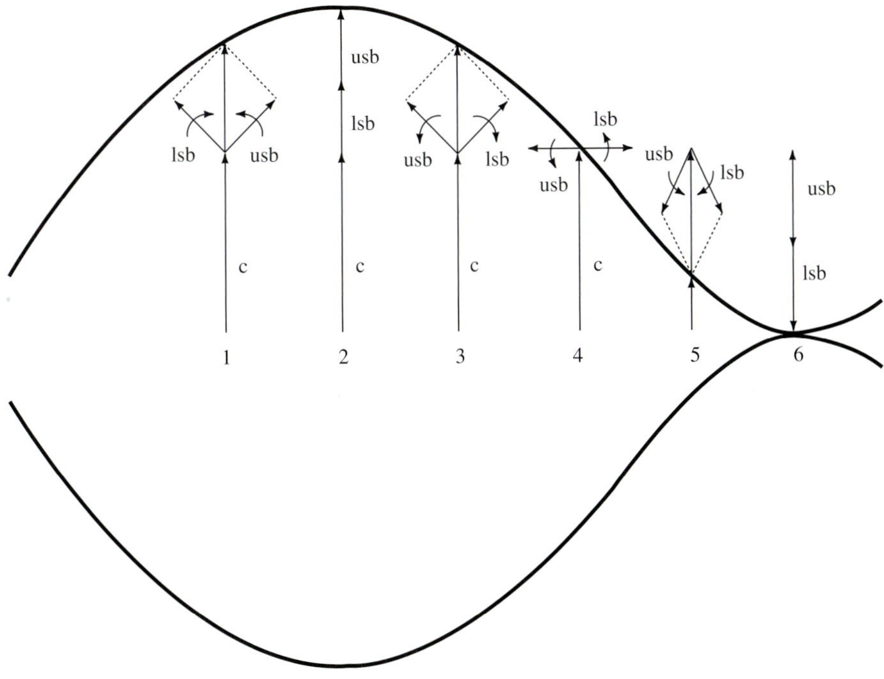

FIGURE 2-7 AM representation using vector addition of phasors.

2. A phasor rotating at a constant rate will generate a sine wave. One full revolution of the phasor corresponds to the full 360° of one sine-wave cycle. The rate of phasor rotation is called angular velocity (ω) and is related to sine-wave frequency ($\omega = 2\pi f$).
3. The sideband phasors' angular velocity is greater and less than the carriers by the modulating signal's angular velocity. This means they are just slightly different from the carriers because the modulating signal is such a low frequency compared to the carrier. You can think of the usb as always slightly gaining on the carrier and the lsb as slightly losing angular velocity with respect to the carrier.
4. If we let the carrier phasor be the reference (stationary with respect to the sidebands) the representation shown in Fig. 2-7 can be studied. Think of the usb phasor as rotating counterclockwise and the lsb phasor rotating clockwise with respect to the "stationary" carrier phasor.
5. The instantaneous amplitude of the AM waveform in Fig. 2-7 is the vector sum of the phasors we have been discussing. At the peak value of the AM signal (point 2) the carrier and sidebands are all in phase, giving a sum of carrier + usb + lsb or twice the carrier amplitude, since each sideband is one-half the carrier amplitude.
6. At point 1 the vector sum of the usb and lsb are added to the carrier and result in an instantaneous value that is also equal to the value at point 3. Notice, however, that the position of the usb and lsb phasors are interchanged at points 1 and 3.
7. At point 4 the vector sum of the three phasors equals the carrier since the sidebands cancel each other. At point 6 the sidebands combine to equal the opposite (negative) of the carrier, resulting in the zero amplitude AM signal that theoretically occurs with exactly 100 percent modulation.

The phasor addition concept helps in understanding how a carrier and sidebands combine to form the AM waveform. It is also helpful in analyzing other communication concepts.

2-3 PERCENTAGE MODULATION

In Sec. 2-2 it was determined that an increase in intelligence amplitude resulted in an AM signal with larger maximums and smaller minimums. It is helpful to have a mathematical relationship between the relative amplitude of the carrier and intelligence signals. The **percentage modulation** provides this, and it is a measure of the extent to which a carrier voltage is varied by the intelligence. The percentage modulation is also referred to as **modulation index** or **modulation factor,** and they are symbolized by *m*.

Percentage Modulation
measure of the extent to which a carrier voltage is varied by the intelligence for AM systems

Modulation Index
another name for percentage modulation; represented as a decimal quantity between 0 and 1 for AM transmitters

Modulation Factor
another name for modulation index

Figure 2-8 illustrates the two most common methods for determining the percentage modulation when modulating with sine waves. Notice that when the intelligence signal is zero, the carrier is unmodulated and has a peak amplitude labeled as E_c. When the intelligence reaches its first peak value (point *w*), the AM signal reaches a peak value labeled E_i (the increase from E_c). Percentage modulation is then given as

$$\%m = \frac{E_i}{E_c} \times 100\% \qquad (2\text{-}3)$$

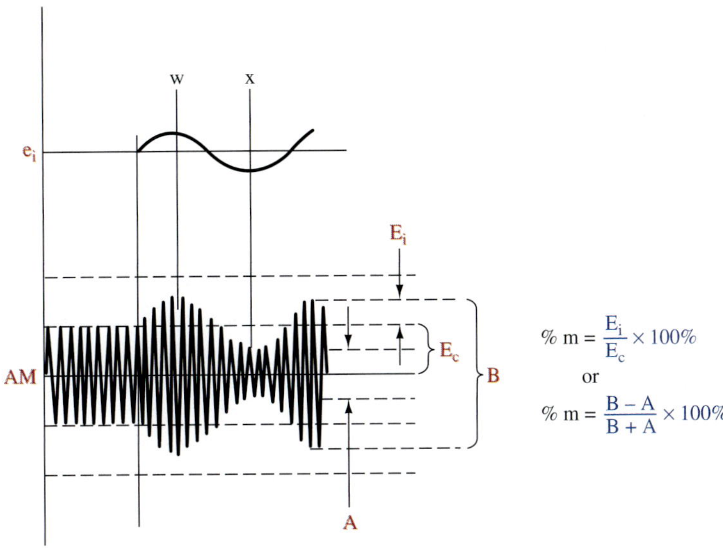

FIGURE 2-8 Percentage modulation determination.

or expressed simply by a ratio:

$$m = \frac{E_i}{E_c} \tag{2-4}$$

The same result can be obtained by utilizing the maximum peak-to-peak value of the AM waveform (point *w*), which is shown as *B*, and the minimum peak-to-peak value (point *x*), which is *A* in the following equation:

$$\%m = \frac{B - A}{B + A} \times 100\% \tag{2-5}$$

This method is usually more convenient in graphical (oscilloscope) solutions.

Overmodulation

If the AM waveform's minimum value *A* falls to zero as a result of an increase in the intelligence amplitude, the percentage modulation becomes

$$\%m = \frac{B - A}{B + A} \times 100\% = \frac{B - O}{B + O} \times 100\% = 100\%$$

This is the maximum possible degree of modulation. In this situation the carrier is being varied between zero and double its unmodulated value. Any further increase in the intelligence amplitude will cause a condition known as **overmodulation** to occur. If this does occur, the modulated carrier will go to more than double its unmodulated value but will fall to zero for an interval of time, as shown in Fig. 2-9. This "gap" produces distortion termed **sideband splatter,** which results in the transmission of frequencies outside a station's normal allocated range. This is an unacceptable condition because it causes severe interference to other stations and causes a loud splattering sound to be heard at the receiver.

Overmodulation when an excessive intelligence signal overdrives an AM modulator producing percentage modulation exceeding 100 percent

Sideband Splatter distortion resulting in an overmodulated AM transmission creating excessive bandwidths

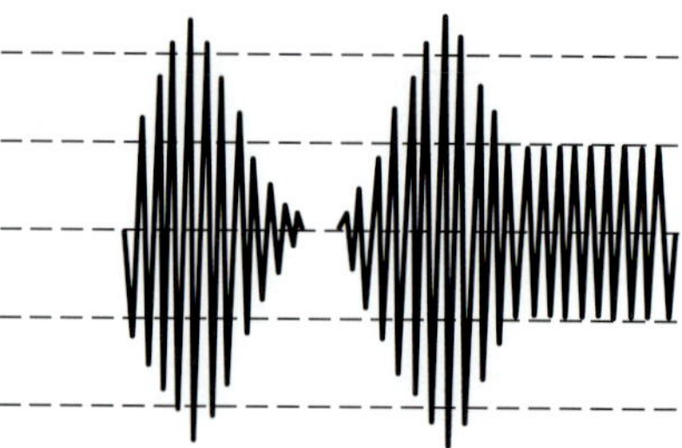

FIGURE 2-9 Overmodulation.

EXAMPLE 2-2

Determine the %m for the following conditions for an unmodulated carrier of 80 V peak-to-peak (p-p).

	Maximum p-p carrier (V)	Minimum p-p carrier (V)
(a)	100	60
(b)	125	35
(c)	160	0
(d)	180	0
(e)	135	35

SOLUTION

(a)
$$\%m = \frac{B - A}{B + A} \times 100\% \tag{2-5}$$
$$= \frac{100 - 60}{100 + 60} \times 100\% = 25\%$$

(b)
$$\%m = \frac{125 - 35}{125 + 35} \times 100\% = 56.25\%$$

(c)
$$\%m = \frac{160 - 0}{160 + 0} \times 100\% = 100\%$$

(d) This is a case of overmodulation since the modulated carrier reaches a value more than twice its unmodulated value.

(e) The increase is greater than the decrease in the carrier's amplitude. This is a distorted AM wave.

2-4 AM ANALYSIS

The instantaneous value of the AM waveform can be developed as follows. The equation for the amplitude of an AM waveform can be written as the carrier peak amplitude, E_c, plus the intelligence signal, e_i. Thus, the amplitude E is

$$E = E_c + e_i$$

but $e_i = E_i \sin \omega_i t$, so that

$$E = E_c + E_i \sin \omega_i t$$

From Eq. (2-2), $E_i = mE_c$, so that

$$\begin{aligned} E &= E_c + mE_c \sin \omega_i t \\ &= E_c(1 + m \sin \omega_i t) \end{aligned}$$

The instantaneous value of the AM wave is the amplitude term E just developed times $\sin \omega_c t$. Thus,

$$\begin{aligned} e &= E \sin \omega_c t \\ &= E_c(1 + m \sin \omega_i t) \sin \omega_c t \end{aligned}$$

Notice that the AM wave (e) is the result of the product of two sine waves. As defined by Eq. 2-2, this product can be expanded with the help of the trigonometric relation $\sin x \sin y = \frac{1}{2}[\cos(x - y) - \cos(x + y)]$. Therefore,

$$e = \overbrace{E_c \sin \omega_c t}^{①} + \overbrace{\frac{mE_c}{2} \cos(\omega_c - \omega_i)t}^{②} - \overbrace{\frac{mE_c}{2} \cos(\omega_c + \omega_i)t}^{③}$$

The preceding equation proves that the AM wave contains the three terms previously listed: the carrier ①, the upper sideband at $f_c + f_i$ ③, and the lower sideband at $f_c - f_i$ ②. It also proves that the instantaneous amplitude of the side frequencies is $mE_c/2$. It shows conclusively that the bandwidth required for AM transmission is twice the highest intelligence frequency.

In the case where a carrier is modulated by a pure sine wave, it can be shown that at 100 percent modulation, the upper- and lower-side frequencies are one-half the amplitude of the carrier. In general, as just developed,

$$E_{SF} = \frac{mE_c}{2} \qquad \textbf{(2-6)}$$

where E_{SF} = side-frequency amplitude
m = modulation index
E_c = carrier amplitude

In an AM transmission, the carrier amplitude and frequency always remain constant, while the sidebands are usually changing in amplitude and frequency. The carrier contains no information since it never changes. However, it does contain the most power since its amplitude is always at least double (when $m = 100\%$) the sideband's amplitude. It is the sidebands that contain the information.

EXAMPLE 2-3

Determine the maximum sideband power if the carrier output is 1 kW and calculate the total maximum transmitted power.

Solution

Since

$$E_{SF} = \frac{mE_c}{2} \tag{2-6}$$

it is obvious that the maximum sideband power occurs when $m = 1$ or 100 percent. At that percentage modulation, each side frequency is $\frac{1}{2}$ the carrier amplitude. Since power is proportional to the square of voltage, each sideband has $\frac{1}{4}$ of the carrier power or $\frac{1}{4} \times 1$ kW, or 250 W. Therefore, the total sideband power is 250 W $\times$ 2 = 500 W and the total transmitted power is 1 kW + 500 W, or 1.5 kW.

Importance of High-Percentage Modulation

It is important to use as high a percentage modulation as possible while ensuring that overmodulation does not occur. The sidebands contain the information and have maximum power at 100 percent modulation. For example, if 50 percent modulation were used in Ex. 2-3, the sideband amplitudes are $\frac{1}{4}$ the carrier amplitude, and since power is proportional to E^2, we have $(\frac{1}{4})^2$, or $\frac{1}{16}$ the carrier power. Thus, total sideband power is now $\frac{1}{16} \times 1$ kW $\times$ 2, or 125 W. The actual transmitted intelligence is thus only $\frac{1}{4}$ of the 500 W sideband power transmitted at full 100 percent modulation. These results are summarized in Table 2-1. Even though the total transmitted power has only fallen from 1.5 kW to 1.125 kW, the effective transmission has only $\frac{1}{4}$ the strength at 50 percent modulation as compared to 100 percent. Because of these considerations, most AM transmitters attempt to maintain between 90 and 95 percent modulation as a compromise between efficiency and the chance of drifting into overmodulation.

A valuable relationship for many AM calculations is

$$P_t = P_c\left(1 + \frac{m^2}{2}\right) \tag{2-7}$$

where P_t = total transmitted power (sidebands and carrier)
P_c = carrier power
m = modulation index

Equation (2-7) can be manipulated to utilize current instead of power. This is a useful relationship since current is often the most easily measured parameter of a transmitter's output to the antenna.

$$I_t = I_c\sqrt{1 + \frac{m^2}{2}} \tag{2-8}$$

Table 2-1 Effective Transmission at 50% versus 100% Modulation

Modulation Index, m	Carrier power (kW)	Power in One Sideband (W)	Total Sideband Power (W)	Total Transmitted Power, P_t (kW)
1.0	1	250	500	1.5
0.5	1	62.5	125	1.125

where I_t = total transmitted current
I_c = carrier current
m = modulation index

Equation (2-8) can also be used with E substituted for I $\left(E_t = E_c\sqrt{1 + m^2/2}\right)$.

Example 2-4

A 500-W carrier is to be modulated to a 90 percent level. Determine the total transmitted power.

Solution

$$P_t = P_c\left(1 + \frac{m^2}{2}\right) \quad \textbf{(2-7)}$$

$$P_t = 500\text{ W}\left(1 + \frac{0.9^2}{2}\right) = 702.5\text{ W}$$

Example 2-5

An AM broadcast station operates at its maximum allowed total output of 50 kW and at 95 percent modulation. How much of its transmitted power is intelligence (sidebands)?

Solution

$$P_t = P_c\left(1 + \frac{m^2}{2}\right) \quad \textbf{(2-7)}$$

$$50\text{ kW} = P_c\left(1 + \frac{0.95^2}{2}\right)$$

$$P_c = \frac{50\text{ kW}}{1 + (0.95^2/2)} = 34.5\text{ kW}$$

Therefore, the total intelligence signal is

$$P_i = P_t - P_c = 50\text{ kW} - 34.5\text{ kW} = 15.5\text{ kW}$$

Example 2-6

The antenna current of an AM transmitter is 12 A when unmodulated but increases to 13 A when modulated. Calculate %m.

Solution

$$I_t = I_c\sqrt{1 + \frac{m^2}{2}} \quad \textbf{(2-8)}$$

$$13\text{ A} = 12\text{ A}\sqrt{1 + \frac{m^2}{2}}$$

$$1 + \frac{m^2}{2} = \left(\frac{13}{12}\right)^2$$

$$m^2 = 2\left[\left(\frac{13}{12}\right)^2 - 1\right] = 0.34$$

$$m = 0.59$$

$$\%m = 0.59 \times 100\% = 59\%$$

EXAMPLE 2-7

An intelligence signal is amplified by a 70% efficient amplifier before being combined with a 10-kW carrier to generate the AM signal. If you want to operate at 100 percent modulation, what is the dc input power to the final intelligence amplifier?

SOLUTION

You may recall that the efficiency of an amplifier is the ratio of ac output power to dc input power. To modulate a 10-kW carrier fully requires 5 kW of intelligence. Therefore, to provide 5 kW of sideband (intelligence) power through a 70 percent efficient amplifier requires a dc input of

$$\frac{5\text{ kW}}{0.70} = 7.14\text{ kW}$$

If a carrier is modulated by more than a single sine wave, the effective modulation index is given by

$$m_{\text{eff}} = \sqrt{m_1^2 + m_2^2 + m_3^2 + \cdots} \qquad \textbf{(2-9)}$$

The total effective modulation index must not exceed 1 or distortion (as with a single sine wave) will result. The term m_{eff} can be used in all previously developed equations using m.

EXAMPLE 2-8

A transmitter with a 10-kW carrier transmits 11.2 kW when modulated with a single sine wave. Calculate the modulation index. If the carrier is simultaneously modulated with another sine wave at 50 percent modulation, calculate the total transmitted power.

SOLUTION

$$P_t = P_c\left(1 + \frac{m^2}{2}\right) \qquad \textbf{(2-7)}$$

$$11.2\text{ kW} = 10\text{ kW}\left(1 + \frac{m^2}{2}\right)$$

$$m = 0.49$$

$$
\begin{aligned}
m_{\text{eff}} &= \sqrt{m_1^2 + m_2^2} \\
&= \sqrt{0.49^2 + 0.5^2} \\
&= 0.7
\end{aligned}
\tag{2-9}
$$

$$
\begin{aligned}
P_t &= P_c\left(1 + \frac{m^2}{2}\right) \\
&= 10\text{ kW}\left(1 + \frac{0.7^2}{2}\right) \\
&= 12.45\text{ kW}
\end{aligned}
$$

2-5 CIRCUITS FOR AM GENERATION

Amplitude modulation is generated by combining carrier and intelligence frequencies through a nonlinear device. Diodes have nonlinear areas, but they are not often used because, being passive devices, they offer no gain. Transistors offer nonlinear operation (if properly biased) and provide amplification, thus making them ideal for this application. Figure 2-10(a) shows an input/output relationship for a typical bipolar junction transistor (BJT). Notice that at both low and high values of current, nonlinear areas exist. Between these two extremes is the linear area that should be used for normal amplification. One of the nonlinear areas must be used to generate AM.

Figure 2-10(b) shows a very simple transistor modulator. It operates with no base bias and thus depends on the positive peaks of e_c and e_i to bias it into the first nonlinear area shown in Fig. 2-10(a). Proper adjustment of the levels of e_c and e_i is necessary for good operation. Their levels must be low to stay in the first nonlinear area, and the intelligence power must be one-half the carrier power (or less) for 100 percent modulation (or less). In the collector a parallel resonant circuit, tuned to the carrier frequency, is used to tune into the three desired frequencies—the upper and lower sidebands and the carrier. The resonant circuit presents a high impedance to the carrier (and any other close frequencies such as the sidebands) and thus allows a high output to those components, but its very low impedance to all other frequencies

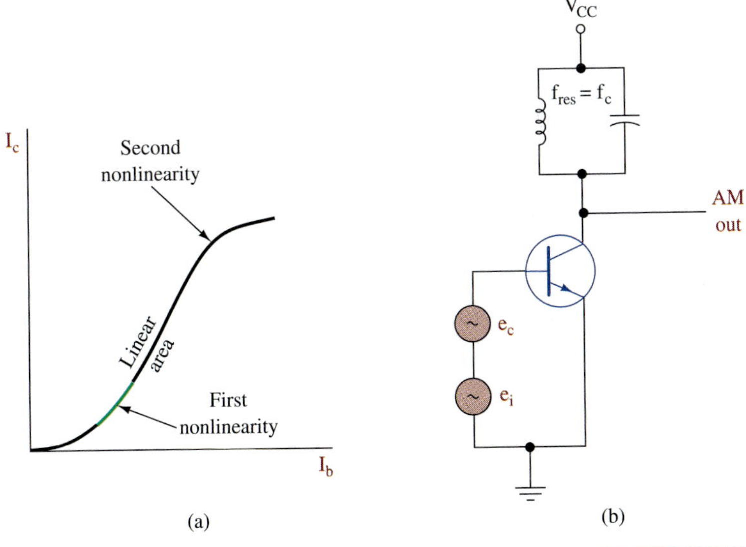

FIGURE 2-10 Simple transistor modulator.

effectively shorts them out. Recall that the mixing of two frequencies through a nonlinear device generates more than just the desired AM components, as illustrated in Fig. 2-2. The tuned circuit then "sorts" out the three desired AM components and serves to provide good sinusoidal components by the flywheel effect.

Base Modulation
a modulation system in which the intelligence is injected into the base of a transistor

In practice, amplitude modulation can be obtained in several ways. For descriptive purposes, the point of intelligence injection is utilized. For example, in Fig. 2-10(b) the intelligence is injected into the base, hence it is termed **base modulation.** *Collector* and *emitter modulation* are also used. In previous years, when vacuum tubes were widely used, the most common form was *plate modulation*, but *grid, cathode*, and (for pentodes) *suppressor-grid* and *screen-grid* modulation schemes were also utilized.

High- and Low-Level Modulation

High-Level Modulation
in an AM transmitter, intelligence superimposed on the carrier at the last point before the antenna

Low-Level Modulation
in an AM transmitter, intelligence superimposed on the carrier; then the modulated waveform is amplified before reaching the antenna

Another common designator for modulators involves whether or not the intelligence is injected at the last possible place or not. For example, the plate-modulated circuit shown in Fig. 2-11 has the intelligence added at the last possible point before the transmitting antenna and is termed a **high-level modulation** scheme. If the intelligence was injected at any previous point, such as at a base, emitter, grid, or cathode, or even at a previous stage, it would be termed **low-level modulation.** The designer's choice between high- and low-level systems is made largely on the basis of the required power output. For high-power applications such as standard radio broadcasting, where outputs are measured in terms of kilowatts instead of watts, high-level modulation is the most economical approach. Vacuum tubes are still the best choice for many high-frequency, high-power transmitter outputs. Recall that class C bias (device conduction for less than 180°) allows for the highest possible efficiency. It realistically provides 70 to 80 percent efficiency as compared to about 50 to 60 percent for the next best configuration, a class B (linear) amplifier. However, class C amplification cannot be used for reproduction of the complete AM signal, and hence large amounts of intelligence power must be injected at the final output to provide a high-percentage modulation.

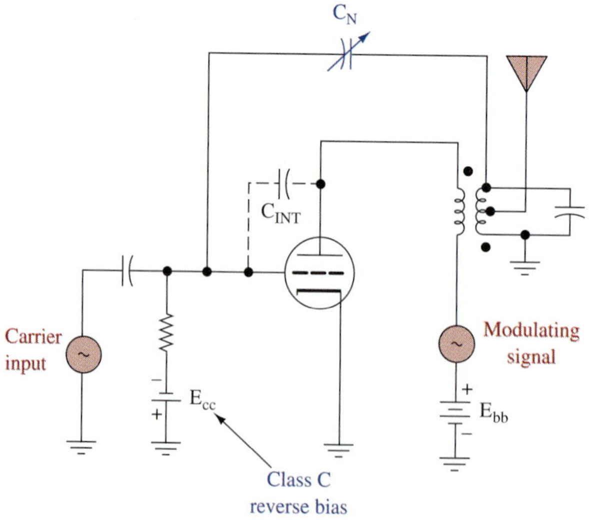

FIGURE 2-11 Plate-modulated class C amplifier.

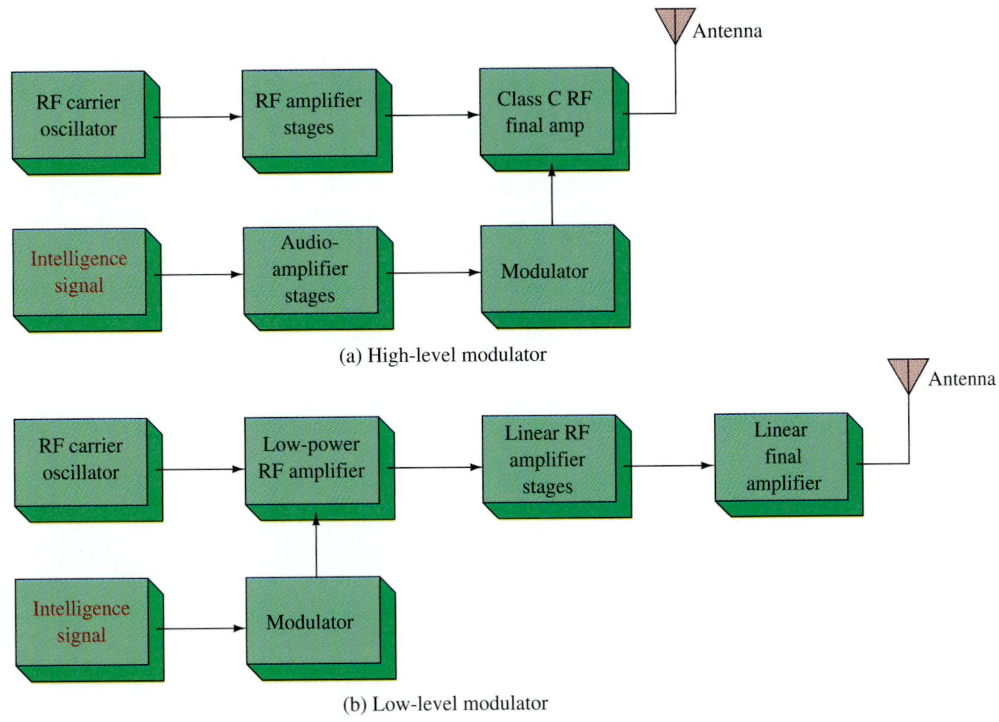

FIGURE 2-12 (a) High- and (b) low-level modulation.

The modulation process is accomplished in a nonlinear device, but all circuitry that follows must be linear. This is required to provide reproduction of the AM signal without distortion. The class C amplifier is not linear but can reproduce (and amplify) the single frequency carrier. However, it would distort the carrier and sidebands combination of the AM signal. This is due to their changing amplitude that would be distorted by the flywheel effect in the class C tank circuit. Block diagrams for typical high- and low-level modulator systems are shown in Fig. 2-12(a) and (b), respectively. Note that in the high-level modulation system [Fig. 2-12(a)] the majority of power amplification takes place in the highly efficient class C amplifier. The low-level modulation scheme has its power amplification take place in the much less efficient linear final amplifier.

In summary, then, high-level modulation requires larger intelligence power to produce modulation but allows extremely efficient amplification of the higher-powered carrier. Low-level schemes allow low-powered intelligence signals to be used, but all subsequent output stages must use less efficient linear (not class C) configurations. Low-level systems usually offer the most economical approach for low-power transmitters.

Neutralization

One of the last remaining applications where tubes offer advantages over solid-state devices is in radio transmitters, where kilowatts of output power are required at high frequencies. Thus, the general configuration shown in Fig. 2-11 is still being utilized. Note the variable capacitor, C_N, connected from the plate tank circuit back

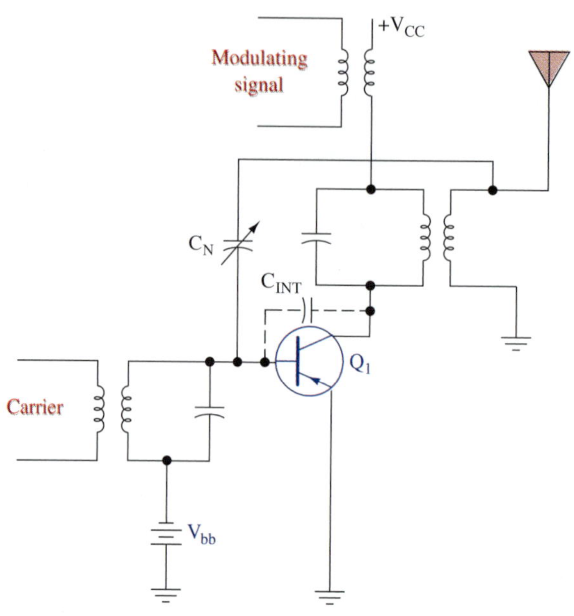

FIGURE 2-13 Collector modulator.

Neutralizing Capacitor
a capacitor that cancels fed-back signals to suppress self-oscillation

to the grid. It is termed the **neutralizing capacitor.** It provides a path for the return of a signal that is 180° out of phase with the signal returned from plate to grid via the internal interelectrode capacitance (C_{INT}) of the tube. C_N is adjusted to cancel the internally fed-back signal to reduce the tendency of self-oscillation. The transformer in the plate is made to introduce a 180° phase shift by appropriate wiring.

Self-oscillation is a problem for all RF amplifiers (both linear and class C). Notice the neutralization capacitor (C_N) shown in the transistor amplifier in Fig. 2-13. The self-oscillation can be at the tuned frequency or at a higher frequency. The higher-frequency self-oscillations are called **parasitic oscillations.** In any event these oscillations are undesirable. At the tuned frequency they prevent amplification from taking place. The parasitic oscillations introduce distortion and reduce desired amplification.

Parasitic Oscillations
higher-frequency self-oscillations in RF amplifiers

Transistor High-Level Modulator

Figure 2-13 shows a transistorized class C, high-level modulation scheme. Class C operation provides an abrupt nonlinearity when the device switches on and off, which allows for the generation of the sum and difference frequencies. This is in contrast to the use of the gradual nonlinearities offered by a transistor at high and low levels of class A bias, as previously shown in Fig. 2-10(a). Generally, the operating point is established to allow half the maximum ac output voltage to be supplied at the collector when the intelligence signal is zero. The V_{bb} supply provides a reverse bias for Q_1 so that it conducts on only the positive peak of the input carrier signal. This, by definition, is class C bias because Q_1 conducts for less than 180° per cycle. The tank circuit in Q_1's collector is tuned to resonate at f_c, and thus the full carrier sine wave is reconstructed there by the flywheel effect at the extremely high efficiency afforded by class C operation.

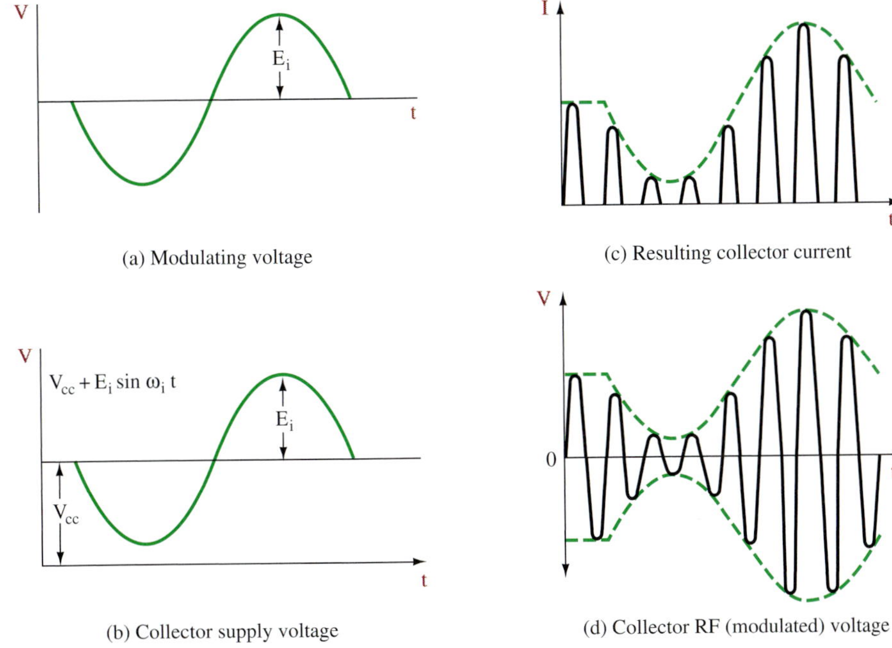

FIGURE 2-14 Collector modulator waveforms.

The intelligence (modulating) signal for the collector modulator of Fig. 2-13 is added directly in series with the collector supply voltage. The net effect of the intelligence signal is to vary the energy available to the tank circuit each time Q_1 conducts on the positive peaks of carrier input. This causes the output to reach a maximum value when the intelligence is at its peak positive value and a minimum value when the intelligence is at its peak negative value. Since the circuit is biased to provide one-half of the maximum possible carrier output when the intelligence is zero, theoretically an intelligence signal level exists where the carrier will swing between twice its static value and zero. This is a fully modulated (100 percent modulation) AM waveform. In practice, however, the collector modulator cannot achieve 100 percent modulation because the transistor's knee in its characteristic curve changes at the intelligence frequency rate. This limits the region over which the collector voltage can vary, and slight collector modulation of the preceding stage is necessary to allow the high modulation indexes that are usually desirable. This is sometimes not a necessary measure in the tube-type high-level modulators.

Figure 2-14(a) shows an intelligence signal for a collector modulator, and Fig. 2-14(b) shows its effect on the collector supply voltage. In Fig. 2-14(c), the resulting collector current variations that are in step with the available supply voltages are shown. Fig. 2-14(d) shows the collector voltage produced by the flywheel effect of the tank circuit as a result of the varying current peaks that are flowing through the tank.

PIN Diode Modulator

Generating AM at frequencies above 100 MHz is expensive since available transistors and ICs are costly. Above 1 GHz (microwave frequencies) it is difficult at any cost with the exception of PIN diodes. Further detail on these devices is provided in the microwave section of Chapter 16.

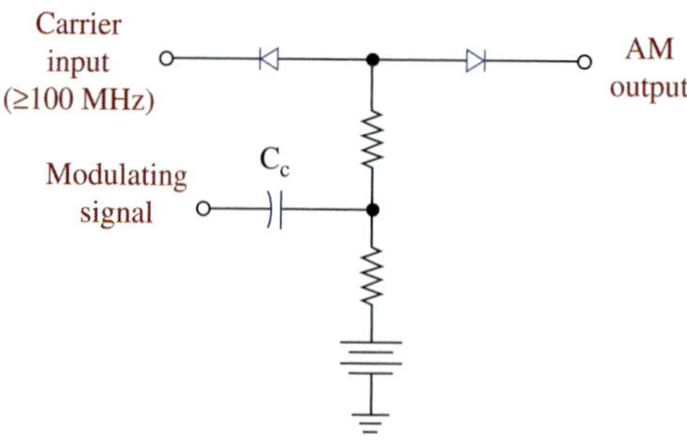

FIGURE 2-15 PIN diode modulator.

PIN diodes are used almost exclusively to generate AM at carrier frequencies above 100 MHz. A basic circuit is shown in Fig. 2-15. The two PIN diodes act as variable resistors when operated above 100 MHz and when forward biased as shown. At lower frequencies they act like regular diodes. The variable resistance when forward biased at high frequencies is quite linear with respect to the level of forward bias. When the modulating signal (applied through C_c) is going positive, it increases the level of forward bias on the two PIN diodes, thereby reducing their resistance and increasing the carrier's output amplitude. The negative-going modulating signal subtracts from the forward bias, thereby decreasing the amount of carrier reaching the output.

The desired AM signal is produced as just described and the quality of the signal is determined by the linearity of the PIN diodes' resistance versus forward bias relationship. The resistance of the PIN diodes attenuates the signal and it must be amplified to bring the AM signal up to a usable level. As stated earlier, PIN diodes are the only practical AM modulators for carriers above about 100 MHz.

Linear-Integrated-Circuit Modulators

The process of generating high-quality AM signals economically is greatly simplified by the availability of low-cost specialty linear integrated circuits (LICs). This is especially true for low-power systems, where low-level modulation schemes are attractive. As an example, the RCA CA3080 operational transconductance amplifier (OTA) can be used to provide AM with an absolute minimum of design considerations. The OTA is similar to conventional operational amplifiers inasmuch as they employ the usual differential input terminals, but its output is best described in terms of the output current, rather than voltage, that it can supply. In addition, it contains an extra control terminal that enhances flexibility for use in a variety of applications, including AM generation.

Figure 2-16(a) shows the CA3080 connected as an amplitude modulator. The gain of the OTA to the input carrier signal is controlled by variation of the amplifier-bias current at pin 5 (I_{ABC}) because the OTA transconductance (and hence gain) is directly proportional to this current. The level of the unmodulated carrier output is determined by the quiescent I_{ABC} current, which is set by the value of R_m. The 100-kΩ potentiometer is adjusted to set the output voltage symmetrically about zero, thus nulling the effects of amplifier input offset voltage. Figure 2-16(b) shows the following:

Top trace—the original intelligence signal (red) superimposed on the upper AM envelope, which gives an indication of the high quality of this AM generator

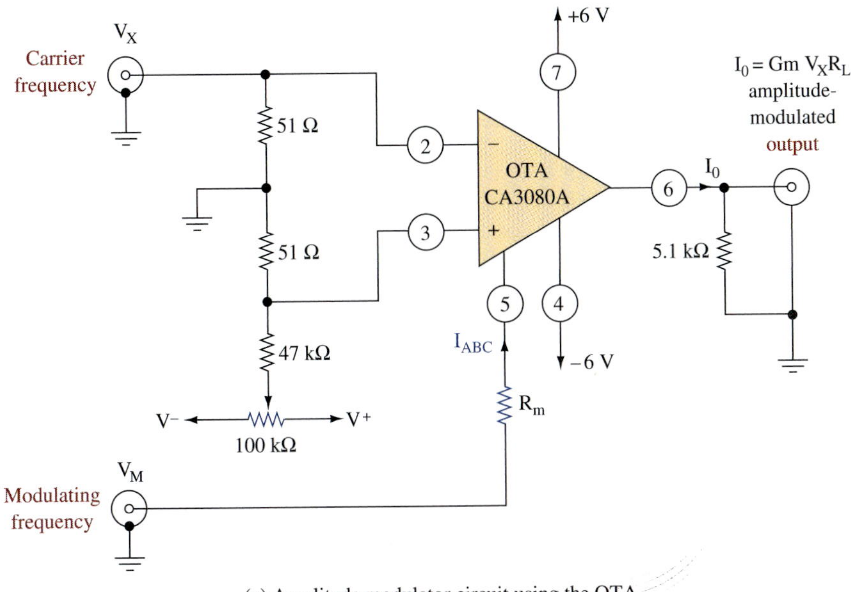

(a) Amplitude modulator circuit using the OTA

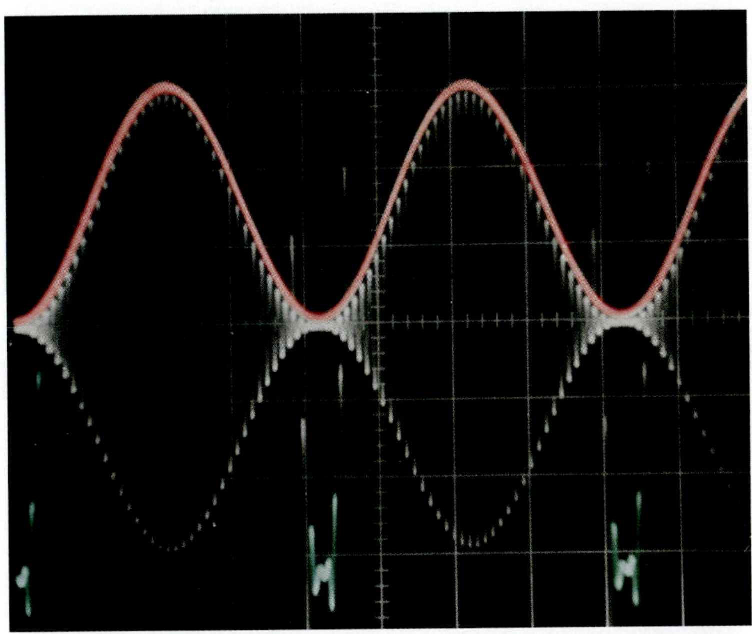

(b) Top trace: modulation frequency input ≈ 20 V p-p and 50 μsec/div
Center trace: amplitude modulation/output 500 mV/div and 50 μsec/div
Bottom trace: expanded output to show depth of modulation 20 mV/div and 50 μsec/div

FIGURE 2-16 LIC amplitude modulator and resulting waveforms. (Courtesy of RCA Solid State Division.)

Center trace—the AM output

Lower trace—the AM output (green) with the scope's vertical sensitivity greatly expanded to show the ability to provide high degrees of modulation (99 percent in this case) with a high degree of quality

Another LIC modulator is shown in Fig. 2-17(a). This circuit uses an HA-2735 programmable dual op amp—half of it to generate the carrier frequency (A_1) and the other half as the AM generator (A_2). In the HA-2735 op amp, the set current (pin 1 for A_1 and pin 13 for A_2) controls the frequency response and gain of each amplifier. This "programmable" function is unnecessary for the oscillator circuit, and thus I_{set1} is fixed by the 147-kΩ resistor. Carrier frequencies up to about 2 MHz can be generated with A_1.

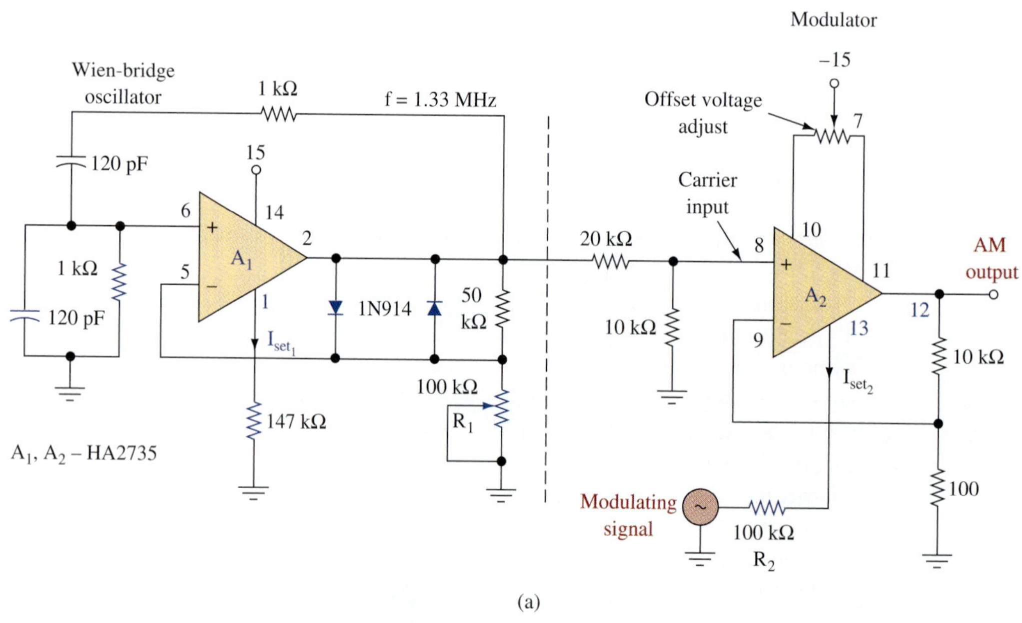

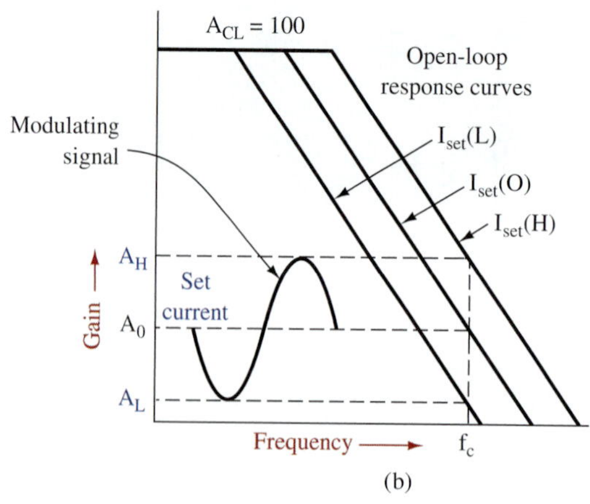

FIGURE 2-17 LIC modulator.

Amplifier A_1 operates as a Wien-bridge oscillator. Amplitude control of the oscillator is achieved with 1N914 clamping diodes in the feedback network. If the output voltage tends to increase, the diodes offer more conductance, which lowers the gain. Resistor R_1 is adjusted to minimize distortion and control the gain, and thus the amount of carrier applied to the modulator. With the components in Fig. 2-17(a), the carrier frequency is approximately 1.33 MHz. This frequency can be changed by selection of different *RC* combinations in the Wien-bridge feedback circuit (1 kΩ and 120 pF).

Amplifier A_2's open-loop response is controlled by the modulating voltage applied to R_2. The percentage of modulation is directly proportional to the modulating voltage. When sinusoidal modulation is applied to R_2, the circuit gain varies from a maximum, A_H, to a minimum, A_L, as A_2's frequency response to the carrier frequency, f_c, is modulated by the set current [Fig. 2-17(b)]. This results in a very distortion-free AM output at pin 12 of A_2. This circuit makes an ideal test generator for troubleshooting AM systems with carrier frequencies to 2 MHz. If a crystal oscillator were used instead of the Wien-bridge circuit, a high-quality AM transmitter could be fabricated with this AM generator.

2-6 AM Transmitter Systems

Section 2-5 dealt with specific circuits to generate AM. Those circuits are only one element of a transmitting system. It is important to obtain a good understanding of a complete transmitting unit, and that is the goal of this section.

Figure 2-18 provides block diagrams of simple high- and low-level AM transmitters. The oscillator that generates the carrier signal will invariably be crystal-controlled to maintain the high accuracy required by the Federal Communications Commission (FCC). The FCC regulates radio and telephone communications in the United States. In Canada the Canadian Radio-Television and Telecommunications Commission performs the same function.

The oscillator is followed by the buffer amplifier, which provides a high-impedance load for the oscillator to minimize drift. It also provides enough gain to drive the modulated amplifier sufficiently. Thus, the buffer amplifier could be a single

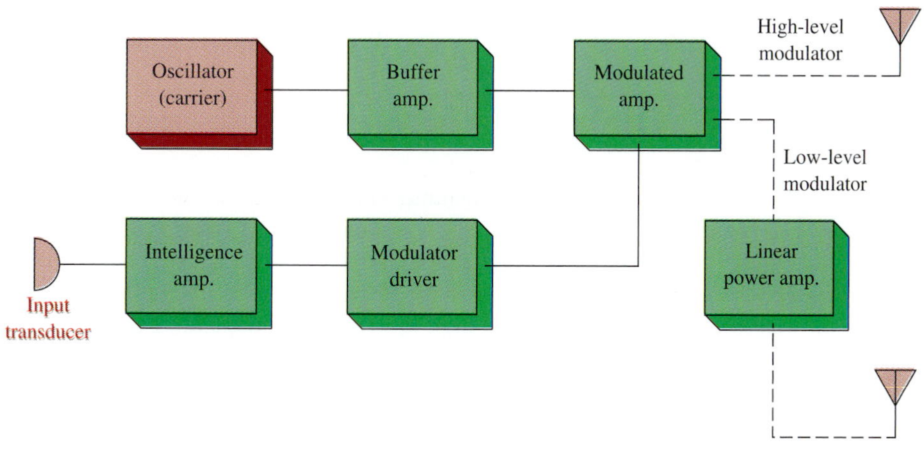

FIGURE 2-18 Simple AM transmitter block diagram.

Typical transmission equipment. (Courtesy of WSNY/WVKO, Columbus, Ohio.)

stage, or however many stages are necessary to drive the following stage, the modulated amplifier.

The intelligence amplifier receives its signal from the input transducer (often a microphone) and contains whatever stages of intelligence amplification are required except for the last one. The last stage of intelligence amplification is called the *modulator*, and its output is mixed in the following stage with the carrier to generate the AM signal. The stage that generates this signal is termed the **modulated amplifier.** This is also the output stage for high-level systems, but low-level systems have whatever number (one or more) of linear power amplifier stages required. Recall that these stages are now amplifying the AM signal and must, therefore, be linear (class A or B), as opposed to the more efficient but nonlinear class C amplifier that can be used as an output stage in high-level schemes.

Modulated Amplifier stage that generates the AM signal

Citizen's Band Transmitter

Figure 2-19 provides a typical AM transmitter configuration for use on the 27-MHz class D citizen's band. It is taken from the Motorola Semiconductor Products Sector Applications Note AN596. It is designed for 13.6-V dc operation, which is the typical voltage level in standard 12-V automotive electrical systems. It employs low-cost plastic transistors and features a novel high-level collector modulation method using two diodes and a double-pi output filter network for matching to the antenna impedance and harmonic suppression.

The first stage uses an MPS 8001 transistor in a common-emitter crystal oscillator configuration. Notice the 27-MHz crystal, which provides excellent frequency stability with respect to temperature and supply voltage variations (well

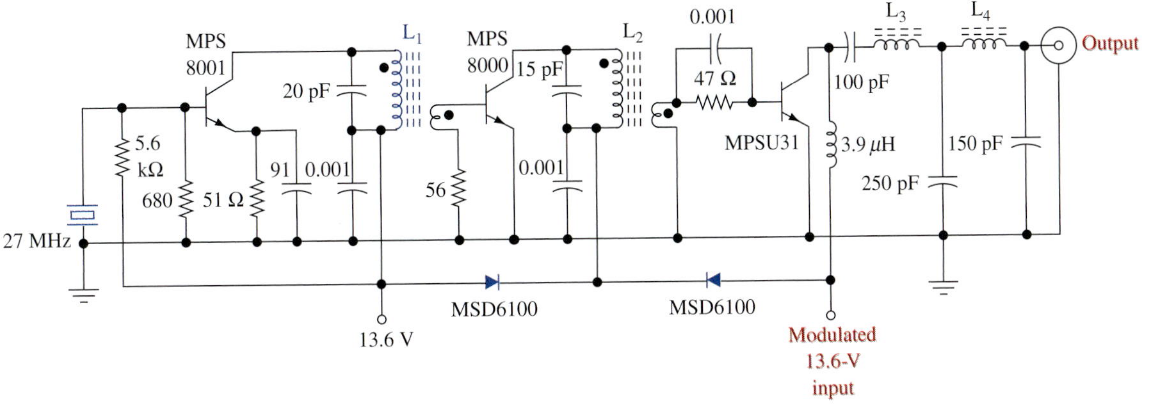

FIGURE 2-19 Class D citizen's band transmitter. (Courtesy of Motorola Semiconductor Products Sector.)

within the 0.005 percent allowance stipulated by FCC regulations for this band). This RF oscillator delivers about 100 mW of 27-MHz carrier power through the L_1 coupling transformer into the buffer (sometimes termed the **driver**) **amplifier,** which uses an MPS8000 transistor in a common-emitter configuration. Information that allows fabrication of the coils used in this transmitter is provided in Fig. 2-20. The use of coils such as these is a necessity in transmitters and allows for required impedance transformations, interstage coupling, and tuning into desired frequencies when combined with the appropriate capacitance to form electrical resonance.

Driver Amplifier
amplifier stage that amplifies a signal prior to reaching the final amplifier stage in a transmitter

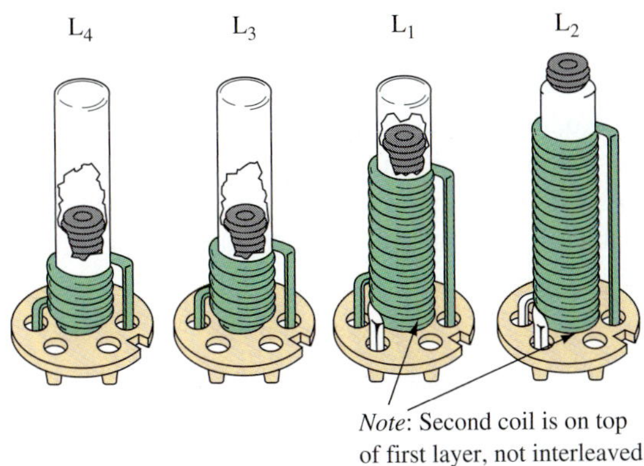

FIGURE 2-20 Coil description for transmitter shown in Fig. 2-19: Conventional transformer coupling is employed between the oscillator and driver stages (L_1) and the driver and final stages (L_2). To obtain good harmonic suppression, a double-pi matching network consisting of (L_3) and (L_4) is utilized to couple the output to the antenna. All coils are wound on standard $\frac{1}{4}$-in. coil forms with No. 22 AWG wire. Carbonyl J $\frac{1}{4}$- × $\frac{3}{8}$-in.-long cores are used in all coils. Secondaries are overwound on the bottom of the primary winding. The cold end of both windings is the start (bottom), and both windings are wound in the same direction. L_1—Primary: 12 turns (close wound); secondary: 2 turns overwound on bottom of primary winding. L_2—Primary: 18 turns (close wound); secondary: 2 turns overwound on bottom of primary winding. L_3—7 turns (close wound). L_4—5 turns (close wound).

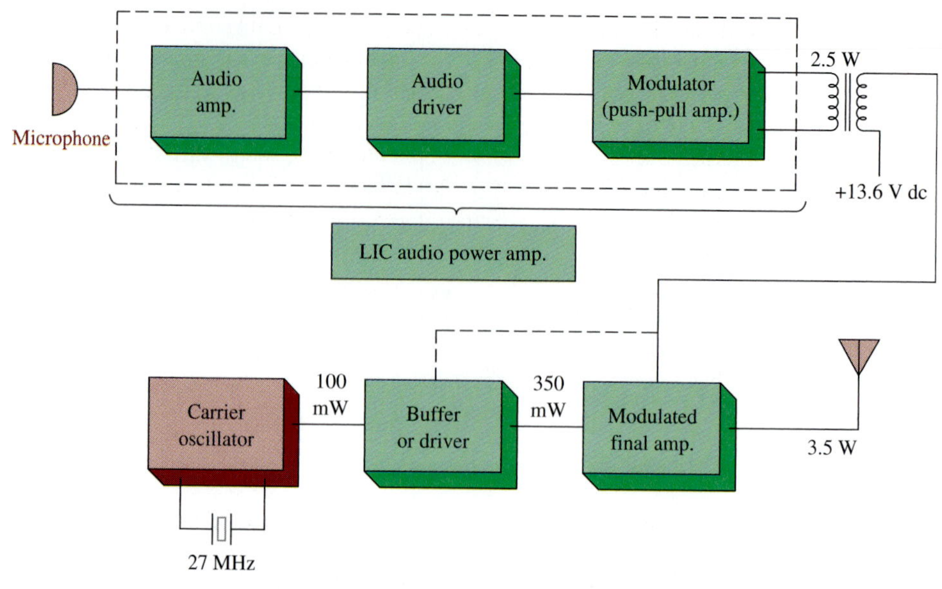

FIGURE 2-21 Citizen's band transmitter block diagram.

The buffer drives about 350 mW into the modulated amplifier. It uses an MPSU31 RF power transistor and can subsequently drive 3.5 W of AM signal into the antenna. This system uses high-level collector modulation on the MPSU31 final transistor, but to obtain a high modulation percentage it is necessary to collector-modulate the previous MPS8000 transistor. This is accomplished with the aid of the MSD6100 dual diode shown in Fig. 2-19. The point labeled "modulated 13.6 V" is the injection point for the intelligence signal riding on the dc supply level of 13.6 V dc. A complete system block diagram for this transmitter is shown in Fig. 2-21. To obtain 100 percent modulation requires about 2.5 W of intelligence power. Thus, the audio amplification blocks between the microphone and the coupling transformer could easily be accomplished by a single low-cost LIC audio power amplifier (shown in dashed lines) capable of 2.5 W of output. The audio output is combined with the dc by the coupling transformer, as is required to modulate the 27-MHz carrier.

Antenna Coupler

Once the 3.5 W of AM signal is obtained at the final stage (MPSU31), it is necessary to "couple" this signal into the antenna. The coupling network for this system comprises L_3, L_4, and the 250- and 150-pF capacitors in Fig. 2-19. This filter configuration is termed a *double-pi network.* To obtain maximum power transfer to the antenna, it is necessary that the transmitter's output impedance be properly matched to the antenna's input impedance. This means equality in the case of a resistive antenna or the complex conjugate in the case of a reactive antenna input. If the transmitter was required to operate at a number of different carrier frequencies, the coupling circuit is usually made variable to obtain maximum transmitted power at each frequency. Coupling circuits are also required to perform some filtering action (to eliminate unwanted frequency components), in addition to their efficient energy transfer function. Conversely, a filter invariably performs a coupling function, and hence the two terms (filter and coupler) are really interchangeable, with what they are called generally governed by the function considered of major importance.

The double-pi network used in the citizen's band transmitter is very effective in suppressing (i.e., filtering out) the second and third harmonics, which would otherwise interfere with communications at 2 × 27 MHz and 3 × 27 MHz. It typically offers 37-dB second harmonic suppression and 55-dB third harmonic suppression. The capacitors and inductors in the double-pi network are resonant to allow frequencies in the 27-MHz region (the carrier and sidebands) to pass, but all other frequencies are severely attenuated. The ratio of the values of the two capacitors determines what part of the total impedance across L_4 is coupled to the antenna, and the value of the 150-pF capacitor has a direct effect on the output impedance.

Transmitter Fabrication and Tuning

The fabrication of high-frequency circuits is much less straightforward than for low frequencies. The minimal inductance of a simple conductor or capacitance between two adjacent conductors can play havoc at high frequencies. Common sense and experimentation generally yield a suitable configuration. The information contained in Fig. 2-22 provides a suggested printed circuit board layout and component mounting photograph for the high-frequency sections (shown schematically in Fig. 2-19).

After assembly it is necessary to go through a tune-up procedure to get the transmitter on the air. Initially, L_1's variable core must be adjusted to get the oscillator to oscillate. This is necessary to get its inductance precisely adjusted so that, in association with its shunt capacitance, it will resonate at the precise 27-MHz resonant frequency of the crystal. The tune-up procedure starts by adjusting the cores of all four coils one-half turn out of the windings. Then turn L_1 clockwise until the oscillator starts and continue for one additional turn. Ensure that the oscillator starts every time by turning the dc on and off (a process termed **keying**) several times. It if does not start reliably every time, turn L_1 clockwise one-quarter turn at a time until it does. Then tune the other coils in order, with the antenna connected, for maximum power output. Apply nearly 100 percent sine-wave intelligence modulation and retune L_2, L_3, and L_4 once again for maximum power output while observing the output on an oscilloscope to ensure that overmodulation and/or distortion do not occur.

Keying
ensuring that an oscillator starts by turning the dc on and off

2-7 Transmitter Measurements

Trapezoid Patterns

Several techniques are available to make operational checks on a transmitter's performance. A standard oscilloscope display of the transmitted AM signal will indicate any gross deficiencies. This technique is all the better if a dual-trace scope is used to allow the intelligence signal to be superimposed on the AM signal, as illustrated in Fig. 2-16. An improvement to this method is known as the *trapezoidal pattern*. It is illustrated in Fig. 2-23. The AM signal is connected to the vertical input and the intelligence signal is applied to the horizontal input with the scope's internal sweep disconnected. The intelligence signal usually must be applied through an adjustable *RC* phase-shifting network, as shown, to ensure that it is exactly in phase with the modulation envelope

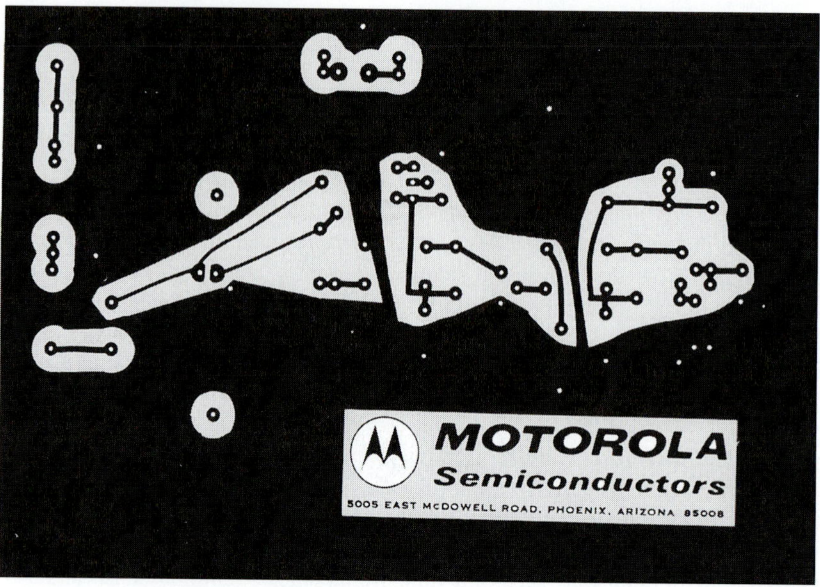

FIGURE 2-22 Citizen's band transmitter PC board layout and complete assembly pictorial. (Courtesy of Motorola Semiconductor Products, Inc.)

of the AM waveform. Figure 2-23(b) shows the resulting scope display with improper phase relationships, and Fig. 2-23(c) shows the proper in-phase trapezoidal pattern for a typical AM signal. It easily allows percentage modulation calculations by applying the *B* and *A* dimensions to the formula presented previously,

$$\%m = \frac{B - A}{B + A} \times 100\% \tag{2-5}$$

In Fig. 2-23(d), the effect of 0 percent modulation (just the carrier) is indicated. The trapezoidal pattern is simply a vertical line because there is no intelligence signal to provide horizontal deflection.

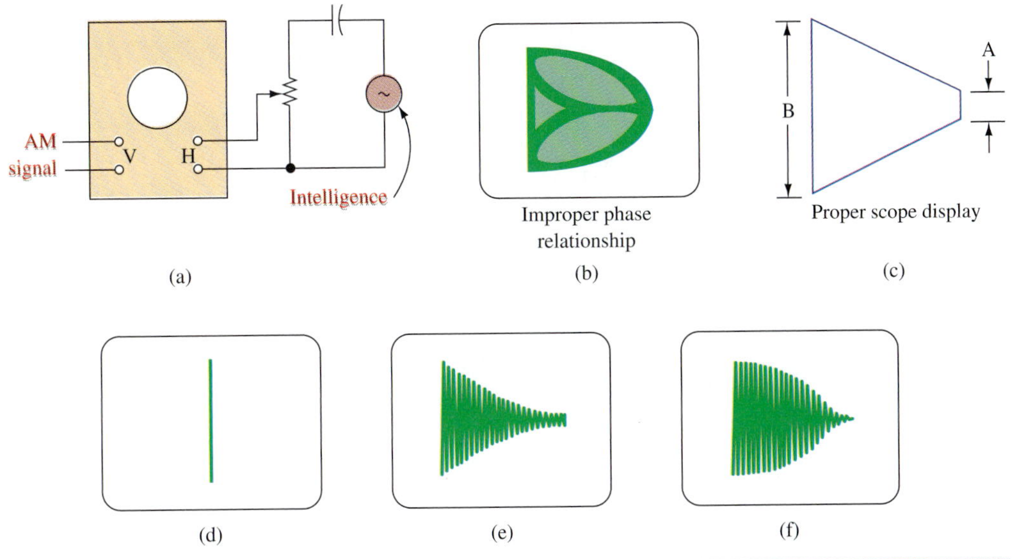

FIGURE 2-23 Trapezoidal pattern connection scheme and displays.

Figures 2-23(e) and (f) show two more trapezoidal displays indicative of some common problems. In both cases the trapezoid's sides are not straight (linear). The concave curvature at (e) indicates poor linearity in the modulation stage, which is often caused by improper neutralization or by stray coupling in a previous stage. The convex curvature at (f) is usually caused by improper bias or low carrier signal power (often termed **low excitation**).

Low Excitation
improper bias or low carrier signal power in an AM modulator

Meter Measurement

It is possible to make some meaningful transmitter checks with a dc ammeter in the collector (or plate) of the modulated stage. If the operation is correct, this current should not change as the intelligence signal is varied between zero and the point where full modulation is attained. This is true because the increase in current during the crest of the modulated wave should be exactly offset by the drop during the trough. A distorted AM signal will usually cause a change in dc current flow. In the case of overmodulation, the current will increase further during the crest but cannot decrease below zero at the trough, and a net increase in dc current will occur. It is also common for this current to decrease as modulation is applied. This malfunction is termed **downward modulation** and is usually the result of insufficient excitation. The current increase during the modulation envelope crest is minimized, but the decrease during the trough is nearly normal.

Downward Modulation
the decrease in dc output current in an AM modulator usually caused by low excitation

Spectrum Analyzers

Spectrum Analyzer
instrument used to measure the harmonic content of a signal by displaying a plot of amplitude versus frequency

The use of spectrum analyzers has become widespread in all fields of electronics, but especially in the communications industry. A **spectrum analyzer** visually displays (on a CRT) the amplitude of the components of a wave as a function of frequency. This can be contrasted with an oscilloscope display, which shows the amplitude of

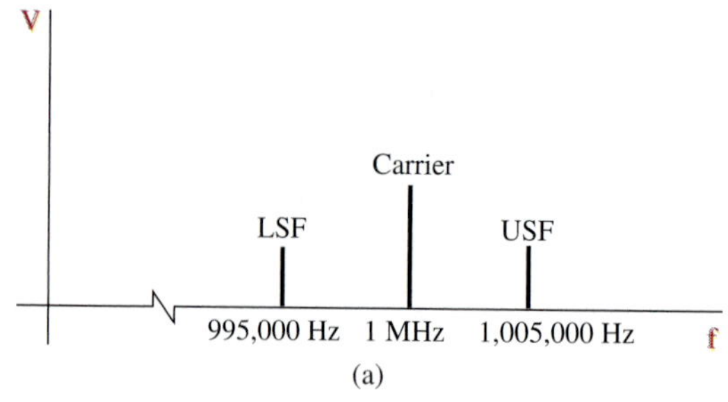

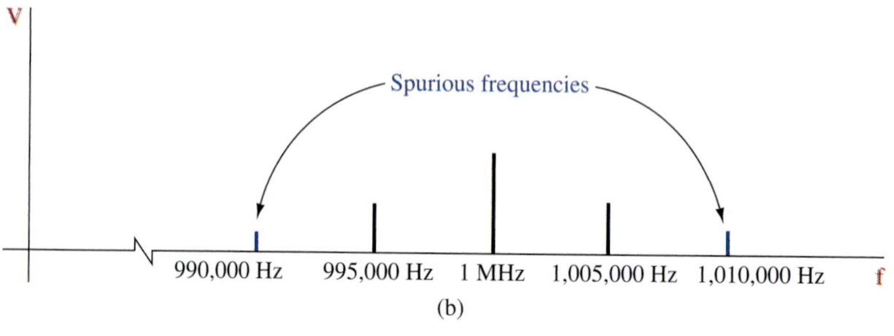

FIGURE 2-24 Spectrum analysis of AM waveforms.

the total wave (all components) versus time. Thus, an oscilloscope shows us the time domain while the spectrum analyzer shows the frequency domain. In Fig. 2-24(a) the frequency domain for a 1-MHz carrier modulated by a 5-kHz intelligence signal is shown. Proper operation is indicated since only the carrier and upper- and lower-side frequencies are present. During malfunctions, and to a lesser extent even under normal conditions, transmitters will often generate **spurious frequencies** as shown in Fig. 2-24(b), where components other than just the three desired are present. These spurious undesired components are usually called **spurs,** and their amplitude is tightly controlled by FCC regulation to minimize interference on adjacent channels. The coupling stage between the transmitter and its antenna is designed to attenuate the spurs, but the transmitter's output stage must also be carefully designed to keep the spurs to a minimum level. The spectrum analyzer is obviously a very handy tool for use in evaluating a transmitter's performance.

Spurious Frequencies
extra frequency components that appear in the spectral display of a signal, signifying distortion

Spurs
undesired frequency components of a signal

The spectrum analyzer is, in effect, an automatic frequency-selective voltmeter that provides both frequency and voltage on its CRT display. It can be thought of as a radio receiver with broad frequency-range coverage and sharp sweep tuning, narrow-bandwidth circuits. The more sophisticated units are calibrated to read signals in dB or dBm. This provides better resolution between low-level sideband signals and the carrier and of course allows direct reading of power levels without resorting to calculation from voltage levels. Most recent spectrum analyzers utilize microprocessor principles (software programming) for ease of operation.

Figure 2-25 shows a typical spectrum analyzer. Available software packages link the waveform analyzer and a computer that enables simplification and automation of complex operations and measurements. A typical display on the computer's CRT is shown in Fig. 2-25. It shows an AM carrier that has four spurious outputs. Notice the noise between the spurs and the carrier. This is commonly referred to as the **noise floor** of the system under test.

Noise Floor
the baseline on a spectrum analyzer display

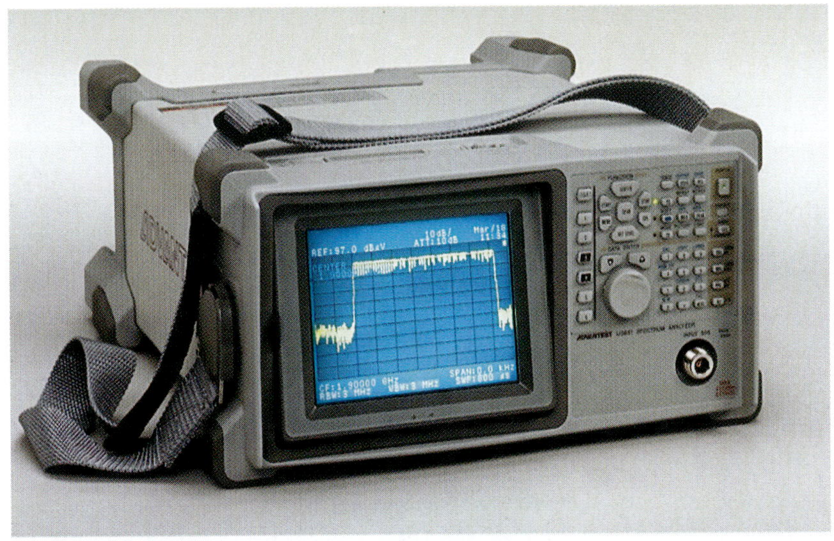

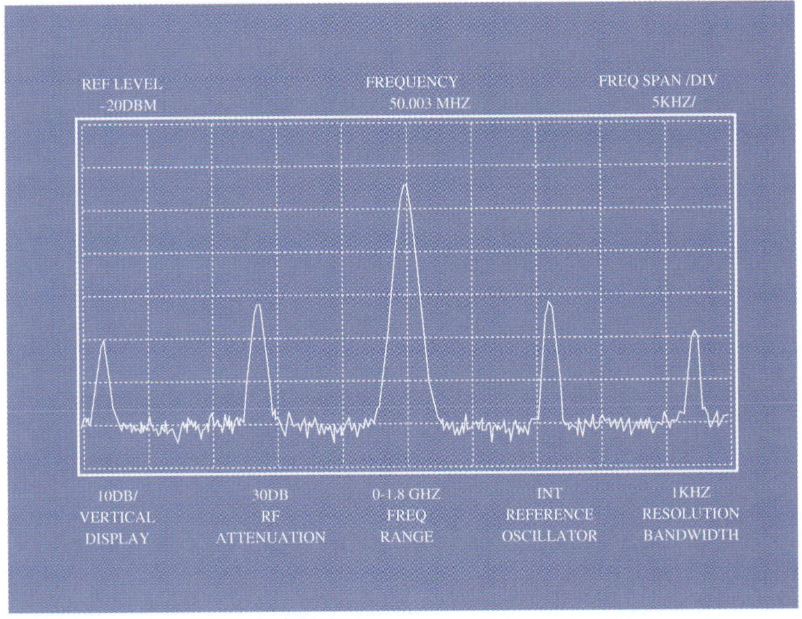

FIGURE 2-25 Spectrum analyzer and typical display. (Courtesy of Tektronix, Inc.)

Harmonic Distortion Measurements

Harmonic distortion measurements can be made easily by applying a spectrally pure signal source to the device under test (DUT). The quality of the measurement is dependent on the harmonic distortion of both the signal source and spectrum analyzer. The source provides a signal to the DUT and the spectrum analyzer is used to monitor the output. Figure 2-26 shows the results of a typical harmonic distortion measurement. The distortion can be specified by expressing the fundamental with respect to the largest harmonic in dB. This is termed the **relative harmonic distortion.**

Relative Harmonic Distortion
expression specifying the fundamental frequency component of a signal with respect to its largest harmonic in dB

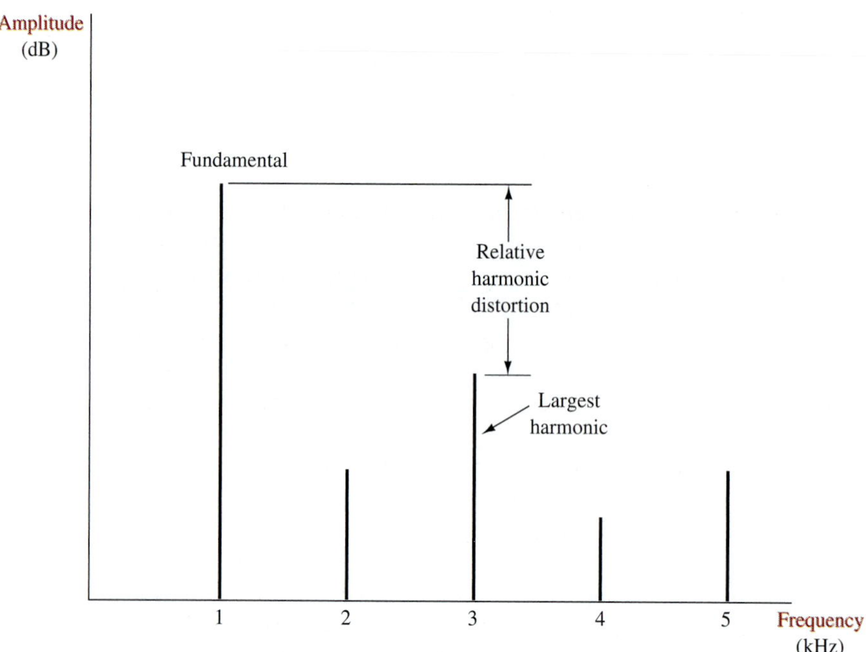

FIGURE 2-26 Relative harmonic distortion.

If the fundamental in Fig. 2-26 is 1 V and the harmonic at 3 kHz (the largest distortion component) is 0.05 V, the relative harmonic distortion is

$$20 \log \frac{1\text{ V}}{0.05\text{ V}} = 26\text{ dB}$$

Total Harmonic Distortion
a measure of distortion that takes all significant harmonics into account

A somewhat more descriptive distortion specification is **total harmonic distortion (THD).** THD takes into account the power in all the significant harmonics:

$$\text{THD} = \sqrt{(V_2^2 + V_3^2 + \cdots)/V_1^2} \qquad \textbf{(2-10)}$$

where V_1 is the rms voltage of the fundamental and V_2, V_3, . . . are the rms voltages of the harmonics. An infinite number of harmonics are theoretically created, but in practice the amplitude falls off for the higher harmonics. Virtually no error is introduced if the calculation does not include harmonics less than one-tenth of the largest harmonic.

Example 2-9

Determine the THD if the spectrum analyzer display in Fig. 2-26 has $V_1 = 1$ V, $V_2 = 0.03$ V, $V_3 = 0.05$ V, $V_4 = 0.02$ V, and $V_5 = 0.04$ V.

Solution

$$\begin{aligned} \text{THD} &= \sqrt{(V_2^2 + V_3^2 + V_4^2 + V_5^2)/V_1^2} \qquad \textbf{(2-10)} \\ &= \sqrt{(0.03^2 + 0.05^2 + 0.02^2 + 0.04^2)/1^2} \\ &= 0.07348 \\ &= 7.35\% \end{aligned}$$

THD calculations are somewhat tedious when a large number of significant harmonics exist. Some spectrum analyzers include an automatic THD function that does all the work and prints out the THD percentage.

Special RF Signal Measurement Precautions

The frequency-domain measurements of the spectrum analyzer provide a more thorough reading of RF frequency signals than does the time-domain oscilloscope. The high cost and additional setup time of the spectrum analyzer dictates the continued use of more standard measurement techniques—mainly voltmeter and oscilloscope usage. Whatever the means of measurement, certain effects must be understood when testing RF signals as compared to the audio frequencies with which you are probably more familiar. These effects are the loading of high-Q parallel-resonant circuits by a relatively low impedance instrument and the frequency response shift that can be caused by test lead and instrument input capacitance.

The consequence of connecting a 50-Ω signal generator into an RF-tuned circuit that has a Z_p in the kilohm region would be a drastically reduced Q and increased bandwidth. The same loading effect would result if a low-impedance detector were used to make RF impedance measurements. This loading is minimized by using resistors, capacitors, or transformers in conjunction with the measurement instrument.

The consequence of test lead or instrument capacitance is to shift the circuit's frequency response. If you were looking at a 10-MHz AM signal that had its resonant circuit shifted to 9.8 MHz by the measurement capacitance, an obvious problem has resulted. Besides some simple attenuation, the equal amplitude relationship between the usb and lsb would be destroyed and waveform distortion results. This effect is minimized by using low-valued series-connected coupling capacitors, or canceled with small series-connected inductors. If more precise readings with inconsequential loading are necessary, specially designed resonant matching networks are required. They can be used as an add-on with the measuring instrument or built into the RF system at convenient test locations.

When testing a transmitter, the use of a dummy antenna is often a necessity. A **dummy antenna** is a resistive load used in place of the regular antenna. It is used to prevent undesired transmissions that may otherwise occur. The dummy antenna (also called a *dummy load*) also prevents damage to the output circuits that may occur under unloaded conditions.

Dummy Antenna
resistive load used in place of an antenna to test a transmitter without radiating the output signal

2-8 Troubleshooting

When traveling uncharted roads, having a map can make the difference between getting lost and finding your way. A plan for troubleshooting, like a map, can help guide you to an equipment malfunction. Made up of a logical sequence of troubleshooting steps, this troubleshooting map can help the technician or engineer hunt down the defect in a piece of electronic communications equipment. By developing and using this strategy, you can become very proficient at locating electronic problems and repairing them.

After completing this section you should be able to

- Describe the purpose of the inspection
- State the sequence of troubleshooting steps

- Troubleshoot an RF amplifier and oscillator
- Check for transmitter operation on proper frequency
- Correct for low transmitter output power

Inspection

The first phase of any repair action is to do a visual inspection of the defective equipment. During this inspection look for broken wires, loose connections, discolored or burned resistors, and exploded capacitors. Burned resistors are easily seen and will give off a distinguishing odor. Equipment that has been dropped may have a broken printed circuit board (PCB). Connectors may have been knocked loose. Look for bad soldering and cold solder joints. Cold solder connections look dull and dingy as opposed to shiny. Intermittent faults are usually the result of cold solder joints. Components hot to the touch after the equipment has been on for a few minutes can indicate shorted components. Listen for unusual sounds when the equipment is turned on. Unusual sounds could lead you to the malfunctioning component. Many defects are found during this inspection phase, and the equipment frequently is fixed without further troubleshooting.

Strategy for Repair

1. Verify That a Problem Does Exist Always verify that the reported problem exists before troubleshooting the equipment. By confirming the problem you will save time that could be wasted looking for a nonexistent defect. If the equipment is not completely dead, try to localize the symptom to a particular stage. Check the service literature for troubleshooting hints. Some manufacturers provide diagnostic charts to help pinpoint malfunctions. The more clues you gather, the more apt you are to associate a fault to a particular circuit function successfully. Another reason for confirming that the problem exists is to rule out operator error. The equipment owner or operator may not know all the equipment's functions and may consider the unit bad if it can't do something it wasn't designed for. Check the operating manual when in doubt about the equipment functions.

2. Isolate the Defective Stage When a problem has been verified in a piece of electronic equipment, begin troubleshooting to isolate the defect to a particular stage. Normally the defective stage can be found by signal tracing or signal injection. The oscilloscope can be used with either the signal tracing or signal injection method. An example of this is in Fig. 2-23, where an AM transmitter's modulation is being monitored by an oscilloscope. As shown, the test pattern appears on the screen and represents over- or undermodulation of the AM transmitted signal. If the transmitted signal were incorrect, the modulator would be adjusted for the proper signal display.

A signal generator or a function generator can be used to supply a test signal at the input of the specific stage under examination. Figure 2-27 illustrates a dual-trace oscilloscope connected to an amplifier stage to show the input test signal and the output signal. The input signal is easily compared to the output signal with this test setup. If the output signal is missing or distorted, then the defective stage has been located.

3. Isolate the Defective Component Once the defective stage has been located, the next step is to find the bad component or components. Voltage and resistance measurements should be used to locate the defective component. Remember, voltage

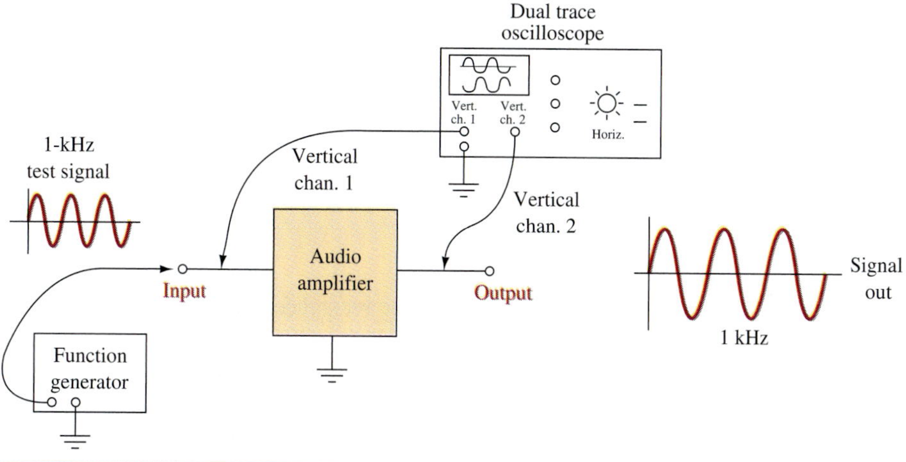

FIGURE 2-27 Comparing input and output signals.

measurements are made with respect to ground, and resistance measurements are done with the power off. The voltage and resistance measurements are compared to specified values stated in the service literature. Incorrect readings usually pinpoint the defective component. It is very possible that there may be more than one bad component in a malfunctioning circuit.

4. **Replace the Defective Component and Hot Check** Before replacing the bad component in a circuit, make sure that another component is not causing it to go bad. For example, a shorted diode or transistor can cause resistors to burn up. Diodes and transistors should be checked for shorted conditions before associated components are replaced. Replace defective components with exact replacement parts when possible. Once the defective component is replaced, ensure that the circuit is operating normally. You may need to do a few more voltage checks to confirm things are back to normal. Burn in (hot check) the equipment by turning it on and letting it operate a number of hours on the test bench. This burn-in is a vital part of equipment repair. If failure is going to reoccur, it will usually show up during this burn-in period.

RF Amplifier Troubleshooting

Bias Supply Many RF amplifiers utilize power from the previous stage to provide dc bias. Figure 2-28 shows how bias for the transistor Q_1 is developed. RF from the previous stage is rectified by the base–emitter junction of Q_1. The current flows

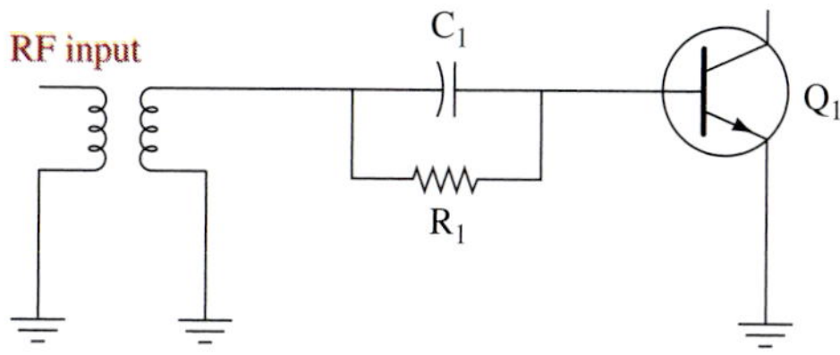

FIGURE 2-28 Self-bias circuit.

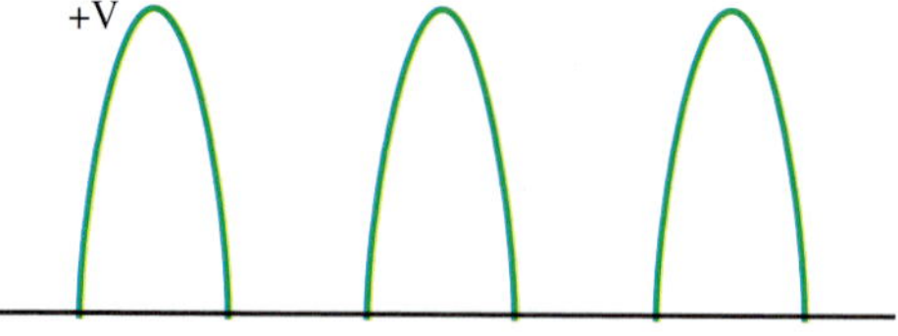

FIGURE 2-29 Voltage at Q_1 base.

through R_1 and the transformer to ground. The reactance of C_1 is low at RF, so the RF bypasses the resistor. C_1 also serves to filter the RF pulses and develop a dc voltage across R_1. At the base of Q_1, this dc voltage is negative with respect to ground. Therefore, Q_1 will be a class C amplifier conducting only on positive RF peaks. Figure 2-29 shows the instantaneous voltage at the base of Q_1 that you can observe with an oscilloscope.

Shorted C_1 If C_1 were to short, excessive drive would reach Q_1. No negative bias for Q_1 could be developed. This would cause Q_1 to draw excessive current and destroy itself. If Q_1 is bad, always check all components ahead of Q_1 before replacing it.

Open C_1 If C_1 were open, the drive reaching Q_1 would be greatly reduced. Bias voltage would be low and Q_1 would not develop full power output.

Open R_1 Resistors in these circuits may overheat and fail to open. C_1 will charge to the negative peak of the RF drive voltage because of the rectifier action of the base–emitter junction. This will cut Q_1 off and there will be no power output.

Output Network Now consider possible faults in components on the output side of Q_1. Common faults are shorted blocking capacitors, overheated tuning capacitors, and open chokes.

Shorted Blocking Capacitor Consider the circuit in Fig. 2-30. Assume that capacitor C_b has shorted. If this amplifier is connected to an antenna that is not dc grounded, there will be no effect at all. C_b is not part of any tuning circuit; its job is to block the dc power supply from the following stage or antenna.

Many antennas show a short circuit to dc. In this case, excessive current would flow through L_1 and L_2, possibly damaging them and the power supply. If a shorted blocking capacitor is found, you should check for damage to the wiring or printed circuit board and the power supply.

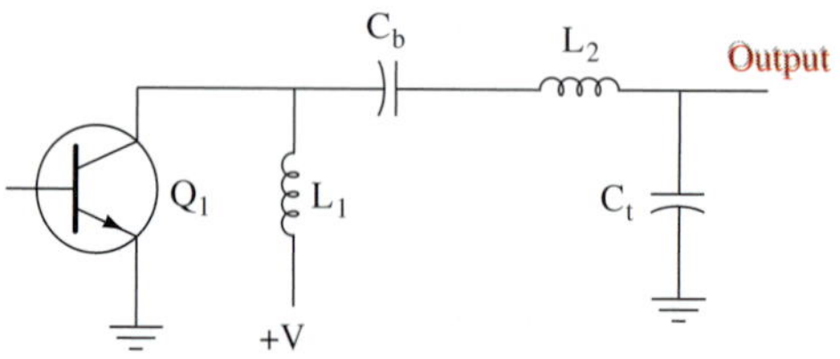

C_b L_2 Output

Q_1 L_1 C_t

+V

FIGURE 2-30 Output components.

Faulty Tuning Capacitor The ac load impedance presented to the transistor Q_1 is dependent on C_t and L_2 forming a tuned circuit that transforms the antenna impedance to the correct value. Assume that C_t is shorted. In this case, the load impedance would probably be too low and Q_1 would draw excessive current. If C_t were open, the opposite would happen. In either case, power output will be very low.

Adjustments Assume that C_t is simply not properly adjusted (it is usually variable). Power output will be too high or too low depending on the direction of error. High-power output will be accompanied by overheating of Q_1 due to excessive collector current.

L_2 is also usually adjustable. You must alternately adjust both C_t and L_2 to obtain the proper impedance match. Look for minimum collector current and maximum power output. Use a spectrum analyzer to be certain the amplifier is not tuned to a harmonic. Some amplifiers will happily tune to the second or third harmonic. Others will break into self-oscillation on many frequencies at once. The spectrum analyzer will reveal many of the bad habits an amplifier might have.

Checking a Transmitter

A word of caution before starting. High-power transmitters (perhaps anything greater than 10 kW) frequently employ vacuum tubes. These are high-voltage devices using voltages of 5 kV or more. *THESE VOLTAGES CAN KILL YOU*. Therefore, troubleshoot with extreme caution. Get experienced help until the following rules become second nature:

1. Before entering the transmitter cabinet, turn off all power switches. Better yet, remove the power plug from the socket. There may be no plug; the transmitter power input lines may be hard-wired to a fuse panel. If so, there will probably be a main switch; turn it off.
2. Even though you have turned off the power, make absolutely certain the high voltage is off before touching any circuits within the transmitter cabinet. The best and only sure test of this is to attach a bare, uninsulated piece of 12-gauge or larger copper wire (no insulation; you must be able to see that the entire length of the wire is intact) to a nonconducting wooden or plastic rod perhaps 2 ft long. Ground one end of this wire; that is, connect it to the metal chassis of the transmitter. Holding the end of the rod farthest from the wire, move the rod so that the ungrounded end of the wire touches those points that would have high voltage on them if the high voltage was still on. You'll see arcing if the voltage is still on.
3. Do not trust automatic switches (interlocks) that are supposed to turn off high-voltage circuits automatically.
4. Remember, charged capacitors—such as those used for power supply filters—can store a lethal charge. Short these units with your bare wire on a rod.

Transmitter Not Operating on Proper Frequency The simplest way to determine a transmitter's operating frequency is to listen to its output signal on a receiver with a calibrated readout that accurately indicates the frequency to which the receiver is tuned. Such receivers may have a built-in crystal oscillator called a crystal calibrator. Calibrator oscillators are specially designed to have rich harmonic output. If, for

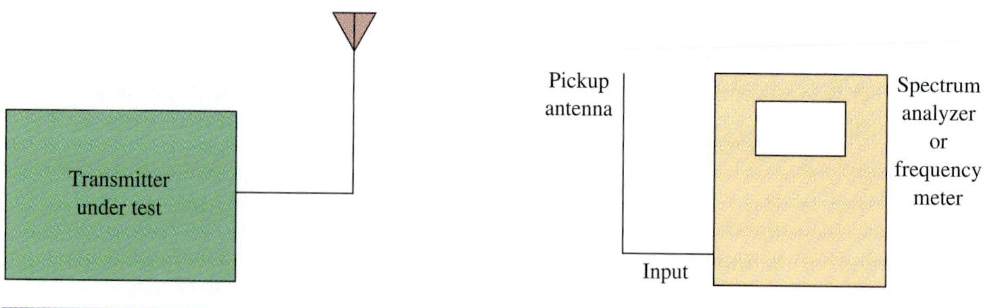

FIGURE 2-31 Determining a transmitter's output frequency.

example, a receiver had a calibrator operating at 100 kHz, signals would be heard on the receiver at harmonics or multiples of 100 kHz over a broad range. Tune the receiver to the 100 kHz multiple nearest the frequency of the transmitter under test. Compare the frequency shown on the receiver's readout to the known multiple of 100 kHz and determine the error between the two. Then, tune to the transmitter frequency and adjust the receiver's readout up or down according to the error. For greater accuracy, the calibrator can be set to the frequency of radio station WWV, operated by the National Institute of Standards and Technology, Fort Collins, Colorado. This station broadcasts on accurate frequencies of 2.5, 5, 10, 15, and 20 MHz on the shortwave bands. Canadian station CHU, broadcasting from Ottawa, can also be used for calibration. It is found at 3330, 7335, and 14,670 kHz on the shortwave bands.

A transmitter's operating frequency can also be determined with a spectrum analyzer or a frequency meter, as shown in Fig. 2-31. It may be necessary to connect a short antenna, perhaps a few feet long, to the input terminal of the measuring equipment if the transmitter has low output power.

If the transmitter is not operating on the correct frequency, adjust the carrier oscillator to the proper frequency. Check the transmitter's maintenance manual for instructions. See additional comments on oscillators in the Chapter 1 troubleshooting section.

Measuring Transmitter Output Power

Figure 2-32 shows the circuit to be used when measuring and troubleshooting the output power of a transmitter. The dummy load acts like an antenna because it absorbs the energy output from the transmitter without allowing that energy to radiate and interfere with other stations. Its input impedance must match (be equal to) the transmitter's output impedance; this is usually 50 Ω.

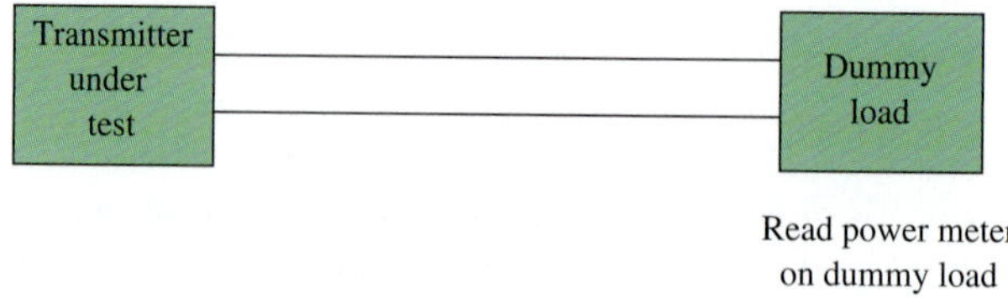

FIGURE 2-32 Checking the output power of an AM transmitter.

Suppose we are checking a low-power commercial transmitter that is rated at 250-W output (the dummy load and wattmeter must be rated for this level). If the output power is greater than the station license allows, the drive control must be adjusted to bring the unit within specs. What if the output is below specs? Let us consider possible causes.

Remember the suggestion in Chapter 1: Do the easy tasks first. Perhaps the easiest thing to do in the case of low-power output is to check the drive control: Is it set correctly? Assuming it is, check the power supply voltage: Is it correct? Observe that voltage on a scope: Is it good, pure dc or has a rectifier shorted or opened, indicated by excessive ripple?

Once the easy tasks have been done, check the tuning of each amplifier stage between the carrier oscillator and the last or final amplifier driving the antenna. If the output power is still too low after peaking the tuning controls, use an oscilloscope to check the output voltage of each stage to see if they are up to specs. Are the signals good sinusoids? If a stage has a clean, undistorted input signal and a distorted output, there may be a defective component in the bias network. Or perhaps the tube/transistor needs checking and/or replacement.

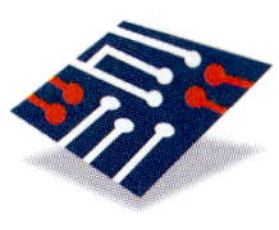

2-9 Troubleshooting with Electronics Workbench™ Multisim

Chapter 2 presented the modulation processes for producing an AM signal. Electronics Workbench™ Multisim can be used to simulate, make measurements, and troubleshoot AM modulator circuits, such as the simple transistor modulator shown in Fig. 2-10. To begin this exercise, open **Fig2-33.ms7 (.msm)** found in your EWB Multisim CD. You should have a circuit that looks like the one shown in Fig. 2-33.

Begin the simulation by clicking on the start simulation switch. View the simulation results by double-clicking on the oscilloscope. You will obtain an image similar to the one shown in Fig. 2-34. Is this an example of amplitude modulation? You can freeze the display by turning the simulation off or pressing pause. A record of the image is continually being saved. You can horizontally shift the displayed screen by adjusting the **Slider,** as shown in Fig. 2-34. Use Eq. 2-5 to determine the modulation index for this simulation. You should obtain a percentage modulation of about 33 percent.

Can you measure the frequency of the carrier from the display? It is easier to make the carrier frequency measurement if you change the timebase settings by clicking on the number in the **Timebase Scale** box. A set of up-down buttons will appear. These buttons are called *spinners*. Use the spinners to change the timebase to 50 μs/Div. Multisim provides a set of cursors on the oscilloscope display that makes it easy to measure the period of the wave. The cursors are located at the top of the oscilloscope. The red cursor is marked number 1 and the blue cursor is marked number 2. The position of the cursors can be moved by clicking on the triangles and dragging them to the desired location. Place cursor 1 at the start of the cycle for a sine wave and cursor 2 at the end of the cycle. The time between the two cursors is shown as **T2 − T1.** The frequency of the sine wave is obtained by inverting the value of the T2 − T1, or $f = 1/(\text{T2} - \text{T1})$. For this example, the measured period is 62.8 μs and $f = 1/62.8\ \mu\text{s} = 15.924$ kHz, which is close to the carrier frequency of the generator (15.9 kHz).

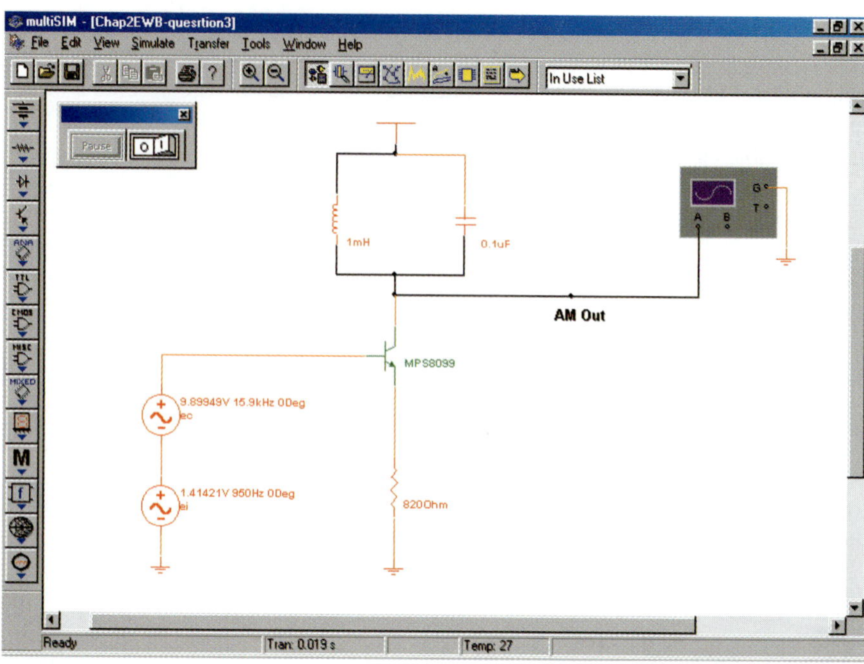

FIGURE 2-33 The Multisim component view for the simple transistor amplitude modulator circuit.

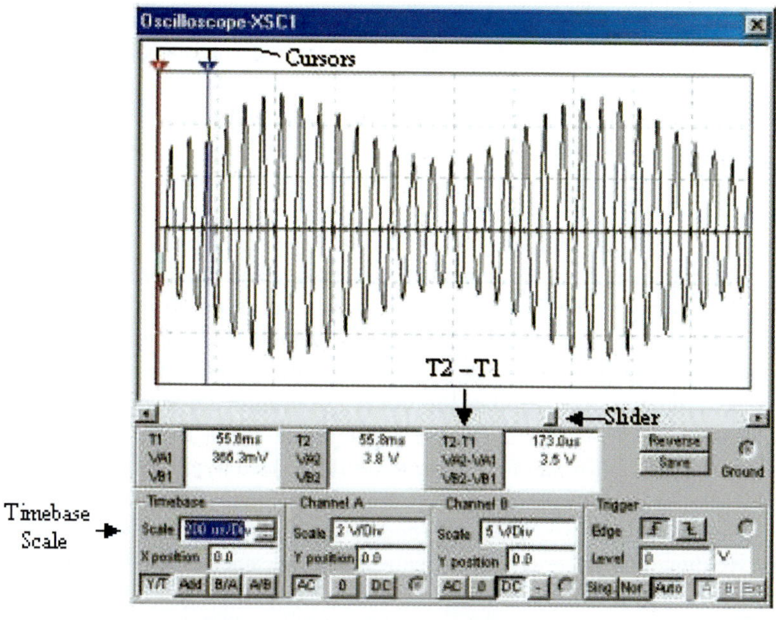

FIGURE 2-34 The Multisim oscilloscope control panel.

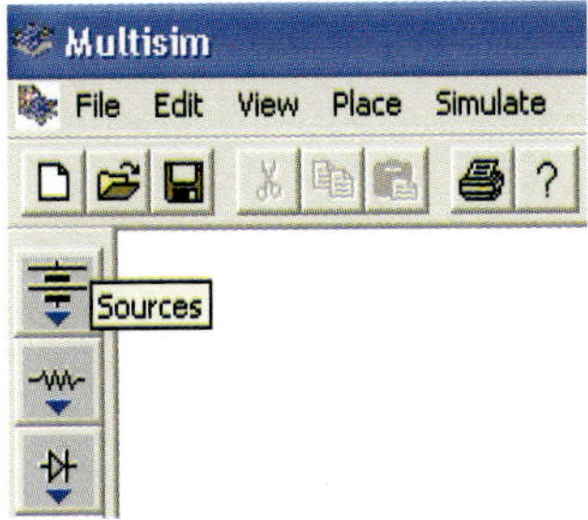

FIGURE 2-35 The Sources icon in Electronics Workbench™ Multisim.

Electronics Workbench™ Multisim also provides an AM source. This feature is convenient when you are learning about the characteristics of an AM signal. The source can be selected by placing the mouse over the Sources icon, as shown in Fig. 2-35. Clicking on the Sources icon provides a list of sources available in Multisim. A partial list of the sources available is shown in Fig. 2-36. A source can be selected from the list by clicking on the Sources icon. The image will open behind the Sources menu. Use your mouse to drag the source to the desired location on the circuit diagram. The parameters for the AM signal can be set by double-clicking on the AM Source. The AM Source and its menu are both shown in Fig. 2-37.

The AM Source menu allows you to set the carrier amplitude, frequency, modulation index, and modulation frequency. Refer to Sections 2-2 through 2-4 for a review of amplitude modulation fundamentals. In this case, the carrier amplitude is set to 1 V, the carrier frequency is 1000 Hz, the modulation index is 1, and the modulation frequency is 100 Hz. The AM waveform produced by the source is shown in Fig. 2-38. The AM Source is used in the Electronics Workbench exercises that follow this section.

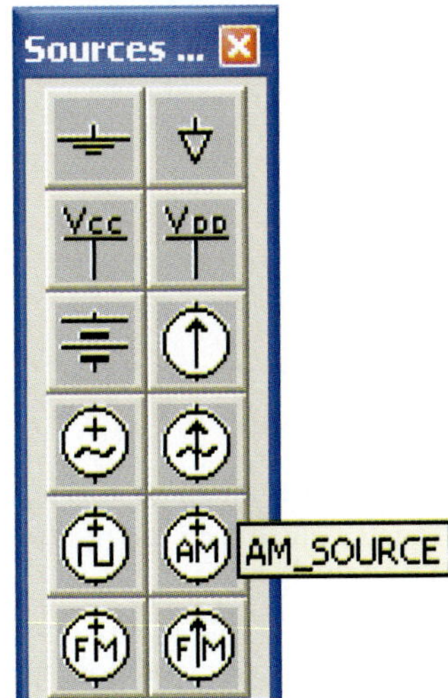

FIGURE 2-36 A partial list of the sources provided by Multisim and the location of the AM Sources icon.

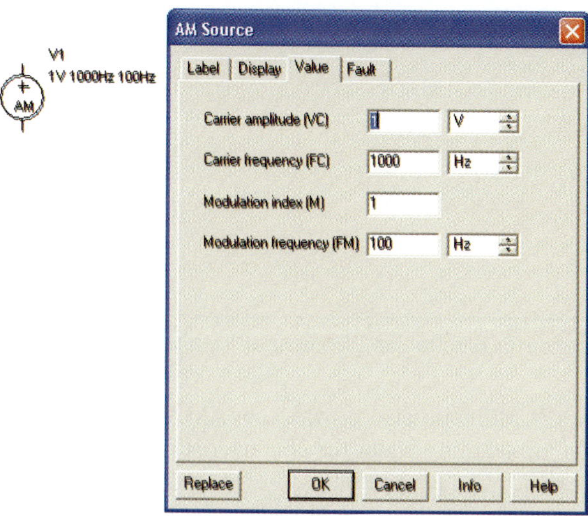

FIGURE 2-37 The AM Source and the menu for setting its parameters.

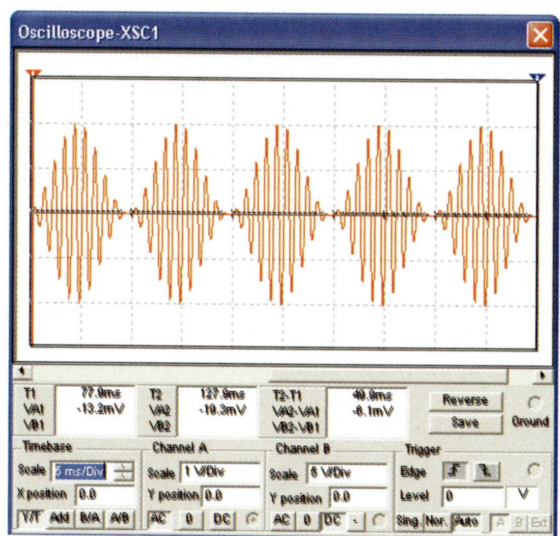

FIGURE 2-38 The output of the AM Source.

Electronics Workbench™ Exercises

1. Open **FigE2-1.ms7 (.msm)** on your EWB CD. Determine the modulation index.
2. Open **FigE2-2.ms7 (.msm)** on your EWB CD. Use the cursors on the oscilloscope to verify that the carrier frequency is 15 kHz.
3. Open **FigE2-3.ms7 (.msm)** on your EWB CD. Add the display of the input signals of the modulating circuit to the channel B input of your oscilloscope so that both the AM signal and the intelligence signal are displayed. Use the scale controls and the Y position so that both traces are easily viewed. Print out the traces displayed on the oscilloscope.

Summary

In Chapter 2 we studied the concept of amplitude modulation as it is specifically utilized in a transmitter. The major topics you should now understand include:

- the fundamental concept of amplitude modulation
- the meaning of modulation index and its use in AM calculations
- the cause of overmodulation and why it must be avoided
- the mathematical analysis of AM and the effect of modulation index on sideband amplitude
- the elements of a simple transistor AM generator and the analysis of its operation
- the understanding of high- and low-level modulation
- the analysis of a high-level transistor modulator
- the analysis and operation of various linear integrated circuit modulators
- the caution necessary when working with high-powered transmitters

Questions and Problems

Section 2-2

1. A 1500-kHz carrier and 2-kHz intelligence signal are combined in a *nonlinear* device. List *all* the frequency components produced.

*2. If a 1500-kHz radio wave is modulated by a 2-kHz sine-wave tone, what frequencies are contained in the modulated wave (the actual AM signal)?

*3. If a carrier is amplitude-modulated, what causes the sideband frequencies?

*4. What determines the bandwidth of emission for an AM transmission?

5. Explain the difference between a sideband and a side frequency.

6. What does the phasor at point 6 in Fig. 2-7 imply about the modulation signal?

7. Explain how the phasor representation can describe the formation of an AM signal.

8. Construct phasor diagrams for the AM signal in Fig. 2-7 midway between points 1 and 2, 3 and 4, and 5 and 6.

Section 2-3

*9. Draw a diagram of a carrier wave envelope when modulated 50 percent by a sinusoidal wave. Indicate on the diagram the dimensions from which the percentage of modulation is determined.

*10. What are some of the possible results of overmodulation?

*11. An unmodulated carrier is 300 V p-p. Calculate $\%m$ when its maximum p-p value reaches 400, 500, and 600 V. (33.3%, 66.7%, 100%)

12. If $A = 200$ V and $B = 60$ V as shown in Fig. 2-8, determine $\%m$. (53.85%)

13. Determine E_c and E_m from Problem 12. ($E_c = 65$ Vpk, $E_m = 35$ Vpk)

*An asterisk preceding a number indicates a question that has been provided by the FCC as a study aid for licensing examinations.

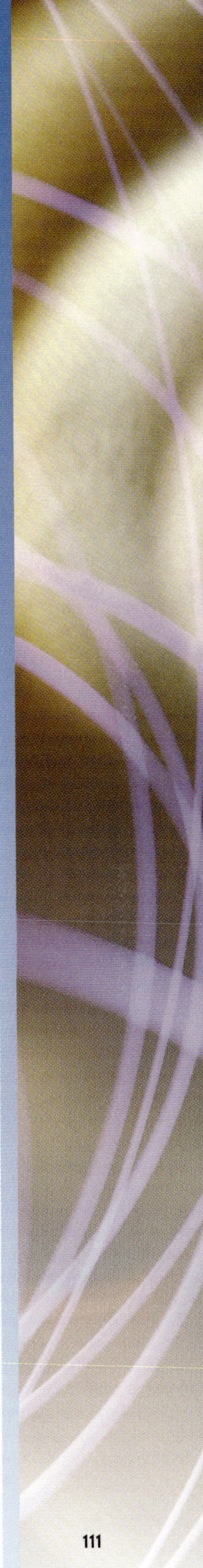

Section 2-4

14. Given that the amplitude of an AM waveform can be expressed as the sum of the carrier peak amplitude and intelligence signal, derive the expression for an AM signal that shows the existence of carrier and side frequencies.
15. A 100-V carrier is modulated by a 1-kHz sine wave. Determine the side-frequency amplitudes when $m = 0.75$. (37.5 V)
16. A 1-MHz, 40-V peak carrier is modulated by a 5-kHz intelligence signal so that $m = 0.7$. This AM signal is fed to a 50-Ω antenna. Calculate the power of each spectral component fed to the antenna. ($P_c = 16$ W, $P_{usb} = P_{lsb} = 1.96$ W)
17. Calculate the carrier and sideband power if the total transmitted power is 500 W in Problem 15. (390 W, 110 W)
18. The ac rms antenna current of an AM transmitter is 6.2 A when unmodulated and rises to 6.7 A when modulated. Calculate $\%m$. (57.9%)

*19. Why is a high percentage of modulation desirable?

*20. During 100 percent modulation, what percentage of the average output power is in the sidebands? (33.3%)

21. An AM transmitter has a 1-kW carrier and is modulated by three different sine waves having equal amplitudes. If $m_{eff} = 0.8$, calculate the individual values of m and the total transmitted power. (0.462, 1.32 kW)
22. A 50-V rms carrier is modulated by a square wave as shown in Table 1-4(c). If only the first four harmonics are considered and $V = 20$ V, calculate m_{eff}. (0.77)

Section 2-5

23. Describe two possible ways that a transistor can be used to generate an AM signal.

*24. What is *low-level modulation*?

*25. What is *high-level modulation*?

26. Explain the relative merits of high- and low-level modulation schemes.

*27. Why must some radio-frequency amplifiers be neutralized?

28. Describe the difference in effect of self-oscillations at a circuit's tuned frequency and parasitic oscillations.
29. Define *parasitic oscillation*.
30. How does self-oscillation occur?
31. Draw a schematic of a class C transistor modulator and explain its operation.

*32. What is the principal advantage of a class C amplifier?

33. Explain the circuit operation of the PIN diode modulator in Fig. 2-15. What type of AM transmitters are likely to use this method of AM generation?

*34. What is the function of a quartz crystal in a radio transmitter?

Section 2-6

*35. Draw a block diagram of an AM transmitter.

*36. What is the purpose of a buffer amplifier stage in a transmitter?

37. Describe the means by which the transmitter shown in Fig. 2-19 is modulated.

*38. Draw a simple schematic diagram showing a method of coupling the radio-frequency output of the final power amplifier stage of a transmitter to an antenna.

39. Describe the functions of an antenna coupler.

*40. A ship radio-telephone transmitter operates on 2738 kHz. At a certain point distant from the transmitter the 2738-kHz signal has a measured field of 147 mV/m. The second harmonic field at the same point is measured as 405 μV/m. To the nearest whole unit in decibels, how much has the harmonic emission been attenuated below the 2738-kHz fundamental? (51.2 dB)

41. What is a *tune-up procedure*?

Section 2-7

*42. Draw a sample sketch of the trapezoidal pattern on a cathode-ray oscilloscope screen indicating low percentage modulation without distortion.

43. Explain the advantages of the trapezoidal display over a standard oscilloscope display of AM signals.

44. A spectrum analyzer display shows that a signal is made up of three components only: 960 kHz at 1 V, 962 kHz at $\frac{1}{2}$ V, 958 kHz at $\frac{1}{2}$ V. What is the signal and how was it generated?

45. Define *spur*.

46. Provide a sketch of the display of a spectrum analyzer for the AM signal described in Problem 63 at both 20 percent and 90 percent modulation. Label the amplitudes in dBm.

47. The spectrum analyzer display shown in Fig. 2-25 is calibrated at 10 dB/vertical division and 5 kHz/horizontal division. The 50.0034-MHz carrier is shown riding on a −20-dBm noise floor. Calculate the carrier power, the frequency, and the power of the spurs. (2.51 W, 50.0149 MHz, 49.9919 MHz, 50.0264 MHz, 49.9804 MHz, 6.3 mW, 1 mW)

*48. What is the purpose of a *dummy antenna*?

49. An amplifier has a spectrally pure sine-wave input of 50 mV. It has a voltage gain of 60. A spectrum analyzer shows harmonics of 0.035 V, 0.027 V, 0.019 V, 0.011 V, and 0.005 V. Calculate the total harmonic distortion (THD) and the relative harmonic distortion. (2.864%, 38.66 dB)

50. An additional harmonic ($V_6 = 0.01$ V) was neglected in the THD calculation shown in Ex. 2-9. Determine the percentage error introduced by this omission. (0.91%)

Section 2-8

51. When troubleshooting, what is the purpose of inspection? Describe the various steps involved in this process.

52. After a repair has been made, how is a hot check accomplished?

53. Discuss the function of C_1 in Fig. 2-28 and explain the effect if it should fail by either shorting or becoming an open circuit.

54. Explain the dangers of working on high-power transmitters and describe the precautions that should be taken.

55. Describe why and how a dummy load is used in checking the output of a transmitter. Why is the impedance of the dummy load an important consideration?

56. There is no output from Q_1 in Fig. 2-28. Briefly describe a plan to isolate the problem.

57. Describe the output of the amplifier in Fig. 2-28 if R_1 is open.

58. Explain the advantages of using a spectrum analyzer.

59. If the inductor L_2 in Fig. 2-30 is shorted, describe its output. Assume the load is an antenna that is not grounded.
60. Describe some of the physical defects of a system that are obvious to the eye.

Questions for Critical Thinking

61. Would the *linear* combination of a low-frequency intelligence signal and a high-frequency carrier signal be effective as a radio transmission? Why or why not?
62. You are analyzing an AM waveform. What significance do the upper and lower envelopes have?
63. An AM transmitter at 27 MHz develops 10 W of carrier power into a 50-Ω load. It is modulated by a 2-kHz sine wave between 20 and 90 percent modulation. Determine:
    (a) Component frequencies in the AM signal.
    (b) Maximum and minimum waveform voltage of the AM signal at 20 percent and 90 percent modulation. (25.3 to 37.9 V peak, 3.14 to 60.1 V peak)
    (c) Sideband signal voltage and power at 20 percent and 90 percent modulation. (2.24 V, 0.1 W, 10.06 V, 2.025 W)
    (d) Load current at 20 percent and 90 percent modulation. (0.451A, 0.530A)
64. Compare the display of an oscilloscope to that of a spectrum analyzer.

Amplitude Modulation: Reception

Chapter Outline

The Grundig Satellit 800 AM/FM/SW radio. (Courtesy of Grundig/Eton Corporation. Reprinted with permission.)

Objectives

- Define the *sensitivity* and *selectivity* of a radio receiver
- Describe the operation of a diode detector in an AM receiver
- Sketch block diagrams for TRF and superheterodyne receivers
- Understand the generation of *image frequencies* and describe how to suppress them
- Recognize and analyze RF and IF amplifiers
- Describe the need for automatic gain control and show how it can be implemented
- Analyze the operation of a complete AM receiver system
- Perform a test analysis on the power levels (dBm) at each stage of an AM receiver system

Key Terms

tuned radio frequency receiver
sensitivity
noise floor
selectivity
envelope detector
product detector
heterodyne detector
first detector
trimmer
padder capacitor
varactor diodes
varicap diodes
VVC diodes
image frequency
double conversion
cross-modulation
converters
first detectors
Schottky diode
self-excited mixer
autodyne mixer
auxiliary AGC diode
dynamic range
signal injection

3-1 Receiver Characteristics

If you were to envision a block diagram for a radio receiver, you would probably go through the following logical thought process:

1. The signal from the antenna is usually very small—therefore, amplification is necessary. This amplifier should have low-noise characteristics and should be tuned to accept only the desired carrier and sideband frequencies to avoid interference from other stations and to minimize the received noise. Recall that noise is proportional to bandwidth.
2. After sufficient amplification, a circuit to detect the intelligence from the radio frequency is required.
3. Following the detection of the intelligence, further amplification is necessary to give it sufficient power to drive a loudspeaker.

Tuned Radio Frequency Receiver
the most elementary receiver design, consisting of RF amplifier stages, a detector, and audio amplifier stages

This logical train of thought leads to the block diagram shown in Fig. 3-1. It consists of an RF amplifier, detector, and audio amplifier. The first radio receivers for broadcast AM took this form and are called **tuned radio frequency** or, more simply, **TRF receivers.** These receivers generally had three stages of RF amplification, with each stage preceded by a separate variable-tuned circuit. You can imagine the frustration experienced by the user when tuning to a new station. The three tuned circuits were all adjusted by separate variable capacitor controls. To receive a station required proper adjustment of all three, and a good deal of time and practice was necessary.

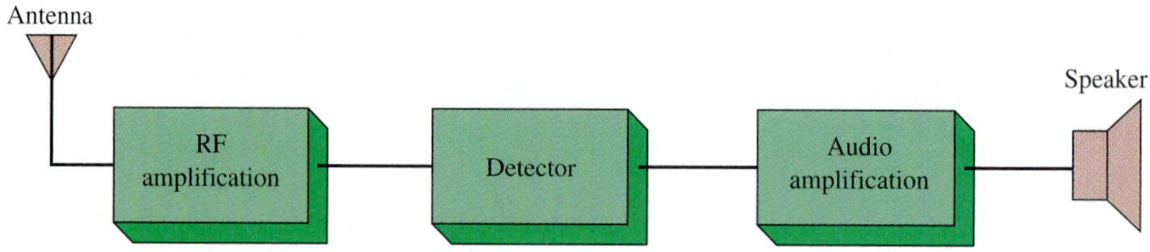

FIGURE 3-1 Simple radio receiver block diagram.

Sensitivity and Selectivity

Sensitivity
the minimum input RF signal to a receiver required to produce a specified audio signal at output

Noise Floor
the baseline on a spectrum analyzer display, representing input noise of the system under test

Two major characteristics of any receiver are its sensitivity and selectivity. A receiver's **sensitivity** may be defined as its ability to drive the output transducer (e.g., speaker) to an acceptable level. A more technical definition is the minimum input signal (usually expressed as a voltage) required to produce a specified output signal or sometimes just to provide a discernible output. The range of sensitivities for communication receivers varies from the millivolt region for low-cost AM receivers down to the nanovolt region for ultrasophisticated units for more exacting applications. In essence, a receiver's sensitivity is determined by the amount of gain provided and, more important, its noise characteristics. In general, the input signal must be somewhat greater than the noise at the receiver's input. This input noise is termed the **noise floor** of the receiver. It is not difficult to insert more gain in a radio, but getting noise figures below a certain level presents a more difficult challenge.

Selectivity may be defined as the extent to which a receiver is capable of differentiating between the desired signal and other frequencies (unwanted radio signals and noise). A receiver can also be overly selective. For instance, on commercial broadcast AM, we have seen that the transmitted signal can include intelligence signals up to about a maximum of 15 kHz, which subsequently generates upper and lower sidebands extending 15 kHz above and below the carrier frequency. The total signal has a 30-kHz bandwidth. Optimum receiver selectivity is thus 30 kHz, but if a 5-kHz bandwidth were selected, the upper and lower sidebands would extend only 2.5 kHz above and below the carrier. The radio's output would suffer from a lack of the full possible fidelity because the output would include intelligence up to a maximum of 2.5 kHz. On the other hand, an excessive selectivity of 50 kHz results in the reception of unwanted adjacent radio signals and the additional external noise that is directly proportional to the bandwidth selected. Unfortunately, TRF receivers did suffer from selectivity problems, which led to their replacement by the superheterodyne receiver.

Selectivity
the extent to which a receiver can differentiate between the desired signal and other signals

As has been stated, broadcast AM can extend to about 30-kHz bandwidth. As a practical matter, however, many stations and receivers use a more limited bandwidth. The lost fidelity is often not detrimental because of the talk-show format of many AM stations. For instance, a 10-kHz bandwidth receiver provides audio output up to 5 kHz, which more than handles the human voice range. Musical reproduction with a 5-kHz maximum frequency is not high fidelity but is certainly adequate for casual listening.

TRF Selectivity

Consider a standard AM broadcast band receiver that spans the frequency range from 550 to 1550 kHz. If the approximate center of 1000 kHz is considered, we can use Eq. (1-25) from Chapter 1 to find that, for a desired 10-kHz BW, a Q of 100 is required.

$$Q = \frac{f_r}{\text{BW}} \tag{1-25}$$

$$= \frac{1000 \text{ kHz}}{10 \text{ kHz}}$$

$$= 100$$

Now, since the Q of a tuned circuit remains fairly constant as its capacitance is varied, a change to 1550 kHz will change the BW to 15.5 kHz.

$$Q = \frac{f_r}{\text{BW}} \tag{1-25}$$

Therefore,

$$\text{BW} = \frac{f_r}{Q}$$

$$= \frac{1550 \text{ kHz}}{100}$$

$$= 15.5 \text{ kHz}$$

The receiver's BW is now too large, and it will suffer from increased noise. On the other hand, the opposite problem is encountered at the lower end of the frequency range. At 550 kHz, the BW is 5.5 kHz.

$$\begin{aligned} \text{BW} &= \frac{f_r}{Q} \\ &= \frac{550\text{ kHz}}{100} \\ &= 5.5\text{ kHz} \end{aligned}$$

The fidelity of reception is now impaired. The maximum intelligence frequency possible is 5.5 kHz/2, or 2.75 kHz, instead of the full 5 kHz transmitted. This selectivity problem led to the general use of the superheterodyne receiver (see Sec. 3-3) in place of TRF designs.

Example 3-1

A TRF receiver is to be designed with a single tuned circuit using a 10-μH inductor.
(a) Calculate the capacitance range of the variable capacitor required to tune from 550 to 1550 kHz.
(b) The ideal 10-kHz BW is to occur at 1100 kHz. Determine the required Q.
(c) Calculate the BW of this receiver at 550 kHz and 1550 kHz.

Solution

(a) At 550 kHz, calculate C using Eq. (1-23) from Chapter 1.

$$f_r = \frac{1}{2\pi\sqrt{LC}} \tag{1-23}$$

$$550\text{ kHz} = \frac{1}{2\pi\sqrt{10\ \mu\text{H} \times C}}$$

$$C = 8.37\text{ nF}$$

At 1550 kHz,

$$1550\text{ kHz} = \frac{1}{2\pi\sqrt{10\ \mu\text{H} \times C}}$$

$$C = 1.06\text{ nF}$$

Therefore, the required range of capacitance is from

1.06 to 8.37 nF

(b)

$$\begin{aligned} Q &= \frac{f_r}{\text{BW}} \\ &= \frac{1100\text{ kHz}}{10\text{ kHz}} \\ &= 110 \end{aligned} \tag{1-25}$$

(c) At 1550 kHz,

$$\begin{aligned} BW &= \frac{f_r}{Q} \\ &= \frac{1550 \text{ kHz}}{110} \\ &= 14.1 \text{ kHz} \end{aligned}$$

At 550 kHz,

$$BW = \frac{550 \text{ kHz}}{110} = 5 \text{ kHz}$$

3-2 AM Detection

The process of detecting the intelligence out of the carrier and sidebands (the AM signal) has thus far been mentioned but not explained. In fact, the detection process can be easily accomplished. Recall our discussions about generating AM. We said if two different frequencies were passed through a nonlinear device, sum and difference components would be generated. The carrier and sidebands of the AM signal are separated in frequency by an amount equal to the intelligence frequency. If the AM signal is passed through a nonlinear device, difference frequencies between the carrier and sidebands will be generated and these frequencies are, in fact, the intelligence. It follows that passing the AM signal through a nonlinear device will provide detection, just as passing the carrier and intelligence through a nonlinear device enables AM generation. The mathematical proof for this was given in Sec. 2-4.

Detection of amplitude-modulated signals requires a nonlinear electrical network. An ideal nonlinear curve for this is one that affects the positive half-cycle of the modulated wave differently than the negative half-cycles. This distorts an applied voltage wave of zero average value so that the average resultant current varies with the intelligence signal amplitude. The curve shown in Fig. 3-2(a) is called an *ideal curve* because it is linear on each side of the operating point P and does not introduce harmonic frequencies.

When the input to an ideal nonlinear device is a carrier and its sidebands, the output contains the following frequencies:

1. The carrier frequency
2. The upper sideband
3. The lower sideband

} original components

4. A dc component
5. A frequency equal to the carrier minus the lower sideband and the upper sideband minus the carrier, which is the original signal frequency

The detector reproduces the signal frequency by producing a distortion of a desirable kind in its output. When the output of the detector is impressed upon a low-pass filter, the radio frequencies are suppressed and only the low-frequency intelligence signal and dc components are left. This is shown as the dashed average current curve in Fig. 3-2(a).

In some practical detector circuits, the nearest approach to the ideal curve is the square-law curve shown in Fig. 3-2(b). The output of a device using this curve contains, in addition to all the frequencies that were listed, the harmonics of each of these

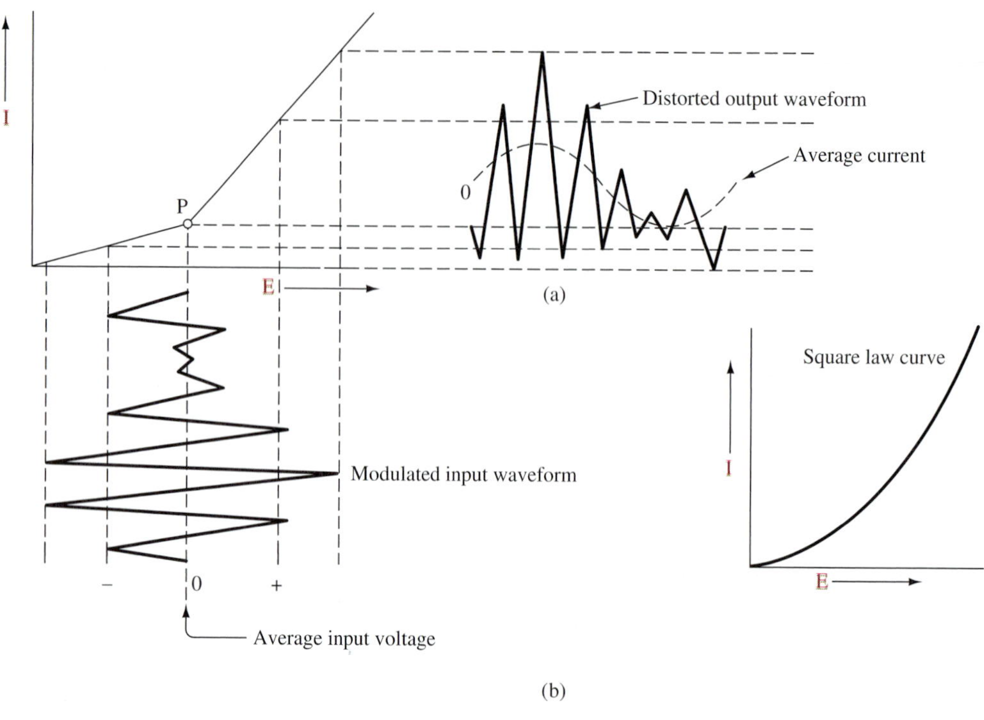

FIGURE 3-2 Nonlinear device used as a detector.

frequencies. The harmonics of radio frequencies can be filtered out, but the harmonics of the sum and difference frequencies, even though they produce an undesirable distortion, may have to be tolerated because they can be in the audio-frequency range.

Diode Detector

One of the simplest and most effective types of detectors, and one with nearly an ideal nonlinear resistance characteristic, is the diode detector circuit shown in Fig. 3-3(a). Notice the *I–V* curve in Fig. 3-3(b). This is the type of curve on which the diode detector at (a) operates. The curved part of its response is the region of low current and indicates that for small signals the output of the detector will follow the square law. For input signals with large amplitudes, however, the output is essentially linear. Therefore, harmonic outputs are limited. The abrupt nonlinearity occurs for the negative half-cycle as shown in Fig. 3-3(b).

The modulated carrier is introduced into the tuned circuit made up of LC_1 in Fig. 3-3(a). The waveshape of the input to the diode is shown in Fig. 3-3(c). Since the diode conducts only during half-cycles, this circuit removes all the negative half-cycles and gives the result shown in Fig. 3-3(d). The average output is shown at (e). Although the average input voltage is zero, the average output voltage across *R* always varies above zero.

The low-pass filter, made up of capacitor C_2 and resistor *R*, removes the RF (carrier frequency), which, so far as the rest of the receiver is concerned, serves no useful purpose. Capacitor C_2 charges rapidly to the peak voltage through the small resistance

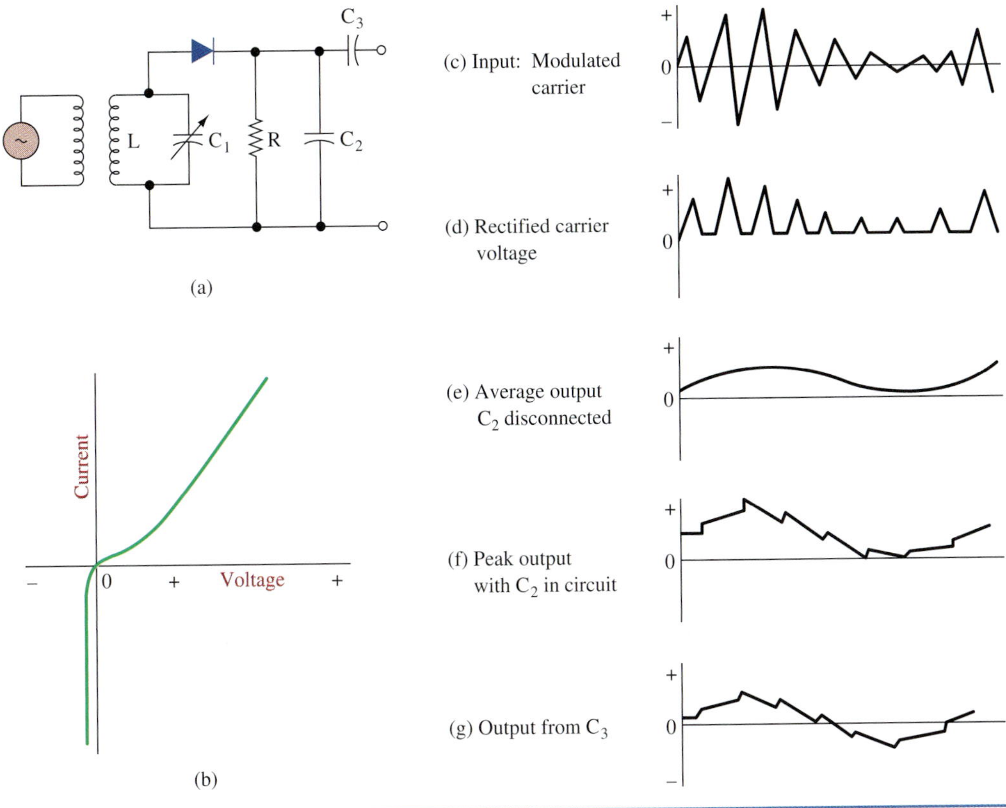

FIGURE 3-3 Diode detector.

of the conducting diode, but discharges slowly through the high resistance of R. The sizes of R and C_2 normally form a rather short time constant at the intelligence (audio) frequency and a very long time constant at the radio frequencies. The resultant output with C_2 in the circuit is a varying voltage that follows the peak variation of the modulated carrier [see Fig. 3-3(f)]. For this reason it is often termed an **envelope detector** circuit. The dc component produced by the detector circuit is removed by capacitor C_3, producing the ac voltage waveshape in Fig. 3-3(g). In communications receivers, the dc component is often used for providing automatic volume (gain) control.

Envelope Detector
another name for diode detector

The advantages of diode detectors are as follows:

1. They can handle relatively high power signals. There is no practical limit to the amplitude of the input signal.
2. Distortion levels are acceptable for most AM applications. Distortion decreases as the amplitude increases.
3. They are highly efficient. When properly designed, 90 percent efficiency is obtainable.
4. They develop a readily usable dc voltage for the automatic gain control circuits.

The disadvantages of the diode detectors are:

1. Power is absorbed from the tuned circuit by the diode circuit. This reduces the Q and selectivity of the tuned input circuit.
2. No amplification occurs in a diode detector circuit.

Detector Diode Types

In noncritical applications, the standard *pn* junction diode can be used. It is usually adequate for the LF, HF, and low VHF bands. Low-cost silicon switching diodes such as the 1N914 and 1N4148 are frequently used. For higher frequencies, point contact diodes are often used. These diodes have the *pn* junction on the surface of the substrate and make contact to the *p*-type material via a small wire. The junction is very small, yielding very low capacitance that makes them useful for microwave operation up to 40 GHz. The commonly used point contact varieties are the 1N21, 1N23, and 1N34.

An important specification for detector diodes is voltage sensitivity. This is a measure of detector output per unit of RF input power. It is specified as V/mW or mV/μW at some specified dc bias current.

Diagonal Clipping

Careful selection of component parts is necessary for obtaining optimum efficiency in diode detector circuits. One very important fact to consider is the value of the time constant RC_2, particularly in the case of pulse modulation. When a carrier modulated by a square pulse [Fig. 3-4(b)] is applied to an ideal diode detector, the waveshape shown in Fig. 3-4(c) is produced. Notice that for clarity, the amplitude of the wave at (c) is exaggerated in comparison to the high frequency carrier shown at (b).

If the time constant of RC_2 is too long compared to the period of the RF wave, several cycles are required to charge C_2, and the leading edge of the output pulse is sloped as shown in Fig. 3-4(d). After the pulse passes by, the capacitor discharges slowly and the trailing edge is exponential rather than square, as desired. This phenomenon is often referred to as diagonal clipping. The diagonal clipping effect from a sine-wave intelligence signal is shown at (e). Notice that the detected sine wave at (e) is distorted. The excessive *RC* time constant did not allow the capacitor voltage to follow the full changes of the sine wave. On the other hand, if the time constant is too short, both the leading and trailing edges can be easily reproduced. However, the capacitor may discharge considerably between carrier cycles. This reduces the average amplitude of the pulse, leaving a sizable component of the carrier frequency in the output, as shown in Fig. 3-4(f).

For these reasons the selection of the time constant is a compromise. The load resistor *R* must be large because the total input voltage is divided across *R* and the internal resistance of the diode when it is conducting. A large value of load resistance ensures that the greater part of this voltage is in the output, where desired. On the other hand, the load resistance must not be so high that capacitor C_2 becomes small enough to approximate the size of C_1 [Fig. 3-4(a)], the internal junction capacitance of the diode. When this occurs, capacitor C_2 will try to discharge through C_1 during the nonconducting periods, which would reduce the amplitude of the detector output.

Synchronous Detection

Diode detectors are used in the vast majority of AM detection schemes. Since high fidelity is usually not an important aspect in AM, the distortion levels of several percentage points or more from a diode detector can be tolerated easily. In applications demanding greater performance, the use of a synchronous detector offers the following advantages:

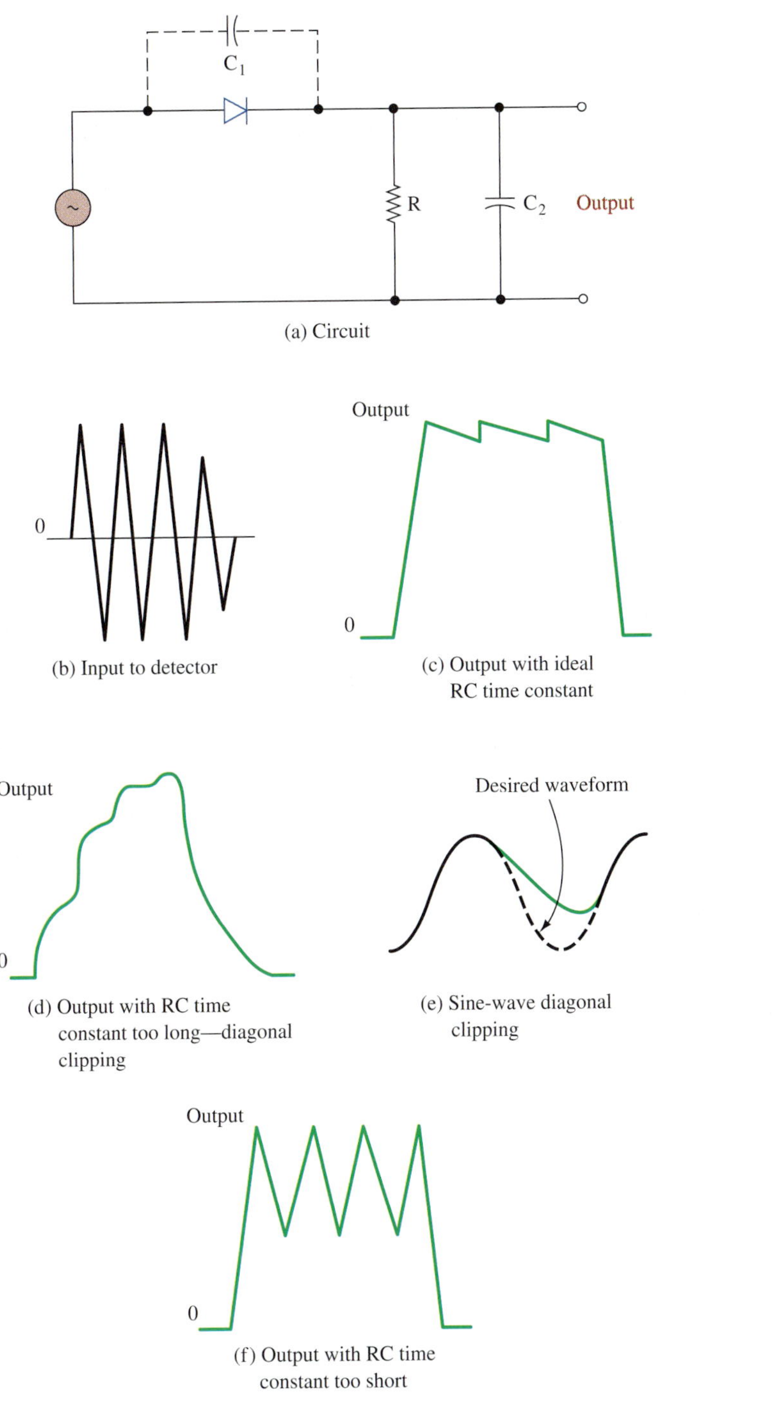

FIGURE 3-4 Diode detector component considerations. Dashed lines indicate average voltage during pulse.

1. Low distortion—well under 1 percent
2. Greater ability to follow fast-modulation waveforms, as in pulse-modulation or high-fidelity applications
3. The ability to provide gain instead of attenuation, as in diode detectors

Product Detector oscillator, mixer, and low-pass filter stage used to obtain the intelligence from an AM signal

Heterodyne Detector another name for synchronous detector or product detector

Synchronous detectors are also called **product** or **heterodyne detectors.** The principle of operation involves mixing in a nonlinear fashion just as in AM generation. Imagine receiving a transmission at 900 kHz. If it contained a 1-kHz tone, the reception consists of three components:

1. The carrier at 900 kHz
2. The usb at 901 kHz
3. The lsb at 899 kHz

If this AM waveform were mixed with an internally generated 900-kHz sine wave through a nonlinear device, a resulting difference frequency is 1 kHz—the desired output intelligence. Of course, a number of much higher sum frequencies are also generated, but they are easily filtered out by a low-pass filter. Detection has been achieved in a completely different fashion from that for the envelope detector (diode detection) discussed previously. A circuit commonly used for product detection is the balanced modulator. It is widely used in single-sideband (SSB) systems and is detailed in Chapter 4.

Since synchronous detectors require rather complex circuitry, the use of an LIC is most appropriate. An LIC offered by RCA (CA3067) or National Semiconductor (LM3067) was designed specifically for use as a TV chroma (color) demodulator. It is easily adapted for use as a synchronous AM detector. Figure 3-5 shows the circuit connection. The tint amplifier provides an initial stage of gain, with the double-balanced

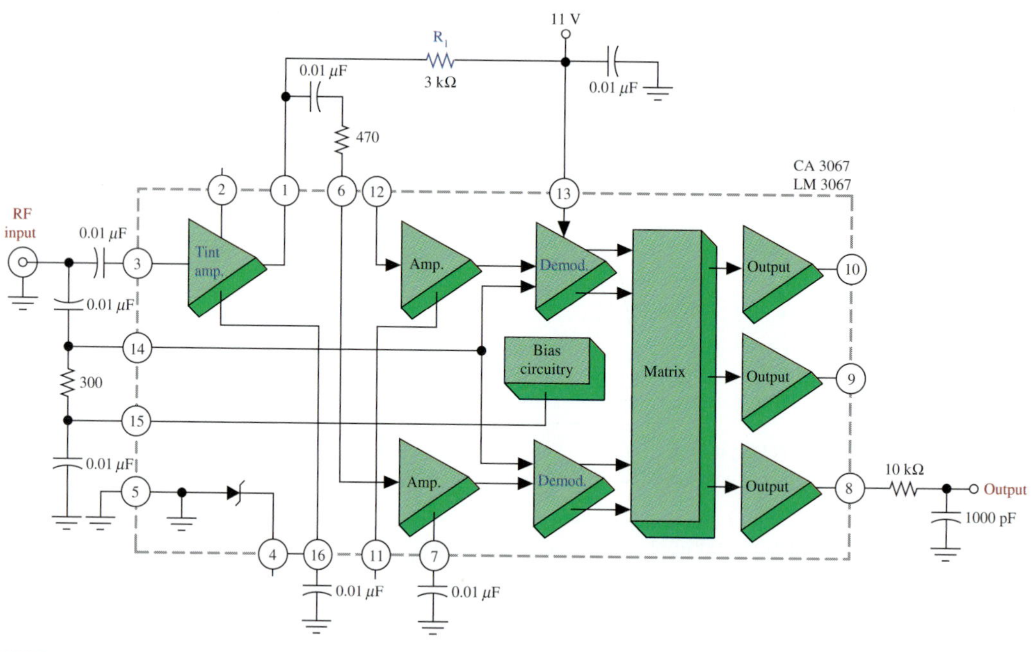

FIGURE 3-5 Synchronous AM detection.

demodulators providing the synchronous detection. With a 35-mV AM signal input, a 450-mV audio output is obtained with less than 0.7 percent distortion at 80 percent modulation. The circuit can be used with carrier frequencies as low as 10 kHz and up to 10 MHz. Higher-frequency operation is made possible by substituting a tuned circuit for R_1, which provides proper adjustment of the carrier phase.

3-3 SUPERHETERODYNE RECEIVERS

The basic variable-selectivity problem in TRF systems led to the development and general usage of the superheterodyne receivers in the early 1930s. This basic receiver configuration is still dominant after all these years, an indication of its utility. A block diagram for a superheterodyne receiver is provided in Fig. 3-6. The first stage is a standard RF amplifier that may or may not be required, depending on factors to be discussed later. The next stage is the mixer, which accepts two inputs, the output of the RF amplifier (or antenna input when an RF amplifier is omitted) and a steady sine wave from the local oscillator (LO). The mixer is yet another nonlinear device utilized in AM. Its function is to mix the AM signal with a sine wave to generate a new set of sum and difference frequencies. Its output, as will be shown, is an AM signal with a constant carrier frequency regardless of the transmitter's frequency. The next stage is the intermediate-frequency (IF) amplifier, which provides the bulk of radio-frequency signal amplification at a *fixed* frequency. This allows for a constant BW over the entire band of the receiver and is the key to the superior selectivity of the superheterodyne receiver. Additionally, since the IF frequency is usually lower than the RF, voltage gain of the signal is more easily attained at the IF frequency. Following the IF amplifiers is the detector, which extracts the intelligence from the radio signal. It is subsequently amplified by the audio amplifiers into the speaker. A dc level proportional to the received signal's strength is extracted from the detector stage and fed back to the IF amplifiers and sometimes to the mixer and/or the RF amplifier. This is

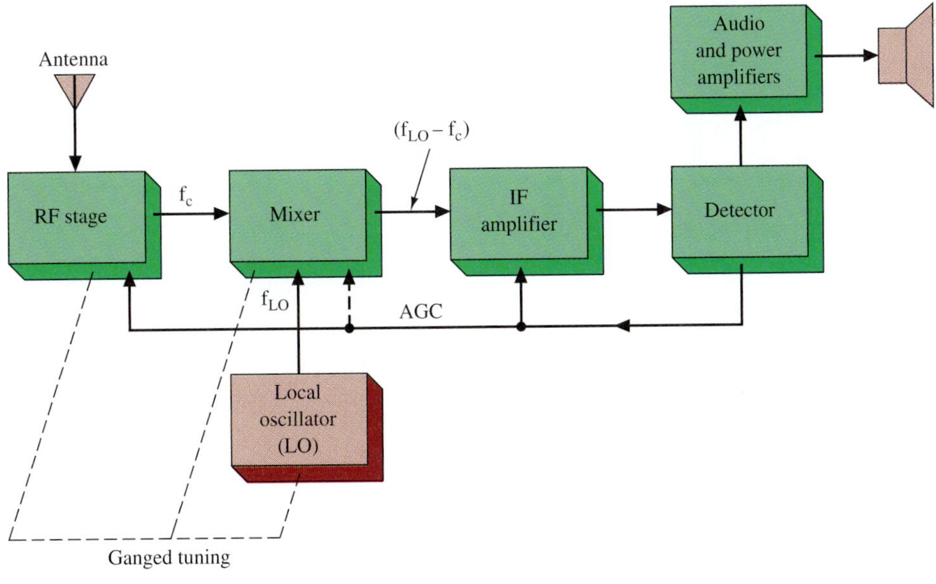

FIGURE 3-6 Superheterodyne receiver block diagram.

the automatic gain control (AGC) level, which allows relatively constant receiver output for widely variable received signals. Detail on AGC is provided in Sec. 3-6.

Frequency Conversion

It has been stated that the mixer performs a frequency conversion process. Consider the situation shown in Fig. 3-7. The AM signal into the mixer is a 1000-kHz carrier that has been modulated by a 1-kHz sine wave, thus producing side frequencies at 999 kHz and 1001 kHz. The LO input is a 1455-kHz sine wave. The mixer, being a nonlinear device, will generate the following components:

1. Frequencies at all of the original inputs: 999 kHz, 1000 kHz, 1001 kHz, and 1455 kHz.
2. Sum and difference components of all the original inputs: 1455 kHz ± (999 kHz, 1000 kHz, and 1001 kHz). This means outputs at 2454 kHz, 2455 kHz, 2456 kHz, 454 kHz, 455 kHz, and 456 kHz.
3. Harmonics of all the frequency components listed in 1 and 2 and a dc component.

The IF amplifier has a tuned circuit that accepts components only near 455 kHz, in this case 454 kHz, 455 kHz, and 456 kHz. Since the mixer maintains the same amplitude proportion that existed with the original AM signal input at 999 kHz, 1000 kHz, and 1001 kHz, the signal now passing through the IF amplifiers is a replica of the original AM signal. The only difference is that now its carrier frequency is 455 kHz. Its envelope is identical to that of the original AM signal. A frequency conversion has occurred that has translated the carrier from 1000 kHz to 455 kHz—a frequency intermediate to the original carrier and intelligence frequencies—which led to the terminology intermediate-frequency amplifier, or IF amplifier. Since the mixer and detector both have nonlinear characteristics, the mixer is often referred to as the **first detector.**

First Detector
the mixer stage in a superheterodyne receiver that mixes the RF signal with a local oscillator signal to form the intermediate frequency signal

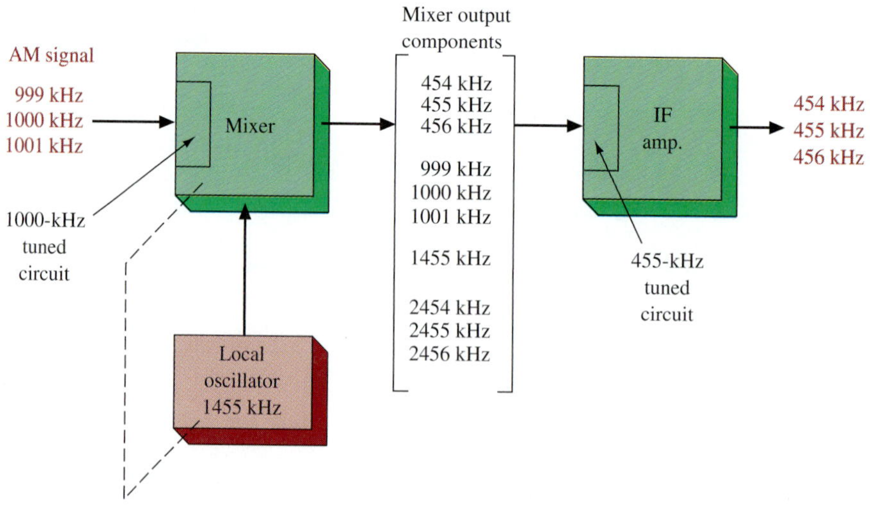

FIGURE 3-7 Frequency conversion process.

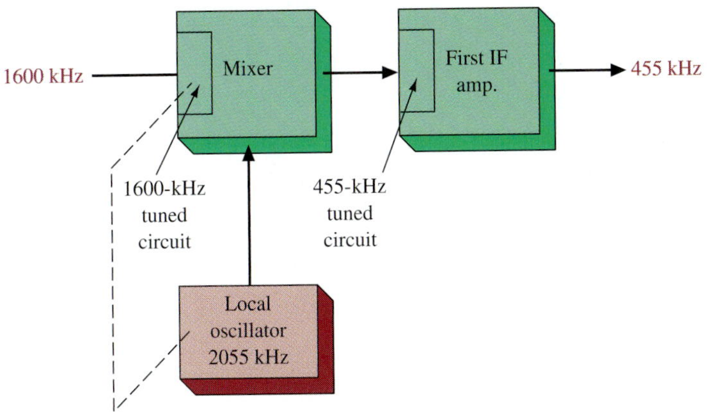

FIGURE 3-8 Frequency conversion.

Tuned-Circuit Adjustment

Now consider the effect of changing the tuned circuit at the front end of the mixer to accept a station at 1600 kHz. This means a reduction in either its inductance or capacitance (usually the latter) to change its center frequency from 1000 kHz to 1600 kHz. If the capacitance in the local oscillator's tuned circuit were simultaneously reduced so that its frequency of oscillation went up by 600 kHz, the situation shown in Fig. 3-8 would now exist. The mixer's output still contains a component at 455 kHz (among others), as in the previous case when we were tuned to a 1000-kHz station. Of course, the other frequency components at the output of the mixer are not accepted by the selective circuits in the IF amplifiers.

Thus, the key to superheterodyne operation is to make the LO frequency track with the circuit or circuits that are tuning the incoming radio signal so that their difference is a constant frequency (the IF). For a 455-kHz IF frequency, the most common case for broadcast AM receivers, this means the LO should always be at a frequency 455 kHz above the incoming carrier frequency. The receiver's front-end tuned circuits are usually made to track together by mechanically linking (ganging) the capacitors in these circuits on a common variable rotor assembly, as shown in Fig. 3-9. Note that this ganged capacitor has three separate capacitor elements.

 3-4 Superheterodyne Tuning

Tracking

It is not possible to make a receiver track perfectly over an entire wide range of frequencies. The perfect situation occurs when the RF amplifier and mixer tuned circuits are exactly together and the LO is above these two by an amount exactly equal to the IF frequency. To obtain a practical degree of tracking, the following steps are employed:

Trimmer
small variable capacitance in parallel with each section of a ganged capacitor

1. A small variable capacitance in parallel with each section of the ganged capacitor, called the **trimmer,** is adjusted for proper operation at the highest frequency. The trimmer capacitors are shown in Fig. 3-9. The highest frequency

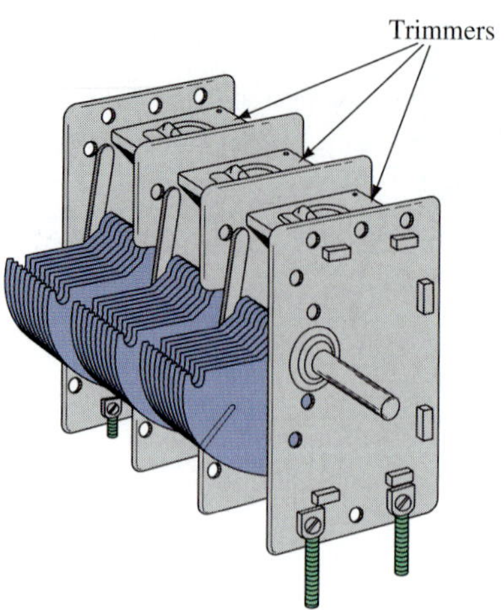

FIGURE 3-9 Variable ganged capacitor.

Padder Capacitor
small variable capacitor in series with each ganged tuning capacitor in a superheterodyne receiver to provide near-perfect tracking at the low end of the tuning range

requires the main capacitor to be at its minimum value (i.e., the plates all the way open). The trimmers are then adjusted to balance out the remaining stray capacitances to provide perfect tracking at the highest frequency.

2. At the lowest frequency, when the ganged capacitors are fully meshed (maximum value), a small variable capacitor known as the **padder capacitor** is put in series with the tank inductor. The padders are adjusted to provide tracking at the low frequency in the band.
3. The final adjustment is made at midfrequency by slight adjustment of the inductance in each tank.

The curve in Fig. 3-10(a) shows that performing the steps above, and then rechecking them once again to allow for interaction effects, provides nearly perfect tracking at three points. The minor imperfections between these points are generally of an acceptable nature. Figure 3-10(b) shows the circuit for each tank circuit and summarizes the adjustment procedure.

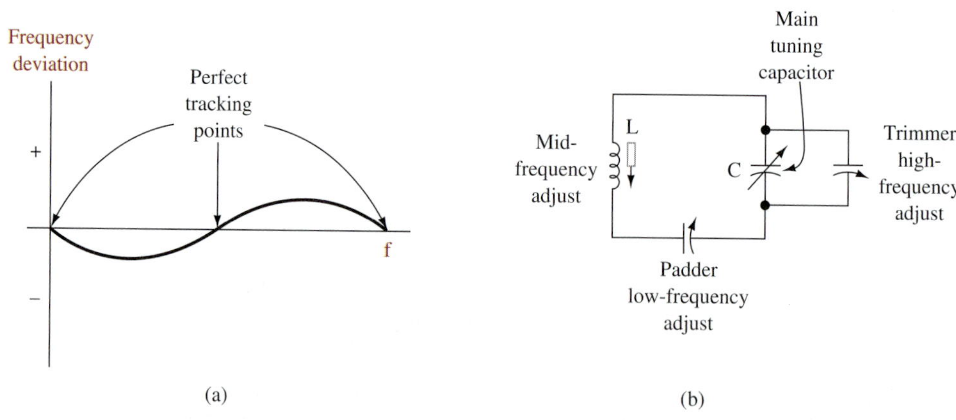

FIGURE 3-10 Tracking considerations.

Trimmer capacitor. (Courtesy of Voltronics Corporation.)

Electronic Tuning

The bulk and cost of ganged capacitors have led to their gradual replacement by a technique loosely called electronic tuning. The majority of new designs use electronic frequency synthesis (see Chapter 7). Another electronic method relies on the capacitance offered by a reverse-biased diode. Since this capacitance varies with the amount of reverse bias, a potentiometer can be used to provide the variable capacitance required for tuning. Diodes that have been specifically fabricated to enhance this variable capacitance versus reverse bias characteristic are referred to as **varactor diodes, varicap diodes,** or **VVC diodes.** Figure 3-11 shows the two generally used symbols for these diodes and a typical capacitance versus reverse bias characteristic.

Varactor Diodes having small internal capacitance that varies as a function of their reverse bias voltage

Varicap Diodes another name for varactor diodes

VVC Diodes another name for varactor or varicap diodes

The amount of capacitance exhibited by a reverse-biased silicon diode, C_d, can be approximated as

$$C_d = \frac{C_0}{(1 + 2|V_R|)^{\frac{1}{2}}} \tag{3-1}$$

where C_0 = diode capacitance at zero bias
V_R = diode reverse bias voltage

The use of Eq. (3-1) allows the designer to determine accurately the amount of reverse bias needed to provide the necessary tuning range. The varactor diode can also be used to generate FM, as explained in Chapter 5.

Figure 3-12 (page 133) shows the front end of a broadcast-band receiver. It does not incorporate an RF amplifier, and Q_1 performs the dual function of mixer and local oscillator. The varactor diode D_1 provides the variable capacitance necessary to

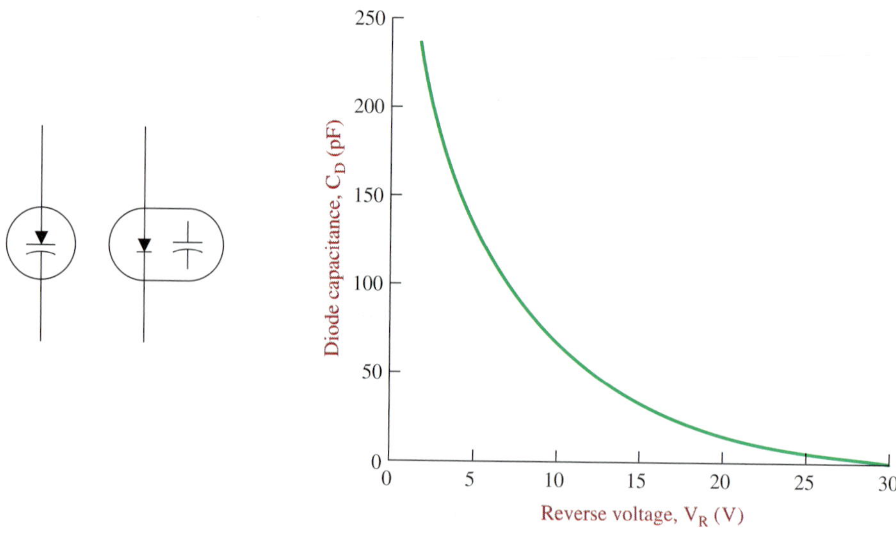

FIGURE 3-11 Varactor diode symbols and *C/V* characteristic.

tune the radio signal from the antenna while D_2 allows for the variable LO frequency. The -1- to -12-V supply comes from the tuning potentiometer and provides the necessary variable reverse voltage for both varactor diodes. The matched diode characteristics required for good tracking often lead to the use of varactor diodes fabricated on a common silicon chip and provided in a single package.

RF inductors. (Courtesy of Coilcraft, Inc., Cary, Illinois.)

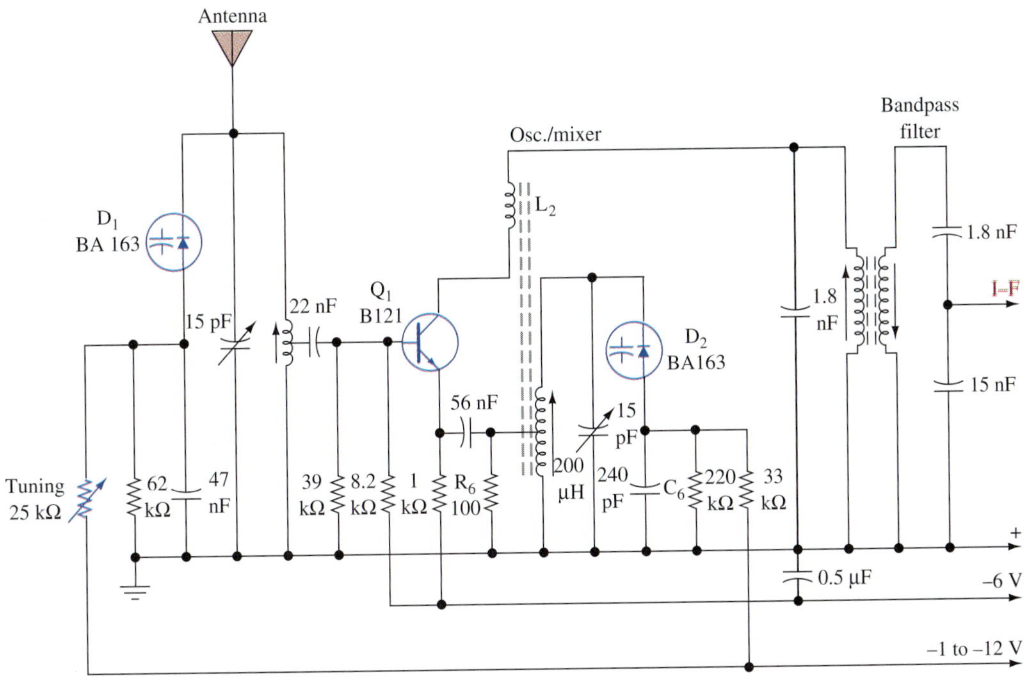

FIGURE 3-12 Broadcast-band AM receiver front end with electronic tuning.

3-5 Superheterodyne Analysis

Image Frequency

The superheterodyne receiver has been shown to have that one great advantage over the TRF—constant selectivity over a wide range of received frequencies. This was shown to be true since the bulk of the amplification in a superheterodyne receiver occurs in the IF amplifiers at a fixed frequency, and this allows for relatively simple and yet highly effective frequency selective circuits. A disadvantage does exist, however, other than the obvious increase in complexity. The frequency conversion process performed by the mixer–oscillator combination sometimes will allow a station other than the desired one to be fed into the IF. Consider a receiver tuned to receive a 20-MHz station that uses a 1-MHz IF. The LO would, in this case, be at 21 MHz to generate a 1-MHz frequency component at the mixer output. This situation is illustrated in Fig. 3-13. If an undesired station at 22 MHz were also on the air, it is possible for it also to get into the mixer. Even though the tuned circuit at the mixer's front end is "selecting" a center frequency of 20 MHz, a look at its response curve in Fig. 3-13 shows that it will not fully attenuate the undesired station at 22 MHz. As soon as the 22-MHz signal is fed into the mixer, we have a problem. It mixes with the 21-MHz LO signal and one of the components produced is 22 MHz − 21 MHz = 1 MHz—the IF frequency! Thus, we now have a desired 20-MHz station and an undesired 22-MHz station, which both look correct to the IF. Depending on the strength of the undesired station, it can either interfere with or even completely override the desired station.

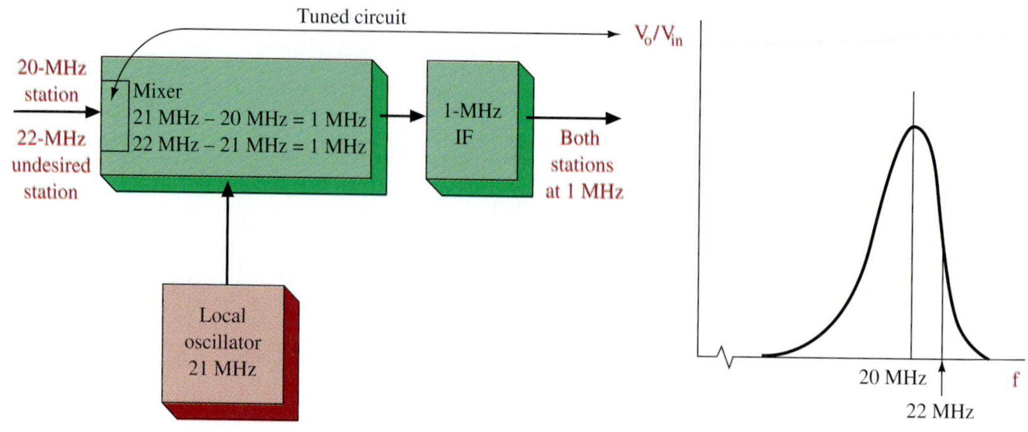

FIGURE 3-13 Image frequency illustration.

Example 3-2

Determine the image frequency for a standard broadcast band receiver using a 455-kHz IF and tuned to a station at 620 kHz.

Solution

The first step is to determine the frequency of the LO. The LO frequency minus the desired station's frequency of 620 kHz should equal the IF of 455 kHz. Hence,

$$\begin{aligned} \text{LO} - 620\ \text{kHz} &= 455\ \text{kHz} \\ \text{LO} &= 620\ \text{kHz} + 455\ \text{kHz} \\ &= 1075\ \text{kHz} \end{aligned}$$

Now determine what other frequency, when mixed with 1075 kHz, yields an output component at 455 kHz.

$$\begin{aligned} X - 1075\ \text{kHz} &= 455\ \text{kHz} \\ X &= 1075\ \text{kHz} + 455\ \text{kHz} \\ &= 1530\ \text{kHz} \end{aligned}$$

Thus, 1530 kHz is the image frequency in this situation.

Image Frequency
undesired input frequency in a superheterodyne receiver that produces the same intermediate frequency as the desired input signal

Double Conversion
superheterodyne receiver design that has two separate mixers, local oscillators, and intermediate frequencies to avoid image frequency problems

In the preceding discussion, the undesired received signal is called the **image frequency.** Designing superheterodyne receivers with a high degree of image frequency rejection is obviously an important consideration.

Image frequency rejection on the standard broadcast band is not a major problem. A glance at Fig. 3-14 serves to illustrate this point. This tuned circuit at the mixer's input comes fairly close to fully attenuating the image frequency, in this case, since 1530 kHz is so far away from the tuned circuit's center frequency of 620 kHz. Unfortunately, this situation is not so easy to attain at the higher frequencies used by many communication receivers. In these cases, a technique known as **double conversion** is employed to solve the image frequency problems. This process is described in Chapter 7.

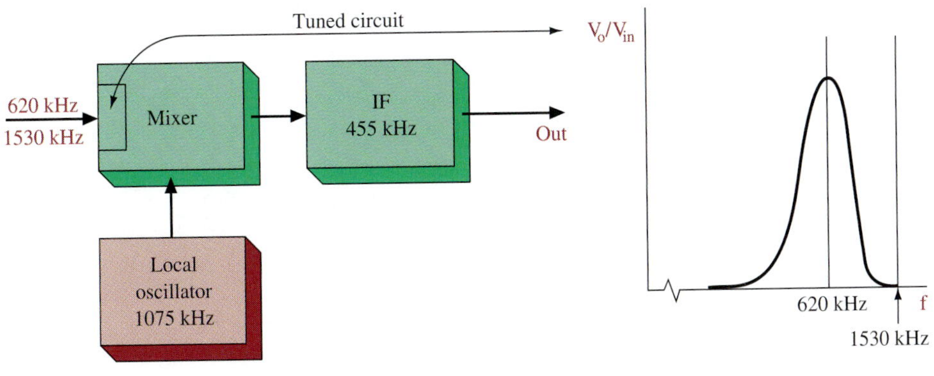

FIGURE 3-14 Image frequency not a problem.

The use of an RF amplifier with its own input tuned circuit also helps to minimize this problem. Now the image frequency must pass through two tuned circuits (tuned to the desired frequency) before it is mixed. These tuned circuits at the input of the RF and mixer stages obviously serve to attenuate the image frequency to a greater extent than can the single tuned circuit in receivers without an RF stage.

RF Amplifiers

The use of RF amplifiers in superheterodyne receivers varies from none in undemanding applications up to three or even four stages in sophisticated communication receivers. Even inexpensive AM broadcast receivers not having an RF amplifier do contain an RF section—the tuned circuit at the mixer's input. The major benefits of using RF amplification are the following:

1. Improved image frequency rejection
2. More gain and thus better sensitivity
3. Improved noise characteristics

The first two are self-explanatory at this point, but the last advantage requires further elaboration. Mixer stages require devices to be operated in a nonlinear area to generate the required difference frequency at their output. This process is inherently more noisy (i.e., higher NF) than normal class A linear bias. The use of RF amplification stages to bring the signal up to appreciable levels minimizes the effect of mixer noise.

The RF amplifier usually employs an FET as its active component. While BJTs certainly can be utilized, the following advantages of FETs have led to their general usage in RF amplifiers.

1. Their high input impedance does not "load" down the Q of the tuned circuit preceding the FET. It thus serves to keep the selectivity at the highest possible level.
2. The availability of dual-gate FETs provides an isolated injection point for the AGC signal.
3. Their input/output square-law relationship allows for lower distortion levels.

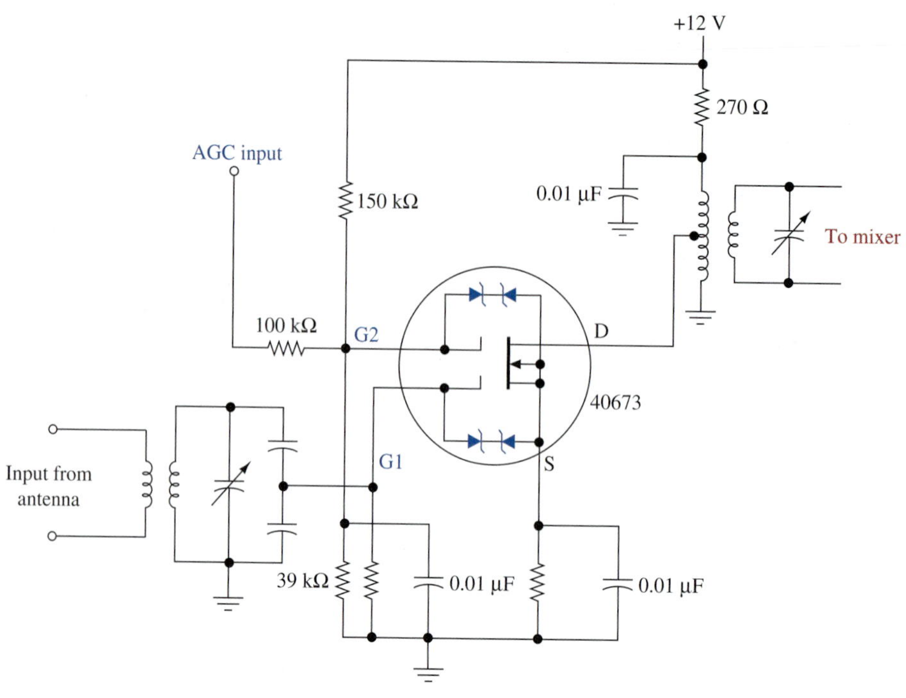

FIGURE 3-15 Dual-gate MOSFET RF amplifier.

Cross-Modulation distortion that results from undesired mixer outputs

The distortion referred to in the last item is called **cross-modulation** and is explained in Sec. 6-2.

A typical MOSFET RF amplifier stage is shown in Fig. 3-15. It is a dual-gate unit, with the AGC level applied to gate 2 to provide for automatically variable gain. The received antenna signal is fed via a tuned coupling circuit to gate 1. The gate 1 and output drain connections are tapped down on their respective coupling networks, which keeps the device from self-oscillation without the need for a neutralizing capacitor. Notice the built-in transient protection shown within the symbol for the 40673 MOSFET. Zener diodes between the gates and source/substrate connections provide protection from up to 10-V p-p transient voltages. This is a valuable safeguard because of the extreme fragility of the MOSFET gate/channel junction. Additional RF amplifier information is included in Sec. 6-2.

Mixer/LO

The frequency conversion accomplished by the mixer/LO combination can be accomplished in a number of ways. The circuits shown in Fig. 3-16 illustrate some of the possibilities. They all make use of a device's nonlinearity to generate sum and difference frequencies between the RF and local oscillator signals. This process generates output components at the IF frequency. Mixers are also referred to as **converters** and **first detectors.**

Converters another name for mixers

First Detectors another name for mixers

The most basic mixer involves using a single diode to provide the required nonlinearity for generating sum and difference frequencies. This is shown in Fig. 3-16(a).

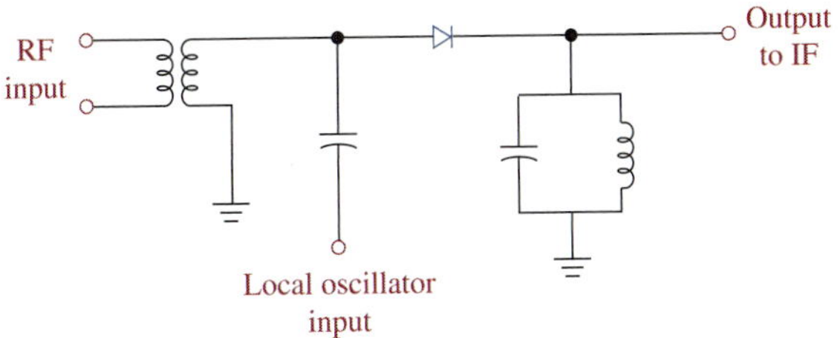

(a)

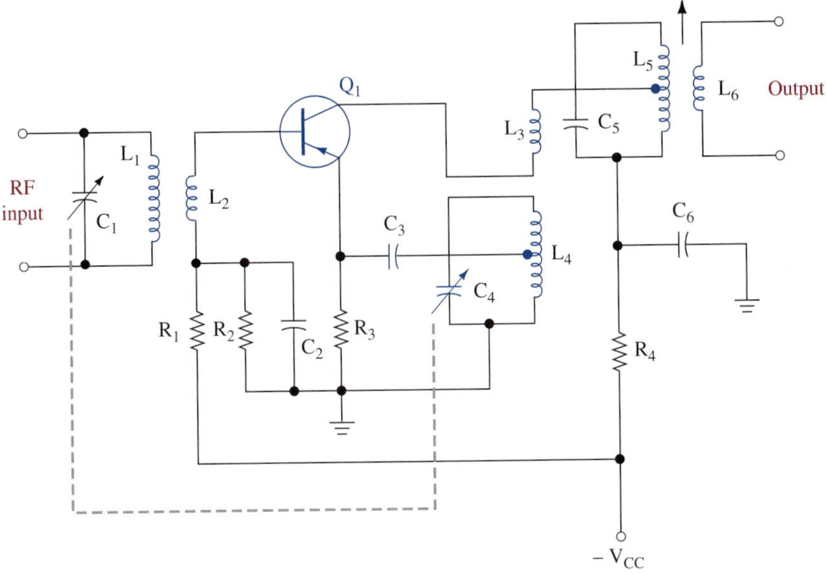

(b)

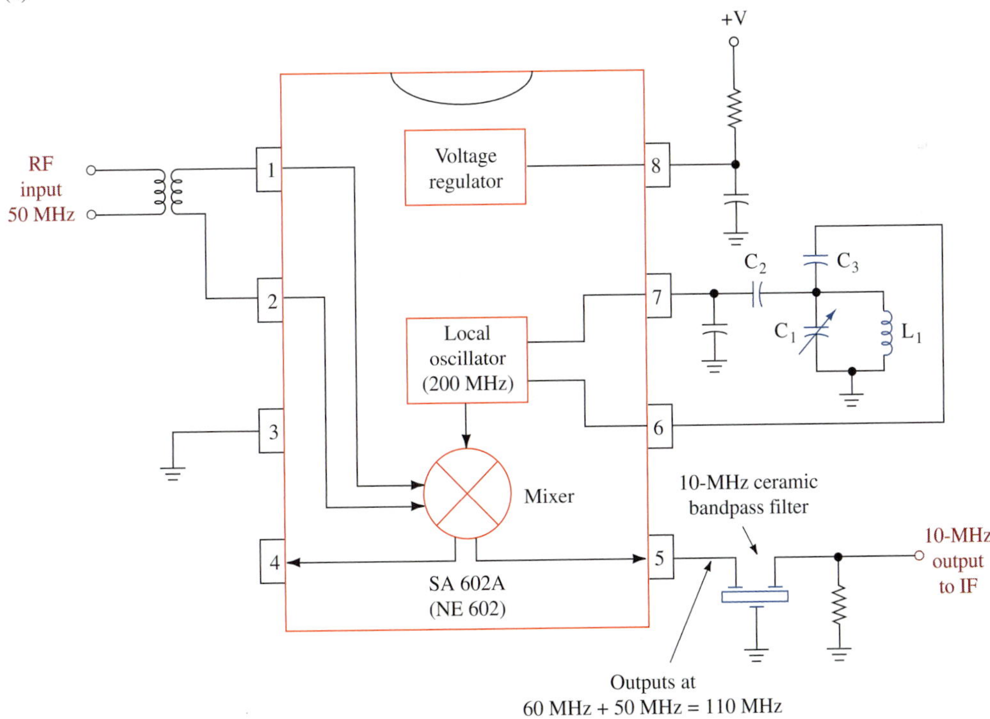

(c)

FIGURE 3-16 Typical mixer circuits.

The *LC* circuit at the output is tuned to the desired IF and attenuates all the other generated frequencies. Diode mixers are especially useful at the higher frequency ranges above 500 MHz. Although they don't offer the gain of transistorized mixers, operation at these frequencies is economically possible by using **Schottky diodes.** The Schottky diode is a specially fabricated device formed from a metal-semiconductor interface. Because the diode is not made from a *pn* junction, it is a majority carrier device and offers an extremely low junction capacitance. This makes it useful up to frequencies of 100 GHz!

Schottky Diode specially fabricated majority carrier device formed from a metal-semiconductor interface, with extremely low junction capacitance

The circuit shown in Fig. 3-16(b) is a **self-excited mixer** because a single device does the mixing and generates the LO frequency. Self-excited mixers are sometimes referred to as **autodyne mixers.** The oscillator-tuned circuit of C_4 and L_4 provides a positive feedback signal to maintain oscillation via coil L_3, which is magnetically coupled to L_4. The oscillator signal is injected into Q_1's emitter via C_3 and the RF signal into its base via L_1–L_2 transformer action. The "mixed" output at Q_1's collector is fed to the C_5–L_5 tank circuit, which tunes in the desired frequency for the IF amplifiers. Recall that mixing signals through a nonlinear device generates many frequency components; the tuned circuit is used to select the desired ones for the IF amplifiers. Further detail on this circuit is provided in the troubleshooting section of this chapter (Sec. 3-8).

Self-Excited Mixer single stage in a superheterodyne receiver that creates the LO signal and mixes it with the applied RF signal to form the IF signal

Autodyne Mixer another name for self-excited mixer

A widely used IC mixer is shown in Fig. 3-16(c). The Philips SA602A (NE602) IC contains a transistorized mixer and an *npn* transistor that generates the local oscillator signal based on the frequency-selective components connected between pins 6 and 7. The L_1, C_1 combination is tuned to the desired LO frequency but a crystal could be substituted if better precision is desired. The C_2, C_3 combination is used to form a Colpitts oscillator with the internal oscillator transistor. The oscillator frequency can be set up to 200 MHz. Notice the output at pin 5 that is fed to a ceramic bandpass filter. This type of filter, which is similar to crystals used as filters, is a commonly used alternative to *LC* filters. Additional detail on these components is provided in Chapter 4. This versatile IC mixer can be used at RF input frequencies up to 500 MHz.

IF Amplifiers

The IF amplifiers provide the bulk of a receiver's gain (and thus are a major influence on its sensitivity) and selectivity characteristics. An IF amplifier is not a whole lot different from an RF stage except it operates at a fixed frequency. This allows the use of fixed double-tuned inductively coupled circuits, which in turn allow for the sharply defined bandpass response characteristic of superheterodyne receivers.

The number of IF stages in any given receiver varies, but from two to four is typical. Some typical IF amplifiers are shown in Fig. 3-17. The circuit at (a) uses the 40673 dual-gate MOSFET while the other two use LICs specially made for IF amplifier applications. Notice the double-tuned LC circuits at the input and output of all three circuits. They are shown within dashed lines to indicate they are one complete assembly. They can be economically purchased for all common IF frequencies and have a variable slug in the transformer core for fine tuning their center frequency. All three of the circuits have provision for the AGC level. Not all receivers utilize AGC to control the gain of mixer and/or RF stages, but they invariably do control the gain of IF stages.

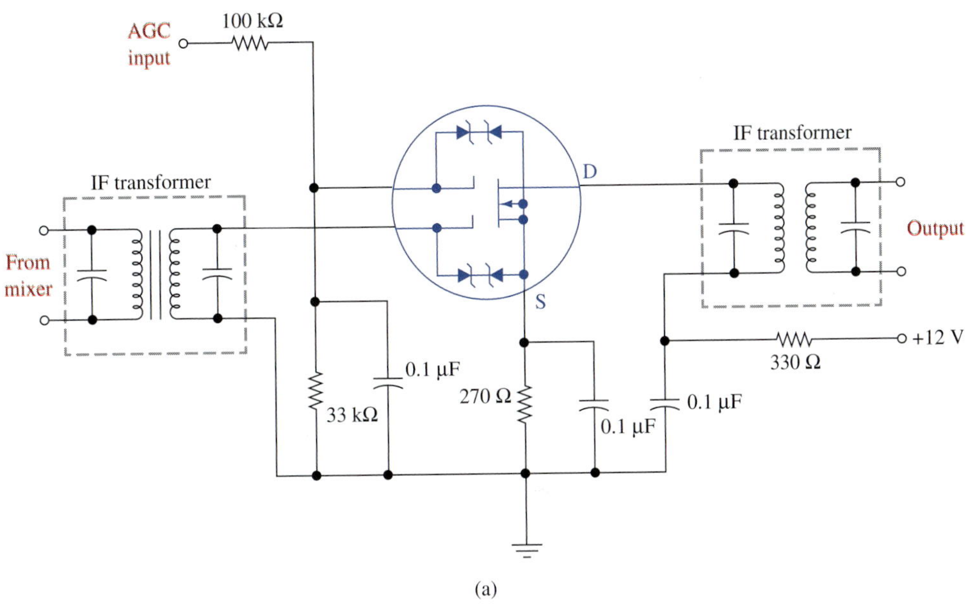

(a)

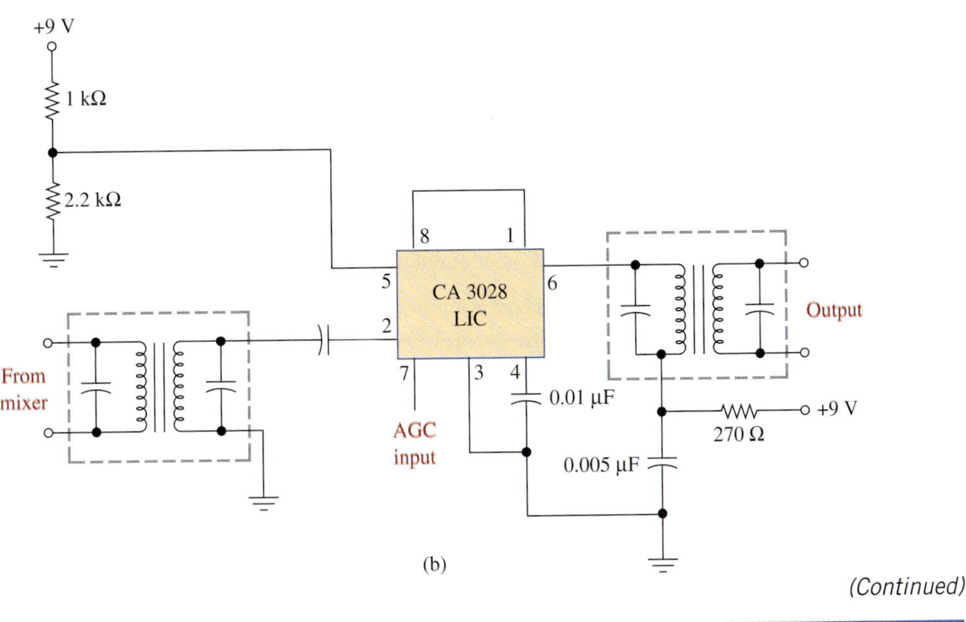

(b)

(Continued)

FIGURE 3-17 Typical IF amplifiers.

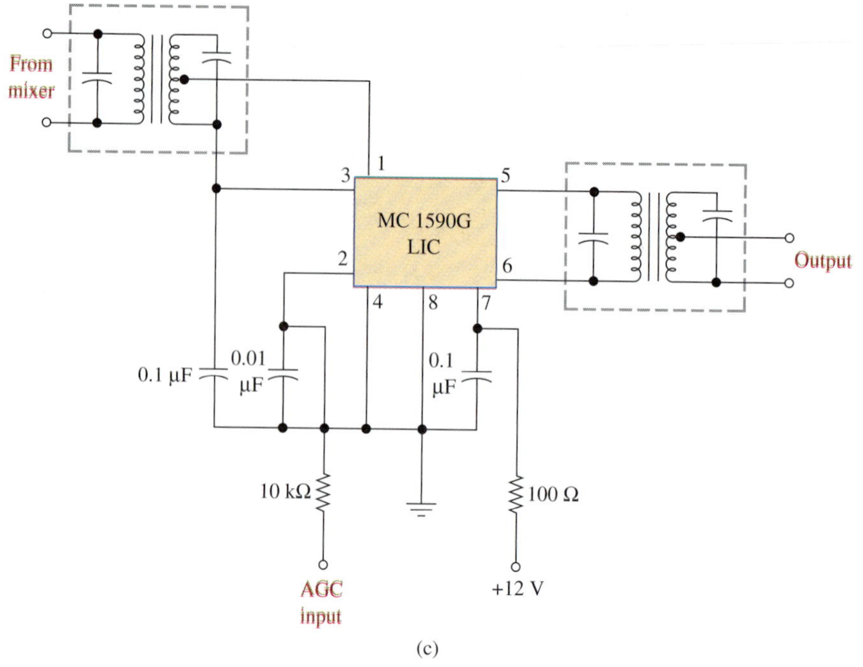

FIGURE 3-17 *(Continued)*

 3-6 Automatic Gain Control

The purpose of automatic gain control (AGC) has already been explained. Without this function, a receiver's usefulness is seriously impaired. The following list gives some of the problems that would be encountered in a receiver without this provision:

1. Tuning the receiver would be a nightmare. To avoid missing the weak stations, you would have the volume control (in the non-AGC set) turned way up. As you tuned to a strong station, you would probably blow out your speaker, whereas a weak station might not be audible.
2. The received signal from any given station is constantly changing as a result of changing weather and ionospheric conditions. The AGC allows you to listen to a station without constantly monitoring the volume control.
3. Many radio receivers are utilized under mobile conditions. For instance, a standard broadcast AM car radio would be almost unusable without a good AGC to compensate for the signal variation in different locations.

Obtaining the AGC Level

Most AGC systems obtain the AGC level just following the detector. Recall that following the detector diode, an *RC* filter removes the high frequency but hopefully leaves the low-frequency envelope intact. By simply increasing that *RC* time constant, a slowly varying dc level is obtained. The dc level changes with variations in the strength of the overall received signal.

Figure 3-18(a) shows the output from a diode detector with no filtering. In this case, the output is simply the AM waveform with the positive portion rectified out for two different levels of received signal into the diode. At (b), the addition of a filter has provided the two different envelope levels while filtering out the high-frequency content. These signals correspond to an undesired change in volume of two different received stations. At (c), a much longer time constant filter has actually

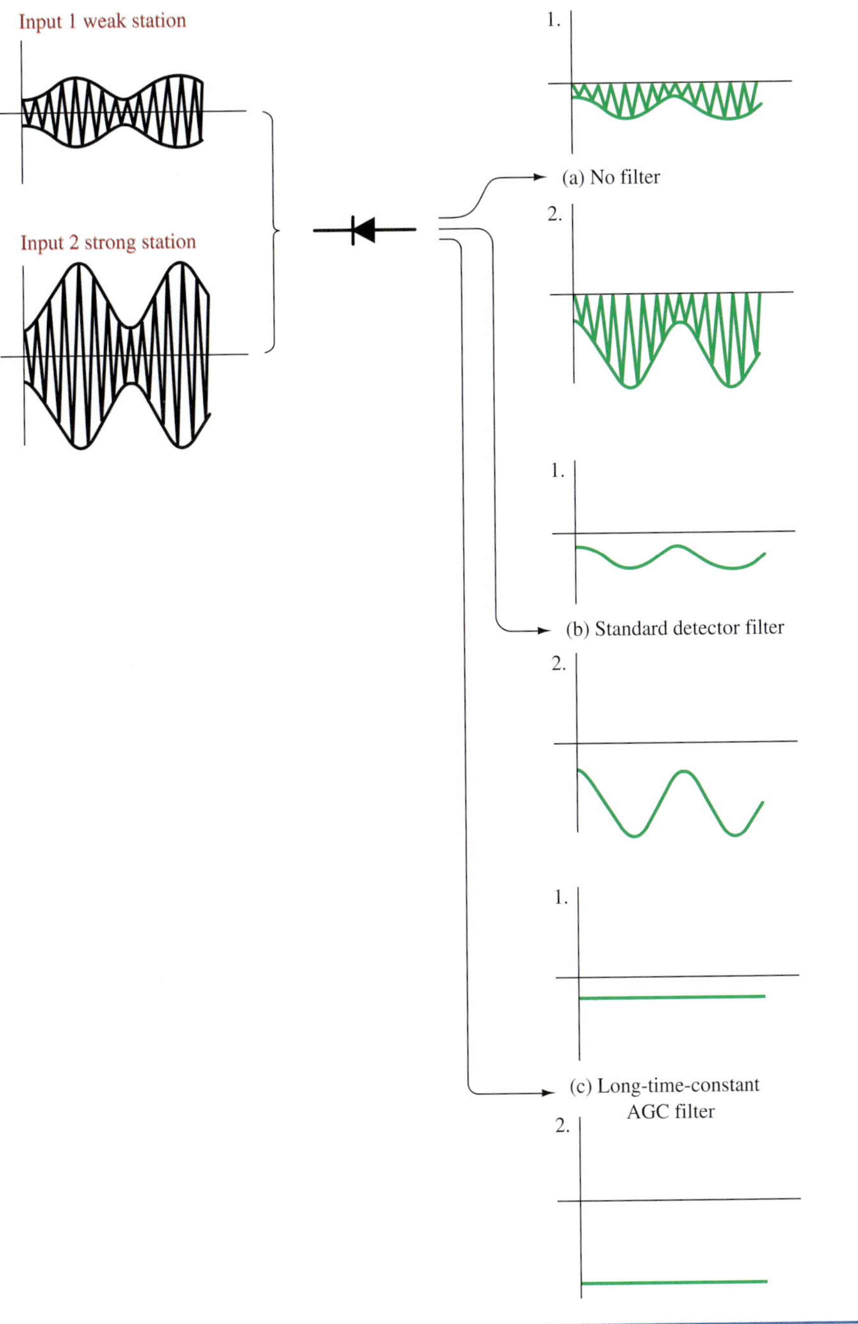

FIGURE 3-18 Development of AGC voltage.

filtered the output into a dc level. Notice that the dc level changes, however, with the two different levels of input signal. This is a typical AGC level that is subsequently fed back to control the gain of IF stages and/or the mixer and RF stages.

In this case, the larger negative dc level at C_2 would cause the receiver's gain to be decreased so that the ultimate speaker output is roughly the same for either the weak or strong station. It is important that the AGC time constant be long enough so that desired radio signal level changes that constantly occur do *not* cause a change in receiver gain. The AGC should respond only to average signal strength changes, and as such usually has a time constant of about a full second.

Controlling the Gain of a Transistor

Figure 3-19 illustrates a method whereby the variable dc AGC level can be used to control the gain of a common emitter (CE) transistor amplifier stage. In the case of a strong received station, the AGC voltage developed across the AGC filter capacitor (C_{AGC}) is a large negative value that subsequently lowers the forward bias on Q_1. It causes more dc current to be drawn through R_2, and hence less is available for the base of Q_1, since R_1, which supplies current for both, can supply only a relatively constant amount. The voltage gain of a CE stage with an emitter bypass capacitor (C_E) is nearly directly proportional to dc bias current, and therefore the strong station reduces the gain of Q_1. The reception of very weak stations would reduce the gain of Q_1 very slightly, if at all. The introduction of AGC in the 1920s marked the first major use of an electronic feedback control system. The AGC feedback path is called the AGC bus because in a full receiver it is usually "bused" back into a number of stages to obtain a large amount of gain control. Some receivers require more elaborate AGC schemes, and they will be examined in Chapter 7.

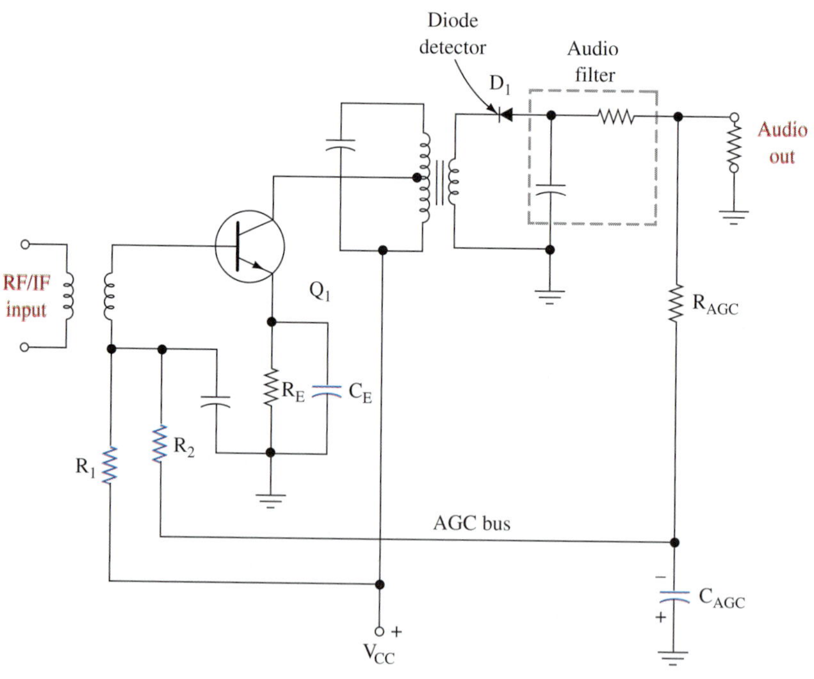

FIGURE 3-19 AGC circuit illustration.

IF/AGC Amplifier

The IF/AGC amplifier shown in Fig. 3-20 operates over an extremely wide input (J1) range of 82 dB. It uses two low-cost transistors (2N3904 and 2N3906) as peak detectors. Q2 functions as a temperature-dependent current source and Q1 as a halfwave detector. Q2 is biased for a collector current of 300 μA at 27°C with a 1 μA/°C temperature coefficient.

The current into capacitor C_{AV} is the difference in the Q1, Q2 collector currents, which is proportional to the output signal at J2. The AGC voltage (V_{AGC}) is the time integral of this difference current. To ensure that V_{AGC} is not sensitive to the short-term output signal changes, the rectified current in Q1 must, on average, balance the current in Q2. If the output of A2 is too small, V_{AGC} will increase, thereby increasing the gain of A1 and A2. This will cause Q1 to conduct further until the current through Q1 balances the current through Q2.

The gain of ICs A1 and A2 is set at 41 dB maximum for a total possible 82-dB gain. They operate sequentially because the gain of A1 goes from minimum to maximum first and then A2's does the same as dictated by the AGC level. The full range of gain occurs from $V_{AGC} \simeq 5$ V (0 dB) to $V_{AGC} \simeq 7$ V (82 dB). This is approximately a linear relationship so that $V_{AGC} = 6$ V would cause a gain of about 41 dB $[(6 - 5)/(7 - 5) \times 82 = 41]$.

The bandwidth exceeds 40 MHz and thereby allows operation at standard IFs such as 455 kHz, 10.7 MHz, or 21.4 MHz. At 10.7 MHz the AGC threshold is 100 μV rms (−67 dBm) and the output is 1.4 V rms (3.9 V p-p). This corresponds to a gain of 83 dB (20 log 1.4 V/100 μV). The output holds steady at 1.4 V rms for inputs from −67 dBm to as high as +15 dBm, giving an 83-dB AGC range. Input signals above 15 dBm overdrive the device. The undesired harmonic outputs are typically at least 34 dB down from the fundamental.

3-7 AM Receiver Systems

We have thus far examined the various sections of AM receivers. It is now time to put it all together and look at the complete system. Figure 3-21 shows the schematic of a widely used circuit for a low-cost AM receiver. In the schematic shown in Fig. 3-21, the push-pull audio power amp, which requires two more transistors, has been omitted.

The L_1–L_2 inductor combination is wound on a powdered-iron (ferrite) core and functions as an antenna as well as an input coupling stage. Ferrite-core loop-stick antennas offer extremely good signal pickup, considering their small size, and are adequate for the strong signal strengths found in urban areas. The RF signal is then fed into Q_1, which functions as the mixer and local oscillator (self-excited). The ganged tuning capacitor, C_1, tunes to the desired incoming station (the *B* section) and adjusts the LO (the *D* section) to its appropriate frequency. The output of Q_1 contains the IF components, which are tuned and coupled to Q_2 by the T_1 package. The IF amplification of Q_2 is coupled via the T_2 IF "can" to the second IF stage, Q_3, whose output is subsequently coupled via T_3 to the diode detector E_2. Of course, T_1, T_2, and T_3 are all providing the very good superheterodyne selectivity characteristics at the standard 455-kHz IF frequency. The E_2 detector diode's output is filtered by C_{11} so that just the intelligence envelope is fed via the R_{12} volume control potentiometer into the Q_4 audio amplifier. The AGC filter, C_4, then allows for a fed-back control level into the base of Q_2.

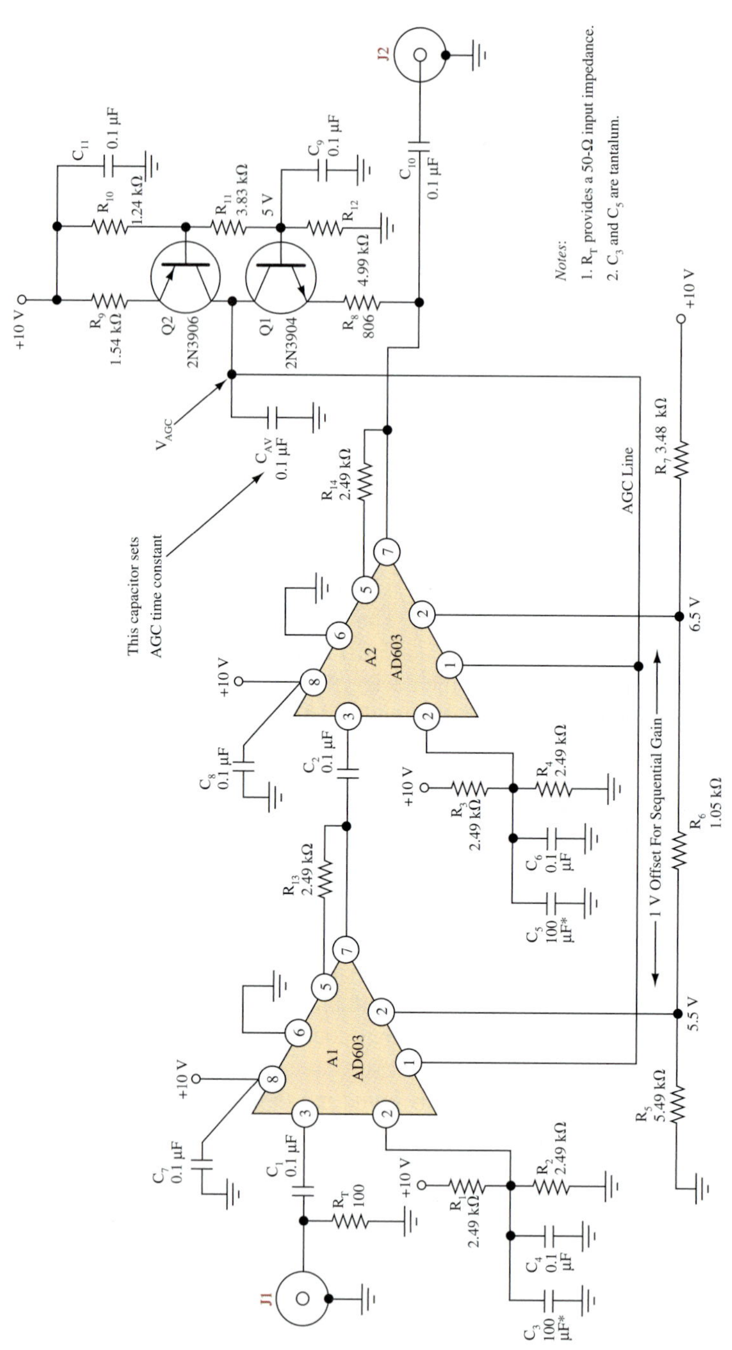

FIGURE 3-20 Wide-range IF/AGC amplifier. (Reprinted with permission from *Electronic Design*, Penton Publishing Company.)

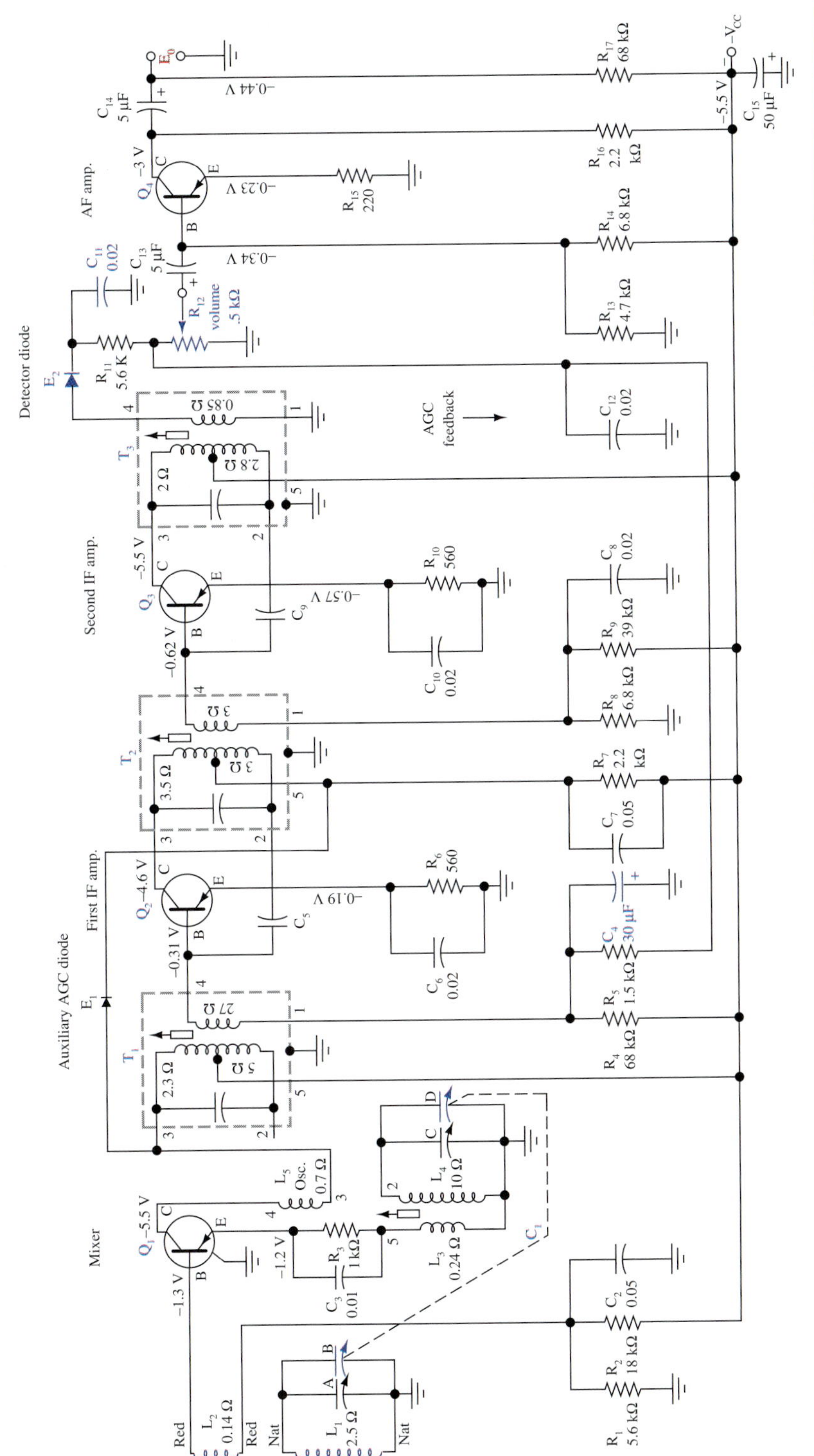

FIGURE 3-21 AM broadcast superheterodyne receiver.

Auxiliary AGC Diode reduces receiver gain for very large signals

This receiver also illustrates the use of an **auxiliary AGC diode** (E_1). Under normal signal conditions, E_1 is reverse biased and has no effect on the operation. At some predetermined high signal level, the regular AGC action causes the dc level at E_1's cathode to decrease to the point where E_1 starts to conduct (forward bias), and it loads down the T_1 tank circuit, thus reducing the signal coupled into Q_2. The auxiliary AGC diode thus furnishes additional gain control for strong signals and enhances the range of signals that can be compensated for by the receiver.

LIC AM Receiver

The complete function of a superheterodyne AM receiver can be accomplished with LICs. The only hitch is that the tuned circuits must be added on externally. Several AM chips are available from the various IC manufacturers. The Philips semiconductor TDA1572T is a typical unit and is shown in Fig. 3-22. Notice that the device uses electronic tuning with variable capacitance diodes.

Even though the use of the LIC greatly reduces component count, the physical size and cost are not appreciably affected because they are mainly determined by the frequency selective circuits. Thus, LIC AM radios are not widely used for low-cost applications but do find their way into higher-quality AM receivers, where certain performance and feature advantages can be realized.

The limiting factor of tuned circuits is the only roadblock to having complete receivers on a chip except for the station selection and volume controls. Alternatives to *LC*-tuned circuits, such as ceramic filters, may be integrable in the future. (See Chapter 7 for additional details on alternative filter circuits.) Another possibility is the use of phase-locked-loop (PLL) technology in providing a nonsuperheterodyne type of receiver. (See Chapter 6 for PLL theory.) Using this approach, it is theoretically possible to fabricate a functional AM broadcast-band receiver using just the chip and two external potentiometers (for volume control and station selection) and the antenna.

AM Stereo

It is known that the reproduction of music with two separate channels can enrich and add to its realism. Broadcast AM has started to move into stereo broadcasts since several schemes were advanced in the late 1970s. Unfortunately, the FCC decided to let the marketplace decide on the best system. This led to confusion and no clear favorite. At this juncture we find that the Motorola system has become the de facto standard. It is no wonder that AM stereo has not become a favorite mode of broadcast as has FM radio, where a single approved system led to essentially total market coverage.

The Motorola C-Quam stereo signal is developed as shown in Fig. 3-23. The carrier is phase-shifted so that essentially two carrier signals are developed. The two audio signals (left and right channels) are used to modulate the two carriers individually. Note that a reference 25-Hz signal also modulates one of the carrier signals. When the receiver detects the 25-Hz tone, it lights up an indicator to indicate stereo reception. The two AM signals are summed out of the modulator for final amplification and transmission. Regular receivers simply detect the left-plus-right signals for normal monaural reception. A specially equipped stereo receiver can differentiate between the two out-of-phase carriers and thereby develop the two separate audio signals.

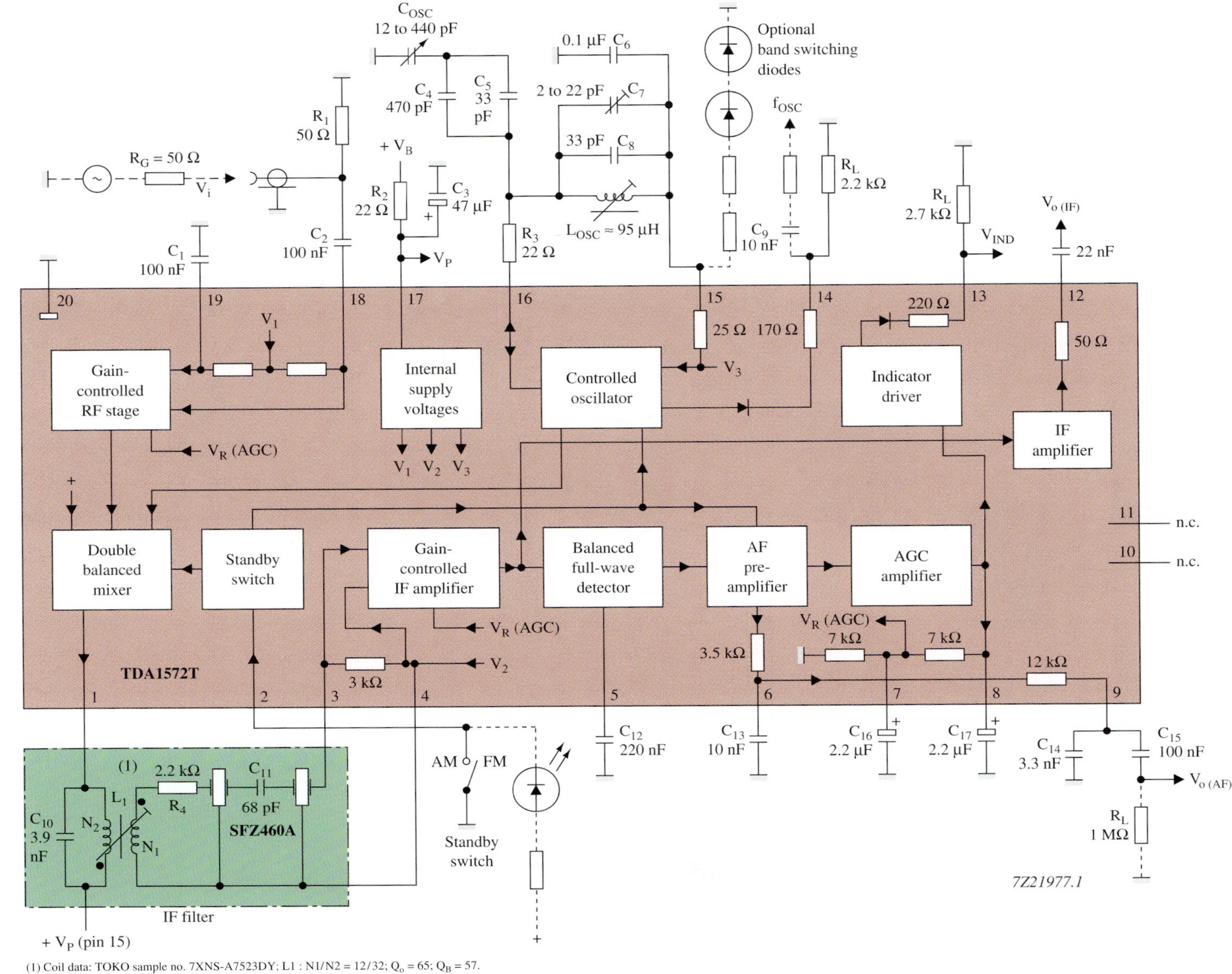

FIGURE 3-22 TDA1572T AM receiver. (Courtesy of Philips Semiconductors.)

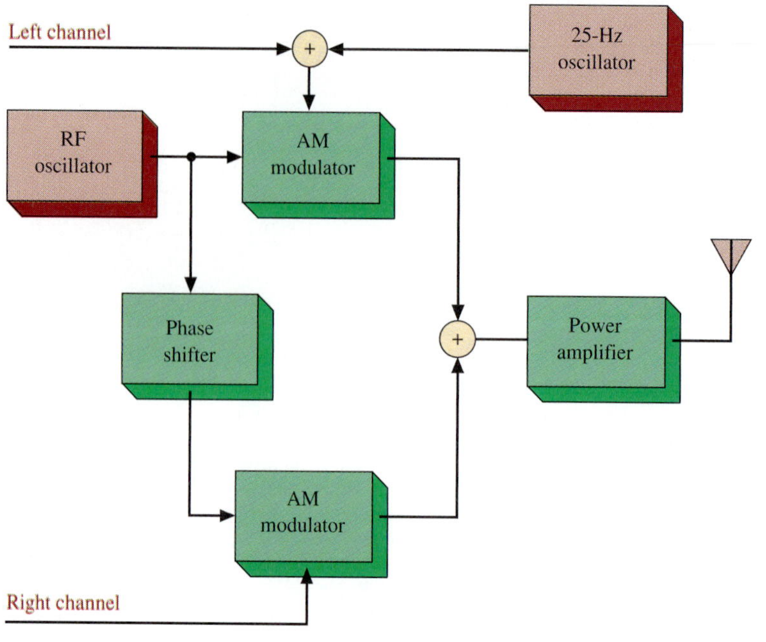

FIGURE 3-23 AM stereo block diagram.

Due to the phase-shifting of the carrier, two sets of sidebands are generated 90° out of phase. Figure 3-24 provides a pictorial representation of this condition. This is an example of combining two separate signals (left and right channels) into one frequency band. Additional information on this concept is provided in subsequent chapters.

A block diagram of a C-Quam AM stereo receiver is shown in Fig. 3-25. An MC13024 IC is the basis of this system and the required "external" components are also shown. This circuit provides the complete receiver function requiring only a stereo audio power amplifier for the left and right channel outputs at pins 23 and 20.

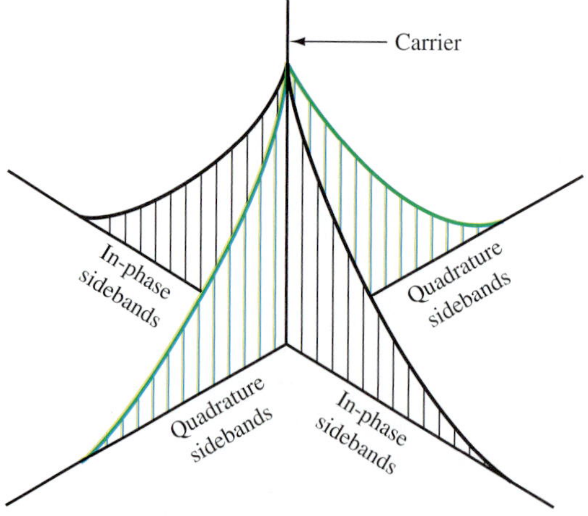

FIGURE 3-24 Phase relationships in AM stereo.

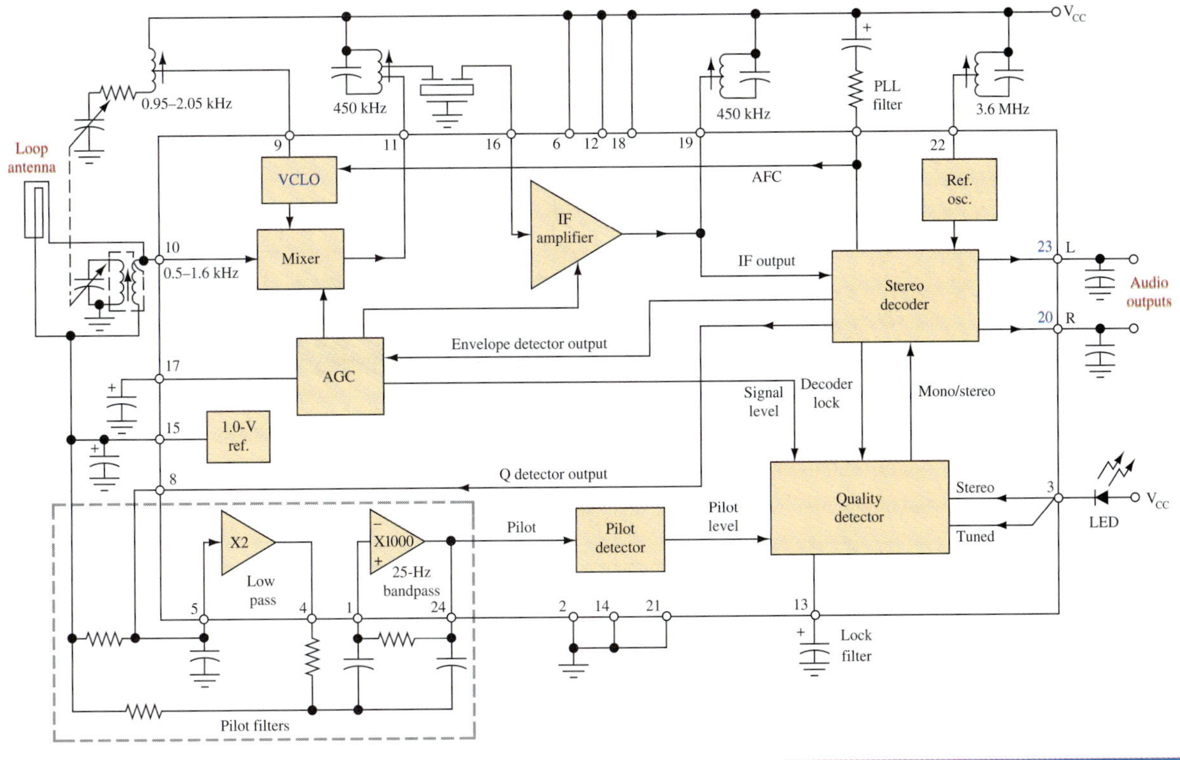

FIGURE 3-25 C-Quam receiver system. (Courtesy of Motorola, Inc.)

As you study this block diagram, you may not understand some of the "blocks." For instance, instead of a local oscillator input to the mixer, a voltage-controlled local oscillator (VCLO) is provided. It is controlled by an automatic frequency control (AFC) signal at pin 7 that is the result of a PLL. All of these devices will be explained in subsequent chapters.

Receiver Analysis

It is convenient to consider power gain or attenuation of various receiver stages in terms of decibels related to a reference power level. The most often used references are with respect to 1 mW (dBm) and 1 W (dBW). In equation form,

$$\text{dBm} = 10 \log_{10} \frac{p}{1 \text{ mW}} \tag{3-2}$$

$$\text{dBW} = 10 \log_{10} \frac{p}{1 \text{ W}} \tag{3-3}$$

A dBm or dBW is an actual amount of power, whereas a dB represents a ratio of power. When dealing with a system that has several stages, the effect of dB and dBm can be dealt with easily. The following example shows this process.

Example 3-3

Consider the radio receiver shown in Fig. 3-26. The antenna receives an 8-μV signal into its 50-Ω input impedance. Calculate the input power in watts, dBm, and dBW. Calculate the power driven into the speaker.

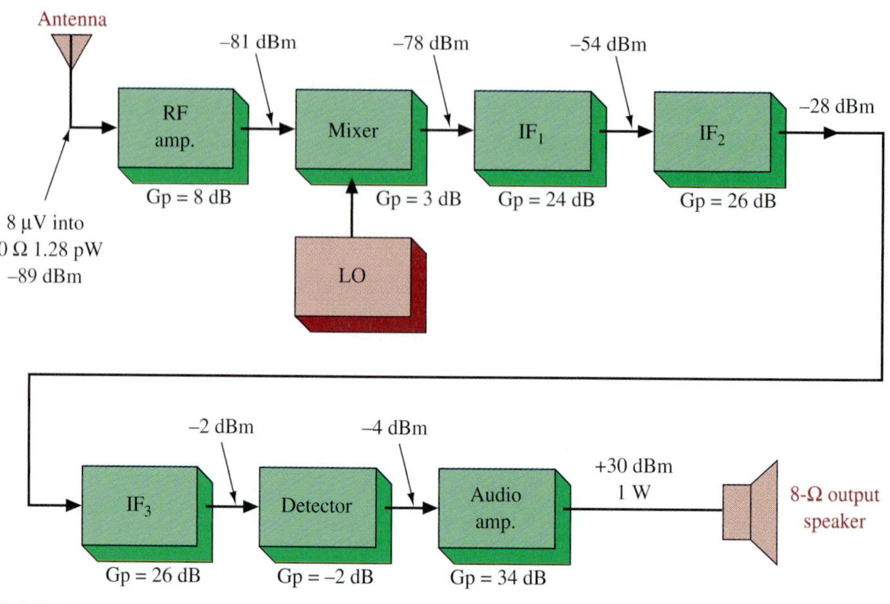

FIGURE 3-26 Receiver block diagram.

Solution

$$P = \frac{V^2}{R} = \frac{(8\ \mu\text{V})^2}{50\ \Omega} = 1.28 \times 10^{-12}\ \text{W}$$

$$\text{dBm} = 10\log_{10}\frac{P}{1\ \text{mW}} \qquad (3\text{-}2)$$

$$= 10\log_{10}\frac{1.28 \times 10^{-12}}{1 \times 10^{-3}} = -89\ \text{dBm}$$

$$\text{dBW} = 10\log_{10}\frac{P}{1\ \text{W}} \qquad (3\text{-}3)$$

$$= 10\log_{10}\frac{1.28 \times 10^{-12}}{1} = -119\ \text{dBW}$$

Notice that dBm and dBW are separated by 30 dB—this is always the case because 30 dB represents a 1000:1 power ratio. To determine the power driven into the speaker, simply add the gains and subtract the losses (in dB) all the way through the system. The −89 dBm at the input is added to the 8-dB gain of the RF stage to give −81 dBm. Notice that the 8-dB gain is simply added to the dBm input to give −81 dBm. This, and all subsequent stages, is shown in Fig. 3-26. Thus,

$$\begin{aligned} P_{\text{out(dBm)}} &= -89\ \text{dBm} + 8\ \text{dB} + 3\ \text{dB} + 24\ \text{dB} + 26\ \text{dB} \\ &\quad + 26\ \text{dB} - 2\ \text{dB} + 34\ \text{dB} \\ &= 30\ \text{dBm into speaker} \end{aligned}$$

$$30\text{ dBm} = 10\log_{10}\frac{P_{out}}{1\text{ mW}}$$

$$3 = \log_{10}\frac{P_{out}}{1\text{ mW}}$$

Therefore,

$$1000 = \frac{P_{out}}{1\text{ mW}}$$

$$P_{out} = 1\text{ W}$$

Example 3-3 assumed that the receiver's AGC was operating at some fixed level based on the input signal's strength. As previously explained, the AGC system will attempt to maintain that same output level over some range of input signal. **Dynamic range** is the decibel difference between the largest tolerable receiver input signal (without causing audible distortion in the output) and its sensitivity (usually the minimum discernible signal). Dynamic ranges of up to about 100 dB represent current state-of-the-art receiver performance.

Dynamic Range
in a receiver, the dB difference between the largest tolerable receiver input level and its sensitivity level

3-8 TROUBLESHOOTING

In this section we will analyze and troubleshoot the AM mixer circuit. The mixer circuit, also known as an autodyne circuit, is a combination of the local oscillator and the mixer in a single stage. We will also discuss power supply and audio amplifier problems in this section.

After completing this section you should be able to

- Troubleshoot an AM mixer circuit
- Identify an open input circuit
- Identify a dead or intermittent local oscillator circuit
- Identify causes for a dead or intermittent local oscillator
- Troubleshoot the receiver's power supply
- Troubleshoot the receiver's audio amplifier

THE MIXER CIRCUIT

In Sec. 3-3 we saw that the local oscillator and the mixer play a very important part in AM reception. Figure 3-27 shows the mixer stage (autodyne circuit) of an AM radio. The received RF input signal is fed into the base of Q_1 from coils L_1 and L_2. The input AM radio signal is selected by tuning C_1; notice also that C_1 and C_4 are ganged. When C_1 is adjusted, C_4 will be adjusted by the same amount. The local oscillator portion of the converter stage is made up of L_3, L_4, and C_4. As C_4 is adjusted, the oscillator frequency changes to maintain a difference frequency of 455 kHz above the received AM signal. The feedback capacitor C_3 sends a portion of the oscillator signal from a tap on L_4 back to the emitter of Q_1. The received RF signal and the

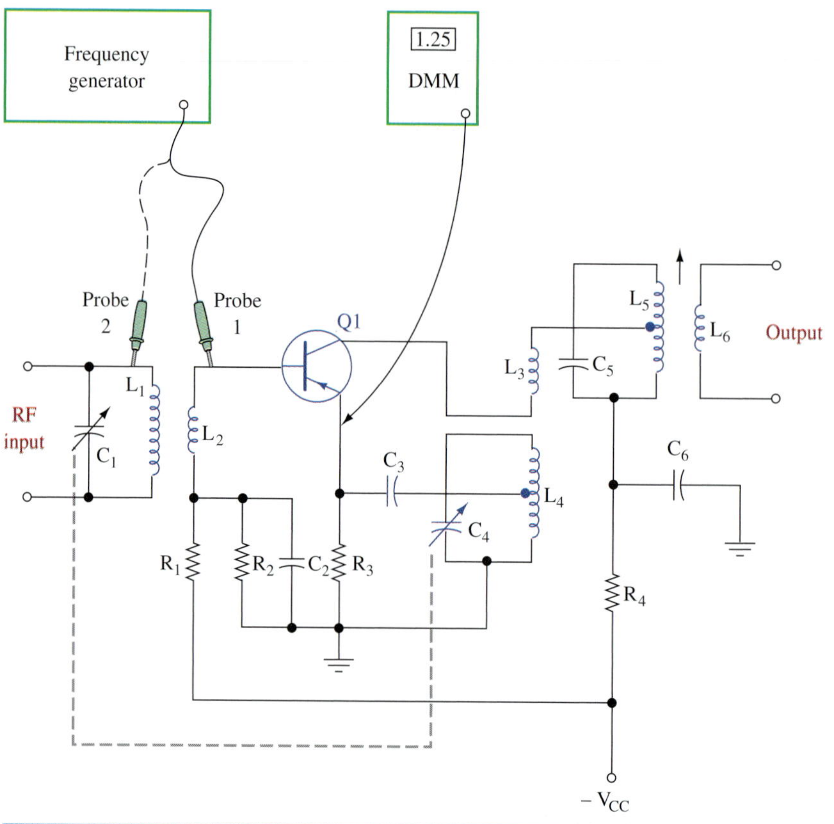

FIGURE 3-27 Troubleshooting a self-excited mixer.

oscillator signal are mixed in Q_1 to produce the IF, which is sent to L_5. All frequencies except the 455-kHz IF signal are filtered out by the tuned circuit's L_5 and C_5. Resistors R_1 and R_2 form a voltage divider network to bias the transistor's base–emitter circuit. Resistor R_3 acts as a dc stabilizer for the emitter circuit. Capacitor C_2 is a decoupling capacitor to keep the IF frequency from being fed back to the base of Q_1. Any IF signal present at the base would be shorted to ground. The transistor's collector dc voltage is supplied by R_4.

No AM RF Signal

If the received AM RF signal does not reach the base of Q_1, no audio will be heard from the speaker. Noise may be heard when the tuning dial is moved across the band, but no stations will come in. An exception might be where a strong AM radio station in close proximity blends through into the converter transistor. A good indication of a working converter stage is to monitor the emitter voltage using a DMM as depicted in Fig. 3-27. As the radio is tuned across the AM band, the voltage reading on DMM will change. An open winding in coil L_1 will cause the received AM signal to be lost. If a test signal were injected at the base of Q_1 (refer to Fig. 3-27, signal generator probe 1), it would be heard from the speaker. If the test signal were applied to L_1

(Fig. 3-27, probe 2), no signal would be heard. If the coil L_2 were open, AM reception would be lost. In addition, an open L_2 will isolate the base from the resistor voltage divider network. As a result, the base–emitter bias would be removed and the transistor would cut off. Coils L_1 and L_2 in most AM receivers are part of the antenna system. The antenna consists of a ferrite metal stick with very fine wires making up the two coils. These fine wires often break at the antenna or come loose from the printed circuit board, causing L_1 or L_2 to become open. Also, the wires from radio transformers usually break at the base of the transformer, where they are connected to the PCB. A dead converter stage can also result from a defective transistor.

Dead Local Oscillator Portion of Converter

Measuring the voltage at the emitter with a DMM and tuning the radio across the AM band is a good indication of oscillator operation. If this voltage changes as the radio is tuned, the oscillator can be assumed to be functioning. An oscilloscope at the emitter of transistor Q_1 will show the oscillator waveform if it is present. If the local oscillator is dead (not operating), the signal will be missing and no voltage change will be detected by the DMM at the emitter of Q_1. An open L_4 will shut down the oscillator operation. The same is true for an open C_4.

Poor AM Reception

A leaky capacitor C_3 can cause erratic operation of the local oscillator circuit. Received radio stations will fade in and fade out as a result of this erratic operation of the oscillator. A station may fade out altogether and the converter quit working from a severely leaking capacitor. This is due to the loading effect on the emitter circuit of Q_1. A local oscillator with poor tracking will affect radio reception at the high end or the low end of the AM band. A faulty C_4 or C_1 is a likely suspect if poor tracking occurs.

Symptoms and Likely Causes

Table 3-1 lists symptoms and the likely circuit components that can cause them. Suspected faulty capacitors should be tested. The best method for testing capacitors is to use a capacitor checker. Some DMMs on the market today have a capacitor

Table 3-1 Mixer Troubleshooting Chart

Symptom	Troubleshooting Checks	Likely Trouble
No reception	Power okay; converter working	No input signal at base of Q_1; L_1 or L_2 open; transistor bad
Stations fade in and out	Q_1's emitter voltage fluctuates	Converter operation erratic; C_3 leaky or open
No stations heard from mid- to low-AM band	DMM voltage changes when radio is tuned	LO not tracking across AM band; C_1 or C_4 faulty
No stations heard from mid- to high-AM band	DMM voltage changes when radio is tuned	LO not tracking across AM band; C_1 or C_4 faulty

check function. The capacitor values are small in the converter circuit and should be tested out of the circuit. Open coils can be found using the ohmmeter setting of the DMM. A good coil will measure a low resistance and an open coil will measure a very high resistance. Coils can usually be measured without removing them from the circuit. If the converter transistor is suspect, test it with a transistor tester. Modern DMMs are equipped with this function. An open or shorted transistor can be tested with the DMM diode check setting or the ohmmeter setting.

Troubleshooting the Power Supply

If the receiver is completely dead, that is, no sound comes from the speaker, you should immediately suspect the power supply. This is one part of a receiver where the technician can often easily find and repair a problem.

Receivers are powered by batteries or a transformer-rectifier supply connected to the 110-V lines. Batteries usually power portable radios. A 9-V battery is most common. To check its output voltage, turn the radio on (to load the battery) and measure the battery's terminal voltage. If it is significantly below 9 V, perhaps 8 V or less, replace the battery and recheck the unit. Also check for corroded terminals.

Some radios employ a group of cells to obtain the necessary voltage. These must be connected in a series-aiding configuration of the positive terminal of one cell to the negative terminal of the next, and so on. The battery compartment has a diagram with battery symbols and plus and minus signs molded into the plastic to help you install the cells properly. Should one cell be placed in the compartment backward, it would cancel the voltage of two cells, thereby dropping the total voltage to the point where the radio would not work. Check for proper installation of all cells. Then perform the loaded test described above for 9-V batteries.

Stereos and communications receivers will most likely use a regulated power supply similar to that shown in Fig. 3-28. Start troubleshooting by checking the output voltage with a DMM connected between point D and ground. If the voltage is correct (per manual specs), your problem lies elsewhere. If not, test the fuse for continuity and be sure the power plug is connected to a "hot" outlet and the switch is on. Next, check the rectifier output waveform at point A with an oscilloscope per the diagram in Fig. 3-29. The waveform should be similar to the one illustrated.

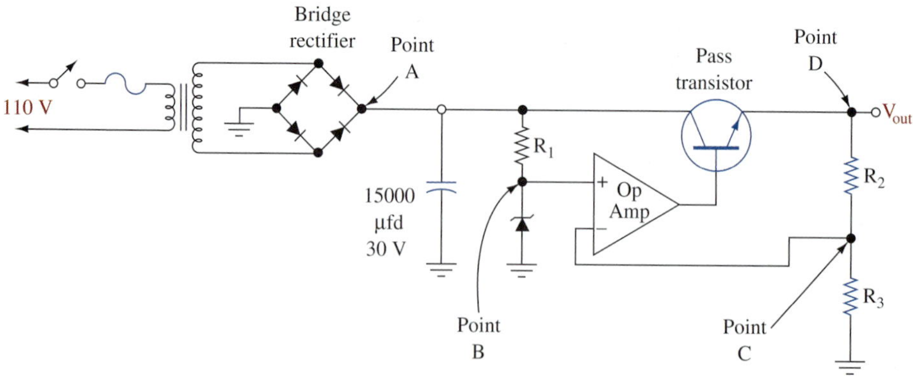

FIGURE 3-28 Regulated power supply.

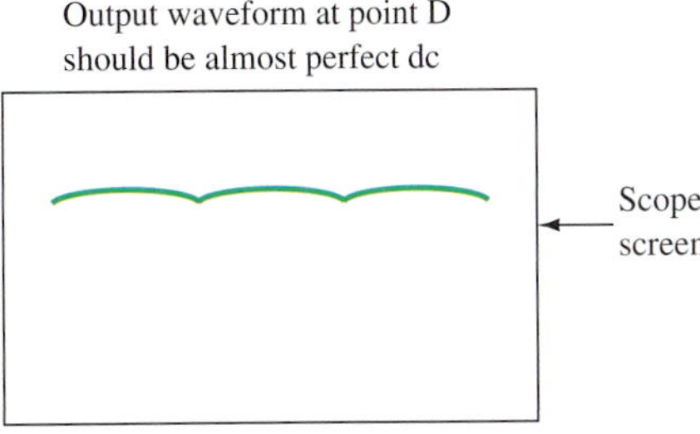

FIGURE 3-29 Bridge rectifier and filter operating properly.

If the rectifier output waveform is not similar to that shown in Fig. 3-29, one or more diodes in the bridge have probably failed. Diodes fail in one of two ways, either by opening or shorting. An open diode changes the bridge rectifier from full-wave to half-wave. As a result, ripple increases dramatically (see Fig. 3-30). A shorted diode causes heavy currents that should blow the fuse or, at the very least, cause overheated components.

Bridge rectifiers are usually encapsulated (you cannot get at the individual diodes). The unit must therefore be replaced should problems be found.

If the filter capacitor (from point A to ground) opens, the bridge output will be unfiltered, making it more difficult for the voltage regulator to eliminate ripple. It's difficult to say exactly what the waveform would look like; check the maintenance manual for details.

A shorted filter capacitor shorts the rectifier output, causing at best a blown fuse, and at worst a burned-out rectifier and/or power transformer. In either case, open or shorted, replace the capacitor.

Assuming the rectifier and filter capacitor pass the tests discussed above, measure the zener reference voltage at point B and compare with the specs per the manual. Measure the voltage at point C, the feedback voltage to the inverting input of the op-amp. It should be within a tenth of a volt or so of the zener voltage. The point C voltage can be calculated using the voltage division formula:

$$V \text{ at point C} = V_{out}(R_3)/(R_2 + R_3)$$

Measure the emitter–collector voltage of the pass transistor. It should be approximately 5 to 7 V depending on power supply load. If this voltage is a few tenths of a volt or less, the transistor is shorted and must be replaced.

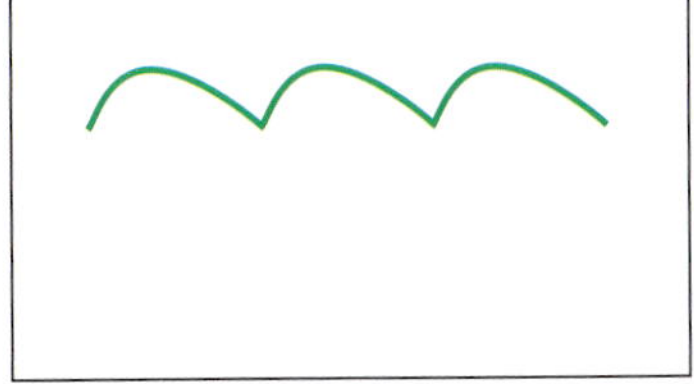

FIGURE 3-30 Ripple increase caused by open diode.

Note: The above comments on power supply troubleshooting apply for any piece of equipment using a regulated power supply, not just superheterodyne receivers.

Troubleshooting the Audio Amplifier

A quick test to determine whether an audio amplifier is working is first find the volume control. It will have three terminals on it. Touch a screwdriver or piece of wire to the center terminal as shown in Fig. 3-31. If the amplifier is working, you should hear a loud 60-Hz hum coming from the loudspeaker.

If the amplifier fails this quick test, do a dc check of voltages throughout the circuit. If nothing shows up, connect an audio generator via a 0.1-μf capacitor to the center terminal of the volume control. Set the generator to approximately 1 kHz at perhaps 50-mV amplitude. Using an oscilloscope, observe the signal at each collector and base between the volume control and loudspeaker. Should the signal be present at one point and not the next, find the defective component and replace it.

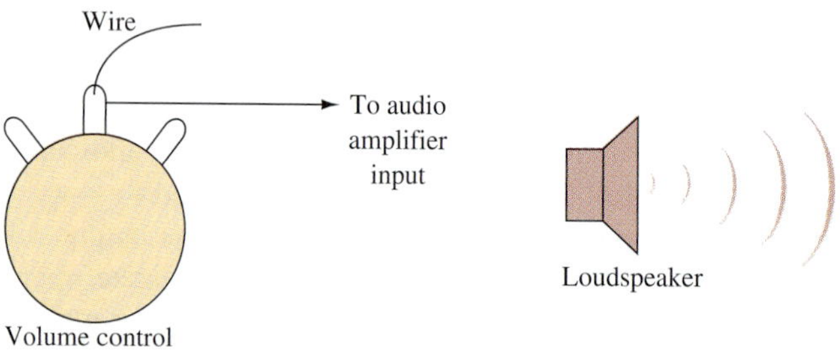

FIGURE 3-31 Testing an audio amplifier.

Troubleshooting the RF Portions of a Superhet Receiver

Signal Injection troubleshooting by injecting an input signal and tracing through the circuit to locate the failed component

In general, troubleshooting a receiver's RF sections is done using the time-tested method of **signal injection** and tracing. The approach is the same discussed for audio amplifiers except that now a high-frequency RF signal, usually modulated, is being used. This signal is connected to or injected into the receiver's antenna input terminals. The signal tracer, which can be either a scope or an RF probe on a voltmeter, is then connected to the inputs and outputs of each amplifier stage, one after the other, until the signal is lost. In this way, the defect is isolated and located with further tests.

3-9 Troubleshooting with Electronics Workbench™ Multisim

This chapter explored the circuits used for receiving and detecting an AM signal. The diode detector shown in Fig. 3-32 is an example of a circuit that can be used to recover the intelligence contained in an AM carrier. **Fig3-32.ms7 (.msm)** found on the EWB Multisim CD is provided to help you investigate further the operation of a diode detector.

Open **Fig3-32.ms7 (.msm)** on your EWB CD. This circuit contains an AM source with a carrier frequency of 100 kHz being modulated by a 1-kHz sinusoid.

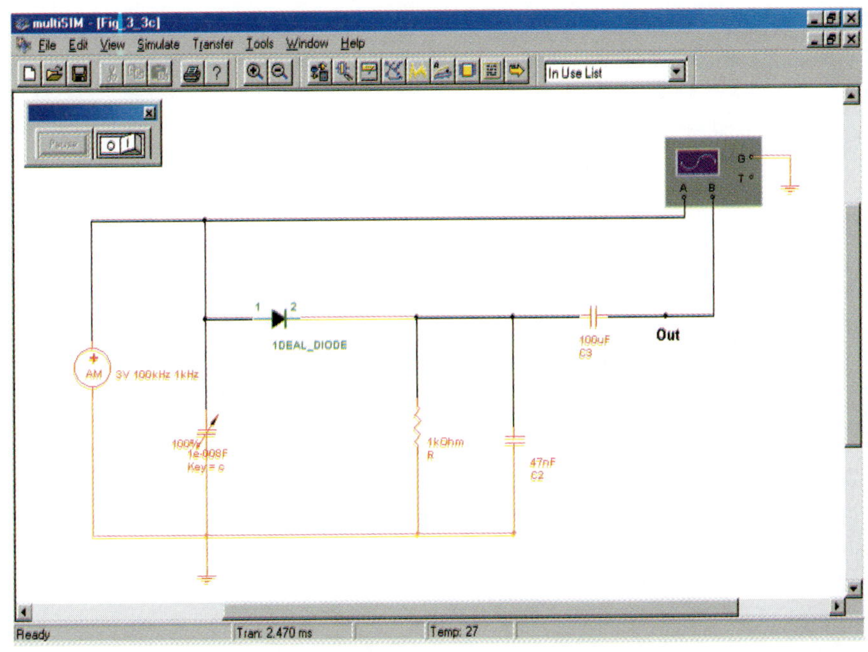

FIGURE 3-32 An AM diode detector circuit as implemented with Electronics Workbench Multisim.

The modulation index is 50 percent. Open the AM source by double-clicking on the **AM** icon. Click on the **value** tab. It should show that the carrier amplitude is 3 V, the carrier frequency is 100 kHz, the modulation index is 0.5 (50 percent), and the modulating frequency is 1 kHz. These values can be changed by the user to meet the needs of a particular simulation. Click the **start simulation** button and observe the traces on the oscilloscope.

The AM source is connected to the channel A input and the output of the detector is connected to the channel B input. The oscilloscope traces from the diode detector are shown in Fig. 3-33. There appears to be a little carrier noise on the recovered 1-kHz sinusoid. How can this noise be removed? (This question is addressed in Exercise 3.) Was a 1-kHz sinusoid recovered? Use the cursors to verify that a 1-kHz signal was recovered. Measure the modulation index of the AM source as shown in Fig. 3-33. Compare your measurement to the expected 50 percent value set in the AM source.

Capacitor 1 in Fig. 3-32 is called a *virtual capacitor*. Double-click on **C1** and select the **value** tab. The information about the virtual capacitor shows that this is a 10-nF capacitor; pressing **c** on the keyboard increases the capacitance value by 5 percent, and pressing **C** decreases the value by 5 percent. Experiment with this adjustment and see if changing the capacitance value affects the recovered signal. Make sure that you click on the schematic window to enable control of the virtual components. Adjustments to the virtual capacitor are not active if another window, such as the oscilloscope window, is currently selected.

Next, open **FigE3-1.ms7 (.msm).** This circuit looks the same as **Fig3-32.ms7 (.msm)** except that this circuit contains a fault. Use the oscilloscope to view the traces in the circuit. Good troubleshooting practice says: *Always perform a visual*

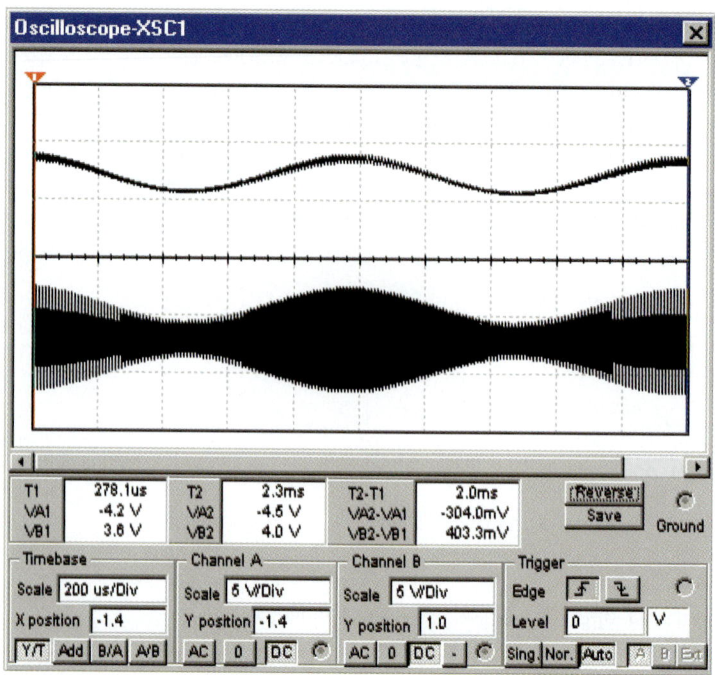

FIGURE 3-33 Oscilloscope output traces from the diode detector.

check of the circuit and check the vital signs. Checking vital signs implies that you must check power-supply voltages and also examine the input signals.

Start the simulation of the circuit and view the output and input traces. Notice that the input AM envelope looks the same, whereas the output is significantly different. This circuit does not show a power supply, but just in case, visually check that the ground connections are in place. The input signal (the AM envelope) and the ground connections are good, so the problem rests with a component. Verify that the output coupling capacitor is allowing the signal to pass properly from the detector to the output. Do this by connecting the oscilloscope A and B channels to each side of C3. You will notice that the signal is the same on both sides, which indicates that C3 is good. Electronics Workbench Multisim provides a feature that allows for the addition of a component fault in a circuit. Double-click on each of the components and check the setting under the **Fault** tab. You will discover that R1 is shorted. Change the fault setting back to **none,** which means no fault, and simulate the circuit again. The circuit should now be operational.

Additional insight into troubleshooting with Electronics Workbench™ Multisim is provided in the EWB exercises below.

Electronics Workbench™ Exercises

1. Open **FigE3-2.ms7 (.msm)** found on your EWB CD. Determine if this circuit is working properly. If it is not, find the fault. Describe why this fault would have caused the output waveform you observed.

2. Open **FigE3-3.ms7 (.msm)** found on your EWB CD. Determine if this circuit is working properly. If it is not, find the fault. Describe why this fault would have caused the output waveform you observed.
3. Open **FigE3-4.ms7 (.msm)** found on your EWB CD. Adjust the virtual capacitors C1 and C3 to provide an output waveform that contains minimal RF noise. This process requires that you adjust C1 and then C3 and keep repeating this sequence until an optimized output is obtained (C1 = 50%, C3 = 70%).

SUMMARY

In Chapter 3 the basics of AM receivers were introduced. The development of receivers from the simplest to superheterodyne systems was discussed. The major topics you should now understand include:

- the basics of a simple radio receiver
- the fundamental concepts of sensitivity and selectivity
- the functional blocks of a tuned radio frequency (TRF) receiver
- the input/output characteristics of a nonlinear device used as an AM detector
- the characteristics, operation, types, and design considerations of diode detectors
- the advantages of synchronous detection over the basic diode detector
- a complete analysis of superheterodyne receiver operation
- the tuning and tracking of a superheterodyne receiver
- an analysis of image frequency and methods for its attenuation
- the operation and typical circuits of the functional blocks in a superheterodyne receiver
- the need for automatic gain control (AGC) in a receiver and the description of a typical circuit and its operation
- the description of various superheterodyne receiver systems with power gain analysis

QUESTIONS AND PROBLEMS

SECTION 3-1

*1. Draw a diagram of a tuned radio-frequency (TRF) radio receiver.
*2. Explain the following: sensitivity of a receiver; selectivity of a receiver. Why are these important characteristics? In what units are they usually expressed?
3. Explain why a receiver can be overly selective.
4. A TRF receiver is to be tuned over the range 550 to 1550 kHz with a 25-μH inductor. Calculate the required capacitance range. Determine the tuned circuit's necessary Q if a 10-kHz bandwidth is desired at 1000 kHz. Calculate the receiver's selectivity at 550 and 1550 kHz. (0.422 to 3.35 nF, 100, 5.5 kHz, 15.5 kHz)

*An asterisk preceding a number indicates a question that has been provided by the FCC as a study aid for licensing examinations.

Section 3-2

*5. Explain the operation of a diode detector.
6. Describe the advantages and disadvantages of a diode detector.
7. Describe diagonal clipping.
8. What association does diagonal clipping have with modulation index?
9. Explain how diagonal clipping occurs in a diode detector.
10. Provide the advantages of a synchronous detector compared to a diode detector. Explain its principle of operation.

Section 3-3

*11. Draw a block diagram of a superheterodyne AM receiver. Assume an incident signal, and explain briefly what occurs in each stage.
*12. What type of radio receivers contains intermediate-frequency transformers?
13. The AM signal into a mixer is a 1.1-MHz carrier that was modulated by a 2-kHz sine wave. The local oscillator is at 1.555 MHz. List all mixer output components and indicate those accepted by the IF amplifier stage.
*14. Explain the purpose and operation of the first detector in a superhet receiver.
15. Explain how the variable tuned circuits in a superheterodyne receiver are adjusted with a single control.

Section 3-4

16. Provide an adjustment procedure whereby adequate tracking characteristics are obtained in a superheterodyne receiver.
17. Draw a schematic that illustrates *electronic* tuning using a varactor diode.
18. A silicon varactor diode exhibits a capacitance of 200 pF at zero bias. If it is in parallel with a 60-pF capacitor and 200-μH inductor, calculate the range of resonant frequency as the diode varies through a reverse bias of 3–15 V. (966 kHz, 1.15 MHz)
19. A varactor diode has C_0 equal to 320 pF. Plot a curve of capacitance versus V_R from 0 to 20 V. The diode is used with a 200-μH coil. Plot the resonant frequency versus V_R from 0 to 20 V and suggest how the response could be linearized.

Section 3-5

*20. If a superheterodyne receiver is tuned to a desired signal at 1000 kHz and its conversion (local) oscillator is operating at 1300 kHz, what would be the frequency of an incoming signal that would possibly cause *image* reception? (1600 kHz)
21. A receiver tunes from 20 to 30 MHz using a 10.7-MHz IF. Calculate the required range of oscillator frequencies and the range of image frequencies.
22. Show why image frequency rejection is not a major problem for the standard AM broadcast band.
*23. What are the advantages to be obtained from adding a tuned radio-frequency amplifier stage ahead of the first detector (converter) stage of a superheterodyne receiver?
*24. If a transistor in the only radio-frequency stage of your receiver shorted out, how could temporary repairs or modifications be made?
25. What advantages do dual-gate MOSFETs have over BJTs for use as RF amplifiers?

*26. What is the *mixer* in a superheterodyne receiver?
27. Describe the advantage of an autodyne mixer over a standard mixer.
28. Why is the bulk of a receiver's gain and selectivity obtained in the IF amplifier stages?

Section 3-6

29. Describe the difficulties in listening to a receiver without AGC.
*30. How is *automatic volume control* accomplished in a radio receiver?
31. Explain how the ac gain of a transistor can be controlled by a dc AGC level.
32. The IF/AGC system in Fig. 3-20 has an AGC level of 5.5 V (V_{AGC} = 5.5 V). Determine the rms output voltage and the gain of the A1, A2 amplifier combination. Calculate the rms input voltage. (1.4 V rms, 20.5 dB, 0.132 V rms)

Section 3-7

33. Describe the function of auxiliary AGC.
34. What is the major limiting function with respect to manufacturing a complete superheterodyne receiver on an LIC chip?
35. A superhet receiver tuned to 1 MHz has the following specifications:
    RF amplifier: P_G = 6.5 dB, R_{in} = 50 Ω *Detector:* 4-dB attenuation
    Mixer: P_G = 3 dB *Audio amplifier*: P_G = 13 dB
    3 IFs: P_G = 24 dB each at 455 kHz
    The antenna delivers a 21-μV signal to the RF amplifier. Calculate the receiver's image frequency and input/output power in watts and dBm. Draw a block diagram of the receiver and label dBm power throughout. (1.91 MHz, 8.82 pW, −80.5 dBm, 10 mW, 10 dBm)
36. A receiver has a dynamic range of 81 dB. It has 0.55 nW sensitivity. Determine the maximum allowable input signal. (0.0692 W)
37. Define *dynamic range.*
38. Describe the C-Quam system of generating broadcast AM stereo. Explain why it hasn't met with widespread acceptance like FM stereo has.
39. Define a *quadrature signal* and explain its use in AM stereo.

Section 3-8

40. You are troubleshooting an AM receiver. You have determined that the RF signal is not reaching Q1's base in the self-excited mixer of Fig. 3-27. Explain possible causes and a procedure to pinpoint the problem.
41. Describe possible problems after it is determined that voltage measurements taken on the emitter of Q1 in Fig. 3-27 show a zero volt reading.
42. Describe operation of the mixer in Fig. 3-27 if the local oscillator stops functioning.
43. Assume the output of the first IF amplifier in Fig. 3-27 is 2455 kHz. What is a probable cause?
44. The regulated power supply in Fig. 3-28 has no output. Describe how you would troubleshoot this circuit.
45. The power supply in Fig. 3-28 provides a 12-V output. Calculate the voltage at point C if R_2 = 330 Ω and R_3 = 470 Ω. (7.05 V)
46. In Fig. 3-28, suppose the 15,000 μfd capacitor was open. Describe the output voltage.

47. Using the block diagram of a receiver (Fig. 3-26), explain how to isolate methodically a problem that lies in the detector stage of the receiver.
48. Describe how a receiver's volume control can be used to determine problems with the audio amplifier.

Questions for Critical Thinking

49. Which of the factors that determine a receiver's sensitivity is more important? Defend your judgment.
50. Would passing an AM signal through a nonlinear device allow recovery of the low-frequency intelligence signal when the AM signal contains only high frequencies? Why or why not?
51. Justify in detail the choice of a superheterodyne receiver in an application that requires constant selectivity for received frequencies.
52. A superheterodyne receiver tunes the band of frequencies from 4 to 10 MHz with an IF of 1.8 MHz. The double-ganged capacitor used has a 325 pF maximum capacitance per section. The tuning capacitors are at the maximum value (325 pF) when the RF frequency is 4 MHz. Calculate the required RF and local oscillator coil inductance and the required tuning capacitor values when the receiver is tuned to receive 4 MHz and 10 MHz. (4.87 μH, 2.32 μH, 52 pF, 78.5 pF)

4 Single-Sideband Communications

Chapter Outline

Motorola Talkabout radios. (Courtesy of Motorola, Inc.)

Objectives

- Describe how an AM generator could be modified to provide SSB
- Discuss the various types of SSB and explain their advantages compared to AM
- Explain circuits that are used to generate SSB in the filter method and describe the filters that can be used
- Analyze the phase-shift method of SSB generation and give its advantages
- Describe several methods used to demodulate SSB systems
- Provide a complete block diagram for an SSB transmitter/receiver
- Determine the frequencies at all points in an SSB receiver when receiving a single audio tone

Key Terms

peak envelope power
pilot carrier
twin-sideband suppressed carrier
independent sideband transmission
balanced modulator
double-sideband suppressed carrier
balanced ring modulator
ring modulator
lattice modulator
surface acoustic wave filter
phasing capacitor
rejection notch
shape factor
peak-to-valley ratio
ripple amplitude
conversion frequency
crystal-lattice filter
continuous wave
compandor
product detector
Butterworth filter
carrier leakthrough

4-1 Single-Sideband Characteristics

The basic concept of single-sideband (SSB) communications was understood as early as 1914. It was first realized through mathematical analysis of an amplitude-modulated RF carrier. Recall that when a carrier is amplitude modulated by a single sine wave, it generates three different frequencies: (1) the original carrier with amplitude unchanged; (2) a frequency equal to the difference between the carrier and the modulating frequencies, with an amplitude up to one-half (at 100% modulation) the modulating signal; and (3) a frequency equal to the sum of the carrier and the modulating frequencies, with an amplitude also equal to a maximum of one-half that of the modulating signal. The two new frequencies, of course, are the side frequencies.

Upon recognition of the fact that sidebands existed, further investigation showed that after the carrier and one of the sidebands were eliminated, the other sideband could be used to transmit the intelligence. Since its amplitude and frequency never change, there is no information contained in the carrier. Further experiments proved that both sidebands could be transmitted, each containing different intelligence, with a suppressed or completely eliminated carrier.

By 1923, the first patent for this system had been granted, and a successful SSB communications system was established between the United States and England. Today, SSB communications play a vital role in radio communications because of their many advantages over standard AM systems. The Federal Communications Commission (FCC) recognizing these advantages, further increased their use by requiring most transmissions in the overcrowded 2- to 30-MHz range to be SSB starting in 1977.

Power Distribution

You should recall that in AM all the intelligence (information) is contained in the sidebands, but two-thirds (or more) of the total power is in the carrier. It would appear that a great amount of power is wasted during transmission. The basic principle of single-sideband transmission is to eliminate or greatly suppress the high-energy RF carrier. This can be accomplished, but accurate tuning is not possible without a carrier and it does affect the fidelity of the music and sounds. However, voice reception is still tolerable.

If a means of suppressing or completely eliminating the carrier is devised, the power that was used for the carrier can be converted into useful power to transmit the intelligence in the sidebands. Since both upper and lower sidebands contain the same intelligence, one of these could also be eliminated, thereby cutting the bandwidth required for transmission in half.

The total power output of a conventional AM transmitter is equal to the carrier power plus the sideband power. Conventional AM transmitters are rated in carrier power output. Consider a low-power AM system operating at 100 percent modulation. The carrier is 4 W and therefore each sideband is 1 W. The total transmitted power at 100 percent modulation is 6 W (4 W + 1 W + 1 W), but the AM transmitter is rated as a 4 W (just the carrier power) transmitter. If this system were converted to SSB, just one sideband at 1 W would be transmitted. This, of course, assumes a sine-wave intelligence signal. SSB systems are most often used for voice communications, which certainly do not generate a sinusoidal waveform.

SSB transmitters (and linear power amplifiers in general) are usually rated in terms of **peak envelope power** (PEP). To calculate PEP, multiply the maximum (peak) envelope voltage by 0.707, square the result, and divide by the load resistance. For instance, an SSB signal with a maximum level (over time) of 150 V p-p driven into a 50-Ω antenna results in a PEP rating of $(150/2 \times 0.707)^2 \div 50\ \Omega = 56.2$ W. This is the same power rating that would be given to the 150-V p-p sine wave, but there is a difference. The 150-V p-p level in the SSB voice transmission may occur only occasionally, while for the sine wave it occurs every cycle. These calculations are valid no matter what type of waveform the transmitter is providing. This could range from a series of short spikes with low average power (perhaps 5 W out of the PEP of 56.2 W) to a sine wave that would yield 56.2 W of average power. With a normal voice signal an SSB transmitter develops an average power of only one-fourth to one-third its PEP rating. Most transmitters cannot deliver an average power output equal to their peak envelope power capability. This is because their power supplies and/or components in the output stage are designed for a lower average power (voice operation) and cannot continuously operate at higher power levels.

Peak Envelope Power
method used to rate the output power of an SSB transmitter

Types of Sideband Transmission

A number of single-sideband systems have been developed. The major types include the following:

1. In the standard single sideband, or simply SSB, system the carrier and one of the sidebands are completely eliminated at the transmitter; only one sideband is transmitted. This is quite popular with amateur radio operators. The chief advantages of this system are maximum transmitted signal range with minimum transmitter power and the elimination of carrier interference.
2. Another system eliminates one sideband and suppresses the carrier to a desired level. The suppressed carrier can then be used at the receiver for a reference, AGC, automatic frequency control (AFC), and, in some cases, demodulation of the intelligence-bearing sideband. This is called a single-sideband suppressed carrier (SSBSC). The suppressed carrier is sometimes called a **pilot carrier.** This system retains fidelity of the received signal and minimizes carrier interference.
3. The type of system often used in military communications is referred to as **twin-sideband suppressed carrier,** or **independent sideband (ISB) transmission.** This system involves the transmission of two independent sidebands, each containing different intelligence, with the carrier suppressed to a desired level.
4. Vestigial sideband is used for television video transmissions. In it, a vestige (trace) of the unwanted sideband and the carrier are included with one full sideband. It is explained with television analysis in Chapter 17.
5. A more recently developed system is called amplitude-compandored single sideband (ACSSB). It is actually a type of SSBSC because a pilot carrier is usually included. In ACSSB the amplitude of the speech signal is compressed at the transmitter and expanded at the receiver. Performance gains of ACSSB systems over SSB are explained in Sec. 4-4.

Pilot Carrier
the suppressed carrier in SSB; the carrier is reduced to a lower level but not removed completely

Twin-Sideband Suppressed Carrier
the transmission of two independent sidebands, containing different intelligence, with the carrier suppressed to a desired level

Independent Sideband Transmission
another name for twin-sideband suppressed carrier transmission

Advantages of SSB

The most important advantage of SSB systems is a more effective utilization of the available frequency spectrum. The bandwidth required for the transmission of one conventional AM signal contains two equivalent SSB transmissions. This type of communications is especially adaptable, therefore, to the already overcrowded high-frequency spectrum.

A second advantage of this system is that it is less subject to the effects of selective fading. In the propagation of conventional AM transmissions, if the upper-sideband frequency strikes the ionosphere and is refracted back to earth at a different phase angle from that of the carrier and lower-sideband frequencies, distortion is introduced at the receiver. Under extremely bad conditions, complete signal cancellation may result. The two sidebands should be identical in phase with respect to the carrier so that when passed through a nonlinear device (i.e., a diode detector), the difference between the sidebands and carrier is identical. That difference is the intelligence and will be distorted in AM systems if the two sidebands have a phase difference.

Another major advantage of SSB is the power saved by not transmitting the carrier and one sideband. The resultant lower power requirements and weight reduction are especially important in mobile communication systems.

The SSB system has a noise advantage over AM due to the bandwidth reduction (one-half). Taking into account the selective fading improvement, noise reduction, and power savings, SSB offers about a 10- to 12-dB advantage over AM.

Obtaining a 12-dB Advantage Given a 50-kW carrier, the peak power is

$$4 \times 50\text{ kW} = 200\text{ kW}$$

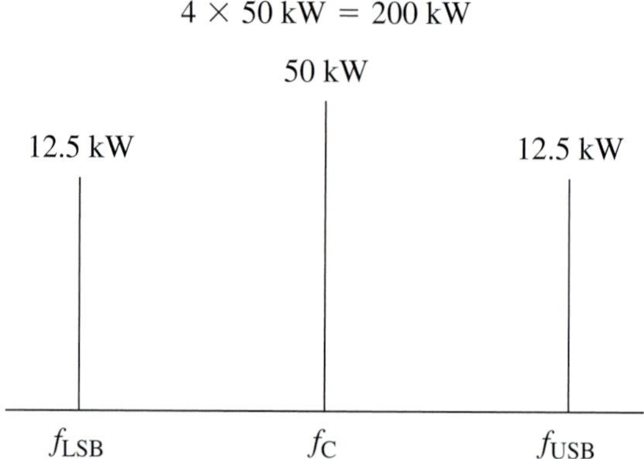

An SSB transmitter with peak power equal to one AM sideband would transmit 12.5 kW.

$$10\log\left(\frac{200}{12.5}\right) = 12\text{ dB}$$

This means that to have the same overall effectiveness, an AM system must transmit 10 to 12 dB more power than SSB. Some controversy exists on this issue because of the many variables that affect the savings. Suffice it to say that a 10-W SSB transmission is at least equivalent to the 100-W AM transmission (10-dB difference).

4-2 Sideband Generation: The Balanced Modulator

The purpose of a **balanced modulator** is to suppress (cancel) the carrier, leaving only the two sidebands. Such a signal is called a DSBSC (**double-sideband suppressed carrier**) signal. A very common balanced modulator is shown in Fig. 4-1. It is sometimes called a **balanced ring modulator** or simply a **ring modulator** and sometimes a **lattice modulator.** Consider the carrier with the instantaneous conventional current flow as indicated by the arrows. The current flow through both halves of L_5 is equal but opposite, and thus the carrier is canceled in the output. This is also true on the carrier's other half-cycle, only now diodes *B* and *C* conduct instead of *A* and *D*.

Considering just the modulating signal, current flow occurs from winding L_2 through diodes *C* and *D* or *A* and *B* but not through L_5. Thus, there is no output of the modulating signal either. Now with both signals applied, but with the carrier amplitude much greater than the modulating signal, the conduction is determined by the polarity of the carrier. The modulating signal either aids or opposes this conduction. When the modulating signal is applied, current will flow from L_2 and diode *D* will conduct more than *A*, and the current balance in winding L_5 is upset. This causes outputs of the desired sidebands but continued suppression of the carrier. This modulator is capable of 60 dB carrier suppression when carefully matched diodes are utilized. It relies on the nonlinearity of the diodes to generate the sum and difference sideband signals.

Balanced Modulator
modulator stage that mixes intelligence with the carrier to produce both sidebands with the carrier eliminated

Double-Sideband Suppressed Carrier
output signal of a balanced modulator

Balanced Ring Modulator
balanced modulator design that connects four matched diodes in a ring configuration

Ring Modulator
another name for balanced ring modulator

Lattice Modulator
another name for balanced ring modulator

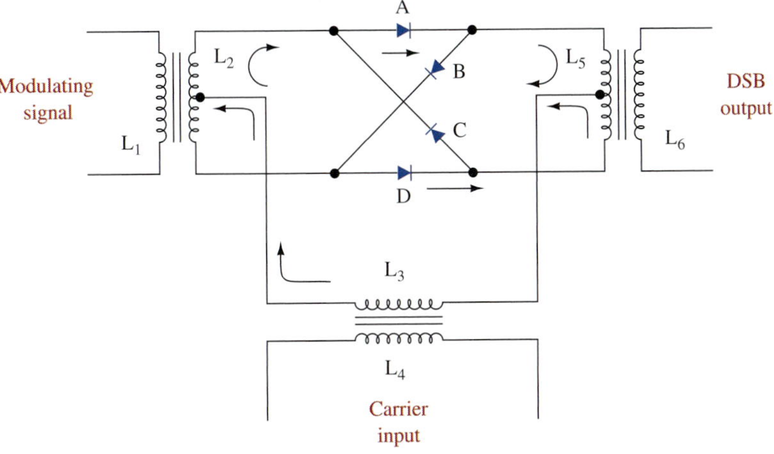

FIGURE 4-1 Balanced ring modulator.

LIC Balanced Modulator

A balanced modulator of the type previously explained requires extremely well matched components to provide good suppression of the carrier (40 or 50 dB suppression is usually adequate). This suggests the use of LICs because of the superior component-matching characteristics obtainable when devices are fabricated on the same silicon chip. A number of devices specially formulated for balanced modulator applications are available. A data sheet for the AD 630 is provided in Fig. 4-2.

Balanced Modulator/Demodulator

AD630

FEATURES
Recovers Signal from +100 dB Noise
2 MHz Channel Bandwidth
45 V/μs Slew Rate
–120 dB Crosstalk @ 1 kHz
Pin Programmable Closed Loop Gains of ±1 and ±2
0.05% Closed Loop Gain Accuracy and Match
100 μV Channel Offset Voltage (AD630BD)
350 kHz Full Power Bandwidth
Chips Available

FUNCTIONAL BLOCK DIAGRAM

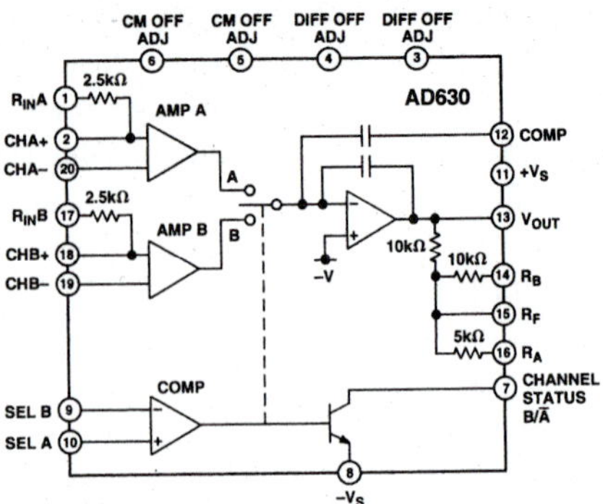

PRODUCT DESCRIPTION

The AD630 is a high precision balanced modulator which combines a flexible commutating architecture with the accuracy and temperature stability afforded by laser wafer trimmed thin-film resistors. Its signal processing applications include balanced modulation and demodulation, synchronous detection, phase detection, quadrature detection, phase sensitive detection, lock-in amplification and square wave multiplication. A network of on-board applications resistors provides precision closed loop gains of ±1 and ±2 with 0.05% accuracy (AD630B). These resistors may also be used to accurately configure multiplexer gains of +1, +2, +3 or +4. Alternatively, external feedback may be employed allowing the designer to implement his own high gain or complex switched feedback topologies.

The AD630 may be thought of as a precision op amp with two independent differential input stages and a precision comparator which is used to select the active front end. The rapid response time of this comparator coupled with the high slew rate and fast settling of the linear amplifiers minimize switching distortion. In addition, the AD630 has extremely low crosstalk between channels of –100 dB @ 10 kHz.

The AD630 is intended for use in precision signal processing and instrumentation applications requiring wide dynamic range. When used as a synchronous demodulator in a lock-in amplifier configuration, it can recover a small signal from 100 dB of interfering noise (see lock-in amplifier application). Although optimized for operation up to 1 kHz, the circuit is useful at frequencies up to several hundred kilohertz.

Other features of the AD630 include pin programmable frequency compensation, optional input bias current compensation resistors, common-mode and differential-offset voltage adjustment, and a channel status output which indicates which of the two differential inputs is active. This device is now available to Standard Military Drawing (DESC) numbers 5962-8980701RA and 5962-89807012A.

PRODUCT HIGHLIGHTS

1. The configuration of the AD630 makes it ideal for signal processing applications such as: balanced modulation and demodulation, lock-in amplification, phase detection, and square wave multiplication.
2. The application flexibility of the AD630 makes it the best choice for many applications requiring precisely fixed gain, switched gain, multiplexing, integrating-switching functions, and high-speed precision amplification.
3. The 100 dB dynamic range of the AD630 exceeds that of any hybrid or IC balanced modulator/demodulator and is comparable to that of costly signal processing instruments.
4. The op-amp format of the AD630 ensures easy implementation of high gain or complex switched feedback functions. The application resistors facilitate the implementation of most common applications with no additional parts.
5. The AD630 can be used as a two channel multiplexer with gains of +1, +2, +3, or +4. The channel separation of 100 dB @ 10 kHz approaches the limit which is achievable with an empty IC package.
6. The AD630 has pin-strappable frequency compensation (no external capacitor required) for stable operation at unity gain without sacrificing dynamic performance at higher gains.
7. Laser trimming of comparator and amplifying channel offsets eliminates the need for external nulling in most cases.

REV. D

One Technology Way, P.O. Box 9106, Norwood, MA 02062-9106, U.S.A.
Tel: 781/329-4700 www.analog.com
Fax: 781/326-8703

FIGURE 4-2 The Analog Devices AD630 balanced modulator/demodulator. (Courtesy of Analog Devices.)

AD630–SPECIFICATIONS (@ 25°C and $\pm V_S = \pm 15$ V unless otherwise noted.)

Model	AD630J/A Min	Typ	Max	AD630K/B Min	Typ	Max	AD630S Min	Typ	Max	Unit
GAIN										
Open Loop Gain	90	110		100	120		90	110		dB
±1, ±2 Closed Loop Gain Error		0.1				0.05		0.1		%
Closed Loop Gain Match		0.1				0.05		0.1		%
Closed Loop Gain Drift		2			2			2		ppm/°C
CHANNEL INPUTS										
V_{IN} Operational Limit[1]	$(-V_S + 4\ V)$ to $(+V_S - 1\ V)$			$(-V_S + 4\ V)$ to $(+V_S - 1\ V)$			$(-V_S + 4\ V)$ to $(+V_S - 1\ V)$			Volts
Input Offset Voltage			500			100			500	µV
Input Offset Voltage T_{MIN} to T_{MAX}			800			160			1000	µV
Input Bias Current		100	300		100	300		100	300	nA
Input Offset Current		10	50		10	50		10	50	nA
Channel Separation @ 10 kHz		100			100			100		dB
COMPARATOR										
V_{IN} Operational Limit[1]	$(-V_S + 3\ V)$ to $(+V_S - 1.5\ V)$			$(-V_S + 3\ V)$ to $(+V_S - 1.5\ V)$			$(-V_S + 3\ V)$ to $(+V_S - 1.3\ V)$			Volts
Switching Window			±1.5			±1.5			±1.5	mV
Switching Window T_{MIN} to T_{MAX}			±2.0			±2.0			±2.5	mV
Input Bias Current		100	300		100	300		100	300	nA
Response Time (–5 mV to +5 mV Step)		200			200			200		ns
Channel Status										
I_{SINK} @ $V_{OL} = -V_S + 0.4$ V[2]	1.6			1.6			1.6			mA
Pull-Up Voltage			$(-V_S + 33\ V)$			$(-V_S + 33\ V)$			$(-V_S + 33\ V)$	Volts
DYNAMIC PERFORMANCE										
Unity Gain Bandwidth		2			2			2		MHz
Slew Rate[3]		45			45			45		V/µs
Settling Time to 0.1% (20 V Step)		3			3			3		µs
OPERATING CHARACTERISTICS										
Common-Mode Rejection	85	105		90	110		90	110		dB
Power Supply Rejection	90	110		90	110		90	110		dB
Supply Voltage Range	±5		±16.5	±5		±16.5	±5		±16.5	Volts
Supply Current		4	5		4	5		4	5	mA
OUTPUT VOLTAGE, @ R_L = 2 kΩ										
T_{MIN} to T_{MAX}	±10			±10			±10			Volts
Output Short Circuit Current		25			25			25		mA
TEMPERATURE RANGES										
Rated Performance-N Package	0		70	0		70		N/A		°C
D Package	–25		+85	–25		+85	–55		+125	°C

NOTES
[1]If one terminal of each differential channel or comparator input is kept within these limits the other terminal may be taken to the positive supply.
[2]I_{SINK} @ $V_{OL} = (-V_S + 1)$ volt is typically 4 mA.
[3]Pin 12 Open. Slew rate with Pins 12 and 13 shorted is typically 35 V/µs.
Specifications subject to change without notice.

AD630

APPLICATIONS: BALANCED MODULATOR

Perhaps the most commonly used configuration of the AD630 is the balanced modulator. The application resistors provide precise symmetric gains of ±1 and ±2. The ±1 arrangement is shown in Figure 9a and the ±2 arrangement is shown in Figure 9b. These cases differ only in the connection of the 10 kΩ feedback resistor (Pin 14) and the compensation capacitor (Pin 12). Note the use of the 2.5 kΩ bias current compensation resistors in these examples. These resistors perform the identical function in the ±1 gain case. Figure 10 demonstrates the performance of the AD630 when used to modulate a 100 kHz square wave carrier with a 10 kHz sinusoid. The result is the double sideband suppressed carrier waveform.

These balanced modulator topologies accept two inputs, a signal (or modulation) input applied to the amplifying channels, and a reference (or carrier) input applied to the comparator.

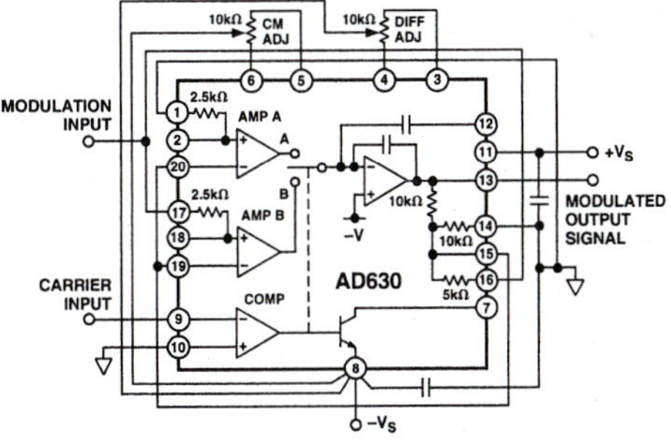

Figure 9b. AD630 Configured as a Gain-of-Two Balanced Modulator

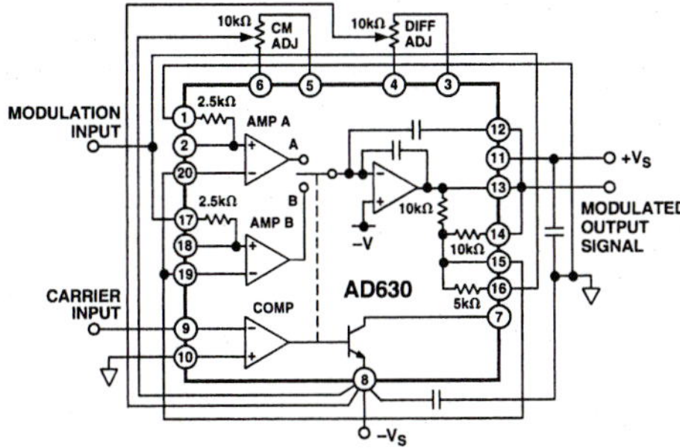

Figure 9a. AD630 Configured as a Gain-of-One Balanced Modulator

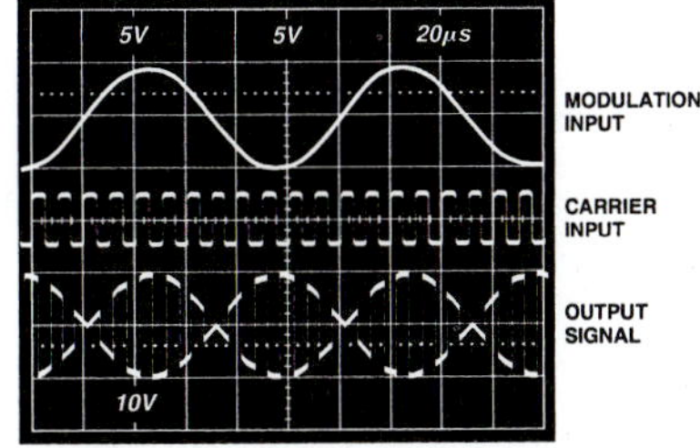

Figure 10. Gain-of-Two Balanced Modulator Sample Waveforms

FIGURE 4-2 *(Continued)*

As shown in the data sheet, this approach does not require the use of transformers or tuned circuits. The balanced modulator function is achieved with matched transistors in the differential amplifiers, with the modulating signal controlling the emitter current of the "diff-amps." The carrier signal is applied to switch the diff-amps' bases, resulting in a mixing process with the mixing product signals out of phase at the collectors. This is an extremely versatile device since it can be used not only as a balanced modulator but also as an amplitude modulator, synchronous detector, FM detector, or frequency doubler.

4-3 SSB Filters

Once the carrier has been eliminated, it is necessary to cancel one of the sidebands without affecting the other one. This requires a sharply defined filter, as Fig. 4-3 helps illustrate. Voice transmission requires audio frequencies from about 100 Hz to 3 kHz. Therefore, the upper and lower sidebands generated by the balanced modulator are separated by 200 Hz, as shown in Fig. 4-3.

The required Q depends on the center or carrier frequency, f_c; the separation between the two sidebands, Δf; and the desired attenuation level of the unwanted sideband. It can be calculated from

$$Q = \frac{f_c(\log^{-1} \text{dB}/20)^{1/2}}{4\Delta f} \tag{4-1}$$

where dB is the suppression of the unwanted sideband.

Example 4-1

Calculate the required Q for the situation depicted in Fig. 4-3 for
(a) A 1-MHz carrier and 80-dB sideband suppression.
(b) A 100-kHz carrier and 80-dB sideband suppression.

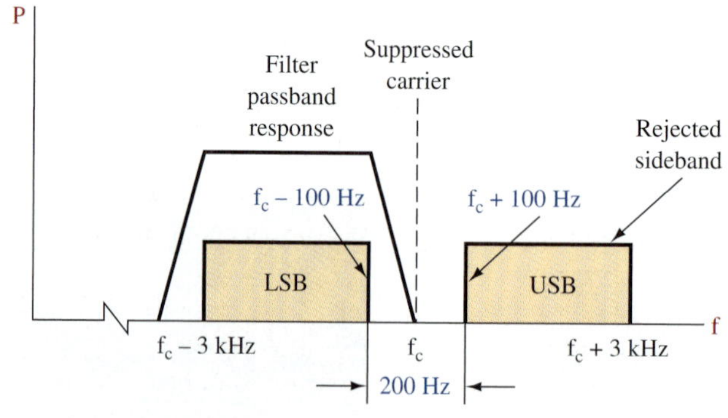

FIGURE 4-3 Sideband suppression.

Solution

(a)
$$Q = \frac{f_c(\log^{-1} \text{dB}/20)^{1/2}}{4\Delta f} \qquad \textbf{(4-1)}$$
$$= \frac{1 \text{ MHz}(\log^{-1} 80/20)^{1/2}}{4 \times 200 \text{ Hz}} = \frac{1 \times 10^6(10^4)^{1/2}}{800}$$
$$= \frac{1 \times 10^8}{8 \times 10^2} = 125{,}000$$

(b)
$$Q = \frac{100 \text{ kHz}(\log^{-1} 80/20)^{1/2}}{4 \times 200 \text{ Hz}}$$
$$= \frac{10^7}{8 \times 10^2} = 12{,}500$$

A practical consequence of the preceding example is that the SSB signal would be generated around the lower 100-kHz carrier in conjunction with a crystal filter. Then, after removing one sideband, an additional frequency translation is usually employed to get the sideband up to the desired frequency range. This is accomplished with a mixer circuit.

Both SSB transmitters and receivers require selective bandpass filters in the region of 100 to 500 kHz. In receivers a high order of adjacent channel rejection is required if channels are to be closely spaced to conserve spectrum space. The filter used, therefore, must have very steep skirt characteristics (fast roll-off) and a flat bandpass characteristic to pass all frequencies in the band equally well. These filter requirements are met by crystal filters, ceramic filters, and mechanical filters. A fourth type of high-Q filter of more recent popularity is the **surface acoustic wave** (SAW) **filter.** It is often used in TV and radar applications and is treated in Chapter 17. It is most applicable to higher frequencies than are typically used in SSB systems.

Surface Acoustic Wave Filter
an extremely high-Q filter often used in TV and radar applications

Crystal Filters

The crystal filter is commonly used in single-sideband systems to attenuate the unwanted sideband. Because of its very high Q, the crystal filter passes a much narrower band of frequencies than the best LC filter. Crystals with a Q up to about 50,000 are available.

The equivalent circuit of the crystal and crystal holder is illustrated in Fig. 4-4(a). Recall that the basics of crystal operation were introduced in Chapter 1. The components L_s, C_s, and R_s represent the series resonant circuit of the crystal itself. C_p represents the parallel capacitance of the crystal holder. The crystal offers a very low-impedance path to the frequency to which it is resonant and a high-impedance path to other frequencies. However, the crystal holder capacitance, C_p, shunts the crystal and offers a path to other frequencies. For the crystal to operate as a bandpass filter, some means must be provided to counteract the shunting effect of the crystal holder. This is accomplished by placing an external variable capacitor in the circuit [C_1 in Fig. 4-4(b)].

In Fig. 4-4(b), a simple bandpass crystal filter is shown. The variable capacitor C_1, called the **phasing capacitor,** counteracts holder capacitance C_p. C_1 can be adjusted so that its capacitance equals the capacitance of C_p. Then both C_p and C_1 pass undesired frequencies equally well. Because of the circuit arrangement, the voltages across C_p and C_1 due to undesired frequencies are equal and 180° out of

Phasing Capacitor
cancels the effect of another capacitance by a 180° phase difference

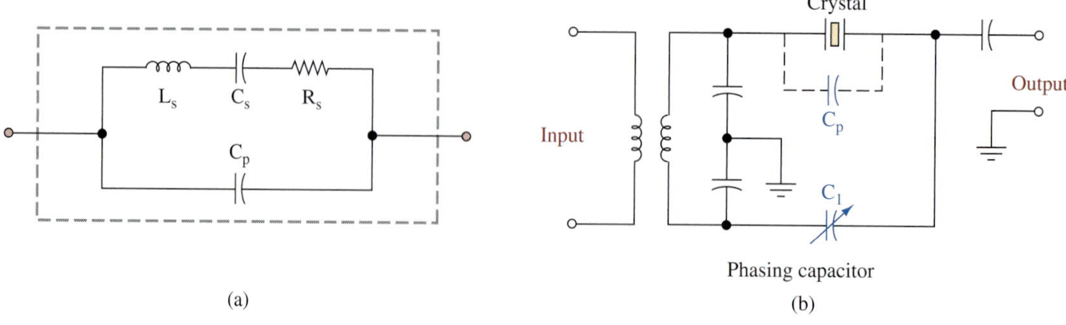

FIGURE 4-4 Crystal equivalent circuit (a) and filter (b).

Rejection Notch a narrow range of frequencies attenuated by a filter, which can be tuned to minimize interference

phase. Therefore, undesirable frequencies are canceled and do not appear in the output. This cancellation effect is called the **rejection notch.**

For circuit operation, assume that a lower sideband with a maximum frequency of 99.9 kHz and an upper sideband with a minimum frequency of 100.1 kHz are applied to the input of the crystal filter in Fig. 4-4(b). Assume that the upper sideband is the unwanted sideband. By selecting a crystal that will provide a low-impedance path (series resonance) at about 99.9 kHz, the lower-sideband frequency will appear in the output. The upper sideband, as well as all other frequencies, will have been attenuated by the crystal filter. Improved performance is possible when two or more crystals are combined in a single filter circuit.

Ceramic Filters

Ceramic filters utilize the piezoelectric effect just as crystals do. However, they are normally constructed from lead zirconate-titanate. While ceramic filters do not offer *Q*s as high as a crystal, they do outperform *LC* filters in that regard. A *Q* of up to 2000 is practical with ceramic filters. They are lower in cost, more rugged, and

Ceramic filters. (Courtesy of Integrated Microwave Corporation, San Diego, CA.)

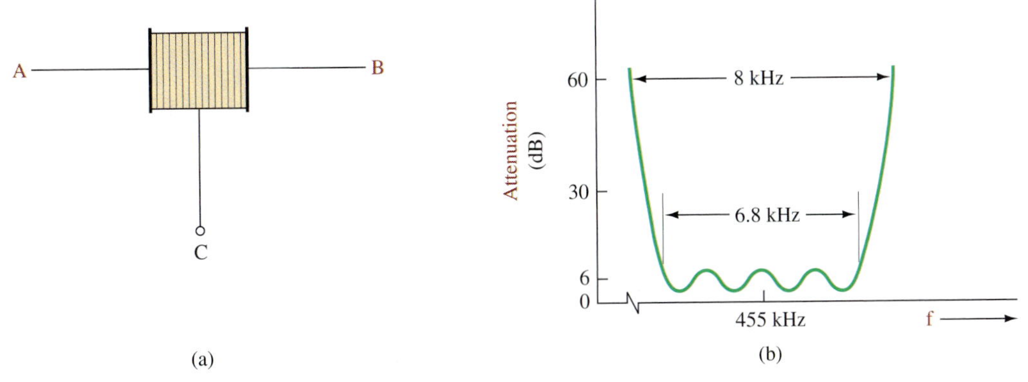

FIGURE 4-5 Ceramic filter and response curve.

smaller in size than crystal filters. They are used not only as sideband filters but also as replacements for the tuned IF transformers for superheterodyne receivers.

The circuit symbol for a ceramic filter is shown in Fig. 4-5(a) and a typical attenuation response curve is shown in Fig. 4-5(b). Note that the bandwidths at 60 dB and at 6 dB are shown. The ratio of these two bandwidths (8 kHz/6.8 kHz = 1.18) is defined as the **shape factor.** The shape factor (60-dB BW divided by a 6-dB BW) provides an indication of the filter's selectivity. The ideal value of 1 would indicate a vertical slope at both frequency extremes. The ideal filter would have a horizontal slope within the passband with zero attenuation. The practical case is shown in Fig. 4-5(b), where a variation is illustrated. This variation is termed the **peak-to-valley ratio** or **ripple amplitude.** The shape factor and ripple amplitude characteristics also apply to the mechanical filters discussed next.

Shape Factor
ratio of the 60-dB and 6-dB bandwidths of a high-Q bandpass filter

Peak-to-Valley Ratio
another name for ripple amplitude

Ripple Amplitude
variation in attenuation of a sharp bandpass filter within its 6-dB bandwidths

Mechanical Filters

Mechanical filters have been used in single-sideband equipment since the 1950s. Some of the advantages of mechanical filters are their excellent rejection characteristics, extreme ruggedness, size small enough to be compatible with the miniaturization of equipment, and a Q in the order of 10,000, which is about 50 times that obtainable with LC filters.

The mechanical filter is a device that is mechanically resonant; it receives electrical energy, converts it to mechanical vibration, then converts this mechanical energy back into electrical energy as the output. Figure 4-6 shows a cutaway view of a typical unit. There are four elements constituting a mechanical filter: (1) an input transducer that converts the electrical energy at the input into mechanical vibrations, (2) metal disks that are manufactured to be mechanically resonant at the desired frequency, (3) rods that couple the metal disks, and (4) an output transducer that converts the mechanical vibrations back into electrical energy.

Not all the disks are shown in the illustration. The shields around the transducer coils have been cut away to show the coil and magnetostrictive driving rods. As you can see by its symmetrical construction, either end of the filter may be used as the input.

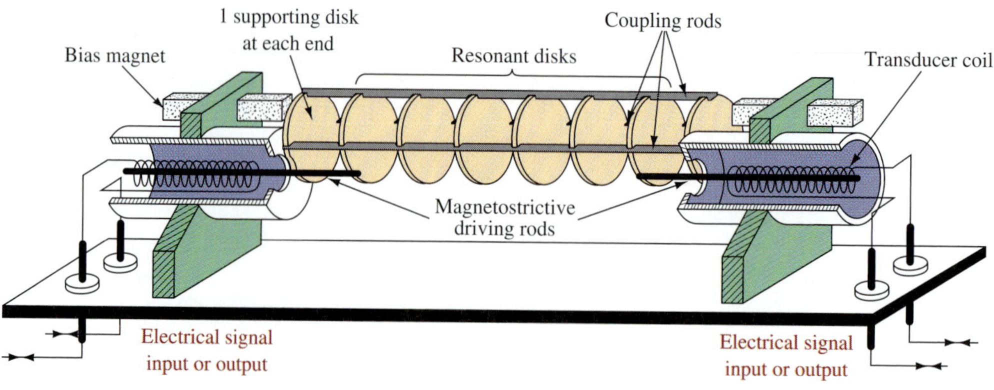

FIGURE 4-6 Mechanical filter.

Figure 4-7 is the electrical equivalent of the mechanical filter. The disks of the mechanical filter are represented by the series resonant circuits L_1C_1 while C_2 represents the coupling rods. The resistance R in both the input and output represents the matching mechanical loads. Phase shift of the input signal is introduced by the L and C components of the mechanical filter. For digital applications, a phase shift can affect the quality of the digital pulse. This can lead to an increase in data errors or bit errors. In analog systems, the voice transmission is not affected as much because the ear is very forgiving of distortion.

Let us assume that the mechanical filter of Fig. 4-6 has disks tuned to pass the frequencies of the desired sideband. The input to the filter contains both sidebands, and the transducer driving rod applies both sidebands to the first disk. The vibration of the disk will be greater at a frequency to which it is tuned (resonant frequency), which is the desired sideband, than at the undesired sideband frequency. The mechanical vibration of the first disk is transferred to the second disk, but a smaller percentage of the unwanted sideband frequency is transferred. Each time the vibrations are transferred from one disk to the next, there is a smaller amount of the unwanted sideband. At the end of the filter there is practically none of the undesired sideband left. The desired sideband frequencies are taken off the transducer coil at the output end of the filter.

Varying the size of C_2 in the electrical equivalent circuit in Fig. 4-7 varies the bandwidth of the filter. Similarly, by varying the mechanical coupling between the disks (Fig. 4-6), that is, by making the coupling rods either larger or smaller,

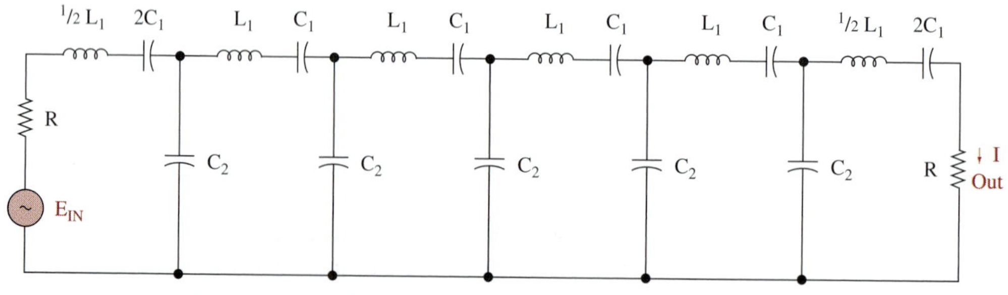

FIGURE 4-7 Electrical analogy of a mechanical filter.

the bandwidth of the mechanical filter is varied. Because the bandwidth varies approximately as the total cross-sectional area of the coupling rods, the bandwidth of the mechanical filter can be increased by using either larger coupling rods or more coupling rods. Mechanical filters with bandwidths as narrow as 500 Hz and as wide as 35 kHz are practical in the range 100 to 500 kHz.

4-4 SSB TRANSMITTERS

Filter Method

Figure 4-8 is a block diagram of a modern single-sideband transmitter using a balanced modulator to generate DSB and the filter method of eliminating one of the sidebands. For illustrative purposes, a single-tone 2000-Hz intelligence signal is used, but it is normally a complex intelligence signal, such as that produced by the human voice.

A 9-MHz crystal frequency is used because of the excellent operating characteristics of monolithic filters at that frequency. The 2-kHz signal is amplified and mixed with a 9-MHz carrier **(conversion frequency)** in the balanced modulator. Remember, neither the carrier nor audio frequencies appear in the output of the balanced modulator; the sum and difference frequencies (9 MHz ± 2 kHz) are its output. As illustrated in Fig. 4-8, the two sidebands from the balanced modulator are applied to the filter. Only the desired upper sideband is passed. The dashed lines show that the carrier and lower sideband have been removed.

Conversion Frequency another name for the carrier in a balanced modulator

The output of the first balanced modulator is filtered and mixed again with a new conversion frequency to adjust the output to the desired transmitter frequency.

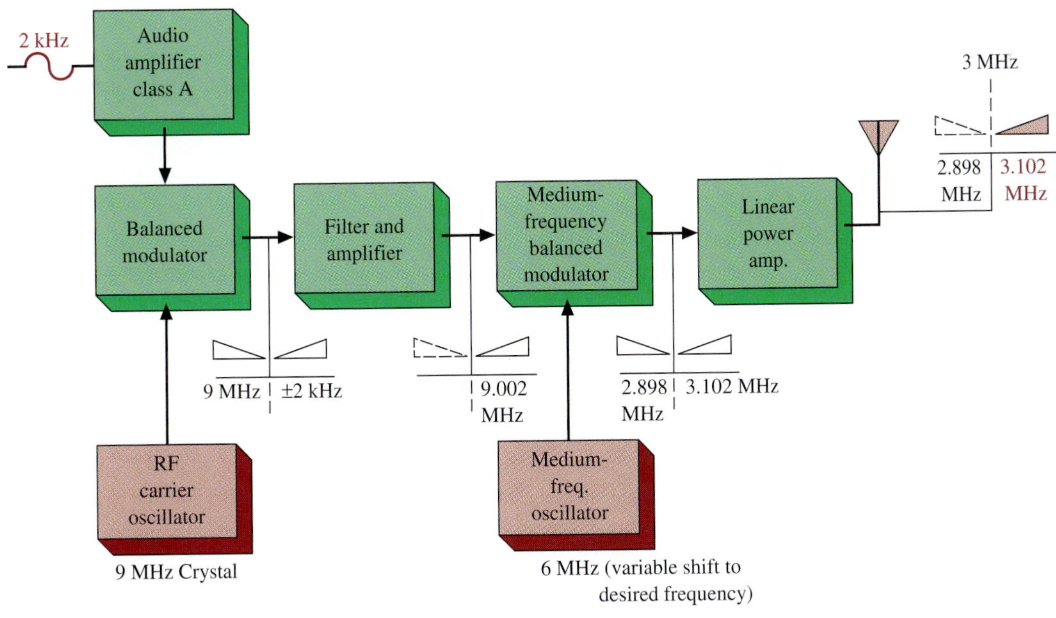

FIGURE 4-8 SSB transmitter block diagram.

After mixing the two inputs to get two new sidebands, the balanced modulator removes the new 3-MHz carrier and applies the two new sidebands (3102 kHz and 2898 kHz) to a tunable linear power amplifier.

Example 4-2

For the transmitter system shown in Fig. 4-8, determine the filter Q required in the linear power amplifier.

Solution

The second balanced modulator created another DSB signal from the SSB signal of the preceding high-Q filter. However, the frequency translation of the second balanced modulator means that a low-quality filter can be used once again to create SSB. The new DSB signal is at about 2.9 MHz and 3.1 MHz. The required filter Q is

$$\frac{3\text{ MHz}}{3.1\text{ MHz} - 2.9\text{ MHz}} = \frac{3\text{ MHz}}{0.2\text{ MHz}} = 15$$

The input and output circuits of the linear power amplifier are tuned to reject one sideband and pass the other to the antenna for transmission. A standard LC filter is now adequate to remove one of the two new sidebands. The new sidebands are about 200 kHz apart ($\simeq$ 3100 kHz $-$ 2900 kHz), so the required Q is quite low. (See Ex. 4-2 for further illustration.) The high-frequency oscillator is variable so that the transmitter output frequency can be varied over a range of transmitting frequencies. Since both the carrier and one sideband have been eliminated, all the transmitted energy is in the single sideband.

Filter SSB Generator

The circuit shown in Fig. 4-9 provides a complete, practical SSB generator. Its output is at 9 MHz and can be heterodyned to any desired frequency. The audio signal is amplified by a 741 op amp (U1). Its output is applied to the gates of Q_1. The balanced modulator is formed by the Q_1–Q_2 combination. The balance for maximum carrier suppression is made by adjusting R_2. The required 180° phase difference for the drains of Q_1 and Q_2 is provided by transformer T_1. It also couples the balanced modulator output into the IF preamplifier, Q_3. The SSB output is filtered by a prepackaged crystal-lattice filter at the collector of Q_3. A **crystal-lattice filter** contains at least two but usually four crystals. It offers a wider possible passband than a single-crystal filter.

Crystal-Lattice Filter
filter containing at least two but usually four crystals

The carrier is generated with either crystal Y_1 (usb) or Y_2 (lsb). The oscillator's output at the drain of Q_5 is amplified by Q_6, which allows 4-V p-p injection into the sources of balanced modulation transistors Q_1 and Q_2. Fine adjustment of carrier frequency is accomplished with trimmer capacitors C_1 or C_2. They are adjusted to just "fit" the desired sideband into the passband of FL_1 and thus attenuate the undesired sideband and any vestige of the carrier that is already suppressed 45 to 50 dB by the balanced modulator.

Continuous Wave
undamped sinusoidal waveform produced by an oscillator in a radio transmitter

Continuous wave (CW) operation is also possible with this system. CW is telegraphy by on–off keying of a carrier. It is the oldest radio modulation system and simply means either to transmit a carrier or not, representing a mark or space

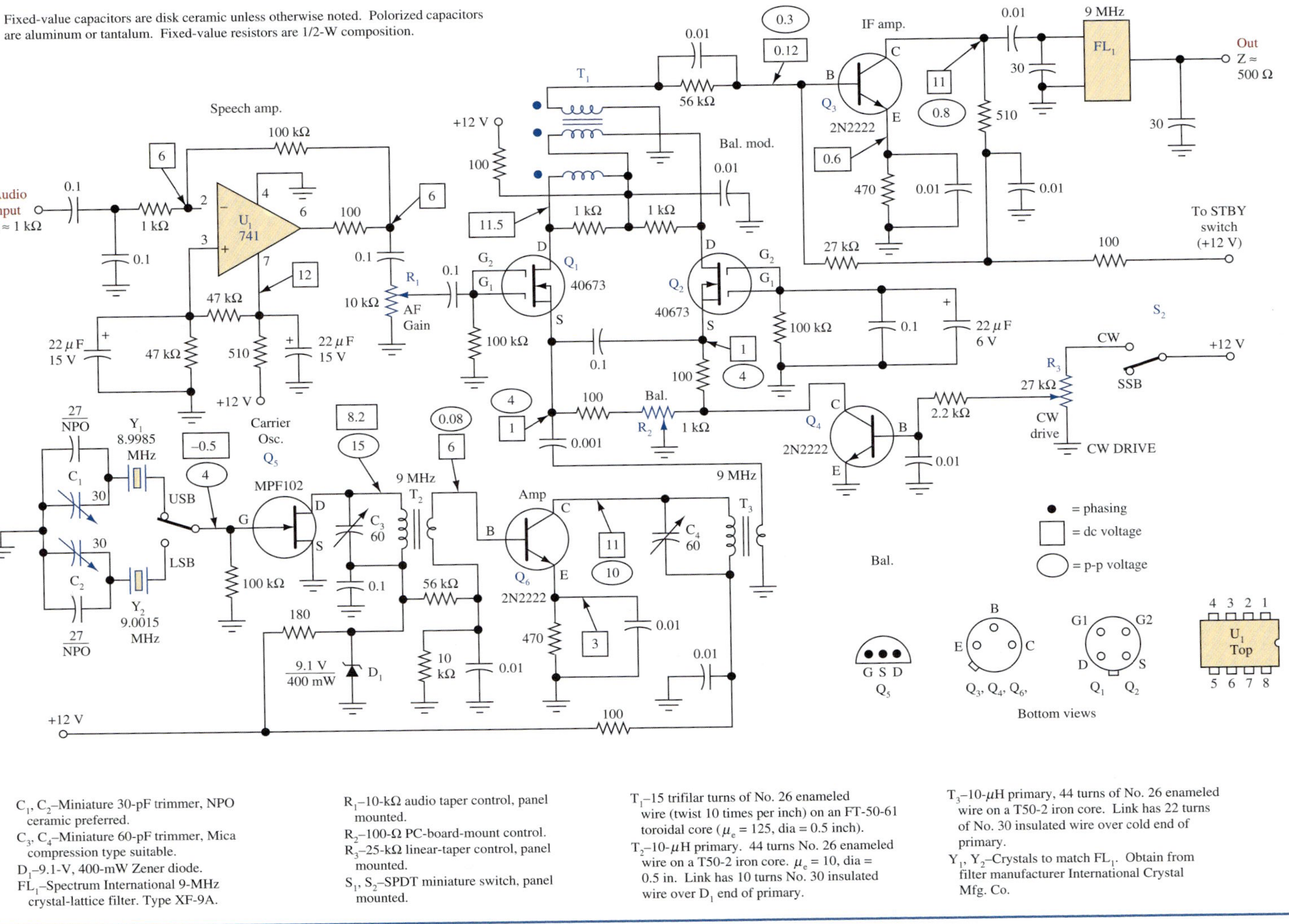

C1, C2–Miniature 30-pF trimmer, NPO ceramic preferred.
C3, C4–Miniature 60-pF trimmer, Mica compression type suitable.
D1–9.1-V, 400-mW Zener diode.
FL1–Spectrum International 9-MHz crystal-lattice filter. Type XF-9A.

R1–10-kΩ audio taper control, panel mounted.
R2–100-Ω PC-board-mount control.
R3–25-kΩ linear-taper control, panel mounted.
S1, S2–SPDT miniature switch, panel mounted.

T1–15 trifilar turns of No. 26 enameled wire (twist 10 times per inch) on an FT-50-61 toroidal core (μ_e = 125, dia = 0.5 inch).
T2–10-μH primary. 44 turns No. 26 enameled wire on a T50-2 iron core. μ_e = 10, dia = 0.5 in. Link has 10 turns No. 30 insulated wire over D1 end of primary.

T3–10-μH primary, 44 turns of No. 26 enameled wire on a T50-2 iron core. Link has 22 turns of No. 30 insulated wire over cold end of primary.
Y1, Y2–Crystals to match FL1. Obtain from filter manufacturer International Crystal Mfg. Co.

FIGURE 4-9 SSB generator-filter method. (From the *ARRL Handbook,* courtesy of the American Radio Relay League.)

in telegraphy. For CW operation, S_2 is switched to the CW position, which activates Q_4 as a variable dc attenuator. R_3 is varied to change the bias of Q_4, which then shifts Q_2's source voltage to permit carrier insertion.

Phase Method

The phase method of SSB generation offers the following advantages over the filter method:

1. There is greater ease in switching from one sideband to the other.
2. SSB can be generated directly at the desired transmitting frequency, which means that intermediate balanced modulators are not necessary.
3. Lower intelligence frequencies can be economically used because a high-Q filter is not necessary.

Despite these advantages, the filter method is rather firmly entrenched for many systems because of adequate performance and the complexity of the phase method. The increased availability of special LICs in the past has increased SSB designs using the phase method.

The phase method of SSB generation relies on the fact that the upper and lower sidebands of an AM signal differ in the sign of their phase angles. This means that phase discrimination may be used to cancel one sideband of the DSB signal.

Consider a modulating signal $f(t)$ to be a pure cosine wave. A resulting balanced modulator output (DSB) can then be written as

$$f_{\text{DSB1}}(t) = (\cos \omega_i t)(\cos \omega_c t) \tag{4-2}$$

where $\cos \omega_i t$ is the intelligence signal and $\cos \omega_c t$ the carrier. The term $\cos A \cos B$ is equal to $\frac{1}{2}[\cos(A + B) + \cos(A - B)]$ by trigonometric identity, and therefore Eq. (4-2) can be rewritten as

$$f_{\text{DSB1}}(t) = \frac{1}{2}[\cos(\omega_c + \omega_i)t + \cos(\omega_c - \omega_i)t] \tag{4-3}$$

If another signal,

$$f_{\text{DSB2}}(t) = \frac{1}{2}[\cos(\omega_c - \omega_i)t - \cos(\omega_c + \omega_i)t] \tag{4-4}$$

were added to Eq. (4-3), the upper sideband would be canceled, leaving just the lower sideband,

$$f_{\text{DSB1}}(t) + f_{\text{DSB2}}(t) = \cos(\omega_c - \omega_i)t$$

Since the signal in Eq. (4-4) is equal to

$$\sin \omega_i t \sin \omega_c t$$

by trigonometric identity, it can be generated by shifting the phase of the carrier and intelligence signal by exactly 90° and then feeding them into a balanced modulator. Recall that sine and cosine waves are identical except for a 90° phase difference.

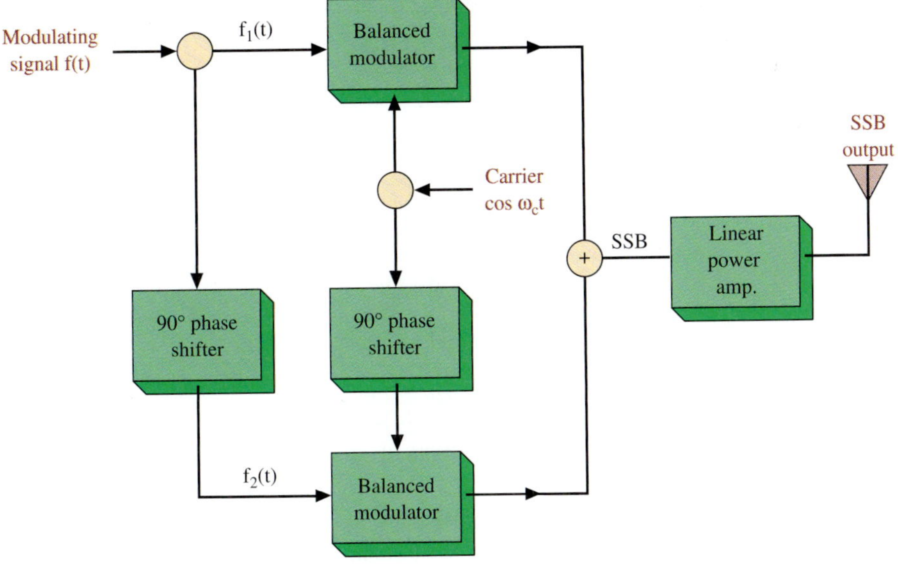

FIGURE 4-10 Phase-shift SSB generator.

A block diagram for the system just described is shown in Fig. 4-10. The upper balanced modulator receives the carrier and intelligence signals directly, while the lower balanced modulator receives both of them shifted in phase by 90°. Thus, combining the outputs of both balanced modulators in the adder results in an SSB output that is subsequently amplified and then driven into the transmitting antenna.

A major disadvantage of this system is the 90° phase-shifting network required for the intelligence signal. The *carrier* 90° phase shift is easily accomplished because of its single-frequency nature, but the audio signal covers a wide range of frequencies. To obtain exactly 90° of phase shift for a complete range of frequencies is difficult. The system is critical inasmuch as an 88° phase shift (2° error) for a given audio frequency results in about 30 dB of unwanted sideband suppression instead of the desired complete suppression obtained at 90° phase shift. The difficulty in obtaining adequate performance of the intelligence phase-shifting network is becoming less of a problem with the newer LICs designed to address this situation.

ACSSB Systems

Amplitude compandoring (**com**pression-ex**pandor**) single-sideband (ACSSB) systems are now allowing narrowband voice communications with the performance of FM systems for the land-mobile communications industry. This equivalent performance is provided with less than one-third the bandwidth of the comparable FM systems. The basis of ACSSB is to compress the audio before modulation and to expand it following demodulation at the receiver. A commonly used method to achieve this is use of the SA571 compandor LIC. It is shown in Fig. 4-11 connected as an expandor. This IC has a unity gain for a 0-dBm input. When used as a compressor, all negative dBm power levels are increased and positive dBm powers are decreased. For example, −40 dBm becomes −20 dBm, +15 dBm becomes +7.5 dBm, and

Compandor
compress/expand; to provide better noise performance, a variable-gain circuit at the transmitter increases its gain for low-level signals; a complementary circuit in the receiver reverses the process to restore the original signal

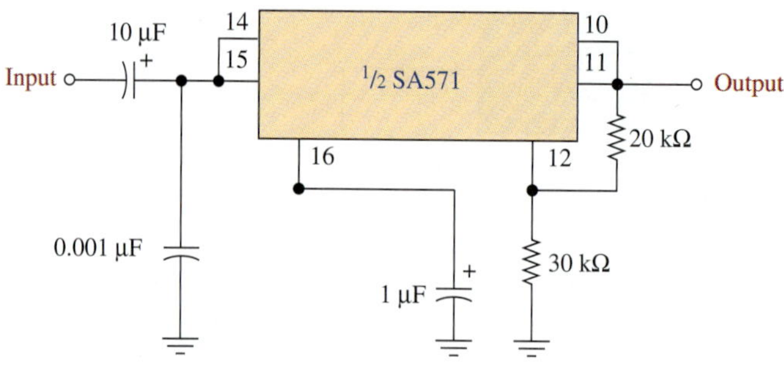

FIGURE 4-11 Amplitude expandor circuit.

so on. The expandor reverses the process to restore the signal's original dynamic range. Thus, a −20 dBm to +7.5 dBm signal becomes −40 dBm to +15 dBm at the expandor's output. The only signals not changed by the 571 IC are those at 0 dBm.

This system significantly cuts down the dynamic range that must be dealt with. It allows the lower-level signals to be transmitted with greater power while remaining within the PEP ratings of the transmitter power amplifier for the highest-level signals. Thus, the *S/N* ratio is significantly improved at the lower end, while the somewhat increased noise for the louder passages (due to their reduced amplitude) is not a problem. At the receiver, the expandor restores the demodulated output to its original dynamic range.

These ACSSB systems also include a pilot carrier signal as illustrated in Fig. 4-12. It is shown added to the audio signal sufficiently separated so that the receiver can ultimately distinguish between the two. The audio passband for voice transmission is fully attenuated by 3 kHz, and a pilot tone at 3.1 kHz above the eliminated carrier is the norm. It is suppressed by 10 dB from the maximum PEP as shown in Fig. 4-12. Thus, the transmitter will have output power of about $\frac{1}{10}$ (−10 dB) of the maximum when there is no voice modulation. At the receiver, the pilot tone is usually compared to a reference oscillator in a phase-locked-loop (PLL) circuit. The PLL difference voltage is used to shift the receiver oscillator until error is eliminated. The pilot tone can also be used for AGC and squelch circuits at the receiver. A complete introduction to the PLL is provided in Chapter 6.

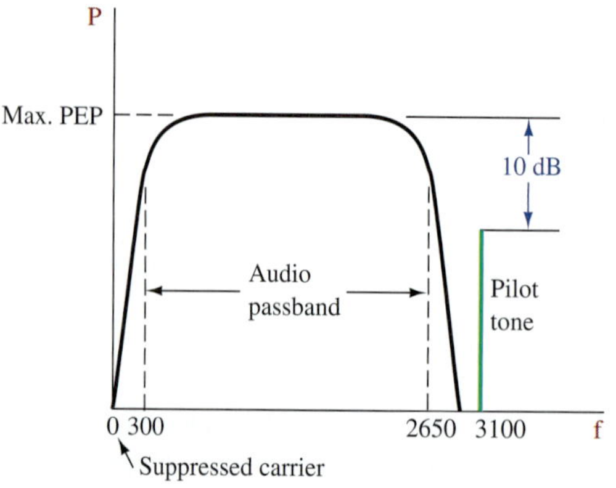

FIGURE 4-12 ACSSB signal.

Transmitter Linear Power Amplifier

Once an SSB signal has been generated, a linear power amplifier is necessary to obtain significant power levels for transmission. The circuit in Fig. 4-13 provides 140 W of PEP nominal output power from 2 to 30 MHz when supplied with about 3 W of signal input. Fairly linear outputs up to 200 W PEP are possible with increased input drive. Its simplicity and use of low-cost components makes it an attractive design for mobile transmitters. It operates on the standard 13.6 V dc available from automotive electrical systems.

The amplifier is a class AB push-pull design. The quiescent current for each MRF454 power transistor (Q_3 and Q_4) is about 500 mA. This amount of bias is

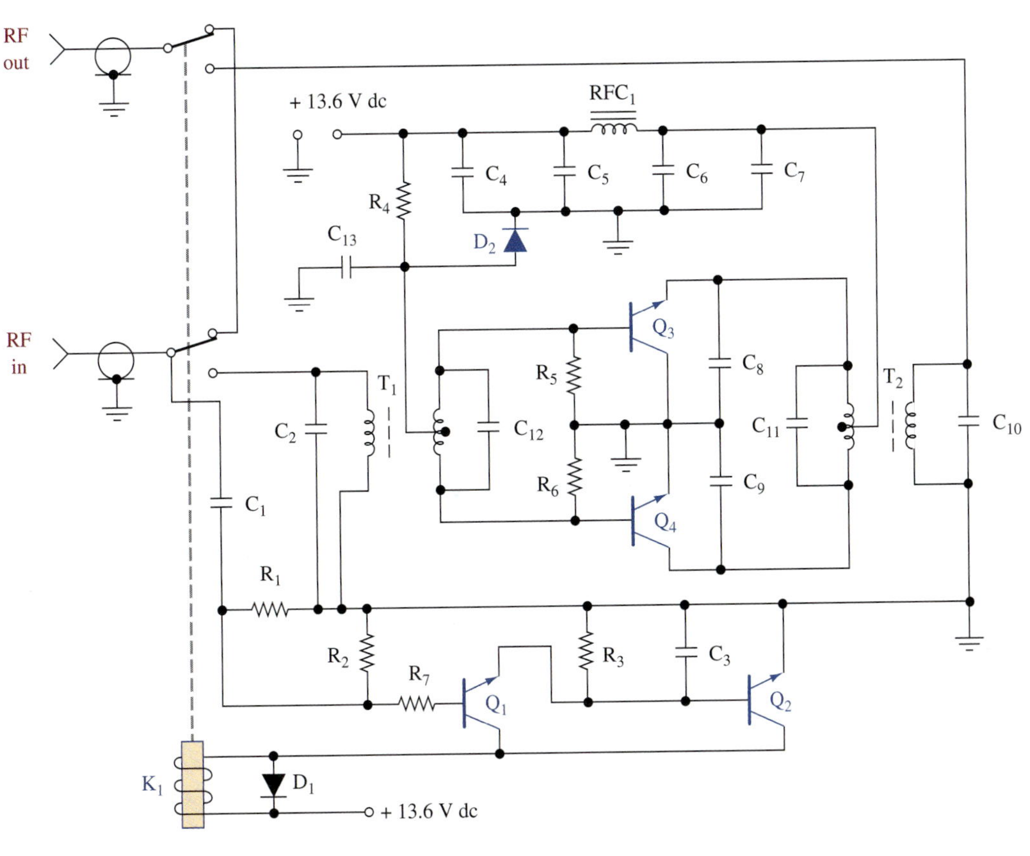

C_1—33 pF dipped mica
C_2—18 pF dipped mica
C_3—10 μF, 35 V dc for AM operation
100 μF, 35 V dc for SSB operation
C_4—0.1 μF Erie
C_5—10 μF, 35 V dc electrolytic
C_6—1 μF tantalum
C_7—0.001 μF Erie disk
C_8, C_9—330 pF dipped mica
C_{10}—24 pF dipped mica
C_{11}—910 pF dipped mica
C_{12}—1100 pF dipped mica
C_{13}—500 μF, 3 V dc electrolytic

R_1—100 kΩ, 0.25 W
R_2—10 kΩ, 0.25 W
R_3—10 kΩ, 0.25 W
R_4—33 Ω, 5 W wirewound
R_5, R_6—10 Ω, 0.5 W
R_7—100 Ω, 0.25 W
RFC_1—9 ferroxcube beads on No. 18 AWG wire
D_1—1N4001
D_2—1N4997
Q_1, Q_2—2N4401
Q_3, Q_4—MRF454
T_1, T_2—16:1 transformers
K_1—Potter & Brumfield KT11A 12 V dc relay or equivalent

FIGURE 4-13 Linear power amplifier. (Courtesy of *Microwaves and RF.*)

needed to prevent crossover distortion under high-output conditions. Diode D_2 is mounted on the same heat sink with the power transistors and "temperature tracks" them to provide bias adjustment with temperature changes. The relay K_1 and associated control circuitry, including Q_1 and Q_2, serve to "engage" the power amplifier only when RF input is present. You will analyze this function in an end-of-chapter question. Further details and circuit construction information can be obtained by requesting engineering bulletin EB63 from Motorola Semiconductor Products, Inc., P.O. Box 20912, Phoenix, AZ 85036.

4-5 SSB Demodulation

One of the major advantages of SSB for voice transmission has been shown to be the elimination of the transmitted carrier. However, this advantage does not apply to music. We have shown that this allows an increase in effective radiated power (erp) because the sidebands contain the information and the never-changing carrier is redundant. Unfortunately, even though the carrier is redundant (contains no information), it *is* needed at the receiver! Recall that the intelligence in an AM system is equal in frequency to the difference of the sideband and carrier frequencies.

Waveforms

Figure 4-14(a) shows three different sine-wave intelligence signals; in Fig. 4-14(b), the resulting AM waveforms are shown, and Fig. 4-14(c) shows the DSB (no carrier) waveform. Notice that the DSB envelope (drawn in for illustrative purposes) looks like a full-wave rectification of the corresponding AM waveform's envelope. It is double the frequency of the AM envelope. In Fig. 4-14(d), the SSB waveforms are simply pure sine waves. This is precisely what is transmitted in the case of a sine-wave modulating signal. These waveforms are either at the carrier plus the intelligence frequency (usb) or carrier minus intelligence frequency (lsb). An SSB receiver would have to somehow "reinsert the carrier" to enable detection of the original audio or intelligence signal. A simple way to form an SSB detector is to use a mixer stage identical to a standard AM receiver mixer. The mixer is a nonlinear device, and the local oscillator input should be equivalent to the desired carrier frequency.

Mixer SSB Demodulator

Figure 4-15 shows this situation pictorially. Consider a 500-kHz carrier frequency that has been modulated by a 1-kHz sine wave. If the upper sideband were transmitted, the receiver's demodulator would see a 501-kHz sine wave at its input. Therefore, a 500-kHz oscillator input will result in a mixer output frequency component of 1 kHz, which is the desired result. If the 500-kHz oscillator is not exactly 500 kHz, the recovered intelligence will not be exactly 1 kHz. If the receiver is to be used on several specific channels, a crystal for each channel will provide the necessary stability. If the receiver is to be used over a complete band of frequencies, the variable frequency oscillator (VFO), often called the beat frequency oscillator (BFO), must have some sort of automatic frequency control (AFC) to provide adequate quality reception. This can be accomplished by including a pilot carrier signal with the transmitted SSB signal. The

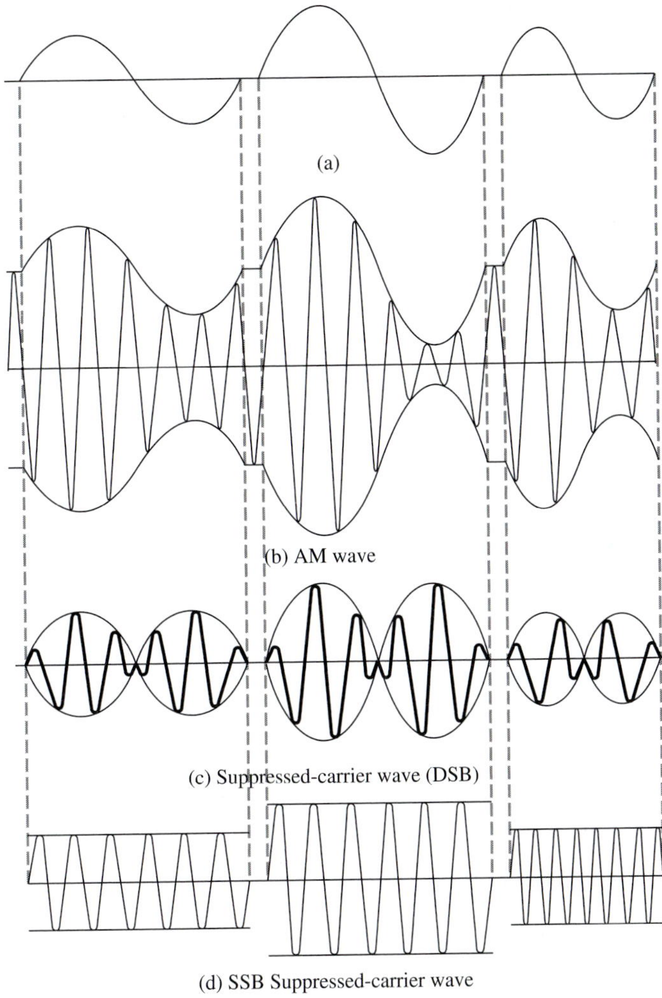

FIGURE 4-14 AM, DSB, and SSB waves from sinusoidal modulating signals.

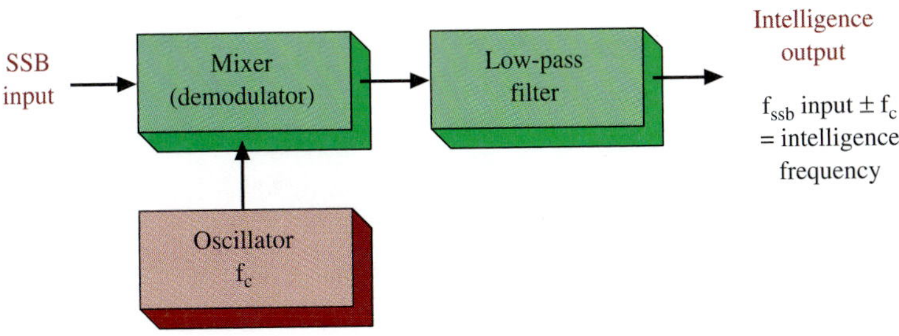

FIGURE 4-15 Mixer used as SSB demodulator.

pilot carrier can then be used to calibrate the receiver's oscillator at periodic intervals. Another approach is to utilize rather elaborate AFC circuits completely at the receiver, and the third possibility is the use of frequency synthesizers. They are covered in Chapter 7.

BFO Drift Effect

In any event, even minor drifts in BFO frequency can cause serious problems in SSB reception. If the oscillator drifts ± 100 Hz, a 1-kHz intelligence signal would be detected either as 1100 Hz or 900 Hz. Speech transmission requires less than a ± 100-Hz shift or the talker starts sounding like Donald Duck and becomes completely unintelligible. Obtaining good-quality SSB reception of music and digital signals requires a carrier.

Example 4-3

At one instant of time, an SSB music transmission consists of a 256-Hz sine wave and its second and fourth harmonics, 512 Hz and 1024 Hz. If the receiver's demodulator oscillator has drifted 5 Hz, determine the resulting speaker output frequencies.

Solution

The 5-Hz oscillator drift means that the detected audio will be 5 Hz in error, either up or down, depending on whether it is a usb or lsb transmission and on the direction of the oscillator's drift. Thus, the output would be either 251, 507, and 1019 Hz or 261, 517, and 1029 Hz. The speaker's output is no longer harmonic (exact frequency multiples), and even though it is just slightly off, the human ear would be offended by the new "music."

Product Detector

Product Detector using a balanced modulator to recover the intelligence in an SSB signal

As we have discussed, to recover the intelligence in an SSB (or DSB) signal, you need to reinsert the carrier. The balanced modulators used to create DSB can also be used to recover the intelligence in an SSB signal. When a balanced modulator is used in this fashion, it is usually called a **product detector.** This is the most common method of detecting an SSB signal.

Figure 4-16 shows another IC balanced modulator being used as a product detector. It is the Plessey Semiconductor SL640C. The capacitor connected to output pin 5 forms the low-pass filter to allow just the audio (low)-frequency component to appear in the output. The simplicity of this demodulator makes its desirability clear.

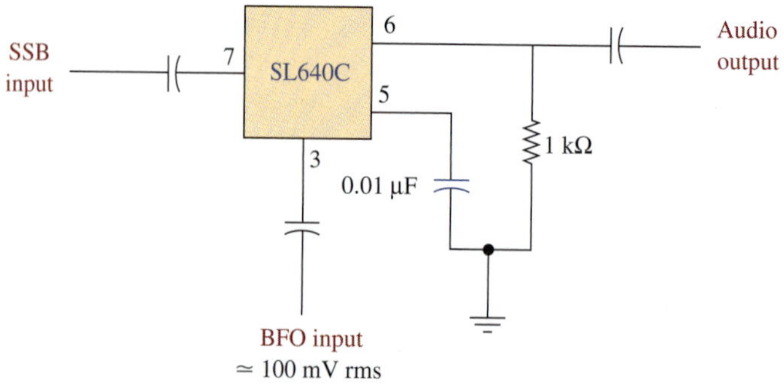

FIGURE 4-16 SL640C SSB detector.

4-6 SSB Receivers

To see the relationship of the parts in a single-sideband receiver, observe the block diagram in Fig. 4-17. Basically, the receiver is similar to an ordinary AM superheterodyne receiver; that is, it has RF and IF amplifiers, a mixer, a detector, and audio amplifiers. To permit satisfactory SSB reception, however, an additional mixer (demodulator) and oscillator must replace the conventional diode detector.

Example 4-4

The SSB receiver in Fig. 4-17 has outputs at 1 kHz and 3 kHz. The carrier used and suppressed at the transmitter was 2 MHz, and the upper sideband was utilized. Determine the exact frequencies at all stages for a 455-kHz IF frequency.

Solution

RF amp and first mixer input	2000 kHz + 1 kHz = 2001 kHz 2000 kHz + 3 kHz = 2003 kHz
Local oscillator	2000 kHz + 455 kHz = 2455 kHz
First mixer output: IF amp and second mixer input (the other components attenuated by tuned circuits)	2455 kHz − 2001 kHz = 454 kHz 2455 kHz − 2003 kHz = 452 kHz
BFO	455 kHz
Second mixer output and audio amp	455 kHz − 454 kHz = 1 kHz 455 kHz − 452 kHz = 3 kHz

As shown before, the carrier frequency was suppressed at the transmitter; thus, for proper intelligence detection, a carrier must be inserted by the receiver. The receiver illustrated in Fig. 4-17 inserts a carrier frequency into the detector, although the carrier frequency may be inserted at any point in the receiver before demodulation.

When the SSB signal is received at the antenna, it is amplified by the RF amplifier and applied to the first mixer. By mixing the output of the local

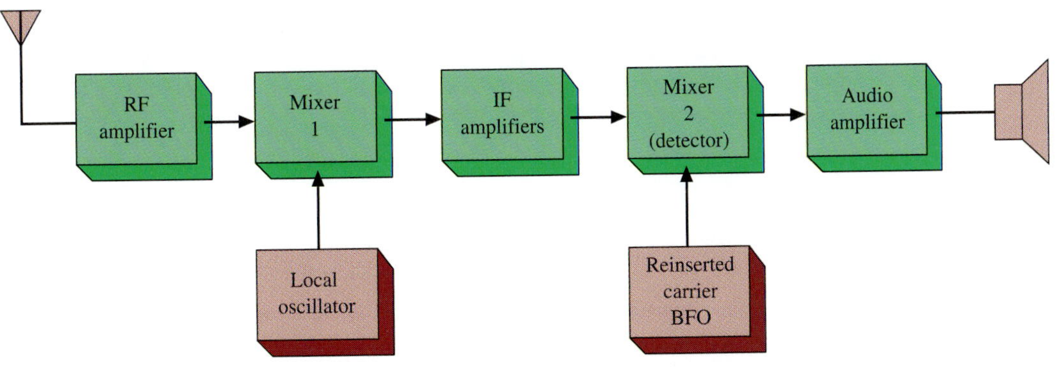

FIGURE 4-17 SSB receiver block diagram.

oscillator with the input signal (heterodyning), a difference frequency, or IF, is obtained. The IF is then amplified by one or more stages. Of course, this is dependent upon the type of receiver. Up to this point it is identical to an AM superheterodyne receiver. The IF output is applied to the second mixer (detector). The detector output is applied to the audio amplifier and then on to the output speaker.

Tuning the sideband receiver is somewhat more difficult than in a regular AM receiver. The carrier injection oscillator must be adjusted precisely to simulate the carrier frequency at all times. As previously explained, any tendency to drift within the oscillator will cause the output intelligence to be distorted.

Basic SSB Receiver

Butterworth Filter
a constant-*k* type of *LC* filter

A basic SSB receiver is shown schematically in Fig. 4-18. This superhet design functions well without an RF amplifier. The input signal comes in to a fixed Butterworth front-end filter (FL_1) that passes 3.75 to 4.0 MHz without tuning. A **Butterworth filter** exhibits a very flat response in the passband and approaches a 6-dB slope per octave. Individual channels in this amateur radio band are tuned by varying the local oscillator (Q_3) frequency. Notice that this stage is labeled with VFO, or *v*ariable *f*requency *o*scillator, and has a range from 4.253 to 4.453 MHz. This signal and the received signal are applied to the two gates of the mixer (Q_1). This 3N211 MOSFET provides high gain ($g_m \simeq 30{,}000\ \mu S$) and its output is applied to a mechanical filter, FL_2. The specified filter has a 2.2-kHz bandwidth at the 3-dB points and has a 5.5-kHz bandwidth at −60 dB.

The mechanical filter output is applied to the IF amplifier, Q_2, which is another 3N211 MOSFET. Its gain, and that of the audio amplifier U_1, are manually controlled by ganged potentiometers R_{1A} and R_{1B}. The bias at gate 2 of Q_2 is varied by R_{1A}. To obtain a wide range of control, it is necessary to have gate 2 a volt or two less than gate 1. This is done by "bootstrapping" this stage with an LED, D_1, that conducts at about 1.5 V. Thus, when R_{1A} has its arm at ground, gate 2 is effectively at −1.5 V and minimum gain for Q_2 occurs.

Another 3N211 device is used as the LO (or VFO if you prefer). Gates 1 and 2 of Q_3 are tied together. The oscillator signal is applied from the gate of Q_3 to gate 2 of the mixer, Q_1. A pure 3-V p-p sine wave is thereby available. D_2 is used as a switching diode to offset the VFO frequency when changing from usb to lsb.

The product detector stage (Q_4) is fed from the IF amp into the source of Q_4. The beat frequency oscillator (Q_5) is a switchable crystal oscillator. S1B selects either crystal Y_1 or Y_2 for lsb or usb, respectively. The product detector output (at Q_4's drain) is applied to a 741 op amp audio amplifier that offers up to 40-dB gain. Its output is sufficient to drive headphones, or an IC power amp could be added if a speaker is needed.

4-7 Troubleshooting

There are two ways to generate SSB signals, but modern manufacturing methods have reduced the cost of filters to the point that nearly all generate the SSB signal with balanced modulators and filters. Most radios even use separate filters to select the upper or lower sideband as desired instead of switching oscillators.

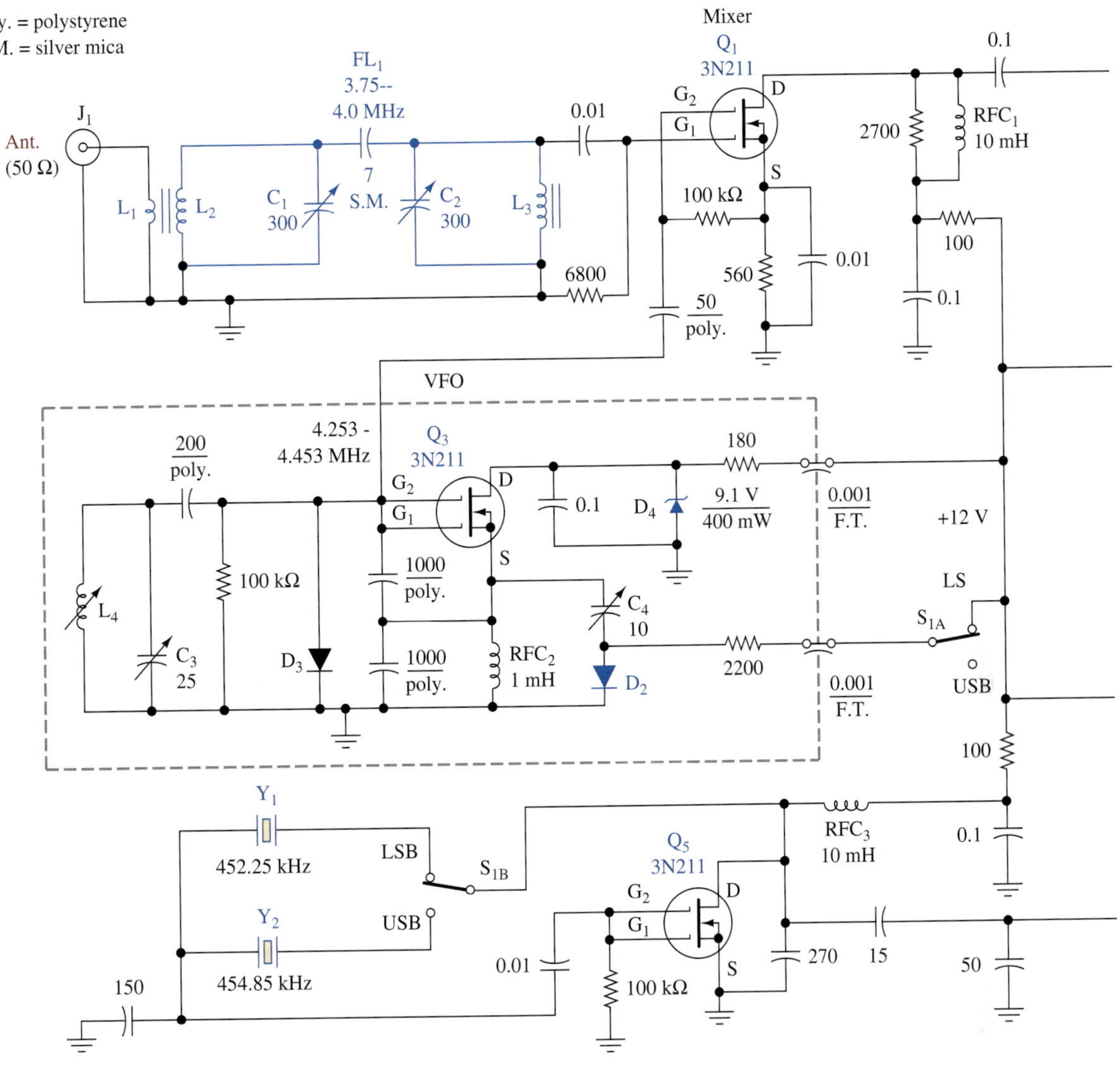

Fixed-value capacitors are disk ceramic unless noted otherwise. Polarized capacitors are electrolytic.
Fixed-value resistors are 1/4- or 1/2-Watt composition.

C_1, C_2—Mica compression trimmer, 300 pF max. Arco 427 or equiv.

C_3—Miniature 25-pF air variable. Hammarlund HF-25 or similar.

C_4—Circuit-board mount subminiature air variable or glass piston trimmer, 10 pF max. NPO miniature ceramic trimmer suitable as second choice.

D_1—LED, any color or size. Used only as 1.5-V reference diode.

D_2, D_3—Silicon switching diode, 1N914 or equiv.

D_4—Polarity-guarding diode. Silicon rectifier, 50 PIV, 1A.

D_5—Zener diode, 9.1 V, 400 mW or 1 Watt.

FL_1—Bandpass filter (see text).

FL_2—Collins Radio CB-type mechanical filter, Rockwell International No. 5269939010, 453.33-kHz center freq.

J_1—SO-239

J_2—Single-hole-mount phono jack.

J_3—Two-circuit phone jack.

L_1—Two turns No. 24 insulated wire over ground end of L2.

(Continued)

FIGURE 4-18 SSB receiver. (From the *ARRL Handbook,* courtesy of the American Radio Relay League.)

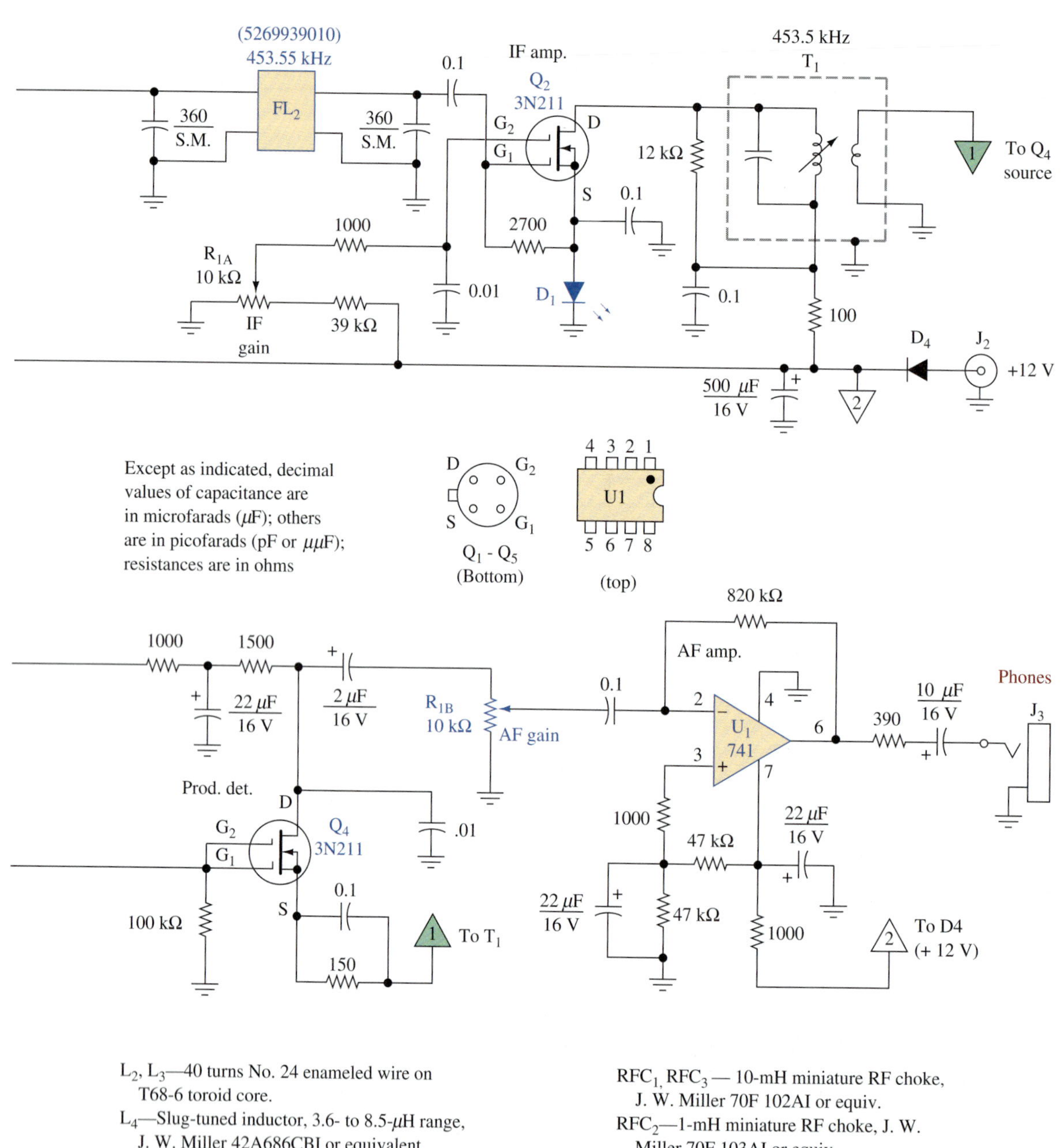

L_2, L_3—40 turns No. 24 enameled wire on T68-6 toroid core.

L_4—Slug-tuned inductor, 3.6- to 8.5-μH range, J. W. Miller 42A686CBI or equivalent suitable. Substitutes should have Q of 100 or greater at 4 MHz and be mechanically rigid.

Q_1–Q_5, incl. —Texas Instruments 3N211 FET,

R_1—Dual control, 10-kΩ per section, linear taper. Allen-Bradley type JD1N200P or similar. Separate controls can be used by providing extra hole in front panel.

RFC_1, RFC_3— 10-mH miniature RF choke, J. W. Miller 70F 102AI or equiv.

RFC_2—1-mH miniature RF choke, J. W. Miller 70F 103AI or equiv.

S_1—Two-pole, two-position phenolic or ceramic wafer switch.

T_1—455-kHz miniature IF transformer (see text). J. W. Miller No. 2067.

U_1—8-pin dual-in-line 741 op amp.

Y_1, Y_2—International Crystal Co. type GP, 30-pF load capacitance, HC-6/U style of holder. LSB 452.25 kHz, and USB 454.85 kHz.

FIGURE 4-18 *(Continued)*

In troubleshooting SSB generators, you will be mainly looking for the presence or absence of various oscillations. A spectrum analyzer is an extremely desirable tool in this regard, but if one is not available, a good general coverage shortwave receiver is the next best tool. It is also desirable to have a frequency counter to measure the exact frequency of the oscillators.

What do we do when faced with a radio receiver that has no reception? Where do we start to look for the trouble? When faced with this kind of problem, how does the technician proceed in formulating a plan of action? This section will show you a popular method used for finding the problem in a receiver with no reception.

After completing this section you should be able to

- Troubleshoot SSB generators and demodulators
- Test for carrier leakthrough with an oscilloscope or spectrum analyzer
- Identify a defective stage in an SSB receiver
- Describe the signal injection method of troubleshooting

Balanced Modulators

What to look for and do:

1. With no audio input, there should be no RF output. An oscilloscope will be helpful here.
2. The voltage from the oscillator must be 6 to 8 times the peak audio voltage. There should be several volts of RF and a few tenths of a volt of audio.
3. The diodes should be well matched. An ohmmeter can be used to select matched pairs or quads.
4. You should be able to null the carrier at the output by adjusting R1 and C1. It may be necessary to adjust each control several times alternately to secure optimum carrier suppression. Further detail on testing for carrier leakthrough is provided in the next few paragraphs.

Testing for Carrier Leakthrough

The purpose of the balanced modulator is to suppress or cancel the carrier. An exactly balanced modulator would totally suppress or remove the carrier. This is an impossibility because there are always imbalances—one diode conducts a little more current than another perhaps. To achieve a circuit's maximum suppression, balanced modulators usually include one or more balance controls, as shown in Fig. 4-19.

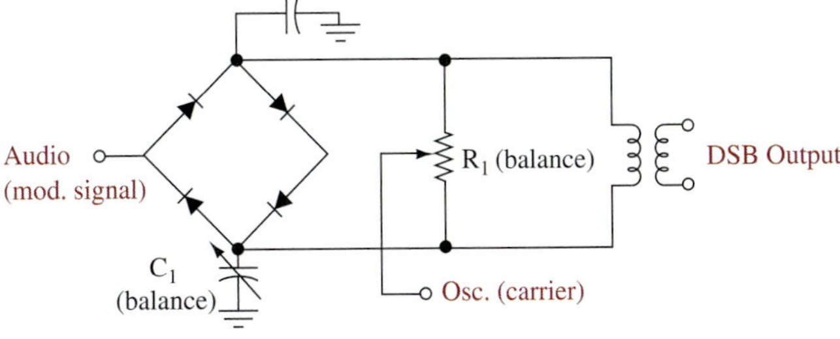

FIGURE 4-19 Balanced modulator.

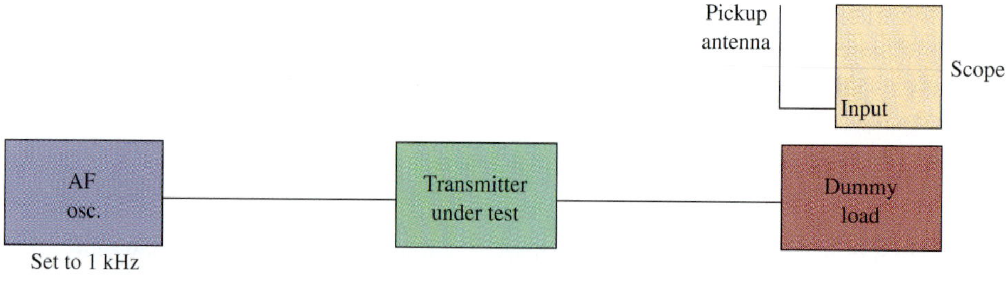

FIGURE 4-20 Checking for carrier leakthrough with an oscilloscope.

Carrier Leakthrough
the amount of carrier not suppressed by the balanced modulator

The first thing to do when troubleshooting a balanced modulator is to check the condition of its balance. This can be done by looking for carrier leakthrough with an oscilloscope. The circuit for this test is shown in Fig. 4-20. **Carrier leakthrough** simply means the amount of carrier not suppressed by the balanced modulator.

The audio signal generator is set to some frequency within the normal audio range of the transmitter, perhaps 1 or 1.5 kHz. Check the manual for the correct RF and audio signal levels into the modulator. (Test points may be included for measuring these.) Typically, the RF input (oscillator output) will be about 4 to 6 times the audio level for proper diode switching. Any DMM can be used to measure the audio, but the meter will require an RF probe for the oscillator output.

Figure 4-21(a) shows the transmitter signal when there is carrier leakthrough; that is, the carrier is not fully suppressed. Note the similarity to a partially modulated AM signal. Figure 4-21(b) shows the signal as it should be, a single tone signal; the carrier is fully suppressed.

One of two conditions could cause carrier leakthrough: either the circuit has become unbalanced or there are defective components. To check for imbalance, adjust the balance control(s) for minimum carrier amplitude. Should there be more than one control, it may be necessary to go back and forth more than once between controls because the setting of one often affects the setting of another.

If there is no balance problem and the input signal levels are correct, there is a defective component, most likely one of the diodes in the bridge assembly. Such bridges are usually a sealed unit; you cannot get at individual diodes. If this is the case, replace the suspect unit with a known good one and recheck for proper operation. Be sure to check balance again with the new unit in place.

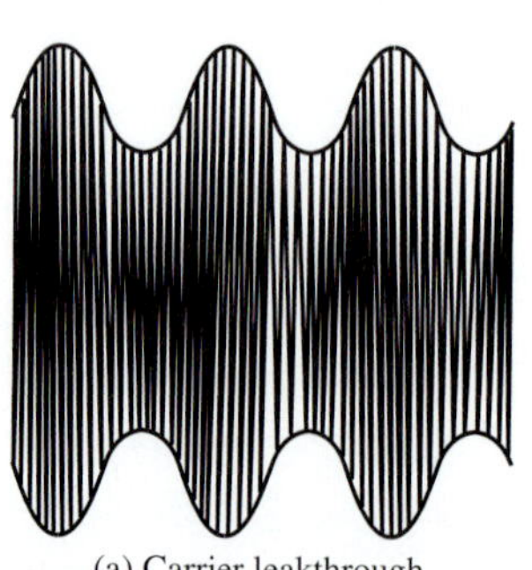

(a) Carrier leakthrough

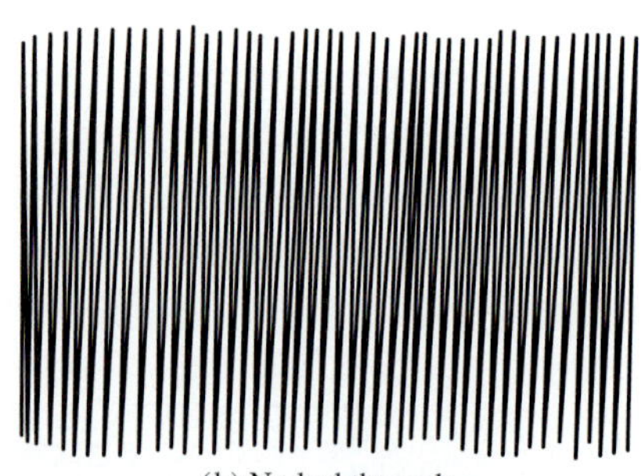

(b) No leakthrough

FIGURE 4-21 Single-sideband signal with and without carrier leakthrough.

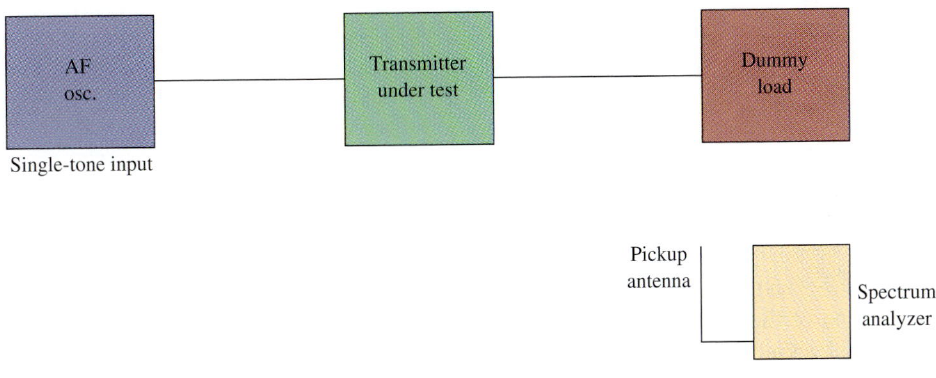

FIGURE 4-22 Checking carrier suppression with a spectrum analyzer.

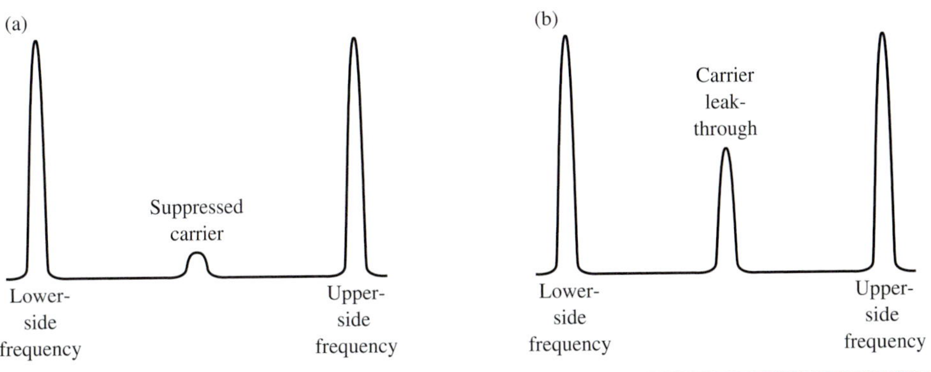

FIGURE 4-23 Carrier suppression as seen on a spectrum analyzer.

Figure 4-22 shows the circuit for checking carrier leakthrough and suppression with a spectrum analyzer.

Having determined that the RF and audio frequencies and levels are correct, observe the screen of the spectrum analyzer. If the modulator is operating correctly, you will see the display in part (a) of Fig. 4-23. Note the location of the suppressed carrier. Part (b) shows the same signal with some carrier leakthrough. As before, adjust the balance controls for minimum carrier amplitude (maximum suppression).

An advantage of the spectrum analyzer is that carrier suppression can be measured in dB directly on the analyzer's log scale. Check the transmitter's manual for the suppression figure; it will be in the neighborhood of −60 to −70 dB. The instruction manual for the spectrum analyzer will tell you how to set the unit's controls for logarithmic measurements.

Testing Filters

Filters designed for SSB service can be ceramic, crystal, or mechanical, but test methods are the same. The filter can have 6 to 10 dB of loss in the passband. Figure 4-24 shows how you would set up the equipment to test a filter.

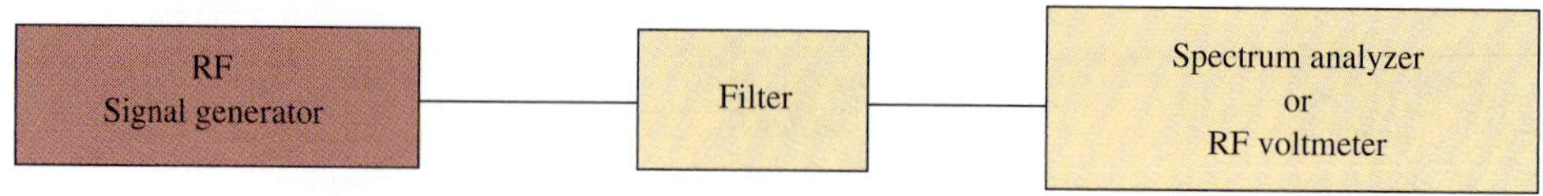

FIGURE 4-24 Filter testing.

The technician should slowly sweep the generator frequency across the passband of the filter. If the sweep speed is too fast, the filter's time delay will cause misleading results. Response of the filter should fall off rapidly at the band edges. Two or three dB of ripple in the passband is normal. If the ripple is as much as 6 to 10 dB, the signal passing through the filter will be badly distorted.

Testing Linear Amplifiers

The two-tone test is generally used to check amplifier linearity. For example, a 400-Hz tone and a 2500-Hz tone are applied to the input of an SSB transmitter. The output is observed with a spectrum analyzer tuned to the transmitter's carrier frequency. Nonlinearity in the amplifier will cause the amplifier to generate mixer-like products. The proper response is shown in Fig. 4-25. The carrier will not be present if the balanced modulator is properly adjusted. In a good linear amplifier, the distortion products will be at least 30 dB below the two desired tones. If the amplifier is not linear, several spurious outputs will appear and may be only a few dB below the desired signals. Amplifier nonlinearity is usually caused by improper bias points in the amplifier.

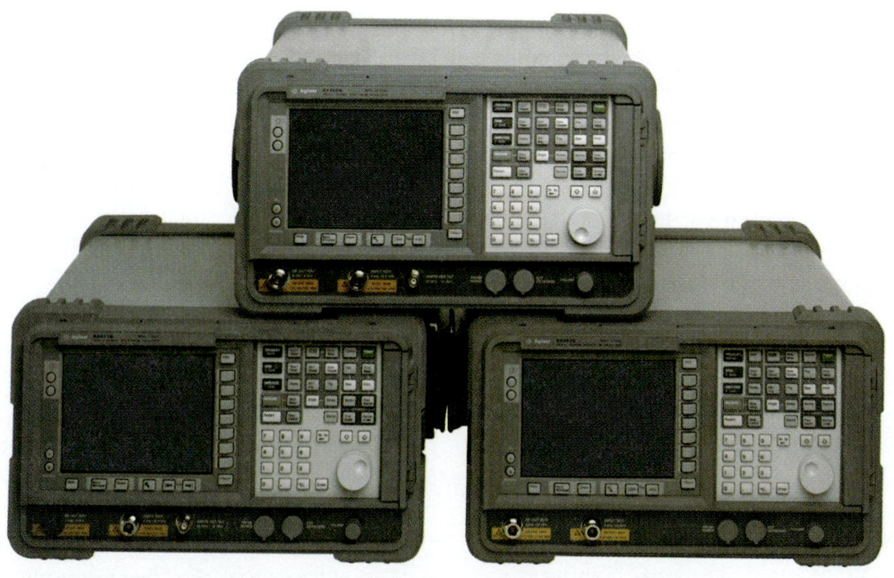

The Agilent family of RF spectrum and signal analyzers. (Courtesy of Agilent Technologies. Reprinted with permission.)

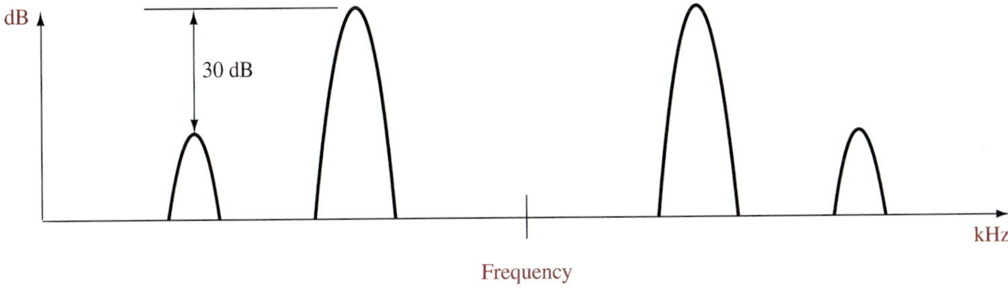

FIGURE 4-25 Two-tone test.

A qualitative check on linearity can be made with the two-tone test by observing the RF output on an oscilloscope. The output should appear as in Fig. 4-14. The output will approach a perfect sine wave if the amplifier is linear. Nonlinearity shows up as flat-topping of the sine wave.

Testing the SSB Receiver System

It's always best to start by having proper servicing material available for the set being repaired. This material includes the service manual, block diagrams, and schematic diagrams. The bare minimum would be the schematic diagrams. These items enhance the troubleshooting job tremendously. With today's complex electronic circuits, it is very difficult to attempt a service job without the service literature.

From the antenna of the receiver illustrated in Fig. 4-26, radio frequency signals are amplified by the RF amplifier. The radio signal can be traced from the antenna to where it is finally heard at the speaker. The RF amplifier boosts the signal before it goes to mixer one. As stated in the Chapter 3 troubleshooting section, a nonworking local oscillator can kill the received signal. So the local oscillator is included in the signal chain as a possible cause of no reception. Next, the converted RF signal is amplified in the IF amplifiers. From the IF strip, the signal is applied to mixer 2, the detector. Mixer 2 is responsible for recovering the original intelligence signal. To recover this original intelligence, the BFO reinserts the missing

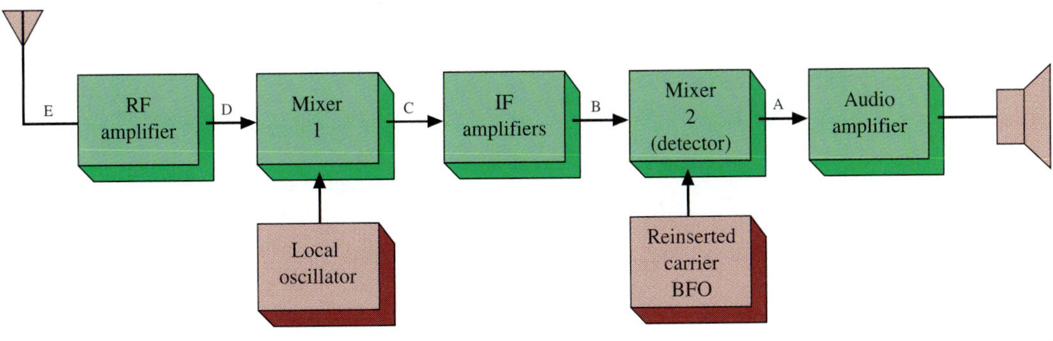

FIGURE 4-26 SSB receiver block diagram.

carrier. If the carrier were not reinserted, the radio would still produce sound at the speaker. The sound is garbled without this reinserted carrier, but the reception is not otherwise hindered. However, a problem in mixer 2 would interfere with the signal reception. The last stage before the speaker is the audio amplifier. In summary, the stages that can interfere with reception are the antenna, the RF amplifier, mixer 1, the local oscillator, the IF amplifiers, mixer 2, the audio amplifier section, the speaker, and the power supply.

The speaker and power supply should be checked first in the sequence of troubleshooting events. Check the power supply voltages using the DMM and compare these measurements to the specified values given in the service literature. If the voltage measurements are correct, the power supply is good. The speaker can be checked by inserting a tone across its terminals. If you hear the tone, then the speaker is working.

Signal Injection

Use a signal generator that is adjusted to a 1-V signal set to the center IF frequency. Modulate this IF with a 1-kHz signal. Inject the IF signal at point B as shown on the block diagram in Fig. 4-26. The 1-kHz test tone should be heard at the speaker output if mixer 2 and the audio amplifier are operating correctly. This would mean that the trouble lies toward the antenna. If you do not hear the tone, inject a pure 1-kHz signal at point A, just ahead of the audio amplifier. A tone from the speaker would now indicate a problem in mixer 2.

If the tone were heard when the signal was injected at point B, we would then move the probe to point C. A tone heard from the speaker indicates the IF amplifier is functioning correctly. Before moving to point D and applying the test signal, readjust the signal generator for a received RF frequency with the modulated 1-kHz signal. Set the generator's output voltage to 20 mV (check the service literature for exact signal levels). If the test tone is heard from the speaker at this point, then the RF amplifier or the antenna can be considered faulty. Decrease the output amplitude of the signal generator to around 2 mV and inject the test signal at point E. If the RF amplifier is bad, no tone will reach the speaker.

Table 4-1 will guide you through the signal-injecting procedure (test points are shown in Fig. 4-26).

Table 4-1

Signal Injected	Test Point	Tone Present	Analysis
1. 1-kHz modulated IF	B	Y (yes)	Mixer 2 and audio amplifier good; go to step 3
		N (no)	Go to step 2
2. 1-kHz signal	A	Y	Mixer 2 stage bad, audio amplifier good
		N	Audio amplifier bad
3. Modulated IF	C	Y	IF amplifier okay; go to step 4
		N	IF amplifier bad
4. Modulated RF	D	Y	Mixer 1 and local oscillator good; go to step 5
		N	Problem resides in mixer 1 or the LO
5. Modulated RF	E	Y	Trouble lies beyond the RF amplifier; check antenna, cabling, connectors, or coupling
		N	RF amplifier bad

Using Table 4-1, you can isolate a defective stage rather quickly. A variation on the above technique is to use an oscilloscope to look at the test signal at the output of each stage instead of relying on hearing the signal.

4-8 Troubleshooting with Electronics Workbench™ Multisim

This section extends understanding of single-sideband systems as implemented in Electronics Workbench™ Multisim. Open **Fig4-27.ms7 (.msm)** on your EWB Multisim CD. This circuit contains two sine-wave generators that are input into a multiplier module. A multiplier circuit produces the sum and difference of the input frequencies on its output. For this example, the sum is 3 MHz + 1 MHz = 4 MHz, and the difference is 3 MHz − 1 MHz = 2 MHz. Start the simulation and observe the inputs to the multiplier and the complex output waveform using the oscilloscope. The circuit is shown in Fig. 4-27.

Next, open the spectrum analyzer, which is connected to the output of the multiplier. The spectrum analyzer shows that frequencies of 2 MHz and 4 MHz are produced. This is shown in Fig. 4-28. This is called a double-sideband spectrum. Use the cursor to determine the frequencies of each spectral component.

It is necessary to remove one of the sidebands in an SSB (single-sideband) system. This can be accomplished by passing the output of the multiplier through a filter. In this example, a five-element Chebyshev high-pass filter has been provided to remove the lower-sideband component. This filter has a 20-dB upper cutoff

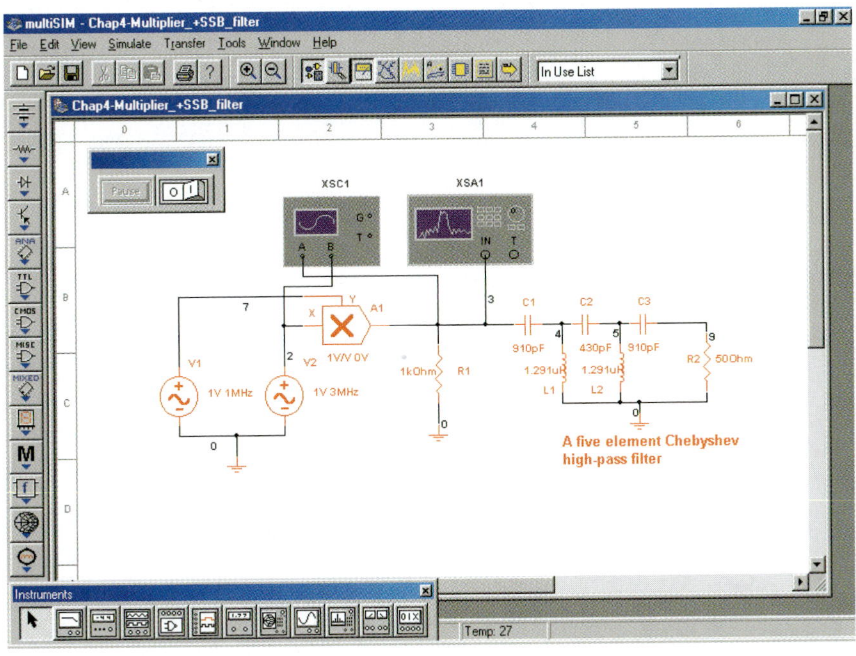

FIGURE 4-27 A multiplier plus SSB filter as implemented with Electronics Workbench™ Multisim.

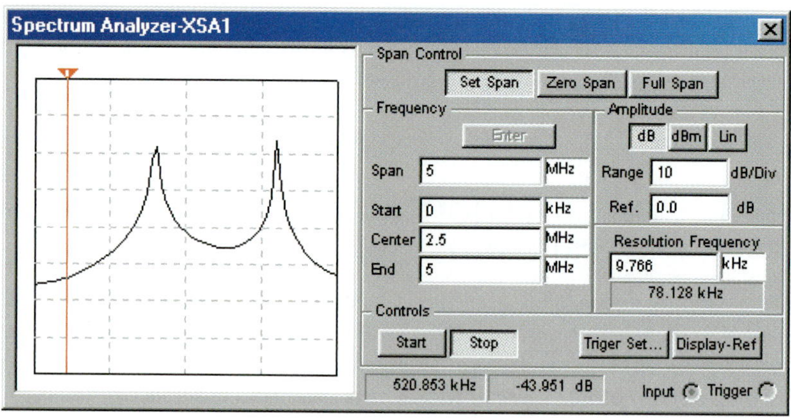

FIGURE 4-28 The double-sideband output spectrum for the multiplier circuit.

frequency of about 2.6 MHz. Therefore, frequencies below 2.6 MHz should be significantly attenuated. Connect the spectrum analyzer to the output of the filter and restart the simulation. Verify that the output contains only the upper-sideband component. The sideband at 2 MHz has been significantly reduced, but the upper sideband is still present. The result is shown in Fig. 4-29.

The next part of this exercise provides you with the opportunity to troubleshoot a filter circuit. Open **FigE4-1.ms7 (.msm)** on your EWB CD. Start the simulation and use the spectrum analyzer to observe the output spectrum. Note that the spectrum analyzer is connected to the output of the filter. Is the circuit working as expected? If not, use the spectrum analyzer and the oscilloscope to check the waveform. Don't forget to perform a visual check of the circuit for potential problems. You will discover that both the upper and lower sidebands are present on the output. What circuit removes the lower sideband? Shouldn't the filter remove the lower sideband? Careful inspection of the filter shows that the ground is missing. Replace the ground and rerun the simulation. The circuit should be functioning properly.

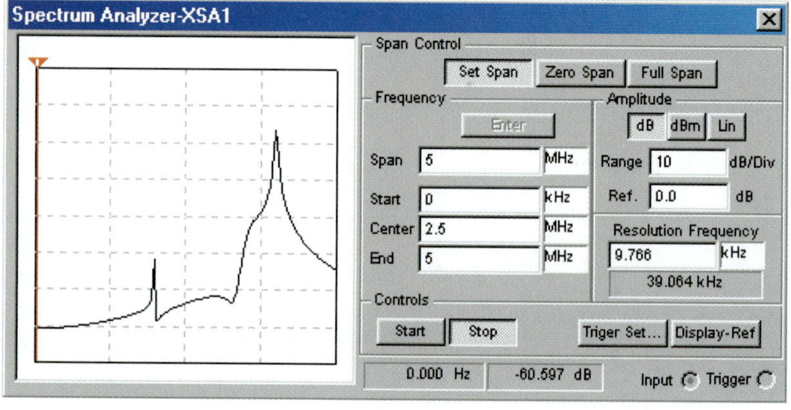

FIGURE 4-29 The multiplier circuit with the lower sideband removed.

Electronics Workbench™ Exercises

1. Open **FigE4-2.ms7 (.msm)** in your EWB CD. This circuit contains a tunable high-pass filter. Use the **(a/A)** and **(b/B) keys** to adjust the inductance values to optimize the performance of the filter. Record your settings and comment on the effect that changing the inductance values has on the output. ($L_1 = 35\%$, $L_2 = 55\%$)
2. Open **FigE4-3.ms7 (.msm)** in your EWB CD. This circuit contains a fault. In this exercise, assume that you have been told that the circuit does not appear to be working properly. You may need to refer back to the waveforms in **Fig4-27.ms7 (.msm)** for a properly functioning circuit to help guide you with your troubleshooting. Confirm that the circuit is not working properly and find the cause of the problem. Once you find the problem, record the fault and the circuit behavior generated by the fault. Correct the fault and rerun the simulation to verify that the problem has been corrected.
3. Open **FigE4-4.ms7 (.msm)** in your EWB CD. This circuit contains a fault. Use your troubleshooting techniques to find the problem. Once you find the problem, record the fault and the circuit behavior observed. Correct the fault and rerun the simulation to verify that the problem has been corrected.

Summary

In Chapter 4 we introduced single-sideband (SSB) systems and explained their various advantages over standard AM systems. The major topics you should now understand include:

- the advantages of SSB systems, including the utilization of available frequency bandwidth, noise reduction, power requirements, and selective fading effects
- the various SSB systems and their general characteristics
- an analysis of the function of a balanced modulator
- the operation of a balanced ring modulator
- the application of linear integrated circuit balanced modulators
- the need for high-Q bandpass filters and the description of mechanical, ceramic, and crystal varieties
- the analysis of SSB transmission systems, including the filter and phase methods
- an understanding of the need for amplitude compandoring and a method of implementation
- the description and operation of a class AB push-pull linear power amplifier
- the analysis of SSB demodulation techniques
- the analysis of a complete SSB receiver

Questions and Problems

Section 4-1

1. An AM transmission of 1000 W is fully modulated. Calculate the power transmitted if it is transmitted as an SSB signal. (167 W)
2. An SSB transmission drives 121 V peak into a 50-Ω antenna. Calculate the PEP. (146 W)

3. Explain the difference between rms and PEP designations.
4. Provide detail on the differences between ACSSB, SSB, SSBSC, and ISB transmissions.
5. List and explain the advantages of SSB over conventional AM transmissions. Are there any disadvantages?
6. A sideband technique called doubled sideband/suppressed carrier (DSBSC) is similar to a regular AM transmission, double sideband full carrier (DSBFC). Using your knowledge of SSBSC, explain the advantage DSBSC has over regular AM.

Section 4-2

7. What are the typical inputs and outputs for a balanced modulator?
8. Briefly describe the operation of a balanced ring modulator.
9. Explain the advantages of using an IC for the four diodes in a balanced ring modulator as compared with four discrete diodes.
10. Referring to the specifications for the AD630 LIC balanced modulator in Fig. 4-2, determine the channel separation at 10 kHz, explain how a gain of +1 and +2 are provided.
11. Explain how to generate an SSBSC signal from the balanced modulator.

Section 4-3

12. Calculate a filter's required Q to convert DSB to SSB, given that the two sidebands are separated by 200 Hz. The suppressed carrier (40 dB) is 2.9 MHz. Explain how this required Q could be greatly reduced. (36,250)

* 13. Draw the approximate equivalent circuit of a quartz crystal.

14. What are the undesired effects of the crystal holder capacitance in a crystal filter, and how are they overcome?

* 15. What crystalline substance is widely used in crystal oscillators (and filters)?

16. Using your library or some other source, provide a schematic for a four-element crystal lattice filter and explain its operation.

* 17. What are the principal advantages of crystal control over tuned circuit oscillators (or filters)?

18. Explain the operation of a ceramic filter. What is the significance of a filter's shape factor?
19. Define *shape factor*. Explain its use.
20. A bandpass filter has a 3-dB ripple amplitude. Explain this specification.
21. Explain the operation and use of mechanical filters.
22. Why are SAW filters not often used in SSB equipment?
23. An SSB signal is generated around a 200-kHz carrier. Before filtering, the upper and lower sidebands are separated by 200 Hz. Calculate the filter Q required to obtain 40-dB suppression. (2500)

---

*An asterisk preceding a number indicates a question that has been provided by the FCC as a study aid for licensing examinations.

Section 4-4

24. Determine the carrier frequency for the transmitter shown in Fig. 4-8. (It is *not* 3 MHz).
25. Draw a detailed block diagram of the SSB generator shown in Fig. 4-9. Label frequencies involved at each stage if the intelligence is a 2-kHz tone and the usb is utilized.
26. The sideband filter (FL_1) in Fig. 4-9 has a 5-dB *insertion loss* (i.e., a 5-dB signal loss from input to output). Calculate the filter's output voltage assuming equal impedances at its input and output. (0.45 V p-p)
27. Calculate the total impedance in the collector of Q_3 in Fig. 4-9. (54.3 Ω)
28. List the advantages of the phase versus filter method of SSB generation. Why isn't the phase method more popular than the filter method?
29. Explain the operation of the phase-shift SSB generator illustrated in Fig. 4-10. Why is the carrier phase shift of 90° not a problem, whereas that for the audio signal is?
30. Explain the operation and need for the control circuitry (K_1, Q_1, Q_2) in the linear power amplifier shown in Fig. 4-13.
31. The PEP transmitted by an ACSSB system is 140 W. It uses an NE571N compandor LIC. Calculate the power transmitted under the no-modulation condition. The audio signal ranges from −28 dBm to +34 dBm before compression. Determine the compressor's output range. (14 W, −14 dBm to +17 dBm)
32. Explain how an ACSSB system can provide improved noise performance compared to a regular SSB system.

Section 4-5

33. List the components of an AM signal at 1 MHz when modulated by a 1-kHz sine wave. What is the component(s) if it is converted to a usb transmission? If the carrier is redundant, why must it be "reinserted" at the receiver?
34. Explain why the BFO in an SSB demodulator has such stringent accuracy requirements.
35. Suppose the modulated signal of an SSBSC transmitter is 5 kHz and the carrier is 400 kHz. At what frequency must the BFO be set?
36. What is a product detector? Explain the need for a low-pass filter at the output of a balanced modulator used as a product detector.
37. Calculate the frequency of a product detector that is fed an SSB signal modulated by 400 Hz and 2 kHz sine waves. The BFO is 1 MHz.

Section 4-6

38. Draw a block diagram for the receiver shown schematically in Fig. 4-18. Suggest a change to the schematic that you feel would improve its performance and explain why.

Section 4-7

39. Describe the output of the modulator in Fig. 4-1 if the L_1 winding was open circuited.
40. When troubleshooting the balanced modulator in Fig. 4-19, provide a detailed procedure to check diode matching by using an ohmmeter. Describe the problem caused with this circuit if the diodes are not matched.

41. Explain the concept of carrier leakthrough and its causes, and provide two methods of testing for it.
42. The two-tone test is used to check amplifier linearity. Explain why a single-tone test will not be as effective as a two-tone test.
43. Explain the effect of injecting a modulated IF signal at point D in Fig. 4-26.
44. Suppose it was determined that there was an output from test point B in Fig. 4-26 and no output from test point A. Explain the possible causes of no output.
45. Describe how signal tracing and signal injection can be used to troubleshoot the SSB receiver in Fig. 4-26.
46. With reference to Table 4-1, explain why doing test 5 before tests 1 through 4 invalidates the analysis.

Questions for Critical Thinking

47. If a carrier and one sideband were eliminated from an AM signal, would the transmission still be usable? Why or why not?
48. Explain the principles involved in a single-sideband, suppressed-carrier (SSBSC) emission. How does its bandwidth of emission and required power compare with that of full carrier and sidebands?
49. You have been asked to provide SSB using a DSB signal, cos $\omega_i t$, cos $\omega_c t$. Can this be done? Provide mathematical proof of your judgment.
50. If, in an emergency, you had to use an AM receiver to receive an SSB broadcast, what modifications to the receiver would you need to make?

5 Frequency Modulation: Transmission

Chapter Outline

Cobra's complete line of two-way radios provides the longest range in the GMRS/FRS category. (Courtesy of Cobra Electronics Corporation. Reprinted with permission.)

Objectives

- Define *angle modulation* and describe the two categories
- Explain a basic capacitor microphone FM generator and the effects of voice amplitude and frequency
- Analyze an FM signal with respect to modulation index, sidebands, and power
- Describe the noise suppression capabilities of FM and how they relate to the capture effect and preemphasis
- Provide various schemes and circuits used to generate FM
- Explain how a PLL can be used to generate FM
- Describe the multiplexing technique used to add stereo to the standard FM broadcast systems

Key Terms

angle modulation
phase modulation
frequency modulation
deviation constant (*k*)
frequency deviation
modulation index
Bessel functions
Carson's rule
guard bands
deviation ratio (DR)
wideband FM
narrowband FM
limiter
capture effect
capture ratio
threshold
preemphasis
deemphasis
Dolby system
varactor diode
reactance modulator
voltage-controlled oscillator
automatic frequency control
Crosby systems
frequency multipliers
exciter
discriminator
Armstrong modulator
pump chain
frequency multiplexing
multiplex operation
frequency-division multiplexing
matrix network

5-1 ANGLE MODULATION

Angle Modulation superimposing the intelligence signal on a high-frequency carrier so that its phase angle or frequency is altered as a function of the intelligence amplitude

Phase Modulation superimposing the intelligence signal on a high-frequency carrier so that the carrier's phase angle departs from its reference value by an amount proportional to the intelligence amplitude

Frequency Modulation superimposing the intelligence signal on a high-frequency carrier so that the carrier's frequency departs from its reference value by an amount proportional to the intelligence amplitude

As has been stated previously, there are three parameters of a sine-wave carrier that can be varied to allow it to carry a low-frequency intelligence signal. They are its amplitude, frequency, and phase. The latter two, frequency and phase, are actually interrelated, as one cannot be changed without changing the other. They both fall under the general category of *angle modulation.* **Angle modulation** is defined as modulation where the angle of a sine-wave carrier is varied from its reference value. Angle modulation has two subcategories, phase modulation and frequency modulation, with the following definitions:

> ***Phase modulation (PM):*** angle modulation where the phase angle of a carrier is caused to depart from its reference value by an amount proportional to the modulating signal amplitude.
>
> ***Frequency modulation (FM):*** angle modulation where the instantaneous frequency of a carrier is caused to vary by an amount proportional to the modulating signal amplitude.

The key difference between these two similar forms of modulation is that in PM the amount of phase change is proportional to intelligence amplitude, while in FM it is the frequency change that is proportional to intelligence amplitude. As it turns out, PM is *not* directly used as the transmitted signal in communications systems but does have importance because it is often used to help generate FM, *and* a knowledge of PM helps us to understand the superior noise characteristics of FM as compared to AM systems. In recent years, it has become fairly common practice to denote angle modulation simply as FM instead of specifically referring to FM and PM.

The concept of FM was first practically postulated as an alternative to AM in 1931. At that point, commercial AM broadcasting had been in existence for over 10 years, and the superheterodyne receivers were just beginning to supplant the TRF designs. The goal of research into an alternative to AM at that time was to develop a system less susceptible to external noise pickup. Major E. H. Armstrong developed the first working FM system in 1936, and in July 1939, he began the first regularly scheduled FM broadcast in Alpine, New Jersey.

Radio Emission Classifications

Table 5-1 gives the codes used to indicate the various types of radio signals. The first letter is A, F, or P to indicate AM, FM, or PM. The next code symbol is one of the numbers 0 through 9 used to indicate the type of transmission. The last code symbol is a subscript. If there is no subscript, it means double-sideband, full carrier. Here are some examples of emission codes:

$A3_a$	SSB, reduced carrier
$A3_i$	SSB, no carrier
F3	FM, double-sideband, full carrier
$A7_i$	SSB, no carrier, multiple sidebands with different messages
10A3	AM, double-sideband, full carrier, 10-kHz bandwidth

Notice the last example. If an emission code is preceded by a number, that number is the bandwidth of the signal in kHz.

Table 5-1 Radio Emission Classifications

Modulation	Type	Subscripts
A Amplitude	0 Carrier on only	None Double-sideband, full carrier
F Frequency	1 Carrier on–off (Morse code, radar)	a Single-sideband, reduced carrier
P Phase	2 Carrier on, keyed tone—on–off	b Two independent sidebands
	3 Telephony, voice, or music	c Vestigial sideband
	4 Facsimile, nonmoving or slow-scan TV	d Pulse amplitude modulation (PAM)
	5 Vestigial sideband, commercial TV	e Pulse width modulation (PWM)
	6 Four-frequency diplex telegraphy	f Pulse position modulation (PPM)
	7 Multiple sidebands, each with different message	g Digital video
		h Single-sideband, full carrier
	8 Unassigned	i Single-sideband, no carrier
	9 General, all other	

5-2 A Simple FM Generator

To gain an intuitive understanding of FM, consider the system illustrated in Fig. 5-1. This is actually a very simple, yet highly instructive, FM transmitting system. It consists of an *LC* tank circuit, which, in conjunction with an oscillator circuit, generates a sine-wave output. The capacitance section of the *LC* tank is not a standard capacitor but is a capacitor microphone. This popular type of microphone is often referred to as a condenser mike and is, in fact, a variable capacitor. When no sound waves reach its plates, it presents a constant value of capacitance at its two output terminals. When sound waves reach the mike, however, they alternately cause its plates to move in and out. This causes its capacitance to go up and down around its center value. The *rate* of this capacitance change is equal to the frequency of the sound waves striking the mike, *and* the *amount* of capacitance change is proportional to the amplitude of the sound waves.

Because this capacitance value has a direct effect on the oscillator's frequency, the following two *important* conclusions can be made concerning the system's output frequency:

1. The frequency of impinging sound waves determines the *rate* of frequency change.
2. The amplitude of impinging sound waves determines the *amount* of frequency change.

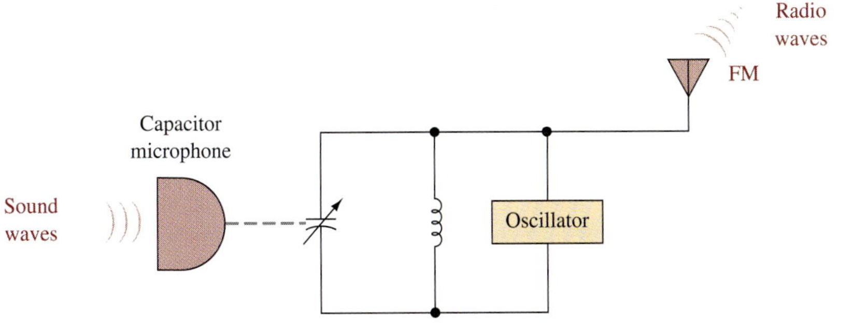

FIGURE 5-1 Capacitor microphone FM generator.

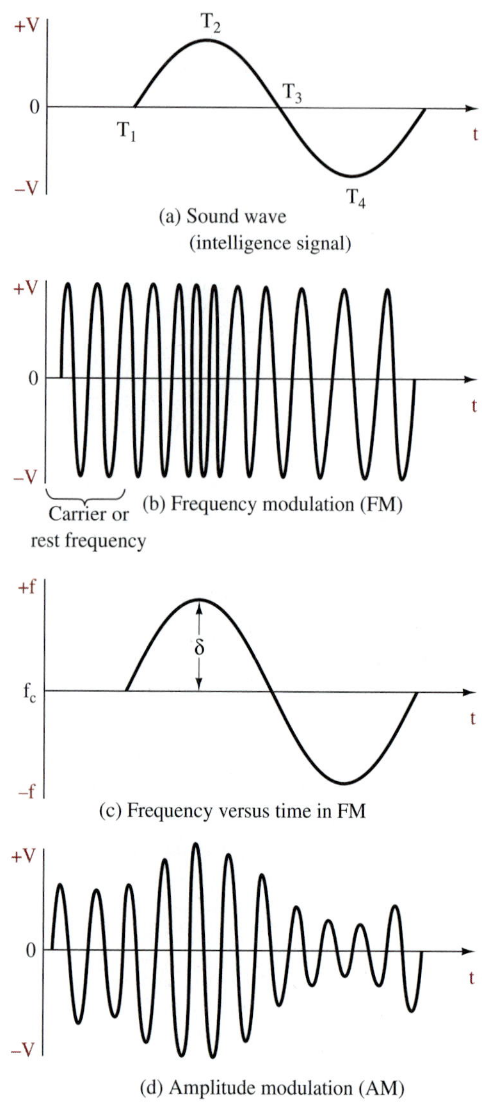

FIGURE 5-2 FM representation.

Consider the case of the sinusoidal sound wave (the intelligence signal) shown in Fig. 5-2(a). Up until time T_1 the oscillator's waveform in Fig. 5-2(b) is a constant frequency with constant amplitude. This corresponds to the carrier frequency (f_c) or *rest* frequency in FM systems. At T_1 the sound wave in Fig. 5-2(a) starts increasing sinusoidally and reaches a maximum positive value at T_2. During this period, the oscillator frequency is gradually increasing and reaches its highest frequency when the sound wave has maximum amplitude at time T_2. From time T_2 to T_4 the sound wave goes from maximum positive to maximum negative and the resulting oscillator frequency goes from a maximum frequency *above* the rest value to a maximum value *below* the rest frequency. At time T_3 the sound wave is passing through zero, and therefore the oscillator output is instantaneously equal to the carrier frequency.

The relationship for the FM signal generated by the capacitor microphone can be written as shown in Eq. (5-1).

$$f_{out} = f_c + ke_i \tag{5-1}$$

where f_{out} = instantaneous output frequency
f_c = output carrier frequency
k = deviation constant [kHz/V]
e_i = modulating (intelligence) input

Equation 5-1 shows that the output carrier frequency (f_c) depends on the amplitude and the frequency of the modulating signal (e_i) and also on the frequency deviation generated by the microphone (k) for a given input level. The unit k is called a deviation constant. This defines how much the carrier frequency will deviate (change) for a given input voltage level. The **deviation constant** (k) is defined in units of kHz/V. For example, if k = 1 kHz/10 mV, then a modulating signal input level of 20 mV will cause a 2-kHz frequency shift. A modulating signal input level of −20 mV will cause a −2-kHz frequency shift. Assuming that the 20-mV level is from a sinusoid (a sinusoid contains both a positive and negative peak), then the rate at which the frequency changes (deviates above and below the original carrier frequency) depends on the frequency of the modulating signal. For example, if a 20-mV, 500-Hz input signal is applied to the microphone, then the carrier frequency will deviate ±2 kHz at a rate of 500 Hz.

Deviation Constant (k)
how much the carrier frequency will deviate for a given modulating input voltage level

Example 5-1

A 25-mV sinusoid at a frequency of 400 Hz is applied to a capacitor microphone FM generator. If the deviation constant for the capacitor microphone FM generator is 750 Hz/10 mV, determine
(a) The frequency deviation generated by an input level of 25 mV.
(b) The rate at which the carrier frequency is being deviated.

Solution

(a) $$\text{positive frequency deviation} = 25\text{ mV} \times \frac{750\text{ Hz}}{10\text{ mV}} = 2250\text{ Hz}, \quad \text{or} \quad 2.25\text{ kHz}$$

$$\text{negative frequency deviation} = -25\text{ mV} \times \frac{750\text{ Hz}}{10\text{ mV}} = -2250\text{ Hz}, \quad \text{or} \quad -2.25\text{ kHz}$$

The total deviation is written as ±2.25 kHz for the given input signal level.

(b) The input frequency (f_i) is 400 Hz; therefore, by Eq. (5-1)

$$f_{out} = f_c + ke_i \tag{5-1}$$

The carrier will deviate ±2.25 kHz at a rate of 400 Hz.

The Two Major Concepts

The amount of oscillator frequency increase and decrease around f_c is called the **frequency deviation,** δ. This deviation is shown in Fig. 5-2(c) as a function of time.

Frequency Deviation
amount of carrier frequency increase or decrease around its center reference value

Notice that this is a graph of frequency versus time—not the usual voltage versus time. It is ideally shown as a sine-wave replica of the original intelligence signal. It shows that the oscillator output is indeed an FM waveform. Recall that FM is defined as a sine-wave carrier that changes in frequency by an *amount* proportional to the instantaneous value of the intelligence wave and at a *rate* equal to the intelligence frequency.

Figure 5-2(d) shows the AM wave resulting from the intelligence signal shown in Fig. 5-2(a). This should help you to see the difference between an AM and FM signal. In the case of AM, the carrier's amplitude is varied (by its sidebands) in step with the intelligence, while in FM, the carrier's frequency is varied in step with the intelligence.

The capacitor microphone FM generation system is seldom used in practical applications; its importance is derived from its relative ease of providing an understanding of FM basics. If the sound-wave intelligence striking the microphone were doubled in frequency from 1 kHz to 2 kHz with constant amplitude, the rate at which the FM output swings above and below the center frequency (f_c) would change from 1 kHz to 2 kHz. Because the intelligence amplitude was not changed, however, the *amount* of frequency deviation (δ) above and below f_c will remain the same. On the other hand, if the 1-kHz intelligence frequency were kept the same but its amplitude were doubled, the *rate* of deviation above and below f_c would remain at 1 kHz, but the *amount* of frequency deviation would double.

As you continue through your study of FM, whenever you start getting bogged down on basic theory, it will often be helpful to review the capacitor mike FM generator. *Remember:*

1. The intelligence amplitude determines the *amount* of carrier frequency deviation.
2. The intelligence frequency (f_i) determines the *rate* of carrier frequency deviation.

Example 5-2

An FM signal has a center frequency of 100 MHz but is swinging between 100.001 MHz and 99.999 MHz at a rate of 100 times per second. Determine:
(a) The intelligence frequency f_i.
(b) The intelligence amplitude.
(c) What happened to the intelligence amplitude if the frequency deviation changed to between 100.002 and 99.998 MHz.

Solution

(a) Because the FM signal is changing frequency at a 100-Hz rate, $f_i = 100$ Hz.
(b) There is no way of determining the actual amplitude of the intelligence signal. Every FM system has a different proportionality constant between the intelligence amplitude and the amount of deviation it causes.
(c) The frequency deviation has now been doubled, which means that the intelligence amplitude is now double whatever it originally was.

5-3 FM Analysis

The complete mathematical analysis of angle modulation requires the use of high-level mathematics. For our purposes, it will suffice simply to give the solutions and discuss them. For phase modulation (PM), the equation for the instantaneous voltage is

$$e = A \sin(\omega_c t + m_p \sin \omega_i t) \quad \textbf{(5-2)}$$

where e = instantaneous voltage
A = peak value of original carrier wave
ω_c = carrier angular velocity ($2\pi f_c$)
m_p = maximum phase shift caused by the intelligence signal (radians)
ω_i = modulating (intelligence) signal angular velocity ($2\pi f_i$)

The maximum phase shift caused by the intelligence signal, m_p, is defined as the **modulation index** for PM.

Modulation Index measure of the extent to which a carrier is varied by the intelligence

The following equation provides the equivalent formula for FM:

$$e = A \sin(\omega_c t + m_f \sin \omega_i t) \quad \textbf{(5-3)}$$

All the terms in Eq. (5-3) are defined as they were for Eq. (5-2), with the exception of the new term, m_f. In fact, the two equations are identical except for that term. It is defined as the modulation index for FM, m_f. It is equal to

$$m_f = \text{FM modulation index} = \frac{\delta}{f_i} \quad \textbf{(5-4)}$$

where δ = maximum frequency shift caused by the intelligence signal (deviation)
f_i = frequency of the intelligence (modulating) signal

Comparison of Eqs. (5-2) and (5-3) points out the only difference between PM and FM. The equation for PM shows that the phase of the carrier varies with the modulating signal amplitude (because m_p is determined by this), and in FM the carrier phase is determined by the ratio of intelligence signal amplitude (which determines δ) to the intelligence frequency (f_i). Thus, FM is *not* sensitive to the modulating signal frequency but PM *is*. The difference between them is subtle—in fact, if the intelligence signal is integrated and then allowed to phase-modulate the carrier, an FM signal is created. This is the method used in the Armstrong indirect FM system, as will be explained in Sec. 5-6. In FM the amount of deviation produced is not dependent on the intelligence frequency, as it is for PM. The amount of deviation is proportional to the intelligence signal amplitude for both PM and FM. These conditions are shown in Fig. 5-3.

FM Mathematical Solution

The FM formula [Eq. (5-3)] is really more complex than it looks because it contains the sine of a sine. To solve for the frequency components of an FM wave requires the use of a high-level mathematical tool, **Bessel functions.** They show that frequency-modulating a carrier with a pure sine wave actually generates an infinite

Bessel Functions mathematical technique for determining the exact bandwidth of an FM signal

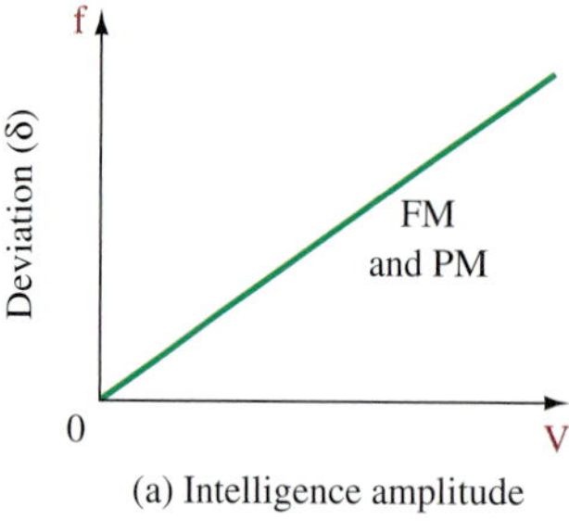

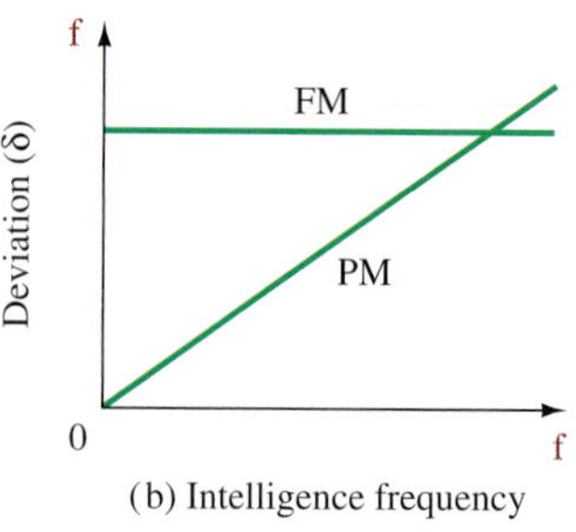

FIGURE 5-3 Deviation effects on FM/PM by intelligence parameters: (a) intelligence amplitude; (b) intelligence frequency.

number of sidebands (components) spaced at multiples of the intelligence frequency, f_i, above and below the carrier! Fortunately, the amplitude of these sidebands approaches a negligible level the farther away they are from the carrier, which allows FM transmission within finite bandwidths. The Bessel function solution to the FM equation is

$$\begin{aligned} f_c(t) = {} & J_0(m_f)\cos\omega_c t - J_1(m_f)[\cos(\omega_c - \omega_i)t - \cos(\omega_c + \omega_i)t] \\ & + J_2(m_f)[\cos(\omega_c - 2\omega_i)t + \cos(\omega_c + 2\omega_i)t] \\ & - J_3(m_f)[\cos(\omega_c - 3\omega_i)t + \cos(\omega_c + 3\omega_i)t] \\ & + \cdots \end{aligned} \tag{5-5}$$

where

$f_c(t)$ = FM frequency components

$J_0(m_f)\cos\omega_c t$ = carrier component

$J_1(m_f)[\cos(\omega_c - \omega_i)t - \cos(\omega_c + \omega_i)t]$ = first set of side frequencies at $\pm f_i$ above and below the carrier

$J_2(m_f)[cos(\omega_c - 2\omega_i)t + \cos(\omega_c + 2\omega_i)]t$ = second set of side frequencies at $\pm 2f_i$ above and below the carrier, etc.

To solve for the amplitude of any side-frequency component, J_n, the following equation should be applied:

$$J_n(m_f) = \left(\frac{m_f}{2}\right)^n \left[\frac{1}{n!} - \frac{(m_f/2)^2}{1!(n+1)!} + \frac{(m_f/2)^4}{2!(n+2)!} - \frac{(m_f/2)^6}{3!(n+3)!} + \cdots\right] \tag{5-6}$$

Thus, solving for these amplitudes is a very tedious process and is strictly dependent on the modulation index, m_f. Note that every component within the brackets is multiplied by the coefficient $(m_f/2)^n$. Table 5-2 gives the solution for several modulation indexes. Notice that for no modulation ($m_f = 0$), the carrier (J_0) is the only frequency present and exists at its full value of 1. However, as the carrier becomes modulated, energy is shifted from the carrier and into the sidebands. For $m_f = 0.25$, the carrier amplitude has dropped to 0.98, and the first side frequencies at $\pm f_i$ around the carrier (J_1) have an amplitude of 0.12. As indicated previously, FM generates an infinite number of sidebands, but in this case, J_2, J_3, J_4, . . . all have negligible value. Thus, an FM transmission with $m_f = 0.25$ requires the same bandwidth ($2f_i$) as an AM broadcast.

Table 5-2 FM Side Frequencies from Bessel Functions

×	(Carrier)	n OR ORDER															
(m_f)	J_0	J_1	J_2	J_3	J_4	J_5	J_6	J_7	J_8	J_9	J_{10}	J_{11}	J_{12}	J_{13}	J_{14}	J_{15}	J_{16}
0.00	1.00	—	—	—	—	—	—	—	—	—	—	—	—	—	—	—	—
0.25	0.98	0.12	—	—	—	—	—	—	—	—	—	—	—	—	—	—	—
0.5	0.94	0.24	0.03	—	—	—	—	—	—	—	—	—	—	—	—	—	—
1.0	0.77	0.44	0.11	0.02	—	—	—	—	—	—	—	—	—	—	—	—	—
1.5	0.51	0.56	0.23	0.06	0.01	—	—	—	—	—	—	—	—	—	—	—	—
2.0	0.22	0.58	0.35	0.13	0.03	—	—	—	—	—	—	—	—	—	—	—	—
2.5	−0.05	0.50	0.45	0.22	0.07	0.02	—	—	—	—	—	—	—	—	—	—	—
3.0	−0.26	0.34	0.49	0.31	0.13	0.04	0.01	—	—	—	—	—	—	—	—	—	—
4.0	−0.40	−0.07	0.36	0.43	0.28	0.13	0.05	0.02	—	—	—	—	—	—	—	—	—
5.0	−0.18	−0.33	0.05	0.36	0.39	0.26	0.13	0.05	0.02	—	—	—	—	—	—	—	—
6.0	0.15	−0.28	−0.24	0.11	0.36	0.36	0.25	0.13	0.06	0.02	—	—	—	—	—	—	—
7.0	0.30	0.00	−0.30	−0.17	0.16	0.35	0.34	0.23	0.13	0.06	0.02	—	—	—	—	—	—
8.0	0.17	0.23	−0.11	−0.29	−0.10	0.19	0.34	0.32	0.22	0.13	0.06	0.03	—	—	—	—	—
9.0	−0.09	0.24	0.14	−0.18	−0.27	−0.06	0.20	0.33	0.30	0.21	0.12	0.06	0.03	0.01	—	—	—
10.0	−0.25	0.04	0.25	0.06	−0.22	−0.23	−0.01	0.22	0.31	0.29	0.20	0.12	0.06	0.03	0.01	—	—
12.0	0.05	−0.22	−0.08	0.20	0.18	−0.07	−0.24	−0.17	0.05	0.23	0.30	0.27	0.20	0.12	0.07	0.03	0.01
15.0	−0.01	0.21	0.04	−0.19	−0.12	0.13	0.21	0.03	−0.17	−0.22	−0.09	0.10	0.24	0.28	0.25	0.18	0.12

Source: E. Cambi, *Bessel Functions,* Dover Publications, Inc., New York, 1948.

Example 5-3

Determine the bandwidth required to transmit an FM signal with $f_i = 10$ kHz and a maximum deviation $\delta = 20$ kHz.

Solution

$$m_f = \frac{\delta}{f_i} = \frac{20\text{ kHz}}{10\text{ kHz}} = 2 \qquad \textbf{(5-4)}$$

From Table 5-2 with $m_f = 2$, the following significant components are obtained:

$$J_0, J_1, J_2, J_3, J_4$$

This means that besides the carrier, J_1 will exist ± 10 kHz around the carrier, J_2 at ± 20 kHz, J_3 at ± 30 kHz, and J_4 at ± 40 kHz. Therefore, the total required bandwidth is 2×40 kHz = 80 kHz.

Example 5-4

Repeat Ex. 5-2 with f_i changed to 5 kHz.

Solution

$$\begin{aligned} m_f &= \frac{\delta}{f_i} \qquad \textbf{(5-4)} \\ &= \frac{20\text{ kHz}}{5\text{ kHz}} \\ &= 4 \end{aligned}$$

In Table 5-2, $m_f = 4$ shows that the highest significant side-frequency component is J_7. Since J_7 will be at $\pm 7 \times 5$ kHz around the carrier, the required BW is 2×35 kHz = 70 kHz.

Examples 5-3 and 5-4 point out a confusing aspect about FM analysis. Deviation and bandwidth are related *but different*. They are related because deviation determines modulation index, which in turn determines significant sideband pairs. The bandwidth, however, is computed by sideband pairs and *not* deviation frequency. Notice in Ex. 5-3 that the maximum deviation was ± 20 kHz, yet the bandwidth was 80 kHz. The deviation is *not* the bandwidth but *does* have an effect on the bandwidth.

Carson's Rule
equation for approximating the bandwidth of an FM signal

An approximation known as **Carson's rule** is often used to predict the bandwidth necessary for an FM signal:

$$\text{BW} \simeq 2(\delta_{\text{max}} + f_{i_{\text{max}}}) \qquad \textbf{(5-7)}$$

This approximation includes about 98 percent of the total power; that is, about 2 percent of the power is in the sidebands outside its predicted BW. Reasonably good fidelity results when limiting the BW to that predicted by Eq. (5-7). Referring back to Ex. 5-3, Carson's rule predicts a BW of 2(20 kHz + 10 kHz) = 60 kHz versus the 80 kHz shown. In Ex. 5-4, BW = 2(20 kHz + 5 kHz) = 50 kHz versus 70 kHz. It should be remembered that even the 70-kHz prediction does not include all of the sidebands created.

Example 5-5

An FM signal, 2000 sin(2π × 10⁸t + 2 sin π × 10⁴t), is applied to a 50-Ω antenna. Determine
(a) The carrier frequency.
(b) The transmitted power.
(c) m_f.
(d) f_i.
(e) BW (by two methods).
(f) Power in the largest and smallest sidebands predicted by Table 5-2.

Solution

(a) By inspection of the FM equation, $f_c = (2\pi \times 10^8)/2\pi = 10^8 = 100$ MHz.
(b) The peak voltage is 2000 V. Thus,

$$P = \frac{(2000/\sqrt{2})^2}{50\ \Omega} = 40\ \text{kW}$$

(c) By inspection of the FM equation, we have

$$m_f = 2 \tag{5-3}$$

(d) The intelligence frequency, f_i, is derived from the sin $\pi\ 10^4 t$ term [Eq. (5-3)]. Thus,

$$f_i = \frac{\pi \times 10^4}{2\pi} = 5\ \text{kHz}$$

(e)

$$m_f = \frac{\delta}{f_i} \tag{5-4}$$

$$2 = \frac{\delta}{5\ \text{kHz}}$$

$$\delta = 10\ \text{kHz}$$

From Table 5-2 with $m_f = 2$, significant sidebands exist to J_4 (4 × 5 kHz = 20 kHz). Thus, BW = 2 × 20 kHz = 40 kHz. Using Carson's rule yields

$$\begin{aligned} \text{BW} &\simeq 2(\delta_{\max} + f_{i\max}) \\ &= 2(10\ \text{kHz} + 5\ \text{kHz}) = 30\ \text{kHz} \end{aligned} \tag{5-7}$$

(f) From Table 5-2, J_1 is the largest sideband at 0.58 times the unmodulated carrier amplitude.

$$p = \frac{(0.58 \times 2000/\sqrt{2})^2}{50\ \Omega} = 13.5\ \text{kW}$$

or 2 × 13.5 kW = 27 kW for the two sidebands at ±5 kHz from the carrier. The smallest sideband, J_4, is 0.03 times the carrier or $(0.03 \times 2000/\sqrt{2})^2/50\ \Omega = 36$ W.

Zero-Carrier Amplitude

Figure 5-4 shows the FM frequency spectrum for various levels of modulation while keeping the modulation frequency constant. The relative amplitude of all components is obtained from Table 5-2. Notice from the table that between $m_f = 2$ and

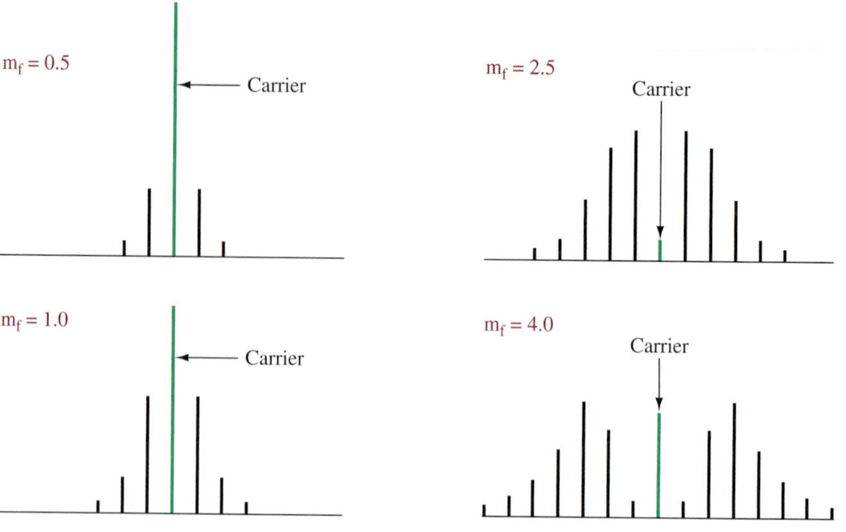

FIGURE 5-4 Frequency spectrum for FM (constant modulating frequency, variable deviation).

$m_f = 2.5$, the carrier goes from a plus to a minus value. The minus sign simply indicates a phase reversal, but when $m_f = 2.4$, the carrier component has zero amplitude and all the energy is contained in the side frequencies. This also occurs when $m_f = 5.5$, 8.65, and between 10 and 12, and 12 and 15.

The zero-carrier condition suggests a convenient means of determining the deviation produced in an FM modulator. A carrier is modulated by a single sine wave at a known frequency. The modulating signal's amplitude is varied while observing the generated FM on a spectrum analyzer. At the point where the carrier amplitude goes to zero, the modulation index, m_f, is determined based on the number of sidebands displayed. If four or five sidebands appear on both sides of the nulled carrier, you can assume that $m_f = 2.4$. The deviation, δ, is then equal to $2.4 \times f_i$. The modulating signal could be increased in amplitude, and the next carrier null should be at $m_f = 5.5$. A check on modulator linearity is thereby possible because the frequency deviation should be directly proportional to the modulating signal's amplitude.

Broadcast FM

Standard broadcast FM uses a 200-kHz bandwidth for each station. This is a very large allocation when one considers that one FM station has a bandwidth that could contain many standard AM stations. Broadcast FM, however, allows for a true high-fidelity modulating signal up to 15 kHz and offers superior noise performance (see Sec. 5-4).

Guard Bands
25-kHz bands at each end of a broadcast FM channel to help minimize interference with adjacent stations

Figure 5-5 shows the FCC allocation for commercial FM stations. The maximum allowed deviation around the carrier is ± 75 kHz, and 25-kHz **guard bands** at the upper and lower ends are also provided. The carrier is required to maintain a ± 2-kHz stability. Recall that an infinite number of side frequencies are generated during frequency modulation, but their amplitude gradually decreases as you move

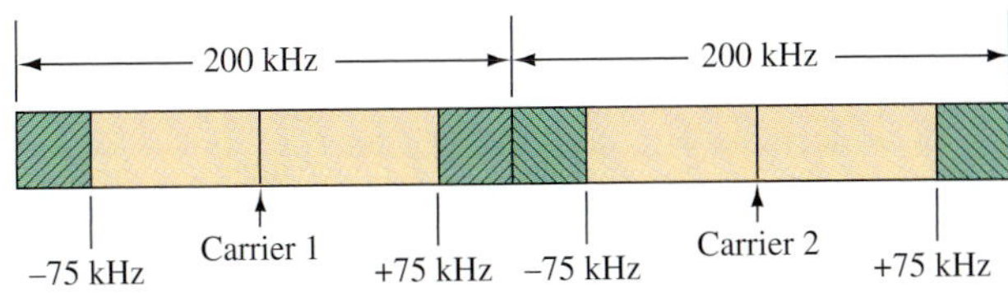

FIGURE 5-5 Commercial FM bandwidth allocations for two adjacent stations.

away from the carrier. In other words, the significant side frequencies exist up to ±75 kHz around the carrier, and the guard bands ensure that adjacent channel interference will not be a problem.

Since full deviation (δ) is 75 kHz, that is 100 percent modulation. By definition, 100 percent modulation in FM is when the deviation is the full permissible amount. Recall that the modulation index, m_f, is

$$m_f = \frac{\delta}{f_i} \qquad \textbf{(5-4)}$$

so that the actual modulation index at 100 percent modulation varies inversely with the intelligence frequency, f_i. This is in contrast with AM, where full or 100 percent modulation means a modulation index of 1 regardless of intelligence frequency.

Deviation Ratio (DR) maximum possible frequency deviation over the maximum input frequency

Another way to describe the modulation index is by **deviation ratio (DR).** Deviation ratio equals the result of dividing the maximum possible frequency deviation by the maximum input frequency, as shown in Eq. (5-8).

$$\text{DR} = \frac{\text{maximum possible frequency deviation}}{\text{maximum input frequency}} = \frac{f_{\text{dev(max)}}}{f_{i(\text{max})}} \qquad \textbf{(5-8)}$$

Deviation ratio is a commonly used term in both television and FM broadcasting. For example, broadcast FM radio permits a maximum carrier frequency deviation, $f_{\text{dev(max)}}$, of ±75 kHz and a maximum audio input frequency, $f_{i(\text{max})}$ of 15 kHz. Therefore, for broadcast FM radio, the deviation ratio (DR) is

$$\text{DR (broadcast FM radio)} = \frac{75 \text{ kHz}}{15 \text{ kHz}} = 5$$

and for broadcast television (NTSC format—see Chapter 17), the maximum frequency deviation of the aural carrier, $f_{\text{dev(max)}}$ is ±25 kHz with a maximum audio input frequency, $f_{i(\text{max})}$, of 15 kHz. Therefore, for broadcast TV (NTSC format), the deviation ratio (DR) is

$$\text{DR (TV NTSC)} = \frac{25 \text{ kHz}}{15 \text{ kHz}} = 1.67$$

Wideband FM a system where the deviation ratio is ≥ 1

FM systems that have a deviation ratio greater than or equal to 1 (DR $\geq$ 1) are considered to be **wideband** systems, whereas FM systems that have a deviation ratio less than 1 (DR $<$ 1) are considered to be narrowband FM systems.

Narrowband FM

Narrowband FM
FM signals used for voice transmissions such as public service communication systems

Frequency modulation is also widely used in communication (i.e., not to entertain) systems such as those used by police, aircraft, taxicabs, weather service, and private industry networks. These applications are often voice transmissions, which means that intelligence frequency maximums of 3 kHz are the norm. These are **narrowband FM** systems because Federal Communications Commission (FCC) bandwidth allocations of 10 to 30 kHz are provided. Narrowband FM (NBFM) systems operate with a modulation index of 0.5 to 1.0. A glance at the Bessel functions in Table 5-2 shows that at these index values, only the first set (J_1) of side frequencies has a significant amplitude; the second (J_2) and third (J_3) lose amplitude quickly. Thus, we see that NBFM has a bandwidth no wider than an AM signal.

Example 5-6

(a) Determine the permissible range in maximum modulation index for commercial FM that has 30-Hz to 15-kHz modulating frequencies.
(b) Repeat for a narrowband system that allows a maximum deviation of 1-kHz and 100-Hz to 2-kHz modulating frequencies.
(c) Determine the deviation ratio for the system in part (b).

Solution

(a) The maximum deviation in broadcast FM is 75 kHz.

$$m_f = \frac{\delta}{f_i} \tag{5-4}$$

$$= \frac{75\text{ kHz}}{30\text{ Hz}} = 2500$$

For $f_i = 15$ kHz:

$$m_f = \frac{75\text{ kHz}}{15\text{ kHz}} = 5$$

(b)

$$m_f = \frac{\delta}{f_i} = \frac{1\text{ kHz}}{100\text{ Hz}} = 10$$

For $f_i = 2$ kHz:

$$m_f = \frac{1\text{ kHz}}{2\text{ kHz}} = 0.5$$

(c)

$$\text{DR} = \frac{f_{\text{dev(max)}}}{f_{i\text{(max)}}} = \frac{1\text{ kHz}}{2\text{ kHz}} = 0.5 \tag{5-8}$$

Example 5-7

Determine the relative total power of the carrier and side frequencies when $m_f = 0.25$ for a 10-kW FM transmitter.

Solution

For $m_f = 0.25$, the carrier is equal to 0.98 times its unmodulated amplitude and the only significant sideband is J_1, with a relative amplitude of 0.12 (from Table 5-2). Therefore, because power is proportional to the voltage squared, the carrier power is

$$(0.98)^2 \times 10 \text{ kW} = 9.604 \text{ kW}$$

and the power of each sideband is

$$(0.12)^2 \times 10 \text{ kW} = 144 \text{ W}$$

The total power is

$$\begin{aligned} 9604 \text{ W} + 144 \text{ W} + 144 \text{ W} &= 9.892 \text{ kW} \\ &\cong 10 \text{ kW} \end{aligned}$$

The result of Ex. 5-7 is predictable. In FM, the transmitted waveform never varies in amplitude, just frequency. Therefore, the total transmitted power must remain constant regardless of the level of modulation. It is thus seen that whatever energy is contained in the side frequencies has been obtained from the carrier. No additional energy is added during the modulation process. The carrier in FM is not redundant as in AM because its (the carrier's) amplitude is dependent on the intelligence signal.

The final power stage of a 20-kW amplifier uses a tube as the active element. (Courtesy of ETO, Inc.—An ASTEX Company.)

5-4 Noise Suppression

The most important advantage of FM over AM is the superior noise characteristics. You are probably aware that static noise is rarely heard on FM, although it is quite common in AM reception. You may be able to guess a reason for this improvement. The addition of noise to a received signal causes a change in its amplitude. Since the amplitude changes in AM contain the intelligence, any attempt to get rid of the noise adversely affects the received signal. However, in FM, the intelligence is *not* carried by amplitude changes but instead by frequency changes. The spikes of external noise picked up during transmission are clipped off by a **limiter** circuit and/or through the use of detector circuits that are insensitive to amplitude changes. Chapter 6 provides more detailed information on these FM receiver circuits.

Limiter
stage in an FM receiver that removes any amplitude variations of the received FM signal before it reaches the discriminator

Figure 5-6(a) shows the noise removal action of an FM limiter circuit, while in Fig. 5-6(b) the noise spike feeds right through to the speaker in an AM system. The advantage for FM is clearly evident; in fact, you may think that the limiter removes all the effects of this noise spike. While it is possible to clip the noise spike off, it still causes an undesired phase shift and thus frequency shift of the FM signal, and this frequency shift *cannot* be removed.

The noise signal frequency will be close to the frequency of the desired FM signal due to the selective effect of the tuned circuits in a receiver. In other words, if you are tuned to an FM station at 96 MHz, the receiver's selectivity provides gain only for frequencies near 96 MHz. The noise that will affect this reception must, therefore, also be around 96 MHz because all other frequencies will be greatly attenuated. The effect of adding the desired *and* noise signals will give a resultant signal with a different phase angle than the desired FM signal alone. Therefore, the noise signal, even though it is clipped off in amplitude, will cause phase modulation (PM), which indirectly causes undesired FM. The amount of frequency deviation (FM) caused by PM is

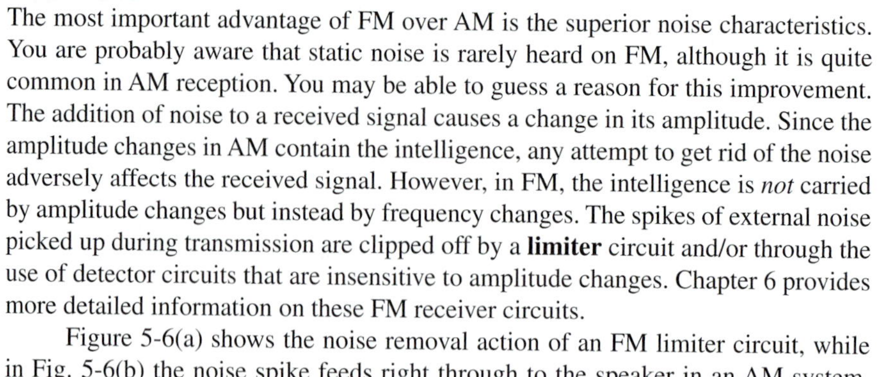

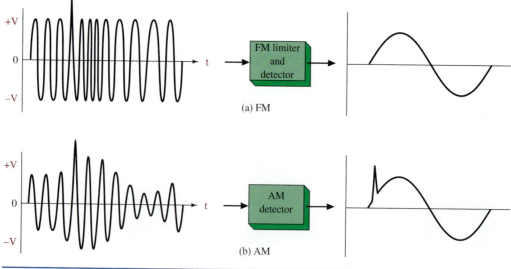

FIGURE 5-6 FM, AM noise comparison.

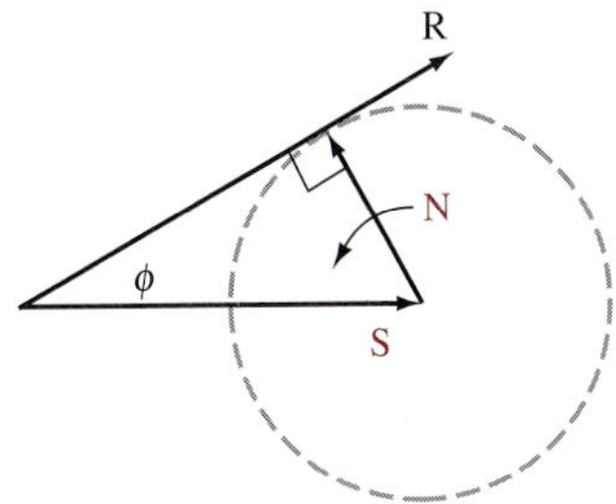

FIGURE 5-7 Phase shift (ϕ) as a result of noise.

$$\delta = \phi \times f_i \tag{5-9}$$

where δ = frequency deviation
ϕ = phase shift (radians)
f_i = frequency of intelligence signal

FM Noise Analysis

The phase shift caused by the noise signal results in a frequency deviation that is predicted by Eq. (5-9). Consider the situation illustrated in Fig. 5-7. Here the noise signal is one-half the desired signal amplitude, which provides a voltage *S/N* ratio of 2:1. This is an intolerable situation in AM but, as the following analysis will show, is not so bad in FM.

Because the noise (*N*) and desired signal (*S*) are at different frequencies (but in the same range, as dictated by a receiver's tuned circuits), the noise is shown as a rotating vector using the *S* signal as a reference. The phase shift of the resultant (*R*) is maximum when *R* and *N* are at right angles to one another. At this worst-case condition

$$\phi = \sin^{-1}\frac{N}{S} = \sin^{-1}\frac{1}{2}$$
$$= 30°$$

or 30°/(57.3° per radian) = 0.52 rad, or about $\frac{1}{2}$ rad.

If the intelligence frequency, f_i, were known, then the deviation (δ) caused by this severe noise condition could now be calculated using Eq. (5-9). Since $\delta = \phi \times f_i$, the worst-case deviation occurs for the maximum intelligence frequency. Assuming an f_i maximum of 15 kHz, the absolute worst case δ due to this severe noise signal is

$$\delta = \phi \times f_i = 0.5 \times 15 \text{ kHz} = 7.5 \text{ kHz}$$

In standard broadcast FM, the maximum modulating frequency is 15 kHz and the maximum allowed deviation is 75 kHz above and below the carrier. Thus, a 75-kHz deviation corresponds to maximum modulating signal amplitude and full volume at the receiver's output. The 7.5-kHz worst-case deviation output due to the $S/N = 2$ condition is

$$\frac{7.5 \text{ kHz}}{75 \text{ kHz}} = \frac{1}{10}$$

and, therefore, the 2:1 signal-to-noise ratio results in an output signal-to-noise ratio of 10:1. This result assumes that the receiver's internal noise is negligible. Thus, FM is seen to exhibit a very strong capability to nullify the effects of noise! In AM, a 2:1 signal-to-noise ratio at the input essentially results in the same ratio at the output. Thus, FM is seen to have an inherent noise reduction capability not possible with AM.

Example 5-8

Determine the worst-case output S/N for a broadcast FM program that has a maximum intelligence frequency of 5 kHz. The input S/N is 2.

Solution

The input $S/N = 2$ means that the worst-case deviation is about $\frac{1}{2}$ rad (see the preceding paragraphs). Therefore.

$$\begin{aligned} \delta &= \phi \times f_i \\ &= 0.5 \times 5 \text{ kHz} = 2.5 \text{ kHz} \end{aligned} \tag{5-9}$$

Because full volume in broadcast FM corresponds to a 75-kHz deviation, this 2.5-kHz worst-case noise deviation means that the output S/N is

$$\frac{75 \text{ kHz}}{2.5 \text{ kHz}} = 30$$

Example 5-8 shows that the inherent noise reduction capability of FM is improved when the maximum intelligence (modulating) frequency is reduced. A little thought shows that this capability can also be improved by increasing the maximum allowed frequency deviation from the standard 75-kHz value. An increase in allowed deviation means that increased bandwidths for each station would be necessary, however. In fact, many FM systems utilized as communication links operate with decreased bandwidths—narrowband FM systems. It is typical for them to operate with a 10-kHz maximum deviation. The inherent noise reduction of these systems is reduced by the lower allowed δ but is somewhat offset by the lower maximum modulating frequency of 3 kHz usually used for voice transmissions.

Example 5-9

Determine the worst-case output S/N for a narrowband FM receiver with $\delta_{max} = 10$ kHz and a maximum intelligence frequency of 3 kHz. The S/N input is 3:1.

Solution

The worst-case phase shift (ϕ) due to the noise occurs when $\phi = \sin^{-1}(N/S)$.

$$\phi = \sin^{-1}\frac{1}{3} = 19.5°, \text{ or } 0.34 \text{ rad}$$

and

$$\begin{aligned} \delta &= \phi \times f_i \\ &= 0.34 \times 3 \text{ kHz} \cong 1 \text{ kHz} \end{aligned} \tag{5-9}$$

The *S/N* output will be

$$\frac{10 \text{ kHz}}{1 \text{ kHz}} = 10$$

and thus the input signal-to-noise ratio of 3 is transformed to 10 or higher at the output.

Capture Effect

This inherent ability of FM to minimize the effect of undesired signals (noise in the preceding paragraphs) also applies to the reception of an undesired station operating at the same or nearly the same frequency as the desired station. This is known as the **capture effect.** You may have noticed when riding in a car that an FM station is suddenly replaced by a different one. You may even find that the receiver alternates abruptly back and forth between the two. This occurs because the two stations are presenting a variable signal as you drive. The capture effect causes the receiver to lock on the stronger signal by suppressing the weaker but can fluctuate back and forth when the two are nearly equal. When they are not nearly equal, however, the inherent FM noise suppression action is very effective in preventing the interference of an unwanted (but weaker) station. The weaker station is suppressed just as noise was in the preceding noise discussion. FM receivers typically have a **capture ratio** of 1 dB—this means suppression of a 1-dB (or more) weaker station is accomplished. In AM, it is not uncommon to hear two separate broadcasts at the same time, but this is certainly a rare occurrence with FM.

Capture Effect
an FM receiver phenomenon that involves locking onto the stronger of two received signals of the same frequency and suppressing the weaker signal

Capture Ratio
the necessary difference (in dB) of signal strength to allow suppression of a weaker signal from a stronger one

The capture effect can also be illustrated by Fig. 5-8. Notice that the *S/N* before and after demodulation for SSB and AM is linear. Assuming noiseless demodulation schemes, SSB (and DSB) has the same *S/N* at the detector's input and output. The degradation shown for AM is due to so much of the signal's power being wasted in the redundant carrier. FM systems with m_f greater than 1 show an actual improvement in *S/N*, as illustrated in Exs. 5-8 and 5-9. For example, consider $m_f = 5$ in Fig. 5-8. When *S/N* before demodulation is 20, the *S/N* after demodulation is about 38—a significant improvement.

Insight into the capture effect is provided by consideration of the inflection point (often termed **threshold**) shown in Fig. 5-8. Notice that a rapid degradation

Threshold
in FM, the point where *S/N* in the output rapidly degrades as *S/N* of received signal is degrading

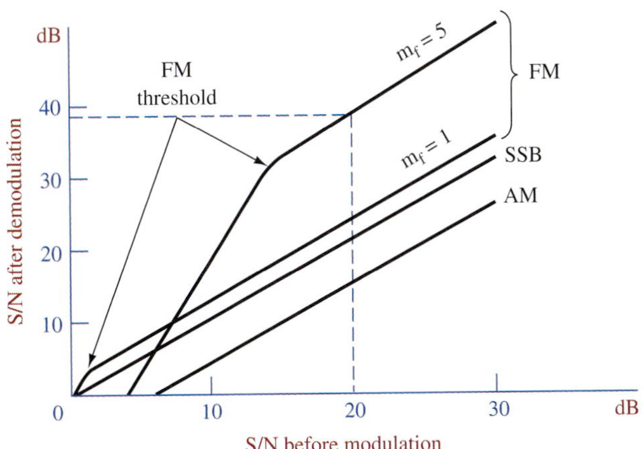

FIGURE 5-8 *S/N* for basic modulation schemes.

in *S/N* after demodulation results when the noise approaches the same level as the desired signal. This threshold situation is noticeable when driving in a large city. The fluttering noise often heard is caused when the FM signal is reflecting off various structures. The signal strength fluctuates widely due to the additive or subtractive effects on the total received signal. The effect can cause the output to blank out totally and resume at a rapid rate as the *S/N* before demodulation moves back and forth through the threshold level.

Preemphasis

The noise suppression ability of FM has been shown to decrease with higher intelligence frequencies. This is unfortunate because the higher intelligence frequencies tend to be of lower amplitude than the low frequencies. Thus, a high-pitched violin note that the human ear may perceive as the same "sound" level as the crash of a bass drum may have only half the electrical amplitude as the low-frequency drum signal. In FM, half the amplitude means half the deviation and, subsequently, half the noise reduction capability. To counteract this effect, almost all FM transmissions provide an artificial boost to the electrical amplitude of the higher frequencies. This process is termed *preemphasis*.

Preemphasis
process in an FM transmitter that amplifies high-frequency more than low-frequency audio signals to reduce the effect of noise

Deemphasis
process in an FM receiver that reduces the amplitudes of high-frequency audio signals down to their original values to counteract the effect of the preemphasis network in the transmitter

By definition, **preemphasis** involves increasing the relative strength of the high-frequency components of the audio signal before it is fed to the modulator. Thus, the relationship between the high-frequency intelligence components and the noise is altered. While the noise remains the same, the desired signal strength is increased.

A potential disadvantage, however, is that the natural balance between high- and low-frequency tones at the receiver would be altered. A **deemphasis** circuit in the receiver, however, corrects this defect by reducing the high-frequency audio by the same amount as the preemphasis circuit increased it, thus regaining the original tonal balance. In addition, the deemphasis network operates on both the high-frequency signal and the high-frequency noise; therefore, there is no change in the improved signal-to-noise ratio. The main reason for the preemphasis network, then, is to prevent the high-frequency components of the transmitted intelligence from being degraded by noise that would otherwise have more effect on the higher than on the lower intelligence frequencies.

The deemphasis network is normally inserted between the detector and the audio amplifier in the receiver. This ensures that the audio frequencies are returned to their original relative level before amplification. The preemphasis characteristic curve is flat up to 500 Hz, as shown in Fig. 5-9. From 500 to 15,000 Hz, there is a sharp increase in gain up to approximately 17 dB. The gain at these frequencies is necessary to maintain the signal-to-noise ratio at high audio frequencies. The frequency characteristic of the deemphasis network is directly opposite to that of the preemphasis network. The high-frequency response decreases in proportion to its increase in the preemphasis network. The characteristic curve of the deemphasis circuit should be a mirror image of the preemphasis characteristic curve. Figure 5-9 shows the pre- and deemphasis curves as used by standard FM broadcasts in the United States. As shown, the 3-dB points occur at 2120 Hz, as predicted by the *RC* time constant (τ) of 75 μs used to generate them.

$$f = \frac{1}{2\pi RC} = \frac{1}{2\pi \times 75\ \mu\text{s}} = 2120\text{ Hz}$$

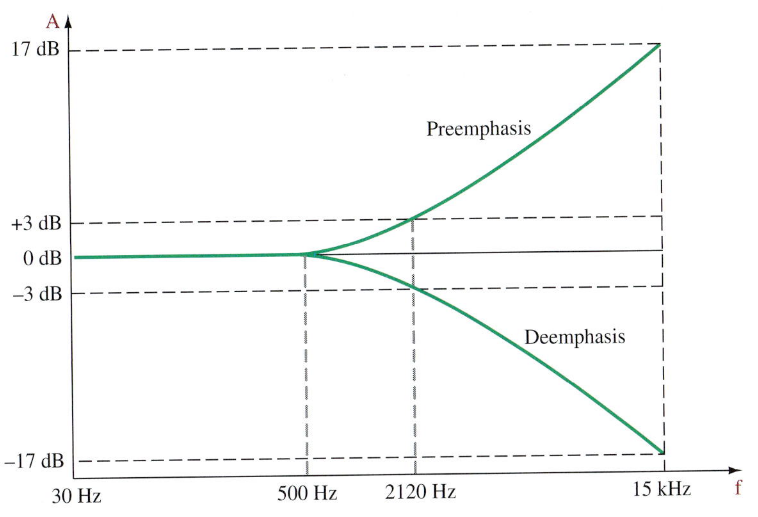

FIGURE 5-9 Emphasis curves ($\tau = 75\ \mu$s).

Figure 5-10(a) shows a typical preemphasis circuit. The impedance to the audio voltage is mainly that of the parallel circuit of C and R_1 because the effect of R_2 is small in comparison to that of either C or R_1. Since capacitive reactance is inversely proportional to frequency, audio frequency increases cause the reactance of C to decrease. This decrease of X_C provides an easier path for high frequencies as compared to R. Thus, with an increase of audio frequency, there is an increase in signal voltage. The result is a larger voltage drop across R_2 (the amplifier's input) at the higher frequencies and thus greater output.

Figure 5-10(b) depicts a typical deemphasis network. Note the physical position of R and C in relation to the base of the transistor. As the frequency of the audio signal increases, the reactance of capacitor C decreases. The voltage division

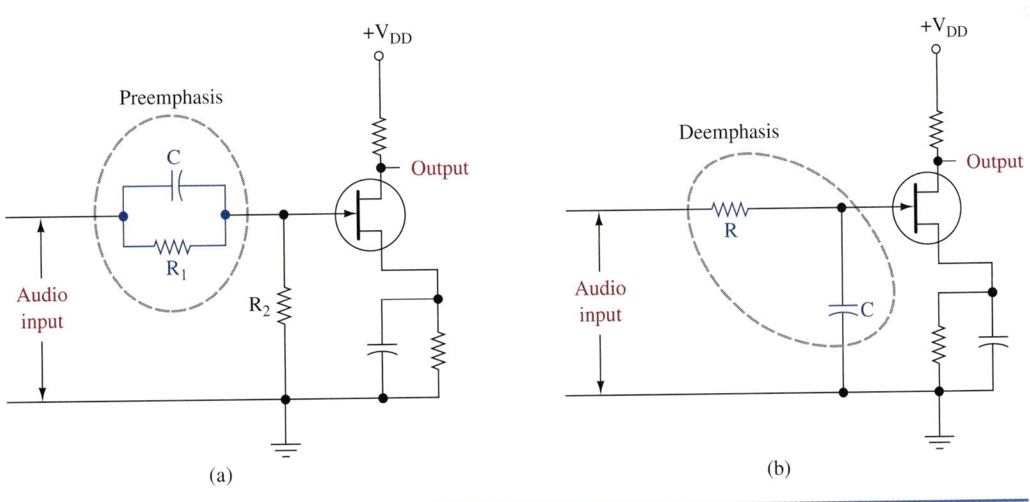

FIGURE 5-10 Emphasis circuits.

between R and C now provides a smaller drop across C. The audio voltage applied to the base decreases; therefore, a reverse of the preemphasis circuit is accomplished. For the signal to be exactly the same as before preemphasis and deemphasis, the time constants of the two circuits must be equal to each other.

Dolby System

Dolby System
advanced noise reduction system in FM systems in which the preemphasis and deemphasis networks work in a dynamic manner

The FCC has ruled that FM broadcast stations can, if desired, use a 25-μs time constant. They then must use the **Dolby system,** which works like preemphasis but in a dynamic fashion. The amount of preemphasis (and subsequent deemphasis in a Dolby receiver) varies depending on the loudness level at any instant. Maximum noise reduction of this system is realized by listeners in fringe areas where noise effects are most detrimental. Those people with regular 75-μs, non-Dolby receivers do not seem to be adversely affected but will usually turn down their treble control somewhat.

The Dolby system used by FM transmitters is not concerned primarily with noise reduction, however. Standard 75-μs preemphasis provides the same amount of boost to both weak and strong high-frequency signals. This created no real problem years ago because few, if any, high frequencies were strong anyway. The better-quality recordings of today do offer some fairly strong high frequencies. Their strength, combined with 75-μs preemphasis, now causes excessive bandwidth of the transmitted FM signal, and thus a station is forced to lower the strength of all signals. To counteract this effect, the transmitting station will artificially reduce the strength of high-frequency signals, resulting in a less-than-true signal reproduction (and less dynamic range) at the receiver.

The Dolby solution to this problem is to provide varying degrees of high-frequency boost, as depicted in Fig. 5-11. The Dolby receiver must reverse these characteristics and thus requires some relatively complex circuitry. Notice how the weak high frequencies are given a significantly greater boost than the stronger ones. The overall result of this system is a stronger received FM signal with greater

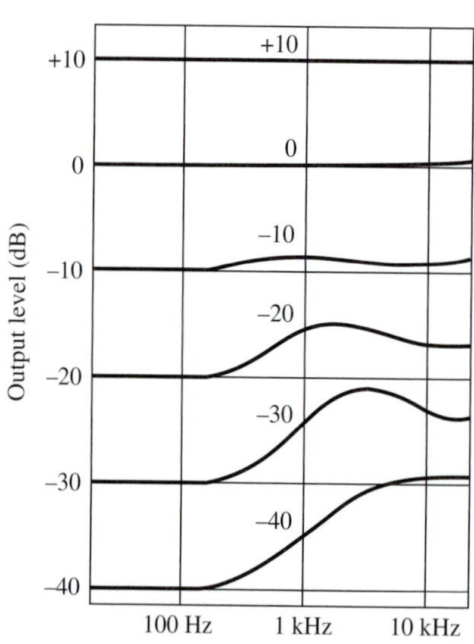

FIGURE 5-11 Dolby dynamic preemphasis.

dynamic range as compared to a broadcast signal, which unnaturally attenuates the high frequencies. The Dolby system is also highly effective in minimizing high-frequency tape hiss noise in tape players.

5-5 Direct FM Generation

The capacitance microphone system explained in Sec. 5-2 can be used to generate FM directly. Recall that the capacitance of the microphone varies in step with the sound wave striking it. Figure 5-1 showed that if the amplitude of sound striking it is increased, the amount of deviation of the oscillator's frequency is increased. If the frequency of sound waves is increased, the rate of the oscillator's deviation is increased. This system is useful in explaining the principles of an FM signal but is not normally used to generate FM in practical systems. It is not able to produce enough deviation as required in actual applications.

Varactor Diode

A varactor diode may be used to generate FM directly. All reverse-biased diodes exhibit a junction capacitance that varies inversely with the amount of reverse bias, as was shown in Fig. 3-11. A diode that is physically constructed to enhance this characteristic is termed a **varactor diode.** Figure 5-12 shows a schematic of a varactor diode modulator. With no intelligence signal (E_i) applied, the parallel combination of C_1, L_1, and D_1's capacitance forms the resonant carrier frequency. The diode D_1 is effectively in parallel with L_1 and C_1 because the $-V_{CC}$ supply appears as a short circuit to the ac signal. The coupling capacitor, C_C, isolates the dc levels and intelligence signal while looking like a short to the high-frequency carrier.

Varactor Diode
diode with a small internal capacitance that varies as a function of its reverse bias voltage

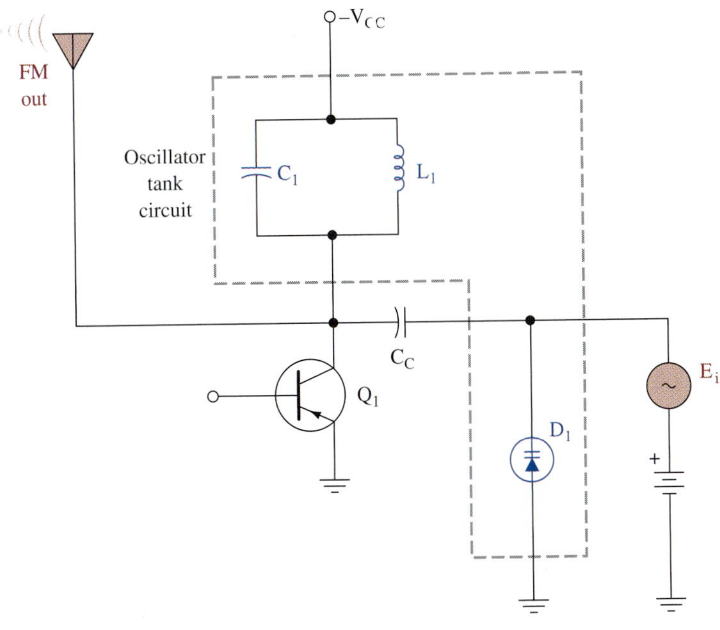

FIGURE 5-12 Varactor diode modulator.

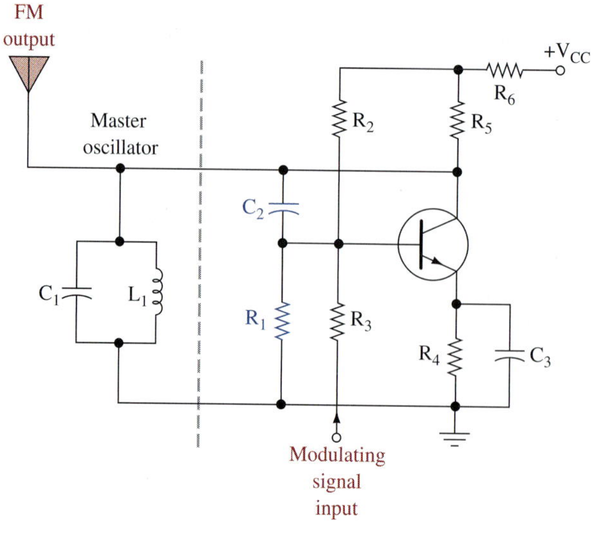

FIGURE 5-13 Reactance modulator.

When the intelligence signal, E_i, is applied to the varactor diode, its reverse bias is varied, which causes the diode's junction capacitance to vary in step with E_i. The oscillator frequency is subsequently varied as required for FM, and the FM signal is available at Q_1's collector. For simplicity, the dc bias and oscillator feedback circuitry is not shown in Fig. 5-12. While the varactor diode modulator can be called a **reactance modulator,** the term is usually applied to those in which an active device is made to look like a variable reactance. The reactance modulator (see Fig. 5-13) is a very popular means of FM generation. The troubleshooting section at the end of this chapter provides details on the operation and repair of a reactance modulator.

Reactance Modulator
amplifier designed so that its input impedance has a reactance that varies as a function of the amplitude of the applied input voltage

LIC VCO FM Generation

Voltage-Controlled Oscillator
designed so that its output voltage varies as a function of the amplitude of the applied input voltage

A **voltage-controlled oscillator** (VCO) produces an output frequency that is directly proportional to a control voltage level. The circuitry necessary to produce such an oscillator with a high degree of linearity between control voltage and frequency was formerly prohibitive on a discrete component basis, but now that low-cost monolithic LIC VCOs are available, they make FM generation extremely simple. Fig. 5-14 provides the specifications for the Philips Semiconductors NE/SE566 VCO. The circuit shown at the end of the specifications provides a high-quality FM generator with the modulating voltage applied to C_2. The FM output can be taken at pin 4 (triangle wave) or pin 3 (square wave). Feeding either of these two outputs into an *LC* tank circuit resonant at the VCO center frequency (i.e., carrier) will subsequently provide a standard sinusoidal FM signal by the flywheel effect.

Crosby Modulator

Now that three practical methods of FM generation have been shown—varactor diode, reactance modulator, and the VCO—it is time to consider the weakness of these methods. Notice that in no case was a crystal oscillator used as the basic reference or carrier frequency. The stability of the carrier frequency is very tightly

Function generator NE/SE566

DESCRIPTION

The NE/SE566 Function Generator is a voltage-controlled oscillator of exceptional linearity with buffered square wave and triangle wave outputs. The frequency of oscillation is determined by an external resistor and capacitor and the voltage applied to the control terminal. The oscillator can be programmed over a ten-to-one frequency range by proper selection of an external resistance and modulated over a ten-to-one range by the control voltage, with exceptional linearity.

FEATURES

- Wide range of operating voltage (up to 24V; single or dual)
- High linearity of modulation
- Highly stable center frequency (200ppm/•C typical)
- Highly linear triangle wave output
- Frequency programming by means of a resistor or capacitor, voltage or current
- Frequency adjustable over 10-to-1 range with same capacitor

PIN CONFIGURATIONS

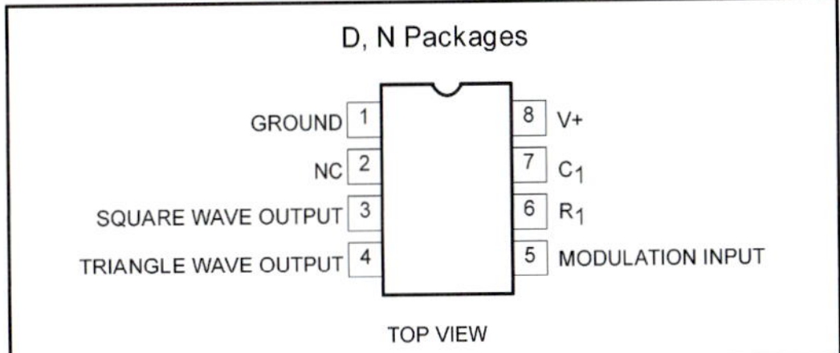

APPLICATIONS

- Tone generators
- Frequency shift keying
- FM modulators
- Clock generators
- Signal generators
- Function generators

ORDERING INFORMATION

DESCRIPTION	TEMPERATURE RANGE	ORDER CODE	DWG #
8-Pin Plastic Small Outline (SO) Package	0 to +70•C	NE566D	0174C
14-Pin Ceramic Dual In-Line Package (CERDIP)	0 to +70•C	NE566F	0581B
8-Pin Plastic Dual In-Line Package (DIP)	0 to +70•C	NE566N	0404B
8-Pin Plastic Dual In-Line Package (DIP)	-55•C to +125•C	SE566N	0404B

BLOCK DIAGRAM

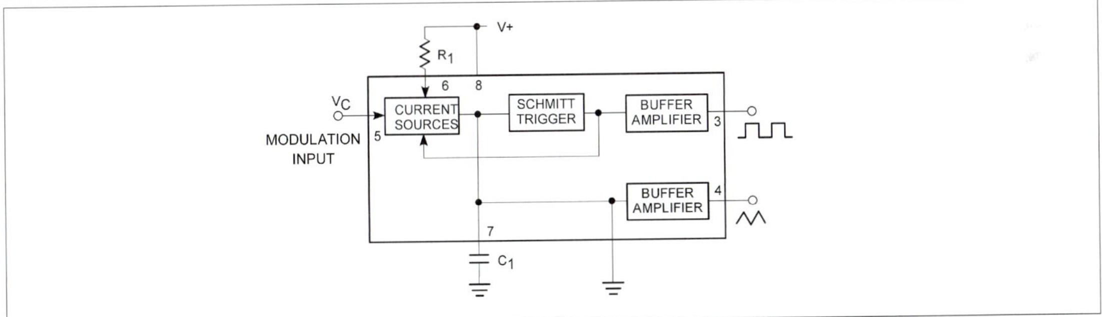

(Continued)

FIGURE 5-14 The NE/SE566 VCO specifications. (Courtesy of Philips Semiconductors.)

EQUIVALENT SCHEMATIC

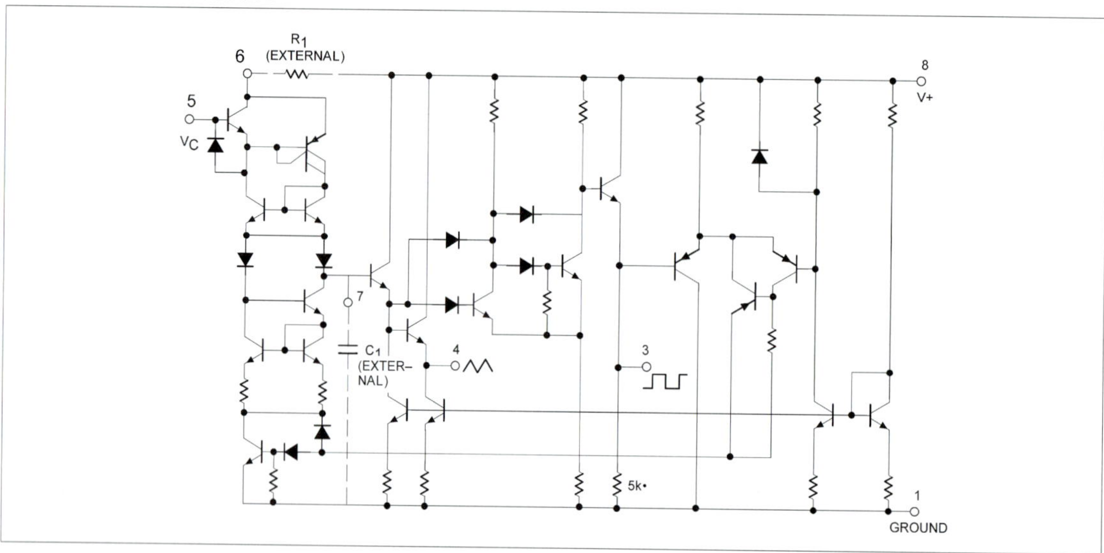

ABSOLUTE MAXIMUM RATINGS

SYMBOL	PARAMETER	RATING	UNIT
V+	Maximum operating voltage	26	V
V_{IN}, V_C	Input voltage	3	V_{P-P}
T_{STG}	Storage temperature range	-65 to +150	°C
T_A	Operating ambient temperature range		
	NE566	0 to +70	°C
	SE566	-55 to +125	°C
P_D	Power dissipation	300	mW

TYPICAL PERFORMANCE CHARACTERISTICS

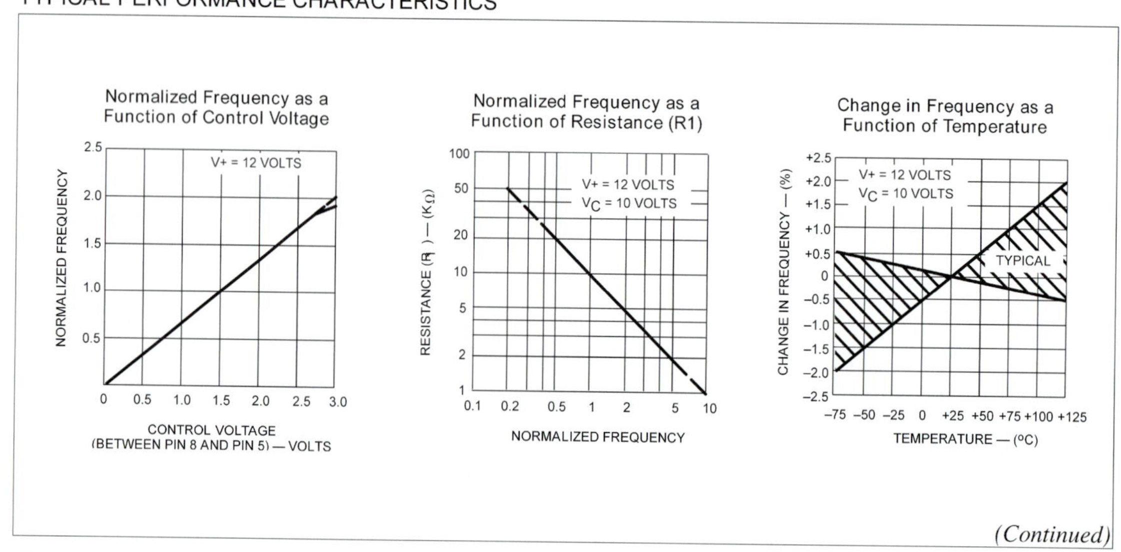

(Continued)

FIGURE 5-14 *(Continued)*

TYPICAL PERFORMANCE CHARACTERISTICS *(Continued)*

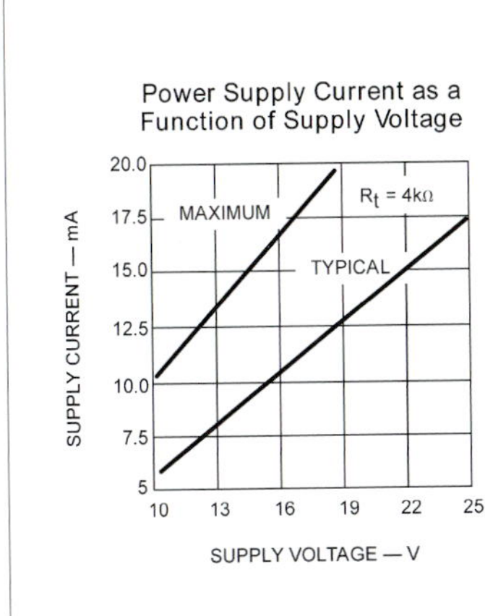

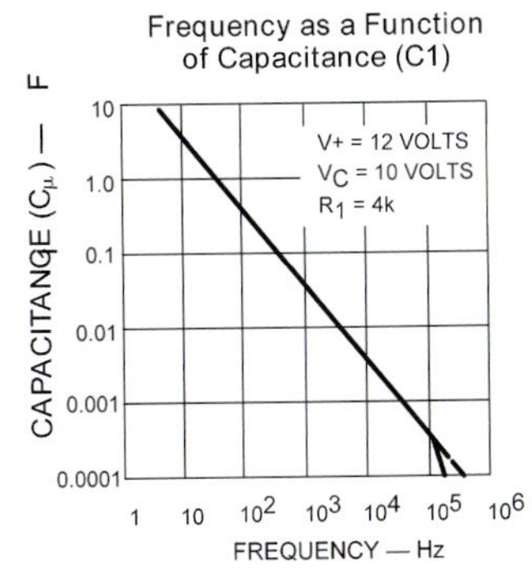

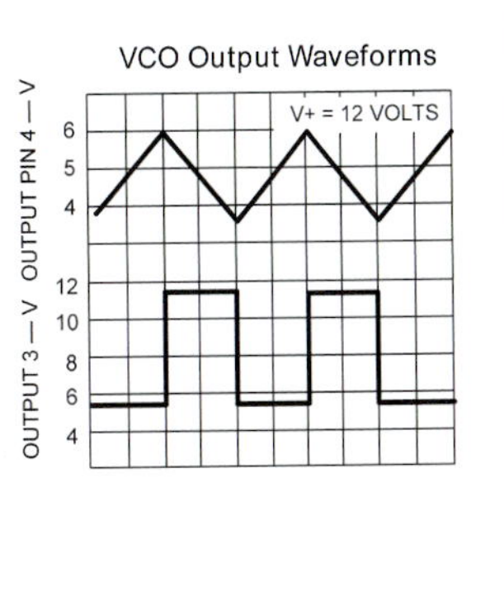

OPERATING INSTRUCTIONS

The NE/SE566 Function Generator is a general purpose voltage-controlled oscillator designed for highly linear frequency modulation. The circuit provides simultaneous square wave and triangle wave outputs at frequencies up to 1MHz. A typical connection diagram is shown in Figure 1. The control terminal (Pin 5) must be biased externally with a voltage (V_C) in the range

$$V+ \leq V_C \leq V+$$

where V_{CC} is the total supply voltage. In Figure 1, the control voltage is set by the voltage divider formed with R_2 and R_3. The modulating signal is then AC coupled with the capacitor C_2. The modulating signal can be direct coupled as well, if the appropriate DC bias voltage is applied to the control terminal. The frequency is given approximately by

$$f_O = \frac{2\,[(V+) - (V_C)]}{R_1\,C_1\,V+}$$

and R_1 should be in the range 2kΩ< R_1<20kΩ.

A small capacitor (typically 0.001μF) should be connected between Pins 5 and 6 to eliminate possible oscillation in the control current source.

If the VCO is to be used to drive standard logic circuitry, it may be desirable to use a dual supply as shown in Figure 2. In this case the square wave output has the proper DC levels for logic circuitry. RTL can be driven directly from Pin 3. For DTL or TTL gates, which require a current sink of more than 1mA, it is usually necessary to connect a 5kΩ resistor between Pin 3 and negative supply. This increases the current sinking capability to 2mA. The third type of interface shown uses a saturated transistor between the 566 and the logic circuitry. This scheme is used primarily for TTL circuitry which requires a fast fall time (<50ns) and a large current sinking capability.

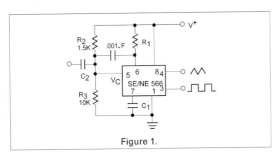

Figure 1.

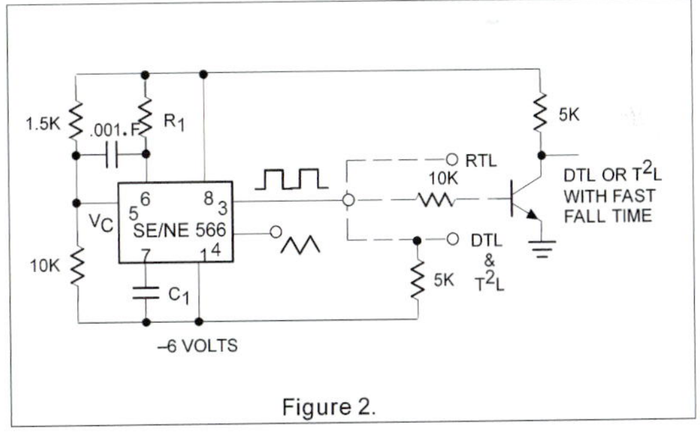

Figure 2.

FIGURE 5-14 *(Continued)*

controlled by the FCC, and that stability is not attained by any of the methods described thus far. Because of the high Q of crystal oscillators, it is not possible to frequency-modulate them directly—their frequency cannot be made to deviate sufficiently to provide workable wideband FM systems. It is possible to modulate directly a crystal oscillator in some narrowband applications. If a crystal is modulated to a deviation of ±50 Hz around a 5-MHz center frequency and both are

Automatic Frequency Control
negative feedback control system used in Crosby FM transmitters to achieve high-frequency stability of the carrier

Crosby Systems
FM systems using direct FM modulation with AFC to control for carrier drift

Frequency Multipliers
amplifiers designed so that the output signal's frequency is an integer multiple of the input frequency

multiplied by 100, a narrowband system with a 500-MHz carrier ±5-kHz deviation results. One method of circumventing this dilemma for wideband systems is to provide some means of **automatic frequency control** (AFC) to correct any carrier drift by comparing it to a reference crystal oscillator.

FM systems utilizing direct generation with AFC are called **Crosby systems.** A Crosby direct FM transmitter for a standard broadcast station at 90 MHz is shown in Fig. 5-15. Notice that the reactance modulator starts at an initial center frequency of 5 MHz and has a maximum deviation of ±4.167 kHz. This is a typical situation because reactance modulators cannot provide deviations exceeding about ±5 kHz and still offer a high degree of linearity (i.e., Δf directly proportional to the modulating voltage amplitude). Consequently, **frequency multipliers** are utilized to provide a ×18 multiplication up to a carrier frequency of 90 MHz (18 × 5 MHz) with a ±75-kHz (18 × 4.167 kHz) deviation. Notice that both the carrier and deviation are multiplied by the multiplier.

Frequency multiplication is normally obtained in steps of ×2 or ×3 (doublers or triplers). The principle involved is to feed a frequency rich in harmonic distortion (i.e., from a class C amplifier) into an *LC* tank circuit tuned to two or three

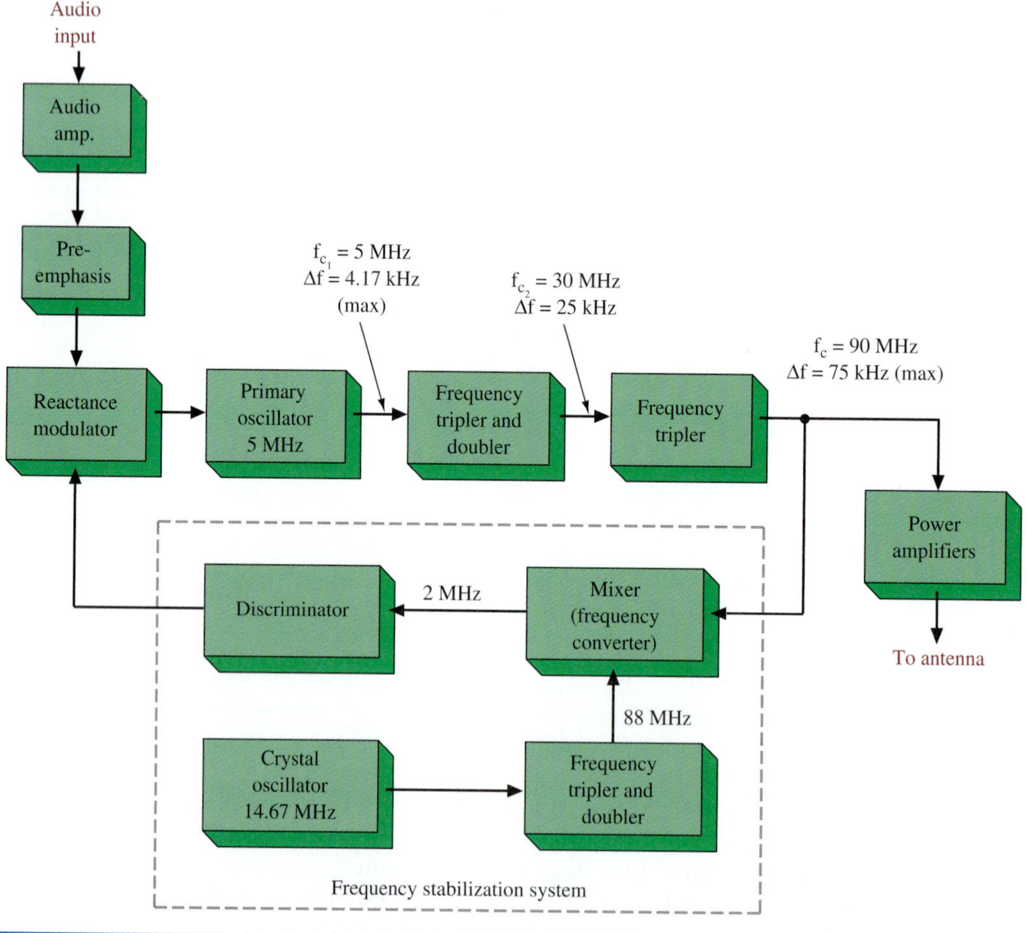

FIGURE 5-15 Crosby direct FM transmitter.

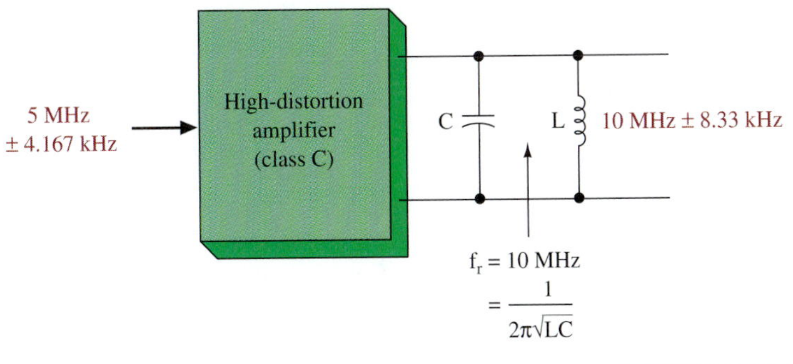

FIGURE 5-16 Frequency multiplication (doubler).

times the input frequency. The harmonic is then the only significant output, as illustrated in Fig. 5-16.

After the ×18 multiplication (3 × 2 × 3) shown in Fig. 5-15, the FM *exciter* function is complete. The term **exciter** is often used to denote the circuitry that generates the modulated signal. The excited output goes to the power amplifiers for transmission *and* to the frequency stabilization system. The purpose of this system is to provide a control voltage to the reactance modulator whenever it drifts from its desired 5-MHz value. The control (AFC) voltage then varies the reactance of the primary 5-MHz oscillator slightly to bring it back on frequency.

Exciter
stages necessary in a transmitter to create the modulated signal before subsequent amplification

The mixer in Fig. 5-15 has the 90-MHz carrier and 88-MHz crystal oscillator signal as inputs. The mixer output accepts only the difference component of 2 MHz, which is fed to the discriminator. A **discriminator** is the opposite of a VCO, because it provides a dc level output based on the frequency input. The discriminator output in Fig. 5-15 will be zero if it has an input of exactly 2 MHz, which occurs when the transmitter is at precisely 90 MHz. Any carrier drift up or down causes the discriminator output to go positive or negative, resulting in the appropriate primary oscillator readjustment. Further detail on discriminator circuits is provided in Chapter 6.

Discriminator
stage in an FM receiver that creates an output dc level that varies as a function of its input frequency

5-6 Indirect FM Generation

If the phase of a crystal oscillator's output is varied, phase modulation (PM) will result. As discussed previously, changing the phase of a signal indirectly causes its frequency to be changed. We thus find that direct modulation of a crystal is possible via PM, which indirectly creates FM. This indirect method of FM generation is usually referred to as the **Armstrong** type, after its originator, E. H. Armstrong. It permits modulation of a stable crystal oscillator without the need for the cumbersome AFC circuitry and also provides carrier accuracies identical to the crystal accuracy, as opposed to the slight degradation of the Crosby system's accuracy.

Armstrong Modulator
FM transmitter that uses a phase modulator that feeds the intelligence signal through a low-pass filter integrator network to convert PM to FM

A simple Armstrong modulator is depicted in Fig. 5-17. The JFET is biased in the ohmic region by keeping V_{DS} low. In that way it presents a resistance from drain to source that is made variable by the gate voltage (the modulating signal). In the ohmic region, the drain-to-source resistance for a JFET transistor behaves like a voltage-controlled resistance (a variable resistor). The resistance value is controlled by the gate voltage (V_{GS}), where a change in the gate voltage will create a

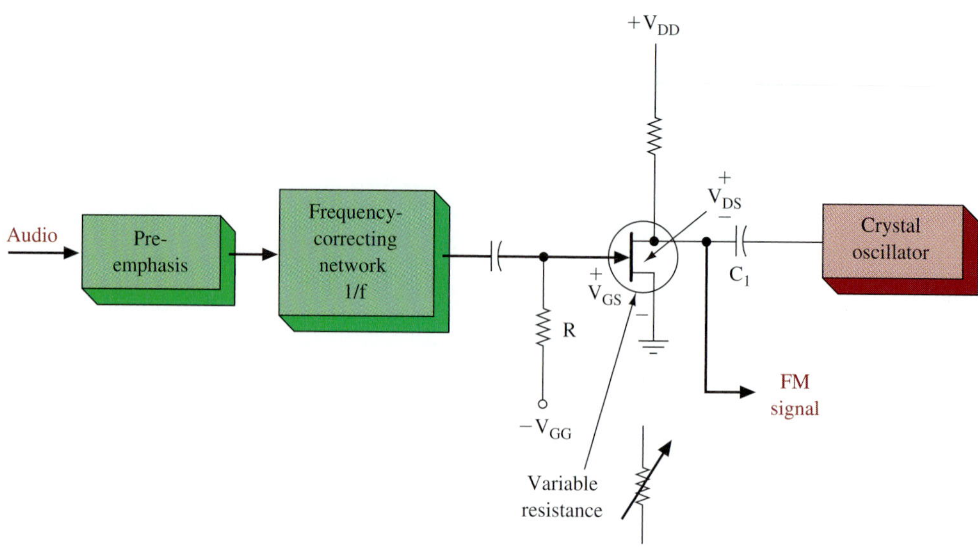

FIGURE 5-17 Indirect FM via PM (Armstrong modulator).

change in the drain-to-source resistance. Notice that the modulating signal is first given the standard preemphasis and then applied to a frequency-correcting network. This network is a low-pass *RC* circuit (an integrator) that makes the audio output amplitude inversely proportional to its frequency. This is necessary because in phase modulation, the frequency deviation created is not only proportional to modulating signal amplitude (as desired for FM) but also to the modulating signal frequency (undesired for FM). Thus, in PM if a 1-V, 1-kHz modulating signal caused a 100-Hz deviation, a 1-V, 2-kHz signal would cause a 200-Hz deviation instead of the same deviation of 100 Hz if that signal were applied to the 1/*f* network.

In summary, the Armstrong modulator of Fig. 5-17 indirectly generates FM by changing the phase of a crystal oscillator's output. That phase change is accomplished by varying the phase angle of an *RC* network (C_1 and the JFET's resistance), in step with the frequency-corrected modulating signal.

Obtaining Wideband Deviation

The indirectly created FM is not capable of much frequency deviation. A typical deviation is 50 Hz out of 1 MHz (50 ppm). Thus, even with a ×90 frequency multiplication, a 90-MHz station would have a deviation of 90 × 50 Hz = 4.5 kHz. This may be adequate for narrowband communication FM but falls far short of the 75-kHz deviation required for broadcast FM. A complete Armstrong FM system providing a 75-kHz deviation is shown in Fig. 5-18. It uses a balanced modulator and 90° phase shifter to phase-modulate a crystal oscillator. Sufficient deviation is obtained by a combination of multipliers and mixing. The ×81 multipliers (3 × 3 × 3 × 3) raise the initial 400-kHz ± 14.47-Hz signal to 32.4 MHz ± 1172 Hz. The carrier *and* deviation are multiplied by 81. Applying this signal to the mixer, which also has a crystal oscillator signal input of 33.81 MHz, provides an output component (among others) of 33.81 MHz − (32.4 MHz ± 1172 Hz), or 1.41 MHz ± 1172 Hz. Notice that the mixer output changes the center frequency *without* changing the

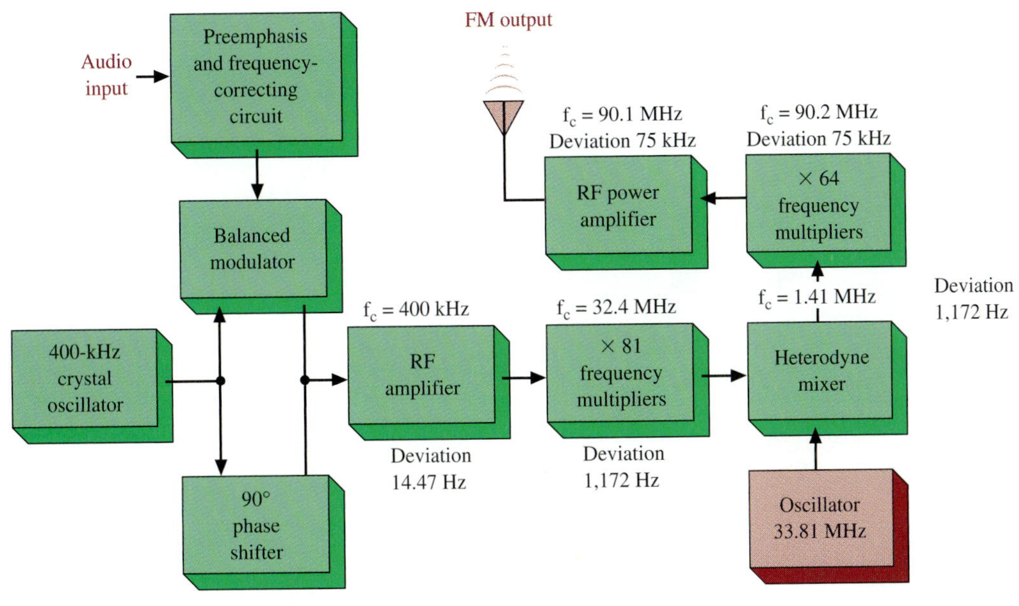

FIGURE 5-18 Wideband Armstrong FM.

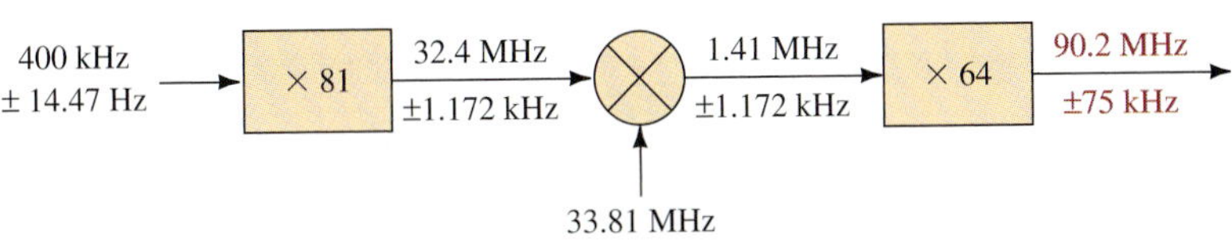

FIGURE 5-19 The pump chain for the wideband Armstrong FM system.

deviation. Following the mixer, the ×64 multipliers accept only the mixer difference output component of 1.41 MHz ± 1172 Hz and raise that to (64 × 1.41 MHz) ± (64 × 1172 Hz), or the desired 90.2 MHz ± 75 kHz.

The electronic circuitry used to increase the operating frequency of a transmitter up to a specified value is called the **pump chain.** A block diagram of the pump chain for the wideband Armstrong FM system is shown in Fig. 5-19.

Pump Chain
electronic circuitry used to increase the operating frequency up to a specified value

5-7 PHASE-LOCKED-LOOP FM TRANSMITTER

Block Diagram

The block diagram shown in Fig. 5-20 provides a very practical way to fabricate an FM transmitter. The amplified audio signal is used to frequency-modulate a crystal oscillator. The crystal frequency is pulled slightly by the variable capacitance exhibited by the varactor diode. The approximate ±200-Hz deviation possible in this fashion is adequate for narrowband systems. The FM output from the crystal oscillator is then divided by 2 and applied as one of the inputs to the phase detector of a phase-locked-loop (PLL) system. As indicated in Fig. 5-20, the other input to the phase detector is the same, and its output is therefore (in this case) the original audio

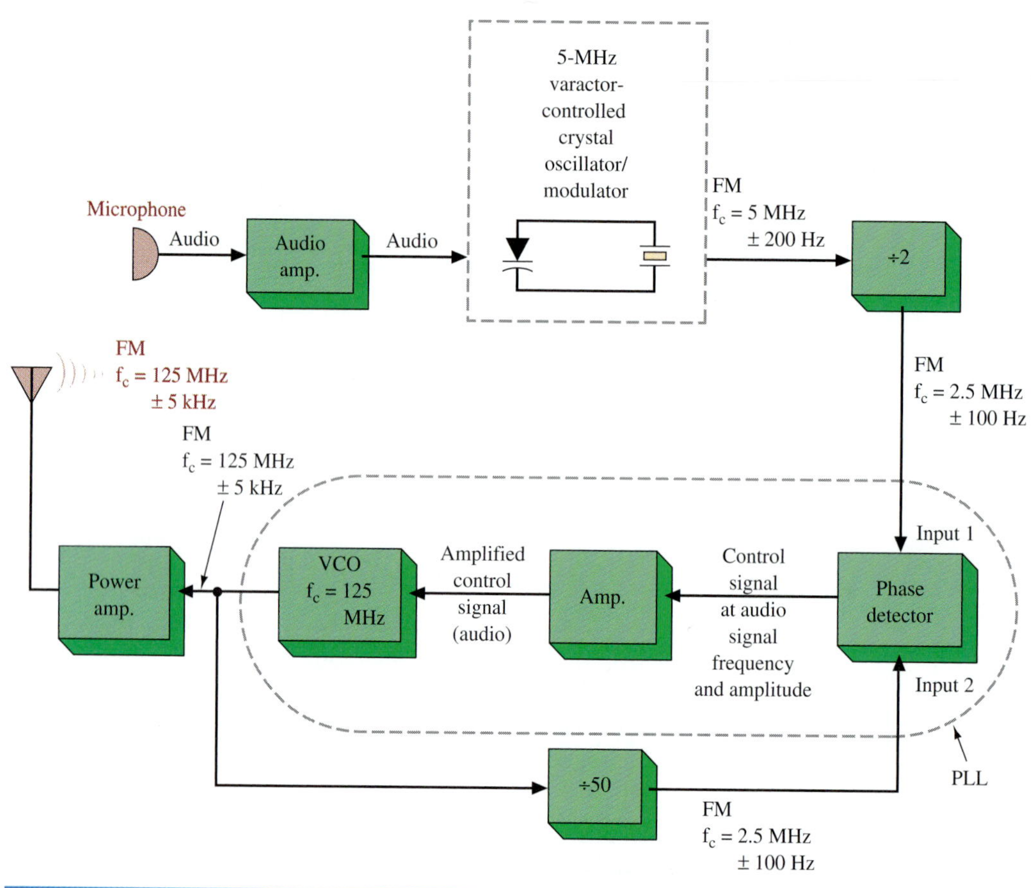

FIGURE 5-20 PLL FM transmitter block diagram.

signal. The input control signal to the VCO is therefore the same audio signal, and its output will be its free-running value of 125 MHz ±5 kHz, which is set up to be exactly 50 times the 2.5-MHz value of the divided-by-2 crystal frequency of 5 MHz.

The FM output signal from the VCO is given power amplification and then driven into the transmitting antenna. This output is also sampled by a ÷50 network, which provides the other FM signal input to the phase detector. The PLL system effectively provides the required ×50 multiplication but, more important, provides the transmitter's required frequency stability. Any drift in the VCO center frequency causes an input to the phase detector (input 2 in Fig. 5-20) that is slightly different from the exact 2.5-MHz crystal reference value. The phase detector output therefore develops an error signal that corrects the VCO center frequency output back to exactly 125 MHz. This dynamic action of the phase detector/VCO and feedback path is the basis of a PLL. More detail on PLL action is provided in Chapter 6.

Circuit Description

The circuit schematic shown in Fig. 5-21 is a practical working system for the block diagram of Fig. 5-20. The crystal frequency is labeled 5 MHz to help you correlate to the block diagram in Fig. 5-20. The actual crystal frequency is to be 5.76–5.92 MHz

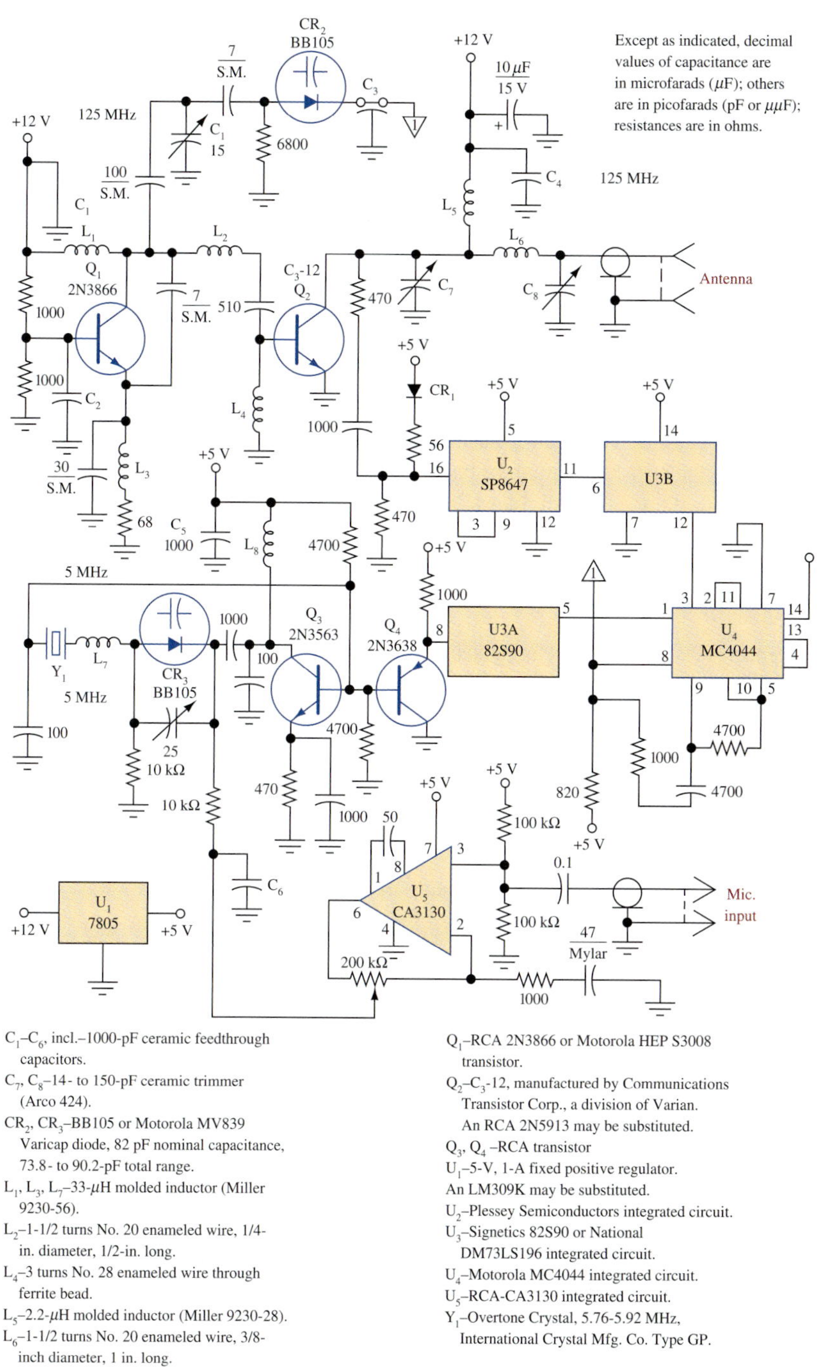

C_1–C_6, incl.–1000-pF ceramic feedthrough capacitors.

C_7, C_8–14- to 150-pF ceramic trimmer (Arco 424).

CR_2, CR_3–BB105 or Motorola MV839 Varicap diode, 82 pF nominal capacitance, 73.8- to 90.2-pF total range.

L_1, L_3, L_7–33-μH molded inductor (Miller 9230-56).

L_2–1-1/2 turns No. 20 enameled wire, 1/4-in. diameter, 1/2-in. long.

L_4–3 turns No. 28 enameled wire through ferrite bead.

L_5–2.2-μH molded inductor (Miller 9230-28).

L_6–1-1/2 turns No. 20 enameled wire, 3/8-inch diameter, 1 in. long.

L_8–100-μF molded inductor (Miller 9230-68).

Q_1–RCA 2N3866 or Motorola HEP S3008 transistor.

Q_2–C_3-12, manufactured by Communications Transistor Corp., a division of Varian. An RCA 2N5913 may be substituted.

Q_3, Q_4 –RCA transistor

U_1–5-V, 1-A fixed positive regulator. An LM309K may be substituted.

U_2–Plessey Semiconductors integrated circuit.

U_3–Signetics 82S90 or National DM73LS196 integrated circuit.

U_4–Motorola MC4044 integrated circuit.

U_5–RCA-CA3130 integrated circuit.

Y_1–Overtone Crystal, 5.76-5.92 MHz, International Crystal Mfg. Co. Type GP.

FIGURE 5-21 PLL FM transmitter schematic.

for the circuit values shown in Fig. 5-21. Similarly, the VCO frequency is really 146 MHz, rather than the 125 MHz shown. This circuitry approach eliminates the cumbersome oscillator–multiplier chain approach and allows for a very compact transmitter. The microphone input is amplified by U_5, which is an RCA CA3130 IC. Its output is used to pull the varactor-controlled crystal oscillator package comprising Y_1, CR_3, and Q_3. Its output is amplified by Q_4 and then divided by 2 by the U_{3A} IC. Refer back to the block diagram in Fig. 5-20 as an aid in this circuit description. The output of the U_{3A} is applied as one of the inputs to the U_4 phase detector amplifier IC. The U_4 output at pin 8 is applied to the VCO made up of varactor diode CR_2 and Q_1. Its output (about 200 mW) is applied to the power amplifier stage (Q_2), which provides about 2 W into the antenna. Its output is sampled by the U_2 IC, which is an emitter-coupled logic (ECL) device that provides a $\div 10$ function at frequencies up to 250 MHz. Its output is then $\div 5$ by the TTL U_{3B} IC and applied as the other input to the U_4 phase detector/amplifier IC. The values in this schematic are selected for operation on the 146-MHz amateur band, but operation can be accomplished at up to 250 MHz, with the U_2 ECL divider IC being the limiting upper frequency factor.

Alignment and Operation

The reference crystal frequency is determined by dividing the desired operating frequency by 25. Varactor voltage is monitored with a DMM or oscilloscope while C_1 is varied through its range. If the loop is locked, the varactor voltage will vary with adjustment of C_1 and should be adjusted to 2.5 V. The transmitter should be terminated in a nonreactive 50-Ω load and the RF amplifier adjusted for maximum power output. Some means of determining deviation will be necessary, and the transmitter will then be ready for use.

Operation on Other Bands

A transmitter may be constructed for use on the 200-MHz band by redesigning the oscillator and RF amplifier tuned circuits to resonate in that band. Q_1 and Q_2 will operate efficiently at frequencies up to 400 MHz. Crystal frequency is determined in the manner indicated previously. If separate oscillator–amplifier modules are constructed for 144 and 200 MHz, or perhaps even 50 MHz, and switched electronically with ECL gates, it is possible to operate on several bands with the same phase-locked-loop components, at a considerable cost savings. It is also possible to select a low-power oscillator and an unmodulated crystal oscillator to generate the LO signal for a receiver. ECL dividers are available that allow application of this circuit at higher frequencies, but a frequency division of more than 50 is required so that the maximum operating frequency of U_4 is not exceeded.

Single-Chip FM Transmitter

An FM transmitter circuit using the Maxim 2606 is shown in Fig. 5-22. This is a low-power FM transmitter operating in the FM broadcast band (88–108 MHz). The circuit runs off a supply voltage from 3 to 5 V connected to pin 5. The left and right audio channels are conncted to resistors R3 and R4, with potentiometer R2 serving as the volume control. Potentiometer R1 is used to select the channel by adjusting

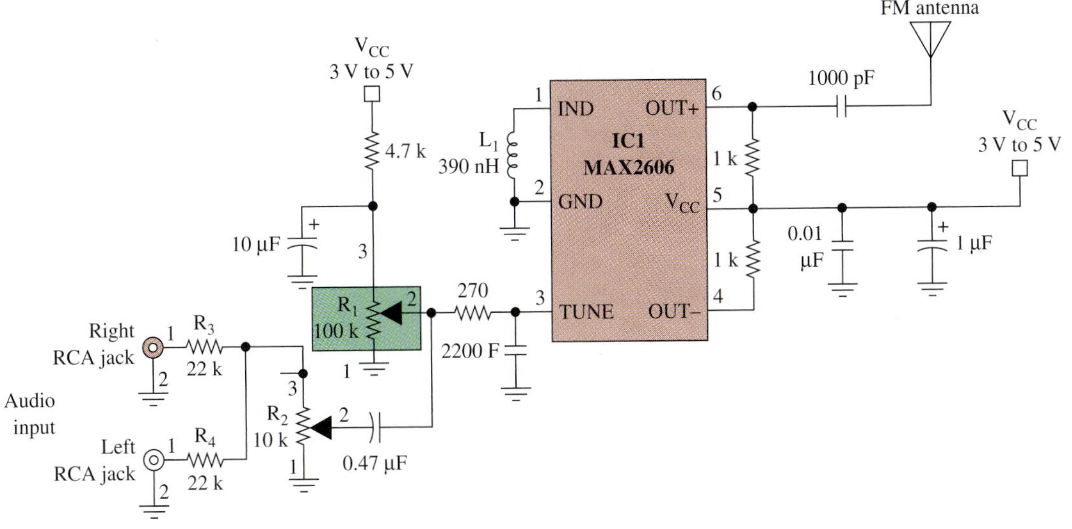

FIGURE 5-22 The MAX 2606 single-chip FM transmitter.

the tuning voltage applied to the TUNE input (pin 3) to the MAX2606 IC. The TUNE input drives a voltage-controlled oscillator with an integrated varactor. An FM antenna connects to pin 6 through the 1000 pF capacitor.

5-8 STEREO FM

The advent of stereo records and tapes and the associated high-fidelity playback equipment in the 1950s led to the development of stereo FM transmissions as authorized by the FCC in 1961. Stereo systems involve generating two separate signals, as from the left and right sides of a concert hall performance. When played back on left and right speakers, the listener gains greater spatial dimension or directivity.

A stereo radio broadcast requires that two separate 30-Hz to 15-kHz signals be used to modulate the carrier so that the receiver can extract the left and right channel information and amplify them separately into their respective speakers. In essence, then, the amount of information to be transmitted is doubled in a stereo broadcast. Hartley's law (Chapter 1) tells us that either the bandwidth or time of transmission must therefore be doubled, but this is not practical. The problem was solved by making more efficient use of the available bandwidth (200 kHz) by **frequency multiplexing** the two required modulating signals. **Multiplex operation** is the simultaneous transmission of two or more signals on one carrier.

Frequency Multiplexing
process of combining signals that are at slightly different frequencies to allow transmission over a single medium

Modulating Signal

The system approved by the FCC is *compatible* because a stereo broadcast received by a normal FM receiver will provide an output equal to the sum of the left plus right channels (L + R), while a stereo receiver can provide separate left and right channel signals. The stereo transmitter has a modulating signal, as shown in Fig. 5-23. Notice that the sum of the L + R modulating signal extends from 30 Hz to 15 kHz as does

Multiplex Operation
simultaneous transmission of two or more signals in a single medium

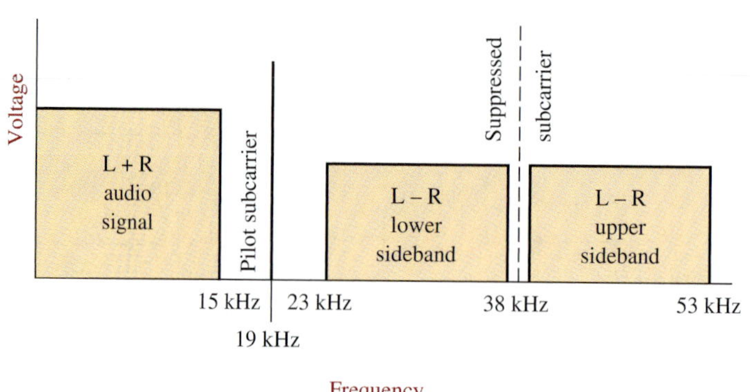

FIGURE 5-23 Composite modulating signals.

the full audio signal used to modulate the carrier in standard FM broadcasts. However, a signal corresponding to the left channel minus right channel (L − R) extends from 23 to 53 kHz. In addition, a 19-kHz pilot subcarrier is included in the composite stereo modulating signal.

The reasons for the peculiar arrangement of the stereo modulating signal will become more apparent when the receiver portion of stereo FM is discussed in Chapter 6. For now, suffice it to say that two different signals (L + R and L − R) are used to modulate the carrier. The signal is an example of **frequency-division multiplexing** because two different signals are multiplexed together by having them exist in two different frequency ranges.

Frequency-Division Multiplexing
simultaneous transmission of two or more signals on one carrier, each on its own separate frequency range, also called frequency multiplexing

FM Stereo Generation

The block diagram in Fig. 5-24 shows the method whereby the composite modulating signal is generated and applied to the FM modulator for subsequent transmission. The left and right channels are picked up by their respective microphones and individually preemphasized. They are then applied to a **matrix network** that inverts the right channel, giving a −R signal, and then combines (adds) L and R to provide an (L + R) signal and also combines L and −R to provide the (L − R) signal. The two outputs *are still* 30-Hz to 15-kHz audio signals at this point. The (L − R) signal and a 38-kHz carrier signal are then applied to a balanced modulator that suppresses the carrier but provides a double-sideband (DSB) signal at its output. The upper and lower sidebands extend from 30 Hz to 15 kHz above and below the suppressed 38-kHz carrier and therefore range from 23 kHz (38 kHz − 15 kHz) up to 53 kHz (38 kHz + 15 kHz). Thus, the (L − R) signal has been translated from audio up to a higher frequency to keep it separate from the 30-Hz to 15-kHz (L + R) signal. The (L + R) signal is given a slight delay so that both signals are applied to the FM modulator in time phase due to the slight delay encountered by the (L − R) signal in the balanced modulator. The 19-kHz master oscillator in Fig. 5-24 is applied directly to the FM modulator and also doubled in frequency, to 38 kHz, for the balanced modulator carrier input.

Matrix Network
adds and/or subtracts and/or inverts electrical signals

Stereo FM is more prone to noise than are monophonic broadcasts. The (L − R) signal is weaker than the (L + R) signal, as shown in Fig. 5-23. The (L − R) signal is also at a higher modulating frequency (23 to 53 kHz), and both of these effects

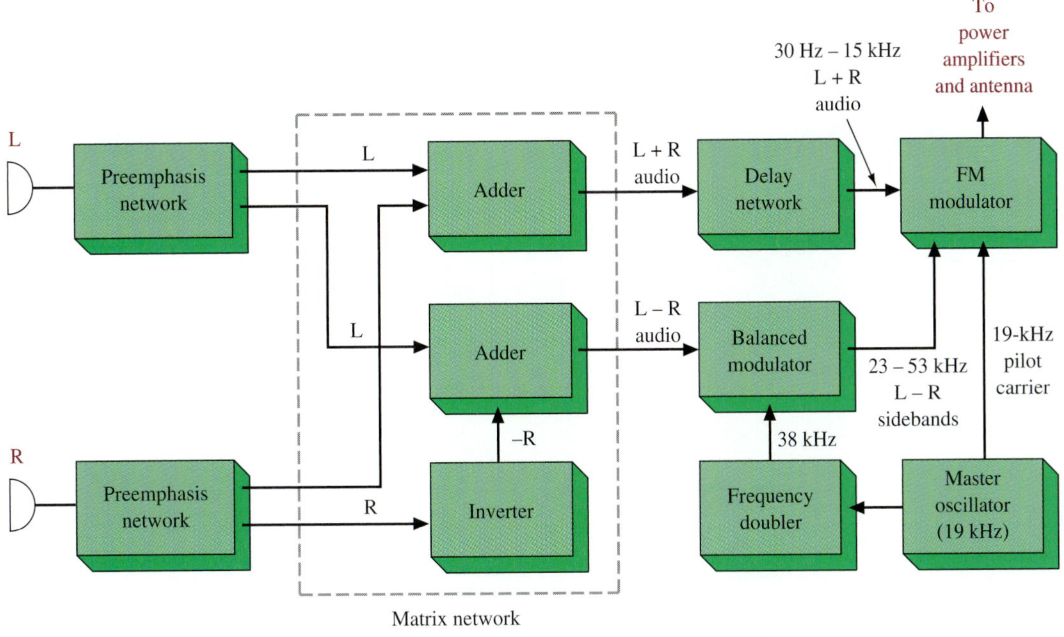

FIGURE 5-24 Stereo FM transmitter.

cause poorer noise performance. The net result to the receiver is an *S/N* of about 20 dB less than the monophonic signal. Because of this, some receivers have a mono/stereo switch so that a noisy (weak) stereo signal can be changed to monophonic for improved reception. A stereo signal received by a monophonic receiver is only about 1 dB worse (*S/N*) than an equivalent monophonic broadcast because of the presence of the 19-kHz pilot carrier. You may wish to refer to Sec. 6-6 at this time to continue this stereo FM discussion into the receiver section.

 5-9 FM TRANSMISSIONS

FM is used in five major categories:

1. Noncommercial broadcast at 88 to 90 MHz
2. Commercial broadcast with 200-kHz channel bandwidths from 90 to 108 MHz
3. Television audio signals with 50-kHz channel bandwidths at 54 to 88 MHz, 174 to 216 MHz, and 470 to 806 MHz
4. Narrowband public service channels from 108 to 174 MHz and in excess of 806 MHz
5. Narrowband amateur radio channels at 29.6 MHz, 52 to 53 MHz, 144 to 147.99 MHz, 440 to 450 MHz, and in excess of 902 MHz

The output powers range from milliwatt levels for the amateurs up to 100 kW for broadcast FM. Note that FM is not used at frequencies below about 30 MHz because of the phase distortion introduced to FM signals by the earth's ionosphere at these frequencies. Frequencies above 30 MHz are transmitted line-of-sight and are not significantly affected by the ionosphere. The limited range (normally 70 to 80 mi) for

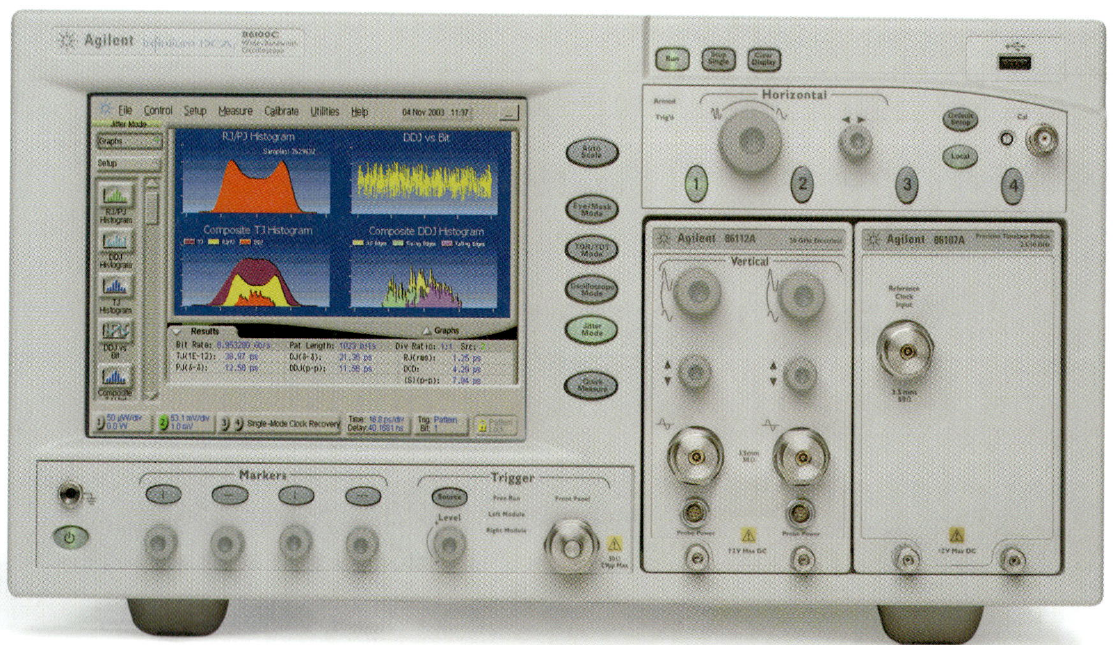

(a)

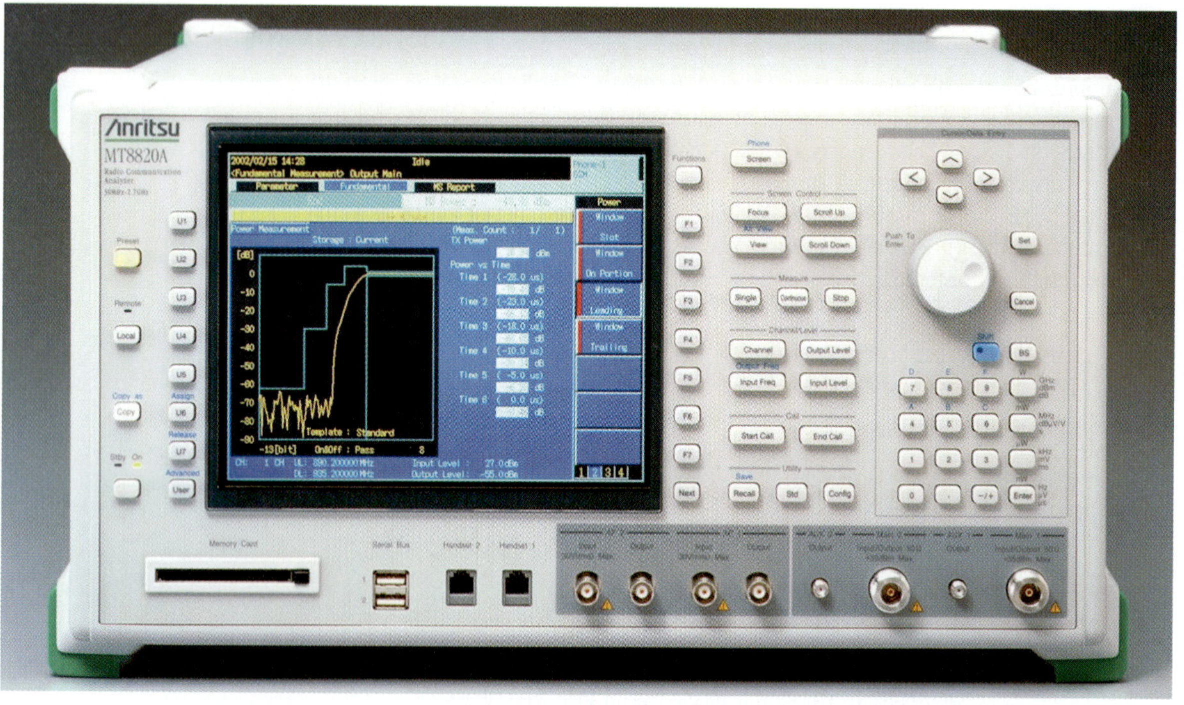

(b)

(a) The 86100C digital communications analyzer with jitter analysis offers breakthrough speed, accuracy, and affordability. (Courtesy of Agilent Technologies. Reprinted with permission.) (b) The MT8802A radio communications analyzer was designed to support the test needs of the manufacturing, R&D, and maintenance markets. (Courtesy of Anritsu Company. Reprinted with permission.)

FM transmission is due to the earth's curvature. See Chapter 13 for a more complete discussion of these effects.

Another advantage that FM has over SSB and AM, other than superior noise performance, is the fact that *low-level modulation* (see Sec. 2-5) can be used with subsequent highly efficient class C power amplifiers. Since the FM waveform does not vary in amplitude, the intelligence is not lost by class C power amplification as it is for AM and SSB. Recall that a class C amplifier tends to provide a constant output amplitude due to the *LC* tank circuit flywheel effect. Thus, there is no need for high-power audio amplifiers in an FM transmitter and, more important, all the power amplification takes place at about 90 percent efficiency (class C), as compared to a maximum of about 70 percent for linear power amplifiers.

5-10 Troubleshooting

The most likely types of FM radio a technician will be called to service are an automotive mobile, a fixed base station, or a handheld portable. Either the mobile or the fixed base station can have power outputs as high as 150 W. Some of these transmitters can be damaged if they are not operated into the proper load impedance, usually 50 Ω. Some contain automatic circuitry to shut the transmitter off if it is not connected to a proper load. Dummy loads are made for this purpose, so make sure you have one rated for sufficient power.

In an FM transmitter, an oscillator is typically controlled so that its frequency changes when an intelligence signal changes. This control is provided by a modulating circuit. Recall from Sec. 5-5 that there are several ways this modulator can work. In this section, we will learn some troubleshooting techniques for a reactance modulator circuit.

Frequency-modulated transmitters can be roughly divided into two categories. The first is wideband FM. More popular than AM broadcasting, it is the FM we listen to on our car radios and home stereos (it also produces the audio portion of a TV signal). The second type of FM transmitters is called narrowband FM. It is used in police and fire department radios as well as taxicabs and VHF boat radios and the popular handheld transceivers called handy talkies, or walkie-talkies. We'll also look at testing a wideband FM generator in this section.

After completing this section you should be able to

- Troubleshoot FM transmitter systems
- Describe the operation of the reactance modulator
- Locate the master oscillator section
- Locate the reactance circuit section
- Recognize the difference between no modulator output, low output, or oscillator output without FM
- Troubleshoot a stereo/SCA FM generator
- Measure an FM transmitter's carrier frequency and deviation

FM Transmitter Systems

1. No RF Output In this case, one should first verify that the oscillator is running. Most of these transmitters multiply the oscillator by 12 or 18 to obtain the output frequency.

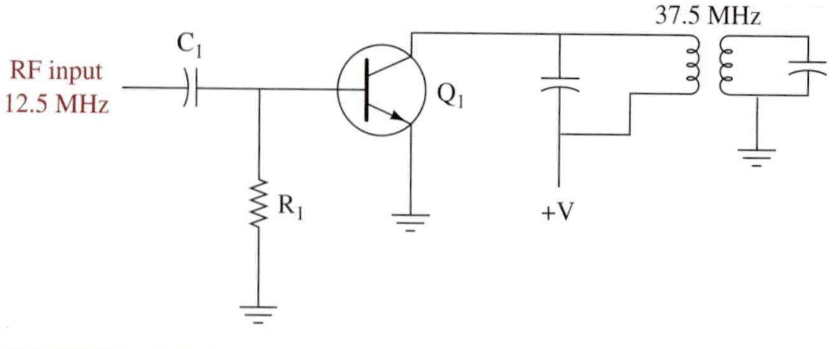

FIGURE 5-25 Multiplier stage.

The service manual will tell you what the multiplier is. It is best to have a spectrum analyzer, but a shortwave receiver will do.

Hints on oscillator problems were given in Chapter 1, so let's assume that the oscillator is running and move on to the first multiplier stage. Figure 5-25 shows a simplified schematic of a multiplier stage. If the base–emitter junction is good and there is sufficient input drive, you will find a negative voltage at the base of Q_1. This is because the RF input is rectified by the junction and the current flows through R_1. When the RF is rectified by the base–emitter junction, current pulses rich in harmonics are amplified and filtered by the tuned circuit in the collector of the transistor.

If a spectrum analyzer is available, the technician can verify that the stage is producing the proper multiple of the input frequency by loosely coupling the analyzer to the output coils. A two- or three-turn coil one-half inch in diameter connected to the analyzer will do.

Either the coils or the capacitors in the multiplier stage will be adjustable. You should be able to peak the output on the proper frequency with these adjustments. If not, check the capacitors and the inductors.

If you do not have a spectrum analyzer, simply measuring the bias voltage on the next stage may be sufficient. You can be reasonably sure that the multiplier is working if you can peak up the drive to the next stage with the adjustments. However, there is some possibility of tuning to the wrong harmonic. If the adjustments are all the way to one end, you may have done this or some component has failed. Another indication of improper tuning is that you will probably not be able to tune the next stage.

A typical transmitter will have three multiplier stages. At some point in the chain you will be able to find enough signal to run a frequency counter. Be careful; too much input to the counter will damage it. At the output of the transmitter, you can use a high-power attenuator.

2. **Incorrect Frequency** Most of these transmitters will have trimmer capacitors to adjust the frequency, while some will have inductors. If the oscillator is off frequency, check the voltage first, then the capacitors. Intermittent capacitors are the hardest to find. Try cooling the capacitor with an aerosol spray sold for this purpose.

3. **Incorrect Deviation** There are usually two adjustments here, one for microphone gain and one on a limiter. The limiter prevents the user from overmodulating the transmitter. When adjusting deviation, make sure the limiter is adjusted so that it doesn't affect the main adjustment. Refer to Fig. 5-26.

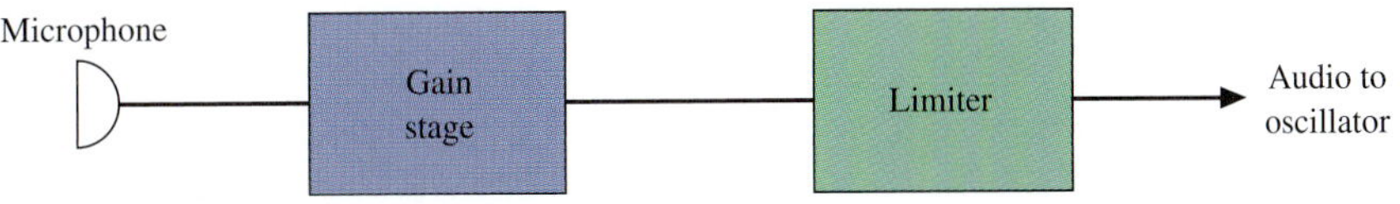

FIGURE 5-26 Audio chain.

Mobile radios are usually set for a peak deviation of 5 kHz. A good service shop will have a deviation meter. If you don't have one, a fair job can be done by simply comparing a known good transmitter with the unit under test by listening to both with any receiver.

If a spectrum analyzer is available, recall Carson's rule and speak into the microphone while adjusting the bandwidth of the transmitted signal to about 16 kHz.

One can also use the zero carrier amplitude method shown in Fig. 5-4. While viewing the transmitter output on the spectrum analyzer, apply a 2-kHz tone to the microphone input. Adjust the gain from zero up until the carrier is null and you have 5-kHz peak deviation.

Reactance Modulator Circuit Operation

The reactance modulator is efficient and provides a large deviation. It is popular and used often in FM transmitters. Figure 5-27 illustrates a typical reactance modulator circuit. Refer to this figure throughout the following discussion. The circuit consists of the reactance circuit and the master oscillator. The reactance circuit operates on the master oscillator to cause its resonant frequency to shift up or shift down depending on the modulating signal being applied. The reactance circuit appears capacitive to the master oscillator. In this case, the reactance looks like a variable capacitor in the oscillator's tank circuit.

Transistor Q_1 makes up the reactance modulator circuit. Resistors R_2 and R_3 establish a voltage divider network that biases Q_1. Resistor R_4 furnishes emitter feedback to thermally stabilize Q_1. Capacitor C_3 is a bypass component that prevents ac input signal degeneration. Capacitor C_1 interacts with transistor Q_1's interelectrode capacitance to cause a varying capacitive reactance directly influenced by the input modulating signal.

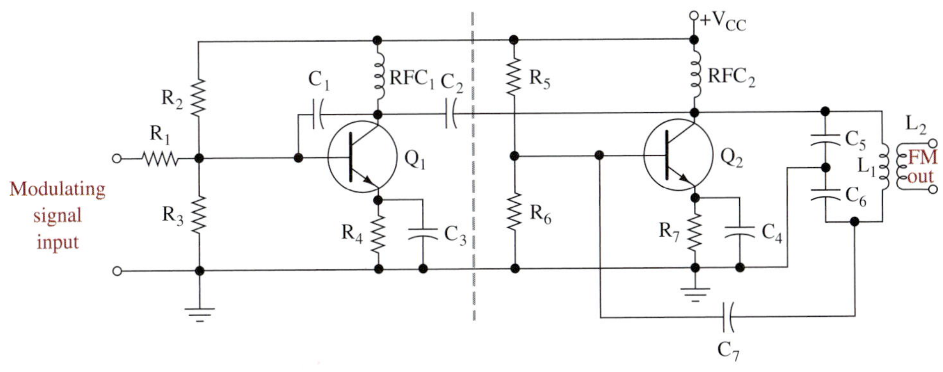

FIGURE 5-27 Reactance modulator.

The master oscillator is a Colpitts oscillator built around transistor Q_2. Coil L_1, capacitor C_5, and capacitor C_6 make up the resonant tank circuit. Capacitor C_7 provides the required regenerative feedback to cause the circuit to oscillate. Q_1 and Q_2 are impedance coupled, and capacitor C_2 effectively couples the changes at Q_1's collector to the tank circuit of transistor Q_2 while blocking dc voltages.

When a modulating signal is applied to the base of transistor Q_1 via resistor R_1, the reactance of the transistor changes in relation to that signal. If the modulating voltage goes up, the capacitance of Q_1 goes down, and if the modulating voltage goes down, the reactance of Q_1 goes up. This change in reactance is felt on Q_1's collector and also at the tank circuit of the Colpitts oscillator transistor Q_2. As capacitive reactance at Q_1 goes up, the resonant frequency of the master oscillator, Q_2, decreases. Conversely, if Q_1's capacitive reactance goes down, the master oscillator resonant frequency increases.

Troubleshooting the Reactance Modulator

The FM output signal from coil L_2 can be lost due to open bias resistors, open RF chokes RFC_1 and RFC_2, or an open winding at coil L_1. In addition, a leaky coupling capacitor C_2 may shift the collector voltage of Q_2 of the master oscillator, causing a low FM output signal to be present at coil L_2. Low collector voltages and weak transistors may produce a low FM signal output condition, as well as changes in bias resistors R_2 and R_3 in transistor Q_1's circuit and resistors R_5 and R_6 in Q_2's circuit. Changes in the emitter resistors of both transistor circuits will lessen the FM output and possibly shut the modulator down.

The master oscillator may operate without being influenced by the reactance circuit. The oscillator output signal at L_2 would not be FM. This situation could occur if the modulating signal were missing from the base of transistor Q_1. An open R_1 would block the modulating signal from getting to the base of Q_1. Without the modulating signal present, Q_1's reactance would not change. A leaky or shorted C_1 could also kill the reactance response of Q_1. Transistor Q_1 may still be operating perfectly but the reactance changes might not be passed to Q_2's tank circuit due to an open C_2.

Table 5-3 is a symptom guide to help you troubleshoot the reactance modulator circuit.

Table 5-3

Symptom	Problem	Probable Cause
No signal out from L_2	No FM modulator output	Open bias resistors in Q_1 and Q_2 circuits; open RFC_1 or RFC_2; C_2 open or leaky; feedback capacitor C_7 open
No FM output at L_2, master oscillator output only	No FM modulation taking place	Q_1 not functioning, check C_1, RFC_1, and R_4; resistor R_1 may be open
Amplitude of FM output low	Low modulator output	Changes in bias resistor values, check R_2, R_3, R_5, and R_6; change in emitter resistors, check R_4 and R_7; Q_2's gain has decreased

Check the resistors with your DMM for proper values. Resistors in the reactance modulator circuit will be precision types with close tolerances. For low FM signal outputs, check capacitors C_1, C_2, C_5, C_6, and C_7 with a capacitor checker.

Any one of these capacitors could cause low output. If any of these capacitors open or become leaky, the FM output may cease altogether. Check coils for open windings using the DMM's continuity function. Look for cold solder joints by observing a dull appearance of the solder connection. The DMM will give a high resistance indication for a cold solder joint.

The Spectrum of a Wideband FM Signal

Figure 5-28 illustrates the frequency spectrum of the modulating signal for a standard wideband FM stereo broadcast system. The signal has two major parts. The first part extends from the carrier up to 53 kHz. It consists of the three components making up the stereo (stereophonic) portion of the signal: the left-plus-right audio channel extending from 0.05 to 15 kHz, the pilot carrier at 19 kHz, and the left-minus-right audio channel from 23 to 53 kHz.

The second part of our FM modulating signal spectrum is the SCA (Subsidiary Communications Authorization) signal extending from 60 to 74 kHz above the carrier. Because the station's voice ID (which by law must be broadcast at regular intervals) is carried on the stereo channel, the SCA channel can legally transmit, for example, weather announcements or perhaps only music. That music, however, is at such a high frequency (note it is centered around 67 kHz) it cannot be heard on regular radios. Instead, the radio station leases special equipment that drops the frequencies to normal hearing range to supply background music to grocery stores and dentists' offices. Figure 5-29 shows the block diagram of a system that generates a stereo/SCA FM signal.

Troubleshooting the stereo/SCA generator is best done with a spectrum analyzer connected as shown in Fig. 5-30. Note first the (L + R) signals from 0.05 to 15 kHz in Fig. 5-28. They will be present whether the station is transmitting monophonic or stereo. The 19-kHz pilot carrier and (L − R) channels, however, are only present during stereo broadcasts. Should the oscillator generating the 19-kHz carrier fail, both that signal as well as the (L − R) channel from 23 to 53 kHz will disappear from the analyzer display. The oscillator will probably be contained within an IC that develops the entire stereo signal. If so, the chip must be replaced

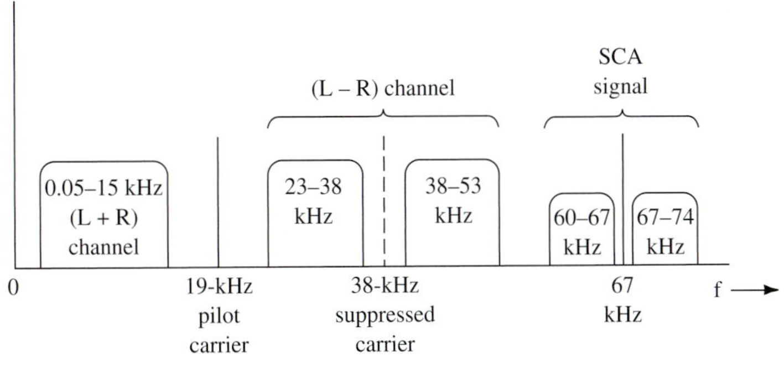

FIGURE 5-28 Spectrum of a wideband FM modulating signal.

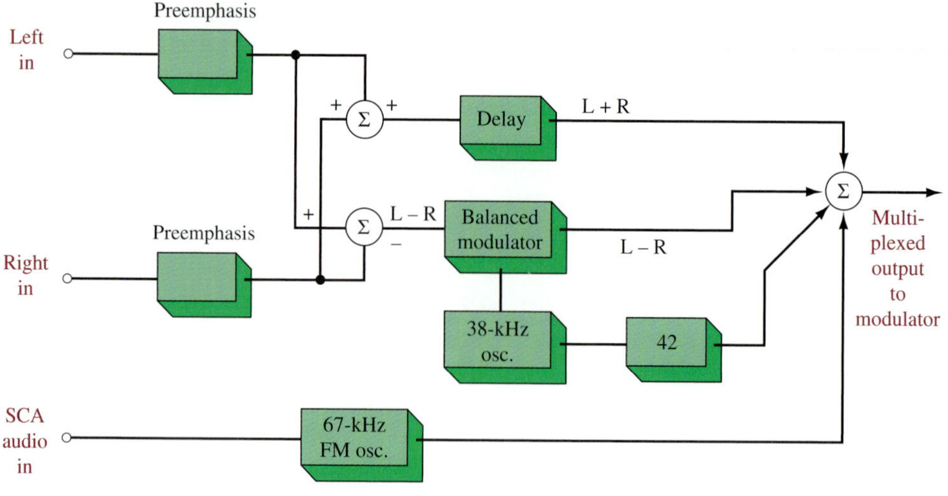

FIGURE 5-29 Generating a stereo/SCA FM modulating signal.

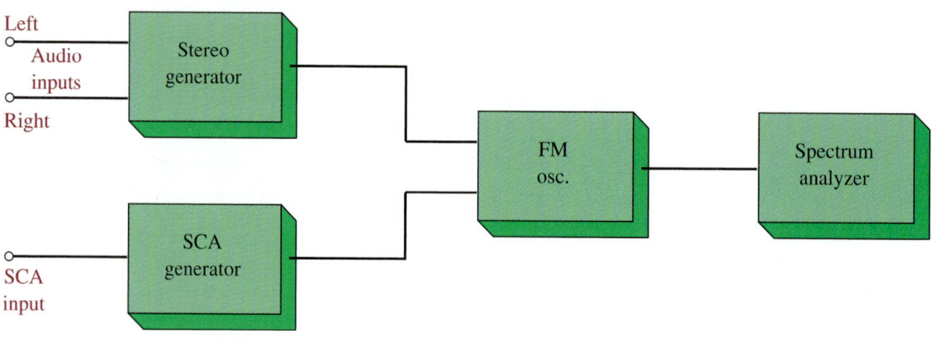

FIGURE 5-30 Troubleshooting the stereo/SCA generator.

(assuming you've determined that V_{CC} and any other required inputs are present). Older units, however, might have a circuit constructed of separate components. In that case, use your regular troubleshooting techniques of checking for proper voltages, resistances, etc., to locate the problem.

The (L − R) channel is a double-sideband suppressed carrier (DSBSC) signal generated by a balanced modulator. The carrier it suppresses is the second harmonic (two times or 38-kHz) of the 19-kHz pilot carrier. Troubleshooting and balance adjustment of balanced modulators were discussed in Chapter 4.

Wideband FM Transmitters' Frequency and Deviation Test

Wideband FM transmissions are defined as those having a modulation index greater than 1. By FCC regulations, these signals are allowed a deviation no greater than ±75 kHz. Because they are modulated by audio signals from 30 Hz to 15 kHz, calculations tell us the modulation index can range from 5 to 2500.

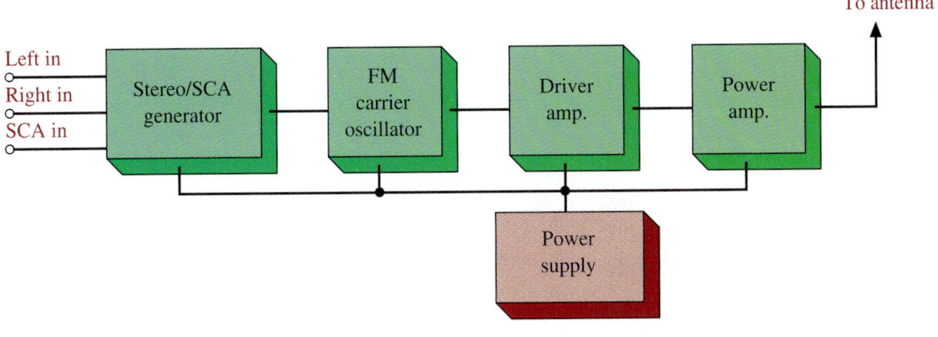

FIGURE 5-31 The wideband FM transmitter.

We will discuss two important tests that are regularly made on FM transmitters. These are carrier frequency and deviation tests.

The block diagram of a wideband FM transmitter is shown in Fig. 5-31. Note the input to the carrier oscillator. It is the output of the stereo/SCA generator discussed above. We say that the stereo/SCA generator frequency modulates the transmitter's carrier oscillator.

The FCC requires an FM broadcast station to hold its carrier frequency accurate to within ±2000 Hz. To achieve this stability, the transmitter employs an automatic frequency control (AFC) circuit. A reference crystal oscillator drives the AFC unit. Its frequency is measured very accurately. To hold this frequency stable against temperature changes, the entire unit is contained in a thermostatically controlled oven.

The AFC control compares the carrier oscillator frequency against that of the reference oscillator. Should there be an error between the two (in other words, should the carrier oscillator change frequency or drift), the AFC circuitry shifts the carrier frequency to return the error to zero.

The test setup of Fig. 5-32 is used to measure a station's carrier frequency. Note the dummy load. It dissipates the energy developed by the transmitter as heat

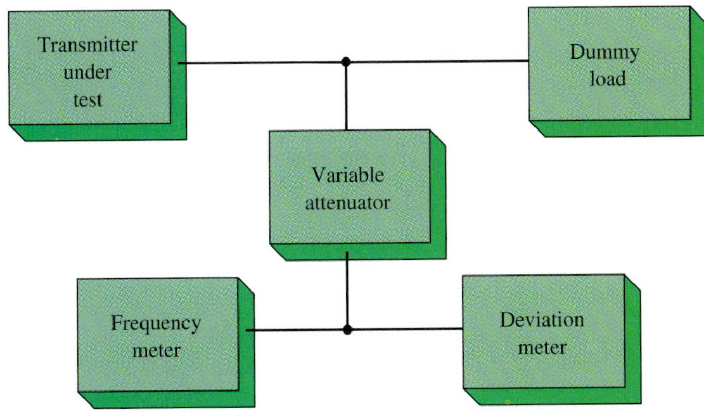

FIGURE 5-32 Measuring the FM transmitter's carrier frequency and deviation.

so no signal goes out over the air during testing and troubleshooting. The input impedance of the dummy load must match (be the same as) the impedance of the transmission line carrying the transmitter's output signal to the antenna.

The frequency meter or counter used to measure carrier frequency in Fig. 5-32 must be accurate, especially if the readings are being taken for FCC certification. Should the frequency be out of specs, check the reference oscillator for proper output with the same frequency meter. Look for problems in the AFC circuitry. The modulating signal should be turned off before you make any carrier measurements.

Carrier deviation can also be measured using the deviation meter in the circuit of Fig. 5-32. As stated above, deviation must not exceed ±75 kHz for wideband FM broadcasting transmitters.

5-11 Troubleshooting with Electronics Workbench™ Multisim

The concept of generating a frequency-modulated signal was introduced in this chapter. This exercise has been developed to help you better understand the concept of generating and analyzing a frequency modulated signal. To begin this exercise, open **Fig5-33.ms7 (.msm),** which can be found on your EWB Multisim CD. This circuit, shown in Fig. 5-33, contains a voltage-controlled oscillator (VCO) that is being driven by a 1-V, 10-kHz triangle wave. The triangle wave is being generated by the

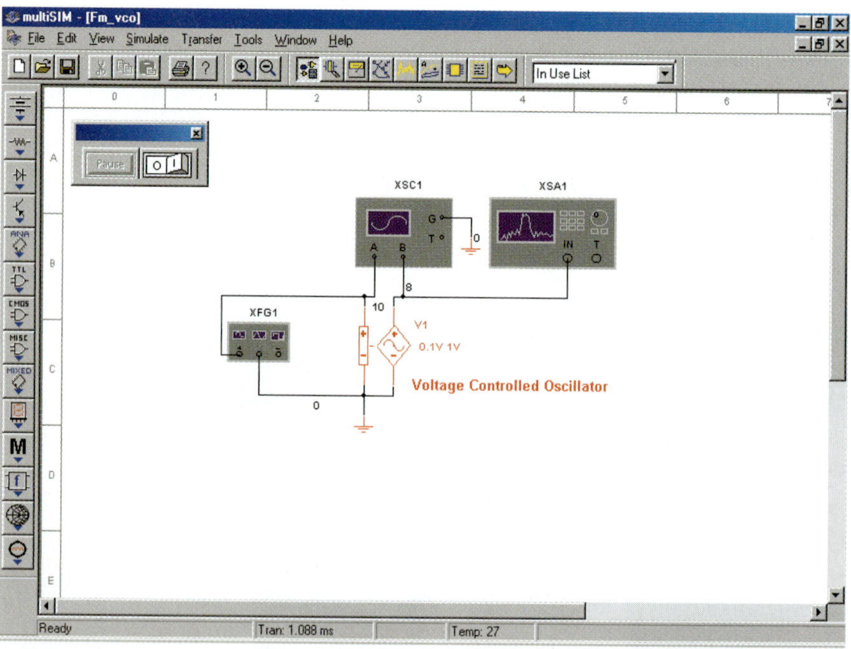

FIGURE 5-33 A frequency-modulation circuit using a voltage-controlled oscillator, as implemented in an Electronics Workbench™ Multisim circuit.

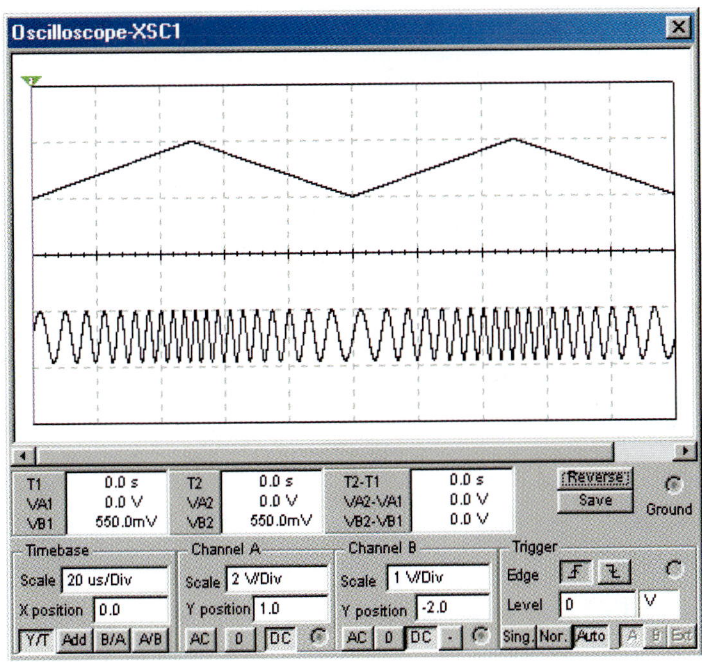

FIGURE 5-34 The oscilloscope display of the 10-kHz triangle wave and the output of the VCO.

function generator. Double-click on the function generator to see the settings. The function generator can produce a sinusoid, triangle, or square wave. The frequency, duty cycle, and offset voltage are adjustable.

Double-click on the VCO to check its settings. The control and frequency arrays are used to specify the ranges for the VCO. In this example, an input voltage of 0 V produces a 100-kHz sinusoid, whereas a 1-V input produces a 200-kHz sinusoid.

The output of the VCO, as viewed with an oscilloscope, is provided in Fig. 5-34. Channel A (top) is the 10-kHz triangle waveform and channel B (bottom) is the VCO output. Notice that as the voltage increases, the VCO frequency increases. Experiment with the circuit and see how the VCO output is affected by inputting a square wave. A square-wave input produces a frequency shift keying (FSK) output.

Close **Fig5-33.ms7 (.msm)** and open the **FigE5-1.ms7 (.msm)** file on your EWB CD. The VCO has been replaced with an FM source. Double-click on the FM source to see the settings. The carrier frequency is 150 kHz, the modulation index is 5, and the signal frequency is 10 kHz. Start the simulation and observe the display on the spectrum analyzer. You should see a picture similar to Fig. 5-35. Notice the spectral width of the signal. Recall from Sec. 5-3 that the bandwidth of an FM signal can be estimated by Carson's rule.

For this example, the input frequency is 10 kHz and the modulation index is 5. The estimated BW is 2(50 kHz + 10 kHz) = 120 kHz. This is a noisy signal, but a quick estimate shows that the 3-dB bandwidth is approximately 120 kHz, which agrees with the estimate using Carson's rule.

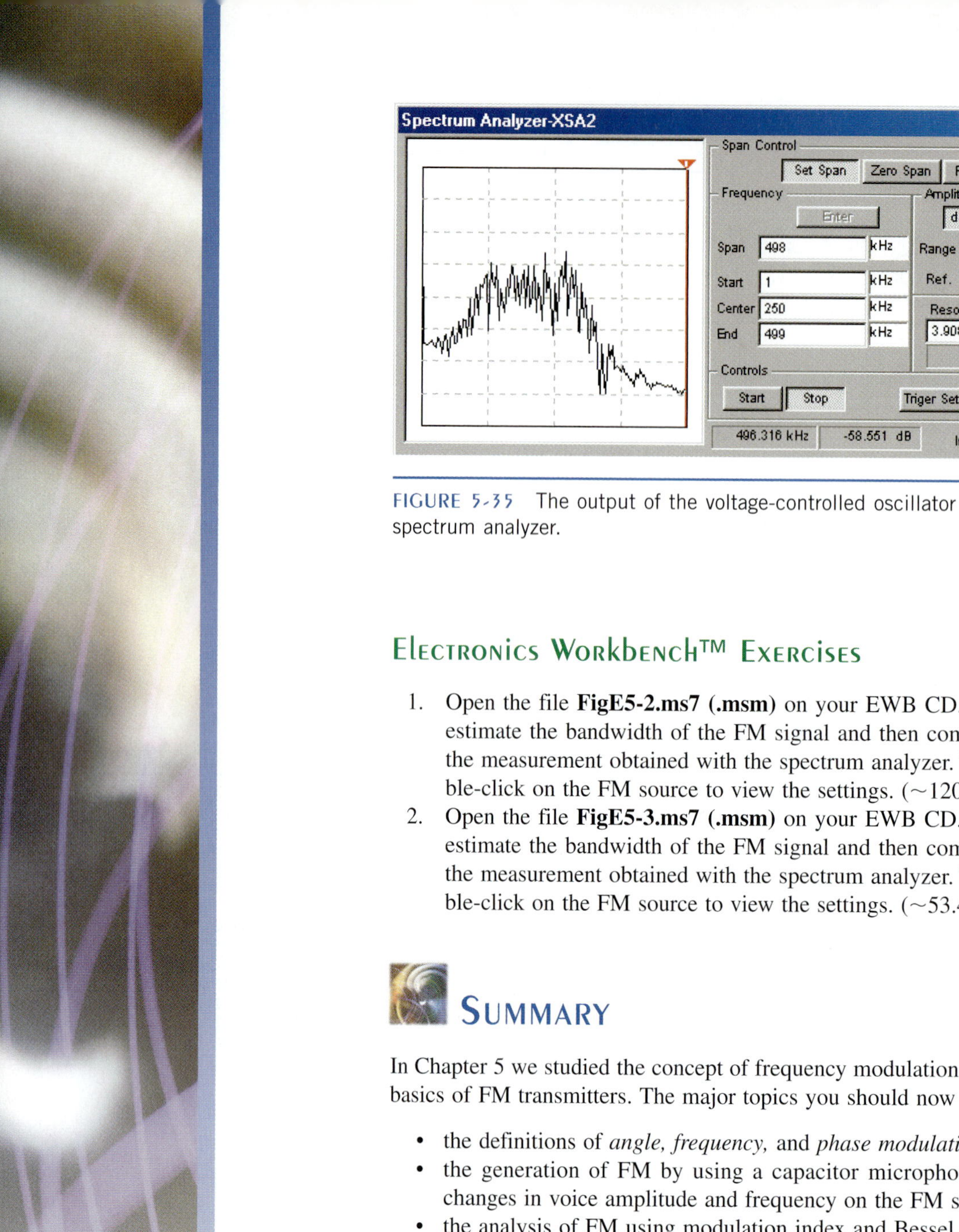

FIGURE 5-35 The output of the voltage-controlled oscillator as viewed with a spectrum analyzer.

Electronics Workbench™ Exercises

1. Open the file **FigE5-2.ms7 (.msm)** on your EWB CD. Use Carson's rule to estimate the bandwidth of the FM signal and then compare your result with the measurement obtained with the spectrum analyzer. You will need to double-click on the FM source to view the settings. (~120 kHz)
2. Open the file **FigE5-3.ms7 (.msm)** on your EWB CD. Use Carson's rule to estimate the bandwidth of the FM signal and then compare your result with the measurement obtained with the spectrum analyzer. You will need to double-click on the FM source to view the settings. (~53.4 kHz)

Summary

In Chapter 5 we studied the concept of frequency modulation (FM) and learned the basics of FM transmitters. The major topics you should now understand include:

- the definitions of *angle, frequency,* and *phase modulation*
- the generation of FM by using a capacitor microphone and the effects of changes in voice amplitude and frequency on the FM signal
- the analysis of FM using modulation index and Bessel functions
- the determination of FM deviation using the zero-carrier condition
- the analysis of noise suppression by limiter circuits and by using phasors and the signal-to-noise ratio (*S/N*)
- the analysis of direct generation FM circuits, including varactor diodes, the reactance modulator, the LIC VCO, and the Crosby modulator
- the operation of an indirect FM system using the Armstrong modulator
- the generation of FM using a phase-locked loop (PLL)
- the changes made to a standard FM transmitter to enable broadcast stereo operation
- the advantage of FM versus SSB or AM

Questions and Problems

Section 5-1

1. Define *angle modulation* and list its subcategories.

*2. What is the difference between frequency and phase modulation?

3. Even though PM is not actually transmitted, provide two reasons that make it important in the study of FM.

4. A radio transmission is classified as $3A1_i$. Describe this signal as fully as possible.

Section 5-2

5. Explain how a condenser microphone can be used very easily to generate FM.

6. Define *deviation constant.*

7. A 50-mV sinusoid, at a frequency of 1 kHz, is applied to a capacitor microphone FM generator. If the deviation constant for the capacitor microphone FM generator is 500 Hz/20 mV, determine:
   (a) The total frequency deviation. (±1.25 kHz)
   (b) The rate at which the carrier frequency is being deviated. (1 kHz)

8. Explain how the intelligence signal modulates the carrier.

9. In an FM transmitter, the output is changing between 90.001 and 89.999 MHz 1000 times a second. The intelligence signal amplitude is 3 V. Determine the carrier frequency and intelligence signal frequency. If the output deviation changes to between 90.0015 and 89.9985 MHz, calculate the intelligence signal amplitude. (90 MHz, 1 kHz, 4.5 V)

*10. What determines the rate of frequency swing for an FM broadcast transmitter?

11. Without knowledge of Sec. 5-3 and using Fig. 5-1, write an equation that expresses the output frequency, f, of the FM generator. *Hint:* If there is no input into the microphone, then $f = f_c$, where f_c is the oscillator's output frequency.

Section 5-3

12. Define *modulation index* (m_f) as applied to an FM system.

*13. What characteristic(s) of an audio tone determines the percentage of modulation of an FM broadcast transmitter?

14. Explain what happens to the carrier in FM as m_f goes from 0 up to 15.

15. Calculate the bandwidth of an FM system (using Table 5-2) when the maximum deviation (δ) is 15 kHz and $f_i = 3$ kHz. Repeat for $f_i = 2.5$ and 5 kHz. (48 kHz, 45 kHz, 60 kHz)

16. Explain the purpose of the *guard bands* for broadcast FM. How wide is an FM broadcast channel?

*17. What frequency swing is defined as 100 percent modulation for an FM broadcast station?

---

*An asterisk preceding a number indicates a question that has been provided by the FCC as a study aid for licensing examinations.

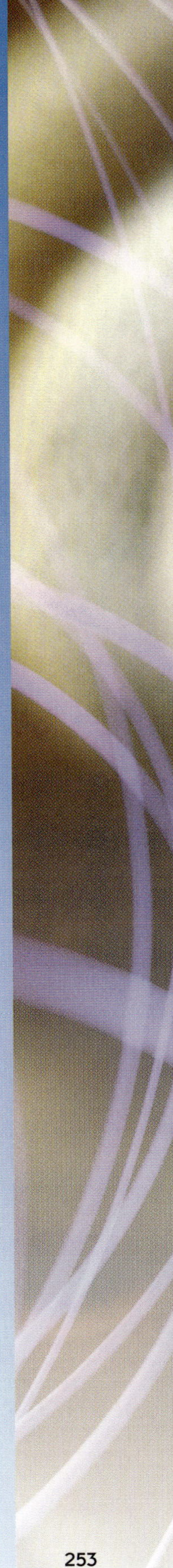

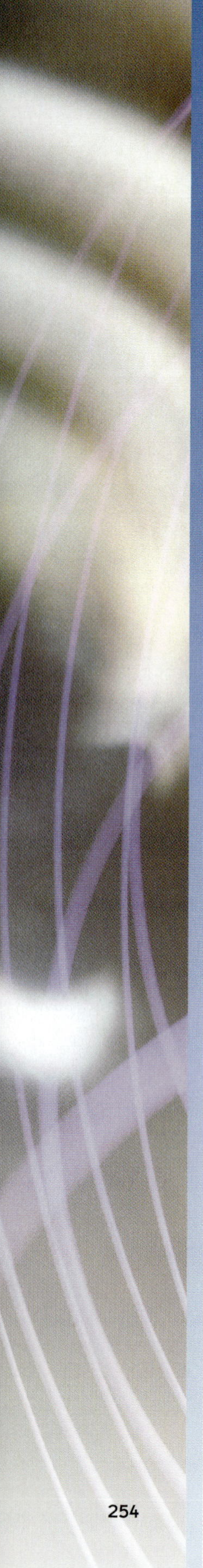

*18. What is the meaning of the term *center frequency* in reference to FM broadcasting?

*19. What is the meaning of the term *frequency swing* in reference to FM broadcast stations?

*20. What is the frequency swing of an FM broadcast transmitter when modulated 60 percent? ($\pm$45 kHz)

*21. An FM broadcast transmitter is modulated 40 percent by a 5-kHz test tone. When the percentage of modulation is doubled, what is the frequency swing of the transmitter?

*22. An FM broadcast transmitter is modulated 50 percent by a 7-kHz test tone. When the frequency of the test tone is changed to 5 kHz and the percentage of modulation is unchanged, what is the transmitter frequency swing?

*23. If the output current of an FM broadcast transmitter is 8.5 A without modulation, what is the output current when the modulation is 90 percent?

24. An FM transmitter delivers, to a 75-Ω antenna, a signal of $v = 1000 \sin(10^9 t + 4 \sin 10^4 t)$. Calculate the carrier and intelligence frequencies, power, modulation index, deviation, and bandwidth. (159 MHz, 1.59 kHz, 6.67 kW, 4, 6.37 kHz, $\sim$ 16 kHz)

25. Assuming that the 9.892-kW result of Ex. 5-7 is exactly correct, determine the total power in the J_2 sidebands and higher. (171 W)

26. Determine the deviation ratio for an FM system that has a maximum possible deviation of 5 kHz and the maximum input frequency is 3 kHz. Is this narrow- or wideband FM? (1.67, wideband)

Section 5-4

*27. What types of radio receivers do not respond to static interference?

*28. What is the purpose of a limiter stage in an FM broadcast receiver?

29. Explain why the limiter does *not* eliminate all noise effects in an FM system.

30. Calculate the amount of frequency deviation caused by a limited noise spike that still causes an undesired phase shift of 35° when f_i is 5 kHz. (3.05 kHz)

31. In a broadcast FM system, the input $S/N = 4$. Calculate the worst-case S/N at the output if the receiver's internal noise effect is negligible. (19.8 : 1)

32. Explain why narrowband FM systems have poorer noise performance than wideband systems.

33. Explain the *capture effect* in FM, and include the link between it and FM's inherent noise reduction capability.

*34. Why is narrowband FM rather than wideband FM used in radio communications systems?

*35. What is the purpose of preemphasis in an FM broadcast transmitter? Of deemphasis in an FM receiver? Draw a circuit diagram of a method of obtaining preemphasis.

*36. Discuss the following for frequency modulation systems:
   (a) The production of sidebands.
   (b) The relationship between the number of sidebands and the modulating frequency.
   (c) The relationship between the number of sidebands and the amplitude of the modulating voltage.
   (d) The relationship between percentage modulation and the number of sidebands.

(e) The relationship between modulation index or deviation ratio and the number of sidebands.
(f) The relationship between the spacing of the sidebands and the modulating frequency.
(g) The relationship between the number of sidebands and the bandwidth of emissions.
(h) The criteria for determining the bandwidth of emission.
(i) Reasons for preemphasis.

Section 5-5

37. Draw a schematic diagram of a varactor diode FM generator and explain its operation.
*38. Draw a schematic diagram of a frequency-modulated oscillator using a reactance modulator. Explain its principle of operation.
39. Using the specifications in Fig. 5-14, draw a schematic of an FM generator using the SE/NE 566 LIC function generator VCO. The center frequency is to be 500 kHz, and the output is to be a sine wave. Show all component values. How much center frequency drift can be expected from a temperature rise of 50°C?
40. Explain the principles of a Crosby-type modulator.
*41. How is good stability of a reactance modulator achieved?
*42. If an FM transmitter employs one doubler, one tripler, and one quadrupler, what is the carrier frequency swing when the oscillator frequency swing is 2 kHz? (48 kHz)
43. Draw a block diagram of a broadcast-band Crosby-type FM transmitter operating at 100 MHz, and label all frequencies in the diagram.
44. Explain the function of a discriminator.

Section 5-6

*45. Draw a block diagram of an Armstrong-type FM broadcast transmitter complete from the microphone input to the antenna output. State the purpose of each stage, and explain briefly the overall operation of the transmitter.
46. Explain the difference in the amount of deviation when passing an FM signal through a mixer as compared to a multiplier.
47. What type of circuits are used to increase a narrowband to a wideband?

Section 5-7

48. Explain the operation of the PLL FM transmitter shown in Fig. 5-20.

Section 5-8

49. Draw a block diagram of a stereo multiplex FM broadcast transmitter complete from the microphone inputs to the antenna output. State the purpose of each stage, and explain briefly the overall operation of the transmitter.
50. Explain how stereo FM can effectively transmit twice the information of a standard FM broadcast while still using the same bandwidth. How is the *S/N* at the receiver affected by a stereo transmission as opposed to monophonic?

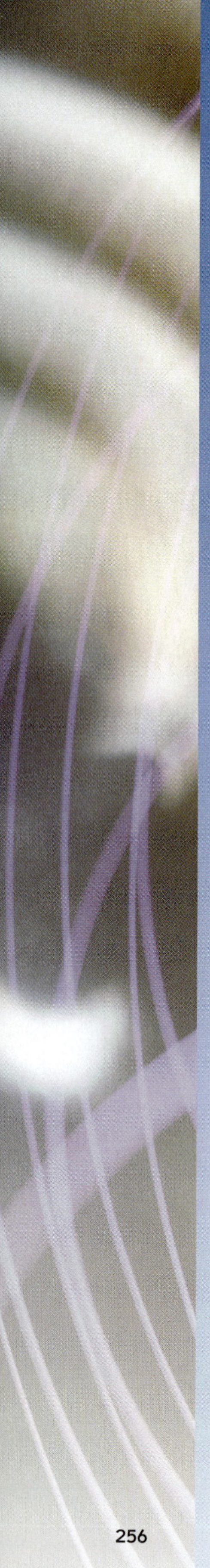

51. Define *frequency-division multiplexing.*
52. Describe the type of modulation used in the L − R signal of Fig. 5-23.
53. Explain the function of a matrix network as it relates to the generation of an FM stereo signal.
54. What difference in noise performance exists between FM stereo and mono broadcasts? Explain what might be done if an FM stereo signal is experiencing noise problems at the receiver.

Section 5-9

*55. What are the merits of an FM communications system compared to an AM system?
*56. Why is FM undesirable in the standard AM broadcast band?
57. What advantage does FM have over AM and SSB?

Section 5-10

58. What is the function of a multiplier stage in an FM transmission system? Explain how to troubleshoot a multiplier stage.
59. Briefly explain the function/operation of the reactance modulator in Fig. 5-27.
60. Resistor R5 in Fig. 5-27 has changed value. Describe the effect on this circuit and how to troubleshoot this situation.
61. What is the SCA signal? Give a technique to troubleshoot a transmitter that is not properly transmitting the SCA signal.
62. Describe the procedure to check an FM transmitter carrier frequency.
63. Describe the FM output if R5 was shorted in Fig. 5-27.
64. Explain what happens if R1 was shorted in Fig. 5-27.
65. In Fig. 5-29, explain what the output would be if the 38-kHz subcarrier were gone.
66. If the balanced modulator in Fig. 5-29 failed, describe the output.

Questions for Critical Thinking

67. Analyze the effect of an intelligence signal's amplitude and frequency when it frequency-modulates a carrier.
68. Contrast the modulation indexes for PM verses FM. Given this difference, could you modify a modulating signal so that allowing it to phase-modulate a carrier would result in FM? Explain your answer.
69. Does the maximum deviation directly determine the bandwidth of an FM system? If not, explain how bandwidth and deviation are related.
70. An FM transmitter puts out 1 kW of power. When $m_f = 2$, analyze the distribution of power in the carrier and all significant sidebands. Use Bessell functions to verify that the sum of these powers is 1 kW.
71. Why is the FCC concerned if an FM broadcast station overmodulates (deviation exceeds ± 75 kHz)?

6 Frequency Modulation: Reception

Chapter Outline

The Motorola cellular telephone contains a full set of features for cellular operation. (Courtesy of Motorola, Inc. Reprinted with permission.)

Objectives

- Describe the operation of an FM receiving system and highlight the difference compared to AM
- Sketch a slope detector schematic and explain how it can provide the required response to the modulating signal amplitude and frequency
- Provide various techniques and related circuits used in FM discriminators
- Explain the operation of the PLL and describe how it can be utilized as an FM discriminator
- Provide the block diagram of a complete stereo broadcast band receiver and explain its operation
- Analyze the operation of an LIC used as a stereo decoder
- Analyze and understand a complete FM receiver schematic

Key Terms

discriminator
automatic frequency control
local oscillator reradiation
cross-modulation
intermodulation distortion
dynamic range
sensitivity
quieting voltage
threshold voltage
limiting knee voltage
quadrature
phase-locked loop
phase comparator
phase detector
capture state
locked
free-running frequency
loop gain
hold-in range
subsidiary communication authorization

6-1 BLOCK DIAGRAM

The basic FM receiver uses the superheterodyne principle. In block diagram form, it has many similarities to the receivers covered in previous chapters. In Fig. 6-1, the only apparent differences are the use of the word *discriminator* in place of *detector,* the addition of a deemphasis network, and the fact that AGC may or may not be used as indicated by the dashed lines.

Discriminator
stage in an FM receiver that creates an output level that varies as a function of its input frequency; recovers the intelligence signal

The **discriminator** extracts the intelligence from the high-frequency carrier and can also be called the detector, as in AM receivers. By definition, however, a discriminator is a device in which amplitude variations are derived in response to frequency or phase variations, and it is the preferred term for describing an FM demodulator.

The deemphasis network following demodulation is required to bring the high-frequency intelligence back to the proper amplitude relationship with the lower frequencies. Recall that the high frequencies were preemphasized at the transmitter to provide them with greater noise immunity, as explained in Sec. 5-4.

Automatic Frequency Control
negative feedback control system in FM receivers used to achieve stability of the local oscillator

The fact that AGC is optional in an FM receiver may be surprising to you. From your understanding of AM receivers, you know that AGC is essential to their satisfactory operation. However, the use of limiters in FM receivers essentially provides an AGC function, as will be explained in Sec. 6-3. Many older FM receivers also included an **automatic frequency control** (AFC) function. This is a circuit that provides a slight automatic control over the local oscillator circuit. It compensates for drift in LO frequency that would otherwise cause a station to become detuned. It was necessary because it had not yet been figured out how to make an economical *LC* oscillator at 100 MHz with sufficient frequency stability. The AFC system is not needed in new designs.

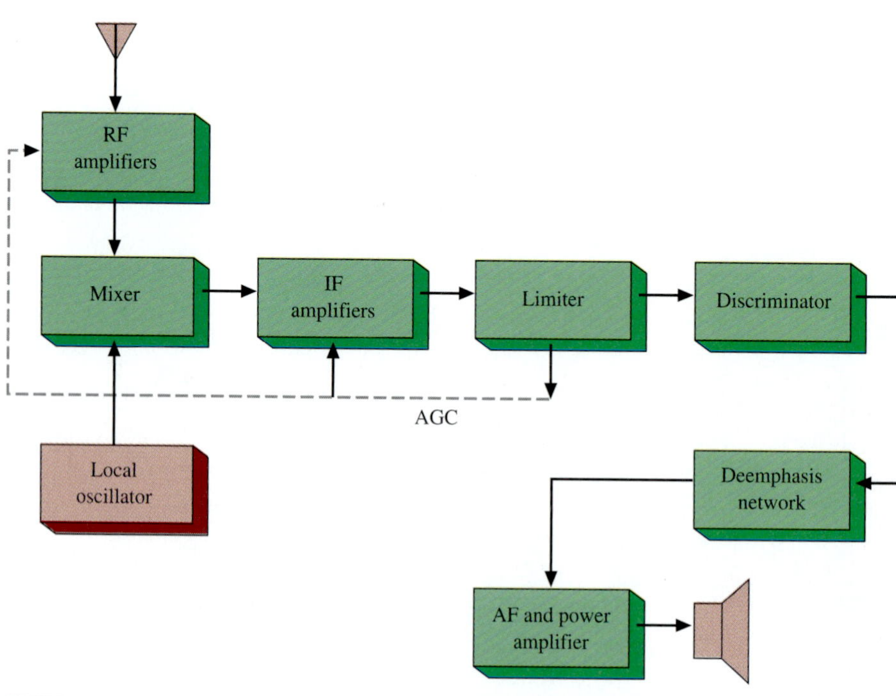

FIGURE 6-1 FM receiver block diagram.

The mixer, local oscillator, and IF amplifiers are basically similar to those discussed for AM receivers and do not require further elaboration. It should be noted that higher frequencies are usually involved, however, because of the fact that FM systems generally function at higher frequencies. The universally standard IF frequency for FM is 10.7 MHz, as opposed to 455 kHz for AM. Because of significant differences in all the other portions of the block diagram shown in Fig. 6-1, they are discussed in the following sections.

 6-2 RF Amplifiers

Broadcast AM receivers normally operate quite satisfactorily without any RF amplifier. This is rarely the case with FM receivers, however, except for frequencies in excess of 1000 MHz (1 GHz), when it becomes preferable to omit it. The essence of the problem is that FM receivers can function with weaker received signals than AM or SSB receivers because of their inherent noise reduction capability. This means that FM receivers can function with a lower sensitivity, and are called upon to deal with input signals of 1 μV or less as compared with perhaps a 30-μV minimum input for AM. If a 1-μV signal is fed directly into a mixer, the inherently high noise factor of an active mixer stage destroys the intelligibility of the 1-μV signal. Therefore, it is necessary to amplify the 1-μV level in an RF stage to get the signal up to at least 10 to 20 μV before mixing occurs. The FM system can tolerate 1 μV of noise from a mixer on a 20-μV signal but obviously cannot cope with 1 μV of noise with a 1-μV signal.

This reasoning also explains the abandonment of RF stages for the ever-increasing FM systems at the 1-GHz-and-above region. At these frequencies, transistor noise is increasing while gain is decreasing. The frequency is reached where it is advantageous to feed the incoming FM signal directly into a diode mixer to step it down immediately to a lower frequency for subsequent amplification. Diode (passive) mixers are less noisy than active mixers.

Of course, the use of an RF amplifier reduces the image frequency problem, as explained in Chapter 3. Another benefit is the reduction in **local oscillator reradiation** effects. Without an RF amp, the local oscillator signal can get coupled back more easily into the receiving antenna and transmit interference.

Local Oscillator Reradiation
undesired radiation of the local oscillator signal through a receiver's antenna

FET RF Amplifiers

Almost all RF amps used in quality FM receivers utilize FETs as the active element. You may think that this is done because of their high input impedance, but this is *not* the reason. In fact, their input impedance at the high frequency of FM signals is greatly reduced because of their input capacitance. The fact that FETs do not offer any significant impedance advantage over other devices at high frequencies is not a deterrent, however, because the impedance that an RF stage works from (the antenna) is only several hundred ohms or less anyway.

The major advantage is that FETs have an input/output square-law relationship while vacuum tubes have a $\frac{3}{2}$-power relationship and BJTs have a diode-type exponential characteristic. A square-law device has an output signal at the input frequency and a smaller distortion component at two times the input frequency, whereas the other devices mentioned have many more distortion components, with some of them occurring at frequencies close to the desired signal. The use of an

FET at the critical small signal level in a receiver means that the device distortion components are filtered out easily by its tuned circuits because the closest distortion component is two times the frequency of the desired signal. This becomes an extreme factor when you tune to a weak station that has a very strong adjacent signal. If the high-level adjacent signal gets through the input tuned circuit, even though greatly attenuated, it would probably generate distortion components at the desired signal frequency by a nonsquare-law device, and the result is audible noise in the speaker output. This form of receiver noise is called **cross-modulation.** This is similar to **intermodulation distortion,** which is characterized by the mixing of *two* undesired signals, resulting in an output component that is equal to the desired signal's frequency. The possibility of intermodulation distortion is also greatly minimized by the use of FET RF amplifiers. Additional discussion of intermodulation distortion is included in Sec. 7-4.

Cross-Modulation
form of distortion resulting in overdriven mixer stages

Intermodulation Distortion
undesired mixing of two signals in a receiver resulting in an output frequency component equal to that of the desired signal

MOSFET RF Amplifiers

A dual-gate, common-source MOSFET RF amplifier is shown in Fig. 6-2. The use of a dual-gate device allows a convenient isolated input for an AGC level to control device gain. The MOSFETs also offer the advantage of increased **dynamic range** over JFETs. That is, a wider range of input signal can be tolerated by the MOSFET while still offering the desired square-law input/output relationship. A similar arrangement is often utilized in mixers because the extra gate allows for a convenient injection point for the local oscillator signal. The accompanying chart in Fig. 6-2 provides component values for operation at 100-MHz and 400-MHz center frequencies. The antenna input signal is coupled into gate 1

Dynamic Range
decibel difference between the largest tolerable receiver input level and its sensitivity

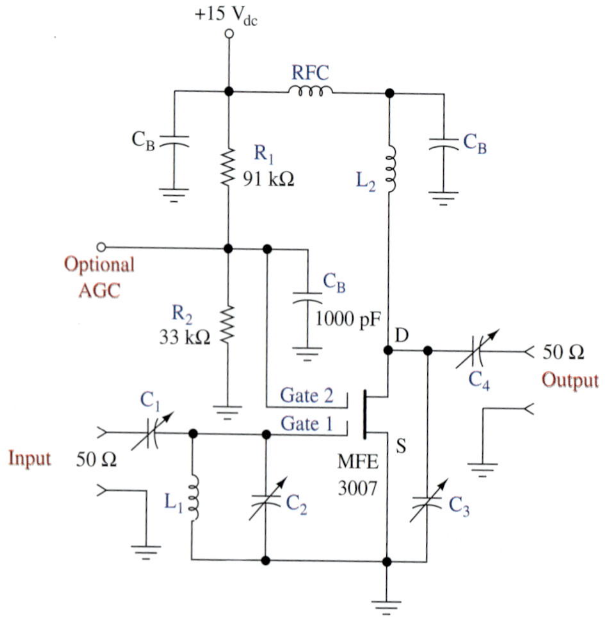

VHF Amplifier
The following component values are used for the different frequencies:

Component Values	100 MHz	400 MHz
C_1	8.4 pF	4.5 pF
C_2	2.5 pF	1.5 pF
C_3	1.9 pF	2.8 pF
C_4	4.2 pF	1.2 pF
L_1	150 nH	16 nH
L_2	280 nH	22 nH
C_B	1000 pF	250 pF

FIGURE 6-2 MOSFET RF amplifier. (Courtesy of Motorola Semiconductor Products, Inc.)

via the coupling/tuning network comprised of C_1, L_1, and C_2. The output signal is taken at the drain, which is coupled to the next stage by the L_2–C_3–C_4 combination. The bypass capacitor C_B next to L_2 and the radio-frequency choke (RFC) ensure that the signal frequency is not applied to the dc power supply. The RFC acts as an open to the signal while appearing as a short to dc, and the bypass capacitor acts in the inverse fashion. These precautions are necessary to RF frequencies because while power supply impedance is very low at low frequencies and dc, it looks like a high impedance to RF and can cause appreciable signal power loss. The bypass capacitor from gate 2 to ground provides a short to any high-frequency signal that may get to that point. It is necessary to maintain the bias stability set up by R_1 and R_2. The MFE 3007 MOSFET used in this circuit provides a minimum power gain of 18 dB at 200 MHz.

6-3 LIMITERS

A limiter is a circuit whose output is a constant amplitude for all inputs above a critical value. Its function in an FM receiver is to remove any residual (unwanted) amplitude modulation and the amplitude variations due to noise. Both of these variations would have an undesirable effect if carried through to the speaker. In addition, the limiting function also provides AGC action because signals from the critical minimum value up to some maximum value provide a constant input level to the detector. By definition, the discriminator (detector) ideally would not respond to amplitude variations anyway because the information is contained in the amount of frequency deviation and the rate at which it deviates back and forth around its center frequency.

A transistor limiter is shown in Fig. 6-3. Notice the dropping resistor, R_C, which limits the dc collector supply voltage. This provides a low dc collector voltage, which makes this stage very easily overdriven. This is the desired result. As

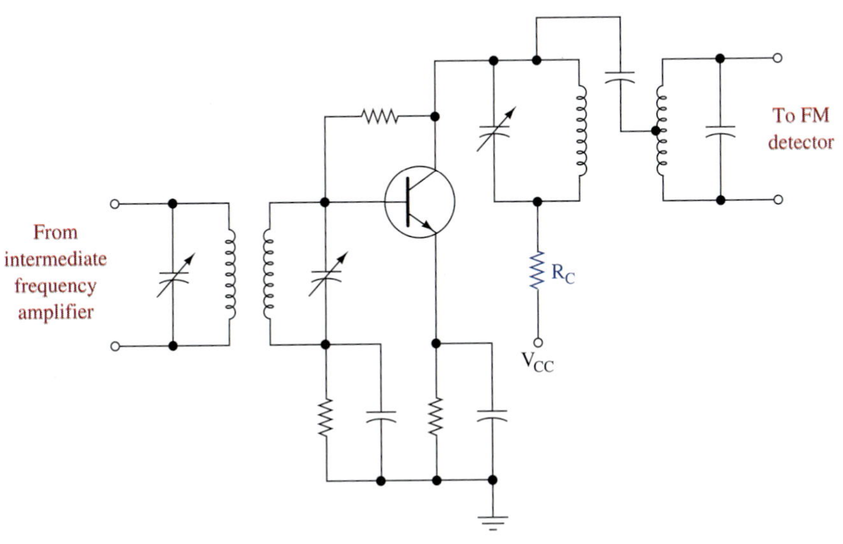

FIGURE 6-3 Transistor limiting circuit.

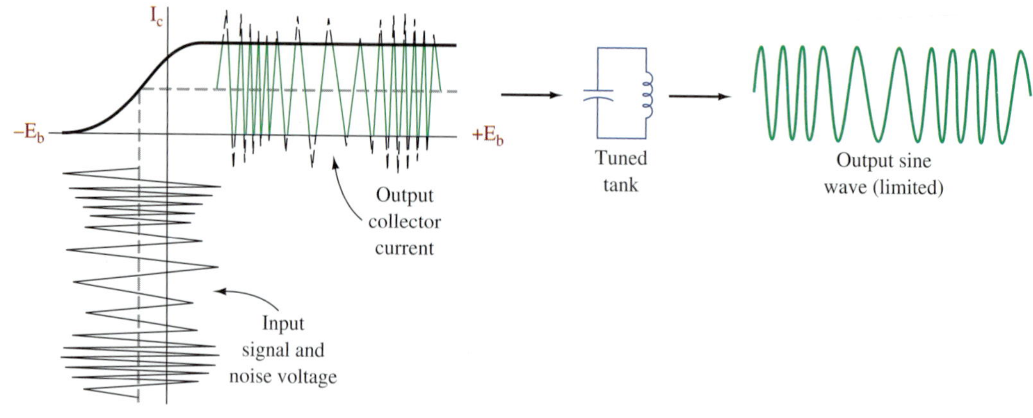

FIGURE 6-4 Limiter input/output and flywheel effects.

soon as the input is large enough to cause clipping at both extremes of collector current, the critical limiting voltage has been attained and limiting action has started.

The input/output characteristic for the limiter is shown in Fig. 6-4, and it shows the desired clipping action and the effects of feeding the limited (clipped) signal into an *LC* tank circuit tuned to the signal's center frequency. The natural flywheel effect of the tank removes all frequencies not near the center frequency and thus provides a sinusoidal output signal as shown. The omission of an *LC* circuit at the limiter output is desirable for some demodulator circuits. The quadrature detector (Sec. 6-4) uses the square-wave-like waveform that results.

Limiting and Sensitivity

Sensitivity
minimum input RF signal to a receiver that is required to produce a specified audio signal at its output

Quieting Voltage
the minimum FM receiver input signal that begins the limiting process

Threshold Voltage
another term for quieting voltage

Limiting Knee Voltage
another term for quieting voltage

A limiter, such as the one shown in Fig. 6-3, requires about 1 V of signal to begin limiting. Much amplification of the received signal is therefore needed prior to limiting, which explains its position following the IF stages. When enough signal arrives at the receiver to start limiting action, the set *quiets*, which means that background noise disappears. The **sensitivity** of an FM receiver is defined in terms of how much input signal is required to produce a specific level of quieting, normally 30 dB. This means that a good-quality receiver with a rated 1.5-μV sensitivity will have background noise 30 dB down from the desired input signal that has a 1.5-μV level.

The minimum required voltage for limiting is called the **quieting, threshold,** or **limiting knee voltage.** The limiter then provides a constant-amplitude output up to some maximum value that prescribes the limiting range. Going above the maximum value results either in a reduced and/or a distorted output. It is possible that a single-stage limiter will not allow for adequate range, thereby requiring a double limiter or the development of AGC control on the RF and IF amplifiers to minimize the possible limiter input range.

It is most common for today's FM receivers to use IC IF amplification. In these cases, the ICs have a built-in limiting action of very high quality (i.e., wide dynamic range). Section 6-7 provides an example of these ICs.

Example 6-1

A certain FM receiver provides a voltage gain of 200,000 (106 dB) prior to its limiter. The limiter's quieting voltage is 200 mV. Determine the receiver's sensitivity.

Solution

To reach quieting, the input must be

$$\frac{200\text{ mV}}{200{,}000} = 1\ \mu\text{V}$$

The receiver's sensitivity is therefore 1 μV.

6-4 Discriminators

The FM discriminator (detector) extracts the intelligence that has been modulated onto the carrier via frequency variations. It should provide an intelligence signal whose amplitude is dependent on instantaneous carrier frequency deviation and whose frequency is dependent on the carrier's rate of frequency deviation. A desired output amplitude versus input frequency characteristic for a broadcast FM discriminator is provided in Fig. 6-5. Notice that the response is linear in the allowed area of frequency deviation and that the output amplitude is directly proportional to carrier frequency deviation. Keep in mind, however, that FM detection takes place following the IF amplifiers, which means that the ±75-kHz deviation is intact but that carrier frequency translation (usually to 10.7 MHz) has occurred.

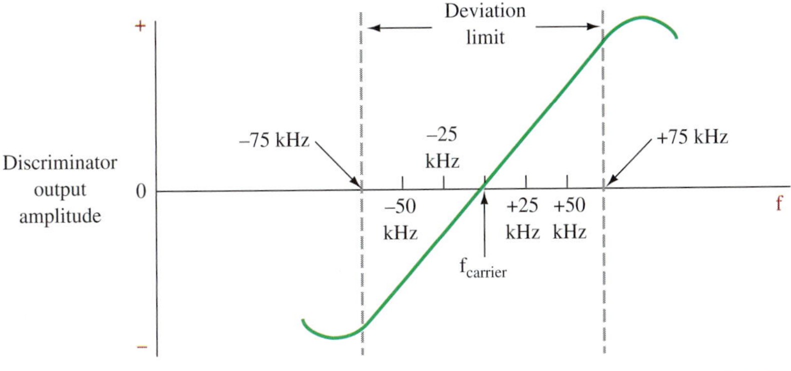

FIGURE 6-5 FM discriminator characteristic.

Slope Detector

The easiest FM discriminator to understand is the slope detector in Fig. 6-6. The *LC* tank circuit that follows the IF amplifiers and limiter is detuned from the carrier frequency so that f_c falls in the middle of the most linear region of the response curve.

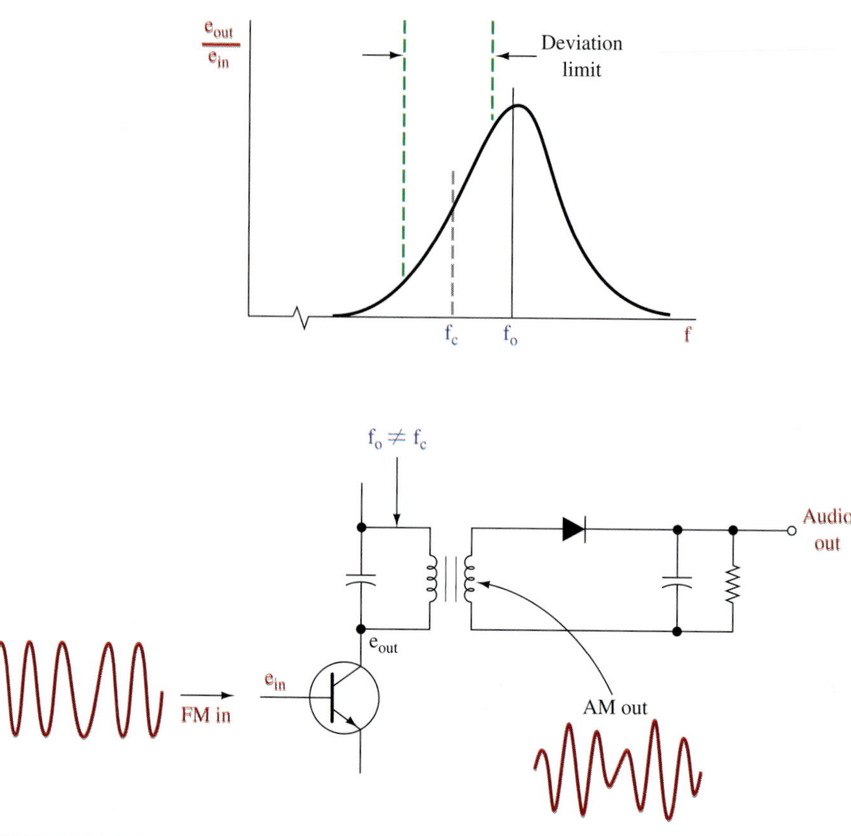

FIGURE 6-6 Slope detection.

When the FM signal rises in frequency above f_c, the output amplitude increases while deviations below f_c cause a smaller output. The slope detector thereby changes FM into AM, and a simple diode detector then recovers the intelligence contained in the AM waveform's envelope. In an emergency, an AM receiver can be used to receive FM by detuning the tank circuit feeding the diode detector. Slope detection is not widely used in FM receivers because the slope characteristic of a tank circuit is not very linear, especially for the large-frequency deviations of wideband FM.

Foster–Seely Discriminator

The two classical means of FM detection are the Foster–Seely discriminator and the ratio detector. While their once widespread use is now diminishing because of new techniques afforded by ICs, they remain a popular means of discrimination using a minimum of circuitry. A typical Foster–Seely discriminator circuit is shown in Fig. 6-7. In it, the two tank circuits [L_1C_1 and $(L_2 + L_3)C_2$] are tuned exactly to the carrier frequency. Capacitors C_c, C_4, and C_5 are shorts to the carrier frequency. The following analysis applies to an unmodulated carrier input:

1. The carrier voltage e_1 appears directly across L_4 because C_c and C_4 are shorts to the carrier frequency.

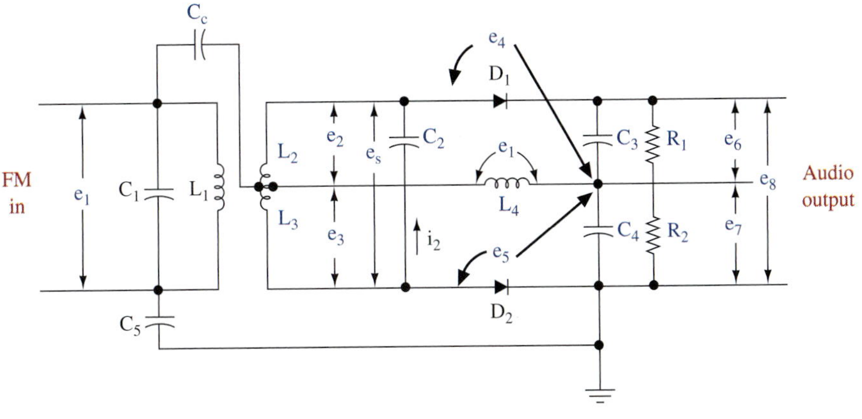

FIGURE 6-7 Foster–Seely discriminator.

2. The voltage e_s across the transformer secondary (L_2 in series with L_3) is 180° out of phase with e_1 by transformer action, as shown in Fig. 6-8(a). The circulating $L_2L_3C_2$ tank current, i_s, is in phase with e_s because the tank is resonant.
3. The current i_s, flowing through inductance L_2L_3, produces a voltage drop that lags i_s by 90°. The individual components of this voltage, e_2 and e_3, are thus displaced by 90° from i_s, as shown in Fig. 6-8(a), and are 180° out of phase with each other because they are the voltage from the ends of a center-tapped winding.
4. The voltage e_4 applied to the diode D_1, C_3, and R_1 network will be the vector sum of e_1 and e_2 [Fig. 6-8(a)]. Similarly, the voltage e_5 is the sum of e_1 and e_3. The magnitude of e_6 is proportional to e_4 while e_7 is proportional to e_5.
5. The output voltage, e_8, is equal to the sum of e_6 and e_7 and is zero because the diodes D_1 and D_2 will be conducting current equally (because $e_4 = e_5$) but in opposite directions through the R_1C_3 and R_2C_4 networks.

The discriminator output is zero with no modulation (zero frequency deviation), as is desired. The following discussion now considers circuit action at some instant when the input signal e_1 is above the carrier frequency. The phasor diagram of Fig. 6-8(b) is used to illustrate this condition:

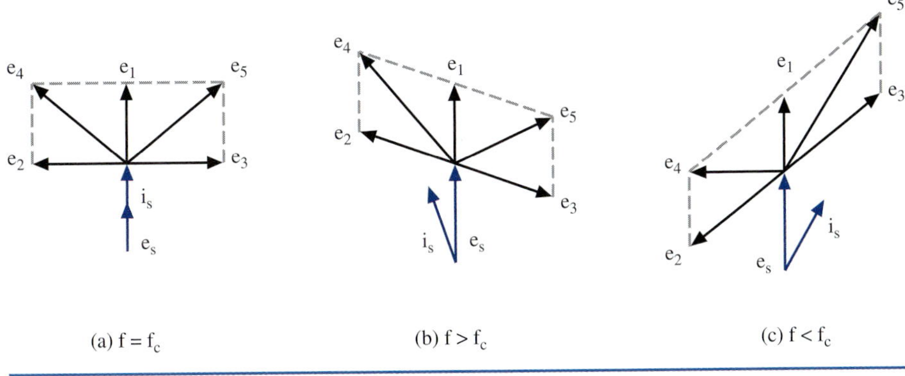

FIGURE 6-8 Discriminator phase relations.

1. Voltages e_1 and e_s are as before, but e_s now sees an inductive reactance because the tank circuit is above resonance. Therefore, the circulating tank current, i_s, lags e_s.
2. The voltages e_2 and e_3 must remain 90° out of phase with i_s, as shown in Fig. 6-8(b). The new vector sums of $e_2 + e_1$ and $e_3 + e_1$ are no longer equal, so e_4 causes a heavier conduction of D_1 than exists for D_2.
3. The output, e_8, which is the sum of e_6 and e_7, will go positive because the current down through R_1C_3 is greater than the current up through R_2C_4 (e_4 is greater than e_5).

The output for frequencies above resonance (f_c) is therefore positive, while the phasor diagram in Fig. 6-8(c) shows that at frequencies below resonance the output goes negative. The amount of output is determined by the amount of frequency deviation, while the frequency of the output is determined by the rate at which the FM input signal varies around its carrier or center value.

Ratio Detector

While the Foster–Seely discriminator just described offers excellent linear response to wideband FM signals, it also responds to any undesired input amplitude variations. The *ratio detector* does not respond to amplitude variations and thereby minimizes the required limiting before detection.

The ratio detector, shown in Fig. 6-9, is a circuit designed to respond only to frequency changes of the input signal. Amplitude changes in the input have no effect upon the output. The input circuit of the ratio detector is identical to that of the Foster–Seely discriminator circuit. The most immediately obvious difference is the reversal of one of the diodes.

The ratio detector circuit operation is similar to the Foster–Seely. A detailed analysis will therefore not be given. Notice the large electrolytic capacitor, C_5, across the R_1–R_2 combination. This maintains a constant voltage that is equal to the peak voltage across the diode input. This feature eliminates variations in the FM signal, thus providing amplitude limiting. The sudden changes in the input signal's amplitude are suppressed by the large capacitor. The Foster–Seely discriminator does not provide amplitude limiting. The voltage E_s is

$$E_s = e_1 + e_2$$

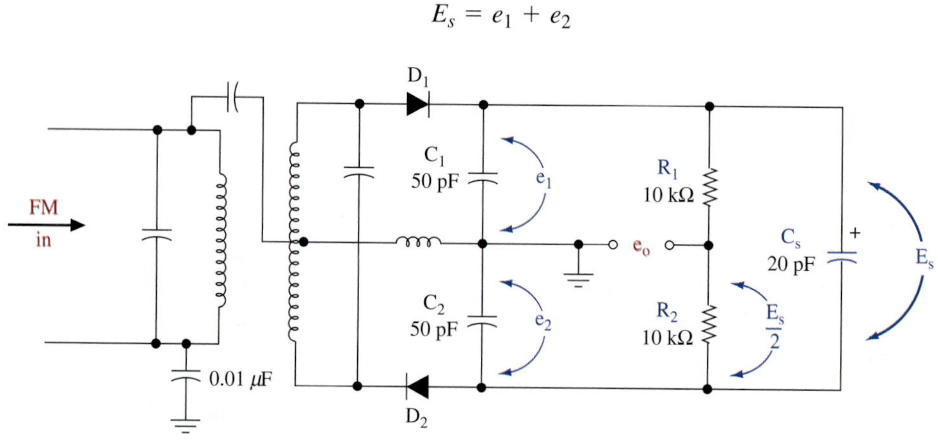

FIGURE 6-9 Ratio detector.

and

$$e_0 = \frac{E_s}{2} - e_2 = \frac{e_1 + e_2}{2} - e_2$$
$$= \frac{e_1 - e_2}{2}$$

When $f_{in} = f_c$, $e_1 = e_2$ and hence the desired zero output occurs. When $f_{in} > f_c$, $e_1 > e_2$, and when $f_{in} < f_c$, $e_1 < e_2$. The desired frequency-dependent output characteristic results.

The component values shown in Fig. 6-9 are typical for a 10.7-MHz IF FM input signal. The output level of the ratio detector is one-half that for the Foster–Seely circuit.

Quadrature Detector

The Foster–Seely and ratio detector circuits do not lend themselves to integration on a single chip due to the transformer required. This has led to increased usage of the quadrature detector and phase-locked loop (PLL). The PLL is introduced in the next section.

Quadrature detectors derive their name from use of the FM signal in phase and 90° out of phase. The two signals are said to be in **quadrature**—at a 90° angle. The circuit in Fig. 6-10 shows an FM quadrature detector using an exclusive-OR gate. The limited IF output is applied directly to one input and the phase-shifted signal to the other. Notice that this circuit uses the limited signal that has not been changed back to a sine wave. The *L*, *C*, and *R* values used at the circuit's input are chosen to provide a 90° phase shift at the carrier frequency to the signal 2 input. The signal 2 input is a sine wave due to the *LC* circuit effects. The upward and downward frequency deviation of the FM signal results in a corresponding higher or lower phase shift. With one input to the gate shifted, the gate output will be a series of pulses with a width proportional to the phase difference. The low-pass *RC* filter at the gate output sums the output, giving an average value that is the intelligence signal. The gate output for three different phase conditions is shown at Fig. 6-10(b). The *RC* circuit output level for each case is shown with dashed lines. This corresponds to the intelligence level at those particular conditions.

Quadrature
signals at a 90° angle

An analog quadrature detector is possible using a differential amplifier configuration, as shown in Fig. 6-11. A limited FM signal switches the transistor current source (Q_1) of the differential pair $Q_2 + Q_3$. L_1 and C_2 should be resonant at the IF frequency. The L_1–C_2–C_1 combination causes the desired frequency-dependent phase shift between the two signals applied to Q_2 and Q_1. The conduction through Q_3 depends on the coincident phase relationships of these two signals. The pulses generated at Q_3's collector are summed by the R_1–C_3 low-pass filter, and the resulting intelligence signal is taken at Q_4's emitter. R_2 is adjusted to yield the desired zero-volt output when an undeviating FM carrier is the circuit's input signal.

The popular 3089 LIC shown in Sec. 6-7 uses the analog quadrature detection technique. It provides an excellent total harmonic distortion (THD) specification of 0.1 percent (typically) for a 10.7-MHz IF and ±75-kHz deviation.

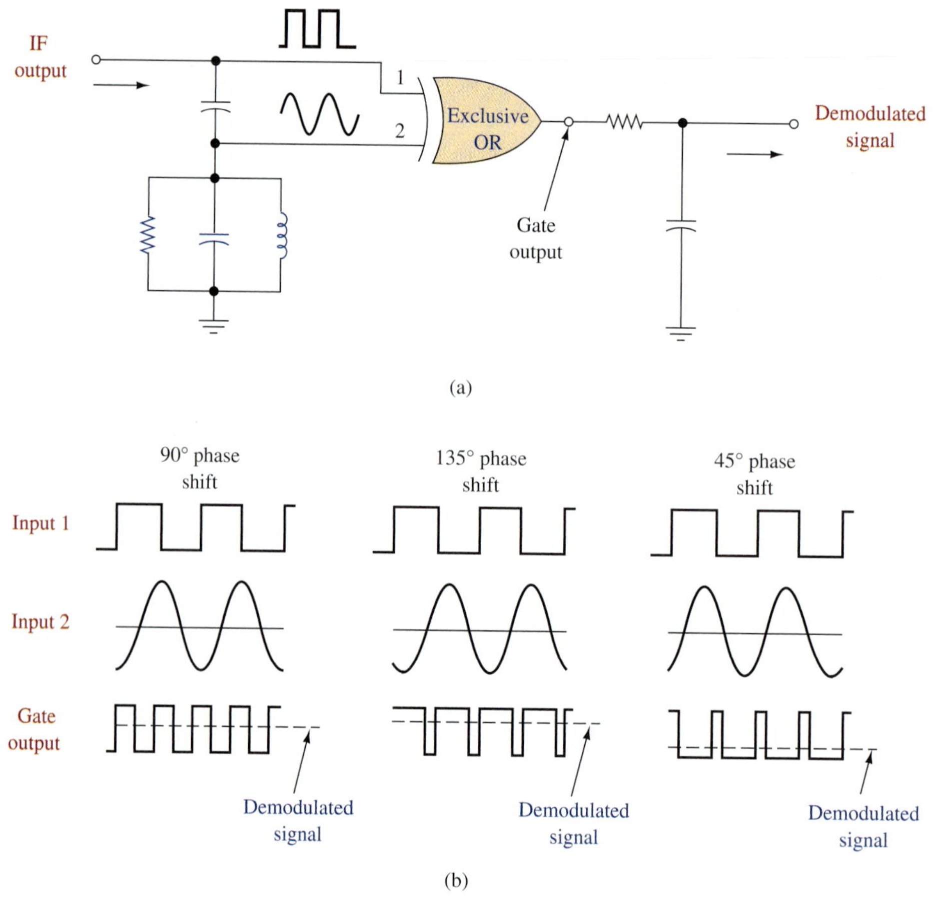

FIGURE 6-10 Quadrature detection.

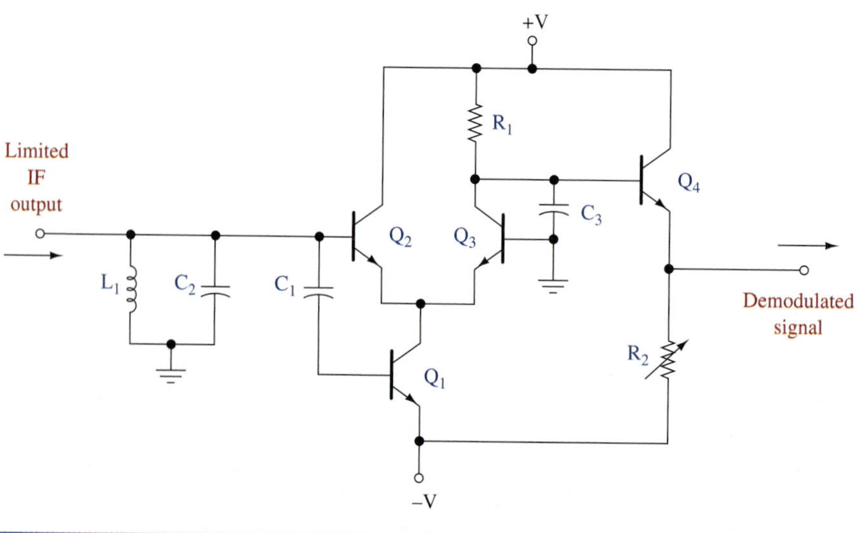

FIGURE 6-11 Analog quadrature detector.

6-5 Phase-Locked Loop

The **phase-locked loop** (PLL) has become increasingly popular as a means of FM demodulation in recent years. It eliminates the need for the intricate coil adjustments of the previously discussed discriminators and has many other uses in the field of electronics. It is an example of an old idea, originated in 1932, that was given a new life by integrated circuit technology. Prior to its availability in a single IC package in 1970, its complexity in discrete circuitry form made it economically unfeasible for most applications.

Phase-Locked Loop
closed-loop control system that uses negative feedback to maintain constant output frequency

The PLL is an electronic feedback control system as represented by the block diagram in Fig. 6-12. This input is to the **phase comparator,** or **phase detector** as it is also called. The VCO within the PLL generates the other signal applied to the comparator.

Phase Comparator
circuit that provides an output proportional to the phase difference of two inputs

Phase Detector
another term for phase comparator

The comparator compares the input signal and the output of the VCO and develops an error signal proportional to the difference between the two. This error signal drives the VCO to change frequency so that the error is reduced to zero. If the VCO frequency equals the input frequency, the PLL has achieved lock and the control voltage will be constant for as long as the PLL input frequency remains constant.

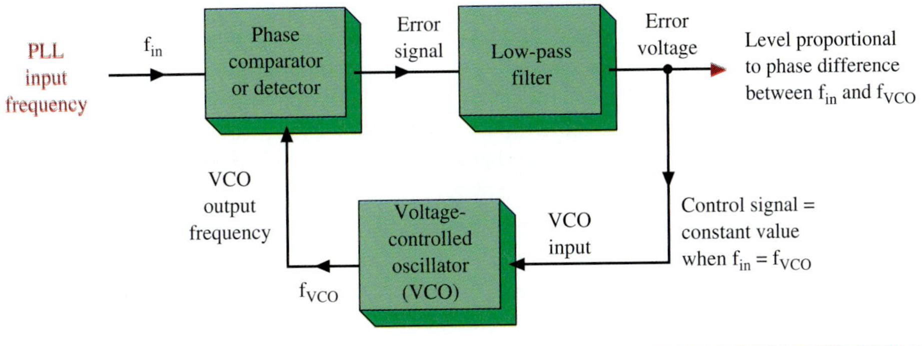

FIGURE 6-12 PLL block diagram.

PLL Capture and Lock

If the VCO starts to change frequency, it is in the **capture state.** It then continues to change frequency until its output is the same frequency as the input. At that point, the PLL is **locked;** the VCO frequency now equals that of the input signal. The PLL has three possible states of operation:

Capture State
when the phase comparator of a PLL generates a signal that forces the VCO to equal the input frequency

Locked
a PLL in the capture state

1. Free-running
2. Capture
3. Locked or tracking

If the input and VCO frequency are too far apart, the PLL free-runs at the nominal VCO frequency, which is determined by an external timing capacitor. This is not a normally used mode of operation. If the VCO and input frequency are close enough, the capture process begins and continues until the locked condition is reached. Once tracking (lock) begins, the VCO can remain locked over a wider input-frequency-range variation than was necessary to achieve capture. The tracking and capture ranges are a function of external resistors and/or capacitors selected by the user.

Example 6-2

A PLL is set up so that its VCO free-runs at 10 MHz. The VCO does not change frequency until the input is within 50 kHz of 10 MHz. After that condition, the VCO follows the input to ±200 kHz of 10 MHz before the VCO starts to free-run again. Determine the lock and capture ranges of the PLL.

Solution

The capture occurred at 50 kHz from the free-running VCO frequency. Assume symmetrical operation, which implies a capture range of 50 kHz × 2 = 100 kHz. Once captured, the VCO follows the input to a 200-kHz deviation, implying a lock range of 200 kHz × 2 = 400 kHz.

PLL FM Demodulator

If the PLL input is an FM signal, the low-pass filter output (error voltage) is the demodulated signal. The modulated FM carrier changes frequency according to the modulating signal. The function of the phase-locked loop is to hold the VCO frequency in step with this changing carrier. If the carrier frequency increases, for example, the error voltage developed by the phase comparator and the low-pass filter rises to make the VCO frequency rise. Let the carrier frequency fall and the error voltage output drops to decrease the VCO frequency. Thus, we see that the error voltage matches the modulating signal back at the transmitter; the error signal is the demodulated output.

The VCO input control signal (demodulated FM) causes the VCO output to match the FM signal applied to the PLL (comparator input). If the FM carrier (center) frequency drifts because of local oscillator drift, the PLL readjusts itself and no realignment is necessary. In a conventional FM discriminator, any shift in the FM carrier frequency results in a distorted output because the *LC* detector circuits are then untuned. The PLL FM discriminator requires no tuned circuits nor their associated adjustments and adjusts itself to any carrier frequency drifts caused by LO or transmitted carrier drift. In addition, the PLL normally has large amounts of internal amplification, which allows the input signal to vary from the microvolt region up to several volts. Since the phase comparator responds only to phase changes and not to amplitudes, the PLL is seen to provide a limiting function of extremely wide range. The use of PLL FM detectors is widespread in current designs.

LM 565 PLL

The circuit shown in Fig. 6-13 demonstrates how a PLL can be used to demodulate a frequency-modulated signal. This circuit consists of an FM source, generated by an LM566 voltage-controlled oscillator (VCO), which is being input into an LM565 PLL. The circuit provides a simple test for PLL operation. The input signal is applied to V_{in} (pin 5 of the VCO). The 1-μf capacitor couples the ac input signal into the VCO input. The VCO is prebiased to a center operating frequency by the voltage divider formed by resistor R_1, R_2, and R_3. Potentiometer R_2 provides adjustment of the center frequency. The output of the VCO is input into the 565 PLL through coupling capacitor C_c. The **free-running frequency** of the PLL is set to

Free-Running Frequency
frequency at which the PLL runs with the input signal removed

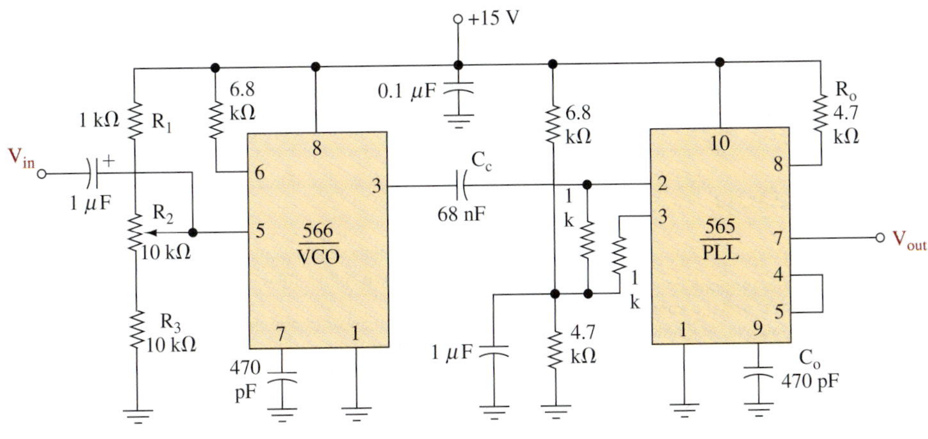

FIGURE 6-13 An example of an FM receiver using the LM565 PLL.

the center frequency of the VCO by the timing capacitor and resistor, C_o (470 pF) and R_o (4.7 kΩ).

The specifications for the LM565 PLL are given in the data sheets in Fig. 6-14. The LM565 can be used in many applications, including data synchronization, modems, low-frequency FSK demodulation, low-frequency FM demodulation, and frequency synthesis. The pin configuration for the device is shown on page 1 of the data sheets.

The formulas used for component calculations for the LM565 are provided in the manufacturers' specifications and application sheets. The following calculations are for the PLL shown in Fig. 6-13.

1. *Free-Running Frequency*

$$f_o \simeq \frac{0.3}{(R_o\, C_o)} = \frac{0.3}{(4.7\ \text{k}\Omega)(470 \times 10^{-12})} = 135.8\ \text{kHz}$$

2. *Loop Gain* $(K_o K_D)$ The **loop gain** is

Loop Gain
total gain of all the internal blocks inside the PLL

$$K_o\, K_D = \frac{33.6 f_o}{V_c} = \frac{(33.6)(135.8\ \text{kHz})}{15} = 3.04 \times 10^5$$

Expressed in dB, the loop gain is 10 log$(3.04 \times 10^5) = 54.8$ dB.

3. *Hold-In Range*

$$f_H = \pm\frac{8 f_o}{V_c} = \frac{(8)(135.8\ \text{k}\Omega)}{15} = \pm 72.43\ \text{k}\Omega$$

Note: The **hold-in range** is the frequency band through which the PLL will remain locked.

Hold-In Range
range of frequencies in which the PLL will remain locked

The PLL used in a frequency synthesizer is detailed in Chapter 7.

National Semiconductor

May 1999

LM565/LM565C Phase Locked Loop

LM565/LM565C Phase Locked Loop

General Description

The LM565 and LM565C are general purpose phase locked loops containing a stable, highly linear voltage controlled oscillator for low distortion FM demodulation, and a double balanced phase detector with good carrier suppression. The VCO frequency is set with an external resistor and capacitor, and a tuning range of 10:1 can be obtained with the same capacitor. The characteristics of the closed loop system — bandwidth, response speed, capture and pull in range — may be adjusted over a wide range with an external resistor and capacitor. The loop may be broken between the VCO and the phase detector for insertion of a digital frequency divider to obtain frequency multiplication.

The LM565H is specified for operation over the −55°C to +125°C military temperature range. The LM565CN is specified for operation over the 0°C to +70°C temperature range.

Features

- 200 ppm/°C frequency stability of the VCO
- Power supply range of ±5 to ±12 volts with 100 ppm/% typical
- 0.2% linearity of demodulated output
- Linear triangle wave with in phase zero crossings available
- TTL and DTL compatible phase detector input and square wave output
- Adjustable hold in range from ±1% to > ±60%

Applications

- Data and tape synchronization
- Modems
- FSK demodulation
- FM demodulation
- Frequency synthesizer
- Tone decoding
- Frequency multiplication and division
- SCA demodulators
- Telemetry receivers
- Signal regeneration
- Coherent demodulators

Connection Diagrams

Metal Can Package

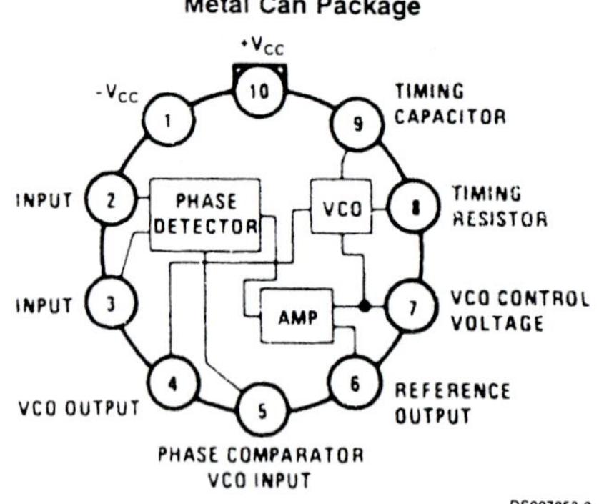

Order Number LM565H
See NS Package Number H10C

Dual-in-Line Package

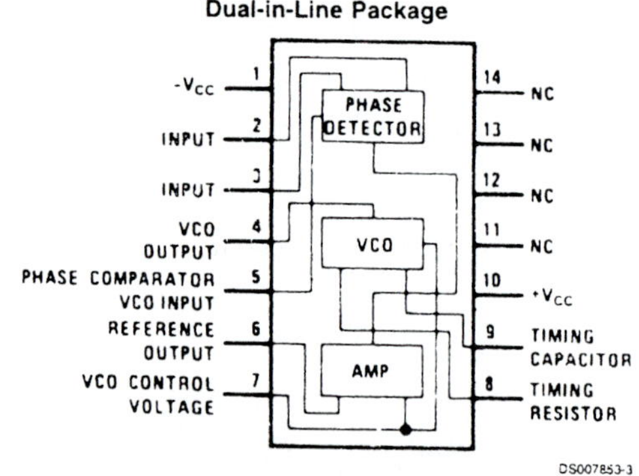

Order Number LM565CN
See NS Package Number N14A

www.national.com

FIGURE 6-14 The LM565 phase-locked-loop data sheets. (Reprinted with permission of National Semiconductor Corporation.)

Absolute Maximum Ratings (Note 1)

If Military/Aerospace specified devices are required, please contact the National Semiconductor Sales Office/ Distributors for availability and specifications.

Supply Voltage	±12V
Power Dissipation (Note 2)	1400 mW
Differential Input Voltage	±1V
Operating Temperature Range	
LM565H	−55°C to +125°C
LM565CN	0°C to +70°C
Storage Temperature Range	−65°C to +150°C
Lead Temperature (Soldering, 10 sec.)	260°C

Electrical Characteristics

AC Test Circuit, $T_A = 25°C$, $V_{CC} = \pm 6V$

Parameter	Conditions	LM565			LM565C			Units
		Min	Typ	Max	Min	Typ	Max	
Power Supply Current			8.0	12.5		8.0	12.5	mA
Input Impedance (Pins 2, 3)	$-4V < V_2, V_3 < 0V$	7	10			5		kΩ
VCO Maximum Operating Frequency	$C_o = 2.7$ pF	300	500		250	500		kHz
VCO Free-Running Frequency	$C_o = 1.5$ nF $R_o = 20$ kΩ $f_o = 10$ kHz	−10	0	+10	−30	0	+30	%
Operating Frequency Temperature Coefficient			−100			−200		ppm/°C
Frequency Drift with Supply Voltage			0.1	1.0		0.2	1.5	%/V
Triangle Wave Output Voltage		2	2.4	3	2	2.4	3	V_{p-p}
Triangle Wave Output Linearity			0.2			0.5		%
Square Wave Output Level		4.7	5.4		4.7	5.4		V_{p-p}
Output Impedance (Pin 4)			5			5		kΩ
Square Wave Duty Cycle		45	50	55	40	50	60	%
Square Wave Rise Time			20			20		ns
Square Wave Fall Time			50			50		ns
Output Current Sink (Pin 4)		0.6	1		0.6	1		mA
VCO Sensitivity	$f_o = 10$ kHz		6600			6600		Hz/V
Demodulated Output Voltage (Pin 7)	±10% Frequency Deviation	250	300	400	200	300	450	mV_{p-p}
Total Harmonic Distortion	±10% Frequency Deviation		0.2	0.75		0.2	1.5	%
Output Impedance (Pin 7)			3.5			3.5		kΩ
DC Level (Pin 7)		4.25	4.5	4.75	4.0	4.5	5.0	V
Output Offset Voltage $\lvert V_7 - V_6 \rvert$			30	100		50	200	mV
Temperature Drift of $\lvert V_7 - V_6 \rvert$			500			500		μV/°C
AM Rejection		30	40			40		dB
Phase Detector Sensitivity K_D			0.68			0.68		V/radian

Note 1: Absolute Maximum Ratings indicate limits beyond which damage to the device may occur. Operating Ratings indicate conditions for which the device is functional, but do not guarantee specific performance limits. Electrical Characteristics state DC and AC electrical specifications under particular test conditions which guarantee specific performance limits. This assumes that the device is within the Operating Ratings. Specifications are not guaranteed for parameters where no limit is given, however, the typical value is a good indication of device performance.

Note 2: The maximum junction temperature of the LM565 and LM565C is +150°C. For operation at elevated temperatures, devices in the TO-5 package must be derated based on a thermal resistance of +150°C/W junction to ambient or +45°C/W junction to case. Thermal resistance of the dual-in-line package is +85°C/W.

www.national.com

FIGURE 6-14 *(Continued)*

AC Test Circuit

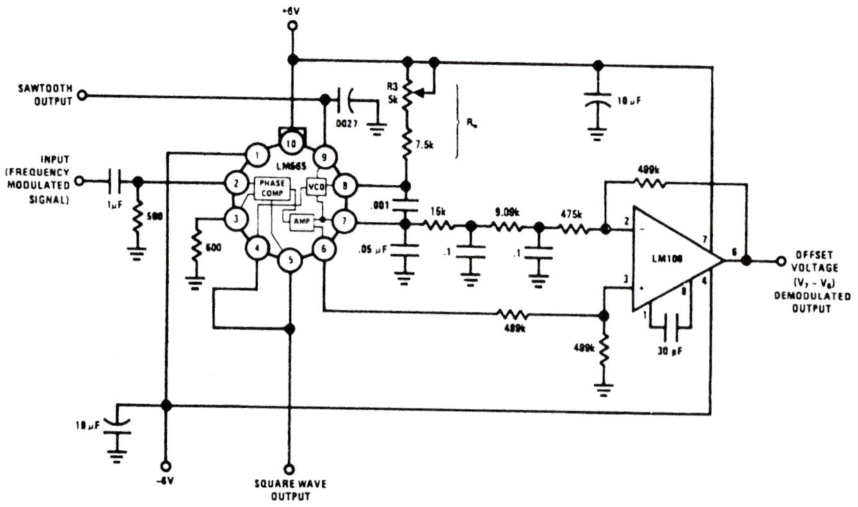

DS007853-5

Note: S_1 open for output offset voltage ($V_7 - V_6$) measurement.

Typical Applications

2400 Hz Synchronous AM Demodulator

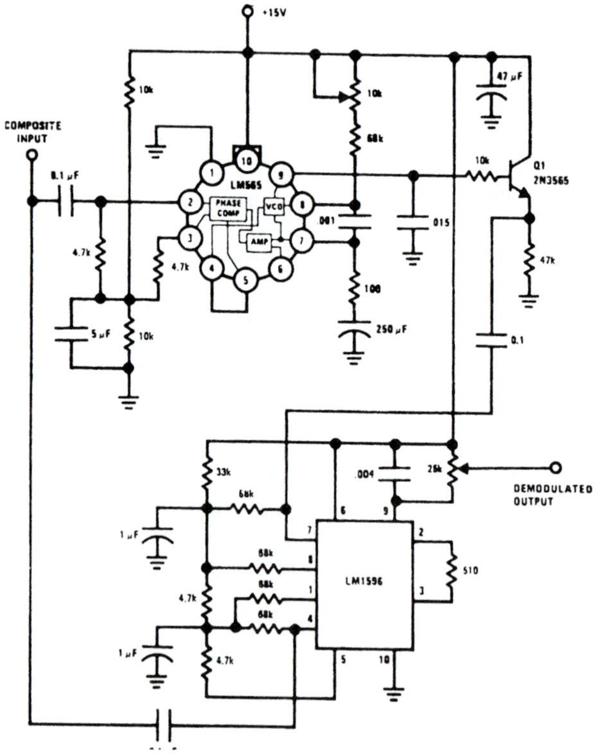

www.national.com

FIGURE 6-14 *(Continued)*

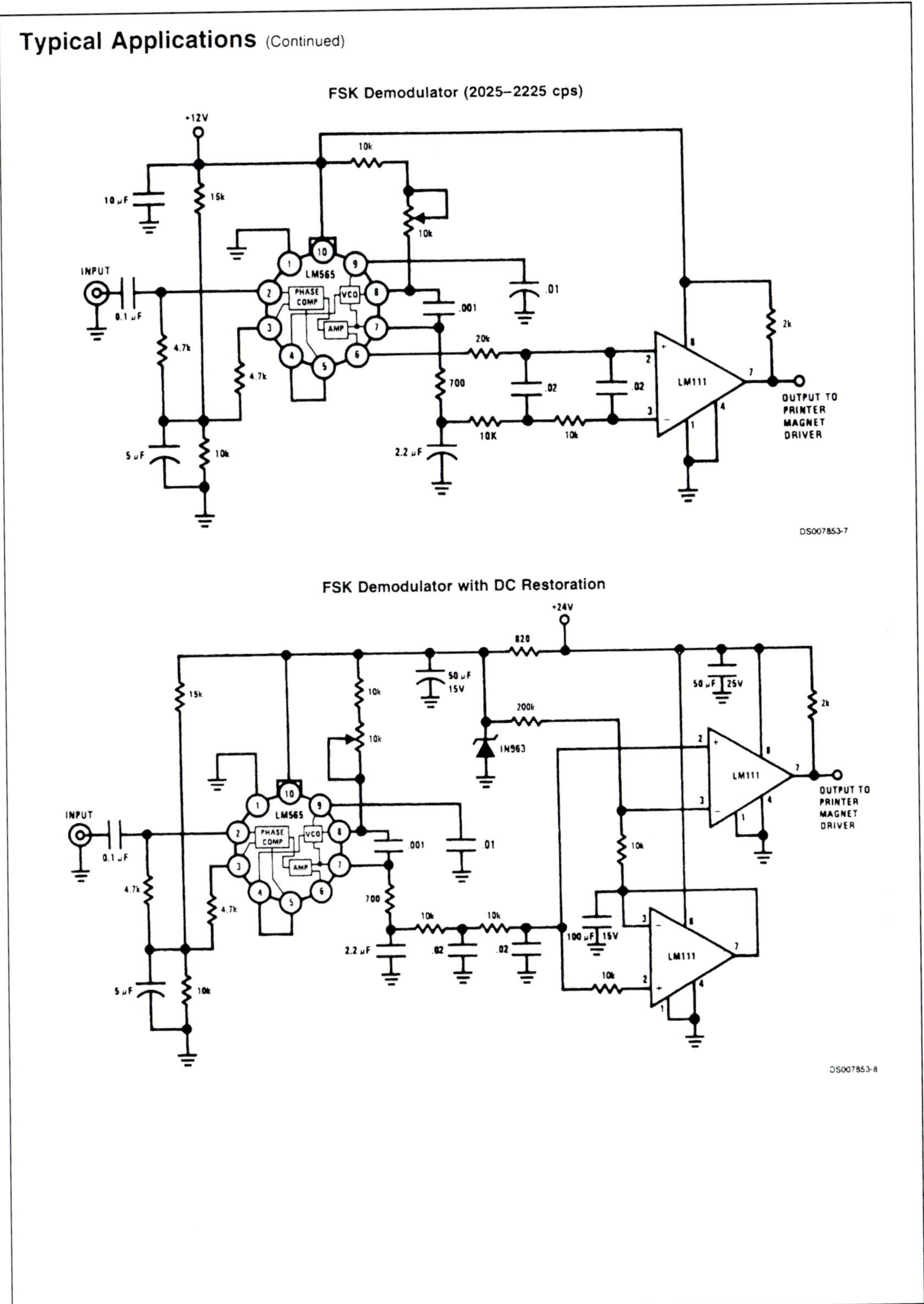

Typical Applications (Continued)

www.national.com

FIGURE 6-14 *(Continued)*

Applications Information

In designing with phase locked loops such as the LM565, the important parameters of interest are:

FREE RUNNING FREQUENCY

$$f_o \cong \frac{0.3}{R_o C_o}$$

LOOP GAIN: relates the amount of phase change between the input signal and the VCO signal for a shift in input signal frequency (assuming the loop remains in lock). In servo theory, this is called the "velocity error coefficient."

$$\text{Loop gain} = K_o K_D \left(\frac{1}{\text{sec}}\right)$$

$$K_o = \text{oscillator sensitivity} \left(\frac{\text{radians/sec}}{\text{volt}}\right)$$

$$K_D = \text{phase detector sensitivity} \left(\frac{\text{volts}}{\text{radian}}\right)$$

The loop gain of the LM565 is dependent on supply voltage, and may be found from:

$$K_o K_D = \frac{33.6 f_o}{V_C}$$

f_o = VCO frequency in Hz

V_c = total supply voltage to circuit

Loop gain may be reduced by connecting a resistor between pins 6 and 7; this reduces the load impedance on the output amplifier and hence the loop gain.

HOLD IN RANGE: the range of frequencies that the loop will remain in lock after initially being locked.

$$f_H = \pm \frac{8 f_o}{V_C}$$

f_o= free running frequency of VCO

V_c= total supply voltage to the circuit

THE LOOP FILTER

In almost all applications, it will be desirable to filter the signal at the output of the phase detector (pin 7); this filter may take one of two forms:

Simple Lead Filter

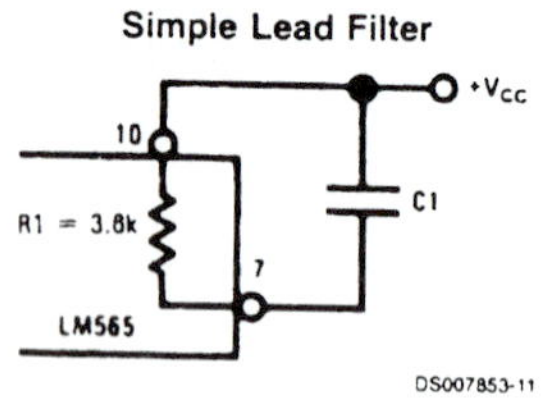

Lag-Lead Filter

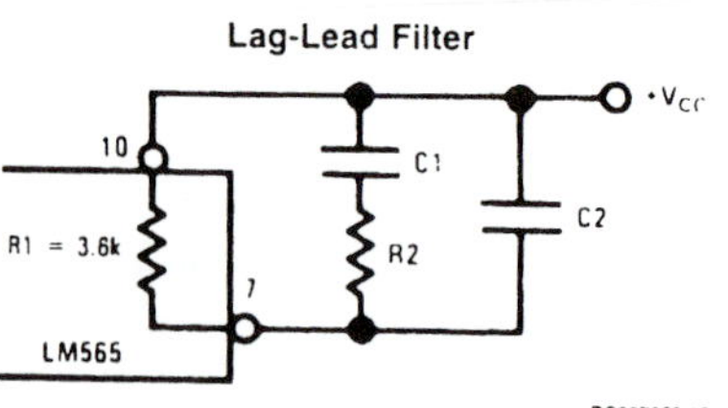

A simple lag filter may be used for wide closed loop bandwidth applications such as modulation following where the frequency deviation of the carrier is fairly high (greater than 10%), or where wideband modulating signals must be followed.

The natural bandwidth of the closed loop response may be found from:

$$f_n = \frac{1}{2\pi}\sqrt{\frac{K_o K_D}{R_1 C_1}}$$

Associated with this is a damping factor:

$$\delta = \frac{1}{2}\sqrt{\frac{1}{R_1 C_1 K_o K_D}}$$

For narrow band applications where a narrow noise bandwidth is desired, such as applications involving tracking a slowly varying carrier, a lead lag filter should be used. In general, if $1/R_1C_1 < K_o K_D$, the damping factor for the loop becomes quite small resulting in large overshoot and possible instability in the transient response of the loop. In this case, the natural frequency of the loop may be found from

$$f_n = \frac{1}{2\pi}\sqrt{\frac{K_o K_D}{\tau_1 + \tau_2}}$$

$$\tau_1 + \tau_2 = (R_1 + R_2) C_1$$

R_2 is selected to produce a desired damping factor δ, usually between 0.5 and 1.0. The damping factor is found from the approximation:

$$\delta \cong \pi \tau_2 f_n$$

These two equations are plotted for convenience.

Filter Time Constant vs Natural Frequency

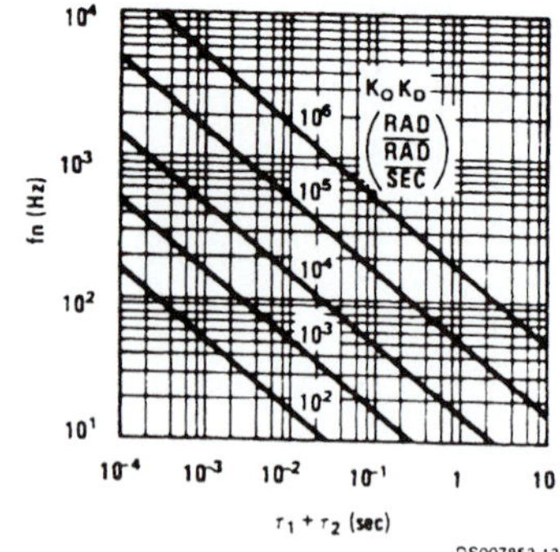

www.national.com

FIGURE 6-14 *(Continued)*

6-6 Stereo Demodulation

FM stereo receivers are identical to standard receivers up to the discriminator output. At this point, however, the discriminator output contains the 30-Hz to 15-kHz (L + R) signal *and* the 19-kHz subcarrier *and* the 23- to 53-kHz (L − R) signal. If a nonstereo (monaural) receiver is tuned to a stereo station, its discriminator output may contain the additional frequencies, but even the 19-kHz subcarrier is above the normal audible range, and its audio amplifiers and speaker would probably not pass it anyway. Thus, the nonstereo receiver reproduces the 30-Hz to 15-kHz (L + R) signal (a full monophonic broadcast) and is not affected by the other frequencies. This effect is illustrated in Fig. 6-15.

The stereo receiver block diagram becomes more complex after the discriminator. At this point, the three signals are separated by filtering action. The (L + R) signal is obtained through a low-pass filter and given a delay so that it reaches the matrix

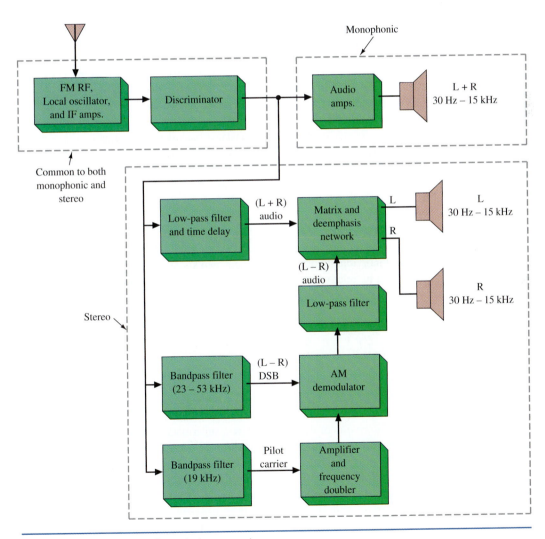

FIGURE 6-15 Monophonic and stereo receivers.

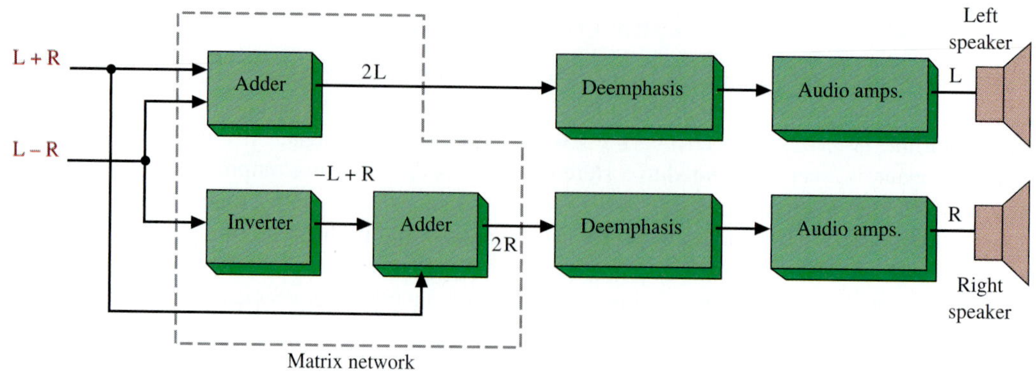

FIGURE 6-16 Stereo signal processing.

network in step with the (L − R) signal. A 23- to 53-kHz bandpass filter selects the (L − R) double sideband signal. A 19-kHz bandpass filter takes the pilot carrier and is multiplied by 2 to 38 kHz, which is the precise carrier frequency of the DSB suppressed carrier 23- to 53-kHz (L − R) signal. Combining the 38-kHz and (L − R) signals through the nonlinear device of an AM detector generates sum and difference outputs of which the 30-Hz to 15-kHz (L − R) components are selected by a low-pass filter. The (L − R) signal is thereby retranslated back down to the audio range and it and the (L + R) signal are applied to the matrix network for further processing.

Figure 6-16 illustrates the matrix function and completes the stereo receiver block diagram of Fig. 6-15. The (L + R) and (L − R) signals are combined in an adder that cancels R because (L + R) + (L − R) = 2L. The (L − R) signal is also applied to an inverter, providing −(L − R) = (−L + R), which is subsequently applied to another adder along with (L + R), which produces (−L + R) + (L + R) = 2R. The two individual signals for the right and left channels are then deemphasized and individually amplified to their own speaker. The process of FM stereo is ingenious in its relative simplicity and effectiveness in providing complete compatibility and doubling the amount of transmitted information through the use of multiplexing.

SCA Decoder

Subsidiary Communication Authorization
an additional channel of multiplexed information authorized by the FCC for stereo FM radio stations to feed services to selected customers

The FCC has also authorized FM stations to broadcast an additional signal on their carrier. It may be a voice communication or other signal for any nonbroadcast-type use. It is often used to transmit music programming that is usually commercial-free but paid for by subscription of department stores, supermarkets, and the like. It is termed the **subsidiary communication authorization** (SCA). It is frequency-multiplexed on the FM modulating signal, usually with a 67-kHz carrier and ±7.5-kHz (narrowband) deviation, as shown in Fig. 6-17. An SCA decoder circuit using the 565 PLL is provided in Fig. 6-18. A resistive voltage divider is used to establish a bias voltage for the input (pins 2 and 3). The demodulated FM signal is fed to the input through a two-stage high-pass filter (510 pF, 4.7 kΩ, 510 pF, 4.7 kΩ), both to allow capacitive coupling and to attenuate the stronger level of the stereo signals. The PLL is tuned to

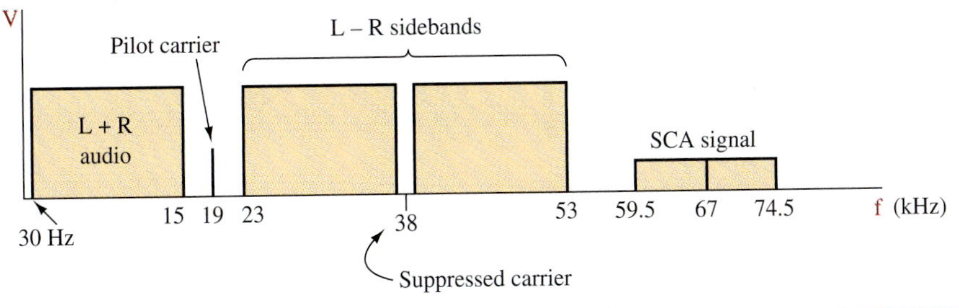

FIGURE 6-17 Composite stereo and SCA modulating signal.

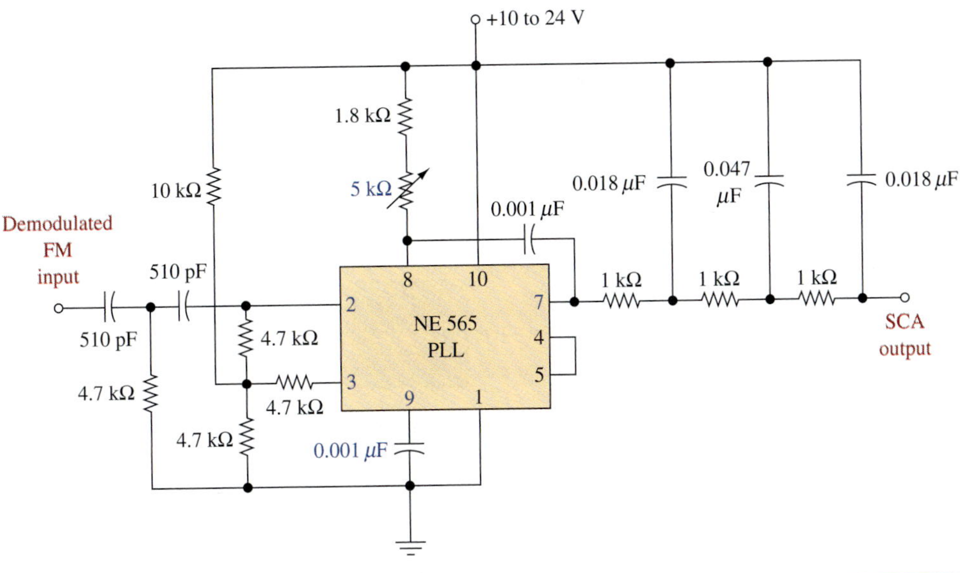

FIGURE 6-18 SCA PLL decoder.

about 67 kHz, with the 0.001-μF capacitor from pin 9 to ground and the 5-kΩ potentiometer providing fine adjustment. The demodulated output at pin 7 is fed through a three-stage low-pass filter to provide deemphasis and attenuate the high-frequency noise that often accompanies SCA transmission.

LIC Stereo Decoder

The decoding of the stereo signals is normally accomplished via special function ICs. The RCA CA3090 is such a device; a functional block diagram is provided in Fig. 6-19. The input signal from the detector is amplified by a low-distortion preamplifier and simultaneously applied to both the 19- and 38-kHz synchronous detectors (see Sec. 3-2). A 76-kHz signal, generated by a local voltage-controlled oscillator (VCO), is counted down by two frequency dividers to a 38-kHz signal and to two 19-kHz signals in phase quadrature. The 19-kHz pilot tone supplied by the FM detector is compared to the locally generated 19-kHz signal in the synchronous detector. The

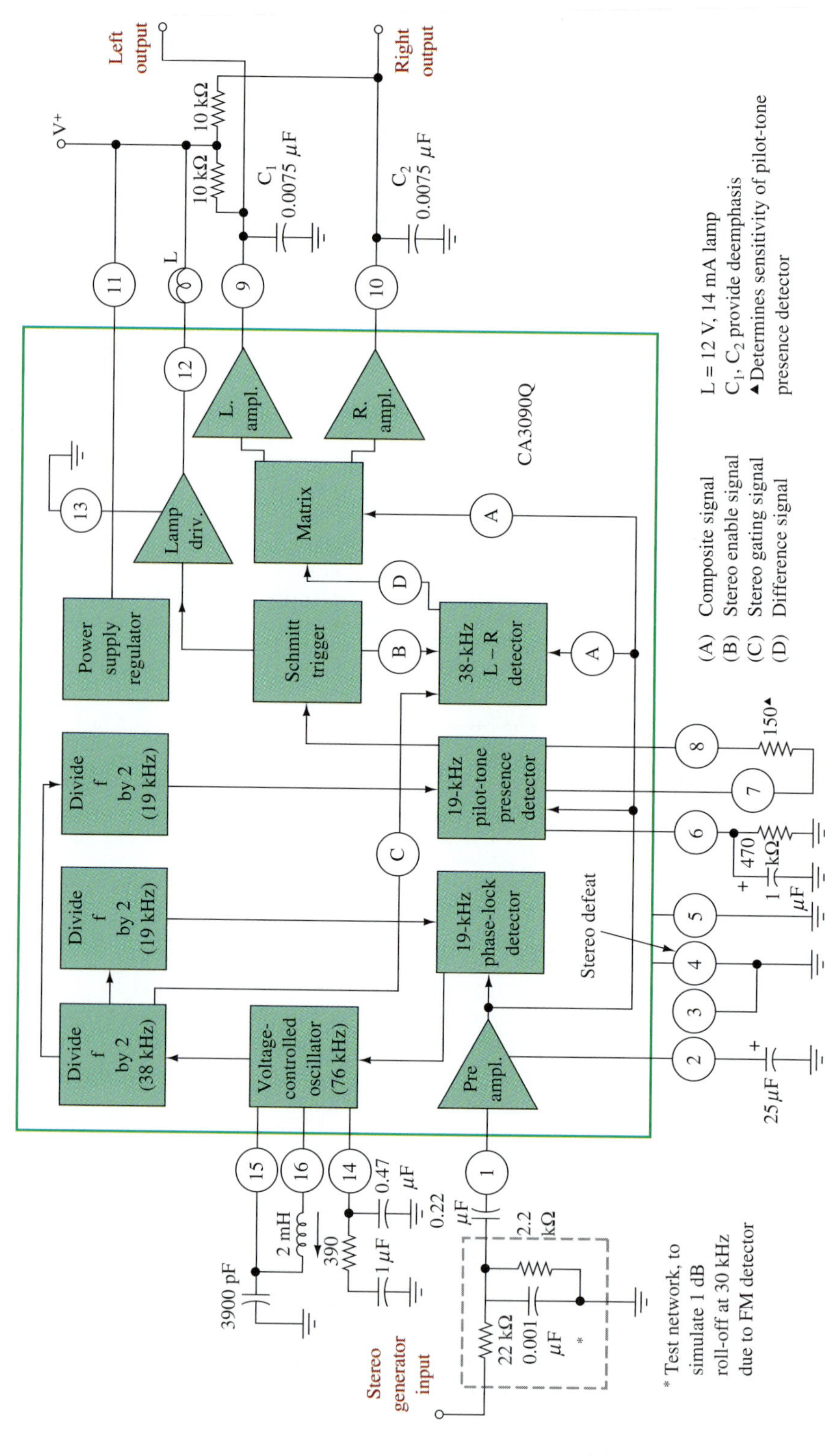

FIGURE 6-19 CA3090 stereo decoder. (Courtesy of RCA.)

resultant signal controls the voltage-controlled oscillator so that it produces an output signal to phase-lock the stereo decoder with the pilot tone. A second synchronous detector compares the locally generated 19-kHz signal with the 19-kHz pilot tone. If the pilot tone exceeds an externally adjustable threshold voltage, a Schmitt trigger circuit is energized. The signal from the Schmitt trigger turns on the small light we see on the panel of our stereos that indicates stereo reception. It also enables the 38-kHz synchronous detector and automatically switches the CA3090 from monaural to stereo operation. The output signal from the 38-kHz detector and the composite signal from the preamplifier are applied to a matrixing circuit, from which emerge the resultant left and right channel audio signals. These signals are applied to their respective left and right channel amplifiers for amplification to a level sufficient to drive most audio power amplifiers.

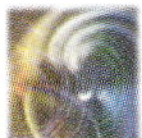

6-7 FM RECEIVERS

A modern approach to FM stereo reception is shown in Fig. 6-20, which is a circuit diagram for the Philips Semiconductors TEA5767/68 single-chip FM stereo radio. The architecture of this IC has dramatically reduced the number of external components. Compare this circuit to the older style of FM receiver shown in Fig. 6-21. In Fig. 6-20, the RF signal connects to pins 35 and 37 on the TEA5767/68. An RF AGC circuit prevents overloading and limits intermodulation problems created by strong adjacent channels.

The circuit in Fig. 6-20 has separate digital (pin 6) and analog (pin 33) grounds and requires a typical analog (pin 34) and digital (pin 7) supply voltage of 3.0 V. The stereo audio appears at pins 22 and 23. The circuit uses a crystal reference frequency to facilitate PLL tuning. External clock frequencies are 32.768 kHz or 13 MHz (pin 16) or 6.5 MHz (pin 17). The crystal oscillator frequencies are used for reference by the

- Frequency divider for the synthesizer PLL.
- IF counter timing.
- Stereo decoder free-running frequency adjustment.
- Center frequency adjustment of the IF filters.

Full microprocessor control of the TEA5767/68 including channel searching and selection audio control, and power on reset are facilitated via the chip's bus connections (pins 8, 9, 11, 12, 13, 14, and 15).

A typical older-style FM receiver involves use of discrete MOSFET RF and mixer stages with a separately excited bipolar transistor local oscillator, as shown in Fig. 6-21 (on pages 285–286). The antenna input signal is applied through the tuning circuit L_1, C_{1A} to the gate of the 40822 MOSFET RF amplifier. Its output at the drain is coupled to the lower gate of the 40823 mixer MOSFET through the C_{1B}–L_2 tuned circuit. The 40244 BJT oscillator signal is applied to the upper gate of the mixer stage. The local oscillator tuned circuit that includes C_{1C} uses a tapped inductor indicating a Hartley oscillator configuration. The tuning capacitor, C_1, has three separate ganged capacitors that vary the tuning range of the RF amp and mixer tuned circuits from 88 to 108 MHz while varying the local oscillator frequency from 98.7 to 118.7 MHz to generate a 10.7-MHz IF signal at the output of the mixer. The mixer output is applied to the commercially available 10.7-MHz double-tuned circuit T_1.

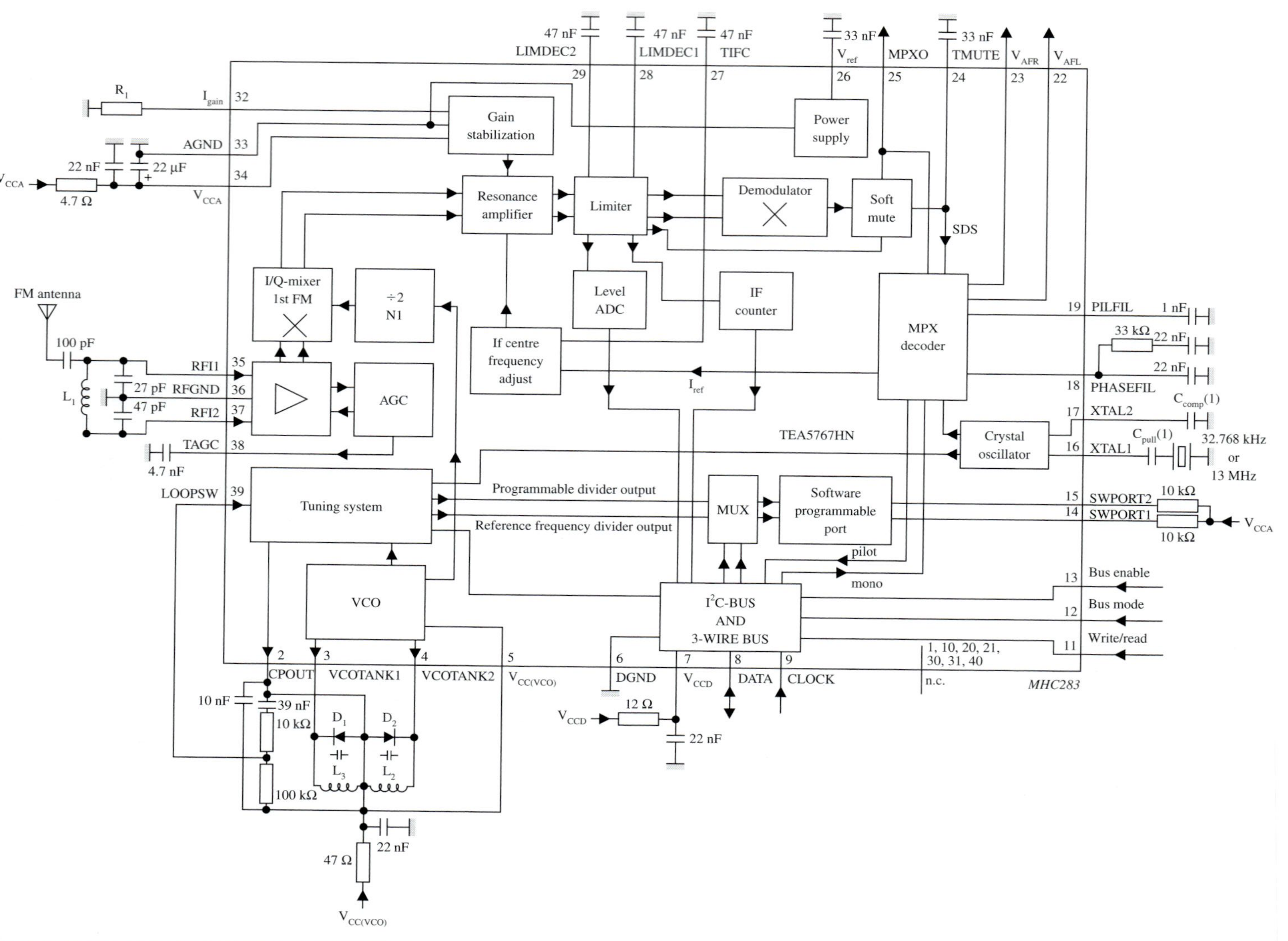

FIGURE 6-20 Block diagram of the Philips Semiconductors TEA5767 single-chip FM stereo radio.

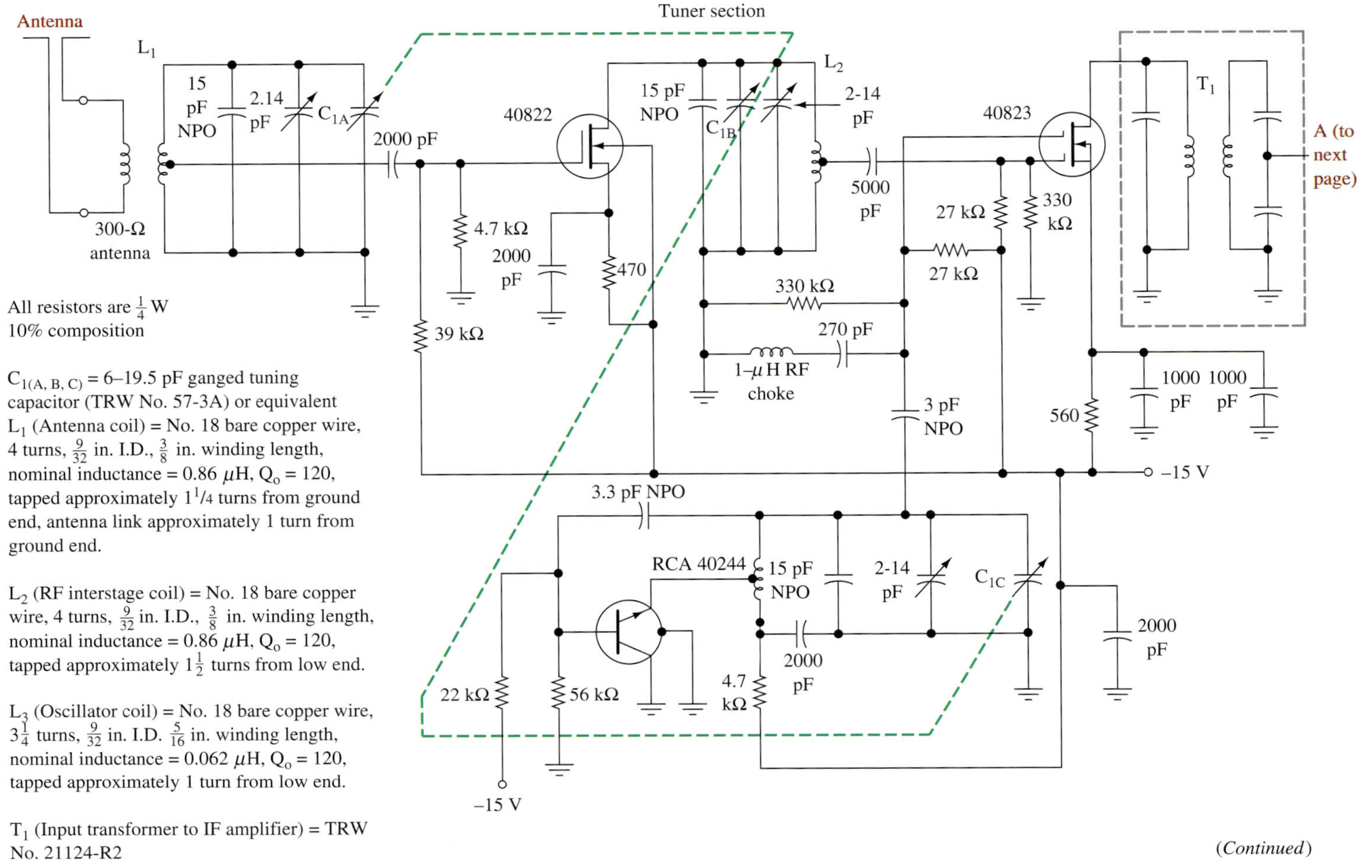

All resistors are $\frac{1}{4}$ W
10% composition

$C_{1(A,\,B,\,C)}$ = 6–19.5 pF ganged tuning capacitor (TRW No. 57-3A) or equivalent
L_1 (Antenna coil) = No. 18 bare copper wire, 4 turns, $\frac{9}{32}$ in. I.D., $\frac{3}{8}$ in. winding length, nominal inductance = 0.86 μH, Q_o = 120, tapped approximately $1^1/4$ turns from ground end, antenna link approximately 1 turn from ground end.

L_2 (RF interstage coil) = No. 18 bare copper wire, 4 turns, $\frac{9}{32}$ in. I.D., $\frac{3}{8}$ in. winding length, nominal inductance = 0.86 μH, Q_o = 120, tapped approximately $1\frac{1}{2}$ turns from low end.

L_3 (Oscillator coil) = No. 18 bare copper wire, $3\frac{1}{4}$ turns, $\frac{9}{32}$ in. I.D. $\frac{5}{16}$ in. winding length, nominal inductance = 0.062 μH, Q_o = 120, tapped approximately 1 turn from low end.

T_1 (Input transformer to IF amplifier) = TRW No. 21124-R2

(Continued)

FIGURE 6-21 Complete 88- to 108-MHz stereo receiver.

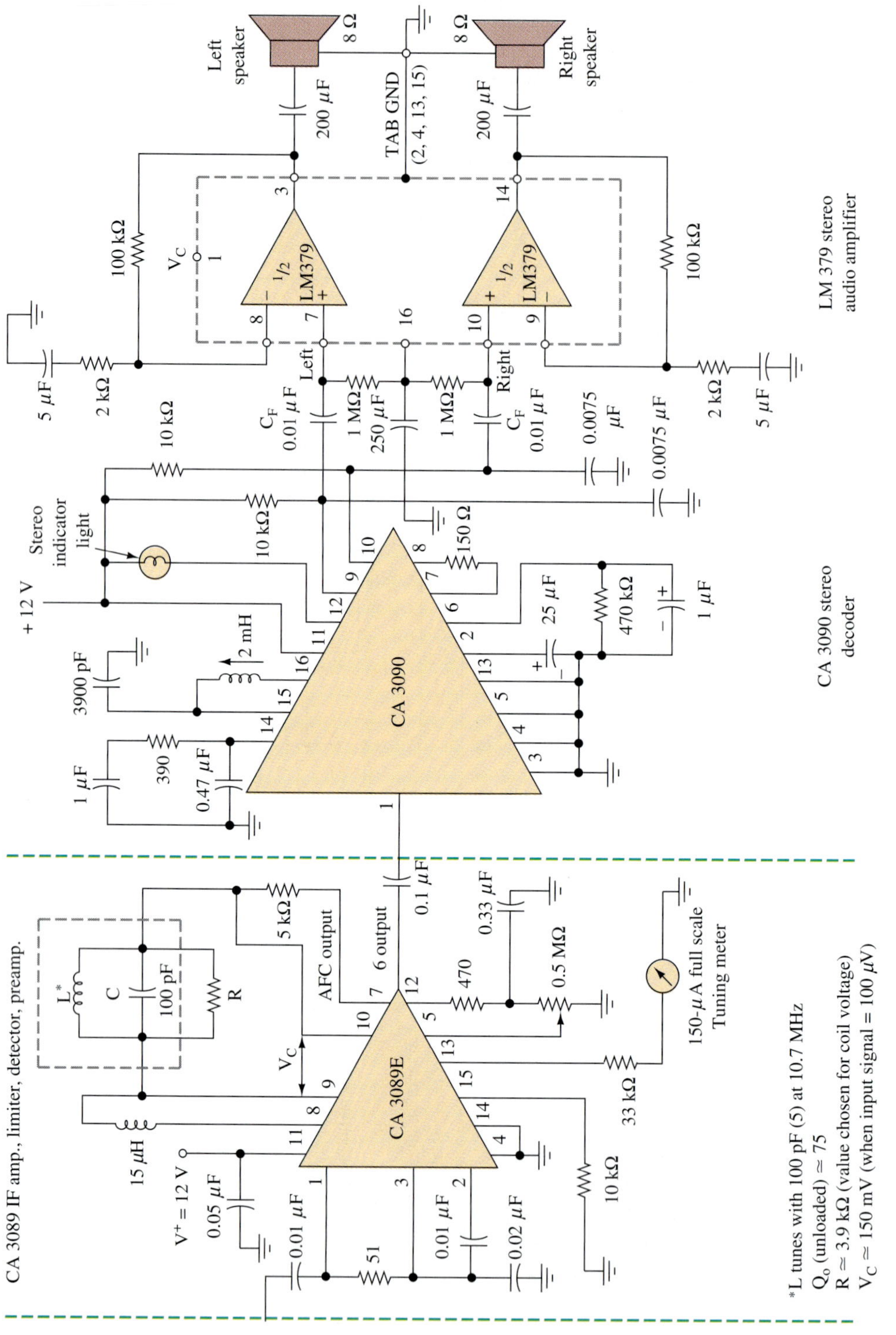

*L tunes with 100 pF (5) at 10.7 MHz
Q_o (unloaded) $\simeq$ 75
R $\simeq$ 3.9 kΩ (value chosen for coil voltage)
$V_C \simeq$ 150 mV (when input signal = 100 μV)

FIGURE 6-21 (*Continued*)

MOSFET receiver front ends offer superior cross-modulation and intermodulation performance as compared to other types, as explained in Sec. 6-2. The Institute of High Fidelity Manufacturers (IHFM) sensitivity for this front end is about 1.75 μV. It is defined as the minimum 100 percent modulated input signal that reduces the total receiver noise and distortion to 30 dB below the output signal. In other words, a 1.75-μV input signal produces 30-dB quieting.

The front-end output through T_1 in Fig. 6-21 is applied to a CA3089 IC. The CA3089 provides three stages of IF amplification—limiting, demodulation, and audio preamplification. It provides demodulation with an analog quadrature detector circuit (Sec. 6-4). It also provides a signal to drive a tuning meter and an AFC output for direct control of a varactor tuner. Its audio output includes the 30-Hz to 15-kHz (L + R) signal, 19-kHz pilot carrier, and 23- to 53-kHz (L − R) signal, which are then applied to the FM stereo decoder IC, the CA3090. The CA3090 was explained in Sec. 6-6 and provides the separate left and right channel outputs as well as a signal to light a stereo indicator light.

The CA3090 audio outputs are then applied to a ganged volume control potentiometer (not shown) and then to an LM379 dual 6-W audio amplifier. It has two separate audio amplifiers in one 16-lead IC and has a minimum input impedance of 2 MΩ per channel. It typically provides a voltage gain of 34 dB, total harmonic distortion (THD) of 0.07 percent at 1-W output, and 70 dB of channel separation.

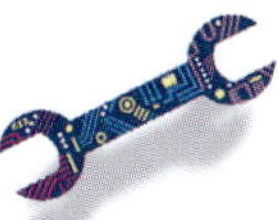

6-8 Troubleshooting

The basic approach to troubleshooting the FM receiver is similar to that of an AM receiver. The FM radio is a superheterodyne receiver like the AM receiver with a few differences. The methods you learn in this section will teach you how to isolate defects in the FM receiver.

After completing this section you should be able to

- Identify defective stages in an FM receiver
- Describe test setups for checking each receiver stage
- Troubleshoot a quadrature detector
- Test a semiconductor diode junction

The FM Receiver

Figure 6-22 represents a typical FM radio receiver. Troubleshooting begins at the limiter stage. Feed in a test signal at the input of the limiter stage, point A in Fig. 6-22. Based on the results of this test, the signal injection point will move toward the antenna or toward the audio section. We will assume, for the sake of this discussion, that the trouble complaint is a dead broadcast band FM receiver and the power supply is good. The signal generator used for this procedure must be capable of producing a test signal at the operating frequencies of the FM receiver and at its IF. In addition, an audio signal is needed to modulate the FM test signal.

Locating a Defective Stage

Feed a modulated test signal at 10.7 MHz (the FM receiver's IF frequency) into the limiter stage. Use a 400- to 1000-Hz signal as the modulating signal. Set the output

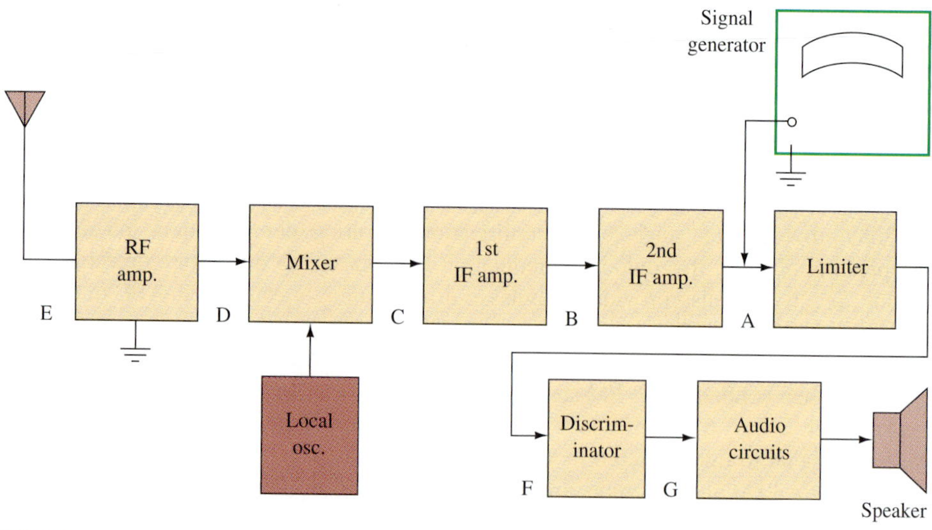

FIGURE 6-22 Typical FM receiver.

signal amplitude of the signal generator to approximately 4 V peak-to-peak (consult the service literature for the exact signal levels to inject for each stage being checked). Wobble the frequency control to each side of the IF center frequency of 10.7 MHz (see Fig. 6-23). If the limiter, discriminator, and audio amplifier circuits are operating correctly, then the wobbled signal will be heard at the speaker as a sound that goes from loud to low as the dial is changed on and off the IF frequency.

With the limiter, discriminator, and audio circuits working properly, move the injected signal to the input of the last IF amplifier, point B in Fig. 6-22. Lower the signal in amplitude and feed it into the base of the IF transistor circuit. For IF circuits composed of IC chips, connect the test signal to the proper input pin on the IC. Consult the service literature for proper signal input points and signal strength. Failure to hear the tone indicates that the IF stage being tested is bad. If the tone is heard at the speaker output, then move the test signal to point C. Continue in this fashion until the defective stage is located. The stage where the test tone is no longer heard is the defective circuit. If the local oscillator or the mixer is determined to be at fault, then troubleshoot these sections as described in the Chapter 3 troubleshooting section.

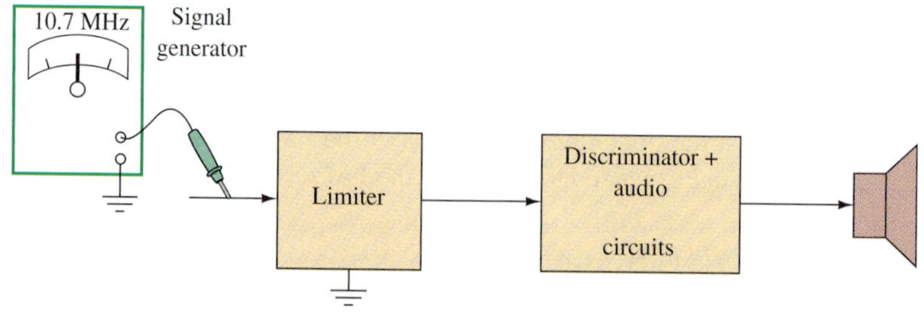

FIGURE 6-23 Testing by wobbling the signal.

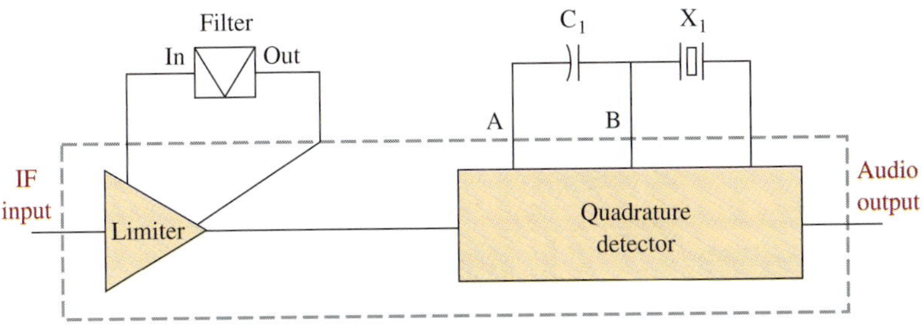

FIGURE 6-24 Quadrature detector.

Quadrature Detectors

Many FM receivers use quadrature detectors because no adjustment is required if the IF is aligned to the detector. If quadrature detectors are found in a communications receiver, they will usually be operated at 455 kHz, and if they appear in a television receiver, the frequency will be 4.5 MHz. The quadrature detector is usually included in an integrated circuit that combines the IF amplifier and the demodulator (see Fig. 6-24). Although the quadrature detector is usually a narrowband device, some home broadcast receivers use them at 10.7 MHz. In that case, C1 might actually be an inductor.

The crystal-like component X_1 is not a quartz crystal. It will be either a ceramic resonator or an *LC* tuned circuit. The filter will probably be a ceramic type, too. Remember, in a communications receiver this type of IF system usually runs at 455 kHz, so an oscilloscope can be used to measure all voltages. The IF input will be less than 1 mV and will be difficult to see. Recovered audio should be around 0.25 V.

A modulated signal generator can be used to provide a test signal. These are narrowband circuits, so the modulating signal should be about 1 kHz and the deviation set to 5 kHz.

First look for the IF signal at the filter's input and output. An RF voltmeter such as the Boonton 91D (an old model but still around) may be necessary because the voltage will be in the order of millivolts. The signal at the filter output will be about one-half the input. Sweep the signal generator $\pm$ 20 kHz from the center frequency and verify that the filter attenuates signals more than 7.5 kHz from the center frequency.

At the center frequency, the voltage at point B should be 90° out of phase with the voltage at point A. This can be observed with your oscilloscope. If you have a single-channel oscilloscope, connect point A to the external sync. Note the position of the trace when looking at point A. The voltage at point B should be delayed from A by one-quarter of a sine wave.

If you do not have an oscilloscope, you can at least observe a peak in the voltage at point B when the signal generator is tuned to center frequency.

Typical problems are:

1. C1 is open or its value has changed. This can cause no output or distorted audio output.
2. Improper alignment. The center frequency of the IF amplifier must be the same as the tuned circuit, X1. One or both may be adjustable. Look for open coils or capacitors that have changed value.
3. Quadrature detectors sometimes have a resistor in the tuned circuit to lower the *Q*. If the resistor is missing or open, the *Q* will be too high, causing distorted audio.

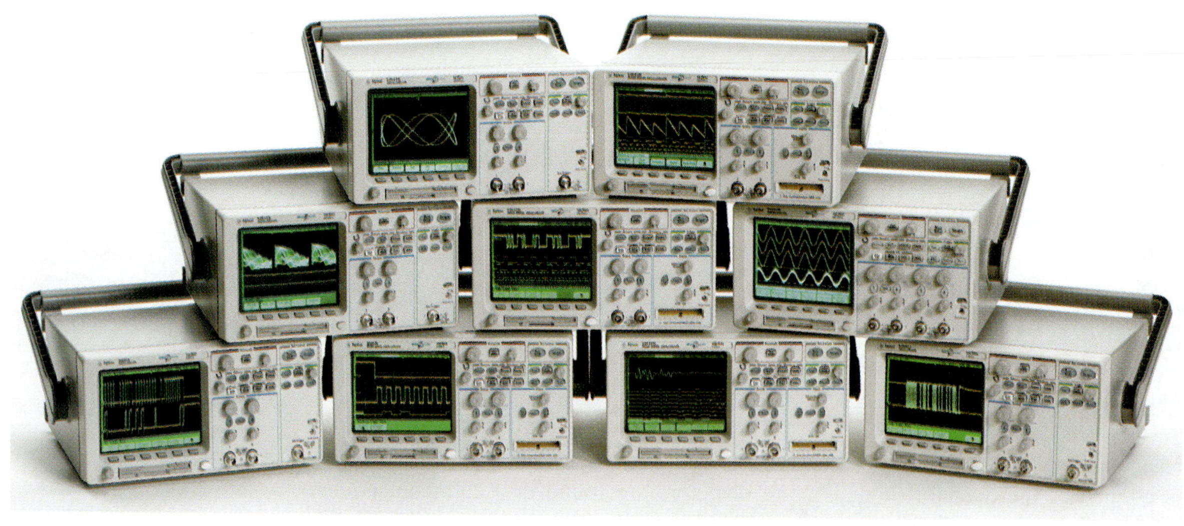

Agilent's 54600 series offers low-cost portable oscilloscopes for analog and digital systems. (Courtesy of Agilent Technologies. Reprinted with permission.)

Discriminators

Foster–Seely discriminators or ratio detectors (Figs. 6-7 and 6-9) are used in older FM receivers. These circuits may drift out of alignment over time and coils or capacitors may short or open. All cause distorted or no audio. Alignment is best done with a sweep generator, but in a pinch apply a fully modulated signal to the receiver and adjust the tuning for minimum distortion.

Stereo Demodulator

When troubleshooting a stereo demodulator (probably an IC), the first thing to look for is proper signal input level. An acceptable input is about 100-mV rms.

Make sure the 76-kHz oscillator is present and on frequency. Look for an open tuning capacitor or inductor if the oscillator is not running or cannot be properly adjusted. Check dc voltages on bypass capacitors. If these capacitors leak, the internal IC bias will be wrong and the IC will not work.

Testing Diodes and Transistors

To test a diode that you suspect is defective, set your DMM to the diode test function, as shown in Fig. 6-25. The procedure for testing diodes and transistors is illustrated in Figs. 6-26(a) and (b). With the DMM in the diode test position, make a reading, then switch the test leads to the opposite ends of the diode and make another reading. Judge the diode based on the following criteria: If one reading shows a value and the other reading shows over range, then the device is good. If both readings result in over range, the device is open. If the two readings are zero or very low, the device is shorted. Refer to Table 6-1 for typical readings. The first and second readings may be in reverse order.

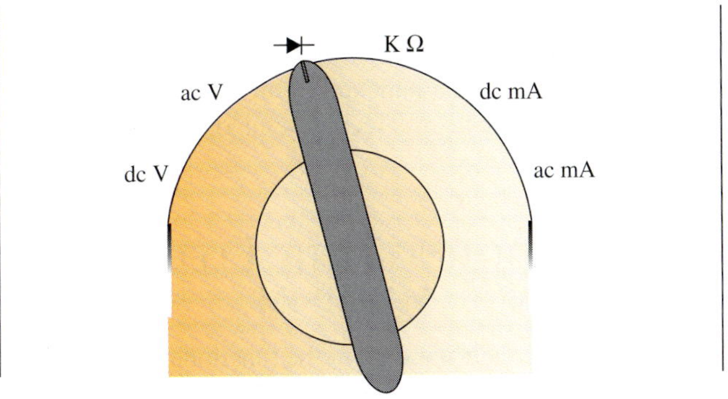

FIGURE 6-25 Diode test range.

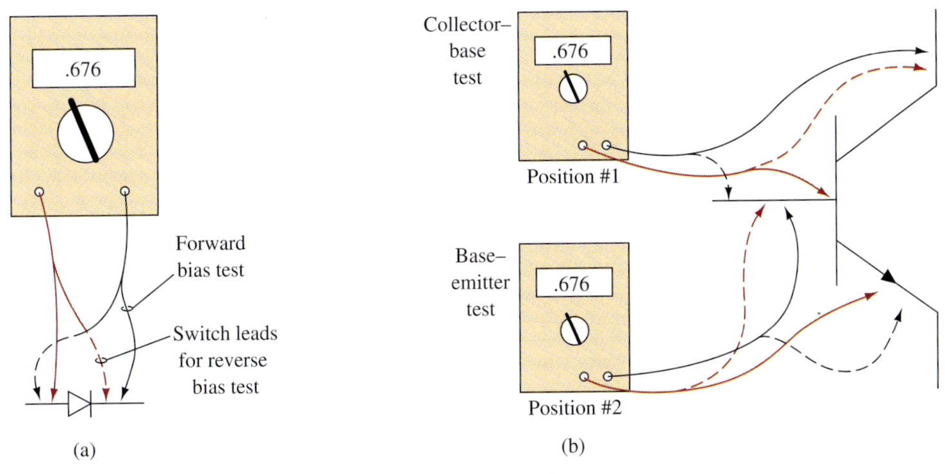

FIGURE 6-26 Diode and transistor testing.

Testing the transistor is very similar to testing the diode if we think of the transistor as two diodes back-to-back. The transistor's collector–base junction is tested as if it were one diode, and the base–emitter junction is tested as another diode, shown in Fig. 6-26(b). Apply the DMM's test leads across the transistor's collector and base connections and make the first reading. Reverse the test leads and make the second reading. Compare your results to the readings in Table 6-1. Next, apply the DMM's test leads to the transistor's base and emitter connections. Make the first reading, then reverse the test leads and make the second reading. Compare the results to the readings in Table 6-1. Good readings should be fairly close to those in

Table 6-1 Displayed Values for a Diode Test

First Reading	Second Reading	Conclusion
Approximately 0.4 V	Over range	Good germanium type
Approximately 0.6 V	Over range	Good silicon type
Over range	Over range	Device is open
Very small or zero	Very small or zero	Device is shorted

the table. For example, if your readings were 0.676 V and over range, this would indicate a good semiconductor junction. Consult the booklet that came with your DMM for the particular values of readings you may get. Remember, two over-range readings or two small-value readings indicate a defective semiconductor junction.

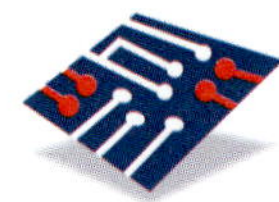

6-9 TROUBLESHOOTING WITH ELECTRONICS WORKBENCH™ MULTISIM

This section gives you a more thorough understanding of the functional blocks within an FM receiver and additional experience troubleshooting electronic communications circuits. Open the file **Fig6-27.ms7 (.msm)** on your EWB Multisim CD. This is an implementation of an FM receiver using Multisim. Each of the building blocks for the FM receiver is identifed. A picture of the Multisim circuit is provided in Fig. 6-27.

A 100-kHz FM signal is being generated by the FM source. The FM carrier is being modulated by a 1-kHz signal and the modulation index is 5. The voltage level has been set to 5 μV to simulate the RF input level that might be received.

The second stage is the RF amplifier, which has a gain of 20,000 V/V. The huge gain is required to provide enough signal voltage for the next stage. The mixer stage follows the RF amplifier and is used to down-convert the 100-kHz frequency to 10 kHz, which is the IF frequency for this circuit. The local oscillator frequency has been set to 110 kHz. The outputs of the mixer are 10 kHz and 210 kHz, which are the difference and sum of the input and local oscillator frequencies. The next stage is the IF amplifier, which includes a bandpass filter that passes the 10-kHz difference and rejects the 210-kHz sum. C_1 and L_1 are used to create the bandpass filter. You can view the Bode plot of the bandpass filter by opening the file **FigE6-1.ms7 (.msm)** on your

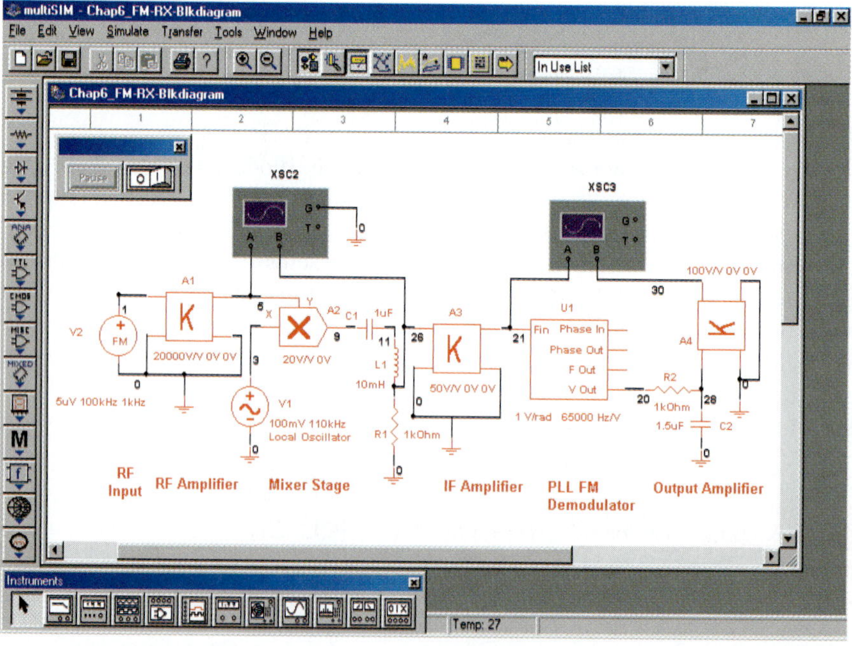

FIGURE 6-27 An implementation of an FM receiver using Multisim.

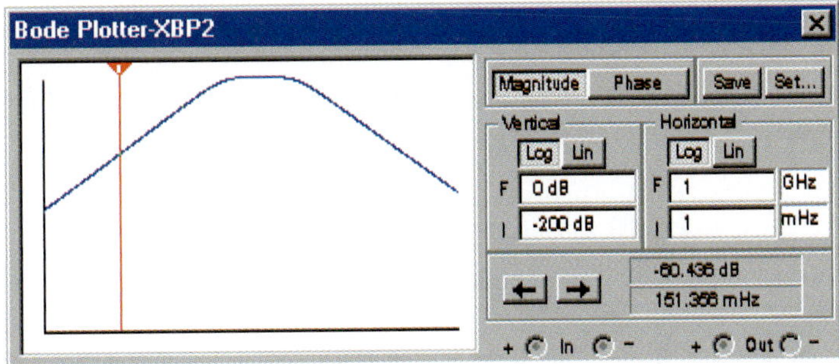

FIGURE 6-28 The frequency plot of the bandpass filter.

EWB CD. Start the simulation and open the Bode plotter. You should see a plot similar to the file shown in Fig. 6-28.

This filter provides a 3-dB bandpass of 140 Hz to 22 kHz. The 210-kHz signal is approximately 80 dB down. You can verify this by sliding the cursor along the frequency plot response of the bandpass filter. The frequency and corresponding dB value is shown at the bottom of the Bode plotter window. Close the Bode plot file and reopen the **Fig6-27.ms7 (.msm)** file.

The output of the bandpass filter is next amplified within the IF amplifier with a gain of 50 V/V. The output of the IF amplifier connects to the PLL FM demodulator. The PLL has been set up to lock to the 10-kHz frequency-modulated signal. The output of the PLL FM demodulator circuit is taken from V_{out}, which is the error voltage output for the PLL. This output is amplified by the output amplifier stage, which has a gain of 100 V/V.

Start the simulation and observe the output. You will notice that a 1-kHz sinusoid is produced, which is the original modulating signal. This simulation may be a little slow and you must let the simulation run for about 10 ms of simulation time for the circuit to stabilize. Become familiar with this circuit; you will next be asked to troubleshoot a faulty version of it.

Open the file **FigE6-2.ms7 (.msm)** on your EWB CD. Start the simulation and observe the output on the oscilloscope. Use your oscilloscope to follow the signal through the circuit. You will find that resistor R2, which connects the output of the PLL to the output amplifier, is open. Double-click on the resistor, R2, and correct the fault by clicking on the **fault** tab and resetting the fault to none. Rerun the simulation to verify that the problem has been corrected.

Electronics Workbench™ Exercises

1. Open the **FigE6-3.ms7 (.msm)** file on your EWB CD. Start the simulation and determine if the circuit is working properly. If it is not working, troubleshoot the circuit and correct the fault(s). Explain your findings.
2. Open the **FigE6-4.ms7 (.msm)** file on your EWB CD. Start the simulation and observe the output signals. Explain how the circuit limits the output level to ±5 V.
3. Open the **FigE6-5.ms7 (.msm)** file on your EWB CD. Describe the purpose of this circuit block in an FM receiver. What is the expected output frequency of the mixer? Verify that the correct frequency is being output by using the trigonometric identity $(\sin A)(\sin B) = 0.5\cos(A - B) - 0.5\cos(A + B)$.

Summary

In Chapter 6 we discussed the basis of an FM receiver and showed the similarities and differences compared to an AM receiver. The major topics you should now understand include:

- the operation of an FM receiver using a block diagram as a guide, including complete descriptions of the discriminator, the deemphasis network, and the limiter functioning as AGC
- the benefits of RF amplifiers, including image frequency attenuation and local oscillator reradiation effects
- the detailed functioning of a transistor limiter circuit
- the description and comparison of slope detector, Foster–Seely discriminator, ratio detector, and quadrature detector circuits
- the description and operation of a phase-locked-loop (PLL) FM demodulator, including its three possible states
- the analysis of a stereo FM demodulation process using a block diagram
- the operation of the subsidiary communication authorization (SCA) decoder operation
- the operation of a complete 88–108-MHz stereo FM receiver by analysis of the schematic

Questions and Problems

Section 6-1

*1. What is the purpose of a discriminator in an FM broadcast receiver?
2. Explain why the automatic frequency control (AFC) function is usually not necessary in today's FM receivers.
*3. Draw a block diagram of a superheterodyne receiver designed for reception of FM signals.
4. The local FM stereo rock station is at 96.5 MHz. Calculate the local oscillator frequency and the image frequency for a 10.7-MHz IF receiver. (107.2 MHz, 117.9 MHz)

Section 6-2

5. Explain the desirability of an RF amplifier stage in FM receivers as compared to AM receivers. Why is this not generally true at frequencies over 1 GHz?
6. Describe the meaning of *local oscillator reradiation,* and explain how an RF stage helps to prevent it.
7. Why is a square-law device preferred over other devices as elements in an RF amplifier?
8. Why are FETs preferred over other devices as the active elements for RF amplifiers?

*An asterisk preceding a number indicates a question that has been provided by the FCC as a study aid for licensing examinations.

9. List two advantages of using a dual-gate MOSFET over a JFET in RF amplifiers.
10. Explain the need for the radio-frequency choke (RFC) in the RF amplifier shown in Fig. 6-2.

Section 6-3

*11. What is the purpose of a limiter stage in an FM broadcast receiver?
*12. Draw a diagram of a limiter stage in an FM broadcast receiver.
13. Explain fully the circuit operation of the limiter shown in Fig. 6-3.
14. What is the relationship among limiting, sensitivity, and quieting for an FM receiver?
15. An FM receiver provides 100 dB of voltage gain prior to the limiter. Calculate the receiver's sensitivity if the limiter's quieting voltage is 300 mV. (3 μV)

Section 6-4

16. Draw a schematic of an FM slope detector and explain its operation. Why is this method not often used in practice?
17. Draw a schematic of a Foster–Seely discriminator, and provide a step-by-step explanation of what happens when the input frequency is below the carrier frequency. Include a phase diagram in your explanation.
*18. Draw a diagram of an FM broadcast receiver detector circuit.
*19. Draw a diagram of a ratio detector and explain its operation.
20. Explain the relative merits of the Foster–Seely and ratio detector circuits.
*21. Draw a schematic diagram of each of the following stages of a superheterodyne FM receiver:
    (a) Mixer with injected oscillator frequency.
    (b) IF amplifier.
    (c) Limiter.
    (d) Discriminator.
    Explain the principles of operation. Label adjacent stages.
22. Describe the process of quadrature detection.

Section 6-5

23. Draw a block diagram of a phase-locked loop (PLL) and briefly explain its operation.
24. Explain in detail how a PLL is used as an FM demodulator.
25. List the three possible states of operation for a PLL and explain each one.
26. A PLL's VCO free-runs at 7 MHz. The VCO does not change frequency until the input is within 20 kHz of 7 MHz. After that condition, the VCO follows the input to ±150 kHz of 7 MHz before the VCO starts to free-run again. Determine the PLL's lock and capture ranges. (300 kHz, 40 kHz)

Section 6-6

27. Explain how separate left and right channels are obtained from the (L + R) and (L − R) signals.
*28. What is SCA? What are some possible uses of SCA?
29. Determine the maximum reproduced audio signal frequency in an SCA system. Why does SCA cause less FM carrier deviation, and why is it thus less noise resistant than standard FM? (*Hint:* Refer to Fig. 6-17.) (7.5 kHz)

30. Explain the principle of operation for the CA3090 stereo decoder.

Section 6-7

31. The receiver front end in Fig. 6-21 is rated to have noise below the signal by 30 dB in the output with a 1.75-μV input. Calculate its output *S/N* ratio with a 1.75-μV input signal. (31.6 to 1)
32. The LIC dual audio amplifiers in Fig. 6-21 are rated to provide 70 dB of channel separation. If the left channel has 1 W of output power, calculate the wattage of the right channel that is included. (0.1 μW)

Section 6-8

33. Explain why you would wobble the IF signal fed into point A of Fig. 6-22.
34. The quadrature detector troubleshooting circuit shown in Fig. 6-24 includes a device labeled X_1. Explain the various options for circuitry at that point.
35. Describe the method for checking a diode with a DMM. Extend that description into a technique for testing a transistor.
36. Describe the operation of the quadrature detector in Fig. 6-24 if C_1 is shorted.
37. Describe possible causes if the Foster–Seely discriminator of Fig. 6-7 has a peak output voltage much less than calculated.
38. If L_2 is shorted in Fig. 6-7, explain what happens to the output voltage.
39. Describe the demodulated signal of Fig. 6-11 if C_3 is shorted.
40. Explain how a low beta (<40) on Q_4 in Fig. 6-11 would affect the circuit's performance.

Questions for Critical Thinking

41. If you were concerned with the sensitivity rating of a communications system, would noise reduction capability be a major factor in your decision-making? Why or why not?
42. Explain why a limiter minimizes or eliminates the need for the AGC function.
43. Draw a schematic of the LM 565 PLL in Fig 6-14 if it is used as an FM demodulator. Pick C_o and R_o so that the free-running frequency is 455 kHz.
44. Draw a block diagram of an FM stereo demodulator. Explain the function of the AM demodulator and the matrix network so nontechnical users can understand. Add a circuit that energizes a light to indicate reception of a stereo station.

7 Communications Techniques

Chapter Outline

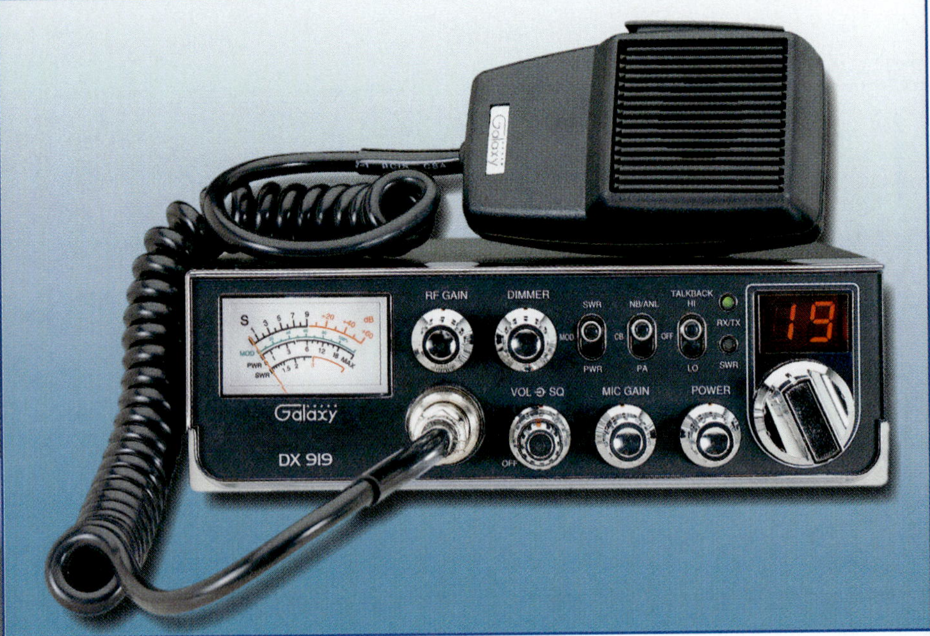

The Galaxy DX 919 40 channel CB radio. (Courtesy of galaxyradios.com. Reprinted with permission.)

Objectives

- Describe double conversion and up-conversion and explain their advantages
- Analyze the advantages of delayed AGC and auxiliary AGC
- Explain the features and their operation that a high-quality receiver may include as compared to a basic receiver
- Analyze and explain the relationships among noise, receiver sensitivity, dynamic range, and the third-order intercept
- Troubleshoot an amplifier suspected of excessive IMD
- Explain the operation of a frequency synthesizer
- Describe the operation of a DDS system and provide advantages and drawbacks compared to analog synthesizers

Key Terms

transceiver
double conversion
up-conversion
preselector
delayed AGC
variable bandwidth tuning
electromagnetic interference
S meter
muting
quieting
noise floor
dynamic range
intermod
third-order intercept point
input intercept
error voltage
direct digital synthesis
phase noise
channel guard

7-1 Introduction

Communications equipment may be loosely defined as that which is *not* used for entertainment. Because much of this equipment is of a vital nature, it is not surprising that communications equipment contains more sophistication than a standard broadcast receiver. In addition, because communication tends to require two-way capabilities, the use of transceivers is prevalent. A **transceiver** is simply a transmitter *and* receiver in a single package. Besides sharing a single package, some circuits are used for both the transmitter and receiver functions. Some examples of shared functions include oscillators, power supplies, and audio amplifiers.

Transceiver
transmitter and receiver sharing a single package and some circuits

Many individuals experience the enjoyment of transceiver contact around the world as radio amateurs. These hams also get into the technology of transceivers. In the early 1980s they developed the first crude model of the Internet. For more information on amateur radio, contact:

American Radio Relay League
225 Main Street
Newington, CT 06111
1-860-594-0200
www.arrl.org

7-2 Frequency Conversion

Double Conversion

One of the most likely areas of change from broadcast receivers to communications receivers is in the mixing process. The two major differences are the widespread use of *double conversion* and the increasing popularity of *up-conversion* in communications equipment. Both of these refinements have as a major goal the minimization of image frequency problems (refer to Chapter 3 for a review of these phenomena).

Double conversion is the process of stepping down the RF signal to a first, relatively high IF frequency and then mixing down again to a second, lower, final IF frequency. Figure 7-1 provides a block diagram for a typical double-conversion system.

Double Conversion
superheterodyne receiver design with two separate mixers, local oscillators, and intermediate frequencies

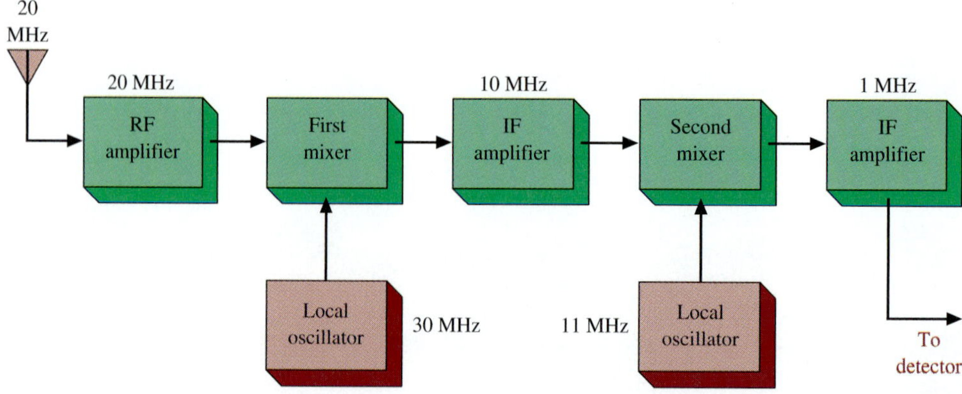

FIGURE 7-1 Double-conversion block diagram.

Notice that the first local oscillator is variable to allow a constant 10-MHz frequency for the first IF amplifier. Now the input into the second mixer is a constant 10 MHz, which allows the second local oscillator to be a fixed 11-MHz crystal oscillator. The difference component (11 MHz − 10 MHz = 1 MHz) out of the second mixer is accepted by the second IF amplifier, which is operating at 1 MHz. The following example illustrates the ability of double conversion to eliminate image frequency problems. Example 7-1 shows that in this case the image frequency is double the desired signal (40 MHz versus 20 MHz), and even the relatively broadband tuned circuits of the RF and mixer stages will almost totally suppress the image frequency. On the other hand, if this receiver uses a single conversion directly to the final 1-MHz IF frequency, it is found that the image frequency will not be fully suppressed.

Example 7-1

Determine the image frequency for the receiver illustrated in Fig. 7-1.

Solution

The image frequency is the one that, when mixed with the 30-MHz first local oscillator signal, will produce a first mixer output frequency of 10 MHz. The desired frequency of 20 MHz mixed with 30 MHz yields a 10-MHz component, of course, but what *other* frequency provides a 10-MHz output? A little thought shows that if a 40-MHz input signal mixes with a 30-MHz local oscillator signal, an output of 40 MHz − 30 MHz = 10 MHz is also produced. Thus, the image frequency is 40 MHz.

The 22-MHz image frequency of Ex. 7-2 is very close to the desired 20-MHz signal. The RF and mixer tuned circuits will certainly provide attenuation to the 22-MHz image, but if it is a strong signal, it will certainly get into the IF stages and will not be removed from that point on. The graph of RF and mixer tuned circuit response in Fig. 7-2 serves to illustrate the tremendous image frequency response rejection provided by the double-conversion scheme.

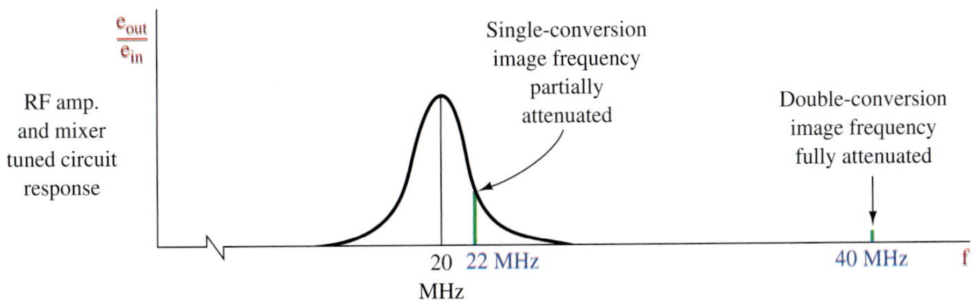

FIGURE 7-2 Image frequency rejection.

Example 7-2

Determine the image frequency for the receiver illustrated in Fig. 7-3.

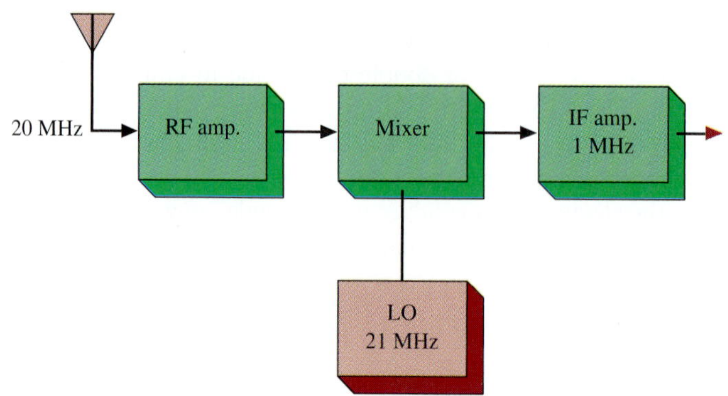

FIGURE 7-3 System for Ex. 7-2.

Solution

If a 22-MHz signal mixes with the 21-MHz local oscillator, a difference component of 1 MHz is produced just as when the desired 20-MHz signal mixes with 21 MHz. Thus, the image frequency is 22 MHz.

Image frequencies are not a major problem for low-frequency carriers, say, for frequencies below 4 MHz. For example, a single-conversion setup for a 4-MHz carrier and a 1-MHz IF means that a 5-MHz local oscillator will be used. The image frequency is 6 MHz, which is far enough away from the 4-MHz carrier that it won't present a problem. At higher frequencies, where images are a problem, the situation is aggravated by the enormous number of transmissions taking place in our crowded communications bands.

Example 7-3

Why do you suppose that images tend to be somewhat less of a problem in FM versus AM or SSB communications?

Solution

Recall the concept of the *capture* effect in FM systems (Chapter 5). It was shown that if a desired and undesired station are picked up simultaneously, the stronger one tends to be "captured" by inherent suppression of the weaker signal. Thus, a 2:1 signal-to-undesired-signal ratio may result in a 10:1 ratio at the output. This contrasts with AM systems (SSB included), where the 2:1 ratio is carried through to the output.

Up-Conversion

Until recently, the double-conversion scheme, with the lower IF frequency (often the familiar 455 kHz) providing most of the receiver's selectivity, has been standard practice because the components available made it easy to achieve the necessary selectivity at low IF frequencies. However, now that VHF crystal filters (30 to 120 MHz) are available for IF circuitry, conversion to a higher IF than RF frequency is popular in sophisticated communications receivers. As an example, consider a receiver tuned to a

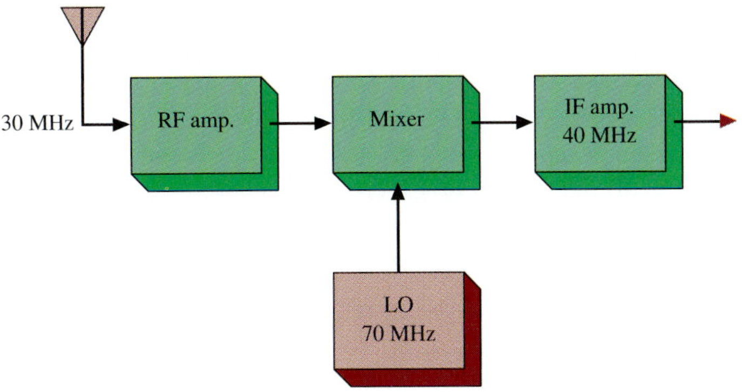

FIGURE 7-4 Up-conversion system.

30-MHz station and using a 40-MHz IF frequency, as illustrated in Fig. 7-4. This represents an **up-conversion** system because the IF is a higher frequency than the received signal. The 70-MHz local oscillator mixes with the 30-MHz signal to produce the desired 40-MHz IF. Sufficient IF selectivity at 40 MHz is possible with a crystal filter.

Up-Conversion
mixing the received RF signal with an LO signal to produce an IF signal higher in frequency than the original RF signal

Example 7-4

Determine the image frequency for the system of Fig. 7-4.

Solution

If a 110-MHz signal mixes with the 70-MHz local oscillator, a 40-MHz output component results. The image frequency is therefore 110 MHz.

Example 7-4 shows the superiority of up-conversion. It is highly unlikely that the 110-MHz image could get through the RF amplifier tuned to 30 MHz. There is no need for double conversion and all its necessary extra circuitry. The only disadvantage to up-conversion is the need for a higher-Q IF filter and better high-frequency response IF transistors. The current state-of-the-art in these areas now makes up-conversion economically attractive. Additional advantages over double conversion include better image suppression and less tuning range requirements for the oscillator. The smaller tuning range for up-conversion is illustrated in the following example and minimizes the tracking difficulties of a widely variable local oscillator.

Example 7-5

Determine the local oscillator tuning range for the systems illustrated in Figs. 7-1 (shown on page 300) and 7-4 if the receivers must tune from 20 to 30 MHz.

Solution

The double-conversion local oscillator in Fig. 7-1 is at 30 MHz for a received 20-MHz signal. Providing the same 10-MHz IF frequency for a 30-MHz signal means that the local oscillator must be at 40 MHz. Its tuning range is from 30 to 40 MHz or 40 MHz/30 MHz = 1.33. The up-conversion scheme of Fig. 7-4 has a 70-MHz local oscillator for a 30-MHz input and requires a 60-MHz oscillator for a 20-MHz input. Its tuning ratio is then 70 MHz/ 60 MHz, or a very low 1.17.

Preselector
the tuned circuits prior to the mixer in a superheterodyne receiver

The tuned circuit(s) prior to the mixer is often referred to as the **preselector.** The preselector is responsible for the image frequency rejection characteristics of the receiver. If an image frequency rejection is the result of a single tuned circuit of known Q, the amount of image frequency rejection can be calculated. The following equation predicts the amount of suppression in dB. If suppression is due to more than a single tuned circuit, the dB of suppression for each is calculated individually and the results added to provide the total suppression:

$$\text{image rejection (dB)} \cong 20 \log\left[\left(\frac{f_i}{f_s} - \frac{f_s}{f_i}\right)Q\right] \tag{7-1}$$

where f_i = image frequency
f_s = desired signal frequency
Q = tuned circuit's Q

Example 7-6

An AM broadcast receiver has two identical tuned circuits prior to the IF stage. The Q of these circuits is 60 and the IF frequency is 455 kHz, and the receiver is tuned to a station at 680 kHz. Calculate the amount of image frequency rejection.

Solution

Using Eq. (7-1), the amount of image frequency rejection per stage is calculated:

$$\text{image rejection (dB)} \cong 20 \log\left[\left(\frac{f_i}{f_s} - \frac{f_s}{f_i}\right)Q\right] \tag{7-1}$$

The image frequency is 680 kHz + (2 × 455 kHz) = 1590 kHz. Thus,

$$20 \log\left[\left(\frac{1590\text{ kHz}}{680\text{ kHz}} - \frac{680\text{ kHz}}{1590\text{ kHz}}\right)60\right] = 20 \log 114.6$$
$$= 41\text{ dB}$$

Thus, the total suppression is 41 dB plus 41 dB, or 82 dB. This is more than enough to provide excellent image frequency rejection.

7-3 Special Techniques

Delayed AGC

Delayed AGC
an AGC that does not provide gain reduction until an arbitrary signal level is attained.

The simple automatic gain control (AGC) discussed in Chapter 3 has a minor disadvantage. It provides some gain reduction even to very weak signals. This is illustrated in Fig. 7-5. As soon as even a weak received signal is tuned, simple AGC provides some gain reduction. Because communications equipment is often dealing with marginal (weak) signals, it is usually advantageous to add some additional circuitry to provide a **delayed AGC,** that is, an AGC that does not provide any gain reduction until some arbitrary signal level is attained and therefore has no gain reduction for weak signals. This characteristic is also shown in Fig. 7-5. It is important for you to understand that delayed AGC does not mean delayed in time.

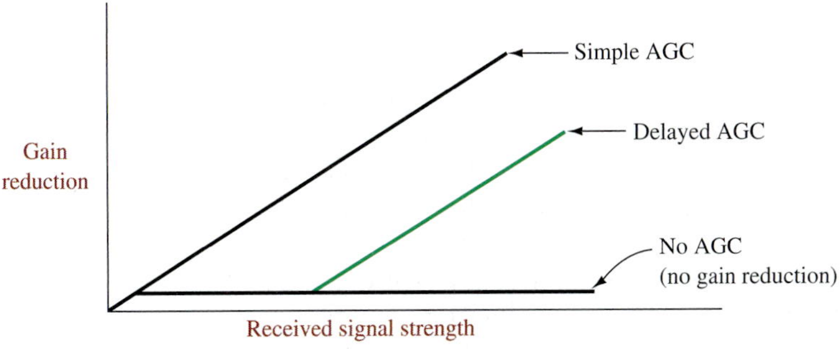

FIGURE 7-5 AGC characteristics.

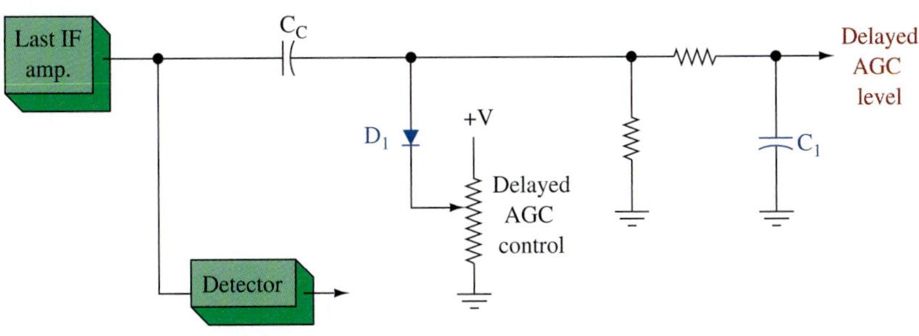

FIGURE 7-6 Delayed AGC configuration.

A simple means of providing delayed AGC is shown in Fig. 7-6. A reverse bias is applied to the cathode of D_1. Thus, the diode looks like an open circuit to the ac signal from the last IF amplifier unless that signal reaches some predetermined instantaneous level. For small IF outputs when D_1 is open, the capacitor C_1 sees a pure ac signal, and thus no dc AGC level is sent back to previous stages to reduce their gain. If the IF output increases, eventually a point is reached where D_1 will conduct on its peak positive levels. This will effectively *short* out the positive peaks of IF output, and C_1 will therefore see a more negative than positive signal and filter it into a relatively constant negative level used to reduce the gain of previous stages. The amplitude of IF output required to start feedback of the "delayed" AGC signal is adjustable by the delayed AGC control potentiometer of Fig. 7-6. This may be an external control so that the user can adjust the amount of delay to suit conditions. For instance, if mostly weak signals are being received, the control might be set so that no AGC signal is developed except for very strong stations. This means the delay interval shown in Fig. 7-5 is increased.

Auxiliary AGC

Auxiliary AGC is used (even on some broadcast receivers) to cause a step reduction in receiver gain at some arbitrarily high value of received signal. It then has the effect of preventing very strong signals from overloading a receiver and thereby causing a distorted, unintelligible output. A simple means of accomplishing the auxiliary AGC function is illustrated in Fig. 7-7(a). Notice the auxiliary AGC diode connected between the collectors of the mixer and first IF transistors. Under normal signal conditions, the dc level at each collector is such that the diode is reverse-biased. In this

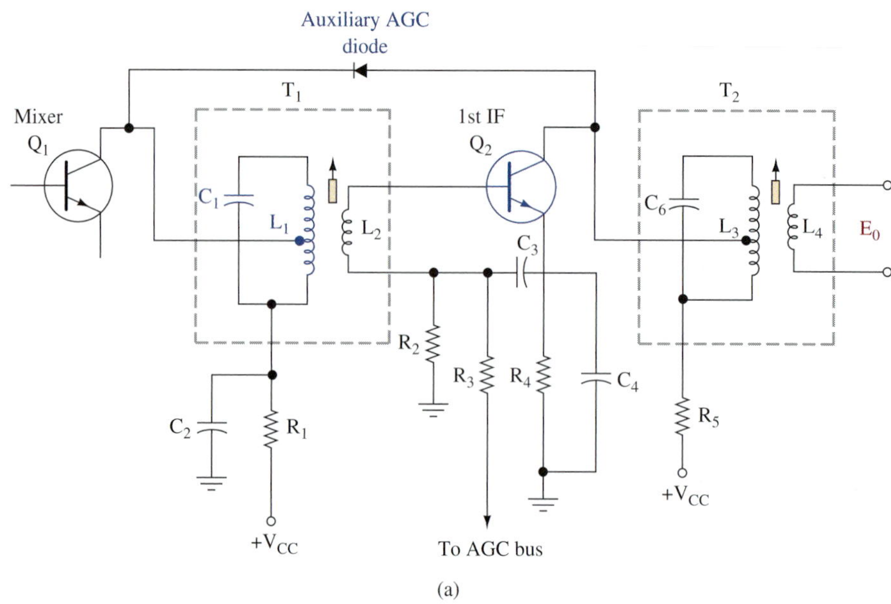

(a)

(b)

(Continued)

FIGURE 7-7 (a) Auxiliary AGC; (b) the Analog Devices AD8369 variable gain amplifier IC;

condition the diode has a very high resistance and has no effect on circuit action. The potential at the mixer's collector is constant because it is not controlled by the normal AGC. However, the AGC control on the first IF transistor, for very strong signals, causes its dc base current to decrease, and hence the collector current also decreases. Thus, its collector voltage becomes more positive, and the diode starts to conduct. The diode resistance goes low, and it loads down the mixer tank (L_1 C_1) and thereby produces a step reduction of the signal coupled into the first IF stage. The dynamic AGC range has thereby been substantially increased.

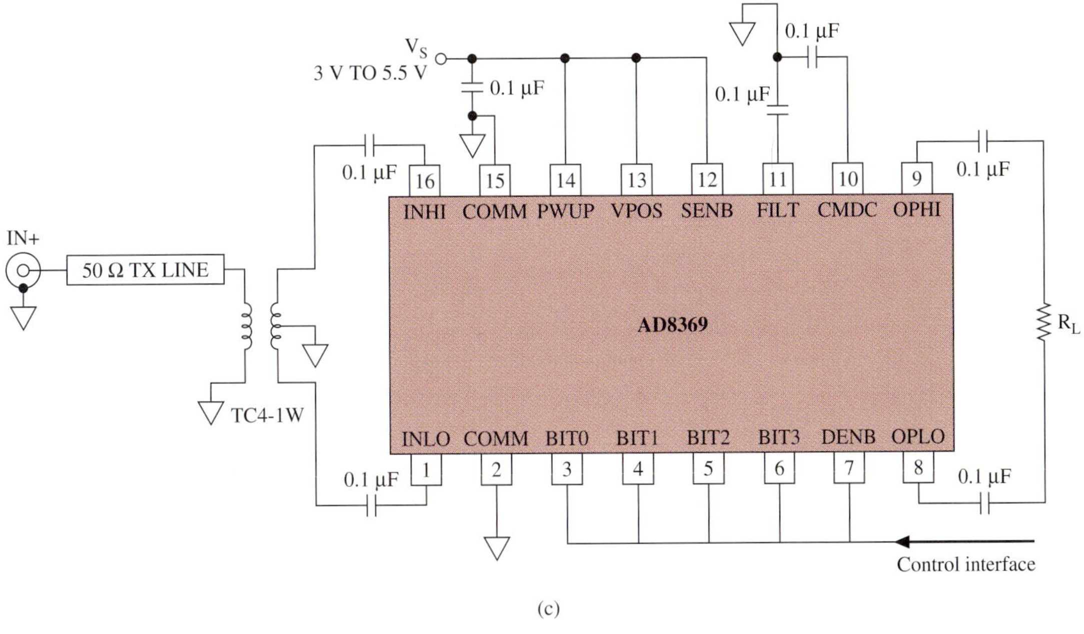

FIGURE 7-7 (*Continued*) (c) basic connections.

An example of a modern digitally controlled variable gain amplifier, the Analog Devices AD8369, is shown in Fig. 7-7(b). It provides a −5 dB to +40 dB digitally adjustable gain in 3-dB increments. The AD8369 is specified for use in RF receive-path AGC loops in cellular receivers. The minimum basic connections for the AD8369 are also shown in Fig. 7-7(c). A balanced RF input connects to pins 16 and 1. Gain control of the device is provided via pins 3, 4, 5, 6, and 7. A 3 V to +5.5 V connects to pins 12, 13, and 14. The balanced differential output appears at pins 8 and 9.

Variable Sensitivity

Despite the increased dynamic range provided by delayed AGC and auxiliary AGC, it is often advantageous for a receiver also to include a variable sensitivity control. This is a manual AGC control because the user controls the receiver gain (and thus sensitivity) to suit the requirement. A communications receiver may be called upon to deal with signals over a 100,000 : 1 ratio, and even the most advanced AGC system does not afford that amount of range. Receivers that are designed to provide high sensitivity and that can handle very large input signals incorporate a manual sensitivity control that controls the RF and/or IF gain.

Variable Selectivity

Many communications receivers provide detection to more than one kind of transmission. They may detect code transmissions, SSB, AM, and FM all in one receiver. The required bandwidth to avoid picking up adjacent channels may well vary from 1 kHz for code up to 30 kHz for narrowband FM.

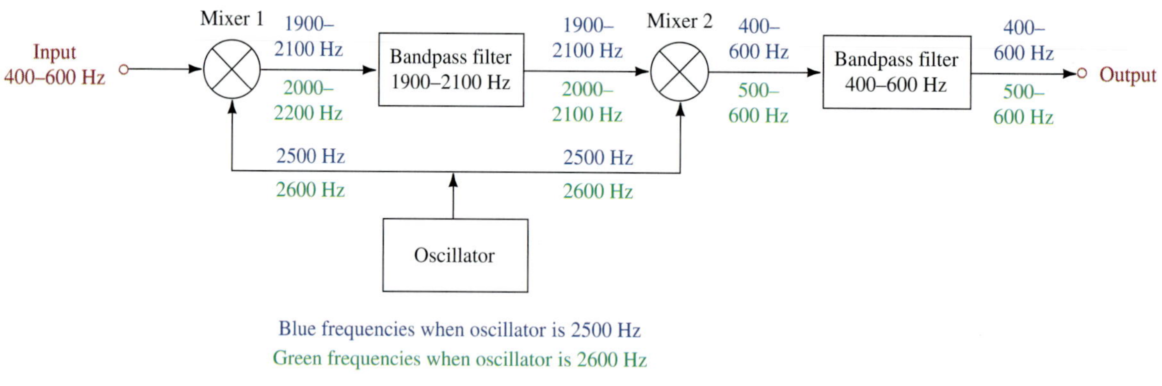

FIGURE 7-8 Variable bandwidth tuning (VBT).

Variable Bandwidth Tuning
technique to obtain variable selectivity to accommodate reception of variable bandwidth signals

Most modern receivers use a technique called **variable bandwidth tuning** (VBT) to obtain variable selectivity. Consider the block diagram shown in Fig. 7-8. The input signal at 500 Hz has half-power frequencies at 400 and 600 Hz, a 200-Hz bandwidth. This is mixed with a 2500-Hz local oscillator output to develop signals from 1900 to 2100 Hz. These drive the bandpass filter, which passes signals from 1900 to 2100 Hz. The filter output is mixed with the LO output producing signals from 400 to 600 Hz. These are fed to another bandpass filter, which passes signals between 400 and 600 Hz. The system output, therefore, is the original 400- to 600-Hz range with the original bandwidth of 200 Hz.

Keeping the same 500-Hz input signal, let us now increase the LO frequency to 2600 Hz. The first mixer output now covers the range from 2000 to 2200 Hz, but only that portion from 2000 to 2100 Hz is passed by the first BPF. These are mixed to produce signals from 500 to 600 Hz, which are passed by the second BPF. The output bandwidth has been reduced to 100 Hz. Increase the LO to 2650 Hz and the system bandwidth drops to 50 Hz. In other words, the bandwidth is now a function of the variable LO frequency.

Noise Limiter

Manufactured sources of external noise are extremely troublesome to highly sensitive communications receivers as well as to any other electronics equipment that is dealing with signals in the microvolt region or less. The interference created by these human-made sources, such as ignition systems, motor communications systems, and switching of high current loads, is a form of **electromagnetic interference** (EMI). Reception of EMI by a receiver creates undesired amplitude modulation, sometimes of such large magnitude as to affect FM reception adversely, to say nothing of the complete havoc created in AM systems. While these noise impulses are usually of short duration, it is not uncommon for them to have amplitudes up to 1000 times that of the desired signal. A noise limiter circuit is employed to silence the receiver for the duration of a noise pulse, which is preferable to a very loud crash from the speaker. These circuits are sometimes referred to as automatic noise limiter (ANL) circuits.

Electromagnetic Interference
unwanted signals produced by devices that produce excessive electromagnetic radiation

A common type of circuit for providing noise limiting is shown in Fig. 7-9. It uses a diode, D_2, that conducts the detected signal to the audio amplifier as long

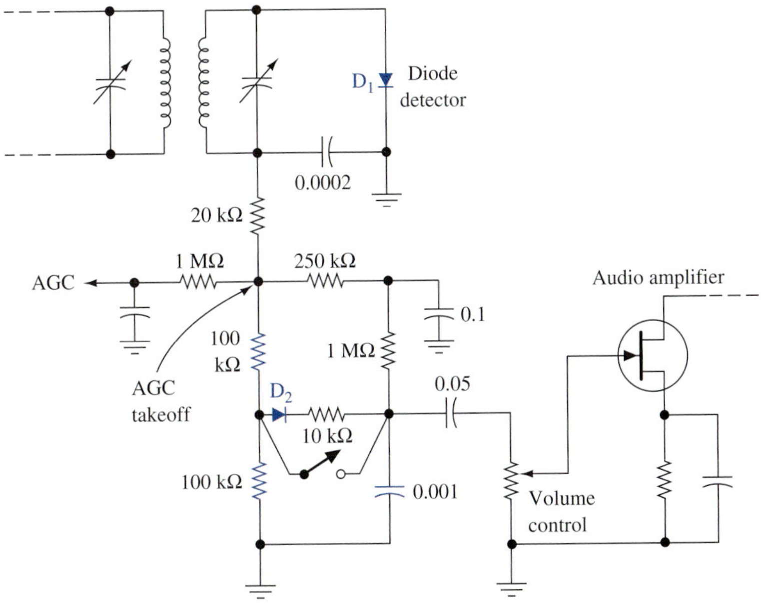

FIGURE 7-9 Automatic noise limiter.

as it is not greater than some prescribed limit. Greater amplitudes cause D_2 to stop conducting until the noise impulse has decreased or ended. The varying audio signal from the diode detector, D_1, is developed across the two 100-kΩ resistors. If the received carrier is producing a −10-V level at the AGC takeoff, the anode of D_2 is at −5 V and the cathode is at −10 V. The diode is on and conducts the audio into the audio amplifier. Impulse noise will cause the AGC takeoff voltage to increase instantaneously, which means the anode of D_2 also does. However, its cathode potential does not change instantaneously since the voltage from cathode to ground is across a 0.001-μF capacitor. Remember that the voltage across a capacitance cannot change instantaneously. Therefore, the cathode stays at −10 V, and as the anode approaches −10 V, D_2 turns off and the detected audio is blocked from entering the audio amplifier. The receiver is silenced for the duration of the noise pulse.

The switch across D_2 allows the noise limiter action to be disabled by the user. This may be necessary when a marginal (noisy) signal is being received and the set output is being turned off excessively by the ANL.

Metering

Many communications receivers are equipped with a meter that provides a visual indication of received signal strength. It is known as the **S meter** and often is found in the emitter leg of an AGC controlled amplifier stage (RF or IF). It reads dc current, which is usually inversely proportional to received signal strength. With no received signal, there is no AGC bias level, which causes maximum dc emitter current flow and therefore maximum stage voltage gain. As the AGC level increases, indicating an increasing received signal, the dc emitter current goes down, which thereby reduces gain. The S meter can thus be used as an aid to accurate tuning as well as providing a relative guide to signal strength. Modern designs use LED bar graphs instead of an electrical meter. This reduces cost and improves reliability.

S Meter
signal strength meter that responds to the received signal level

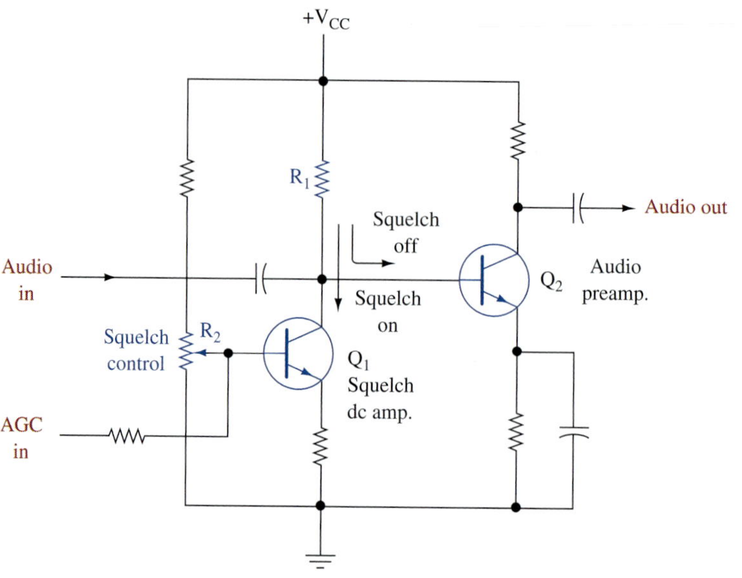

FIGURE 7-10 Squelch circuit.

In some receivers, the S meter can be electrically switched into different areas to aid in troubleshooting a malfunction. In those cases, the operator's manual provides a troubleshooting guide on the basis of meter readings obtained in different areas of the receiver.

Squelch

Muting
the squelch capability of better-quality broadcast FM receivers

Quieting
the tendency for an FM receiver's audio output signal to turn off as the detector responds to a low input carrier level or no carrier input

When a sensitive receiver receives no carrier, the AGC action causes maximum system gain, which results in a high degree of noise output. This sounds like a hissing or rushing noise. This occurs in many communications applications where the user is constantly monitoring a transmission, such as in police service, but there is no transmission most of the time. Without a squelch system to cut off the receiver's output during transmission lulls, the noise output would cause the user severe aggravation. Squelch circuitry is also useful in minimizing noise output that occurs when tuning between stations. In fact, even better-quality broadcast FM receivers provide a squelch capability, but in these applications it is usually termed **muting.** Squelch circuitry is also referred to as the **quieting** or simply the Q circuit.

Figure 7-10 shows a squelch configuration that causes the audio amplifier stage to be cut off whenever no carrier (or an extremely weak station) is being received. In that case the AGC level is zero, which causes the dc amplifier stage, Q_1, to be on. This draws the current being supplied by R_1 into Q_1's collector away from Q_2's base, which means the audio preamp transistor, Q_2, is cut off. Thus, no ac signal is available from Q_2's collector for subsequent power amplification to the speaker. When a signal is picked up by the receiver, the AGC level goes to some negative dc value, which causes Q_1 to turn off. This allows the dc current being supplied by R_1 to enter Q_2's base, which biases it on so that audio preamplification can take place.

User adjustment of the squelch control (R_2) allows the cut-in point of quieting to be varied. This is necessary so that a very weak station, one that generates a small AGC level, will not cause the receiver's output to be squelched.

Squelch Techniques*

The squelch circuitry in a receiver is employed to mute the audio output when the matching transmitter is turned off, or when signal conditions are too poor to produce a usable signal to noise ratio. Several different methods are used:

1. Fixed RF level threshold.
2. Variable level controlled by HF audio noise.
3. Pilot tone control signal.
4. Digital code control signal.
5. Microprocessor controlled algorithm (SmartSquelch™).

Two opposite conditions require different squelch activity:

1. Close operating range with a strong average RF level.
2. Distant operating range with a weak average RF level.

At a close operating range with a generally strong RF level, an ideal squelch would be aggressive and mute noise caused by multipath dropouts without allowing any noise to be produced in the audio signal. The problem with this approach is that an aggressive squelch will reduce operating range significantly. At longer operating distances with a lower average RF signal level, an ideal squelch would be less aggressive and allow the RF signal to dip closer to the noise floor and thus extend the operating range. This approach, however, can allow brief noise-ups at close range caused by multipath signal conditions.

Fixed RF-level squelch systems monitor only the incoming signal level to determine the need to squelch. While the squelch threshold is often adjustable in this type of design, determining an optimum setting is difficult at best because the average RF level in any given situation is almost impossible to predict. The receiver can also be falsely triggered by interference when the matching transmitter is turned off.

A squelch system that utilizes HF audio noise to control the squelch threshold is effective at muting the receiver when the transmitter is turned off. This approach also assumes that a dropout is preceded by a buildup of high frequency audio noise. While this type of squelch is fairly effective in most cases, it can also be fooled by audio containing a large amount of high frequency content such as jingling car keys or coins.

A pilot tone controlled squelch system normally uses a continuous supersonic audio signal generated in the transmitter to control the audio output of the receiver. The receiver must be more sensitive to the pilot tone signal than the RF carrier to avoid inadvertent squelching when the carrier is weak, yet still strong enough to produce usable audio. This approach is highly reliable in muting the receiver when the transmitter is turned off, but it does not address the issue of strong and weak RF signal conditions when the transmitter is close or at a distance.

A digital code squelch technique utilizes a supersonic audio signal containing a unique 8-bit code generated in the transmitter to signal the receiver to open the audio output when the transmitter is turned on. The code is repeated several times at turn-on to ensure that it is picked up by the receiver. At turn-off, the transmitter first emits another code to signal the receiver to mute the audio, then after a brief delay, shuts down the transmitter power. A different code is used in every system to avoid conflicts in multichannel wireless systems. This approach is highly effective at keeping

*(Courtesy of Lectrosonics Inc. Reprinted with permission.)

the receiver quiet when the transmitter is off, and it eliminates noise at turn-on and turn-off, but it does not address the issue of strong and weak signal conditions.

A unique technique called SmartSquelch™ is employed in some Lectrosonics receivers. This is a microprocessor-controlled technique that automatically controls the squelch activity by monitoring the RF level, the audio level, and the recent squelching history over a time period of several seconds. The system provides aggressive squelch activity during strong RF signal conditions to eliminate completely the noise caused by multipath conditions at close range. During weak RF signal conditions, the system provides less aggressive squelch activity to allow maximum operating range by taking advantage of audio masking to bury background noise.

7-4 RECEIVER NOISE, SENSITIVITY, AND DYNAMIC RANGE RELATIONSHIPS

Now that you have become more knowledgeable about receivers, it is appropriate to expand on the noise considerations discussed in Chapter 1. As you will see, there are various trade-offs and relationships among noise figure, sensitivity, and dynamic range when dealing with high-quality receiver systems.

To understand fully these relationships for a receiver, it is first necessary to recognize the factors limiting sensitivity. In one word, the factor most directly limiting sensitivity is noise. Without noise, it would be necessary only to provide enough amplification to receive any signal, no matter how small. Unfortunately, noise is always present and must be understood and controlled as much as possible.

As explained in Chapter 1, there are many sources of noise. The overwhelming effect in a receiver is thermal noise caused by electron activity in a resistance. From Chapter 1 the noise power is

$$P_n = kT\,\Delta f \tag{1-10}$$

For a 1-Hz bandwidth (Δf) and at 290 K:

$$\begin{aligned} P &= 1.38 \times 10^{-23}\ \text{J/K} \times 290\ \text{K} \times 1\ \text{Hz} \\ &= 4 \times 10^{-21}\ \text{W} = -174\ \text{dBm} \end{aligned}$$

For a 1-Hz, 1-K system:

$$P = 1.38 \times 10^{-23}\ \text{W} = -198\ \text{dBm}$$

The preceding shows the temperature variable is of interest because it is possible to lower the circuit temperature and decrease noise without changing other system parameters. At 0 K no noise is generated. It is very expensive and difficult, however, to operate systems anywhere near 0 K. Most receiving systems are operated at ambient temperature. The other possible means to lower thermal noise is to reduce the bandwidth. However, the designer has limited capability in this regard.

Noise and Receiver Sensitivity

What is the sensitivity of a receiver? This question cannot be answered directly without making certain assumptions or knowing certain facts that will have an effect on the result. Examination of the following formula illustrates the dependent factors in determining sensitivity.

$$S = \text{sensitivity} = -174 \text{ dBm} + \text{NF} + 10 \log_{10} \Delta f + \text{desired } S/N \quad \textbf{(7-2)}$$

where −174 dBm is the thermal noise power at room temperature (290 K) in a 1-Hz bandwidth. It is the performance obtainable at room temperature if no other degrading factors are involved. The $10 \log_{10} \Delta f$ factor in Eq. (7-2) represents the change in noise power due to a change above a 1-Hz bandwidth. The wider the bandwidth, the greater the noise power and the higher the noise floor. *S/N* is the desired signal-to-noise ratio in dB. It can be determined for the signal level, which is barely detectable, or it may be regarded as the level allowing an output at various ratings of fidelity. Often, a 0-dB *S/N* is used, which means that the signal and noise power at the output are equal. The signal can therefore also be said to be equal to the **noise floor** of the receiver. The receiver noise floor and the receiver output noise are one and the same thing.

Noise Floor
the baseline on a spectrum analyzer display, equal to the output noise of the device under test

Consider a receiver that has a 1-MHz bandwidth and a 20-dB noise figure. If an *S/N* of 10 dB is desired, the sensitivity (S) is

$$\begin{aligned} S &= -174 + 20 + 10 \log(1{,}000{,}000) + 10 \\ &= -84 \text{ dBm} \end{aligned}$$

You can see from this computation that if a lower *S/N* is required, better receiver sensitivity is necessary. If a 0-dB *S/N* is used, the sensitivity would become −94 dBm. The −94-dBm figure is the level at which the signal power equals noise power in the receiver's bandwidth. If the bandwidth were reduced to 100 kHz while maintaining the same input signal level, the output *S/N* would be increased to 10 dB due to noise power reduction.

Dynamic Range

The **dynamic range** of an amplifier or receiver is the input power range over which it provides a useful output. It should be stressed that a receiver's dynamic and AGC ranges are usually two different quantities. The low-power limit is essentially the sensitivity specification discussed in the preceding paragraphs. It is a function of the noise. The upper limit has to do with the point at which the system no longer provides the same linear increase as related to the input increase. It also has to do with certain distortion components and their degree of effect.

Dynamic Range
for a receiver, the decibel difference between the largest tolerable receiver input level and its sensitivity (smallest useful input level)

When testing a receiver (or amplifier) for the upper dynamic range limit, it is common to apply a single test frequency and determine the *1-dB compression point*. As shown in Fig. 7-11, this is the point in the input/output relationship where the output has just reached a level where it is 1 dB down from the ideal linear response. The input power at that point is then specified as the upper power limit determination of dynamic range.

When two frequencies (f_1 and f_2) are amplified, the second-order distortion products are generally out of the system passband and are therefore not a problem. They occur at $2f_1$, $2f_2$, $f_1 + f_2$, and $f_1 - f_2$. Unfortunately, the third-order products at $2f_1 + f_2$, $2f_1 - f_2$, $2f_2 - f_1$, and $2f_2 + f_1$ usually have components in the system bandwidth. The distortion thereby introduced, *intermodulation distortion* (IMD), was mentioned in Chapter 6. It is often referred to simply as **intermod.** Recall that

Intermod
intermodulation distortion

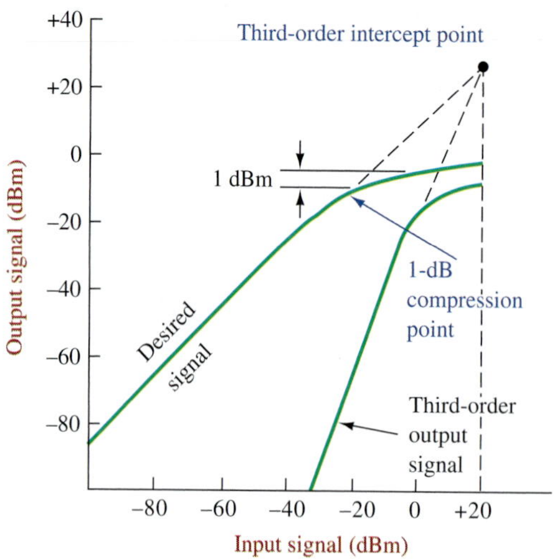

FIGURE 7-11 Third-order intercept and compression point illustration.

the use of MOSFETs at the critical RF and mixer stages is helpful in minimizing these third-order effects. Intermodulation effects have such a major influence on the upper dynamic range of a receiver (or amplifier) that they are often specified via the **third-order intercept point** (or **input intercept**). This is illustrated in Fig. 7-11. It is the input power at the point where straight-line extensions of desired and third-order input/output relationships meet. It is about 20 dBm in Fig. 7-11. It is used only as a figure of merit. The better a system is with respect to intermodulation distortion, the higher will be its input intercept.

Third-Order Intercept Point
receiver figure of merit describing how well it rejects intermodulation distortion from third-order products resulting at the mixer output

Input Intercept
another name for third-order intercept point

The dynamic range of a system is usually approximated as

$$\text{dynamic range (dB)} \simeq \frac{2}{3}\,(\text{input intercept} - \text{noise floor}) \qquad \textbf{(7-3)}$$

Poor dynamic range causes problems, such as undesired interference and distortion, when a strong signal is received. The current state of the art is a dynamic range of about 100 dB.

EXAMPLE 7-7

A receiver has a 20-dB noise figure (NF), a 1-MHz bandwidth, a +5-dBm third-order intercept point, and a 0-dB S/N. Determine its sensitivity and dynamic range.

SOLUTION

$$S = -174\text{ dBm} + \text{NF} + 10\log_{10}\Delta f + \frac{S}{N} \qquad \textbf{(7-2)}$$

$$= -174\text{ dBm} + 20\text{ dB} + 10\log_{10}10^6 + 0 = -94\text{ dBm}$$

$$\text{dynamic range} \simeq \frac{2}{3}\,(\text{input intercept} - \text{noise floor}) \qquad \textbf{(7-3)}$$

$$= \frac{2}{3}[5\text{ dBm} - (-94\text{ dBm})]$$

$$= 66\text{ dB}$$

Example 7-8

The receiver from Ex. 7-7 has a preamplifier at its input. The preamp has a 24-dB gain and a 5-dB NF. Calculate the new sensitivity and dynamic range.

Solution

The first step is to determine the overall system noise ratio (NR). Recall from Chapter 1 that

$$\text{NR} = \log^{-1}\frac{\text{NF}}{10}$$

Letting NR_1 represent the preamp and NR_2 the receiver, we have

$$\text{NR}_1 = \log^{-1}\frac{5\text{ dB}}{10} = 3.16$$

$$\text{NR}_2 = \log^{-1}\frac{20\text{ dB}}{10} = 100$$

The overall NR is

$$\text{NR} = \text{NR}_1 + \frac{\text{NR}_2 - 1}{P_{G_1}} \tag{1-16}$$

and

$$P_{G1} = \log^{-1}\frac{24\text{ dB}}{10} = 251$$

$$\text{NR} = 3.16 + \frac{100 - 1}{251} = 3.55$$

$$\text{NF} = 10\log_{10} 3.55 = 5.5\text{ dB}$$

$$= \text{total system NF}$$

$$S = -174\text{ dBm} + 5.5\text{ dB} + 60\text{ dB} = -108.5\text{ dBm}$$

The third-order intercept point of the receiver alone had been +5 dBm but is now preceded by the preamp with 24-dB gain. Assuming that the preamp can deliver 5 dBm to the receiver without any appreciable intermodulation distortion, the system's third-order intercept point is +5 dBm −24 dB = −19 dBm. Thus,

$$\text{dynamic range} \simeq \frac{2}{3}[-19\text{ dBm} - (-108.5\text{ dBm})]$$

$$= 59.7\text{ dB}$$

Example 7-9

The 24-dB gain preamp in Ex. 7-8 is replaced with a 10-dB gain preamp with the same 5-dB NF. What are the system's sensitivity and dynamic range?

Solution

$$\text{NR} = 3.16 + \frac{100 - 1}{10} = 13.1$$

$$\text{NF} = 10 \log_{10} 13.1 = 11.2 \text{ dB}$$

$$S = -174 \text{ dBm} + 11.2 \text{ dB} + 60 \text{ dB} = -102.8 \text{ dBm}$$

$$\text{dynamic range} \simeq \frac{2}{3}[-5 \text{ dBm} - (-102.8 \text{ dB})] = 65.2 \text{ dB}$$

The results of Exs. 7-7 to 7-9 are summarized as follows:

	Receiver Only	Receiver and 10-dB Preamp	Receiver and 24-dB Preamp
NF (dB)	20	11.2	5.5
Sensitivity (dBm)	−94	−102.8	−108.5
Third-order intercept point (dBm)	+5	−5	−19
Dynamic range (dB)	66	65.2	59.7

An analysis of the examples and these data shows that the greatest sensitivity can be realized by using a preamplifier with the lowest noise figure and highest available gain to mask the higher NF of the receiver. Remember that as gain increases, so does the chance of spurious signals and intermodulation distortion components. A preamplifier used prior to a receiver input has the effect of decreasing the third-order intercept proportionally to the gain of the amplifier, while the increase in sensitivity is less than the gain of the amplifier. Therefore, to maintain a high dynamic range, it is best to use only the amplification needed to obtain the desired noise figure. It is not helpful in an overall sense to use excessive gain. The data in the chart show that adding the 10-dB gain preamplifier improved sensitivity by 8.8 dB and decreased dynamic range by only 0.8 dB. The 24-dB gain preamp improved sensitivity by 14.5 dB but decreased the dynamic range by 6.3 dB.

Intermodulation Distortion Testing

It is common to test an amplifier for its intermodulation distortion (IMD) by comparing two test frequencies to the level of a specific IMD product. As previously mentioned, the second-order products are usually outside the frequency range of concern. This is generally true for all the even-order products and is illustrated in Fig. 7-12. It shows some second-order products that would be outside the bandwidth of interest for most systems.

The odd products are of interest because some of them can be quite close to the test frequencies f_1 and f_2 shown in Fig. 7-12. The third-order products shown ($2f_2 - f_1$ and $2f_1 - f_2$) have the most effect, but even the fifth-order products ($3f_2 - 2f_1$ and $3f_1 - 2f_2$) can be troublesome. Figure 7-13(a) shows a typical spectrum analyzer display when two test signals are applied to a mixer or small-signal amplifier. Notice that the third-order products are shown 80 dB down from the test signals, while the fifth-order products are more than 90 dB down.

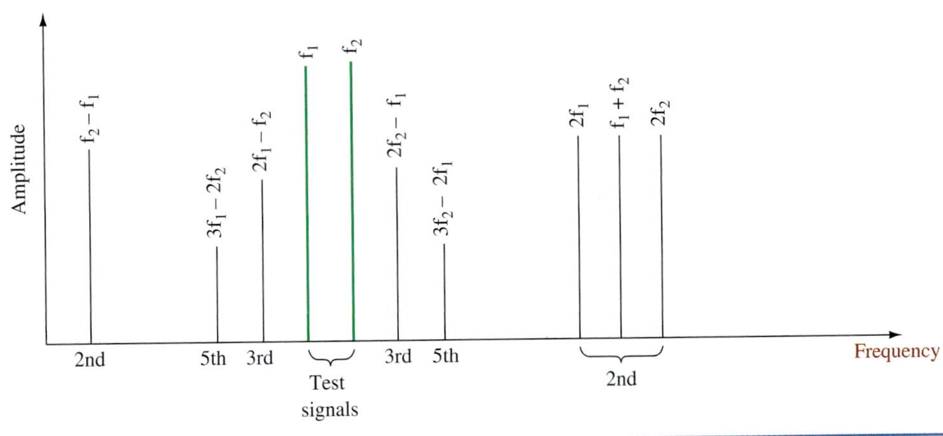

FIGURE 7-12 IMD products (second-, third-, and fifth-order for two test signals).

Figure 7-13(b) shows the IMD testing result when applying two frequencies to a typical Class AB linear power amplifier. The higher odd-order products (up to the eleventh in this case) are significant for the power amplifier. Fortunately, these effects are less critical in power amplifiers than in the sensitive front end of a radio receiver.

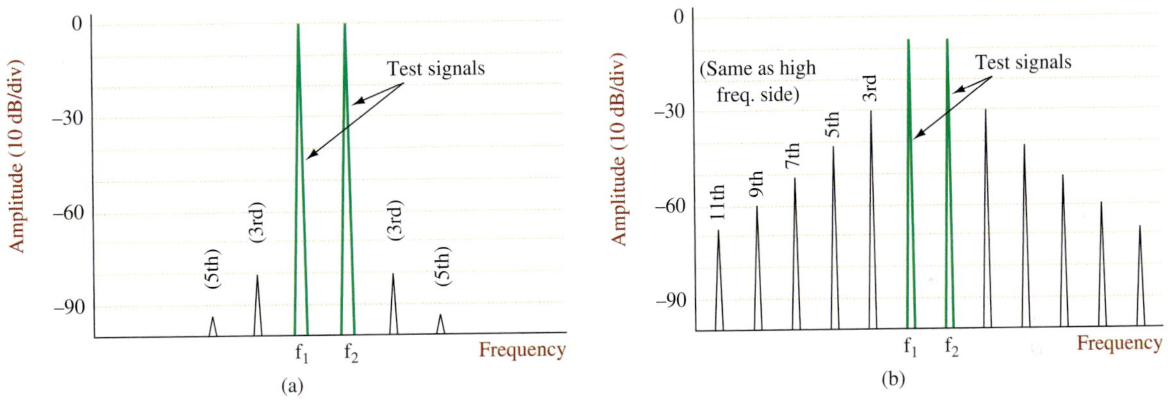

FIGURE 7-13 IMD testing: (a) mixer; (b) Class AB linear power amplifier.

 7-5 Frequency Synthesis

Most transceiver designs use frequency synthesizers to generate the highly accurate frequencies used for the transmitter carrier and receiver local oscillator. They are also widely used in signal generators and instrumentation systems such as spectrum analyzers and modulation analyzers. The concept of frequency synthesis has been around since the 1930s, but the cost of the circuitry necessary was prohibitive for most designs until integrated circuit technology started offering the phased-locked loop (PLL) in a single, low-cost chip. Recall that basic PLL theory was provided in Chapter 6.

A basic frequency synthesizer is shown in Fig. 7-14. Besides the PLL, the synthesizer includes a very stable crystal oscillator and the divide-by-*N* programmable

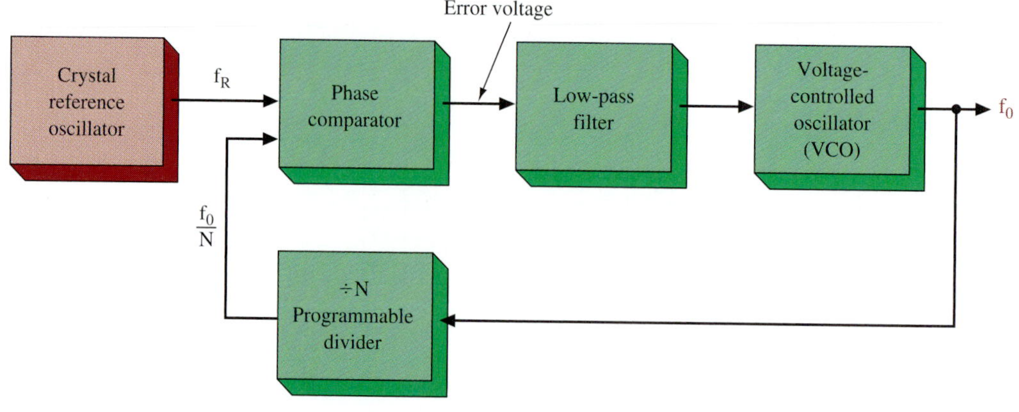

FIGURE 7-14 Basic frequency synthesizer.

Error Voltage output of the phase comparator in a PLL

divider. The output frequency of the voltage-controlled oscillator (VCO) is a function of the applied control voltage.

The output of the phase comparator is a voltage termed the **error voltage,** which is proportional to the phase difference between the signals at its two inputs. This output controls the frequency of the VCO so that the phase comparator input from the VCO via the variable divider ($\div N$) remains at a constant phase difference with the reference input, f_R, so that the frequencies are equal. The VCO frequency is thus maintained at Nf_R. Such a synthesizer will produce a number of frequencies separated by f_R and is the most basic form of phase-locked synthesizer. Its stability is directly governed by the stability of the reference input f_R, although it is also related to noise in the phase comparator, noise in any dc amplifier between the phase comparator and the VCO, and the characteristics of the low-pass filter usually placed between the phase comparator and the VCO.

Consider the case where the programmable divider in Fig. 7-14 can divide by integers from N equals 1 to 10. If the reference frequency is 100 kHz and $N = 1$, then the output should be 100 kHz. If $N = 2$, f_0 must equal 200 kHz to provide a constant phase difference for the phase comparator. Similarly, for $N = 5$, $f_0 =$ 500 kHz. The pattern should be apparent to you now. A synthesizer with outputs of 100 kHz, 200 kHz, 300 kHz, and so on, is not useful for most applications. Much smaller spacing between output frequencies is necessary, and means to attain that condition are discussed next.

The design of frequency synthesizers using the principle described above involves the design of various subsystems, including the VCO, the phase comparator, any low-pass filters in the feedback path, and the programmable dividers.

Programmable Division

A typical programmable divider is shown in Fig. 7-15. It consists of three stages with division ratios K_1, K_2, and K_3, which may be programmed by inputs P_1, P_2, and P_3, respectively. Each stage divides by K_n except during the first cycle after the program input P_n is loaded, when it divides by P (which may have any integral value from n to K). Hence the counter illustrated divides by $P_3 \times (K_1K_2) + P_2K_1 + P_1$, and when an output pulse occurs, the program inputs are reloaded. The counter will divide by any integer between 1 and $(K_1K_2K_3 - 1)$.

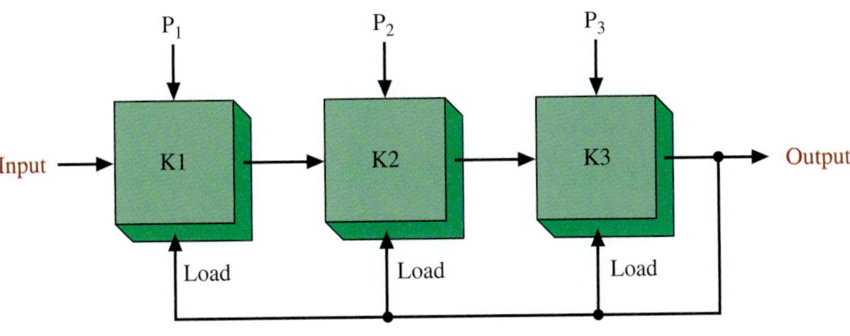

FIGURE 7-15 Typical programmable divider.

The most common programmable dividers are either decades or divide-by-16 counters. These are readily available in various logic families, including CMOS and TTL. CMOS devices are preferred when power consumption is a consideration. Using such a package, one can program a value of N from about 3 to 9999. The theoretical minimum count of 1 is not possible because of the effects of circuit propagation delays. The use of such counters permits the design of frequency synthesizers, which are programmed with decimal thumbwheel switches and use a minimum number of components. If a synthesizer is required with an output of nonconsecutive frequencies and steps, a custom programmable counter may be made using some custom devices such as programmable logic devices (PLDs) or programmable divider ICs. This is the case for the citizen's band synthesizer discussed later in this section.

The maximum input frequency of a programmable divider is limited by the speed of the logic used, and more particularly by the time taken to load the programmed count. The power consumption of high-frequency digital circuitry can be an issue in low-power applications (e.g., cell phones). The output frequency of the simple synthesizer in Fig. 7-14 is limited, of course, to the maximum frequency of the programmable divider.

There are many ways of overcoming this limitation on synthesizer frequency. The VCO output may be mixed with the output of a crystal oscillator and the resulting difference frequency fed to the programmable divider, or the VCO output may be multiplied from a low value in the operating range of the programmable divider to the required high output frequency. Alternatively, a fixed ratio divider capable of operating at a high frequency may be interposed between the VCO and the programmable divider. These methods are shown in Figs. 7-16(a), (b), and (c), respectively.

All the methods discussed above have their problems, although all have been used and will doubtless continue to be used in some applications. Method (a) is the most useful technique because it allows narrower channel spacing or high reference frequencies (hence faster lock times and less loop-generated jitter) than the other two, but it has the drawback that, because the crystal oscillator and the mixer are within the loop, any crystal oscillator noise or mixer noise appears in the synthesizer output. Nevertheless, this technique has much to recommend it.

The other two techniques are less useful. Frequency multiplication introduces noise, and both techniques must either use a very low reference or rather wide channel spacing. What is needed is a programmable divider that operates at the VCO frequency—one can then discard the techniques described above and synthesize directly at whatever frequency is required.

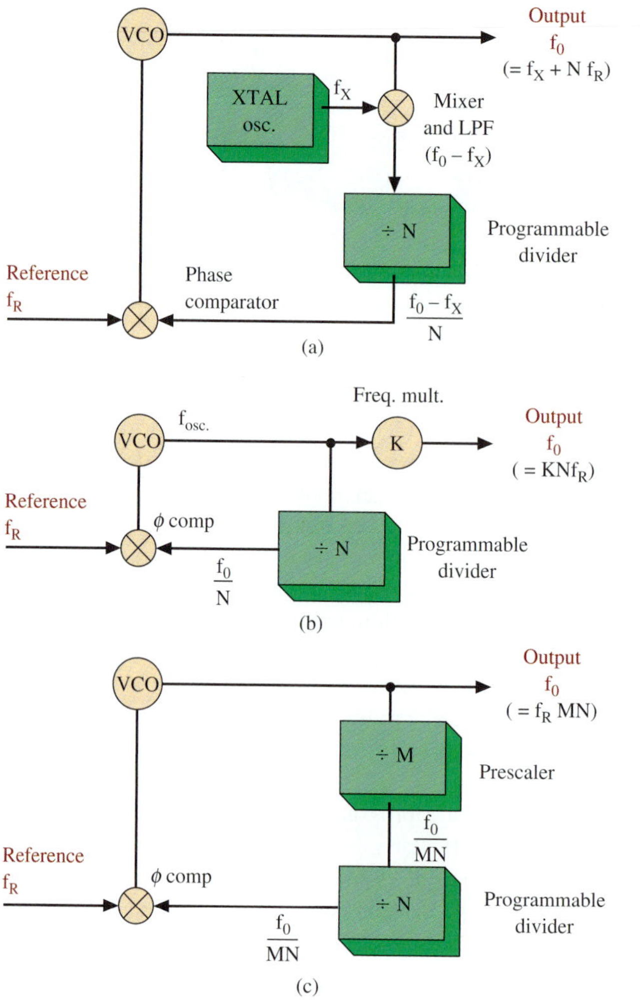

FIGURE 7-16 Synthesizer alternatives.

Two-Modulus Dividers

Considerations of speed and power make it impractical to design programmable counters of the type described above, even using ECL, at frequencies much into the VHF band (30 to 300 MHz) or above. A different technique exists, however, using two-modulus dividers; that is, in one mode it divides by N and in the other mode, by $N + 1$.

Figure 7-17 shows a divider using a two-modulus prescaler. The system is similar to the one shown in Fig. 7-16(c), but in this case the prescaler divides by either N or $N + 1$, depending on the logic state of the control input. The output of the prescaler feeds two normal programmable counters. Counter 1 controls the two-modulus prescaler and has a division ratio A. Counter 2, which drives the output, has a division ratio M. In operation the $N/(N + 1)$ prescaler (Fig. 7-17) divides by $N + 1$ until the count in programmable counter 1 reaches A and then divides by N until the count in programmable counter 2 reaches M, when both counters are reloaded, a pulse passes to output, and the cycle restarts. The division ratio of the entire system is $A(N + 1) + N(M - A)$, which equals $NM + A$. There is only one constraint on the system—because the two-modulus prescaler does not change modulus until counter 1 reaches A, the count in counter 2 (M) must never be less than A. This limits the minimum count the system may reach to $A(N + 1)$, where A is the maximum possible value of count in counter 1.

The use of this system entirely overcomes the problems of high-speed programmable division mentioned earlier. A number of ÷ 10/11 counters working at frequencies of up to 500 MHz and also ÷ 5/6, ÷ 6/7, and ÷ 8/9 counters working up to 500 MHz are now readily available. There is also a pair of circuits intended to allow ÷ 10/11 counters to be used in ÷ 40/41 and ÷ 80/81 counters in 25-kHz and 12.5-kHz channel VHF synthesizers. It is not necessary for two-modulus prescalers to divide by $N/(N + 1)$. The same principles apply to $\div N/(N + Q)$ counters (where Q is any integer), but $\div N/(N + 1)$ tends to be most useful.

Citizen's Band Synthesizer The availability of low-cost PLL and programmable divider ICs has led to the use of synthesizers in almost all channelized transceivers. This is true even for the very low-cost systems used on the 40-channel citizen's band. A typical CB transceiver is shown in Fig. 7-18.

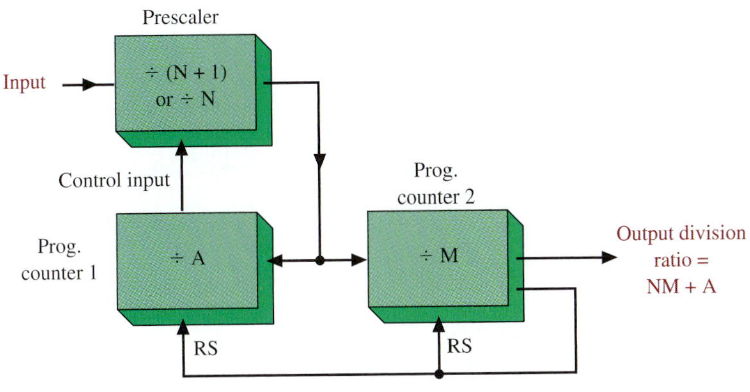

FIGURE 7-17 Divider system with two-modulus prescaler.

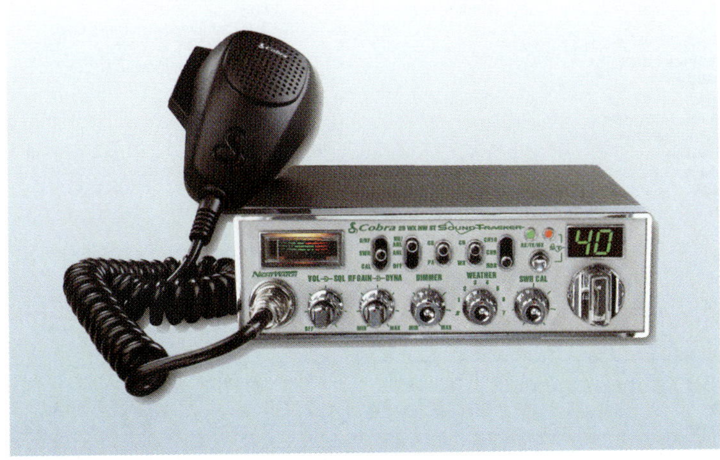

FIGURE 7-18 CB transceiver. (Courtesy of Cobra Electronics/Golin Harris. Reprinted with permission.)

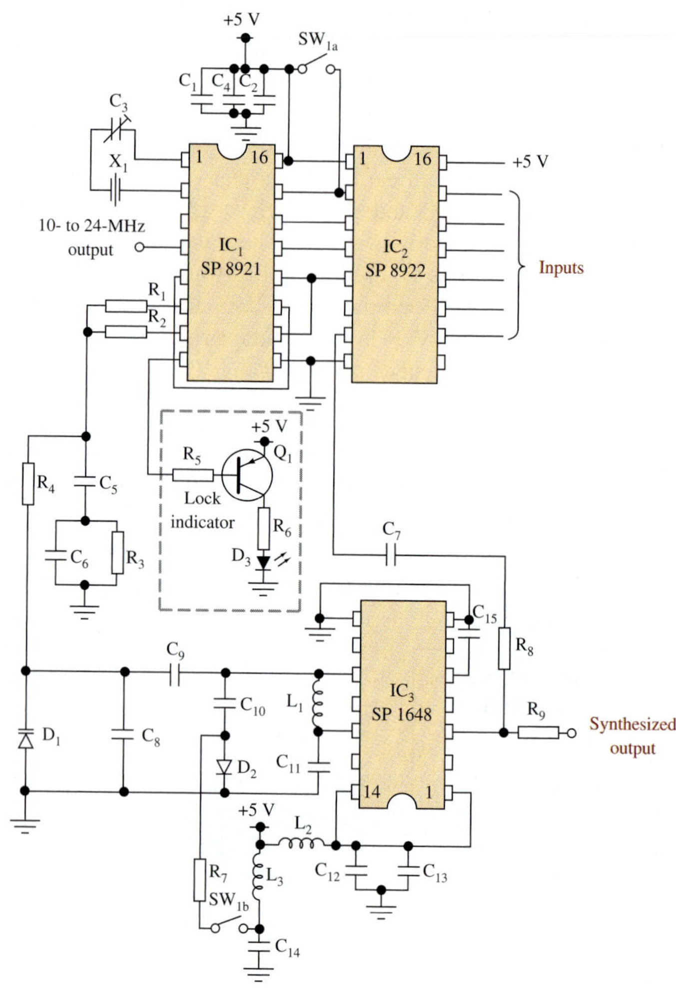

All resistors are $\frac{1}{8}$ W ± 10% unless otherwise stated. Capacitor values are in microfarads unless otherwise stated.

IC_1	SP8921
IC_2	SP8922/SP8923
IC_3	SP1648
Q_1	2N 3906
D_1	ZC822, Ferranti varactor diode
D_2	1N4148 Silicon diode
D_3	LED lock indicator
X_1	10.240-MHz crystal, series mode
L_1	11 turns 30-gauge cotton-covered wire on Neosid A7 assembly
L_2	100-μH RF choke
L_3	100-μH RF choke
R_1	1.0 kΩ ± 5%
R_2	1.0 kΩ ± 5%
R_3	8.2 kΩ ± 5%
R_4	33 kΩ
R_5	10 kΩ
R_6	150 (adjust for LED brightness)

R_7	1 kΩ
R_8	1 kΩ
R_9	470 (adjust for required output level)
C_1	0.1
C_2	100, 10 V solid tantalum
C_3	2-22 pF variable
C_4	100 10 V solid tantalum
C_5	10 V solid tantalum
C_6	0.1
C_7	1000 pF
C_8	22 pF +10%
C_9	0.01
C_{10}	100 pF ± 10%
C_{11}	0.01
C_{12}	0.1
C_{13}	10 solid tantalum
C_{14}	0.1
C_{15}	1000 pF
SW_1	2-pole, 1-way switch, (receive/transmit)

FIGURE 7-19 CB synthesizer circuit.

The synthesizer circuit shown in Fig. 7-19 allows all necessary frequencies for a CB transceiver to be generated by using a single-crystal oscillator. The 40 channels are spaced at 10-kHz intervals (with some gaps) between 26.965 and 27.405 MHz. Local oscillator frequencies for the reception of these channels with intermediate frequencies of 455 kHz, 10.240 MHz, and 10.700 MHz are also synthesized. Table 7-1 shows the relationship between the program input and the channel selected. By using an input code other than one of the 40 given, other frequencies may be selected—in fact, there are 64 channels at 10-kHz separation available from 26.895 to 27.525 MHz and programming starts at all zeros on input A through F for 26.895, and each increase of one bit to the binary number on these inputs increases the channel frequency by 10 kHz until all 1s give 27.535 MHz. The A input is the least significant bit, F the most significant. The programming input on pin 16 of the SP8922 IC in Fig. 7-19 is normally kept high, but making it low increases the programmed frequency by 5 kHz. Table 7-2 shows the programming required to obtain various offsets. This synthesizer is intended for use in double-conversion receivers with IFs of 10.695 MHz and 455 kHz and generates either the frequency programmed or the frequency programmed less 10.695 MHz.

If other offsets are programmed in connections to pin 15 of the SP8921 IC and pin 2 of the SP8922 IC, they must be altered according to Table 7-2. The synthesizer consists of the SP8921 and the SP8922 plus an SP1648 voltage-controlled oscillator. The programming inputs to the SP8922 are as shown in Table 7-1. Logic 1 is +3 V or more; logic 0 is either ground or an open circuit.

The crystal oscillator in the SP8921 (Fig. 7-19) is trimmed by a small variable capacitor, C_3, which must be set up during alignment of the synthesizer so that the output frequency on pin 4 is 10.240000 MHz. The only other adjustment is to set the core of L_1 so that the varicap control voltage (D_1) is 2.85 V when the synthesizer is set to channel 30 transmit. Because the difference between transmit and receive frequencies is over 10 MHz, it is not possible to tune both with the same tuned circuit, and an extra capacitor is switched by means of a diode during reception.

The phase/frequency comparator of the SP8921 can have an output swing from 0.5 V to 3.8 V, but it is better to work in the range 1.5 to 3.0 V because the phase-error output voltage is more linear in this region. The ZC822 tuning diode specified for this synthesizer may be replaced by any other tuning diode provided that it will tune the VCO over the required range, or a little more, as the control voltage goes from 1.5 V to 3.0 V. With slight coil changes, the MV2105 has been used successfully in this synthesizer.

The low-pass filter of the PLL consists of C_5, C_6, and R_3. If faster lock (at the expense of larger noise and reference sidebands) is required, the filter may be redesigned. If the synthesizer is used in a scanning receiver, a switched filter should be used to give fast lock during scanning but a slower lock and cleaner signal during normal operation. A scanning receiver automatically "searches" a number of channels until it finds a good received signal. It then stays tuned to that channel until "nudged" by the operator and/or the reception is lost. The lock output on pin 8 of the SP8921 is used to light an indicator when the loop is not locked and should also be used, in a transmitter or transceiver, to prevent transmission when the loop is unlocked.

Table 7-1 Input Code for CB Synthesizer of Fig. 7-19

Channel Number	Input Code						Output Frequency with $R/T = 0$ (MHz)
	F	E	D	C	B	A	
1	0	0	0	1	1	1	26.965
2	0	0	1	0	0	0	26.975
3	0	0	1	0	0	1	26.985
4	0	0	1	0	1	1	27.005
5	0	0	1	1	0	0	27.015
6	0	0	1	1	0	1	27.025
7	0	0	1	1	1	0	27.035
8	0	1	0	0	0	0	27.055
9	0	1	0	0	0	1	27.065
10	0	1	0	0	1	0	27.075
11	0	1	0	0	1	1	27.085
12	0	1	0	1	0	1	27.105
13	0	1	0	1	1	0	27.115
14	0	1	0	1	1	1	27.125
15	0	1	1	0	0	0	27.135
16	0	1	1	0	1	0	27.155
17	0	1	1	0	1	1	27.165
18	0	1	1	1	0	0	27.175
19	0	1	1	1	0	1	27.185
20	0	1	1	1	1	1	27.205
21	1	0	0	0	0	0	27.215
22	1	0	0	0	0	1	27.225
23	1	0	0	1	0	0	27.255
24	1	0	0	0	1	0	27.235
25	1	0	0	0	1	1	27.245
26	1	0	0	1	0	1	27.265
27	1	0	0	1	1	0	27.275
28	1	0	0	1	1	1	27.285
29	1	0	1	0	0	0	27.295
30	1	0	1	0	0	1	27.305
31	1	0	1	0	1	0	27.315
32	1	0	1	0	1	1	27.325
33	1	0	1	1	0	0	27.335
34	1	0	1	1	0	1	27.345
35	1	0	1	1	1	0	27.355
36	1	0	1	1	1	1	27.365
37	1	1	0	0	0	0	27.375
38	1	1	0	0	0	1	27.385
39	1	1	0	0	1	0	27.395
40	1	1	0	0	1	1	27.405

Table 7-2 Offset Codes for Synthesizer of Fig. 7-19

Offset	SP8921	SP8922
0	0	0
455 kHz	0	1
10.240 MHz	1	0
10.695 MHz	1	1

(a)

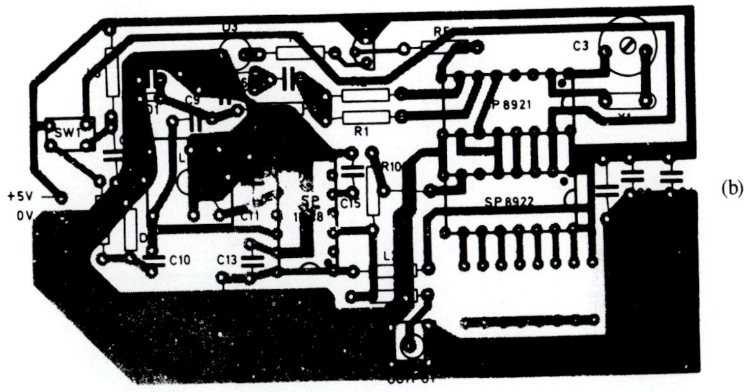

(b)

FIGURE 7-20 Printed circuit board details: (a) printed circuit layout for CB synthesizer; (b) component layout for CB synthesizer.

Figure 7-20 shows the circuit board layout and component location of this synthesizer. It requires a single +5-V supply and draws about 60 mA. The performance is improved if a double-sided board is used with a ground plane on one side. A small additional improvement would come from the use of a grounded screening can over the whole system to prevent stray noise pickup.

The synthesizer has reference frequency sidebands 50 dB down at 1.25 kHz from the carrier. All output over 5 kHz from the carrier is over 70 dB down. Lock time for a change from channel 0 to channel 40 (a frequency change of 440 kHz) is around 35 ms. Stepping from transmit to receive, or vice versa, takes somewhat longer because of the much larger change of frequency but is generally complete within 75 ms.

A block diagram of a UHF receiver incorporating a PLL frequency synthesizer circuit is provided in Fig. 7-21. The receiver features microprocessor control of 256 frequencies with 0.025-MHz spacing. The unit can also be built to provide 13 different frequency blocks from 537.6 MHz to 865.0 MHz. The schematic of the UHF multifrequency receiver is provided in Fig. 7-22 (page 327).

The signal at the antenna is fed through an RF filter and amplifier. The amplified RF signal is then mixed with the frequency, generated by the Philips Semiconductor SA7025 1-GHz low-voltage fractional-*N* synthesizer to down-convert the receive signal to the IF frequency. A Philips TDA7021T FM radio circuit for MTS provides the circuitry for the FM reception. The IC uses a frequency-locked loop (FLL) technology with an intermediate frequency of 76 kHz.

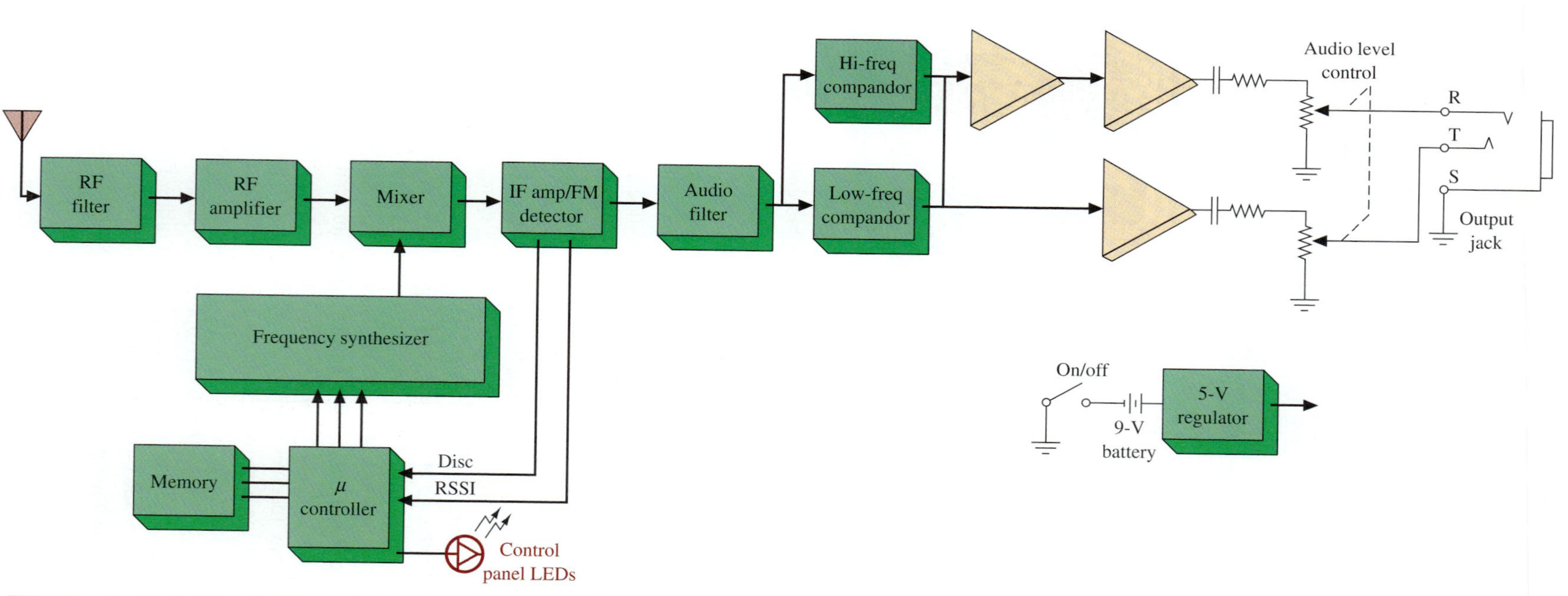

FIGURE 7-21 UCR110 block diagram. (Courtesy of Lectrosonics, Inc.)

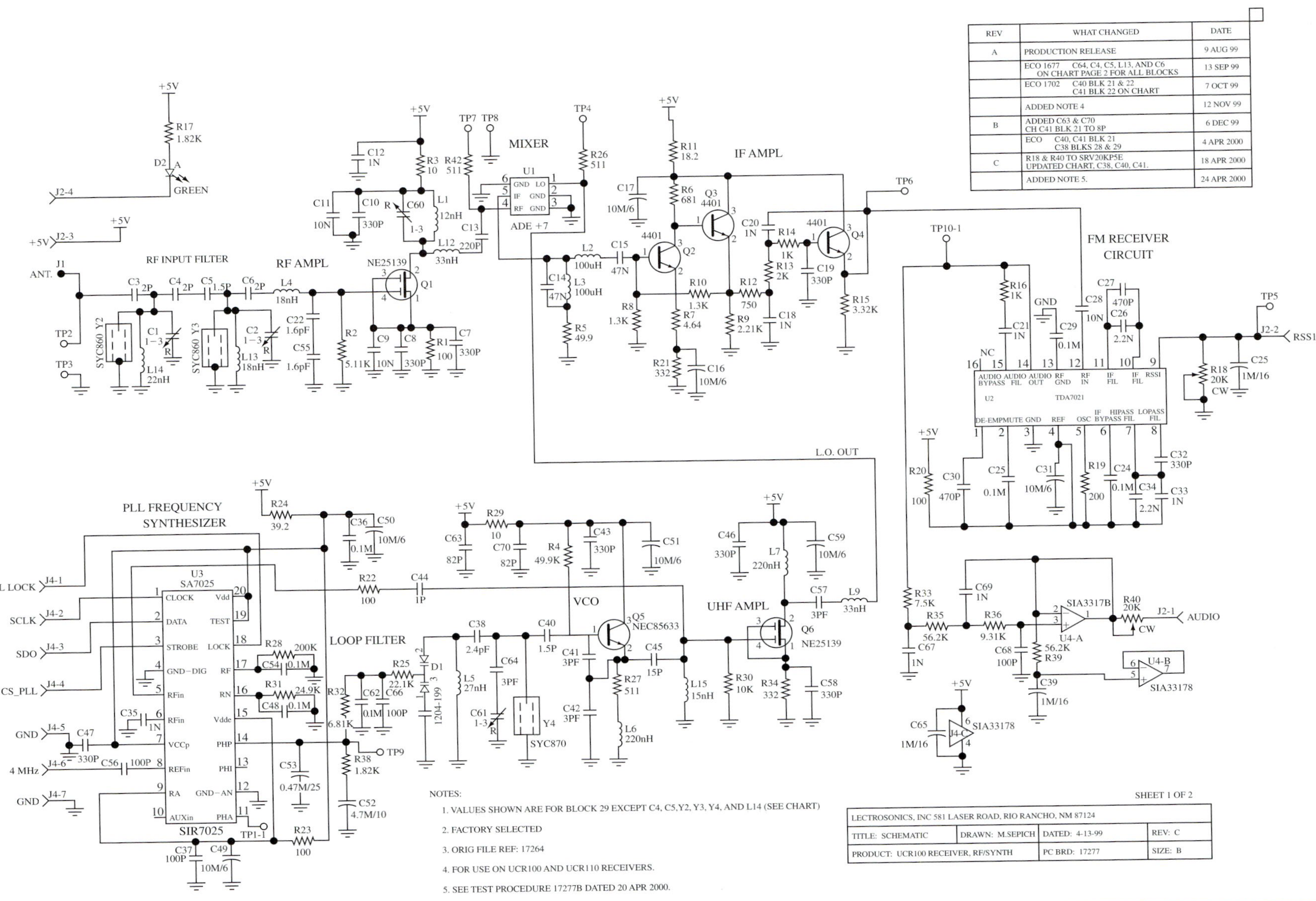

FIGURE 7-22 The schematic of a UHF multifrequency receiver. (Courtesy of Lectrosonics, Inc.)

7-6 Direct Digital Synthesis

Direct Digital Synthesis frequency synthesizer design that has better repeatability and less drifting but limited maximum output frequencies, greater phase noise, and greater complexity and cost

Direct digital synthesis (DDS) systems became economically feasible in the late 1980s. They offer some advantages over the analog synthesizers discussed in the previous section but generally tend to be somewhat more complex and expensive. They are useful, however, for some applications. The digital logic used can improve on the repeatability and drift problems of analog units that often require select-by-test components. These advantages also apply to digital filters that have replaced some standard analog filters in recent years. The disadvantages of DDS (and digital filters) are the relatively limited maximum output frequency and greater complexity/cost considerations.

A block diagram for a basic DDS system is provided in Fig. 7-23. The numerically controlled oscillator (NCO) contains the phase accumulator and read-only memory (ROM) look-up table. The NCO provides the updated information to the digital-to-analog converter (DAC) to generate the RF output.

The phase accumulator generates a phase increment of the output waveform based on its input (Δ phase in Fig. 7-23). The input (Δ phase) is a digital word that, in conjunction with the reference oscillator (f_{CLK}), determines the frequency of the output waveform. The output of the phase accumulator serves as a variable-frequency oscillator generating a digital ramp. The frequency of the signal is defined by the Δ phase as

$$f_{out} = \frac{(\Delta \text{ phase}) f_{CLK}}{2^N} \tag{7-4}$$

for an N-bit phase accumulator.

Translating phase information from the phase accumulator into amplitude data is accomplished by means of the look-up table stored in memory. Its digital output (amplitude data) is converted into an analog signal by the DAC. The low-pass filter provides a spectrally pure sine-wave output.

The final output frequency is typically limited to about 40 percent of f_{CLK}. The phase accumulator size is chosen based on the desired frequency resolution, which is equal to $f_{CLK} \div 2^N$ for an N-bit accumulator.

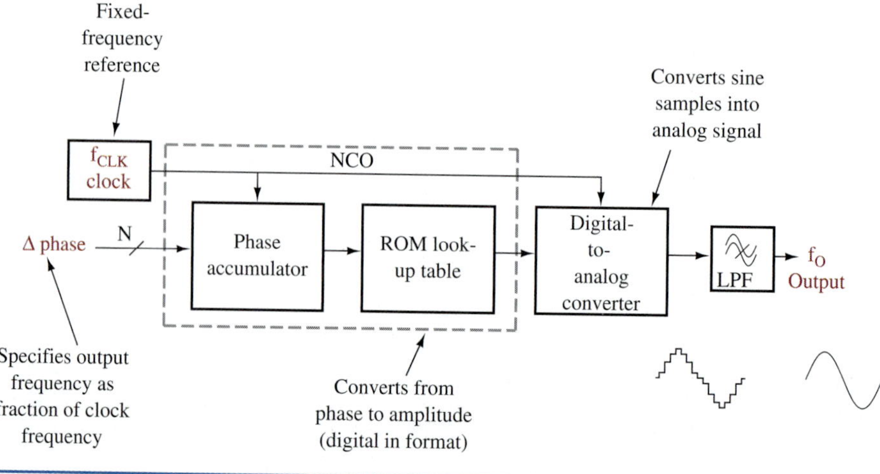

FIGURE 7-23 DDS block diagram.

EXAMPLE 7-10

Calculate the maximum output frequency and frequency resolution for a DDS when operated at $f_{CLK\,MAX}$.

Solution

The maximum output frequency is approximately 40 percent of $f_{CLK\,MAX}$:

$$= 0.40 \times 100 \text{ MHz}$$
$$= 40 \text{ MHz}$$

The frequency resolution is given by $f_{CLK} \div 2^N$:

$$= \frac{100 \text{ MHz}}{2^{32}} \cong 0.023 \text{ Hz}$$

The preceding example shows that DDS offers the possibility for extremely small frequency increments. This is one of the advantages offered by DDS over the analog synthesizers described in Sec. 7-5. Another DDS advantage is the ability to shift frequencies quickly. This characteristic is useful for the spread-spectrum systems described in Sec. 7-8.

The disadvantages of DDS include the limit on maximum output frequency and higher **phase noise.** Spurious changes in phase of the synthesizer's output result in energy at frequencies other than the desired one. This phase noise is often specified for all types of oscillators and synthesizers. It is usually specified in dB/$\sqrt{\text{Hz}}$ at a particular offset from center frequency. A specification of -90 dB/$\sqrt{\text{Hz}}$ at a 10-kHz offset means that noise energy at a 1-Hz bandwidth 10 kHz

Phase Noise
spurious changes in the phase of a frequency synthesizer's output that produce frequencies other than the desired one

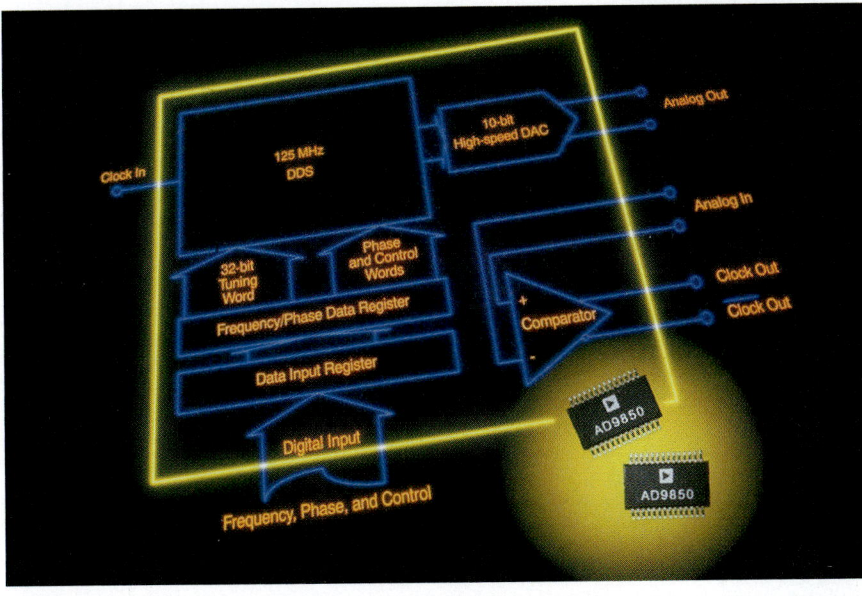

AD 9850 DDS block diagram. (Courtesy of Analog Devices, Inc.)

away from the center frequency should be 90 dB lower than the center frequency output. In a sensitive receiver, phase noise will mask out a weak signal that would otherwise be detected.

 7-7 FM Communications Transceivers

A large amount of two-way communication takes place in the 150- to 174-MHz band via frequency modulation. In fact, this band is so heavily utilized by business and industry that the overflow is going to allocations at 450 or 470 MHz and around 900 MHz. The radio described in this section is a basic design selected to show you, in detailed fashion, complete circuit and troubleshooting information.

The unit shown in Fig. 7-24 is the General Electric Phoenix model. It is a basic FM transceiver available at one specific frequency of operation in the range 150 to 174 or 450 to 470 MHz. We shall detail a 150- to 174-MHz version, but it is similar to the higher-frequency model. These units provide reliable communication up to about 40 miles using transmitter output powers of 25 W (150 to 174 MHz) or 20 W (450 to 470 MHz).

Channel Guard
in a transceiver, causing a specific audio frequency to be encoded onto the carrier together with the audio

The only major feature of this transceiver that has not previously been described is the **channel guard** function. It causes a specific audio frequency to be encoded onto the carrier together with the audio. If the intended receiver does not detect this tone, it inhibits reception of the call. This feature minimizes unwanted reception and allows for a certain degree of privacy in these extremely crowded channels.

The detailed block diagram for this transceiver is shown in Fig. 7-25 (pages 332–333). The detailed circuit analysis to follow will be easier to understand if you take a few minutes now to familiarize yourself with the functional block operation.

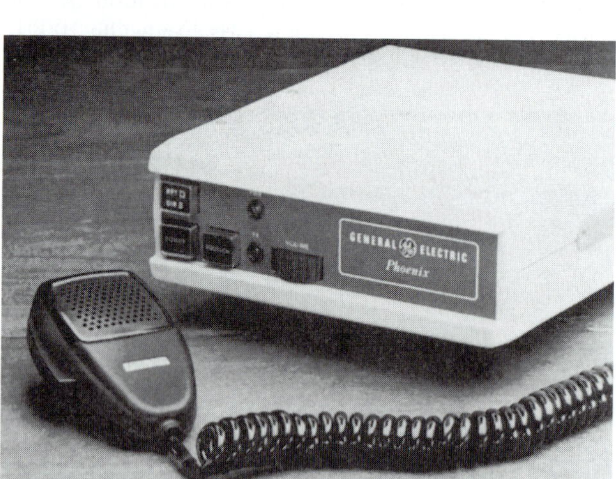

FIGURE 7-24 Basic communications transceiver. (Courtesy of General Electric Company.)

Circuit Analysis

Transmitter The transmitter (Fig. 7-25) utilizes a crystal-controlled, frequency-modulated exciter for two-frequency operation in the 150- to 174-MHz frequency band. The transmitter consists of audio processor U_{101}, oscillator Q_{151}, buffer Q_{153}, exciter stages Q_{201} through Q_{203}, PA amplifier Q_{251} and Q_{252}, and power control circuit Q_{254} through Q_{257}. The exciter provides approximately 300 to 350 mW of modulated RF to the power amplifier, which provides rated output power. Figure 7-25 is a block diagram of the radio showing both the transmitter and receiver.

Microphone Preamplifier A preamplifier stage (Q_{901} and associated circuitry) is provided for the standard electret microphone without a built-in preamplifier (Fig. 7-26, page 334). The preamplifier circuit is located on the IOC board.

With this microphone, MIC HI is coupled through $J_{911\text{-}5}$ to the preamplifier stage. The $J_{911\text{-}5}$ refers to the J_{911} connector (on the IOC board in Fig. 7-26) and indicates pin 5 of that connector. The amplified output is coupled through R_{908} and C_{905} to the audio processor. For optional microphones with a built-in preamplifier, audio is coupled through $J_{911\text{-}4}$, bypassing MIC preamp Q_{901}.

Audio Processor U_{101} The audio processor (Fig. 7-26) provides audio preemphasis with amplitude limiting and postlimiter filtering. A total gain of approximately 24 dB is realized through the audio processor. Twenty dB is provided by U_{101B} and 4 dB by U_{101A}.

The 8.5-V regulator powers the audio processor and applies regulated +8.5 V through $P_{903\text{-}2}$ to a voltage divider consisting of R_{101}, R_{111}, R_{110}, and R_{109}. The +4.25-V output from the voltage divider establishes the operating reference point for both operational amplifiers. C_{106} provides an ac ground at the summing input of both operational amplifiers (U_{101A} and U_{101B}).

Resistors R_{109}, R_{110}, and R_{111} and diodes D_{101} and D_{102} provide limiting for U_{101B}. Diodes D_{101} and D_{102} are reverse-biased at +1.7 V dc. Voltage-divider network R_{109}, R_{110}, and R_{111} provides +5.9 V dc at the cathode of D_{101} and +2.6 V dc at the anode of D_{102}. The voltage at the junction of D_{101} and D_{102} is 4.25 V dc. C_{104} and C_{108} permit a dc-level change between $U_{101B\text{-}7}$ and the voltage-divider network for diode biasing.

When the input signal to $U_{101B\text{-}6}$ is of a magnitude such that the amplifier output at $U_{101B\text{-}7}$ does not exceed 4 V p-p, the amplifier provides a nominal 20-dB gain. When the audio signal level at $U_{101B\text{-}7}$ exceeds 4 V p-p, diodes D_{101} and D_{102} conduct on the positive and negative half-cycles, providing 100 percent negative feedback to reduce the amplifier gain to 1. This limits the audio amplitude at $U_{101B\text{-}7}$ to 5 V p-p.

Resistors R_{105}, R_{106}, R_{107}, and C_{107} comprise the audio preemphasis network that enhances the signal-to-noise ratio (see Sec. 5-4). R_{107} and C_{107} control the preemphasis curve below limiting. R_{106} and C_{107} control the cutoff point for high-frequency preemphasis. As high frequencies are attenuated, the gain of U_{101} is increased.

Audio from the preamplifier or microphone is coupled to the input of operational amplifier $U_{101B\text{-}6}$. The amplified output of U_{101B} is coupled through R_{114}, R_{112}, R_{104}, and R_{117} to a second operational amplifier, U_{101A}.

The channel guard (CG) tone input is applied to $U_{101A\text{-}2}$. The CG tone is then combined with the microphone audio.

A postlimiter filter consisting of U_{101A}, R_{112}–R_{114}, C_{108}, and C_{110} provides 12 dB/octave roll-off. R_{104} and C_{102} provide an additional 6 dB/octave roll-off, for a total of 18 dB.

The output of the postlimiter filter is coupled through C_{110} to the temperature-compensated transmitter oscillators Q_{151} and Q_{152}.

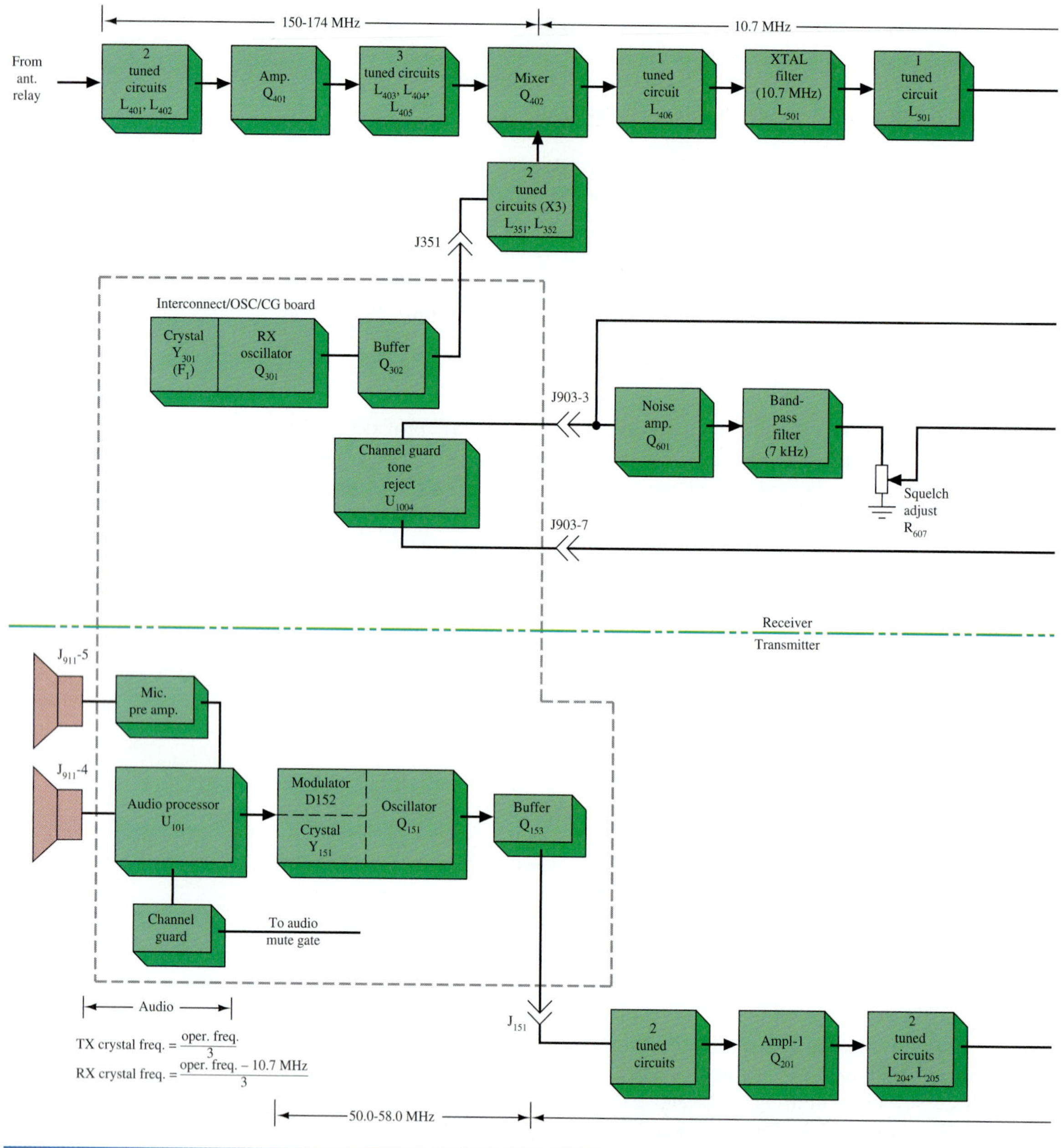

FIGURE 7-25 Block diagram. (Courtesy of General Electric Company.)

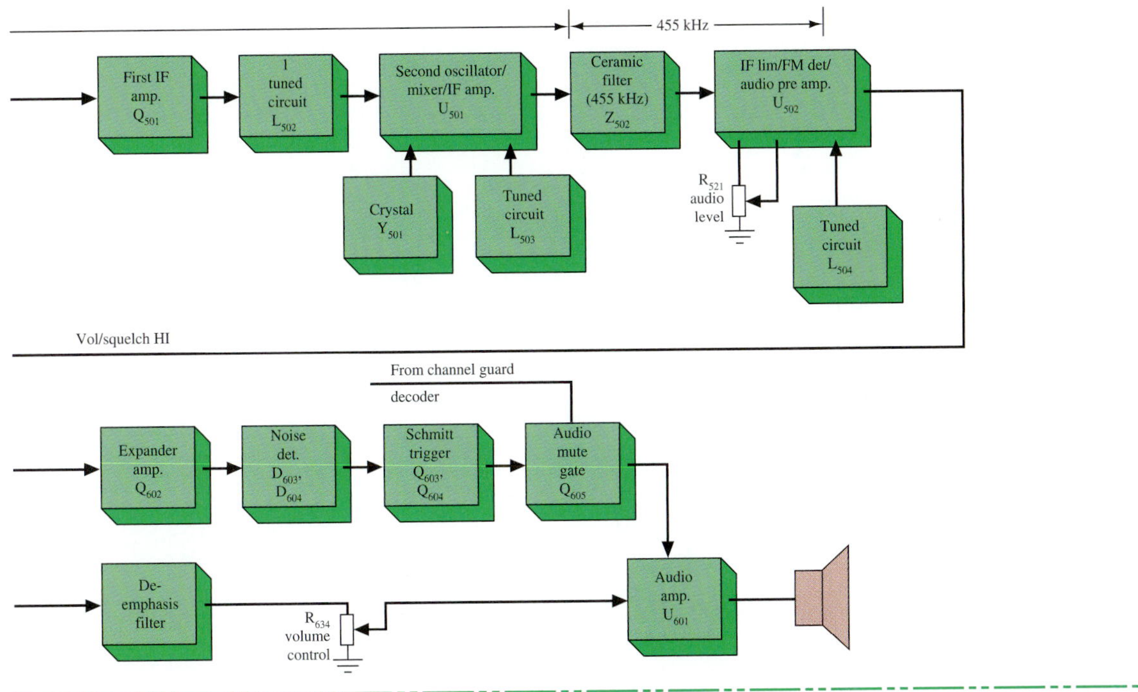

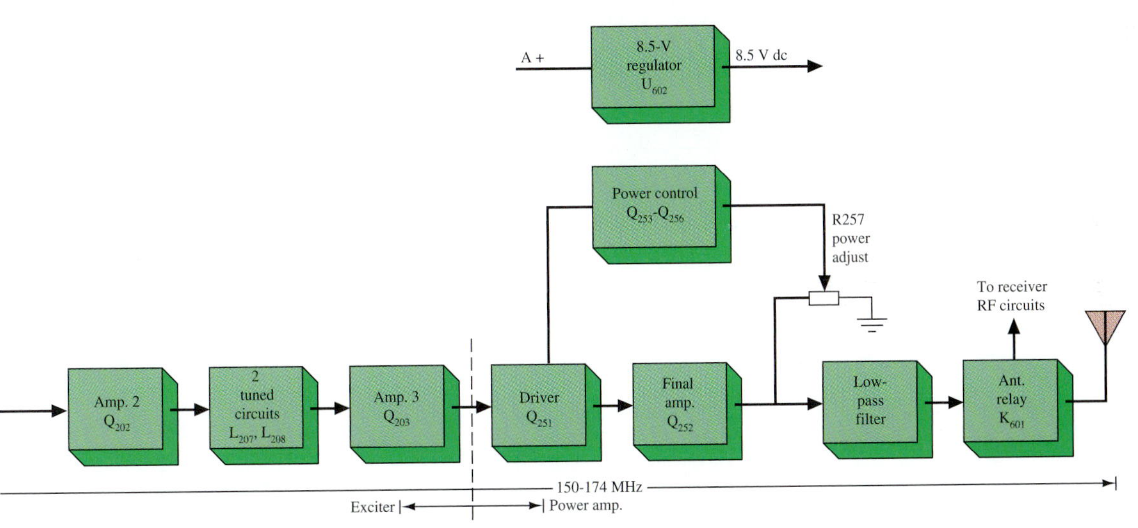

FIGURE 7-25 *(Continued)*

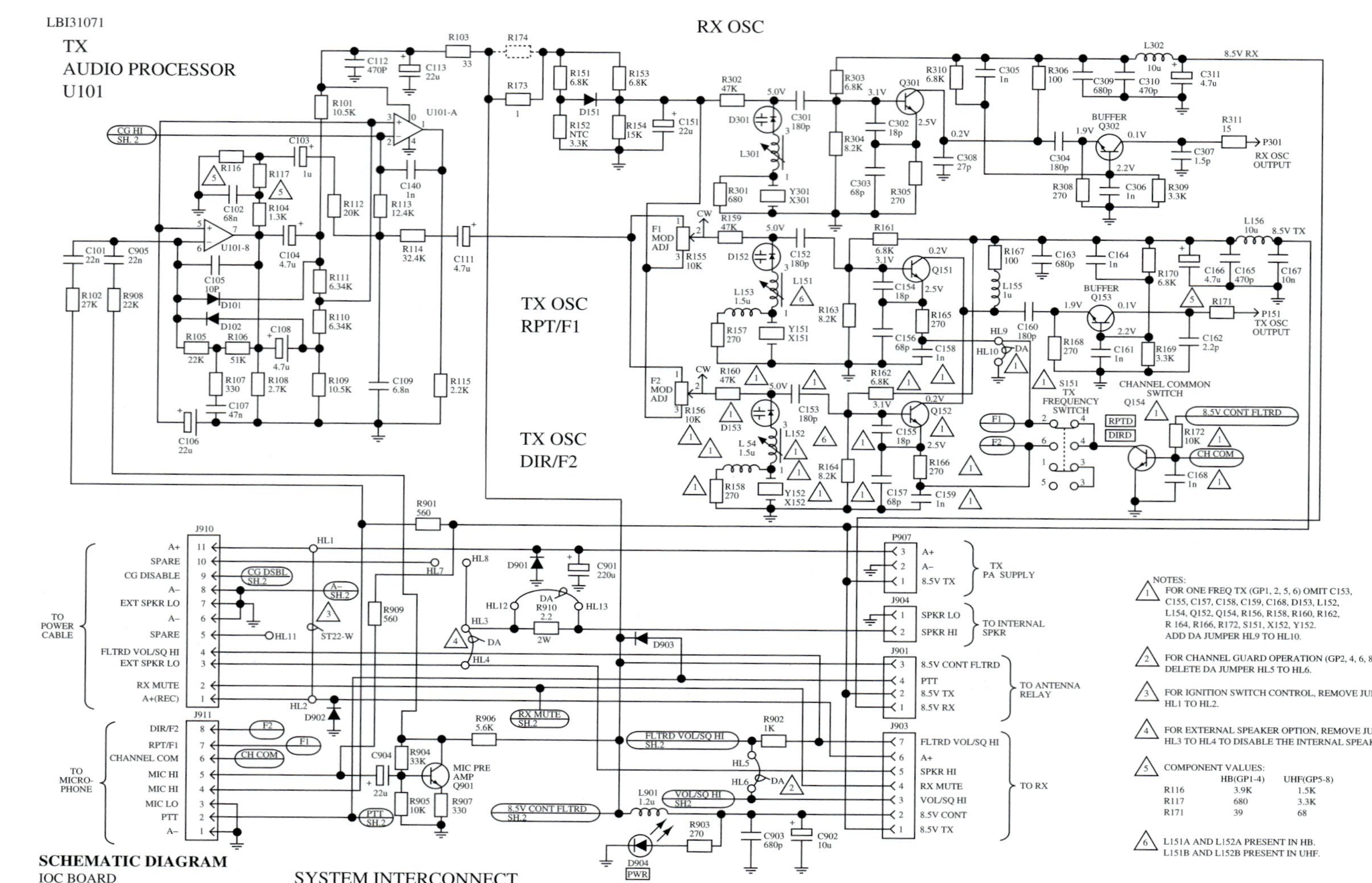

FIGURE 7-26 Audio, IOC, and oscillator schematic. (Courtesy of General Electric Company.)

Transmitter Oscillator A temperature-compensating network consisting of R_{151}, R_{152}, R_{153}, R_{154}, D_{151}, and C_{151} maintains oscillator frequency over a temperature range of -30 to $+60$°C (Fig. 7-26). The temperature-compensating dc voltage and audio are applied to FM modulator D_{152} through MOD ADJ control R_{155} and R_{159}. Modulator varactor D_{151} varies the transmit frequency at the audio rate applied from the audio processor.

Q_{151}, Y_{151}, and associated circuitry comprise a Colpitts oscillator that generates a frequency that is one-third of the desired carrier. The transmit oscillator is adjusted to the assigned operating frequency by L_{151}. The oscillator output is applied to buffer Q_{153} and then coupled through P_{151} to the exciter circuitry on the transmitter/receiver (Tx/Rx) board.

Exciter The exciter (Fig. 7-27) consists of three amplifier stages that provide a minimum of 300 mW to the PA section of the transmitter. In addition to providing approximately 22 dB of gain, the exciter contains filters that determine the bandwidth and spurious characteristics of the transmitter.

The output of the oscillator and buffer stages is coupled to the input of the exciter via J_{151}, which is connected to a tap on L_{201}. This tap also supplies voltage to buffer transistor Q_{153} (Fig. 7-26). L_{201} and L_{202} select the third harmonic (150 MHz), which is present at J_{151}.

C_{206} and C_{207} match the output of the two-pole filter to the base of Q_{201}. Q_{201} provides approximately 10 dB of gain. C_{213} and C_{214} match the collector of Q_{201} to the input of a second two-pole filter, consisting of L_{204} and L_{205}. The emitter voltage on Q_{201} (AMPL-1) can be monitored at TP_{201} and is typically $+0.3$ V.

The collector of Q_{201} is matched to the second two-pole filter, consisting of L_{204} and L_{205}. The output of this filter (approximately 10 to 15 mW) is matched to the base of amplifier Q_{202}. The collector of this transistor in turn provides the signal to the third two-pole filter, consisting of L_{207} and L_{208}. The emitter voltage on Q_{202} is monitored at TP_{202} and is typically $+0.5$ V.

The output of the third two-pole filter (approximately 60 to 80 mW) is applied to amplifier Q_{203}, which supplies 300 to 350 mW (typically) to the PA driver. An "L" network (L_{209}, L_{210}, and C_{232}) matches the output impedance of Q_{203} to 50 Ω. A 50-Ω microstrip (W_{251}) applies the 300- to 350-mW signal to the PA circuitry. (See Chapter 15 for an explanation of microstrip circuit connections.) An additional metering point is available on this microstrip to monitor power at the exciter-PA interface. TP_{251} consists of two pins that will accept a special RF detector probe that can be used with any multimeter.

Power Amplifier The two-stage power amplifier (Fig. 7-27) consists of driver Q_{251}, power amplifier Q_{252}, and associated circuitry. Collector voltage for driver Q_{251} is applied from A+ through pass transistor Q_{256} and L_{253}, L_{252}, and R_{252}. The collector voltage for Q_{251} is a result of the output power setting and voltage variations at any given time. The output of driver Q_{251} is coupled to the base of power amplifier Q_{252} through an impedance-matching network consisting of C_{258}, C_{259}, L_{254}–L_{256}, C_{261}, C_{262}, and R_{253}. Collector voltage for Q_{252} is provided from A+ through L_{258}, L_{257}, and R_{255}.

The 25-W output of the power amplifier is connected to the low-pass filter by W_{252}–W_{254} and then to antenna relay K_{601}. The output power adjust circuit allows the transmitter to be set to rated output power. The power adjustment is attained by

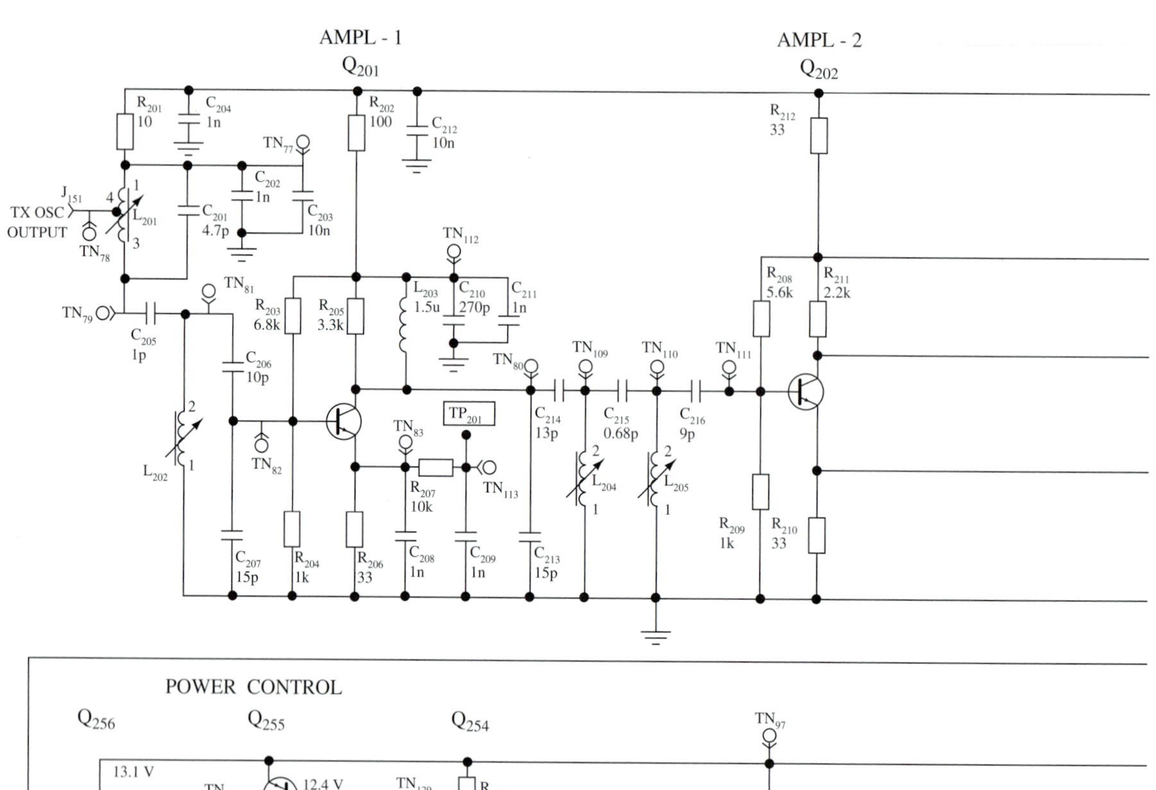

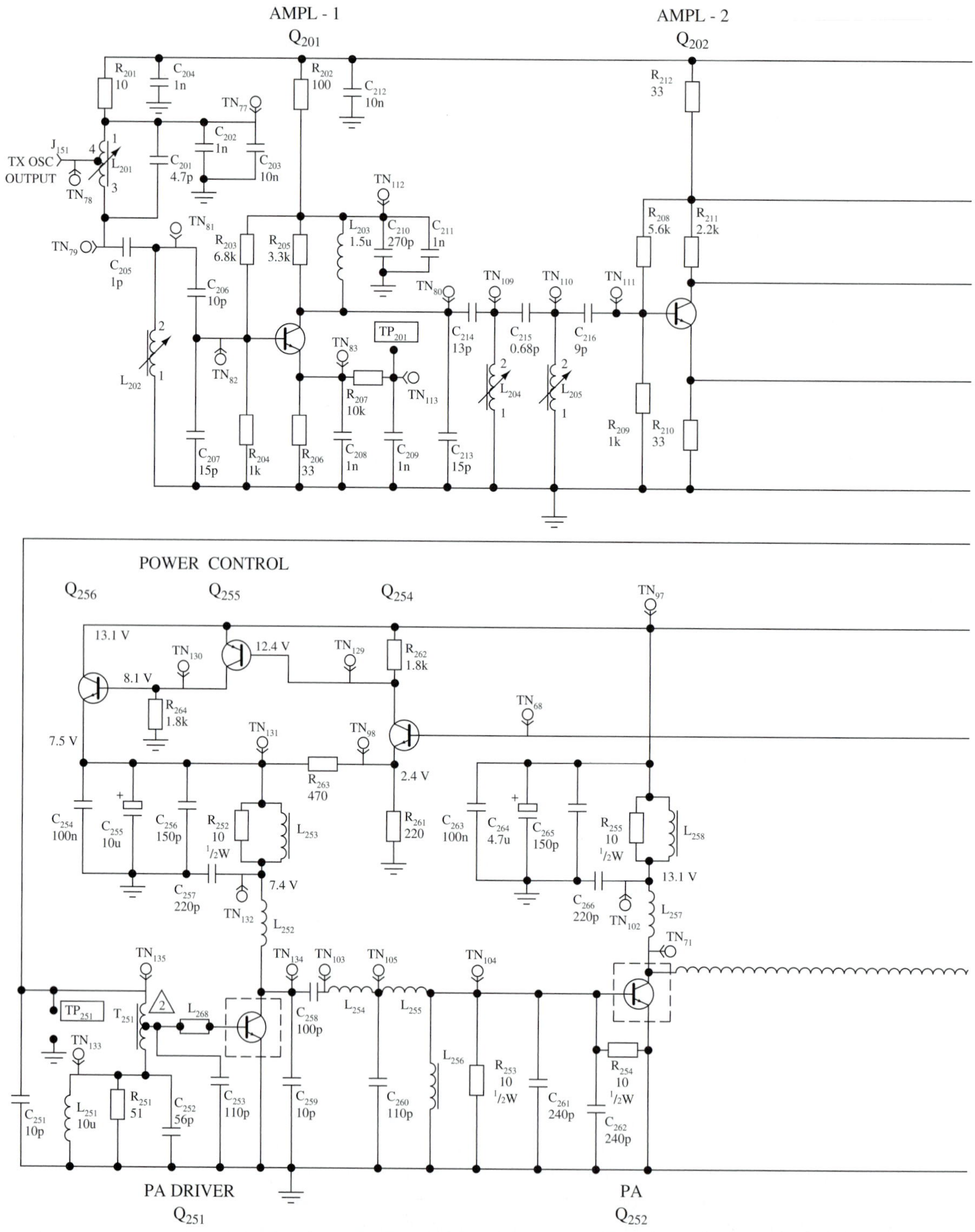

FIGURE 7-27 Transmitter schematic. (Courtesy of General Electric Company.)

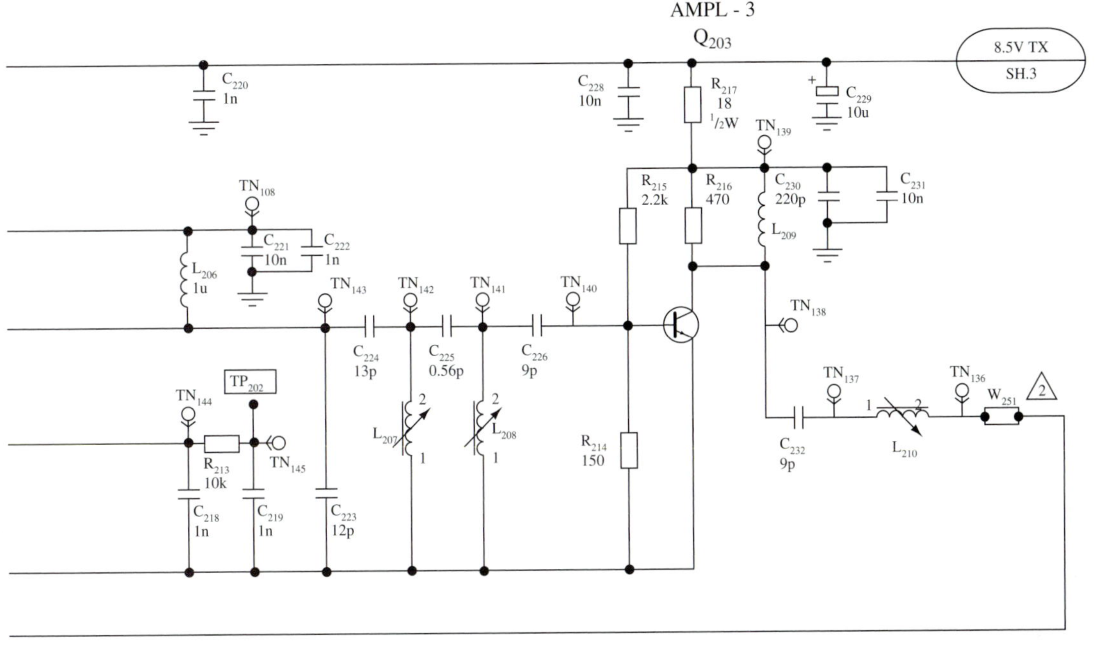

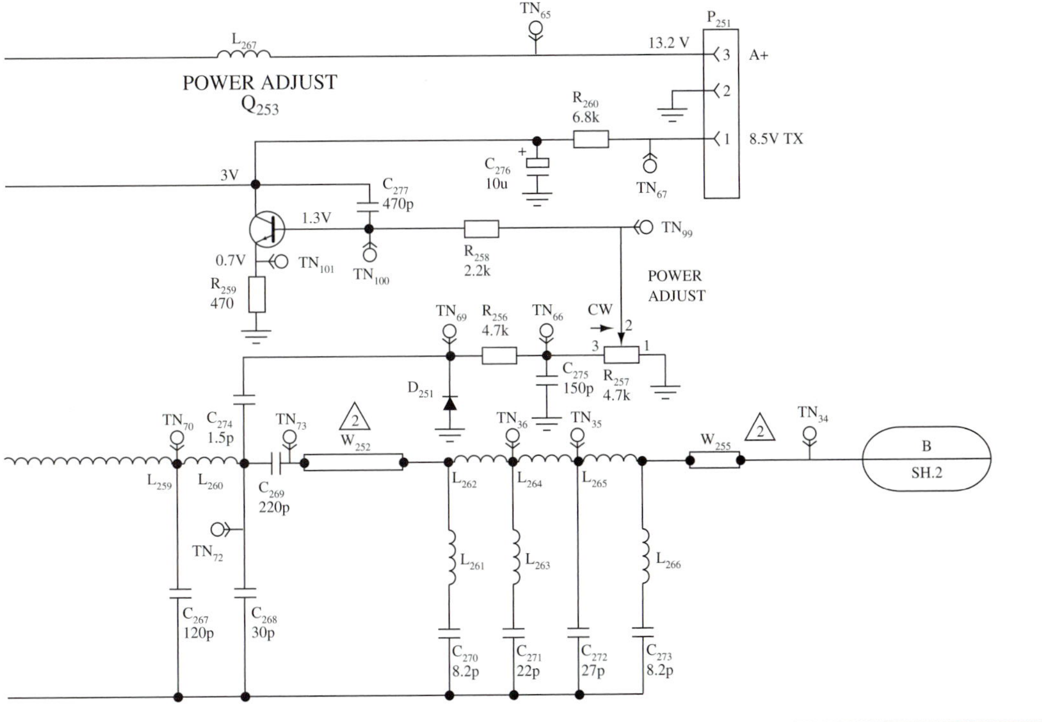

FIGURE 7-27 *(Continued)*

controlling the dc collector voltage to drive Q_{251} through pass transistor Q_{256}. The pass transistor is controlled by a feedback loop consisting of Q_{253} through Q_{256}. The power is set by potentiometer R_{257}.

A change in output power is sensed by D_{251}, causing the base voltage of Q_{253} to change accordingly. For example, if the output power increases, the base of Q_{253} goes more positive, causing it to increase conduction, which lowers its collector voltage. Q_{253} controls Q_{254}; therefore, as Q_{253} increases conduction, Q_{254} decreases conduction. This raises the voltage applied to the base of Q_{255}. The conduction of Q_{255} decreases proportionally, lowering the base voltage of pass transistor Q_{256}. The resulting decrease in conduction of Q_{256} lowers the collector voltage of driver Q_{251}, thereby lowering the output power in proportion to the excessive power originally sensed by the base circuit of Q_{253}.

Receiver The receiver (Fig. 7-25) is dual-conversion, superheterodyne FM designed for one-frequency operation in the frequency range 150 to 174 MHz. A regulated 8.5 V is used for all receiver stages except for the audio PA IC, which operates from the A+ supply.

The receiver has intermediate frequencies of 10.7 MHz and 455 kHz. Adjacent channel selectivity is obtained by using two bandpass filters: a 10.7-MHz crystal filter and a 455-kHz ceramic filter.

All of the receiver circuitry except the oscillator is mounted on the transmitter/receiver (Tx/Rx) board (Fig. 7-28, pages 340–343). The receiver consists of:

Receiver front end
10.7-MHz first IF circuitry
First and second oscillators
455-kHz second IF circuitry with FM detector
Audio PA circuitry
Squelch circuitry

Receiver Front End As shown in Fig. 7-28, an RF signal from the antenna is coupled through antenna relay K_{601} and two tuned circuits (L_{401}, C_{401}, C_{402}; and L_{402}, C_{404}, C_{405}) to the emitter of RF amplifier Q_{401}. The output of Q_{401} is coupled through three additional tuned circuits (L_{403}, C_{408}; L_{404}, C_{411}; and L_{405}, C_{413}) to the gate of first mixer Q_{402}. The front-end selectivity is provided by these five tuned circuits.

Oscillator and Multiplier Q_{301}, Y_{301}, and associated circuitry make up a Colpitts oscillator (Fig. 7-26). The frequency is controlled by a third-mode crystal operated at one-third of the required injection frequency. Voltage-variable capacitor D_{301}, L_{301}, and Y_{301} are connected in series to provide compensation capability. The compensation voltage used to control the transmitter oscillators is applied to D_{301} to maintain stability. L_{301} is adjustable to set the oscillator frequency. R_{301} is in parallel with Y_{301} to ensure operation on the third overtone of the crystal.

The output of Q_{301} is coupled through C_{304} to the emitter of buffer Q_{302}. The output of Q_{302} is coupled through P_{301} to two tuned circuits (L_{351} and L_{352}) on the Tx/Rx board. L_{351} and L_{352} are tuned to the third harmonic of the oscillator frequency, which is applied to the source input of mixer Q_{402}.

The RF frequency from the oscillator/multiplier chain and input level to the mixer can be measured at TP_{401}. The meter reading at TP_{401} is typically 2 to 4 V.

First Mixer The first mixer (Fig. 7-28) uses a junction FET (Q_{402}) as the active device. The FET mixer provides a high-input impedance, high-power gain, and an output relatively free of intermodulation products.

In the mixer stage, RF from the front-end filter is applied to the gate of the mixer. Injection voltage from the multiplier stages is applied to the source of the mixer. The 10.7-MHz mixer first IF output signal is coupled from the drain of Q_{402} through an impedance-matching network (L_{406}, C_{415} and C_{416}) to crystal filter Z_{501}.

The highly selective crystal filter provides the first portion of the receiver IF selectivity. The output of the filter is coupled through impedance-matching network L_{501} to the first IF amplifier.

First and Second IF and Detector Stages First IF amplifier Q_{501} is a dual-gate MOSFET (Fig. 7-28). The crystal filter output is applied to gate 1 of the amplifier, and the amplified signal is taken from the drain. The biasing on gate 2 and the drain load determines the gain of the stage. The amplifier provides approximately 20 dB of IF gain. The output of Q_{501} is coupled through an impedance-matching network (L_{502}) that matches the amplifier output to the input of IC U_{501}.

U_{501} and associated circuitry consist of the second oscillator, mixer, and second IF amplifier. The crystal for the oscillator is Y_{501}, and the oscillator operates at 10.245 MHz. This frequency is mixed with the 10.7-MHz input. The output of the mixer is limited by D_{501} and D_{502}. L_{503} is tuned for the 455-kHz second IF frequency.

The output of U_{501} is coupled through ceramic filter Z_{502}, which provides the 455-kHz selectivity, and applied to U_{502}. Test point TP_{501} is used in aligning the receiver and can be used to check the output of U_{501}.

U_{502} and associated circuitry consist of a 455-kHz limiter, a quadrature-type FM detector, and an audio preamplifier. L_{504} is the quadrature detector coil. Audio-level potentiometer R_{521} is used to set the audio output level to the audio amplifier.

Audio and Squelch Circuits Audio (Fig. 7-28) is applied to the channel guard tone reject filter through $P_{903\text{-}3}$ and back to the deemphasis network through $P_{903\text{-}7}$. The audio passes through the deemphasis network (R_{633}, C_{608}, and C_{609}) to volume control R_{634}.

Audio amplifier IC U_{601} drives the speaker at the desired audio level (up to 3 W). The feedback loop containing R_{637}, R_{638}, and C_{611} determines the amplifier's closed-loop gain. R_{636} and C_{613} provide the audio high-frequency roll-off above 6 kHz.

The audio amplifier can be muted by a dc voltage from 8.5 V Tx, or from receiver mute gate Q_{605}. The two logic inputs for Q_{605} are a squelch signal and Rx MUTE.

The squelch circuit (Fig. 7-28) operates on the noise components contained in the FM detector output. The output of U_{502} is applied to frequency-selective noise amplifier Q_{601}, which has a resonant circuit (L_{601}, R_{604}, and C_{602}) as the collector load. The output is noise in a band around 7 kHz.

The noise output is coupled through squelch control R_{607} to expander amplifier Q_{602}, which improves the level discrimination characteristics of the circuit. The output of Q_{602} is applied to a passive voltage-doubler circuit (D_{603} and D_{604}). This circuit has a high source impedance and operates as an average-value rectifier.

Following the voltage doubler is a Schmitt trigger (Q_{603}–Q_{604}). The trigger provides the necessary hysteresis and a well-defined output signal for Rx MUTE gate Q_{605}.

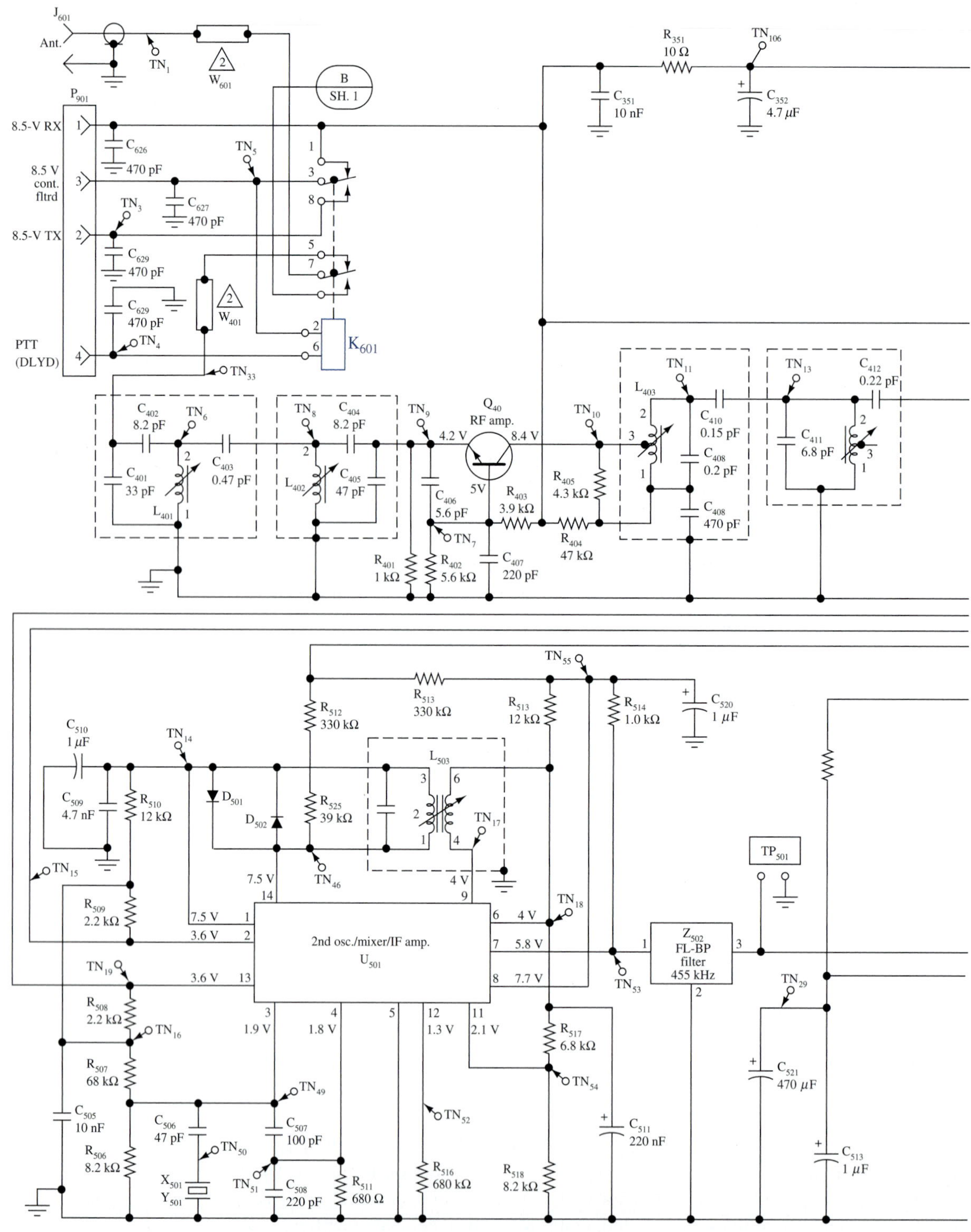

FIGURE 7-28 Receiver schematic. (Courtesy of General Electric Company.)

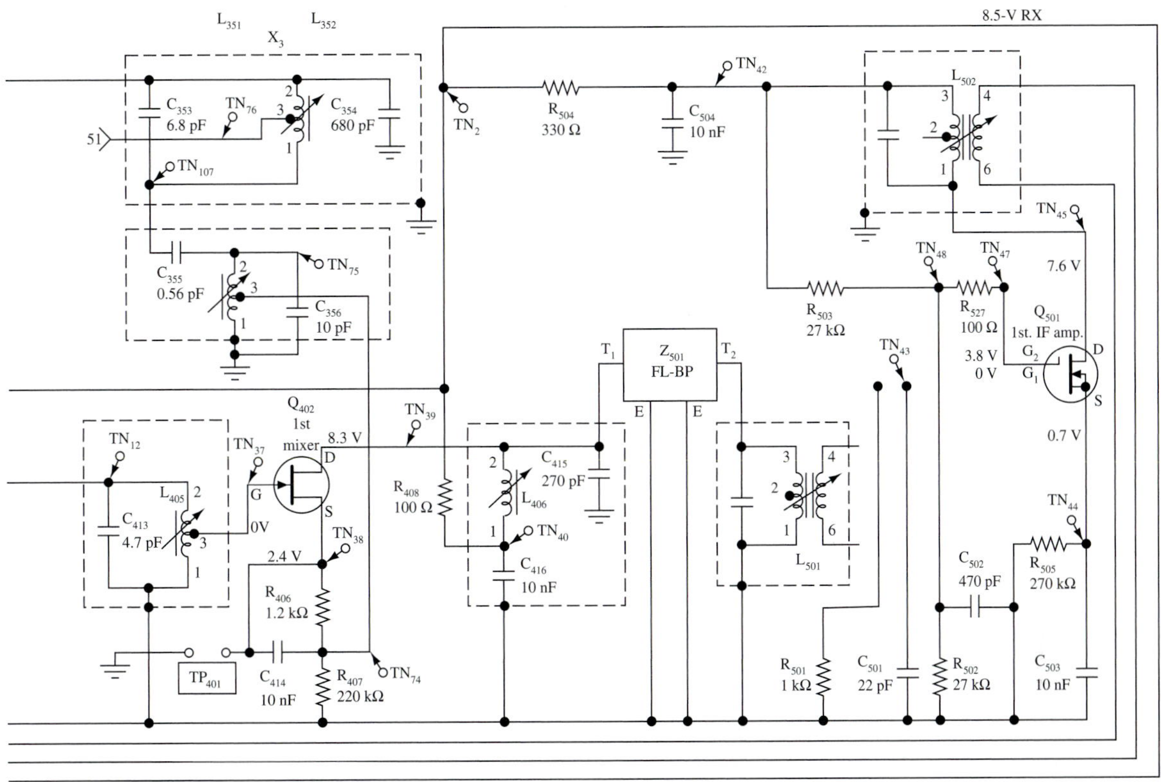

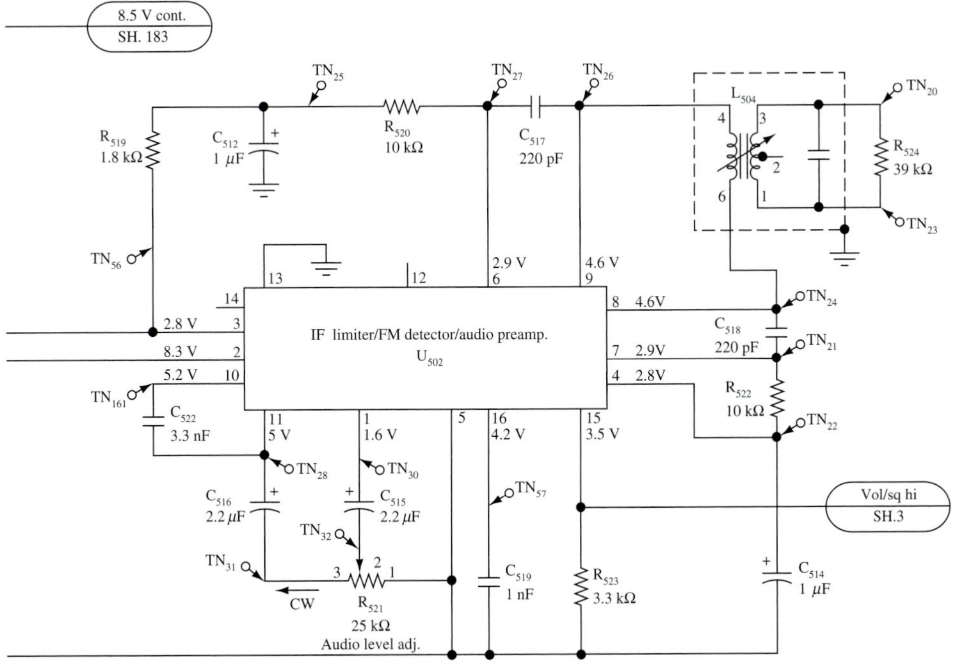

FIGURE 7-28 *(Continued)*

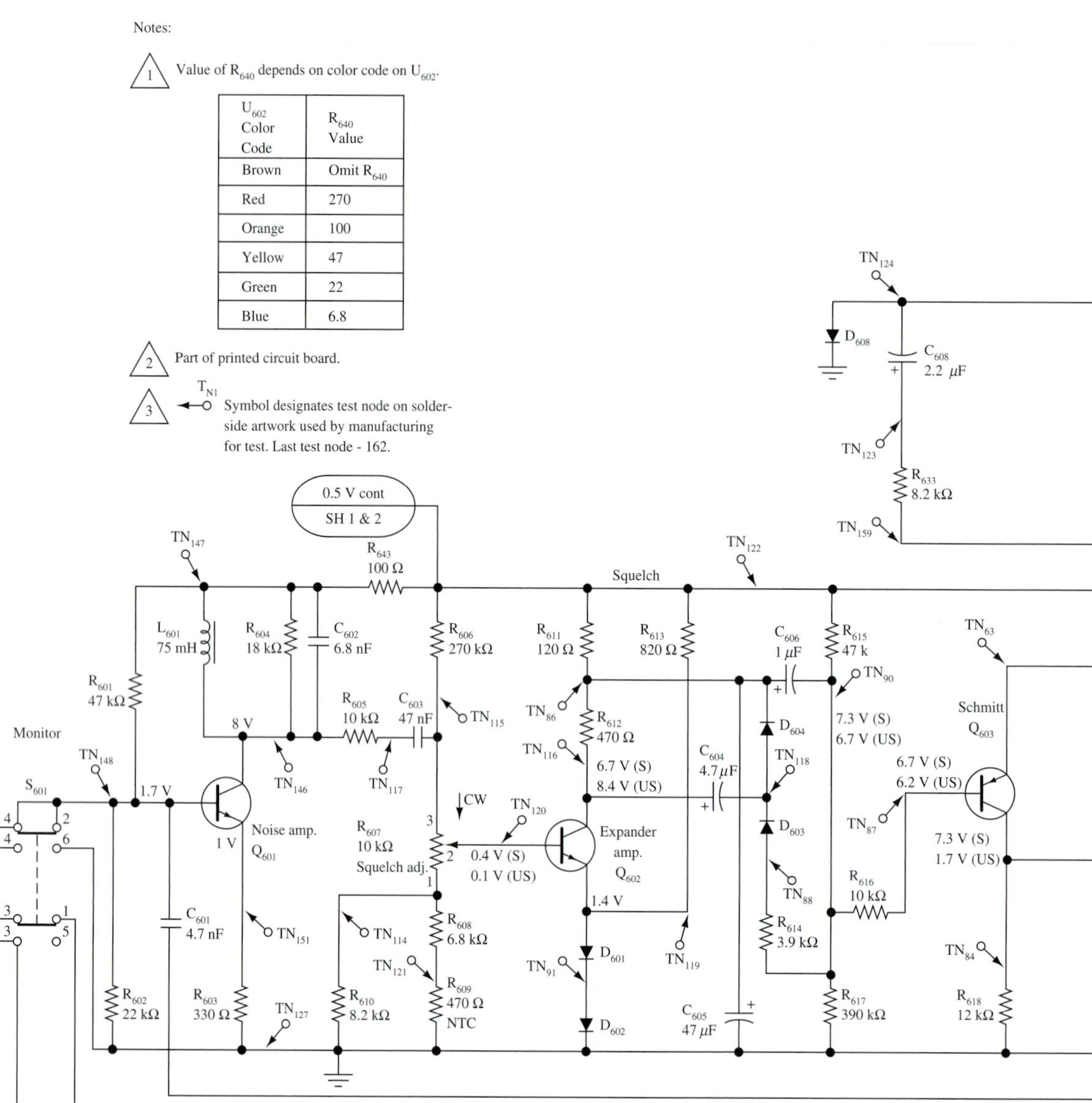

Notes:

1. Value of R_{640} depends on color code on U_{602}.

U_{602} Color Code	R_{640} Value
Brown	Omit R_{640}
Red	270
Orange	100
Yellow	47
Green	22
Blue	6.8

2. Part of printed circuit board.

3. T_{N1} Symbol designates test node on solder-side artwork used by manufacturing for test. Last test node - 162.

All Resistors are 1/4 W unless otherwise specified.

Voltage Readings

Voltage readings are typical readings measured to system negative with a 20,000-Ω/V dc voltmeter under the following conditions:

1. No signal input
2. Volume control (R_{634}) set to minimum
3. Monitor switch (S_{601}) in out position
4. Unsquelched (US)-squelch adjust (R_{607}) set to minimum (CW)
5. Squelched (S)-squelch adjust (R_{607}) set to maximum (CCW)

FIGURE 7-28 *(Continued)*

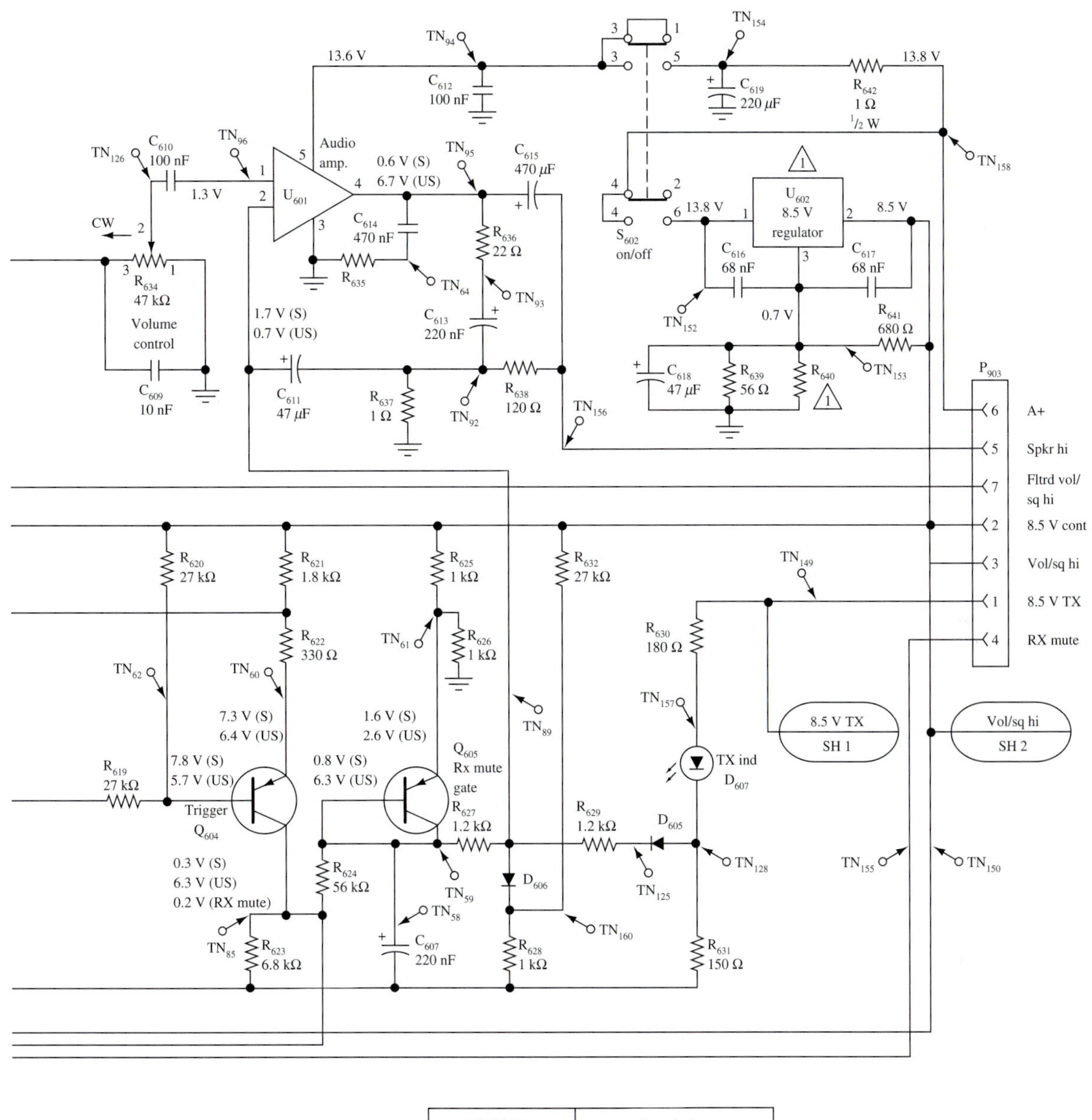

Model No.	Description
19D900599G1 19D900602G1	TX/RX BD assembly Chassis assembly

FIGURE 7-28 *(Continued)*

With no RF signal present, the detected noise at the voltage-doubler output turns on Q_{603}, turning off Q_{604}. This causes Q_{605} to turn on, applying +1 V to pin 2 of audio amplifier U_{601}. This voltage turns off U_{601} and mutes the receiver.

When an RF signal is received, the noise at the output of Q_{601} decreases and drive to Q_{603} is removed. This turns off Q_{603} and allows Q_{604} to turn on. With Q_{604} turned on, Rx MUTE gate Q_{605} turns off. This turns on U_{601} so that audio is heard at the speaker.

The squelch sensitivity is adjusted by R_{607} in the base circuit of expander amplifier Q_{602}.

Pressing the MONITOR pushbutton on the front of the radio opens the Rx MUTE to disable the channel guard. It also grounds the base of Q_{601} and disables the squelch function.

Channel Guard The channel guard (Fig. 7-29, pages 346–347) is a continuous-tone encoder/ decoder for operation on tone frequencies in the range 71.9 to 210.7 Hz. The encoder provides tone-coded modulation to the transmitter. The decoder operates in conjunction with the receiver to inhibit all calls that are not tone-coded with the proper channel guard frequency.

The channel guard circuitry consists of discrete components for the encode disable, PTT switch, and receiver mute switch; and four thick-film integrated circuit modules consisting of decode module U_{1001}, encode module U_{1002}, frequency-switchable selective amplifier (FSSA) U_{1003}, plug-in Versatone network Z_{1001}, and IC U_{1004} in the tone reject filter.

For a functional diagram of the channel guard encoder/decoder, refer to the troubleshooting procedures on pages 345–351.

FSSA Frequency-switchable selective amplifier (FSSA) U_{1003} (Fig. 7-29) is a highly stable active bandpass filter for the frequency range 71.9 to 210.7 Hz. The selectivity of the filter is shifted across the bandpass frequency range by switching Versatone networks in the filter circuit.

The gain of the FSSA is a function of the tone frequency (Table 7-3). The tone frequency is determined by the tone network connected in the FSSA circuit. Versatone network Z_{1001} is a precision resistor network.

Encode When the PTT switch is operated, the channel guard encode tone (Fig. 7-29) is generated by coupling the output or FSSA bandpass filter U_{1003} back to its input through a phase-inverting amplifier circuit and a limiter circuit. The output of the FSSA is coupled from U_{1003} to the input of the phase-inverting amplifier at $U_{1002\text{-}9}$.

Table 7-3 Standard Tone Frequencies (Hz)

71.9	88.5	107.2	131.8	162.2
74.4	91.5	110.9	136.5	167.9
77.0	94.8	114.8	141.3	173.8
79.7	97.4	118.8	146.2	179.9
82.5	100.0	123.0	151.4	186.2
85.4	103.5	127.3	156.7	192.8
			203.5	
			210.7	

An amplifier provides 180° phase shift of the tone frequency at the output. The output of the phase-inverting amplifier circuit is coupled from $U_{1002\text{-}6}$ to the input of the limiter circuit at $U_{1002\text{-}5}$. A limiting network sets the tone output coupled from $U_{1002\text{-}4}$ to the input of the FSSA ($U_{1003\text{-}12}$) at 53 mV p-p.

The limiter circuit is also used as an encode switch. Keying the transmitter applies +5.4 V to $U_{1002\text{-}2}$. This starts the circuit oscillating. The tone frequency is determined by the tone network connected in the FSSA circuit.

The tone output of the encoder circuit is taken from $U_{1002\text{-}7}$ and coupled through tone output amplifier Q_{1002} and modulation adjustment R_{1010} to the audio processor on the transmitter/receiver board.

Decode Audio from volume/squelch high that contains the correct frequency is coupled to pin 1 of decode module U_{1001} (Fig. 7-29). Pin 1 of U_{1001} is the input of an active, three-stage, low-pass filter. The low-pass filter attenuates frequencies over 210.7 Hz. The output of the low-pass filter at $U_{1001\text{-}15}$ is applied to $U_{1001\text{-}14}$. $U_{1001\text{-}14}$ is the input of a limiter circuit, limiting the output at $U_{1001\text{-}13}$ to 55 mV p-p.

The output from the limiter is coupled to pin 12 of FSSA U_{1003}. Because the tone is the proper frequency, the FSSA will allow it to pass. The output of the FSSA is coupled to $U_{1001\text{-}3}$. $U_{1001\text{-}3}$ is the input to an amplifier circuit. The output of the amplifier at $U_{1001\text{-}4}$ is coupled to the input of a threshold detector at $U_{1001\text{-}6}$.

In the mute mode, when the tone decoder in U_{1001} detects the channel guard frequency, Q_{1005} turns Q_{1006} off. This unmutes the receiver audio. In the squelch mode, Q_{1006} is operating, grounding the Rx MUTE lead and muting the receiver audio.

Audio from VOL/SQ HI is applied to the tone reject filter. The tone reject filter is an active filter consisting of U_{1004} and associated circuitry. All frequencies from 80 to 210.7 Hz are rejected by the filter while passing all other audio frequencies back to the receiver audio circuits (filtered VOL/SQ HI).

Encode Disable The encode disable circuit (Fig. 7-29) consists of Q_{1003} and Q_{1004}. To disable the encode circuit, a positive voltage (+8.5 to 14 V dc) is applied to connector $P_{910\text{-}9}$ at the rear of the radio. This is accomplished by temporarily jumpering $P_{910\text{-}9}$ (GC DISABLE) to $P_{910\text{-}11}$ (A+). This positive voltage is applied to the base of Q_{1003}, turning on both Q_{1003} and Q_{1004}. When turned on, Q_{1004} applies +8.5 V dc to the base of the PTT switch Q_{1001}, forcing it off. With Q_{1001} off, the operating voltage for the encoder IC U_{1002} and encode tone output stage transistor Q_{1002} is removed, preventing any tone output.

The encode disable circuit has been incorporated as a maintenance aid for the service technician. This circuit disables the channel guard encode circuit and allows for transmitter distortion and modulation checks without removing the cover from the radio.

Troubleshooting

Troubleshooting charts are provided in Figs. 7-30 (pages 348–349) and 7-31 (pages 350–351) for all major areas of the transceiver. Refer now to these self-explanatory guides. Note that the receiver troubleshooting information (Fig. 7-31) takes the form of a flowchart path to solution. The transmitter information (Fig. 7-30) takes the more conventional approach of listing symptoms, probable causes, and troubleshooting procedures.

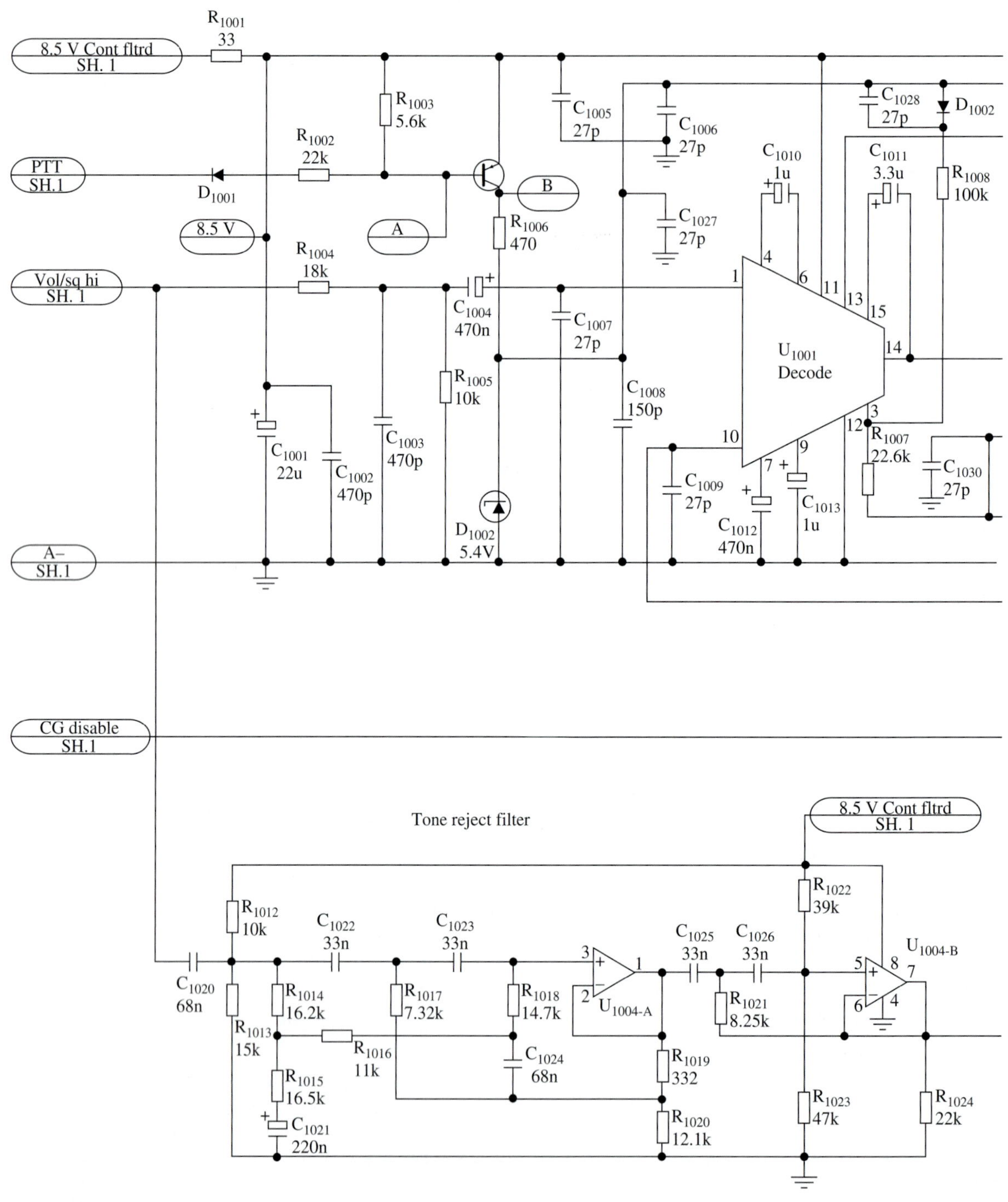

FIGURE 7-29 Channel guard schematic. (Courtesy of General Electric Company.)

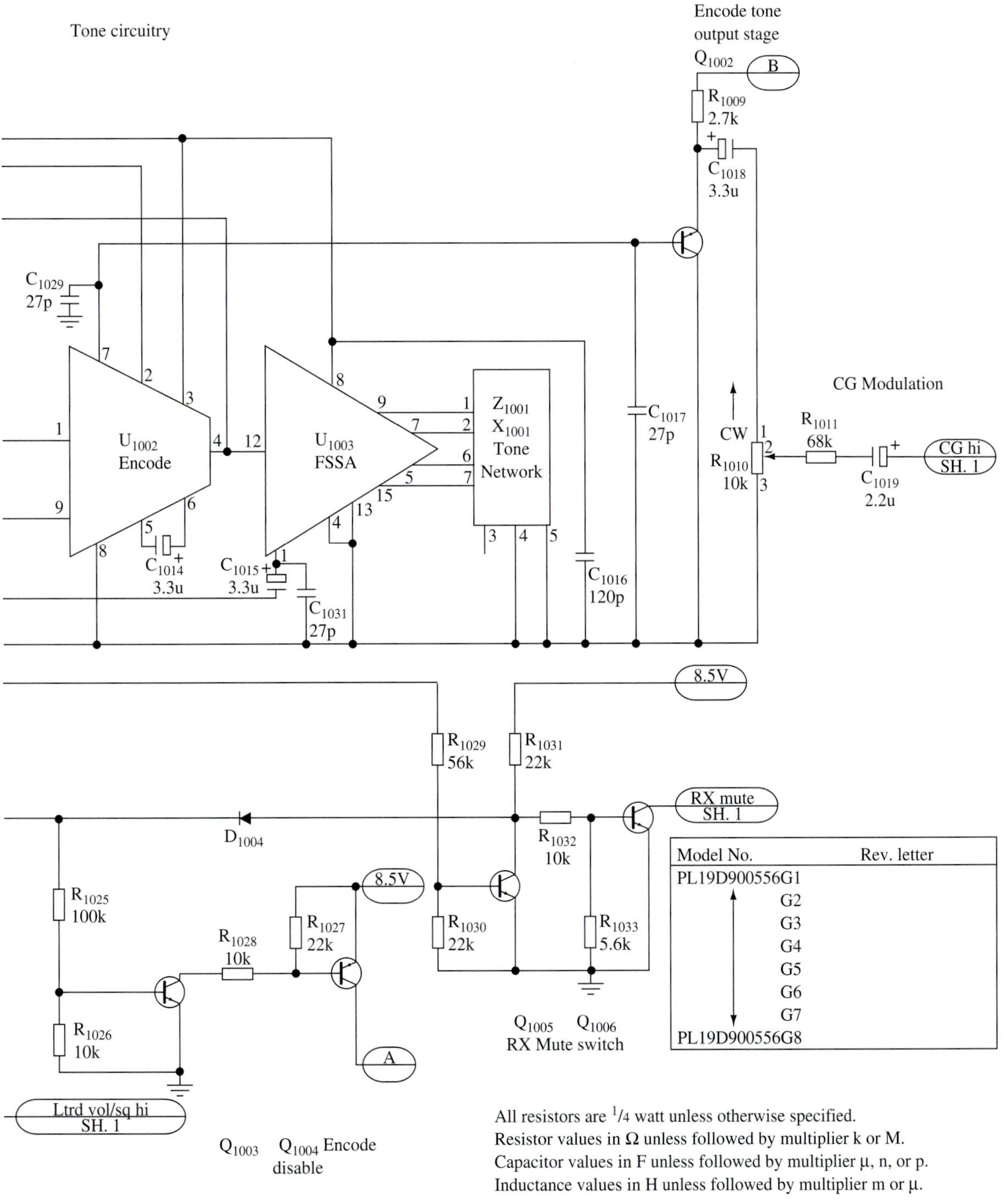
Tone circuitry
Encode tone
output stage
Q1002
B
R1009
2.7k
C1018
3.3u
C1029
27p
U1002
Encode
U1003
FSSA
Z1001
X1001
Tone
Network
C1017
27p
CW
R1010
10k
CG Modulation
R1011
68k
C1019
2.2u
CG hi
SH. 1
C1014
3.3u
C1015
3.3u
C1031
27p
C1016
120p
8.5V
R1029
56k
R1031
22k
RX mute
SH. 1
D1004
R1032
10k
Model No.
Rev. letter
PL19D900556G1
G2
G3
G4
G5
G6
G7
PL19D900556G8
R1025
100k
R1028
10k
R1027
22k
8.5V
R1030
22k
R1033
5.6k
R1026
10k
A
Q1005 Q1006
RX Mute switch
Ltrd vol/sq hi
SH. 1
Q1003 Q1004 Encode
disable
All resistors are 1/4 watt unless otherwise specified.
Resistor values in Ω unless followed by multiplier k or M.
Capacitor values in F unless followed by multiplier μ, n, or p.
Inductance values in H unless followed by multiplier m or μ.

FIGURE 7-29 *(Continued)*

PA Troubleshooting Chart			
Symptom	**Typical Voltage/Signal**	**Probable Cause**	**Troubleshooting Procedure**
No Power	0 V at TP_{203}	No exciter output	Repair exciter.
	0.65 V at TP_{203} but 0 V at TP_{251}	C_{251} stage inoperative	Check for A+ at collector of Q_{251} and check associated circuitry.
	0.8 V at TP_{251} but 0 V on collector Q_{252}	Power control circuitry faulty	Repair power control circuit (see circuit analysis for operation).
	Radio draws 4 to 5 A	Open circuit after Q_{253} output	Check U_{256}, filter, relay, etc., for open circuit.
	Radio current less than 2 A but 0.8 V on TP_{251} and 12 V collector Q_{251}	Q_{252} or Q_{253} stages inoperative	If Q_{252} draws around 0.75 A Problem is Q_{253} or associated circuitry. If Q_{253} draws zero current, problem is Q_{253} or associated circuitry.
Low Power	Significantly less than 0.65 V at TP_{203}	Less exciter output	Repair exciter.
	Significantly less than 0.8 V at TP_{251}	Low output Q_{251} stage	Problem is Q_{251} or associated circuitry.
	Less than 11 V on collector Q_{252}	Power control	Adjust, then if necessary, repair power control.
	12 V on collector Q_{252}, radio draws approx. 5A	Antenna filter	Tune filter.
	Radio current less than 4 A	Q_{252} or Q_{253} stages faulty	If Q_{252} draws around 0.75 A, problem is Q_{253} or associated circuitry. If Q_{252} draws significantly less than 0.75 A, problem is Q_{252} or associated circuitry.

Troubleshooting	
Symptom	**Procedure**
Unit does not decode	1. Place squelch switch in the unsquelched position and check for proper receiver operation. 2. If the receiver operates properly, set squelch to the "out" position. Apply the proper channel guard tone to the radio and check for 8.5 V dc at position $U_{1001-10}$. 3. If reading is not correct, check voltage readings on connections between the tone network FL_{1001} and U_{1005}. 4. If the readings between the tone network and U_{1003} are incorrect, ensure good contact between the tone network and the network socket. 5. If readings are correct, check voltage readings at all other points identified.
Unit does not encode	1. Check for 3.1 V dc at U_{1002-7}. 2. If reading is correct, check mod. adj. R_{1001}, then check the transmitter oscillator module. 3. If reading is not correct, check voltage readings on connections between the tone network FL_{1001} and U_{1003}. 4. If the readings between the tone network and U_{1003} are incorrect, ensure good contact between the tone network and the network socket. 5. If readings are correct, check voltage readings at all other points identified.

FIGURE 7-30 Transmitter troubleshooting. (Courtesy of General Electric Company.)

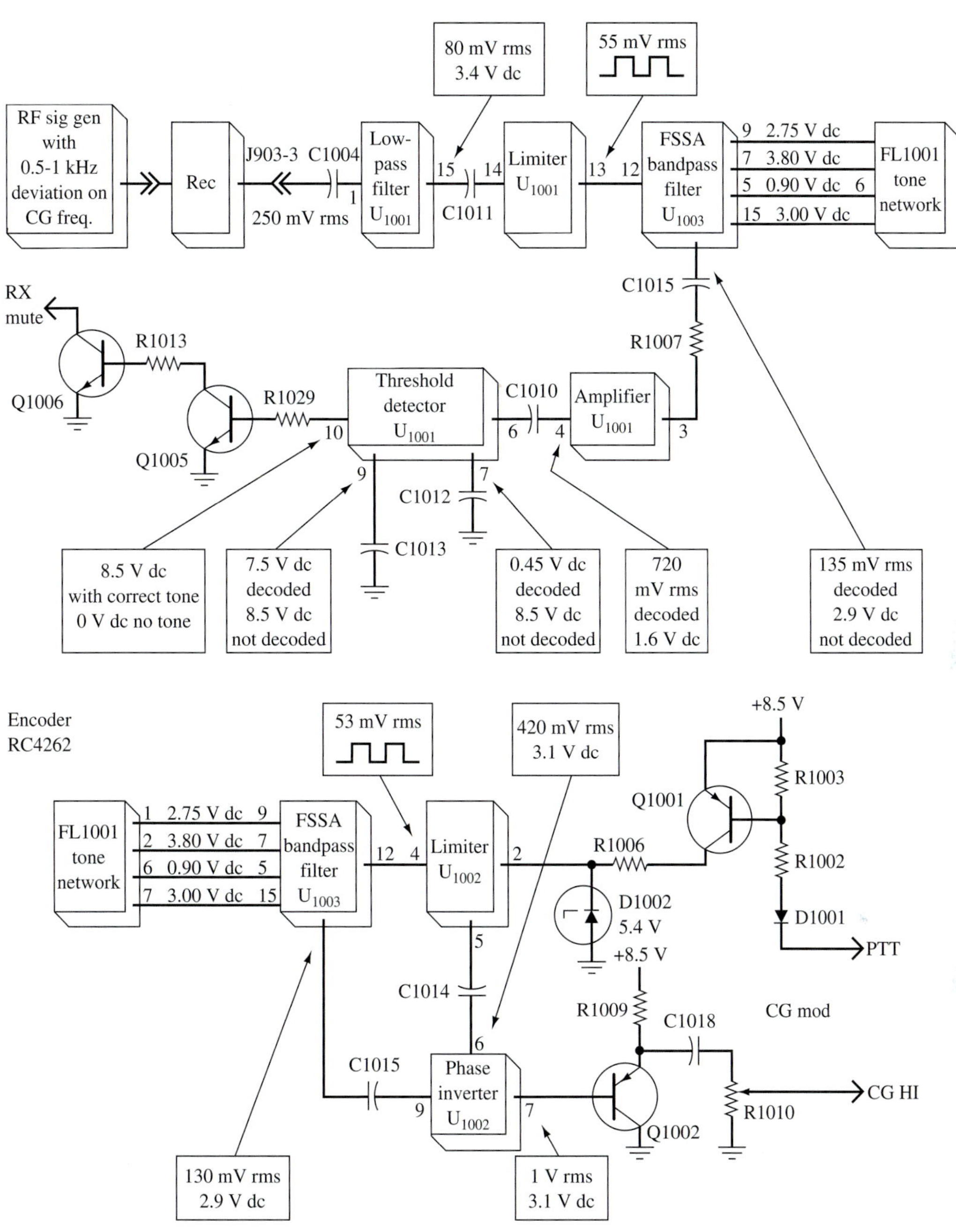

FIGURE 7-30 *(Continued)*

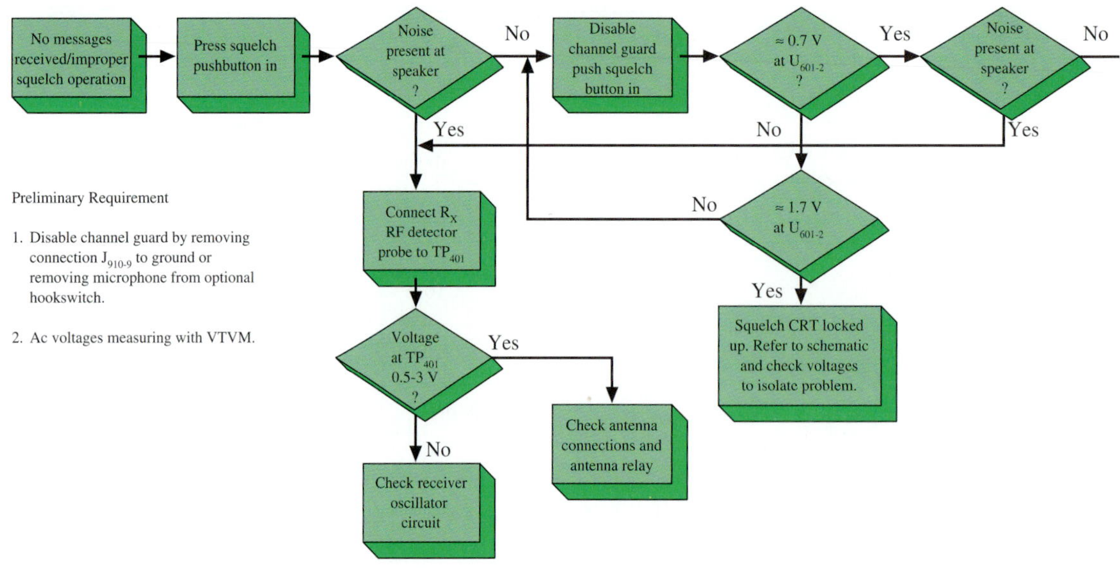

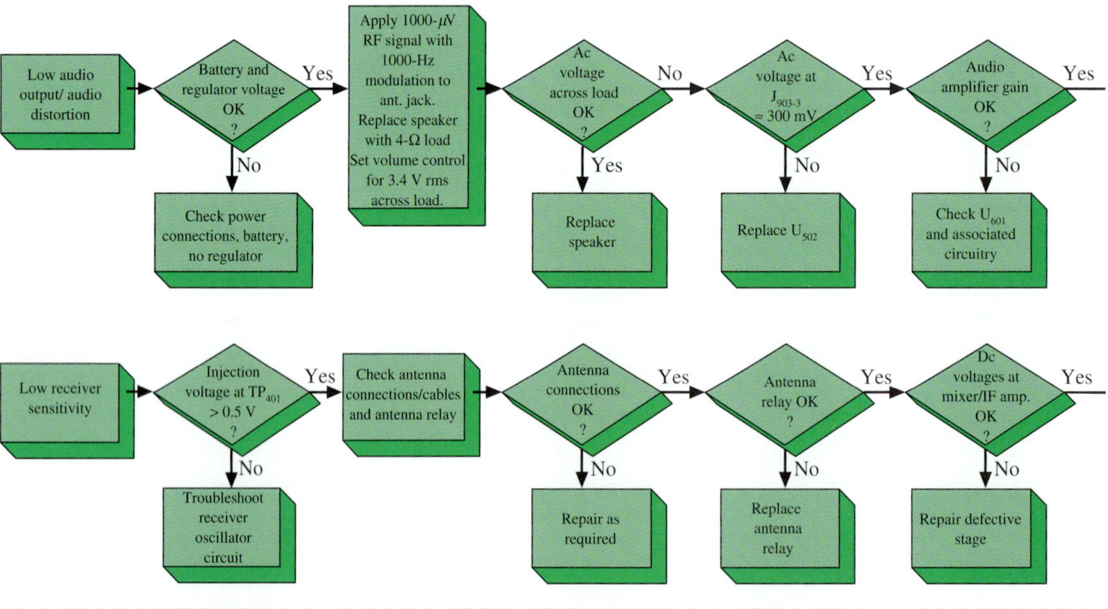

FIGURE 7-31 Receiver troubleshooting. (Courtesy of General Electric Company.)

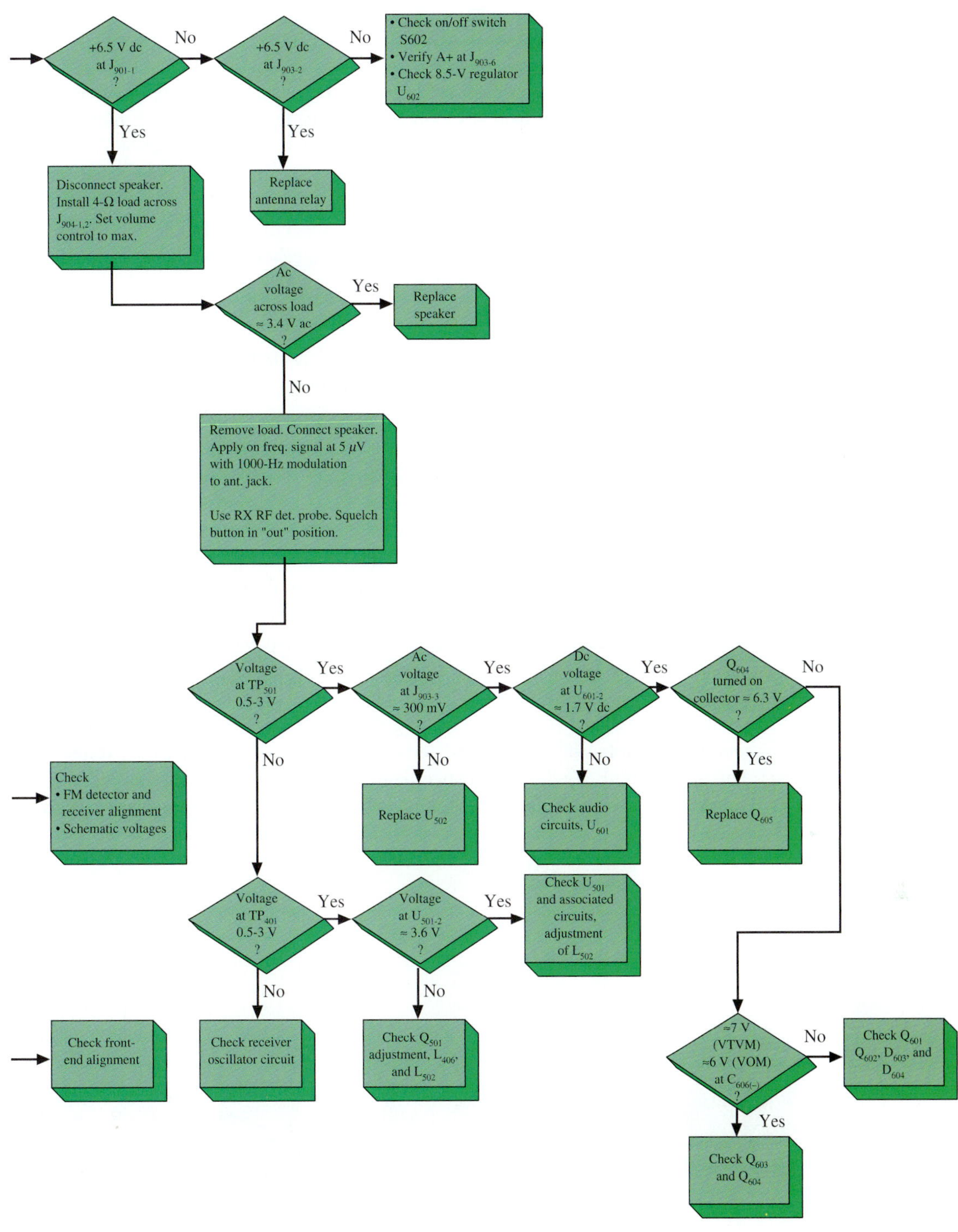

FIGURE 7-31 *(Continued)*

7-8 Troubleshooting

Transceivers, or two-way radios, are found in many commercial applications. In this section we will look at troubleshooting the transmitter portion of a mobile transceiver. General troubleshooting techniques are presented in this section. You should always consult the service manual before disassembling a transceiver and making any adjustments or repairs on it.

Today's communication equipment usually includes digital logic circuits to control various functions. We will learn to troubleshoot some basic logic circuits. We'll also consider troubleshooting a frequency synthesizer.

After completing this section you should be able to

- Describe the signal flow in a mobile FM transmitter circuit
- Describe common mobile transmitter failures
- Troubleshoot basic logic circuits
- Troubleshoot a frequency synthesizer

Transceiver Transmitter

The block diagram in Fig. 7-32 depicts the transmitter portion of a mobile transceiver. Mobile transmitters may differ somewhat in design. For example, this particular transmitter uses several frequency multiplier circuits in the exciter stage to step up the frequency to the necessary operating frequency. A press-to-talk microphone feeds the voice signal into an audio amplifier. The voice signal is amplified and sent to the phase modulator. The phase modulator is also fed by a crystal-controlled oscillator. The signal driving the power amplifier is FM from the phase modulator that has been amplified and multiplied in frequency by the exciter stage. The power amplifier delivers a specified output power to the antenna via the harmonic filter and the antenna-switching relay. Typical output power ratings are 20 to 25 W.

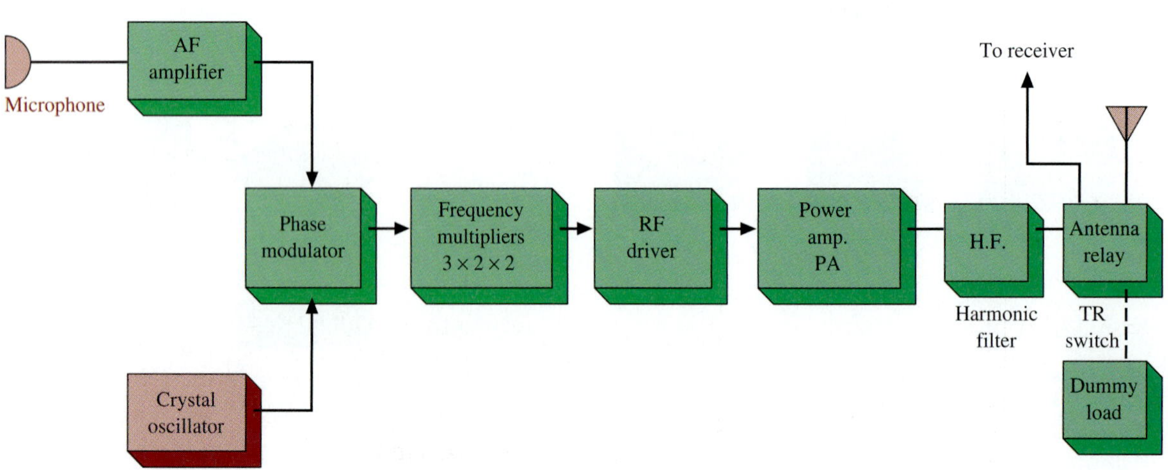

FIGURE 7-32 Block diagram of a mobile FM transceiver, transmitter portion.

Transmitter Troubles

Transmitter troubles fall into several categories: low-power output, exciter troubles, oscillator troubles, power amplifier troubles, transmit-receive (TR) antenna-switching relay problems, modulator troubles, and microphone-associated problems. We will look at some of these common transmitter troubles in the following discussion.

1. **Microphone and Audio Troubles** Microphone failures are more often the problem in mobile radios than audio amplifiers. This is true because of the extensive use the microphones get. In addition, these radios are exposed to temperature extremes and moisture. Being subject to these weather conditions will cause the microphone's cords to become stiff and brittle. Flexing the cord may damage the insulation or cause the cord to break at its plug. Solder connections can become poor conductors over a period of time and weathering. Press-to-talk switches become contaminated and corroded from dirt and moisture. These switches get a tremendous amount of use, and often the internal contacts fail to make good connection. An inoperative microphone and its cord should be replaced.

2. **Modulator Troubles** A common modulator problem is low modulation. This low modulation reduces the transmitter's operational range. Leaky coupling capacitors, open bypass capacitors, and off-tolerance resistors are likely candidates in the modulator circuits that can reduce the modulation level. Check bias resistors in the transistor circuits of the modulator. Also, review the Chapter 5 troubleshooting section for the reactance modulator.

3. **TR Switching Troubles** The TR switch is responsible for switching the antenna between the transmitter and the receiver in the transceiver radio. By this switching action a common antenna is shared by both the receiver and the transmitter. Absence of RF power at the antenna could be due to contact trouble in the antenna-switching relay. Figure 7-28 shows a typical antenna-switching relay, K601. Notice the contact switching positions. A defective relay can usually be checked by pressing on the contacts using a plastic wand while keying the transmitter. If the relay works with pressure applied, then it should be replaced. Cleaning the contacts may work if the relay points are not pitted.

4. **PA Troubles** The presence of a strong exciter output, but the absence of a signal or a weak signal at the power-amplifier output, indicates power-amplifier troubles. A very common cause for this missing output is blown output transistors. Output transistors can blow from not being terminated and from impedance mismatches with the antenna or dummy load. Never key a transmitter unless it is connected to the antenna or the dummy load. For no- or low-output power check the transistors. If the transistors are not bad, look for passive components that might be defective and for changes in resistor values.

5. **Oscillator Troubles** Improper tuning of the oscillator stage may cause unstable transmitter operation. Sometimes a tank inductance is varied in the oscillator circuit. In other circuits, the tank is fixed and a capacitor may be used to trim the frequency. Improper tuning in either case can put the oscillator at the edge of stable or unstable operation. Weak crystals will cause the oscillator's output to decrease. The weak crystal may even shut the oscillator down completely. Verify the oscillator's operation by monitoring the transmitter output or by measuring the voltage at a frequency

multiplier test point. Note any changes in frequency at the transmitter's output or any voltage variations at the test point. Consult the service manual for specific steps to repair, adjust, and replace crystals.

Logic Problems

Today's receivers and transmitters usually utilize a good bit of digital logic. When troubleshooting any digital circuit, the technician will first look to see if minimum logic levels are being met. The voltage levels do vary for many of today's logic circuits, but we will assume that the parts are operating from a +5-V supply. The part is guaranteed to output at least 2.4 V (logic one) and less than 0.4 V for a logic zero. Current ratings go with the voltages that guarantee the number of loads a gate can drive.

The part is guaranteed to accept any voltage more than 2.0 V as a logic one on its input, point A or B. The gate is guaranteed to accept any input voltage less than 0.8 V as a logic zero.

Note that the part is guaranteed to give more than it requires, 2.4 V versus 2.0 V. This is done so that there is a guaranteed margin to allow for age and noise. It sounds simple.

Now the technician has to understand what the circuit should do functionally. The best sources of help are a good manual, a set of schematics, and experience. Modern logic circuits are challenging, but a good block diagram, example logic diagrams, and good test gear can help you solve the problem.

Logic Zero Incorrect To place a logic zero on the TTL gate's input, we must draw a small amount of current from the gate. If the part driving the gate cannot do this, the voltage will not fall in the guaranteed region and the gate will not know what to do.

It is also possible that the gate input has shorted to the supply. In this case it is impossible for any output stage to drive the bad input. You will have to change one or the other or both. Which one? See if other gates in that IC are working. Is one hot? It might be the bad one.

Logic One Incorrect This is exactly the opposite of the above problem. TTL is guaranteed to supply 40 μA. If one of the diodes in the following gate is shorted, the gate will not be able to output the required 2 V the next gate requires to function. Again, it is difficult to tell which gate is malfunctioning, input or output. You will have to make an educated guess and change one.

Synthesizer Problems

Figure 7-14 is a block diagram of a synthesizer found in many communications receivers. The output of the VCO is connected to the first mixer to set the receiver frequency. Here are some typical troubles.

1. *Small frequency error:* Check the frequency of the oscillator. Remember, error at the oscillator is multiplied by *N*.
2. *Large frequency error:* Is the loop locked? Probably not. Look at the output of the phase comparator. If the loop is not locked, you will see a waveform that is the difference in the divider output and the reference oscillator. Check components in the VCO and look at logic levels in the divider. Check to see that the division ratio is correct.

3. *No output at all:* A failure in the system may drive the varactor in the VCO to some condition that will not allow oscillations. You might isolate the VCO from the low-pass filter. Check all VCO components.

7-9 Troubleshooting with Electronics Workbench™ Multisim

In this section, we first review the concept of a multiplier circuit or, as it is sometimes called, a mixer. Open **Fig7-33.ms7 (.msm)** in your EWB Multisim CD. In this circuit, 20- and 21-MHz sine waves are being input into a mixer stage. The circuit is shown in Fig. 7-33. Recall that when two frequencies are mixed together, we should see the sum and differences of the A and B frequencies at the output of the mixer stage.

$$\sin A \times \sin B = 0.5 \cos (A - B) - 0.5 \cos (A + B)$$

Start the simulation and examine the output of the mixer with the oscilloscope. You should see a complex signal containing the product of the A and B frequencies. The oscilloscope trace of the complex signal is shown in Fig. 7-34. The figure shows a high-frequency sine wave riding on top of a lower-frequency sine wave. Use the oscilloscope to measure the period and determine the frequency of each sine-wave component. You should find that this complex signal contains a 1-MHz and a 41-MHz component.

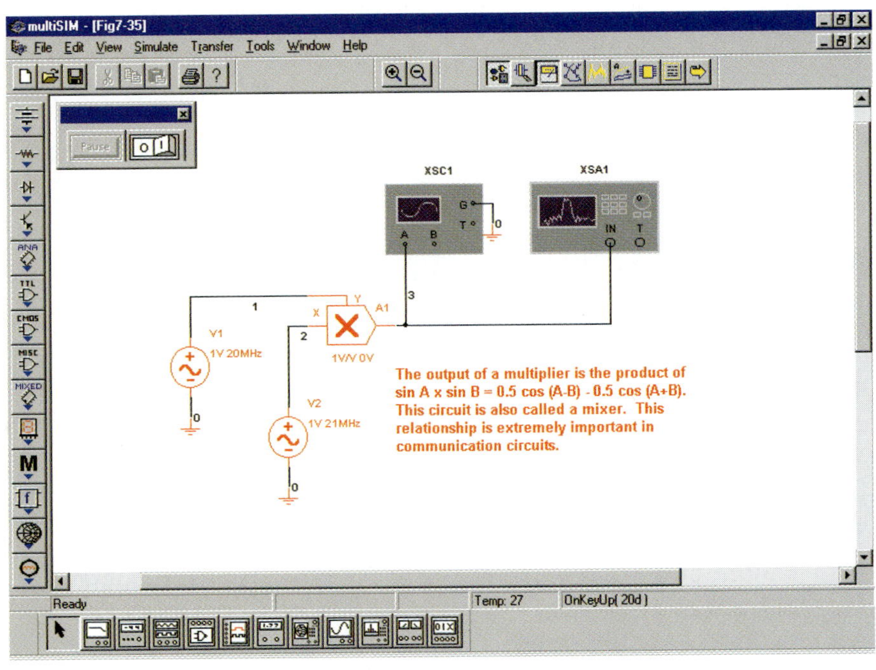

FIGURE 7-33 The mixer circuit as implemented with Multisim.

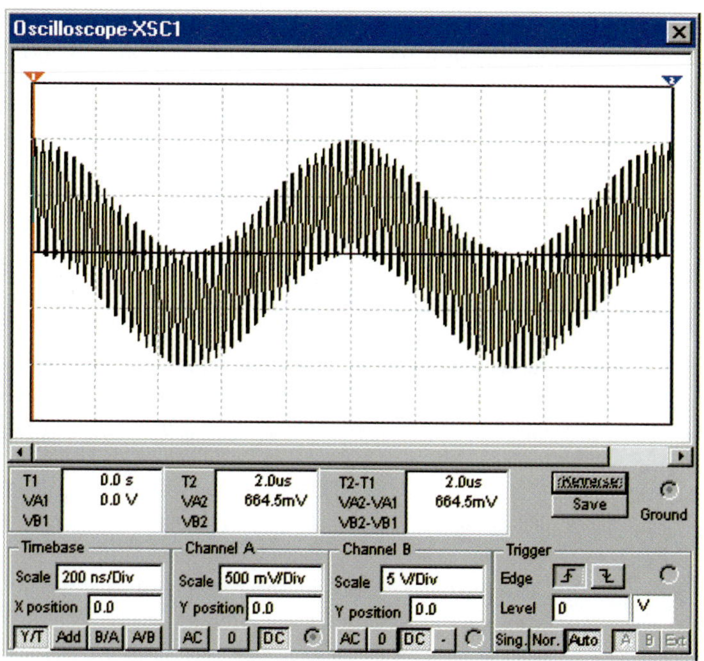

FIGURE 7-34 The output of the mixer as viewed with an oscilloscope. The input frequencies to the mixer are 20 and 21 MHz.

You can use the spectrum analyzer to verify the frequency components at the output of the mixer. Start the simulation and double-click on the spectrum analyzer module. You should see a spectral display similar to that shown in Fig. 7-35.

Use the cursor on the spectrum analyzer to measure the frequency components. You will find that the spectral output of the mixer stage contains a 1-MHz and a 41-MHz component, as predicted by the sin $A \times \sin B$ equations and by the analysis with the oscilloscope.

Change the input frequencies so that both inputs are 20 MHz and the amplitudes of each sine wave are equal. What trace do you expect to see on the

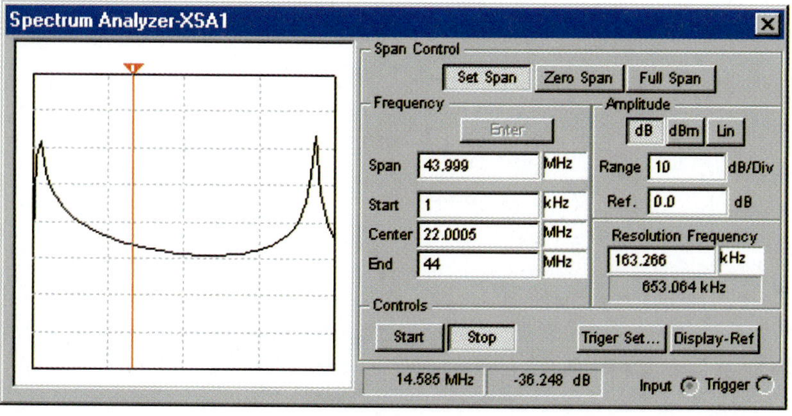

FIGURE 7-35 The output of the mixer as viewed by a spectrum analyzer.

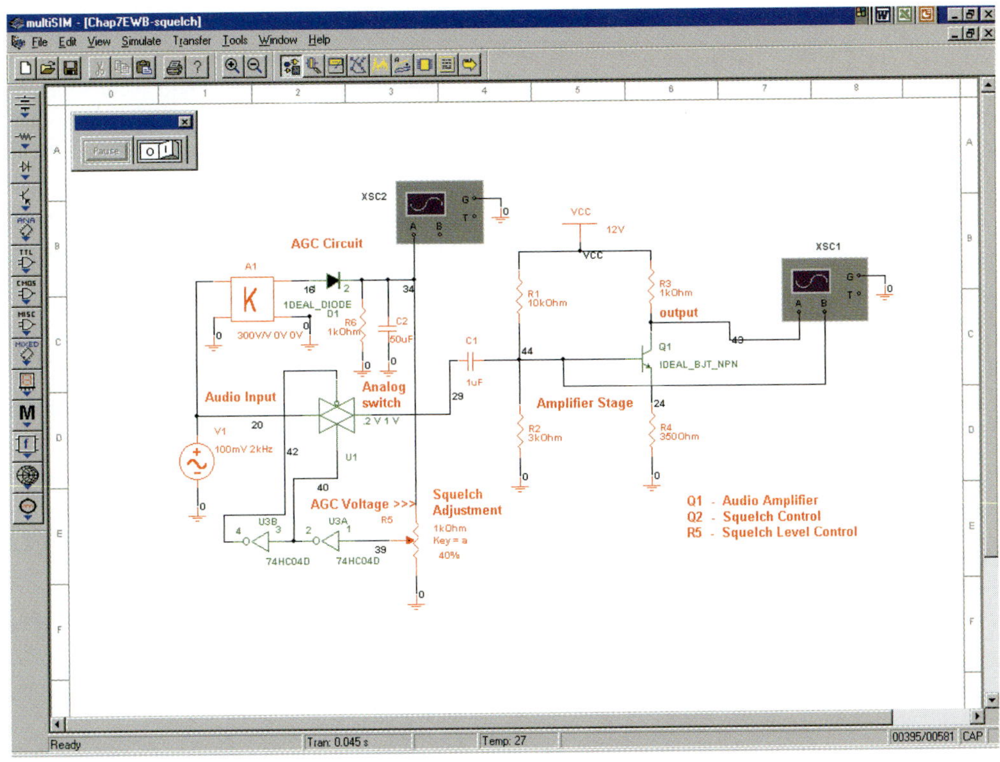

FIGURE 7-36 An example of a squelch circuit as implemented with Multisim.

oscilloscope and on the spectrum analyzer? Start the simulation and see if you are correct. You should find that the oscilloscope shows a 40-MHz sine wave and the spectrum analyzer shows a 40-MHz frequency component. We are seeing the $A + B$ frequency term, whereas the $A - B$ frequency term is zero.

The next exercise provides an opportunity to experiment with a squelch circuit, as implemented with Multisim. Squelch circuits are commonly used in communication receivers to turn off (squelch) the output signal when the signal strength of the received signal is low and noisy or not present. This minimizes annoying noise problems when listening to radio transmissions as they are keyed on and off. Open **Fig7-36.ms7 (.msm)** in your EWB CD to experiment with the squelch circuit.

The input audio signal for the squelch circuit, shown in Fig 7-36, is provided by a 2-kHz sine-wave generator. This signal feeds an amplifier stage and an automatic gain control (AGC) circuit. The AGC circuit uses a simple half-wave rectifier and an RC filter to provide the AGC voltage. The AGC voltage is fed to a 1-kΩ potentiometer that provides squelch adjustment. The signal from the potentiometer feeds two inverters whose outputs are connected to an analog switch, which connects the audio signal to the output amplifier. If sufficient AGC voltage is present (indicating a strong signal), the analog switch will be turned on and the 2-kHz audio signal is passed through to the output amplifier. If the AGC voltage is low or not present (indicating a poor-quality signal or no carrier), the analog switch is turned off. This disconnects the 2-kHz audio signal from the output amplifier.

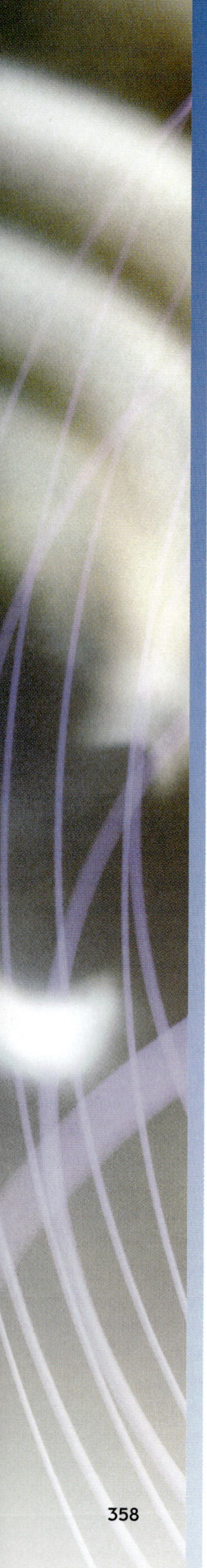

Start the simulation and double-click on the oscilloscope (XSC1) to view the amplifier output. You should see a sine wave on both the input (Ch. B) and output (Ch. A) of the amplifier stage. A virtual potentiometer has been provided for squelch-level adjustment. It should be set to 40 percent. You can squelch the 100-mV sine wave by adjusting the squelch control to 0 percent. To adjust the squelch level, press *a* to decrease and *A* to increase the resistance of the potentiometer.

Return the squelch setting to 40 percent, stop the simulation and change the sine-wave input level from 100 mV to 10 mV. This change simulates a weak receive signal. A squelch control setting of 40 percent will reject weak signals. Verify this by starting the simulation and viewing the signals on the oscilloscope. The oscilloscope will show a flat line or no signal with the control set to 40 percent. Change the squelch control to 50 percent, and the signals should return. Returning the squelch control to 40 percent will reject the signal. In other words, you now have adjustable squelch control.

Experiment with this circuit. Use the multimeter to examine voltage levels throughout the circuit. View the output of the AGC circuit with the oscilloscope. View both the AC and DC components to understand the AGC signal better.

Electronics Workbench™ Exercises

1. Open the file **FigE7-1.ms7 (.msm)** in your EWB CD. You have discovered that this squelch circuit is not working. Troubleshoot the squelch circuit to determine the cause of the failure. Correct the fault and rerun the simulation. Report on your findings.
2. Open the file **FigE7-2.ms7 (.msm)** in your EWB CD. You have been told that this squelch circuit is working but does not have the range of control that it previously had. Troubleshoot the squelch circuit to determine the cause of the failure. Correct the fault and rerun the simulation. Report on your findings.
3. Open the file **FigE7-3.ms7 (.msm)** in your EWB CD. You discovered that the squelch circuit is not working. You suspect that an employee at the facility dropped a screwdriver into the receiver cabinet while power was on. Troubleshoot the squelch circuit to determine the cause of the failure. Correct the fault(s) and rerun the simulation. Report on your findings.

Summary

In Chapter 7 we described various improvements to receiver design and discussed some of the more complicated specifications used in high-quality receivers. The concept of spread-spectrum communications was also introduced. The major topics you should now understand include:

- the analysis of advanced techniques for image frequency reduction, including double conversion and up-conversion
- the description and explanation of special techniques for improving receiver operation, including delayed AGC, auxiliary AGC, manual sensitivity control, variable notch filters, ANL circuits, and squelch control
- the analysis of the relationship among noise, sensitivity, and dynamic range in a high-quality receiver

- the analysis of intermodulation distortion (IMD) testing
- the description and analysis of various frequency synthesizers
- the method used to obtain direct digital synthesis (DDS) systems
- the analysis of spread-spectrum techniques, including description of CDMA, frequency hopping, time hopping, and direct sequence

Questions and Problems

Section 7-2

1. Explain the difference between an FM stereo receiver and a communications transceiver.
2. Draw a block diagram for a double-conversion receiver when tuned to a 27-MHz broadcast using a 10.7-MHz first IF and 1-MHz second IF. List all pertinent frequencies for each block. Explain the superior image frequency characteristics as compared to a single-conversion receiver with a 1-MHz IF, and provide the image frequency in both cases.
3. Draw block diagrams and label pertinent frequencies for a double-conversion *and* up-conversion system for receiving a 40-MHz signal. Discuss the economic merits of each system and the effectiveness of image frequency rejection.
4. A receiver tunes the HF band (3 to 30 MHz), utilizes up-conversion with an intermediate frequency of 40.525 MHz, and uses high-side injection. Calculate the required range of local oscillator frequencies. (43.5 to 70.5 MHz)
5. An AM broadcast receiver's preselector has a total effective Q of 90 to a received signal at 1180 kHz and uses an IF of 455 kHz. Calculate the image frequency and its dB of suppression. (2090 kHz, 40.7 dB)

Section 7-3

6. Discuss the advantages of delayed AGC over normal AGC and explain how it may be attained.
7. Explain the function of auxiliary AGC and give a means of providing it.
8. Explain the need for variable sensitivity and show with a schematic how it could be provided.
9. Explain the need for variable selectivity. Describe how VBT is accomplished if the oscillator in Fig. 7-8 is changed to 2650 Hz.
10. What is the need for a noise limiter circuit? Explain the circuit operation of the noise limiter shown in Fig. 7-9.
11. List some possible applications for *metering* on a communications transceiver.

*12. What is the purpose of a squelch circuit in a radio communications receiver?

13. List two other names for a squelch circuit. Provide a schematic of a squelch circuit and explain its operation. List five different squelch methods.
14. Describe the effects of EMI on a receiver.
15. Describe the operation of an automatic noise limiter (ANL).

---

*An asterisk preceding a number indicates a question that has been provided by the FCC as a study aid for licensing examinations.

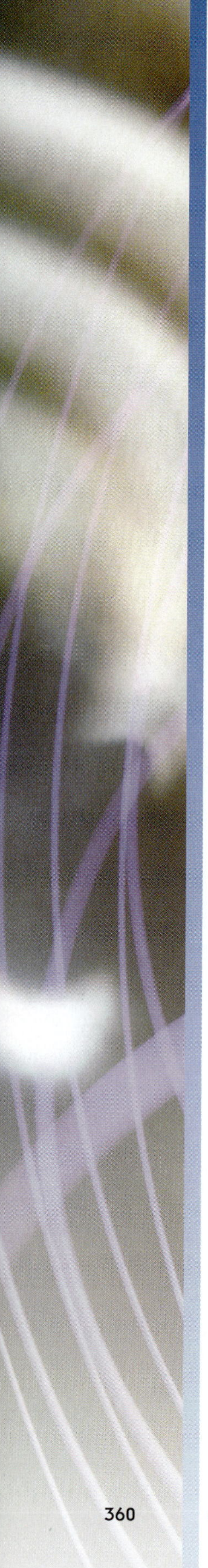

Section 7-4

16. We want to operate a receiver with NF = 8 dB at $S/N = 15$ dB over a 200-kHz bandwidth at ambient temperature. Calculate the receiver's sensitivity. (−98 dBm)
17. Explain the significance of a receiver's 1-dB compression point. For the receiver represented in Fig. 7-11, determine the 1-dB compression point. (≅10 dBm)
18. Determine the third-order intercept for the receiver illustrated in Fig. 7-11. (≅ +20 dBm)
19. The receiver described in Problem 16 has the input/output relationship shown in Fig. 7-11. Calculate its dynamic range. (78.7 dB)
20. A receiver with a 10-MHz bandwidth has an S/N of 5 dB and a sensitivity of −96 dBm. Find the required NF. (3 dB)

Section 7-5

21. Explain the operation of a basic frequency synthesizer as illustrated in Fig. 7-14. Calculate f_0 if $f_R = 1$ MHz and $N = 61$. (61 MHz)
22. Discuss the relative merits of the synthesizers shown in Figs. 7-16(a), (b), and (c) as compared to the one in Fig. 7-14.
23. Describe the operation of the synthesizer divider in Fig. 7-17. What basic problem does it overcome with respect to the varieties shown in Figs. 7-14 and 7-16?
24. Calculate the output frequency of a synthesizer using the divider technique shown in Fig. 7-17 when the reference frequency is 1 MHz, $A = 26$, $M = 28$, and $N = 4$. (138 MHz)
25. Determine the output frequency for the synthesizer of Fig. 7-19 when the input code is 100011. (27.245 MHz)
26. Explain the operation of the UHF multifrequency receiver of Fig. 7-22.

Section 7-6

27. Briefly explain DDS operation based on the block diagram shown in Fig. 7-23.
28. A DDS system has $f_{CLK\,MAX} = 60$ MHz and a 28-bit phase accumulator. Calculate its approximate maximum output frequency and frequency resolution when operated at $f_{CLK\,MAX}$. (24 MHz, 0.223 Hz)

Section 7-7

29. Describe the function of the channel guard function in a communication transceiver.
30. Analyze the system block diagram in Fig. 7-25. Redraw it more simply to explain its operation to technicians who are not experts in the field of electronic communication.
31. The relay (K601) in Fig. 7-28 is malfunctioning. The contacts are not releasing when the coil is deenergized. Describe the effect this would have on the transceiver's performance.

Section 7-8

32. Explain how the technician would know that the AF amplifier in Fig. 7-32 lost its gain.
33. Describe a possible output that leads the technician to suspect that the problem is in the frequency multipliers of Fig. 7-32.
34. What are some of the problems that can occur in the FM transceiver of Fig. 7-32 from a bad oscillator?
35. Describe the output if the harmonic filter in Fig. 7-32 was leaky.

Questions for Critical Thinking

36. Describe the process of up-conversion. Explain its advantages and disadvantages compared to double conversion.
37. You have been asked to extend the dynamic range of a receiver. Can this be done? What factors determine the limits of dynamic range? Can they be changed? Explain.
38. In evaluating a receiver, how important is its ability to handle intermodulation distortion? Explain the process you would use to analyze a receiver's ability to handle this distortion. Include the concept of third-order intercept point in your explanation.
39. The receiver in Problem 19 has a 6-dB NF preamp (gain = 20 dB) added to its input. Calculate the system's sensitivity and dynamic range. (−99.94 dBm, 66.96 dB)

Digital Communications: Coding Techniques

Chapter Outline

A digital test set using the Agilent 71612A 12 Gbps pattern generator and detector. (Court of Agilent Technologies. Reprinted with permission.)

Objectives

- Describe the quantization process in a PCM system in terms of how it is created, how to determine the Nyquist sampling frequency, and how to define quantization levels
- Determine the dynamic range and signal-to-noise ratio of a PCM system
- Describe the common digital signal encoding formats
- Understand the concept of Hamming distance as applied to the technique of error detection and correction
- Describe the various techniques for code error detection and correction, including parity, block check character, cyclic redundancy check, Hamming code, and Reed–Solomon codes.

Key TERMS

regeneration
digital signal processing
algorithms
ASCII
parity
EBCDIC
Baudot code
Gray code
acquisition time
aperture time
natural sampling
flat-top sampling
Nyquist rate
aliasing, or foldover distortion
antialiasing filter
quantization
quantile
quantile interval
quantization levels
quantizing error
quantizing noise
dynamic range
uniform quantization level
linear quantization level
nonlinear coding
nonuniform coding
idle channel noise
amplitude companding
codec
multilevel binary
Hamming distance
minimum distance (D_{min})
symbol substitution
block check character (BCC)
longitudinal redundancy check (LRC)
cyclic redundancy check
BCC
systematic code
(n, k) cyclic code
generating polynomial
syndrome
forward error-correcting
Hamming code
interleaving

8-1 INTRODUCTION

The field of digital and data communications has experienced explosive growth in recent years. In general, this field includes the transfer of analog signals using digital techniques and the transfer of digital data using digital and/or analog techniques. It is difficult to separate the two topics totally because of their interrelationships.

Digital communications is the transfer of information in digital form. As shown in Fig. 8-1, if the information is analog (voice in this case), it is converted to digital for transmission. At the receiver, it is reconverted to analog. Figure 8-1 also shows a digital computer signal transmitted to another computer. Notice that this is shown to represent both digital and data communications. The third system in Fig. 8-1 shows a computer's digital signal converted to analog for transmission and then reconverted to digital by the modem. We look at modems in detail in Chapter 11. The reason that various techniques are used boils down to performance and cost, which will be apparent as we take a close look at the systems involved.

The move to digital and/or data communications is due to several factors. It is occurring despite the increased complexity and bandwidth necessary for transmission. Noise performance is one of two major advantages. Consider an analog signal with an instantaneous received value of 1 mV. If at the same time an instantaneous 0.1-mV noise spike changes the received value to 1.1 mV, there is normally no way of knowing the correct value of the signal. In a digital system, however, the received signal may ultimately be changed to either a logical 0- or 1-V level. Now the received noise of 0.1 mV may still be there but certainly would not cause an error.

The digital system can re-create the original signal by having circuits that change any signal below 0.5 V into the 0-V level and any signal above 0.5 V into

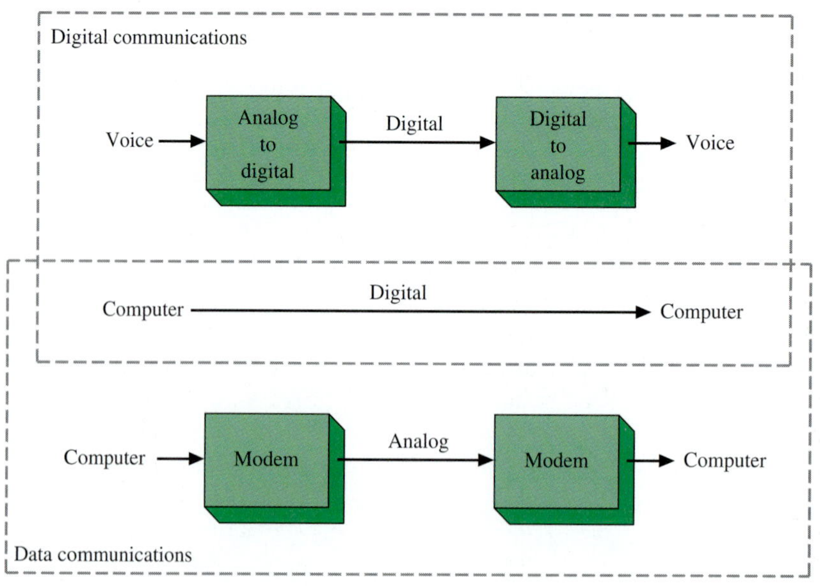

FIGURE 8-1 Digital/data communications.

the 1-V level. This ability to restore a noise-corrupted signal to its original value is called **regeneration.** Obviously, if the noise is so great as to cause the 0-V level to be seen as a 0.6-V level, an error will occur.

Regeneration
restoring a noise-corrupted signal to its original condition

Another advantage of using a digital format involves the ability to process the signal at the transmitter (preprocessing) and/or the receiver (postprocessing). Both of these operations are termed **digital signal processing.** Signals in digital format can be stored in computer memory and be easily manipulated by **algorithms**—a plan or set of instructions followed to achieve a specific goal. They can be implemented by digital circuitry, which is a *hardware* solution. Increasingly, they are implemented by *software* instructions (via a *computer program*) that instruct a microprocessor how to perform specific manipulations. Many communications systems use *microcontrollers* that are microprocessors programmed to do one basic task only.

Digital Signal Processing
using programming techniques to process a signal while in digital form

Algorithms
a plan or set of instructions to achieve a specific goal

8-2 Alphanumeric Codes

The most common alphanumeric coding scheme for binary data is the American Standard Code for Information Interchange (ASCII). The Extended Binary-Coded Decimal Interchange Code (EBCDIC) is still used in large computing systems but sees little use in digital communication systems.

The ASCII Code

ASCII is a 7-bit code used for representing alphanumeric symbols with a distinctive code word. The ASCII code was developed by a committee of the American National Standards Institute (ANSI) for the purpose of coding binary data. ASCII-77 is the adopted international standard. Figure 8-2 provides a list of the codes.

ASCII
standardized coding scheme for alphanumeric symbols

There are 2^7 (128) possible 7-bit code words available with an ASCII system. The binary codes are ordered sequentially, which simplifies the grouping and sorting of the characters. The 7-bit words are ordered with the least significant bit (lsb) given as bit 1 (b_1), while the most significant bit (msb) is bit 7 (b_7). Notice that a binary value is not specified by the code for bit 8 (b_8). Usually the bit 8 (b_8) position is used for parity checking. **Parity** is an error detection scheme that identifies whether an even or odd number of logical ones are present in the code word. This concept is discussed in greater detail in Sec. 8-6. For ASCII data used in a serial transmission system b_1, the lsb bit, is transmitted first.

Parity
a common method of error detection, adding an extra bit to each code representation to give the word either an even or odd number of 1s

The ASCII system is based on the binary-coded-decimal (BCD) code in the last 4 bits. The first 3 bits indicate whether a number, letter, or character is being specified. Notice that 0110001 represents "1," while 1000001 represents "A" and 1100001 represents "a." It uses the standard binary progression (i.e., 0110010 represents "2"), and this makes mathematical operations possible. Because the letters are also represented with the binary progression, alphabetizing is also achieved via binary mathematical procedures. You should also be aware that analog waveform coding is accomplished simply by using the BCD code for PCM systems covered in Sec. 8-3.

In some systems the actual transmission of these codes includes an extra pulse at the beginning (start) and ending (stop) for each character. When start/stop pulses are used in the coding of signals, it is called an *asynchronous* (nonsynchronous) transmission. A synchronous transmission (without start/stop pulses)

Bit Positions:											
7 →				0	0	0	0	1	1	1	1
6 →				0	0	1	1	0	0	1	1
5 →				0	1	0	1	0	1	0	1
4 ↓	3 ↓	2 ↓	1								
0	0	0	0	**NUL**	**DLE**	SP	0	@	P	'	p
0	0	0	1	**SOH**	**DC1**	!	1	A	Q	a	q
0	0	1	0	**STX**	**DC2**	"	2	B	R	b	r
0	0	1	1	**ETX**	**DC3**	#	3	C	S	c	s
0	1	0	0	**EOT**	**DC4**	$	4	D	T	d	t
0	1	0	1	**ENQ**	**NAK**	%	5	E	U	e	u
0	1	1	0	**ACK**	**SYN**	&	6	F	V	f	v
0	1	1	1	**BEL**	**ETB**	'	7	G	W	g	w
1	0	0	0	**BS**	**CAN**	(	8	H	X	h	x
1	0	0	1	**HT**	**EM**	)	9	I	Y	i	y
1	0	1	0	**LF**	**SUB**	*	:	J	Z	j	z
1	0	1	1	**VT**	**ESC**	+	;	K	[	k	{
1	1	0	0	**FF**	**FS**	,	<	L		l	¦
1	1	0	1	**CR**	**GS**	–	=	M	]	m	}
1	1	1	0	**SO**	**RS**	.	>	N	^	n	~
1	1	1	1	**SI**	**US**	/	?	O	—	o	DEL

Sample of Control Characters (Bold)
STX = Start of text
EOT = End of transmission
CR = Carriage return
HT = Horizontal tabulation

Examples:
1000011 = C
0110011 = 3
1010000 = P
0110000 = 0 (Zero)
0100000 = SP (space)

FIGURE 8-2 American Standard Code for Information Interchange (ASCII).

allows more characters to be transmitted within a given sequence of bits. The transmission of information between various computer installations may require the less efficient asynchronous transmitting mode depending on computer characteristics.

The EBCDIC Code

EBCDIC
standardized coding scheme for alphanumeric symbols

The Extended Binary-Coded Decimal Interchange Code **(EBCDIC)** is an 8-bit alphanumeric code. The term *binary-coded decimal* is used in the name because of the structure present in the coding scheme, which uses only the 0–9 positions. A list of the code words for the EBCDIC system is given in Fig. 8-3, and the acronyms for the control characters are listed in Table 8-1 on page 368.

The Baudot Code

Baudot Code
fairly obsolete coding scheme for alphanumeric symbols

Another interesting code presented for historical reasons is the **Baudot code.** The Baudot code was developed in the days of teletype machines such as the ASR-33

EBCDIC CODES

Bit Positions 4, 5, 6, 7	Second Hexadecimal Digit																	
		00				01				10				11				Bit Positions 0,1
		00	01	10	11	00	01	10	11	00	01	10	11	00	01	10	11	Bit Positions 2,3
		0	1	2	3	4	5	6	7	8	9	A	B	C	D	E	F	First Hexadecimal Digit
0000	0	NUL	DLE	DS		SP	&	-						(	)	\	0	
0001	1	SOH	DC1	SOS		RSP		/		a	j	–		A	J	NSP	1	
0010	2	STX	DC2	FS	SYN					b	k	s		B	K	S	2	
0011	3	ETX	DC3	WUS	IR					c	l	t		C	L	T	3	
0100	4	SEL	RES/ ENP	BYP/ INP	PP					d	m	u		D	M	U	4	
0101	5	HT	NL	LF	TRN					e	n	v		E	N	V	5	
0110	6	RNL	BS	ETB	NBS					f	o	w		F	O	W	6	
0111	7	DEL	POC	ESC	BOT					g	p	x		G	P	X	7	
1000	8	GE	CAN	SA	SBS					h	q	y		H	Q	Y	8	
1001	9	SPS	EM	SPE	IT				▲	i	r	z		I	R	Z	9	
1010	A	RPT	UBS	SM/ SW	RFF	¢	!	|	:					SHY				
1011	B	VT	CU1	CSP	CU3	.	$	,	#									
1100	C	FF	IFS	MFA	DC4	<	*	%	@									
1101	D	CR	IGS	ENQ	NAK	(	)	–	▲									
1110	E	SO	IRS	ACK		+	;	>	=									
1111	F	SI	IUS/ ITB	BEL	SUB	¦	¬	?	"								BO	

FIGURE 8-3 The Extended Binary-Coded Decimal Interchange Code.

Teletype terminal. Baudot is an alphanumeric code based on five binary values. The Baudot code is not very powerful, but it does have its place in communications history. The Baudot code is provided in Fig. 8-4.

The alphabet has 26 letters, and there is an almost equal number of commonly used symbols and numbers. The 5-bit Baudot code is capable of handling these possibilities. A 5-bit code can have only 2^5 or 32 bits of information but actually provides 26×2 bits by transmitting a 11111 to indicate all following items are "letters" until a 11011 transmission occurs, indicating "figures." Notice that no provision for lowercase letters is provided.

Figure 8-5(a) shows an example of the Baudot code to transmit "YANKEES 4 REDSOX 3." Be sure to work out the code in Fig. 8-5(b) on your own; it is the only "X-rated" part of this book that we were allowed to include.

Table 8-1 The EBCDIC Code—List of Abbreviations

ACK	Acknowledge	ETB	End of Transmission	RFF	Required Form Feed
BEL	Bell	ETX	End of Text	RNL	Required New Line
BS	Backspace	FF	Form Feed	RPT	Repeat
BYP/INP	Bypass/Inhibit Presentation	FS	Field Separator	SA	Set Attribute
CAN	Cancel	GE	Graphic Escape	SBS	Subscript
CR	Carriage Return	HT	Horizontal Tab	SEL	Select
CSP	Control Sequence Prefix	IFS	Interchange File Sep.	SFE	Start Field Extend
CU1	Customer Use 1	IGS	Interchange Group Sep.	SI	Shift In
CU3	Customer Use 3	IR	Index Return	SM/SW	Set Mode/Switch
DC1	Device Control 1	IRS	Interchange Record Sep.	SO	Shift Out
DC2	Device Control 2	IT	Indent Tab	SOH	Start of Heading
DC3	Device Control 3	IUS/ITB	Interchange Unit Sep./Intermediate Text Block	SOS	Start of Significance
DC4	Device Control 4	LF	Line Feed	SPS	Superscript
DEL	Delete	MFA	Modify Field Attribute	STX	Start of Text
DLE	Data Link Escape	NAK	Negative Acknowledge	SUB	Substitute
DS	Digit Select	NBS	Numeric Backspace	SYN	Synchronous Idle
EM	End of Medium	NL	New Line	TRN	Transparent
ENQ	Enquiry	NUL	Null	UBS	Unit Backspace
EO	Eight Ones	POC	Program-Operator Comm.	VT	Vertical Tab
EOT	End of Transmission	PP	Presentation Position	WUS	Word Underscore
ESC	Escape	RES/NEP	Restore/Enable Presentation		

FIGURE 8-4 The Baudot code.

Character Shift		Binary Code
		BIT
Letter	Figure	4 3 2 1 0
A	-	1 1 0 0 0
B	?	1 0 0 1 1
C	:	0 1 1 1 0
D	$	1 0 0 1 0
E	3	1 0 0 0 0
F	!	1 0 1 1 0
G	&	0 1 0 1 1
H	#	0 0 1 0 1
I	8	0 1 1 0 0
J	'	1 1 0 1 0
K	(	1 1 1 1 0
L	)	0 1 0 0 1
M	.	0 0 1 1 1
N	,	0 0 1 1 0
O	9	0 0 0 1 1
P	0	0 1 1 0 1
Q	1	1 1 1 0 1
R	4	0 1 0 1 0
S	BEL	1 0 1 0 0
T	5	0 0 0 0 1
U	7	1 1 1 0 0
V	;	0 1 1 1 1
W	2	1 1 0 0 1
X	/	1 0 1 1 1
Y	6	1 0 1 0 1
Z	"	1 0 0 0 1
Figure Shift		1 1 1 1 1
Letter Shift		1 1 0 1 1
Space		0 0 1 0 0
Line Feed		0 1 0 0 0
Null		0 0 0 0 0

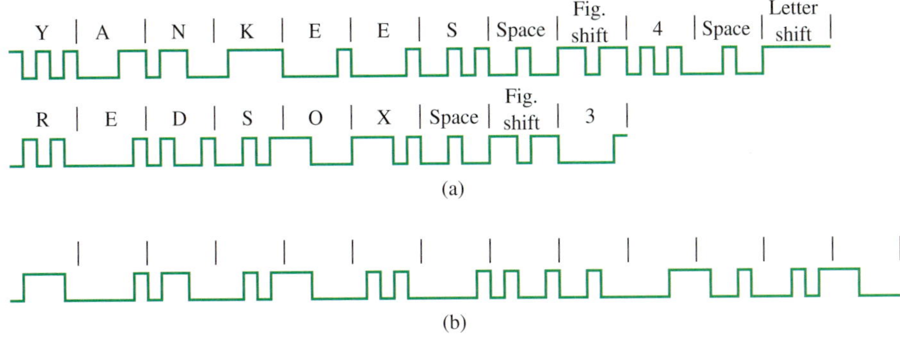

FIGURE 8-5 Baudot code examples.

The Gray Code

The last alphanumeric code we will look at is the **Gray code.** The Gray code is a numeric code for representing the decimal values from 0 to 9. It is based on the relationship that only one bit in a binary word changes for each binary step. For example, the code for 7 is 0010 while the code for 8 is 0011. Notice that only one binary bit changes when the decimal value changes from 7 to 8. This is true for all of the numbers (0–9). The Gray code is shown in Fig. 8-6.

Gray Code
numeric code for representing decimal values from 0 to 9

The Gray code is used most commonly in telemetry systems that have slowly changing data or in communication links that have a low probability of bit error. This coding scheme works well for detecting errors in slowly changing outputs, such as data from a temperature sensor (thermocouple). If more than one change is detected when words are decoded, then the receiving circuitry assumes that an error is present.

 8-3 Pulse-Code Modulation

Pulse-code modulation (PCM) is the most common technique used today in digital communications for representing an analog signal by a digital word. PCM is used in many applications, such as your telephone system, digital audio recording (DAT or digital audio tape), CD laser disks, digitized video special effects, voice mail, digital video, and many other applications. PCM techniques and applications are a primary building block for many of today's advanced communications systems.

Pulse-code modulation is a technique for converting the analog signals into a digital representation. The PCM architecture consists of a sample-and-hold (S/H) circuit and a system for converting the sampled signal into a representative binary format. First, the analog signal is input into a sample-and-hold circuit. At

Binary	#
0000	0
1000	1
1100	2
0100	3
0110	4
1110	5
1010	6
0010	7
0011	8
1011	9

1-bit change for each step value.

FIGURE 8-6 The Gray code.

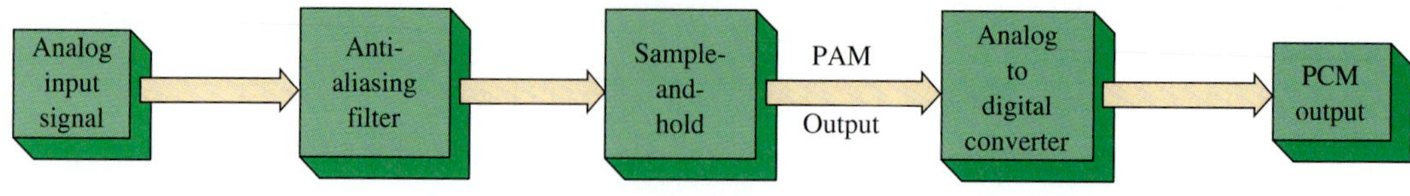

FIGURE 8-7 Block diagram of the PCM process.

fixed time intervals, the analog signal is sampled and held at a fixed voltage level until the circuitry inside the A/D converter has time to complete the conversion process of generating a binary value. A block diagram of the process is shown in Fig. 8-7.

The Sample-and-Hold Circuit

Most A/D integrated circuits come with sample-and-hold (S/H) circuits integrated into the system, but it is still important for the user to have a good understanding of the capabilities and the limitations of the S/H circuit. A typical S/H circuit is shown in Fig. 8-8. The analog signal is typically input into a buffer circuit. The purpose of the buffer circuit is to isolate the input signal from the S/H circuit and to provide proper impedance matching as well as drive capability to the hold circuit. Many times the buffer circuit is also used as a current source to charge the hold capacitor. The output of the buffer is fed to an analog switch, which is typically the drain of a junction field-effect transistor (JFET) or a metal-oxide semiconductor field-effect transistor (MOSFET). The JFET or MOSFET is wired as an analog switch, which is controlled at the gate by a sample pulse generated by the sample clock. When the JFET or MOSFET transistor is turned on, the switch will short the analog signal from drain to source. This connects the buffered input signal to a hold capacitor. The capacitor begins to charge to the input voltage level at a time constant determined by the hold capacitor's capacitance and the analog switch's and buffer circuit's "on" channel resistance.

When the analog switch is turned off, the sampled analog signal voltage level is held by the "hold" capacitor. Figure 8-9(a) shows a picture of a sinusoid on the input of the S/H circuit. The sample times are indicated by the vertical dotted lines.

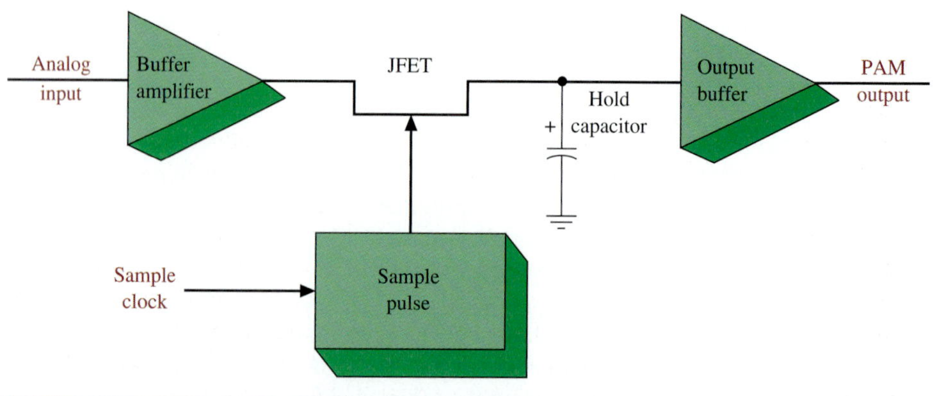

FIGURE 8-8 A sample-and-hold circuit.

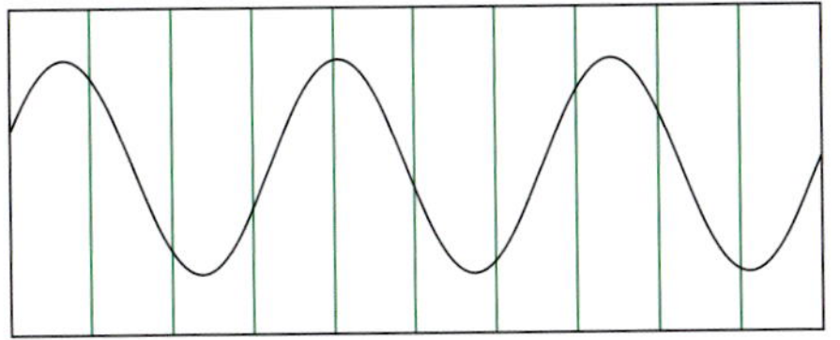

(a) Sample intervals for an input sinusoid.

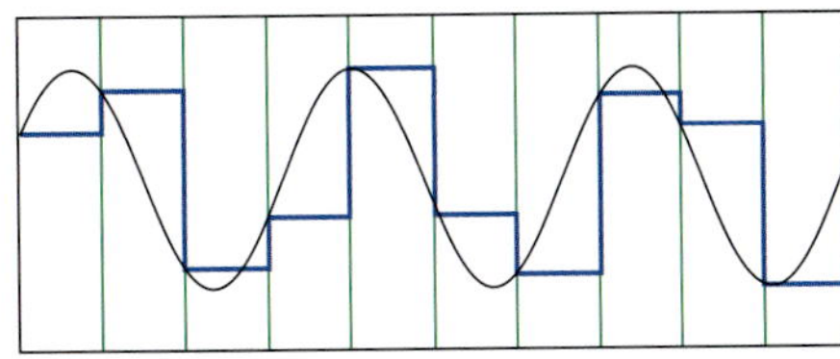

(b) A pulse-amplitude-modulated (PAM) signal.

FIGURE 8-9 Generation of PAM.

In Fig. 8-9(b) the sinusoid is redrawn as a sampled signal. Note that the sampled signal maintains a fixed voltage level between samples. The region where the voltage level remains relatively constant is called the *hold time*. The resulting waveform, shown in Fig. 8-9(b), is called a pulse-amplitude-modulated (PAM) signal. The S/H circuit is designed so that the sampled signal is held long enough for the signal to be converted by the A/D circuitry into a binary representation.

The time required for an S/H circuit to complete a sample is based partly on the acquisition and aperture time. The **acquisition time** is the amount of time it takes for the hold circuit to reach its final value. (During this time the analog switch connects the input signal to the hold capacitor.)

Acquisition Time
amount of time it takes for the hold capacitor to reach its final value

The acquisition time is controlled by the sample pulse. The **aperture time** is the time that the S/H circuit must hold the sampled voltage. The aperture and acquisition times limit the maximum input frequency that the S/H circuit can accurately process.

Aperture Time
the time that the S/H circuit must hold the sampled voltage

To provide a good-quality S/H circuit, a couple of design considerations must be met. The analog switch "on" resistance must be small. The output impedance of the input buffer must also be small. By keeping the input resistance minimal, the overall time constant for sampling the analog signal can be controlled by the selection of an appropriate hold capacitor. Ideally a minimal-size hold capacitor should be selected so that a fast charging time is possible, but a small capacitor will have trouble holding a charge for a very long period. A 1-nF hold capacitor is a popular choice for many circuit designers. It is important too that the hold capacitor be of high quality. High-quality capacitors have polyethylene, polycarbonate, or teflon dielectrics. These types of dielectrics minimize voltage variations due to capacitor characteristics.

Pulse-Amplitude Modulation

The concept of pulse-amplitude modulation (PAM) has already been introduced in this chapter, but there are a few specifics regarding the creation of a pulse-amplitude-modulated signal at the output of a sample-and-hold circuit that necessitate discussion.

Two basic sampling techniques are used to create a PAM signal. The first is called **natural sampling.** Natural sampling occurs when the tops of the sampled waveform (the sampled analog input signal) retain their natural shape. An example of natural sampling is shown in Fig. 8-10(a). Notice that one side of the analog switch is connected to ground. When the transistor is turned on, the JFET will short the signal to ground, but it will pass the unaltered signal to the output when the transistor is off. Note, too, that there is not a hold capacitor present in the circuit.

Natural Sampling
sampling in which the tops of the sampled waveforms retain their natural shape

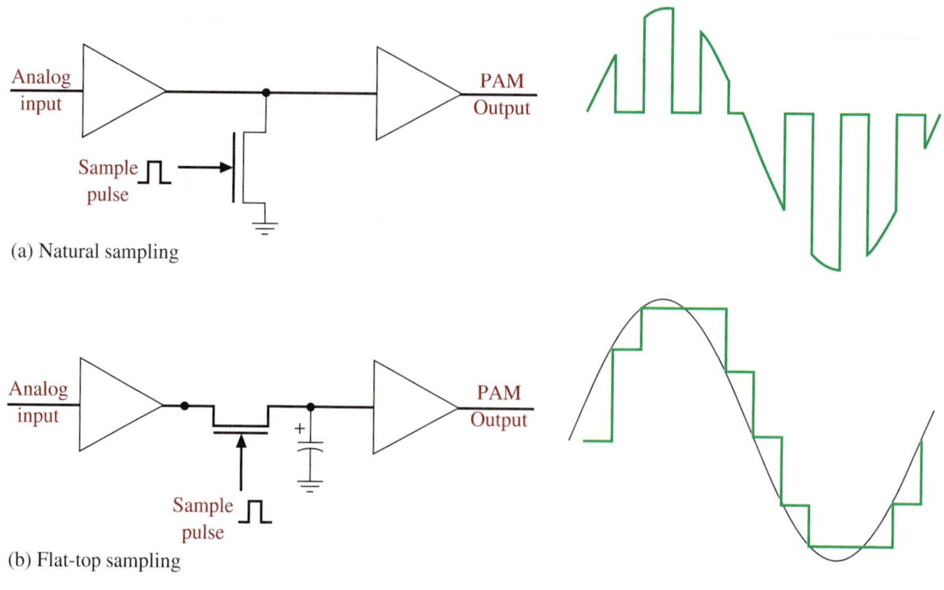

FIGURE 8-10 (a) Natural sampling; (b) flat-top sampling.

Flat-Top Sampling
sampling in which the signal voltage is held constant during samples, creating a staircase that tracks the changing input waveform

Probably the most popular type of sampling used in PCM systems is called **flat-top sampling.** In flat-top sampling, the sample signal voltage is held constant between samples. The method of sampling creates a staircase that tracks the changing input signal. This method is popular because it provides a constant voltage during a window of time for the binary conversion of the input signal to be completed. An example of flat-top sampling is shown in Fig. 8-10(b). Note that this is the same type of waveform as shown in Fig. 8-9(b). With flat-top sampling, the analog switch connects the input signal to the hold capacitor.

The Sample Frequency

Nyquist Rate
states that the sample frequency must be at least twice the highest input frequency

One of the most critical specifications in a PCM system is the selection of the sample frequency. The sample frequency is governed by the **Nyquist rate.** The Nyquist rate states that the sample frequency (f_s) must be at least twice the highest input frequency (f_a).

$$f_s \geq 2f_a \tag{8-1}$$

Sampling a signal gives many of the same properties that a "mixer" circuit in RF communications possesses. The mathematical relationship for a mixer circuit and a sampling circuit is expressed by the trigonometric identity

$$\sin A \times \sin B = 0.5 \cos (A - B) - 0.5 \cos (A + B) \tag{8-2}$$

From Eq. (8-2) it is evident that if the A frequency (f_s) is not twice the B frequency (f_b), then the $A - B$ ($f_s - f_a$) term will produce a signal whose frequency is less than B's (f_a). This created signal will appear within the original frequency bandwidth. Figure 8-11 graphically depicts the relationship of the sample frequency to the input frequency.

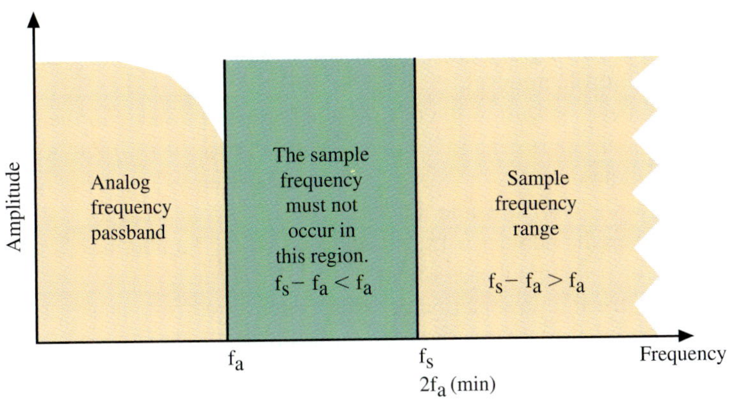

FIGURE 8-11 The sample frequency and input frequency relationship.

The phenomenon associated with the generation of an erroneously created signal in the sampling process is called **aliasing,** or **foldover distortion.** These error signals can be minimized by incorporating an **antialiasing filter** (i.e., a low-pass filter) on the input to the S/H circuit. The antialiasing filter bandlimits the input frequencies so that foldover distortion, or aliasing, is eliminated or minimized.

Aliasing, or Foldover Distortion
the phenomenon associated with the generation of error signals in the sampling process

For example, a voice channel on a telephone system is band-limited to a maximum of 4 kHz. The sample rate for the telephone system is 8 kHz, twice the highest input frequency. The input frequency is band-limited by either an active or a passive low-pass filter circuit. Therefore, the difference component ($A - B$) term will create a signal above the band-limited range of 4 kHz. Keep in mind that the input signal will seldom be a pure sinusoid so there will be some harmonic content to the signal (refer back to Chapter 1, FFT). The harmonics, if not filtered, can lead to aliasing, or foldover distortion, problems.

Antialiasing Filter
a filter that bandlimits the input frequencies to $\frac{1}{2}$ the sampling frequency so that foldover distortion, or aliasing, is prevented

Example 8-1

A CD audio laser-disk system has a frequency bandwidth of 20 Hz to 20 kHz. What is the minimum sample rate required to satisfy the Nyquist sampling rate?

Solution

$$f_s \geq 2f_a \qquad \textbf{(8-1)}$$

$$f_s \geq 2 \times 20 \text{ kHz}$$

$$f_s \geq 40 \text{ kHz}$$

Note: The sample rate for CD audio players is 44.1 kHz.

Quantization

Once an analog signal has been properly sampled, the process of converting the sampled signal to a binary value can begin. In PCM systems the sampled signal is segmented into different voltage levels, with each level corresponding to a different

Quantization
process of segmenting a sampled signal in a PCM system into different voltage levels, each level corresponding to a different binary number

Quantile
a quantization level step-size

Quantile Interval
another name for quantile

Quantization Levels
another name for quantile

Quantizing Error
an error resulting from the quantization process

Quantizing Noise
another name for quantizing error

binary number. This process is called **quantization.** The quantization levels also determine the resolution of the digitizing system. Each quantization level step-size is called a **quantile,** or **quantile interval.**

Analog signals are quantized to the closest binary value provided in the digitizing system. This is an approximation process. For example, if our numbering system is the set of whole numbers 1, 2, 3, . . . , and the number 1.4 must be converted (rounded off) to the closest whole number, then 1.4 is translated to 1. If the input number is 1.6, then the number is translated to a 2. If the number is 1.5, then we have the same error if the number is rounded off to a 1 or a 2.

In PCM, the electrical representation of voice is converted from analog form to digital form. This process of encoding is shown in Fig. 8-12. There are a set of amplitude levels and sampling times. The amplitude levels are termed **quantization levels,** and 12 such levels are shown. At each sampling interval, the analog amplitude is quantized into the closest available quantization level, and the analog-to-digital converter (ADC) puts out a series of pulses representing that level in the binary code.

For example, at time t_2 in Fig. 8-12, voice waveform S_1 is closest to level q_8, and thus the coded output at that time is the binary code 1000, which represents 8 in binary code. Note that the quantizing process resulted in an error, which is termed the **quantizing error,** or **quantizing noise.** The maximum

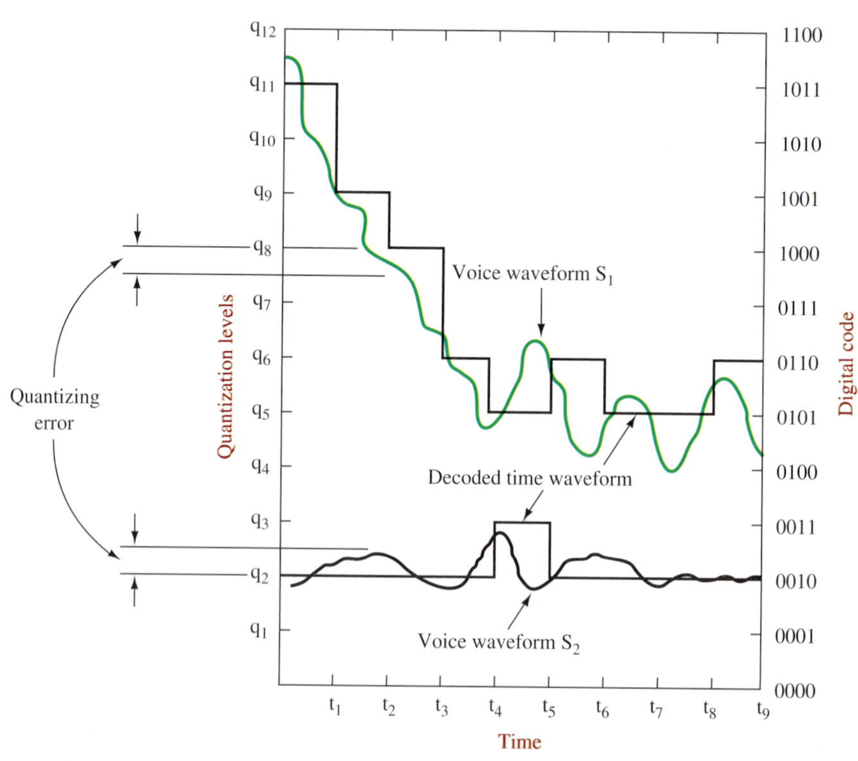

FIGURE 8-12 PCM encoding. (From the November 1972 issue of the *Electronic Engineer,* with the permission of the publisher.)

voltage of the quantization error is one-half the voltage of the minimum step-size $V_{LSB/2}$. Voice waveform S_2 provides a 0010 code at time t_2, and its quantizing error is also shown in Fig. 8-12. The amount of this error can be minimized by increasing the number of quantizing levels, which of course lessens the space between each one. The 4-bit code shown in Fig. 8-12 allows for a maximum of 16 levels because $2^4 = 16$. The use of a higher-bit code decreases the error at the expense of transmission time and/or bandwidth because, for example, a 5-bit code (32 levels) means transmitting 5 high or low pulses instead of 4 for each sampled point. The sampling rate is also critical and must be greater than twice the highest significant frequency, as previously described. It should be noted that the sampling rate in Fig. 8-12 is lower than the highest-frequency component of the information. This is not a practical situation but was done for illustrative purposes.

While a 4- or 5-bit code may be adequate for voice transmission, it is not adequate for transmission of television signals. Figure 8-13 provides an example of TV pictures for 5-bit and 8-bit (256 levels) PCM transmissions, each with 10-MHz sampling rates. In the first (5-bit) picture, contouring in the forehead and cheek areas is very pronounced. The 8-bit resolution results in an excellent-fidelity TV signal that is not discernibly different from a standard continuous modulation transmission.

Notice in Fig. 8-14 that at the sample intervals, the closest quantization level is selected for representing the sine-wave signal. The resulting waveform has poor resolution with respect to the sine-wave input. *Resolution* with respect to a digitizing system refers to the accuracy of the digitizing system in representing a sampled signal. It is the smallest analog voltage change that can be distinguished by the converter. For example, the analog input to our PCM system has a minimum voltage of 0.0 V and a maximum of 1.0 V. Then

$$q = \frac{V_{\max}}{2^n} = \frac{V_{FS}}{2^n}$$

where q = the resolution
n = number of bits
V_{FS} = full-scale voltage

If a 2-bit system is used for quantizing a signal, then 2^2, or 4, quantized levels are used. Referring to Fig. 8-14 we see that the quantized levels (quantile intervals) are each 0.25 V in magnitude. Typically it is stated that this system has 2-bit resolution. This follows from the equation just presented.

To increase the resolution of a digitizing system requires that the number of quantization levels be increased. To increase the number of quantization levels requires that the number of binary bits representing each voltage level be increased. If the resolution of the example in Fig. 8-14 is increased to 3 bits, then the input signal will be converted to 1 of 8 possible values. The 3-bit example with improved resolution is shown in Fig. 8-15.

Another way of improving the accuracy of the quantized signal is to increase the sample rate. Figure 8-16 shows the sample rate doubled but still using a 3-bit system. The resultant signal shown in Fig. 8-16 is dramatically improved compared to the quantized waveform shown in Fig. 8-15 by this change in sampling rate.

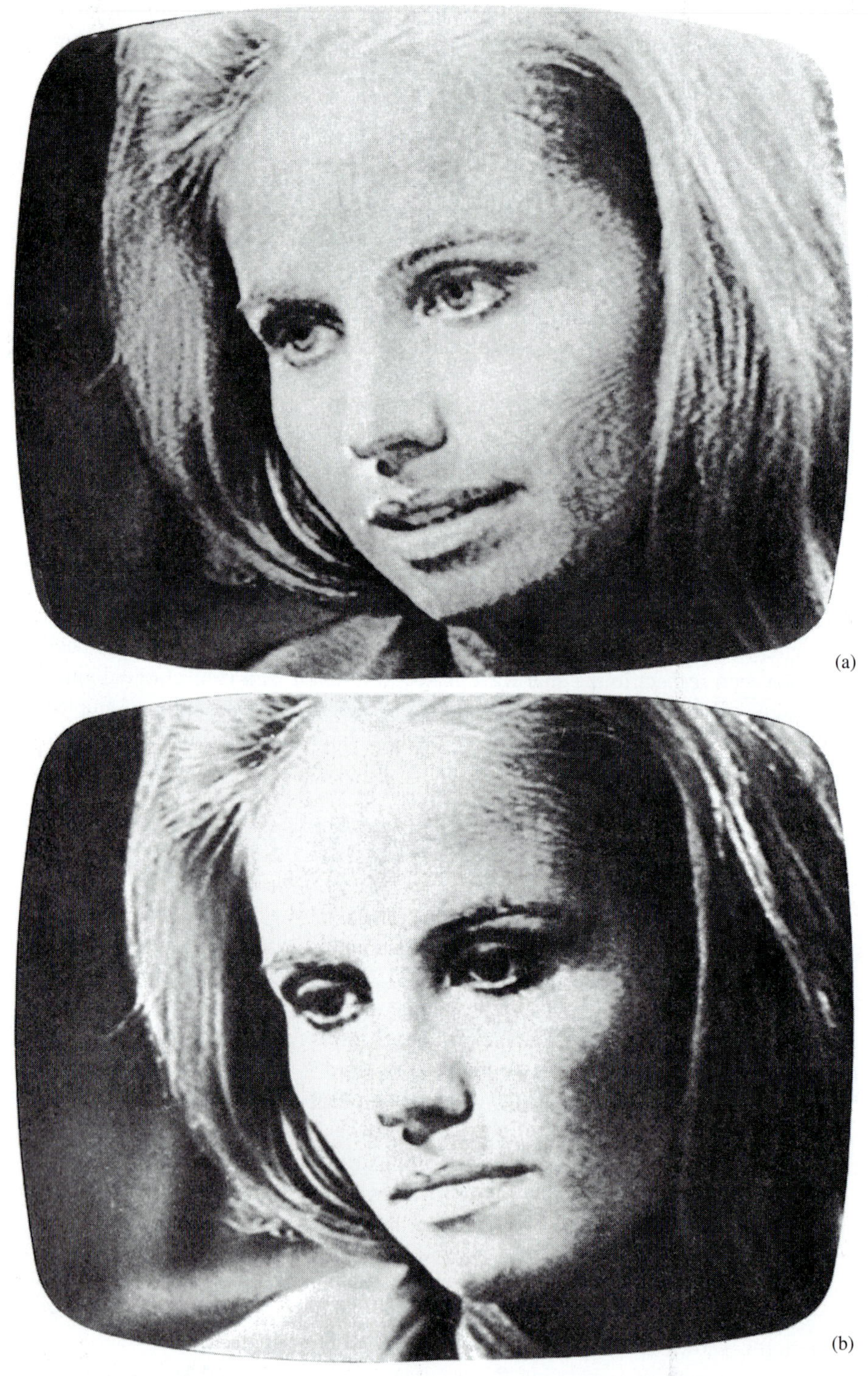

FIGURE 8-13 PCM TV transmission: (a) 5-bit resolution; (b) 8-bit resolution.

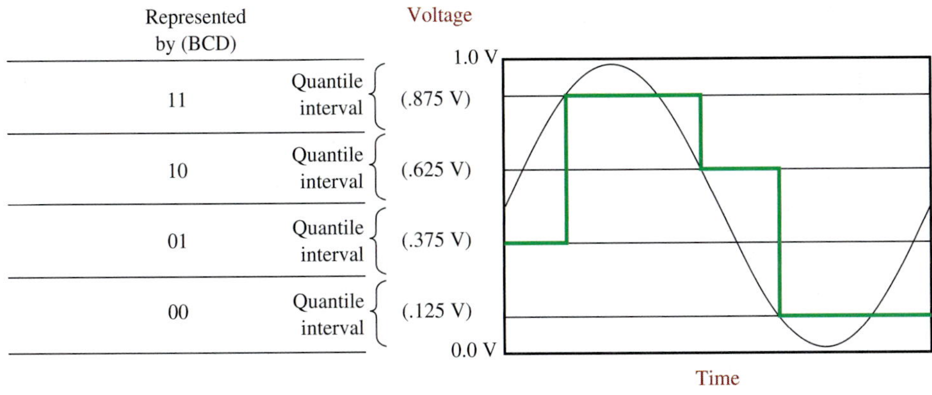

FIGURE 8-14 Voltage levels for a quantized signal.

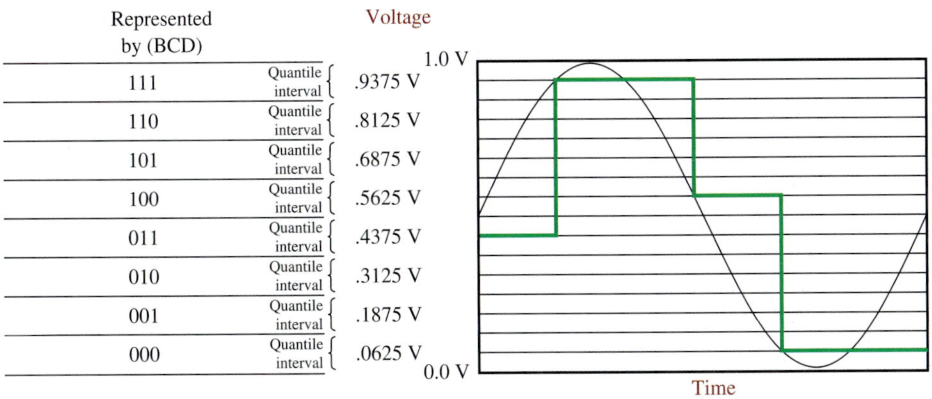

FIGURE 8-15 An example of 3-bit quantization.

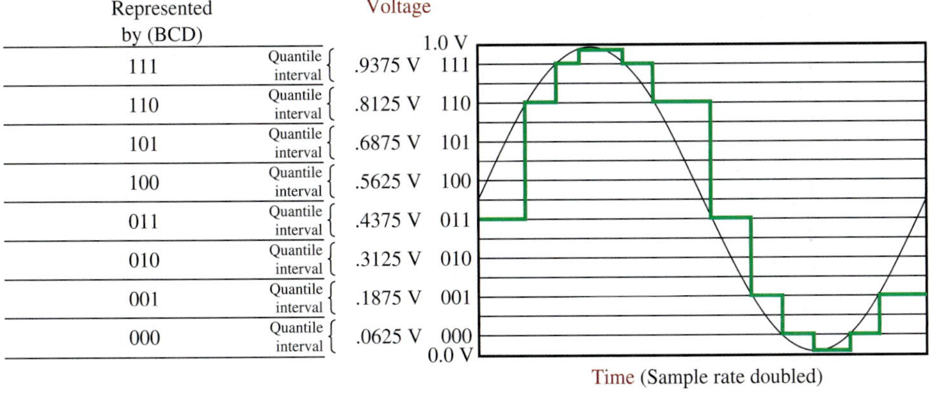

FIGURE 8-16 An example of 3-bit quantization with increased sample rate.

Dynamic Range and Signal-to-Noise Calculations

Dynamic Range
in a PCM system, the ratio of the maximum input or output voltage level to the smallest voltage level that can be quantized and/or reproduced by the converters

Dynamic range (DR) for a PCM system is defined as the ratio of the maximum input or output voltage level to the smallest voltage level that can be quantized and/or reproduced by the converters. It is the same as the converter's parameters:

$$\frac{V_{FS}}{q}, \frac{\text{full-scale voltage}}{\text{resolution}}$$

This value is expressed as:

$$\text{DR} = \frac{V_{\max}}{V_{\min}} = 2^n \qquad \textbf{(8-3)}$$

Dynamic range is typically expressed in terms of decibels. For a binary system, each bit can have two logic levels, either a logical low or logical high. Therefore the dynamic range for a single-bit binary system can be expressed logarithmically, in terms of dB, by the expression:

$$\text{DR}_{\text{dB}} = 20 \log \frac{V_{\max}}{V_{\min}}$$

$$\text{DR}_{\text{dB}} = 20 \log 2^n \qquad \textbf{(8-4)}$$

where $n =$ number of bits in the digital word.

The dynamic range (DR) for a binary system is expressed as 6.02 dB/bit or $6.02 \times n$, where n represents the number of quantizing bits. This value comes from $20 \log 2 = 6.02$ dB, where the 2 represents the two possible states of one binary bit. To calculate the dynamic range for a multiple-bit system, simply multiply the number of quantizing bits (n) times 6.02 dB per bit. For example, an 8-bit system will have a dynamic range (expressed in dB) of:

$$(8 \text{ bits})(6.02 \text{ dB/bit}) = 48.16 \text{ dB}$$

The *signal-to-noise ratio (S/N)* for a digitizing system is written as:

$$S/N = [1.76 + 6.02n] \qquad \textbf{(8-5)}$$

where $n =$ the number of bits used for quantizing the signal
$S/N =$ the signal-to-noise ratio in dB

This relationship is based on the ratio of the rms quantity of the maximum input signal to the rms quantization noise.

Another way of measuring digitized or quantized signals is the *signal-to-quantization-noise level* $(S/N)_q$. This relationship is expressed mathematically, in dB, as

$$(S/N)_{q(\text{dB})} = 10 \log 3L^2 \qquad \textbf{(8-6)}$$

where $L =$ number of quantization levels
$L = 2^n$, where $n =$ the number of bits used for sampling

Example 8-2 shows how Eqs. (8-4), (8-5), and (8-6) can be used to obtain the number of quantizing bits required to satisfy a specified dynamic range and determine the signal-to-noise ratio for a digitizing system.

Example 8-2

A digitizing system specifies 55 dB of dynamic range. How many bits are required to satisfy the dynamic range specification? What is the signal-to-noise ratio for the system? What is $(S/N)_q$ for the system?

Solution

First solve for the number of bits required to satisfy a dynamic range (DR) of 55 dB.

$$\text{DR} = 6.02 \text{ dB/bit } (n) \tag{8-4}$$

$$55 \text{ dB} = 6.02 \text{ dB/bit } (n)$$

$$n = \frac{55}{6.02} = 9.136$$

Therefore, 10 bits are required to achieve 55 dB of dynamic range. Nine bits will provide a dynamic range of only 54.18 dB. The tenth bit is required to meet the 55 dB of required dynamic range. Ten bits provides a dynamic range of 60.2 dB. To determine the signal-to-noise (S/N) ratio for the digitizing system:

$$S/N = [1.76 + 6.02n] \text{ dB} \tag{8-5}$$

$$S/N = [1.76 + (6.02)10] \text{ dB}$$

$$S/N = 61.96 \text{ dB}$$

Therefore, the system will have a signal-to-noise ratio of 61.96 dB. For this example, 10 sample bits are required; therefore, $L = 2^{10} = 1024$ and

$$(S/N)_{q\text{ (dB)}} = 10 \log 3L^2 = 10 \log 3(1024)^2 = 64.97 \text{ dB} \tag{8-6}$$

For Ex. 8-2, the dynamic range is 60.2 dB, $S/N = 61.96$ dB, and $(S/N)_q = 64.97$ dB. The differences result from the assumptions made about the sampled signal and the quantization process. For practical purposes, the 60.2-dB value is a good estimate, and it is easy to remember that each quantizing bit provides about 6 dB of dynamic range.

Companding

Up to this point our discussion and analysis of PCM systems have been developed around **uniform** or **linear quantization levels.** In linear (uniform) quantization systems each quantile interval is the same step-size. An alternative to linear PCM systems is **nonlinear,** or **nonuniform, coding** in which each quantile interval step-size may vary in magnitude.

Uniform Quantization Level
each quantile interval is the same step-size

Linear Quantization Level
another name for uniform quantization level

Nonlinear Coding
each quantile interval step-size may vary in magnitude

Nonuniform Coding
another name for nonlinear coding

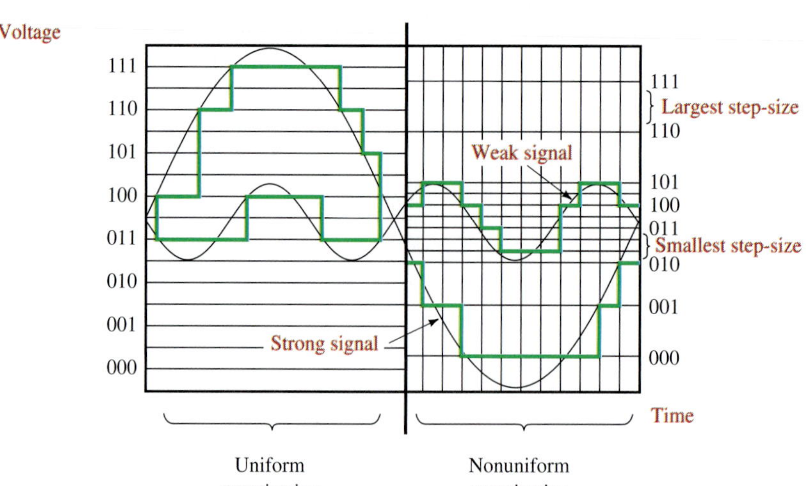

FIGURE 8-17 Uniform (left) and nonuniform (right) quantization.

It is quite possible for the amplitude of an analog signal to vary throughout its full range. In fact, this is expected for systems exhibiting a wide dynamic range. The signal will change from a very strong signal (maximum amplitude) to a weak signal (minimum amplitude—V_{1sb} for quantized systems). For the system to exhibit good signal-to-noise characteristics, either the input amplitude must be increased with reference to the quantizing error or the quantizing error must be reduced.

The justification for the use of a nonuniform quantization system will be presented, but let's discuss some general considerations before proceeding. How can the quantization error be modified in a nonuniform PCM system so that an improved *S/N* results? The answer can be obtained by first examining a waveform that has uniform quantile intervals as shown in Fig. 8-17. Notice that poor resolution is present in the weak signal regions, yet the strong signal regions exhibit a reasonable facsimile of the original signal. Figure 8-17 also shows how the quantile intervals can be changed to provide smaller step-sizes within the area of the weak signal. This will result in an improved *S/N* ratio for the weak signal.

What is the price paid for incorporating a change such as this in a PCM system? The answer is that the large amplitude signals will have a slightly degraded *S/N*, but this is an acceptable situation if the goal is improving the weak signal's *S/N*.

Idle Channel Noise

Idle Channel Noise small-amplitude signal that exists due to the noise in the system

Digital communications systems will typically have some noise in the electronics and the transmission systems. This is true of even the most sophisticated technologies currently available. One of the noise signals present is called **idle channel noise.** Simply put, this is a noise source of small amplitude that exists in the channel independent of the analog input signal and that can be quantized by the A/D converter. One method of eliminating the noise source in the quantization process is to incorporate a quantization procedure that does not recognize the idle channel noise as large enough to be quantized. This usually involves increasing the quantile interval step-size in the noise region to a large-enough value so that the noise signal can no longer be quantized.

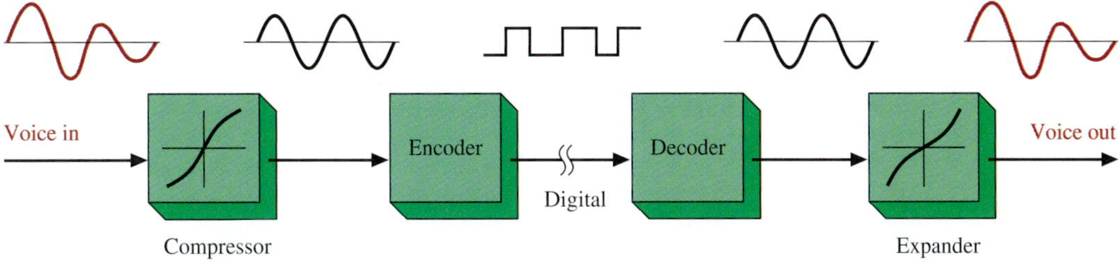

FIGURE 8-18 Companding process.

Amplitude Companding

Amplitude Companding process of volume compression before transmission and volume expansion after detection

The other form of companding is called **amplitude companding.** Amplitude companding involves the process of volume compression before transmission and expansion after detection. This is illustrated in Fig. 8-18. Notice how the weak portion of the input is made nearly equal to the strong portion by the compressor but restored to the proper level by the expander. Companding is essential to quality transmission using PCM and the delta modulation technique introduced in Chapter 9.

The use of time-division-multiplexed (TDM) PCM transmission for telephone transmissions has proven its ability to cram more messages into short-haul cables than frequency-division-multiplexed (FDM) analog transmission. This concept is explored fully in Chapters 9 and 11. The TDM PCM methods were started by Bell Telephone in 1962 and are now the only methods used in new designs except for delta modulation schemes (see Chapter 9). Once digitized, these voice signals can be electronically switched and restored without degradation. The standard PCM system in U.S. and Japanese telephony uses μ-*law* companding. In Europe, the CCITT* specifies A-law companding. The μ-*law* companded signal is predicted by

$$V_{\text{out}} = \frac{V_{\max} \times \ln(1 + \mu\, V_{\text{in}}/V_{\max})}{\ln(1 + \mu)} \qquad \textbf{(8-7)}$$

The μ parameter defines the amount of compression. For example, $\mu = 0$ indicates no compression and the voltage gain curve is linear. Higher values of μ yield nonlinear curves. The early Bell systems have $\mu = 100$ and a 7-bit PCM code. An example of μ-law companding is provided in Fig. 8-19. This figure shows the encoder transfer characteristic.

Digital-to-Analog Converters

As we saw in Fig. 8-7, the analog-to-digital converter (ADC) is used to convert the information signal to a digital format. This process is known as *digitizing*. A block diagram of a PCM system (transmitter and receiver) is shown in Fig. 8-20. The ADC is shown in the transmitting section and the DAC in the receiver section.

The function of the DAC is to convert a digital (binary) bit stream to an analog signal. The DAC accepts a parallel bit stream and converts it to its analog equivalent. Figure 8-21 illustrates this point.

*Consultative Committee on International Telephone & Telegraph.

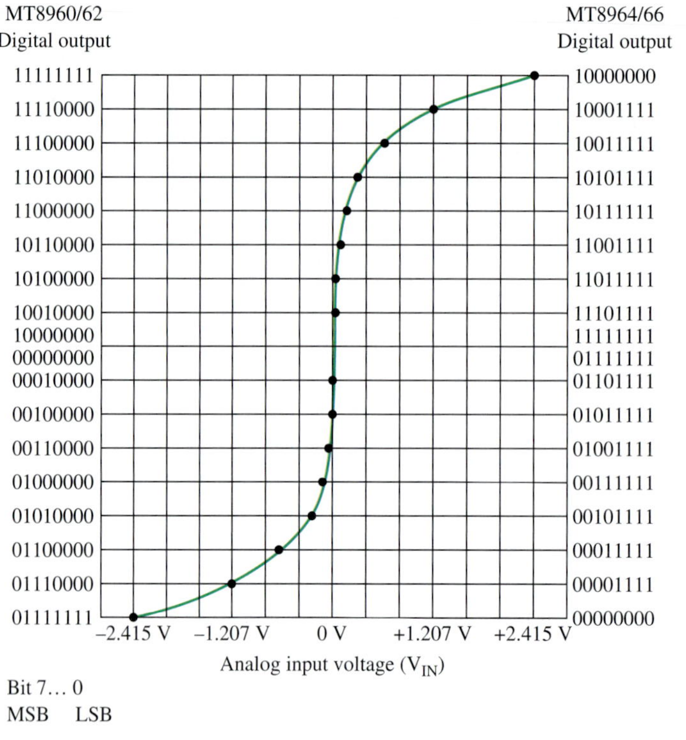

FIGURE 8-19 μ-law encoder transfer characteristic.

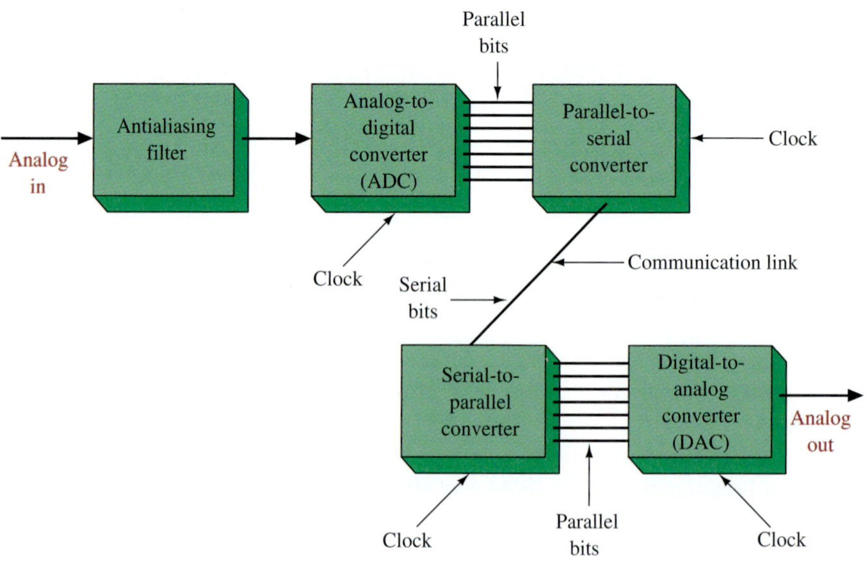

FIGURE 8-20 PCM communication system.

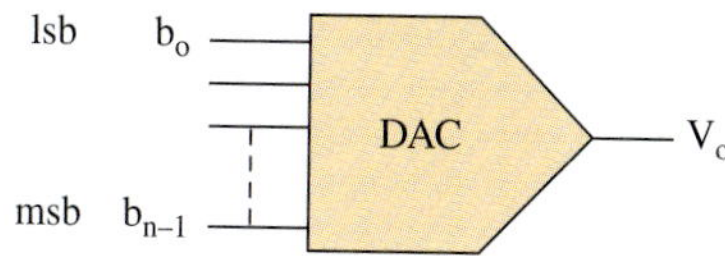

FIGURE 8-21 DAC input/output.

The least significant bit (lsb) is called b_0 and the most significant bit (msb) is called b_{n-1}. The resolution of a DAC is the smallest change in the output that can be caused by a change of the input. This is the step-size of the converter. It is determined by the least significant bit (lsb). The full-scale voltage (V_{FS}) is the largest voltage the converter can produce. In a digital-to-analog converter, the step-size or resolution q is given as

$$q = \frac{V_{FS}}{2^n} \tag{8-8}$$

where n is the number of binary digits.

A binary-weighted resistor DAC is shown in Fig. 8-22. It is one of the more simple DACs to analyze. For simplicity we have used four bits of data. Note that the value of the resistor is divided by the binary weight for that bit position. For example, in bit position 2^0, which has a value of 1, the entire value of R is used. This is also the lsb. Because this is a summing amp, the voltages are added to give the output voltage.

The output voltage is given as

$$V_o = -V_{Ref}\left(\frac{b_1 R_f}{R/2^0} + \frac{b_2 R_f}{R/2^1} + \cdots + \frac{b_{n-1} R_f}{R/2^{n-1}}\right) \tag{8-9}$$

An R-$2R$ ladder-type DAC is shown in Fig. 8-23. This is one of the more popular DACs and is widely used. Note that each switch is activated by a parallel data stream and is summed by the amp. We show a 4-bit R-$2R$ circuit for simplicity.

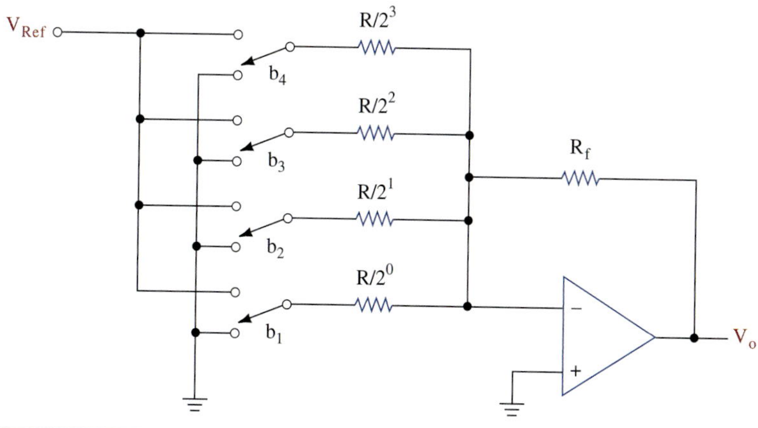

FIGURE 8-22 Binary-weighted resistor DAC.

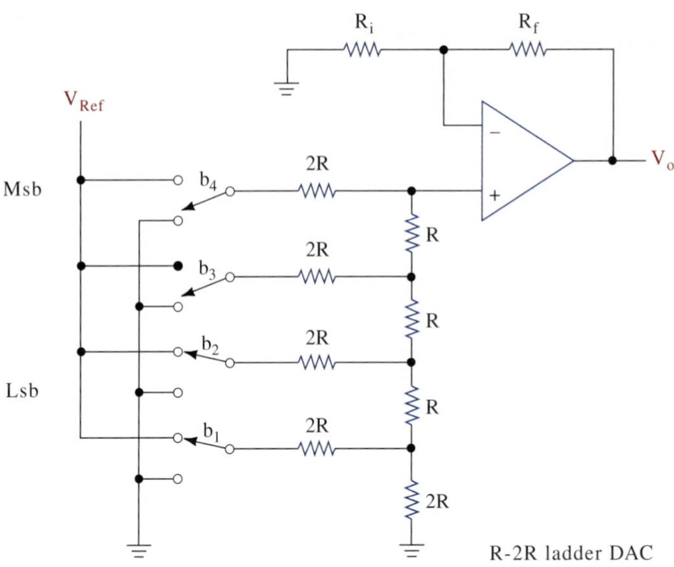

FIGURE 8-23 R-2R ladder DAC.

The output voltage is given as

$$V_o = V_{Ref}\left(1 + \frac{R_f}{R}\right)\left(\frac{b_n}{2^1} + \frac{b_{n-1}}{2^2} + \cdots + \frac{b_1}{2^n}\right) \quad \textbf{(8-10)}$$

where b is either 0 or 1, depending on the digital word being decoded.

Example 8-3

Assume the circuit in Fig. 8-22 has the following values: $R = 100\ k\Omega$ and $R_f = 10\ k\Omega$. Assume $V_{Ref} = -10\ V$. Determine the step-size, or resolution, and the output voltage if all switches are closed.

Solution

The step-size is determined by leaving all switches open and closing the lsb. Thus,

$$V_o = -(-10\text{ V})\,(R_f/R) = 10\text{ V}\left(\frac{10\text{ k}\Omega}{100\text{ k}\Omega}\right) = 1.0$$

The resolution is 1.0. If all switches are closed, a logic 1 is input. So, using Eq. (8-9), we have

$$\begin{aligned} V_o &= -(-10\text{ V})\left(\frac{10\text{ k}\Omega}{100\text{ k}\Omega} + \frac{10\text{ k}\Omega}{50\text{ k}\Omega} + \frac{10\text{ k}\Omega}{25\text{ k}\Omega} + \frac{10\text{ k}\Omega}{12.5\text{ k}\Omega}\right) \\ &= (10\text{ V})(0.1 + 0.2 + 0.4 + 0.8) \\ &= (10\text{ V})(1.5) = 15\text{ V} \end{aligned}$$

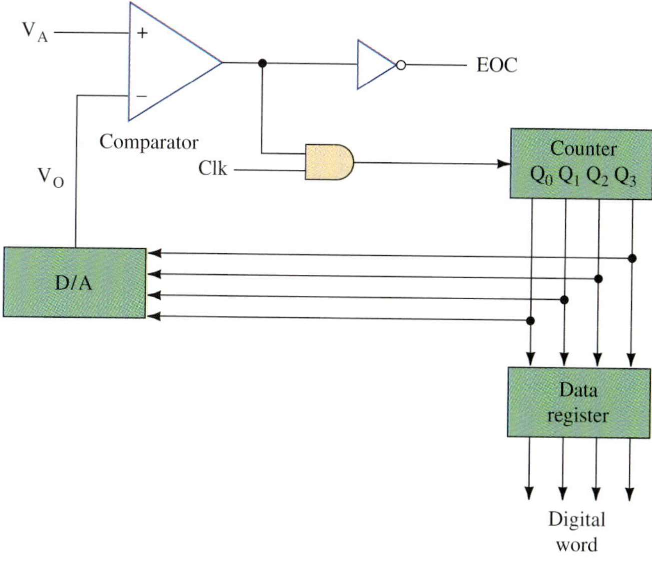

FIGURE 8-24 4-bit ramp analog-to-digital converter.

Analog-to-Digital Converters

Figure 8-24 shows a simple 4-bit ramp ADC. The analog information goes into the comparator. The output is ANDed with the clock to cause the counter to begin counting. When the counter's digital output reaches the analog equivalent, the AND gate is low and the counter stops counting. The end of conversion (EOC) signal is used to latch data into the registers and reset the counter. Some delay must be used before resetting the counter, otherwise the data would not be latched into the register. This time is longer than the time it takes the register to latch the data.

Other types of analog-to-digital converters are the successive-approximation ADC and the dual-slope ADC. The successive approximation ADC is more widely used. This is illustrated by its use in the coder-decoder circuits for telephone operations.

Codec

The A/D circuitry in PCM systems is often referred to as the encoder. The D/A circuitry at the receiver is correspondingly termed the decoder. These functions are often combined in a single LSI chip termed a **codec** (coder-decoder). The block diagram for a typical codec is provided in Fig. 8-25. These devices are widely used in the telephone industry to allow voice transmission to be accomplished in digital form. Basic telephone operation will be explained in Chapter 11.

Codec
a single LSI chip containing both the ADC and DAC circuitry

Figure 8-25 shows the functional block diagram of the MT8960–67. These devices provide the conversion interface between the voiceband analog signals of a telephone subscriber loop and the digital signals required in a digital pulse code

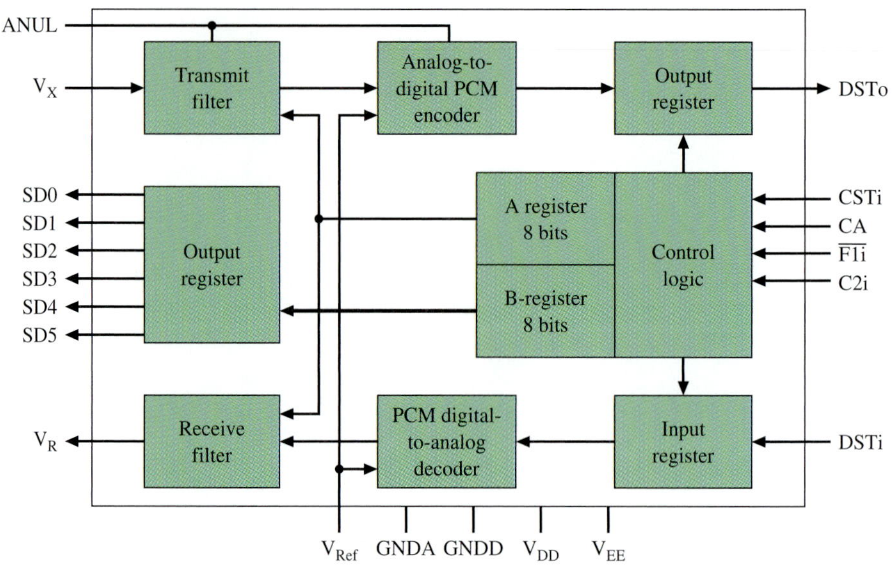

FIGURE 8-25 Codec block diagram.

modulation (PCM) switching system. Analog (voiceband) signals in the transmit path enter the chip at V_X and are sampled at 8 kHz. The samples are quantized and assigned 8-bit digital values defined by logarithmic PCM encoding laws. Analog signals in the receive path leave the chip at V_R after reconstruction from digital 8-bit words.

Separate switched capacitor filter sections are used for bandlimiting prior to digital encoding in the transmit path and after digital decoding in the receive path. Eight-bit PCM encoded digital data enter and leave the chip serially on DSTi and DSTo pins, respectively.

 8-4 Digital Signal Encoding Formats

Transmission of digital data using a binary format (+5 V—hi, 0.0 V—low) is usually limited to short distances such as a computer to a printer interface. Typically the binary data are transmitted serially over a single wire, fiber, or RF link. This requires that the binary data be encoded so that the highs and lows can be detected easily. The transmission systems are typically serial, either asynchronous or synchronous. This requires the addition of clocking information in the data for synchronous systems.

The digital signal encoding formats presented in this section are the most commonly used PCM waveforms. (Note that we are identifying the encoding format as a pulse-code-modulated waveform.) The waveforms are classified as one of four encoding groups:

1. NRZ—nonreturn-to-zero
2. RZ—return-to-zero
3. Phase-encoded and delay modulation
4. Multilevel binary

The encoding formats described are of the *baseband* type. A baseband signal is one that is not modulated. These waveforms are still in a binary or pseudo-binary format, so therefore they are classified as baseband.

The NRZ Group

The NRZ group is a popular method for encoding binary data. NRZ codes are also one of the easiest to implement. NRZ codes get their name from the fact that the data signal does not return to zero during an interval. In other words, NRZ codes remain constant during an interval. Because of this feature, the code has a dc component in the waveform. For example, a data stream containing a chain of 1s or 0s will appear as a dc signal at the receive side. Look at the waveform for the NRZ-L code that is shown in Fig. 8-26. Notice that the code remains constant for several clock cycles for a series of zeros or ones.

Another important factor to consider is that the NRZ codes do not contain any self-synchronizing capability. NRZ codes will require the use of start bits or some kind of synchronizing data pattern to keep the transmitted binary data synchronized. There are three coding schemes in the NRZ group: NRZ-L (level), NRZ-M (mark),

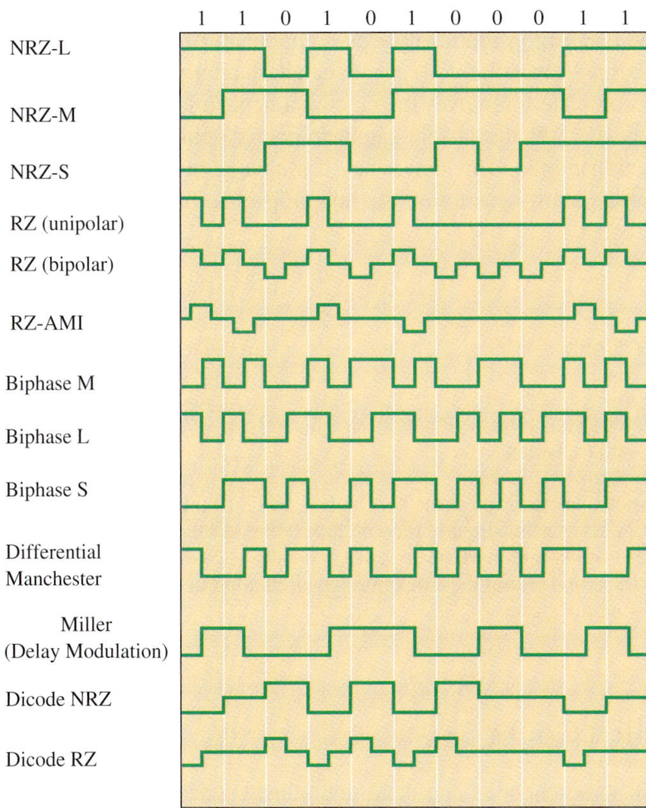

FIGURE 8-26 Digital signal encoding formats.

and NRZ-S (space). The waveforms for these formats are provided in Fig. 8-26. The NRZ code descriptions are provided in Table 8-2.

Table 8-2 NRZ Codes

NRZ-L	(nonreturn-to-zero—level)
	1 (hi)—high level
	0 (low)—low level
NRZ-M	(nonreturn-to-zero—mark)
	1 (hi)—transition at the beginning of the interval
	0 (low)—no transition
NRZ-S	(nonreturn-to-zero—space)
	1 (hi)—no transition
	0 (low)—transition at the beginning of the interval

The RZ Codes

The *RZ-unipolar* code shown in Fig. 8-26 has the same limitations and disadvantages as the NRZ group. A dc level appears on the data stream for a series of 1s or 0s. Synchronizing capabilities are also limited. These deficiencies are overcome by modifications in the coding scheme, which include using bipolar signals and alternating pulses. The *RZ-bipolar* code provides a transition at each clock cycle, and a bipolar pulse technique is used to minimize the dc component. Another RZ code is *RZ-AMI*. The alternate-mark-inversion code provides alternating pulses for the 1s. This technique almost completely removes the dc component from the data stream, but since a data value of 0 is 0 V, the system can have poor synchronizing capabilities if a series of 0s is transmitted. This deficiency can also be overcome by transmission of the appropriate start, synchronizing, and stop bits. Table 8-3 provides descriptions of the RZ codes.

Table 8-3 RZ Codes

RZ (unipolar) (return-to-zero)	
	1 (hi)—transition at the beginning of the interval
	0 (low)—no transition
RZ (bipolar) (return-to-zero)	
	1 (hi)—positive transition in the first half of the clock interval
	0 (low)—negative transition in the first half of the clock interval
RZ-AMI (return-to-zero—alternate-mark inversion)	
	1 (hi)—transition within the clock interval alternating in direction
	0 (low)—no transition

Biphase and Miller Codes

The popular names for phase-encoded and delay-modulated codes are *biphase* and *Miller* codes. Biphase codes are popular for use in optical systems, satellite telemetry links, and magnetic recording systems. *Biphase M* is used for encoding Society of Motion Picture and Television Engineers (SMPTE) time-code data for recording on videotapes. The biphase code is an excellent choice for this type of media because the code does not have a dc component to it. Another important benefit is that the code is *self-synchronizing*, or *self-clocking*. This feature allows the data stream speed to vary (tape shuttle in fast and slow search modes) while still providing the receiver with clocking information.

The *biphase L* code is commonly known as *Manchester coding*. This code is used on the *Ethernet* standard IEEE 802.3 for local area networks (LANs). Chapter 11 provides more detail on Ethernet and LANs. Figure 8-26 provides examples of these codes and Table 8-4 summarizes their characteristics.

Table 8-4 Phase-Encoded and Delay-Modulation (Miller) Codes

Biphase M (biphase-mark)
- 1 (hi)—transition in the middle of the clock interval
- 0 (low)—no transition in the middle of the clock interval
- *Note:* There is always a transition at the beginning of the clock interval.

Biphase L (biphase-level/manchester)
- 1 (hi)—transition from high-to-low in the middle of the clock interval
- 0 (low)—transition from low-to-high in the middle of the clock interval

Biphase S (biphase-space)
- 1 (hi)—no transition in the middle of the clock interval
- 0 (low)—transition in the middle of the clock interval
- *Note:* There is always a transition at the beginning of the clock interval.

Differential Manchester
- 1 (hi)—transition in the middle of the clock interval
- 0 (low)—transition at the beginning of the clock interval

Miller/delay modulation
- 1 (hi)—transition in the middle of the clock interval
- 0 (low)—no transition at the end of the clock interval unless followed by a zero

Multilevel Binary Codes

Codes that have more than two levels representing the data are called **multilevel binary** codes. In many cases the codes will have three levels. We have already examined two of these codes in the RZ group: RZ (bipolar) and RZ-AMI. Also included in this group are *dicode NRZ* and *dicode RZ*. Table 8-5 summarizes the multilevel binary codes.

Multilevel Binary
codes that have more than two levels representing the data

Table 8-5 Multilevel Binary Codes

Dicode NRZ
- One-to-zero and zero-to-one data transitions change the signal polarity. If the data remain constant, then a zero-level is output.

Dicode RZ
- One-to-zero and zero-to-one data transitions change the signal polarity in half-step voltage increments. If the data don't change, then a zero-voltage level is output.

8-5 Coding Principles

An ideal digital communications system is error-free. Unfortunately, a digital transmission will occasionally have an error. Modifications to the data can provide an increase in the receive system's capability to detect and possibly correct the error. Suppose that we transmit a zero (0) or a one (1). If either data value changes, then we have an error. How can the chance of an error be decreased? The process of decreasing an error depends on the transmission system being used and the digital encoding and modulation techniques employed. Even if the chances of detecting and correcting the error are increased, there is still some probability of receiving a data-bit error in the received message, but if the error can be corrected, then our message is still usable. Methods for improving the likelihood that the data-bit error can be both detected and corrected are presented next.

Our discussion on coding principles begins with a fundamental look at the basic coding techniques and establishing some rules for correcting errors. Let's first look at a very simple data system that has only two possible states. For this example let's assume that the data is just a single binary value with status indications for zero (0) and one (1). The system also requires that data-bit errors be corrected at the receiver without the need for retransmitting the data. If only a single zero (0) and a single one (1) are transmitted for each state, then the receiver will not be able to distinguish a correct bit from an error. This is the case because all the possible data values map directly to a valid value (0 or 1).

What can be done to the representative data value for each state so that an error can be detected? What if the number of binary bits representing each state are altered so that a logical zero is defined to be (00) and a logical one is defined to be (11)? Adding a data bit to each state effectively increases the distance between each code word to two. The distance can be visually shown by listing the possible binary states for the 2-bit words. The *distance* between each defined state is called the **Hamming distance,** also known as the **minimum distance, (D_{min}).** This relationship is shown in Fig. 8-27.

Hamming Distance the logical distance between defined states

Minimum Distance (D_{min}) another name for the Hamming distance

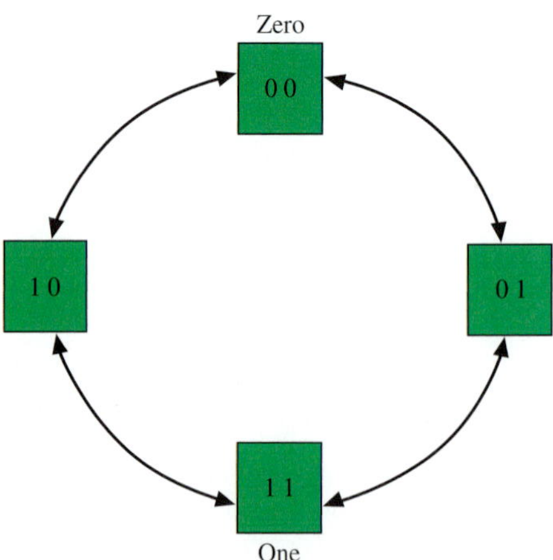

FIGURE 8-27 Coding for a zero (00) and a one (11) for a minimum distance, D_{min}, of 2.

If either 01 or 10 is received, then a data-bit error has occurred. The receive system has no trouble detecting a 1-bit error. As for correcting the bit error, each possibility, (01) and (10), can represent an error in a 0 or a 1. Therefore, increasing the number of binary bits representing each state to a minimum distance, D_{min}, of 2 improves the capability of the receive system to detect the error but does not improve the system's capability to correct the error.

Let's once again increase the number of binary bits for each state so that a logical zero is now defined to be 000 and a logical one is 111. The minimum distance between each code word is now 3. This relationship is depicted in Fig. 8-28.

If an error does occur in a data bit, can a coding system with $D_{min} = 3$ correct the error? This question can be answered by looking at an example. Let's assume that the data word 011 is received. Because this data word is neither a 111 nor a 000, it can be assumed that a data bit error has occurred. Look at Fig. 8-28 and see if you can determine which code, 000 or 111, that the received error code, 011, is *most likely* to belong to. The answer is the 111 code. The distance from 011 to 111 is 1 and the distance from 011 to 000 is 2. Therefore, it can be stated that, for a coding system with a minimum distance between each code of 3 or greater, the errors in the received code can be detected and corrected. The minimum distance between each code will determine the number of data-bit errors that can be corrected. The relationships for the detection and correction of data-bit errors to minimum distance is provided in Table 8-6.

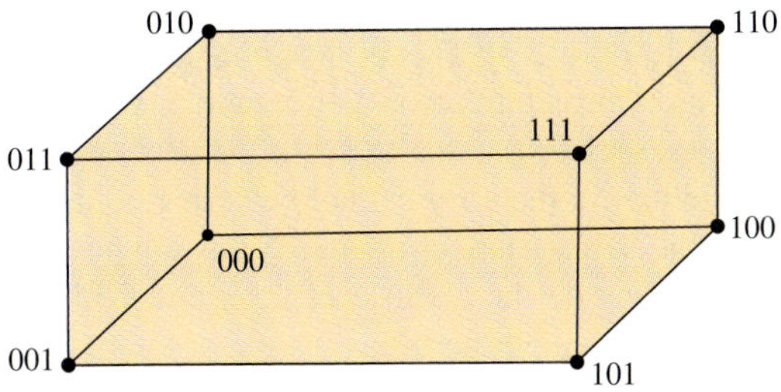

FIGURE 8-28 A code message for a zero (000) and a one (111) with a minimum distance, D_{min}, of 3.

Table 8-6 ERROR DETECTION AND CORRECTION BASED ON D_{MIN}

Error Detection

For a minimum distance, D_{min}, between code words, $(D_{min} - 1)$ errors can be detected.

Error Correction

If D_{min} is even, then $[(D_{min}/2) - 1]$ errors can be corrected.

If D_{min} is odd, then $\frac{1}{2}(D_{min} - 1)$ errors can be corrected.

Example 8-4

Determine the number of errors that can be detected and corrected for the distances
(a) 2.
(b) 3.
(c) 4.

Solution

(a) $D_{min} = 2$; the number of errors detected is $(D_{min} - 1) = 2 - 1 = 1$. D_{min} is *even*; therefore, the number of errors corrected equals

$$(D_{min}/2) - 1 = (\tfrac{2}{2}) - 1 = 0$$

(b) $D_{min} = 3$; the number of errors detected equals $(D_{min} - 1) = 3 - 1 = 2$. D_{min} is *odd*; therefore, the number of errors corrected equals

$$\tfrac{1}{2}(D_{min} - 1) = \tfrac{1}{2}(3 - 1) = 1$$

(c) $D_{min} = 4$; the number of errors detected equals $(D_{min} - 1) = 4 - 1 = 3$. D_{min} is *even*; therefore, the number of errors corrected equals

$$(D_{min}/2) - 1 = (\tfrac{4}{2}) - 1 = 1$$

What would have to be done if all of the eight possible states shown in Fig. 8-28 were to be transmitted and a minimum distance of 2 were to be required? To create an eight-level code with a minimum distance of 2 requires 4 bits [3 bits to provide eight levels (2^3) and 1 bit to provide a D_{min} of 2 (2^1)]. This is shown in Table 8-7.

A code with D_{min} equal to 2 cannot correct an error. For example, what if a (1 1 0 1) is received? Is the correct word (1 1 0 0) or is it (1 1 1 1)? Without sufficient overhead bits creating the required D_{min}, the error cannot be corrected. If the eight-level code is changed so that a 1-bit error can be corrected, 5 bits are now required to represent each word [3 bits to provide the eight levels (2^3) and 2 bits to provide a D_{min} of 3]. The 5-bit code is shown in Table 8-8.

To see how the code can correct, let's assume that the received code word is (0 0 1 1 1). To determine the distance of this code word to any of the possible receive codes, simply XOR the received code word with any of the eight possible valid codes. The XOR operation that yields the smallest result tells us which is the correct code. This process is demonstrated in Example 8-5.

Table 8-7 Eight-Level Code with a Distance of 2

0 0 0 0 (0 0 0)	1 1 0 0 (1 0 0)
0 0 0 1	1 1 0 1
0 0 1 1 (0 0 1)	1 1 1 1 (1 0 1)
0 0 1 0	1 1 1 0
0 1 1 0 (0 1 0)	1 0 1 0 (1 1 0)
0 1 1 1	1 0 1 1
0 1 0 1 (0 1 1)	1 0 0 1 (1 1 1)
0 1 0 0	1 0 0 0

Table 8-8 Eight-Level Code with Distance 3

0 0 0 0 0 (0 0 0)	0 1 0 1 1 (1 0 0)
0 0 0 0 1	0 1 0 1 0
0 0 0 1 1	0 1 0 0 0
0 0 0 1 0 (0 0 1)	1 1 0 0 0 (1 0 1)
0 0 1 1 0	1 1 0 0 1
0 0 1 1 1	1 1 0 1 1
0 0 1 0 1 (0 1 0)	1 1 1 1 1 (1 1 0)
0 0 1 0 0	1 1 1 0 1
0 1 1 0 0	1 1 1 1 0
0 1 1 0 1 (0 1 1)	1 0 1 1 0 (1 1 1)
0 1 1 1 1	
0 1 1 1 0	

Example 8-5

Determine the distance for a received code of (0 0 1 1 1), shown in Table 8-7, to all the possible correct codes by XORing the received code with all the possible correct codes. The result with the least number of bit-position differences is most likely the correct code. Then state which is most probably the correct code based on the minimum distance.

Solution

```
0 0 1 1 1            0 0 1 1 1            0 0 1 1 1
0 0 0 0 0            0 0 0 1 0            0 0 1 0 1
---------            ---------            ---------
0 0 1 1 1   (3)      0 0 1 0 1   (2)      0 0 0 1 0   (1)

0 0 1 1 1            0 0 1 1 1            0 0 1 1 1
0 1 1 0 1            0 1 0 1 1            1 1 0 0 0
---------            ---------            ---------
0 1 0 1 0   (2)      0 1 1 0 0   (2)      1 1 1 1 1   (5)

0 0 1 1 1            0 0 1 1 1
1 1 1 1 1            1 0 1 1 0
---------            ---------
1 1 0 0 0   (2)      1 0 0 0 1   (2)
```

Therefore, based on comparing the received code with all the possible correct codes, the most likely code is (0 0 1 0 1), which is the code for (0 1 0).

8-6 Code Error Detection and Correction

Codes and raw digital data are being transmitted with increasing volume every year. Unless some means of error detection is used, it is not possible to know when errors have occurred. These errors are caused by noise and transmission system impairments. In contrast, it is obvious when a voice transmission has been impaired by noise or equipment problems.

Redundancy is used as the means of error detection when codes and digital data are transmitted. A basic redundancy system transmits everything twice to make sure that exact correlation exists. Transmitting redundant data uses bandwidth, which slows the transfer of data. Fortunately, schemes have been developed that do not require such a high degree of redundancy and provide for a more efficient use of available bandwidth.

Parity

The most common method of error detection is the use of parity. A single bit called the *parity bit* is added to each code representation. If it makes the total number of 1s even, it is termed *even parity*, and an odd number of 1s is *odd parity*. For example, if the ASCII code for A is to be generated, the code is P1000001 and P is the parity bit. Odd parity would be 11000001 because the number of 1s is now 3 (see Fig. 8-29). The receiver checks for parity. If an even number of 1s occurs in a character grouping (digital word), an error is indicated and the receiver usually requests a retransmission. Unfortunately, if two errors (an even number) occur, parity systems will not indicate an error. In many systems a burst of noise causes two or more errors, so that more elaborate error-detection schemes may be required.

Many circuits are used as parity generators and/or checkers. A simple technique is shown in Fig. 8-30. If there are n bits per word, $n - 1$ exclusive-OR (XOR) gates are needed. The first two bits are applied to the first gate and the remaining individual bits to each subsequent gate. The output of this circuit will always be a 1 if there is an odd number of 1s and a 0 for an even number of 1s. If odd parity is desired, the output is fed through an inverter. When used as a parity checker, the word and parity bit is applied to the inputs. If no errors are detected, the output is low for even parity and high for odd parity.

When an error is detected there are two basic system alternatives:

1. An automatic request for retransmission (ARQ)

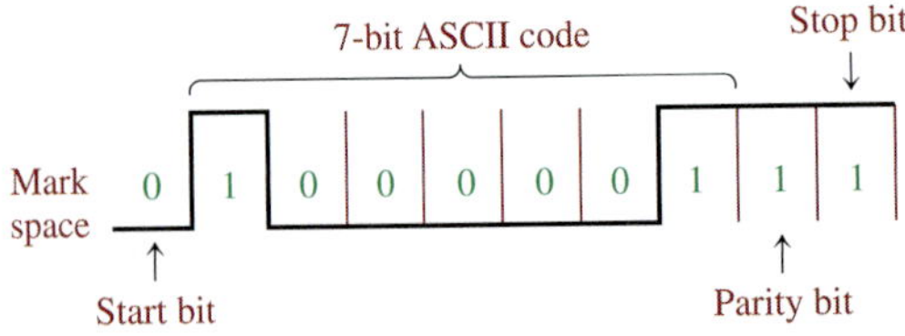

FIGURE 8-29 ASCII code for A with odd parity. Note that lsb b_1 is the first bit of the digital word transmitted.

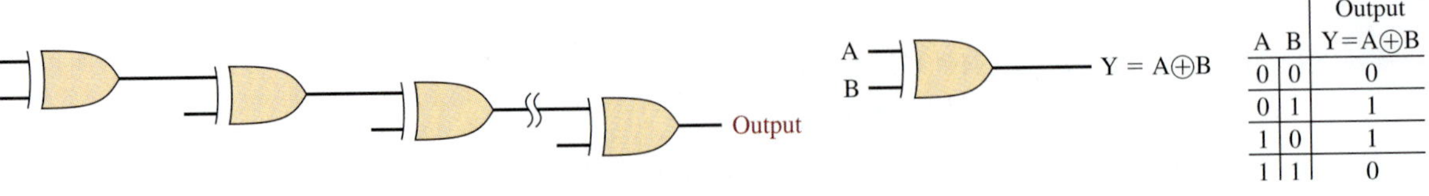

FIGURE 8-30 Serial parity generator and/or checker.

2. Display of an unused symbol for the character with a parity error (called **symbol substitution**)

Symbol Substitution displaying an unused symbol for the character with a parity error

Most systems use a request for retransmission. If a block of data is transmitted and no error is detected, a positive acknowledgment (ACK) is sent back to the transmitter. If a parity error is detected, a negative acknowledgment (NAK) is made and the transmitter repeats that block of data.

Block Check Character

A more sophisticated method of error detection than simple parity is needed in higher data-rate systems. At higher data speeds, telephone data transmission is usually synchronous and blocked. A block is a group of characters transmitted with no time gap between them. It is followed by an *end-of-message* (EOM) indicator and then a **block check character (BCC).** A block size is typically 256 characters. The transmitter uses a predefined algorithm to compute the BCC. The same algorithm is used at the receiver based on the block of data received. The two BCCs are compared and, if identical, the next block of data is transmitted.

Block Check Character (BCC) method of error detection involving sending a block of data, then an end of message indicator, then a block check character representing characteristics of the data that was sent

There are many algorithms used to generate a BCC. The most elementary one is an extension of parity into two dimensions, called the **longitudinal redundancy check (LRC).** This method is illustrated with the help of Fig. 8-31. Shown is a small block of 4-bit characters using odd parity. The BCC is formed as an odd-parity bit for each vertical column. Now suppose that a double error occurred in character 2 as shown in Fig. 8-31(b). With odd parity, the third and fourth bits from the left in the BCC should be zeros; instead they are 1s. As described previously, simple parity would not pick up this error. With the BCC, however, the error is detected. If a single error occurs [Fig. 8-31(c)], the erroneous bit can be pinpointed as the intersection of the row and column containing the error. Correction is achieved by inverting the bad bit. If a double error occurs in a column, the scheme is defeated.

Longitudinal Redundancy Check (LRC) extending parity into two dimensions

The error location process just described is not usually utilized. Rather, the receiver checks for character and LRC errors and if either (or both) occur, a

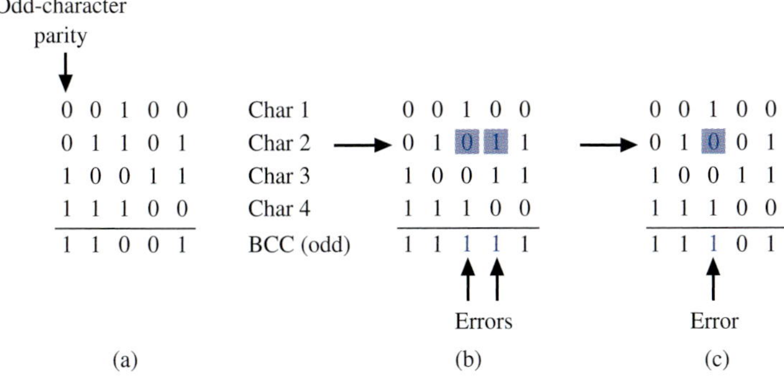

FIGURE 8-31 LRC error detection.

retransmission is requested. Occasionally, an error will occur in the BCC. This is unavoidable, but the only negative consequence is an occasional unnecessary retransmission. This scheme is useful in low noise environments.

Cyclic Redundancy Check

Cyclic Redundancy Check
method of error detection involving performing repetitive binary division on each block of data and checking the remainders

One of the most powerful error-detection schemes in common use is the **cyclic redundancy check** (CRC). The CRC is a mathematical technique that is used in synchronous data transmission. It can effectively catch 99.95 percent of transmission errors.

In the CRC technique, each string of bits is represented by a polynomial function. The technique is done by division. It is illustrated as follows:

$$\frac{M(x)}{G(x)} = Q(x) + R(x) \qquad \textbf{(8-11)}$$

where $M(x)$ is the binary data, called the message function, and $G(x)$ is a special code for which the message function is divided, called the generating function. The process yields a quotient function, $Q(x)$, and a remainder function, $R(x)$. The quotient is not used, and the remainder, which is the CRC block check code **(BCC),** is attached to the end of the message. This is called a **systematic code,** where the BCC and the message are transmitted as separate parts within the transmitted code. At the receiver, the message and CRC check character pass through its block check register BCR. If the register's content is zero, then the message contains no errors.

BCC
block check code, the code generated when creating the CRC transmit code

Systematic Code
a code in which the message and block check code are transmitted as separate parts within the same transmitted code

Cyclic codes are popular not only because of their capability to detect errors but also because they can be used in high-speed systems. They are easy to implement, requiring only the use of shift registers, EXOR gates, and feedback paths. Cyclic block codes are expressed as **(n, k) cyclic codes,** where

n = length of the transmitted code
k = length of the message

(n, k) Cyclic Code
nomenclature used to identify cyclic codes in terms of their transmitted code length (n) and message length (k)

For example, a (7, 4) cyclic code simply means that the bit length of the transmitted code is 7 bits ($n = 7$) and the message length is 4 bits ($k = 4$). This information also tells us the length or number of bits in the block check code (BCC), which is the code transmitted with each word. The number of bits in the BCC is $(n - k)$. This relationship is expressed as

$$\text{BCC length} = n - k \qquad \textbf{(8-12)}$$

This code is combined with the message to generate the complete transmit binary code and is used at the receiver to determine if the transmitted message contains an error. The general form for systematic (n, k) CRC code generation is shown in Fig. 8-32.

The number of shift registers required in the CRC generating circuit is the length of the block check code (BCC), which is $(n - k)$. For a (7, 4) cyclic code, the number of shift registers required to generate the BCC is $7 - 4 = 3$. Note that the highest order of the generating polynomial described next (3 in this case, from x^3) is also the number of shift registers required.

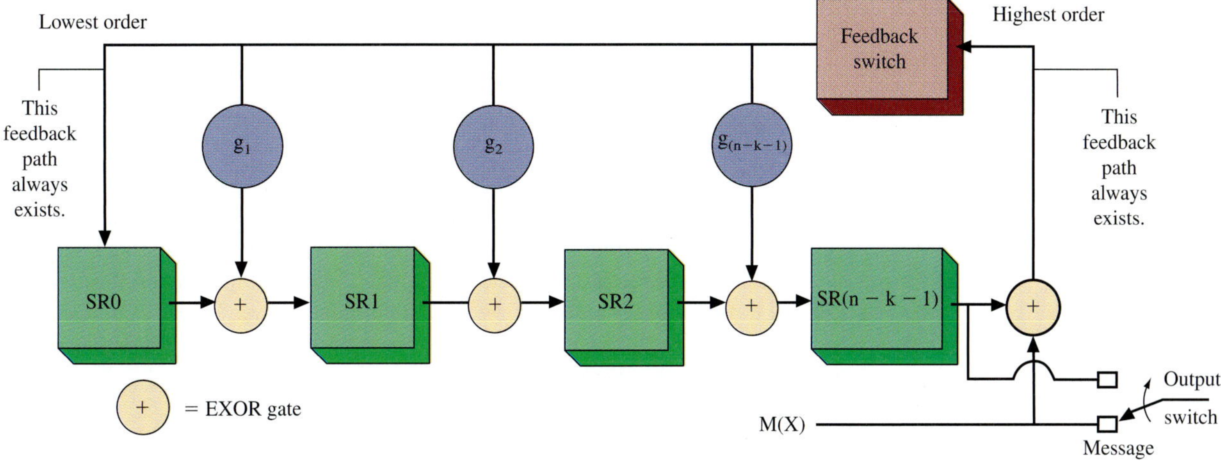

FIGURE 8-32 CRC code generator for an (n, k) cyclic code.

Construction of a CRC generating circuit is guided by the **generating polynomial,** $G(x)$. The feedback paths to each XOR gate are determined by the coefficients for the generating polynomial, $G(x)$. If the coefficient for a variable is 1, then a feedback path exists. If the coefficient is 0, then there isn't a feedback path. The general form for the CRC generator with respect to $G(x)$ is

Generating Polynomial defines the feedback paths to be used in the CRC generating circuit

$$G(x) = 1 + g_1 x + g_2 x^2 + \cdots + g_{n-k-1} x^{n-k-1} + x^{n-k} \quad \textbf{(8-13)}$$

The lowest-order value is $x^0 = 1$; therefore, the feedback always exists. This is indicated in Fig. 8-32.

The highest-order value, which varies, also has a coefficient of 1. In CRC circuits, the lowest- and highest-order polynomial values have a coefficient of 1. This feedback must always exist because of the cyclic nature of the circuit. For example, if the generating polynomial expression is $G(x) = 1 + x + x^3$, then the feedback paths are provided by 1 (x^0), x, and x^3. The coefficient for x^2 is 0; therefore, no feedback is specified.

In addition to the EXOR gates and shift registers for the circuit shown in Fig. 8-32, the CRC generator contains two switches for controlling the shifting of the message and code data to the output. The procedure for CRC code generation using this circuit is as follows:

CRC Code Generation: Procedure for Using Fig. 8-32

1. Load the message bits serially into the shift registers. This requires:
   Output switch connected to the message input $M(x)$,
   Feedback switch closed.
   Note: k shifts are performed; k is obtained from the (n, k) specification. The k shifts load the message into the shift registers and at the same time the message data are serially sent to the output and the BCC is generated.
2. After completing the transmission of the kth bit, the feedback switch is opened and the output switch is changed to select the shift registers.

3. The contents of the shift registers containing the BCC are shifted out using $(n - k)$ shifts. *Note:* The total number of shifts to the CRC generator circuit is n, which is the length of the transmit code.

The operation of this circuit can be treated mathematically as follows. The message polynomial, $M(x)$, is being multiplied by $x^{(n-k)}$. This results in $(n - k)$ zeros being added to the end of the message polynomial, $M(x)$. The number of zeros is equal to the binary size of the block check code (BCC). The generator polynomial is then modulo 2 divided into the modified $M(x)$ message. The remainder left from the division is the block check code (BCC). In systematic form, the BCC is appended to the original message, $M(x)$. The completed code is ready for transmission. An example of this process and an example of implementing a cyclic code using Fig. 8-32 is given in Ex. 8-6.

EXAMPLE 8-6

For a (7, 4) cyclic code and given a message polynomial $M(x) = (1\,1\,0\,0)$ and a generator polynomial $G(x) = x^3 + x + 1$, determine the BCC (a) mathematically and (b) using the circuit provided in Fig. 8-32.

SOLUTION

(a) The code message $M(x)$ defines the length of the message (4 bits). The number of shift registers required to generate the block check code is determined from the highest order in the generating polynomial $G(x)$, which is x^3. This indicates that three shift registers will be required, so we will pad the message (1 1 0 0) with three zeros to get (1 1 0 0 0 0 0). Remember, a (7, 4) code indicates that the total length of the transmit CRC code is 7 and the

$$\text{BCC length} = n - k \tag{8-12}$$

The modified message is next divided by the generator polynomial, $G(x)$. This is called modulo-2 division.

$$G(x)\overline{)M(x) \cdot x^{n-k}}$$

```
         1110
 1011)1100000
      1011
       1110
       1011
        1010
        1011
         010
```

The remainder 010 is attached to $M(x)$ to complete the transmit code, $C(x)$. The transmit code word is 1100010.

(b) Next, we use the circuit in Fig. 8-32 to generate the same cyclic code for the given $M(x)$ and $G(x)$. The CRC generating circuit can be produced from the general form provided in Fig. 8-32. $G(x) = x^3 + x + 1$ for the (7, 4) cyclic code. This information tells us the feedback paths (determined by the coefficients of the generating polynomial expression equal to 1) as well as the number of shift registers required, $(n - k)$. The CRC generating circuit is provided in Fig. 8-33. The serial output sequence and the shift register contents for each shift are shown in Table 8-9.

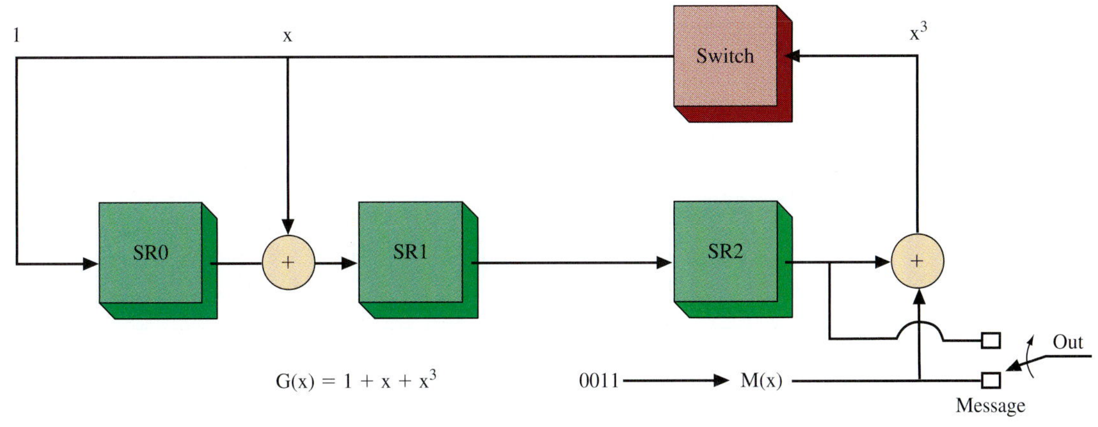

FIGURE 8-33 CRC code generator for a (7, 4) cyclic code using a generator polynomial $G(x) = x^3 + x + 1$.

Table 8-9 THE SHIFT SEQUENCE FOR EX. 8-6

Data [$M(x)$]					Register Contents			
x^0	x^1	x^2	x^3	Shift Number	SR0	SR1	SR2	Output
0	0	1	1	0	0	0	0	
	0	0	1	1	1	1	0	1
		0	0	2	1	0	1	1
			0	3	1	0	0	0
			–	4	0	1	0	0

The result in both cases for Ex. 8-8 is a transmitted code of 1100010, where 1100 is the message $M(x)$ and 010 is the BCC. The division taking place in part (a) of Ex. 8-8 is modulo-2 division. Remember, modulo-2 is just an XORing of the values. The division of $G(x)$ into $M(x)$ requires only that the divisor has the same number of binary positions as the value being divided (the most significant position must be a 1). The value being divided does not have to be larger than the divisor.

CRC Code-Dividing Circuit

At the receive side, the received code is verified by feeding the received serial CRC code into a *CRC dividing circuit*. The dividing circuit has $(n - k)$ shift registers. The general form for a CRC divide circuit is given in Fig. 8-34.

The arrangement of the feedback circuit and the shift registers depends on the generator polynomial, $G(x)$, and the coefficients for each expression. If the coefficient for X is 1, then a feedback path to an EXOR is provided for that shift register. If the coefficient is 0, then there is not a feedback path to the input of the respective shift register. The number of shift registers required in the circuit is still $(n - k)$. The circuit requires k shifts to check the received data for errors. The result (or remainder) of the shifting (division) should be all 0s. The remainder is

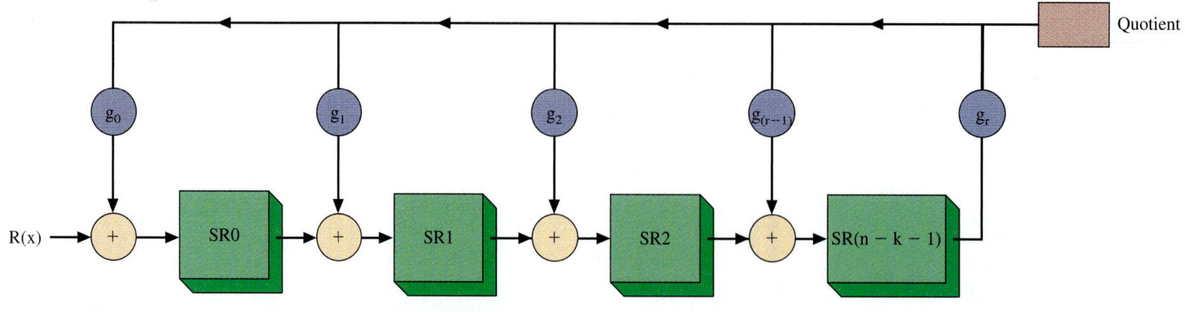

FIGURE 8-34 A CRC divide circuit for the general form of $G(x) = g_0 + g_1x + g_2x^2 + \cdots + g_rx^r$.

Syndrome
the value left in the CRC dividing circuit after all data have been shifted in

called the **syndrome.** If the syndrome contains all 0s, then it is assumed that the received data are correct. If the syndrome is a nonzero value, then bit error(s) have been detected. In most cases the receive system will request a retransmission of the message. This is the case for ethernet computer networks. In limited cases, the code *can* then be corrected by adding the syndrome vector to the received vector. This method of correction requires the use of a look-up table to determine the correct word. Example 8-7 demonstrates the use of the CRC detection circuit using the information from Ex. 8-6.

In Ex. 8-7 the transmitted code length is seven, ($n = 7$); the message length is four, ($k = 4$); and the code length is three, $(n - k) = (7 - 4) = 3$.

Example 8-7

The serial data stream 0 1 0 0 0 1 1 was generated by a (7, 4) cyclic coding system using the generator polynomial $G(x) = 1 + x + x^3$ (see Ex. 8-6). Develop a circuit that will check the received data stream for bit errors. Verify your answer both (a) mathematically and (b) by shifting the data through a CRC divide circuit.

Solution

(a) To verify the data mathematically requires that the *received data* be divided by the coefficients defining generator polynomial, $G(x)$. This is similar to the procedure used in Ex. 8-6 except the data being divided contains the complete code.

```
           1110
    1011)1100010
         1011
          1110
          1011
           1011
           1011
            000
```

The syndrome (remainder) is zero; therefore, the received data do not contain any errors.

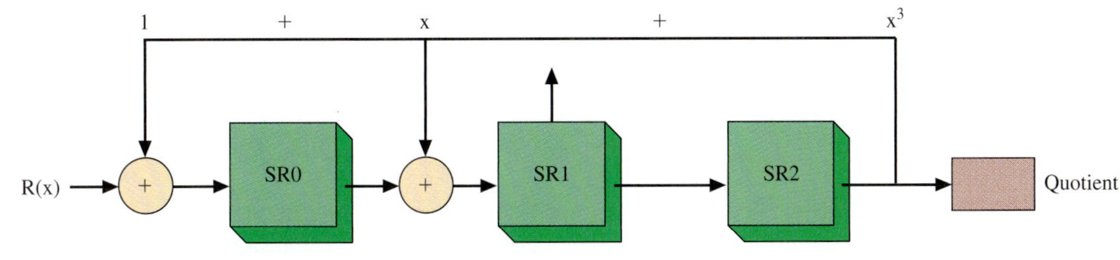

FIGURE 8-35 A CRC divide circuit for $G(x) = 1 + x + x^3$.

(b) The CRC divide circuit is generated from the CRC generator polynomial expression, $G(x)$. $G(x)$ for this system is $1 + x + x^3$. The circuit is created using the form shown in Fig. 8-35. The shift register contents are first reset to zero. Next, the received data are input to $R(x)$. The data are shifted serially into the circuit. The results of the shifting are provided in Table 8-10. A total of n shifts are required.

The result of the shifting in the data produced all 0s remaining in the shift registers. Therefore, the syndrome is 0, which indicates that the received data do not contain any bit errors.

Table 8-10 The Shift Sequence for Ex. 8-7

Code lsb msb	Data lsb msb	Shift Number	Register Contents SR0	SR1	SR2	Output
0 1 0 0	0 1 1		0	0	0	
0 1 0	0 0 1	1	1	0	0	1
0 1	0 0 0	2	1	1	0	1
0	1 0 0	3	0	1	1	0
	0 1 0	4	1	1	1	0
	0 1	5	1	0	1	0
	0	6	0	0	0	1
	–	7	0	0	0	0

Hamming Code

The error-detection schemes thus far presented require retransmission if errors occur. Techniques that allow correction at the receiver are called **forward error-correcting** (FEC) codes. The basic requirement of such codes is for sufficient redundancy to allow error correction without further input from the transmitter. The **Hamming code** is an example of an FEC code named for R. W. Hamming, an early developer of error-detection/correction systems.

Forward Error-Correcting error-checking techniques that permit correction at the receiver, rather than retransmitting the data

Hamming Code a forward error-checking technique named for R. W. Hamming

If m represents the number of bits in a data string and n represents the number of bits in the Hamming code, n must be the smallest number such that

$$2^n \geq m + n + 1 \tag{8-14}$$

Consider a 4-bit data word 1101. The minimum number of parity bits to be used is 3 when Eq. (8-14) is referenced. A possible setup, then, is

P_1	P_2	1	P_3	1	0	1	
1	2	3	4	5	6	7	bit location

We'll let the first parity bit, P_1, provide even parity for bit locations 3, 5, and 7. P_2 does the same for 3, 6, and 7, while P_3 checks 5, 6, and 7. The resulting word, then, is

1	0	1	0	1	0	1	
1	2	3	4	5	6	7	bit location
P_1	P_2		P_3				

When checked, a 1 is assigned to incorrect parity bits, while a 0 represents a correct parity bit. If an error occurs so that bit location 5 becomes a 0, the following process takes place. P_1 is a 1 and indicates an error. It is given a value of 1 at the receiver. P_2 is not concerned with bit location 5 and is correct and therefore given a value of 0. P_3 is incorrect and is therefore assigned a value of 1. These three values result in the binary word 101. Its decimal value is 5, and this means that bit location 5 has the wrong value and the receiver has pinpointed the error without a retransmission. It then changes the value of bit location 5 and transmission continues. The Hamming code is not able to detect multiple errors in a single data block. More complex (and more redundant) codes are available if necessary.

Reed–Solomon Codes

Reed–Solomon (RS) codes are also forward error-correcting codes (FEC) like the Hamming code. They belong to the family of BCH (Bose–Chaudhuri–Hocquenghem) codes. Unlike the Hamming code, which can detect only a single error, RS codes can detect multiple errors, which makes RS codes very attractive for use in CD players because a CD can contain scratches in its surface, and in mobile communications, where a burst of errors is expected.

Interleaving
a technique used to rearrange the data into a nonlinear ordering scheme to improve the chance of data correction

RS codes often employ a technique called **interleaving** to enable the system to correct multiple data-bit errors. Interleaving is a technique used to rearrange the data into a nonlinear ordering scheme to improve the chance of correcting data errors. The benefit of this technique is that burst errors can be corrected because the likelihood is that all the data message won't be destroyed. For example, a large scratch on a CD disk will cause the loss of large amounts of data in a track; however, if the data bits representing the message are distributed over several tracks in different blocks, then the likelihood that they can be recovered is increased. The same is also true for a message transmitted over a communications channel. Burst errors will cause the loss of only a portion of the message data, and if the Hamming distance is large enough, the message can be recovered.

The primary disadvantage of implementing the RS coding scheme is the time delay for encoding (interleaving the data) at the transmitter and decoding (deinterleaving the data) at the receiver. In applications such as a CD ROM, data size and the delay are not a problem, but implementing RS codes in mobile communications could create significant delay problems in real-time applications and possibly a highly inefficient use of data bandwidth.

8-7 Troubleshooting

Digital communications provide vital links to transfer information in today's world. Digital data are used in every aspect of electronics in one form or another. In this section we will look at digital pulses and the effects that noise, impedance, and frequency have on them. Digital communications troubleshooting requires that the technician be able to recognize digital pulse distortion and to identify what causes it.

After completing this section you should be able to

- Identify a good pulse waveform
- Identify frequency distortion
- Describe effects of incorrect impedance on the square wave
- Identify noise on a digital waveform

The Digital Waveform

A square wave signal is a digital waveform and is illustrated in Fig. 8-36. The square wave shown is a periodic wave that continually repeats itself. It is made up of a positive alternation and a negative alternation. The ideal square wave will have sides that are vertical. These sides represent the high-frequency components. The flat-top and bottom lines represent the low-frequency components. A square wave is composed of a fundamental frequency and an infinite number of odd harmonics, as described in Chapter 1 in the FFT analysis section.

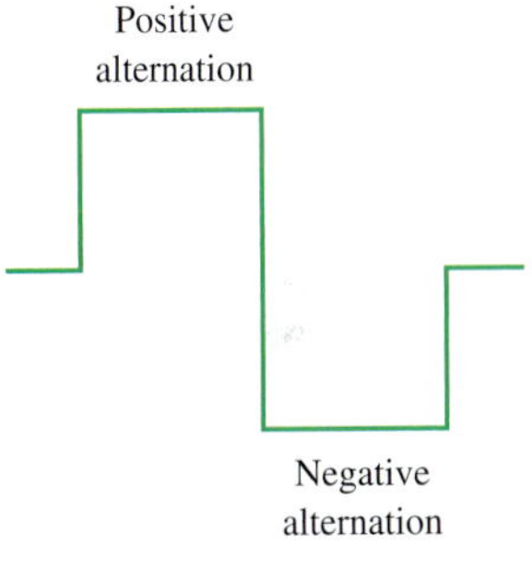

FIGURE 8-36 Ideal square wave with 50 percent duty cycle.

Figure 8-37 shows this same square wave stretched out to illustrate a more true representation of it. Notice the sides are not ideally vertical but have a slight slope to them. The edges are rounded off because transition time is required for the low pulse to go high and back to low again. Figure 8-37 shows the positive alternation as a positive pulse and the negative alternation as a negative pulse. Rise time refers to the pulse's low-to-high transition and is normally measured from the 10 percent point to the 90 percent point on the waveform. Fall time represents the high-to-low transition and is measured from the 90 percent and 10 percent points. From the pulse's maximum low point to its maximum high point is the amplitude measured in volts. By observing the pulse waveform in response to circuit conditions that it may encounter, much can be determined about the circuit.

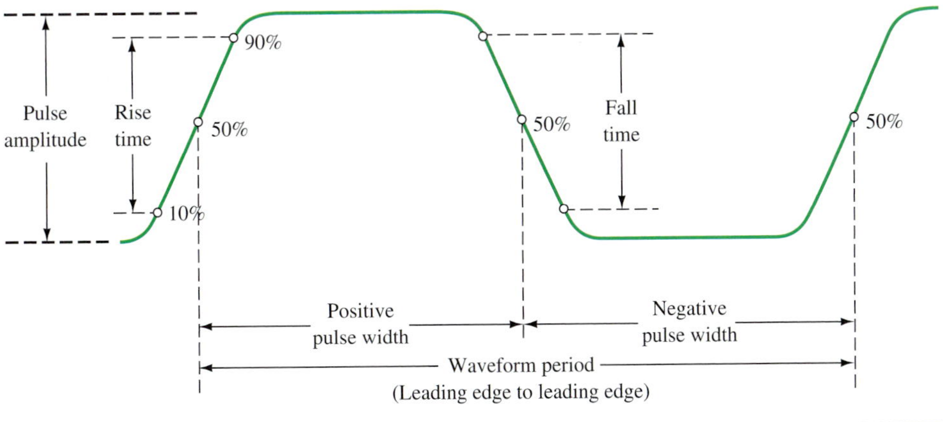

FIGURE 8-37 Illustration of noise on a pulse.

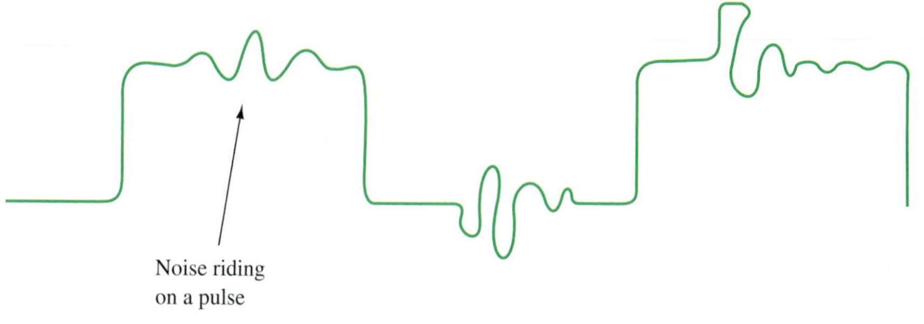

FIGURE 8-38 Nonideal square wave.

Effects of Noise on the Pulse

From previous discussions about noise you learned that noise has an additive effect on a signal. A signal's amplitude is changed by noise adding to it or subtracting from it. This concept is depicted in Figure 8-38. The positive pulse and the negative pulse have changed from the ideal as a direct result of encountering noise. The noise has changed the true amplitudes of the pulses. If the noise becomes too severe, the positive and negative noise excursions might be mistaken by logic circuits as high and low pulses. Proper noise compensating techniques, shielding, and proper grounds help to reduce noise. If the digital waveforms under test show signs of deterioration due to noise, troubleshoot by checking compensation circuits, shielding, and ensuring that proper grounds are made.

Effects of Impedance on the Pulse

The square wave pulse can show the effects of impedance mismatches, as seen in Fig. 8-39. The pulse in Fig. 8-39(a) is severely distorted by an impedance that is below that required. For example, if RG-58/AU coaxial cable, carrying data pulses, became shorted or if one of its connectors developed a low-resistance leakage path to ground, then the data pulses would suffer low-impedance distortion. Figure 8-39(b) shows a type of distortion on the top and bottom of the square wave called ringing. Ringing is the result of the effects of high impedance on the pulse. If a transmission line is improperly terminated or develops a high resistance for some reason, then ringing can occur. A tank circuit with too high of a Q will cause ringing. By adjusting the tank's Q or by adding proper termination to a transmission line, the waveform can be brought back to normal, as shown in Fig. 8-39(c). When working with data lines, ensure that proper terminations are made. The effect of impedance loading is reduced in communications equipment like transmitters, receivers, and data handling circuits when proper repair and alignment maintenance techniques are used.

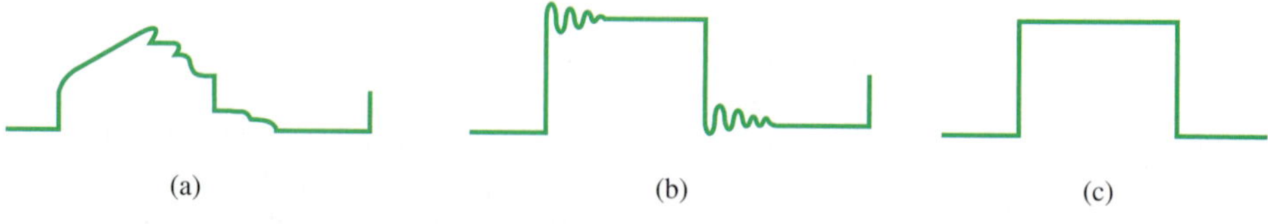

FIGURE 8-39 Effects of impedance mismatches: (a) impedance is too low; (b) impedance is too high (ringing); (c) impedance is matched.

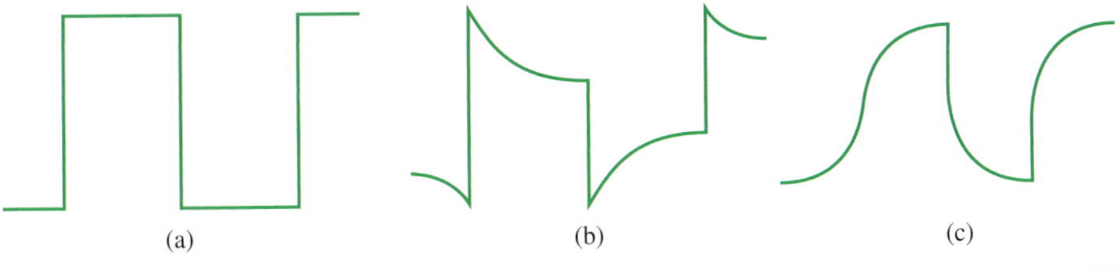

FIGURE 8-40 Effects of frequency on data pulses: (a) good pulses; (b) low-frequency attenuation; (c) high-frequency attenuation.

Effects of Frequency on the Pulse

Digital pulses will not be distorted when passing through an amplifier with a sufficient bandwidth or a transmission line with a sufficient bandwidth. Figure 8-40(a) shows that the pulse does not become distorted in any fashion. The dip at the top and bottom of the waveform in Fig. 8-40(b) represents low-frequency attenuation. The high-frequency harmonic components pass without being distorted, but the low-frequency components are attenuated. Coupling capacitors in communications circuits can cause low-frequency distortion. Low-frequency compensating network malfunctions will also attenuate the low-frequency components of the pulse waveform.

High-frequency distortion occurs when the high-frequency harmonic components of a digital square wave are lost. The edges become rounded off, as seen in Fig. 8-40(c). The high harmonic frequencies are lost when the media bandwidth becomes too narrow to pass the pulse in its entirety. A circuit's bandwidth can change when the value of circuit components such as coils, capacitors, and resistors increases or decreases. Open and shorted capacitors found in tank circuits can also cause high-frequency losses.

8-8 Troubleshooting with Electronics Workbench™ Multisim

Proper sampling of the input signal is an extremely important process when converting an analog signal to a digital format. This Electronics Workbench™ Multisim exercise has been developed to advance your understanding of the sampling process and to reinforce the importance of properly selecting the sample frequency. Examples are provided to demonstrate the properly sampled signal and the components generated by aliasing when the sample frequency is inadequate.

Start this exercise by opening the file **Fig8-41.msm** found on your EWB Multisim CD. This circuit contains a function generator that supplies the sample frequency (f_s = 5 kHz) and a sine-wave generator that provides the analog signal (f_i = 1 kHz) being sampled. The Multisim circuit is shown in Fig. 8-41.

Start the simulation and observe the traces on the oscilloscope. Trace A is the input sine wave and trace B is the sampled signal (also called a pulse-amplitude-modulated [PAM] signal). The oscilloscope traces are shown in Fig. 8-42.

Experiment with the sample frequency. Double-click on the function generator and set the sample frequency to 12 kHz. Restart the simulation and notice the improvement in the sampled signal due to the significant increase in the sample

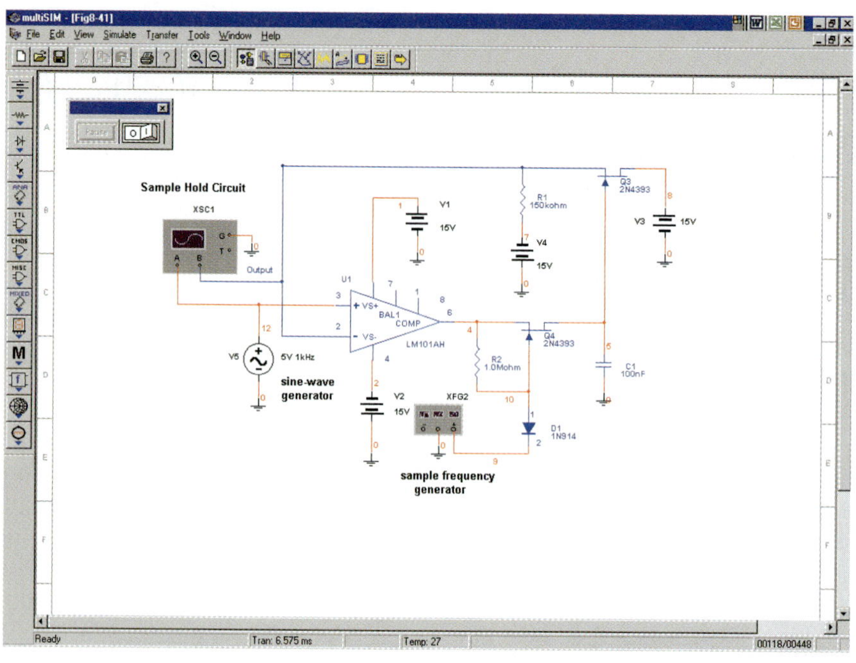

FIGURE 8-41 A sample-and-hold circuit as implemented in Electronics Workbench™ Multisim.

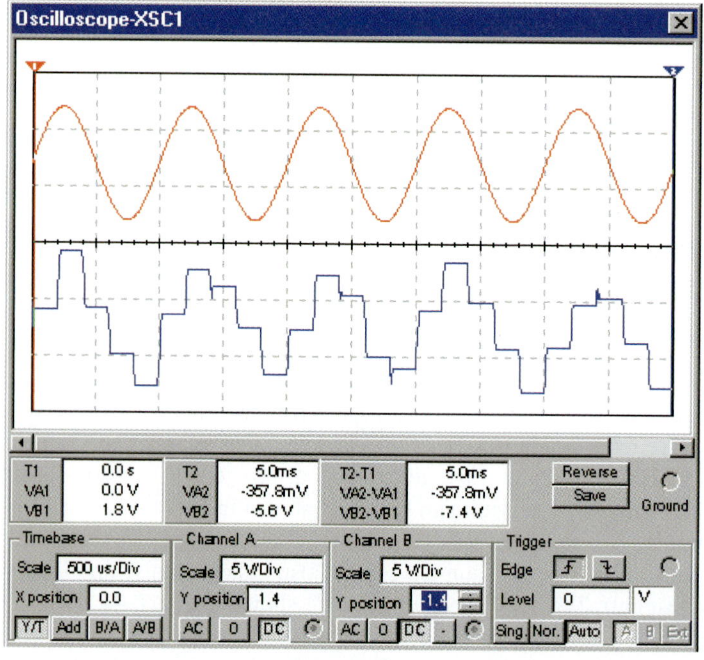

FIGURE 8-42 The oscilloscope traces for the sample-and-hold circuit.

frequency. The Nyquist sampling theorem states that the sample frequency (f_s) must be at least twice the highest input frequency. What happens if the sample frequency does not satisfy the Nyquist criteria? Open the file **FigE8-1.ms7 (.msm)** on your EWB CD. Notice that a spectrum analyzer has been connected to the output. The sample frequency (f_s) has been set to 6 kHz and the input frequency (f_i) is 4 kHz. Start the simulation and observe the display of frequencies on the spectrum analyzer. The signal you should see on the spectrum analyzer is shown in Fig. 8-43.

Use the cursor to determine the frequencies displayed. You will see the following: 2 kHz, 4 kHz, 6 kHz, 8 kHz, and 10 kHz. Where did these frequencies come from? The sample frequency is 6 kHz. The input frequency is 4 kHz. Therefore, you have the following.

Frequency	**Origination**
(6 − 4) kHz = 2 kHz	$f_s - f_i$
4 kHz	Input signal
6 kHz	Sample frequency
(6 + 2) kHz = 8 kHz	Sum of the sample frequency (f_s) and the 2-kHz ($f_s - f_i$) frequency
(6 + 4) kHz = 10 kHz	Sum of the sample frequency (f_s) and the input frequency (f_i)

This example demonstrates the importance of properly selecting the sample frequency. The input signals in communications systems are typically more complex than a simple sine wave and contain many harmonic frequencies. This example underscores the importance of incorporating antialiasing filters in sampling circuits to ensure that the maximum input frequency never exceeds $f_s/2$. If the Nyquist criteria is not met, aliasing frequencies will be generated, leading to a decrease in system performance.

One of the Electronics Workbench™ exercises that follows further investigates the sample-and-hold circuit and provides you with the opportunity to examine the

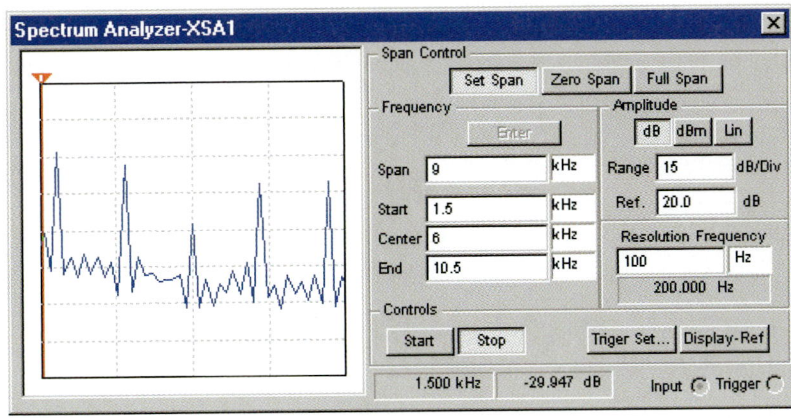

FIGURE 8-43 The spectrum of a signal that contains aliased frequencies due to improper selection of the sampling frequency.

spectral content of a signal generated by a sample-and-hold circuit. Additionally, two of the exercises have faults incorporated in the circuit and provide you with the opportunity to troubleshoot the circuit.

Electronics Workbench™ Exercises

1. Open the file **FigE8-2.ms7 (.msm)** on your EWB CD. Use the spectrum analyzer to determine the frequencies generated in the sampled signal. The input frequency is 5 kHz. Where did the 3-kHz signal come from?
2. Open the file **FigE8-3.ms7 (.msm)** on your EWB CD. This circuit contains a fault. Start the simulation and use the oscilloscope to isolate the fault. When you discover the fault, double-click on the component, click on the **fault** tab, and set the faults to none. Restart the simulation and see if you have repaired the circuit. Specifiy what component(s) you repaired.
3. Open the file **FigE8-4.ms7 (.msm)** on your EWB CD. This circuit contains a fault. Start the simulation and use the oscilloscope to isolate the fault. When you discover the fault, double-click on the component, click on the **fault** tab, and set the faults to none. Restart the simulation and see if you have repaired the circuit. Specifiy what component(s) you repaired.

Summary

In Chapter 8 we studied the coding techniques commonly used in digital communications. The techniques behind pulse-code modulation were explored, as were digital signal encoding formats, coding principles, and error correction and detection. The major topics you should now understand include:

- the definition of the Nyquist sampling frequency
- aliasing and foldover distortion
- the definition of the dynamic range of a sampled signal
- identification of the major digital encoding formats including the NRZ, RZ, phase-encoded, and multilevel binary formats
- the concept of Hamming distance
- the explanation of the various codes' error detection and correction methods, including parity, cyclic redundancy check (CRC) codes, Hamming code, and Reed–Solomon codes

Questions and Problems

Section 8-1

1. With the assistance of Fig. 8-1, describe a transmission that is
   (a) Digital but not data.
   (b) Both digital and data.
   (c) Data but not digital.
2. Define *digital signal processing*. Provide an example that is not described in this book.

Section 8-2

3. What do the abbreviations ASCII and EBCDIC stand for?
4. Provide the ASCII code for 5, a, A, and STX.
5. Provide the EBCDIC code for 5, a, A, and STX.
6. Describe the Gray code.
7. Provide an application of the Gray code.

Section 8-3

8. Define *acquisition time* for a sample-and-hold circuit.
9. Define *aperture time* for a sample-and-hold circuit.
10. What is the typical capacitance value for a sample-and-hold circuit?
11. Draw the PAM signal for a sinusoid using
    (a) Natural sampling.
    (b) Flat-top sampling.
12. An audio signal is band-limited to 15 kHz. What is the minimum sample frequency if this signal is to be digitized?
13. A sample circuit behaves like what other circuit used in radio-frequency communications?
14. What is the dynamic range (in dB) for a 12-bit PCM system?
15. Define the resolution of a PCM system. Provide two ways that resolution can be improved.
16. Calculate the number of bits required to satisfy a dynamic range of 48 dB.
17. What is meant by quantization?
18. Explain the difference in linear PCM and nonlinear PCM.
19. Explain the process of companding and the benefit it provides.
20. A μ-law companding system with $\mu = 100$ is used to compand a 0- to 10-V signal. Calculate the system output for inputs of 0, 0.1, 1, 2.5, 5, 7.5, and 10 V. (0, 1.5, 5.2, 7.06, 8.52, 9.38, 10)

Section 8-4

21. Describe the characteristics of the four basic encoding groups: NRZ, RZ, phase-encode binary, and multilevel binary.
22. Sketch the data waveforms for 1 1 0 1 0 using NRZ-L, biphase M, differential Manchester, and dicode RZ.
23. What are the differences between bipolar codes and unipolar codes?
24. Which coding format is considered to be self-clocking? Explain the process.

Section 8-5

25. What is D_{min}?
26. How can the minimum distance between two data values be increased?
27. Determine the number of errors that can be detected and corrected for data values with a Hamming distance of
    (a) Two.
    (b) Five.
28. Determine the distance between the following two digital values:

    1 1 0 0 0 1 0 1 0
    0 1 0 0 0 0 0 1 0

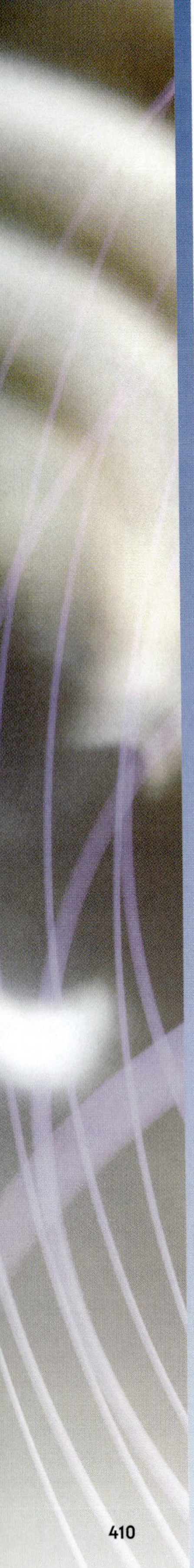

29. What is the minimum distance between data words if
    (a) Five errors are to be detected?
    (b) Eight errors are to be detected?

Section 8-6

30. CRC codes are commonly used in what computer networking protocol?
31. Provide a definition of systematic codes.
32. With regard to (n, k) cyclic codes, define n and k.
33. What is a block check code?
34. How are the feedback paths determined in a CRC code-generating circuit?
35. Draw the CRC generating circuit if $G(x) = x^4 + x^2 + x + 1$.
36. Given that the message value is 1 0 1 0 0 1 and $G(x) = 1\ 1\ 0\ 1$, perform modulo-2 division to determine the block check code (BCC). (111)
37. What is a syndrome?
38. What does it mean if the syndrome value is
    (a) All zeros?
    (b) Not equal to zero?
39. What does it mean for a code to be forward error-correcting?
40. What popular commercial application uses Reed–Solomon coding and why?

Questions for Critical Thinking

41. Explain why a technique such as CRC coding is used. Can you achieve the same results using parity checking?
42. A PCM system requires 72 dB of dynamic range. The input frequency is 10 kHz. Determine the number of sample bits required to meet the dynamic range requirement and specify the minimum sample frequency to satisfy the Nyquist sampling frequency. (12 bits, 20 kHz)
43. The voltage range being input to a PCM system is 0 to 1 V. A 3-bit A/D converter is used to convert the analog signal to digital values. How many quantization levels are provided? What is the resolution of each level? What is the value of the quantization error for this system? (8, 0.125, 0.0625)

9 Wired Digital Communications

Chapter Outline

The 16900 Se logic analyzers take advantag advanced com puter and netw technology to achieve fastes measurement trol and analys in the industry (Courtesy of Ag lent Technolog Reprinted with permission.)

Objectives

- Describe six combinations for transmitting analog or digital signals using either an analog or digital channel
- Explain the concepts of error probability and bit error rate
- Describe the basics of digital links and protocols
- Detail the bandwidth issues of a digital communications link
- Describe the fundamentals of TDMA and how it is used to transport digital data
- Describe the delta and pulse modulation techniques
- Explain the fundamentals of data transmission, including T1, T3, packet switching, frame relay, and ATM
- Describe the current computer serial communication standards, facsimile, and computer bus standards

Key Terms

wired digital communications
baseband
coding
mark, space
dit
error probability (P_e)
bit error rate (BER)
Energy per bit (E_b), or bit energy
E_b/N_o
synchronous
asynchronous
handshaking
protocols
framing
line control
multipoint circuits
flow control
sequence control
character insertion
character stuffing
bit stuffing
transparency
bps
baud rate
fractional T1
point of presence
CSU/DSU
D4 framing
ESF
loopback
AMI
bipolar coding
B8ZS
minimum ones density
bipolar violation
HDLC
PPP
data encapsulation
packets
statistical concentration
frame relay
public data network
Telco
X.25
committed information rate
bursty
committed burst information rate (CBIR)
ATM
packet switching
payload
virtual path connection (VPC)
virtual channel connection (VCC)
SVC
VPI
VCI
time-division multiple access (TDMA)
time slot
demultiplexer (DMUX)
guard times
intersymbol interference (ISI)
delta modulation
slope modulation
tracking ADC
slope overload
continuously variable slope delta
pulse modulation
time-division multiplexing
Nyquist rate
pulse-amplitude modulation
pulse-width modulation
pulse-position modulation
pulse-time modulation
pulse-duration modulation
pulse-length modulation
continuous wave
interrupted continuous wave
asynchronous system
start bit, stop bit
synchronous system
universal serial bus (USB)
hot-swappable
Type A connector
Type B connector
Firewire (IEEE 1394)
RS-232
null modem
data terminal equipment
data communications equipment
RS-422, RS-485
balanced mode
facsimile
fax

9-1 INTRODUCTION

Wired Digital Communications
digital communications over a wired connection

In this chapter and Chapter 10, we will consider the important aspects of the transmission of signals as they relate to digital communications. Chapter 9 focuses on digital communications over a wired link, which is called **wired digital communications.** Chapter 10 focuses on the techniques associated with wireless digital communications. Before we continue, it would be helpful to consider different transmission schemes for digital and analog signals. This will provide a perspective on the various methods presented in this chapter and help you to compare them with some other transmission systems.

Six possible methods for transmitting information are provided in Fig. 9-1. In Fig. 9-1(a) the analog signal is transmitted directly. An example of this is a basic intercom system where a microphone creates an electrical analog of your voice, which is transmitted directly via a pair of wires. Of course, that signal is usually amplified at both the input and output sides of the channel (pair of wires). Notice that the channel is labeled an analog baseband channel in Fig. 9-1(a). **Baseband** means the signal is transmitted at its base frequencies, and no modulation to another frequency range has occurred. Figure 9-1(b) shows a standard analog modulation scheme. We spent Chapters 2–8 studying these systems.

Baseband
the signal is transmitted at its base frequency with no modulation to another frequency range

When a computer transmits digital data to another computer or a computer peripheral such as a printer, it may transmit that signal directly. This situation is illustrated in Fig. 9-1(c). Notice that a coder and decoder are shown in this system. We'll consider these types of communications in Sec. 9-7.

The transmission of digital signals via an analog channel is shown in Fig. 9-1(d). A modem allows the digital signal to be transmitted on an analog channel. This is the scheme used by most personal computers to transmit their information over telephone lines. We'll look at this in Chapter 11.

You may wonder, why go to the complexities shown in Figs. 9-1(e) and (f) just to transmit an analog signal? In Fig. 9-1(e) the signal is converted to digital form by the analog-to-digital converter and transmitted as digital pulses on the digital channel. In Fig. 9-1(f) the same processing occurs, only this channel cannot carry the digital pulses, so the modem is used to allow an analog channel to carry digital data. As you will see, the digitization of analog signals offers advantages when multiplexing more than one signal in a transmission. There can also be advantages when noise is a significant problem.

9-2 BACKGROUND MATERIAL FOR DIGITAL COMMUNICATIONS

This section introduces some of the basic digital communications concepts needed to understand and appreciate fully the material presented in this chapter. The basic principles of binary coding are first reviewed, followed by a discussion on code noise immunity. The concept of bit-error rate (BER) and the probability of a bit error (P_e) are presented next. The section concludes with definitions for the fundamental communication concepts of simplex, half-duplex, full duplex, synchronous, and asynchronous data communications.

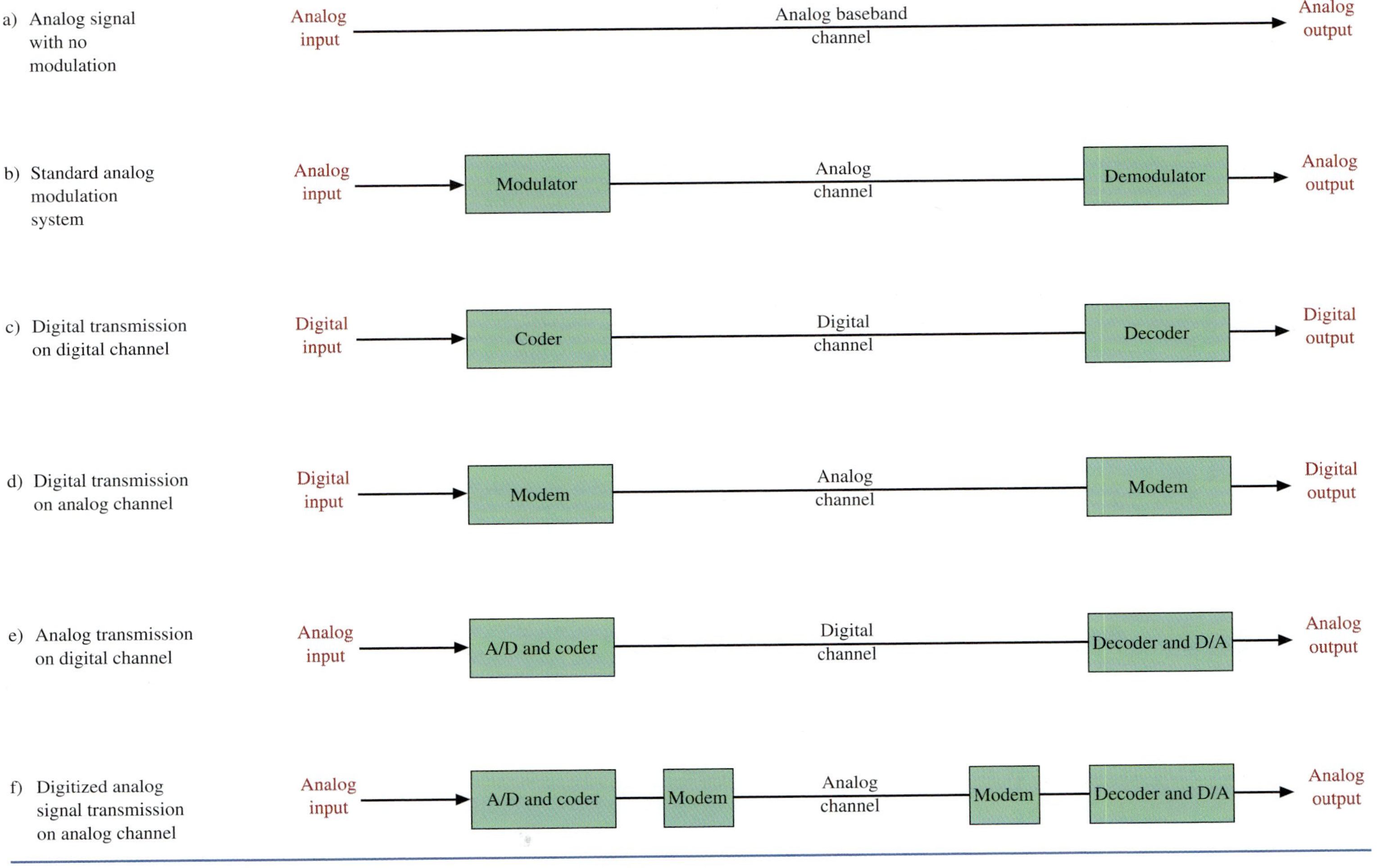

FIGURE 9-1 Transmission schemes for analog and digital signals.

Binary Coding

Some of the earliest forms of electrical communications used coding to send messages rather than direct transmission of *voice*. The telegraph demonstration by Samuel Morse in 1843 is an example. It is ironic that digital communications (as telegraphy is) now promise to help ease the problems of overcrowded voice transmission facilities. The future will certainly bring more and more coded speech, transmitted in digital format because of the following advantages:

1. Less sensitive to noise
2. Less crosstalk (cochannel interference)
3. Lower distortion levels
4. Faded signals more easily re-created
5. Greater transmission efficiency

What at first seemed a barrier to digital transmission—that is, the need to encode (convert) an analog signal into digital form—is proving to be an advantage. Coding now allows speech to be compressed to its minimum essential content and therefore permits the greatest possible efficiency in transmission.

Coding
transforming messages or signals in accordance with a definite set of rules

Coding may be defined as the process of transforming messages or signals in accordance with a definite set of rules. Many different codes are available for use, but one thing they share universally is the use of two levels. We can then refer to this as a binary system. In such a system, the next signal will either be *high* or *low* and should have an equal (50:50) chance of being one or the other. A *bit* is a unit of information required to allow proper selection of one out of two equally probable events. For example, assuming that heads or tails when flipping a coin are equally probable, let a high condition represent heads and a low condition represent tails. You must have 1 bit of information to predict the result of the coin toss. If that 1 bit of information is a high condition (usually termed the 1 level), then you can correctly predict that heads came up.

Mark, Space
analog signal representations of digital high or low states, respectively; usually sine waves of specific frequencies

The high and low conditions are referred to as 1 and 0, or **mark** and **space,** respectively. With respect to an electrical signal used to represent these conditions, the relationship is shown in Fig. 9-2. In this figure, 7 bits of information are

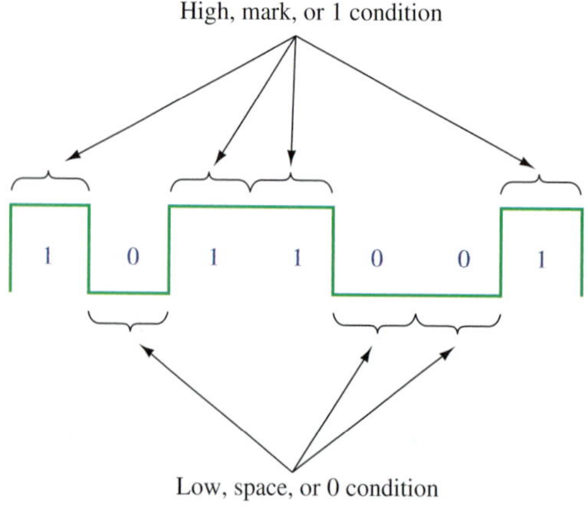

FIGURE 9-2 Binary information.

provided with the following sequence: 1 0 1 1 0 0 1. With 7 bits available, it is possible to select the correct result out of 128 equiprobable events because the 7 bits can be arranged in 128 different configurations, and each one may be allowed to represent 1 of 128 equiprobable events in a code. Because 2 raised to the seventh power equals 128, we can infer the following relationship:

$$n = \log_2 M \tag{9-1}$$

where n is the number of bits of information required to predict one out of M equiprobable events and by the logarithm definition, $2^n = M$.

If it were required to use a binary code to represent the 26 letters in the alphabet, Eq. (9-1) shows that

$$n = \log_2 26 = 4.7$$

Because the number of bits required must always be an integer, 5 bits are necessary. Because 5 bits means that a choice out of 32 different events ($2^5 = 32$) is possible, not all the capability of a 5-bit system is utilized. The coding efficiency is given as

$$\mu = \frac{\text{\# bits required}}{\text{\# bits used}} \tag{9-2}$$

Thus,

$$\mu = \frac{4.7}{5} \times 100\% = 94\%$$

It is not absolutely necessary to use a binary system for coding, but it is easy to show why it is used almost exclusively. A decimal system or 10 different levels is sometimes utilized. In it, the **dit** (decimal digit) is the basic unit of information. If it were used to represent our alphabet, the number of required dits would be $\log_{10} 26 = 1.415$ dits. Thus, 2 dits would be required to represent the 26 different letters, which means the efficiency of the decimal system in this case would be

Dit
decimal digit

$$\mu = \frac{1.415}{2} \times 100\% = 71\%$$

It is a fact that binary coding is more efficient than other coding except in isolated instances. It is also simpler to implement than other coding schemes.

Example 9-1

Determine the number of bits required for a binary code to represent 110 different possibilities, and compare its efficiency with a decimal system to accomplish the same goal.

Solution

In a binary system,

$$\begin{aligned} n &= \log_2 M \qquad (9\text{-}1)\\ &= \log_2 110 = 6.78 \end{aligned}$$

Solution of the equation above can be accomplished by taking the log of 110 divided by the log of 2.

$$\frac{\log_{10} 110}{\log_{10} 2} = 6.78$$

and therefore $2^{6.78} = 110$.
Thus, 7 bits are required, and the efficiency is

$$\mu = \frac{6.78}{7} \times 100\% = 97\% \quad \textbf{(9-2)}$$

In a decimal system, the number of dits required is $\log_{10} 110 = 2.04$, or a total of 3 dits. The efficiency is

$$\mu = \frac{2.04}{3} \times 100\% = 68\% \quad \textbf{(9-2)}$$

Code Noise Immunity

We have seen that binary coding systems are generally more efficient than other systems. Perhaps an even greater advantage is their superior noise immunity. Consider a binary system where 0 V represents 0 and 9 V represents 1. In such a system, it would take a noise level of roughly 5 V or more to cause an error in output intelligibility. In a decimally coded system of the same total output power, 0 is represented by 0 V, 1 by 1 V, 2 by 2 V, etc., up to 9 by 9 V. Thus, the 10 discrete levels are 1 V apart, and a 0.5-V noise level can impair intelligibility. If the output levels in this decimal system were 0, 10 V, 20 V, . . . , 90 V, then the 9-V binary and 90-V decimal system would have comparable noise immunities. Because power is proportional to the square of voltage, the decimal system requires 10^2 or 100 times the power of the binary system to offer the same noise immunity. By now in your study of communications, you may have come to the correct conclusion that noise is the single most important consideration. The reasoning just concluded regarding efficiency and noise immunity is an elementary example of the work of an information theory specialist. Recall from Chapter 1 that *information theory* is the branch of learning concerned with optimization of transmitted information.

Error Probability (P_e)
in a digital system, the number of errors per total number of bits received

Errors occur as a result of noise. There are many sources of noise, as explained in Chapter 1. The **error probability** (P_e) in a digital system is the number of errors per total number of bits received. For instance, if 1 error bit per 100,000 bits occurs, the error probability expressed as P_e is 1/100,000, or 10^{-5}. The acceptable error probability in communications systems ranges from about 10^{-5} up to 10^{-12} in more demanding applications. The average number of errors in a transmission m bits long can be calculated as

$$\text{average number of errors} = m \times \text{error probability } (P_e) \quad \textbf{(9-3)}$$

Bit Error Rate (BER)
the number of bit errors that occur for a given number of bits transmitted

The most common method of referring to the quality of a digital communications systems is its **bit error rate (BER).** The BER is the number of bit errors that occur for a given number of bits transmitted. The bit error rate is related to the error probability, P_e, because it is the ratio of bit errors to bits transmitted.

Example 9-2

A digital transmission has an error probability, P_e, of 10^{-6}, and 10^7 bits are received. Calculate the expected number of errors.

Solution

$$\begin{aligned}\text{average number of errors} &= m \times \text{error probability } (P_e) \qquad \textbf{(9-3)}\\ &= 10^7 \times 10^{-6}\\ &= 10\end{aligned}$$

or 10 expected bit errors if 10 million bits are received.

Energy per Bit

The **energy per bit** (E_b), or **bit energy,** is the amount of power in a digital bit for a given amount of time. The equation for calculating the bit energy is provided in Eq. (9-4).

Energy Per Bit (E_b), or Bit Energy
amount of power in a digital bit for a given amount of time for that bit

$$E_b = P_t \cdot T_b \qquad \textbf{(9-4)}$$

where E_b = energy per bit (in joules/bit)
P_t = total carrier power (watts)
T_b = 1/(bit frequency)

Example 9-3

The transmit power for a cellular phone is 0.8 W. The phone is transferring digital data at a rate of 9600 bps. Determine the energy per bit, E_b, for the data transmission.

Solution

$$\text{bit frequency} = 1/9600 = 1.042 \times 10^{-4}$$
$$E_b = P_t \cdot T_b = 0.8 \cdot 1.042 \times 10^{-4} = 8.336\ 10^{-5}\ \text{J} \quad (\text{or } 83.36\ \mu\text{J}) \qquad \textbf{(9-4)}$$

The value for the bit energy (E_b) is typically provided as a relative measure to the total system noise (N_o). Recall from Chapter 1 that the electronics in a communications system generate noise and that the primary noise contribution is thermal noise. The relationship for the bit-energy-to-noise value is stated as $\boldsymbol{E_b/N_o}$. In digital communication systems, the probability of a bit error (P_e) is a function of two factors, the bit-energy-to-noise ratio (E_b/N_o) and the method used to modulate the data digitally (e.g., QPSK; see Sec. 10-2). Basically, if the bit energy increases (within operational limits), then the probability of a bit error, P_e, decreases, and if the bit energy is low, then the probability of a bit error will be high.

E_b/N_o
the bit-energy-to-noise ratio

Communication Links and Protocols

Transmission of information in a communications link is defined by three basic protocol techniques: simplex, half-duplex, and full duplex. An example of simplex operation is a radio station transmitter. The ratio station is sending transmissions to you, but there is typically not a return communications link. Most handitalkies operate in the half-duplex mode. When one unit is transmitting, the other unit must be in the receive mode. A cellular telephone is an example of a communications link that is operating full duplex. In full-duplex mode, the units can be transmitting and receiving at the same time. The following terms apply to both analog and digital communications links.

Simplex	Communication is in one direction only.
Half-duplex	Communication is in both directions but only one can talk at a time.
Full-duplex	Both parties can talk at the same time.

Synchronous
this mode of operation implies that the transmit and receiver data clocks are locked together

Asynchronous
this mode of operation implies that the transmit and receiver data clocks are not locked together and the data must provide start and stop information to lock the systems together temporarily

The communications link can also be classified according to whether it is a **synchronous** or an **asynchronous** channel. Synchronous operation implies that the transmit and receive data clocks are locked together. In many communications systems, this requires that the data contain clocking information (called self-clocking data). Some of the data protocols presented in Sec. 8-4 meet this criteria. For example, biphase codes contain clocking information. The other option for transmitting data is to operate asynchronously, which means that the clocks on the transmitter and receiver are not locked together. That is, the data do not contain clocking information and typically contain start and stop bits to lock the systems together temporarily. NRZ-L is an example of a data format that requires the use of start and stop bits to lock the systems temporarily.

Protocols

Handshaking
procedures allowing for orderly exchange of information between a central computer and remote sites

Protocols
set of rules to make devices sharing a channel observe orderly communication procedures

Framing
separation of blocks into information and control sections

The vast number of digital data facilities now in existence require complex networks to allow equipment to "talk" to one another. To maintain order during the interchange of data, rules to control the process are necessary. Initially, procedures allowing orderly interchange between a central computer and remote sites were needed. These rules and procedures were called **handshaking.** As the complexity of data communications systems increased, the need grew for something more than a "handshake." Thus, sets of rules and regulations called **protocols** were developed. A protocol may be defined as a set of rules designed to force the devices sharing a channel to observe orderly communications procedures.

Protocols have four major functions:

1. *Framing:* Data are normally transmitted in blocks. The **framing** function deals with the separation of blocks into the information (text) and control sections. A maximum block size is dictated by the protocol. Each block will normally contain control information such as an *address field* to indicate the intended recipient(s) of the data and the block check character (BCC) for error detection. The protocol also prescribes the process for error correction when one is detected.

2. *Line control:* **Line control** is the procedure used to decide which device has permission to transmit at any given time. In a simple full-duplex system with just two devices, line control is obviously not necessary. However, systems with three or more devices **(multipoint circuits)** require line control.
3. *Flow control:* Often there is a limit on the rate at which a receiving device can accept data. A computer printer is a prime example of this condition. **Flow control** is the protocol process used to monitor and control these rates.
4. *Sequence control:* This is necessary for complex systems where a message must pass through numerous links before it reaches its final destination. **Sequence control** keeps message blocks from being lost or duplicated and ensures that they are received in the proper sequence. This has become an especially important consideration in packet-switching systems. They are introduced later in this chapter.

Line Control
procedure that decides which device has permission to transmit at a given time

Multipoint Circuits
systems with three or more devices

Flow Control
protocol used to monitor and control rates at which receiving devices receive data

Sequence Control
keeps message blocks from being lost or duplicated and ensures that they are received in the proper sequence

Protocols are responsible for integration of control characters within the data stream. Control characters are indicated by specific bit patterns that can also occur in the data stream. To circumvent this problem, the protocol can use a process termed **character insertion,** also called **character stuffing** or **bit stuffing.** If a control character sequence is detected in the data, the protocol causes an insertion of a bit or character that allows the receiving device to view the preceding sequence as valid data. This method of control character recognition is called **transparency.**

Character Insertion
insertion of a bit or character so that a data stream is not mistaken for a control character

Character Stuffing
another name for character insertion

Bit Stuffing
another name for character insertion

Transparency
control character recognition by using the character insertion process

Protocols are classified according to their framing technique. Character-oriented protocols (COPs) use specific binary characters to separate segments of the transmitted information frame. These protocols are slow and bandwidth-inefficient and are seldom used anymore. Bit-oriented protocols (BOPs) use frames made up of well-defined fields between 8-bit start and stop flags. A flag is fixed in both length and pattern. The BOPs include high-level data link control (HDLC) and synchronous data link control (SDLC). Synchronous data link control is considered a subset of HDLC. The byte-count protocol used in Digital Equipment Corporation's digital data communications message protocol (DDCMP) uses a header followed by a count of the characters that will follow. They all have similar frame structures and are therefore independent of codes, line configurations, and peripheral devices. There are many protocol variations.

To illustrate the use of BOPs, we will discuss synchronous data link communication (SDLC). It was developed by IBM. Data link control information is transferred on a bit-by-bit basis. Figure 9-3 illustrates the SDLC BOP format.

The start and stop flag for the SDLC frame is 01111110 ($7E_H$). The SDLC protocol requires that there be six consecutive ones (1s) within the flag. The transmitter will insert a zero after any five consecutive ones. The receiver must remove the zero when it detects five consecutive 1s followed by a zero. The communication equipment synchronizes on the flag pattern.

Start flag	Address	Control	Message	Frame check sequence	Stop flag
8 bits	8 bits	8 bits	Multiples of 8 bits	8 bits	8 bits

FIGURE 9-3 Bit-oriented protocol format, SDLC frame format.

9-3 Bandwidth Considerations

This section begins with a discussion on the capacity limits of a digital system, then we study the bandwidth considerations of the digital pulse stream. The system capacity and bit rate for a digital system are not unlimited. The relationship for bit rate to channel bandwidth is defined by the Shannon–Hartley theorem. The theorem relates the capacity of a channel when its bandwidth and noise are known.

$$C = \text{BW} \log_2(1 + S/N) \tag{9-5}$$

where C = channel capacity (bps)
BW = bandwidth of the system
S/N = signal-to-noise ratio

An example using Eq. (9-5) is provided in Ex. 9-4.

Example 9-4

Calculate the capacity of a telephone channel that has an S/N of 1023 (60 dB).

Solution

The telephone channel has a bandwidth of about 3 kHz. Thus,

$$\begin{aligned} C &= \text{BW} \log_2(1 + S/N) \qquad \textbf{(9-5)} \\ &= 3 \times 10^3 \log_2(1 + 1023) \\ &= 3 \times 10^3 \log_2(1024) \\ &= 3 \times 10^3 \times 10 \\ &= 30{,}000 \text{ bits per second} \end{aligned}$$

Example 9-4 shows that a telephone channel could theoretically handle 30,000 b/s if an S/N power ratio of 1023 (60 dB) exists on the line. The Shannon–Hartley theorem represents a fundamental limitation. The only consequence of exceeding it is a very high bit error rate. Generally, an acceptable bit error rate (BER) of 10^{-5} or better requires significant reductions from the Shannon–Hartley theorem prediction.

Bandwidth Considerations

bps
bits per second

Baud Rate
a measure of the symbols per second

Bandwidth considerations for digital systems are much like those of an analog system. Technical specifications such as the frequency response, available frequency spectrum, etc., are still of great importance. The available bandwidth dictates the amount of data that can be transmitted over time (i.e., bits per second, **bps**) and the bandwidth (in hertz) influences the choice of the digital modulation techniques. For clarification purposes, bit rate and **baud rate** are two different terms. Bit rate is measured in bits per second, but baud rate is a measure of symbols per second. For binary (two-level) data transmission, however, the bit rate and the baud rate are the same.

The first consideration for digital data transmission is that the bandwidth of our transmission system will not be infinite and therefore the transmitted pulse will

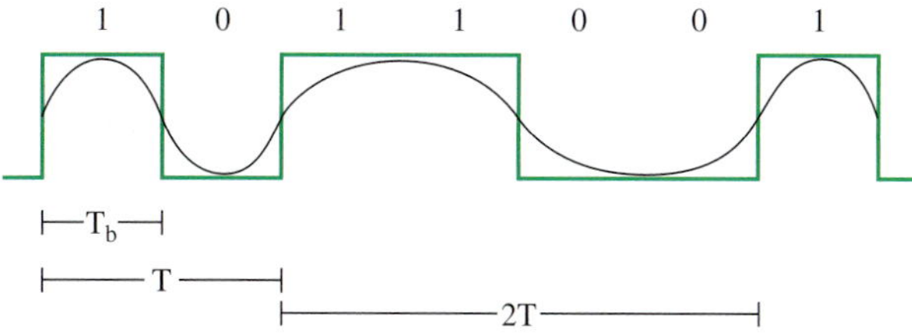

FIGURE 9-4 A digital pulse stream of frequency $1/T$ and a pulse width of T_b.

not be a perfect rectangle. Remember from Chapter 1 that the square wave is made up of a summation of sinusoids of harmonically related frequencies. It is important to note that the bandwidth requirements for a digital data stream depend on the encoding method used to transmit the data. Figure 9-4 shows the pulses in an NRZ-L-encoded digital data stream. (Refer to Sec. 8-4 for a review of digital encoding schemes.)

If an NRZ-L code is being used, the highest bit frequency requirement rate will be for an alternating 0 1 pattern. Note that the sinusoid component for the 1 1 0 0 sequence has a period twice that of the 1 0 pattern. Therefore, the 1 1 0 0 sequence will not be a limiting factor for the minimum bandwidth. The minimum bandwidth, BW_{min}, is determined by the sinusoidal component contained in the data stream with the shortest period of T (in seconds), which is the alternating 1 0 pattern. The equation for calculating the minimum bandwidth is

$$BW_{min} = \frac{1}{2T_b} = \frac{1}{T} \quad [\text{Hz}] \tag{9-6}$$

EXAMPLE 9-5

An 8-kbps NRZ-L encoded data stream is used in a digital communication link. Determine the minimum bandwidth required for the communications link.

SOLUTION

Using Eq. (9-6), the minimum bandwidth is

$$T_b = \frac{1}{8 \text{ kbps}} = 125 \ \mu\text{s}$$

$$BW_{min} = \frac{1}{2\,T_b} = \frac{1}{2 \cdot 125 \ \mu\text{s}} = 4 \text{ kHz} \tag{9-6}$$

9-4 DATA TRANSMISSION

This section introduces the basic concepts of high-speed serial data transmission. Topics in this section include an introduction to the data standards currently being used in data communications and the data formats being used. These data standards include T1 to T3, DS-1 to DS-3, and E1, E3. The OC (optical carrier) transmission data rates are discussed in Chapter 18. An overview of packet switching, frame relay, and the asynchronous transfer protocol (ATM) and SDLC/HDLC are also presented.

Data Channels

The common communication data rates (for end users) are T1 to T3 and DS-1 to DS-3. T1/DS-1 and T3/DS-3 are actually the same data rates. Data for the T-level carriers are listed in Table 9-1.

Table 9-1 Data Rates for the T and DS Carriers

Designation	Data Rate
T1 (DS-1)	1.544 Mbps
T2 (DS-2)	6.312 Mbps
T3 (DS-3)	44.736 Mbps
T4 (DS-4)	274.176 Mbps

The T1 line is capable of carrying 24 time-division-multiplexed telephone calls. Each phone call requires 64 kbps.

$$8 \text{ bits/sample} \times 8000 \text{ samples/second} = 64{,}000 \text{ bits per second} \quad (64 \text{ kbps})$$

Framing bits are added to the multiplexed voice data at a rate of 8 kbps to maintain data flow. This brings the total data rate for a T1 channel to:

$$\begin{array}{rr} 24 \text{ channels} \times 64 \text{ kbps/channel} = & 1.536 \text{ Mbps} \\ + \qquad \text{framing bits} = & 8 \text{ kbps} \\ \hline \text{total bit rate} = & 1.544 \text{ Mbps} \end{array}$$

T1 data lines are not restricted to carrying voice data. The data lines can be leased from a communications carrier for carrying any type of data, including voice, data, and video. It is important to note that when you lease a T1 line for providing a data connection from point A to point B, the communications carrier does not provide you with your own private physical connection. What this means is the communication carrier is providing you with sufficient bandwidth in their system to carry your T1 (1.544-Mbps) data rate. Your data will most likely be multiplexed with hundreds or even thousands of other T1 data channels.

A data facility may need only a portion of the data capability of the T1 bandwidth. **Fractional T1** (FT1) is the term used to indicate that only a portion of the T1 bandwidth is being used. Fractional T1 data rates up to 772 kbps are available. The most common fractional T1 data rate is 56 kbps. This used to be called a DS-0 line. The 56-kbps lines are provided as 24-hour dedicated leased lines or through a dial-up connection called switch-56. The user then pays for the data bandwidth only when it is needed.

Fractional T1
a term used to indicate that only a portion of the data bandwidth of a T1 line is being used

Two other designations for data rates are E1 and E3. These designations are used throughout the world where the T-carrier designation is not used. For example, the E1, E3 designations are primarily used in Europe. The data rates for E1 and E3 are listed in Table 9-2.

Table 9-2 E1 and E3 Data-Transmission Rates

Designation	Date Rate
E1	2.028 Mbps
E3	34.368 Mbps

Point of Presence

The point where the communication carrier brings in service to a facility is called the **point of presence.** This is the point where users connect their data to the communications carrier. The link to the communications carrier can be copper, fiber, digital microwave, or digital satellite. The communications carriers will also require that data be connected through a **CSU/DSU** (channel service unit/data service unit). The CSU/DSU provides the data interface to the communications carrier, which includes adding the framing information for maintaining the data flow, storing performance data, and providing management of the line. An example of inserting the CSU/DSU in the connection to the Telco cloud is shown in Fig. 9-5. The CSU/DSU also has three alarm modes for advising the user about problems on the link. The alarms are red, yellow, and blue. The conditions for each alarm are defined in Table 9-3.

Point of Presence
the point where users connect their data to the communications carrier

CSU/DSU
channel service unit/data service unit, which provides the data interface to the communications carrier doing framing and line management

FIGURE 9-5 Insertion of the CSU/DSU in the connection to the Telco cloud.

Table 9-3 The CSU/DSU Alarms

Red alarm	A local equipment alarm that indicates that the incoming signal has been corrupted.
Yellow alarm	Indicates that a failure in the link has been detected.
Blue alarm	Indicates a total loss of incoming signal.

T1 Framing

The original framing for the data in T1 circuits is based on **D4 framing.** The D4 frame consists of 24 voice channel (8 kbps) × 8 bits/channel plus one framing bit, for a total of 193 bits. The D4 system uses a 12-bit framing sequence of

1 0 0 0 1 1 0 1 1 1 0 0

to maintain synchronization of the receiving equipment. The D4 12-bit framing sequence is generated from 12 D4 frames.

D4 Framing
the original data framing used in T1 circuits

ESF
extended superframe framing

Extended superframe framing **(ESF)** is an improvement in data performance over D4 framing. ESF extends the frame length to 24 frames compared at D4, which uses 12 frames. The extended frame length creates 24 ESF framing bits. The use of the 24 bits in the ESF frame are listed in Table 9-4.

Table 9-4 The Function of the 24 ESF Framing Bits

6 bits	Frame synchronization
6 bits	Error detection
12 bits	Communications link control and maintenance

ESF uses only 6 bits for frame synchronization compared to the 12 synchronizing bits used in D4 framing. ESF uses 6 bits for computing an error check code. The code is used to verify the data transmission was received without errors. Twelve bits of the ESF frame are used for maintenance and control of the communications link. Examples of the use of the maintenance and control bits include obtaining performance data from the link and configuring loopbacks for testing the link. A **loopback** is when the data is routed back to the sender.

Loopback
data is routed back to the sender

Three loopback tests are shown in Fig. 9-6. The loopback test marked A is used to test the cable connecting the router to the CSU/DSU. Loopback test B is used to test the link through the CSU/DSU. Loopback test C tests the CSU and the link to Telco.

Line Coding Formats

The data connection to the communications carrier requires that the proper data encoding format be selected for the CSU/DSU. The data are encoded so that timing information of the binary stream is maintained and the logical 1s and 0s can still be detected.

A fundamental coding scheme that was developed for transmission over T1 circuits is alternate mark inversion **(AMI)**. The AMI code provides for alternating

AMI
alternate mark inversion

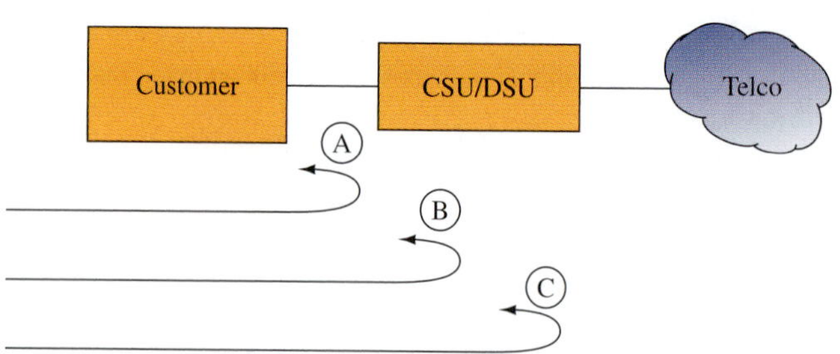

(A) Tests the cable connecting the router to the CSU/DSU

(B) Tests the link through the CSU/DSU

(C) Tests the CSU/DSU and the link to Telco

FIGURE 9-6 Three loopback configurations used to test the serial communications link.

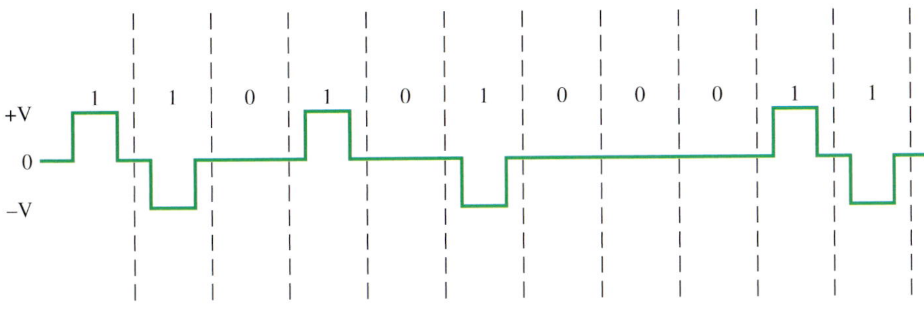

FIGURE 9-7 An example of the AMI data encoding format.

voltage level pulses (+V and −V) for representing the ones. This technique removes the DC component of the data stream almost completely, which helps to maintain synchronization. An example of the AMI coded waveform is shown in Fig. 9-7. Notice that successive 1s are represented by pulses in the opposite direction (+V and −V). This is called **bipolar coding.**

Bipolar Coding
successive ones are represented by pulses in the opposite voltage direction

The 0s have a voltage level of 0 V. Notice that when there are successive 0s a straight line at 0 V is generated. A flat line generated by a long string of 0s can produce a loss of timing and synchronization. This deficiency can be overcome by the transmission of the appropriate start, stop, and synchronizing bits, but this comes at the price of adding overhead bits to the data transmission and consuming a portion of the data communication channel's bandwidth.

B8ZS
bipolar 8 zero substitution

The bipolar 8 zero substitution **(B8ZS)** data encoding format was developed to improve data transmission over T1 circuits. T1 circuits require that a minimum ones density level be met so that the timing and synchronization of the data link are maintained. Maintaining a **minimum ones density** means that a pulse is intentionally sent in the data stream, even if the data being transmitted is a series of 0s. Intentionally inserting the pulses in the data stream helps to maintain the timing and synchronization of the data stream. The inserted data pulses include two **bipolar violations,** which means that the pulse is in the same voltage direction as the previous pulse in the data stream.

Minimum Ones Density
a pulse is intentionally sent in the data stream even if the data being transmitted is a series of 0s only

Bipolar Violation
the pulse is in the same voltage direction as the previous pulse

In B8ZS encoding, eight consecutive 0s are replaced with an 8-bit sequence that contains two intentional bipolar violations. An example of B8ZS encoding is shown in Fig. 9-8 for both cases of bipolar swing prior to the transmission of eight consecutive 0s. Fig. 9-8(a) shows the bipolar swing starting with a +V, while Fig. 9-8(b) shows the bipolar swing starting with −V. The receiver detects the bipolar violations in the data stream and replaces the inserted byte (8 bits) with all 0s to recover the original data stream. The result is that timing is maintained without corrupting the data. The advantage of using the B8ZS encoded fromat is that the bipolar violations enable the timing of the data transmission to remain synchronized without the need for overhead bits. This provides the use of the full data capacity of the channel.

HDLC
high-level data link control; a synchronous proprietary protocol

PPP
point-to-point protocol

Two other serial line protocols commonly used in wide area networking are high-level data link control **(HDLC)** and point-to-point protocol **(PPP).** Both protocols are used to carry data over a serial line connection, typically over direct connections such as with T1. The hardware at each end of the data connection must be configured with the proper data encapsulation. **Data encapsulation** means that the data is packaged properly for transport over a serial communications line. The type

Data Encapsulation
properly formatting the data for transport over a serial communications line

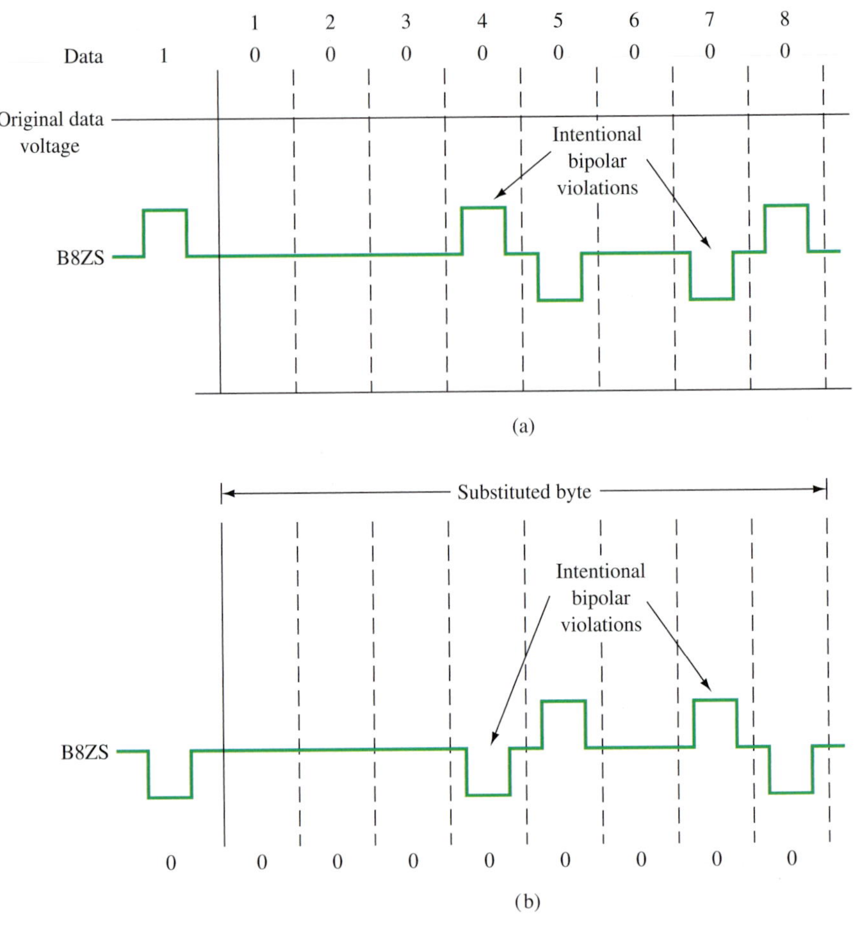

FIGURE 9-8 Intentional bipolar violations in B8ZS encoding.

of encapsulation depends on the hardware used to make the connection. The HDLC data encapsulation formats are implemented differently by some vendors, and some equipment is not always interoperable with other equipment, even though they both have specified the HDLC encapsulation. In that case, another encapsulation format such as PPP can be used to make the direct connection.

Packet Switching

Packets
segments of data

Statistical Concentration
processors at switching centers directing packets so that a network is used most efficiently

In this data technique, data are divided into small segments called **packets.** A typical packet size is 1000 bits. The individual packets of an overall message do not necessarily take the same path to a destination. They are held for very short periods of time at switching centers and are therefore transmitted in near real time. The processors at the switching centers monitor the packets continuously from the standpoint of source, destination, and priority. The processors then direct each packet so the network is used most efficiently. The process is termed **statistical concentration.** The price paid for the very high efficiency afforded and near-real-time transmission is the need for very complex protocols and switching arrangements. As packet switching technology progresses, the user may not need to choose in advance between packet and circuit switching; the techniques may merge. This could be the result of "fast" packet switches. The development of packet switch systems operating at millions of packets per second could eliminate the need for circuit switching.

Frame Relay

A **frame relay** is a packet switching network designed to carry data traffic over a **public data network** (PDN) (e.g., **Telco,** the local telephone company). Frame relay is an extension of the X.25 packet switching system; however, frame relay does not provide the error-checking and data-flow control of X.25. **X.25** was designed for data transmission over analog lines, and frame relay is operated over higher-quality, more reliable digital data lines.

Frame relay operates on the premise that the data channels will not introduce bit errors or worst-case minimal bit errors. Without the overhead bits for error checking and data-flow control, the transfer of data in a frame relay system is greatly improved. If an error is detected, then the receiver system corrects the error. The frame relay protocol enables calls or connections to be made within a data network. Each data frame contains a connection number that identifies the source and destination addresses.

The commercial carrier (telco) provides the switch for the frame relay network. Telco provides a guaranteed data rate, or **committed information rate (CIR),** based on the service and bandwidth requested by the user. For example, the user may request a T1 data service with a CIR of 768 kbps. A T1 connection allows a maximum data rate of 1.544 Mbps. The network (Telco) allows for bursty data transmissions and will allow the data transfer to go up to the T1 connection carrier bandwidth, even though the CIR is 768 kbps. **Bursty** means that the data rates can momentarily exceed the leased CIR data rate of the service. Communication carriers use what is called a **committed burst information rate (CBIR),** which enables subscribers to exceed the committed information rate (CIR) during times of heavy traffic. Note: The bursty data transmission rate can never exceed the data rate of the physical connection. For example, the maximum data rate for a T1 data service is 1.544 Mbps, and the bursty data rate can never exceed this rate.

Frame Relay
a packet switching network designed to carry data traffic over a public data network

Public Data Network
a local telephone company or a communications carrier

Telco
the local telephone company

X.25
a packet-switched protocol designed for data transmission over analog lines

Committed Information Rate
guaranteed data rate or bandwidth to be used in the frame relay connection

Bursty
a state in which the data rates can momentarily exceed the leased data rate of the service

Committed Burst Information Rate (CBIR)
enables subscribers to exceed the committed information rate (CIR) during times of heavy traffic

Asynchronous Transfer Mode (ATM)

The asynchronous transfer mode **(ATM)** is a cell relay technique designed for voice, data, and video traffic. Cell relay is considered to be an evolution of **packet switching** because packets or cells are processed at switching centers and directed to the best network for delivery. The stations connected to an ATM network transmit octets (8 bits of data) in a cell that is 53 octets (bytes) long. Forty-eight bytes of the cell are used for data (or **payload**) and five bytes are used for the cell header. The ATM cell header contains the data bits used for error checking, virtual circuit identification, and payload type.

All ATM stations are always transmitting cells, but the empty cells are discarded at the ATM switch. This technique provides more efficient use of the available bandwidth and allows for bursty traffic. The station is also guaranteed access to the network with a specified data frame size. This is not true for IP networks,

ATM
asynchronous transfer mode; a cell relay network designed for voice, data, and video traffic

Packet Switching
packets are processed at switching centers and directed to the best network for delivery

Payload
another name for the data being transported

where heavy traffic can bring the system to a crawl. The ATM protocol was designed for use in high-speed multimedia networking, including operation in high-speed data transmission from T1 up to T3, E3, and SONET. SONET is the synchronous optical network and is covered in Chapter 18. The standard data rate for ATM is 155 Mbps, although the data rates for ATM are continually evolving.

Virtual Path Connection (VPC)
used to connect the end users

Virtual Channel Connection (VCC)
carries the ATM cell from user to user

SVC
switched virtual circuit

ATM is connection oriented and uses two different types of connections, a **virtual path connection (VPC)** and a **virtual channel connection (VCC).** A virtual channel connection is used to carry the ATM cell data from user to user. The virtual channels are combined to create a virtual path connection that is used to connect the end users. Virtual circuits can be configured as permanent virtual connections (PVCs) or they can be configured as switched virtual circuits **(SVCs).**

Five classes of services are available with ATM. These classes are based on the needs of the user. In some applications, the users needs a constant bit rate for applications such as teleconferencing. In another application, the user may need only limited periods of higher bandwidth to handle bursty data traffic. The five ATM service classes are provided in Table 9-5.

Table 9-5 The Five ATM Service Classes

ATM Service Class	Abbreviation	Description	Typical Use
Constant bit rate	CBR	Cell rate is constant	Telephone, videoconferencing, television
Variable bit rate/ not real time	VBR-NRT	Cell rate is variable	Email
Variable bit rate/ real time	VBR-RT	Cell rate is variable but can be constant on demand	Voice traffic
Available bit rate	ABR	Users are allowed to specify a minimum cell rate	File transfers/ email
Unspecified bit rate	UBR		TCP/IP

VPI
virtual path identifier

VCI
virtual channel identifier

ATM uses an 8-bit virtual path identifier **(VPI)** to identify the virtual circuits used to deliver cells in the ATM network. A 16-bit virtual circuit identifier **(VCI)** is used to identify the connection between the two ATM stations. The VPI and VCI numbers are provided by Telco. Together, the numbers are used to create an ATM permanent virtual circuit (PVC) through the ATM cloud.

9-5 Time-Division Multiple Access (TDMA)

Time-Division Multiple Access (TDMA)
a technique used to transport data from multiple users over the same data channel

Time-division multiple access (TDMA) is a technique used to transport data from multiple sources over the same serial data channel. TDMA techniques are used for transporting data over wired systems and are also used for wireless communication (e.g., some cellular telephones). An example of generating a TDMA output is provided in Fig. 9-9. A1–A4, B1–B4, and C1–C4 represent digital data from three different users. A multiplexer (MUX) is used to combine the source data into one serial data stream. A key to successful data multiplexing is that the multiplexing

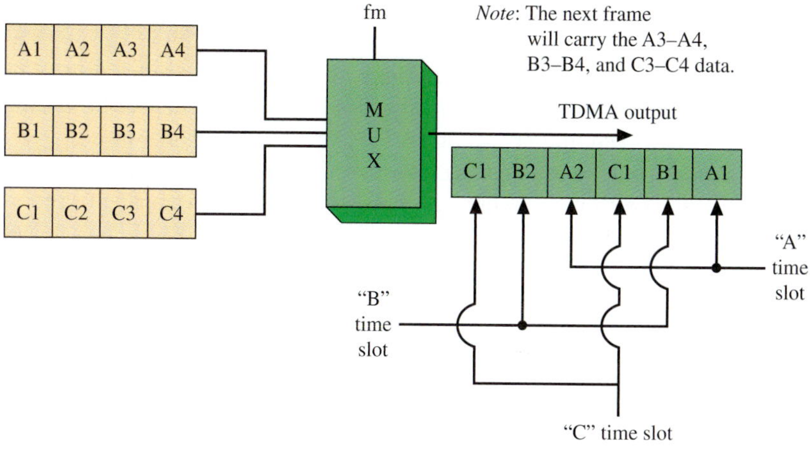

FIGURE 9-9 An example of generating a TDMA output.

frequency (f_m) must be fast enough so that the throughputs of the A, B, and C data are output fast enough, so that the system does not become congested and none of the data are lost.

The TDMA output data stream (Fig. 9-9) shows the placement of the A, B, and C data in the TDMA frame. Each block represents a **time slot** within the frame. The time slot provides a fixed location (relative in time to the start of a data frame) for each group of data so at the receiver, the data can be recovered easily. The process of recovering the data is shown in Fig. 9-10. The serial data is input into a **demultiplexer (DMUX),** which recovers the A, B, and C group data. If only the information in the A time slots is to be recovered, then the data in the B and C time slots are ignored.

Time Slot
provides a fixed location (relative in time to the start of a data frame) for each group of data

Demultiplexer (DMUX)
recovers the individual groups of data from the TDMA serial data stream

In wireless systems the arrival of the TDMA data can be an issue because of potential multipath RF propagation problems with stationary systems and the added RF propagation path problems introduced by mobile communications. To compensate for the variation in data arrival times, **guard times** are added to the TDMA frame. If the data arrive too closely together, then potential **intersymbol interference (ISI)** due to data overlapping can occur, resulting in an increase in the bit error rate (BER). The guard times provide an additional margin of error, thereby minimizing intersymbol interference and bit errors.

Guard Times
time added to the TDMA frame to allow for the variation in data arrival

Intersymbol Interference (ISI)
the overlapping of data, which can increase the bit error rate

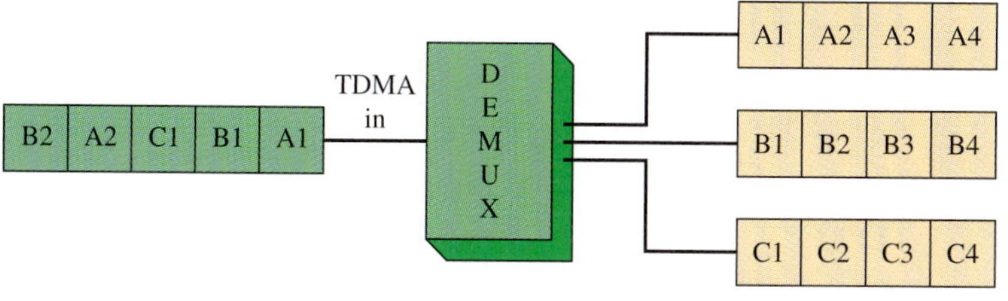

FIGURE 9-10 An example of recovering TDMA data.

9-6 Delta and Pulse Modulation

Delta Modulation digital modulation technique in which the encoder transmits information regarding whether the analog information increases or decreases in amplitude

Slope Modulation another name for delta modulation

Tracking ADC an ADC whose output indicates input changes rather than absolute values of input

Slope Overload in delta modulators, when the analog signal has a high rate of amplitude change, the encoder can produce a distorted analog signal

Continuously Variable Slope Delta increasing the step-size in a delta modulation system when three or more consecutive ones or zeros occur

Delta modulation (DM), sometimes called **slope modulation,** is another truly digital system (as is PCM). It transmits information only to indicate whether the analog signal it encodes is to "go up" or "go down." This process is shown in Fig. 9-11. Note that the encoder outputs are highs or lows that "instruct" whether to go up or go down, respectively. The relative simplicity of this system is readily apparent as compared to PCM. Delta modulation takes advantage of the fact that voice signals do not change abruptly and there is generally only a small change in level from one sample to the next. On the other hand, PCM can respond to very abrupt level changes between samples, such as analog signals that have been time-domain multiplexed before encoding.

A schematic for a delta modulator is shown in Fig. 9-12. A demodulator would consist of an integrator (just like the one in Fig. 9-12) followed by a sharp-cut-off low-pass filter. The integrator output would look like waveform B in Fig. 9-12. The filter smooths it out to provide the final analog signal. (You will be asked to describe/explain fully the modulator's operation in a question at the end of the chapter.) Further detail on the demodulation process is provided in Fig. 9-13.

The major advantage of delta modulation is simplicity. The delta modulator is called a **tracking ADC** because it follows the contours of the input and provides output that shows input changes rather than exact values. It does not require the synchronization of PCM systems and inherently provides a serial stream of bits, so that there is no need for a parallel-to-serial converter.

A difficulty faced by these systems is **slope overload.** When the analog signal has a high rate of change, the delta modulator can "fall behind" and a distorted output results. An increased sample rate could be used, but this necessitates a higher bandwidth for transmission of the signal. In systems where this is a problem, a technique called **continuously variable slope delta** (CVSD) modulation is used. A typical CVSD scheme is to increase the step-size whenever there is a longer run than three successive 1s or 0s. When the modulator catches up (as indicated by a change from 1 to 0 or 0 to 1), the step-size returns to normal. CVSD modulation is

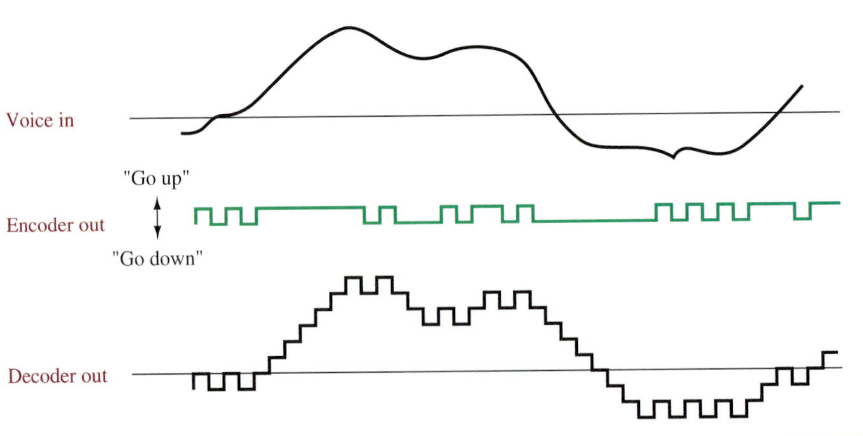

FIGURE 9-11 Linear delta modulation.

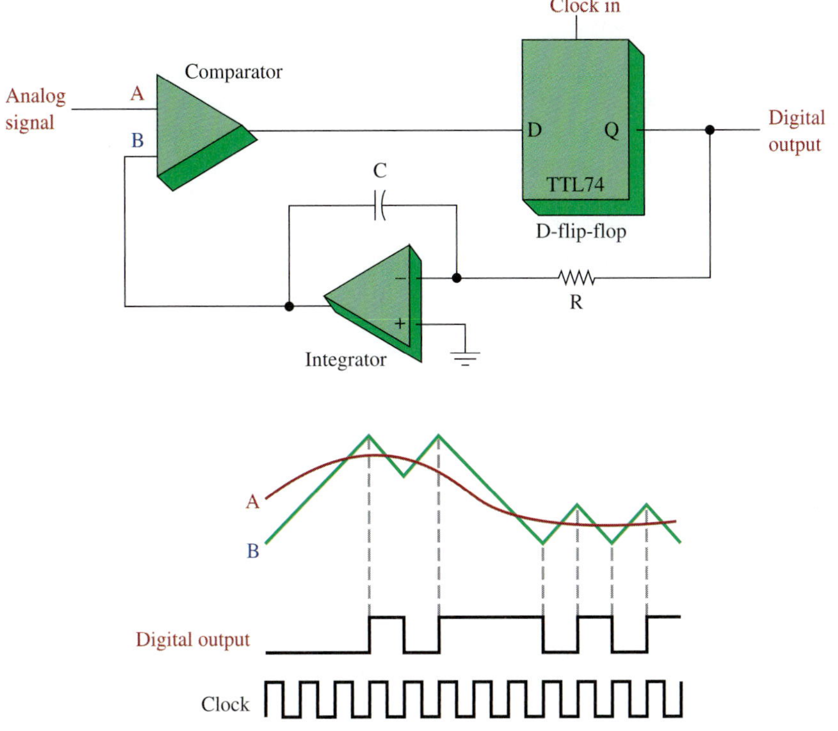

FIGURE 9-12 Delta modulator (linear).

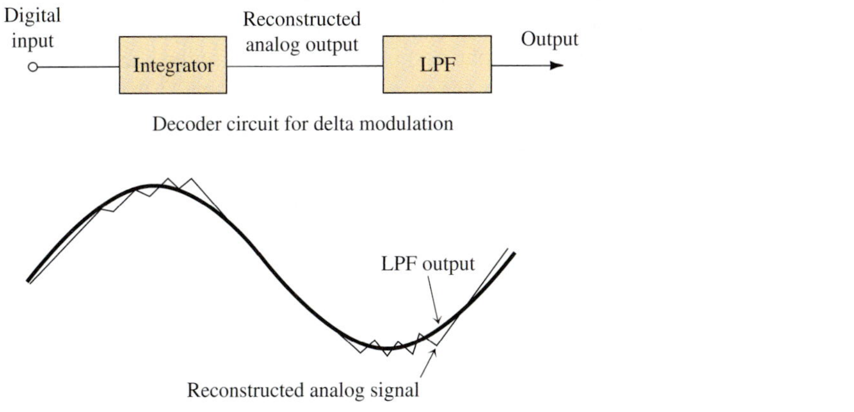

FIGURE 9-13 Demodulation circuit for delta modulation.

accomplished by using decision circuitry to count the number of 1s and 0s from the output of the D-flip-flop and a variable gain amplifier.

CVSD systems are typically implemented using a single LSI chip for both the transmitter and receiver sections. Similarly, the algorithm used for transmitting is used to reverse the process for reception. The net overall effect of CVSD can be compared to companding, except that CVSD depends on past values and adapts, while companding is a fixed algorithm.

Pulse Modulation

You have undoubtedly drawn graphs of *continuous* curves many times during your education. To do that, you took data at some finite number of discrete points, plotted each point, and then drew the curve. Drawing the curve may have resulted in a very accurate replica of the desired function even though you did not look at every possible point. In effect, you took *samples* and guessed where the curve went between the samples. If the samples had sufficiently close spacing, the result is adequately described. It is possible to apply this line of thought to the transmission of an electrical signal, that is, to transmit only the samples and let the receiver reconstruct the total signal with a high degree of accuracy. This is termed **pulse modulation.**

Pulse Modulation
the process of using some characteristic of a pulse (amplitude, width, position) to carry an analog signal

The key distinction between pulse modulation and normal AM or FM is that in AM or FM some parameter of the modulated wave varies continuously with the message, whereas in pulse modulation some parameter of a sample pulse is varied by each sample value of the message. The pulses are usually of very short duration so that a pulse modulated wave is "off" most of the time. This factor is the main reason for using pulse modulation because it allows

1. Transmitters to operate on a very low duty cycle ("off" more than "on"), as is desirable for certain microwave devices and lasers
2. The time intervals between pulses to be filled with samples of other messages

The latter reason conveniently allows several different messages to be transmitted on the same channel. This is the form of multiplexing known as **time-division multiplexing** (TDM). It is analogous to computer time sharing, where several users utilize a computer simultaneously.

Time-Division Multiplexing
two or more intelligence signals are sequentially sampled to modulate the carrier in a continuous, repeating fashion

Nyquist Rate
the sampling frequency must be at least twice the highest frequency of the intelligence signal or there will be distortion that cannot be corrected by the receiver

It was shown in Chapter 8 that a signal sampled at twice the rate of its highest significant frequency component can be reconstructed fully at the receiver to a high degree of accuracy. Stated inversely, a given bandwidth can carry pulse signals of half its high-frequency cutoff. This is known as the **Nyquist rate.** In the case of voice transmission, the standard sampling rate is 8 kHz, it being just slightly more than twice the highest significant frequency component. This implies a pulse rate of 8 kHz or a 125-μs period. Because a pulse duration of 1 μs may be adequate, it is easy to see that several different messages could be multiplexed (TDM) on the channel, or alternatively it would allow a high peak transmitted power with a much lower (1/125) average power. The high peak power can provide a very high signal-to-noise ratio or a greater transmission range.

Note that a price must be paid for system gains obtained by pulse modulation schemes. More important than the greater equipment complexity is the requirement for greater channel (bandwidth) size. If a maximum 3-kHz signal directly amplitude-modulates a carrier, a 6-kHz bandwidth is required. If a 1-μs pulse does the modulating, just allowing its fundamental component of 1/1 μs or 1 MHz to do the modulating means a 2-MHz bandwidth is required in AM. In spite of the large bandwidth required, TDM is still preferable (if not the only possible way) to using 100 different transmitters, antennas or transmission lines, and receivers in cases where large numbers of messages must be conveyed simultaneously.

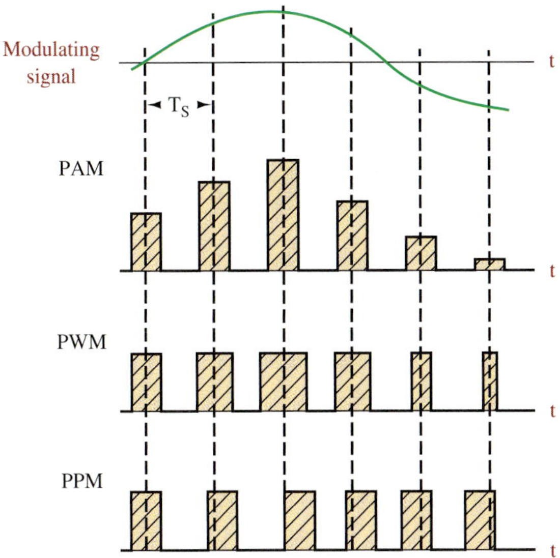

FIGURE 9-14 Types of pulse modulation.

In its strictest sense, pulse modulation is not modulation but rather a *message-processing* technique. The message to be transmitted is sampled by the pulse, and the pulse is subsequently used to either amplitude- or frequency-modulate the carrier. The three basic forms of pulse modulation are illustrated in Fig. 9-14. There are numerous varieties of pulse modulation, and no standard terminology has yet evolved. The three types we shall consider here are usually termed **pulse-amplitude modulation** (PAM), **pulse-width modulation** (PWM), and **pulse-position modulation** (PPM). For the sake of clarity, the illustration of these modulation schemes has greatly exaggerated the pulse widths. Because a major application of pulse modulation occurs when TDM is to be used, shorter pulse durations, leaving room for more multiplexed signals, are obviously desirable. As shown in Fig. 9-14, the pulse parameter that is varied in step with the analog signal is varied in direct step with the signal's value at each sampling interval. Notice that the pulse amplitude in PAM and pulse width in PWM are not zero when the signal is minimum. This is done to allow a constant pulse rate and is important in maintaining synchronization in TDM systems.

Pulse-Amplitude Modulation
sampling short pulses of the intelligence signal; the resulting pulse amplitude is directly proportional to the intelligence signal's amplitude

Pulse-Width Modulation
sampling short pulses of the intelligence signal; the resulting pulse width is directly proportional to the intelligence signal's amplitude

Pulse-Position Modulation
sampling short pulses of the intelligence signal; the resulting position of the pulses is directly proportional to the intelligence signal's amplitude

Pulse-Amplitude Modulation

In pulse-amplitude modulation (PAM), the pulse amplitude is made proportional to the modulating signal's amplitude. This is the simplest pulse modulation to create because a simple sampling of the modulating signal at a periodic rate can be used to generate the pulses, which are subsequently used to modulate a high-frequency carrier. An eight-channel TDM PAM system is illustrated in Fig. 9-15. At the transmitter, the eight signals to be transmitted are periodically sampled. The sampler illustrated is a rotating machine making periodic brush contact with each signal. A similar rotating machine at the receiver is used to distribute the eight separate signals, and it must be synchronized to the transmitter. A mechanical

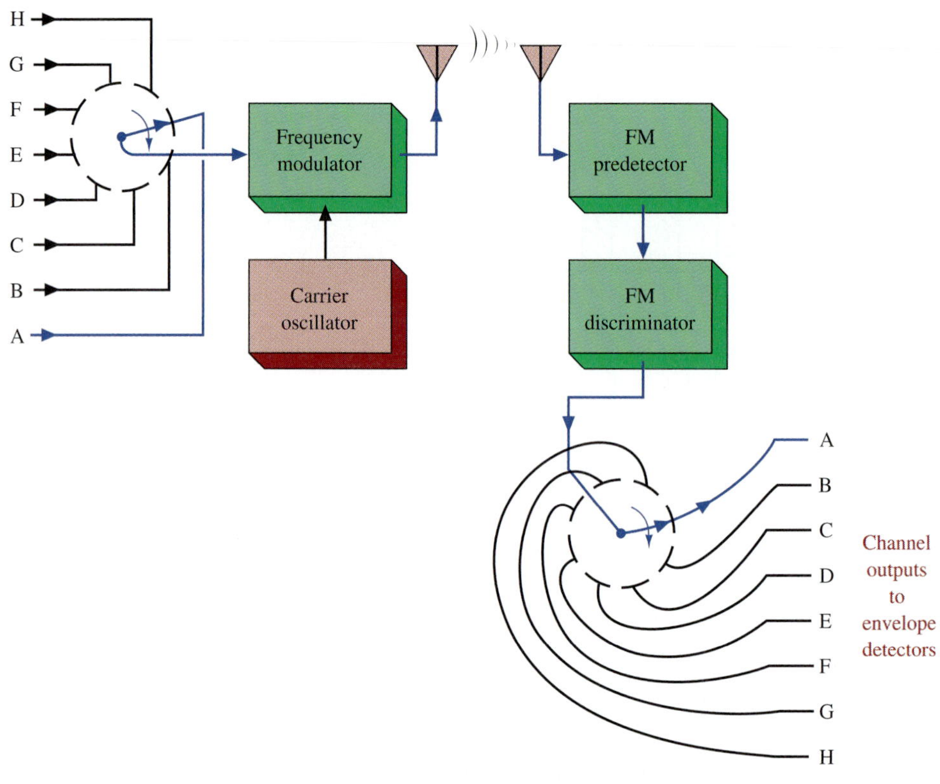

FIGURE 9-15 Eight-channel TDM PAM system.

Pulse-Time Modulation
modulation schemes that vary the timing (not the amplitude) of pulses

Pulse-Duration Modulation
another name for pulse-width modulation

Pulse-Length Modulation
another name for pulse-width modulation

sampling system such as this may be suitable for the low sampling rates like those encountered in some telemetry systems but it would not be adequate for the 8-kHz rate required for voice transmissions. In that case, an electronic switching system would be incorporated.

At the transmitter, the variable amplitude pulses are used to frequency-modulate a carrier. A rather standard FM receiver re-creates the pulses, which are then applied to the electromechanical *distributor* going to the eight individual channels. This distributor is virtually analogous to the distributor in a car that delivers high voltage to eight spark plugs in a periodic fashion. The pulses applied to each line go into an envelope detector that serves to re-create the original signal. This can be a simple low-pass *RC* filter such as that used following the detection diode in a standard AM receiver.

While PAM finds some use due to its simplicity, PWM and PPM use constant amplitude pulses and provide superior noise performance. The PWM and PPM systems fall into a general category termed **pulse-time modulation** (PTM) because their timing, and not amplitude, is the varied parameter.

Pulse-Width Modulation

Pulse-width modulation (PWM), a form of PTM, is also known as **pulse-duration modulation** (PDM) and **pulse-length modulation** (PLM). A simple means of

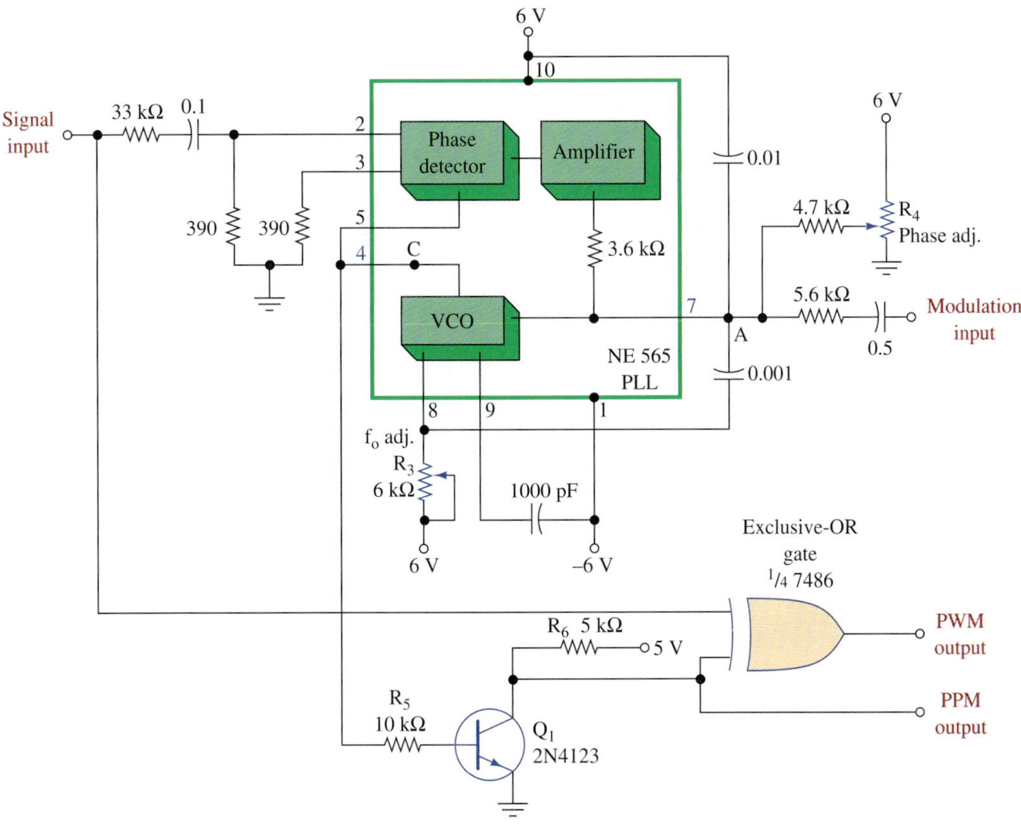

FIGURE 9-16 PLL generation of PWM and PPM.

PWM generation is provided in Fig. 9-16 using a 565 PLL. It actually creates PPM at the VCO output (pin 4), but by applying it and the input pulses to an exclusive-OR gate, PWM is also created. For the phase-locked loop (PLL) to remain locked, its VCO input (pin 7) must remain constant. The presence of an external modulating signal upsets the equilibrium. This causes the phase detector output to go up or down to maintain the VCO input (control) voltage. However, a change in the phase detector output also means a change in phase difference between the input signal and the VCO signal. Thus, the VCO output has a phase shift proportional to the modulating signal amplitude. This PPM output is amplified by Q_1 in Fig. 9-16 just prior to the output. The exclusive-OR circuit provides a high output only when just one of its two inputs is high. Any other input condition produces a low output. By comparing the PPM signal and the original pulse input signal as inputs to the exclusive-OR circuit, the output is a PWM signal at twice the frequency of the original input pulses.

Adjustment of R_3 varies the center frequency of the VCO. The R_4 potentiometer may be adjusted to set up the quiescent PWM duty cycle. The outputs (PPM or PWM) of this circuit may then be used to modulate a carrier for subsequent transmission.

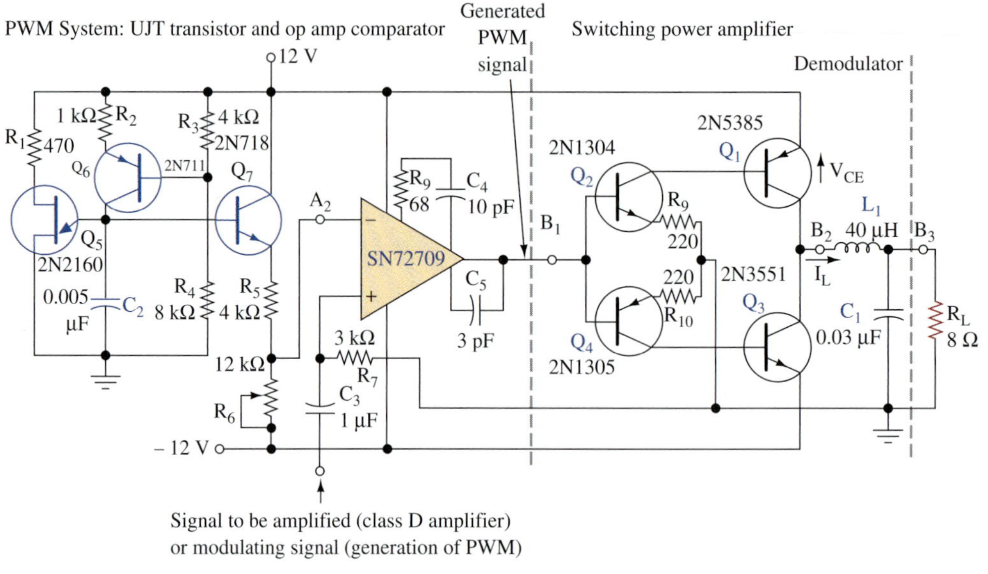

FIGURE 9-17 PWM generator and class D power amplifier.

Class D Amplifier and PWM Generator PWM forms the basis for a very efficient form of power amplification. The circuit in Fig. 9-17 is a so-called class D amplifier because the actual power amplification is provided to the PWM signal, and because it is of constant amplitude, the transistors used can function between cutoff and saturation. This allows for maximum efficiency (in excess of 90 percent) and is the reason for the increasing popularity of class D amplifiers as a means of amplifying any analog signal.

The circuit of Fig. 9-17 illustrates another common method for generation of PWM and also illustrates class D amplification. The Q_6 transistor generates a constant current to provide a linear charging rate to capacitor C_2. The unijunction transistor, Q_5, discharges C_2 when its voltage reaches Q_5's firing voltage. At this time C_2 starts to charge again. Thus, the signal applied to Q_7's base is a linear saw tooth as shown at A in Fig. 9-18. That sawtooth following amplification by the Q_7 emitter-follower in Fig. 9-17 is applied to the op amp's inverting input. The modulating signal or signal to be amplified is applied to its noninverting input, which causes the op amp to act as a comparator. When the sawtooth waveform at A in Fig. 9-18 is less than the modulating signal B, the comparator's output (C) is high. At the instant A becomes greater than B, C goes low. The comparator (op amp) output is therefore a PWM signal. It is applied to a push-pull amplifier (Q_1, Q_2, Q_3, Q_4) in Fig. 9-17, which is a highly efficient switching amplifier. The output of this power amp is then applied to a low-pass LC circuit (L_1, C_1) that converts back to the original signal (B) by integrating the PWM signal at C, as shown at D in Fig. 9-18. The output of the op amp in Fig. 9-17 would be used to modulate a carrier in a communications system, while a simple integrating filter would be used at the receiver as the detector to convert from pulses to the original analog modulating signal.

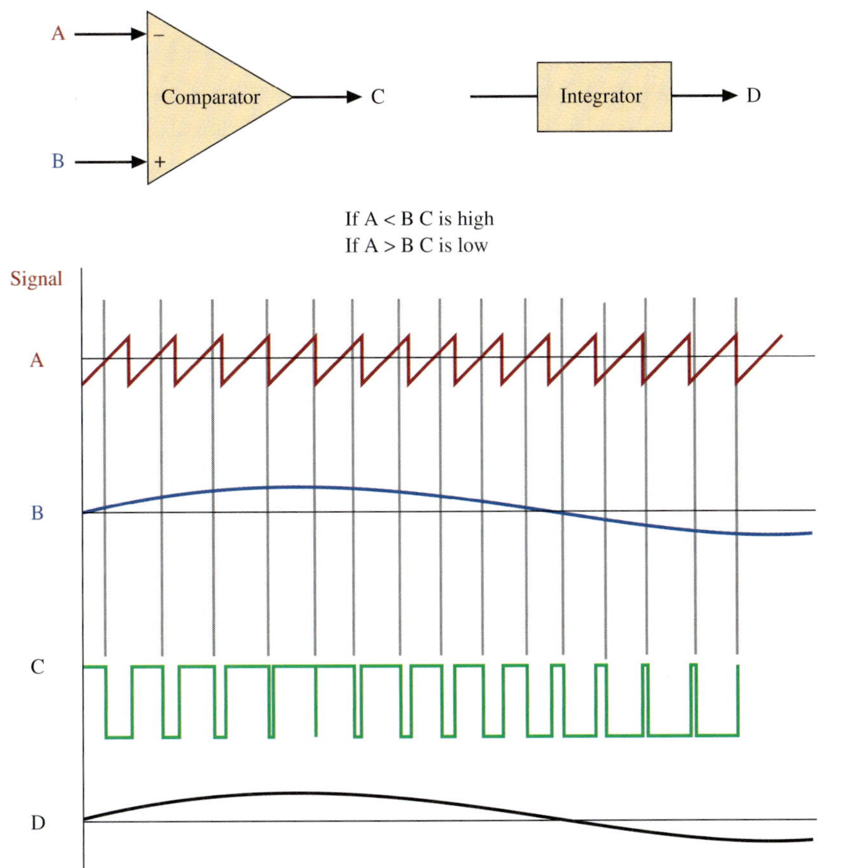

FIGURE 9-18 PWM generation waveforms.

Pulse-Position Modulation

PWM and pulse-position modulation (PPM) are very similar, a fact that is underscored in Fig. 9-19, which shows PPM being generated from PWM. Because PPM has superior noise characteristics, it turns out that the major use for PWM is to generate PPM. By inverting the PWM pulses in Fig. 9-19 and then differentiating them, the positive and negative spikes shown are created. By applying them to a Schmitt trigger sensitive only to positive levels, a constant amplitude and constant pulse-width signal is formed. However, the position of these pulses is variable and now proportional to the original modulating signal, and the desired PPM signal has been generated. The information content is *not* contained in either the pulse amplitude or width as in PAM and PWM, which means the signal now has a greater resistance to any error caused by noise. In addition, when PPM modulation is used to amplitude-modulate a carrier, a power savings results because the pulse width can be made very small.

At the receiver, the detected PPM pulses are usually converted to PWM first and then converted to the original analog signal by integrating as previously described. Conversion from PPM to PWM can be accomplished by feeding the

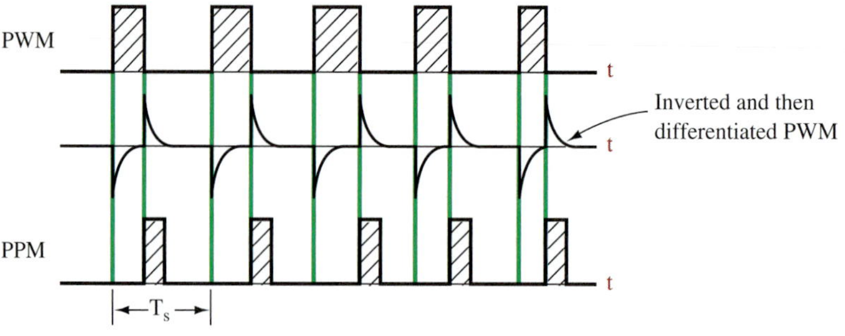

FIGURE 9-19 PPM generation.

PPM signal into the base of one transistor in a flip-flop. The other base is fed from synchronizing pulses at the original (transmitter) sampling rate. The period of time that the PPM-fed transistor's collector is low depends on the difference in the two inputs, and it is therefore the desired PWM signal.

This detection process illustrates the one disadvantage of PPM compared to PAM and PWM. It requires a pulse generator synchronized from the transmitter. However, its improved noise and transmitted power characteristics make it the most desirable pulse modulation scheme.

Demodulation

The process of demodulation is one of reproducing the original analog signal. We have noticed that the PAM signals contained harmonics of higher frequencies. Reshaping the original information signal would necessarily require removal of the higher frequencies. The use of a low-pass filter will accomplish this task. The upper cutoff frequency is selected to eliminate the highest frequency contained in the information. Figure 9-20 shows a block diagram of a PAM demodulator.

Demodulating a PWM signal is simple. The approach is similar to demodulation of the PAM signal. A low-pass filter can be used along with some waveshaping circuit. By using an RS flip-flop, the PPM signal can be converted to PWM and then demodulated using the technique for demodulating PWM.

FIGURE 9-20 Block diagram of a PAM demodulator.

Data Transmission via AM

The earliest form of data transmission using amplitude modulation occurred in the very first days of radio about 100 years ago. The International Morse Code was transmitted by simply turning a carrier on and off. The Morse Code is not a true bi-

A	·—	N	—·	1	————
B	—···	O	———	2	··———
C	—·—·	P	·——·	3	···——
D	—··	Q	——·—	4	····—
E	·	R	·—·	5	·····
F	··—·	S	···	6	—···
G	——·	T	—	7	——···
H	····	U	··—	8	———··
I	··	V	···—	9	————·
J	·———	W	·——	0	—————
K	—·—	X	—··—		
L	·—··	Y	—·——		
M	——	Z	——··		

. (period)	·—·—·—
, (comma)	——··——
? (question mark) ($\overline{\text{IMI}}$)	··——··
/ (fraction bar)	—··—·
: (colon)	———···
; (semicolon)	—·—·—·
((parenthesis)	—·——·—
) (parenthesis)	—·——·
' (apostrophe)	·————·
- (hyphen or dash)	—····—
$ (dollar sign)	···—··—
" (quotation marks)	·—··—

FIGURE 9-21 International Morse Code.

nary code because it not only includes marks and spaces but also differentiates between the duration of these conditions. The Morse Code is still used in amateur radio-telegraphic communications. A human skilled at code reception can provide highly accurate decoding. The International Morse Code is shown in Fig. 9-21. It consists of dots (short mark), dashes (long mark), and spaces. A *dot* is made by pressing the telegraph key down and allowing it to spring back rapidly. The length of a dot is one basic time unit. The *dash* is made by holding the key down (keying) for three basic time units. The spacing between dots and dashes in one letter is one basic time unit and between letters is three units. The spacing between words is seven units.

The most elementary form of transmitting highs and lows is simply to key a transmitter's carrier on and off. Figure 9-22(a) shows a dot, dash, dot waveform, while Fig. 9-22(b) shows the resulting transmitter output if the mark allows the carrier to be transmitted and space cuts off transmission. Thus, the carrier is conveying intelligence by simply turning it on or off according to a prearranged code. This type of transmission is called **continuous wave** (CW); however, because the wave is periodically interrupted, it might more appropriately be called an interrupted continuous wave. As a concession to the CW misnomer, it is sometimes called **interrupted continuous wave** (ICW).

Continuous Wave
a type of transmission where a continuous sinusoidal waveform is interrupted to convey information

Interrupted Continuous Wave
a more accurate name for a continuous wave transmission

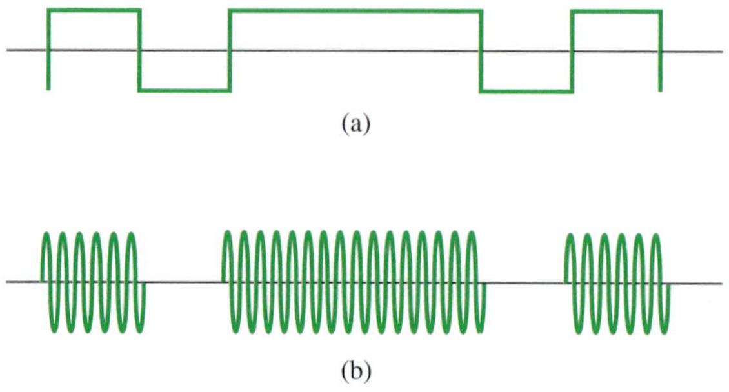

FIGURE 9-22 CW waveforms.

Whether the CW shown in Fig. 9-22(b) is created by a hand-operated key, a remote-controlled relay, or an automatic system such as a computer, the rapid rise and fall of the carrier presents a problem. The steep sides of the waveform are rich in harmonic content, which means the channel bandwidth for transmission would have to be extremely wide or else adjacent channel interference would occur. This is a severe problem because a major advantage of coded transmission versus direct voice transmission is narrow bandwidth channels. The situation is remedied by use of an *LC* filter, as shown in Fig. 9-23. The inductor L_3 slows down the rise time of the carrier, while the capacitor C_2 slows down the decay. This filter is known as a *keying filter* and is also effective in blocking the radio frequency interference (RFI), created by arcing of the key contacts, from being transmitted. This is accomplished by the L_1, L_2 RF chokes and capacitor C_1 that form a low-pass filter.

CW is a form of AM and therefore suffers from noise to a much greater extent than FM systems. The space condition (no carrier) is also troublesome to a receiver because at that time the receiver's gain is increased by AGC action to make received noise a problem. Manual receiver gain control helps but not if the received signal is fading between high and low levels, as is often the case. The simplicity

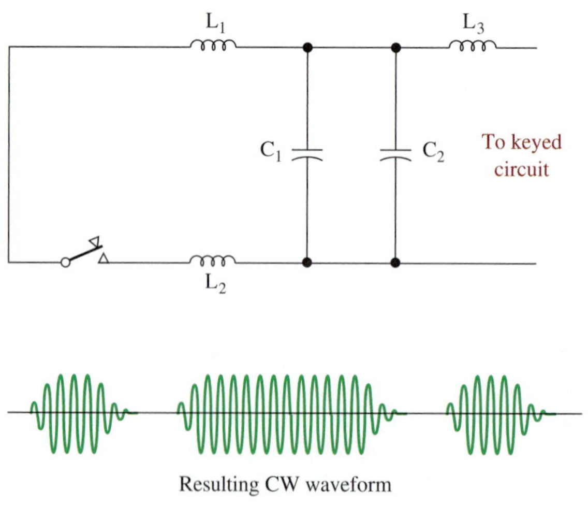

FIGURE 9-23 Keying filter and resulting waveform.

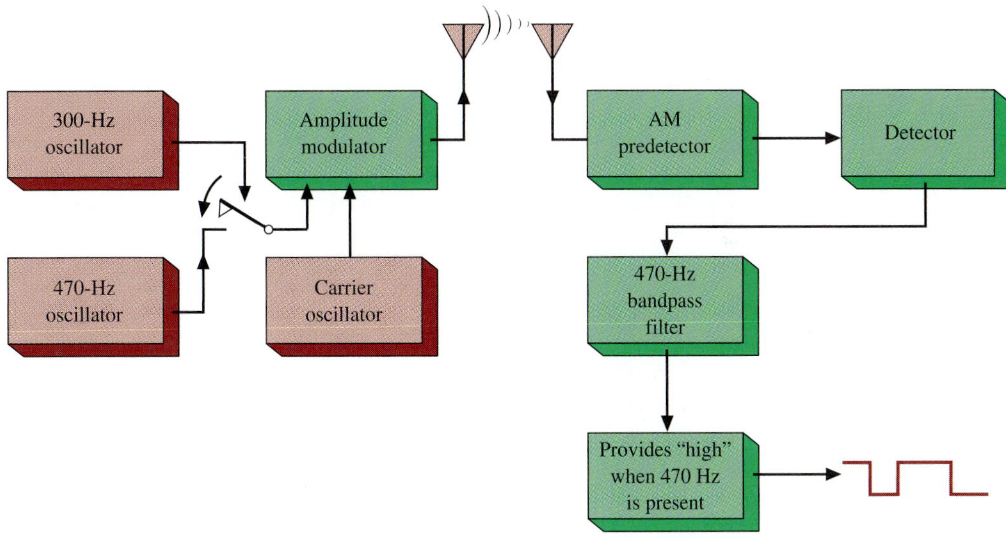

FIGURE 9-24 Two-tone modulation system (AM).

and narrow bandwidth of CW make it attractive to radio amateurs, but its major value today is to show the historical development of data transmission.

Two-Tone Modulation Two-tone modulation is a form of AM, but in it the carrier is always transmitted. Instead of simply turning the carrier on and off, the carrier is amplitude-modulated by two different frequencies representing either a one or zero. The two frequencies are usually separated by 170 Hz. An example of such a telegraphy system is provided in Fig. 9-24. When the transmitter is keyed, the carrier is modulated by a 470-Hz signal (1 condition); it is modulated by a 300-Hz signal for the 0 condition. At the receiver, after detection, either 300- or 470-Hz signals are present. A 470-Hz bandpass filter provides an output for the 1 condition that makes the output high whenever 470 Hz is present and low otherwise.

Example 9-6

The two-tone modulation system shown in Fig. 9-24 operates with a 10-MHz carrier. Determine all possible transmitted frequencies and the required bandwidth for this system.

Solution

This is an amplitude modulation system; therefore, when the carrier is modulated by 300 Hz, the output frequencies will be 10 MHz and 10 MHz $\pm$ 300 Hz. Similarly, when modulated by 470 Hz, the output frequencies will be 10 MHz and 10 MHz $\pm$ 470 Hz. Those are all possible outputs for this system. The bandwidth required is therefore 470 Hz $\times$ 2 = 940 Hz, which means that a 1-kHz channel would be adequate.

Example 9-6 shows that two-tone modulation systems are very effective with respect to bandwidth utilized. One hundred 1-kHz channels could be sandwiched in the frequency spectrum from 10 MHz to 10.1 MHz. The fact that a carrier is always transmitted eliminates the receiver gain control problems previously mentioned, and the fact that three different frequencies (a carrier and two side frequencies) are always being transmitted is another advantage over CW systems. In CW either one frequency, the carrier, or none is transmitted. Single-frequency transmissions are much more subject to ionospheric fading conditions than multifrequency transmissions. This phenomenon will be elaborated on in Chapter 13.

9-7 Computer Communication

The data communication that takes place between computers and peripheral equipment is of two basic types—serial and parallel—and uses ASCII format. In addition, the data that are sent in serial form (i.e., one bit after another on a single pair of wires) may be classified as either synchronous or asynchronous.

Asynchronous System the transmitter and receiver clocks free-run at approximately the same speed

Start Bit, Stop Bit used to precede and follow each transmitted data word

Synchronous System the transmitter and receiver clocks run at exactly the same frequency

In an **asynchronous system,** the transmit and receive clocks free-run at approximately the same speed. Each computer word is preceded by a **start bit** and followed by at least one **stop bit** to frame the word. In a **synchronous system** both sender and receiver are exactly synchronized to the same clock frequency. This is most often accomplished by having the receiver derive its clock signal from the received data stream.

Many choices are available today for selecting a serial communications interface. Most often the choice is dictated by the computer and the electronic equipment. For example, a spectrum analyzer might have to contain a GPIB and RS-232 interface, while a digital camera might contain a USB or Firewire interface. Industrial equipment might be interconnected using the RS422 or RS485 standard. This section addresses the key standards that the user may find on computer and electronic communications equipment, which includes the following:

- USB
- Firewire
- RS232
- RS485
- RS422
- GPIB

Universal Serial Bus (USB)

Universal Serial Bus (USB) a hot-swappable, high-speed serial communications interface

Hot-swappable a term used to describe that an external device can be plugged in or unplugged at any time

The **universal serial bus (USB)** port is becoming a very popular choice for high-speed serial communications interfaces. The reasons are simple:

- Almost all computer peripherals (mouse, printers, scanners, etc.) are now available in a USB version.
- The USB devices are **hot-swappable,** which means the external devices can be plugged in or unplugged at any time.
- USB devices are detected automatically once they are connected to the computer.
- The USB 2 supports a maximum data rate of 480 Mbps; USB 1.1 supports 12 Mbps.
- A total of 127 peripherals can be connected to one USB port.

The USB cable used to interconnect the peripheral to the computer consists of four wires inside a shielded jacket. The function of the four wires and their wire colors are listed in Table 9-6.

Table 9-6 The USB Wire Colors and Functions

Color	Function
Red	+5 V
Brown	Ground
Yellow	Data
Blue	Data

Two type of connectors are used with USB ports and cables, **Type A** and **Type B.** They are shown in Fig. 9-25. The Type A connector is the upstream connection that connects to the computer. The Type B connector is for the downstream connection to the peripheral.

Type A Connector the USB upstream connection that connects to the computer

Type B Connector the USB downstream connection to the peripheral

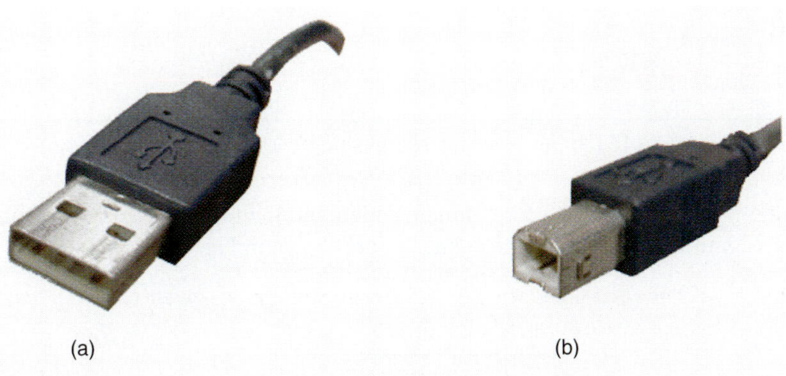

FIGURE 9-25 The USB Type A and Type B connectors.

An example of a USB connection using the MAX3451 transceiver is shown in Fig. 9-26.

This diagram shows the MAX3451 being used to interface the peripheral device through the IC to the PC. The SPD input is used to select the data transfer rate of 1.5 Mbps (SPD = low) or 12 Mbps (SPD = +V). The D+ and D− pins are bidirectional bus connections. The OE pin is used to control the data flow. To transmit data from the peripheral to the USB side, bring OE low and SUS low. Receiving data requires OE to be high and SUS to be low. VP and VM terminals function as receiver outputs when OE is high (transmit mode) and duplicate D+ and D− when OE is low (receiver mode). The supply voltage range for the device is +1.65 V to +3.6 V.

Firewire (IEEE 1394)

Firewire is another high-speed serial connection available for computers and peripherals. **Firewire A (IEEE 1394a)** supports data transfers up to 400 Mbps, while **Firewire B (IEEE 1394b)** supports 800 Mbps, data speeds in the gigabits are planned. The firewire cable uses a shielded twisted-pair cable with 3 pairs (6 wires).

Firewire A (IEEE 1394a) a high-speed serial connection that supports data transfers up to 400 Mbps

Firewire B (IEEE 1394b) a high-speed serial connection that supports data transfers up to 800 Mbps

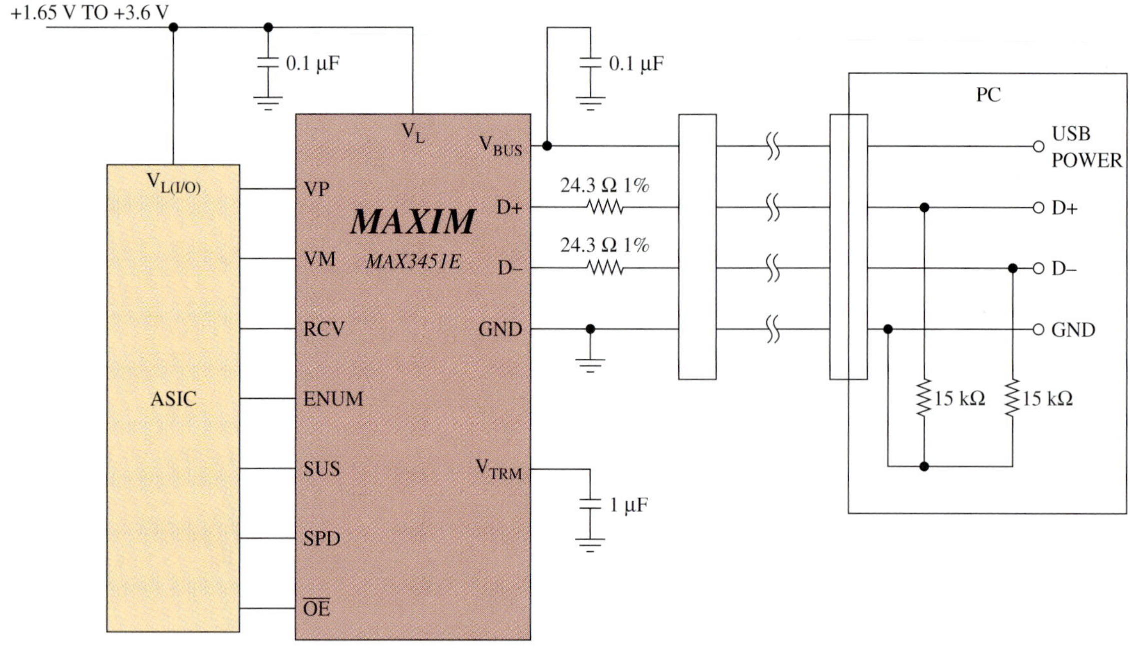

FIGURE 9-26 An example of using the MAX3451 transceiver for establishing a USB connection.

Two of the wire pairs are used for communication and the third is used for power. The pin assignments for the firewire connector are shown in Fig. 9-27.

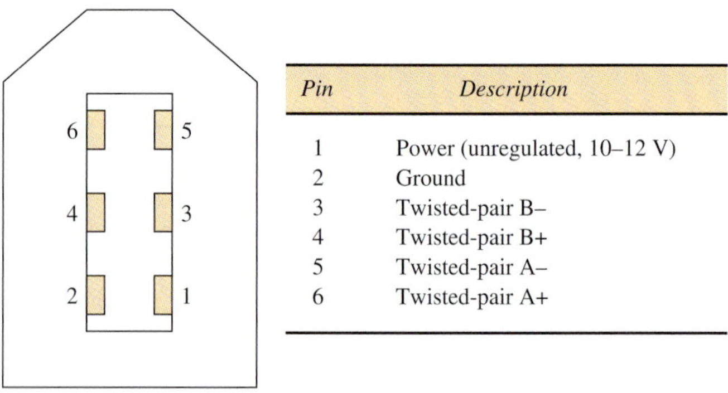

Pin	*Description*
1	Power (unregulated, 10–12 V)
2	Ground
3	Twisted-pair B–
4	Twisted-pair B+
5	Twisted-pair A–
6	Twisted-pair A+

FIGURE 9-27 The firewire connector and pin assignments.

RS-232 Standard

RS-232
a standard of voltage levels, timing, and connector pin assignments for serial data transmission

Older serial data communications follow a standard called **RS-232,** or more correctly, RS-232 C. Usually everyone is referring to the "C" version of RS-232 because that is what is currently in use, but often the "C" is omitted. Be aware that even though we may refer to the standard as RS-232, we really mean RS-232 C. The RS-232 C standard is set by the Electronics Industry Association (EIA).

In addition to setting a standard of voltages, timing, etc., standard connectors have also been developed. This normally consists of what is called a DB-25 con-

RS-232 "D-type" connector—front view.

Pin	*Name*	*Abbreviation*
1	Frame ground	FG
2	Transmit data	TD
3	Receive data	RD
4	Request to send	RTS
5	Clear to send	CTS
6	Data set ready	DSR
7	Signal ground	SG
8	Data carrier detect	DCD
20	Data terminal ready	DTR

Most popular pins implemented in RS-232 connections.

FIGURE 9-28 DB-25 connector.

nector, which is a connector with two rows of pins, arranged so that there are 13 in one row and 12 in the other. A diagram of this connector is provided in Fig. 9-28.

It should be noted that even though the DB-25 connector is usually used, the actual RS-232 C standard specifications do not define the actual connector. In the last fifteen years or so, another connector has also been used for RS-232. This is the DB-9 connector, which has become a sort of quasi-standard for use on IBM compatible personal computers. See Fig. 9-29 for a diagram of the DB-9 connector. You may ask, "How can 25 pins from a DB-25 connector all fit into the 9 pins of a DB-9 connector?" As we will see, all 25 pins of the DB-25 connector are not used, and, in fact, 9 pins are enough to do the job.

The original purpose of RS-232 was to provide a means of interfacing a computer with a modem. The computer in this case was likely a mainframe computer because personal computers, at least as we know them today, had not yet been developed. Modems were always external, so it was necessary that some means of connection between the modem and the computer be made. It would be even nicer if this connection would be made standard so that all computers, and all modems, could be connected interchangeably. This was the original purpose of RS-232. However, it has evolved into being many other things.

Today an RS-232 interface is used to interface a mouse to a personal computer, to interface a printer to a personal computer, and probably to interface about

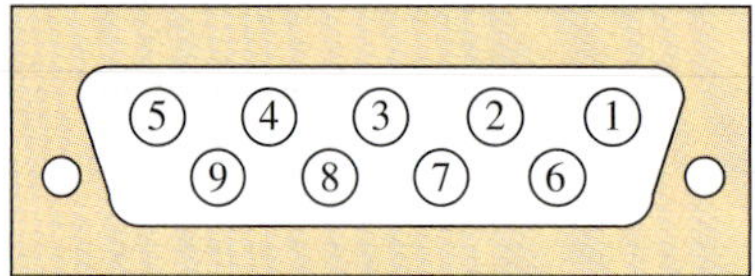

Front view

Pin			
	1	DCD	(8)
	2	RD	(3)
	3	TD	(2)
	4	DTR	(20)
	5	SG	(7)
	6	DSR	(6)
	7	RTS	(4)
	8	CTS	(5)
	9	RI	(22)

FIGURE 9-29 DB-9 connector.

as many other things as one can think of to a personal computer. In many cases, it is used to interface instrumentation to a PC. This means that the standard has evolved, and in the process of this evolution has changed in terms of the real world.

The standard did not define the connector but did define signal levels and different lines that could be used. The signal levels that are defined are very broad. The voltage levels are to be between 3 and 25 V. A minus voltage indicates a "1" and a plus voltage indicates a "0." Although the definition covers 3 to 25 V, the real signal levels are usually a nominal 12 to 15 V. Many chips available today will not respond to the 3-V levels, so from a real-world practical point of view, the signal level is between 5 and 15 V. This is still a very broad range and can obviously allow for a lot of loss in a cable.

In addition to the signal levels, the RS-232 standard specifies that the maximum distance for a cable is 50 ft, and the capacitance of the cable cannot exceed 2500 pF. In reality, it is the capacitance of the cable that limits the distance. In fact, distances that far exceed 50 ft are commonly used today for serial transmission.

The RS-232 standard also contains another interesting statement. It says that if any two pins are shorted together, the equipment should not be damaged. This obviously requires a good buffer. This buffering is normally provided. It should be noted that the standard says only that the equipment will not be damaged. It does not say that the equipment will work in that condition. In other words, if you short the pins on an RS-232 connector, there should be no smoke, but it might not work!

Perhaps the most important part of the standard from a technical point of view is that it defines the way the computer should "talk" to the modem; the timing involved, including the sequence of signals; and how each is to respond.

RS-232 Line Descriptions

Now that we know a little about what RS-232 is designed to do, it is time to look at the actual signal lines involved and see what they do. As explained earlier, the DB-25 connector definition is not a part of the original standard, but because it has become a de facto standard, we will use it in our discussion. Refer to Fig. 9-28 as a reference in this description. The complete signal description chart in Fig. 9-30 will also be helpful.

PIN NO.	EIA CKT.	CCITT CKT.	Signal description	Common abbrev.	From DCE	To DCE
1	AA	101	Protective (chassis) ground	GND		
2	BA	103	Transmitted data	TD		X
3	BB	104	Received data	RD	X	
4	CA	105	Request to send	RTS		X
5	CB	106	Clear to send	CTS	X	
6	CC	107	Data set ready	DSR	X	
7	AB	102	Signal ground/common return	SG	X	X
8	CF	109	Received line signal detector	DCD	X	
9			Reserved			
10			Reserved			
11			Unassigned			
12	SCF	122	Secondary received line signal detector		X	
13	SCB	121	Secondary clear to send		X	
14	SBA	118	Secondary transmitted data			X
15	DB	114	Transmitter signal element timing (DCE)		X	
16	SBB	119	Secondary received data		X	
17	DD	115	Receiver signal element timing		X	
18			Unassigned			
19	SCA	120	Secondary request to send			X
20	CD	108/2	Data terminal ready	DTR		X
21	CG	110	Signal quality detector	SQ	X	
22	CE	125	Ring indicator	RI	X	
23	CH	111	Data signal rate selector (DTE)			X
23	CI	112	Data signal rate selector (DCE)		X	
24	DA	113	Transmitter signal element timing (DTE)			X
25			Unassigned			

FIGURE 9-30 Signal description for DB-25.

1. Ground pins: Actually, there are two ground pins in RS-232. They are *not* the same, however, and serve very different purposes.

   Pin 1 is the *protective ground* (GND). It is connected to the chassis ground and is there simply to make certain that no potential difference exists between the chassis of the computer and the chassis of the peripheral equipment. This is *not* the signal ground. The protective ground functions much the same as the third prong of a 115-V ac 3-prong outlet. The circuit will work without a connection to this pin, but the operator will also lose protection. In other words, make certain this pin is connected.

   Pin 7 is the *signal ground* (SG). This is the pin used for the ground return of all the other signal lines. Look at the location of the signal ground, pin 7. One of the problems often associated with RS-232 and especially DB-25 connectors is knowing if you are looking from the front or back and which end you should start counting from. Many connectors have the pin numbers printed on them, but this printing is usually so small that you cannot read it. Note that regardless of which end you count from, pin 7 always comes out in the same place!
2. Data signal pins: Data can be sent from both the computer and peripheral equipment. Therefore a bidirectional path is necessary.

   Pin 2 is *transmit data* (TD). In theory, this pin will contain the actual data flowing from the computer to the peripheral equipment.

Pin 3 is *receive data* (RD). In theory, this pin has the actual data flowing from the peripheral equipment to the computer. Note that it is the same as pin 2 listed above, but in the opposite direction.

The problem is that theory and reality are not always the same. What if you want to link computers together? Which one is sending data, and which one is receiving data? It would seem that there should be an easy answer to this question, but such is not the case. Which end of the cable are you looking at? In this instance, when one is transmitting data, those same data become receive data to the other computer. What this really means is that it is not easy to define which is receive and which is transmit.

Because of this problem, what is usually called a null modem cable has been developed. It has pins 2 and 3 crisscrossed, i.e., pin 2 at one end is connected to pin 3 at the other end, and vice versa. Of course, this problem does not exist if we were to use RS-232 as originally intended, i.e., for the computer to talk to a modem.

Another way of completing an RS-232 connection, a null modem connection, is shown in Fig. 9-31. The RTS-CTS and DSR-DTR pins are connected together as shown. This tricks the RS-232 device into believing a DCE is connected and enables the TX and RX lines to function properly.

3. Handshaking pins: Pins 4 and 5 are used for handshaking or flow control. More correctly, pin 4 is called the *request to send* (RTS) pin, while pin 5 is the *clear to send* pin (CTS). These two lines work together to determine that everything is fine for data to flow. Originally, this was used to turn on the modem's carrier, but today it is more often used to check for buffer overflow. Almost all modern modems (and other serial devices) have some sort of buffer that is used when receiving and sending data. It would not be satisfactory for

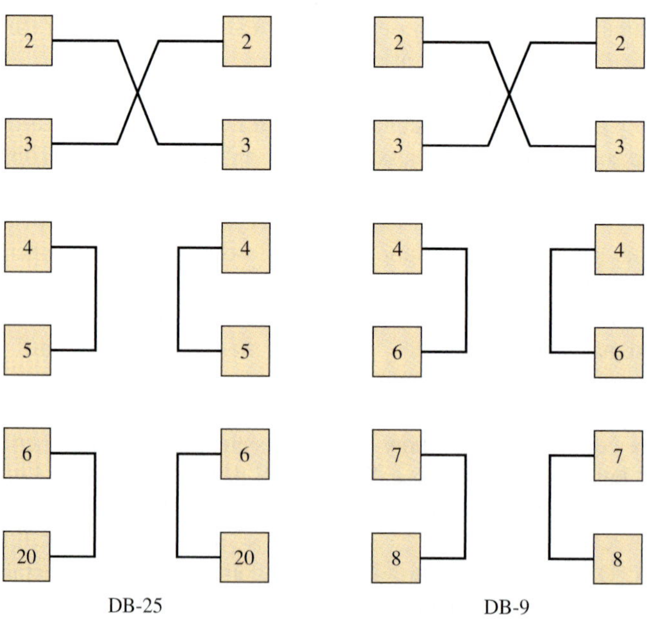

FIGURE 9-31 The RS-232 null modem connection.

that buffer to be sending more information than it could actually hold. Pins 4 and 5 may be used to handle this problem. It should be noted that if one computer is talking to another one via a null modem cable, these pins need to be crisscrossed just like pins 2 and 3.

In many cases, software flow control is used, and pins 4 and 5 do not serve any useful purpose at all. In this case, pins 4 and 5 need to be jumpered at the DB-25 connector. You may wonder why it would be necessary to jumper these pins if they are not actually used. This is because many serial ports will check these pins before they will send any data. Many serial interface cards are configured so that this signal is required for any data transfer to take place.

4. Equipment ready pins: Pins 6 and 20 are complementary pairs much as pins 4 and 5 are. Actually, pin 6 is called *data set ready* (DSR), while pin 20 is called *data terminal ready* (DTR). The original purpose of these pins was simply to make certain that the power of the external modem was turned on and that the modem was ready to go to work. In some instances, these pins were also used to indicate if the phone was off hook or on hook.

   Today, these pins may be used for any number of purposes, such as paper out indicators on printers. The two pins work together and normally should be jumpered on the DB-25 connector like pins 4 and 5 if their use is not expected.

   It should be noted that, up to this point, the complementary pairs of pins have been adjacent to one another. Obviously, pins 6 and 20 are not in consecutive order. An inspection of the DB-25 connector will reveal, however, that pins 6 and 20 are almost directly above and below each other in their respective rows.

5. Signal detect pin: Pin 8 is usually called the *data carrier detect* (DCD) or sometimes simply the *carrier detect* (CD) line. In an actual modem, it is used to indicate that a carrier (or signal) is present. It may also indicate that the signal-to-noise ratio is such that data transmission may take place.

   Many computer interface cards require that this signal be present before they will communicate. If the RS-232 connection is not a modem requiring this signal, it is often tied together with pins 6 and 20 (DSR and DTR). As has been the case with other pins, this pin is probably not used for its original purpose. What it really is used for is highly dependent on the device in use.

6. Ring indicator pin: Pin 22 is known as the *ring indicator* (RI). Its original purpose was to do just what its name indicates—to let the computer know when the phone was ringing. Most modems are equipped with automatic answer capabilities. Obviously, the modem needed to tell the computer that someone with data to send was calling!

   Pin 22 is forced true only when the ringing voltage is present. This means that this signal will go on and off in rhythm with the actual telephone ring. As is true with other signals, this pin may not be used. Often it is tied together with pins 6, 20, and 8. In other cases, it is simply ignored. Again, the actual use of this pin today will vary widely with the equipment connected.

7. Other pins: Up to this point, we have discussed the 10 pins used most often. In reality, the chassis ground is usually accomplished by grounding each piece of equipment separately and is not necessary with battery-powered computer

equipment. These considerations have led to the 9-pin connector usage shown in Fig. 9-29. When the DB-25 connector is used, the remaining pins are occasionally used for special situations, as warranted by the equipment connected.

It should be noted that the terms *computer, modem*, and *peripheral equipment* have been used in the discussion thus far. Two more technically correct terms may be used in literature, texts, and diagrams. The term **data terminal equipment** (DTE) is used to indicate a computer, computer terminal, personal computer, etc. The term **data communications equipment** (DCE) is used to indicate the peripheral equipment (such as the modem, printer, mouse, etc.). Notice that the DCE label is used in Fig. 9-30.

Data Terminal Equipment
refers to a computer, terminal, personal computer, etc.

Data Communications Equipment
refers to peripheral computer equipment such as a modem, printer, mouse, etc.

Many communication standards available on today's computers complement the use of RS-232. Table 9-7 provides a brief overview of many of the standards currently being used.

Table 9-7 Overview of Current Serial Computer Communication Standards

RS-232	The common serial data connection for computer modems and the mouse interface.
RS-422	A balanced serial communications link that can support data speeds up to 10 Mbps at a distance of 4000 ft.
RS-449	Intended as a replacement for the RS-232 standard, but the two standards are not completely compatible either mechanically or electrically. Data speeds are from 9600 bps to 10 Mbps, and they depend on the length and type of cable used.
RS-485	A balanced differential output that allows multiple data connections. Data speeds of 30 Mbps are supported.
RS-530	A standard replacing RS-449 and also complementing the RS-232 standard. This standard will also interface the balanced systems RS-422 and 423.

Source: www.blackbox.com.

RS-422, RS-485

The RS-232 output is a single-ended signal, which means that a signal line and a ground line are used to carry the data. This type of arrangement can be susceptible to noise and ground bounce. The **RS-422** and **RS-485** use a differential technique that provides a significant improvement in performance and thus yields greater distances and the capacity to support higher data rates. In a differential technique, the signals on the wires are set up for a high (+) and low (−) signal line. The (+) indicates that the phase relationship of the signal on the wire is positive and the (−) indicates that the phase of the signal on the wire is negative; both signals are relative to a virtual ground. This is called a **balanced mode** of operation. In a balanced mode of operation, the balanced operation of the two wire pairs helps to maintain the required level of performance in terms of crosstalk and noise rejection.

RS-422, RS-485
balanced mode serial communications standards that support multidrop applications

Balanced Mode
neither wire in the wire pairs connects to ground

RS-422 and RS-485 also support multidrop applications, which means that the standard supports multiple drivers and receivers connected to the same data line. RS-422 is not a true multidrop network because it can drive ten receivers but only allows one transmitter to be connected. RS-422 supports a data rate of 10 Mbps over a distance of 4000 feet. The RS-485 standard allows a true multiport connection. The standard supports 32 drivers and 32 receivers on a single two-wire bus. It supports data rates up to 30 Mbps over a distance of 4000 feet.

Computer interface cards are available that provide the functions for virtually any communications test set or gear. For example, spectrum analyzers and data and protocol analyzer interface cards are available on a PC interface card. Table 9-8 outlines some of the common computer bus interfaces available today. This information is helpful for the user in specifying the proper bus or an interface card for a computer. This list is not complete but it does provide some of the more common computer bus interfaces.

Table 9-8 STANDARD COMPUTER BUS INTERFACES

Interface	Description
PCI (Peripheral Component Interconnect)	PCI is the best bus choice for current computers. PCI supports 32- and 64-bit implementations.
ISA (Industry Standard Architecture)	Allows for 16-bit data transfers between the motherboard and the expansion board.
EISA (Extended Industry Standard Architecture)	EISA extends the ISA bus to 32-bit data transfers.
MCA (Micro Channel Architecture)	
VLB [VESA (Video Electronics Standards Association) Local Bus]	Supports 32-bit data transfers at 50 MHz.
USB (Universal Serial Bus)	Does not require that the computer be turned off or rebooted to activate the connection. Supports data rates up to 12 Mbps. Two type of USB connectors are shown in Figs. 9-25(a) and (b).
IEEE 1394 (Firewire, i-Link)	A high-speed, low-cost interconnection standard that supports data speeds of 100 to 400 Mbps (see Fig. 9-27).
SCSI (Small Computer System Interface)	Consists of an SCSI host adapter; SCSI devices such as hard drives; DVD; CD-ROM; and internal or external SCSI cables, terminators, and adapters. SCSI technology supports data transfer speeds up to 160 Mbps.
IDE (Integrated Drive Electronics)	Standard electronic interface between the computer's motherboard and storage devices such as hard drives. The data transfer speeds are slower than SCSI.
AGP (Accelerated Graphics Port)	Used in high-speed 3D graphics applications.

FACSIMILE

Facsimile is the process whereby fixed graphic material such as pictures, drawings, or written material is scanned and converted to an electrical signal, transmitted, and after reception used to produce a likeness (facsimile) of the subject copy. It is roughly comparable to the transmission of a single TV frame except the output is reproduced on paper rather than on the face of a CRT. Facsimile has been

Facsimile
system of transmitting images in which the image is scanned at the transmitter, reconstructed at the receiving end, and duplicated on paper

used for years for the rapid transmission of photos to local newspapers by news services (e.g., AP Wirephoto) and aboard ships for reception of up-to-date weather charts or maps.

Fax
abbreviation for facsimile

In recent years facsimile **(fax)** equipment has been designed to appeal to the industrial world. This equipment generally uses standard telephone lines as the communications link. This enables industry to send important business papers rapidly across town or around the world. As the cost and problems concerned with standard mail delivery escalate, it is not far-fetched to envision facsimile postal service.

A facsimile unit is shown in Fig. 9-32. The copy to be transmitted is scanned by a laser light source. Light reflected from the light and dark elements on each page is gathered by a solid-state photoreceptor. This signal is then used to frequency-modulate a low-frequency carrier before it is applied to the phone line by using a modem.

Standard phone lines have a very limited bandwidth of less than 3 kHz. Recall Hartley's law (Chapter 1), which states that the information transmitted is proportional to bandwidth times the time of transmission. Today's fax machines use *compression* techniques to shorten document transmission time. The receiver is programmed to reverse the data compression scheme. The basic compression technique

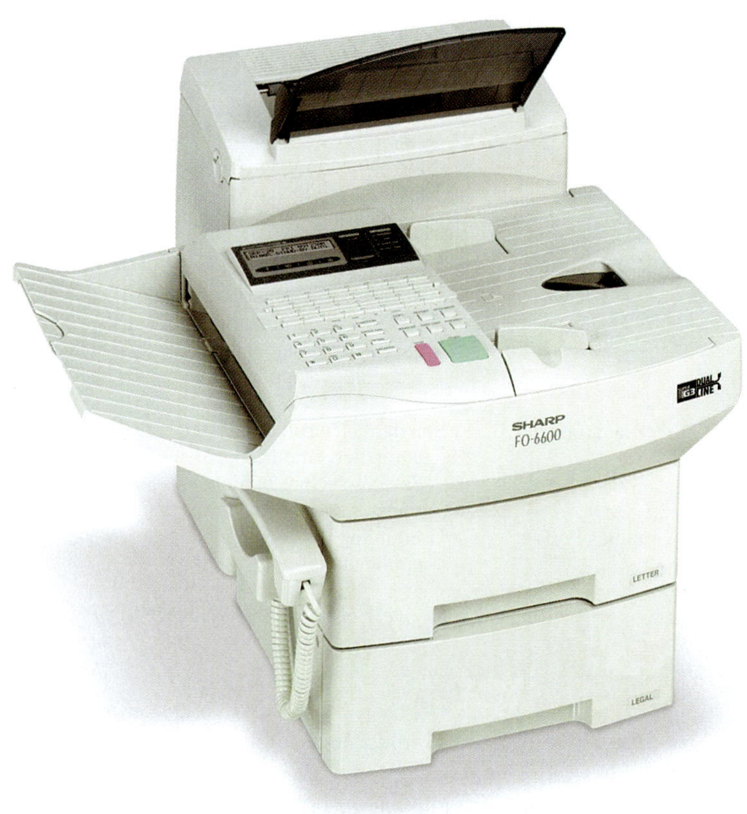

FIGURE 9-32 Facsimile unit. (Photo courtesy of Sharp Electronics Corporation.)

uses the concept that documents often contain a large amount of blank (white) area. The fax transmits information that indicates how many consecutive "white" areas occur in the scanned line, thereby greatly reducing the time for transmission. This involves usage of computer memory at both transmitter and receiver. More advanced (faster) fax machines rely on the fact that adjacent scan lines are usually very similar. These machines transmit the difference between one line and the next to reduce transmission time even further.

9-8 TROUBLESHOOTING

Telemetry systems have become big business in recent years. In addition to the military applications, many commercial uses are being developed. Today's telemetry market ranges from remote reading of gas and electric meters to credit-card validation, security monitoring, token-free highway toll systems, and remote inventory control.

After completing this section you should be able to

- Troubleshoot a radio-telemetry receiver
- Troubleshoot a radio-telemetry transmitter

RECEIVER PROBLEMS

Let's start by looking at some possible receiver problems in Fig. 9-33. Because this receiver has a low-frequency IF amplifier, an oscilloscope can be used to check signal levels and adjust the tuned circuits. However, a typical oscilloscope probe presents a capacitive load to the circuit you connect it to. The best probe for this work will be an X10 type. These probes divide the signal to the oscilloscope by 10 but shunt the circuit under test with only 8 to 10 pF, whereas an X1 probe would have a shunt capacitance of 100 pF. Sometimes even 10 pF is significant, but it is usually not a problem if you are aware of the effect.

NO OUTPUT Assume you know that a good transmitted signal is present. Check the easy items first: battery or power supply and the antenna connections. Refer to Fig. 9-33 in this discussion.

1. Disconnect the receiver from the decoder. It is possible the decoder is shorted, therefore killing the receiver.
2. Check the local oscillator. You can check the LO signal at pin 2 with an oscilloscope by using the low-capacitance probe. If you don't have one, just use a small capacitor, 10 pF or less, in series with your probe. The capacitor forms a voltage divider with the probe capacitance. If the capacitance of the added capacitor is small compared to that of the probe, the loss is approximately the ratio of the capacitances.

   If the oscillator is not running, try adjusting the inductor, L1. After the oscillator runs, you should be able to find a peak amplitude followed by a point where it quits. Adjust the coil away from the quitting point and just below the output peak.

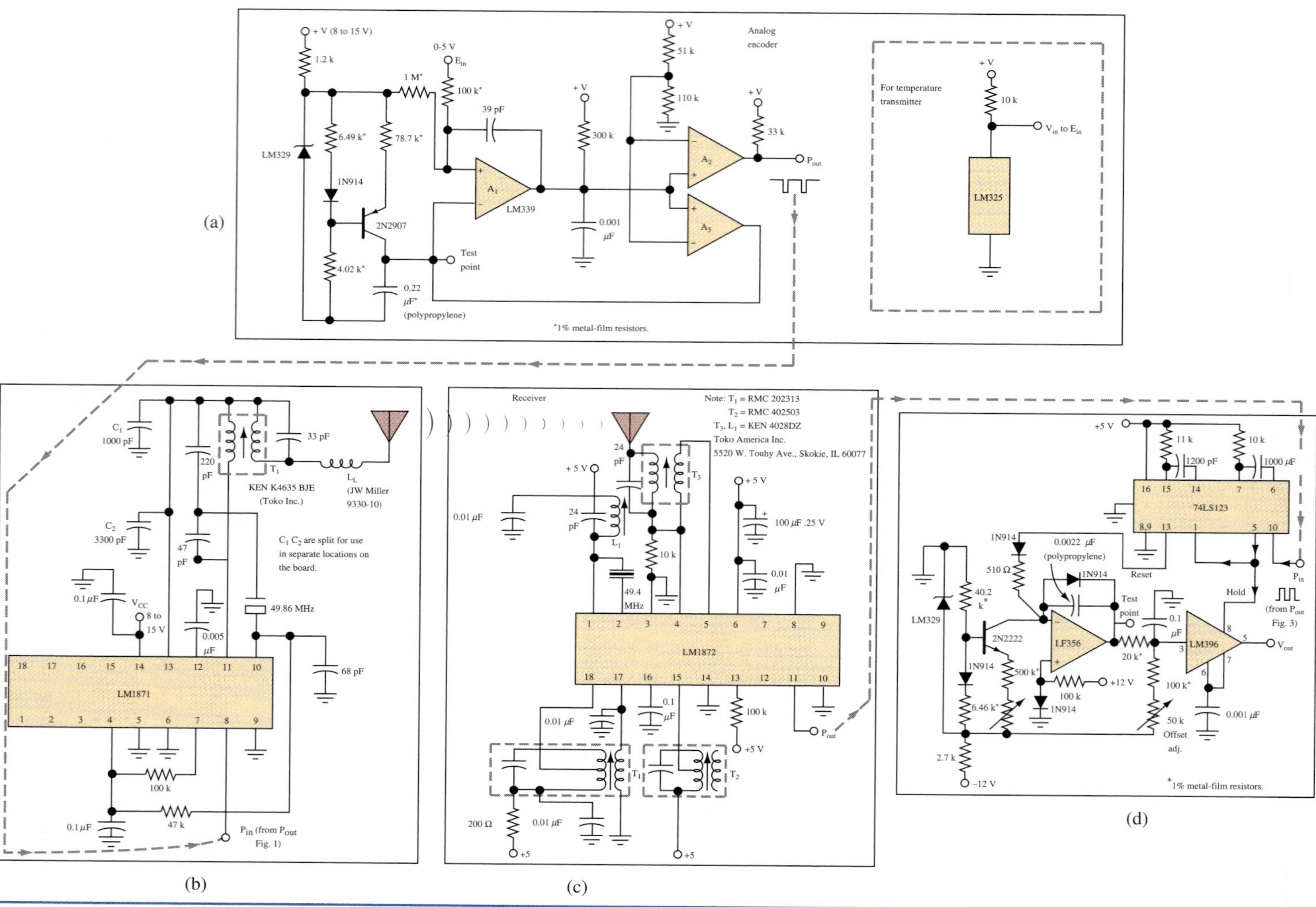

FIGURE 9-33 Complete radio-telemetry system: (a) encoder; (b) transmitter; (c) receiver; (d) decoder. (Reprinted with permission from *Electronic Design*, Vol. 29, No. 10; copyright Hayden Publishing Co., Inc.)

If adjustment is fruitless, check the crystal as described in Chapter 1 or substitute another. Don't forget the inductor and its tuning capacitor. Integrated circuits are usually more reliable than the rest of the parts and harder to unsolder.

3. Check for signal in the IF amplifier stages. You have no way to adjust the oscillator frequency, so the IF must be aligned to the transmitter's frequency. The manufacturer recommends killing the AGC by grounding pin 16. The secondary of T3 is a good test point because it isolates the IF amplifier from the oscilloscope probe. The signal amplitude will be down by a factor of 8 according to the data sheet, but we are looking only for a peak. Adjust T1, T2, and T3 for maximum output. After some signal is obtained, you may find that one coil will not tune properly. Check the inductance and its associated capacitor.

Transmitter Problems

In this case, you will need some means of monitoring the transmitter's RF output without actually making a physical connection to the transmitter. A spectrum analyzer is efficient, but almost any receiver capable of operating on the proper frequency will do.

1. No output. Check for modulation at the input, pin 8, and at the internal modulator's output, pin 13. The modulator simply turns the oscillator's power supply on and off. The off-on rate will be in the low audio frequency region. If these are present and there is no RF output, it's time to check the crystal and the tuned circuits.
2. Low output. This can be caused by low modulator output. Check the modulation voltage peak amplitude at pin 13. For the LM1871, you should have 4.5 V.

Monitor the signal strength with a receiver or the spectrum analyzer while adjusting the oscillator transformer, T1. Look for a definite peak. If none can be found, check the associated 220-pF and 47-pF capacitors.

Encoder/Decoder Problems

The current source is the heart of both the encoder and the decoder. Both circuits also depend on a high-quality capacitor to integrate the current provided by the current source.

First, a short discussion of the current source is necessary. The LM329 is an active 6.9-V reference. This voltage is divided by the 6.49- and the 4.02-kΩ resistors to provide a fixed base–emitter voltage on the 2N2907. Because this voltage is fixed, the voltage across the emitter resistor is fixed. Therefore, the collector current is constant. Now, let's look at some typical problems.

1. Assume the 2N2907 has a base-to-collector short. Now the current is only limited by the emitter resistor. The capacitor will charge rapidly. Therefore, a lower signal voltage will cause the comparator (A1) to trip at a lower voltage, and the gain will appear to be too high. Because the voltage across the capacitor is no longer linear, the system output will also be nonlinear.
2. Assume the integrating capacitor has developed a high-resistance leak. Now part of the current is not building charge on the capacitor. The gain will appear to be lower and some nonlinearity may appear. This applies to both the encoder and decoder.

3. Assume a failure of the 74LS123 in the decoder. It must provide the proper logic levels to reset the integrator and trigger the sample-and-hold IC. Incorrect logic levels result in no output.
4. Assume the sample-and-hold capacitor (0.001 μF) has developed a leak. The operational amplifier may have enough drive to charge the capacitor, but voltage will immediately begin to fall. Looking at the output with an oscilloscope, you will see a sawtooth waveform, and the output voltage will be too low.

9-9 Troubleshooting with Electronics Workbench™ Multisim

In this section, we examine a circuit used to detect a unique binary data sequence. This circuit is called a sequence detector. Once the unique binary sequence is detected, an output signal is triggered. Circuits such as this one are often used in communications to detect the beginning of a serial data stream. These type of circuits are frequently discussed in digital design textbooks, and you are encouraged to refer to these texts for the steps to implement sequence detectors.

This section examines the method for testing the sample sequence generator shown in Fig. 9-34. This circuit has been designed to detect a sequence of three consecutive 1s (1 1 1). The objective of this section is to demonstrate how to test the sequence detector circuit using the Multisim Word Generator and Logic Analyzer.

To begin, open the Multisim file **Fig9-34.ms7 (.msm)** using Electronics Workbench™ Multisim. The circuit should look similar to Fig. 9-34 except the word generator is labeled XWG1 and the logic analyzer is labeled XLA1. Double-click on the word generator to open its control panel. An example of the window displayed is provided in Fig. 9-35. The Multisim word generator provides for 32-bit words to be programmed into the generator. Over 16,000 32-bit numbers can be

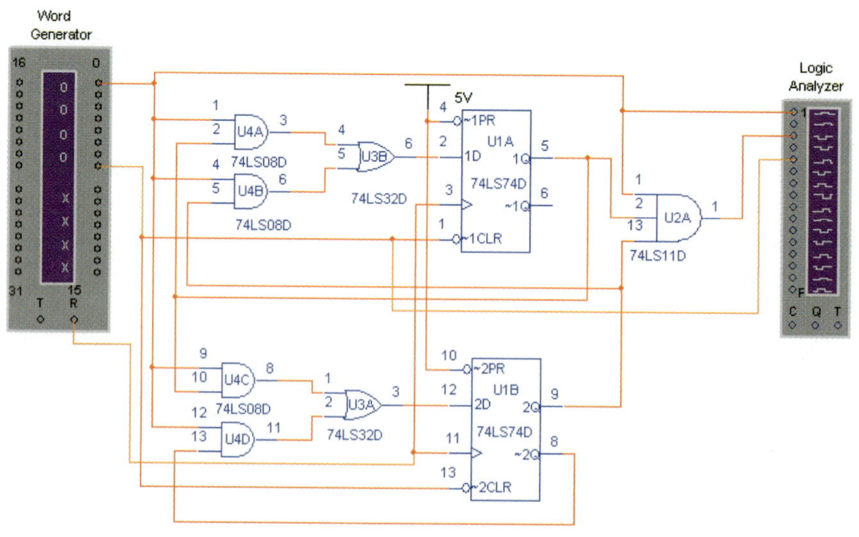

FIGURE 9-34 A "1 1 1" sequence detector circuit as implemented in Electronics Workbench™ Multisim.

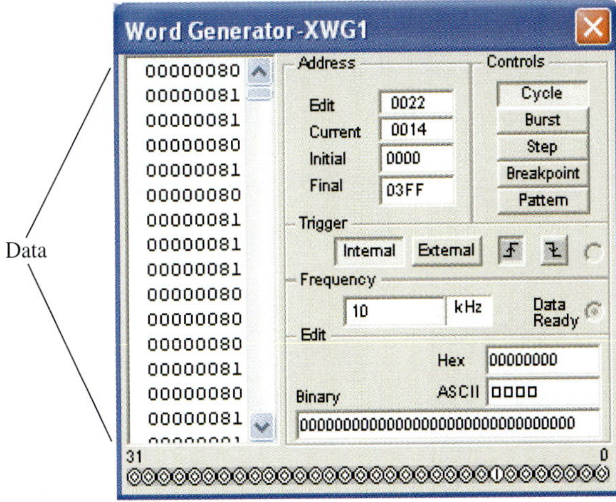

FIGURE 9-35 The control panel for the Multisim word generator.

entered. In this case we will be using only the first data bit (D0) for the simulation of a serial data stream so the entries shown will have data values only in the D0 position. The data values displayed are in hexadecimal. The 80_H translates to 1000000 in binary. The 81_H translates to 10000001.

There are several options for running the simulation. The option selected in Fig. 9-35 is the cycle. This mode will run the simulation through the entire set of more than 16,000 data values. Another useful option is the step. In this mode, the simulation will proceed one clock cycle at a time. In this mode, clicking on the Multisim pause button will let the simulation proceed for one more clock cycle.

Double-click on the logic analyzer icon to display the selected traces from the sequence detector. Start the simulation and let the simulation fill the logic analyzer's screen. You should see an image similar to that shown in Fig. 9-36. You should also

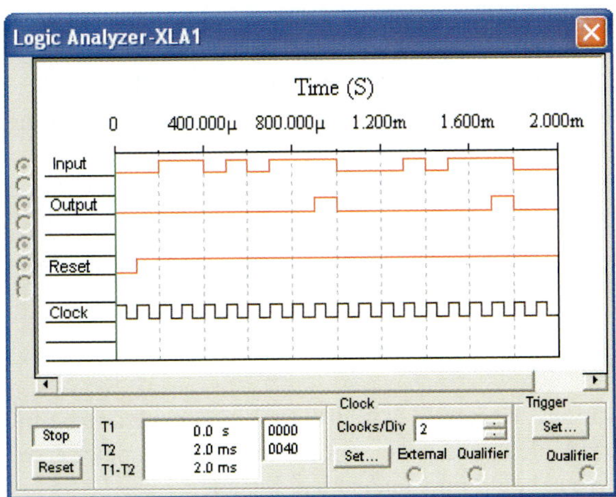

FIGURE 9-36 The simulation results as displayed by the Multisim logic analyzer.

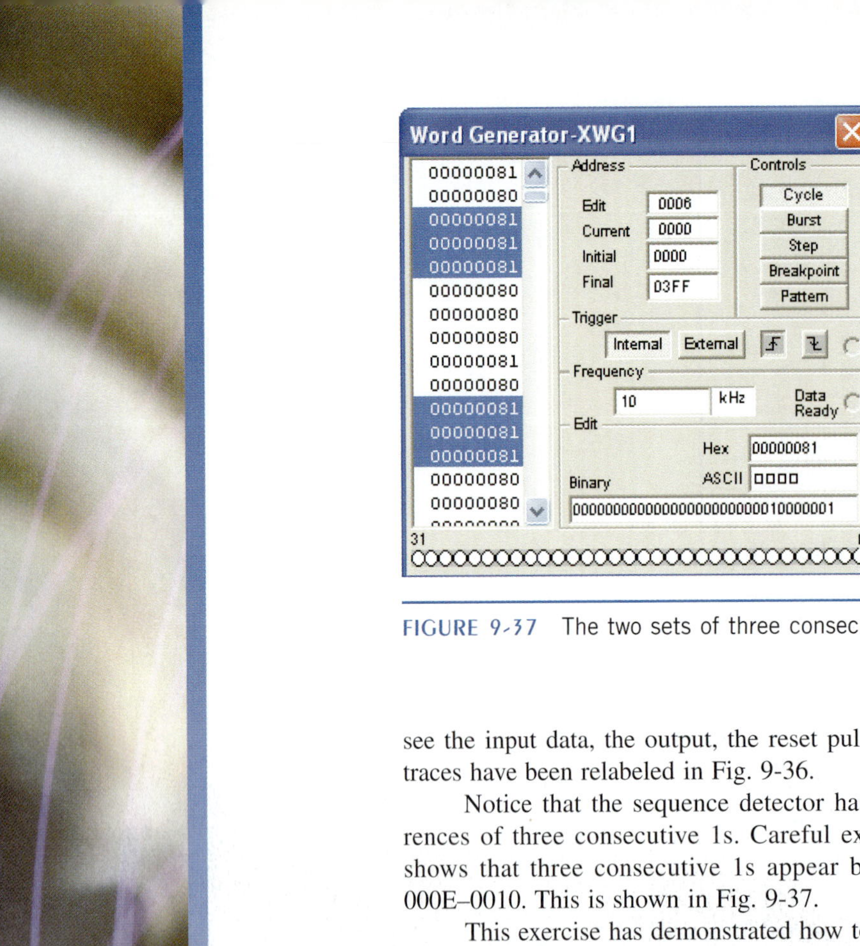

FIGURE 9-37 The two sets of three consecutive 1s in the Multisim simulation.

see the input data, the output, the reset pulse, and the internal clock (clock). The traces have been relabeled in Fig. 9-36.

Notice that the sequence detector has detected the presence of two occurrences of three consecutive 1s. Careful examination of the word generator file shows that three consecutive 1s appear beginning at 0006–0008 and again at 000E–0010. This is shown in Fig. 9-37.

This exercise has demonstrated how to use the Multisim word generator and the logic analyzer for examining the traces from a sequence detector circuit. The following Electronic Workbench™ Multisim exercises will also investigate your understanding of the material presented in this section.

Electronic Workbench™ Exercises

1. Open the Multisim file **Fig9-34.ms7 (.msm)** and determine if another sequence of three consecutive 1s appears in the simulation. (data sequence 0025–0027)
2. Open the Multisim file **FigE9-1.ms7 (.msm)** and determine if the circuit is working properly. If it is not, determine which circuit has a fault. Correct the fault and verify that the circuit is working properly.
3. Open the Multisim file **FigE9-2.ms7 (.msm)** and determine if the circuit is working properly. If it is not, find and correct the fault.

Summary

In Chapter 9 we examined various aspects of wired digital communications, including bandwidth issues, TDMA, delta and pulse modulation, and computer communication. The major topics you should now understand include:

- the concepts of error probability and bit error rate
- the basic concept of establishing a data link and the associated protocol

- the bandwidth issues in wired digital communications
- the fundamentals of TDMA
- the techniques used in delta and pulse modulation
- the fundamentals of data transmission, including T1, T3, packet switching, frame relay, and ATM
- computer serial communications, facsimile, and computer bus standards

Questions and Problems

Section 9-1

1. Explain what is meant by a baseband transmission.
2. Which transmission scheme of the six shown in Fig. 9-1 is used by a broadcast FM radio station?
3. Provide some possible reasons why an analog signal is digitized when an analog output is desired. A block diagram of such a system is shown in Fig. 9-1(f).
4. How are analog signals converted to digital?

Section 9-2

5. In what ways is coded voice transmission advantageous over direct transmission? What are some possible disadvantages?
6. Define *coding* and *bit*.
7. Sketch a voltage waveform that could be used to represent the binary number 11010100100.
8. Explain the noise immunity advantages of the binary code over any other code.
9. Calculate the error probability in a system that produces 7 error bits out of 5,700,000 total bits. (1.23×10^{-6})
10. How many bits are required to obtain an efficiency of 90 percent if 4 bits are used? (3.6)
11. The transmit power for a digital communications device is 1 W. The data rate is 28.8 kbps. Determine the energy per bit. (34.7 μJ)
12. Describe the differences between an asynchronous and a synchronous communications channel. Provide examples for each.
13. Describe the purpose of a communication protocol.

Section 9-3

14. Calculate the channel capacity (bits per second) of a standard phone line that has an *S/N* of 511. (27 kbps)
15. A 56-kbps NRZ-L encoded data stream is used in a digital communications link. Determine the minimum bandwidth required for the communications link. (28 kHz)

Section 9-4

16. List the data rates for the T and DS carriers.
17. How many telephone calls can be carried over a T1 line? How is the total of 1.544 Mbps obtained for a T1 line?

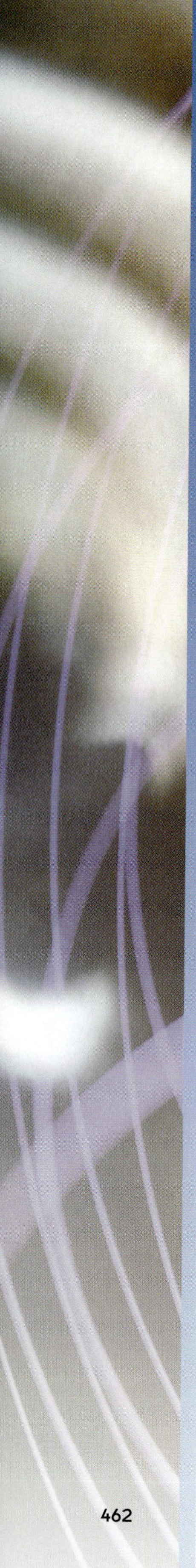

18. Define *fractional T1.*
19. Define *point of presence.*
20. What is the purpose of a CSU/DSU?
21. Describe how packet switching works. What is statistical concentration?
22. Explain the operation of a frame relay network. What technique does it use to increase data throughput and why is this a viable approach?
23. What is ATM and how are data moved quickly with this technique?

Section 9-5

24. Describe the TDMA technique for transmitting data over a serial communications channel.
25. What is the purpose of guard time and why is it necessary to provide this in a TDMA system?
26. Draw a block diagram of a TDMA system. Are there any special requirements for the multiplexing frequency? Describe what is required.

Section 9-6

27. Explain the process of delta modulation.
28. Provide a detailed explanation of the delta modulator shown in Fig. 9-12.
29. Explain how slope overload occurs in DM and how the CVSD feature helps to overcome it.
30. Compare CVSD to companding.
31. List two advantages of pulse modulation and explain their significance.
32. With a sketch similar to Fig. 9-14, explain the basics of PAM, PWM, and PPM.
33. Describe a means of generating and detecting PWM.
34. Describe a means of generating and detecting PPM.
35. Draw a diagram showing demodulation of a PWM signal.
36. Define *continuous-wave transmission.* In what ways is this an appropriate name?
37. Explain the AGC difficulties encountered in the reception of CW. What is two-tone modulation, and how does it remedy the receiver AGCC problem of CW?
38. Calculate all possible transmitted frequencies for a two-tone modulation system using a 21-MHZ carrier with 300- and 700-Hz modulating signals to represent mark and space. Calculate the channel bandwidth required.

Section 9-7

39. Describe the origin of the RS-232 C standard and discuss its current status as a standard.
40. What are the USB wire functions and colors?
41. The USB Type A connector is used for upstream or downstream connection?
42. What is handshaking?
43. What data rates do USB and Firewire support?
44. Describe the difference in USB Type A and B connectors.

Questions for Critical Thinking

45. Explain the difference between committed information rate and committed burst information rate.
46. Describe the three CSU/DSU alarms and three loopback tests for a CSU/DSU. Why are loopback tests important?

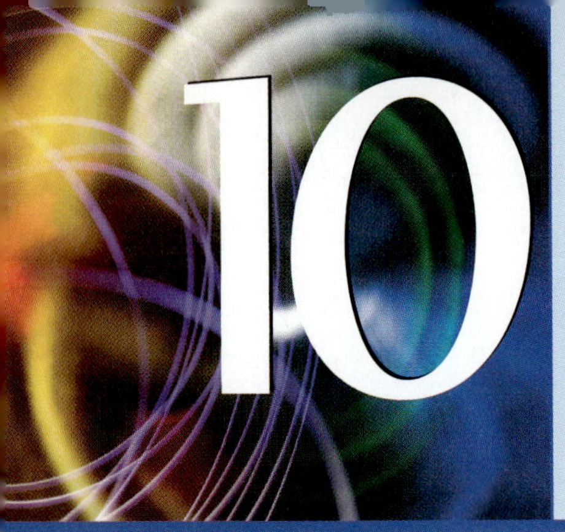

Wireless Digital Communications

Chapter Outline

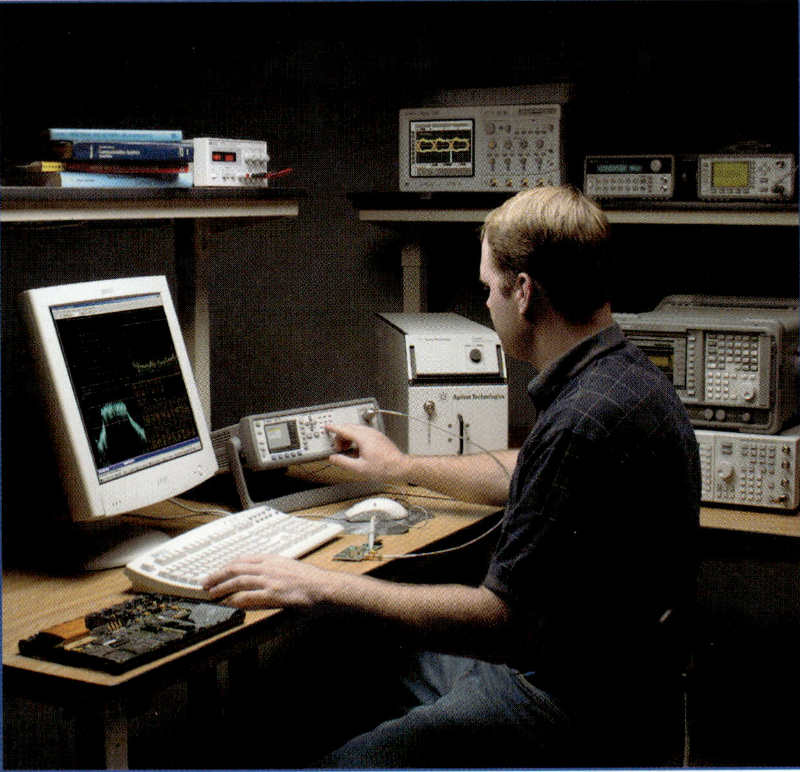

Bluetooth™ and WLAN test equipment allows for a broad range of signal measurements to evaluate wireless connectivity formats. (Courtesy of Agilent Technologies. Reprinted with permission.)

Objectives

- Describe the basics of a wireless digital communications link
- Provide detail on the various schemes used to transmit digital signals, including FSK, PSK, BPSK, QPSK, DPSK, and QAM
- Describe the generation of eye patterns and explain their use
- Describe the OFDM technique and explain why it is used
- Detail the operation of a complete radio-telemetry system
- Understand the basic steps for troubleshooting cell phone problems

Key Terms

wireless digital communications
wireless
frequency shift keying
phase shift keying
data bandwidth compression
quadrature amplitude modulation
constellation pattern
loopback
eye patterns
pseudonoise (PN) codes
spread
PN sequence length
maximal length
frequency hopping spread spectrum
dwell time
DSSS
chips
code division multiple access (CDMA)
multiple access
hit
signature sequence
despread
orthogonal frequency division multiplexing (OFDM)
multitone modulation
orthogonal
cyclic prefix
flash OFDM
telemetry
radio telemetry
water mark sticker
preferred roaming list (PRL)
OTA
RF shield box

10-1 INTRODUCTION

wireless digital communications the transport of digital data over a wireless medium

wireless the term used today to describe telecommunications technology that uses radio waves, rather than cables, to carry the signal

This chapter focuses on the physical layer technologies used in **wireless digital communications.** This area of communications has evolved rapidly over the past ten years, and it is safe to say that the evolution is just beginning. The term **wireless** is used today to describe telecommunications technologies that use radio waves, rather than cables, to carry the signal. Examples of wireless digital equipment and systems used today include cellular phones, wireless networking, cordless devices (for example, Bluetooth), digital television, satellite TV, satellite radio, global positioning systems (GPSs), pagers, garage door openers, and telemetry systems. Wireless is not new; Chapters 1 to 7 addressed the fundamental techniques that have been used for many years to transport both analog and digital signals over radio waves. In fact, digital data has been transported over wireless systems for many years, but the recent evolution of wireless digital services (for example, personal communication services [PCS], third-generation [3G] services, and mobile Internet) is behind the wireless explosion. The objective of this chapter is to present the communications techniques used to transport digital data over wireless networks.

Wireless technologies can be divided into the following three groups:

1. Fixed wireless: both transmit and receive units are at fixed locations (for example, wireless local area networks [LANs]).
2. Mobile wireless: the transmit and/or receive units are moving (for example, cellular and PCS phones).
3. Infrared (IR) wireless: the transmit and receive units use infrared light sources and detectors to provide the communications link (for example, building-to-building communications links)

The basic digital modulation techniques used in wireless communications are first examined in Sec. 10-2. These techniques include frequency-shift keying, phase-shift keying, BPSK, QPSK, and QAM. Spread-spectrum techniques are introduced in Sec. 10-3, which thoroughly explains CDMA and how spread-spectrum communication works. The section also includes a demonstration of the spreading and despreading of the BPSK signal using the Electronics Workbench™ Multisim software.

Orthogonal frequency division multiplexing (OFDM) and the technique used to create an OFDM signal are examined in Sec. 10-4. An overview of flash OFDM is presented in the same section, which also reviews the spread-spectrum version of OFDM. Radio telemetry is examined in Sec. 10-5. The techniques of remote metering are introduced to give you an overview of how the data is gathered remotely.

A section on troubleshooting cell phone problems is provided in Sec. 10-6. The basic troubleshooting steps are examined, as is an example of running full diagnostics of a cell phone using a CDMA test set. The chapter concludes with the use of Electronics Workbench™ Multisim to simulate a BPSK transmit-receive circuit.

10-2 DIGITAL MODULATION TECHNIQUES

The transmission of digital data via frequency modulation (FM) or phase modulation (PM) offers some of the same advantages over amplitude modulation that occur in standard analog systems. In this section we'll examine the common FM/PM schemes used to transmit digital data. Keep in mind that these schemes are used for transmission of any type of binary code.

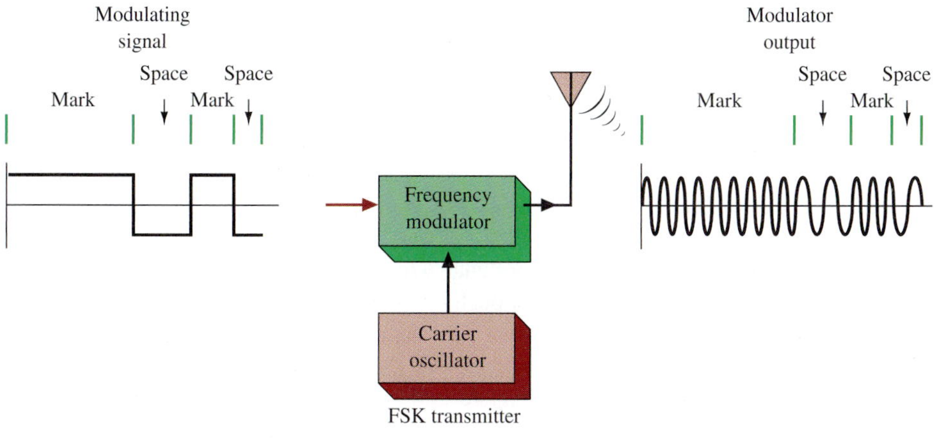

FIGURE 10-1 FSK transmitter.

Frequency Shift Keying

Frequency shift keying (FSK) is a form of frequency modulation in which the modulating wave shifts the output between two predetermined frequencies—usually termed the mark and space frequencies. It may be considered as an FM system in which the carrier frequency is midway between the mark and space frequencies and is modulated by a rectangular wave, as shown in Fig. 10-1. The mark condition causes the carrier frequency to increase by 42.5 Hz, while the space condition results in a 42.5-Hz downward shift. Thus, the transmitter frequency is constantly changing by 85 Hz as it is keyed. This 85-Hz shift is the standard for narrowband FSK, while an 850-Hz shift is the standard for wideband FSK systems.

Frequency Shift Keying
a form of data transmission in which the modulating wave shifts the output between two predetermined frequencies

Example 10-1

Determine the channel bandwidth required for the narrowband and wideband FSK systems.

Solution

The fact that narrowband FSK shifts a total of 85 Hz does *not* mean the bandwidth is 85 Hz. While shifting 85 Hz, it creates an infinite number of sidebands, with the extent of significant sidebands determined by the modulation index. If this is difficult for you to accept, it would be wise to review the basics of FM in Chapter 5.

In practice, most narrowband FSK systems utilize a channel of several kilohertz, while wideband FSK uses 10 to 20 kHz. Because of the narrow bandwidths involved, FSK systems offer only slightly improved noise performance over the AM two-tone modulation scheme. However, the greater number of sidebands transmitted in FSK allows better ionospheric fading characteristics than do two-tone AM modulation schemes.

FSK Generation

The generation of FSK can be accomplished easily by switching an additional capacitor into the tank circuit of an oscillator when the transmitter is keyed. In narrowband FSK it is often possible to get the required frequency shift by shunting the capacitance directly across a crystal, especially if frequency multipliers follow the oscillator, as is usually the case. FSK can also be generated by applying the rectangular wave modulating signal to a voltage-controlled oscillator (VCO). Such a system is shown in Fig. 10-2. The VCO output is the desired FSK signal, which is then transmitted to an FM receiver. The receiver is a standard unit up through the IF amps. At that point a 565 phase-locked loop (PLL) is used for detecting the original modulating signal. As the IF output signal appears at the PLL input, the loop locks to the input frequency and tracks it between the two frequencies with a corresponding dc shift at its output, pin 7. The loop filter capacitor C_2 is chosen to set the proper overshoot on the output, and the three-stage ladder filter is used to remove the sum frequency component. The PLL output signal is a rounded-off version of the original binary modulating signal and is therefore applied to the comparator circuit (the 5710 in Fig. 10-2) to make it logic compatible.

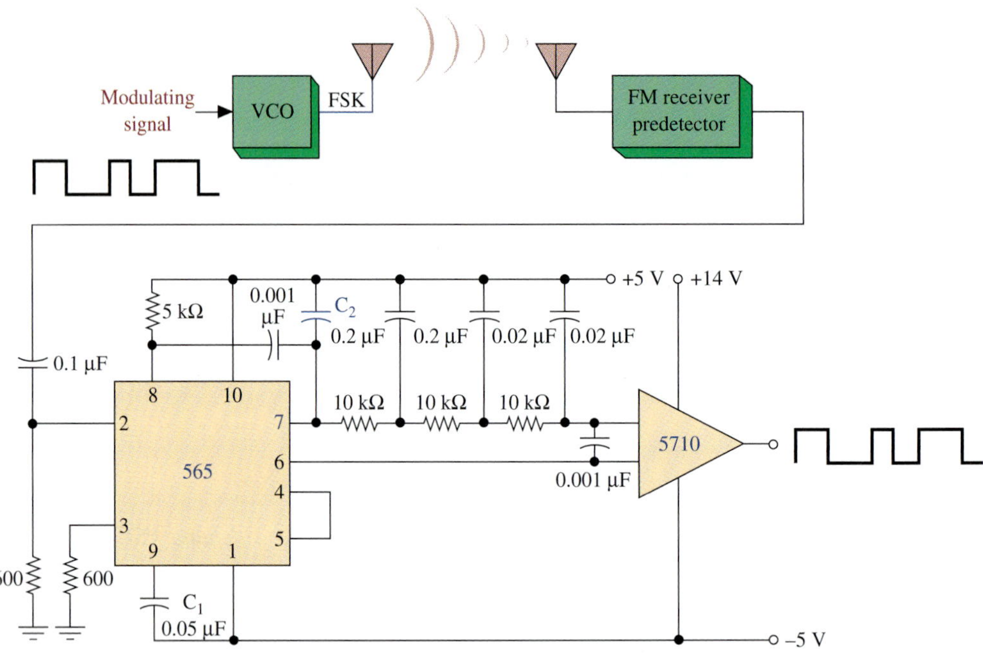

FIGURE 10-2 Complete FSK system.

Phase Shift Keying

Phase Shift Keying method of data transmission in which data causes the phase of the carrier to shift by a predefined amount

One of the most efficient methods for data modulation is **phase shift keying** (PSK). PSK systems provide a low probability of error. The incoming data cause the phase of the carrier to phase-shift a defined amount. This relationship is expressed as:

$$V_o(t) = V \sin\left[\omega_c(t) + \frac{2\pi(i-1)}{M}\right] \quad \textbf{(10-1)}$$

where $i = 1, 2, \ldots, M$

$M = 2^n$, number of allowable phase states

n = the number of data bits needed to specify the phase state

ω_c = angular velocity of carrier

There are many versions of the PSK signal. Three common versions are shown in Table 10-1.

With M (the number of allowable phase states) greater than four, the systems are referred to as *M-ary* systems, and the output signal is called a *constellation.* In a BPSK signal, the phase of the carrier will shift by 180° (i.e., $\pm\sin\omega_c t$). A diagram showing a BPSK constellation is provided in Fig. 10-3. For a QPSK signal, the phase changes 90° for each possible state.

Table 10-1 COMMON PSK SYSTEMS

Binary phase shift keying—BPSK	$M = 2$	$n = 1$
Quadrature phase shift keying—QPSK	$M = 4$	$n = 2$
8PSK	$M = 8$	$n = 3$

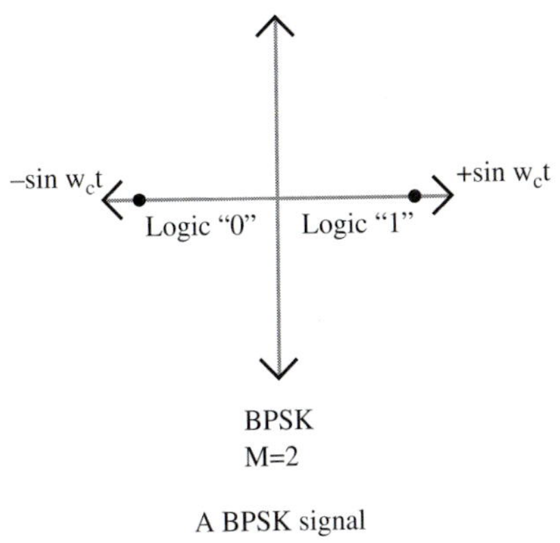

FIGURE 10-3 Illustration of a BPSK constellation.

Binary Phase Shift Keying

For a BPSK signal, $M = 2$ and $n = 1$ (see Table 10-1). The $+\sin(\omega_c t)$ vector provides the logical "1" and the $-\sin(\omega_c t)$ vector provides the logical "0," as shown in Fig. 10-3. The BPSK signal does not require that the frequency of the carrier be shifted, as with the FSK system. Instead, the carrier is directly phase modulated, meaning that the phase of the carrier is shifted by the incoming binary data. This relationship is shown in Fig. 10-4 for an alternating pattern of 1s and 0s.

Generation of the BPSK signal can be accomplished in many ways. A block diagram of a simple method is shown in Fig. 10-5. The carrier frequency $[+\sin(\omega_c t)]$ is phase-shifted 180°. The + and − values are then fed to a 1 of 2 selector circuit, which is driven by the binary data. If the binary data is a 1, then the output is $+\sin(\omega_c t)$. If the binary-input data is a 0, then the $-\sin(\omega_c t)$ signal is selected for the output. The actual devices selected for performing this operation are dependent on the binary input data rate and the transmit carrier frequency.

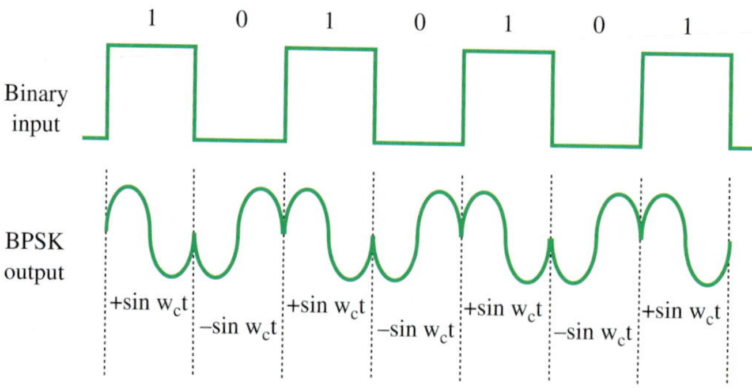

FIGURE 10-4 The output of a BPSK modulator circuit for a 1010101 input.

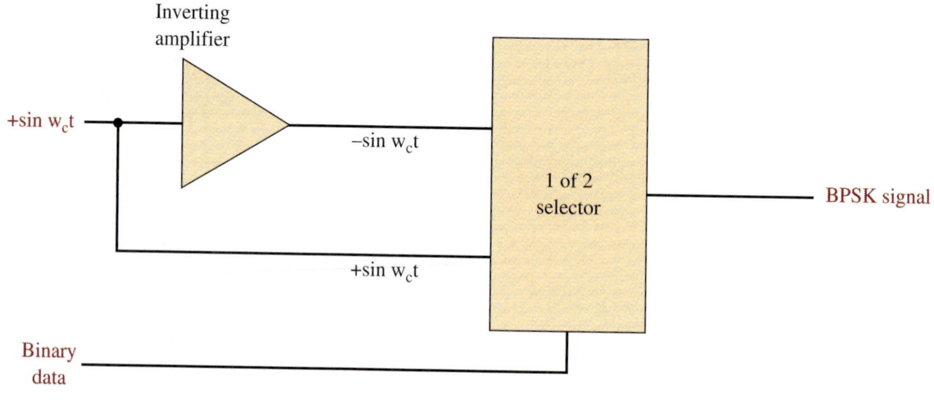

FIGURE 10-5 A circuit for generating a BPSK signal.

A BPSK receiver detects the phase shift in the received signal. One possible way of constructing a BPSK receiver is by using a mixer circuit. The received BPSK signal is fed into the mixer circuit. The other input to the mixer circuit is driven by a reference oscillator synchronized to $\sin(\omega_c t)$. This is referred to as *coherent carrier recovery*. The recovered carrier frequency is mixed with the BPSK input signal to provide the demodulated binary output data. A block diagram of the receive circuit is provided in Fig. 10-6. Mathematically, the BPSK receiver shown in Fig. 10-6 can provide a 1 and a 0 as shown:

$$\text{“1” output} = [\sin(\omega_c t)][\sin(\omega_c t)] = \sin^2(\omega_c t)$$

and by trig identity $\sin^2 \mathrm{A} = \frac{1}{2}[1 - \cos 2A]$. Therefore

$$\text{“1” output} = \tfrac{1}{2} - \tfrac{1}{2}[\cos(2\omega_c t)]$$

The $\frac{1}{2}[cos(2\omega_c t)]$ term is filtered out by the low-pass filter shown in Fig. 10-6. This leaves

$$\text{“1” output} = \tfrac{1}{2}$$

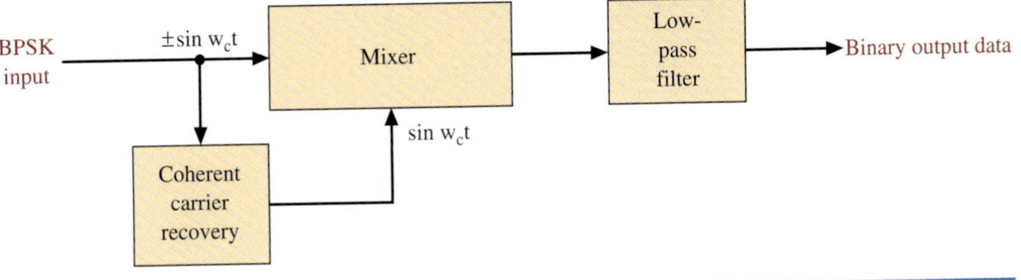

FIGURE 10-6 A BPSK receiver using coherent carrier recovery and a mixer circuit.

Similar analysis will show

$$\begin{aligned}\text{"0" output} &= [-\sin(\omega_c t)][\sin(\omega_c t)] \\ &= -\sin^2(\omega_c t) \\ &= -\tfrac{1}{2}\end{aligned}$$

The $\pm\frac{1}{2}$ represent dc values that correspond to the 1 and 0 binary values. The $\pm\frac{1}{2}$ values can be conditioned for the appropriate input level for the receive digital system.

The QPSK System

The QPSK constellation provides four vectors for representing the binary data. The savings by transmitting data this way are in the reduced bandwidth requirement. The QPSK system uses two data channels, identified as the *I* and *Q* channels. Each channel contributes to the direction of the vector within the phase constellation. The four possible values for I and Q are shown in Fig. 10-7. The BPSK signal requires a $BW_{min} = f_b$, where the QPSK signal will require only $f_b/2$ for each channel. The quantity f_b refers to the frequency of each of the original bits in the data. With QPSK transmission, a **data bandwidth compression** is realized. This means that more data are being compressed into the same available bandwidth. This relationship is shown pictorially in Fig. 10-8.

Data Bandwidth Compression
in QPSK transmission, more data are being compressed into the same available bandwidth as compared to BPSK

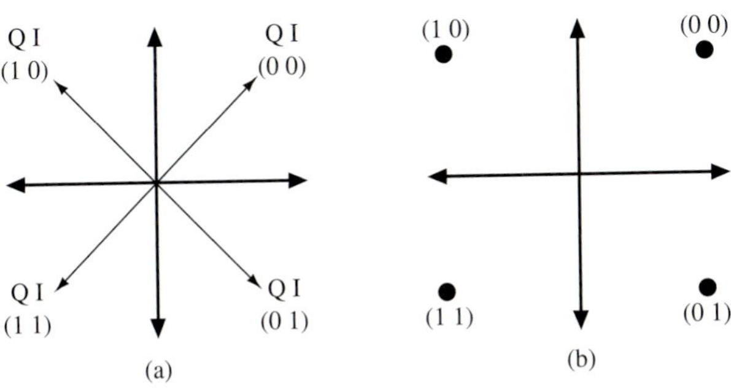

FIGURE 10-7 The QPSK phase constellation: (a) vector representation; (b) the data points.

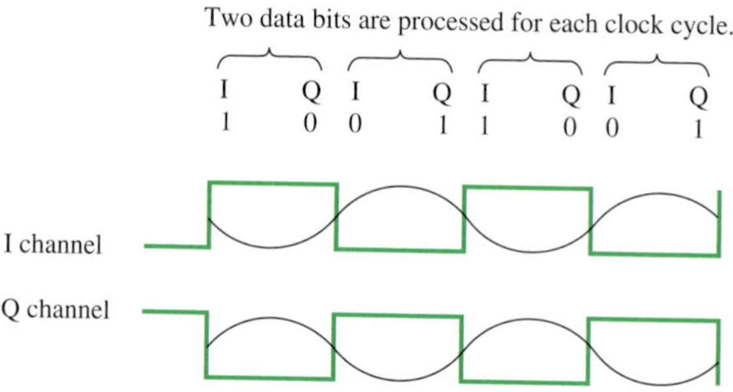

FIGURE 10-8 The I and Q data channels for a QPSK signal.

A block diagram of a QPSK demodulating circuit is shown in Fig. 10-9. A carrier recover circuit is used to generate a local clock frequency, which is locked to the QPSK input carrier frequency (sin $\omega_c t$). This frequency is then used to drive the phase-detector circuits for recovering the I and Q data. The phase detector is basically a mixer circuit where sin $\omega_c t$ (the recovered carrier frequency) is mixed with the QPSK input (expressed as sin $\omega_c t + \phi_d$). The ϕ_d term indicates that the carrier frequency has been shifted in phase, which is to be expected for phase shift keying data. The output of the phase detector (v_{pd}) is then determined as follows:

$$
\begin{aligned}
v_{pd} &= (\sin \omega_c t)(\sin \omega_c t + \phi_d) \\
&= 0.5A \cos(0 + \phi_d) - 0.5A \cos 2\omega_c t
\end{aligned}
$$

A low-pass filter (LPF) removes the $2\omega_c$ high-frequency component, leaving

$$v_{pd} = 0.5A \cos \phi_d$$

The remaining ϕ_d value is the dc voltage representing the direction of the data vector along the I-axis. The Q data are recovered in the same way, and together the I

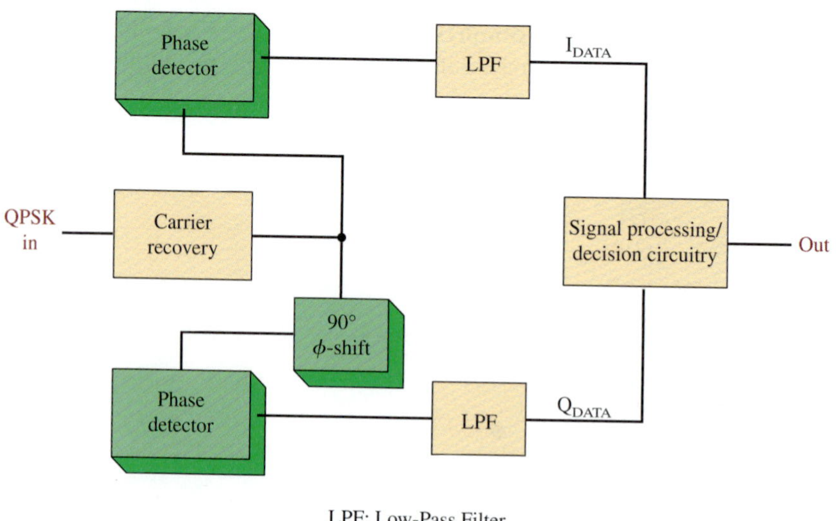

FIGURE 10-9 A block diagram of a QPSK demodulating circuit.

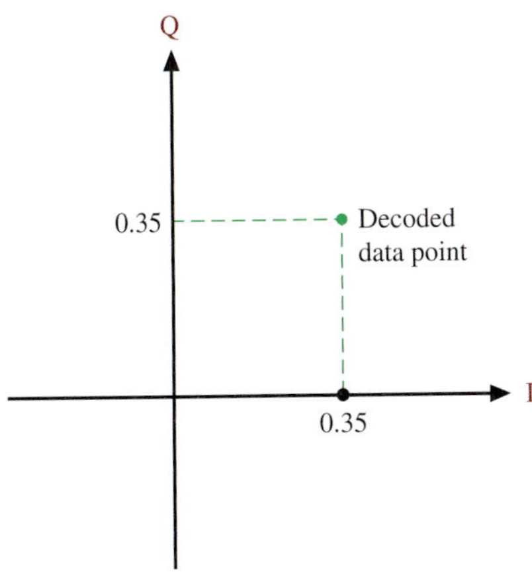

FIGURE 10-10 The decoded data point.

and Q values define the direction of the data vector. For a data value of (0, 0), the data point is at a vector of 45°. Therefore, v_{pd} = 0.5/cos (45°) = (0.5)(0.707) = 0.35 V. The decoded data point is shown in Fig. 10-10. The amplitude of the data and the phase, ϕ, will vary due to noise introduced in the transmission and data recovery. Each contributes to a noisy constellation, as shown in Fig. 10-11. This is an example of what the decoded QPSK digital data can look like at the output of a digital receiver. The values in each quadrant are no longer single dots, as in the ideal case [Fig. 10-7 (b)]. During transmission, noise is added to the data, and the result is a cluster of data in each quadrant of the QPSK constellation.

Once the I and Q data are decoded, signal processing and a decision circuit are used to analyze the resulting data and assign a digital value to it (e.g., 00, 10, 11, 01). The shaded areas in Fig. 10-11 represent the decision boundaries in each quadrant. The data points, defined by the I and Q signals, will fall in one of the boundary regions and are assigned a digital value based on which quadrant the point is within.

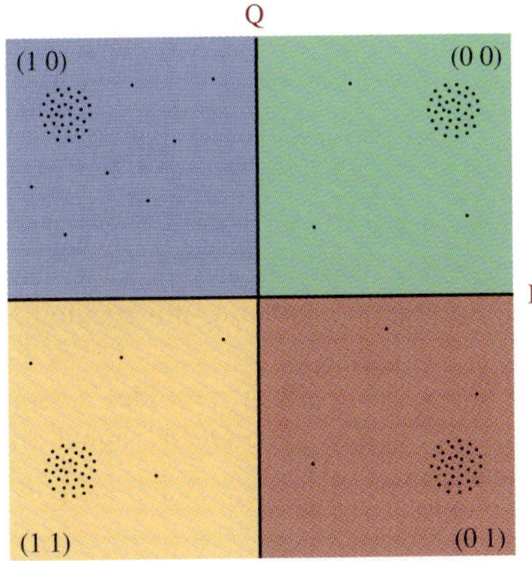

FIGURE 10-11 The QPSK constellation with noise at the receiver. The decision boundaries are shown by the different shaded areas.

Differential Phase Shift Keying

Differential phase shift keying (DPSK) uses the BPSK vector relationship for generating an output. The logical "0" and "1" are determined by comparing the phase of two successive data bits. In BPSK, $\pm\sin \omega_c t$ represent a 1 and 0, respectively. This is also true for a DPSK system. Generation of the 1 and 0 outputs is accomplished by initially comparing the first bit (1) transmitted with a reference value (0). The two values are EXNORed to generate the DPSK output (0). The "0" output is used to drive a PSK circuit shown in Fig. 10-5. A 0 generates $-\sin \omega_c t$, and a 1 generates $+\sin \omega_c t$.

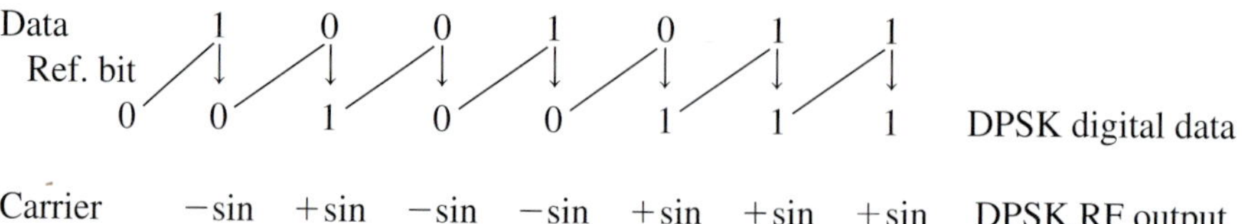

A simplified block diagram of a DPSK transmitter is shown in Fig. 10-12.

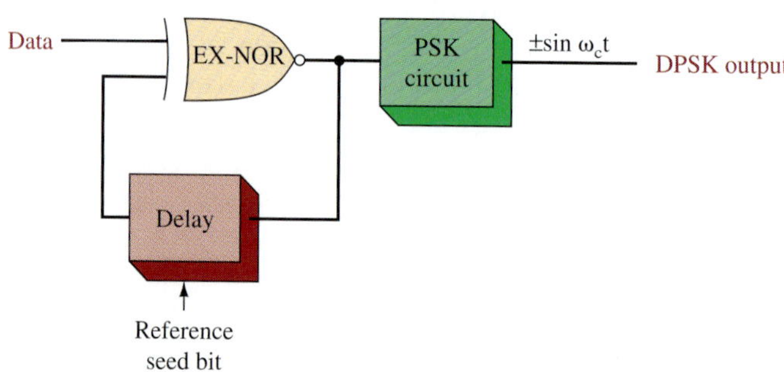

FIGURE 10-12 The simplified block diagram of a DPSK transmitter.

At the DPSK receiver, the incoming data is EXNORed with a 1-bit delay of itself. For the previous example, the data will be shifted as shown:

Rx carrier	−sin	+sin	−sin	−sin	+sin	+sin	+sin	
	(ref)							
Shifted Rx carrier	−sin	−sin	+sin	−sin	−sin	+sin	+sin	+sin
Recovered data	1	0	0	1	0	1	1	

The advantage of a DPSK system is that carrier recovery circuitry is not required. The disadvantage of DPSK is that it requires a good signal-to-noise ratio for it to achieve a bit error rate equivalent to QPSK.

A DPSK receiver is shown in Fig. 10-13. The delay circuit provides a 1-bit delay of the DPSK $\pm\sin \omega_c t$ signal. The delayed signal and the DPSK signal are mixed together. The possible outputs of the DPSK receiver are

$$(\sin \omega_c t)(\sin \omega_c t) = 0.5\cos(0) - 0.5\cos 2w_c t$$
$$(\sin \omega_c t)(-\sin \omega_c t) = -0.5\cos(0) + 0.5\cos 2w_c t$$
$$(-\sin \omega_c t)(-\sin \omega_c t) = 0.5\cos(0) - 0.5\cos 2w_c t$$

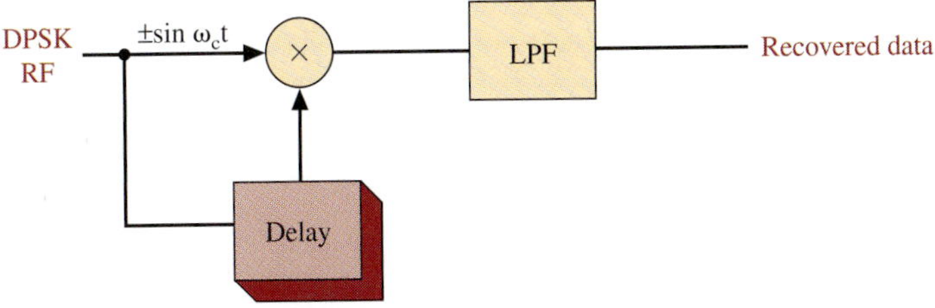

FIGURE 10-13 A DPSK receiver.

The higher-frequency component 2ω is removed by the low-pass filter, leaving the $\pm 0.5\cos(0)$ dc term. Because $\cos(0)$ equals 1, this term reduces to ± 0.5, where $+0.5$ represents a logical 1 and -0.5 represents a logical 0.

Quadrature Amplitude Modulation

Several special modulation techniques are used, over and above those previously described, to transmit digital signals. The most popular system to achieve high data rates in limited bandwidth channels is called **quadrature amplitude modulation** (QAM).

Quadrature Amplitude Modulation
method of achieving high data rates in limited bandwidth channels, characterized by two data signals that are 90° out of phase with each other

The block diagram in Fig. 10-14 shows a QAM transmitter. The binary data are first fed to a ÷ data block that essentially produces two data signals at half the original bit rate by feeding every other data bit to its two outputs. These two-level outputs are then converted to four-level baseband streams. The resultant four-level symbol streams of the I and Q channels are then applied to the modulators in Fig. 10-14. Notice the carrier for the Q-channel modulator. It is shifted 90° from the

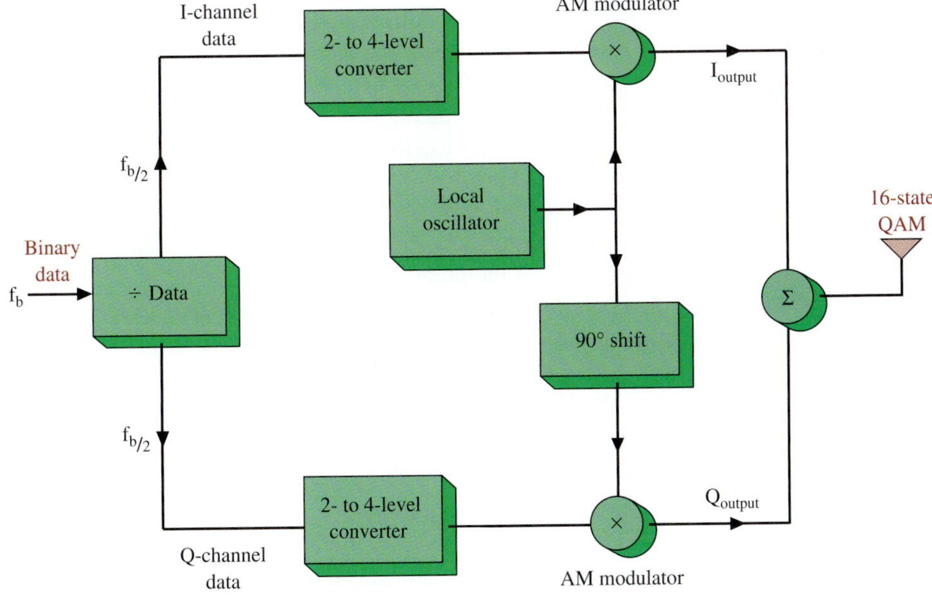

FIGURE 10-14 16-QAM transmitter.

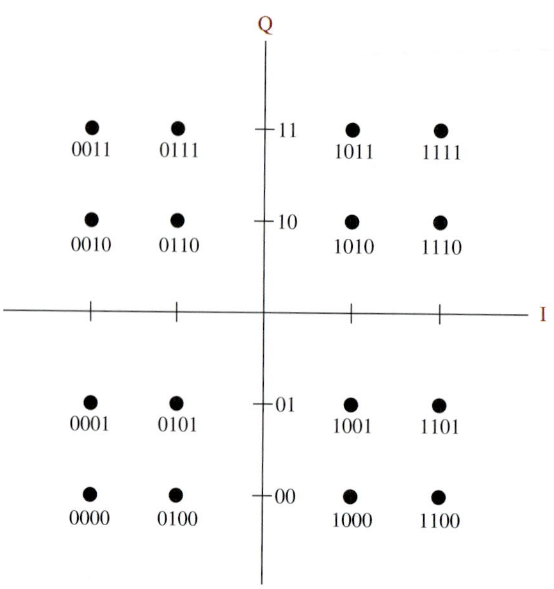

FIGURE 10-15 16-QAM (4 × 4) constellation pattern.

I channel and is said to be in quadrature. This explains its Q designation, while the I channel is so named because it is the in-phase channel. The net result of this is the ability to transmit large amounts of digital data through limited bandwidth channels.

The QAM demodulator reverses the modulator process, thereby providing the original binary data signal. Some insight into QAM systems is provided by feeding the demodulated I signal into the horizontal input and the Q signal into the vertical input of an oscilloscope. The result (shown in Fig. 10-15) is called a **constellation pattern** because of its resemblance to stars. The gain and position of each channel must be properly adjusted and a signal must also be applied to the scope's Z-axis input to kill the scope intensity during the digital state transition times.

Constellation Pattern
display used to monitor QAM data signals on an oscilloscope to provide information on linearity and noise

The constellation pattern can be used to analyze the system's linearity and noise performance. Because QAM involves AM, linearity of the transmitter's power amplifiers can be a cause of system error. Linearity problems are indicated by unequal spacing in the constellation pattern. Noise problems are indicated by excessive blurring and spreading out of the points.

QAM digital transmission systems have become dominant in recent years. Besides the 4 × 4 system introduced here, 2 × 2 and 8 × 8 systems are also commonly used.

Loopbacks

Loopback
test configuration for a data link; the receiver takes the data and sends it back to the transmitter, where it is compared with the original data to indicate system performance

Many digital modulation systems include a **loopback** capability. The receiver takes the received data and sends them back to the transmitter. The data are then compared with the originally transmitted data to provide an indication of system performance. Bit errors can occur both in the original transmission and in the loopback, so the bit error rate cannot be pinpointed because it will not be known where the error

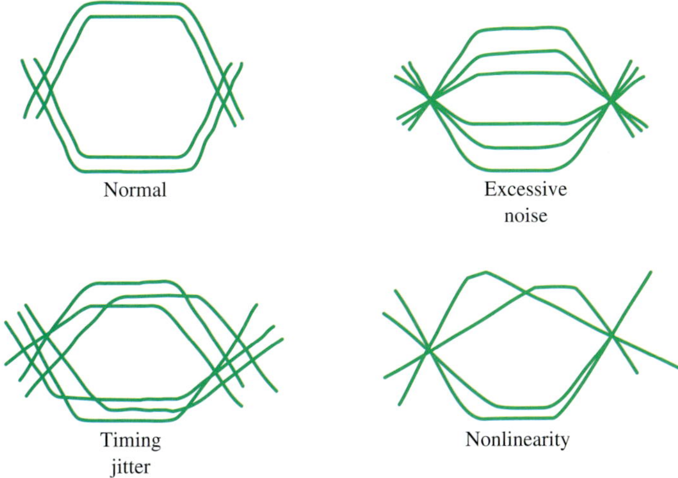

FIGURE 10-16 Eye patterns.

occurred. Nonetheless, the loopback test is very helpful in diagnosing the basic system performance.

Eye Patterns

Another technique that is extremely helpful in diagnosing the performance of a digital modulation system is the generation of **eye patterns.** They are generated by "overlaying" on an oscilloscope all the digital bit signals received. Ideally, this would result in a rectangular display because of the persistence of the CRT phosphor. The effects of transmission cause various rounding effects, which result in a display resembling an eye.

Eye Patterns
using the oscilloscope to display overlayed received data bits that provide information on noise, jitter, and linearity

Refer to Fig. 10-16 for the various possible patterns. The opening of the eye represents the voltage difference between a 1 or a 0. A large variation indicates noise problems, while nonsymmetrical shapes indicate system distortion. Jitter and undesired phase shifting are also discernible on the eye pattern. The eye pattern can be viewed while making adjustments to the system. This allows for the immediate observation of the effects of filter, circuit, or antenna adjustments.

10-3 Spread-Spectrum Techniques

The first spread-spectrum systems were utilized by the U.S. government toward the end of World War II. These techniques allow transmissions that cannot be jammed (i.e., rendered useless by a counter "noise" signal at the same frequency) nor detected by the enemy. Military applications are still widespread, but as explained subsequently, commercial use is increasing due to performance advantages.

Spread spectrum started to gain serious commercial attention in 1985 when the Federal Communications Commission (FCC) opened the industrial, scientific, and medical (ISM) band for unlicensed operation of devices under FCC technical regulations 15.247. The FCC permits spread-spectrum modulation at a maximum transmitter power of 1 W in three bands: 902–928 MHz, 2400–2483.5 MHz, and 5725–5850 MHz.

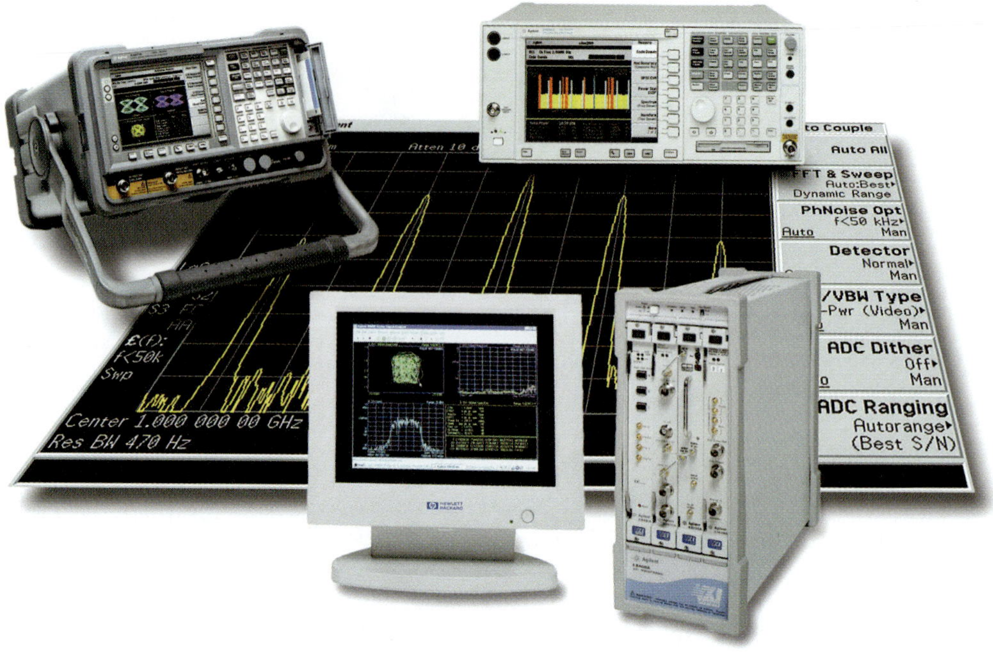

Agilent has a wide range of bench-top or portable spectrum analyzers. (Courtesy of Agilent Technologies. Reprinted with permission.)

Spread spectrum is becoming the technology of choice in many wireless applications, including cellular telephones, mobile Internet, wireless local-area networks (WLANs), and automated data collection systems using portable scanners of universal product code (UPC) codes. Other applications include remote heart monitoring, industrial security systems, and very small aperture satellite terminals (VSAT). Multiple spread-spectrum users can coexist in the same bandwidth if each user is assigned a different "spreading code," which will be explained subsequently.

As we have seen, regular modulation schemes tend to fully utilize a single band of frequencies. Noise in that band will obviously degrade the signal, and therefore it is vulnerable to jamming. This single-frequency band also allows detection by undesired recipients, who may track down the signal source via direction-finding (DF) techniques. An antenna commonly used in DF applications is described in Chapter 14. The spread-spectrum solution to these problems takes one of two forms, frequency hopping and direct sequence. Both of these spread spectrum techniques use a pseudonoise technique for spreading the signal. The concept of pseudonoise codes is presented first. That discussion is followed by a look at frequency hopping and direct sequence spread spectrum.

Pseudonoise (PN) Codes
digital codes with pseudorandom output data streams that appear to be noiselike

Spread
the RF signal is spread randomly over a range of frequencies in a noiselike manner

Pseudonoise (PN) Codes

Pseudonoise (PN) codes are popular in spread-spectrum communications because the randomness of the serial output data stream appears to be noiselike. These random data streams are used to **spread** the RF signal so that it appears to be noise, which means that the RF signal is randomly spread over a range of frequencies in a noiselike manner. The randomness of the signal is due to the fact that the 1 0 data

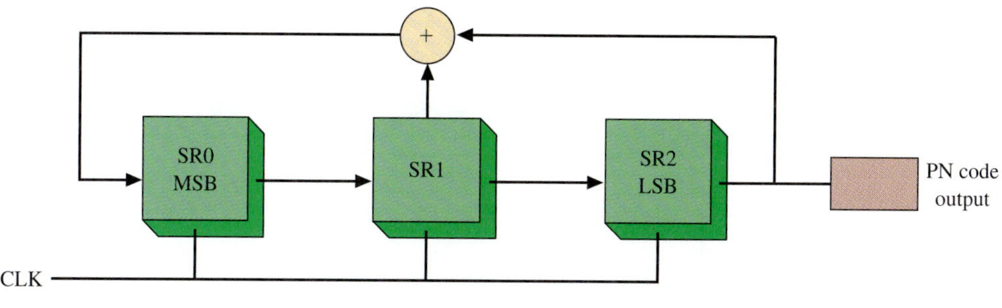

FIGURE 10-17 A 7-bit sequence generator.

pattern does not appear to repeat. PN sequences do repeat, however, but only after a long time. How often the pattern repeats depends on the PN sequence generating circuit. The application of PN codes in spread-spectrum communications will be explained in this section. The operation of a PN sequence generator circuit is explained first.

The method for implementing PN codes is quite simple. It requires the use of shift registers, feedback paths, and EXOR gates. A random pattern of 1s and 0s are output as the circuit is clocked. The data pattern will repeat if the circuit is clocked a sufficient number of times. For example, the circuit for a 7-bit PN sequence generator is shown in Fig. 10-17. The 7-bit PN sequence generator circuit contains three shift registers, an EXOR gate, and feedback paths. The circuit structure and the number shift registers in the circuit can be used to determine how many times the circuit must be clocked before repeating. This is called the **PN sequence length.** The equation for calculating the length of the PN sequence is provide in Eq. (10-2).

PN Sequence Length
the number of times a PN generating circuit must be clocked before repeating the output data sequence

$$\text{PN sequence length} = 2^n - 1 \qquad \textbf{(10-2)}$$

where $n =$ the number of shift registers in the circuit. A PN sequence that is $2^n - 1$ in length is said to be of **maximal length.** For example, the PN sequence generator shown in Fig. 10-17 contains three shift registers ($n = 3$) and a maximal length of $2^3 - 1 = 7$. An example of using Eq. (10-2) to determine the PN sequence length is provided in Ex. 10-2.

Maximal Length
indicates that the PN code has a length of $2^n - 1$

EXAMPLE 10-2

Determine the sequence length of a properly connected PN sequence generator containing
(a) 3 shift registers ($n = 3$).
(b) 7 shift registers ($n = 7$).

SOLUTION

(a) $n = 3$, PN sequence length $= 2^3 - 1 - 8 - 1 = 7$ **(10-2)**
(b) $n = 7$, PN sequence length $= 2^7 - 1 = 127$ **(10-2)**

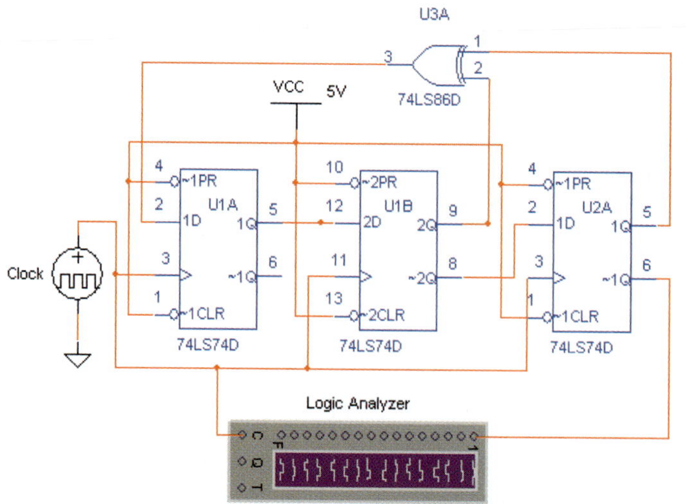

FIGURE 10-18 An implementation of a 7-bit PN sequence generator using Electronics Workbench™ Multisim.

An implementation of a 7-bit PN sequence generator circuit using Electronics Workbench™ Multisim is shown in Fig. 10-18. The circuit consists of three shift registers using 74LS74 D-type flip-flops (U1A, U1B, and U2A). Each shift register in the PN sequence generator shown in Fig. 10-18 is driven by the system clock. The Q outputs of U1B and U2A are fed to the input of the EXOR gate U3A (74LS86). The circuit is clocked $2^n - 1$ times (7 times) before repeating. The data is output serially, as shown in Fig. 10-19. Notice that the PN code output data stream repeats after seven cycles. The reason the PN circuit repeats after $2^n - 1$ times is that an all zero condition is not allowed for the shift register because such a condition will not propagate any changes in the shift-register states when the system is clocked.

The circuits required to implement PN sequences are well known. Modern spread-spectrum circuits have the PN sequence generator integrated into the system

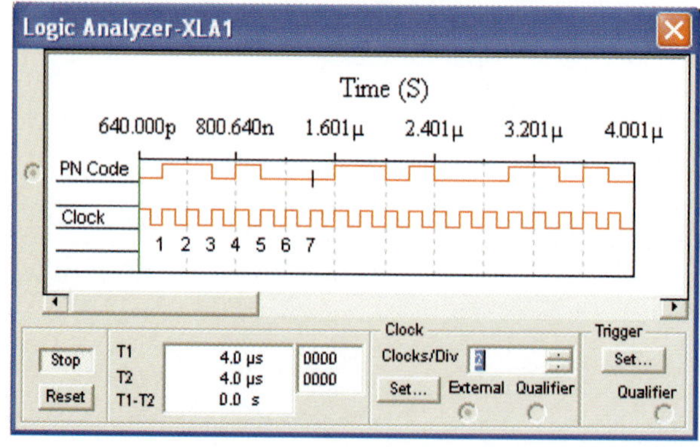

FIGURE 10-19 The serial data output stream for the 7-bit PN sequence generator.

Table 10-2 CONNECTIONS FOR CREATING PN SEQUENCE GENERATORS

Number of Shift Registers (n)	Sequence Length	EXOR Inputs
2	3	1, 2
3	7	2, 3
4	15	3, 4
5	31	3, 5
6	63	5, 6
7	127	6, 7
9	511	5, 9
25	33,554,431	22, 25
31	2,147,483,647	28, 31

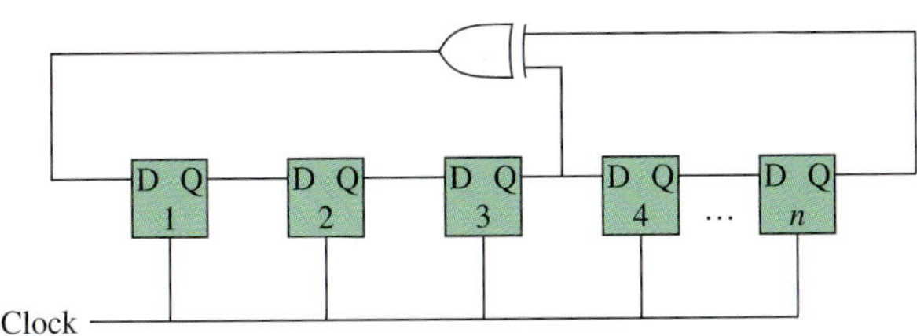

FIGURE 10-20 The PN sequence generator circuit structure for the connections listed in Table 10-2. The circuit shown is connected with five shift registers ($n = 5$) and a maximal length of 31.

IC. Table 10-2 lists the connections required to create different sequence lengths for the circuit structure provided in Fig. 10-20. The table shows only a partial listing of some of the shorter PN sequence lengths. Many spread-spectrum systems use a PN sequence length that is much longer. The PN circuit shown in Fig. 10-20 contains five shift registers ($n = 5$). Table 10-2 shows the EXOR inputs for the circuit that come from the Q outputs of the shift registers.

FREQUENCY HOPPING

In **frequency hopping spread spectrum,** the data is transmitted by a carrier that is switched in frequency in a pseudorandom fashion. *Pseudorandom* implies a sequence that can be re-created (e. g., at the receiver) but has the properties of randomness. The time of each carrier block is called **dwell time.** Dwell times are usually less than 10 ms. The receiver knows the order of the frequency switching, picks up the successive blocks, and assembles them into the original message.

Frequency Hopping Spread Spectrum
transmitting data by a carrier that is switched in frequency in a pseudorandom fashion

Dwell Time
the time each carrier spends at a specific frequency

A block diagram for a frequency hopping system is provided in Fig. 10-21. Essentially identical programmable frequency synthesizers and hopping sequence generators are the basis for these systems. The receiver must synchronize itself to the transmitter's hopping sequence, which is done via the synchronization logic. A spread-spectrum system obviously has an extra degree of complexity compared to conventional systems and is more and more feasible economically due to the advantages offered. Of course, this has always been true in many military applications.

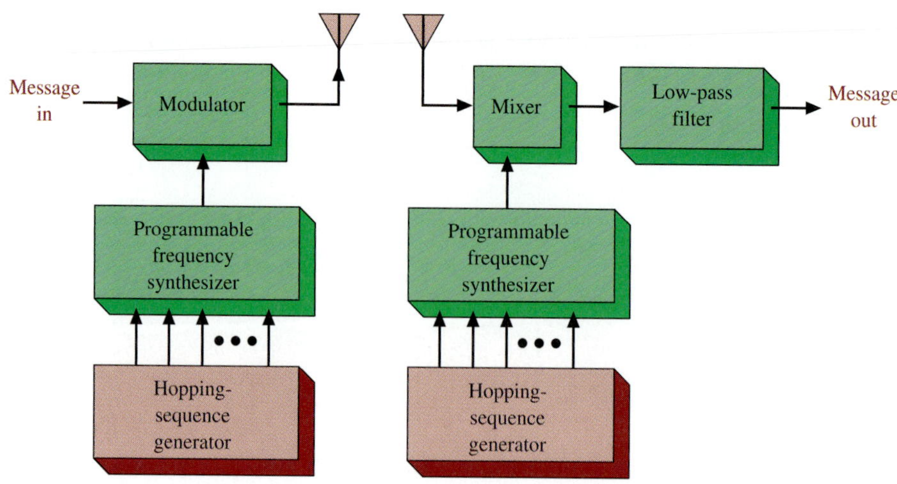

FIGURE 10-21 Frequency hopping spread spectrum.

A simplified picture of the RF spectrum for a frequency hopping signal is provided in Fig. 10-22. This image shows the shifting of the carrier frequency over a 600 kHz range. Note that seven frequencies are displayed. The pattern for the shifting of the frequencies is generated by a PN code sequence generator. In this case, the 3-bit contents of the PN sequence generator shown in Fig. 10-17 were used to generate the frequency hopping. The current state of the shift registers (SR0–SR2) randomly select each frequency, which requires that the PN sequence generator be initiated with a seed value.

The first step for configuring the PN sequence generator is to initialize the shift registers with a known value. In this case, the sequence generator is seeded with a 1 0 0 value. This step is shown in Table 10-3, which lists the shift register contents as the generator is clocked. At shift 0, the 1 0 0 contents select frequency D. At shift 1, the shift register contents of 0 1 0 select frequency B. This repeats for the length of the PN sequence generator and produces the seven unique states and frequencies A–G. The frequencies are randomly selected until the sequence begins repeating on shift 7, as shown with the shift register contents of 1 0 0 and the selection of frequency D.

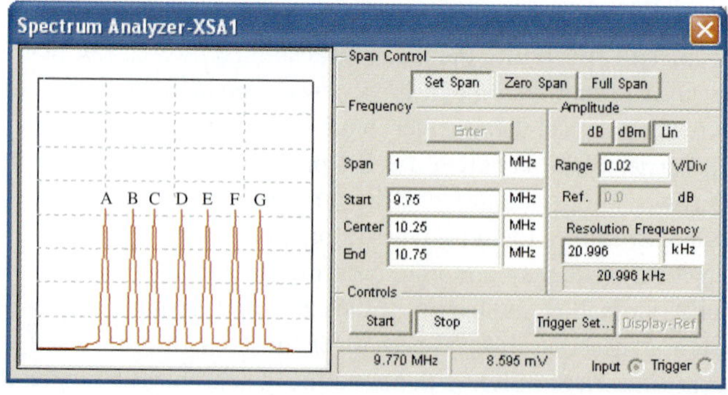

FIGURE 10-22 A simplified RE spectrum for a frequency hopping spread-spectrum signal.

Table 10-3 The Shift Register Contents for a PN Sequence Generator with $N = 3$ and a Seed Value of 1 0 0

Shift Number	SR0	SR1	SR2	Freqency
0	1	0	0	D
1	0	1	0	B
2	1	0	1	E
3	1	1	0	F
4	1	1	1	G
5	0	1	1	C
6	0	0	1	A
7	1	0	0	D

Direct Sequence Spread Spectrum (DSSS)

Direct sequence spread spectrum (DSSS) is often used in the transmission of digital bits in modern wireless digital systems. Its pseudorandom sequence uses pulses that are shorter than the message bit, called **chips.** The chips successively modulate fractions of the bit that typically phase-shift the carrier. The receiver multiplies the incoming signal by the same chip signal sequence to recover the original modulated digital signal. Systems utilizing this technique are called **code-division multiple-access (CDMA)** systems. **Multiple access (MA)** applies to any method used to multiplex many signals in one communications channel.

DSSS
direct sequence spread spectrum

Chips
in the transmission of digital bits, pulses shorter than the message bits

Code-Division Multiple-Access (CDMA)
communications system in which spread-spectrum techniques are used to multiplex more than one signal within a single channel

Multiple Access
any method of multiplexing many channels in one communications channel

These spread-spectrum techniques utilize a far wider bandwidth than that required by conventional modulation schemes. Besides the two advantages (detection and the prevention of jamming) for the military already described, other advantages have led to nonmilitary applications. Spread spectrum permits many transmitters to operate over the same channel with minimal interference because the pseudorandom sequences only rarely coincide. This coincidence, termed a **hit,** adds only a low-level noise to the overall receptions. The net result is a more efficient channel utilization (more channels), especially under conditions where continuous transmissions are not being made. In fact, spread-spectrum communications are now being added to bands fully used by conventional communications techniques with minimal interference. Another advantage of spread spectrum is the reduction in fading it affords compared to conventional (narrowband) systems. Different parts of the frequency spectrum tend to fade at variable rates. Thus, each hop of a frequency-hopping signal has an independent characteristic. The ratio of maximum to minimum received signal strength (fading) is typically 2 to 3 dB, compared with 20 to 30 dB for conventional transmission.

Hit
when two spread-spectrum transmitters momentarily transmit at the same frequency; they coincide at that instant

Spread-spectrum techniques are also widely used in radar systems (see Chapter 16). The spread carrier, as determined by the pseudorandom sequence, allows the receiver to determine accurately the time the transmitter sent its energy with minimal effects from noise, and so on. The distance to a transmitter (ranging) or object via a reflection can thereby be made accurately with high reliability.

Figure 10-23 shows the basic format of a direct-sequence spread-spectrum system. These systems are also referred to as pseudonoise (PN) spread-spectrum signaling systems. The spreading signal is a carrier modulated with a binary sequence having a pseudorandom (pseudonoise or PN) nature generated by a PN sequence generator circuit. The sequence generated is called the **signature sequence.** The clock rate of the shift register is very high compared with the data rate to be transmitted. One

Signature Sequence
the pseudorandom digital sequence used to spread the signal

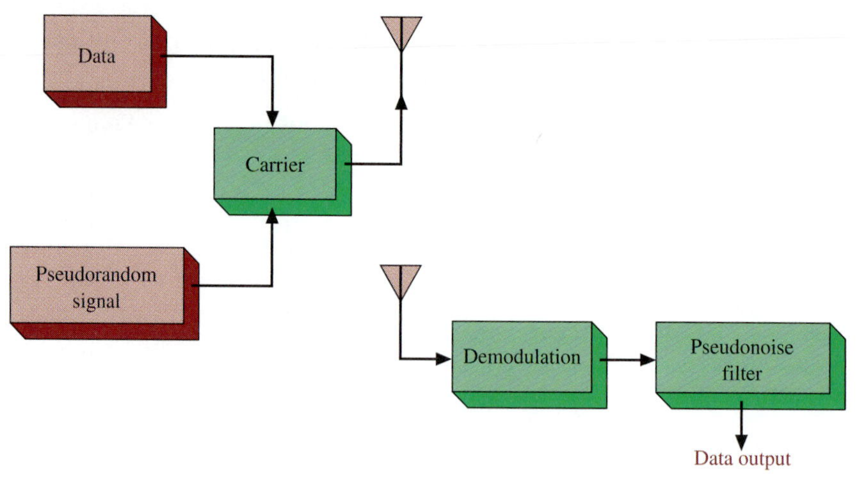

FIGURE 10-23 Direct sequence spread spectrum.

period of the shift-register sequence contains thousands of binary transitions call chips, and they are modulated with one bit of binary data. The receiver uses a PN sequence generator identically programmed (via logic connections) to the transmitter's PN sequence. The positions of the two shift registers along with the pseudorandom sequence are synchronized by analysis of the received waveform, with a decision circuit picking the signal with the highest correlation level. Other users in the band may have similar transmitters and receivers but use different logic connections in their PN sequence generators so they can present a different signature. Receivers do not recognize other signatures; they see them as background noise.

An example of implementing a DSSS circuit using Electronics Workbench™ Multisim is provided next. This material demonstrates how a BPSK signal is spread using the PN code and how the original data is recovered. The circuit used in this example is shown in Fig. 10-24. This DSSS circuit uses a 7-bit PN sequence to spread the signal. The PN sequence generating circuit is not shown in this schematic but is shown in Fig. 10-18. An alternating 1 0 data pattern is fed into a BPSK circuit. The BPSK circuit is considered the modulating circuit and the 1 0 data pattern is the input data. The waveforms for the BPSK output and the PN code are shown in Fig. 10-25. Notice the regular phase shift of the BPSK output signal due to the input data 1 0 pattern. The PN code repeats after 7 clock cycles (also shown in Fig. 10-25). The BPSK output and the PN code are then mixed together to create the DSSS output, as shown in the schematic of the circuit. This is indicated as mixer-TX in Fig. 10-24. The resulting output (DSSS output) is provided in Fig. 10-26. Notice the shape of the BPSK signal in the time domain. What is shown are the chips of data that are being transmitted.

The circuit shows that the DSSS output is connected to the receiver circuit. In practice the DSSS signal is transmitted to the receiver via RF. The receiver filters and amplifies the signal, which is the same basic concept that applies to all receiver front-ends. The received DSSS signal is fed into a receive mixer circuit (mixer-RX), which mixes the DSSS signal with the receiver's PN code. The PN code must be locked (synchronized) to the receiver's data stream, a necessary condition for the DSSS signal to be **despread.** Despreading the signal means that the signal is returned to its original modulated format. In this case, the original format

Despread
return the DSSS signal back to its original modulated format

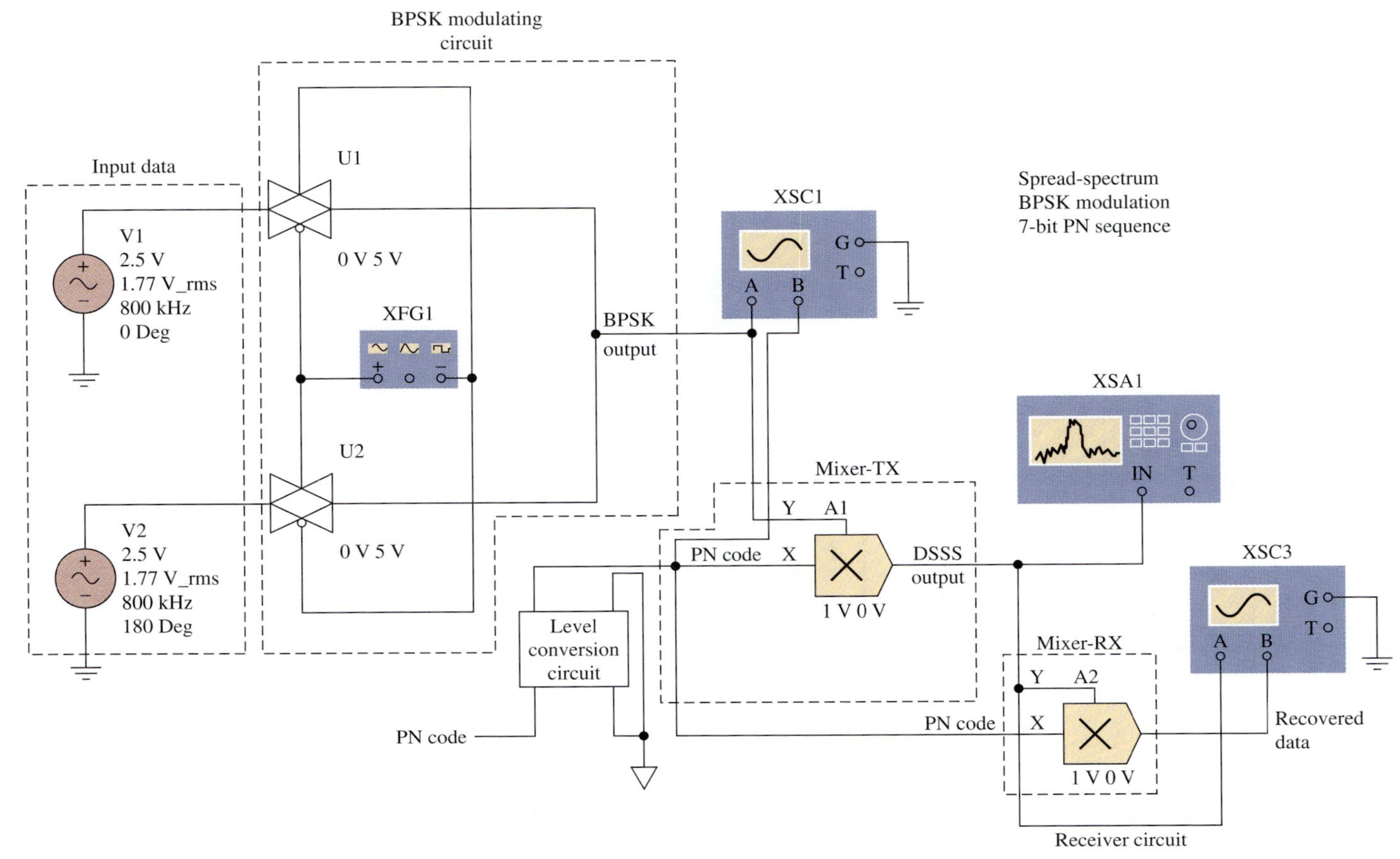

FIGURE 10-24 An implementation of a DSSS circuit using Electronics Workbench™ Multisim.

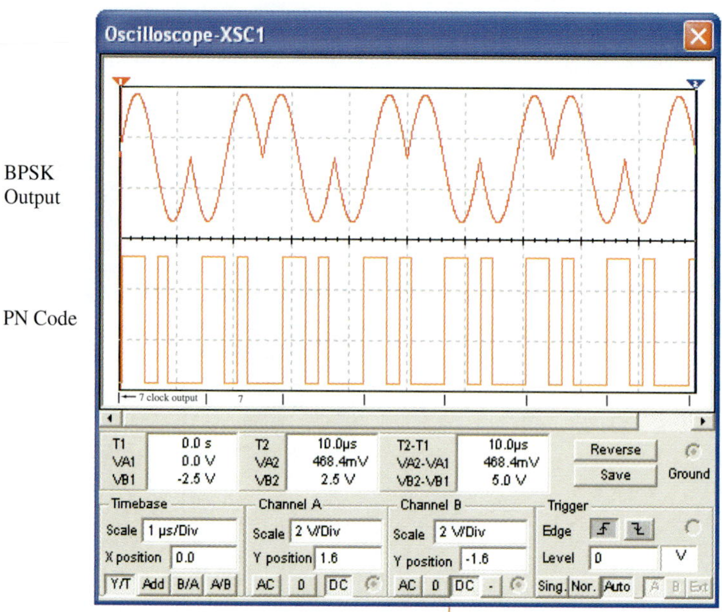

FIGURE 10-25 The waveforms for the BPSK output and the PN code.

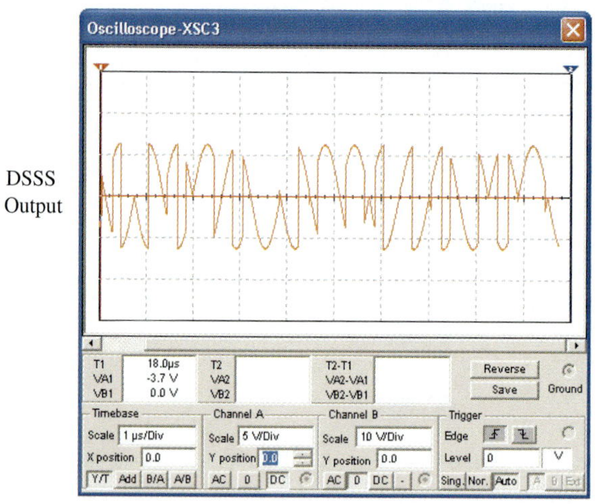

FIGURE 10-26 The DSSS output signal generated by mixing the BPSK output and the PN code.

is the BPSK signal. The receiver must have the same PN code as the transmitter to recover the data. The DSSS output and the recovered data are provided in Fig. 10-27. Notice that the recovered data is the same as the original BPSK data stream shown in Fig. 10-25.

The first example using Electronics Workbench™ Multisim demonstrated how the DSSS signal was generated and how the original BPSK data was recovered. The

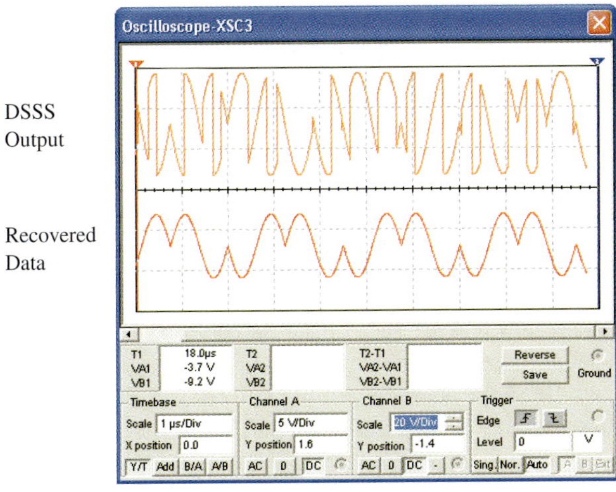

FIGURE 10-27 The DSSS signal input to the receiver and the recovered BPSK data.

next example demonstrates the spreading of the modulated digital signal. The modulation technique is BPSK, and the spectrum for the BPSK signal is shown in Fig. 10-28. Notice that the center of the two peaks is at about 2.8 MHz and that there are two peaks at about ± 100 kHz from the center frequency (2.8 MHz). The following guidelines were used to set the circuit parameters to spread the BPSK signal properly.

BPSK modulation rate	Lower frequency
Chip rate	Medium frequency (approximately five to seven times the modulation rate)
Carrier frequency	High frequency (two to five times the chip rate)

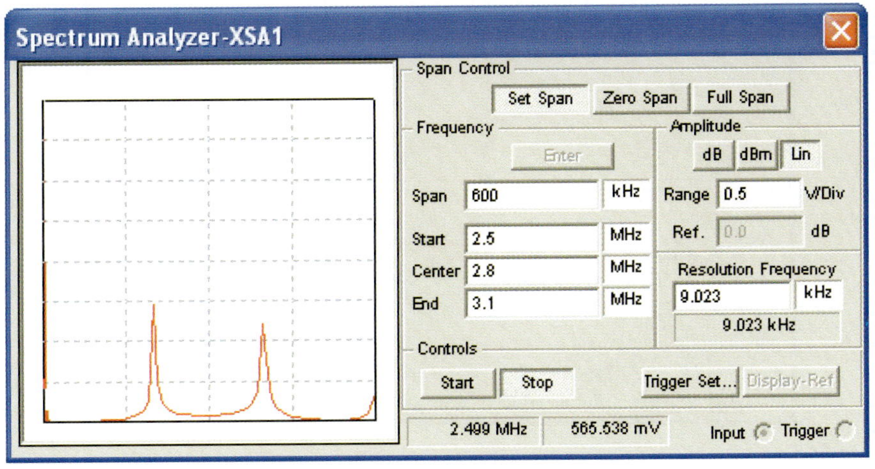

FIGURE 10-28 The spectrum of the BPSK signal.

Table 10-4 The Circuit Settings for Spreading the Digital Signal

BPSK modulation rate	100 kHz
Chip rate	700.280 kHz
Carrier frequency	2.8 MHz

The BPSK modulation rate is the digital bit rate. The chip rate is the rate of the bit pattern from the PN sequence generator circuit. The carrier frequency has been set to 2.8 MHz. For this example, the parameters are set as listed in Table 10-4.

Notice from Table 10-4 that the BPSK modulation rate of 100 kHz is the ± frequency separation of the two BPSK peaks shown in Fig. 10-28. The chip rate has been set to about seven times the BPSK modulation rate. This will significantly change the time domain signals for the BPSK and PN code. An example is shown in Fig. 10-29. The BPSK signal now repeats several cycles before changing phase. Compare to Fig. 10-25, where the phase changes after each cycle. The circuit parameters for Fig. 10-29 have been changed to satisfy the requirements set forth in Table 10-4.

Now examine the signal after it has been spread by the DSSS circuit. The result is shown in Fig. 10-30. Notice that the signal is now spread about 700 kHz, which was accomplished with a 7-bit PN with a chip rate of about 700 kHz. Compare the spectrum of this waveform to that shown in Fig. 10-28. There is a significant change

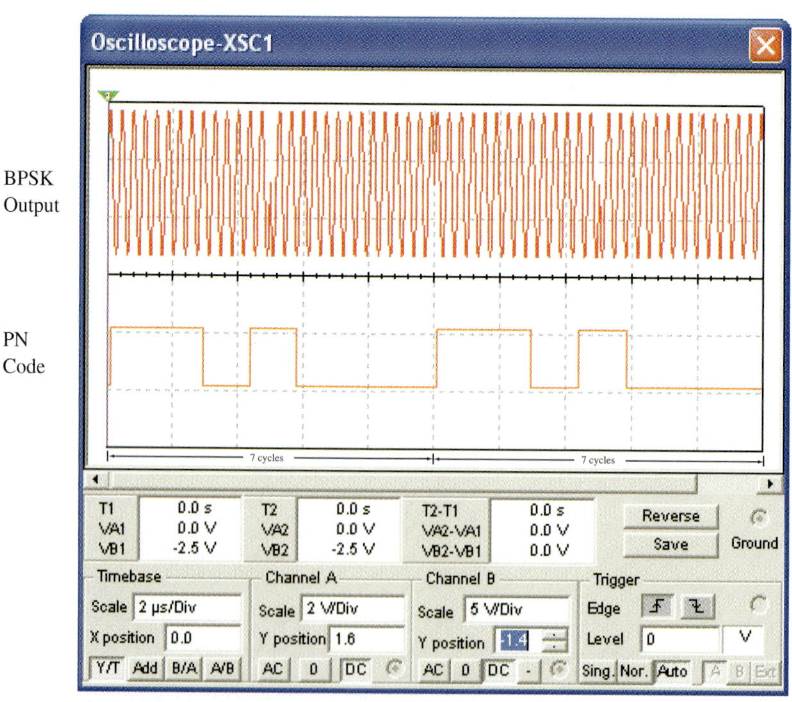

FIGURE 10-29 The time domain waveforms for the BPSK and PN code outputs using the proper chip–modulation rate relationship.

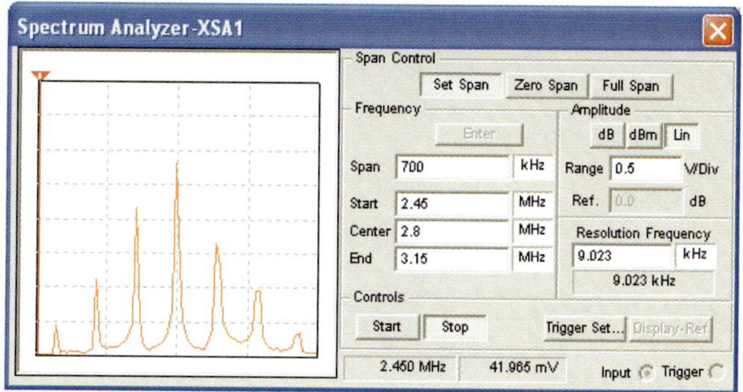

FIGURE 10-30 The spectrum of the signal after spreading.

in the spectral content. The spreading of the DSSS signal can be calculated using Eq. (10-3). Examples of applying Eq. (10-3) are provided in Example 10-3.

$$\text{Spreading} = \frac{\text{RC}}{\text{RB}} \tag{10-3}$$

where RC is the chip rate and RB is the modulation bit rate.

Example 10-3

Determine the spreading of a DSSS signal given the following parameters.

(a) Modulation bit rate: 56 kbps
Chip rate: 560 kbps

(b) Modulation bit rate: 256 kbps
Chip rate: 1792 kbps

Solution

(a)

$$\text{Spreading} = \frac{560\,\text{kbps}}{56\,\text{kbps}} = 10 \tag{10-3}$$

(b)

$$\text{Spreading} = \frac{1792\,\text{kbps}}{256\,\text{kbps}} = 7 \tag{10-3}$$

10-4 Orthogonal Frequency Division Multiplexing (OFDM)

Orthogonal frequency division multiplexing (OFDM), also called **multitone modulation,** is a digital communications technique implemented today in many wireless applications. These applications include the 802.11a wireless LANs (see Chapter 11), some DSL and cable modems, and North American digital radio. The OFDM concept originated in the 1960s but has only recently been implemented because of the availability of affordable digital signal processing.

Orthogonal Frequency Division Multiplexing (OFDM) a technique used in digital communications to transmit the data on multiple carriers over a single communications channel

Multitone Modulation another name for orthogonal frequency division multiplexing

In OFDM, the data is divided into multiple data streams and is transmitted using multiple subcarriers at fixed frequencies over the same wired or wireless communications channel. The data streams over the subcarriers can be of any digital

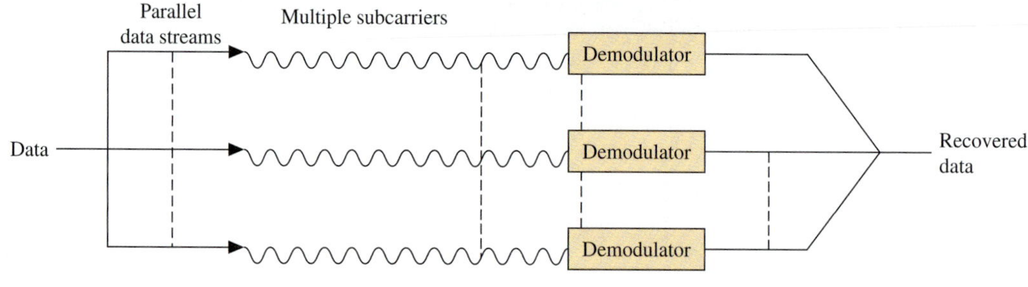

FIGURE 10-31 Block diagram of OFDM transmission.

modulation method. This technique enables the receiver's demodulators to see only one carrier. The benefits include

- Improved immunity to RF interference
- Low multipath distortion

Orthogonal
two signals are orthogonal if the signals can be sent over the same medium without interference

The OFDM technique increases the symbol length, which provides separation in the bits in the data streams, thereby minimizing the possibility of intersymbol interference (ISI). Thus, the demodulators for each channel do not see an interfering signal from an adjacent channel. (Note: Two signals are **orthogonal** if the signals can be sent over the same medium without interference.) This concept is demonstrated in Fig. 10-31.

An example of an OFDM transmission of a data sequence is provided next. In this example, the data sequence is the ASCII message "BUY." The ASCII data value is transmitted in parallel over eight BPSK carries, which is shown in Fig. 10-32. The

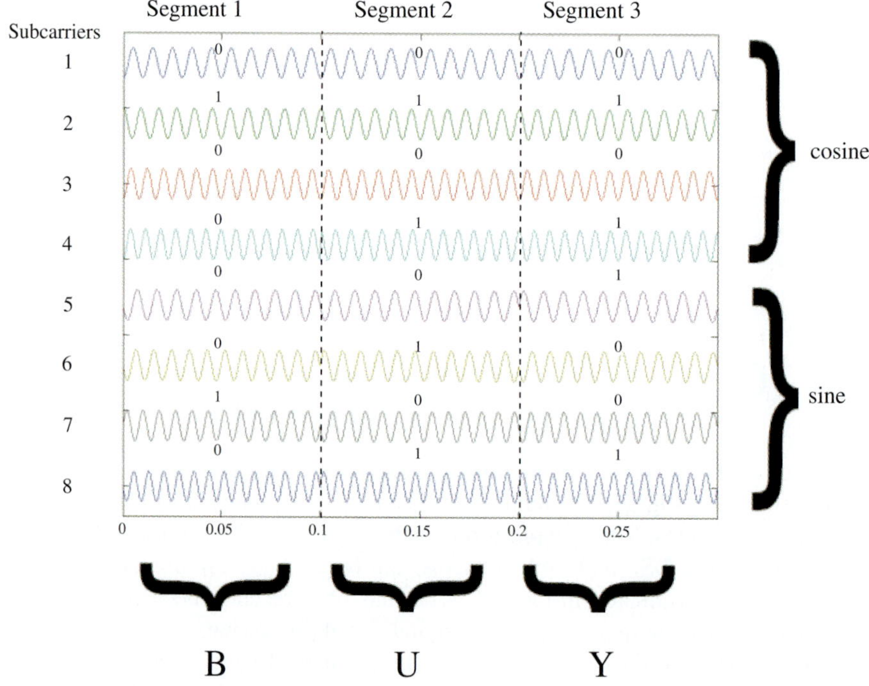

FIGURE 10-32 Example of the OFDM transmission of the ASCII characters B, U, and Y.

Table 10-5 The ASCII Values for the Letters B, U, and Y

Letter	8-Bit ASCII Value
B	0100 0010
U	0101 0101
Y	0101 1001

binary values for each carrier are listed. Subcarriers 1–4 are cosine waves, while subcarriers 5–8 are sine waves. Figure 10-32 shows three segments for the multiple BPSK carriers. Segment 1 shows the ASCII code for "B." Segment 2 shows the ASCII code for "U," and segment 3 shows the ASCII code for "Y." The ASCII values for BUY are provided in Table 10-5. Note that each subcarrier is being used to carry just one bit of information per segment. To prevent intersymbol interference in OFDM systems, a small guard time is usually inserted (not shown in Fig. 10-32). This guard time may be left as a gap in transmission or it may be filled with a **cyclic prefix**, which means that the end of a symbol is copied to the beginning of the data stream, thus leaving no gap. This modification makes the symbol easier to demodulate.

Cyclic Prefix
the end of a symbol is copied to the beginning of the data stream, thereby increasing its overall length and thus removing any gaps in the data transmission

Another OFDM technique used in digital communication is **flash OFDM,** which is considered a spread-spectrum technology. A fast hopping technique is used to transmit each symbol over a different frequency. The frequencies are selected in a pseudorandom manner. In other words, the frequency selected appears to be totally random. This technique provides the advantages of CDMA but has the additional benefit of frequency diversity, which means that the OFDM signal is less susceptible to fading because the entire signal is spread over multiple subcarriers on a wide frequency range.

Flash OFDM
a spread-spectrum version of OFDM

10-5 Telemetry

Telemetry may be defined as remote metering. It is the process of gathering data on some particular phenomenon without the presence of human monitors. The gathered data may be recorded on chart recorders, tape recorders, or computer memory and then picked up at some convenient time. If the data are transmitted as a radio wave, it is called **radio telemetry.** This process started during World War II when telemetry systems were developed to obtain flight data from aircraft and missiles. It offered an alternative to having human observers on board when that was impractical or considered too dangerous. Besides the military applications, many new commercial uses are being developed. Today's telemetry markets range from remote reading of gas and electric meters to credit card validation, security monitoring, token-free highway toll systems, and remote inventory control.

Telemetry
remote metering; gathering data on some phenomenon without the presence of human monitors

Radio Telemetry
gathering data on some phenomenon without the presence of human monitors and transmitting the data to another site via radio

Since situations to be remotely metered invariably involve more than one measurement, the different signals are always multiplexed. This allows the use of a single transmitter/receiver and in fact was the first major use of multiplexing techniques. It was also the beginning of pulse modulation techniques, described in Chapter 9, because of the ease with which they can be multiplexed.

Telemetry Block Diagram

A radio telemetry system block diagram is shown in Fig. 10-33. The process begins with the system to be monitored. Transducers (sensors) convert from the entity to be

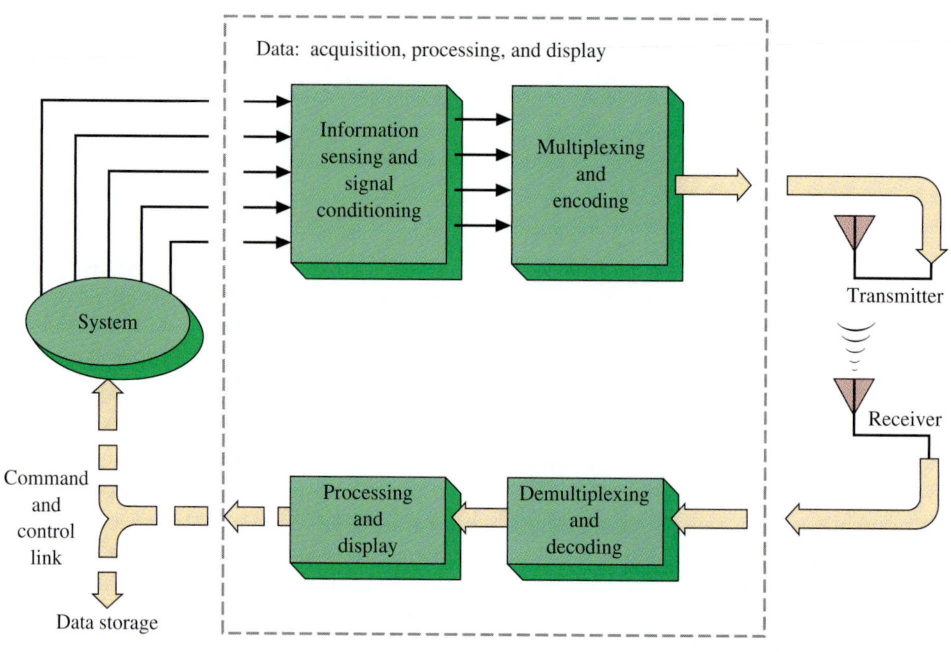

FIGURE 10-33 Radio-telemetry block diagram.

measured to an electrical signal. Five transducer outputs are shown in Fig. 10-33, but sophisticated systems, such as a Mars exploration probe, may include hundreds of transmitted measurements. In any measurement system, stimuli such as temperature, pressure, movement, or acceleration must be converted to a form that can be processed electrically. While a mercury thermometer produces a good visual output of temperature, a thermistor or thermocouple transducer converts temperature to an electrical signal usable by a telemetry system.

Following the sensing, Fig. 10-33 shows a signal-conditioning function. The outputs from the transducers will be electrical in nature but may not have much else in common. The outputs may be variable resistance, capacitance, inductance, voltage, or current of many different magnitudes. The conditioning circuits turn the raw data into uniform digestible information. Following conditioning, the signals are applied to the multiplexing and encoding block. They are then transmitted to the receiver for demultiplexing and decoding, followed by the processing and display function. Complex telemetry systems incorporate a computer for maximum efficiency in data processing. From this block a command and control link is shown in Fig. 10-33 for those systems that send control signals back to the system under measurement. This may be done to change the performance to a desired level. This "completes the loop" except for the path to data storage shown in Fig. 10-33.

Typical Telemetry System

Telemetry systems may use FDM or TDM and in some cases both, as shown in Fig. 10-34. In this instance, three wideband channels and six narrowband channels are provided. Many times there are some conditions that must be monitored more

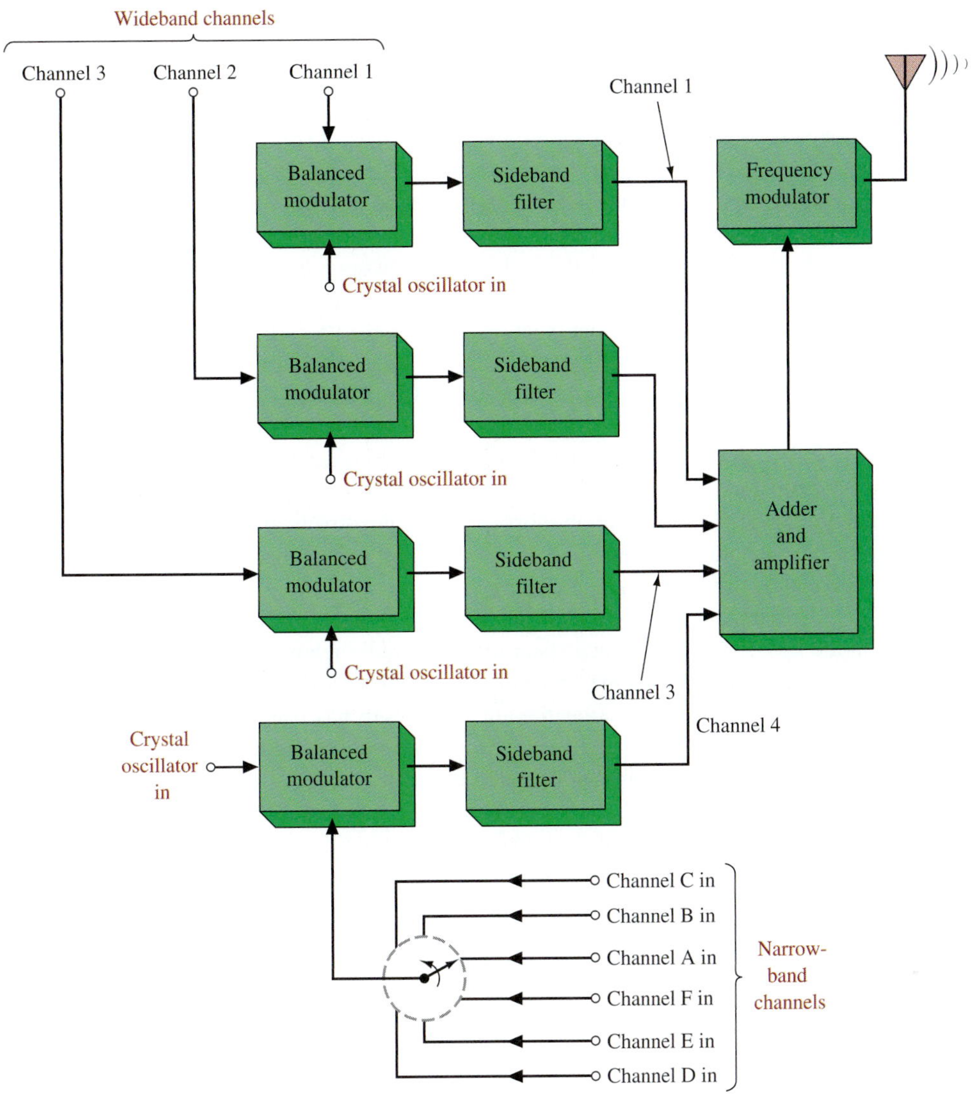

FIGURE 10-34 TDM and FDM telemetry transmitter system.

often than others. They are represented by channels 1, 2, and 3 in Fig. 10-34. The six conditions that do not require rapid monitoring (subchannels *A* through *F*) are time-division-multiplexed (TDM) onto channel 4. They are referred to as subchannels because 6 of them make up one of the frequency-division-multiplexed (FDM) main channels. There are countless modulation combinations possible in radio telemetry. In the system illustrated in Fig. 10-34, the subchannels are PCM-encoded to amplitude-modulate a subcarrier that is converted into SSB, which is used to frequency-modulate the main carrier. The three wideband channels use SSB directly to frequency-modulate the main carrier. This is termed a PCM/SSB/FM system. However, any other method is possible (and has probably been used), such as PAM/FM/FM, PWM/AM/FM, PPM/AM/AM, and so on.

Radio-Telemetry System

A complete radio-telemetry system is shown in Fig. 10-35. Part (a) of the figure shows the analog encoder that converts a 0- to 5-V analog signal into a digital pulse stream. The LM 325 IC shown within dashed lines illustrates a simple way this system can be used in a temperature-monitoring telemetry system. The LM 325 IC contains the temperature transducer and circuitry to provide the linear 0- to 5-V signal representing a range of temperatures.

In the encoder [Fig. 10-35 (a)], a current-source-fed integrating capacitor at the inverting input of A_1 (0.22 μF) generates a linear ramp that is compared with the input voltage by A_1, one section of a quad comparator. An offset voltage is summed with the input so that the circuit will generate a transmittable, nonzero signal for a ground-level input to A_2 and A_3. Comparator A_3 discharges the 0.22-μF integrating capacitor while A_2 provides a separate buffered output. When the ramp is reset, A_1 goes high. Because A_1 has an open-collector output, A_2 and A_3 do not see the change until C_2 has charged to 6 V. This delay assures that C_1 will be discharged on reset. The delay is short so that timing errors are not induced in the system. The encoder's output appears as a string of pulses that have cycle periods proportional to the input voltage (PWM).

The transmitter portion of the system is shown in Fig. 10-35 (b). It is based on the LM 1871 IC designed specifically for radio telemetry. The analog encoder's output signal modulates the transmitter's 49-MHz carrier. Coil T_1 must be trimmed for maximum RF output after the antenna length has been selected. A length of 2 ft is sufficient; shorter lengths will work, but with a reduced range. The transmit range is about 150 m but can be extended with additional power amplification.

The receiver [Fig. 10-35 (c)] contains a local oscillator, a mixer, a 455-kHz IF section, and digital detection and decoding circuitry. Receiver alignment is simple and does not require special equipment. The local-oscillator coil (L_1) is tuned while that section's output signal (pin 2) is viewed with a 10-pF or less oscilloscope probe. The coil is trimmed to the point just before that at which the waveform amplitude peaks and then disappears.

The antenna input transformer (T_3), mixer (T_1), and IF transformer (T_2) are adjusted by means of the companion transmitter and the intended receiving antenna. The internal automatic-frequency-control (AFC) circuitry is disabled by grounding pin 16. The 455-kHz IF output (pin 15) is adjusted for peak amplitude. The amplitude must remain below 400 mV p-p throughout the adjustment procedures to prevent clipping of the IF waveform. To control the amplitude, you must remove the transmitter's antenna and separate the transmitter and receiver to obtain a usable IF output level.

Once the first adjustment is complete, the scope should be connected to the unused T_2 secondary to prevent detuning. Transformers T_1, T_2, and T_3 should be trimmed for maximum waveform amplitude. This LM 1871 and 1872 system can handle more than one channel of telemetry data, as explained in the manufacturer's specifications and application notes.

In the analog decoder [Fig. 10-35 (d)] a reference-current source combines with an operational amplifier/integrator (LF356) to generate a second linear ramp. This signal is put into a sample-and-hold (S/H) circuit (LM396), which "looks" for a time specified by the incoming pulse string. A negative offset is also summed into the S/H input to correct for the voltage added during coding. Output pulses from the receiver trigger a dual one-shot (74LS123), which generates two successive

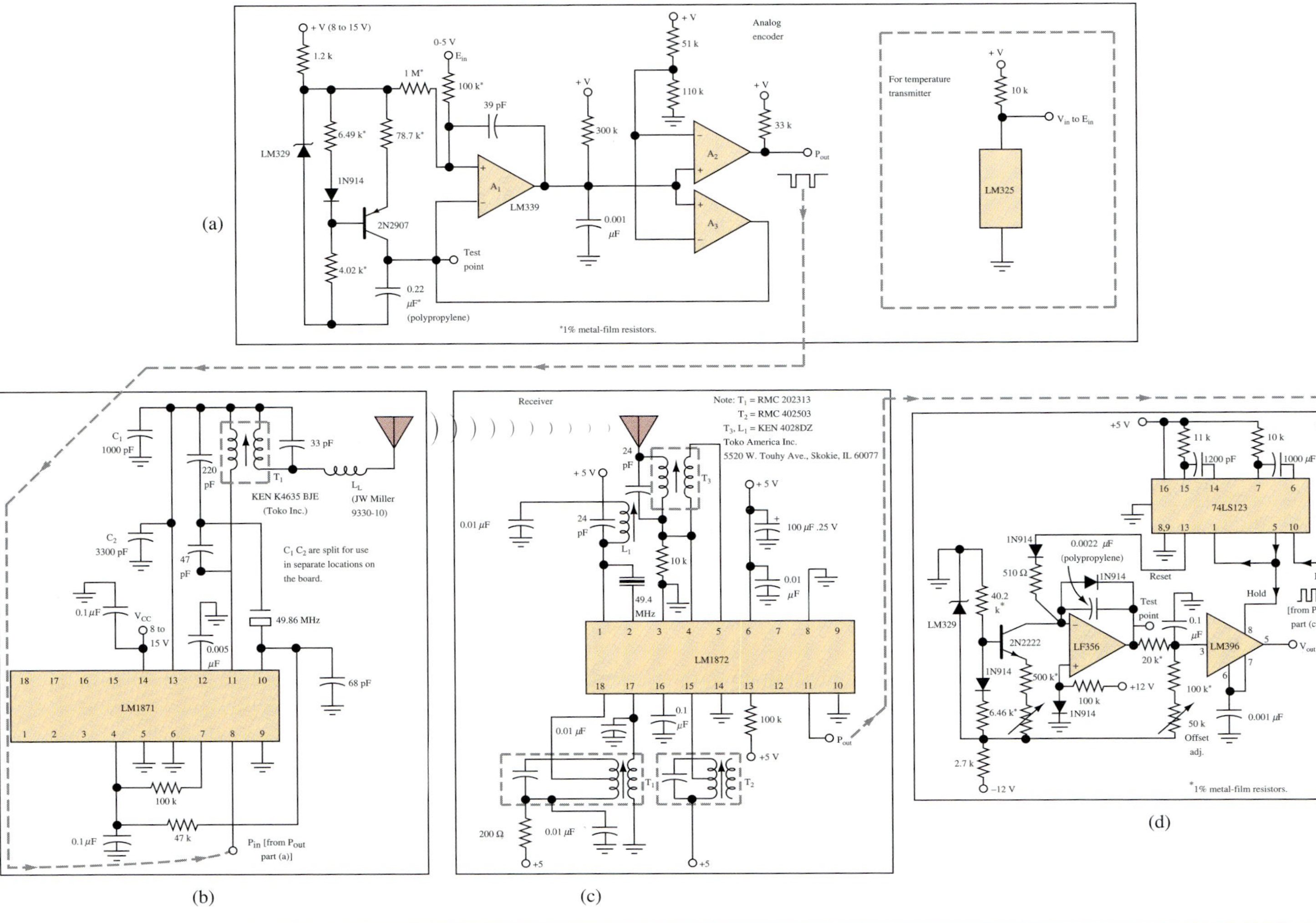

FIGURE 10-35 Complete radio-telemetry system: (a) encoder; (b) transmitter; (c) receiver; (d) decoder. (Reprinted with permission from *Electronic Design*, Vol. 29, No. 10; copyright Hayden Publishing Co., Inc.)

pulses of approximately 5-μs duration. The first pulse triggers the S/H, and the second resets the integrator. The LM396 output is therefore a signal that is proportional to the pulse width from the receiver, as is desired. The gain and offset adjustments are somewhat interactive. With a two-point calibration, however, no more than three iterations should be needed.

With the addition of a 10-kΩ resistor and a two-terminal temperature sensor, the circuit can serve as a remote temperature transmitter, with the voltage output equaling 10 mV/K. The most significant contributor to circuit inaccuracy is the temperature coefficient of the integrating capacitors in the encoder and decoder circuitry. The circuits in Figs. 10-35 (a) and (d) use polypropylene capacitors with a temperature coefficient of approximately -150 ppm/°C. By paralleling capacitors with opposite temperature coefficients, the designer can reduce the temperature sensitivity of the circuitry by a factor of 5. A combination of silver–mica and polystyrene in a 3:1 ratio would have such an effect.

For analog data acquisition, RF telemetry presents an attractive alternative to the usual hard-wired approaches. The most common complications in analog data acquisition—source inaccessibility, the lack of practical line routes, and the need for nonstationary sensors—can be sidestepped. In addition, the RF route takes care of problems such as ground loops and wire losses.

A wide variety of uses with this system is possible, including the transmission of physiological data from human and animal subjects without the need for their confinement, the collection of data from rotating or otherwise moving machines without the need for brushes or slip rings, and wireless outdoor-to-indoor links for the collection of weather and other information.

10-6 Troubleshooting

The focus of this chapter has been on digital wireless communications. Probably one of the most visible and most widely used digital wireless communications devices today is the wireless telephone. This section focuses on the steps used by technicians to troubleshoot cellular telephone problems.

The first step when troubleshooting a cell phone problem is to collect information about the phone and the type of problems the customer is having. This is typically prepared in the form of a trouble ticket. For example, the customer may be losing calls or not getting messages. In addition to the customer's problems, information collected from the cell phone, which may be required, can include the following:

- Electronic serial number (ESN): a unique number assigned to a cell phone at the factory.
- Mobile identification number (MIN): a ten-digit number assigned by the seller. It identifies the cell phone within the cellular service provider's wireless network.
- Master subsidy lock (MSL): a proprietary code used to program a MIN into the phone. It is used when the phone is being serviced. Every ESN is assigned a unique MSL.

The MIN is programmed into the cell phone, using the MSL, when purchased. Access to viewing the codes varies for each cell phone manufacturer, but it is typically available via the cell phone's menu.

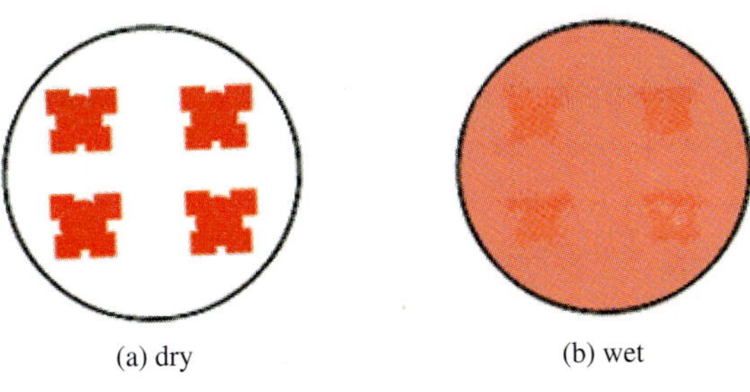

FIGURE 10-36 The cell phone's water mark sticker: (a) without water damage; (b) discolored due to water damage.

It is common for cell phones to experience physical problems. Typical physical problems are LCD screen problems, button problems, evidence of something spilt on the cell phone, damage caused by dropping the phone, and water damage. Most of these problems are detected easily by a quick visual inspection of the cell phone. Water damage is also detected easily by the technician via a **water mark sticker** located in the battery compartment area. An example of a water mark sticker is provided in Fig. 10-36. Figure 10-36(a) shows an undamaged water mark sticker, and Fig. 10-36(b) shows a discolored water mark sticker due to water damage. An important note: the cell phone's warranty is typically voided if the water mark sticker has been removed.

Water Mark Sticker
a sticker used to detect water damage

Assuming the unit passes the physical inspection, the cell phone is checked to verify that the phone has the latest roaming list of available cell towers. This list is the **preferred roaming list (PRL).** A cell phone with an out-of-date PRL is subject to lost calls. This PRL should be upgraded anytime a phone is in for maintenance. The upgrade requires the new PRL file to be uploaded from a computer file into the cell phone. This upgrade is accomplished via a USB or an RS-232 serial interface connection and takes less than a minute. PRLs can also be updated over the air **(OTA)** if enabled by the service provider.

Preferred Roaming List (PRL)
the roaming list of available cell towers

OTA
over the air

In some cases, full diagnostics of a cell phone may be required. A CDMA tester is used to exercise fully the cell phone's operation. A block diagram of the CDMA test setup is provided in Fig. 10-37. The CDMA tester provides the tests recommended by the cell phone manufacturer to test the operation. The mobile unit under test is inserted into the RF shield box and is connected to the tester. The **RF shield box** isolates the mobile unit under test (the cell phone being tested) from any possible RF interference from nearby cell towers or other interfering RF sources.

RF Shield Box
isolates the mobile unit under test from any possible interference from nearby towers

A CDMA tester typically provides support for a CDMA base station simulator, which is used to verify the functional aspects of the phone. The tester simulates call initiating and disconnect and voice quality checks (echo testing). The CDMA tester tests tuned channel power measurements that enable the system to verify maximum power and frequency error. The CDMA tester also verifies the mobile unit's

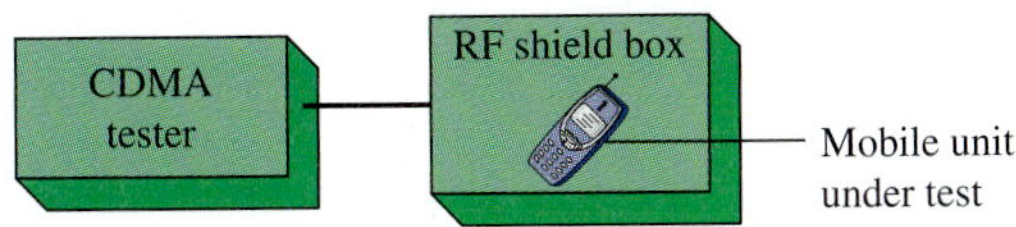

FIGURE 10-37 The test setup for the full CDMA diagnostics.

Test	Measured	Lower Limit	Upper Limit	Pass/Fail
PCS US traffic channel = 325				
Called number 555-5555				
Origination successful	Yes			
Page successful	Yes			
CPD voice quality	1 P/F	1 P/F	1 P/F	Pass
CDMA rho	1.0	.94	1	Pass
CDMA frequency error	4.0 Hz	−150 Hz	150 Hz	Pass
CDMA static timing offset	−0.46 μs	−1 μs	1 μs	Pass
CDMA sensitivity FER at −102 dBm	0.0%		.5 %	Pass
CDMA max RF output power	24.3 dBm	21 dBm	30 dBm	Pass

FIGURE 10-38 Example of Test Data Generated by a CDMA Test

frame error rate (FER). This is a measure of the phone's ability to demodulate the received signal in a high noise environment. The test shown in Fig. 10-38 shows a measured FER of 0.0 percent with an upper limit of .5 percent. The test also shows a maximum RF output power of 24.3 dBm.

10-7 Troubleshooting with Electronics Workbench™ Multisim

This chapter presented techniques for encoding digital data for transmission. In this exercise, you will have the opportunity to use Electronics Workbench™ Multisim to gain a better understanding of several important communication building blocks, including generation of a BPSK signal, a coherent carrier recovery circuit, a mixer, and recovering a BPSK encoded signal. The circuit used in this example is shown in Fig. 10-39.

To begin this exercise, open the file **Fig10-39.ms7 (.msm)** on your EWB Multisim CD. This circuit contains a BPSK generating circuit and a receiver based on the block diagrams provided in Figs. 10-5 and 10-6.

A 1-kHz sine wave is used to simulate the carrier frequency. The 1-kHz sine wave is fed to an inverting operational amplifier with a gain of −1. The op amp provides a 180° phase shift of the sine wave. This result is shown in Fig. 10-40.

Both phases of the sine-wave signal are fed into a 1-of-2 selector constructed with two analog switches. The control of the analog switches is provided by a 500-Hz square wave, which is used to represent the digital data. In this case, the digital data is an alternating 1 0 pattern. The square wave is inverted by the 74HC04 inverter, and the inverted and noninverted signals are used to select which phase is being output. Start the simulation and view the output of the BPSK generating circuit with oscilloscope XCS1. Notice that the phase is alternating, based on which analog switch is selected.

The BPSK generated signal is next input into a BPSK receiver that consists of a coherent carrier recovery circuit, a mixer or multiplier stage, and a low-pass filter. This circuit is discussed in Sec. 10-2. The BPSK received signal is connected

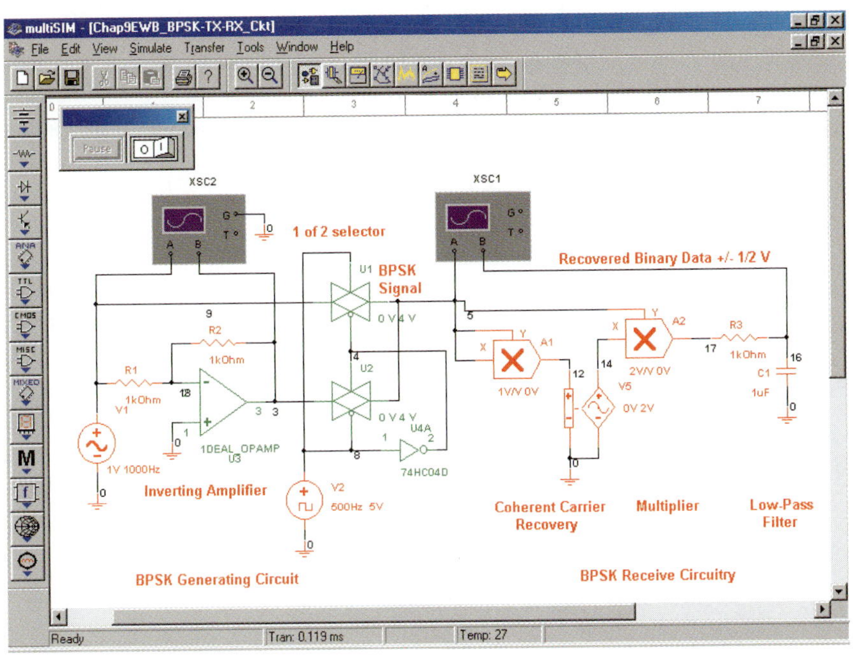

FIGURE 10-39 A BPSK transmit-receive circuit as implemented in Electronics Workbench™ Multisim.

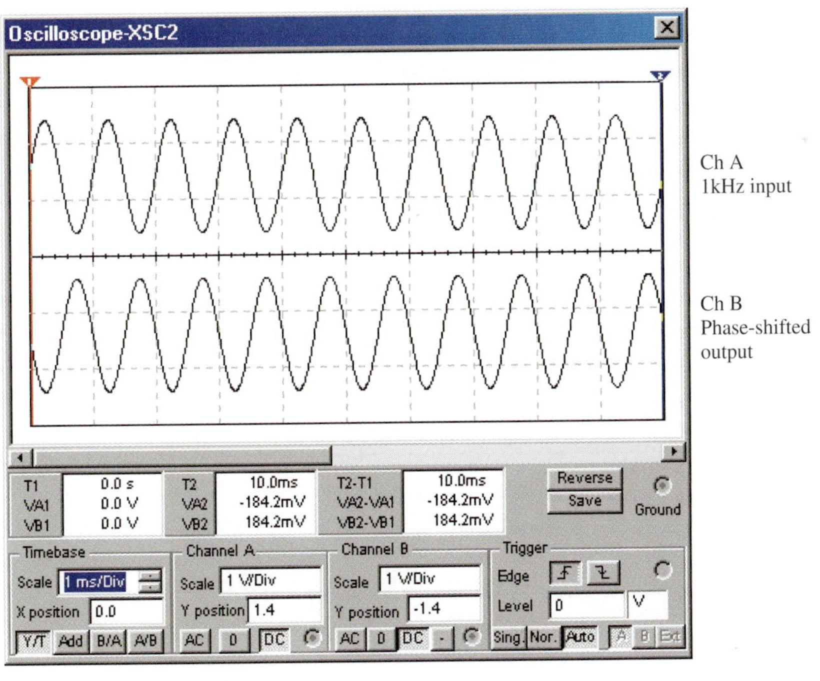

FIGURE 10-40 The phase relationship for the input and output of an inverting amplifier.

to both inputs of a multiplier stage. The purpose of this circuit is to square the input signal so that only a $+\sin 2\omega t$ term is created. The $+\sin 2\omega t$ term indicates that the new signal, which is twice the original frequency, has been created and the new signal does not alternate-phase. Connect the oscilloscope to the output of the multiplier and you will observe that a 2-kHz signal has been generated. Why was a 2-kHz signal created? Recall that $(\sin A) \times (\sin B) = 0.5 \cos(A - B) - 0.5 \cos(A + B)$. For this example, the $A - B$ term (1 kHz − 1 kHz) will go to zero, whereas the $A + B$ term will equal 2 kHz.

The output of the multiplier is fed to a voltage-controlled oscillator, which has been set to output a 1-kHz signal. This circuit is being used to simulate a phase-locked-loop circuit, where the internal VCO on the PLL phase-locks to the 2-kHz signal and the output is preset to generate a 1-kHz sine wave.

The recovered carrier and the BPSK signal are both input into a multiplier. Once again, this is a $(\sin A) \times (\sin B)$ circuit. As mentioned in Sec. 10-2, the output of this circuit will be either

$$\sin A \times \sin A = 0.5 - 0.5 \cos(2A)$$

or

$$\sin A \times -\sin A = -0.5 + 0.5 \cos(2A)$$

The low-pass filter consisting of R_3 and C_1 (Fig. 10-39) removes the high-frequency term, leaving only the ±0.5-V term. This result can be viewed by connecting the oscilloscope (XSC1) to the output of the recovery circuit. Notice in Fig. 10-41 that the oscilloscope is showing a ±0.5-V output, which is exactly what was predicted. The output is changing with the change in phase of the BPSK signal.

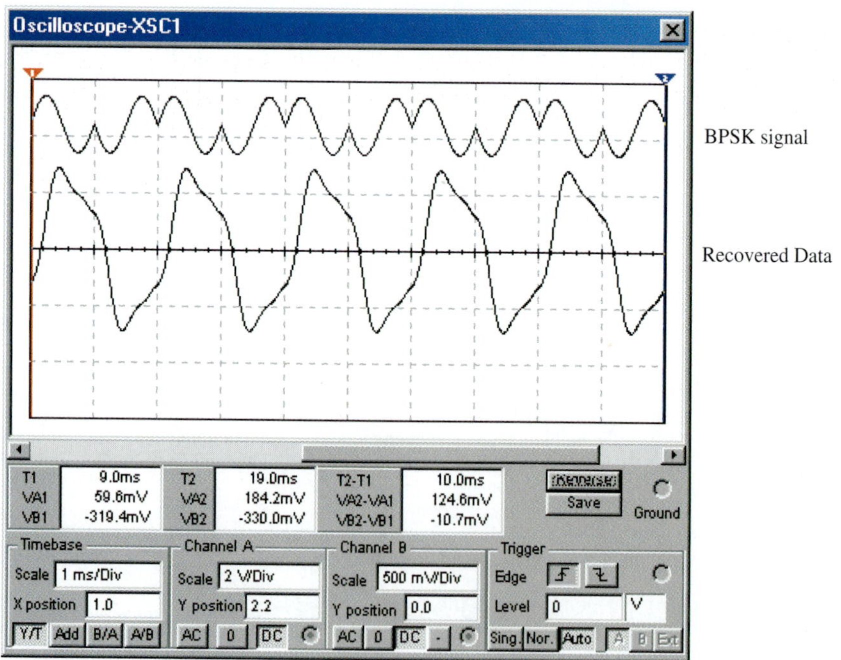

FIGURE 10-41 The oscilloscope view of the BPSK signal and the recovered data.

This exercise has demonstrated how a BPSK digital communications block can be constructed and analyzed using Electronics Workbench™ Multisim. Be sure to use the oscilloscope to verify that you have gained a complete understanding of the signal path for both the generating and receiver sides of a BPSK digital communications system. You will need a thorough understanding of this circuit to complete the Electronics Workbench™ Multisim exercises in this chapter.

Electronics Workbench™ Exercises

1. Open the file **FigE10-1.ms7 (.msm)** on your EWB CD. This circuit contains a fault. Troubleshoot the circuit to find the fault. When you find the fault, correct it and rerun the simulation to determine if the circuit is now working properly. If it is not, then continue troubleshooting the circuit. After you are done, explain your findings.
2. Open the file **FigE10-2.ms7 (.msm)** on your EWB CD. This circuit contains a fault. Troubleshoot the circuit to find the fault. When you find the fault, correct it and rerun the simulation to determine if the circuit is now working properly. After you are done, explain your findings.
3. Use the hysteresis block provided in Electronics Workbench™ Multisim to provide level conversion of the output of the BPSK circuit provided in the file **Fig10-39.ms7 (.msm).** Your modification should enable the receiver to output +5.0 V and 0.0 V for the ±0.2-V level provided from the output of the RC (R_3 and C_1) filter. (The solution can be found in **Fig10-39-solution.ms7 (.msm).**)

Summary

In Chapter 10, we examined various aspects of wireless digital communications transmission, including FSK, PSK, BPSK, QPSK, DPSK, frequency hopping, and direct sequence spread spectrum. The techniques of OFDM and radio telemetry were also examined. The major topics you should now understand include:

- the different digital modulation techniques used
- the analysis of data transmission techniques using FM/PM, including frequency and phase-shift keying
- the analysis of quadrature amplitude modulation (QAM), including constellation patterns, loopback, and eye tests
- the issues of PN sequence generation
- frequency hopping and direct sequence spread spectrum
- the spreading of the spread-spectrum signal
- the issues of OFDM transmission
- the basic concept of telemetry and a description of a complete radio-telemetry system

Questions and Problems

Section 10-1

1. Explain what is meant by the term *wireless*.

Section 10-2

2. What is a frequency-shift-keying system? Describe two methods of generating FSK.
3. Explain how the FSK signal is detected.
4. Describe the PSK process.
5. What do the M and n represent in Table 10-1?
6. Explain a method used to generate BPSK using Fig. 10-5 as a basis.
7. Describe a method of detecting the binary output for a BPSK signal using Fig. 10-6 as a basis.
8. What is coherent carrier recovery?
9. Describe the recovery of QPSK using Fig. 10-9 as a basis.
10. Provide a brief description of the QAM system. Explain why it is an efficient user of frequency spectrum.
11. Describe how to generate a constellation.
12. What is the purpose of any eye pattern?
13. The input data to a DPSK transmitter is 1 1 0 1. Determine the DPSK digital data stream and the DPSK RF output.

Section 10-3

14. What is a PN code and why is it noiselike?
15. What does it mean to spread the RF signal?
16. What is the PN sequence length of a PN sequence generator with each of the following?
    (a) 4 shift registers
    (b) 9 shift registers
    (c) 23 shift registers
17. What does it mean for a PN sequence to be of maximal length?
18. Draw the circuit for a PN sequence generator with $n = 5$.
19. Explain the concept of frequency hopping spread spectrum.
20. Define *dwell time.*
21. Generate the shift-register contents for a PN sequence generator with $n = 3$ and a seed value of 1 0 1.
22. Define *code division multiple access.*
23. Define the term *hit* relative to spread spectrum.
24. What is a signature sequence?
25. Draw the block diagram of a DSSS transmit-receive system.
26. Determine the spreading of a DSSS signal if the chip rate is 1 Mbps and the modulation rate is 56 kbps.

Section 10-4

27. Draw the block diagram of an OFDM transmission.
28. How does flash OFDM differ from OFDM?
29. Determine the received OFDM signal for the traces shown in Fig. 10-42.

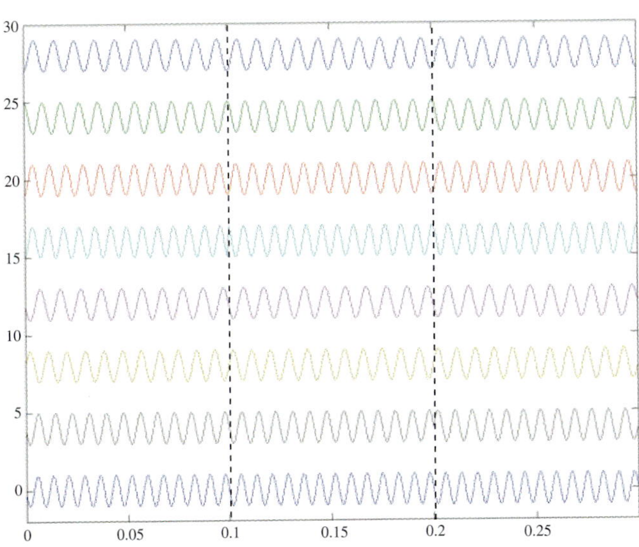

FIGURE 10-42 The OFDM traces for Problem 29.

Section 10-5

30. What is the purpose of a telemetry system?
31. Explain why telemetry systems invariably use multiplexing and pulse modulation techniques.

Questions for Critical Thinking

32. Why is OFDM not considered a true spread-spectrum system?
33. Determine the 7-bit code generated for the PN generator shown in Fig. 10-17 if a seed value of 1 1 1 is used. What happens if a seed value of 0 0 0 is used?
34. Explain how it is possible for multiple CDMA communications devices to share the same RF spectrum.

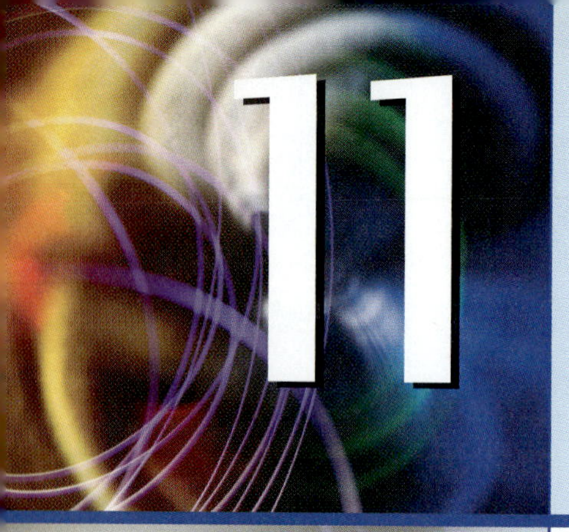

Network Communications

Chapter Outline

The Cisco voice-over IP communications system. (Courtesy of Cisco Systems. Reprinted with permission.)

Objectives

- Describe the basic telephone operation, including tip, ring, local loop, and DTMF
- Discuss the concept of telephone-line quality and explain an attenuation distortion diagram
- Describe the block diagram of a telephone system
- Describe telephone traffic units, congestion, traffic observation, and measurement
- Explain the operation of cellular and PCS phone systems
- Explain the various LAN topologies and the operation of the Ethernet protocol
- Describe the methodology required to construct an office and a building LAN
- Describe the operation of wireless LANs
- Discuss the evolution of the Internet, the details of IP addressing, and IP telephony
- Describe the new developments in modem technologies, including V.92, xDSL, and cable modems
- Describe the ISDN network, including the roles of the R, S, and T interfaces

Key Terms

tip
ring
local loop
trunk
T3
OC-1
loaded cable
attenuation distortion
delay distortion
delay equalizer
cell sites
frequency reuse
cell splitting
advanced mobile phone service (AMPS)
handoff
NAMPS (narrowband analog mobile phone service)
DAMPS (digital advanced mobile phone service)
Rayleigh fading
PCS (personal communications services)
GSM
dual-band
microcellular systems
DCCH
air-interface
3G
IMT-2000
W-CDMA
FDD
TDD
CDMA-2000
local area network
topology
protocol
CSMA/CD
network interface card (NIC)
MAC address
broadcast address
100BaseT
RJ-45
CSMA/CA
DSSS
FHSS
ISM
pseudorandom
hopping sequence
U-NII
metropolitan area network
wide area network
open systems interconnection
IANA
IP telephony (voice-over IP)
quality of service (QoS)
V.44 (V.34)
V.92 (V.90)
asymmetric operation
cable modems
ranging
xDSL
DSL
ADSL (asymmetric DSL)
discrete multitone (DMT)
latency
wireless markup language (WML)
WMLScript
microbrowser

11-1 Introduction

A network is an interconnection of users that allows communication among them. The most extensive existing network is the worldwide telephone grid. It allows direct connection between two users simply by dialing an access code. Behind that apparent simplicity is an extremely complex system, which we look at in Sec. 11-2. Because the telephone network is so convenient and inexpensive, it is often used to allow one computer to "speak" with another. The mushrooming cellular and PCS phone network is explored in Sec. 11-3. The techniques that allow transmission of computer bits over telephone lines are the topics of Sec. 11-9.

Computers often need to communicate with more than one other computer or terminal device. In fact, the proliferation of low-cost computer systems has spawned the growth of networks that include hundreds of computers. In these networks, one computer may wish to send data to all the others or to just a few specified units. Networks that allow this kind of communication form the basis of this chapter.

Telephone and computer networks have blended together in their functionality. Computer networks now have their own telephone systems (IP telephony) and wireless telephones now provide Internet access and browsing capabilities. This chapter introduces both network technologies, their new developments, and how they are interfacing together.

11-2 Basic Telephone Operation

The Greek word *tele* means "far" and *phone* means "sound." The telephone system represents a worldwide grid of connections that enables point-to-point communications between the many subscribers. The early systems used mechanical switches to provide routing of a call. Initially, the *Strowger stepping switch* was used, and subsequently *crossbar switching* was used. Today's systems utilize solid-state electronics for switching under computer control to determine and select the best routing possibilities. The possible routes for local calls include hard-wired paths or fiber-optic systems. Long-distance calls are routed using these same paths, but they can also use radio transmission via satellite or microwave transmission paths. It is possible for a call to use all these paths in getting from its source to its destination. These multipath transmissions are especially tough on digital transmissions, as we will later see.

Tip
the grounded wire in two-wire phone service

Ring
the nongrounded wire in two-wire phone service

Local Loop
the twisted-pair telephone line from the subscriber to the central phone office

The telephone company (telco) provides two-wire service to each subscriber. One wire is designated the **tip** and the other the **ring.** The telco provides −48 V dc on the ring and grounds the tip, as shown in Fig. 11-1. The telephone circuitry must work with three signal levels: the received voice signal, which could be as low as a few millivolts; the transmitted voice signal of 1 to 2 V rms; and an incoming ringing signal of 90 V rms. They also accept the dc power of −48 V at 15 to 80 mA. Until the phone is removed from the hook, only the ring circuits are connected to the line. The subscriber's telephone line is usually either AWG 22, 24, or 26 twisted-pair wire, which handles the 300-Hz to 3-kHz audio voice signal but in reality can work up to several megahertz. The twisted-pair line, or **local loop** as it is often called, runs up to a few miles to a central phone office, or in a business setting to a PBX (private business exchange). When a subscriber lifts the handset, a switch is closed that indicates a dc loop circuit between the tip and the ring line through the phone's microphone. The handset earpiece is transformer or electronically coupled into this circuit also. The telco senses the off-hook condition and responds with an audible dial tone. The central office/PBX function shown in Fig. 11-1 will be detailed later in this section.

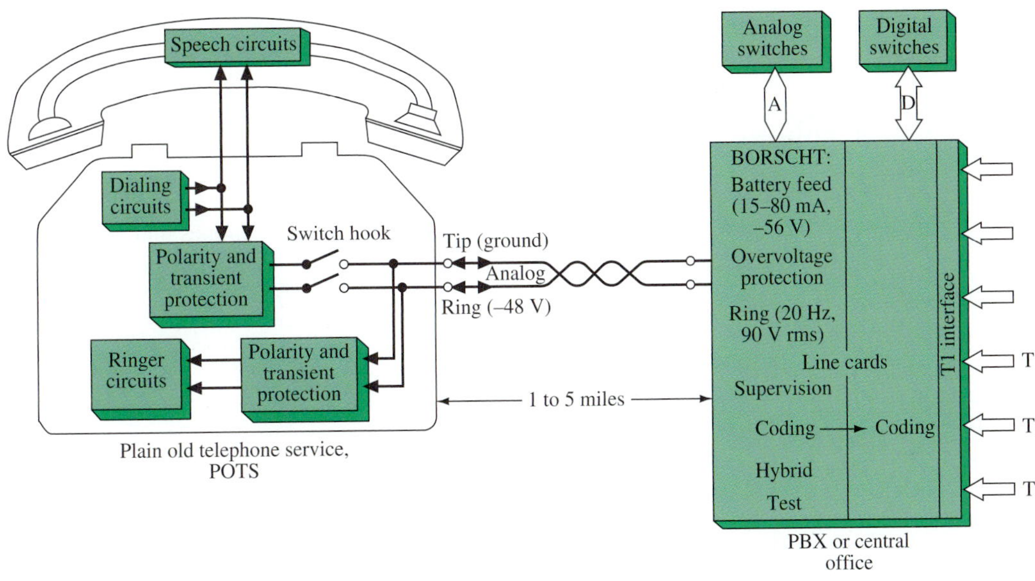

FIGURE 11-1 Telephone representation.

At this point the subscriber dials or keys in the desired number. Dial pulsing is the interruption of the dc loop circuit according to the number dialed. Dialing 2 interrupts the circuit twice and an 8 interrupts it eight times. Tone-dialing systems utilize a dual-tone multifrequency (DTMF) electronic oscillator to provide this information. Figure 11-2 shows the arrangement of this system. When selecting the digit 8, a dual-frequency tone of 852 Hz and 1336 Hz is transmitted. When the telco receives the entire number selected, its central computer makes a path selection. It then either sends the destination a 90-V ac ringer signal or sends the originator a busy signal if the destination is already in use.

The connection paths for telephone service were all initially designed for voice transmission. As such, the band of frequencies of interest were about 300 to 3000 Hz. To meet the increasing demands of data transmission, the telco now provides special dedicated lines with enhanced performance and bandwidths up to 30 MHz for high-speed applications. The characteristics of these lines have a major

697	1	2	3
770	4	5	6
852	7	8	9
941	*	0	#
Frequency (Hz)	1209	1336	1477

FIGURE 11-2 Touch-tone dialing.

effect on their usefulness for data transmission applications. Line-quality considerations are explored later in this section.

Telephone Systems

A complete telephone system block diagram is shown in Fig. 11-3. On the left, three subscribers are shown. The top one is an office with three phones and a computer (PC). The digital signal of the PC is converted to analog by the modem. The office

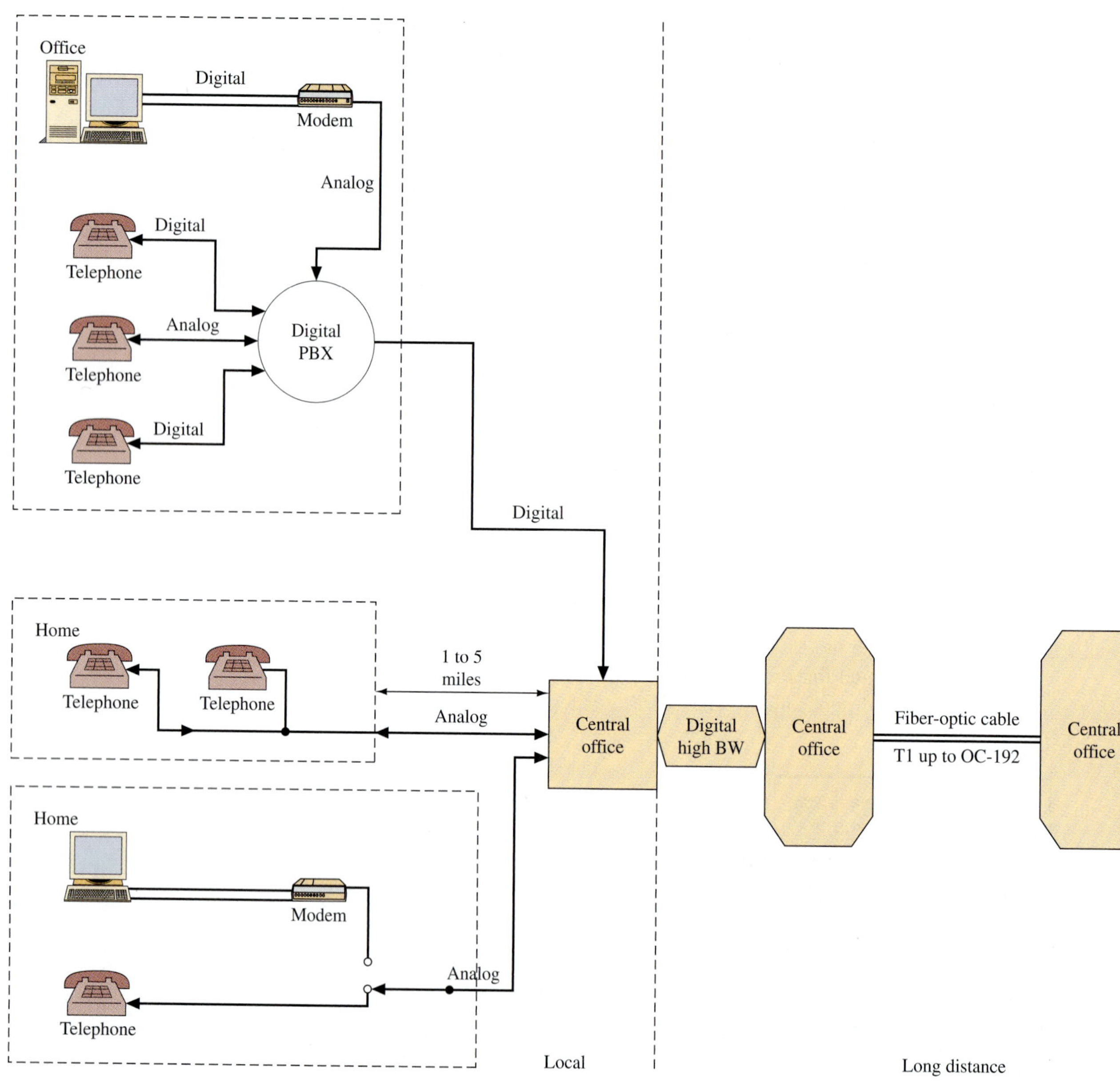

FIGURE 11-3 Telephone system block diagram.

system is internally tied together via its private branch exchange (PBX). The PBX also connects it to the outside world, as shown. The next subscriber in Fig. 11-3 is a home with two phones, while the third subscriber is a home with a phone and a PC. Notice the switch used for voice/data communications.

The primary function of the PBX and central office is the same: switching one telephone line to another. In addition, most central offices multiplex many conversations onto one line. The multiplexing may be based on time or frequency division, and the transmitted signals may be analog or digital; however, in current transmission practice, multiplexing is almost universally PCM digital.

Before switching at the PBX or central office, circuitry residing on the line cards handles the so-called BORSCHT function for their line. These interface circuits are also called the *subscriber loop interface circuit* (SLIC). BORSCHT is an acronym generated as follows:

***B**attery feeding:* supplying the −48 V at 15- to 80-mA power required by the telephone service. Batteries under continuous charge at the central office allow phone service even during power failures.

***O**vervoltage protection:* guarding against induced lightning strike transients and other electrical pickups.

***R**inging:* producing the 90-V rms signal shared by all lines; a relay or high-voltage solid-state switch connects the ring generator to the line.

***S**upervision:* alerting the central office to on- and off-hook conditions (dial tone, ringing, operator requests, busy signals); also the office's way of auditing the line for billing.

***C**oding:* if the central office uses digital switching or is connected to TDM/PCM digital lines, there are codes and their associated filters on the line card.

***H**ybrid:* separating the two-wire subscriber loops into two-wire pairs for transmitting and receiving. The phone system is obviously a full-duplex system. Other than the local loop, four wires are used by the phone company, two for transmitting and two for receiving. This arrangement is shown in Fig. 11-4. The hybrid was a transformer-based circuit, but today an electronic circuit contained within an IC provides the BORSCHT functions.

***T**esting:* enabling the central office to test the subscriber's line.

Historically, analog switches ranged from stepping switches and banks of relays to crossbar switches and crosspoint arrays of solid-state switches. They were slow, had limited bandwidth, ate up space, consumed lots of power, and were difficult to control with microprocessors or even CPUs. To overcome the limitations inherent in analog switching and to enable central offices to use TDM/PCM links in urban areas, telephone companies have converted to digital transmission using

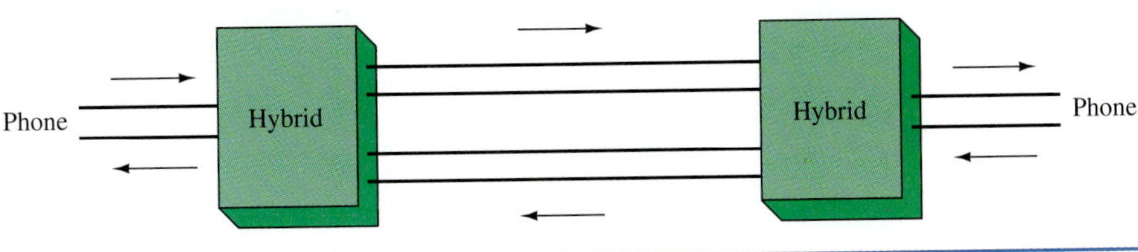

FIGURE 11-4 Two- to four-wire conversion.

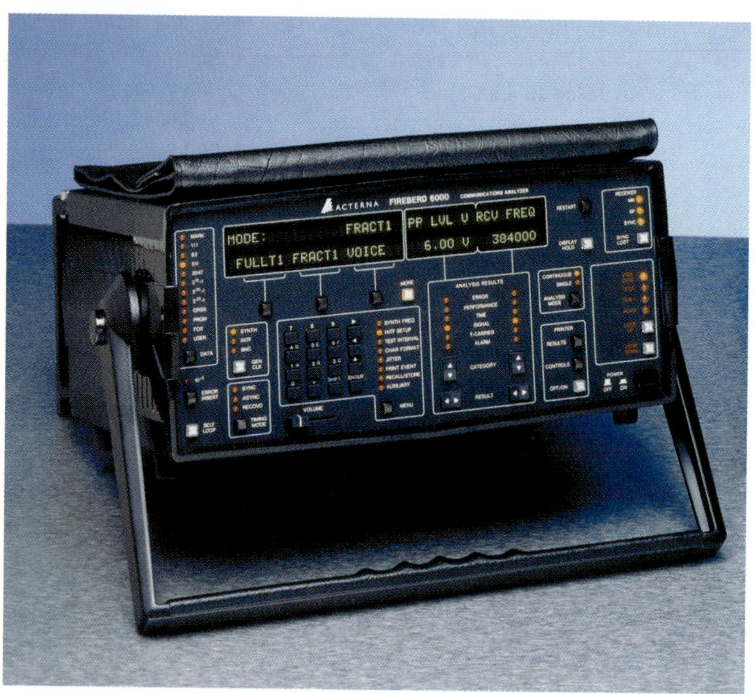

The Fireberd 6000 communications analyzer provides test solutions for frame relay and asynchronous transfer mode. (Courtesy of Acterna, Inc. Reprinted with permission.)

8-bit words. [See Chapter 8 for a discussion of pulse-code modulation (PCM).] The conversions were almost complete by the late 1990s.

On the other hand, if the digital words are multiplexed onto a T1 carrier (see Fig. 11-3) en route to another office, the serial bit stream is converted to the equivalent of 24 voice channels plus supervisory, signaling, and framing bits. Frames are transmitted every 125 μs (an 8-kHz rate). The circuit connecting one central office to another is called the **trunk** line.

Trunk
the circuit connecting one central office to another

Bundles of T1 lines carry most voice channels between central offices in densely populated areas. If the lines stretch more than 6000 ft or so, a repeater amplifies the signal and regenerates its timing. When the number of T1s becomes large, they are further combined into a **T3** signal, which contains a total of 28 T1s and operates at a data rate of 44.736 Mbps. At higher channel densities, it is necessary to convert to optical signals. The T3 then becomes known as an **OC-1,** or optical carrier level 1, operating at 51.84 Mbps. On very high density routes, levels of OC-12 and OC-192 are common, and on long domestic transcontinental routes, OC-48 and OC-192 are in service. Optical fibers have replaced most of the copper wire used in telephone systems, especially in urban areas. Refer to Chapter 18 for details on these fiber-optic systems.

T3
a signal with a digital data rate of 44.736 Mbps

OC-1
optical carrier level 1, which operates at 51.84 Mbps

Line Quality Considerations

An ideal telephone line would transmit a perfect replica of a signal to the receiver. Ideally, this would be true for a basic analog voice signal, an analog version of a digital signal, or a pure digital signal. Unfortunately, as we're sure you would expect, the ideal does not occur. In many cases, the existing cable infrastructure in the United States is more than 30 years old and will be in use for many more years. We now look at the various reasons that a signal received via telephone lines is less than perfect (i.e., is distorted).

Attenuation Distortion

The local loop for almost all telephone transmissions is a two-wire twisted-pair cable. This rudimentary transmission line is usually made up of copper conductors surrounded by polyethylene plastic for insulation. The transmission characteristics of this line are dependent on the wire diameter, conductor spacing, and dielectric constant of the insulation. The resistance of the copper causes signal attenuation. Unfortunately, as explained in Chapter 12, transmission lines have inductance and capacitance that have a frequency-dependent effect. A frequency versus attenuation curve for a typical twisted pair cable is shown in Fig. 11-5. The higher frequencies are obviously attenuated much more than the lower ones. This distortion can be very troublesome to digital signals because the pulses become very rounded and data errors (is it a 1 or a 0?) occur. This distortion is also troublesome to analog signals.

This higher-frequency attenuation can be greatly curtailed by adding inductance in series with the cable. The dashed curve in Fig. 11-5 is a typical frequency versus attenuation response for a cable that has had 88 mH added in series every 6000 ft. Notice that the response is nearly flat below 2 kHz. This type of cable, termed **loaded cable,** is universally used by the phone company to extend the range of a connection. Loaded cable is denoted by the letters H, D, or B to indicate added inductance every 6000, 4500, or 3000 ft, respectively. Standard values of added inductance are 44, 66, or 88 mH. A twisted-pair cable with a 26 D 88 label indicates a 26-gauge wire with 88 mH added every 4500 ft.

Loaded Cable
cable with added inductance every 6000, 4500, or 3000 feet

Attenuation distortion is the difference in gain or loss at some frequency with respect to a reference tone of 1004 Hz. The basic telephone line (called a 3002 channel) specification is illustrated graphically in Fig. 11-6. As can be seen, from 500 to 2500 Hz, the signal cannot be more than 2 dB above the 1004-Hz level or

Attenuation Distortion
in telephone lines, the difference in gain at some frequency with respect to a reference tone of 1004 Hz

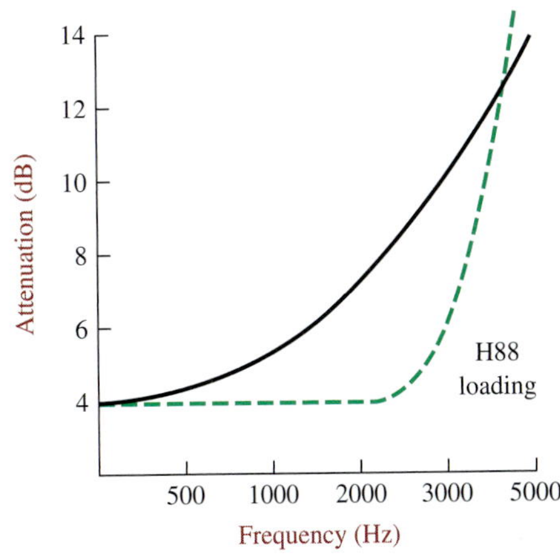

FIGURE 11-5 Attenuation for 12,000 ft of 26-gauge wire.

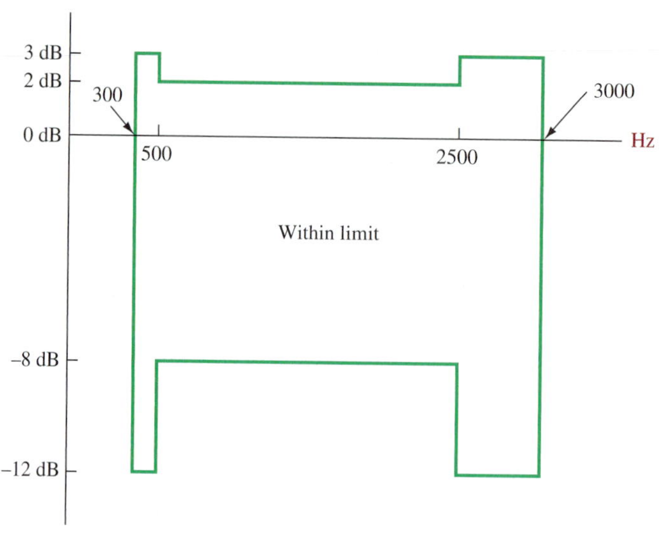

FIGURE 11-6 Attenuation distortion limit for 3002 channel.

8 dB below the 1004-Hz level. From 300 to 500 Hz and 2500 to 3000 Hz, the allowable limits are +3 dB and −12 dB. A subscriber can sometimes lease a better line if necessary. A commonly encountered "improved" line is designated as a C2 line. It has limits of +1 dB and −3 dB between 500 and 2800 Hz. From 300 to 500 Hz and 2800 to 3000 Hz, the C2 limits are +2 dB and −6 dB.

Delay Distortion

Delay Distortion
when various frequency components of a signal are delayed different amounts during transmission

A signal traveling down a transmission line experiences some delay from input to output. That is not normally a problem. Unfortunately, not all frequencies experience the same amount of delay. The **delay distortion** can be quite troublesome to data transmissions. The basic 3002 channel is given an *envelope delay* specification by the Federal Communications Commission (FCC) of 1750 μs between 800 and 2600 Hz. This specifies that the delay between any two frequencies cannot exceed 1750 μs. The improved C2 channel is specified to be better than 500 μs from 1000 to 2600 Hz, 1500 μs from 600 to 1000 Hz, and 3000 μs from 500 to 600 and 2600 to 2800 Hz.

Delay Equalizer
an *LC* filter that removes delay distortion from signals on phone lines by providing increased delay to those frequencies least delayed by the line, so that all frequencies arrive at nearly the same time

The delay versus frequency characteristic for a typical phone line is shown with dashed lines in Fig. 11-7, while the characteristics after delay equalization are also provided. The **delay equalizer** is a complex *LC* filter that provides increased delay to those frequencies least delayed by the phone line, so that all frequencies arrive at nearly the same time. Delay, or phase, equalizers typically have several sections, one for each small group of frequencies across the band. Also, they may be fixed or adjustable.

Telephone Traffic

Let us take a typical workday. Every motorist knows what rush-hour traffic is all about. These are the periods of the day when plenty of vehicles are lined up in every available lane of traffic: in the morning before office hours, at lunch, and after office hours. Motorists take into account delays caused by rush-hour traffic to arrive at their

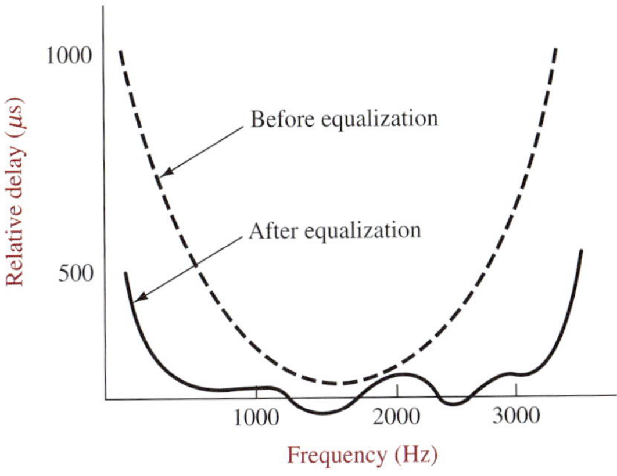

FIGURE 11-7 Delay equalization.

destinations in a timely manner. Those who are unfamiliar with traffic flow patterns are likely to miss appointments (how they wish there were always enough lanes and no intersections on the way to their destinations).

Normal telephone traffic is not so different from vehicular traffic. There are two times in a typical workday when telephone traffic intensifies: between 9:00 and 11:00 in the morning and between 2:00 and 4:00 in the afternoon. The morning traffic is the heaviest, because of the volume of business calls, followed by a lesser volume of calls in the afternoon. That period of one hour when traffic is heaviest is referred to as the busy hour. An example of a busy hour period is from 9:45 A.M. to 10:45 A.M. The busy hour traffic intensity on Mondays is mostly the highest, and it tapers off toward Wednesdays. It starts to pick up again on Thursdays and even more on Fridays (especially on pay-day-Fridays).

The Internet and home computers have changed traditional telephone traffic patterns. This is particularly true in the evening, when computer modem connections to the Internet tie up telephone connections for many hours at a time. The result is that traditional studies for predicting telephone traffic no longer apply to evening telephone use.

The Unit of Traffic

In telephony, one way to define traffic is the average number of calls in progress during a specific period of time, often one hour. A circuit or path that carries its usage for one traffic call at a time is referred to as a trunk. Telephone traffic, although it is a dimensionless quantity, is expressed either in erlang (named after Agner Krarup Erlang, a Danish pioneer of traffic theory) or in hundred-call-seconds (CCS), the latter being commonly used in North America. One erlang equals 36 CCS. A simple way to understand this unit is to look at a bicycle. In a period of one hour the most that you can ride this bicycle is 36 hundred seconds. Two or more persons can ride, or occupy, or hold this bicycle one at a time, but they cannot exceed a total of 36 hundred seconds (or 36 hundred-ride-seconds) in a period of one hour. A group of ten bicycles for rent, while it has a theoretical capacity of 360 hundred-ride-seconds, may sometimes be used only for an average of 36 hundred-ride-seconds on a Monday morning, but on a Sunday this same group of ten bicycles may not be enough, so "congestion" occurs. Some lose interest, others wait for the next available bicycle, and some try again later.

Telephone trunks are arranged in groups. The traffic capacity of a group of trunks depends on the nature or distribution of call durations or holding time, Widely distributed call holding times tend to reduce traffic handling capacity. Traffic carried by a group of trunks may, therefore, be stated as follows:

$$A = C \times \frac{H}{T}$$

where A = traffic in erlangs
C = average number of calls in progress during a period of time
H = average holding time of each call
T = 3,600 seconds (1 hour)

Congestion

A situation in a telephone switching office when calls are unable to reach their destination is referred to as congestion. Getting a busy tone instead of a ringback tone because the called subscriber station is busy is not part of the congestion by definition. Equipment installed in an office provides terminal access for all the subscribers it serves, but not for all the calls they make during the busy hour period. Some calls are allowed to be lost during this time to meet the economic objectives of providing service. It is highly prohibitive to provide sufficient trunks to carry all traffic offered to a system with no calls lost at all. The measure of calls lost during a busy hour period is known as grade of service. In the dimensioning of a telephone network, traffic engineers look at grade of service in the following ways:

1. The probability that a call will be lost due to congestion
2. The probability of congestion
3. The proportion of time during which congestion occurs

Grade of service, *B*, is designed into the system by traffic engineers. After the system is put into service, it is observed and verified using traffic scanning devices. Traffic observation and measurement show how many calls are offered, carried, and lost in the system. Grade of service, *B*, is then determined as:

$$B = \frac{\text{number of calls lost}}{\text{number of calls offered}}$$

or

$$B = \frac{\text{traffic lost}}{\text{traffic offered}}$$

The lower this number, the higher the grade of service.

Traffic Observation and Measurement

Just as the traffic control center of a city would like to ensure a smooth flow of vehicular traffic, telephone operating companies consider traffic management to be the most important function in providing reliable and efficient public telephone service.

Continuous traffic measurement is done to detect and resolve potential sources of congestion. Calls that are either delayed or lost due to congestion problems mean customer dissatisfaction and ultimately lost revenues. The traffic measurement studies are made to determine customer calling patterns, which serve as a basis for discounted toll rates. While operating companies provide contingencies for their network failures, telephone systems are never engineered to handle all calls without congestion. It is also from studying traffic that operating companies forecast future demands and project capital expenditures for their expansion programs.

11-3 Cellular Phone, PCS Systems, and 3G Wireless

Mobile telephone service originated in the late 1940s. It was never a widely used system because of its limited frequency spectrum allocation and the high cost of the required equipment. Additional frequency spectrum was afforded by the FCC's re-allocation of the 800- to 900-MHz band away from UHF TV in the mid-1970s. This allowed for the hundreds of voice channels required for large-scale mobile service, while semiconductor advances and ingenious system design were able to bring costs under control.

The Bell Telephone Company developed the current system called advanced mobile phone service (AMPS). It is commonly referred to as cellular mobile radio for reasons that will soon be obvious. The cellular concept involves an essentially regular array of transmitter–receiver stations called **cell sites.** Figure 11-8 shows a cellular system with 21 "cells" being served by seven different channel groups. The cells cover the entire geographic area to be served by the system. The hexagon shape was chosen after detailed studies showed it led to the most cost-efficient and easily managed system. The two major concepts of cellular systems are frequency reuse and cell splitting.

Cell Sites
a regular array of transmitter–receiver stations

Frequency Reuse

Frequency reuse is the process of using the same carrier frequency (channel) in different cells that are geographically separated. Power levels are kept low enough so that cochannel interference is not objectionable. Thus, cell sites A_1 and A_2 in Fig. 11-8 use channels at the same frequency but have enough separation that they do not interfere with each other. Actually, this process is used in most radio services but not on the shrunken geographic scale of cellular telephone. Instead of covering an entire metropolitan area

Frequency Reuse
in cellular phones, the process of using the same carrier frequency in different cells that are geographically separated

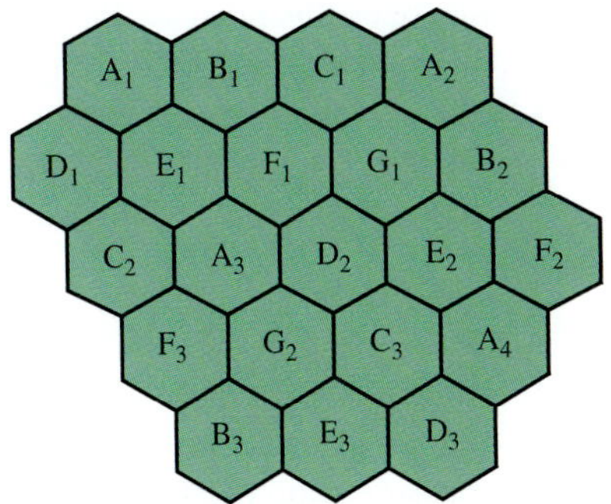

FIGURE 11-8 Cellular phone system layout.

A cellular telephone antenna site. (Courtesy of Agilent Technologies. Reprinted with permission.)

from a single transmitter site with high-power transceivers as the previous mobile phone systems had, the service provider distributes moderate-power systems at each cell site. Through frequency reuse, a cellular system can handle several simultaneous calls, greatly exceeding the number of allocated channels. The multiplier by which the system capacity exceeds allocated channels depends primarily on the number of cell sites.

Cell Splitting

Cell Splitting
for cellular phones, if all traffic in a given cell increases beyond a reasonable capacity, the cell is split into smaller coverage areas

Several frequency channels are assigned to each cell in the system. This is called a channel set. Obviously, one channel is required for each phone call taking place at any one instant of time. If all traffic in a cell increases beyond reasonable capacity, a process called **cell splitting** is utilized. Figure 11-9 illustrates this process. An area from Fig. 11-8 is split into a number of cells that perhaps corresponds to the city's downtown area where phone traffic is heaviest. Successive stages of cell splitting would further increase the available call traffic, if necessary. The techniques of frequency reuse and cell splitting permit service to a large and growing area while using a relatively small frequency spectrum allocation.

System Operation

Advanced Mobile Phone Service (AMPS)
800- to 900-MHz frequency band for mobile telephones

The **advanced mobile phone service (AMPS)** system operates in the 800- to 900-MHz frequency band and uses frequency modulation with a peak deviation of 12-kHz and 30-kHz channel spacing. A duplex phone conversation requires a 30-kHz channel for transmitting and one for receiving. A typical metropolitan system with 666 duplex channels thereby requires a 40-MHz (30 kHz $\times$ 2 $\times$ 666) spectrum allocation.

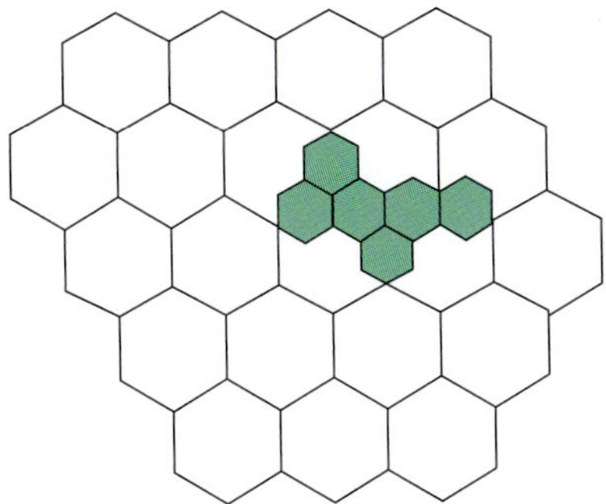

FIGURE 11-9 Cell splitting.

The system includes a mobile telephone switching office (MTSO), the call sites, and the mobile units. The MTSO central processor controls the switching equipment needed to interconnect mobile users with the land telephone network. It also controls cell-site actions and many of the mobile unit actions through commands relayed to them by the cell sites. An 11-cell system is shown in Fig. 11-10, with the MTSO providing a link with two mobile users.

The MTSO is linked to each cell site with land telephone connections, over which the cell sites exchange information necessary for processing calls. Each cell site contains one transceiver for each of its assigned voice channels and the transmitting and receiving antennas for those channels. The cell site also includes signal-level monitoring equipment and a setup radio, as explained subsequently.

The mobile equipment includes a control unit, logic unit, transceiver, and antenna. The control unit contains the user interfaces—handset, dialer, and indicator lights. The transceiver includes a synthesizer to tune all allocated channels. The logic unit interprets customer actions and various system commands. Subsequently, it controls the transceiver and control units. One of the antennas is used for transmission, and both antennas are used for reception in a space-diversity configuration, as explained later in this section.

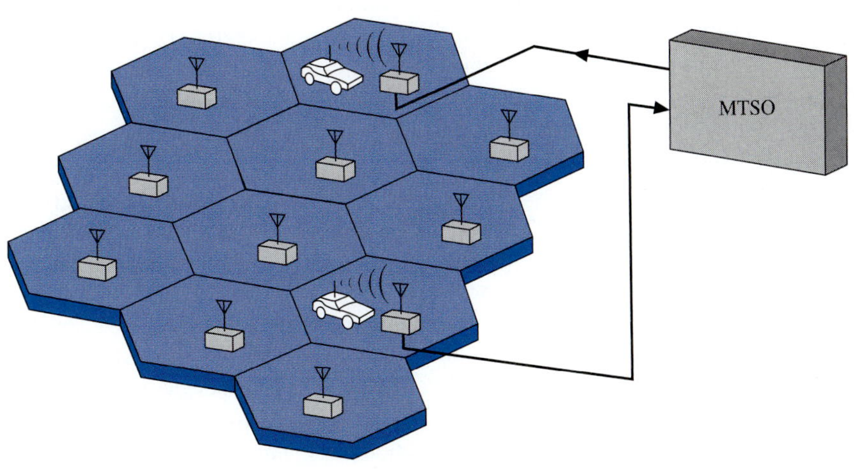

FIGURE 11-10 MTSO linking two mobile users.

A few of the radio channels are used for setup and allow exchange of information needed to establish (set up) calls. Whenever a mobile unit is turned on but not in use, it is monitoring the setup channels. The mobile unit samples all received setup channels and selects the strongest one. Its current cell location has thereby been determined. It then synchronizes with the data stream being transmitted and interprets it. The setup channel data include identification numbers of mobile units to which calls are currently being directed.

When a mobile unit determines that it is being called, it samples the signal strength of all received setup channels again and responds through the cell site offering the strongest signal. It transmits its choice to that cell site, and that cell transmits back the voice channel assignment. The mobile unit tunes to the assigned channel and then receives a command to alert the mobile user (i.e., the telephone rings). The sequence of events is similar but reversed when the mobile user originates a call.

The system examines the call being received every few seconds at the cell site. When necessary, the system looks for another site to serve the call. If need for another site is determined, the system sends a command to the user to retune to a channel allocated to the new cell site. This process of changing channels is called **handoff.** This procedure causes only a brief interruption of the conversation—typically 50 ms. This handoff is not noticed by the user. A command signal from the base station causes the frequency synthesizer (under microprocessor control) in the phone to switch automatically to the carrier frequency of the new cell.

Handoff
the process of changing channels for a new cell site

The power output is from 0.7 to 3 W. Its level is controlled by a power up/down signal from the base station in seven 1-dB steps. This is done to reduce interference with other phones and to minimize overloading of the base station receiver.

The means of speech transmission is standard (analog) narrowband FM using a 12.5- to 30-kHz bandwidth. Frequency-division multiple access separates the signals of different terminals. The second-generation cellular systems use digital speech transmission.

NAMPS

The popularity of cellular phone service created a traffic congestion problem. During the day, the system can reach full capacity, which prevents a cell phone from connecting to the network. To help address the potential congestion problem, a new system called **NAMPS (narrowband analog mobile phone service)** was developed to triple the capacity of the AMP systems. NAMPS divides the 30-kHz single channel into three 10-kHz channels, which increases the capacity of the system and thus enables more air traffic. But the increased number of channels also increases the possibility of adjacent channel interference. The NAMPS system is already operational in some U.S. and overseas markets.

NAMPS (Narrowband Analog Mobile Phone Service)
a system that triples the capacity of the AMPS system

DAMPS (Digital Advanced Mobile Phone Service)
the digital system for mobile phone service

Another answer for the increased demand of cellular service is **DAMPS (digital advanced mobile phone service).** DAMPS provides for increased capacity and security due to its digital data format. The DAMPS system uses the same AMPS network; however, it provides more channels by incorporating TDMA and CDMA technologies. These technologies allow for a greater number of mobile phone calls within the same bandwidth. In fact, up to six digital channels can occupy one AMPS 30-kHz channel using TDMA. CDMA systems are claiming 8 to 15 times the capacity of an AMPS system. CDMA systems all share the same bandwidth. Isolation of each CDMA transmission is provided by a unique identification code, which keeps its transmission isolated. However, CDMA systems require careful control of the power in each mobile phone's transmission to maximize the number of allocated channels. Minimal power levels that still meet the received signal level are the best case for transmission.

Rayleigh Fading

The FM capture effect (Chapter 5) is very helpful in minimizing cochannel interference effects in cellular systems. Unfortunately, this benefit is considerably degraded by **Rayleigh fading**—a rapid variation in signal strength received by mobile units in urban environments. To maintain adequate signal strength during the fades, transmitter power must be increased by the fading margin of up to 20 dB.

Rayleigh Fading
rapid variation in signal strength received by mobile units in urban environments

The signal received by the mobile user can take many paths in an urban area. The signal reflects off buildings and many other obstructions. This multipath reception causes the signal to contain components from many different path lengths. The fading results because, in some relative positions, phases of the signals arriving from the various paths interfere constructively, while in other positions the phases add destructively. The received signal in cellular systems varies by as much as 30 dB (Rayleigh fading) at a very rapid rate because the signal can go from one extreme to the other during a half-wavelength movement by the receiving vehicle. This does not take long for operation in the range 800 to 900 MHz, where a wavelength is about one-third of a meter!

PCS: Personal Communications Services

Another development in mobile communications is **PCS (personal communications services),** which is the North American implementation of the global system for mobile communications, **GSM.** It is also called the second generation wireless (2G). In addition to phone service, PCS systems provide enhanced features such as messaging, paging, and data services, including access to the Internet. The data rate, over the air, for GSM is slow, 9600 bps, but there are websites designed for the limited access speed of the PCS system. The PCS system operates in the 1900-MHz range, although some phones can operate in both the 800- and 1900-MHz bands. This is called **dual-band** technology. A comparison of the cellular and PCS spectrum is provided in Fig. 11-11. PCS radios are intended to be small enough to fit in a shirt pocket. The transmitted power is low and therefore more base stations are required than for cellular service. PCS systems are also called **microcellular systems.**

PCS (Personal Communications Services)
a 1900-MHz mobile phone service with enhanced features such as messaging, paging, and data service

GSM
global system for mobile communications

Dual-Band
indicates that a phone can operate in two different bands

Microcellular Systems
another name for PCS systems

Each PCS channel has two frequencies associated with it, one for uplink (phone to the system) and one for downlink (system to the phone). This provides for the simultaneous transmission of information. A **DCCH** (digital control channel) provides PCS systems with improved functionality over cellular phones. PCS incorporates a multilayered protocol of four layers, similar to computer networking, to manage the **air-interface.** The four layers are shown in Fig. 11-12.

DCCH
digital control channel

Air-Interface
used by PCS systems to manage the transfer of information

1. *Physical layer:* manages the radio interface
2. *Data link layer:* manages the data packaging and transfer, including error correction
3. *Message layer:* manages the transfer of messages transmitted and received
4. *Upper applications layer:* the actual service being used

For example, if you are using a message transfer, the upper application will be the user interface on your PCS phone for conducting the message transfer. The upper applications layer next passes the information to the message layer, which

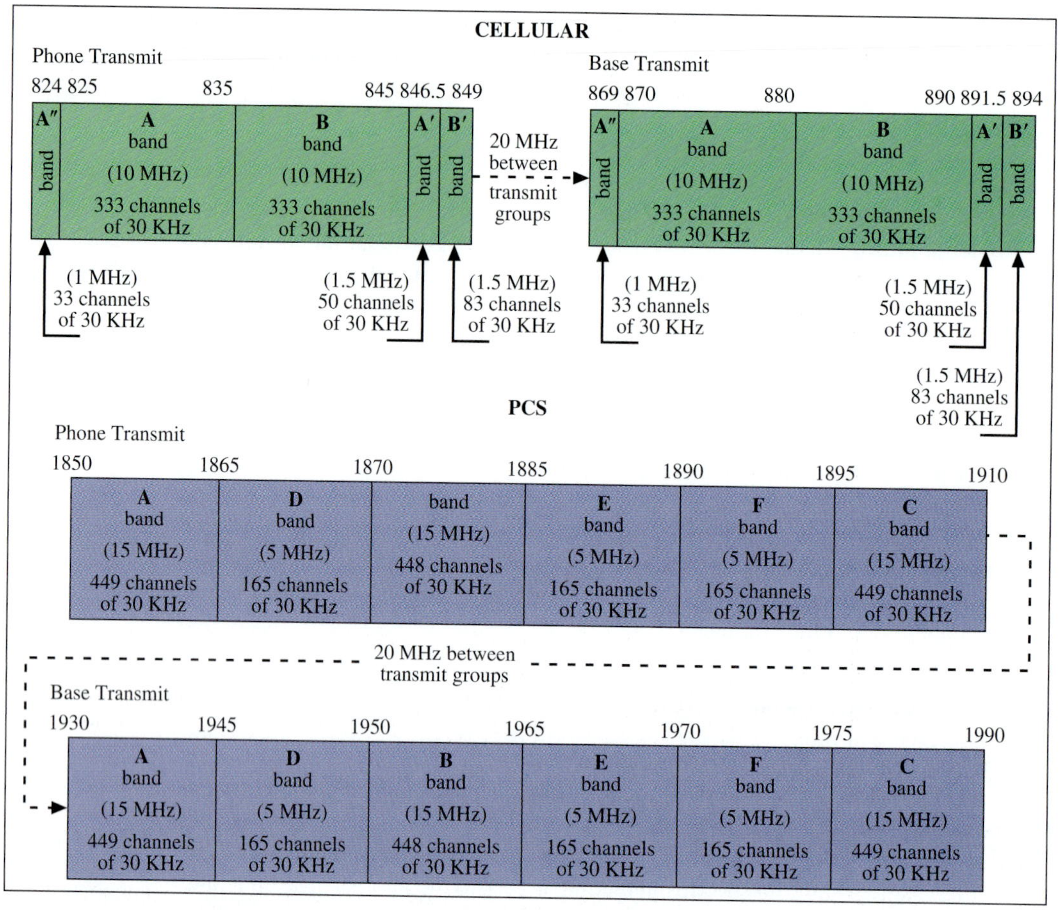

CELLULAR
Phone Transmit
824 825
835
845 846.5 849
A″ band
A band (10 MHz) 333 channels of 30 KHz
B band (10 MHz) 333 channels of 30 KHz
A′ band
B′ band
20 MHz between transmit groups
Base Transmit
869 870
880
890 891.5 894
A″ band
A band (10 MHz) 333 channels of 30 KHz
B band (10 MHz) 333 channels of 30 KHz
A′ band
B′ band
(1 MHz) 33 channels of 30 KHz
(1.5 MHz) 50 channels of 30 KHz
(1.5 MHz) 83 channels of 30 KHz
(1 MHz) 33 channels of 30 KHz
(1.5 MHz) 50 channels of 30 KHz
(1.5 MHz) 83 channels of 30 KHz
PCS
Phone Transmit
1850
1865
1870
1885
1890
1895
1910
A band (15 MHz) 449 channels of 30 KHz
D band (5 MHz) 165 channels of 30 KHz
band (15 MHz) 448 channels of 30 KHz
E band (5 MHz) 165 channels of 30 KHz
F band (5 MHz) 165 channels of 30 KHz
C band (15 MHz) 449 channels of 30 KHz
20 MHz between transmit groups
Base Transmit
1930
1945
1950
1965
1970
1975
1990
A band (15 MHz) 449 channels of 30 KHz
D band (5 MHz) 165 channels of 30 KHz
B band (15 MHz) 448 channels of 30 KHz
E band (5 MHz) 165 channels of 30 KHz
F band (5 MHz) 165 channels of 30 KHz
C band (15 MHz) 449 channels of 30 KHz

FIGURE 11-11 A comparison of the cellular and PCS spectrums. (Courtesy of the International Engineering Consortium and the IEC Web Pro Forums, http://www.iec.org/tutorials/)

Physical layer
Data link layer
Message layer
Upper applications layer

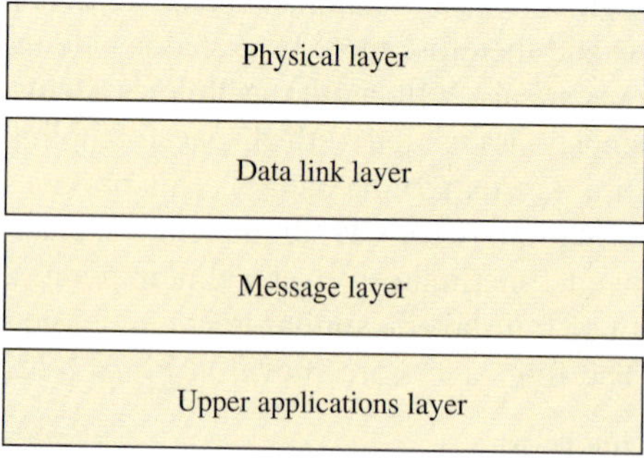

FIGURE 11-12 The four layers of the PCS air interface.

incorporates whatever management and control information it may require. Next, the message layer passes the information to the data link layer, which packages the data for transmission. Finally, the packaged data are passed to the physical layer for transmission over the RF channel. The reverse of this process is incorporated when a message is received. Another example of an upper applications layer is the PCS Internet browser.

The layer of protocols provides for the addition of more advanced future applications while maintaining the same operation of the lower layers. This is comparable to the OSI layers in computer networking.

3G Wireless

The **3G** (third generation) wireless system is considered to be the next development in wireless connectivity. 3G was developed to provide broadband network (broadband wireless) services with expected data rates exceeding 2 Mbps. The standard defining 3G wireless is called international mobile telecommunications, or **IMT-2000.** It includes the following features:

- Worldwide use
- High data rates (up to 2 Mbps)
- Spectrum efficiency
- Support for both packet-switched and circuit-switched data transmission
- Support for all mobile applications

3G
the third generation in wireless connectivity

IMT-2000
international mobile telecommunications; the standard defining 3G wireless

The most significant development from IMT-2000 has been wideband code division multiple access **(W-CDMA).** This is also called the universal mobile telecommunication system (UMTS), which has been identified as the successor to GSM (2G wireless). The W-CDMA technology provides for a peak throughout of 1.246 Mbps. It also provides for backward compatibility with GSM (2G wireless).

There are two modes of operation for W-CDMA: the **FDD** mode (frequency division duplex, FDD, W-CDMA) and the **TDD** mode (time division duplex, TDD, W-CDMA). It is not clear which one is best. Both technologies are being implemented and it appears that both will continue to be supported by wireless vendors. The frequency assignments for 3G wireless are not fully defined for the United States. However, the uplink/downlink frequency ranges for Europe have been defined and are listed in Table 11-1.

W-CDMA
wideband code division multiple access

FDD
frequency division duplex

TDD
time division duplex

Table 11-1 The 3G Wireless Frequencies (Europe)

Frequency (MHz)	Application
1920–1980	FDD, W-CDMA uplink
2110–2170	FDD, W-CDMA downlink
1900–1920	TDD, W-CDMA uplink
2010–2025	TDD, W-CDMA downlink

Another development for 3G wireless is **CDMA-2000,** an air interface that is popular in the United States. CDMA-2000 is divided into three phases based on bandwidth support.

CDMA2000 1X: supports data rates up to 614 kbps

CDMA2000 1X EV: supports data rates up to 2.8 Mbps

CDMA2000 3X: supports triple the bandwidth of CDMA2000 1X, data rates in excess of 2 Mbps

CDMA2000
a 3G wireless development popular in the United States

11-4 LOCAL AREA NETWORKS

The dramatic decrease in computer system cost and increase in availability have led to an explosion in computer usage. Organizations such as corporations, colleges, and government agencies have acquired large numbers of single-user computer systems. They may be dedicated to word processing, scientific computation, process control, etc. A need to interconnect these locally distributed computer networks soon became apparent. Interconnection allows the users to send messages to the other network members. It also allows resource sharing of expensive equipment such as high-quality graphics printers or access to a large mainframe computer to run programs too complicated for the local computer. The local computer is usually a personal microcomputer-type system. The network used to accomplish this is called a **local area network** (LAN). LANs are typically limited to separations of a mile or two and to several hundred users, but are usually smaller in scope.

Local Area Network
network of users that share computers in a limited area

Topology
architecture of a network

Protocol
a means for a user to gain control of the network to allow transmission

Local area networks are defined in terms of the **topology** (architecture) used to interconnect the networking equipment and the **protocol** used for accessing the network. The most common architectures for local area networks (LANs) are shown in Fig. 11-13. Two of the networking protocols in common use are carrier sense multiple access with collision detection (CSMA/CD), which is associated with the bus and star topologies, and token passing, which is associated with the token-ring topology.

The token-ring topology is shown in Fig. 11-14. The token-passing technique is well suited to the ring network topology. An electrical token is placed in the channel and circulates around the ring. If a user wishes to transmit, the station must wait until possession of the token exists. Each station is assured access for transmission of its messages. A disadvantage of this system is that if an error changes the token pattern, it causes the token to stop circulating. Also, ring networks rely on each system to relay all data to the next user. A failed station causes data traffic to cease. Another

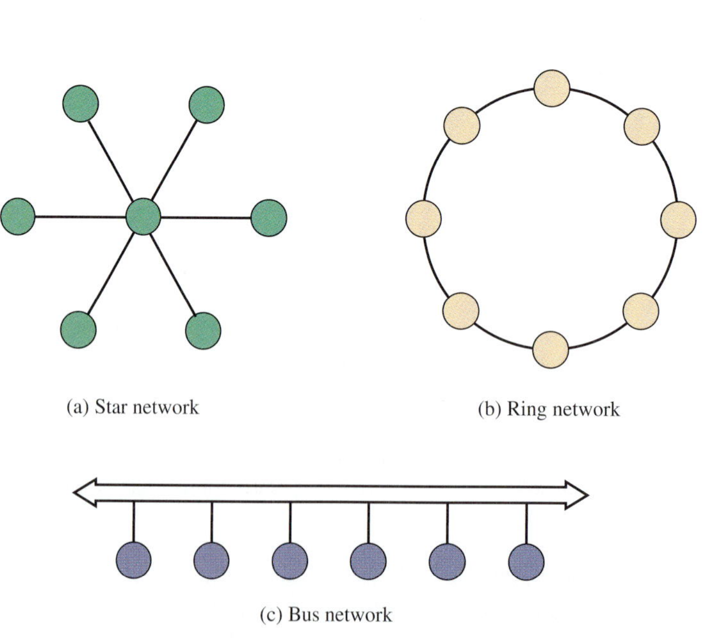

FIGURE 11-13 Network topologies.

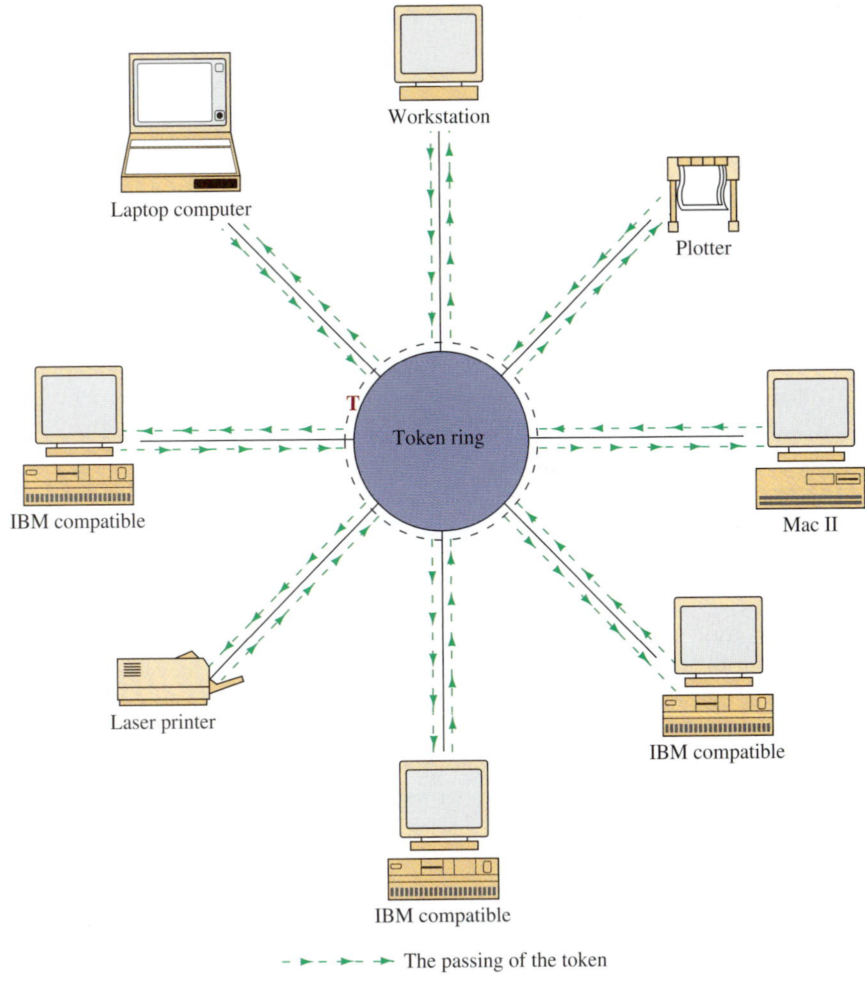

FIGURE 11-14 The token-ring topology.

approach for the token-ring technique is to attach all the computers to a central token-ring hub. Such a device manages the passing of the token rather than assigning this task to individual computers, which improves the reliability of the network.

An extensive number of LAN systems are currently available. Many are applicable to a specific manufacturer's equipment only. The Institute of Electrical and Electronic Engineers (IEEE) standards board approved LAN standards in 1983. The following IEEE 802 standards provided impetus for different manufacturers to use the same codes, signal levels, etc.: IEEE 802.3 CSMA/CD IEEE 802.5 Token-ring.

The bus network topology is shown in Fig. 11-15. A bus network shares the media for data transmission. This means that while one computer is talking on the LAN, the other network devices (e.g., other computers) must wait until the transmission is complete. For example, if computer 1 is printing a large file, the line of communication will be between computer 1 and the printer, and this will tie up the

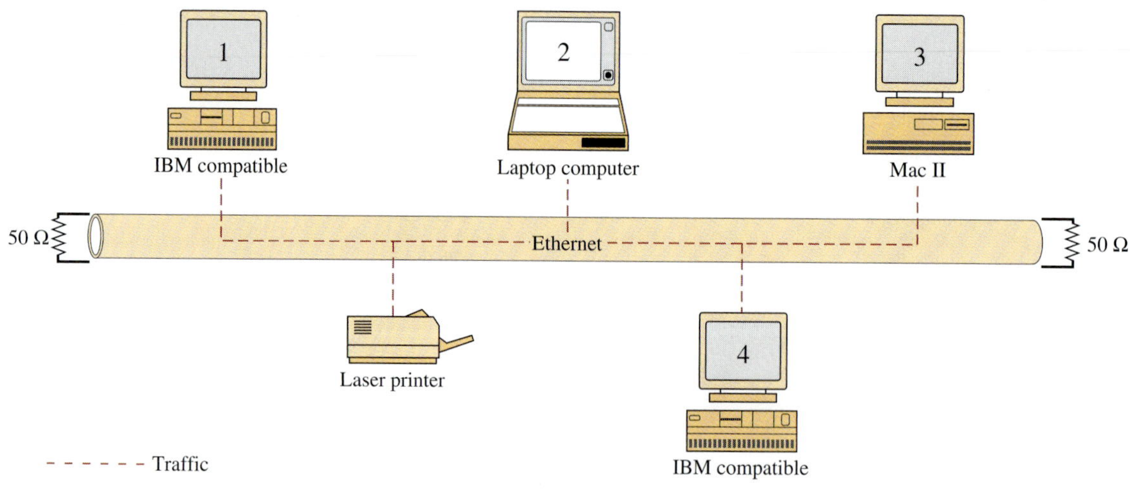

FIGURE 11-15 The bus topology.

network bus for a good portion of the time. All network devices on the bus see the data traffic from computer 1 to the printer, and the other network devices must wait for pauses in transmission or until the transmission is complete before they can assume control of the bus and initiate transmission. This means that bus topologies are not very efficient. This is one reason—but not the only reason—that bus topologies are seldom used in modern computer networks.

The star topology, shown in Fig. 11-16, is the most common in today's LANs. At the center of a star network is either a hub or a switch. The hub or the switch is used to connect the network devices together and facilitate the transfer of data. For example, if computer 1 wants to send data to the network printer, the hub or switch provides the network connection. Actually, either a switch or a hub can be used, but there is a significant advantage to using a switch. In a hub environment, the hub will rebroadcast the message to all computers connected to the star network. This is very similar to the bus topology because all data traffic on the LAN is being seen by all computers. However, if a switch is used instead of a hub, then the message is transmitted directly from computer 1 to the printer. This greatly improves the efficiency of the available bandwidth. It also permits additional devices to communicate without tying up the network. For example, while computer 1 is printing a large file, computers 5 and 6 can communicate with each other.

Ethernet LAN

Ethernet is a baseband CSMA/CD protocol local area network system. It originated in 1972, and the full specification was provided via a joint effort among Xerox, Digital Equipment Corporation, and Intel in 1980. **CSMA/CD** stands for carrier sense multiple access with collision detection.

CSMA/CD
the Ethernet LAN protocol, carrier sense multiple access with collision detection

Basically, for a computer to talk on the Ethernet network, it first listens to see if there is any data traffic (carrier sense). This means that any computer on the LAN can be listening for data traffic and any of the computers on the LAN can access the network (multiple access). There is a chance that two or more computers will

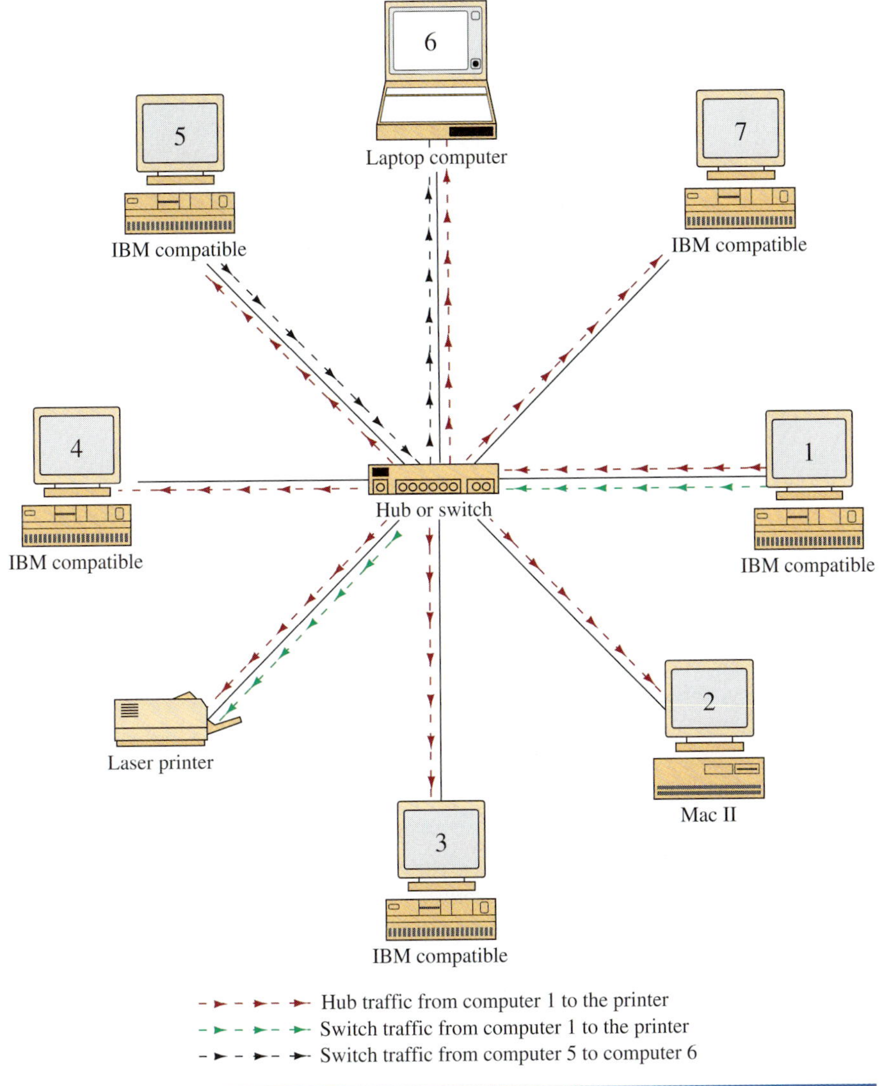

FIGURE 11-16 The star topology.

attempt to broadcast a message at the same time; therefore, Ethernet systems have the capability for detecting data collisions (collision detection).

How is the destination for the data determined? The Ethernet protocol provides information regarding the source and destination addresses. The structure for the Ethernet frame is shown and described in Fig. 11-17.

Preamble	Start frame delimiter	Destination MAC address	Source MAC address	Length type	Data	Pad	Frame check sequence

FIGURE 11-17 The data structure for the Ethernet frame.

An Etherlink 10/100 Mbps network interface card. (Courtesy of 3COM Corporation. Reprinted with permission.)

Network Interface Card (NIC)
the electronic hardware used to interface the computer to the network

MAC Address
a unique 6-byte address assigned by the vendor of the network interface card

Broadcast Address
setting the destination MAC address to all 1s broadcasts the message to all computers on the LAN

Preamble: an alternating pattern of 1s and 0s used for synchronization.

Start frame delimiter: A binary sequence of 1 0 1 0 1 0 1 1 that indicates the start of the frame.

Destination MAC address and source MAC address: Each Ethernet **network interface card (NIC)** has a unique media access control (MAC) address associated with it. The **MAC address** is 6 bytes in length. The first 3 bytes are used to indicate the vendor, and the last 3 bytes are unique numbers assigned by the vendor. This is the information that ultimately enables the data to reach a destination in a LAN. This is also how computer 1 and the printer communicated directly in the star topology example using the switch (Fig. 11-16). The switch used the MAC address information to redirect the data from computer 1 directly to the printer. *Note:* If the destination MAC address is all 1s, then this is called a **broadcast address,** and the message is sent to all stations on the network. The following are examples of a MAC address and a broadcast address; the addresses shown are in hexadecimal code (base 16).

	Vendor	NIC Card ID
MAC address	0 0 A A 0 0	B 6 7 A 5 7
Broadcast address	F F F F F F	F F F F F F

Length/type: an indication of the number of bytes in the data field if this value is less than 1500. If this number is greater than 1500, it indicates the type of data format, for example, IP and IPX.

Data: the data being transferred from the source to the destination.

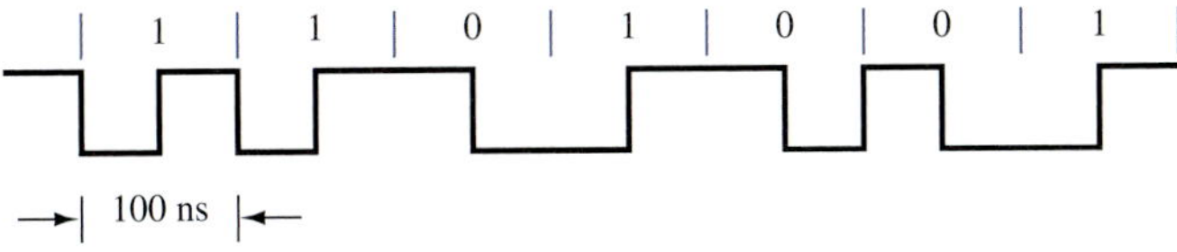

FIGURE 11-18 Manchester encoding.

Pad: a field used to bring the total number of bytes up to the minimum of 46 if the data file is less than 46 bytes.

Frame check sequence: a 4-byte cyclic redundancy check (CRC) value used for error detection. The CRC check is performed on the characters from the destination MAC address through the pad fields. If an error is detected, then the system requests a retransmission.

The minimum length of the Ethernet frame is 64 bytes from the destination MAC address through the frame check sequence. The maximum Ethernet frame length is 1522 bytes. The 0s and 1s in the Ethernet frame are formatted using Manchester encoding (biphase-L, see Sec. 8-4). An example of Manchester encoding is shown in Fig. 11-18.

11-5 Assembling a LAN

This section presents two examples of assembling LANs. The examples demonstrate a technique that can be used to assemble an office LAN and a building LAN. These examples are presented from the point of view of assembling the hardware necessary for establishing network communications between the computers and ancillary network devices. Many possible configurations can be used to solve these problems; this is one solution. Note that all computer networks require some type of networking software to run the LAN. Networking software is available with Microsoft Windows, Windows NT, NT server, Novell, and the Macintosh operating system, to name a few.

The Office LAN Example

Our example of an office LAN consists of 10 computers, 2 printers, and 1 server. The layout for the office LAN is shown in Fig. 11-19. Each computer, the printers, and

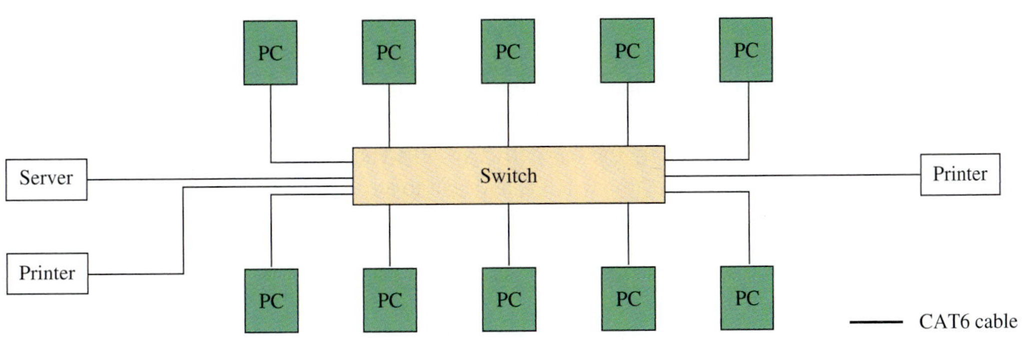

FIGURE 11-19 An example of an office LAN.

the server on the LAN are all connected to a common switch. The connection from each unit in the network to the switch is provided by a CAT6 (category 6) twisted-pair cable. CAT6 cables are capable of carrying 1000 Mbps of data up to a length of 100 meters. Twisted-pair cables and the various category specifications are discussed in Sec. 12-2. If the network hardware and software are properly set up, all computers will be able to access the server, the printer, and other computers.

The media used for transporting data in a modern computer network are either twisted-pair or fiber cables. Fiber cables, optical LANs, and their numerics are discussed in Chapter 18. Table 11-2 lists the common numerics used to describe the data rates for the copper coaxial cable and twisted-pair media being used in a LAN.

Table 11-2 Common Numerics for LAN Cabling

Numeric	Description
10Base2	10 Mbps over coax up to 185 m, also called ThinNet (seldom used anymore)
10Base5	10 Mbps over coax up to 500 m, also called ThickNet (seldom used anymore)
10BaseT	10 Mbps over twisted pair
100BaseT	100 Mbps over twisted pair
100BaseFX	100 Mbps over fiber
1000BaseT	1 Gbps over twisted pair
1000BaseFX	1 Gbps over fiber
10GBase—	The family of fiber products supporting 10-gigabit Ethernet (10GbE)

Assembling a Building LAN

A building LAN describes a network where multiple LANs within a building are connected together. An example of assembling a building LAN is provided in Fig. 11-20. For this example, three switches were required because the distance from the computers to a central switch exceeded the 100-m maximum distance for CAT6 twisted-pair cable. To meet the 100-m maximum-distance requirement, it was decided to place a switch inside a closet in each of the three wings. The connection from each device is **100BaseT.** This means that the data rate is 100-Mbps baseband, and CAT6 twisted-pair cable was used. All network devices in each wing were routed to their respective switches, located in either closet A, B, or C. Each closet has **RJ-45** patch panels for routing the cables. RJ-45 connectors are 8-pin modular types used to connectorize CAT6 twisted-pair cable. The patch panels were included to maximize the flexibility of the network, including future wiring changes needed to accommodate changes in the network. The number of computers, printers, workstations, and servers input to each switch is listed by the respective switch. The fiber optic feeds from switches A and B are combined with switch C and sent to a router. The router provides a connection to the Internet and routes data traffic back to closets A, B, and C.

100BaseT
indicates 100-Mbps data baseband over twisted-pair cable

RJ-45
the 8-pin modular connector used to connectorize CAT6 cable

Wireless LANs

A typical computer network uses twisted-pair and fiber-optic cable for the data links in modern computer networks. Another medium competing for use in higher data-rate

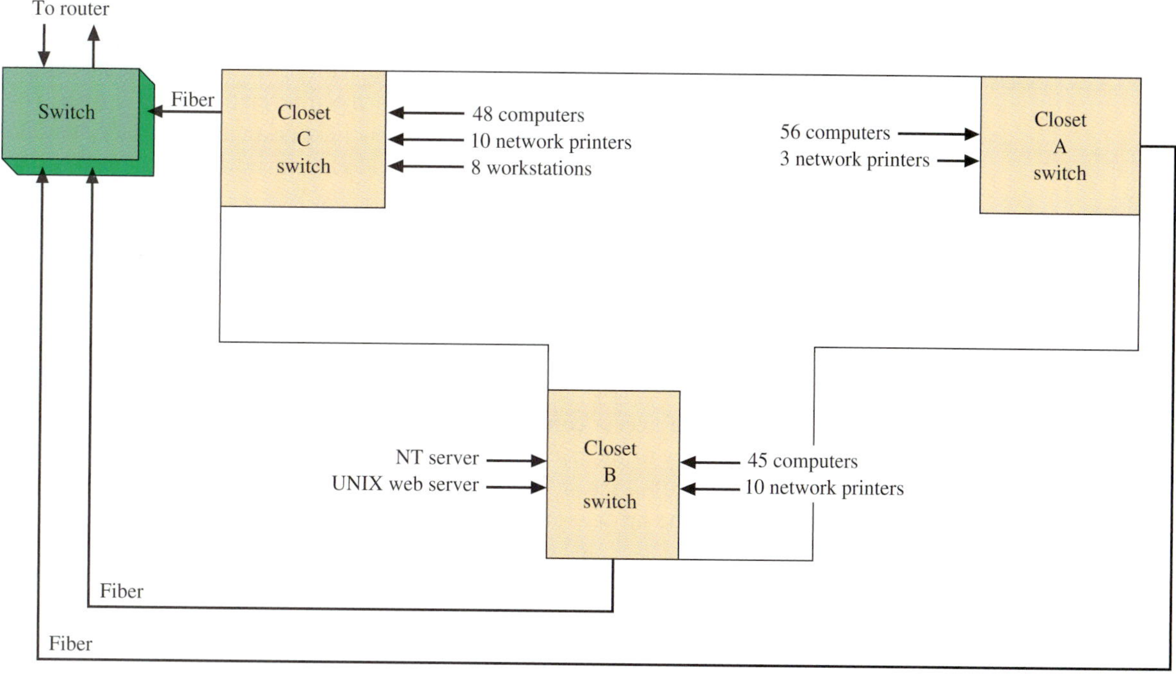

FIGURE 11-20 An example of a building LAN.

LANs is wireless, based on the IEEE 802.11 wireless standard. The advantages of wireless include:

- User mobility in the workplace
- Cost-effective for use in areas that are difficult or too costly to wire

The Cisco catalyst 2926 10/100 autosensing switch. (Courtesy of Cisco Systems. Reprinted with permission.)

The concept of user mobility in the workplace opens the door to many opportunities to provide more flexibility in the workplace. Workers can potentially access the network from almost any location within the workplace. Accessing information from the network is as easy as if the information is on a disk. This section examines the fundamentals of wireless networking, the 802.11 standard and its family (802.11b, 802.11a, and 802.11g), and how wireless LANs are configured.

The IEEE 802.11 wireless LAN standard defines the physical (PHY) layer, the medium access control (MAC) layer, and the MAC management protocols and services.

The PHY (physical) layer defines

- The method of transmitting the data: RF and infrared

The MAC (media access control) layer defines

- the reliability of the data service
- access control to the shared wireless medium
- protecting the privacy of the transmitted data

The wireless management protocols and services are

- authentication
- association
- data delivery
- privacy

DSSS
direct sequence spread spectrum

FHSS
frequency hopping spread spectrum

ISM
industrial, scientific, and medical

Pseudorandom
the number sequence appears random but actually repeats

Four physical layer technologies are currently used in 802.11 wireless networking: direct sequence spread spectrum **(DSSS),** frequency hopping spread spectrum **(FHSS),** infrared, and orthogonal frequency division multiplexing (OFDM). DSSS is used most often in 802.11b wireless networks, and OFDM is used in 802.11a. (Refer to Chapter 10 for the discussion on wireless digital communications.)

DSSS uses three nonoverlapping 22 MHz channels in the 2.4 GHz industrial, scientific, and medical **(ISM)** band. The frequency channels used in North America are listed in Table 11-3. An example of the frequency spectrum for three channels in DSS is shown in Fig. 11-21.

In frequency hopping spread spectrum (FHSS), the transmit signal frequency changes based on a pseudorandom sequence. **Pseudorandom** means that the

Table 11-3 The DSSS Channels

Channel Number	Frequency (GHz)
1	2.412
2	2.417
3	2.422
4	2.427
5	2.432
6	2.437
7	2.442
8	2.447
9	4.452
10	2.457
11	2.462

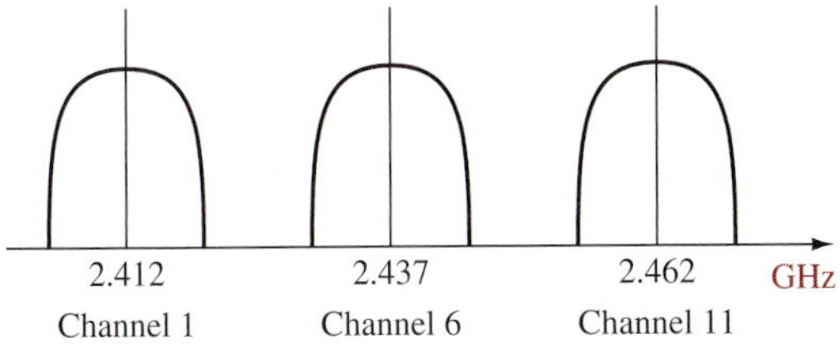

FIGURE 11-21 An example of the three channels in the DSSS spectrum.

sequence appears to be random but it does repeat, typically after some lengthy period of time. (Refer to Chapter 10 for the discussion on PN circuits.) FHSS uses 79 channels, each 1 MHz wide, in the ISM 2.4 GHz band. FHSS requires that the transmit and receive units know the **hopping sequence** (the order of frequency changes) so that a communications link can be established and synchronized. FHSS data rates are typically 1 and 2 Mbps.

Hopping Sequence
the order of frequency changes

The maximum transmit power of 802.11b wireless devices is 1000 mW; however, the nominal transmit power level is 100 mW. The 2.4 GHz frequency range used by 802.11b is shared by many technologies, including Bluetooth, cordless telephones, and other wireless LANs. The RF signals emitted from these technologies in the 2.4-GHz range contribute noise and can effect wireless data reception. A significant improvement in wireless performance is available with the IEEE 802.11a standard. The 802.11a equipment operates in the 5-GHz range, as opposed to the 2.4-GHz range for 802.11b, and 802.11a provides significant improvement with RF interference.

The 802.11a standard uses a technique called orthogonal frequency division multiplexing (OFDM) to transport the data over 12 possible channels in the unlicensed national information infrastructure **(U-NII).** U-NII was set aside by the Federal Communications Commission (FCC) to support short-range high-speed wireless data communications. The operating frequencies for 802.11a are listed in Table 11-4. The transmit power levels for 802.11a are provided in Table 11-5.

U-NII
unlicensed national information infrastructure

Table 11-4 The IEEE 802.11a Channels and Operating Frequencies

	Channel	Center Frequency (GHz)
Lower Band	36	5.180
	40	5.20
	44	5.22
	48	5.24
Middle Band	52	5.26
	56	5.28
	60	5.30
	64	5.32
Upper Band	149	5.745
	153	5.765
	157	5.785
	161	5.805

Table 11-5 Maximum Transmit Power Levels for 802.11a with a 6 dBi Antenna Gain

Band	Power Level
Lower	40 mW
Middle	200 mW
Upper	800 mW

IEEE 802.11a equipment is not compatible with 802.11b. One advantage of this fact is that the 802.11a equipment will not interfere with 802.11b. Therefore, 802.11a and 802.11b links can run next to each other without causing any interference. The downside of 802.11a is the increased cost of the equipment and the increased power consumption because of the OFDM technology. These factors are of particular concern with mobile users and the effect it can have on battery life. Also, the maximum usable distance (RF range) is about half that for the 802.11b.

Another IEEE 802.11 wireless standard is the IEEE 802.11g. The 802.11g standard supports the higher data transmission rates of 54 Mbps but operates in the same 2.4-GHz range as 802.11b. The 802.11g equipment is also backward compatible with 802.11b equipment, which means that 802.11b wireless clients can communicate with the 802.11g access points and the 802.11g wireless client equipment can communicate with the 802.11b access points.

The obvious advantage of this is that companies with an existing 802.11b wireless network can migrate to the higher data rates provided by 802.11g without sacrificing network compatibility. In fact, some manufacturers support both the 2.4 GHz and 5GHz standards.

CSMA/CA
carrier sense multiple access with collision avoidance—the protocol used by wireless LANs

The 802.11b uses a modified version of the CSMA/CD protocol called carrier sense multiple access with collision avoidance **(CSMA/CA).** The CSMA/CA protocol avoids collisions by waiting for an acknowledgment (ACK) that a packet has arrived intact before initiating another transmission. If an ACK is not received by the sender, then it is assumed that the packet was not received intact and the sender retransmits the packet.

An example of an 802.11b wireless Ethernet office LAN is shown in Fig. 11-22. Each PC has a module connected to it called a wireless LAN adapter (WLA). This unit connects to the Ethernet port on the PC's network interface card or is plugged into the motherboard. The wireless LAN adapters use a low-gain (2.2 dBi) dipole antenna. The wireless LAN adapters communicate directly with another wireless device called an *access point.* The access point serves as an interface between the wireless LAN and the wired network.

Another wireless technology, called Bluetooth, and its associated technologies have been developed to replace the cable connecting computers, mobile phones, handheld devices, portable computers, and fixed electronic devices. The information normally carried by a cable is transmitted over the 2.4 GHz ISM frequency band using a frequency-hopping technique. Communication between portable computers is similar to computers connected via a wired LAN but distances are limited to less than 100 meters.

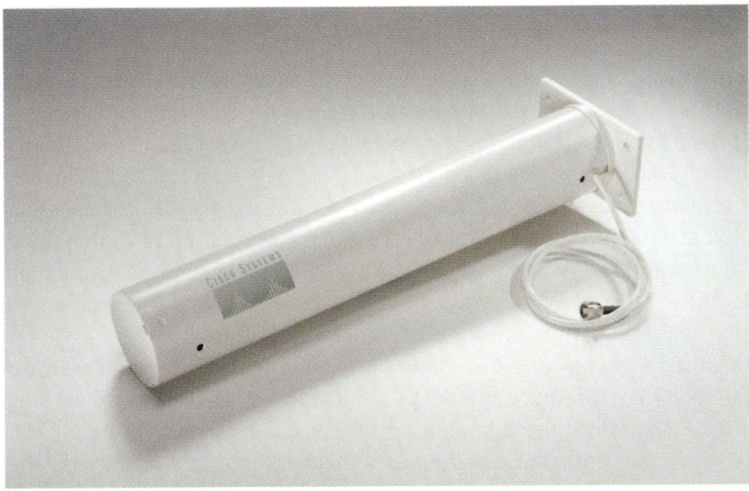

The Cisco Aironet wireless Yagi antenna. (Courtesy of Cisco Systems. Reprinted with permission.)

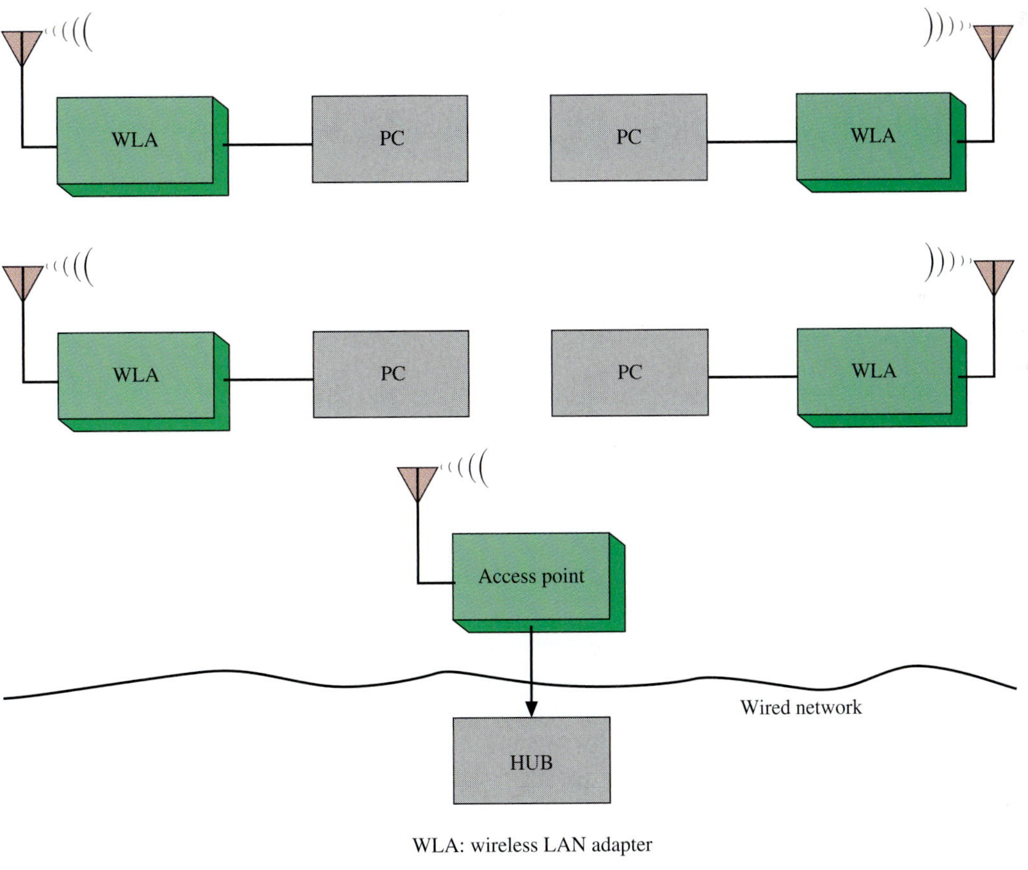

FIGURE 11-22 An example of a wireless LAN and its interface to a wired network.

11-6 LAN Interconnection

The utility of LANs led to the desire to connect two (or more) networks together. For instance, a large corporation may have had separate networks for its research and engineering and for its manufacturing units. Typically, these two systems used totally different technologies, but it was deemed necessary to "tie" them together. This led to **metropolitan area networks** (MANs)—two or more LANs linked together within a limited geographical area. Once the techniques were in place to do this, it was decided that it would be helpful to link the MAN with the marketing division on the other side of the country. Now two or more LANs were linked together over a wide geographical area, resulting in a **wide area network** (WAN).

Metropolitan Area Network
two or more LANs linked together over a limited geographical area

Wide Area Network
two or more LANs linked together over a wide geographical area

Open Systems Interconnection
reference model to allow different types of networks to be linked together

To allow different types of networks to be linked together, an **open systems interconnection** (OSI) reference model was developed by the International Organization for Standardization. It contains seven layers, as shown in Fig. 11-23. It provides for everything from the actual physical network interface to software applications interfaces.

7. Application
6. Presentation
5. Session
4. Transport
3. Network
2. Data link
1. Physical

FIGURE 11-23 OSI reference model.

1. *Physical layer:* provides the electrical connection to the network. It doesn't speak to the modulation or physical medium used.
2. *Data link layer:* handles error recovery, flow control (synchronization), and sequencing (which terminals are sending and which are receiving). It is considered the "media access control layer."
3. *Network layer:* accepts outgoing messages and combines messages or segments into packets, adding a header that includes routing information. It acts as the network controller.
4. *Transport layer:* is concerned with message integrity between the source and destination. It also segments/reassembles (the packets) and handles flow control.
5. *Session layer:* provides the control functions necessary to establish, manage, and terminate the connections as required to satisfy the user request.
6. *Presentation layer:* accepts and structures the messages for the application. It translates the message from one code to another if necessary.
7. *Application layer:* logs the message in, interprets the request, and determines what information is needed to support the request.

Interconnecting LANs

The interconnection of two or more LANs (into a MAN or WAN) is accomplished in several ways, depending on the LAN similarities.

Bridges: Bridges use only the bottom two OSI layers to link LANs that usually have identical protocols at the physical and data link layers. This is illustrated in Fig. 11-24.

Routers: Routers interconnect LANs by using the bottom three OSI layers, as shown in Fig. 11-25. They manage traffic congestion by employing a flow-control mechanism to direct traffic to alternative paths and can provide protocol conversion when needed.

Gateways: This is an older term used to describe a device that encompasses all seven OSI layers. It interconnects two networks that use different protocols and formats. In the capacity shown in Fig. 11-26, the gateway is being used to perform protocol conversion at the applications layer. Today, routers manage all the functions previously handled by a gateway.

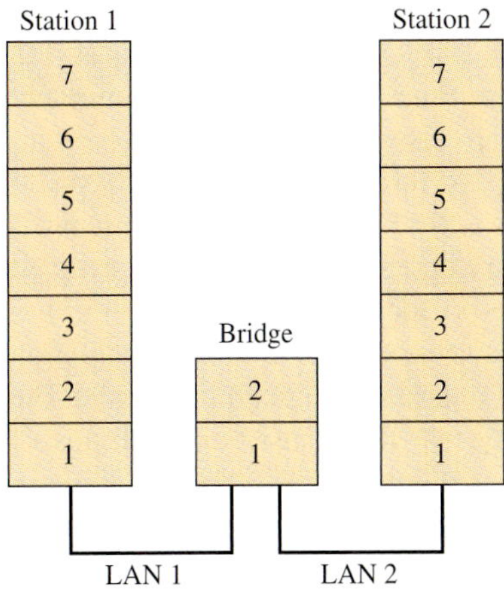

FIGURE 11-24 Bridge connecting two LANs.

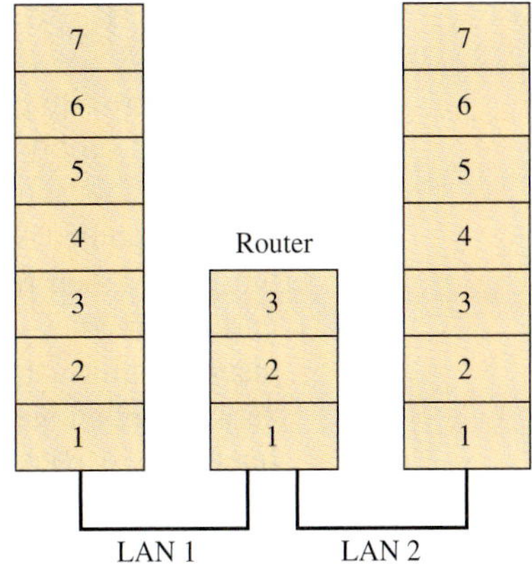

FIGURE 11-25 Router connecting two LANs.

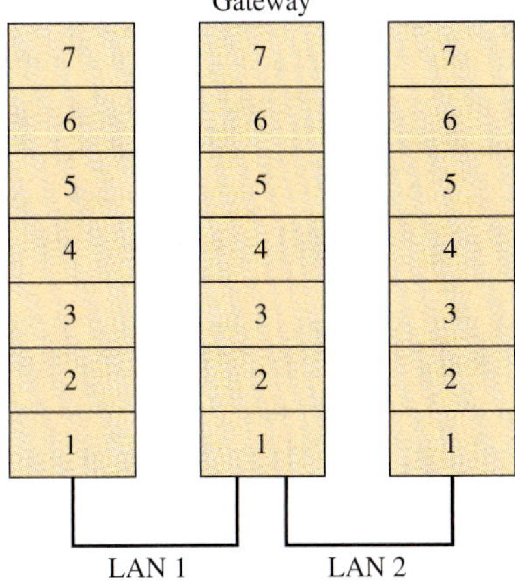

FIGURE 11-26 Gateway connecting two LANs.

The Cisco 12000 series gigabit switch router. (Courtesy of Cisco Systems. Reprinted with permission.)

11-7 INTERNET

The most exciting network development in recent times is the Internet (commonly referred to simply as the Net). It allows for the interconnection of LANs and individuals. In a few short years it has become a pervasive force in our society. A system originally called ARPANET was developed in the early 1970s to link academic institutions and their researchers involved with military defense activities. It has evolved into the Internet—a network that has worldwide broadcasting capability, a mechanism for information dissemination, and a medium for collaboration and interaction between individuals and their computers without regard to geographic location. The Internet is a packet-switched, global network system that consists of millions of local area networks and computers (hosts).

A definition of the term *Internet* has been provided by the Federal Networking Council (FNC): "Internet refers to the global information system that—(i) is logically linked together by a globally unique address space based on the Internet Protocol (IP) or its subsequent extensions/follow-ons; (ii) is able to support communications using the Transmission Control Protocol/Internet Protocol (TCP/IP) suite or its subsequent extensions/follow-ons, and/or other IP-compatible protocols; and (iii) provides, uses or makes accessible, either publicly or privately, high level services layered on the communications and related infrastructure described herein."

The Internet was designed before LANs existed but has accommodated this network technology. It was envisioned as supporting a range of functions including file sharing, remote login, and resource sharing/collaboration. Over the years it has also enabled electronic mail and the World Wide Web (WWW). The Internet has not finished evolving as evidenced by the Internet telephone that is coming on-line, to be followed by Internet television.

In 1992, the WWW software of Tim Lee was released to the public. It is termed a hypertext system—one that gives the ability to link documents together. The major breakthrough came in June 1993, with the release of the Mosaic browser for Windows, which dramatically improved its look and interface. It was created by the National Center for Supercomputing Applications. The initial versions of the Mosaic are very similar to the browsers we use today. Web popularity is shown by the fact that it became the dominant Internet use one year later in 1994.

The Web is a method (and system) that provides users the opportunity to create and disseminate information on a global basis. It unleashes the power of individual creativity and allows a cross-connection of all people of the world. The growth of the Web has been rapid and is now becoming very commercial as product marketing and pay-for-access sites become common. There is a risk that it will lose its diversity and democratic nature with increased commercialism. One of the great appeals of the Internet is that ordinary people with limited resources can publish and/or gather material just as do large corporations and organizations. If this capability is lost, the Internet may regress to just another passive medium like television.

IP (INTERNET PROTOCOL) ADDRESSING

Moving data across the country and even moving data through routers in LANs requires a better addressing scheme than the MAC address. The MAC address provides the physical address for the network interface card, but where is it located? On what LAN, or which building, or what city, or even what country? IP addressing provides a solution to worldwide addressing through incorporating a unique address that tells on which network the computer is located.

IP network numbers are assigned by **IANA** (Internet Assigned Numbers Authority), an agency that assigns IP addresses to computer networks and makes sure no two different networks are assigned the same IP address. IP addresses are issued based on the class of the network. Examples of the three classes of IP networks are provided in Table 11-6.

IANA
the agency that assigns the computer network IP addresses

Table 11-6 The Three Classes of IP Networks

Class	Description	Example IP Numbers
Class A	Governments, very large networks	44.*.*.*
Class B	Midsize companies, universities, etc.	128.123.*.*
Class C	Small networks	192.168.1.*

Sample numbers of network addresses are also shown in Table 11-6. The numbers indicate the network portion of the IP address for each class. This provides sufficient information for routing the data to the appropriate destination network. The destination network then uses the remaining information (the * portion) to direct the packet to the destination computer. The * portion of the address is typically assigned by the local network system administrator or is dynamically assigned when users need access outside their local networks. For example, your Internet service provider (ISP) dynamically assigns an IP address to your computer when you log on to the Internet.

11-8 IP Telephony

IP telephony (voice-over IP) is the telephone system for computer networks. It incorporates technologies comparable to PBX (private branch exchange) telephone systems while maintaining the flexibility of computer networks. In fact, the IP telephone system is called NBX (the abbreviation for "network branch exchange") by 3COM Corporation. The NBX easily enables the user to incorporate telephone systems within their facility. Features that you would expect of a traditional PBX are provided with the NBX, including internal calls, message forwarding, speed dialing, voice mail, and access to the local PSTN (public switch telephone network). The access to the PSTN is provided through traditional telephone-line connections. Long-distance calls can be placed through the access to the PSTN or through Internet IP delivery via the NBX. Installation and management of the IP telephone system can typically be done by the computer networking staff. The cable requirements and terminations are the same as CAT5e/6 and RJ-45 cabling. 3COM's NBX 100 uses an Internet browser, which has been set up for a direct connection to access its internal management features.

IP Telephony (Voice-Over IP)
the telephone system for computer networks

Each telephone in the NBX system is assigned an internal extension number, much in the same way as in a PBX phone system. In addition to the phone number, each telephone has its own MAC address, which is used to deliver the voice data traffic within the LAN. The telephone can also be assigned an IP number and a gateway address so that phone calls can be routed outside the immediate LAN for long-distance calling over the Internet or over leased corporate computer data links.

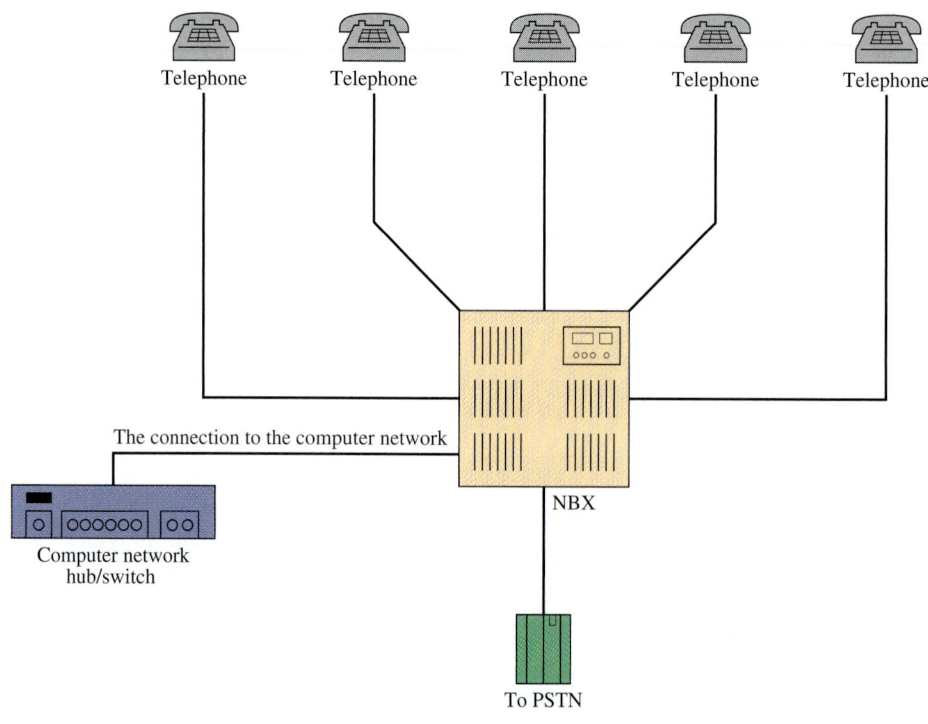

FIGURE 11-27 An IP telephone network.

Quality of Service (QoS)
the expected quality of the service

The phones in the NBX system are connected in a star topology to a central switch so that the **quality of service (QoS)** for voice traffic is not affected by computer network usage. Traditional telephone systems are very reliable, and the public expects a high quality of service. IP telephone systems must adhere to this implied measure of quality for the public to accept them as a viable alternative to the traditional PSTN. The computer network can experience very heavy data traffic and the NBX system can experience heavy voice traffic without any loss in system performance because the network switch provides the direct connection between the parties participating in the telephone call.

An example of an IP telephone LAN is provided in Fig. 11-27. This network looks very similar to the star topology network (see Fig. 11-16). Integration into the computer network LAN is provided through a connection to a switch or hub.

11-9 Interfacing the Networks

Sections 11-1 to 11-8 introduced the basics of both telephone and computer networks. Not too long ago telephone and computer networks were considered totally separate technologies. Today, both sides of these networks are developing technologies and applications that will help integrate their technologies and capabilities into a total information network framework. This section addresses the current issues of interfacing the networks and discusses the latest in modem technologies and standards (including V.92), cable modems, traditional data connections such as ISDN, and the latest in data connection (xDSL). Also, a summary is provided of the new protocols being developed that help facilitate the integration of the networks.

Modem Technologies

The voice frequency channels of the public switched telephone network are used extensively for the transmission of digital data. To use these channels, the data must be converted to an analog form that can be sent over the bandwidth-limited line. In voice-grade telephone lines, the transformers, carrier systems, and loading considerations attenuate all signals below 300 Hz and above 3400 Hz. While the bandwidth from 300 to 3400 Hz is suitable for voice transmission, it is not appropriate for digital data transmission because the digital pulse contains harmonics well outside this range. To transmit data via a phone requires the conversion of a signal totally within the 300- to 3400-Hz range. This conversion is provided by a modem.

There are currently two major standards for providing high-speed modem connections to an analog telephone line. These standards are **V.44 (V.34),** which is totally analog and which provides data rates up to 33.6 Kbps, and **V.92 (V.90),** which is a combination of digital and analog and provides data rates up to 56 Kbps. The V.92 (V.90) modem connection requires a V.92 (V.90) compatible modem and a service provider who has a digital line service back to the phone company. The data transfer with V.92 (V.90) is called **asymmetric operation** because the data-rate connection to the service provider is typically at V.44 (V.34) speeds, whereas the data rate connection from the service provider is at the V.92 (V.90) speed. The difference in the data rates in asymmetric operation is due to the noise introduced by the analog-to-digital conversion process. The modem link from your computer to the PSTN (your telephone connection) is typically analog. This analog signal is then converted to digital at the phone company's central office. If the Internet service provider (ISP) has a digital connection to the phone company, then an analog-to-digital conversion is not required. The signal from the ISP through the phone company is converted back to analog for reception by your modem. However, the digital-to-analog process does not typically introduce enough noise to affect the data rate.

V.44 (V.34)
the standard for an all-analog modem connection with a maximum data rate of up to 34 Kbps

V.92 (V.90)
the standard for a combination analog and digital modem connection with a maximum data rate up to 56 Kbps

Asymmetric Operation
a term used to describe the modem connection when the data-transfer rates to and from the service provider differ

Cable Modems

Cable modems provide an alternative way of accessing a service provider. Cable modems capitalize on their high-bandwidth network to deliver high-speed, two-way data. Data rates range from 128 kbps to 10 Mbps upstream (computer to the cable-head end) and 10 to 30 Mbps downstream (cable-head end back to the computer). The cable modem connections can also be one-way when the television service implemented on the cable system precludes two-way communications. In this case, the subscriber connects to the service provider via the traditional telephone and receives the return data via the cable modem. The data service does not impair the delivery of the cable television programming. Currently the cable systems are using Ethernet protocol for transferring the data over the network. Many subscribers use the same upstream connection. This leads to potential collision problems, so a technique called **ranging** is used, where each cable modem determines the amount of time needed for its data to travel to the cable-head end. This technique minimizes collision rate, keeping it less than 25 percent.

Cable Modems
modems that use the high bandwidth of a cable television system to deliver high-speed data to and from the service provider

Ranging
a technique used by the modems to determine the time it takes for data to travel to the cable-head end

The ISDN

The integrated services digital network (ISDN) is an established data communications link for both voice and data using a set of standardized interfaces. For business, the primary attractions will be increased capability, flexibility, and decreased cost. If one type of service—say, facsimile—is required in the morning and another in the afternoon—perhaps teleconferencing or computer links—it can easily shift back and forth. At present, the hookup for a given service might take a substantial amount of time to complete. With ISDN, new capacities will be available just by asking for them through a terminal.

The ISDN contains four major interface points as shown in Fig. 11-28. The R, S, T, and U interface partitions allow for a variety of equipment to be connected into the system. Type 1, or TE1, equipment includes digital telephones and terminals that comply with ISDN recommendations. Type 2, or TE2, gear is not compatible with ISDN specifications. It needs a terminal adapter to change the data to the ISDN's 64-kbps *B* channel rate. The TE2 equipment interfaces the network via the *R* reference point.

The ISDN standards also define two network termination (NT) points. NT_1 represents the telephone companies' network termination as viewed by the customer. NT_2 represents the termination of items such as local area networks (see Sec. 11-4) and private branch exchanges (PBXs).

The customer ties in with the ISDN's NT_1 point with the *S* interface. If an NT_2 termination also exists, an additional *T* reference point linking both NT_2 and NT_1 terminations will act as an interface. Otherwise, the *S* and *T* reference points are identical. The ITU-T recommendations call for both *S* and *T* reference points to be four-wire synchronous interfaces that operate at a basic access rate of 192 kbps. They are called the local loop. Reference point *U* links NT_1 points on either side of a pair of users over a two-wire 192-kbps span. The two termination points are essentially the central office switches.

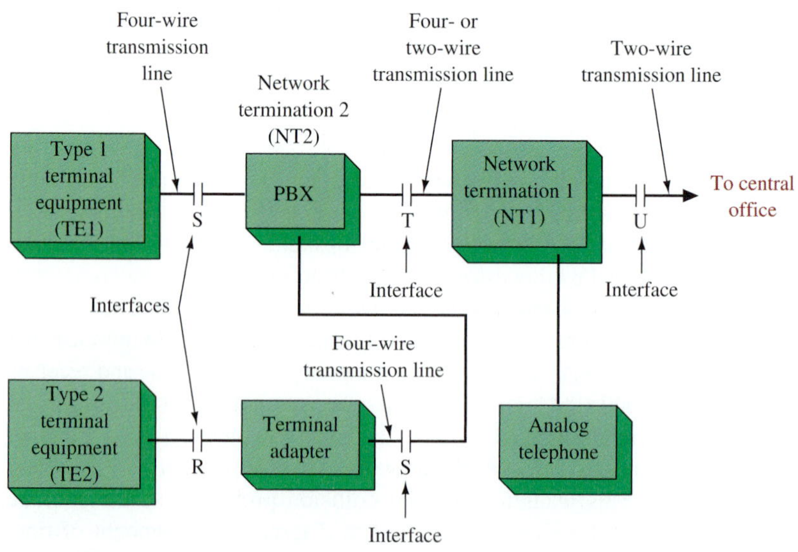

FIGURE 11-28 ISDN setup illustration of R, S, T, and U interfaces.

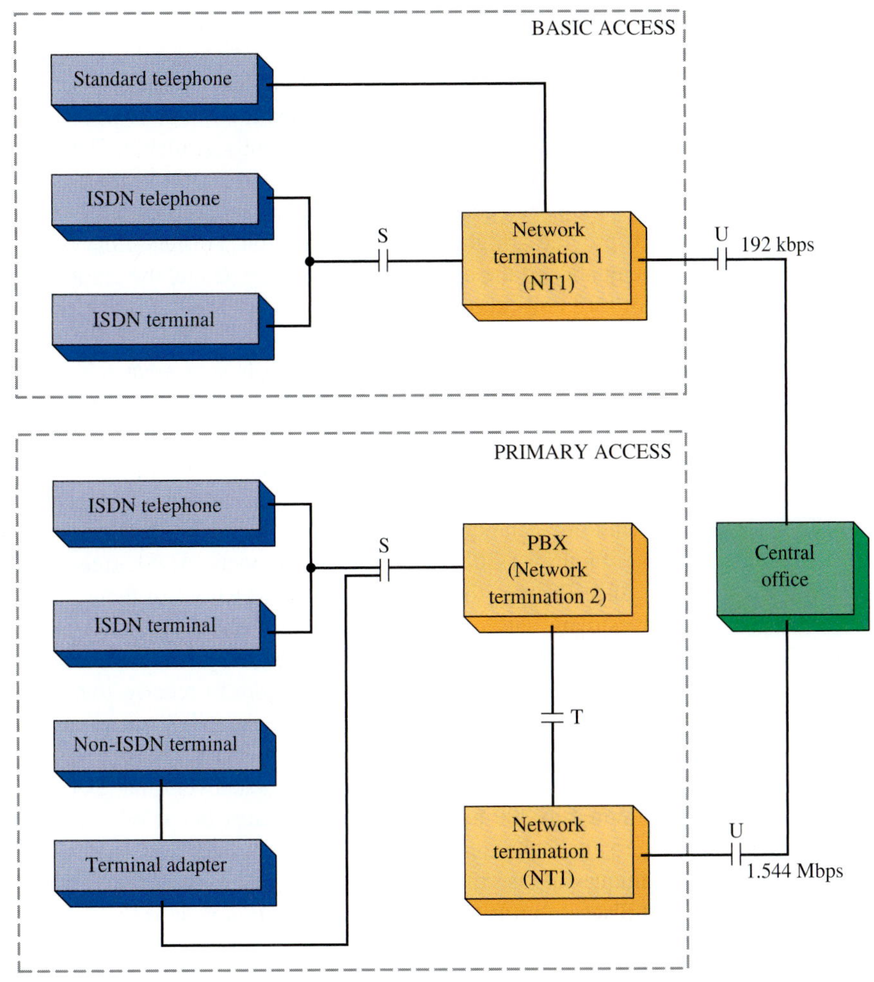

FIGURE 11-29 Basic and primary access ISDN system.

The ISDN specifications spell out a basic system as two *B* channels and one *D* channel (2*B* + *D*). The two *B* channels operate at 64 kbps each, while the *D* channel is at 16 kbps, for a total of 144 kbps. The 48-kbps difference between the basic 192-kbps access rate and the 2*B* + *D* rate of 144 kbps is mainly for containment of protocol signaling. The 2*B* + *D* channels are what the *S* and *T* four-wire reference points see. The *B* channels carry voice and data while the *D* channel handles signaling, low-rate packet data, and low-speed telemetry transmissions.

The ITU-T defines two types of communication channels from the ISDN central office to the user. They are shown in Fig. 11-29. The basic access service is the 192-kbps channel already discussed and serves small installations. The primary access channel has a total overall data rate of 1.544 Mbps and serves installations with large data rates. This channel contains 23 64-kbps *B* channels plus a 64-kbps *D* channel. From any angle, the potential for ISDN is enormous. The capabilities of worldwide communications are now taking a quantum leap forward.

xDSL Modems

xDSL
a generic representation of the various DSL technologies that are available

DSL
digital subscriber line

The **xDSL** modem is considered to be the next generation of high-speed Internet access technology. **DSL** stands for digital subscriber line, and the "x" generically represents the various types of DSL technologies that are currently available. The DSL technology uses the existing copper telephone lines for carrying the data. Copper telephone lines can carry high-speed data over limited distances, and the DSL technologies use this trait to provide a high-data-rate connection. However, the actual data rate depends on the quality of the copper cable, the wire gauge, the amount of crosstalk, the presence of load coils, the bridge taps, and the distance of the connection from the phone service's central office.

DSL is the base technology in the xDSL services. It is somewhat related to the ISDN service; however, the DSL technologies provide a significant increase in bandwidth and DSL is a point-to-point technology. ISDN is a switch technology and can experience traffic congestion at the phone service's central office. The available xDSL services and their projected data rates are provided in Table 11-7.

ADSL (Asymmetric DSL)
a service providing up to 1.544 Mbps from the user to the service provider and up to 8 Mbps back to the user from the service provider

DSL services use filtering techniques to enable the transport of data and voice traffic on the same cable. Figure 11-30 shows an example of the ADSL frequency spectrum. Note that the voice channel, the upstream data connection (from the home computer), and the downstream data connection (from the service provider) each occupy their own portion of the frequency spectrum. **ADSL (asymmetric DSL)** is based on the assumption that the user needs more bandwidth to receive transmissions (downstream link) than for transmission (upstream link). ADSL can provide data rates up to 1.544 Mbps upstream and 1.5 to 8 Mbps downstream.

Discrete Multitone (DMT)
an industry standard data-modulation technique used by ADSL that uses multiple subchannel frequencies to carry the data

It was stated earlier in the section on interfacing to the network that a copper telephone line is band-limited to 300 to 3400 Hz. This is true, but xDSL services use special signal-processing techniques for recovering the received data and a unique modulation technique for inserting the data on the line. For ADSL, a multi-carrier technique called **discrete multitone (DMT)** modulation is used to carry the

Table 11-7 xDSL Services and Their Projected Data Rates

Technology	Data Rate	Distance Limitation
ADSL	1.5–8 Mbps downstream Up to 1.544 Mbps upstream	18,000 ft
IDSL	Up to 144 kbps full-duplex	18,000 ft
HDSL	1.544 Mbps full-duplex	12,000 to 15,000 ft
SDSL	1.544 Mbps full-duplex	10,000 ft
VDSL	13–52 Mbps downstream 1.5–2.3 Mbps upstream	1,000 to 4,500 ft

Upstream: computer user to the service provider
Downstream: from the service provider back to the computer user
ADSL: Asymmetric digital subscriber line
IDSL: ISDN digital subscriber line
HDSL: High-bit-rate digital subscriber line
SDSL: Single-line digital subscriber line
VDSL: Very high bit rate digital subscriber line

Source: "xDSL Local Loop Access Technology, Delivering Broadband over Copper Wires," ©3COM Technical Paper. Reproduced with permission of 3COM Corporation.

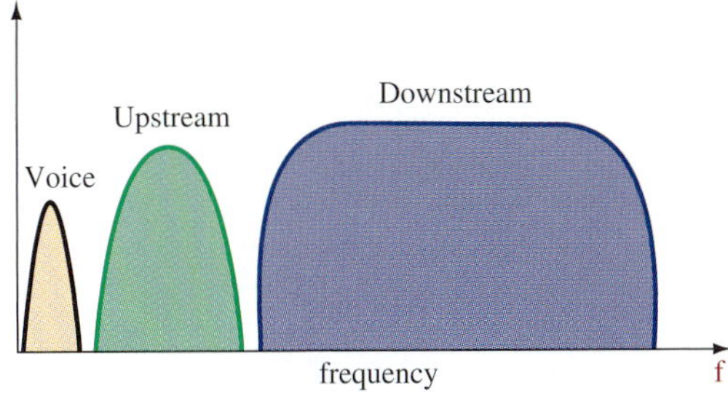

FIGURE 11-30 The ADSL frequency spectrum.

data over the copper lines. It is well understood that the performance of copper lines can vary from site to site. DMT uses a technique to optimize the performance of each site's copper telephone lines. The DMT modem can use up to 256 subchannel frequencies for carrying the data over the copper telephone lines. A test is initiated at start-up to determine which of the 256 subchannel frequencies should be used to carry the data. The system then selects the best subchannels and splits the data over the available subchannels for transmission.

ADSL is receiving the most attention because its data-modulation technique, DMT, is already an industry standard. An example of an xDSL network is shown in Fig. 11-31. The ADSL system requires an ADSL modem, which must be compatible with the service provider. Additionally, a POTS splitter is required to separate the voice and data transmission.

WAP Protocol

The *wireless application protocol* (WAP) is a world standard that has been developed to bridge the gaps between mobile communications, the Internet, and corporate intranets. WAP was developed to address the issues and standardize the solutions for providing web-based services and wireless Internet access. The WAP standards address key delivery problems with wireless data networks. These include limited bandwidth, **latency,** connection stability, and availability. Latency is the time delay from the request for information until a response is obtained.

The WAP specification includes several key specifications:

Wireless markup language (WML), which optimizes hypertext methodologies for a wireless environment, including graphics, and **WMLScript,** which is the WML comparable version of Javascript

A specification for a **microbrowser** that is web-browser adapted for the wireless environment

The framework for wireless telephony applications (WTA) within the WML environment.

Some features of WAP include:

A wireless transaction protocol (WTP), used to manage the data transfer in the WAP protocol. The WTP is analogous to the TCP layer in computer networks. WTP provides the minimum information required to handle each request/response transaction.

Latency
time delay from the request for information until a response is obtained

Wireless Markup Language (WML)
the hypertext language for the wireless environment

WMLScript
the WML-comparable version of Javascript

Microbrowser
analogous to a web browser that has been adapted for the wireless environment

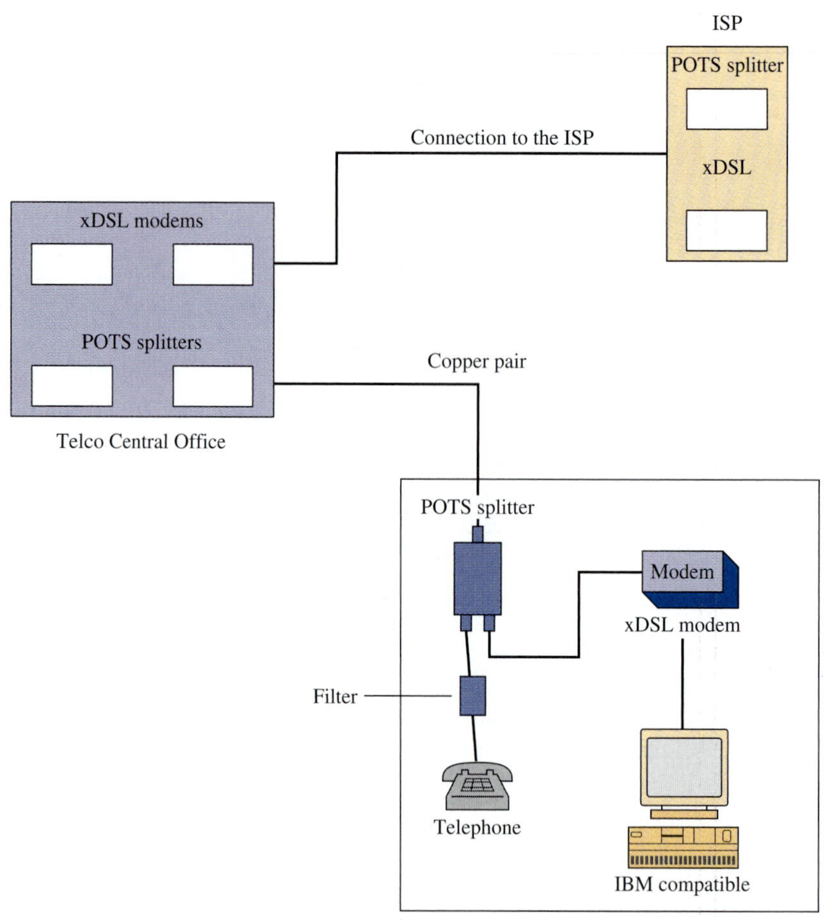

FIGURE 11-31 An xDSL connection to an ISP.

A wireless transport layer security (WTLS), which provides for a more secure wireless connection. This layer is comparable to the computer networking industry standard transport layer security (TLS) protocol, which used to be called the secure socket layer.

WAP provides enormous possibilities for growth in the wireless environment. The standard addresses the key limiting issues of wireless phone communications but provides the adaptability for supporting existing information sources such as the World Wide Web and provides a methodology for incorporating a multitude of new uses. The majority of the world's mobile telephone companies now support WAP.

11-10 TROUBLESHOOTING

Local area networks (LANs) are finding their way into every kind of business at an ever-increasing rate. From small offices to very large government agencies, LANs are becoming an indispensable part of business communications. Computer data and audio and video information are shared on LANs every day. In this section, you will be introduced to a typical LAN configuration and to some common LAN problems. Opportunities abound in LAN technology for those who are willing to prepare with specialized training. LAN seminars, community college classes, and hands-on training will help prepare you for this technology.

After completing this section you should be able to

- Define near-end crosstalk interference
- Describe two common problems when using twisted pair
- Name the two types of modular eight connectors and their proper use

Troubleshooting a LAN

The most common maintenance situation for larger LAN installations is to set up a help desk. LAN users who experience problems call the help desk. Usually a technician is dispatched if the problem can't be resolved over the phone. Let's take a look at some of the problems the technician may encounter when dispatched.

Some preliminary checks should be made first. Ensure that the workstation is plugged into electrical power. Is it turned on? Is the CRT brightness turned down? These are obvious things that are easily overlooked and warrant a check. The network interface card should be checked for proper installation in the PC. Check the hub-to-workstation connection. Is the user's account set up properly on the server? Sometimes passwords and accounts get deleted by accident. Check the user's LAN connection software for proper boot-up. This software can become corrupted and may need to be reinstalled on the workstation.

The following two commands are useful for troubleshooting computer networks. The first is ping, which provides a way to verify the operation of the network. The command structure is as follows:

```
Usage: ping[-t][-a][-n count][-l size][-f][-i TTL][-v TOS]
       [-r count][-s count][[-j host-list]|[-k host-list]
       [-w timeout] destination-list
```

Options

```
-t            Ping the specified host until stopped
              To see statistics and continue, type Control-
              Break
              To stop, type Control-C
-a            Resolve addresses to host-names
-n count      Number of echo requests to send
-l size       Send buffer size
-f            Set Don't Fragment flag in packet
-i TTL        Time To Live
-v TOS        Type Of Service
-r count      Record route for count hops
-s count      Timestamp for count hops
-j host-list  Loose source route along host-list
-k host-list  Strict source route along host-list
-w timeout    Timeout in milliseconds to wait for each reply
```

The following is an example of pinging a network site using its IP address. You can try this if your computer is connected to the Internet. The site is a computer called invincible.nmsu.edu and has an IP address of 128.123.24.123. This computer has been set up for outside users to experiment with, so feel free to ping this site.

```
Pinging 128.123.24.123 with 32 bytes of data:

      Reply from 128.123.24.123: bytes=32 time=3ms TTL=253
      Reply from 128.123.24.123: bytes=32 time=2ms TTL=253
      Reply from 128.123.24.123: bytes=32 time=2ms TTL=253
      Reply from 128.123.24.123: bytes=32 time=3ms TTL=253

Ping statistics for 128.123.24.123:
      packets: sent=4, received=4, lost=0 (0% loss)

Approximate round trip times in milliseconds:
      minimum=2 ms, maximum=3 ms, average=2 ms
```

You will get a timed-out message if the site does not respond.

Another command that is useful for troubleshooting networks is tracert, which traces the route of the data through the network

```
      Usage: tracert [-d] [-h maximum_hops][-j host-list] [-w
             timeout]
      target_name
```

Options

```
-d                  Do not resolve addresses to hostnames.
-h maximum_hops     Maximum number of hops to search for target
-j host-list        Loose source route along host-list
-w timeout          Wait timeout milliseconds for each reply
```

The following is an example of tracing the route to 128.123.24.123.

```
Tracing route to pc-ee205b-8.NMSU.Edu [128.123.24.123] over a
maximum of 30 hops:

1 1 ms  1 ms 1 ms jett-gate-e2.NMSU.Edu [128.123.83.1]
2 3 ms  2 ms 2 ms r101-2.NMSU.Edu [128.123.101.2]
3 2 ms  2 ms 3 ms pc-ee205b-8.NMSU.Edu [128.123.24.123]
```

The trace is complete. The tracert command provides the path the data traveled.

Troubleshooting Unshielded Twisted-Pair Networks Twisted-pair networks represent a different challenge to the troubleshooter than coaxial cable. Two common problems that occur with unshielded twisted-pair wiring are that the pairs become crossed or split up. Both conditions produce data signal degeneration. Near-end crosstalk (NEXT) is generated from split pairs. NEXT stems from interference between the twisted pairs. Let's use the example of a signal being transmitted from the workstation to the hub. The signal is smallest (maximum attenuation) at the hub. A transmitted signal originating at the hub—a strong signal—will feed over into the attenuated weak signal. Preventing crossed and split pairs is the best insurance against NEXT. Crossed pairs are not difficult to find but usually require a certain amount of wire tracing and continuity checking. Split pairs are more difficult to find, and special test instruments should be used to track down the splits. A LAN cable meter has several specialized test functions, and the miswired feature is one of them.

Another common problem that happens with unshielded twisted-pair wiring is that the wrong kind of connector is used. Stranded copper conductors need a piercing-type modular eight connector, and solid core conductors use a connector that straddles the wire. Both types of connectors look alike, so they can easily be mixed up and often are. When the wrong connector is used, the result is an open or an intermittent connection. Use care in replacing connectors. Keep these connectors separate in clearly labeled bins.

Some Cabling Tips With a little extra precaution, many LAN problems can be eliminated. Be sure to keep wiring links short without stretching the wire tight. Twisted pair should never be placed near ac power lines or other noise sources. Install your cable base carefully. Wiring runs should be dry. Moisture causes corrosion over a period of time. Use only good-quality connectors. Never untwist more twists than necessary when making twisted-pair connections. The rule of thumb is untwist a maximum of $\frac{1}{2}$ in. Finally, keep good-quality wiring diagrams of the network installation.

11-11 Troubleshooting with Electronics Workbench™ Multisim

This exercise introduces the techniques for making audio-signal-level and distortion measurements using Electronics Workbench™ Multisim simulations. Obtaining signal-level measurements and measuring signal-path performance are common maintenance, installation, and troubleshooting practices in all areas of network communications. Communication networks require that proper signal levels are maintained to ensure minimum line distortion and crosstalk. This section examines the techniques for making dB (decibel) and THD (total harmonic distortion) measurements.

To begin the execise, open the file **Fig11-32.ms7 (.msm)** on your Electronics Workbench™ Multisim CD. The circuit is shown in Fig. 11-32. This circuit contains an ac signal source, a 600-Ω load, and two multimeters. Start the simulation and double-click on both multimeters. The top multimeter is measuring the dB level, and the bottom multimeter is used to measure the voltage across the load. Recall from Sec. 1-2 that 0.774 V across a 600-Ω load represents 0 dBm. The example provided in Fig. 11-32 shows a 0-dBm measurement. Note the voltage value specified on the ac voltage source. If we are measuring 0 dBm, then why isn't the value 0.774 V? Why is the level from the 0-dBm signal source set to 2.188 V? Careful examination of the circuit shows that the output impedance of the signal generator is 600 Ω. The load resistance is also 600 Ω, and the combination of the two resistors forms a voltage divider of two equal-value resistors; therefore, only half of the original signal generator voltage ($\approx$ 1.094 V) appears across the load. This still is not the expected value of 0.774 V. Why the difference? Recall that power measurements, including dB measurements, require that the ac signal be expressed in terms of its rms value. If you multiply 1.084 V by 0.707, then you will obtain the expected 0.774 V measured across the 600-Ω load, and this explains how the -0.017-($\approx$0) dB value is obtained.

Total harmonic distortion (THD) measurements provide a measure of distortion that takes all significant harmonics into account. Electronics Workbench™ Multisim provides an instrument that measures THD. Open the file **FigE11-1.ms7 (.msm)** on your EWB CD. Double-click on the distortion analyzer. You will see an image similar to the one shown in Fig. 11-33.

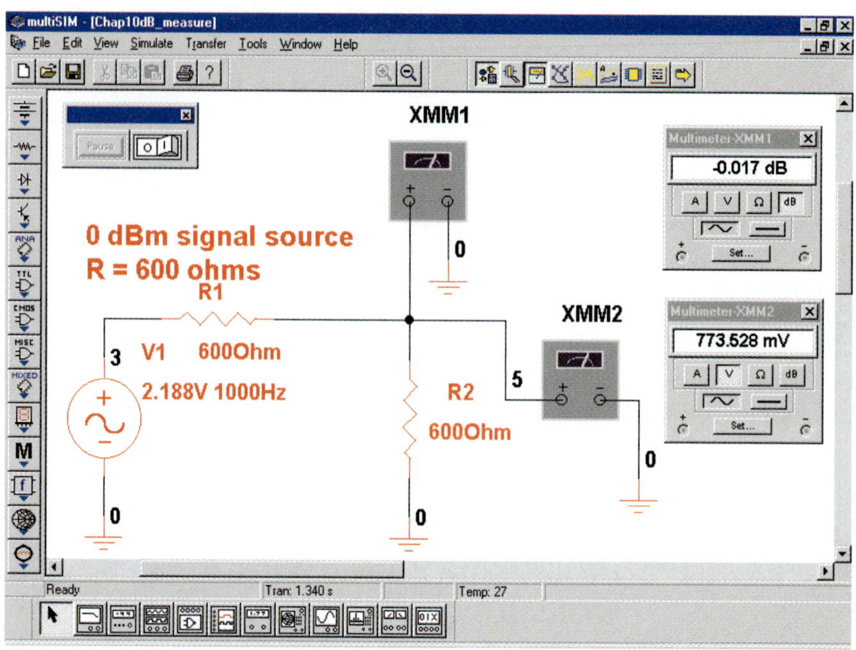

FIGURE 11-32 The Multisim circuit used to demonstrate the dB measurements.

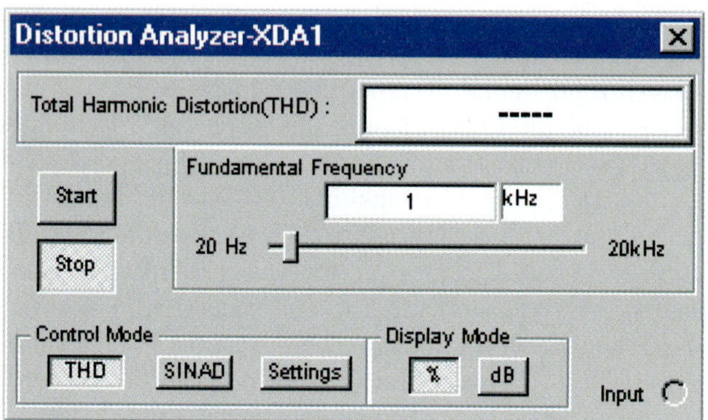

FIGURE 11-33 The display control panel for the Multisim distortion analyzer.

The distortion analyzer provides modes for measuring THD and SINAD (see Sec. 1.4 for a discussion on SINAD). This exercise focuses on the THD measurement. Notice that the control panel allows the user to specify the fundamental frequency of the signal being measured. For example, if you are measuring the THD of a 1-kHz signal, then you would set the fundamental frequency to 1 kHz. The instrument also provides for the range and number of harmonics being measured. Click on the settings button. For this example, the range is 20 to 20 kHz and the number of harmonics being measured is 10.

Double-click on the waveform generator and verify that the frequency is 1 kHz and a sinusoid has been selected. Start the simulation and observe the value of the THD. You should see 0.000 percent. This is possible with an ideal sine wave that produces only the single fundamental 1-kHz frequency, but ideal function generators do not exist. Stop the simulation and change the waveform generator so that it outputs a triangle wave. Leave the frequency at 1 kHz and start the simulation. You should get a THD of 12.049 percent. A high value for THD is expected because a triangle wave contains multiple harmonics of the fundamental frequency.

What happens if a THD measurement is taken on a 2-kHz sine wave but the instrument is set to measure a fundamental frequency of 1 kHz? Change the waveform generator back to a sinusoid but with a frequency of 2 kHz. Start the simulation and obtain a THD measurement. The instrument display will show **–E–,** which indicates a measurement error.

The following Electronics Workbench™ exercises provide additional opportunities to explore the use of dB and distortion measurements to help you become more familiar with a distortion analyzer.

Electronics Workbench™ Exercises

1. Open the file **FigE11-2.ms7 (.msm)** in your EWB CD. This circuit contains a 600-Ω audio source, a T-type attenuator, and a 600-Ω load. Measure the dB level at the output and the input and determine the amount of attenuation being provided by the attenuator.
2. Open the file **FigE11-3.ms7 (.msm)** in your EWB CD. Measure the THD introduced by the simple transmission line (0.044 %).
3. Open the file **FigE11-4.ms7 (.msm)** in your EWB CD. Measure the THD introduced by the sampling process. The simulation will have to run 100 ms of simulation time for the software to calculate all harmonic contributions. (6.33%).

Summary

In Chapter 11 we studied the various networks encountered in digital and analog communications. These include the telephone and cellular networks and local area networks. The major topics you should now understand include:

- the basis of telephone operation, including definitions of tip, ring, trunk, PBX, DTMF, BORSCHT function, and T1 line
- the line quality considerations, including the effects of attenuation and signal delay distortion
- the analysis of cellular and PCS telephone systems, including frequency reuse, cell splitting, and Rayleigh fading
- the description of telephone network structure
- the description of telephone traffic, traffic units, and congestion
- the operation and methodology of implementing local area networks
- an understanding of the various modem technologies available for interfacing computer networks with the telephone networks
- the explanation of the integrated-services digital network (ISDN)
- the description of local area network (LAN) topologies

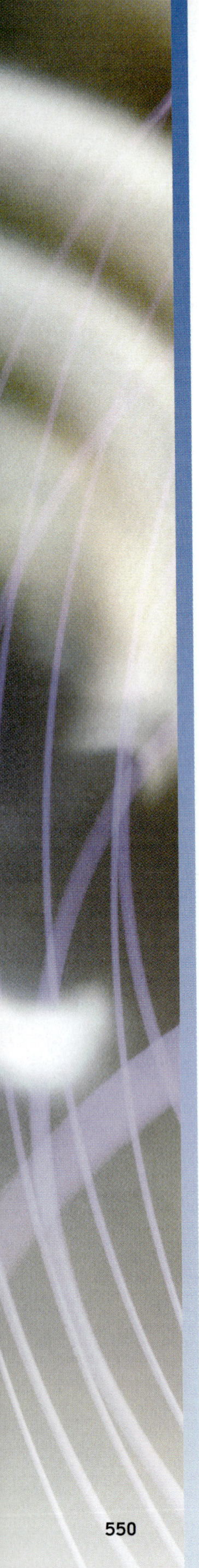

- the description of the OSI seven-layer reference model and definitions of networking devices such as bridges and routers
- the evolution of the Internet, including development of the World Wide Web

Questions and Problems

Section 11-1

1. Describe the basic limitation of using the telephone system in computer communication.

Section 11-2

2. List the three signal levels that telephone circuitry must work with.
3. Describe the sequence of events taking place when a telephone call is initiated through to its completion.
4. Transcribe your phone number into the two possible electrical signals commonly used in the phone system.
5. What is a PBX and what is its function?
6. List the BORSCHT functions.
7. Explain the causes of *attenuation distortion.*
8. Define *loaded cable.*
9. What is a C2 line? List its specification.
10. What are repeaters and when are they used on a T1 line?
11. Define *attenuation distortion.*
12. Define *delay distortion* and explain its causes.
13. Describe the envelope delay specification for a 3002 channel.
14. Why are telephone lines widely used for transmission of digital data? Explain the problems involved with their use.
15. What is telephone traffic? What is busy-hour telephone traffic?
16. What are the two units used to express telephone traffic?
17. What is congestion?
18. What is grade of service? How do traffic engineers interpret or look at grade of service?
19. How is grade of service measured? What does it mean when this figure is very low?
20. What is the significance of continuously observing and measuring traffic?

Section 11-3

21. Redesign the cellular system in Fig. 11-8 so that only six different channel groups (A, B, C, D, E, F) are used instead of the seven shown.
22. Describe the concepts of frequency reuse and cell splitting.
23. Design the split-cell system in Fig. 11-9 so that the minimum number of channel groups are used.
24. Describe the sequence of events when a mobile user makes a call to a land-based phone. Include in your description the handoff once the call has been made.
25. The cellular system in Fig. 11-10 requires splitting of the two cells surrounded by other cells. They need to be split into five cells (from the original

two). Determine the minimum number of channel sets that can serve the original and new systems.

26. Describe Rayleigh fading and explain how to minimize its effects.
27. A cellular system operates at 840 MHz. Calculate the Rayleigh fading rate for a mobile user traveling at 40 mi/h (100 fades per second).
28. A mobile user is transmitting two steps above minimum power. Calculate its power output. (1.11 W)
29. Explain TDMA and CDMA as related to future cellular systems.
30. Describe the personal communications services (PCS).
31. Describe the operation of the NAMPS and DAMPS mobile phone service.
32. Define the concept of dual-band technology in mobile phone communications.
33. Detail the differences between the cellular and PCS spectrums.
34. List the four layers for the PCS air interface.

Section 11-4

35. Provide a general description of a local area network.
36. List the basic topologies available for LANs and explain them.
37. Describe the operation of the Ethernet protocol.
38. Describe the Ethernet frame structure, including a description of MAC level addressing.
39. Define the concept of a broadcast address in an Ethernet frame.

Section 11-5

40. Discuss the layout, interconnection of the devices, and issues in implementing an office LAN.
41. Discuss the layout, interconnection of the devices, and issues in implementing a building LAN.
42. List the common numerics used for LAN cabling.
43. What is the current data rate for the IEEE 802.11 wireless LAN protocol? What protocol is used with wireless communications?

Section 11-6

44. Describe the differences among LANs, MANs, and WANs.
45. Provide a brief description of the functions addressed by the OSI reference model.
46. Explain the functions of bridges and routers.

Section 11-7

47. Describe the evolution of the Internet.
48. Describe the concept of IP addressing.
49. Find four Internet sites that provide technical tutorials on cellular communications.

Section 11-8

50. Discuss the operation of IP telephony.
51. Discuss the issue of quality of service (QoS).

Section 11-9

52. Describe the operation of ADSL.
53. Discuss the implementation of cable modems for interfacing the networks.

54. Describe how V.92 can achieve such high data rates over the analog phone lines.
55. Explain the objective of ISDN and briefly explain its organization with the help of Figs. 11-28 and 11-29.
56. List five xDSL services and indicate their projected data rates.
57. Describe the use of DMT (discrete multitone) operation in ADSL systems.
58. Discuss the key issues of the WAP protocol.

Questions for Critical Thinking

59. You are at a company where phone lines are being used for signal transmission without delay equalization. Predict the results of unequal delays to the different frequency components of a received signal and justify the need for delay equalization.
60. You hear someone refer to the "handshaking protocol." Is this an accurate use of terms? Why or why not?
61. Discuss the issues of connecting a building network (LAN) to a T1 connection that has been brought into the building. What information do you need to know to complete the job?
62. You are asked to install a wireless LAN in a building. What are the issues that should be examined before you start this task?

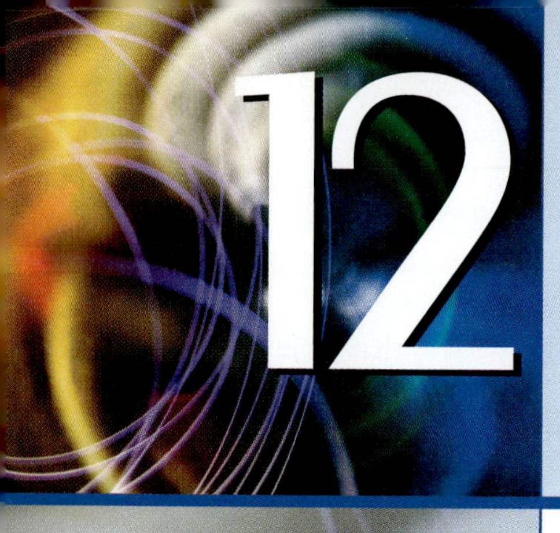

Transmission Lines

Chapter Outline

Category 6 twisted-pair cable. (Courtesy of Blackbox Corp. Reprinted with permission.)

Objectives

- Describe the operational characteristics of twisted-pair cable and its testing considerations
- Describe the physical characteristics of standard transmission lines and calculate Z_0
- Calculate the velocity of propagation and the delay factor
- Analyze wave propagation and reflection for various line configurations
- Describe how standing waves are produced and calculate the standing wave ratio
- Use the Smith chart to find input impedance and match loads to a line with matching sections and single-stub tuners
- Explain the use of line sections to simulate discrete circuitry
- Troubleshoot the location of a line break using TDR concepts

Key Terms

transmission line
CAT6/5e
RJ-45
attenuation
near-end crosstalk (NEXT)
crosstalk
ACR
delay skew
power-sum NEXT testing (PSNEXT)
return loss
unbalanced line
balanced line
common mode rejection
baluns
characteristic impedance
surge impedance
skin effect
velocity of propagation
delay line
velocity constant
velocity factor
wavelength
nonresonant line
traveling waves
resonant line
reflection
standing wave
voltage standing wave ratio
standing wave ratio
flat line
quarter-wavelength matching transformer
electrical length
Smith chart
normalizing
single-stub tuner
double-stub tuner
slotted line

12-1 Introduction

In previous chapters we have been concerned with the generation and reception of communications signals. In Chapters 12 to 15 we shall learn of the methods of getting these signals from transmitter to receiver. Transmission may take place via transmission lines, antennas, waveguides, or optical fibers. Sometimes a combination such as transmission line from transmitter to its antenna, to receiving antenna, to transmission line, and to receiver is used. A **transmission line** may be defined as the conductive connections between system elements that carry signal power. You may be thinking that if the wire connection between two points is a transmission line, why is a whole chapter of study required? It turns out that at very high frequencies, even simple wire connections start behaving in a peculiar fashion. What appears to be a short circuit may no longer be one. Or energy sent down the wire is reflected back. These phenomena and others form the basis of this chapter. A good understanding of the material presented in this chapter is a necessary prerequisite for the antenna and waveguide chapters to follow.

Transmission Line
the conductive connections between system elements that carry signal power

12-2 Types of Transmission Lines

Two-Wire Open Line

One type of parallel line is the two-wire open line illustrated in Fig. 12-1. This line consists of two wires that are generally spaced from $\frac{1}{4}$ to 6 in. apart. It is sometimes used as a transmission line between antenna and transmitter or antenna and receiver. An advantage of this type of line is its simple construction. Another type of parallel line is the twin lead or two-wire ribbon type. This line is illustrated in Fig. 12-2. This line is essentially the same as the two-wire open line, except that uniform spacing is assured by embedding the two wires in a low-loss dielectric, usually polyethylene. The dielectric space between conductors is partly air and partly polyethylene.

Twisted Pair

The twisted-pair transmission line is illustrated in Fig. 12-3. As the name implies, the line consists of two insulated wires twisted to form a flexible line without the use of spacers. It is not used for high frequencies because of the high losses that occur in the rubber insulation. When the line is wet, the losses increase greatly. Local area networks (LANs) are often wired using unshielded twisted pair.

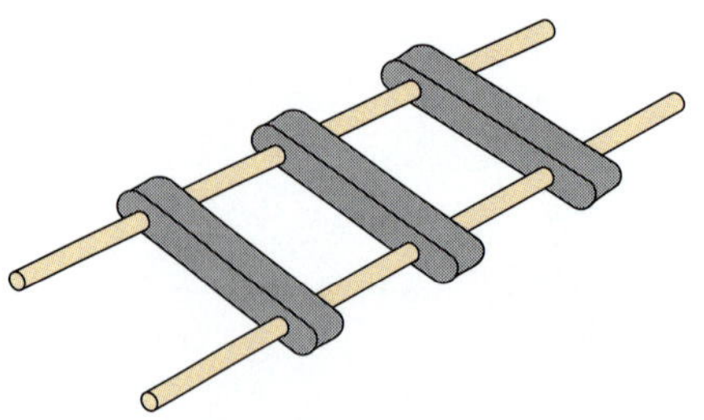

FIGURE 12-1 Parallel two-wire line.

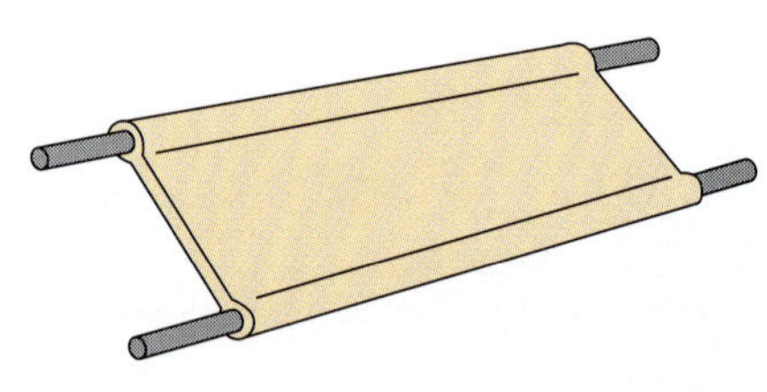

FIGURE 12-2 Two-wire ribbon-type lines.

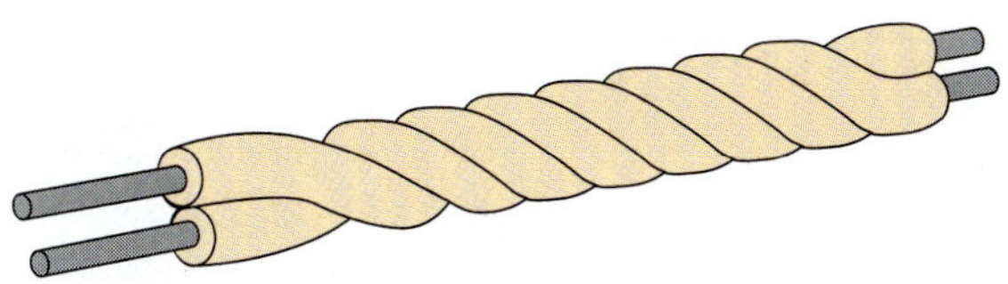

FIGURE 12-3 Twisted pair.

Unshielded Twisted Pair (UTP)

Unshielded twisted-pair (UTP) cable is playing an increasingly important role in computer networking. The most common twisted-pair cable standard used for computer networking is UTP category 6 **(CAT6)** and 5e **(CAT5e),** which are cable-tested to provide the transmission of data rates up to 1000 Mbps for a maximum length of 100 m. CAT6/5e cable consists of four color-coded pairs of 22- or 24-gauge wires terminated with an **RJ-45** connector. The precise manner in which the twist of the cables is maintained, even at the terminations, provides a significant increase in signal-transmission performance. CAT5e standards allow 0.5 in. of untwisted conductors at the termination; CAT6 recommends about $3/8^{11}$ untwisted. The balanced operation of the two wires per pair help to maintain the required level of performance in terms of crosstalk and noise rejection.

CAT6/5e
category 5e computer networking cable capable of handling a 1000 MHz bandwidth up to a length of 100 m

RJ-45
the four-pair termination commonly used for terminating CAT6/5e cable

The need for increased data rates is pushing the technology of twisted-pair cable to even greater performance requirements. The CAT6/5e designation is simply a minimum performance measurement of the cables. The cable must satisfy minimum **attenuation** loss and **near-end crosstalk (NEXT)** for a minimum frequency of 100 MHz. Attenuation loss defines the amount of loss in signal strength as it propagates down the wire. When current travels in a wire, an electromagnetic field is created. This field can induce a voltage in adjacent wires resulting in crosstalk.

Attenuation
the amount of loss in the signal strength as it propagates down a wire

Near-End Crosstalk (NEXT)
a measure of the level of crosstalk or signal coupling within the cable, with a high NEXT (dB) value being desirable

Crosstalk
unwanted coupling caused by overlapping electric and magnetic fields

Crosstalk is what you occasionally hear on the telephone when you can faintly hear another conversation. NEXT is a measure of the level of crosstalk, or signal coupling, within the cable. The measurement is called near-end testing because the receiver is more likely to pick up the crosstalk from the transmit to the receiver wire pairs at the ends. The transmit-signal levels at each end are strong, and the cable is more susceptible to crosstalk at this point. Additionally, the receive-signal levels have been attenuated due to normal cable path loss and are significantly weaker than the transmit signal. A high NEXT (dB) value is desirable. Near-end crosstalk is graphically depicted in Fig. 12-4.

Manufacturers combine the two measurements of attenuation and crosstalk on data sheets and list this combined measurement as the attenuation-to-crosstalk ratio

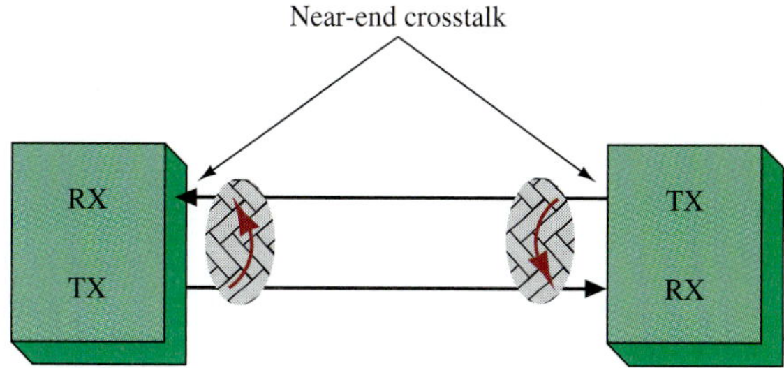

FIGURE 12-4 A graphical illustration of near-end crosstalk.

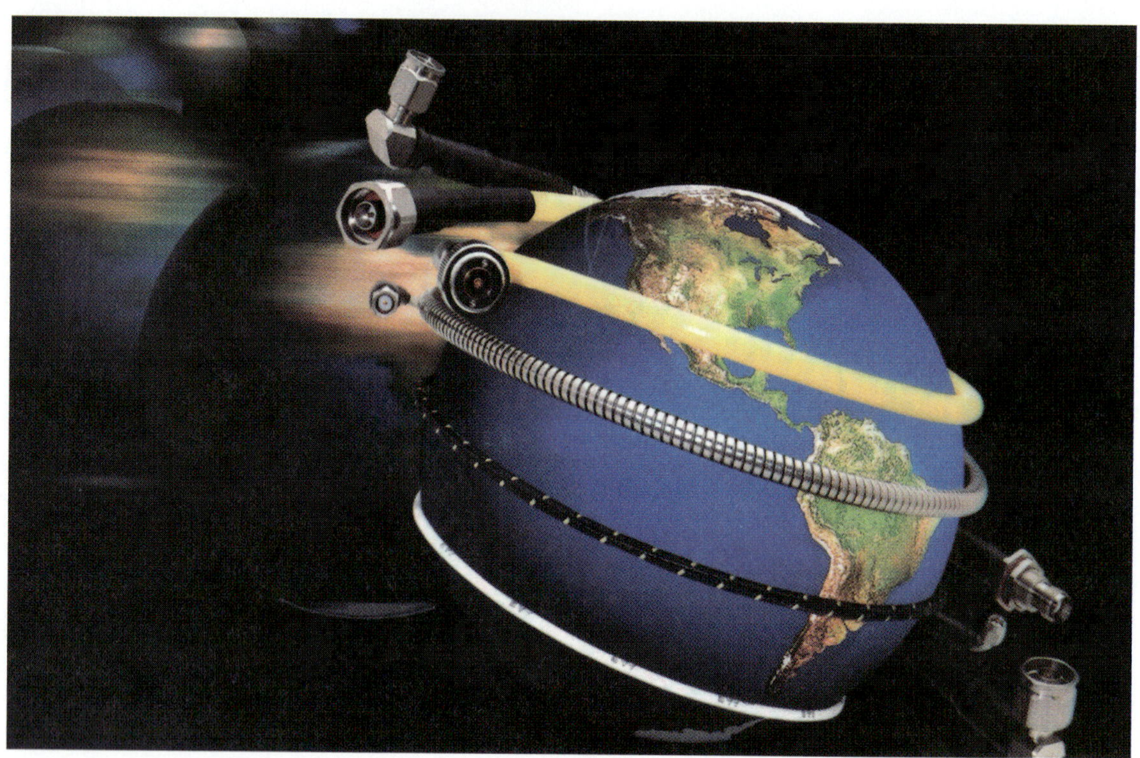

Transmission lines and connectors. (Courtesy of Insulated Wire Inc. Reprinted with permission.)

ACR
combined measurement of attenuation and crosstalk; a large ACR indicates greater bandwidth

(ACR). A larger ACR indicates that the cable has a greater data capacity. Essentially, ACR is a figure of merit for twisted-pair cable, where figure of merit implies a measure of the quality of the cable.

Wiring of the RJ-45 connector for RJ-45 CAT6/5e cable is defined by the Telecommunications Industry Association standard TIA568B. Within the TIA568B standard are the wiring guidelines T568A and T568B. These are shown in Table 12-1.

Table 12-2 lists the different categories, a description, and bandwidth for twisted-pair cable. Notice that CAT1, CAT2, and CAT4 are not listed. There never were CAT1 and CAT2 cable specifications, although you see them in many textbooks, handbooks, and on-line sources. The first CAT or category specification was

Table 12-1 T568A/T568B Wiring

Wire Color	Pair	Pin No. T568A	Pin No. T568B
White/blue	1	5	5
Blue/white	1	4	4
White/orange	2	3	1
Orange/white	2	6	2
White/green	3	1	3
Green/white	3	2	6
White/brown	4	7	7
Brown/white	4	8	8

The OMNIScanner2 tests, certifies, and documents high-speed copper and fiber networks with superior accuracy and repeatability. (Courtesy of Fluke Networks. Reprinted with permission.)

for CAT3. The CAT4 specification was removed from the TIA568B standard because it is an obsolete specification. CAT3 is still listed and is still used to a limited extent in telephone installations; however, modern telephone installations use CAT6/5e.

Enhanced data capabilities of twisted-pair cable include new specifications for testing **delay skew.** It is critical in high-speed data transmission that the data on the wire pair arrive at the other end at the same time. If the wire lengths of different wire pairs are significantly different, then the data on one wire pair will take longer to

Delay Skew
measure of the difference in time for the fastest to the slowest wire pair in a UTP cable

Table 12-2 Different Categories for Twisted-Pair Cable, Based on TIA568B

Category	Description	Bandwidth/Data Rate
Category 3 (CAT3)	Telephone installations Class C	Up to 16 Mbps
Category 5 (CAT5)	Computer networks Class D	Up to 100 MHz/100 Mbps for a 100-m length
Enhanced CAT5 (CAT5e)	Computer networks	100-MHz/1000 Mbps applications with improved noise performance
Category 6 (CAT6)	Current standard for higher-speed computer networks Class E	Up to 250 MHz/1000 Mbps
Category 7 (CAT7)	Proposed standard for higher-speed computer networks Class F	Up to 600 MHz

Power-Sum NEXT Testing (PSNEXT)
measures the total crosstalk of all cable pairs to ensure that the cable can carry data traffic on all four pairs at the same time with minimal interference.

Return Loss
a measure of the ratio of power transmitted into a cable to the amount of power returned or reflected

propagate along the wire, hence arriving at the receiver at a different time and potentially creating distortion of the data. Therefore, delay skew is a measure of the difference in time between the fastest and the slowest wire pair in a UTP cable. Additionally, the enhanced twisted-pair cable must meet four-pair NEXT requirements, which is called **power-sum NEXT (PSNEXT)** testing. Basically, power-sum testing measures the total crosstalk of all cable pairs. This test ensures that the cable can carry data traffic on all four pairs at the same time with minimal interference.

It has also been indicated that twisted pair can handle gigabit Ethernet networks at 100 m. The proposed gigabit data-rate capability of twisted pair requires the use of all four wire pairs in the cable, with each pair handling 250 Mbps of data. The total bit rate is 4×250 Mbps, or 1 Gbps.

An equally important twisted-pair cable measurement is **return loss.** This provides a measure of the ratio of power transmitted into a cable to the amount of power returned or reflected. The signal reflection is due to impedance changes in the cable link and the impedance changes contributing to cable loss. Cables are not perfect, so there will always be some reflection. Examples of the causes for impedance changes are nonuniformity in impedance throughout the cable, the diameter of the copper, cable handling, and dielectric differences. Return-loss tests are now specified for qualifying CAT5/class D, CAT5e, and CAT6/class E twisted-pair links.

Shielded Pair

The shielded pair, shown in Fig. 12-5, consists of parallel conductors separated from each other, and surrounded by a solid dielectric. The conductors are contained within a copper braid tubing that acts as a shield. The assembly is covered with a rubber or flexible composition coating to protect the line from moisture or mechanical damage.

The principal advantage of the shielded pair is that the conductors are balanced to ground; that is, the capacitance between the cables is uniform throughout the length of the line. This balance is due to the grounded shield that surrounds the conductors with a uniform spacing along their entire length. The copper braid shield isolates the conductors from external noise pickup. It also prevents the signal on the shielded-pair cable from radiating to and interfering with other systems.

Coaxial Lines

There are two types of coaxial lines: the rigid or air coaxial line, and the flexible or solid coaxial line. The electrical configuration of both types is the same; each contains two concentric conductors.

The rigid air coaxial line consists of a wire mounted inside, and coaxially with, a tubular outer conductor. This line is shown in Fig. 12-6. In some applications

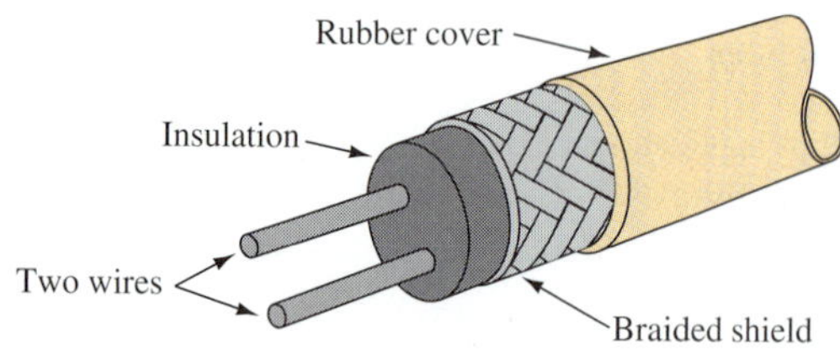

FIGURE 12-5 Shielded pair.

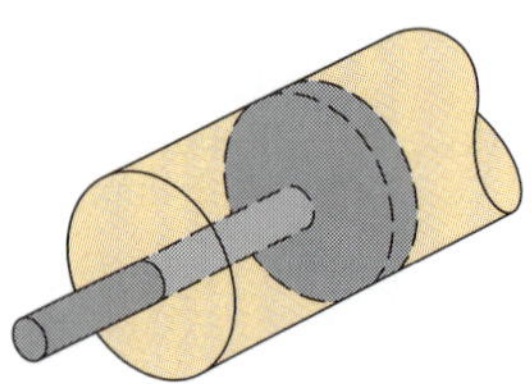

FIGURE 12-6 Air coaxial: cable with washer insulator.

Connectors for coaxial transmission lines. (Courtesy of KMW, Inc. Reprinted with permission.)

the inner conductor is also tubular. The inner conductor is insulated from the outer conductor by insulating spacers, or beads, at regular intervals. The spacers are made of Pyrex, polystyrene, or some other material possessing good insulating characteristics and low loss at high frequencies.

The chief advantage of this type of line is its ability to minimize radiation losses. The electric and magnetic fields in the two-wire parallel line extend into space for relatively great distances, and radiation losses occur. No electric or magnetic fields extend outside the outer (grounded) conductor in a coaxial line. The fields are confined to the space between the two conductors; thus, the coaxial line is a perfectly shielded line. It is important, however, to have the connectors properly installed to avoid leakage. Noise pickup from other lines is also prevented.

This line has several disadvantages: it is expensive to construct, it must be kept dry to prevent excessive leakage between the two conductors, and although high-frequency losses are somewhat less than in previously mentioned lines, they are still excessive enough to limit the practical length of the line.

The condensation of moisture is prevented in some applications by the use of an inert gas, such as nitrogen, helium, or argon, pumped into the line at a pressure of from 3 to 35 psi. The inert gas is used to dry the line when it is first installed, and a pressure is maintained to ensure that no moisture enters the line.

Concentric cables are also made, with the inner conductor consisting of flexible wire insulated from the outer conductor by a solid, continuous insulating material. Flexibility may be gained if the outer conductor is made of braided wire, although this reduces the level of shielding. Early attempts at obtaining flexibility

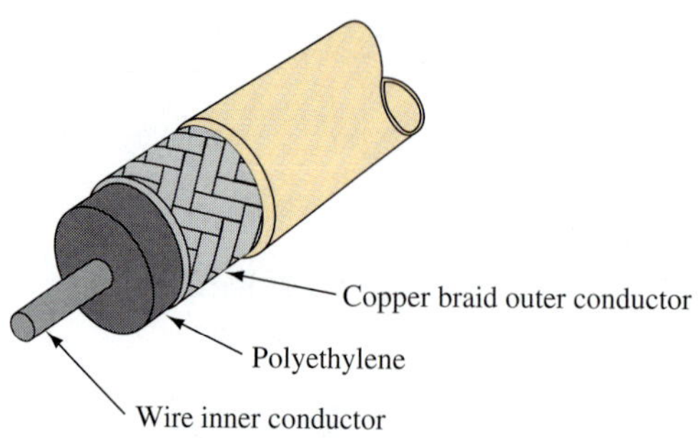

FIGURE 12-7 Flexible coaxial.

employed the use of rubber insulators between the two conductors. The use of rubber insulators caused excessive losses at high frequencies and allowed moisture-carrying air to enter the line, resulting in high leakage current and arc-over when high voltages were applied. These problems were solved by the development of polyethylene plastic, a solid substance that remains flexible over a wide range of temperatures. A coaxial line with a polyethylene spacer is shown in Fig. 12-7. Polyethylene is unaffected by seawater, gasoline, oils, and liquids. High-frequency losses due to the use of polyethylene, although greater than the losses would be if air were used, are lower than the losses resulting from the use of most other practical solid dielectric material. Solid flexible coaxial transmission lines are the most frequently used type of transmission line. Teflon is also commonly used as the dielectric in transmission lines.

Balanced/Unbalanced Lines

Unbalanced Line
the electrical signal in a coaxial line is carried by the center conductor with respect to the grounded outer conductor

Balanced Line
the same current flows in each wire but 180° out of phase

Common Mode Rejection
when signals that are 180° out of phase cancel each other

Baluns
circuits that convert between balanced and unbalanced operation

The electrical signal in a coaxial line is carried by the center conductor with respect to the grounded outer conductor. This is called an **unbalanced line.** Operation with the other types (two-wire open, twisted pair, shielded pair) is usually done with what is termed a **balanced line.** In it, the same current flows in each wire but 180° out of phase. Figure 12-8 shows the common technique for converting between unbalanced and balanced signals using a center-tapped transformer.

Any noise or unwanted signal picked up by the balanced line is picked up by both wires. Because these signals are 180° out of phase, they ideally cancel each other at the output center-tapped transformer. This is called **common mode rejection** (CMR). Practical common mode rejection ratio (CMRR) figures are 40–70 dB. In other words, the undesired noise or signal picked up by a two-wire balanced line is attenuated by 40–70 dB.

Circuits that convert between balanced and unbalanced operation are called **baluns.** The center-tapped transformers in Fig. 12-8 are therefore baluns. In Sec. 12-9, a balun using just a section of transmission line will be introduced.

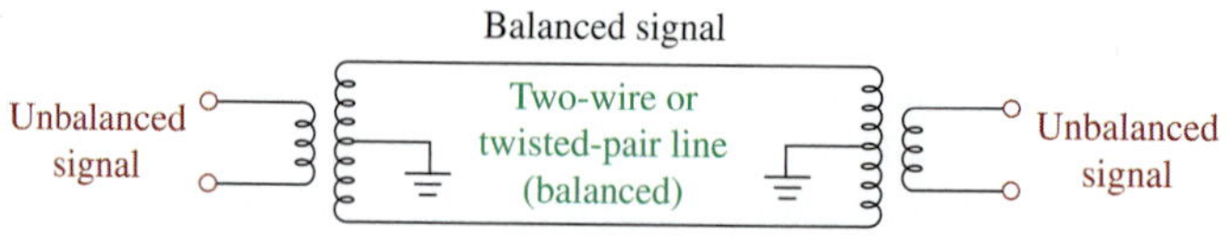

FIGURE 12-8 Balanced/unbalanced conversion.

12-3 Electrical Characteristics of Transmission Lines

Two-Wire Transmission Line

The end of a two-wire transmission line that is connected to a source is ordinarily called the *generator end* or *input end.* The other end of the line, if connected to a load, is called the *load end* or *receiving end.*

The electrical characteristics of the two-wire transmission line depend primarily on the construction of the line. Because the two-wire line can be viewed as a long capacitor, the change of its capacitive reactance is noticeable as the frequency applied to it is changed. Because the long conductors have a magnetic field about them when electrical energy is being passed through them, the properties of inductance are also observed. The values of the inductance and capacitance present depend on various physical factors, and the effects of the line's associated reactances also depend on the frequency applied. No dielectric is perfect (electrons manage to move from one conductor to the other through the dielectric), so there is a conductance value for each type of two-wire transmission line. This conductance value will represent the value of current flow that may be expected through the insulation. If the line is uniform (all values equal at each unit length), one small section of the line may be represented as shown in Fig. 12-9. Such a diagram may represent several feet of line.

In many applications, the values of conductance and resistance are insignificant and may be neglected. If they are neglected, the circuit appears as shown in Fig. 12-10. Notice that this network is *terminated* with a resistance that represents the impedance of the infinite number of sections exactly like the section of line under consideration. The termination is considered to be a load connected to the line.

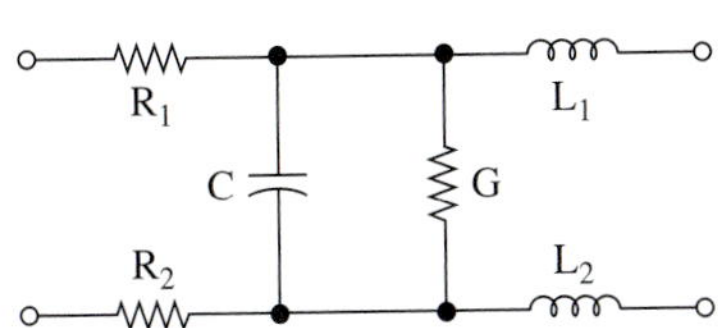

L_1 = inductance of top wire
L_2 = inductance of bottom wire
R_1 = resistance of top wire
R_2 = resistance of bottom wire
G = conductance between wires
C = capacitance between wires

FIGURE 12-9 Equivalent circuit for a two-wire transmission line.

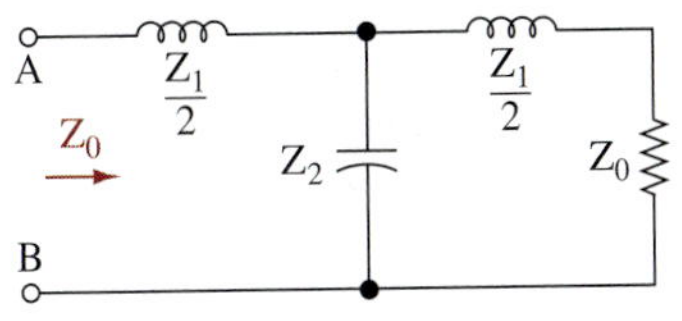

FIGURE 12-10 Simplified circuit terminated with its characteristic impedance.

Characteristic Impedance

A line infinitely long can be represented by an infinite number of inductors and capacitors. If a voltage is applied to the input terminals of the line, current would begin to flow. Because there are an infinite number of these sections of line, the current would flow indefinitely. If the infinite line were uniform, the impedance of each section would be the same as the impedance offered to the circuit by any other section of line of the same unit length. Therefore, the current would be of some finite value. If the current flowing in the line and the voltage applied across it are known, the impedance of the infinite line could be determined by using Ohm's law. This impedance is called the **characteristic impedance** of the line. The symbol used to represent the characteristic impedance is Z_0. If the characteristic impedance of the line could be measured at any point on the line, it would be found to be the same. The characteristic impedance is sometimes called the **surge impedance.**

Characteristic Impedance
the input impedance of a transmission line either infinitely long or terminated in a pure resistance exactly equal to its characteristic impedance

Surge Impedance
another name for characteristic impedance

In Fig. 12-10 the distributed inductance of the line is divided equally into two parts in the horizontal arms of the T. The distributed capacitance is lumped and shown connected in the central leg of the T. The line is terminated in a resistance equal to that of the characteristic impedance of the line as seen from terminals *AB*. The reasons for using this value of resistive termination will be fully explained in Sec. 12-6. Because the circuit in Fig. 12-10 is nothing more than a series–parallel *LCR* circuit, the impedance of the network may be determined by the formula that will now be developed.

The impedance, Z_0, looking into terminals *AB* of Fig. 12-10 is

$$Z_0 = \frac{Z_1}{2} + \frac{Z_2[(Z_1/2) + Z_0]}{Z_2 + (Z_1/2) + Z_0} \tag{12-1}$$

Simplifying yields

$$Z_0 = \frac{Z_1}{2} + \frac{(Z_1Z_2/2) + Z_0Z_2}{Z_2 + (Z_1/2) + Z_0} \tag{12-2}$$

Expressing the right-hand member in terms of the least common denominator, we obtain

$$Z_0 = \frac{Z_1Z_2 + (Z_1^2/2) + Z_1Z_0 + (2Z_1Z_2/2) + 2Z_0Z_2}{2[Z_2 + (Z_1/2) + Z_0]} \tag{12-3}$$

If both sides of this equation are multiplied by the denominator of the right-hand member, the result is

$$2Z_2Z_0 + \frac{2Z_1Z_0}{2} + 2Z_0^2 = Z_1Z_2 + \frac{Z_1^2}{2} + Z_1Z_0 + \frac{2Z_1Z_2}{2} + 2Z_0Z_2 \tag{12-4}$$

Simplifying gives us

$$2Z_0^2 = 2Z_1Z_2 + \frac{Z_1^2}{2} \tag{12-5}$$

or

$$Z_0^2 = Z_1Z_2 + \left(\frac{Z_1}{2}\right)^2 \tag{12-6}$$

If the transmission is to be accurately represented by an equivalent network, the T-network section of Fig. 12-10 must be replaced by an infinite number of similar sections. Thus, the distributed inductance in the line will be divided into n sections, instead of the number (2) as indicated in the last term of Eq. (12-6). As the number of sections approaches infinity, the last term Z_1/n will approach zero. Therefore,

$$Z_0 = \sqrt{Z_1 Z_2} \tag{12-7}$$

Because the term Z_1 represents the inductive reactance and the term Z_2 represents the capacitive reactance,

$$Z_0 = \sqrt{2\pi f L \times \frac{1}{2\pi f C}}$$

and

$$Z_0 = \sqrt{\frac{L}{C}} \tag{12-8}$$

The derivation resulting in Eq. (12-8) shows that a line's characteristic impedance depends on its inductance and capacitance.

Example 12-1

A commonly used coaxial cable, RG-8A/U, has a capacitance of 29.5 pF/ft and inductance of 73.75 nH/ft. Determine its characteristic impedance for a 1-ft section and for a length of 1 mi.

Solution

For the 1-ft section,

$$Z_0 = \sqrt{\frac{L}{C}} \tag{12-8}$$

$$= \sqrt{\frac{73.75 \times 10^{-9}}{29.5 \times 10^{-12}}} = \sqrt{2500} = 50\ \Omega$$

For the 1-mi section,

$$Z_0 = \sqrt{\frac{5280 \times 73.75 \times 10^{-9}}{5280 \times 29.5 \times 10^{-12}}} = \sqrt{\frac{5280}{5280} \times 2500} = 50\ \Omega$$

Example 12-1 shows that the line's characteristic impedance is independent of length and is in fact a *characteristic* of the line. The value of Z_0 depends on the ratio of the distributed inductance and the capacitance in the line. An increase in the separation of the wires increases the inductance and decreases the capacitance. This effect takes place because the effective inductance is proportional to the flux that may be established between the two wires. If the two wires carrying current in opposite directions are placed farther apart, more magnetic flux is included between them (they cannot cancel their magnetic effects as completely as if the wires were closer together), and the distributed inductance is increased. The capacitance is lowered if the plates of the capacitor (the plates are the two conducting wires) are more widely spaced.

Thus, the effect of increasing the spacing of the two wires is to increase the characteristic impedance because the L/C ratio is increased. Similarly, a reduction in the diameter of the wires also increases the characteristic impedance. The reduction in the size of the wire affects the capacitance more than the inductance because the effect is equivalent to decreasing the size of the plates of a capacitor to decrease the capacitance. Any change in the dielectric material between the two wires also changes the characteristic impedance. If a change in the dielectric material increases the capacitance between the wires, the characteristic impedance, by Eq. (12-8), is reduced.

The characteristic impedance of a two-wire line may be obtained from the formula

$$Z_0 \simeq \frac{276}{\sqrt{\epsilon}} \log_{10} \frac{2D}{d} \qquad \textbf{(12-9)}$$

where D = spacing between the wires (center to center)
d = diameter of one of the conductors
ϵ = dielectric constant of the insulating material relative to air

The characteristic impedance of a concentric or coaxial line also varies with L and C. However, because the difference in construction of the two lines causes L and C to vary in a slightly different manner, the following formula must be used to determine the characteristic impedance of the coaxial line:

$$Z_0 \simeq \frac{138}{\sqrt{\epsilon}} \log_{10} \frac{D}{d} \qquad \textbf{(12-10)}$$

where D = inner diameter of the outer conductor
d = outer diameter of the inner conductor
ϵ = dielectric constant of the insulating material relative to air

The relative dielectric constant of air is 1; polyethylene, 2.3; and teflon, 2.1.

EXAMPLE 12-2

Determine the characteristic impedance of
(a) A parallel wire line with $D/d = 2$ with air dielectric.
(b) An air dielectric coaxial line with $D/d = 2.35$.
(c) RG-8A/U coaxial cable with $D = 0.285$ in. and $d = 0.08$ in. It uses a polyethylene dielectric.

SOLUTION

(a)

$$Z_0 \simeq \frac{276}{\sqrt{\epsilon}} \log_{10} \frac{2D}{d} \qquad \textbf{(12-9)}$$

$$\simeq \frac{276}{1} \log_{10} 4$$

$$\simeq 166\ \Omega$$

(b) $$Z_0 \simeq \frac{138}{\sqrt{\epsilon}} \log_{10} \frac{D}{d} \quad \textbf{(12-10)}$$

$$\simeq \frac{138}{1} \log_{10} 2.35$$

$$\simeq 51.2\ \Omega$$

(c) $$Z_0 \simeq \frac{138}{\sqrt{2.3}} \log_{10} \frac{0.285}{0.08}$$

$$\simeq 50\ \Omega$$

It's important that coaxial cable not be stepped on, crimped, or bent in too small a radius. Any of these conditions changes the *D/d* ratio and therefore changes Z_0. As you'll see from following discussions, this causes problems with respect to circuit operation.

Transmission Line Losses

Whenever the electrical characteristics of lines are explained, the lines are often thought of as being loss-free. Although this allows for simple and more readily understood explanations, the losses in practical lines cannot be ignored. There are three major losses that occur in transmission lines: copper losses, dielectric losses, and radiation or induction losses.

The resistance of any conductor is never zero. When current flows through a transmission line, energy is dissipated in the form of I^2R losses. A reduction in resistance will minimize the power loss in the line. The resistance is indirectly proportional to the cross-sectional area. Keeping the line as short as possible will decrease the resistance and the I^2R loss. The use of a wire with a large cross-sectional area is also desirable; however, this method has its limitations due to the resulting cost increases.

At high frequencies the I^2R loss is mainly due to the **skin effect.** When a dc current flows through a conductor, the movement of electrons through its cross section is uniform. The situation is somewhat different when ac is applied.

Skin Effect
the tendency for high-frequency electric current to flow mostly near the surface of the conductive material

The flux density at the center of a conductor is greater than at the outer edge. Therefore, the inductance and inductive reactance are greater, causing lower current in the center and more along the outer edge. This effect increases with frequency. Forcing current to the edge effectively reduces the conductor's cross-sectional area in which current can flow and, because resistance is inversely proportional to cross-sectional area, the resistance increases. This increase in resistance is called skin effect. At very high frequencies, skin effect is so great that wires can no longer be used to carry current and we must resort to waveguides; these are discussed in Chapter 15.

Dielectric losses are proportional to the voltage across the dielectric. They increase with frequency and, when coupled with skin effect losses, limit most practical operation to a maximum frequency of about 18 GHz. These losses are lowest when air dielectric lines are used. In many cases the use of a solid dielectric is required, for example, in the flexible coaxial cable, and if losses are to be minimized, an insulation with a low dielectric constant is used. Polyethylene allows the construction of a flexible cable whose dielectric losses, though higher than air, are still much lower than the losses that occur with other types of low-cost dielectrics. Because I^2R losses and dielectric losses are proportional to length, they are usually

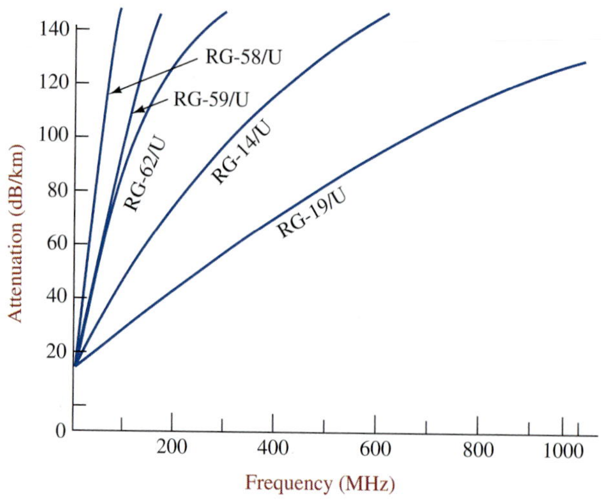

FIGURE 12-11 Line attenuation characteristics.

lumped together and expressed in decibels of loss per meter. The loss versus frequency effects for some common lines are shown in Fig. 12-11.

The electrostatic and electromagnetic fields that surround a conductor also cause losses in transmission lines. The action of the electrostatic fields is to charge neighboring objects, while the changing magnetic field induces an electromagnetic force (EMF) in nearby conductors. In either case, energy is lost.

Radiation and induction losses may be greatly reduced by terminating the line with a resistive load equal to the line's characteristic impedance and by proper shielding of the line. Proper shielding can be accomplished by the use of coaxial cables with the outer conductor grounded. The problem of radiation loss is of consequence, therefore, only for parallel wire transmission lines.

12-4 Propagation of DC Voltage down a Line

Physical Explanation of Propagation

To understand the characteristics of a transmission line with an ac voltage applied, the infinitely long transmission line will first be analyzed with a dc voltage applied. This will be accomplished using a circuit like the one illustrated in Fig. 12-12. In this circuit the resistance of the line is not shown. The line is assumed to be loss-free.

Considering only the capacitor C_1 and the inductor L_1 as a series circuit, when voltage is applied to the network, capacitor C_1 has the ability to charge through inductor L_1. It is characteristic of an inductor that at the first instant of time when voltage is applied, a maximum voltage is developed across it and minimum current is permitted to pass through it. At the same time, the capacitor has a minimum of voltage across it and passes a maximum current. The maximum current is not permitted to flow at the first instant because of the action of the inductor, which is in the charge path of the capacitor. At this instant the voltage across points c and d is zero. Because the remaining portion of the line is connected to points c and d, 0 V is developed across it at the first instant of time. The voltage across the rest of the

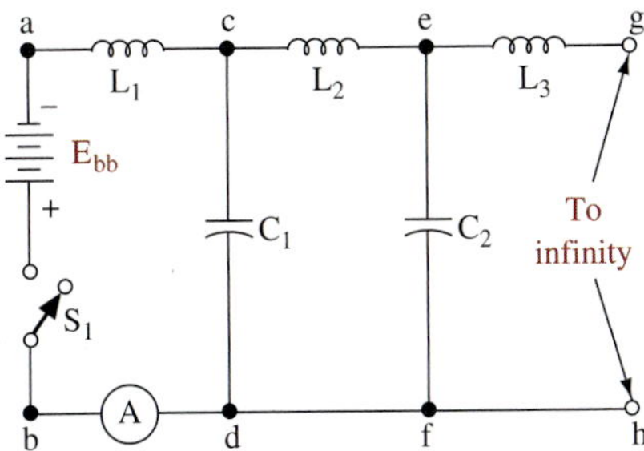

FIGURE 12-12 DC voltage applied to a transmission line.

line is dependent on the charging action of the capacitor, C_1. Some finite amount of time is required for capacitor C_1 to charge through inductor L_1. As capacitor C_1 is charging, the ammeter records the changing current. When C_1 charges to a voltage that is near the value of the applied voltage, capacitor C_2 begins to charge through inductors L_1 and L_2. The charging of capacitor C_2 again requires time. In fact, the time required for the voltage to reach points *e* and *f* from points *c* and *d* is the same time as was necessary for the original voltage to reach points *c* and *d*. This is true because the line is uniform, and the values of the reactive components are the same throughout its entire length. This action continues in the same manner until all of the capacitors in the line are charged. Because the number of capacitors in an infinite line is infinite, the time required to charge the entire line would be an infinite amount of time. It is important to note that current is flowing continuously in the line and that it has some finite value.

Velocity of Propagation

When a current is moving down the line, its associated electric and magnetic fields are said to be *propagated* down the line. Time is required to charge each unit section of the line, and if the line were infinitely long, it would require an infinitely long time to charge. The time for a field to be propagated from one point on a line to another may be computed because if the time and the length of the line are known, the **velocity of propagation** may be determined. The network shown in Fig. 12-13 is the circuit that will be used to compute the time required for the voltage wave-front to pass a section of line of specified length. The total charge (Q) in coulombs on capacitor C_1 is determined by the relationship

Velocity of Propagation
the speed at which an electrical signal moves through a conductor

$$Q = Ce \tag{12-11}$$

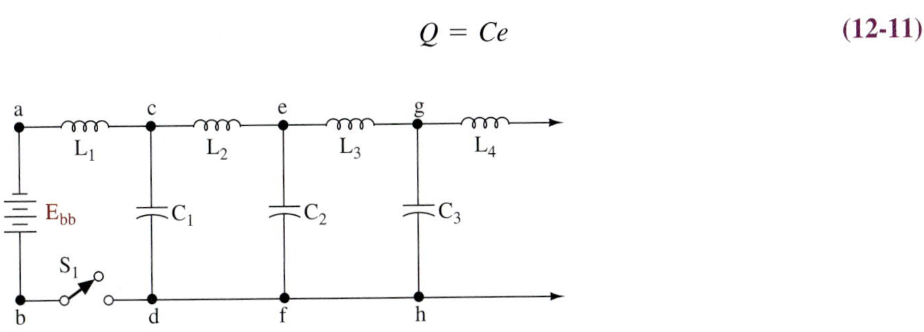

FIGURE 12-13 Circuit for computing time of travel.

Because the charge on the capacitor in the line had its source at the battery, the total amount of charge removed from the battery will be equal to

$$Q = it \tag{12-12}$$

Because these charges are equal, they may be equated:

$$Ce = it \tag{12-13}$$

As the capacitor C_1 charges, capacitor C_2 contains a zero charge. Because capacitor C_1's voltage is distributed across C_2 and L_2, at the same time the charge on C_2 is practically zero, the voltage across C_1 (points c and d) must be, by Kirchhoff's law, entirely across L_2. The value of the voltage across the inductor is given by

$$e = L\frac{\Delta i}{\Delta t} \tag{12-14}$$

Because current and time start at zero, the change in time and the change in current are equal to the final current and the final time. Equation (12-14) becomes

$$et = Li \tag{12-15}$$

Solving the equation for i, we have

$$i = \frac{et}{L} \tag{12-16}$$

Solving the equation that was a statement of the equivalency of the charges for current [Eq. (12-13)], we obtain

$$i = \frac{Ce}{t} \tag{12-17}$$

Equating both of these expressions yields

$$\frac{et}{L} = \frac{Ce}{t} \tag{12-18}$$

Solving the equation for t gives us

$$t^2 = LC$$

or

$$t = \sqrt{LC} \tag{12-19}$$

Because velocity is a function of both time and distance ($V = d/t$), the formula for computing propagation velocity is

$$V_P = \frac{d}{\sqrt{LC}} \tag{12-20}$$

where V_p = velocity of propagation
d = distance of travel
$\sqrt{LC}$ = time (t)

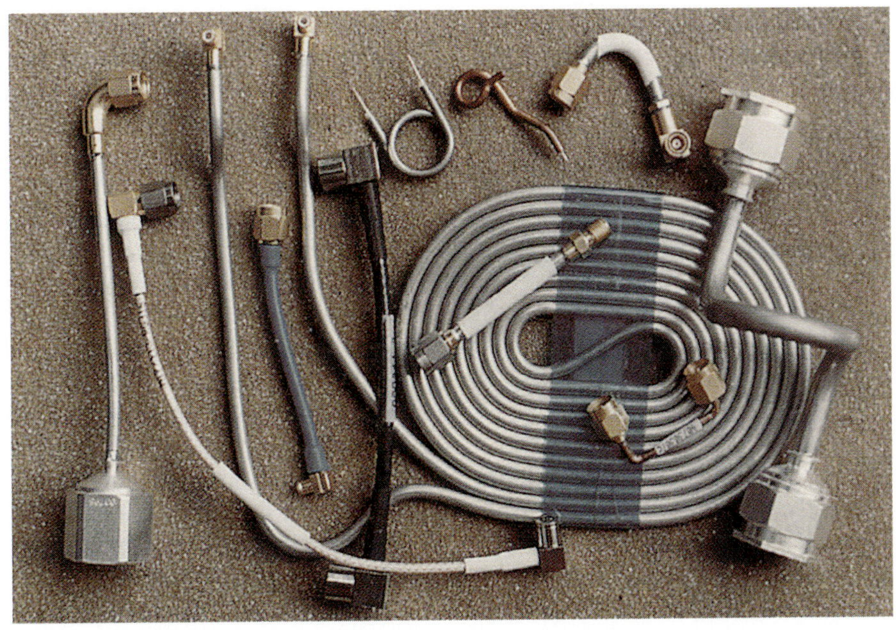

Delay lines. (Courtesy of E/Z From Cable Corp. Reprinted with permission.)

It should again be noted that the time required for a wave to traverse a transmission line segment depends on the value of L and C and that these values will be different, depending on the type of the transmission line considered.

Delay Line

The velocity of electromagnetic waves through a vacuum is the speed of light, or 3×10^8 m/s. It is just slightly reduced for travel through air. It has just been shown that a transmission line decreases this velocity because of its inductance and capacitance. This property is put to practical use when we want to delay a signal by some specific amount of time. A transmission line used for this purpose is called a **delay line.**

Delay Line
a length of a transmission line designed to delay a signal from reaching a point by a specific amount of time

Example 12-3

Determine the amount of delay and the velocity of propagation introduced by a 1-ft section of RG-8A/U coaxial cable used as a delay line.

Solution

From Ex. 12-1, we know that this cable has a capacitance of 29.5 pF/ft and inductance of 73.75 nH/ft. The delay introduced by 1 ft of this line is

$$t = \sqrt{LC} \qquad \textbf{(12-19)}$$

$$= \sqrt{73.75 \times 10^{-9} \times 29.5 \times 10^{-12}}$$

$$= 1.475 \times 10^{-9}\,\text{s} \qquad \text{or} \qquad 1.457\text{ ns}$$

The velocity of propagation is

$$V_p = \frac{d}{\sqrt{LC}} \tag{12-20}$$

$$= \frac{1 \text{ ft}}{1.475 \text{ ns}} = 6.78 \times 10^8 \text{ ft/s} \qquad \text{or} \qquad 2.07 \times 10^8 \text{ m/s}$$

Velocity Constant ratio of actual velocity to velocity in free space

Velocity Factor another name for velocity constant

Example 12-3 showed that the energy velocity for RG-8A/U cable is roughly two-thirds the velocity of light. This ratio of actual velocity to the velocity in free space is termed the **velocity constant** or **velocity factor** of a line. It can range from about 0.55 up to 0.97, depending on the type of line, the *D/d* ratio, and the type of dielectric. As an approximation for non-air-dielectric coaxial lines, the velocity factor, v_f, is

$$v_f \simeq \frac{1}{\sqrt{\epsilon_r}} \tag{12-21}$$

where v_f = velocity factor

ε_r = relative dielectric constant

$$\epsilon_r\,(\text{RG} - 8\text{A/U}) \simeq 2.3$$

Example 12-4

Determine the velocity factor for RG-8A/U cable by using the results of Ex. 12-3 and also by using Eq. (12-21).

Solution

From Ex. 12-3, the velocity was 2.07×10^8 m/s. Therefore,

$$v_f = \frac{2.07 \times 10^8 \text{ m/s}}{3 \times 10^8 \text{ m/s}} = 0.69$$

Using Eq. (12-21), we obtain

$$v_f \simeq \frac{1}{\sqrt{\epsilon_r}} = \frac{1}{\sqrt{2.3}} = 0.66$$

Wavelength

Wavelength the distance traveled by a wave during a period of one cycle

A wave that is radiated through space travels at about the speed of light, or 186,000 mi/s (3×10^8 m/s). The velocity of this wave is constant regardless of frequency, so that the distance traveled by the wave during a period of one cycle (called one **wavelength**) can be found by the formula

$$\lambda = \frac{c}{f} \tag{12-22}$$

where λ (the Greek lowercase letter lambda, used to symbolize wavelength) is the distance in meters from the crest of one wave to the crest of the next, f is the frequency, and c is the velocity of the radio wave in meters per second. It should be noted that the wavefront travels more slowly on a wire than it does in free space.

Example 12-5

Determine the wavelength (λ) of a 100-MHz signal in free space and while traveling through an RG-8A/U coaxial cable.

Solution

In free space, the wave's velocity (c) is 3×10^8 m/s. Therefore,

$$\lambda = \frac{c}{f} \qquad (12\text{-}22)$$
$$= \frac{3 \times 10^8 \text{ m/s}}{1 \times 10^8 \text{ Hz}} = 3 \text{ m}$$

Note: The free-space wavelength is typically labeled λ_0. In RG-8A/U cable we found in Ex. 12-3 that the velocity of propagation is 2.07×10^8 m/s. Therefore,

$$\lambda = \frac{c}{f} \qquad (12\text{-}22)$$
$$= \frac{2.07 \times 10^8 \text{ m/s}}{1 \times 10^8 \text{ Hz}} = 2.07 \text{ m}$$

Example 12-5 shows that line wavelength is less than free-space wavelength for any given frequency signal.

12-5 Nonresonant Line

Traveling DC Waves

A **nonresonant** line is defined as one of infinite length or that is terminated with a resistive load equal in ohmic value to the characteristic impedance of the line. In a nonresonant line, all of the energy transferred down the line is absorbed by the load resistance and any inherent resistance in the line. The voltage and current waves are called **traveling waves** and move in phase with one another from the source to the load.

Nonresonant Line one of infinite length or that is terminated with a resistive load equal in ohmic value to its characteristic impedance

Traveling Waves voltage and current waves moving through a transmission line

Because the nonresonant line may be either an infinite line or one terminated in its characteristic impedance, the physical length is not critical. In the resonant line that will be discussed shortly, the physical length of the line is quite important.

The circuit in Fig. 12-14 shows a line terminated with a resistance equal to its characteristic impedance. The charging process and the ultimate development of a voltage across the load resistance will now be described. At the instant switch S_1 is closed, the total applied voltage is felt across inductor L_1. After a very short time has elapsed, capacitor C_1 begins to assume a charge. C_2 cannot charge at this time because all the voltage felt between points c and d is developed across inductor L_2 in the same way the initial voltage was developed across inductor L_1. Capacitor C_2 is unable to charge until the charge on C_1 approaches the amplitude of the supply voltage. When this happens, the voltage charge on capacitor C_2 begins to rise. The voltage across capacitor C_2 will be felt between points e and f. Because the load resistor is also effectively connected between points e and f, the voltage across the resistor is equal to the voltage appearing across C_2. The voltage input has been

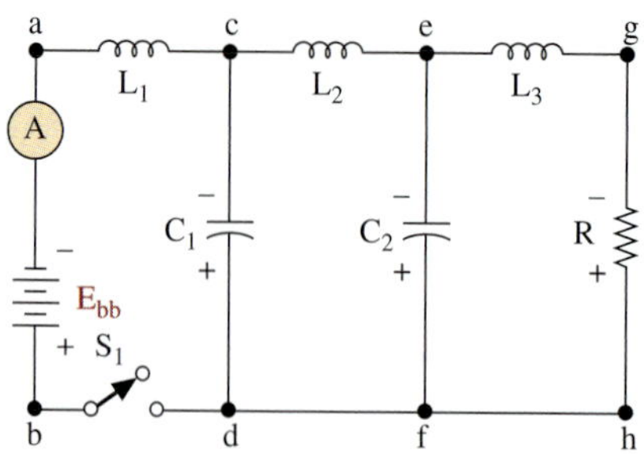

FIGURE 12-14 Charged nonresonant line.

transferred from the input to the load resistor. While the capacitors were charging, the ammeter recorded a current flow. After all of the capacitors are charged, the ammeter will continue to indicate the load current that will be flowing through the dc resistance of the inductors and the load resistor. The current will continue to flow as long as switch S_1 is closed. When it is opened, the capacitors discharge through the load resistor in much the same way as filter capacitors discharge through a bleeder resistor.

Traveling AC Waves

There is little difference between the charging of the line when an ac voltage is applied to it and when a dc voltage is applied. The charging sequence of the line with an ac voltage applied will now be discussed. Refer to the circuit and waveform diagrams in Fig. 12-15.

As the applied voltage begins to go positive, the voltage wave begins traveling down the line. At time t_3, the first small change in voltage arrives at point A, and the voltage at that point starts increasing in the positive direction. At time t_5, the same voltage rise arrives at point *B*, and at time t_7, the same voltage rise arrives at the end of the line. The waveform has moved down the line as a wavefront. The time required for the voltage changes to move down the line is the same as the time required for the dc voltage to move down the same line. The time for both of these waves to move down the line for a specified length may be computed by using Eq. (12-19). The following general remarks concerning the ac charging of the line may now be made. All of the instantaneous voltages produced by the generator travel down the line in the order in which they were produced. If the voltage waveform is plotted at any point along the line, the resulting waveform will be a duplicate of the generator waveform. Because the line is terminated with its characteristic impedance, all of the energy produced by the source is absorbed by the load impedance.

 12-6 Resonant Transmission Line

Resonant Line
a transmission line terminated with an impedance that is not equal to its characteristic impedance

A **resonant line** is defined as a transmission line that is terminated with an impedance that is *not* equal to its characteristic impedance. Unlike the nonresonant line, the length of the resonant line is critical. In some applications, the resonant line may be terminated in either an open or a short. When this occurs, some very interesting effects may be observed.

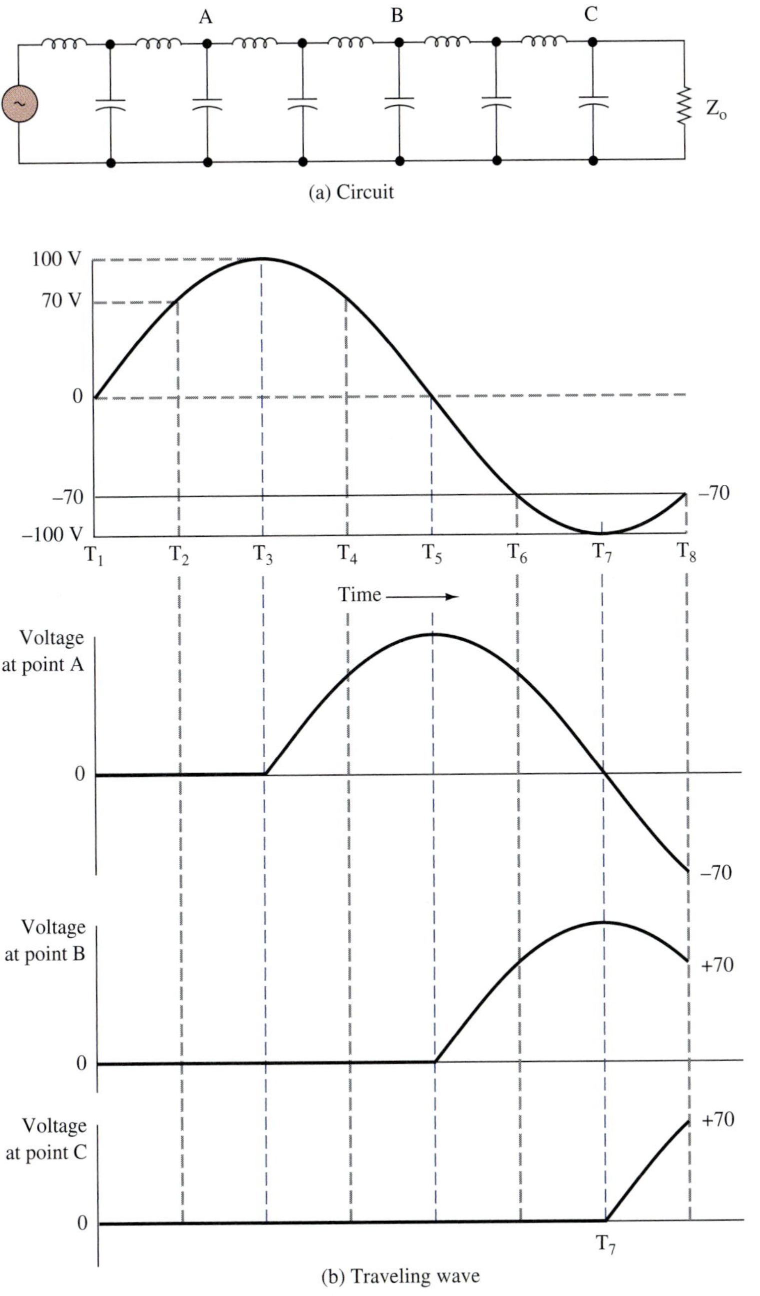

FIGURE 12-15 Alternating-current charging analysis.

DC Applied to an Open-Circuited Line

A transmission line of finite length terminated in an open circuit is illustrated in Fig. 12-16. The characteristic impedance of the line may be assumed to be equal to the internal impedance of the source. Because the impedance of the source is equal to the impedance of the line, the applied voltage is divided equally between the

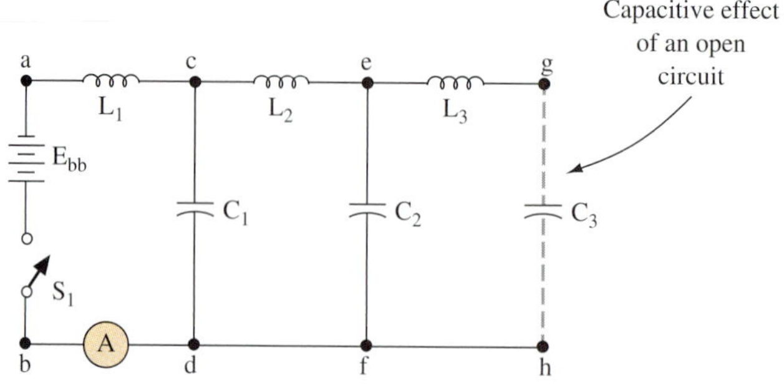

FIGURE 12-16 Open-ended transmission line.

impedance of the source and the impedance of the line. When switch S_1 is closed, current begins to flow as the capacitors begin to charge through the inductors. As each capacitor charges in turn, the voltage will move down the line. As the last capacitor is charged to the same voltage as every other capacitor, there will be no difference in potential between points e and g. This is true because the capacitors possess exactly the same charge. The inductor L_3 is also connected between points e and g. Because there is no difference in potential between points e and g, there can no longer be current flow through the inductor. This means that the magnetic field about the inductor is no longer sustained. The magnetic field must collapse. It is characteristic of the field about an inductor to tend to keep current flowing in the same direction when the magnetic field collapses. This additional current must flow into the capacitive circuit of the open circuit, C_3. Because the energy stored in the magnetic field is equal to that stored in the capacitor, the charge on capacitor C_3 doubles. The voltage on capacitor C_3 is equal to the value of the applied voltage. Because there is no difference of potential between points c and e, the magnetic field about inductor L_2 collapses, forcing the charge on capacitor C_2 to double its value. The field about inductor L_1 also collapses, doubling the voltage on C_1. The combined effect of the collapsing magnetic field about each inductor in turn causes a voltage twice the value of the original apparently to move back toward the source. This voltage movement in the opposite direction caused by the conditions just described is called **reflection.** The reflection of voltage occurred in the same polarity as the original charge. Therefore, it is said of a transmission line terminated in an open that the reflected voltage wave is always of the same polarity and amplitude as the incident voltage wave. When this reflected voltage reaches the source, the action stops because of the cancellation of the voltages. The current, however, is reflected back with an opposite polarity because when the field about the inductor collapsed, the current dropped to zero. As each capacitor is charged, causing the reflection, the current flow in the inductor that caused the additional charge drops to zero. When capacitor C_1 is charged, current flow in the circuit stops, and the line is charged. It may also be said that the line now "sees" that the impedance at the receiving end is an open.

Reflection
abrupt reversal in direction of voltage and current

Incident and Reflected Waves

The situation for a 100-V battery with a 50-Ω source resistance is illustrated in Fig. 12-17. The battery is applied to an open-circuited 50-Ω characteristic impedance line at time $t = 0$. Initially, a 50-V level propagates down the line, and 1 A is

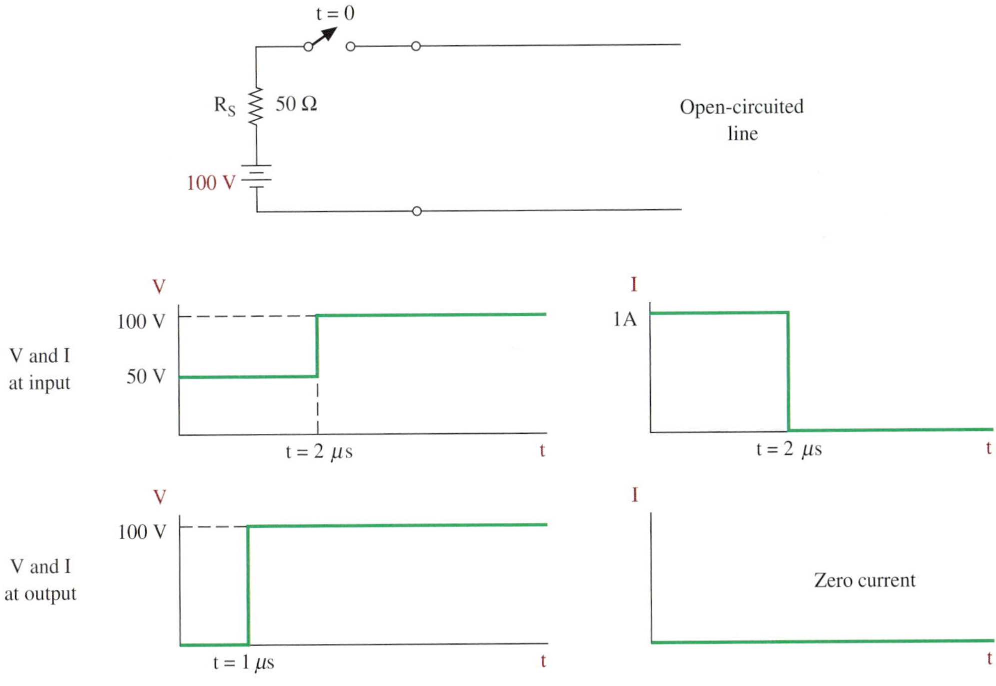

FIGURE 12-17 Direct current applied to an open-circuited line.

drawn from the battery. Notice that 50 V is dropped across R_s at this point in time. The reflected voltage from the open circuit is also 50 V so that the resultant voltage along the line is 50 V + 50 V, or 100 V, once the reflection gets back to the battery. If it takes 1 μs for energy to travel the length of this transmission line, the voltage at the line's input will initially be 50 V until the reflection gets back to the battery at $t = 2\ \mu$s. Remember, it takes 1 μs for the energy to get to the end of the line and then 1 μs additional for the reflection to get back to the battery. Thus, the voltage versus time condition shown in Fig. 12-16 for the line's input is 50 V until $t = 2\ \mu$s, when it changes to 100 V. The voltage at the load is zero until $t = 1\ \mu$s, as shown, when it jumps to 100 V due to the 50 V just reaching it and the resulting 50-V reflection.

The current conditions on the line are also shown in Fig. 12-17. The incident current is 1 A, which is 100 V $\div$ $(R_s + Z_0)$. The reflected current for an open-circuited line is out of phase and, therefore, is -1 A with a resultant of 0 A. The current at the line's input is 1 A until $t = 2\ \mu$s, when the reflected current reaches the source to cancel the incident current. After $t = 2\ \mu$s, the current is therefore zero. The current at the load is always zero because, when the incident 1 A reaches the load, it is immediately canceled by the reflected -1 A.

Thus, after 1 μs at the load and after 2 μs at the source, the results are predicted by standard analysis; that is, the voltage on an open-circuited wire is equal to the source voltage, and the current is zero. Unfortunately, when ac signals are applied to a transmission line, the results are not so simple because the incident and reflected signals occur on a continuous repetitive basis.

DC Applied to a Short-Circuited Line

The condition of applying a 100-V, 50-Ω source resistance battery to a shorted 50-Ω transmission line is illustrated in Fig. 12-18. Once again, assume that it takes 1 μs for energy to travel the length of the line. A complete analysis as presented with the open-circuited line would now be repetitive. Instead, only the differences and final results will be provided.

When the incident current of 1 A [100 V ÷ (R_s + Z_0)] reaches the short-circuited load, the reflected current is 1 A and in phase so that the load current becomes 1 A + 1 A = 2 A. The incident voltage (+50 V) is reflected back out of phase (−50 V) so that the resultant voltage at the short circuit is zero, as it must be across a short. The essential differences here from the open-circuited line are:

1. The voltage reflection from an open circuit is in phase, while from a short circuit it is out of phase.
2. The current reflection from an open circuit is out of phase, while from a short circuit it is in phase.

The resultant load voltage is therefore always zero, as it must be across a short circuit, and is shown in Fig. 12-18. However, the voltage at the line's input is initially +50 V until the reflected out-of-phase level (−50 V) gets back in 2 μs, which causes the resultant to be zero. The load current is zero until t = 1 μs, when the incident and reflected currents of +1 A combine to cause 2 A of current to flow. The

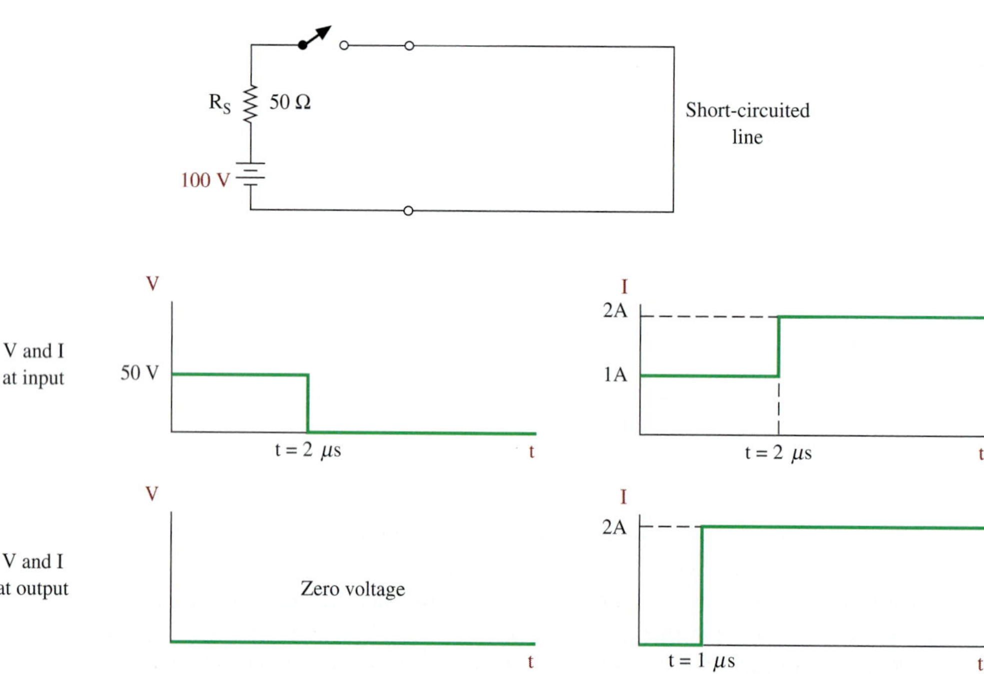

FIGURE 12-18 Direct current applied to a short-circuited line.

current at the line's input is initially 1 A until the reflected current of 1 A arrives at $t = 2\ \mu s$ to cause the total current to be 2 A.

Standing Waves: Open Line

The conditions of an open or shorted transmission line are extraordinary conditions. Reflected waves are, on the whole, highly undesirable. It is known that when the impedance of the source is equal to the characteristic impedance of the line and that line is terminated in its characteristic impedance, there is a maximum transfer of power, complete absorption of energy by the load, and no reflected waves. If the line is not terminated in its characteristic impedance, there will be reflected waves present on the line. The type and amount of reflected waves depend on the type and amount of mismatch. When a mismatch occurs, there is an interaction between the incident and reflected waves. When the applied signal is ac, this interaction results in the creation of a new kind of wave called a **standing wave.** This name is given to these waveforms because they apparently remain in one position, varying only in amplitude. These waves, and the variations in amplitude, are illustrated in Fig. 12-19.

Standing Wave waveforms that apparently seem to remain in one position, varying only in amplitude

The left-hand column in Fig. 12-19 represents the in-phase reflection of the voltage wave on the open line or current wave on a shorted line. Note the dashed line extending past the load in the top left-hand diagram. It is an extension of the incident traveling wave. By simply "folding" it back across the termination point, the reflected wave for in-phase reflection is obtained. All the diagrams of in-phase reflection show the incident, reflected, and resultant waves on a line at various instants of time. They are graphs of a wave versus *position* at some instant of time and *not* waves versus time, as you are accustomed to seeing. The resultant waves for each instant of time are shown with a blue line and are simply the vector sum of the incident and reflected waves.

Note that at positions d_1 and d_3 in Fig. 12-19 the resultant voltage (current) on the line (shown in blue) is always zero. If you stationed yourself at these points with an oscilloscope, no wave would ever be seen once the very first incident wave and then reflected wave had passed by. On the other hand, at position d_2, the generator and, at the load, the resultants have a maximum value. The resultant is truly a *standing* wave. If readings of the rms voltage and current waves for an open line were taken along the line, the result would be as shown in Fig. 12-20 on page 581. This is a conventional picture of standing waves. The voltage at the open is maximum, while the current is zero, as it must be through an open circuit. The positions d_1 through d_6 correspond to those positions shown in Fig. 12-19 with respect to summation of the absolute values of the resultants shown in Fig. 12-19.

Standing Waves: Shorted Line

The right-hand column of Fig. 12-19 shows the out-of-phase reflection that occurs for current on an open line and voltage on a shorted line. The first graph shows that the incident wave is extended (in dashed lines) past the load *180° out of phase* from the incident wave and then folded back to provide the reflected wave. At *b* the reflected wave coincides with the incident wave, providing the maximum possible resultant. At *d* the reflected wave cancels with the incident wave so that the resultant at that instant of time is zero at all points on the line.

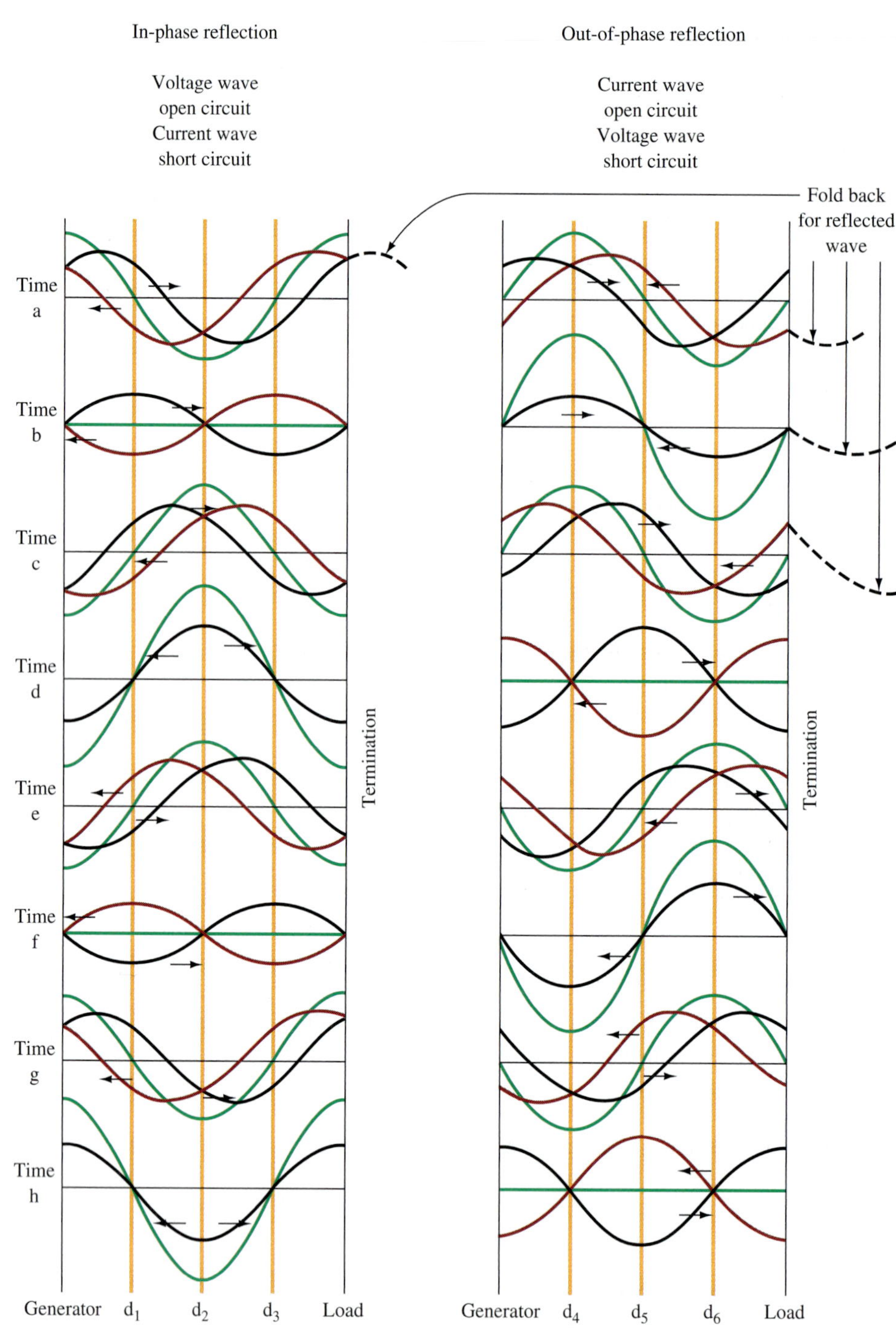

FIGURE 12-19 Development of standing waves.

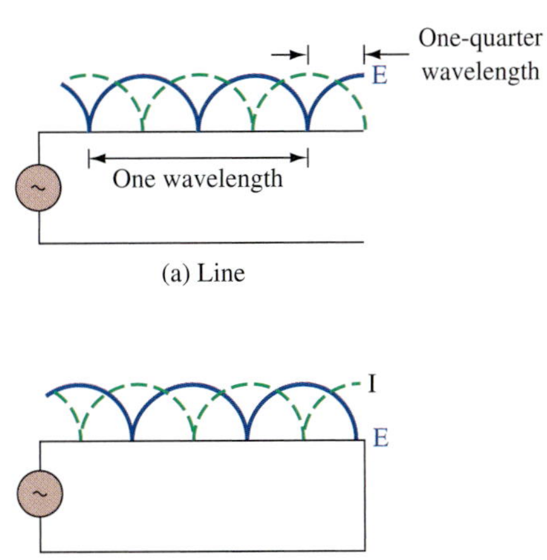

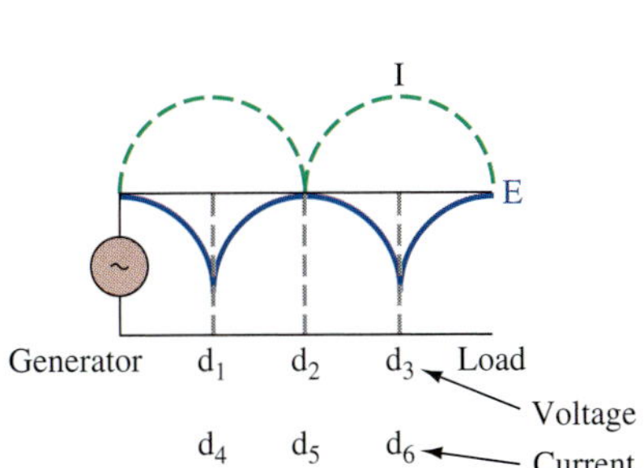

FIGURE 12-20 Conventional picture of standing waves—open line.

FIGURE 12-21 Standing waves of voltage and current.

At the end of a transmission line terminated in an open, the current is zero and the voltage is maximum. This relationship may be stated in terms of phase. The voltage and current at the end of an open-ended transmission line are 90° out of phase. At the end of a transmission line terminated in a short, the current is maximum and the voltage is zero. The voltage and current are again 90° out of phase.

These current–voltage relationships are shown in the diagrams in Fig. 12-21. These phase relationships are important because they indicate how the line acts at different points throughout its length.

A transmission line has points of maximum and minimum voltage as well as points of maximum and minimum current. The position of these points can be predicted accurately if the applied frequency and type of line termination are known.

EXAMPLE 12-6

An open-circuited line is 1.5λ long. Sketch the incident, reflected, and resultant voltage and current waves for this line at the instant the generator is at its peak positive value.

SOLUTION

Recall that to obtain the reflected voltage from an open circuit, the incident wave should be continued past the open (in your mind) and then folded back toward the generator. This process is shown in Fig. 12-22(a). Notice that the reflected wave coincides with the incident wave, making the resultant wave double the amplitude of the generator voltage. The current wave at an open circuit should be continued 180° out of phase, as shown in Fig. 12-22(b), and then folded back to provide the reflected wave. In this case, the incident and reflected current waves cancel one another, leaving a zero resultant all along the line at this instant of time.

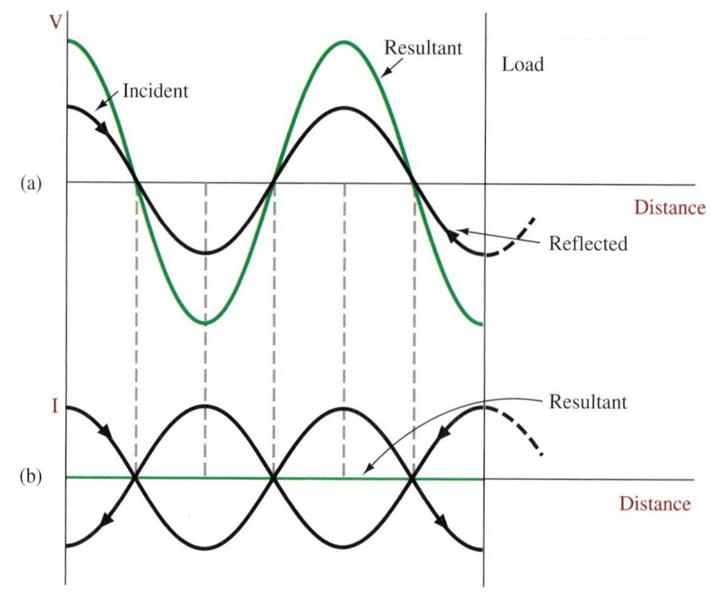

FIGURE 12-22 Diagram for Ex. 12-6.

12-7 Standing Wave Ratio

A standing wave is the result of an incident and reflected wave. The ratio of reflected voltage to incident voltage is called the reflection coefficient, Γ. That is,

$$\Gamma = \frac{E_r}{E_i} \tag{12-23}$$

where Γ is the reflection coefficient, E_r is the reflected voltage, and E_i is the incident voltage. When a line is terminated with a short circuit, open circuit, or purely reactive load, no energy can be absorbed by the load so that total reflection takes place. The reflected wave equals the incident wave so that $|\Gamma| = 1$. When a line is terminated with a resistance equal to the line's Z_0, no reflection occurs, and therefore $\Gamma = 0$. For all other cases of termination, the load absorbs some of the incident energy (but not all), and the reflection coefficient is between 0 and 1.

The reflection coefficient may also be expressed in terms of the load impedance:

$$\Gamma = \frac{Z_L - Z_0}{Z_L + Z_0} \tag{12-24}$$

where Z_L is the load impedance.

Voltage Standing Wave Ratio
ratio of the maximum voltage to minimum on a line

On a lossless line, the voltage maximums and minimums of a standing wave have a constant amplitude. The ratio of the maximum voltage to minimum on a line is called the **voltage standing wave ratio** (VSWR):

$$VSWR = \frac{E_{\max}}{E_{\min}} \tag{12-25}$$

In a more general sense, it is sometimes referred to as simply the **standing wave ratio** (SWR) because it is also equal to the ratio of maximum current to minimum current. Therefore,

Standing Wave Ratio another name for voltage standing wave ratio

$$\text{SWR} = \text{VSWR} = \frac{E_{max}}{E_{min}} = \frac{I_{max}}{I_{min}} \tag{12-26}$$

The VSWR can also be expressed in terms of Γ:

$$\text{VSWR} = \frac{1 + |\Gamma|}{1 - |\Gamma|} \tag{12-27}$$

The VSWR has an infinite value when total reflection occurs because E_{min} is 0 and has a value of 1 when no reflection occurs. No reflection means no standing wave so that the rms voltage along the line is always the same (neglecting losses); thus E_{max} equals E_{min} and the VSWR is 1.

Effect of Mismatch

The perfect condition of no reflection occurs only when the load is purely resistive and equal to Z_0. Such a condition is called a **flat line** and indicates a VSWR of 1. It is highly desirable because all the generator power capability is getting to the load. Also, if reflection occurs, the voltage maximums along the line may exceed the cable's dielectric strength, causing a breakdown. In addition, the existence of a reflected wave means greater I^2R (power) losses, which can be a severe problem when the line is physically long. The energy contained in the reflected wave is absorbed by the generator (except for the I^2R losses) and not "lost" unless the generator is not perfectly matched to the line, in which case a rereflection of the reflected wave occurs at the generator. In addition, a high VSWR also tends to accentuate noise problems and causes "ghosts" to be transmitted with video or data signals. Refer to Chapter 13 for a discussion of ghost problems.

Flat Line condition of no reflection; VSWR is 1

To summarize, then, the disadvantages of not having a perfectly matched (flat line) system are as follows:

1. The full generator power does not reach the load.
2. The cable dielectric may break down as a result of high-value standing waves of voltage (voltage nodes).
3. The existence of reflections (and rereflections) increases the power loss in the form of I^2R heating, especially at the high-value standing waves of current (current nodes).
4. Noise problems are increased by mismatches.
5. "Ghost" signals can be created.

The VSWR can be determined quite easily, if the load is a known value of pure resistance, by the following equation:

$$\text{VSWR} = \frac{Z_0}{R_L} \quad \text{or} \quad \frac{R_L}{Z_0} \quad \text{(whichever is larger)} \tag{12-28}$$

where R_L is the load resistance. Whenever R_L is larger than Z_0, R_L is used in the numerator so that the VSWR will be greater than 1. For instance, on a 100-Ω line

($Z_0 = 100\,\Omega$), a 200-Ω or 50-Ω R_L results in the same VSWR of 2. They both create the same degree of mismatch.

The higher the VSWR, the greater is the mismatch on the line. Thus, a low VSWR is the goal in a transmission line system except when the line is being used to simulate a capacitance, inductance, or tuned circuit. These effects are considered in the following sections.

Example 12-7

A citizen's band transmitter operating at 27 MHz with 4-W output is connected via 10 m of RG-8A/U cable to an antenna that has an input resistance of 300 Ω. Determine

(a) The reflection coefficient.

(b) The electrical length of the cable in wavelengths (λ).

(c) The VSWR.

(d) The amount of the transmitter's 4-W output absorbed by the antenna.

(e) How to increase the power absorbed by the antenna.

Solution

(a)
$$\Gamma = \frac{Z_L - Z_0}{Z_L + Z_0} \tag{12-24}$$
$$= \frac{300\ \Omega - 50\ \Omega}{300\ \Omega + 50\ \Omega} = \frac{5}{7} = 0.71$$

(b)
$$\lambda = \frac{v}{f} \tag{12-22}$$
$$= \frac{2.07 \times 10^8\ \text{m/s}}{27 \times 10^6\ \text{Hz}} = 7.67\ \text{m}$$

Because the cable is 10 m long, its electrical length is

$$\frac{10\ \text{m}}{7.67\ \text{m/wavelength}} = 1.3\lambda$$

(c) Because the load is resistive,

$$\text{VSWR} = \frac{R_L}{Z_0} \tag{12-28}$$
$$= \frac{300\ \Omega}{50\ \Omega} = 6$$

An alternative solution, because Γ is known, is

$$\text{VSWR} = \frac{1 + \Gamma}{1 - \Gamma} \tag{12-27}$$
$$= \frac{1 + \frac{5}{7}}{1 - \frac{5}{7}} = \frac{\frac{12}{7}}{\frac{2}{7}} = 6$$

(d) The reflected voltage is Γ times the incident voltage. Because power is proportional to the square of voltage, the reflected power is $(5/7)^2 \times 4\text{ W} = 2.04\text{ W}$, and the power to the load is

$$\begin{aligned} P_{\text{load}} &= 4\text{ W} - P_{\text{refl}} \\ &= 4\text{ W} - 2.04\text{ W} = 1.96\text{ W} \end{aligned}$$

(e) You do not have enough theory yet to answer this question. Continue reading this chapter and read Chapter 14 on antennas.

Example 12-7 showed the effect of mismatch between line and antenna. The transmitted power of only 1.96 W instead of 4 W would seriously impair the effective range of the transmitter. It is little wonder that great pains are taken to get the VSWR as close to 1 as possible.

Quarter-Wavelength Transformer

One simple way to match a line to a resistive load is by use of a **quarter-wavelength matching transformer.** It is not physically a transformer but does offer the property of impedance transformation. To match a resistive load, R_L, to a line with characteristic impedance Z_0, a λ/4 section of line with characteristic impedance Z_0' should be placed between them. In equation form, the value of Z_0' is

$$Z_0' = \sqrt{Z_0 R_L} \tag{12-29}$$

Quarter-Wavelength Matching Transformer
quarter-wavelength piece of transmission line of specified line impedance used to force a perfect match between a transmission line and its load resistance

The required λ/4 section to match the 50-Ω line to the 300-Ω resistive load of Ex. 12-7 would have $Z_0' = \sqrt{50\ \Omega \times 300\ \Omega} = 122\ \Omega$. This solution is shown pictorially in Fig. 12-23. The input impedance looking into the matching section is 50 Ω. This is the termination on the 50-Ω cable, and thus the line is flat, as desired, from there on back to the generator. Remember that λ/4 matching sections are only effective working into a resistive load. The principle involved is that there will now be two reflected signals, equal but separated by λ/4. Because one of them travels λ/2 farther than the other, this 180° phase difference causes cancellation of the reflections.

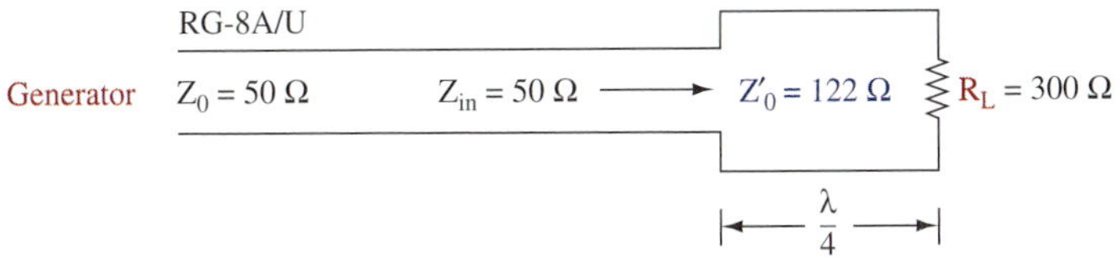

FIGURE 12-23 λ/4 matching section for Ex. 12-7.

Electrical Length

The **electrical length** of a transmission line is of importance to this overall discussion. Recall that when reflections occur, the voltage maximums occur at λ/2 intervals. If the transmission line is only λ/16 long, for example, the reflection still occurs, but the line is so short that almost no voltage variation along the line exists. This situation is illustrated in Fig. 12-24.

Electrical Length
the length of a line in wavelengths, not physical length

Notice that the λ/16 length line has almost no "standing" voltage variations along the line because of its short electrical length, while the line that is one λ long has two complete variations over its length. Because of this, the transmission line

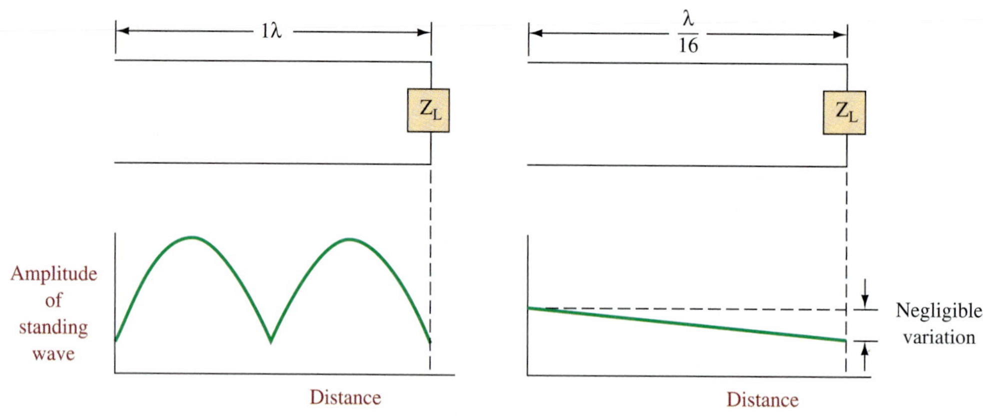

FIGURE 12-24 Effect of line electrical length.

effects we have been discussing are applicable only to electrically long lines—generally only those greater than $\lambda/16$ long.

Be careful that you understand the difference between electrical length and physical length. A line can be miles in length and still be electrically short if the frequency it carries is low. For example, a telephone line carrying a 300-Hz signal has a λ of 621 mi. On the other hand, a 10-GHz signal has a λ of 3 cm so that even extremely small circuit interconnections require application of transmission line theory.

12-8 The Smith Chart

Transmission Line Impedance

Transmission line calculations are very cumbersome from a purely mathematical standpoint. It is often necessary to know the input impedance of a line of a certain length with a given load. The impedance at any point on a line with standing waves is repetitive every half wavelength because the voltage and current waves are similarly repetitive. The impedance is therefore constantly changing along the line and is equal to the ratio of voltage to current at any given point. This impedance can be solved via the following equation for a lossless line:

$$Z_s = Z_0 \frac{Z_L + jZ_0 \tan \beta s}{Z_0 + jZ_L \tan \beta s} \tag{12-30}$$

where Z_s = line impedance at a given point
Z_L = load impedance
Z_0 = line's characteristic impedance
βs = distance from the load to the point where it is desired to know the line impedance (electrical degrees)

Because the tangent function is repetitive every 180°, the result will similarly be repetitive, as we know it should be. If the line is $\frac{3}{4}\lambda$ long (270 electrical degrees), its impedance is the same as at a point $\lambda/4$ from the load.

Smith Chart Introduction

Equation (12-30) can be solved, but the solution using the **Smith chart** is far more convenient and versatile. This impedance chart was developed by P. H. Smith in 1938 and is still widely used for line, antenna, and waveguide calculations in spite of the widespread availability of computers and programmable calculators.

Smith Chart
impedance chart developed by P. H. Smith, useful for transmission line analysis

Figure 12-25 illustrates the Smith chart. It contains two sets of lines. The lines representing constant resistance are circular and are all tangent to each other at the right-hand end of the horizontal line through the center of the chart. The value of resistance along any one of these circles is constant, and its value is indicated just above the horizontal line through the center of the chart.

The second set of lines represents arcs of constant reactance. These arcs are also tangent to one another at the right-hand side of the chart. The values of reactance for

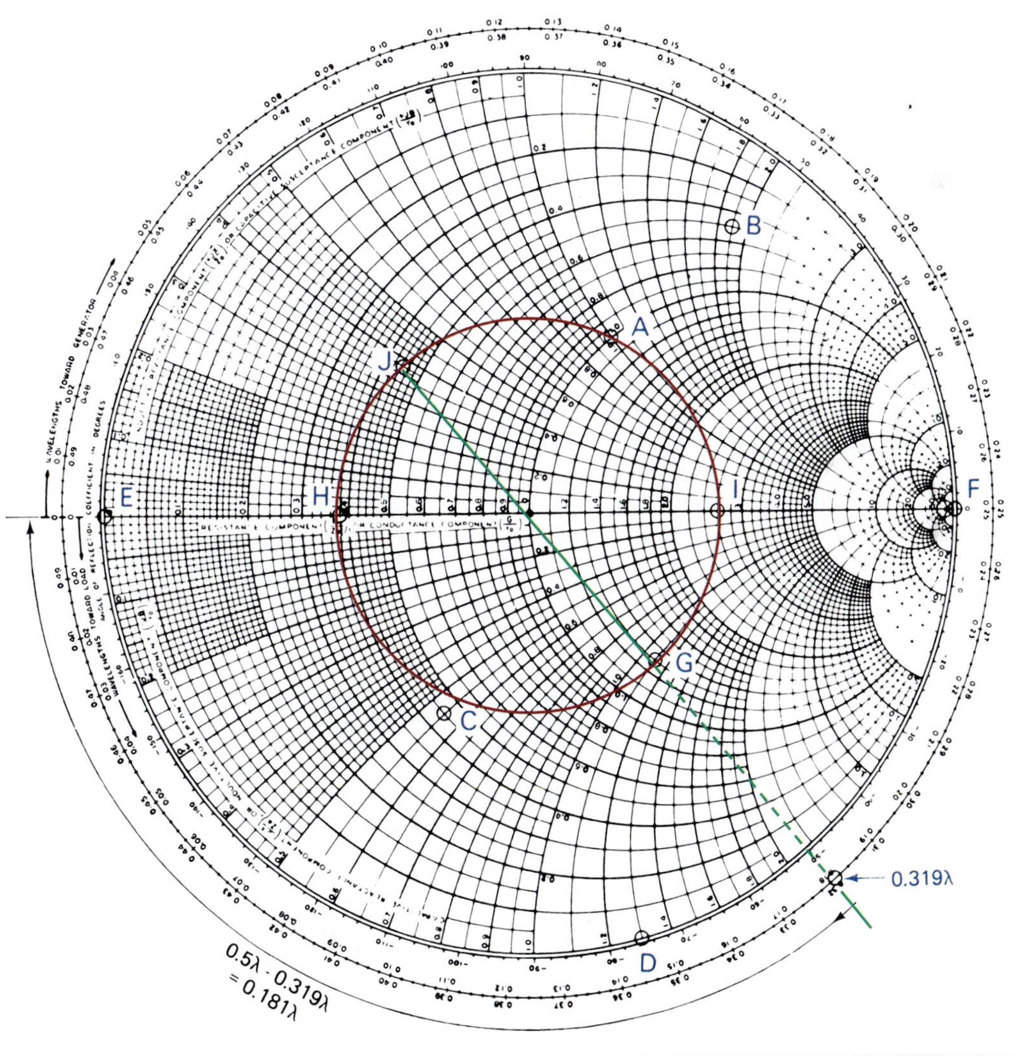

FIGURE 12-25 Smith chart.

each arc are labeled at the circumference of the chart and are positive on the top half and negative on the bottom half.

Several computer-aided design (CAD) programs are available that can accurately predict high-frequency circuit performance. It may therefore seem strange that the pencil-and-paper Smith chart is still utilized. The intuitive and graphical nature of Smith chart design is still desirable, however, because the effect of each design step is clearly apparent.

Using the Smith Chart

Normalizing
dividing impedances by the line's characteristic impedance

So the chart can be used for a wide range of impedance and admittance values, a process of **normalizing** is employed. This involves dividing all impedances by the line's Z_0 so that if the impedance $100 + j50$ were to be plotted for a 50-Ω line, the value $100/50 + j50/50 = 2 + j1$ would be used. The normalized impedance is then represented by a lowercase letter,

$$z = \frac{Z}{Z_0} \tag{12-31}$$

where z is the normalized impedance. When working with admittances,

$$y = \frac{Y}{Y_0} \tag{12-32}$$

where y is the normalized admittance and Y_0 is the line's characteristic admittance.

The point A in Fig. 12-25 is $z = 1 + j1$ or could be $y = 1 + j1$. It is the intersection of the one resistance circle and one reactive arc. Point B is $0.5 + j1.9$, while point C is $0.45 - j0.55$. Be sure that you now understand how to plot points on the Smith chart.

Point D (toward the bottom of the chart) in Fig. 12-25 is $0 - j1.3$. All points on the circumference of the Smith chart, such as point D, represent pure reactance except for the points at either extreme of the horizontal line through the center of the chart. At the left-hand side, point E, a short circuit ($z = 0 + j0$) is represented, while at the right-hand side, point F, an open circuit is represented ($z = \infty$).

Points G and J in Fig. 12-25 illustrate a practical application of the Smith chart. Assume that a 50-Ω transmission line has a load $Z_L = 65 - j55\ \Omega$. That load, when normalized, is

$$z_L = \frac{Z_L}{Z_0} = \frac{65}{50} - j\frac{55}{50} = 1.3 - j1.1$$

Point G is plotted as $z_L = 1.3 - j1.1$. By drawing a circle through point G and using the chart center as the circle's center, the locations of all impedances along the transmission line are described by the circle. Recall that the impedance along a line varies but repeats every half wavelength. A full revolution around the circle corresponds to a half-wavelength (180°) movement on a line. A clockwise (CW) rotation on the chart means you are moving toward the generator, while a counterclockwise (CCW) rotation indicates movement toward the load. The scales on the outer circumference of the chart indicate the amount of movement in wavelengths. For example, moving from the load (point G) toward the generator to point H brings us to a point on the line where the impedance is purely resistive and equal to $z = 0.4$. As shown in Fig. 12-25, that movement is equal to $0.5\lambda - 0.319\lambda = 0.181\lambda$. In other words, at a point 0.181λ from the load, the impedance on the line is purely resistive and has a normalized value of $z = 0.4$. In terms of actual impedance, the impedance is $Z = z \times Z_0 = 0.4 \times 50\ \Omega = 20\ \Omega$.

Moving another λ/4 or halfway around the circle from point *H* in Fig. 12-25 brings you to another point of pure resistance, point *I*. At point *I* on the line, the impedance is $z = 2.6$. That also is the VSWR on the line. Wherever the circle drawn through z_L for a transmission line crosses the right-hand horizontal line through the chart center, that point is the VSWR that exists on the line. For this reason, the circle drawn through a line's load impedance is often called its VSWR circle.

The Smith chart allows for a very simple conversion of impedance to admittance, and vice versa. The value of admittance corresponding to any impedance is always diagonally opposite and the same distance from the center. For example, if an impedance is equal to $1.3 - j1.1$ (point *G*), then the admittance *y* is at point *J* and is equal to $0.45 + j0.38$ as read from the chart. Because $y = 1/z$, you can mathematically show that

$$\frac{1}{1.3 - j1.1} = 0.45 + j0.38$$

but the Smith chart solution is much less tedious. Recall that mathematical *z* to *y* and *y* to *z* transformations require first a rectangular to polar conversion, taking the inverse, and then a polar to rectangular conversion.

Other applications of the Smith chart are most easily explained by solution of actual examples.

Example 12-8

Find the input impedance and VSWR of a transmission line 4.3λ long when $Z_0 = 100\ \Omega$ *and* $Z_L = 200 - j150\ \Omega$.

Solution

First normalize the load impedance:

$$z_L = \frac{Z_L}{Z_0} \tag{12-31}$$

$$= \frac{200 - j150\ \Omega}{100\ \Omega} = 2 - j1.5$$

That point is then plotted on the Smith chart in Fig. 12-26, and its corresponding VSWR circle is also shown. That circle intersects the right-hand half of the horizontal line at 3.3 as shown, and therefore the VSWR = 3.3. Now, to find the line's input impedance, it is required to move from the load toward the generator (CW rotation) 4.3λ. Because each full revolution on the chart represents λ/2, it means that eight full rotations plus 0.3λ is necessary. Thus, just moving 0.3λ from the load in a CW direction will provide z_{in}. The radius extended through z_L intersects the outer wavelength scale at 0.292λ. Moving from there to 0.5λ provides a movement of 0.208λ. Thus, additional movement of 0.092λ (0.3λ − 0.208λ) brings us to the generator and provides z_{in} for a 4.3λ line (or 0.3λ, 0.8λ, 1.3λ, 1.8λ, etc., length line). The line's z_{in} is read as

$$z_{in} = 0.4 + j0.57$$

from the chart, which in ohms is

$$Z_{in} = z_{in} \times Z_0 = (0.4 + j0.57) \times 100\ \Omega = 40\ \Omega + j57\ \Omega$$

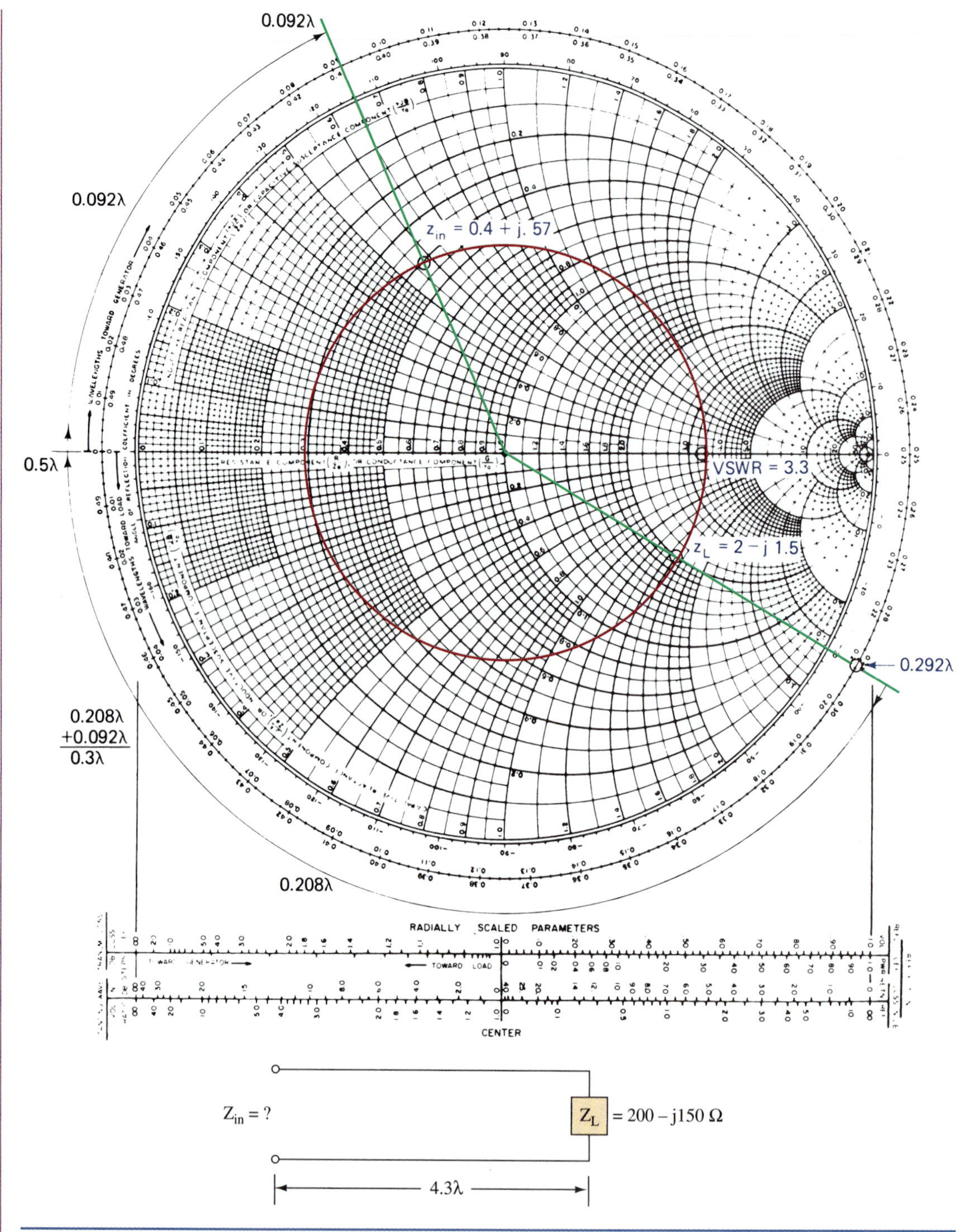

FIGURE 12-26 Smith chart for Ex. 12-8.

Corrections for Transmission Loss

Example 12-8 assumed the ideal condition of a lossless line. If line attenuation cannot be neglected, the incident wave gets weaker as it travels toward the load, and the reflected wave gets weaker as it travels back toward the generator. This causes the VSWR to get weaker as we approach the source end of the line. A true standing wave representation on the Smith chart would be a spiral. The correction for this condition is made using the "transm. loss, 1-dB steps" scale at the bottom left-hand side of the Smith chart. Other scales at the bottom of the chart can be used to provide the VSWR (which can be taken directly off the chart, as previously shown), the voltage and power reflection coefficient, and decibel loss information.

Matching Using the Smith Chart

Many of the calculations involved with transmission lines pertain to matching a load to the line and thus keeping the VSWR as low as possible. The following examples illustrate some of these situations.

Example 12-9

A load of 75 Ω + j50 Ω is to be matched to a 50-Ω transmission line using a λ/4 matching section. Determine the proper location and characteristic impedance of the matching section.

Solution

1. Normalize the load impedance:

$$z_L = \frac{Z_L}{Z_0} = \frac{75\ \Omega + j50\ \Omega}{50\ \Omega} = 1.5 + j1$$

2. Plot z_L on the Smith chart and draw the VSWR circle. This is shown in Fig. 12-27.
3. From z_L move toward the generator (CW) until a point is reached where the line is purely resistive. Recall that the λ/4 matching section works only between a pure resistance and the line.
4. Point A or B in Fig. 12-27 could be used as the point where the matching section is inserted. We shall select point A because it is closest to the load. It is 0.058λ from the load. At that point the line should be cut and the λ/4 section inserted as shown in Fig. 12-27.
5. The normalized impedance at point A is purely resistive and equal to $z = 2.4$. That also is the VSWR on the line with no matching. The actual resistance is 2.4 × 50 Ω = 120 Ω. Therefore, the characteristic impedance of the matching section is

$$Z_0' = \sqrt{Z_0 \times R_L} \tag{12-29}$$
$$= \sqrt{50\ \Omega \times 210\ \Omega} = 77.5\ \Omega$$

Single-Stub Tuner the stub's distance from the load and the location of its short circuit are adjustable to allow a match between line and load

Double-Stub Tuner has fixed stub locations, but the position of the short circuits is adjustable to allow a match between line and load

Stub Tuners

The use of short-circuited stubs is prevalent in matching problems. A **single-stub tuner** is illustrated in Fig. 12-28(a). In it, the stub's distance from the load and location of its short circuit are adjustable. The **double-stub tuner** shown in Fig. 12-28(b) has fixed stub locations, but the position of both short circuits is adjustable. The following example illustrates the procedure for matching with a single-stub tuner.

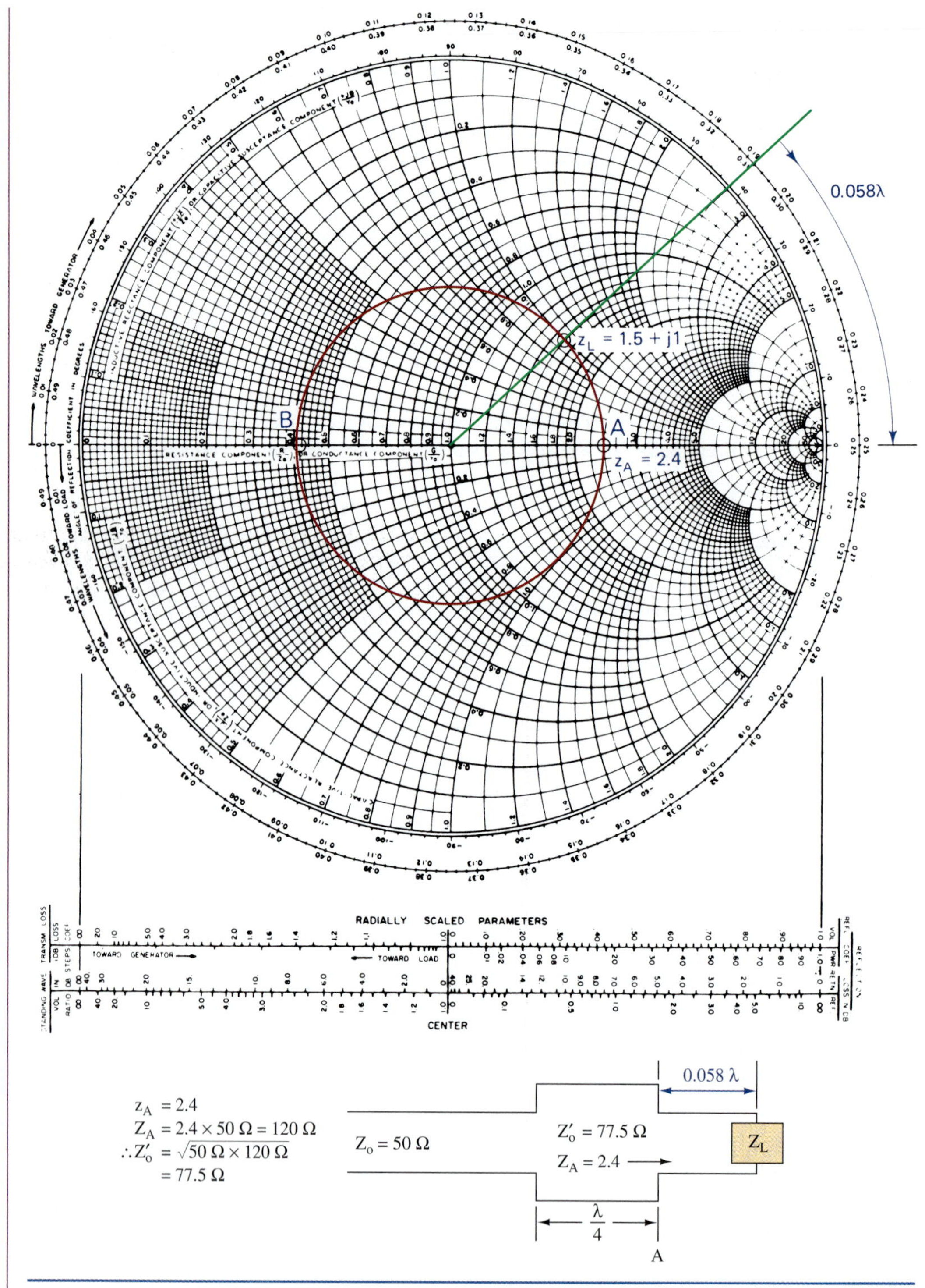

FIGURE 12-27 Smith chart for Ex. 12-9.

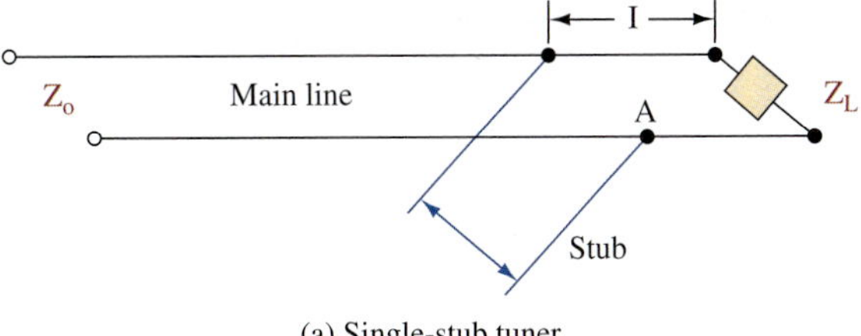

(a) Single-stub tuner

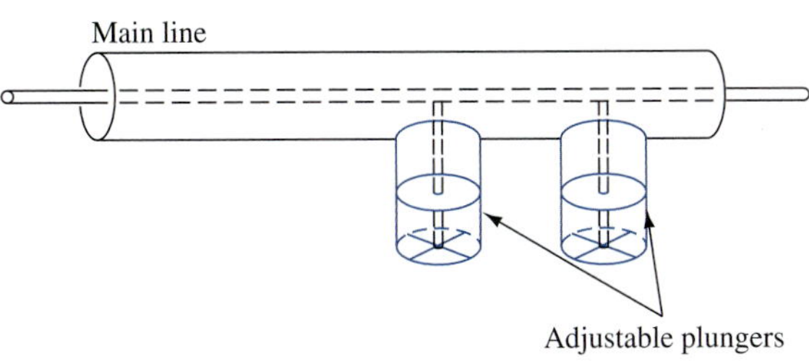

(b) Double-stub tuner

FIGURE 12-28 Stub tuners.

Example 12-10

The antenna load on a 75-Ω line has an impedance of 50 − j100 Ω. Determine the length and position of a short-circuited stub necessary to provide a match.

Solution

1.
$$z_L = \frac{Z_L}{Z_o} = \frac{50 - j100\ \Omega}{75\ \Omega} = 0.67 - j1.33$$

2. Plot z_L on the Smith chart (Fig. 12-29), and draw the VSWR circle.
3. Convert z_L to y_L by going to the diagonally opposite side of the VSWR circle from z_L. Read $y_L = 0.27 + j0.59$.
4. Move from y_L to the point where the admittance is 1 ± whatever j term results. That is point A in Fig. 12-29. Read point A as $y_A = 1 + j1.75$ and note that point A is 0.093λ from the load. The short-circuited stub should be located 0.093λ from the load.
5. Now the $+j1.75$ term at point A must be canceled by the stub admittance. If the stub admittance is $y_s = -j1.75$, the imaginary terms cancel because the total admittance at point A with the parallel stub is $(1 + j1.75 - j1.75) = 1$. Recall that parallel admittances are directly additive.
6. The load admittance of the short-circuited stub is infinite. That is plotted on the Smith chart as point B. From point B, move toward the generator until the stub admittance is $-j1.75$. That corresponds to a distance of 0.083λ and is the required length for the short-circuited stub.
7. A match is now accomplished because the total admittance at the stub location is 1. This means that

$$z = \frac{1}{y} = \frac{1}{1} = 1$$

or $Z = 1 \times 75\ \Omega = 75\ \Omega$, which matches the line to the point where the stub is connected back toward the generator.

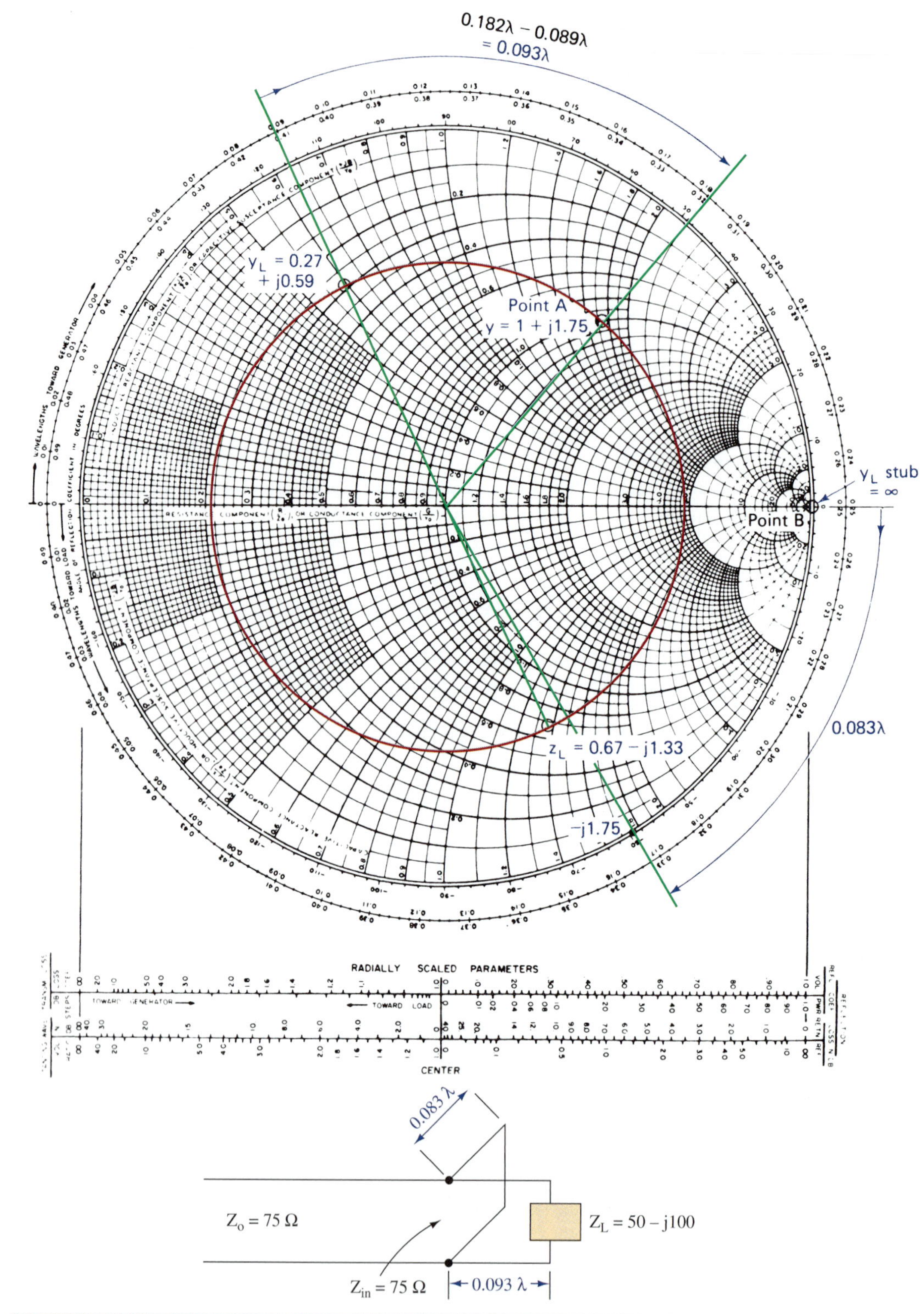

FIGURE 12-29 Smith chart for Ex. 12-10.

The key to matching with the single-stub tuner of the previous example is moving back from the load until the admittance takes on a normalized real term of 1. Whatever j term (imaginary) results can then be canceled with a short-circuited stub made to look like the same j term with opposite polarity. While the Smith chart may at first be perplexing to work with, a bit of practice allows mastery in a short period of time.

12-9 Transmission Line Applications

Discrete Circuit Simulation

Transmission line sections can be used to simulate inductance, capacitance, and LC resonance. Figure 12-30 tabulates these effects for open and shorted sections with lengths equal to, less than, and greater than a quarter wavelength. A shorted quarter-wavelength section looks like a parallel LC circuit or ideally, therefore, like an open circuit. A shorted section less than $\lambda/4$ looks like a pure inductance, and a section greater than $\lambda/4$ looks like a pure capacitance. These effects can be verified on a Smith chart by plotting z_L of 0 for a short-circuited line at the left center of the chart. Moving CW toward the generator less than $\lambda/4$ puts you on the top half of the chart, indicating a $+j$ term and thus inductance. Moving exactly $\lambda/4$ puts you at the center-right point on the chart, indicating the infinite impedance of an ideal tank circuit, and moving past that point provides the $-j$ impedance of capacitance.

While open-circuit sections would seem to provide similar effects, they are seldom used. The open-circuited line tends to radiate a fair amount of energy off the end of the line so that total reflection does not take place. This causes the simulated circuit element to take on a resistive term, which greatly reduces its quality. These losses do not occur with shorted sections, and the simulated circuit has better quality than is possible using discrete inductors and/or capacitors. Short-circuited $\lambda/4$ sections offer Qs of about 10,000 as compared to a maximum of about 1000 using a very high-grade inductor and capacitor.

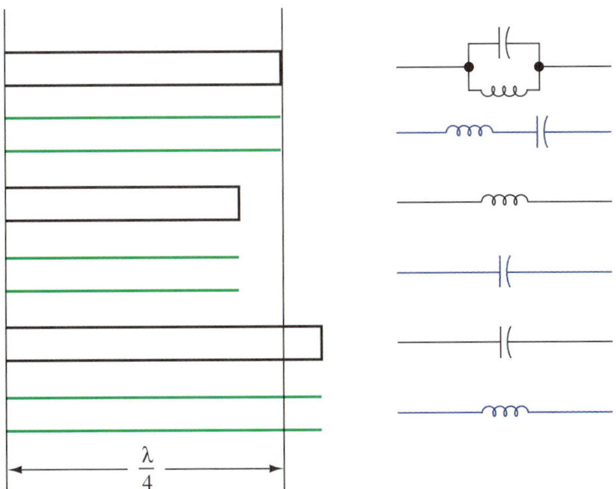

FIGURE 12-30 Transmission line section equivalency.

For this reason, it is normal practice to use transmission line sections to replace inductors and/or capacitors at frequencies above 500 MHz where the line section becomes short enough to be practically used. They are commonly found in the oscillator of UHF tuners for television; the tuners operate from about 500 to 800 MHz.

Baluns

As described in Sec. 12-2, parallel wire line generally carries two equal but 180° out-of-phase signals with respect to ground. Such a line is called a *balanced* line. In a coaxial line, the outer conductor is usually grounded so that the two conductors do not carry signals with the same relationship to ground. The coaxial line is therefore termed an *unbalanced* line.

It is sometimes necessary to change from an unbalanced to a balanced condition such as when a coaxial line feeds a balanced load like a dipole antenna. They cannot be connected directly together because the antenna lead connected to the cable's grounded conductor would allow the shield to become part of the antenna.

This situation is solved through use of an unbalanced-to-balanced transformer, which is usually called a *balun*. It can be a transformer as shown in Fig. 12-31(a) and described in Sec. 12-2 or a special transmission line configuration as shown in Fig. 12-31 (b). The use of the standard transformer is limited due to excessive losses at high frequencies. The balun in Fig. 12-31(b) does not suffer from that disadvantage. The inner conductor of the coaxial line is tapped at 180° ($\lambda/2$) from the end. The tap and the end of the inner conductor provide two equal but 180° out-of-phase signals, neither of which is grounded. Thus, they supply the required signals for the balanced line. These baluns are reversible because they function equally well in going from a balanced to an unbalanced condition.

Transmission Lines as Filters

A quarter-wave section of transmission line can be used as an efficient filter or suppressor of even harmonics. Other types of filters may be used to filter out the odd

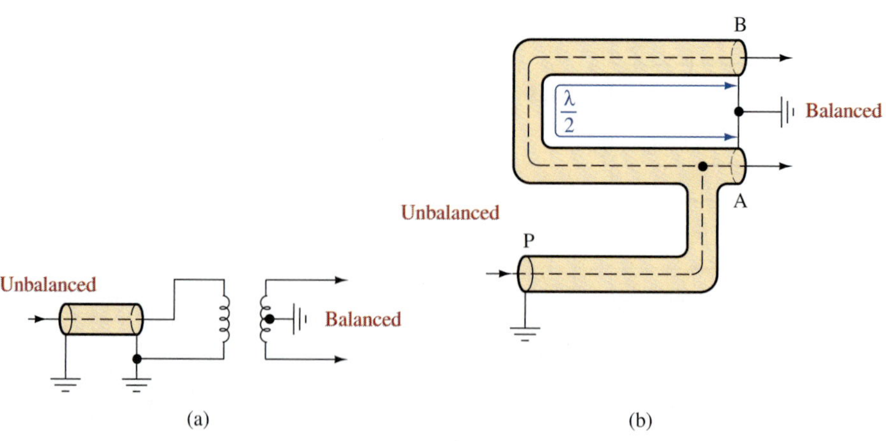

FIGURE 12-31 Baluns.

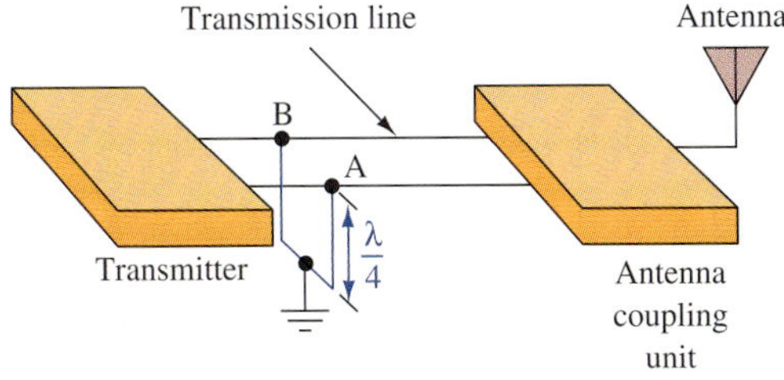

FIGURE 12-32 Quarter-wave filters.

harmonics. In fact, filters may be designed to eliminate the radiation of an entire single sideband of a modulated carrier.

Suppose that a transmitter is operating on a frequency of 5 MHz, and it is found that the transmitter is causing interference at 10 and 20 MHz. A shorted quarter-wave line section may be used to eliminate these undesirable harmonics. A quarter-wave line shorted at one end offers a high impedance at the unshorted end to the fundamental frequency. At a frequency twice the fundamental, such a line is a half-wave line, and at a frequency 4 times the fundamental, the line becomes a full-wave line. A half-wave or full-wave line offers zero impedance when its output is terminated in a short. Therefore, the radiation of even harmonics from the transmitter antenna can be eliminated almost completely by the circuit shown in Fig. 12-32.

The resonant filter line, *AB*, is a quarter wave in length at 5 MHz and offers almost infinite impedance at this frequency. At the second harmonic, 10 MHz, the line *AB* is a half-wave line and offers zero impedance, thereby shorting this frequency to ground. The quarter-wave filter may be inserted anywhere along the nonresonant transmission line with a similar effect.

Slotted Lines

One of the simplest and yet most useful measuring instruments at very high frequencies is the **slotted line.** As its name implies, it is a section of coaxial line with a lengthwise slot cut in the outer conductor. A pickup probe is inserted into the slot, and the magnitude of signal picked up is proportional to the voltage between the conductors at the point of insertion. The probe rides in a carriage along a calibrated scale so that data can be obtained to plot the standing wave pattern as a function of distance. From these data, the following information can be determined:

Slotted Line
section of coaxial line with a lengthwise slot cut in the outer conductor to allow measurement of the standing wave pattern

1. VSWR
2. Generator frequency
3. Unknown load impedance

Time-Domain Reflectometry

Time-domain reflectometry (TDR) is a system whereby a short-duration pulse is transmitted into a line. If monitored with an oscilloscope, the reflection of that pulse provides much information regarding the line. Of course, no reflection at all indicates an infinitely long line or (more likely) a line with a perfectly matched load and no discontinuities.

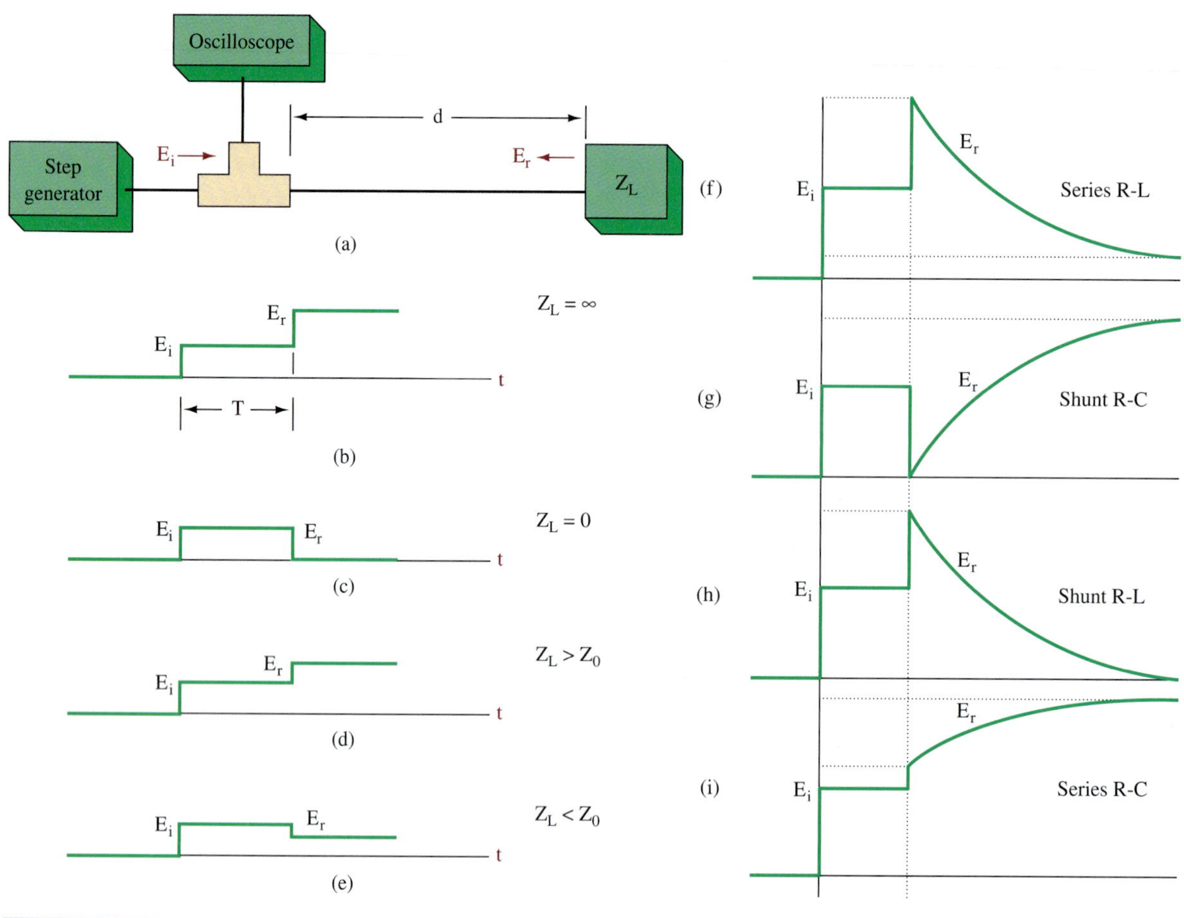

FIGURE 12-33 Time-domain reflectometry.

A very common problem with transmission line communications involves cable failure between communications terminals. These problems are usually due to chemical erosion or a mechanical break. Location of these problems is done by using a time-domain reflectometer. These instruments can pinpoint a fault within several feet at a distance of 10 mi. This is accomplished by simply measuring the time taken for the return pulse and then calculating distance based on the cable's propagation velocity.

TDR is also often used in laboratory testing. In these cases the transmitted signal is sometimes a fast-rise-time step signal. The setup for this condition is shown in Fig. 12-33(a). In Fig. 12-33(b) the case for an open-circuited line is shown. Assuming a lossless line, the reflected wave is in phase and effectively results in a doubling of the incident step voltage. The time T shown is the time of incidence plus reflection and therefore should be divided by 2 to calculate the distance to the open circuit. Thus, the distance is propagation velocity (V_p) times $T/2$.

The case for a shorted line is shown in Fig. 12-33(c). In this case the reflection is out of phase and results in cancellation. The situation of Z_L greater than Z_0 is shown in Fig. 12-33(d), while Fig. 12-33(e) shows the case for Z_L less than Z_0.

The magnitudes shown in the scope display can be used to determine Z_L or the reflection coefficient Γ because

$$\Gamma = \frac{E_r}{E_i} \tag{12-23}$$

and

$$\Gamma = \frac{Z_L - Z_0}{Z_L + Z_0} \tag{12-24}$$

and the final voltage, E_F, is

$$\begin{aligned} E_F &= E_i + E_r \\ &= E_i(1 + \Gamma) \\ &= E_i\left(1 + \frac{Z_L - Z_0}{Z_L + Z_0}\right) \end{aligned}$$

Figures 12-33(f) through (i) displays the TDR signal for complex impedances.

12-10 TROUBLESHOOTING

Vital links of communication data flow over transmission lines daily. Phone conversations, computer data, and audio and video information all depend on the media's ability to pass these signals wholly without any loss of quality. In this section, we will look at a popular application for transmission lines. We will also look at problems encountered in the cabling application. Many opportunities exist today for the technician who can effectively track down and repair cabling troubles.

After completing this section you should be able to

- Identify popular types of cabling available for LANs and other uses
- Describe crosstalk interference and give an example of it
- Describe simple methods for testing cable resistance, insulation, and TDR
- Troubleshoot television antenna lines

COMMON APPLICATION

A very common application for miles and miles of wire today is the local area network (LAN). The most popular kind of wiring used to connect these systems is the unshielded twisted-pair UTP. Twisted pair is used mainly to connect the workstation to the hub/switch, or punch-down block. The vitality of a local area network depends on the cabling base—the transmission media. As a matter of fact, 75 percent of LAN failures are wiring related. Therefore, technicians must learn techniques to track cable problems and repair them effectively.

LOSSES ON TRANSMISSION LINES

From previous discussions in this chapter, transmission line losses can be summed up as energy losses, magnetic field losses, and electric field losses. Energy losses are a result of wire heating and leakage currents in the dielectric material used as the

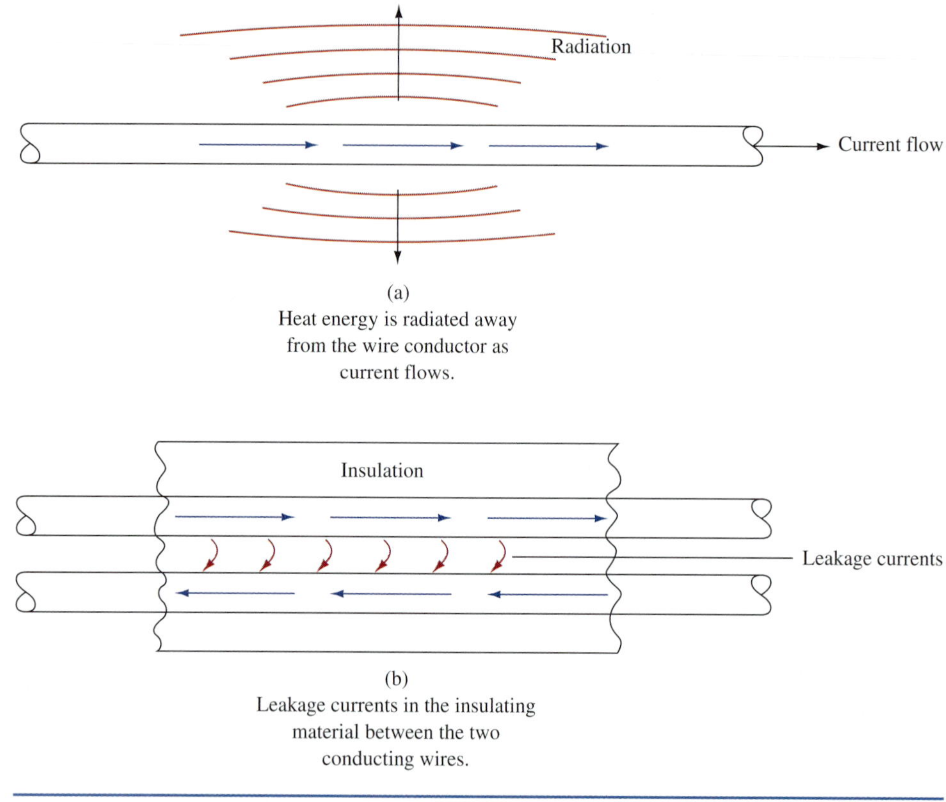
Radiation
Current flow
(a)
Heat energy is radiated away from the wire conductor as current flows.
Insulation
Leakage currents
(b)
Leakage currents in the insulating material between the two conducting wires.

FIGURE 12-34 Heat radiation and leakage losses.

insulator. Heat is produced when current flows in the wire that has resistance. The heat is radiated away from the wire as shown in Fig. 12-34(a). Figure 12-34(b) illustrates the effect of dielectric losses. Voltage potentials develop between the conductors that in turn cause minute currents to flow through the dielectric material. Heating and dielectric losses increase as signal frequency increases.

Magnetic field losses occur when currents are induced in a nearby conductor from the influence of another conductor's magnetic fields. Likewise, electrostatic fields tend to build a charge in a nearby conductor, causing electric field loss. Another effect of magnetic and electrostatic fields is induced interference, called crosstalk.

Interference on Transmission Lines

When UTP is used for the wire base in a LAN, there is a transfer of energy between the wires, causing interference to be induced on wires. This unwanted coupling is caused by overlapping electric and magnetic fields and is called crosstalk. All of us have experienced hearing another party's conversation while having our own on the telephone. This is an example of crosstalk. Crosstalk can be reduced by careful attention to wire separation and twists. Tighter twists create better coupling, which cancels the effects of magnetic and electrostatic fields, thus reducing crosstalk interference. Don't untwist twisted pair, even at the connection points. Crosstalk can be measured by special handheld instruments that directly compare the crosstalk reading against set standards. Crosstalk is also reduced by using STP or coaxial cable, but expense goes up and must be weighed against the system's requirements.

Logic circuits can be fooled into creating bad data from noise. Noise can come from various sources. Electric lights, electric generators, and switching devices generate noise that can become part of the signal if proper precautions are not observed. Do not run sensitive cabling near ac power lines or motors. Avoid running cable near overhead fluorescent lighting. In troubleshooting a system that is bothered by interference, check the wiring; someone may have ignored the above warnings. Shielding and proper cable terminations are also safeguards against noise interference.

Cable Testing

Test instruments have been developed to do sophisticated testing on cable links. Discussing these instruments in detail is beyond the scope of this textbook; however, some instruments will be mentioned. Simple testing techniques can be used to discover faulty cabling. Pin-to-pin continuity testing done from one end of the wire to the opposite end is a simple effective means to detect miswiring and shorted and open links. Resistance measurements can be made on lengths of single wire and compared to the cable manufacturer's specified resistance value. At the far end, the wires to be tested would be shorted together, and the resistance would be measured at the near end. The resistance value would represent the resistance of two wires and must be taken under consideration. Resistance readings above or below the expected value indicates opens or shorts in the wires. Insulation testers or the megohm meter (megometer) give a readout or audible indication of the insulation test results. Normally the megometer is connected to the wires at the near end, and the wires are disconnected at the opposite end. The readout represents the condition of the insulation. Breakdowns of the insulation at any point along the wiring path are readily displayed.

Television Antenna Line Repair

Transmission lines seldom totally fail without human intervention; however, they do age. Sunlight is hard on plastic dielectrics used in ribbon-type lines, sometimes called twin-lead. Where twin-lead is used near salt water, it deteriorates more rapidly. Moisture and salt collect on the dielectric, causing the loss to rise to the point that the twin-lead is not usable.

Coaxial cables are relatively immune to aging due to weather. Older types suffered from contamination due to plasticizer in the outer cover. Water can enter the cable at connectors and breaks in the outer covering. Water has a very high dielectric constant and changes the characteristic impedance and increase the loss. Sometimes coaxial cable is installed with very tight turns around corners. Over time, the center conductor will walk through the dielectric and eventually short the cable.

Suppose you are called to service a television receiver with "snow" almost overwhelming the picture. TV twin-lead is being used and is too cheap to justify much testing. Simply inspect it, looking for cracks in the plastic and discoloration of the plastic. You might take the time to short the far end and check resistance with an ohmmeter; there should be almost no resistance. If the far end is open, the resistance should be infinite. When in doubt, change the twin-lead.

12-11 TROUBLESHOOTING WITH ELECTRONICS WORKBENCH™ MULTISIM

The Smith chart was introduced in Chapter 12. In this exercise, this important impedance-calculating tool is reintroduced using Electronics Workbench™ Multisim. Multisim provides a network analyzer instrument that contains the Smith chart analysis in addition to many other useful analytical tools. A network analyzer is used to measure the parameters commonly used to characterize circuits or elements that operate at high frequencies. This exercise focuses on the Smith chart and the Z-parameter calculations. Z-parameters are the impedance values of a network expressed using its real and imaginary components. Refer to Sec. 12-8 for additional Smith chart examples and a more detailed examination of their function.

Begin the exercise by opening **Fig12-35.ms7 (.msm)** in your Electronics Workbench™ Multisim CD. This circuit, shown in Fig. 12-35, contains a 50-Ω resistor connected to port 1 (P1) of the network analyzer, and port 2 (P2) is terminated with a 50-Ω resistor. The first circuit being examined by the network analyzer is a simple resistive circuit. This example provides a good starting point for understanding the setup for the network analyzer and how to read the simulation results. The first circuit being tested by the network analyzer is shown in Fig. 12-35.

Start the simulation. The impedance calculations performed by the network analyzer are very quick and the start-simulation button resets quickly. Before you look at the test results, predict what you will see. Based on the information you learned in Sec. 12-8 and the fact that you are testing a resistor, you would expect to see a purely resistive result. Double-click on the network analyzer to open the instrument. You should see a Smith chart similar to the one shown in Fig. 12-36.

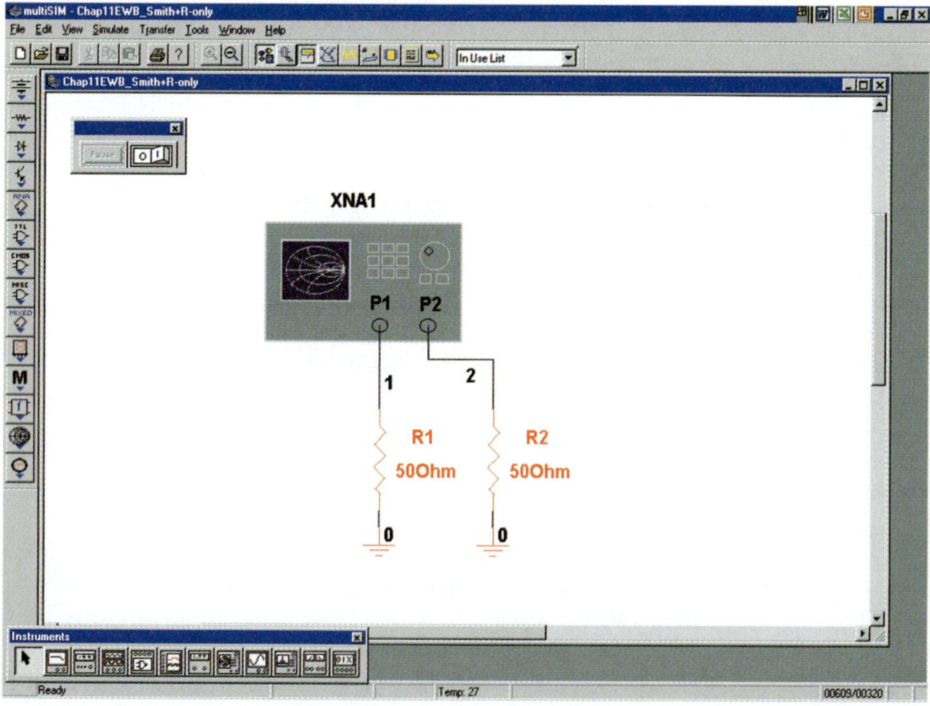

FIGURE 12-35 An example of using the Multisim Network Analyzer to analyze a 50-Ω resistor.

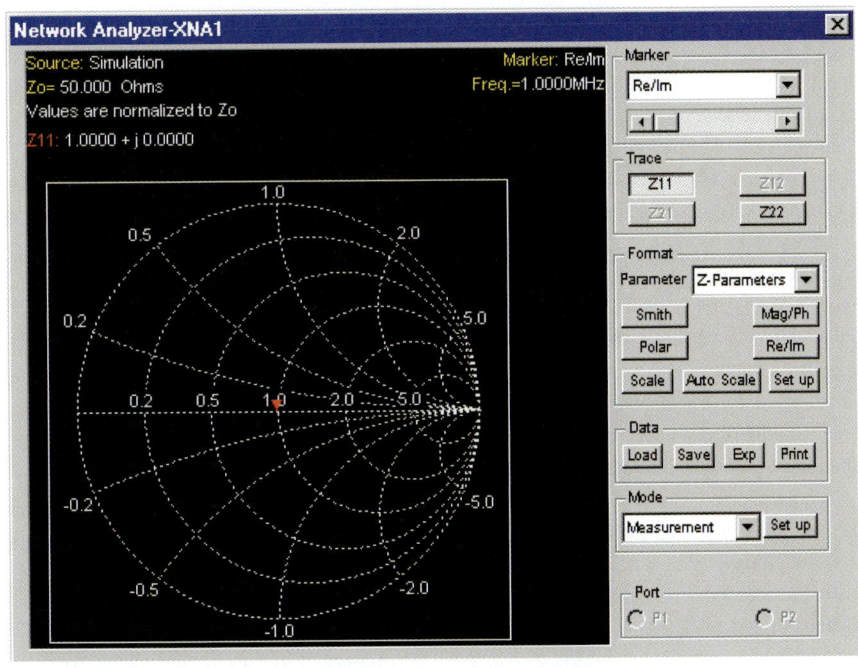

FIGURE 12-36 The Smith chart for the test of the 50-Ω resistor.

The Smith chart indicates the following:

$$Z_o = 50\ \Omega$$

$$Z_{11} = 1 + j0 \qquad \text{Values are normalized to } Z_o$$

The value $Z_{11} = 1 + j0$ indicates that the input impedance for the network being analyzed is purely resistive and its normalized value is 1, which translates to 50 Ω. Recall that the values on a Smith chart are divided by the normalized resistance. Notice the red marker on the Smith chart located at 1.0 on the real axis. The 1.0 translates to 50 Ω, and this value is obtained by multiplying the Smith chart measured resistance of 1 Ω by the characteristic impedance of 50 Ω to obtain the actual resistance measured. In this case, the computed resistance is 50 Ω.

The frequency at which this calculation was made is shown in the upper-right corner of the Smith chart screen. In this case, a frequency of 1.0 MHz was used. Immediately to the right of the frequency is a slider that can be used to select the frequency being used to determine the impedance. Move the slider from one end to the other to see how the impedance values change through the frequency range. The frequency range is set by clicking on the **Set-up** button next to the Measurement mode at the bottom right of the network analyzer screen. Double-click on the **Set-up** button to check the settings. You will notice the following:

Start frequency	1 MHz
Stop frequency	10 GHz
Sweep type	Decade
Number of points per decade	25
Characteristic impedance Z_o	50 Ω

The start and stop frequencies provide control of the frequency range when testing a network. The sweep type can be either decade or linear, but decade is used most of the time. The number of points per decade enables the user to control the resolution of the plotted trace displayed, and the characteristic impedance Z_o provides for user control of the normalizing impedance.

The next two Multisim exercises provide examples of using the Multisim network analyzer to compute the impedances of simple *RC* and *RL* networks. These exercises will help you better understand the Smith chart results when analyzing complex impedances. Open the file **FigE12-1.ms7 (.msm)** in your EWB CD. This example contains a simple *RC* network of $R = 25\ \Omega$ and $C = 6.4$ nF. The network analyzer is set to analyze the frequencies from 1 MHz to 100 MHz. The results of the simulation are shown in Fig. 12-37. At 1 MHz, the normalized input impedance to the *RC* network shows that $Z_{11} = 0.5 - j0.497$. Multiplying these values by the normalized impedance of 50 Ω yields a Z of approximately $25 - j25$, which is the expected value for this *RC* network at 1 MHz.

Next, open the file **FigE12-2.ms7 (.msm)** in your EWB CD. This example contains a simple *RL* network of $R = 25\ \Omega$ and $L = 4\ \mu$H. The network analyzer is set to analyze the frequencies from 1 MHz to 10 GHz. The results of the simulation are shown in Fig. 12-38. At 1 MHz, the normalized input impedance to the *RL* network shows that $Z_{11} = 0.5 + j0.5$. Multiplying these values by the normalized impedance of 50 Ω yields $Z = 25 + j25$, which is the expected value for this *RL* network at 1 MHz.

The following Electronics Workbench™ Multisim exercises provide additional experience with using the Multisim network analyzer and analyzing the results displayed on the Smith chart.

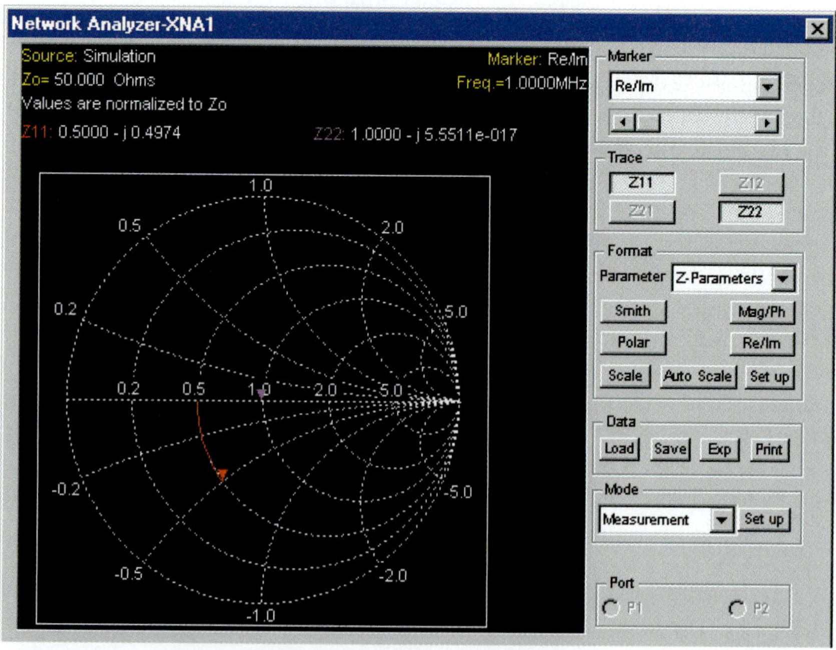

FIGURE 12-37 The Smith chart result for the simple *RC* network.

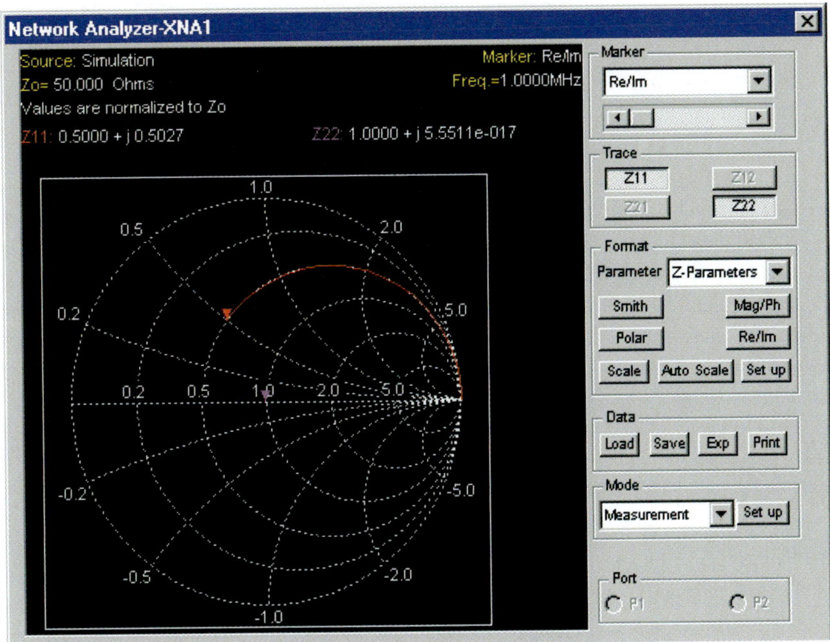

FIGURE 12-38 The Smith chart result for the simple *RL* network.

Electronics Workbench™ Exercises

1. Open the file **FigE12-3.ms7 (.msm)** on your EWB CD. Determine the impedance of this *RC* network at 10 MHz. ($Z = 25 - j1.59 \times 10^{-3}$)
2. Open the file **FigE12-4.ms7 (.msm)** on your EWB CD. Determine the impedance of this *RL* network at 1 GHz and 10 GHz. *Note:* Don't forget to multiply the impedance displayed on the network analyzer by the characteristic impedance. ($Z = 25 + j62830$, $Z = 25.79 + j62.83E \times 10^4$, basically an open circuit)
3. Open the file **FigE12-5.ms7 (.msm)** on your EWB CD. Use the Smith chart to determine the resonant frequency of the network. *Hint:* At resonance, a series *RLC* circuit will be resistive only. You must open the setup on the network analyzer to change the start frequency and improve the resolution of the plotted values by changing the number of points per decade to 50 so you can see at what frequency the impedance curve passes through the real axis. ($f = 50.11$ kHz)

Summary

In Chapter 12 we introduced basic transmission line theory. You learned that a basic wire doesn't just act like a strip of copper when the frequency is high enough. The major topics you should now understand include:

- the definition and types of transmission lines, including two-wire open line, twisted pair, shielded pair, and coaxial cable
- the key aspects for testing UTP

- the discussion of transmission line electrical characteristics, emphasizing characteristic impedance and the various line losses
- the explanation of dc voltage propagation, including velocity, delay, and reflections from both short-circuited and open-circuited lines
- the analysis of standing waves for open and shorted lines
- the definition of voltage standing wave ratio (VSWR), electrical length of a line, effects of a mismatched load, and use of a quarter-wavelength matching transformer
- the description of a Smith chart and its use in matching a load to a line
- the application of transmission lines as discrete circuit components, unbalanced-to-balanced transformers (baluns), and filters
- the testing of transmission lines using slotted lines and time-domain reflectometers

Questions and Problems

Section 12-2

1. Define *transmission line*. If a simple wire connection can be a transmission line, why is an entire chapter of study devoted to it?
2. In general terms, discuss the various types of transmission lines. Include the advantages and disadvantages of each type.

* 3. What would be the considerations in choosing a solid dielectric cable over a hollow pressurized cable for use as a transmission line?

* 4. Why is an inert gas sometimes placed within concentric radio-frequency transmission cables?

5. Where are twisted-pair cables often used?
6. What is meant by the CAT6/5e designation?
7. Define *near-end crosstalk* and *attenuation* relative to testing twisted-pair cable.
8. Describe three additional test considerations for twisted-pair cable with the enhanced data capabilities.
9. Explain how an unbalanced line differs from a balanced line.
10. Explain how the pickup of unwanted signals on a balanced transmission line is attenuated when converted to an unbalanced signal.
11. A balanced transmission line picks up an undesired signal with a 5-mW level. After conversion to an unbalanced signal using a center-tapped transformer, the undesired signal is $0.011 \mu W$. Calculate the common mode rejection ratio (CMRR). (56.6 dB)

Section 12-3

12. Draw an equivalent circuit for a transmission line, and explain the physical significance of each element.
13. Provide a physical explanation for the meaning of a line's characteristic impedance (Z_0).
14. Calculate Z_0 for a line that exhibits an inductance of 4 nH/m and 1.5 pF/m. (51.6 Ω)

---

*An asterisk preceding a number indicates a question that has been provided by the FCC as a study aid for licensing examinations.

15. Calculate the capacitance per meter of a 50-Ω cable that has an inductance of 55 nH/m. (22 pF/m)
16. In detail, explain how an impedance bridge can be used to determine Z_0 for a piece of transmission line.
* 17. If the spacing of the conductors in a two-wire radio-frequency transmission line is doubled, what change takes place in the surge impedance (Z_0) of the line?
* 18. If the conductors in a two-wire radio-frequency transmission are replaced by larger conductors, how is the surge impedance affected, assuming no change in the center-to-center spacing of the conductor?
* 19. What determines the surge impedance of a two-wire radio-frequency transmission line?
20. Determine Z_0 for the following transmission lines:
(a) Parallel wire, air dielectric with $D/d = 3$. (215 Ω)
(b) Coaxial line, air dielectric with $D/d = 1.5$. (24.3 Ω)
(c) Coaxial line, polyethylene dielectric with $D/d = 2.5$. (36.2 Ω)
21. List and explain the various types of transmission line losses.
* 22. A long transmission line delivers 10 kW into an antenna; at the transmitter end, the line current is 5 A, and at the coupling house (load) it is 4.8 A. Assuming the line to be properly terminated and the losses in the coupling system to be negligible, what is the power lost in the line? (850 W)
23. Define *surge impedance*.

Section 12-4

24. Derive the equation for the time required for energy to propagate through a transmission line [Eq. (12-19)].
* 25. What is the velocity of propagation for radio-frequency waves in space?
26. A delay line using RG-8A/U cable is to exhibit a 5-ns delay. Calculate the required length of this cable. (3.39 ft)
27. Explain the significance of the velocity factor for a transmission line.
28. Determine the velocity of propagation of a 20-km line if the LC product is $7.5 \times 10^{-12}\ s^2$. (7.3×10^9 m/s)
29. What is the wavelength of the signal in Problem 28 if the signal's frequency is 500 GHz? (0.0146)
30. If the velocity of propagation of a 20-ft transmission line is 600 ft/s, how long will it take for the signal to get to the end of the line? (33.3 ms)

Section 12-5

31. Define a *nonresonant transmission line,* and explain what its traveling waves are and how they behave.
* 32. An antenna is being fed by a properly terminated two-wire transmission line. The current in the line at the input end is 3 A. The surge impedance of the line is 500 Ω. How much power is being supplied to the line? (4.5 kW)
33. With the help of Fig. 12-15, provide a charging analysis of a nonresonant line with an ac signal applied.

Section 12-6

34. Explain the properties of a resonant transmission line. What happens to the energy reaching the end of a resonant line? Are reflections a generally desired result?

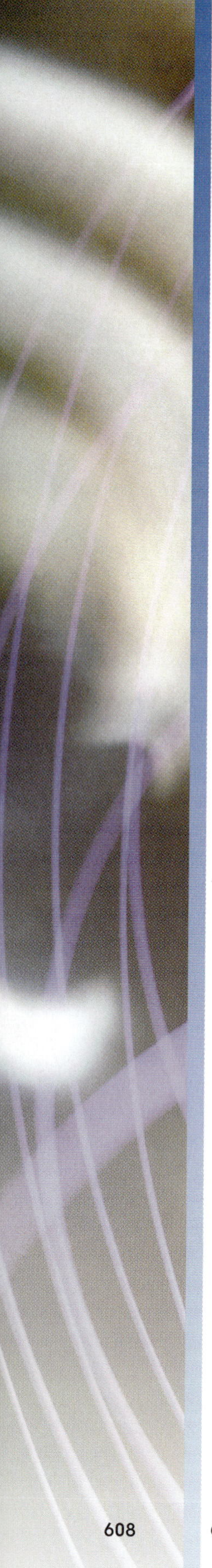

35. A dc voltage from a 20-V battery with $R_s = 75\ \Omega$ is applied to a 75-Ω transmission line at $t = 0$. It takes the battery's energy 10 μs to reach the load, which is an open circuit. Sketch current and voltage waveforms at the line's input and load.
36. Repeat Problem 35 for a short-circuited load.
37. What are *standing waves, standing wave ratio* (SWR), and *characteristic impedance*, in reference to transmission lines? How can standing waves be minimized?
38. With the help of Fig. 12-19, explain how standing waves develop on a resonant line.
* 39. If the period of one complete cycle of a radio wave is 0.000001 s, what is the wavelength? (300 m)
* 40. If the two towers of a 950-kHz antenna are separated by 120 electrical degrees, what is the tower separation in feet? (345 ft)

Section 12-7

41. Define *reflection coefficient,* Γ, in terms of incident and reflected voltage and also in terms of a line's load and characteristic impedances.
42. Express SWR in terms of
    (a) Voltage maximums and minimums.
    (b) Current maximums and minimums.
    (c) The reflection coefficient.
    (d) The line's load resistance and Z_0.
43. Explain the disadvantages of a mismatched transmission line.
* 44. What is the primary reason for terminating a transmission line in an impedance equal to the characteristic impedance of the line?
* 45. What is the ratio between the currents at the opposite ends of a transmission line one-quarter wavelength long and terminated in an impedance equal to its surge impedance?
46. An SSB transmitter at 2.27 MHz and 200 W output is connected to an antenna ($R_{in} = 150\ \Omega$) via 75 ft of RG-8A/U cable. Determine
    (a) The reflection coefficient.
    (b) The electrical cable length in wavelengths.
    (c) The SWR.
    (d) The amount of power absorbed by the antenna.
* 47. What should be the approximate surge impedance of a quarter-wavelength matching line used to match a 600-Ω feeder to a 70-Ω (resistive) antenna?

Section 12-8

48. Calculate the impedance of a line 675 electrical degrees long. $Z_0 = 75\ \Omega$ and $Z_L = 50\ \Omega + j75\ \Omega$. Use the Smith chart *and* Eq. (12-30) as separate solutions, and compare the results.
49. Convert an impedance, $62.5\ \Omega - j90\ \Omega$, to admittance mathematically *and* with the Smith chart. Compare the results.
50. Find the input impedance of a 100-Ω line, 5.35λ long, and with $Z_L = 200\ \Omega + j300\ \Omega$.
* 51. Why is the impedance of a transmission line an important factor with respect to matching "out of a transmitter" into an antenna?
* 52. What is *stub tuning*?

53. The antenna load on a 150-Ω transmission line is 225 Ω − $j300$ Ω. Determine the length and position of a short-circuited stub necessary to provide a match.
54. Repeat Problem 53 for a 50-Ω line and an antenna of 25 Ω + $j75$ Ω.

Section 12-9

55. Calculate the length of a short-circuited 50-Ω line necessary to simulate an inductance of 2 nH at 1 GHz.
56. Calculate the length of a short-circuited 50-Ω line necessary to simulate a capacitance of 50 pF at 500 MHz.
57. Describe two types of baluns, and explain their function.
* 58. How may harmonic radiation of a transmitter be prevented?
* 59. Describe three methods for reducing harmonic emission of a transmitter.
* 60. Draw a simple schematic diagram showing a method of coupling the radio-frequency output of the final power amplifier stage of a transmitter to a two-wire transmission line, with a method of suppression of second and third harmonic energy.
61. Explain the construction of a slotted line and some of its uses.
62. Explain the principle of TDR and some uses for this technique.
63. A pulse is sent down a transmission line that is not functioning properly. It has a propagation velocity of 2.1×10^8 m/s, and an inverted reflected pulse (equal in magnitude to the incident pulse) is returned in 0.731 ms. What is wrong with the line, and how far from the generator does the fault exist?
64. A fast-rise-time 10-V step voltage is applied to a 50-Ω line terminated with an 80-Ω resistive load. Determine Γ, E_F, and E_r. (0.231, 12.3 V, 2.3 V)

Section 12-10

65. Describe some of the causes of crosstalk and list possible solutions.
66. Explain why cabling should not be run close to ac power lines.
67. List some of the causes of magnetic field losses in a cable.
68. Explain the effects of extreme sunlight (heat radiation) on cables.

Questions for Critical Thinking

69. With the help of Fig. 12-12, provide a step-by-step explanation of how a dc voltage propagates through a transmission line.
70. An open-circuited line is 1.75λ. Sketch the incident, reflected, and resultant waveforms for both voltage and current at the instant the generator is at its peak negative value. Sketch and compare the waveforms for a short-circuited line.
71. You are asked to design a line "free of transmission line effects." You design one that is λ/16 long. How would you justify this design?
72. Match a load of 25 Ω + $j75$ Ω to a 50-Ω line using a quarter-wavelength matching section. Determine the proper location and characteristic impedance of the matching section. Repeat this problem for a Z_L = 110 Ω − $j50$ Ω load. Provide *two* separate solutions.

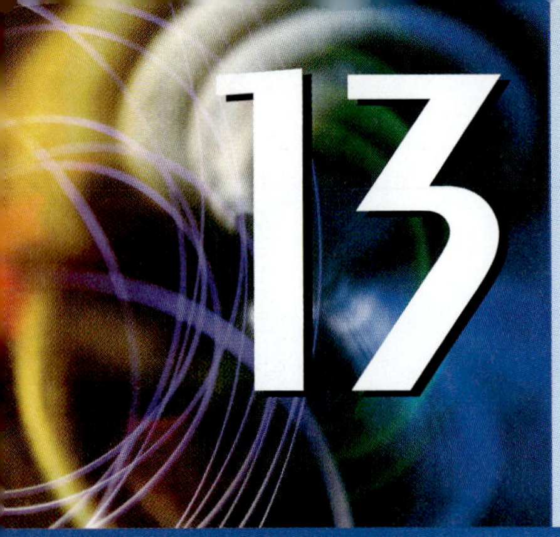

Wave Propagation

Chapter Outline

The very large array (VLA) antenna site in Socorro, New Mexico. (Courtesy of The Stock Market/Lester Lefkowitz. Reprinted with permission.)

Objectives

- Discuss the makeup of an electromagnetic wave and the characteristics of an isotropic point source
- Explain the processes of wave reflection, refraction, and diffraction
- Describe ground- and space-wave propagation and calculate the ghosting effect in TV reception
- Calculate the approximate radio horizon based on antenna height
- Discuss the effects of the ionosphere on sky-wave propagation
- Define the critical angle and skip zone for sky-wave propagation
- Describe the important aspects of satellite communication
- Calculate the power received by a satellite based on a SATCOM power budget analysis

Key Terms

transducer
radio-frequency interference
electromagnetic interference
transverse
polarization
isotropic point source
wavefront
coefficient of reflection
diffraction
shadow zone
ground wave
surface wave
radio horizon
ghosting
sky wave
skipping
isothermal region
critical frequency
critical angle
maximum usable frequency
optimum working frequency
quiet zone
skip zone
fading
diversity reception
synchronous orbit
uplink
downlink
transponder
geosynchronous orbit
differential GPS
frequency-division multiple-access
time-division multiple-access
code-division multiple-access
perigee
apogee

13-1 Electrical to Electromagnetic Conversion

Early radios were often referred to as the "wireless." This new machine could speak without being "wired" to the source like the telegraph and telephone. The transmitter's output is coupled to its surrounding atmosphere and then intercepted by the receiver. We know that the atmosphere is *not* a conductor of electrons like a copper wire—air is, in fact, a very good insulator. Thus, the electrical energy fed into a transmitting antenna must be converted to another form of energy for transmission. In this chapter we shall study the effects of the transformed energy and its propagation.

Transducer
device that converts energy from one form to another

The transmitting antenna converts its input electrical energy into electromagnetic energy. The antenna can thus be thought of as a **transducer**—a device that converts from one form of energy into another. In that respect, a light bulb is very similar to an antenna. The light bulb also converts electrical energy into electromagnetic energy—light. The only difference between light and the radio waves we shall be concerned with is their frequency. Light is an electromagnetic wave at about 5×10^{14} Hz, while the usable radio waves extend from about 1.5×10^4 Hz up to 3×10^{11} Hz. The human eye is responsive to (able to perceive) the very narrow range of light frequencies, and consequently we are blind to the radio waves. Actually, that is an advantage because the great number of radio waves surrounding our earth would otherwise paint a chaotic picture.

The receiving antenna intercepts the transmitted wave and converts it back into electrical energy. An analogous transducer is the photovoltaic cell that also converts a wave (light) into electrical energy. Because a basic knowledge of waves is necessary to your understanding of antennas and radio communications, the following section is presented prior to your study of wave propagation.

13-2 Electromagnetic Waves

Radio-Frequency Interference
undesired radiation from a radio transmitter

Electromagnetic Interference
unwanted signals from devices that produce excessive electromagnetic radiation

Transverse
when the oscillations of a wave are perpendicular to the direction of propagation

Polarization
the direction of the electric field of an electromagnetic wave

Electricity and electromagnetic waves are interrelated. An electromagnetic field consists of an electric field and a magnetic field. These fields exist with all electric circuits because any current-carrying conductor creates a magnetic field around the conductor, and any two points in the circuit with a potential difference (voltage) between them create an electric field. These two fields contain energy, but in circuits, this field energy is usually returned to the circuit when the field collapses. If the field does not fully return its energy to the circuit, it means the wave has been at least partially *radiated,* or set free, from the circuit. This radiated energy is undesired because it may cause interference with other electronic equipment in the vicinity. It is termed **radio-frequency interference** (RFI) if it is undesired radiation from a radio transmitter, and if from another source, it is termed **electromagnetic interference** (EMI) or, more simply, noise.

In the case of a radio transmitter, it is hoped that the antenna efficiently causes the wave energy to be set free. The antenna is designed *not* to allow the electromagnetic wave energy to collapse back into the circuit.

An electromagnetic wave is pictured in Fig. 13-1. In it, $1\frac{1}{2}$ wavelengths of the electric field (E) and the magnetic field (H) are shown. The direction of propagation is shown to be perpendicular to both fields, which are also mutually perpendicular to each other. The wave is said to be **transverse** because the oscillations are perpendicular to the direction of propagation. The **polarization** of an electromagnetic

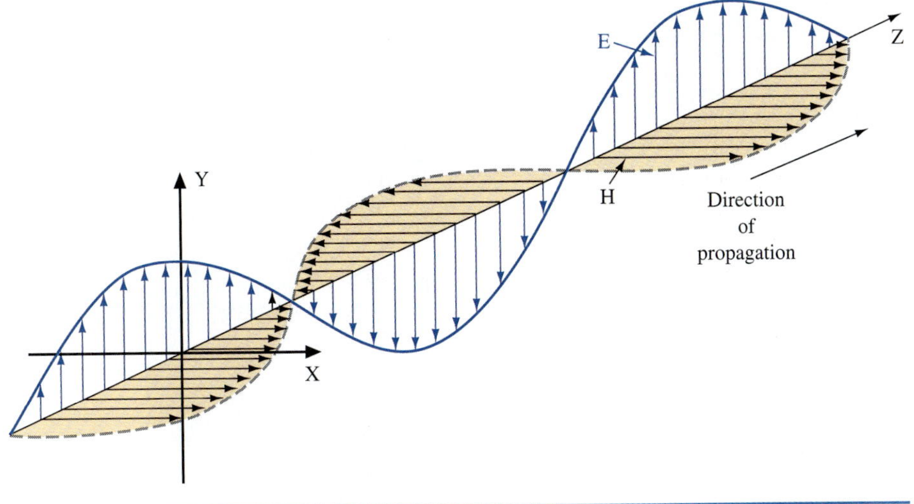

FIGURE 13-1 Electromagnetic wave.

wave is determined by the direction of its *E* field component. In Fig. 13-1 the *E* field is vertical (*y* direction), and the wave is therefore said to be vertically polarized. As we will see, the antenna's orientation determines polarization. A vertical antenna results in a vertically polarized wave.

Wavefronts

If an electromagnetic wave were radiated equally in all directions from a point source in free space, a spherical wavefront would result. Such a source is termed an **isotropic point source.** A **wavefront** may be defined as a surface joining all points of equal phase. Two wavefronts are shown in Fig. 13-2. An *isotropic* source radiates equally in

Isotropic Point Source
a point in space that radiates electromagnetic radiation equally in all directions

Wavefront
a plane joining all points of equal phase in a wave

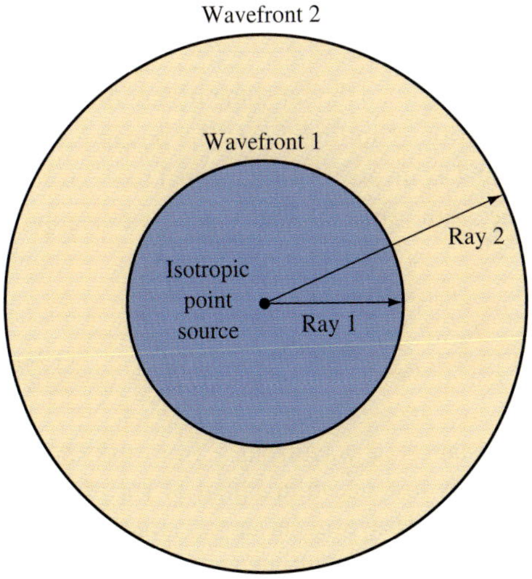

Wavefront 2
Wavefront 1
Ray 2
Isotropic point source
Ray 1

FIGURE 13-2 Antenna wavefronts.

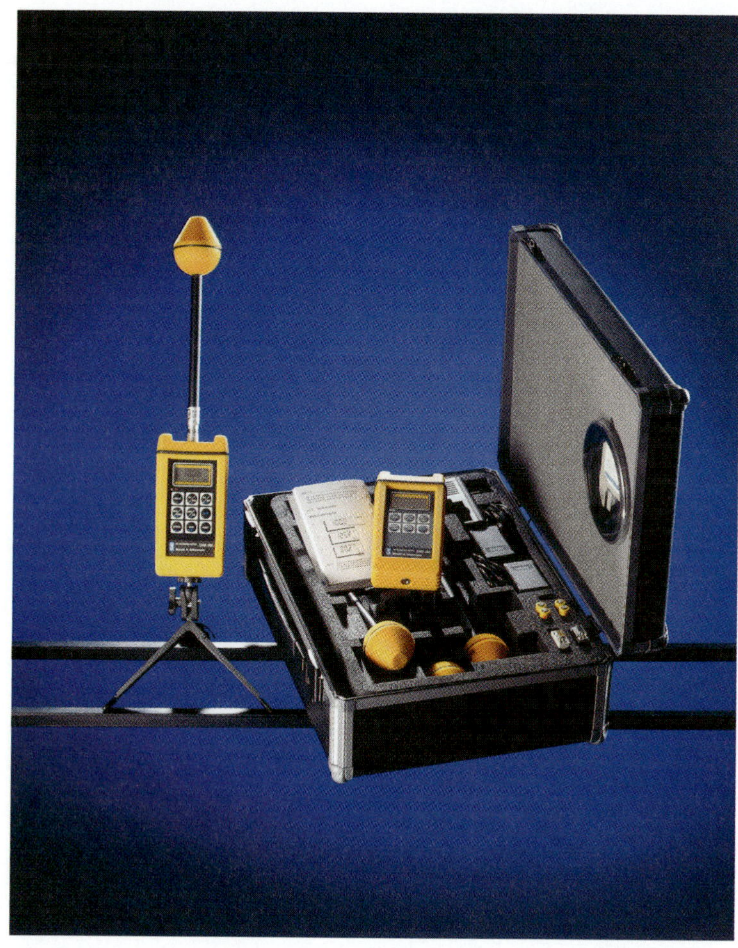

Handheld meters measure electromagnetic fields. (Courtesy of Wandel & Goltermann. Reprinted with permission.)

all directions. The wave travels at the speed of light so that at some instant the energy will have reached the area indicated by wavefront 1 in Fig. 13-2. The power density, $\mathcal{P}$ (in watts per square meter), at wavefront 1 is inversely proportional to the square of its distance, r (in meters), from its source, with respect to the originally transmitted power, P_t. Stated mathematically,

$$\mathcal{P} = \frac{P_t}{4\pi r^2} \tag{13-1}$$

If wavefront 2 in Fig. 13-2 is twice the distance of wavefront 1 from the source, then its power density in watts per unit area is just one-fourth that of wavefront 1. Any section of a wavefront is curved in shape. However, at appreciable distances from the source, small sections are nearly flat. These sections can then be considered as *plane wavefronts*, which simplifies the treatment of their optical properties provided in Sec. 13-3.

Characteristic Impedance of Free Space

The strength of the electric field, $\mathscr{E}$ (in volts per meter), at a distance r from a point source is given by

$$\mathscr{E} = \frac{\sqrt{30P_t}}{r} \tag{13-2}$$

where P_t is the originally transmitted power in watts. This is one of Maxwell's equations, which were postulated in 1873 and allowed mathematical analysis of electromagnetic wave phenomena. They were experimentally verified by Hertz in 1888.

Power density $\mathscr{P}$ and the electric field $\mathscr{E}$ are related to impedance in the same way that power and voltage relate in an electric circuit. Thus,

$$\mathscr{P} = \frac{\mathscr{E}^2}{\mathscr{L}} \tag{13-3}$$

where $\mathscr{L}$ is the characteristic impedance of the medium conducting the wave. For free space, Eqs. (13-1) and (13-2) can be substituted into Eq. (13-3) to give

$$\mathscr{L} = \frac{\mathscr{E}^2}{\mathscr{P}} = \frac{30P_t}{r^2} \div \frac{P_t}{4\pi r^2} = 120\pi = 377\ \Omega \tag{13-4}$$

Thus, you can see that free space has an intrinsic impedance similar to a transmission line.

The characteristic impedance of any electromagnetic wave-conducting medium is provided by

$$\mathscr{L} = \sqrt{\frac{\mu}{\epsilon}} \tag{13-5}$$

where μ is the medium's permeability and ϵ is the medium's permittivity.

For free space, $\mu = 1.26 \times 10^{-6}$ H/m and $\epsilon = 8.85 \times 10^{-12}$ F/m. Substituting in Eq. (13-5) yields

$$\mathscr{L} = \sqrt{\frac{\mu}{\epsilon}} = \sqrt{\frac{1.26 \times 10^{-6}}{8.85 \times 10^{-12}}} = 377\ \Omega$$

which agrees with the result from Eq. (13-4).

13-3 Waves Not in Free Space

Until now, we have discussed the behavior of waves in free space, which is a vacuum or complete void. We now consider the effects of our environment on wave propagation.

Reflection

Just as light waves are reflected by a mirror, radio waves are reflected by any medium such as metal surfaces or the earth's surface. The angle of incidence is equal to the angle of reflection, as shown in Fig. 13-3. Note that there is a change in phase

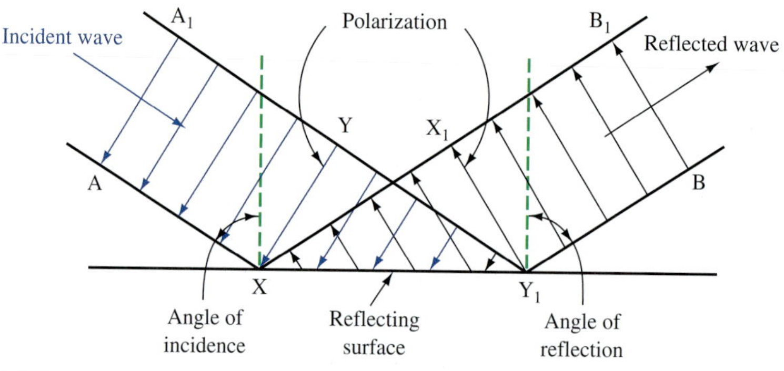

FIGURE 13-3 Reflection of a wavefront.

of the incident and reflected waves, as seen by the difference in the direction of polarization. The incident and reflected waves are 180° out of phase.

Coefficient of Reflection ratio of the reflected electric field intensity divided by the incident intensity

Complete reflection occurs only for a theoretically perfect conductor and when the electric field is perpendicular to the reflecting element. For complete reflection, the **coefficient of reflection** ρ is 1 and is defined as the ratio of the reflected electric field intensity divided by the incident intensity. It is less than 1 in practical situations due to the absorption of energy by the nonperfect conductor and also because some of the energy actually propagates right through it.

The previous discussion is valid when the electric field is *not* normal to the reflecting surface. If it is fully parallel to the reflecting (conductive) surface, the electric field is *shorted* out, and all of the electromagnetic energy is dissipated in the form of generated surface currents in the conductor. If the electric field is partially parallel to the surface, it will be partially shorted out.

If the reflecting surface is curved, as in a parabolic antenna, the wave may be analyzed using the appropriate optical laws with regard to focusing the energy, etc. This is especially true with respect to microwave frequencies, which are discussed in Chapter 16.

Refraction

Refraction of electromagnetic radio waves occurs in a manner akin to the refraction of light. Refraction occurs when waves pass from a medium of one density to another medium with a different density. A good example is the apparent bending of a spoon when it is immersed in water. The bending seems to take place at the water's surface, or exactly at the point where there is a change of density. Obviously, the spoon does not bend from the pressure of the water. The light forming the image of the spoon is bent as it passes from the water, a medium of high density, to the air, a medium of comparatively low density.

The bending (refraction) of an electromagnetic wave (light or radio wave) is shown in Fig. 13-4. Also shown is the reflected wave. Obviously, the coefficient of reflection is less than 1 here because a fair amount of the incident wave's energy is propagated through the water—after refraction has occurred.

The angle of incidence, θ_1, and the angle of refraction, θ_2, are related by the following expression, which is Snell's law:

$$n_1 \sin \theta_1 = n_2 \sin \theta_2 \tag{13-6}$$

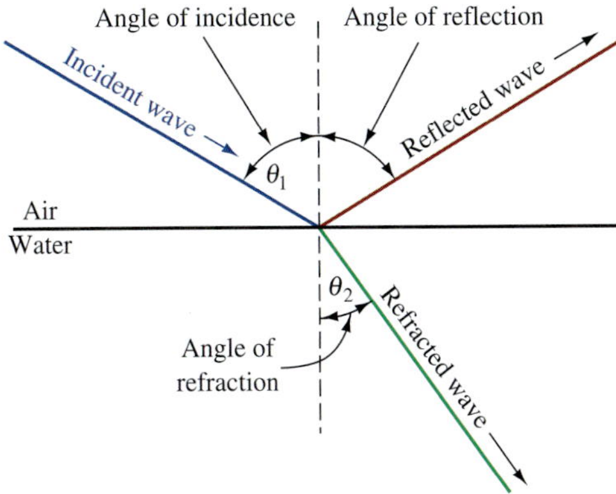

FIGURE 13-4 Wave refraction and reflection.

where n_1 is the refractive index of the incident medium and n_2 is the refractive index of the refractive medium. Recall that the refractive index for a vacuum is exactly 1 and it is approximately 1 for the atmosphere. For glass, it is about 1.5, and for water it is 1.33.

Diffraction

Diffraction is the phenomenon whereby waves traveling in straight paths bend around an obstacle. This effect is the result of Huygens' principle, advanced by the Dutch astronomer Christian Huygens in 1690. The principle states that each point on a spherical wavefront may be considered as the source of a secondary spherical wavefront. This concept is important to us because it explains radio reception behind a mountain or tall building. Figure 13-5 shows the diffraction process allowing reception beyond a mountain in all but a small area, which is called the **shadow zone.** The figure shows that electromagnetic waves are diffracted over the top and around the sides of an obstruction. The direct wavefronts that just get by the obstruction

Diffraction
the phenomenon whereby waves traveling in straight paths bend around an obstacle

Shadow Zone
an area following an obstacle that does not receive a wave by diffraction

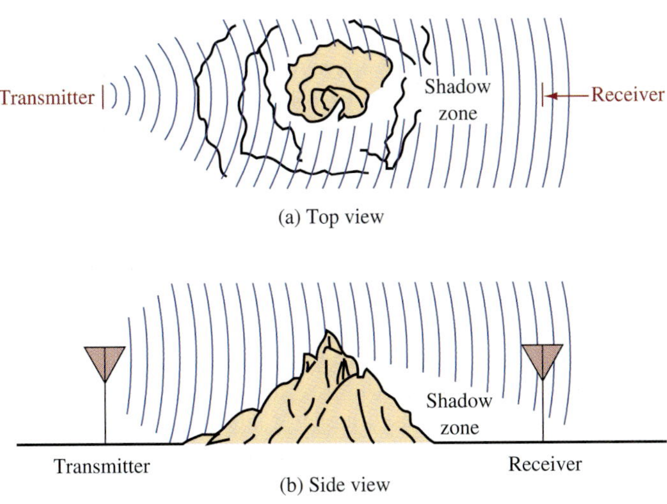

FIGURE 13-5 Diffraction around an object.

become new sources of wavefronts that start filling in the void, making the shadow zone a finite entity. The lower the frequency of the wave, the quicker is this process of diffraction (i.e., the shadow zone is smaller).

13-4 GROUND- AND SPACE-WAVE PROPAGATION

There are four basic modes of getting a radio wave from the transmitting to receiving antenna:

1. Ground wave
2. Space wave
3. Sky wave
4. Satellite communications

As we will see in the following discussions, the frequency of the radio wave is of primary importance in considering the performance of each type of propagation.

Ground-Wave Propagation

Ground Wave radio wave that travels along the earth's surface

Surface Wave another name for ground wave

A **ground wave** is a radio wave that travels along the earth's surface. It is sometimes referred to as a **surface wave.** The ground wave must be vertically polarized (electric field vertical) because the earth would short out the electric field if horizontally polarized. Changes in terrain have a strong effect on ground waves. Attenuation of ground waves is directly related to the surface impedance of the earth. This impedance is a function of conductivity and frequency. If the earth's surface is highly conductive, the absorption of wave energy, and thus its attenuation, will be reduced. Ground-wave propagation is much better over water (especially salt water) than, say, a very dry (poor conductivity) desert terrain.

The ground losses increase rapidly with increasing frequency. For this reason ground waves are not very effective at frequencies above 2 MHz. Ground waves are, however, a very reliable communications link. Reception is not affected by daily or seasonal changes such as with sky-wave propagation.

Ground-wave propagation is the only way to communicate into the ocean with submarines. Extremely low frequency (ELF) propagation is utilized. ELF waves encompass the range 30 to 300 Hz. At a typically used frequency of 100 Hz, the attenuation is about 0.3 dB/m. This attenuation increases steadily with frequency so that, at 1 GHz, a 1000-dB/m loss is sustained! Seawater has little attenuation to ELF signals, so these frequencies can be used to communicate with submerged submarines without their having to surface and be vulnerable to detection.

Space-Wave Propagation

The two types of space waves are shown in Fig. 13-6. They are the direct wave and ground reflected wave. Do not confuse these with the ground wave just discussed. The direct wave is by far the most widely used mode of antenna communications. The propagated wave is direct from transmitting to receiving antenna and does not travel along the ground. The earth's surface, therefore, does not attenuate it.

The direct space wave does have one severe limitation—it is basically limited to so-called *line-of-sight* transmission distances. Thus, the antenna height and the

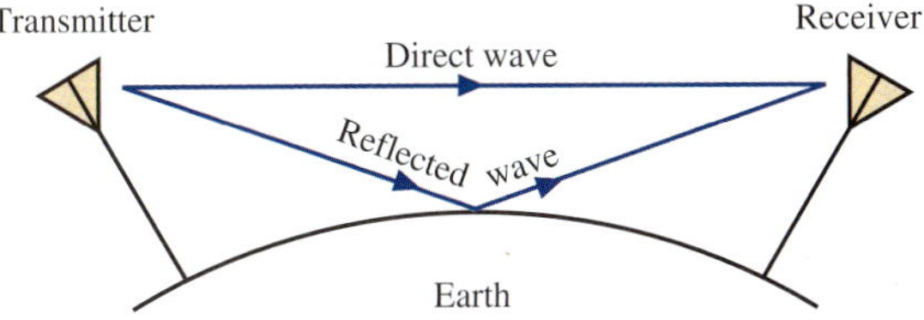

FIGURE 13-6 Direct and ground reflected space waves.

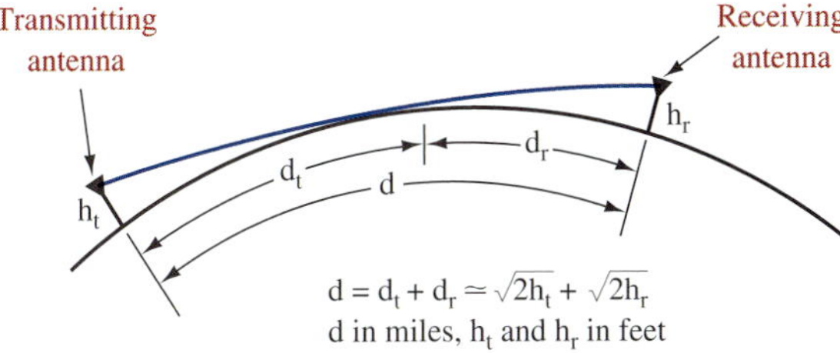

FIGURE 13-7 Radio horizon for direct space waves.

curvature of the earth are the limiting factors. The actual **radio horizon** is about $\frac{4}{3}$ greater than the geometric line of sight because of diffraction effects and is empirically predicted by the following approximation:

Radio Horizon
a distance about $\frac{4}{3}$ greater than line-of-sight, approximate limit for direct space-wave propagation

$$d \simeq \sqrt{2h_t} + \sqrt{2h_r} \qquad \textbf{(13-7)}$$

where d = radio horizon (mi)
h_t = transmitting antenna height (ft)
h_r = receiving antenna height (ft)

The diffraction effects cause the slight wave curvature, as shown in Fig. 13-7. If the transmitting antenna is 1000 ft above ground level and the receiving antenna is 20 ft high, a radio horizon of about 50 mi results. This explains the coverage that typical broadcast FM and TV stations provide because they are utilizing direct space-wave propagation.

The reflected wave shown in Fig. 13-6 can cause reception problems. If the phase of these two received components is not the same, some degree of signal fading and/or distortion will occur. This can also result when both a direct and ground wave are received or when any two or more signal paths exist. A special case involving TV reception is presented next.

Ghosting in TV Reception Any tall or massive objects obstruct space waves. This results in diffraction (and subsequent shadow zones) and reflections. Reflections pose a specific problem because, for example, reception of a TV signal may be the combined result of a direct space wave and a reflected space wave, as shown in Fig. 13-8. This condition results in **ghosting,** which manifests itself in the form of a double-image distortion. This is due to the two signals arriving at the receiver at two different times—the reflected signal has a farther distance to travel. The reflected signal is weaker than the direct signal because of the inverse square-law relationship of signal strength to distance [Eq. (13-1)] and because of losses incurred during reflection.

Ghosting
when the same signal arrives at the TV receiver at two different times; the reflected signal has farther to travel and is weaker than the direct signal, resulting in a double image

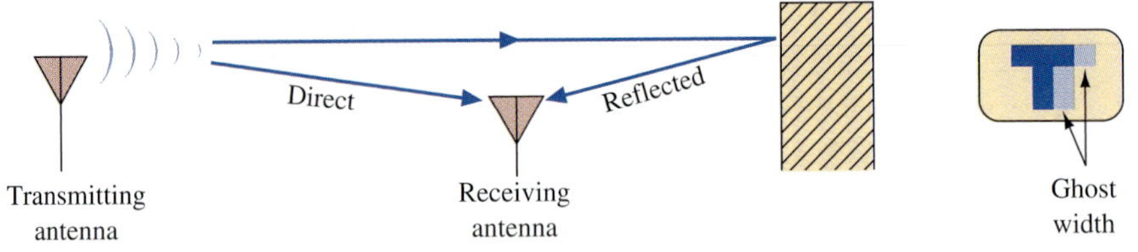

FIGURE 13-8 Ghost interference.

Example 13-1

Determine the ghost width on a TV screen 15 in. wide when a reflected wave results from an object $\frac{1}{2}$ mi "behind" a receiver.

Solution

The reflected wave travels 1 mi farther than the direct wave (2 × 0.5 mi). Each horizontal line on the receiver is 53.5 μs in duration (Chapter 17). Assuming the wave travels at the speed of light, the time of delay between the direct and reflected signal is

$$t = \frac{d}{v} = \frac{1 \text{ mi}}{186{,}000 \text{ mi/s}} = 5.38\ \mu\text{s}$$

The ghost width will therefore be

$$\frac{5.38\ \mu\text{s}}{53.5\ \mu\text{s/trace}} \times 15 \text{ in.} = 1.51 \text{ in.}$$

Ghosting is seen in the image on the left. (Courtesy of Philips Consumer Electronics. Reprinted with permission.)

A possible solution to the ghosting problem shown in Ex. 13-1 is to detune the receiving antenna orientation so that the reflected wave is too weak to be displayed. Of course, the direct wave must exceed the receiver's sensitivity limit because it will also be reduced in level when the antenna is detuned. It should be noted that ghosting can also be caused by transmission line reflections between antenna and set.

 13-5 SKY-WAVE PROPAGATION

One of the most frequently used methods of long-distance transmission is by the use of the **sky wave.** Sky waves are those waves radiated from the transmitting antenna in a direction that produces a large angle with reference to the earth. The sky wave has the ability to strike the ionosphere, be refracted from it to the ground, strike the ground, be reflected back toward the ionosphere, and so on. The refracting and reflecting action of the ionosphere and the ground is called **skipping.** An illustration of this skipping effect is shown in Fig. 13-9.

Sky Wave
those radio waves radiated from the transmitting antenna in a direction toward the ionosphere

Skipping
the alternate refracting and reflecting of a sky wave signal between the ionosphere and the earth's surface

The transmitted wave leaves the antenna at point *A*, is refracted from the ionosphere at point *B*, is reflected from the ground at point *C*, is again refracted from the ionosphere at point *D*, and arrives at the receiving antenna *E*. The critical nature of the sky waves and the requirements for refraction will be discussed thoroughly in this section.

To understand the process of refraction, the composition of the atmosphere and the factors that affect it must be considered. Insofar as electromagnetic radiation is concerned, there are only three layers of the atmosphere: the troposphere, the stratosphere, and the ionosphere. The troposphere extends from the surface of the earth up to approximately 6.5 mi. The next layer, the stratosphere, extends from the upper limit of the troposphere to an approximate elevation of 23 mi. From the upper limit of the stratosphere to a distance of approximately 250 mi lies the region known as the ionosphere. Beyond the ionosphere is free space. The temperature in the stratosphere is considered to be a constant unfluctuating value. Therefore, it is not subject to temperature inversions, nor can it cause significant refractions. The constant temperature stratosphere is also called the **isothermal region.**

Isothermal Region
the stratosphere, considered to have a constant temperature

The ionosphere is appropriately titled because it is composed primarily of ionized particles. The density at the upper extremities of the ionosphere is very low

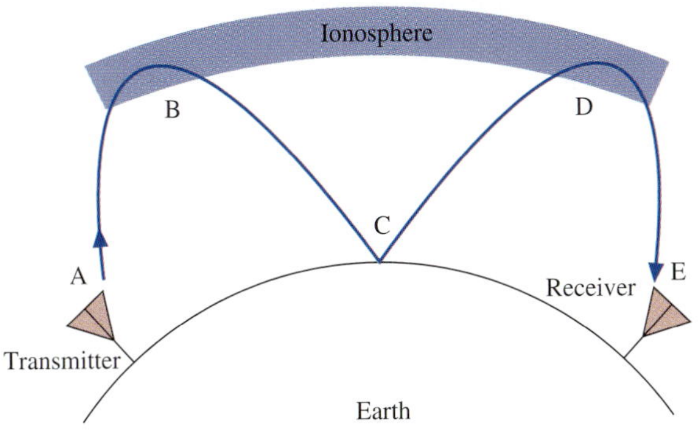

FIGURE 13-9 Sky-wave propagation.

and becomes progressively higher as it extends downward toward the earth. The upper region of the ionosphere is subjected to severe radiation from the sun. Ultraviolet radiation from the sun causes ionization of the air into free electrons, positive ions, and negative ions. Even though the density of the air molecules in the upper ionosphere is small, the radiation particles from space are of such high energy at that point that they cause wide-scale ionization of the air molecules that are present. This ionization extends down through the ionosphere with diminishing intensity. Therefore, the highest degree of ionization occurs at the upper extremities of the ionosphere, while the lowest degree occurs in the lower portion of the ionosphere.

Ionospheric Layers

The ionosphere is composed of three layers designated, respectively, from lowest level to highest level as *D*, *E*, and *F*. The *F* layer is further divided into two layers designated F_1 (the lower layer) and F_2 (the higher layer). The presence or absence of these layers in the ionosphere and their height above the earth vary with the position of the sun. At high noon, radiation from the sun in the ionosphere directly above a given point is greatest, while at night it is minimal. When the radiation is removed, many of the ions that were ionized recombine. The interval of time between these conditions finds the position and number of the ionized layers within the ionosphere changing. Because the position of the sun varies daily, monthly, and yearly with respect to a specified point on earth, the exact characteristics of the layers are extremely difficult to predict. However, the following general statements can be made:

1. The *D* layer ranges from about 25 to 55 mi. Ionization in the *D* layer is low because it is the lowest region of the ionosphere (farthest from the sun). This layer has the ability to refract signals of low frequencies. High frequencies pass right through it but are partially attenuated in so doing. After sunset, the *D* layer disappears because of the rapid recombination of its ions.
2. The *E* layer limits are from approximately 55 to 90 mi high. This layer is also known as the Kennelly–Heaviside layer because these two men were the first to propose its existence. The rate of ionic recombination in this layer is rather rapid after sunset and is almost complete by midnight. This layer has the ability to refract signals of a higher frequency than were refracted by the *D* layer. In fact, the *E* layer can refract signals with frequencies as high as 20 MHz.
3. The *F* layer exists from about 90 to 250 mi. During the daylight hours, the *F* layer separates into two layers, the F_1 and F_2 layers. The ionization level in these layers is quite high and varies widely during the course of a day. At noon, this portion of the atmosphere is closest to the sun, and the degree of ionization is maximum. The atmosphere is rarefied at these heights, so the recombination of the ions occurs slowly after sunset. Therefore, a fairly constant ionized layer is present at all times. The *F* layers are responsible for high-frequency, long-distance transmission due to refraction for frequencies up to 30 MHz.

The relative distribution of the ionospheric layers is shown in Fig. 13-10. With the disappearance of the *D* and *E* layers at night, signals normally refracted by these layers are refracted by the much higher layer, resulting in greater skip distances at night. The layers that form the ionosphere undergo considerable variations in altitude, density, and thickness, due primarily to varying degrees of solar activity. The F_2 layer

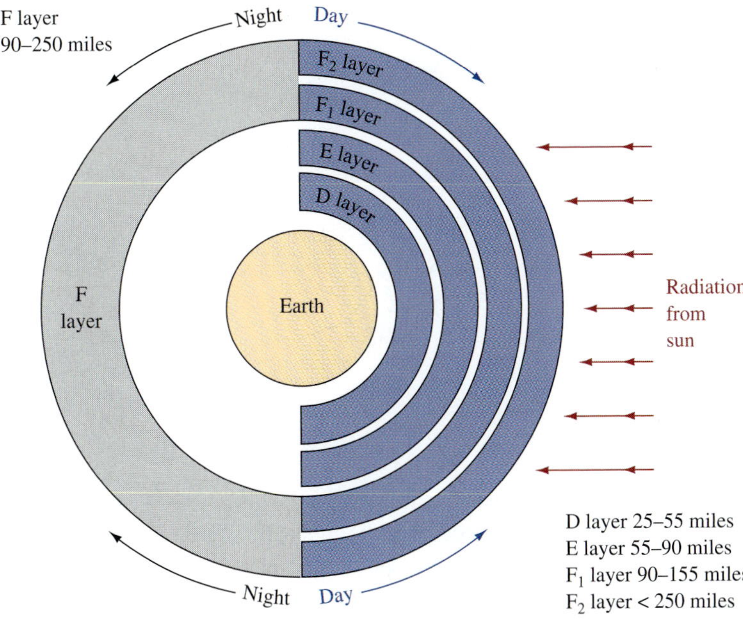

FIGURE 13-10 Layers of the ionosphere.

undergoes the greatest variation due to solar disturbances (sunspot activity). There is a greater concentration of solar radiation in the earth's atmosphere during peak sunspot activity, which recurs in 11-year cycles, as discussed in Chapter 1. During periods of maximum sunspot activity, the *F* layer is more dense and occurs at a higher altitude. During periods of minimum sunspot activity, the lower altitude of the *F* layer returns the sky waves (dashed lines) to points relatively close to the transmitter compared with the higher altitude *F* layer occurring during maximum sunspot activity. Consequently, skip distance is affected by the degree of solar disturbance.

Effects of the Ionosphere on the Sky Wave

The ability of the ionosphere to return a radio wave to the earth depends on the ion density, the frequency of the radio wave, and the angle of transmission. The refractive ability of the ionosphere increases with the degree of ionization. The degree of ionization is greater in summer than in winter and is also greater during the day than at night. As mentioned previously, abnormally high densities occur during times of peak sunspot activity.

Critical Frequency If the frequency of a radio wave being transmitted vertically is gradually increased, a point is reached where the wave is not refracted sufficiently to curve its path back to earth. Instead, these waves continue upward to the next layer, where refraction continues. If the frequency is sufficiently high, the wave penetrates all layers of the ionosphere and continues out into space. The highest frequency that is returned to earth when transmitted vertically under given ionospheric conditions is called the **critical frequency.**

Critical Frequency
the highest frequency that will be returned to the earth when transmitted vertically under given ionospheric conditions

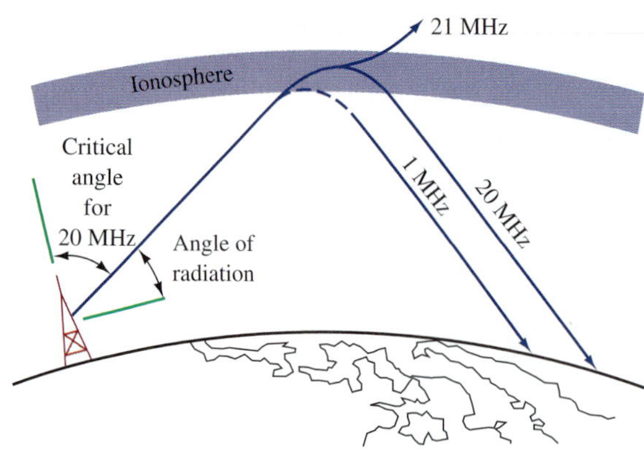

FIGURE 13-11 Relationship of frequency to refraction by the ionosphere.

Critical Angle In general, the lower the frequency, the more easily the signal is refracted; conversely, the higher the frequency, the more difficult is the refracting or bending process. Figure 13-11 illustrates this point. The angle of radiation plays an important part in determining whether a particular frequency is returned to earth by refraction from the ionosphere. Above a certain frequency, waves transmitted vertically continue into space. However, if the angle of propagation is lowered (from the vertical), a portion of the high-frequency waves below the critical frequency is returned to earth. The highest angle at which a wave of a specific frequency can be propagated and still be returned (refracted) from the ionosphere is called the **critical angle** for that particular frequency. The critical angle is the angle that the wavefront path makes with a line extended to the center of the earth. Refer to Fig. 13-11, which shows the critical angle for 20 MHz. Any wave above 20 MHz (e.g., the 21-MHz wave shown) is not refracted back to earth but goes through the ionosphere and into space.

Critical Angle
the highest angle with respect to a vertical line at which a radio wave of a specified frequency can be propagated and still be returned to the earth from the ionosphere

Maximum Usable Frequency (MUF) There is a best frequency for optimum communication between any two points at any specific condition of the ionosphere. As you can see in Fig. 13-12, the distance between the transmitting antenna and the point at which the wave returns to earth depends on the angle of propagation, which in turn is limited by the frequency. The highest frequency that is returned to earth at a given distance is called the **maximum usable frequency** (MUF) and has an average monthly value for any given time of the year. The **optimum working frequency** is the one that provides the most consistent communication and is

Maximum Usable Frequency
the highest frequency that is returned to the earth from the ionosphere between two specific points on earth

Optimum Working Frequency
the frequency that provides for the most consistent communication path via sky waves

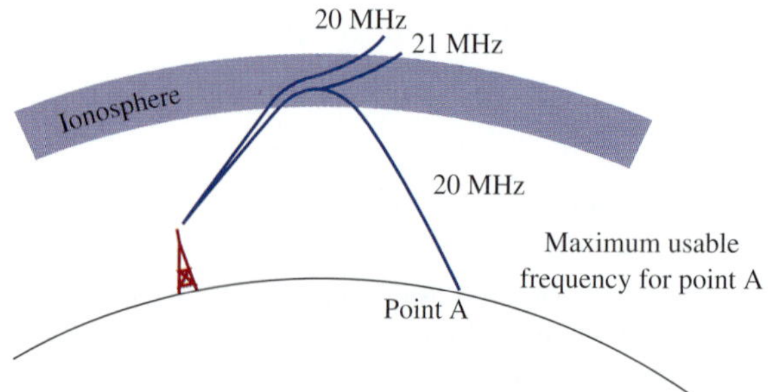

FIGURE 13-12 Relationship of frequency to critical angle.

therefore the best one to use. For transmission using the F_2 layer, the optimum working frequency is about 85 percent of the MUF, while propagation via the *E* layer is consistent, in most cases, if a frequency near the MUF is used. Because ionospheric attenuation of radio waves is inversely proportional to frequency, using the MUF results in maximum signal strength.

Because of this variation in the critical frequency, nomograms and frequency tables are used to predict the maximum usable frequency for every hour of the day for every locality in which transmissions are made. This information is prepared from data obtained experimentally from stations scattered all over the world. All this information is pooled, and the results are tabulated in the form of long-range predictions that remove most of the guess-work from this type of radio communications.

The U.S. government transmits propagation data on a regular basis. The two stations are WWV, Fort Collins, Colorado, at 18 minutes past every hour on frequencies of 2.5, 5, 10, 15 and 20 MHz; and WWVH, Hawaii, on 5, 10 and 15 MHz, 45 minutes past every hour. These stations transmit the A and K indices and the solar flux, which can be used to predict MUF as well as other propagation characteristics. The K index, from 0 to 8, is a measure of the earth's geomagnetic activity. A value above approximately 4 indicates a geomagnetic storm with severe effects on radio communications. The K index is updated every three hours and shows useful "trend" information. The A index is open-ended; that is, it has no maximum value, but readings above about 100 or so are rare. Values of perhaps 10 or lower indicate quiet conditions and good propagation. Based on the K index, the A index is updated every 24 hours at 1800 UT. Solar flux is a measure of sunspot activity. Like the A index, low values indicate good propagation.

Quiet Zone
between the point where the ground wave is completely dissipated and the point where the first sky wave is received

Skip Zone
another name for quiet zone

Skip Zone Between the point where the ground wave is completely dissipated and the point where the first sky wave returns, *no* signal will be heard. This area is called the **quiet** or **skip zone** and is shown in Fig. 13-13. You can see that the skip zone occurs for a given frequency, when propagated at its critical angle. The skip zone is the distance from the end of ground-wave reception to the point of the first sky-wave reception. This occurs for the energy propagated at the critical angle. Similarly, the skip distance is the minimum distance from the transmitter to where the sky wave can be returned to earth and also occurs for energy propagated at the critical angle.

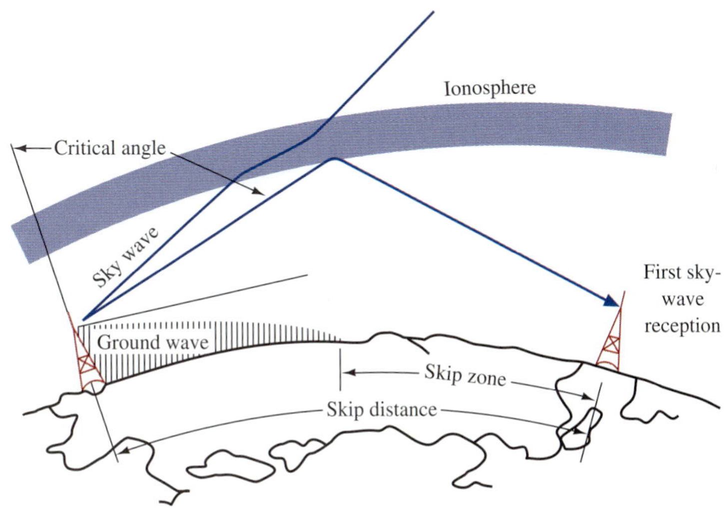

FIGURE 13-13 Skip zone.

Fading
variations in signal strength that may occur at the receiver over a period of time

Fading **Fading** is a term used to describe variations in signal strength that occur at a receiver during the time a signal is being received. Fading may occur at any point where both the ground wave and the sky wave are received, as shown in Fig. 13-14(a). The two waves may arrive out of phase, thus producing a cancellation of the usable signal. This type of fading is encountered in long-range communications over bodies of water where ground-wave propagation extends for a relatively long distance. In areas where sky-wave propagation is prevalent, fading may be caused by two sky waves traveling different distances, thereby arriving at the same point out of phase, as shown in Fig. 13-14(b). Such a condition may be caused by a portion of the transmitted wave being refracted by the E layer while another portion of the wave is refracted by the F layer. A complete cancellation of the signal would occur if the two waves arrived 180° out of phase with equal amplitudes. Usually, one signal is weaker than the other, and therefore a usable signal may be obtained.

Because the ionosphere causes somewhat different effects on different frequencies, a received signal may have phase distortion. As mentioned in Chapter 4, SSB is least susceptible to phase distortion problems. FM is so susceptible to these effects that it is rarely used below 30 MHz (where sky waves are possible). The greater the bandwidth, the greater the problem with phase distortion.

Frequency blackouts are closely related to certain types of fading, some of which are severe enough to blank out the transmission completely. The changing conditions in the ionosphere shortly before sunrise and shortly after sunset may

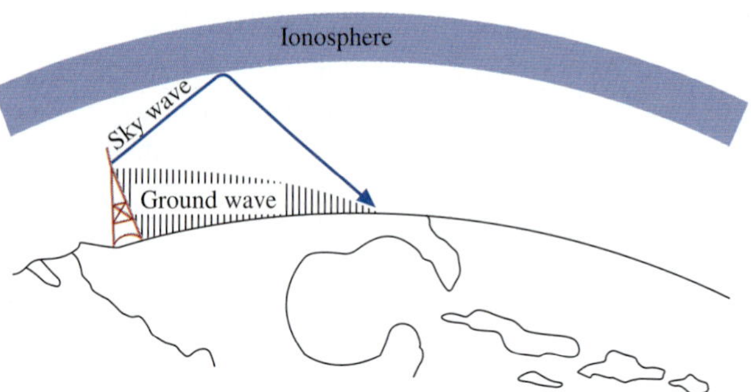

(a) Fading caused by arrival of ground wave and sky wave at the same point out of phase

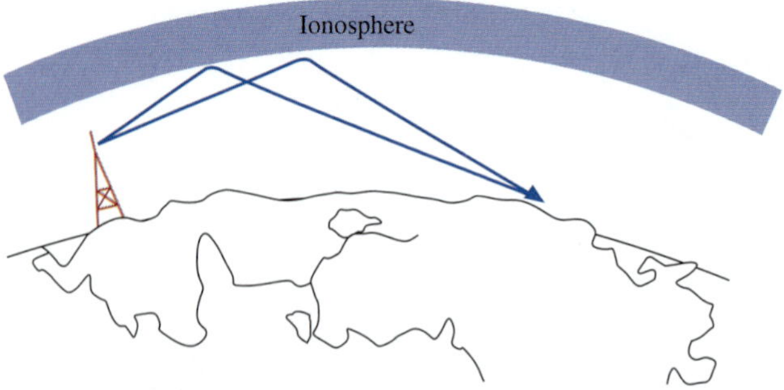

(b) Fading caused by arrival of two sky waves at the same point out of phase

FIGURE 13-14 Fading.

cause complete blackouts at certain frequencies. The higher-frequency signals may then pass through the ionosphere, while the lower-frequency signals are absorbed.

Ionospheric storms (turbulent conditions in the ionosphere) often cause radio communications to become erratic. Some frequencies will be completely blacked out, while others may be reinforced. Sometimes these storms develop in a few minutes, and at other times they require as much as several hours to develop. A storm may last several days.

Tropospheric Scatter

Tropospheric scatter transmission can be considered as a special case of sky-wave propagation. Instead of aiming the signal toward the ionosphere, however, it is aimed at the troposphere. The troposphere ends just 6.5 mi above the earth's surface. Frequencies from about 350 MHz to 10 GHz are commonly used with reliable communications paths of up to 400 mi.

The scattering process is illustrated in Fig. 13-15. As shown, two directional antennas are pointed so that their beams intersect in the troposphere. The great majority of the transmitted energy travels straight up into space. However, by a little-understood process, a small amount of energy is *scattered* in the forward direction. As shown in Fig. 13-15, some energy is also scattered in undesired directions. The best and most widely used frequencies are around 0.9, 2, and 5 GHz. Even then, however, the received signal is only one-millionth to one-billionth of the transmitted power. There is an obvious need for high-powered transmitters and extremely sensitive receivers. In addition, the scattering process is subject to two forms of fading. The first is due to multipath transmissions within the scattering path, with the effect occurring as quickly as several times per minute. Atmospheric changes provide a second, but slower, change in the received signal strength.

To accommodate these severe fading problems, some form of **diversity reception** is always used. This is the process of transmitting and/or receiving several signals and then either adding them all together at the receiver or selecting

Diversity Reception
transmitting and/or receiving several signals and either adding them together at the receiver or selecting the best one at any given instant

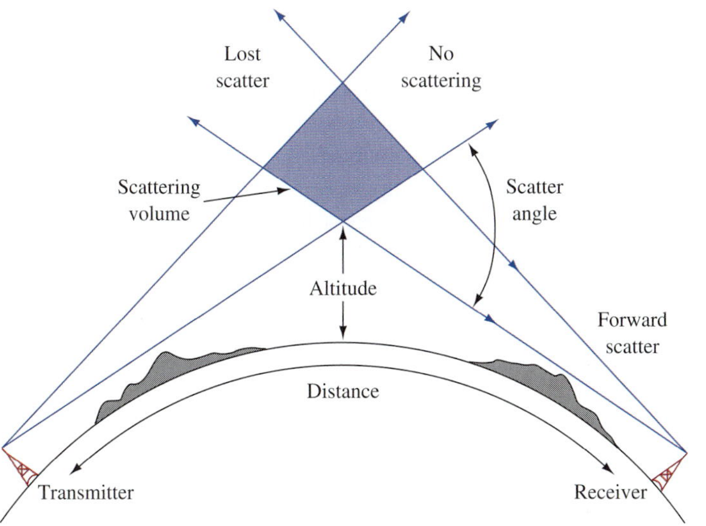

FIGURE 13-15 Tropospheric scatter.

the best one at any given instant. The types of diversity reception utilized include one or combinations of the following:

Space diversity: comprising two or more receiving antennas separated by 50 wavelengths or more. The best received signal at any instant is selected as input for the receiver.

Frequency diversity: transmission of the same information on slightly different frequencies. The different frequencies fade independently even when transmitted and received through the same antennas.

Angle diversity: transmission of information at two or more slightly different angles. This results in two or more paths based on illuminating different scattering volumes in the troposphere.

Polarization diversity: the capability of receiving horizontally and vertically polarized signals.

In spite of the high-power and diversity requirements and the more recent satellite communications, the use of tropospheric scatter continues since its first use in 1955. It provides reliable long-distance communication links in areas such as deserts and mountain regions and between islands. It is used for voice and data links by the military and commercial users.

13-6 SATELLITE COMMUNICATIONS

Synchronous Orbit
when a satellite's position remains fixed with respect to the earth's rotation

A final category of wave propagation is satellite communications (SATCOM). In these systems, a communications satellite is placed into **synchronous orbit**—which means its position remains fixed with respect to the earth's rotation. This is accomplished when the satellite is stationed approximately 22,000 mi above the earth's surface.

The transmitter sends a signal via a highly directional antenna, through the ionosphere, to the satellite's receiving antenna. It is then reamplified within the satellite and transmitted back to earth. The satellite is powered by a bank of batteries whose charge is maintained by panels of solar cells. The service life of communications satellites is from 5 to 10 years. The unit is designed to be extremely reliable because service calls are tough to accomplish!

Uplink
sending signals to a satellite

Downlink
a satellite sending signals to earth

Transponder
electronic system on a satellite that performs reception, frequency translation, and retransmission of received radio signals

Satellite communication allows transoceanic links, and wide bandwidths are utilized to allow the multiplexing of several different signals. Frequencies used are in excess of 1 GHz. At these high frequencies, the effects of ionospheric refraction and attenuation are negligible. The frequencies used range from about 1 GHz up to 40 GHz. The signals received and subsequently retransmitted by the satellite are at different carrier frequencies. For example, the Intelsat III satellite shown in Fig. 13-16 receives signals (the **uplink**) at 5.93 to 6.42 GHz, amplifies, translates down to 3.705 to 4.195 GHz, and then reamplifies via a TWT output stage to a 7-W level for transmission back to earth (the **downlink**). The frequency translation is used to prevent interference between the two signals both at the ground station and satellite. The frequency bands commonly utilized and their designations are shown in Table 13-1. The Intelsat III would be classified as a C/S band satellite because its uplink is nominally in the C-band and its downlink is in the S-band.

An electronic system performing reception, frequency translation, and retransmission is called a **transponder.** The total power consumption for satellite operation is about 150 W. The Intelsat 907 satellite was launched in 2003. It provides enhanced C-band and high-powered Ku-band service.

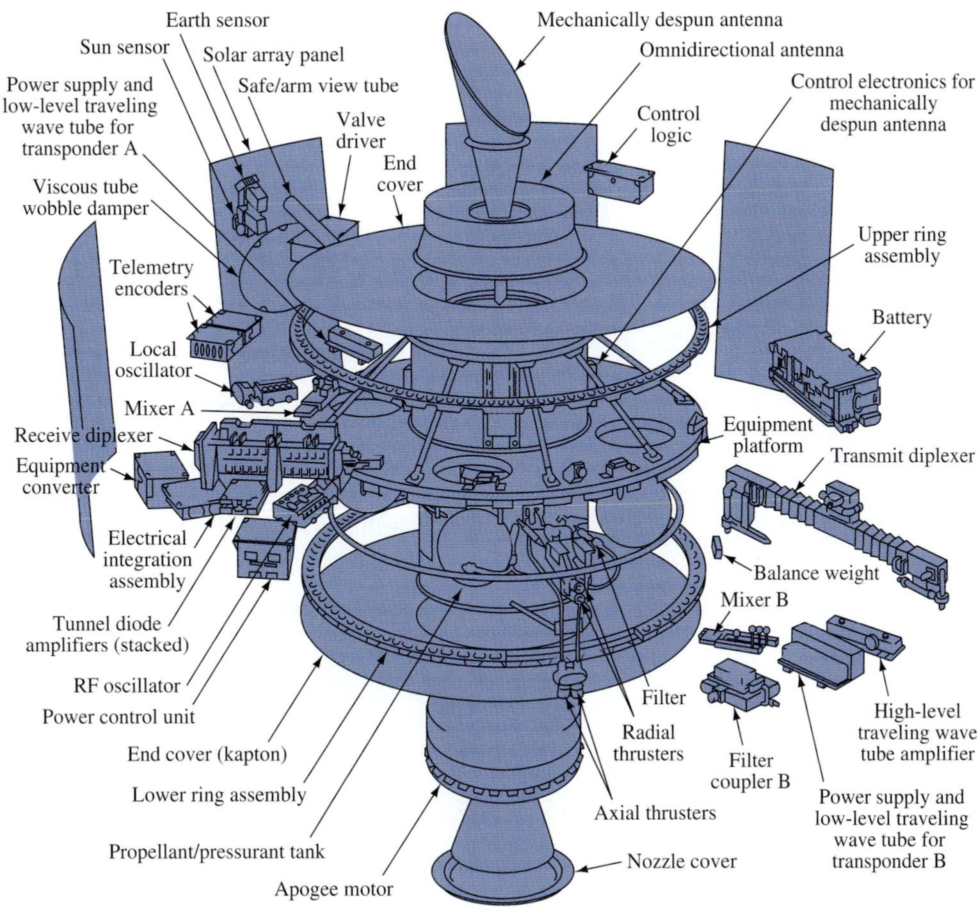

FIGURE 13-16 Intelsat III satellite. (Courtesy of TRW. Reprinted with persmission.)

Table 13-1 Satellite Frequency Ranges and Band Designators

Band Designator	Frequency (GHz)
L	1–2
S	2–4
C	4–8
X	8–12
Ku	12–18
K	18–27
Ka	27–40

The round-trip distance for a satellite relay is typically 90,000 km. The total transmission time is about 300 ms. Thus, in a transoceanic telephone conversation, a 600-ms delay occurs before you hear a reply. Because of this, care is exercised in the routing of international calls to ensure that no more than a single satellite hop is utilized. Special circuitry is also incorporated to reduce delayed echo to reasonable levels.

Satellites in **geosynchronous orbit** (GEO) have become numerous. International regulations limit their spacing to prevent interference. This puts the

Geosynchronous Orbit another name for synchronous orbit

orbital slots over prime real estate, such as North America, Europe, or Japan, at a premium. A recent trend is therefore to use *low earth orbit* (LEO) satellites. At LEO altitudes (250 to 1000 mi), signal time delay shrinks to 5–10 ms and the launch costs drop considerably from that of the GEO satellites. These LEO satellites are not stationary with respect to a specific point on earth. They orbit the earth with periods of about 90 minutes and are visible to an earth station for only 5–20 minutes each 90 minutes. If real-time communication is required, several LEO satellites are necessary. In addition, we must devise some method of handing off subscriber connections between satellites every few minutes, as they appear and disappear over the horizon. This requires a high degree of intelligence within the network, much like a standard cellular telephone system, except in this case, the base stations move while the subscribers stay relatively still.

The global positioning system (GPS) is another application made possible by satellite technology. It provides pinpoint geographic location information. The GPS was originally used by government and law enforcement agencies but the recent availability of low-cost handheld receivers has allowed personal use. You can now obtain your exact location when traveling by car or boat or when hiking.

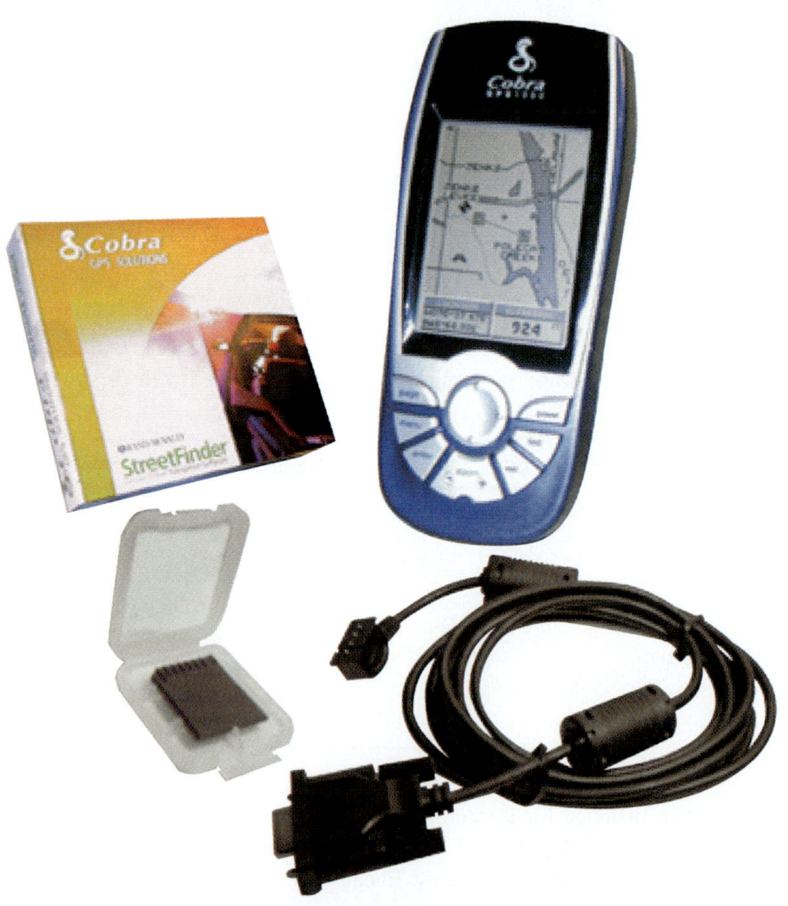

Cobra's GPS handheld line provides street-level detail mapping software, faster acquisition time, and intuitive interfaces. (Courtesy of Cobra Electronics Corporation. Reprinted with permission.)

The GPS satellites transmit position data signals and a GPS receiver processes and computes the time to receive each one. Doing this from four different satellites allows the receiver to determine your exact latitude and longitude.

GPS currently uses a constellation of 28 satellites orbiting above the earth at a distance of 10,900 mi. The GPS satellites complete an orbit about every 12 hours. The satellites transmit two signals, a course acquisition (C/A) signal, transmitted on 1575.42 MHz, which is available for civilian use, and a precision code (P-code), transmitted on 1227.6 MHz and 1575.42 MHz, which are for military use only. GPS receivers measure the time it takes for the satellite signals to travel from the satellites to the receiver; from this information, the receiver can fix our position (locate where we are). It takes three satellites to fix our position in terms of latitude and longitude, whereas it takes four satellites to determine three-dimensional information: latitude, longitude, and elevation.

Civilian receivers can have a position accuracy of about 10 m, but this distance can vary. The position accuracy can be improved by using a technique called **differential GPS.** Receiver accuracy is improved by using a ground receiver at a known location to provide corrections to the satellite civilian signal error, thereby providing increased position accuracy of less than a meter.

Differential GPS
a technique in which GPS satellite clocking corrections are transmitted so that the position error can be minimized

Multiplexing Techniques

A single satellite typically allows simultaneous communications among multiple users. Consider the situation shown in Fig. 13-17. The satellite shown has a *footprint* (coverage area) as indicated. Some satellites use highly directional antennas so that the

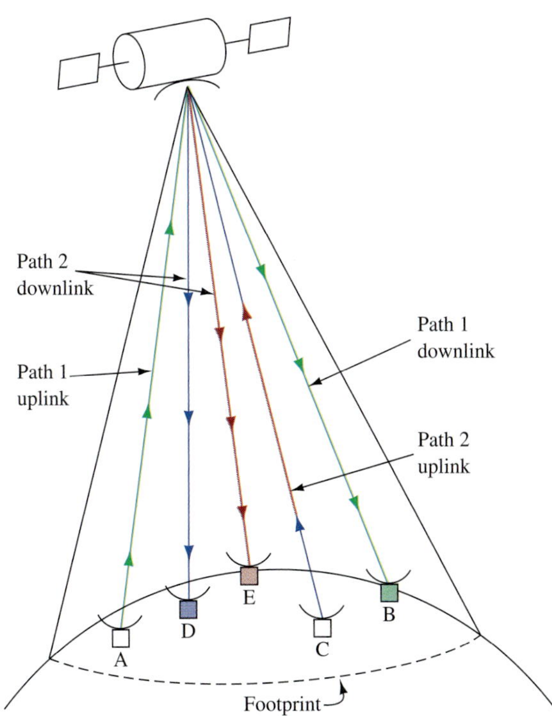

FIGURE 13-17 Satellite footprint and multiple communications.

footprint may include two specific areas. For example, it may be desirable to utilize a satellite between Hawaii and the west coast of the United States. In that case, there is no sense in wasting downlink signal power over a large portion of the Pacific Ocean.

In Fig. 13-17 communication among five earth stations is taking place simultaneously. Station A is transmitting to station B on path 1. Station C is transmitting to stations D and E on path 2. Control signals included with the original transmitted signals are used to allow reception at the appropriate receiver(s).

Frequency-Division Multiple-Access
operating on different frequencies based on which channel is available

Two different multiplexing methods are commonly used to allow multiple transmissions with a single satellite. The early satellite systems all used **frequency-division multiple-access** (FDMA). In these systems the satellite is a wideband receiver/transmitter that includes several frequency channels, much as broadcast radio contains several channels. An earth station that sends a signal indicating a desire to transmit is sent a control signal telling it which available frequency to transmit on. When the transmission is complete, the channel is released back to the "available" pool. In this fashion a *m*ultiple *a*ccess capability for the earth stations is provided—FD*MA*.

Time-Division Multiple-Access
a single satellite can service multiple earth stations simultaneously on the same frequency based on available bursts of time

Most of the newer SATCOM systems use **time-division multiple-access** (TDMA) as a means to allow a single satellite to service multiple earth stations simultaneously. In TDMA all stations use the same carrier frequency, but they transmit one or more traffic bursts in nonoverlapping time frames. This is illustrated in Fig. 13-18, where three earth stations are transmitting simultaneously but never at the same instant of time. The traffic bursts are amplified by the satellite transponder and retransmitted in a downlink beam that is received by the desired station(s). The computer control of these systems is rather elaborate, as you can well imagine.

TDMA offers the following advantages over FDMA systems:

1. A single carrier for the transponder to operate on is a major advantage. Its traveling wave tube (TWT) power amplifier is much less subject to intermodulation problems and can operate at a higher power output when dealing with a smaller range of frequencies.

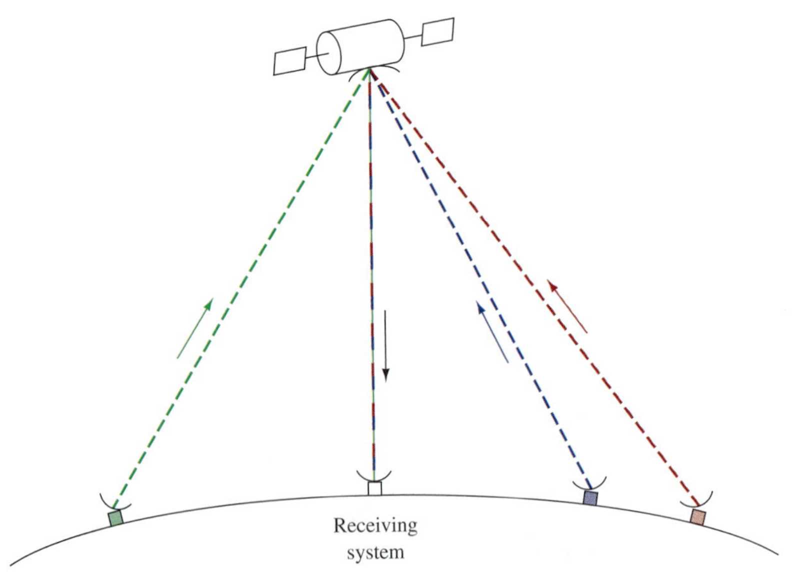

FIGURE 13-18 TDMA illustration.

2. The use of the time domain rather than frequency domain to achieve selectivity is advantageous. In FDMA the earth station must transmit and receive on a multiplicity of frequencies and must provide a large number of frequency-selective up-conversion and down-conversion chains. In TDMA the selectivity is accomplished in time rather than frequency. This is much simpler and less expensive to accomplish.
3. TDMA is ideally suited to digital communications because they are naturally suited to the storage, rate conversions, and time-domain processing used in TDMA implementation. TDMA is also ideally suited to demand-assigned operation, in which the traffic burst durations are adjusted to accommodate demand.

It should be mentioned that a third multiplexing technique is now receiving some attention. **Code-division multiple-access** (CDMA) also allows the use of just one carrier. In it, each station uses a different binary sequence to modulate the carrier. The control computer uses a "correlator" that can separate and "distribute" the various signals to the appropriate downlink station.

Code-Division Multiple-Access
each station uses a different binary sequence to modulate the same carrier

VSAT and MSAT Systems

The two areas of satellite communications showing the greatest growth are (1) very small aperture terminal (VSAT) fixed satellite communication systems, and (2) ultrasmall aperture terminal mobile satellite (MSAT) systems. Technological advances and market demand have driven the development of these new markets. MSAT terminals, which can be called "VSATs on wheels," have several features in common with VSATs. While VSATs take telecommunication services directly to fixed users, MSAT terminals take them to moving vehicles. The most visible VSAT systems are the direct TV and dish network, antennas mounted outside people's homes and apartments. These systems provide entertainment and data (Internet) access.

VSAT systems allow multiple inexpensive stations to be linked to a large central installation. For example, Kmart installed small aperture antenna systems (VSATs) at over 2000 stores and linked them with its central mainframe computer in Michigan. This arrangement allows Kmart to verify checks and credit cards quickly and to convey data, such as what customers are buying and how much inventory is on hand. They can supply each store with the items its customers are buying and speed up the checkout process.

The VSAT dish antenna is typically 1 m in diameter, and a transmitter power of just 2 to 3 W is sufficient. This market is expected to be immune to optical fiber competition for another 20 years or more, until fiber replaces copper cable. When that takes place, the economics may favor the terrestrial fiber transmission systems.

Figure 13-19 provides a pictorial representation of Chrysler's VSAT network. It connects the automaker's headquarters with more than 6000 dealerships and corporate facilities in North America. It is used to assist mechanics with repair and allows salespeople to order and confirm delivery dates for cars from a showroom computer. It also helps maintain proper inventories of automobiles and spare parts. Satellite systems have the advantage of simultaneous delivery of information to and from multiple sites using TDMA techniques. The transmit power requirements are minimal, and all sites can share access.

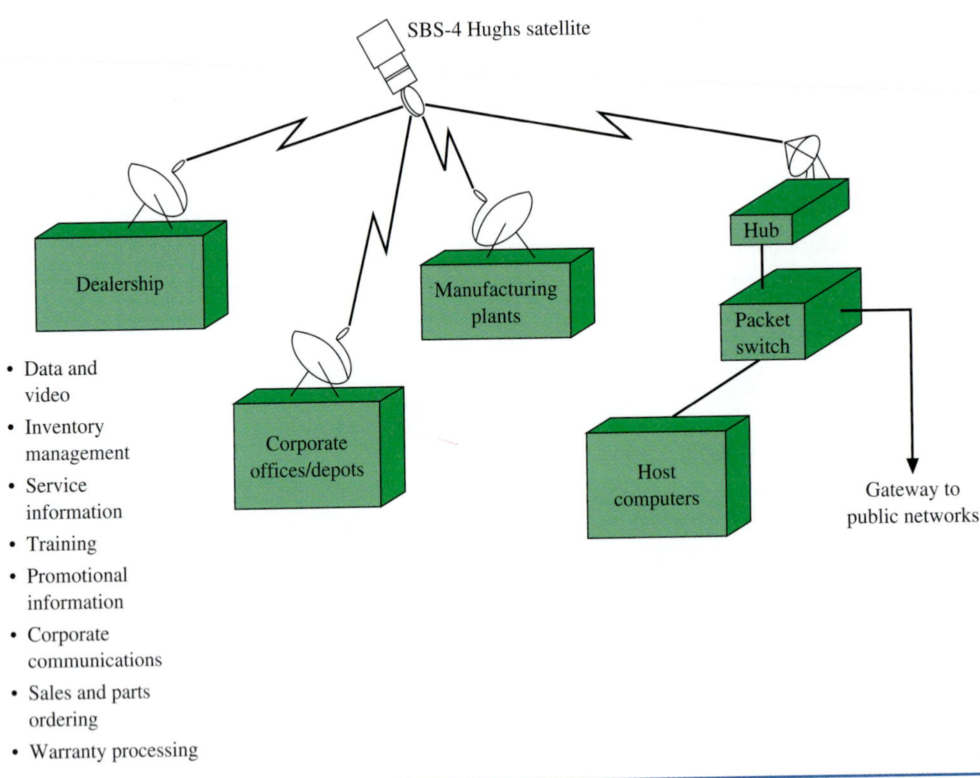

FIGURE 13-19 VSAT network.

The first application of MSAT systems includes large national trucking firms. The system allows the dispatching center to maintain continuous communication with each of its trucks.

Satellite Radio

The Federal Communications Commission (FCC) allocated RF spectrum in the S-band (2.3 GHz) for the Digital Audio Radio Service (DARS) in 1992. Currently there are three satellite radio services: XM Satellite Radio, Sirius Satellite Radio, and WorldSpace. WorldSpace began satellite radio transmission in 1998; XM, in 2001; and Sirius, in 2002. The satellites used by these radio services orbit in geostationary or inclined orbital patterns. Both patterns are shown in Fig. 13-20. The **perigee** is the closest distance of the satellite's orbit to earth, and the **apogee** is the farthest distance of the satellite's orbit from earth. Both are shown in Fig. 13-21.

Perigee
closest distance of a satellite's orbit to earth

Apogee
farthest distance of a satellite's orbit to earth

The XM Satellite Radio service uses two geostationary satellites. These satellites are parked approximately 22,300 miles above the earth at a fixed location. The difference in the distance of the apogee and perigee for geostationary satellites is minimal. The Sirius Satellite Radio service uses three satellites in an inclined orbit. Each satellite is above the continental United States at least 16 hours each day. The orbits of the satellites are arranged so that at least one satellite is always over the United States. The apogee for the Sirius satellites is 29,200 miles above earth and at this point, the satellites are over North America. The perigee for the Sirius satellites is 14,900 miles. WorldSpace provides satellite radio service for areas outside the

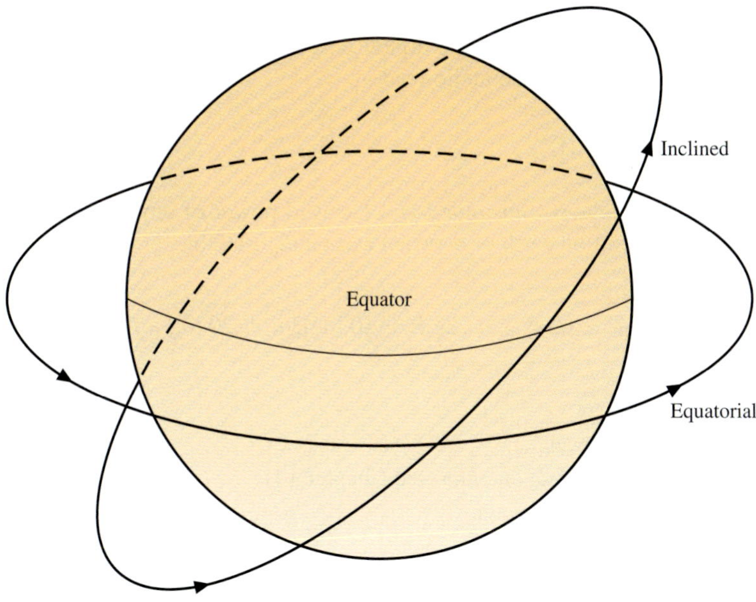

FIGURE 13-20 The geostationary and inclined orbital patterns for satellites.

United States and currently has three satellites parked in geostationary orbit. These satellites transmit satellite radio over the L-band spectrum (1467 to 1492) MHz.

Reception of satellite radio services requires an antenna and custom chip sets to process the received signal. Reception of the Sirius Satellite Radio signal is made possible using diversity receivers (in addition to the special chip sets) and antennas. Receiver diversity is the reception of two signals from two satellites at any given time and the selection of the best one. This process is called spatial diversity. The Sirius satellites transmit the radio signal on three different frequencies in the 12.5 MHz band. Once again, the best received signal is selected. The Sirius system also uses time

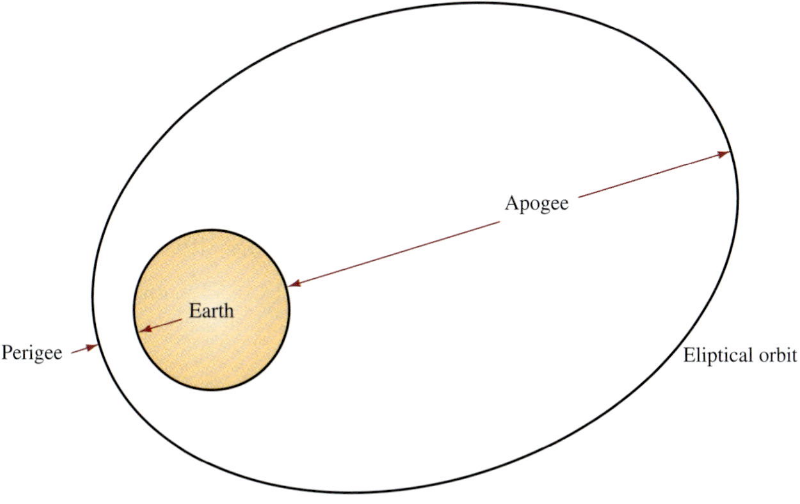

FIGURE 13-21 The apogee and perigee of a satellite's orbit.

diversity, which is provided by delaying the audio by about four seconds. This delay is accomplished by storing the satellite's digital data stream so that a momentary loss of signal does not interrupt the audio feed.

SATCOM Power Budget

The following equation relates transmitted and received power of any space wave, whether it be point to point on earth or between earth and a satellite.

$$\left(\frac{P_R}{P_T}\right)_{\text{dB}} \simeq (G_T)_{\text{dB}} + (G_R)_{\text{dB}} - [32.5 + 20\log_{10} d + 20\log_{10} f]\ \text{dB} \quad \textbf{(13-8)}$$

where P_R = received power
P_T = transmitted power
G_T = gain of transmitting antenna (see Chapter 14)
G_R = gain of receiving antenna
d = distance (km) between antennas
f = frequency (MHz)

Example 13-2 shows the great amount of energy lost in long-distance transmissions. The return link suffers the same type of attenuation, and this whole exercise helps in understanding the need for extremely sophisticated, low-noise, and costly equipment in satellite systems.

Example 13-2

Calculate the power received at a satellite given the following conditions:

1. *The power gain of the transmitting parabolic dish antenna (see Chapter 16) is 30,000.*
2. *The transmitter drives 2 kW of power into the antenna at a carrier frequency of 6.21 GHz.*
3. *The satellite receiving antenna has a power gain of 30.*
4. *The transmission path is 45,000 km.*

Solution

Using Eq. (13-8),

$$\left(\frac{P_R}{2\text{ kW}}\right)_{\text{dB}} \simeq 10\log_{10}30{,}000 + 10\log_{10}30 - (32.5 + 20\log_{10}45{,}000 + 20\log_{10}6210)\text{ dB}$$

$$\left(\frac{P_R}{2\text{ kW}}\right)_{\text{dB}} \simeq 44.8\text{ dB} + 14.8\text{ dB} - (32.5 + 93.1 + 75.9)\text{ dB}$$

$$= 59.6\text{ dB} - 201.5\text{ dB}$$

$$= -141.9\text{ dB}$$

$$\frac{P_R}{2\text{ kW}} = \frac{1}{\text{antilog } 14.19\text{ dB}} = \frac{1}{1.55 \times 10^{14}}$$

$$P_R = \frac{2\text{ kW}}{1.55 \times 10^{14}} = 1.29 \times 10^{-11}\text{ W}$$

$$= 12.9\text{ pW!}$$

13-7 Troubleshooting

A radio communications system transmits a radio frequency signal and depends on a significant amount of that signal being intercepted at the receiver. In earlier chapters, you learned about noise and interference that could affect the transmitted signal. Now we will take a closer look at the problems interference causes in TV and FM radio systems because we all can identify with these. We will also discuss some methods used to resolve interference problems.

After completing this section you should be able to

- Identify different types of interference
- Describe three methods to reduce interference
- Troubleshoot various antenna installation problems

Radio Interference

This discussion will be limited to TV and FM reception because we are most familiar with both. Sources of unwanted signals (noise and interference) happen naturally or are human-made. Review Chapter 1 for sources of noise. Good circuit and antenna design reduces noise to negligible levels. Human-made sources generate most of the interference that usually disturbs the quality of the signal at the receiver site. There are various solutions to remedy interference problems at the receiver. The first step in eliminating an unwanted signal is to pinpoint the source of it. After finding the source of the interference that is causing the disturbance, remove it if possible. When it is not possible to remove the interference source, try increasing the distance of the undesirable source from the receiver. This usually reduces the effects of interference on the receiver and may clear the problem. Using filters is another practical approach for removing unwanted signals. One more method is to protect the receiver by shielding the antenna, the ac input power line, or the whole receiver from the interference signal. The following paragraphs talk about the kinds of interference you may encounter and practical methods to resolve the problems.

Capture and Cochannel Interference Effects In areas of congested radio, TV, and communications channels, receivers are subject to the capture effect and cochannel interference. From the discussion in Chapter 5, the capture effect causes a stronger station to overpower and replace a weaker station at the receiver. The weak station is usually lost completely. Cochannel interference takes the form of two or more broadcasting stations bleeding into each other at the receiver. To the listener, this bleeding-over effect turns into bothersome noise. The best solution for these kinds of interference is to rotate the antenna or obtain a more directional antenna.

EMI and RFI Section 13-2 introduced EMI and RFI. Electromagnetic interference (EMI) shows up on TV as vertical bands of dots moving on the screen. On FM, EMI causes distortion to the audio. Radio frequency interference (RFI) displays itself as several bars or wavy lines on the TV screen. Strong RFI will cause complete loss of the TV's picture. FM audio is affected by RFI with garbled sound, often causing it to be pure gibberish.

The automobile ignition or spark plugs and kitchen appliances like the blender or microwave oven are sources of EMI. In addition, computers and electric motors produce EMI. EMI can enter the receiver through the antenna, lead-in wire, or power line. To decide which is bringing in the EMI, disconnect the lead-in antenna wire to the receiver and short the receiver antenna terminals together. If the interference disappears, then the source was the antenna or lead-in wire. If the interference continues, then the unwanted EMI is coming through the power line. When it is not possible to remove the EMI source or relocate it, use shielding or filtering to remove the interference. When the interference is entering the receiver by way of the antenna, it may be necessary to relocate the antenna.

Ham radio and CB radio transmitters are a common cause of RFI. If the source can be found (towering antennas in the neighborhood are usually a dead giveaway), attempt to contact the owner and let them know the problem exists. Install a high-pass filter between the antenna and receiver input on the lead-in wire to eliminate RFI. However, the best way to eliminate the interference is to remove the source.

Fading Fading is one of the most troublesome hindrances in communications. Fading is the result of the signal arriving at the receiver from two different paths—a direct path and the skyway path. A typical example of this type of problem occurs when an airplane flies over an area where outside TV antennas are used. The airplane causes reflected signals to mix with the direct signal, and a fluttering results in the picture on the TV receiver. Using a high-gain directional antenna will often resolve this kind of interference.

Reflections Figure 13-6 shows a problem that exists between any type of broadcast station and its receiver. In practice, the reflected wave is quite strong, almost as strong as the direct wave, but the path taken by the reflected wave is longer than that of the direct wave. The important fact to remember is that even though the wavelength may be only 1 m and the path is several miles, every time the path difference is equivalent to $\frac{1}{2}$ wavelength or 180°, there will be a null in the signal. Conversely, when the path difference is a multiple of wavelengths, the signals add, potentially doubling the signal strength. Equation (13-9) will enable you to determine where a peak in signal strength might be found.

$$\theta = \frac{1.385 \times 10^{-4} \times H_t \times H_r \times f}{D} \tag{13-9}$$

Remember, every time θ is an odd multiple of 180°, you are in a null. H_t is the transmitter height in feet, H_r is the receiver height in feet, D is the distance in miles, and f is the frequency in MHz.

The point is, when you need more signal, move the antenna. Intuition would tell you to increase the height, but you may actually be able to lower the antenna and find more signal.

> *Diffraction:* Diffraction is much more complicated than reflection, but the solution is the same. As you move away from the mountain shown in Fig. 13-5 you will find hot spots. The technician who finds the hot spot can save a great deal of money on an antenna installation.
>
> *Ghosting in TV Reception:* It may be possible to fight the ghost problem described in Fig. 13-8 with knowledge of your antenna patterns. Most TV antennas have a fairly broad main beam and several null and side lobes (see Fig. 13-22). Additional information on antenna patterns is provided in Chapter 14.

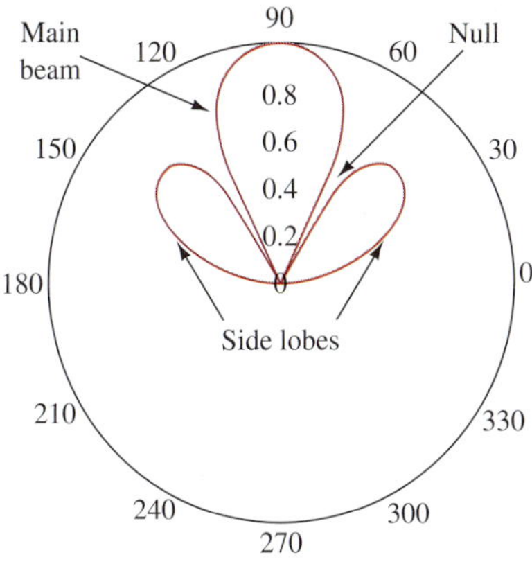

FIGURE 13-22 Antenna pattern.

Try orienting the antenna to place the ghost signal in a null and the desired signal somewhere on the main beam. If this isn't possible, try a single-frequency antenna. Single-frequency antennas have a high rejection of side lobes.

Sky-Wave Propagation

Commercial use of the shortwave frequencies is steadily declining, but it is still the cheapest way to communicate with remote areas of the world. The technician or engineer needs some knowledge of system planning. People interested in short-wave propagation or forecasting should obtain a copy of a computer program called "Ioncap." It was developed by the National Bureau of Standards, now called National Institute of Standards and Technology, and can be purchased from the U.S. Government Printing Office. Several commercial programs are available that use Ioncap as a basis and are easier to use.

Satellite Communications

When servicing or installing a satellite system, your chief problem is aligning the antenna with the satellite. Beams are no more than 2° wide and polarity might be unknown. All are usually adjustable and often need adjustment when performance is not satisfactory.

13-8 Troubleshooting with Electronics Workbench™ Multisim

In this exercise, we investigate the simulation of crystals and crystal oscillators using Electronics Workbench™ Multisim. Crystals are used when greater frequency stability is required than that provided by LC oscillators. Chapter 13 focused primarily on the concepts of radio-wave propagation and the effects propagation has on the different frequencies. This is a good opportunity to explore the components used to generate these different frequencies.

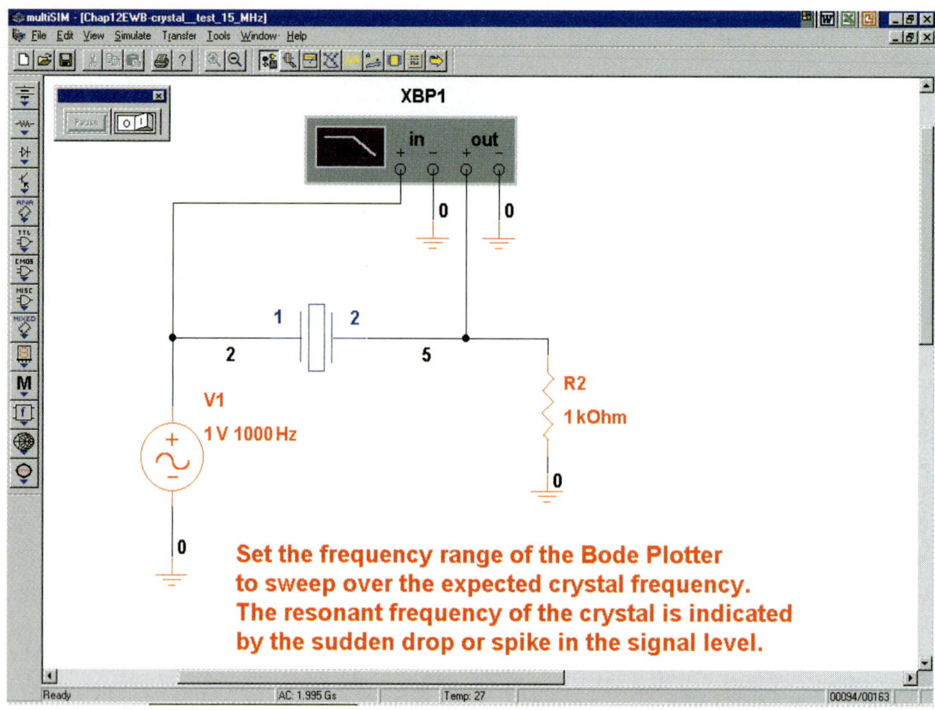

FIGURE 13-23 The test circuit for the crystal oscillator using EWB Multisim.

The first exercise investigates the property of a crystal. Crystals and crystal oscillators were first introduced in Chapter 1, where it was mentioned that a crystal can be modeled as a series *RLC* resonant circuit with a very high Q. Refer to Fig. 1-26 for a drawing of the electrical equivalent circuit of a crystal. Open the file **Fig13-23.ms7 (.msm)** on your EWB Multisim CD. This circuit contains a crystal connected to the Bode plotter, a signal source, and 1-k Ω termination. The circuit is shown in Fig. 13-23.

Recall from Chapter 1 that at the series resonant point, the crystal should have a very low resistance, whereas at other frequencies, the crystal impedance should be quite high. Based on this information, the Bode plotter provided by Multisim can be used to provide a frequency sweep of the crystal using the test circuit provided. At resonance, we should see a change or disturbance in the output response. The frequency of the crystal is not listed on the circuit, so make sure the range for the Bode plotter has been set to sweep a wide frequency range. For this example, the initial (*I*) frequency is 1 kHz and the final (*F*) frequency is 2 GHz.

Start the simulation and view the frequency-sweep results by double-clicking on the Bode plotter instrument. You should see an image similar to that shown in Fig. 13-24. Move the cursor so that you can measure the frequency of the disturbance in the frequency sweep. You will find that the disturbance in the frequency sweep is at about 15 MHz. Double-click on the crystal. You should see that the crystal's frequency value is 15 MHz. Click on **Edit Model** to view how the crystal is being modeled by Multisim. The crystal is modeled by a series *LCR* circuit that is defined by the circuit nodes and the component values consisting of LS 1 3 0.005, CS 3 4 2.2e-014, and RS 4 2 10, and a parallel capacitance CO 1 2 5e-012. You can use these values to verify that the resonant frequency of the model is 15 MHz.

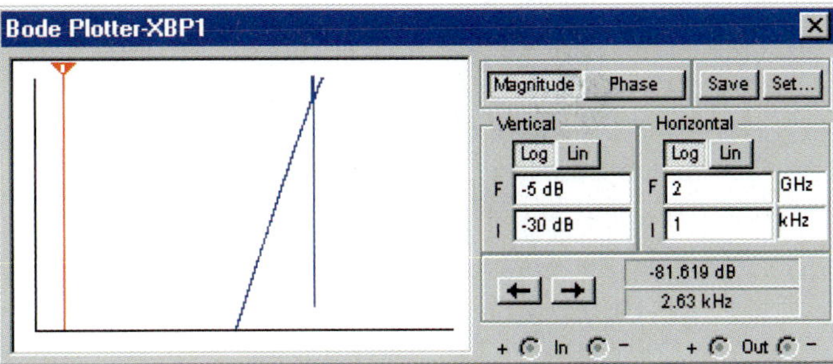

FIGURE 13-24 The frequency sweep of a crystal under test.

Next, open the file **FigE13-1.ms7 (.msm)** in your EWB CD. This circuit is an example of a Pierce crystal oscillator. The crystal frequency is 32.768 kHz, which is a common clock frequency used in digital clocks and watches. This frequency, 32.768 kHz, is equal to 2^{15}, and this value is easily divided by down to 1 pulse per second using digital logic circuits. Start the simulation and check the output signal. You should see a wave with a period of about 30.5 μS, which is the period of a 32.768-kHz signal. The RFC (RF choke) is placed in series with the connection to the power supply to minimize the coupling of oscillator noise to the power supply.

Electronics Workbench™ Exercises

1. Open the file **FigE13-2.ms7 (.msm)** on your EWB CD. Use the technique described in the text to determine the frequency of the crystal. Verify your answer by double-clicking on the crystal and viewing the values.
2. Open the file **FigE13-3.ms7 (.msm)** on your EWB CD. Determine if the Pierce crystal oscillator is working properly. Correct any faults and retest the circuit. Report on your findings.
3. Open the file **FigE13-4.ms7 (.msm)** on your EWB CD. Determine if the Pierce crystal oscillator is working properly. Correct any faults and retest the circuit. Report on your findings.

Summary

In Chapter 13 we studied various considerations of wave propagation. We discovered that electrical energy can be converted to wave energy with many properties in common with light wave propagation. The major topics you should now understand include:

- the definition of an electromagnetic wave, isotropic point source, wavefront, and characteristic impedance of free space
- the understanding of environmental effects on wave propagation, including reflection, refraction, and diffraction
- the explanation of ground- and space-wave propagation
- the description of ionospheric layers and their effects on sky-wave propagation
- the definitions of skipping, critical frequency, critical angle, maximum usable frequency (MUF), skip zone, fading, and tropospheric scatter

- the description and use of satellite communications
- the explanations of multiplexing techniques used in satellite communications, including FDMA, TDMA, and CDMA
- the description of very small aperture terminal (VSAT) and ultrasmall aperture terminal mobile satellite (MSAT) communication
- the power-loss calculations used in satellite communications analysis

Questions and Problems

Section 13-1

1. Explain why an antenna can be thought of as a transducer.
2. List the similarities and dissimilarities between light waves and radio waves.

Section 13-2

3. What are the two components of an electromagnetic wave? How are they created? Explain the two possible things that can happen to the energy in an electromagnetic wave near a conductor.

*4. What is *horizontal and vertical polarization* of a radio wave?

*5. What kinds of fields emanate from a transmitting antenna, and what relationships do they have to each other?

6. Define *wavefront.*
7. Calculate the power density in watts per square meter (on earth) from a 10-W satellite source that is 22,000 mi from earth. (6.35×10^{-16} W/m^2)
8. Calculate the power received from a 20-W transmitter, 22,000 mi from earth, if the receiving antenna has an effective area of 1600 mi^2. (2.03×10^{-12} W)
9. Calculate the electric field intensity, in volts per meter, 20 km from a 1-kW source. How many decibels down will that field intensity be if the distance is an additional 30 km from the source? (8.66 mV/m, 7.96 dB)
10. Calculate the characteristic impedance of free space using two different methods.

*11. How does the field strength of a standard broadcast station vary with distance from the antenna?

12. Define *permeability.*

Section 13-3

13. In detail, explain the process of reflection for an electromagnetic wave.
14. With the aid of Snell's law, fully explain the process of refraction for an electromagnetic wave.
15. What is *diffraction* of electromagnetic waves? Explain the significance of the shadow zone and how it is created.
16. Write the equation for the coefficient of reflection. ($\rho = \epsilon_r/\epsilon_i$)
17. Define *refraction.*
18. Define *shadow zone.*

---

*An asterisk preceding a number indicates a question that has been provided by the FCC as a study aid for licensing examinations.

Section 13-4

19. List the three basic modes whereby an electromagnetic wave propagates from a transmitting to a receiving antenna.
20. Describe ground-wave propagation in detail.
21. Explain why ground-wave propagation is more effective over seawater than desert terrain.
* 22. What is the relationship between operating frequency and ground-wave coverage?
* 23. What are the lowest frequencies useful in radio communications?
24. Fully explain space-wave propagation. Explain the difference between a direct and reflected wave.
25. Explain the phenomenon of *ghosting* in TV reception. What would be the effect if this occurred with a voice transmission?
26. Calculate the ghost width for a 17-in.-wide TV screen when a reflected wave results from an object $\frac{3}{8}$ mi "behind" a receiver. How could this effect be minimized? (1.28 in.)

Section 13-5

27. List the course of events in the process of sky-wave propagation.
28. Provide a detailed discussion of the ionosphere—its makeup, its layers, its variations, and its effect on radio waves.
* 29. What effects do sunspots and the aurora borealis have on radio communications?
30. Define and describe *critical frequency, critical angle,* and *maximum usable frequency* (MUF). Explain their importance to sky-wave communications.
31. What is the optimum working frequency, and what is its relationship to the MUF?
32. What frequencies have substantially straight-line propagation characteristics analogous to those of light waves and are unaffected by the ionosphere?
33. What radio frequencies are useful for long-distance communications requiring continuous operation?
34. In radio transmissions, what bearings do the angle of radiation, density of the ionosphere, and frequency of emission have on the length of the skip zone?
35. Why is it possible for a sky wave to "meet" a ground wave 180° out of phase?
36. What is the process of tropospheric scatter? Explain under what conditions it might be used.
* 37. What is the purpose of a diversity antenna receiving system?
38. List and explain three types of diversity reception schemes.
39. What is skipping?
40. Define *fading.*
41. What happens when a signal is above the critical frequency?

Section 13-6

42. What are *satellite communications*? List reasons for their increasing popularity.
43. Explain the difference between GEO and LEO satellite systems. Describe the advantages and disadvantages of each system.
44. Describe a typical VSAT installation. How does it differ from an MSAT system?

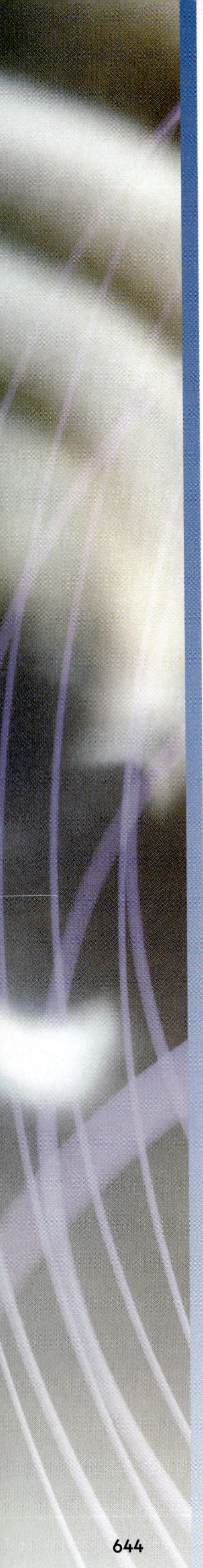

45. The signal received by the satellite in Ex. 13-2 is amplified to a 7-W level at 4 GHz and retransmitted to earth via the same antennas. Calculate the power received by the earth station. (0.108 pW)
46. Explain the methods of multiplexing in SATCOM systems, and provide the advantages of TDMA over FDMA.

Section 13-7

47. Describe the effects of EMI on a receiver.
48. Explain the best methods for reducing EMI.
49. Explain the problems associated with ghosting.
50. Explain the best way to reduce reflection.
51. Your television is exhibiting interference on the picture. How can you determine whether you have an EMI or RFI problem?
52. What are the three ways EMI can be picked up by a receiver? Explain how you can test to determine the source.

Questions for Critical Thinking

53. A user complains about "interference." How can you determine whether this is electromagnetic interference (EMI) or radio-frequency interference (RFI)?
54. Calculate the radio horizon for a 500-ft transmitting antenna and a receiving antenna of 20 ft. Calculate the required height increase for the receiving antenna if a 10 percent increase in radio horizon were required. (37.9 mi, 31.2 ft)
55. In the strictest sense, define *skip distance* and *skip zone.*
56. You will be receiving sky waves. In what ways can you anticipate fading to occur?

Antennas

Chapter Outline

14-1 Basic Antenna Theory
14-2 Half-Wave Dipole Antenna
14-3 Radiation Resistance
14-4 Antenna Feed Lines
14-5 Monopole Antenna
14-6 Antenna Arrays
14-7 Special-Purpose Antennas
14-8 Troubleshooting
14-9 Troubleshooting with Electronics Workbench™ Multisim

A microwave antenna site. (Courtesy of The Stock Market/Lester Lefkowitz.)

Objectives

- Describe the development of the half-wave dipole antenna from transmission line theory
- Define the properties of antenna reciprocity and polarization
- Explain antenna radiation and induction field, radiation pattern, gain, and radiation resistance
- Calculate and define antenna efficiency
- Describe the physical and electrical characteristics of common antenna types and arrays
- Explain the ability to "electronically steer" the radiation pattern of phased arrays
- Differentiate between antenna beamwidth and bandwidth
- Design a log-periodic antenna given the range of frequencies it is to be operated over and its design ratio

Key Terms

reciprocity
polarization
half-wave antenna
dipole antenna
radiation field
induction field
near field
far field
radiation pattern
omnidirectional
directional
beamwidth
antenna gain
dBi
dBd
radiation resistance
corona discharge
feed line
delta match
monopole antenna
image antenna
counterpoise
loading coil
antenna array
parasitic array
reflector
director
lobes
front-to-back ratio
driven array
collinear array
phased array
null
twin lead
grid-dip meter
anechoic chamber

14-1 BASIC ANTENNA THEORY

In this chapter we introduce the fundamentals of antennas and describe the most commonly encountered types. Antennas for use at microwave frequencies are described in Chapter 16.

An antenna is a circuit element that provides a transition from a guided wave on a transmission line to a free space wave and it provides for the collection of electromagnetic energy. In a transmitting system, a radio-frequency signal is developed, amplified, modulated, and applied to the antenna. The RF currents flowing through the antenna produce electromagnetic waves that radiate into the atmosphere. In a receiving system, electromagnetic waves "cutting" through the antenna induce alternating currents for use by the receiver.

To have adequate signal strength at the receiver, either the power transmitted must be extremely high or the efficiency of the transmitting and receiving antennas must be high because of the high losses in wave travel between the transmitter and the receiver.

Any receiving antenna transfers energy from the atmosphere to its terminals with the same efficiency with which it transfers energy from the transmitter into the atmosphere. This property of interchangeability for transmitting and receiving operations is known as antenna **reciprocity.** Antenna reciprocity occurs because antenna characteristics are essentially the same regardless of whether an antenna is sending or receiving electromagnetic energy.

Reciprocity
an antenna's ability to transfer energy from the atmosphere to its receiver with the same efficiency with which it transfers energy from the transmitter into the atmosphere

Because of reciprocity, we will generally treat antennas from the viewpoint of the transmitting antenna, with the understanding that the same principles apply equally well when the antenna is used for receiving electromagnetic energy.

Antennas produce or collect electromagnetic energy and they should do so in an efficient manner. Consequently, antennas are composed of conductors arranged to permit efficient operation. Efficient operation also requires that the receiving antenna be of the same polarization as the transmitting antenna. **Polarization** is the direction of the electric field and is, therefore, the same as the antenna's physical configuration. Thus, a vertical antenna will transmit a vertically polarized wave. The received signal is theoretically zero if a vertical E field cuts through a horizontal receiving antenna.

Polarization
the direction of the electric field of a given electromagnetic radiated signal

The received signal strength of an antenna is usually described in terms of the electric field strength. If a received signal induces a 10-μV signal in an antenna 2 m long, the field strength is 10 μV/2 m, or 5 μV/m. Recall from Chapter 13 that the received field strength is inversely proportional to the distance from the transmitter [Eq. (13-2)].

14-2 HALF-WAVE DIPOLE ANTENNA

Any antenna having a physical length that is one half-wavelength of the applied frequency is called a half-wave dipole antenna. Half-wave dipole antennas are predominantly used with frequencies above 2 MHz. It is unlikely that a half-wave dipole antenna will be found in applications below 2 MHz because at these low-frequencies this antenna is physically too large. Consider a half-wave dipole antenna for a 60-Hz signal.

$$\lambda = \frac{c}{f} = \frac{186{,}000 \text{ mi/s}}{60} = 3100 \text{ mi}$$

A $\frac{1}{2}\lambda$ antenna for 60 Hz is therefore 3100 mi/2, or 1550 mi!

Development of the Half-Wave Dipole Antenna

When the open two-wire transmission line was discussed in Chapter 12, it was found that one of its disadvantages was excessive radiation at high frequencies. Radiation from a transmission line is undesirable since the perfect transmission line would be one that possessed no losses. Although the two-wire transmission line was considered to be an adequate transmission medium at extremely high frequencies, it can become an effective antenna. For this reason, an analysis of the open-ended, quarter-wave transmission line will furnish an excellent introduction for understanding basic antenna theory. The open-ended quarter-wave transmission line segment is shown in Fig. 14-1.

The characteristics of the open-ended line are such that the voltage at the end of the line is maximum, and the current at the end is zero. This is true of an open-ended line regardless of the wavelength of the line. On either the open or shorted line, standing waves will be produced. Because the voltage applied to the line is sinusoidal, the line will constantly be charging and discharging. Current will be flowing in the line continuously. Because the current at the ends of the line is minimum, a quarter-wave back (at the source), the current must be maximum. The impedance at the sending end is low, and the impedance at the open circuit is high. At the open end, E is high and I is very low. This causes the impedance, Z, which is equal to E/I, to be very high. The opposite situation exists at the sending end. The standing waves of current and voltage are shown on the quarter-wave section in Fig. 14-1.

It is desirable to have maximum radiation from an antenna. Under such conditions all energy applied to the antenna would be converted to electromagnetic waves and radiated. This maximum radiation is not possible with the two-wire transmission line because the magnetic field surrounding each conductor of the line is in a direction that opposes the lines of force about the other conductor. Under these conditions, the quarter-wave transmission line proves to be an unsatisfactory antenna; however, with only a slight physical modification, this section of transmission line can be transformed into a relatively efficient antenna. This transformation is accomplished by bending each line outward 90° to form a **half-wave,** or a $\lambda/2$ **dipole,** as shown in Fig. 14-2.

Half-Wave Antenna
an antenna whose receive elements are one half-wavelength in length

Dipole Antenna
straight radiator typically one half-wavelength long, usually separated at the center by an insulator and fed by a balanced transmission line

The antenna shown in Fig. 14-2 is composed of two quarter-wave sections. The electrical distance from the end of one to the end of the other is a half-wavelength. If voltage is applied to the line, the current is maximum at the input and minimum at the ends. The voltage is maximum between the ends, and minimum between the input terminals.

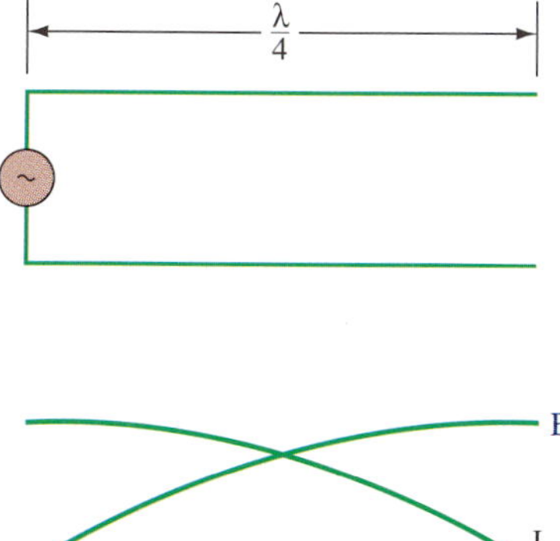

FIGURE 14-1 Quarter-wave transmission line segment (open-ended).

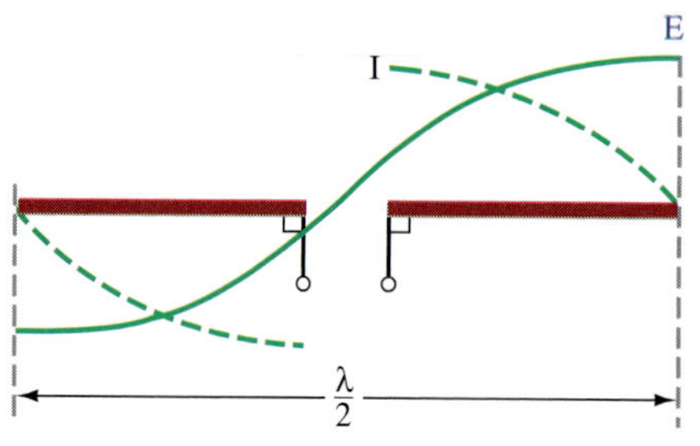

FIGURE 14-2 Basic half-wave dipole antenna.

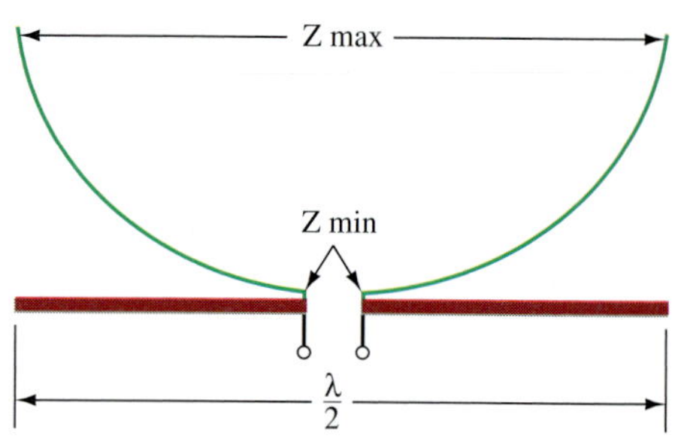

FIGURE 14-3 Impedance along a half-wave antenna.

Half-Wave Dipole Antenna Impedance

An impedance value may be specified for a half-wave antenna thus constructed. Generally, the impedance at the ends is maximum, while that at the input is minimum. Consequently, the impedance value varies from a minimum value at the generator to a maximum value at the open ends. An impedance curve for the half-wave antenna is shown in Fig. 14-3. Notice that the line has different impedance values for different points along its length. The impedance values for half-wave antennas vary from about 2500 Ω at the open ends to 73 Ω at the source ends.

Radiation and Induction Field

Feeding the Hertz antenna at the center results in an input impedance that is purely resistive and equal to 73 Ω. Recall that with an open-circuited $\lambda/4$ transmission line, the input impedance was 0 Ω, and it could therefore not absorb power. By spreading the open $\lambda/4$ transmission line out into a half-wave dipole antenna, its input impedance has taken on a finite resistive value. It can now absorb power, but the question is, how? The answer is that it can now efficiently accept electrical energy and radiate it into space as electromagnetic waves. While the mechanisms of launching a wave from a current-carrying wire are not fully understood, the fields surrounding the antenna do not collapse their energy back into the antenna but rather radiate it out into space. This radiated field is appropriately termed the **radiation field.** Antennas also have an **induction field** associated with them. It is the portion of field energy that *does* collapse back into the antenna and is therefore limited to the zone immediately surrounding the antenna. Its effect becomes negligible at a distance more than about one-half wavelength from the antenna.

Radiation Field
radiation that surrounds an antenna but does not collapse its field back into the antenna

Induction Field
radiation that surrounds an antenna and collapses its field back into the antenna

Near Field
region less than $2D/\lambda$ from the antenna

Far Field
region greater than $2D/\lambda$ from the antenna

Other designators for antenna fields are the **near field** and the **far field.** The far-field region begins when the distance

$$\text{(a)}\ R_{ff} = 1.6\lambda: \quad \frac{D}{\lambda} < 0.32 \qquad \textbf{(14-1a)}$$

$$\text{(b)}\ R_{ff} = 5D: \quad 0.32 < \frac{D}{\lambda} < 2.5 \qquad \textbf{(14-1b)}$$

$$\text{(c)}\ R_{ff} = \frac{2D^2}{\lambda}: \quad \geq 2.5\lambda \qquad \textbf{(14-1c)}$$

where R_{ff} = far field distance from the antenna [meters]
D = dimension of the antenna [meters]
λ = wavelength of the transmitted signal [meters/cycle]

The near-field region is any distance less than R. The effects of the induction field are negligible in the far field.

EXAMPLE 14-1

Determine the distance from a $\lambda/2$ dipole to the boundary of the far field region if the $\lambda/2$ dipole is used in a 150-MHz communications system.

SOLUTION

The wavelength (λ) for a $\lambda/2$ dipole at 150 MHz is approximately

$$\lambda = \frac{3 \times 10^8}{150 \times 10^6} = 2 \frac{\text{m}}{\text{cycle}}$$

Therefore $\lambda/2 = 1$ m, which is the antenna's dimension (D).

$$\frac{D}{\lambda} = \frac{1}{2} = 0.5$$

Therefore select Eq. (14-1b).

$$R_{ff} = 5D = 5(1) = 5 \text{ m}$$

Therefore, the boundary for the far field region is any distance greater than 1 m from the antenna. In this case, the far-field distance is equal to the diameter (D) of the $\lambda/2$ dipole.

EXAMPLE 14-2

Determine the distance from a parabolic reflector with diameter (D) = 4.5 m to the boundary of the far-field region if the parabolic reflector is used for Ku-band transmission of a 12-GHz signal.

SOLUTION

The wavelength (λ) for a 12-GHz signal is approximately

$$\lambda = \frac{3 \times 10^8}{12 \times 10^9} = 0.025 \frac{\text{m}}{\text{cycle}}$$

$D = 4.5$ meter.

$$\frac{D}{\lambda} = \frac{4.5}{.025} = 180$$

Therefore, select Eq. (14-1c).

$$R > \frac{2(4.5)^2}{0.025} = 1620 \text{ m}$$

Therefore, the boundary for the far field region for this parabolic reflector is a distance greater than 1620 m from the antenna. The far-field boundary for high-gain antennas (e.g., a parabolic reflector) will always be greater than for low-gain antennas (e.g., a dipole antenna).

Radiation Pattern

Radiation Pattern
diagram indicating the intensity of radiation from a transmitting antenna or the response of a receiving antenna as a function of direction

Omnidirectional
a spherical radiation pattern

Directional
concentrating antenna energy in certain directions at the expense of lower energy in other directions

Beamwidth
the angular separation between the half-power points on an antenna's radiation pattern

The radiation pattern for the $\lambda/2$ dipole antenna is shown in Fig. 14-4(a). A **radiation pattern** is an indication of radiated field strength around the antenna. The pattern shown in Fig. 14-4(a) shows that maximum field strength for the $\lambda/2$ dipole occurs at right angles to the antenna, while virtually zero energy is launched "off the ends." So if you wish to communicate with someone, the best results would be obtained when he or she is in the direction of A or 180 degrees opposite; the person should not be located off the ends of the antenna. Recall from Chapter 13 that we considered an isotropic source of waves. Its radiation pattern is spherical, or as shown in two dimensions [Fig. 14-4(b)], it is circular or **omnidirectional.** The half-wave dipole antenna is termed **directional** because it concentrates energy in certain directions at the expense of lower energy in other directions.

Another important concept is an antenna's **beamwidth.** It is the angular separation between the half-power points on its radiation pattern. It is shown for the $\lambda/2$ dipole in Fig. 14-4(a). A three-dimensional radiation pattern cross section for a vertically polarized $\lambda/2$ dipole is shown in Fig. 14-5. You can see that it is a doughnut-shaped pattern. If the antenna were mounted close to ground, the pattern would be altered by the effects of ground reflected waves.

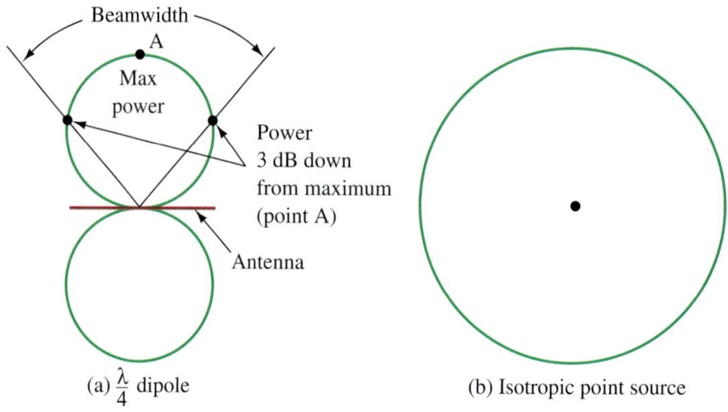

FIGURE 14-4 Radiation patterns.

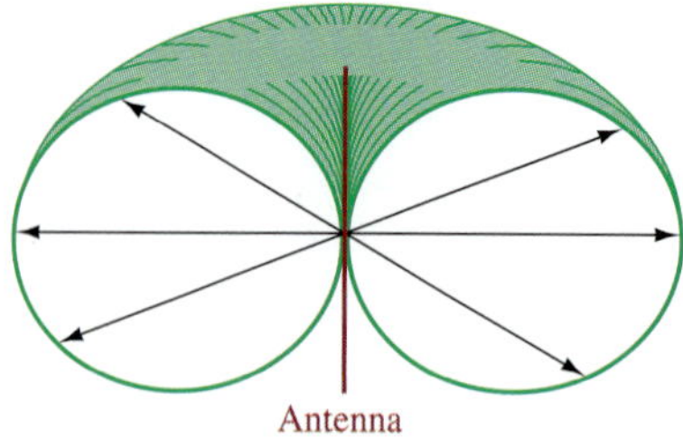

FIGURE 14-5 Three-dimensional radiation pattern for a $\lambda/2$ dipole.

Antenna Gain

Antenna Gain
a measure of how much more power in dB an antenna will radiate in a certain direction with respect to that which would be radiated by a reference antenna, i.e., an isotropic point source or dipole

The half-wave dipole antenna has *gain* with respect to the theoretical isotropic radiator. **Antenna gain** is different from amplifier gain because feeding 50 W into a dipole does not result in more than 50 W of radiated field energy. It is, instead, a gain relative to a reference antenna. The dipole, therefore, has a gain relative to the isotropic radiator in a certain direction. The half-wave dipole antenna has a 2.15-dB gain (at right angles to the antenna) as compared to an isotropic radiator. However, because a perfect isotropic radiator cannot be practically realized, the $\lambda/2$ dipole antenna is sometimes taken as the standard reference to which all other antennas are compared with respect to their *gain*. When the gain of an antenna is multiplied by its power input, the result is termed its effective radiated power (ERP). For instance, an antenna with a gain of 7 and fed with 1 kW has an ERP of 7 kW.

The gain for an antenna whose gain is provided with respect to an isotropic radiator is often expressed as **dBi.** In other words, the half-wave dipole antenna's gain can be expressed as 2.15 dBi. If an antenna's gain is given in decibels with respect to a dipole, it is expressed as **dBd.** This occurs somewhat less often than dBi in antenna literature. The gain of an antenna in dBi is 2.15 dB more than when expressed in dBd. Thus, an antenna with a gain of 3 dBd has a gain of 5.15 dBi (3 dB + 2.15 dB).

dBi
antenna gain relative to an isotropic radiator

dBd
antenna gain relative to a dipole antenna

The amount of power received by an antenna through free space can be predicted by the following:

$$P_r = \frac{P_t G_t G_r \lambda^2}{16\pi^2 d^2} \tag{14-2}$$

where P_r = power received (W)
P_t = power transmitted (W)
G_t = transmitting antenna gain (ratio, not dB) compared to isotropic radiator
G_r = receiving antenna gain (ratio, not dB) compared to isotropic radiator
λ = wavelength (m)
d = distance between antennas (m)

Example 14-3

Two $\lambda/2$ dipoles are separated by 50 km. They are "aligned" for optimum reception. The transmitter feeds its antenna with 10 W at 144 MHz. Calculate the power received.

Solution

The two dipoles have a gain of 2.15 dB. That translates into a gain ratio of $\log^{-1}$ 2.15 dB = 1.64.

$$P_r = \frac{P_t G_t G_r \lambda^2}{16\pi^2 d^2} \tag{14-2}$$

$$= \frac{10\text{ W} \times 1.64 \times 1.64 \times \left(\dfrac{3 \times 10^8\text{ m/s}}{144 \times 10^6}\right)^2}{16\pi^2 \times (50 \times 10^3\text{ m})^2}$$

$$= 2.96 \times 10^{-10}\text{ W}$$

The received signal in Ex. 14-3 would provide a voltage of 147 μV into a matched 73-Ω receiver system [$(P = V^2/R)$, $v = (2.96 \times 10^{-10}\text{ W} \times 73\ \Omega) = 147\ \mu$V]. This is a relatively strong signal because receivers can often provide a usable output with less than a 1-μV signal.

14-3 Radiation Resistance

Radiation Resistance the portion of an antenna's input impedance that results in power radiated into space

The portion of an antenna's input impedance that is the result of power radiated into space is called the **radiation resistance,** R_r. Note that R_r is not the resistance of the conductors that form the antenna. It is simply an effective resistance that is related to the power radiated by the antenna. Since a relationship exists between the power radiated by the antenna and the antenna current, radiation resistance can be mathematically defined as the ratio of total power radiated to the square of the effective value of antenna current, or

$$R_r = \frac{P}{I^2} \tag{14-3}$$

where R_r = radiation resistance (Ω)
I = effective rms value of antenna current at the feed point (A)
P = total power radiated from the antenna

It should be mentioned at this point that not all of the energy absorbed by the antenna is radiated. Power may be dissipated in the actual antenna conductor by high-powered transmitters, by losses in imperfect dielectrics near the antenna, by eddy currents induced in metallic objects within the antenna's induction field, and by arcing effects in high-powered transmitters. These arcing effects are termed **corona discharge.** If these losses are represented by one lumped value of resistance, R_d, and the sum of R_d and R_r is called the antenna's total resistance, R_T, the antenna's efficiency can be expressed as

Corona Discharge luminous discharge of energy by an antenna from ionization of the air around the surface of the conductor

$$\eta = \frac{P_{\text{transmitted}}}{P_{\text{input}}} = \frac{R_r}{R_r + R_d} = \frac{R_r}{R_T} \tag{14-4}$$

Effects of Antenna Length

The radiation resistance varies with antenna length, as shown in Fig. 14-6. For a half-wave antenna, the radiation resistance measured at the current maximum (center of the antenna) is approximately 73 Ω. For a quarter-wave antenna, the radiation resistance measured at its current maximum is approximately 36.6 Ω. These are free-space values, that is, the values of radiation resistance that would exist if the antenna were completely isolated so that its radiation pattern would not be affected by ground or other reflections.

Ground Effects

For practical antenna installations, the height of the antenna above ground affects radiation resistance. Changes in radiation resistance occur because of ground reflections that intercept the antenna and alter the amount of antenna current flowing. Depending on their phase, the reflected waves may increase antenna current or decrease it. The phase of the reflected waves arriving at the antenna, in turn, is a function of antenna height and orientation.

At some antenna heights, it is possible for a reflected wave to induce antenna currents in phase with transmitter current so that total antenna current increases. At other antenna heights, the two currents may be 180° out of phase so that total antenna current is less than if no ground reflection occurred.

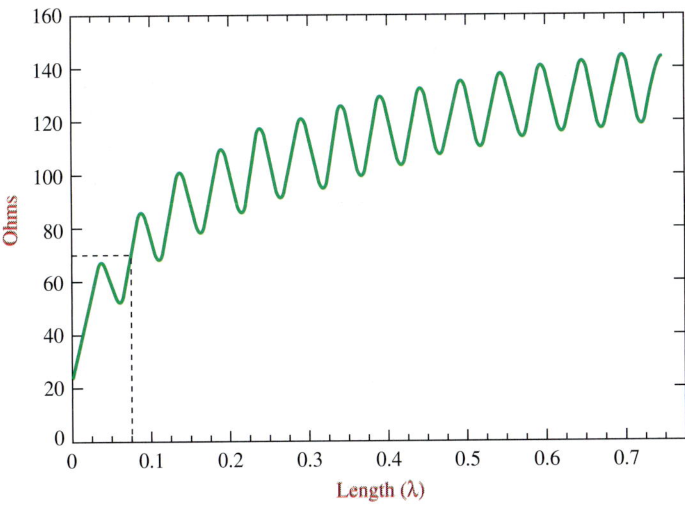

FIGURE 14-6 Radiation resistance of antennas in free space plotted against length.

With a given input power, if antenna current increases, the effect is as if radiation resistance decreases. Similarly, if the antenna height is such that the total antenna current decreases, the radiation resistance is increased. The actual change in radiation resistance of a half-wave antenna at various heights above ground is shown in Fig. 14-7. The radiation resistance of the horizontal antenna rises steadily to a maximum value of 90 Ω at a height of about three-eighths wavelength. The resistance then continues to rise and fall around an average value of 73 Ω, which is the free-space value. As the height is increased, the amount of variation keeps decreasing.

The variation in radiation resistance of a vertical antenna is much less than that of the horizontal antenna. The radiation resistance (dashed line in Fig. 14-7) is a maximum value of 100 Ω when the center of the antenna is a quarter-wavelength above ground. The value falls steadily to a minimum value of 70 Ω at a height of a half-

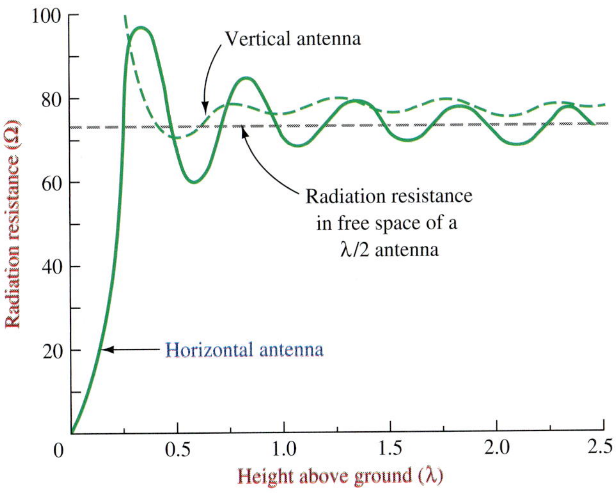

FIGURE 14-7 Radiation resistance of half-wavelength antennas at various heights.

wavelength above ground. The value then rises and falls by several ohms about an average value slightly above the free-space value of a horizontal half-wave antenna.

Since antenna current is affected by antenna height, the field intensity produced by a given antenna also changes. In general, as the radiation resistance is reduced, the field intensity increases, whereas an increase in radiation resistance produces a drop in radiated field intensity.

Electrical versus Physical Length

If an antenna is constructed of very thin wire and is isolated in space, its electrical length corresponds closely to its physical length. In practice, however, an antenna is never isolated completely from surrounding objects. For example, the antenna will be supported by insulators with a dielectric constant greater than 1. The dielectric constant of air is arbitrarily assigned a numerical value equal to 1. Therefore, the velocity of a wave along a conductor is always slightly less than the velocity of the same wave in free space, and the physical length of the antenna is less (by about 5 percent) than the corresponding wavelength in space. The physical length can be approximated as about 95 percent of the calculated electrical length.

Example 14-4

We want to build a $\lambda/2$ dipole to receive a 100-MHz broadcast. Determine the optimum length of the dipole.

Solution

At 100 MHz,

$$\lambda = \frac{c}{f} = \frac{3 \times 10^8 \text{ m/s}}{100 \times 10^6 \text{ Hz}} = 3 \text{ m}$$

Therefore, its electrical length is $\lambda/2$, or 1.5 m. Applying the 95 percent correction factor, the actual optimum physical length of the antenna is

$$0.95 \times 1.5 \text{ m} = 1.43 \text{ m}$$

The result of the preceding example is also obtained by using the following formula:

$$L = \frac{486}{f(\text{MHz})} \qquad \textbf{(14-5)}$$

where L is dipole length in feet. For Ex. 14-4, it would give $L = \frac{468}{100} = 4.68$ ft, which is equal to 1.43 m.

Effects of Nonideal Length

The 95 percent correction factor is an approximation. If ideal results are desired, a trial-and-error procedure is used to find the exact length for optimum antenna performance. If the antenna length is not the optimum value, its input impedance looks like a capacitive circuit or an inductive circuit depending on whether the antenna is shorter or longer than the specified wavelength. A half-wave dipole antenna slightly longer than a half-wavelength acts like an inductive circuit, and an antenna slightly shorter than a half-wavelength appears to the source as a capacitive circuit. Compensation for additional length can be made by cutting the antenna down to proper length or by tuning out the inductive reactance by adding a capacitance in series. This added X_C completely cancels the inductive reactance, and the source then sees a pure resistance, provided the proper size capacitor is used. If an antenna is shorter than the required length, the source end of the line appears capacitive. This condition may be corrected by adding inductance in series with the antenna input.

14-4 Antenna Feed Lines

Two of the most common ways to describe antennas are based on the point at which the transmission line connects to the antenna. Connect the line to the end of the antenna and we have an end-fed antenna; connect it to the center and it is called center fed. If the transmission line joins the antenna at a high-voltage point, the antenna is said to be voltage fed. Conversely, connect to a high-current point and we have a current-fed antenna. All of these types are shown in Fig. 14-8.

It is seldom possible to connect a generator directly to an antenna. It is usually necessary to transfer energy from the generator (transmitter) to the antenna by use of a transmission line (also called an antenna **feed line**). Such lines may be resonant, nonresonant, or a combination of both types.

Feed Line
transmission line that transfers energy from the generator to the antenna

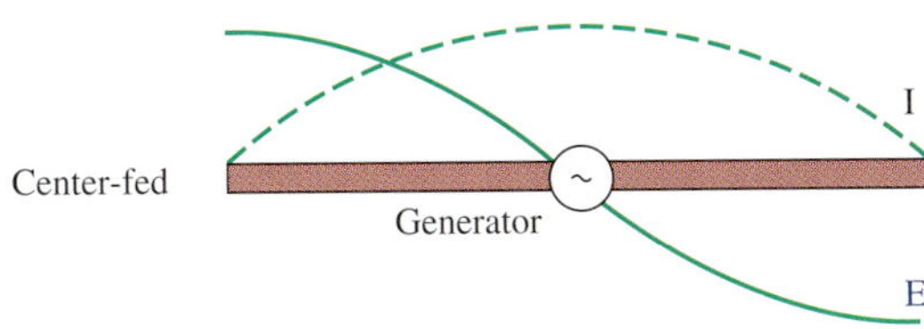

(a) Generator at current maximum means current feed

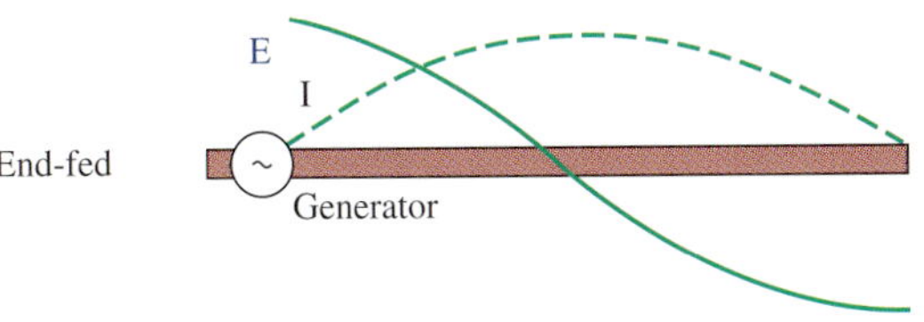

(b) Generator at voltage maximum means voltage feed

FIGURE 14-8 (a) Current feed and (b) voltage feed.

Resonant Feed Line

The resonant transmission line is not widely used as an antenna feed method because it tends to be inefficient and is very critical with respect to its length for a particular operating frequency. In certain high-frequency applications, however, resonant feeders sometimes prove convenient.

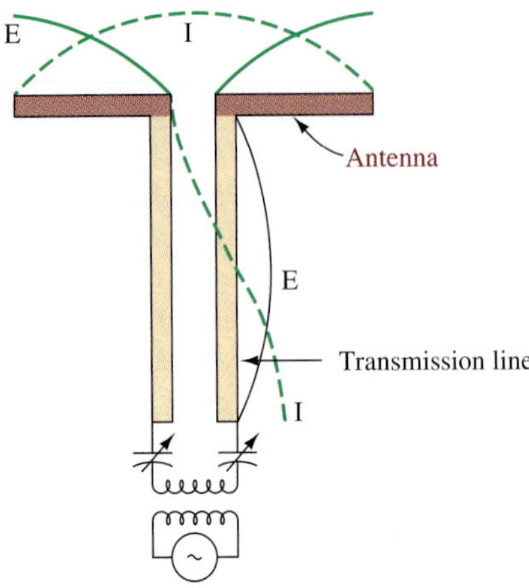

FIGURE 14-9 Current feed with resonant line.

In the current-fed antenna with a resonant line, shown in Fig. 14-9, the transmission line is connected to the center of the antenna. This antenna has a low impedance at the center and, like the voltage-feeding transmission line, has standing waves on it. Constructing it to be exactly a half-wavelength causes the impedance at the sending end to be low. A series resonant circuit is used to develop the high currents needed to excite the line. Adjusting the capacitors at the input compensates for slight irregularities in line and antenna length.

Although this example of an antenna feed system is a simple one, the principles described apply to antennas and to lines of any length provided both are resonant. The line connected to the antenna may be either a two-wire or coaxial line. In high-frequency applications, the coaxial line is preferred due to its lower radiation loss.

One advantage of connecting a resonant transmission line to an antenna is that it makes impedance matching unnecessary. In addition, it makes it possible to compensate for any irregularities in either the line or the antenna by providing the appropriate resonant circuit at the input. Its disadvantages are increased power losses in the line due to high standing waves of current, increased probability of arc-over because of high standing waves of voltage, very critical length, and production of radiation fields by the line due to the standing waves on it.

Nonresonant Feed Line

The nonresonant feed line is the more widely used technique. The open-wire line, the shielded pair, the coaxial line, and the twisted pair may be used as nonresonant lines. This type of line has negligible standing waves if it is properly terminated in its characteristic impedance at the antenna end. It has a great advantage over the resonant line because its operation is practically independent of its length.

The illustrations in Fig. 14-10 show the excitation of a half-wave antenna by nonresonant lines. If the input to the center of the antenna in Fig. 14-10(a) is 73 Ω and if the coaxial line has a characteristic impedance of 73 Ω, a common method of feeding this antenna is accomplished by connecting directly to the center of the

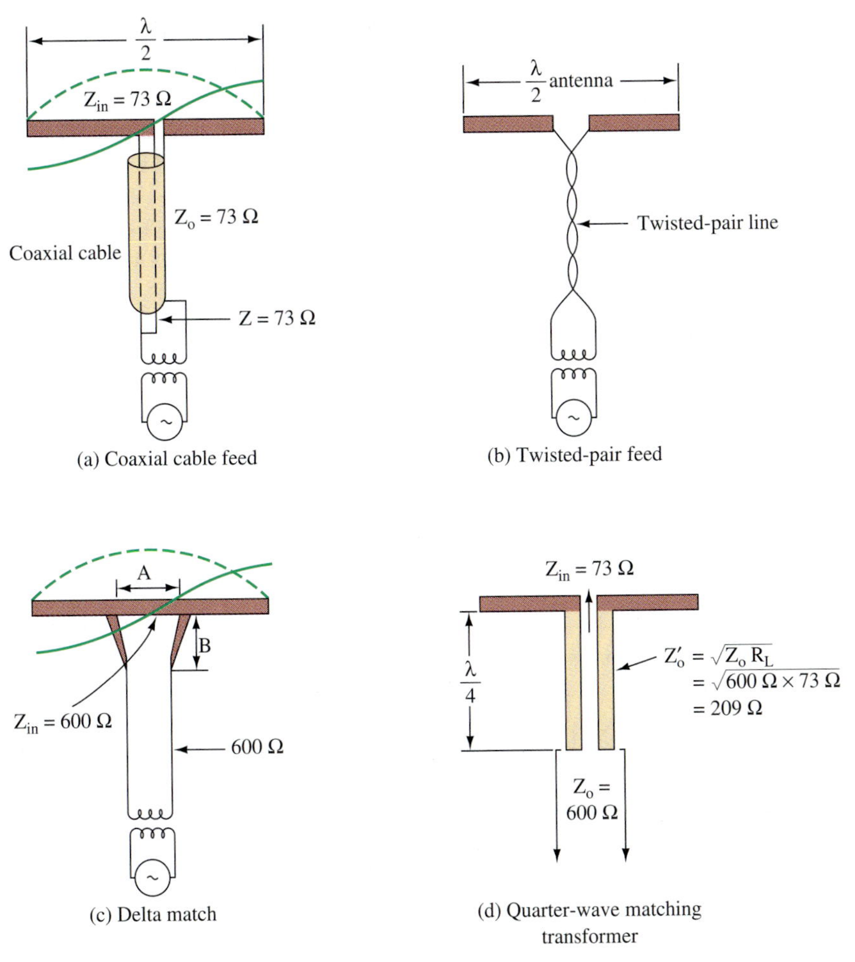

FIGURE 14-10 Feeding antennas with nonresonant lines.

antenna. This method of connection produces no standing waves on the line when the line is matched to a generator. Coupling to a generator is often made through a simple untuned transformer secondary.

Another method of transferring energy to the antenna is through the use of a twisted-pair line, as shown in Fig. 14-10(b). It is used as an untuned line for low frequencies. Due to excessive losses occurring in the insulation, the twisted pair is not used at higher frequencies. The characteristic impedance of such lines is about 70 Ω.

Delta Match

When a line does not match the impedance of the antenna, it is necessary to use special impedance matching techniques such as those discussed with Smith chart applications in Chapter 12. An example of an additional type of impedance matching device is the **delta match,** shown in Fig. 14-10(c). Due to inherent characteristics, the open, two-wire transmission line does not have a characteristic impedance

Delta Match
an impedance matching device that spreads the transmission line as it approaches the antenna

(Z_0) low enough to match a center-fed dipole with $Z_{\text{in}} = 73\ \Omega$. Practical values of Z_0 for such lines lie in the range 300 to 700 Ω. To provide the required impedance match, a delta section [shown in Fig. 14-10(c)] is used. This match is obtained by spreading the transmission line as it approaches the antenna. In the example given, the characteristic impedance of the line is 600 Ω, and the center impedance of the antenna is 73 Ω. As the end of the transmission line is spread, its characteristic impedance increases. Proceeding from the center of the antenna to either end, a point will be reached where the antenna impedance equals the impedance at the output terminals of the delta section. Recall that the antenna impedance increases as you move from its center to the ends. The delta section is then connected at this distance to either side of the antenna center.

The delta section becomes part of the antenna and, consequently, introduces radiation loss (one of its disadvantages). Another disadvantage is that trial-and-error methods are usually required to determine the dimensions of the A and B sections for optimum performance. Both the distance between the delta output terminals (its width) and the length of the delta section are variable, so adjustment of the delta match is difficult.

Quarter-Wave Matching

Still another impedance-matching device is the quarter-wave transformer, or matching transformer, as shown in Fig. 14-10(d). This device is used to match the low impedance of the antenna to the line of higher impedance. Recall from Chapter 12 that the quarter-wave matching section is effective only between a line and purely resistive loads.

To determine the characteristic impedance (Z_0') of the quarter-wave section, the following formula from Chapter 12 is used.

$$Z_0' = \sqrt{Z_0 R_L} \qquad \textbf{(12-29)}$$

where Z_0' = characteristic impedance of the matching line
Z_0 = impedance of the feed line
R_L = resistive impedance of the radiating element

For the example shown, Z_0' has a value slightly over 209 Ω. With this matching device, standing waves will exist on the $\lambda/4$ section but not on the 600-Ω line. Recall from Chapter 12 the use of stub matching techniques as another alternative.

This matching technique is useful for narrowband operation, while the delta section is more broadband in operation.

 14-5 Monopole Antenna

Monopole Antenna usually a quarter-wave grounded antenna

The **monopole antenna** (sometimes called a vertical antenna) is used primarily with frequencies below 2 MHz. The difference between the vertical antenna and the half-wave dipole antenna is that the vertical type requires a conducting path to ground, and the half-wave dipole type does not. The monopole antenna is usually a quarter-wave grounded antenna or any odd multiple of a quarter-wavelength.

Antenna transmission line measurement system. (Courtesy of Anritsu Co.)

Effects of Ground Reflection

A monopole antenna used as a transmitting element is shown in Fig. 14-11. The transmitter is connected between the antenna and ground. The actual length of the antenna is one quarter-wavelength. However, this type of antenna, by virtue of its connection to ground, uses the ground as the other quarter-wavelength, making the antenna electrically a half-wavelength. This is so because the earth is considered to be a good conductor. In fact, there is a reflection from the earth that is equivalent to

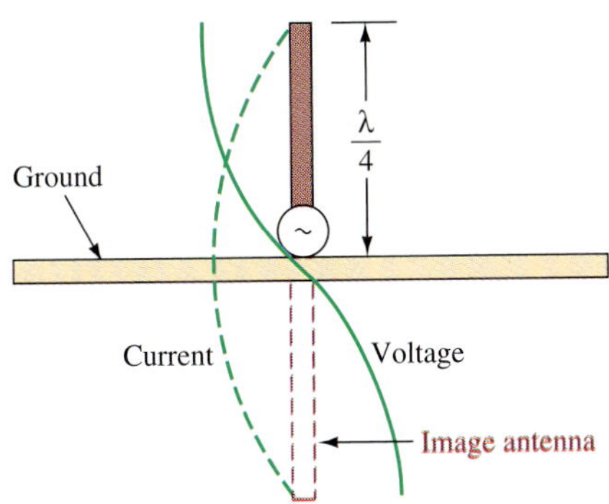

FIGURE 14-11 Grounded monopole antenna.

the radiation that would be realized if another quarter-wave section were used. The reflection from the ground looks as if it is coming from a $\lambda/4$ section beneath the ground. This is known as the **image antenna** and is shown in Fig. 14-11. By use of the monopole antenna, which is a quarter-wave in actual physical length, half-wave operation may be obtained. All of the voltage, current, and impedance relationships characteristic of a half-wave antenna also exist in this antenna. The only exception is the input impedance, which is approximately 36.6 Ω at the base. The effective current in the monopole grounded antenna is maximum at the base and minimum at the top, while voltage is minimum at the bottom and maximum at the top.

Image Antenna
the simulated $\lambda/4$ antenna resulting from the earth's conductivity with a monopole antenna

When the conductivity of the soil in which the monopole antenna is supported is very low, the reflected wave from the ground may be greatly attenuated. A great attenuation of the reflected signal is highly undesirable. To overcome this disadvantage, the site location can be moved to a location where the soil possesses a high conductivity, such as damp areas. If it is impractical to move the site, provisions must be made to improve the reflecting characteristics of the ground by installing a buried ground screen.

The Counterpoise

When an actual ground connection cannot be used because of the high resistance of the soil or a large buried ground screen is impractical, a **counterpoise** may replace the usual direct ground connection. This is required for monopole antennas mounted on the top of tall buildings. The counterpoise consists of a structure made of wire erected a short distance above the ground and *insulated from the ground.* The size of the counterpoise should be at least equal to, and preferably larger than, the size of the antenna.

Counterpoise
reflecting surface of a monopole antenna if the actual earth ground cannot be used; a flat structure of wire or screen placed a short distance above ground with at least a quarter-wavelength radius

The counterpoise and the surface of the ground form a large capacitor. Due to this capacitance, antenna current is collected in the form of charge and discharge currents. The end of the antenna normally connected to ground is connected through the large capacitance formed by the counterpoise. If the counterpoise is not well insulated from ground, the effect is much the same as that of a leaky capacitor, with a resultant loss greater than if no counterpoise were used.

Although the shape and size of the counterpoise are not particularly critical, it should extend for equal distances in all directions. When the antenna is mounted vertically, the counterpoise may have any simple geometric pattern, like those shown in Fig. 14-12. The counterpoise is constructed so that it is nonresonant at the

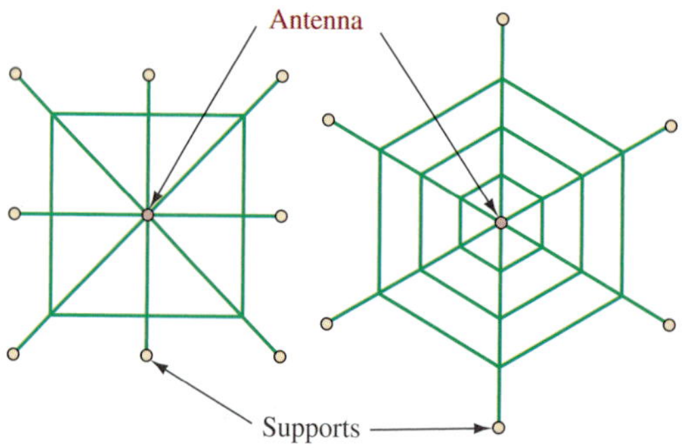

FIGURE 14-12 Counterpoise (top view).

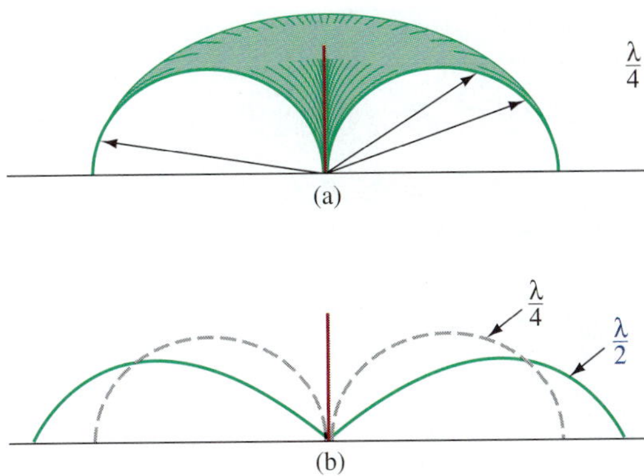

FIGURE 14-13 Monopole antenna radiation patterns.

operating frequency. The operation realized by use of either the well-grounded monopole antenna or the monopole antenna using a counterpoise is the same as that of the half-wave antenna of the same polarization.

Radiation Pattern

The radiation pattern for a monopole antenna is shown in Fig. 14-13(a). It is omnidirectional in the ground plane but falls to zero off the antenna's top. Thus, a large amount of energy is launched as a ground wave, but appreciable sky-wave energy also exists. By increasing the vertical height to $\lambda/2$, the ground-wave strength is increased, as shown in Fig. 14-13(b). The maximum ground-wave strength is obtained by using a length slightly less than $5/8\lambda$. Any greater length produces high-angle radiation of increasing strength, and horizontal radiation is reduced. At a height of 1λ, there is no ground wave.

Loaded Antennas

In many low-frequency applications, it is not practical to use an antenna that is a full quarter-wavelength. This is especially true for mobile transceiver applications. Monopole antennas less than a quarter-wavelength have an input impedance that is highly capacitive, and they become inefficient radiators. The reason for this is that a highly reactive load cannot accept energy from the transmitter. It is reflected and sets up high standing waves on the feeder transmission line. An example of this is a $\lambda/8$ vertical antenna, which exhibits an input impedance of about $8\ \Omega - j500\ \Omega$ at its base.

To remedy this situation, the *effective* height of the antenna should be $\lambda/4$, and this can be accomplished with several different techniques. Figure 14-14 shows a series inductance that is termed a **loading coil.** It is used to tune out the capacitive appearance of the antenna. The coil–antenna combination can thus be made to appear resonant (resistive) so that it can absorb the full transmitter power. The inductor can be variable to allow adjustment for optimum operation over a range of transmitter frequencies. Notice the standing wave of current shown in Fig. 14-14. It has maximum amplitude at the loading coil and thus does not add to the radiated power. This results in heavy I^2R losses in the coil instead of this energy being radiated. However, the transmission line feeding the loading coil/antenna is free of standing waves when the loading coil is properly tuned.

Loading Coil
a series inductance used to tune out the capacitive appearance of an antenna

Satellite antenna. (Courtesy of Scientific Atlanta.)

A more efficient solution is the use of top loading, as shown in Fig. 14-15(a). Notice that the high-current standing wave now exists at the base of the antenna so that maximum possible radiation now occurs. The metallic *spoked wheel* at the top adds shunt capacitance to ground. This additional capacitance reduces the antenna's capacitive reactance because C and X_c are inversely related. The antenna can, therefore, be made nearly resonant with the proper amount of top loading. This does not allow for convenient variable frequency operation as with the loading

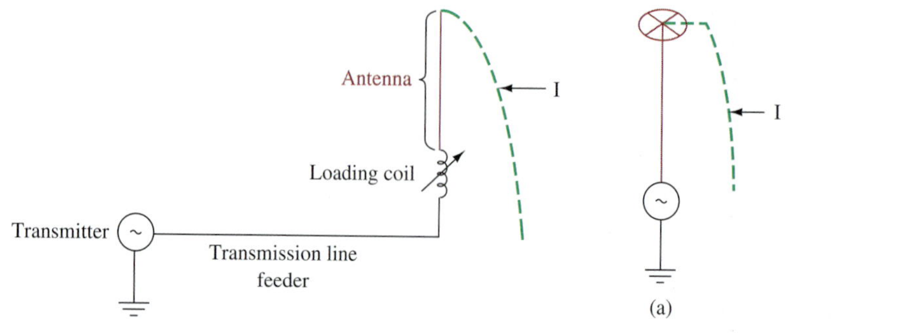

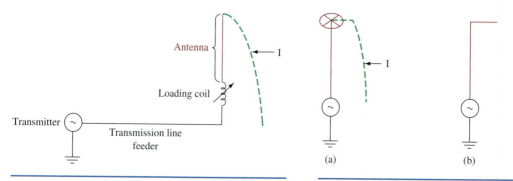

FIGURE 14-14 Monopole antenna with loading coil.

FIGURE 14-15 Top-loaded monopole antennas.

coil, but it is a more efficient radiator. The *inverted L* antenna in Fig. 14-15(b) accomplishes the same goal as the top-loaded antenna but is usually less convenient to construct physically.

14-6 ANTENNA ARRAYS

Half-Wave Dipole Antenna with Parasitic Element

An **antenna array** is one that has more than one element or component. If one or more of the elements is not electrically connected, it is called a **parasitic array.** The most elementary antenna array is shown in Fig. 14-16(a). It consists of a simple half-wave dipole and a nondriven (not electrically connected) half-wave element located a quarter-wavelength behind the dipole. The nondriven element is also termed a *parasitic* element because it is not electrically connected.

Antenna Array
group of antennas or antenna elements arranged to provide the desired directional characteristics

Parasitic Array
when one or more of the elements in an antenna array is not electrically connected

The dipole radiates electromagnetic waves with the usual bidirectional pattern. However, the energy traveling toward the parasitic element, upon reading it, induces voltages and currents but incurs a 180° phase shift in the process. These voltages and currents cause the parasitic element also to radiate a bidirectional wave pattern. However, due to the 180° phase shift, the energy traveling away from the driven element cancels that from the driven element. The energy from the parasitic element traveling toward the driven element reaches it in phase and causes a doubling of energy propagated in that direction. This effect is shown by the radiation pattern in Fig. 14-16(b). The parasitic element is also termed a **reflector** because it effectively "reflects" energy from the driven element. Notice that this simple array has resulted in a more directive antenna and thus exhibits gain with respect to a standard half-wave dipole antenna.

Reflector
the parasitic element that effectively reflects energy from the driven element

Let us consider *why* the energy from the reflector gets back to the driven element in phase and thus reinforces propagation in that direction. Recall that the initial energy from the driven element travels a quarter-wavelength before reaching the reflector. This is equivalent to 90 electrical degrees of phase shift. An additional 180° of phase shift occurs from the induction of voltage and current into the reflector. The reflector's radiated energy back toward the driven element experiences another 90° of phase shift before reaching the driven element. Thus, a total phase shift of 360° (90° + 180° + 90°) results so that the reflector's energy reaches the driven element in phase.

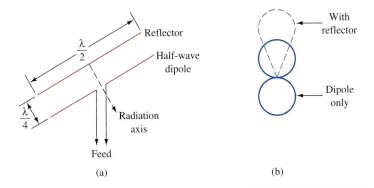

FIGURE 14-16 Elementary antenna array.

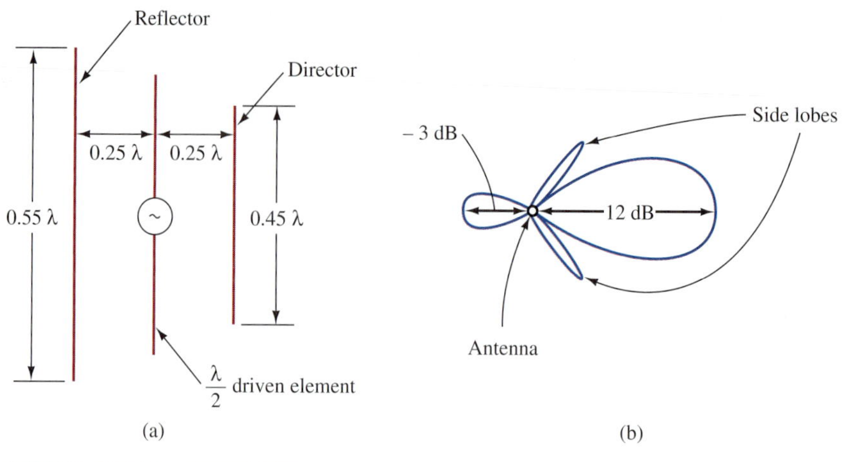

FIGURE 14-17 Yagi–Uda antenna.

Yagi–Uda Antenna

The Yagi–Uda antenna consists of a driven element and two or more parasitic elements. It is named after the two Japanese scientists who were instrumental in its development. The version shown in Fig. 14-17(a) has two parasitic elements: a reflector and a director. A **director** is a parasitic element that serves to "direct" electromagnetic energy because it is in the direction of the propagated energy with respect to the driven element. The radiation pattern is shown in Fig. 14-17(b). Notice the two side **lobes** of radiated energy that result. They are generally undesired, as is the small amount of reverse propagation. The difference in gain from the forward to the reverse direction is defined as the **front-to-back ratio** (*F*/*B* ratio). For example, the pattern in Fig. 14-17(b) has a forward gain of 12 dB and a −3 dB gain (actually, loss, because it is a negative gain) in the reverse direction. Its *F*/*B* ratio is therefore [12 dB − (− 3 dB)], or 15 dB.

Director
the parasitic element that effectively directs energy in the desired direction

Lobes
small amounts of radiation shown on a radiation pattern; generally undesirable

Front-to-Back Ratio
the difference in antenna gain in dB from the forward to the reverse direction

This Yagi–Uda antenna provides about 10 dB of power gain with respect to a half-wavelength dipole reference. This is somewhat better than the approximate 3-dB gain of the simple array shown in Fig. 14-16. In practice, the Yagi–Uda antenna often consists of one reflector and two or more directors to provide even better gain characteristics. They are often used as HF transmitting antennas and as VHF/UHF television receiving antennas.

The analysis of how the radiation patterns of these antennas result is rather complex and cannot be simply accomplished, as was done for the simple array shown in Fig. 14-16. More often than not, the lengths and spacings of the parasitic elements are the result of experiments rather than theoretical calculations.

Driven Collinear Array

Driven Array
multielement antenna in which all the elements are excited through a transmission line

Collinear Array
any combination of half-wave elements in which all the elements are excited by a connected transmission line

A **driven array** is a multielement antenna in which all of the elements are excited through a transmission line. A four-element collinear array is shown in Fig. 14-18(a). A **collinear array** is any combination of half-wave elements in which all the elements are placed end to end to form a straight line. Each element is excited so that their fields are all in phase (additive) for points perpendicular to the array. This is accomplished by the $\lambda/2$ length of transmission line (a $\lambda/4$ twisted pair) between the

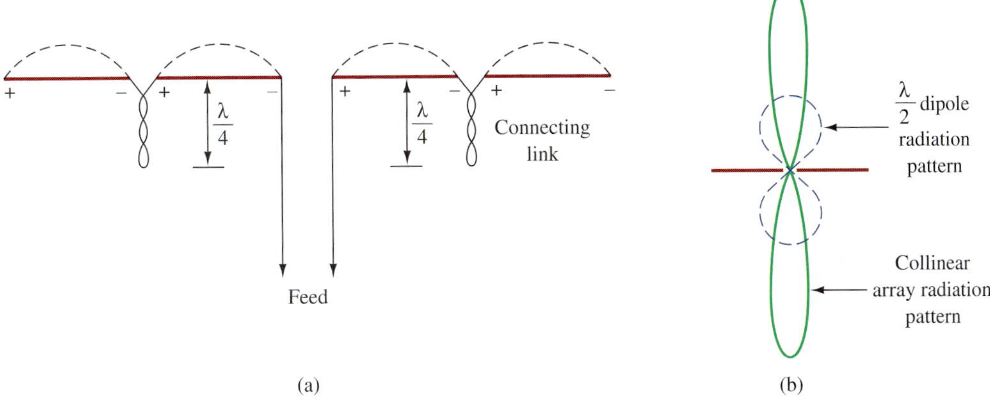

FIGURE 14-18 Four-element collinear array.

elements on both sides of the feed point. They are twisted so that the fields created by the line cancel each other to minimize losses.

The radiation pattern for this antenna is provided in Fig. 14-18(b). Energy off the ends is canceled from the $\lambda/2$ spacing (cancellation) of elements, but reinforcement takes place perpendicular to the antenna. The resulting radiation pattern thus has gain with respect to the standard half-wave dipole antenna radiation pattern shown with dashed lines in Fig. 14-18(b). It has gain at the expense of energy propagated away from the antenna's perpendicular direction. The full three-dimensional pattern for both antennas is obtained by revolving the pattern shown about the antenna axis. This results in the doughnut-shaped pattern for the half-wave dipole antenna and flattened doughnut shape for this collinear array. The array is a more directive antenna (smaller beamwidth). Increased directivity and gain are obtained by adding more collinear elements.

Broadside Array

If a group of half-wave elements is mounted vertically, one over the other, as shown in Fig. 14-19, a broadside array is formed. Such an array provides greater directivity in both the vertical and horizontal planes than the collinear array. With the arrangement shown in Fig. 14-19, the separation between each stack is a half-wavelength. The signal reversal shown in the connecting wires puts the voltage and current in each element of each stack in phase. The net resulting radiation pattern is a directive pattern in the horizontal plane (as with the collinear array) but also a directive pattern in the vertical plane (in contrast to the collinear array).

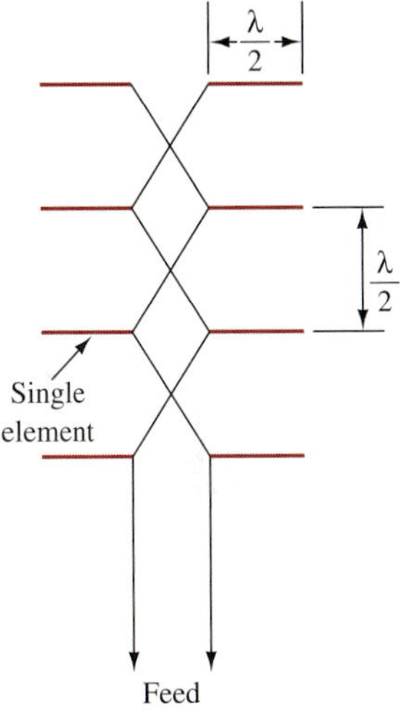

FIGURE 14-19 Eight-element broadside array.

Vertical Array

You have probably noticed that some standard broadcast AM transmitters usually utilize three or more vertical antennas lined up in a row with equal spacing between them. The radiation pattern of a single vertical antenna is omnidirectional in the horizontal plane, which may be undesirable due to interference possibilities with an adjacent channel station or due to geographical population density patterns. For instance, it doesn't make sense for a New York City station to beam half of its energy to the Atlantic Ocean. By properly controlling the phase and power level into each of the towers, almost any

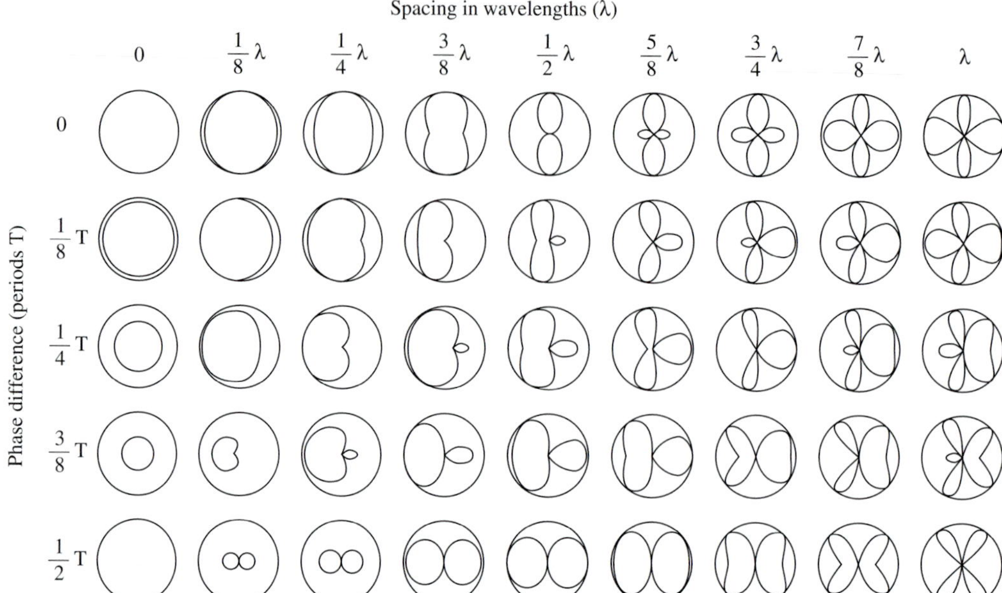

FIGURE 14-20 Phase-array antenna patterns. (From Henry Jaski, Ed., Antenna Engineering Handbook, 1961; courtesy of McGraw-Hill Book Company, New York.)

radiation pattern desired can be obtained. Thus, the energy that would have been wasted over the Atlantic Ocean can be redirected to the areas of maximum population density.

Phased Array
combination of antennas in which there is control of the phase and power of the signal applied at each antenna resulting in a wide variety of possible radiation patterns

This arrangement is called a **phased array** because controlling the phase (and power) to each element results in a wide variety of possible radiation patterns. A station may easily change its pattern at sunrise and sunset because increased sky-wave coverage at night might interfere with a distant station operating at about the same frequency. To give an idea of the countless radiation patterns possible with a phased array, refer to Fig. 14-20. It shows the radiation for just two $\lambda/4$ vertical antennas with variable spacing and input voltage phase. The patterns are simply the vector sum of the instantaneous field strength from each individual antenna.

14-7 SPECIAL-PURPOSE ANTENNAS

Log-Periodic Antenna

The log-periodic antenna is a special case of a driven array. It was first developed in 1957 and has proven so desirable that its many variations now make up an entire class of antennas. It provides reasonably good gain over an extremely wide range of frequencies. Therefore, it is useful for multiband transceiver operation and as a TV receiving unit to cover the entire VHF and UHF bands. It can be termed a wide-bandwidth or broad-band antenna. Bandwidth is not to be confused with beamwidth in this situation.

Antenna *bandwidth* is defined with respect to its design frequency, often termed its *center frequency.* If a 100-MHz (center frequency) log-periodic antenna's transmitted or received power is 3 dB down at 50 MHz and 200 MHz, its bandwidth is 200 MHz − 50 MHz, or 150 MHz. This measurement is made in the direction of highest antenna directivity.

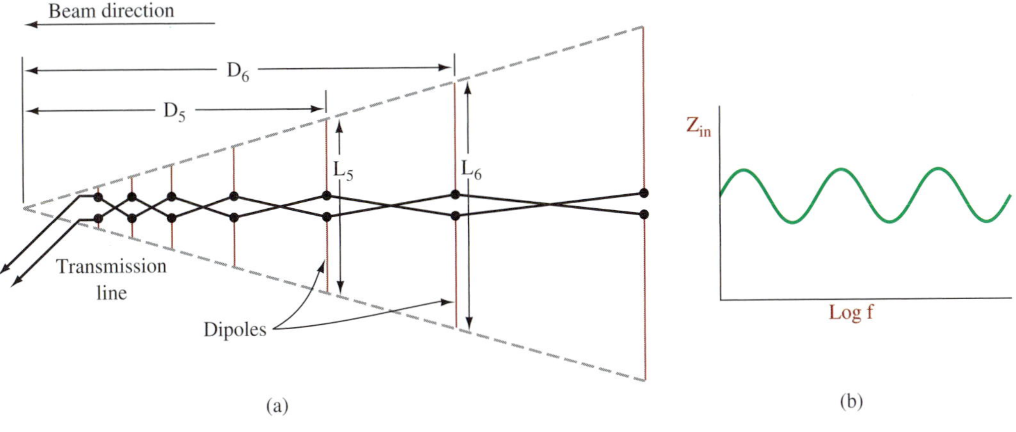

FIGURE 14-21 Log-periodic dipole array.

The most elementary form of log-periodic antenna is shown in Fig. 14-21(a). It is termed a log-periodic dipole array and derives its name from the fact that its important characteristics are periodic with respect to the logarithm of frequency. This is true of its impedance, its SWR with a given feed line, and the strength of its radiation pattern. For instance, its input impedance is shown to be nearly constant (but periodic) as a function of the log of frequency in Fig. 14-21(b).

The log-periodic array in Fig. 14-21(a) consists of several dipoles of different lengths and spacings. The dipole lengths and spacings are related by

$$\frac{D_1}{D_2} = \frac{D_2}{D_3} = \frac{D_3}{D_4} = \frac{D_4}{D_5} \cdots = \tau = \frac{L_1}{L_2} = \frac{L_2}{L_3} = \frac{L_3}{L_4} = \frac{L_4}{L_5} \cdots \qquad \textbf{(14-6)}$$

where τ is called the *design* ratio with a typical value of 0.7. The range of frequencies over which it is useful is determined by the frequencies at which the longest and shortest dipoles are a half-wavelength.

Small-Loop Antenna

A loop antenna is a turn of wire whose dimensions are normally much smaller than 0.1λ. When this condition exists, the current in it may all be considered in phase. This results in a magnetic field that is everywhere perpendicular to the loop. The resulting radiation pattern is sharply bidirectional, as indicated in Fig. 14-22, and is

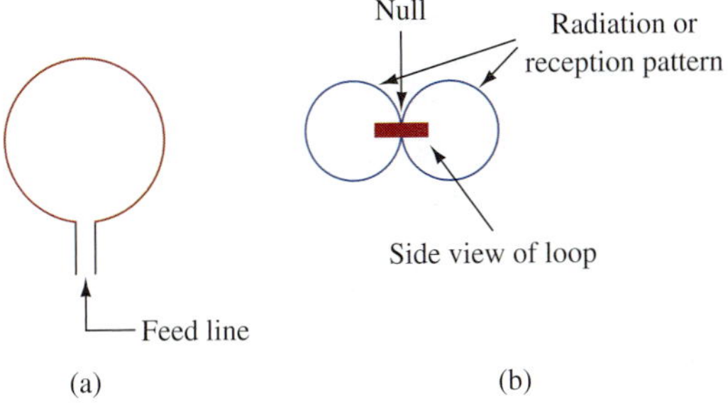

FIGURE 14-22 Loop antenna.

effective over an extremely wide range of frequencies—those for which its diameter is about $\lambda/16$ or less. The antenna is usually circular, but any shape is effective.

Null
a direction in space with minimal signal level

Because of its sharp **null** (see Fig. 14-22), small size, and broad-band characteristics, the loop antenna's major application is in direction-finding (DF) applications. The goal is to determine the direction of some particular radiation. Generally, readings from two different locations are required due to the antenna's bidirectional pattern. If the two locations are far enough apart, the distance and direction of the radiation source can be calculated using trigonometry. Since the signal falls to zero much more sharply than it peaks, the nulls are used in the DF applications.

While other antennas with directional characteristics can be used in DF, the loop's small size seems to outweigh the gain advantages of larger directive antennas.

Ferrite Loop Antenna

The familiar ferrite loop antenna found in most broadcast AM receivers is an extension of the basic loop antenna just discussed. The effect of using a large number of loops wound about a highly magnetic core (usually ferrite) serves to increase greatly the effective diameter of the loops. This forms a highly efficient receiving antenna, considering its small physical size compared to the hundreds of feet required to obtain a quarter-wavelength for the broadcast AM band. The directional characteristics of this antenna are verified by the fact that a portable AM receiver can usually be

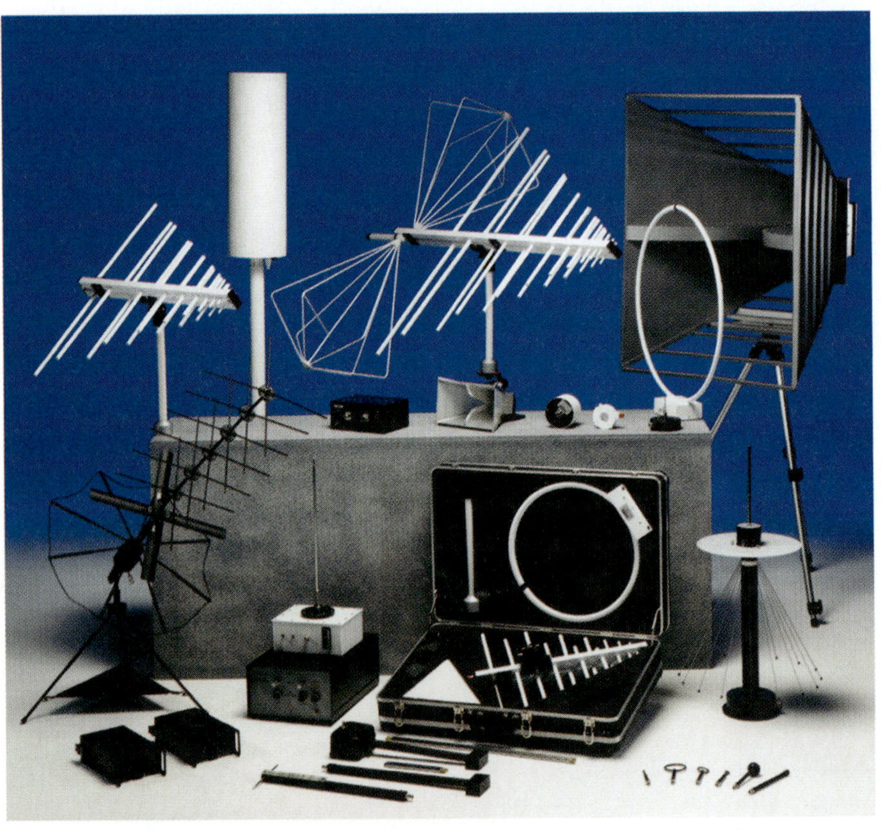

Various antennas available from one manufacturer. (Courtesy of Electro-Metrics, Inc.)

oriented to *null* out reception of a station. You should now be able to determine a line through which that broadcasting station exists when the null is detected.

Folded Dipole Antenna

Recall that the standard half-wavelength dipole antenna has an input impedance of 73 Ω. Recall also that it becomes very inefficient whenever it is not used at the frequency for which its length equals $\lambda/2$ (i.e., it has a narrow bandwidth). The folded dipole antenna shown in Fig. 14-23(a) offers the same radiation pattern as the standard half-wave dipole antenna but has an input impedance of 288 Ω (approximately 4×73 Ω) and offers relatively broad-band operation.

A standard half-wave dipole antenna can provide the same broad-band characteristics as the folded dipole by incorporating a parallel tank circuit, as shown in Fig. 14-23(b). With the tank circuit resonant at the frequency corresponding to the antenna's $\lambda/2$ length, the tank presents a very high resistance in parallel with the antenna's 73 Ω and has no effect. However, as the frequency goes down, the antenna becomes capacitive, while the tank circuit becomes inductive. The net result is a resistive overall input impedance over a relatively wide frequency range.

The folded dipole is a useful receiving antenna for broadcast FM and for VHF TV. Its input impedance matches well with the 300-Ω input impedance terminals common to these receivers. It can be inexpensively fabricated by using a piece of standard 300-Ω parallel wire transmission line, commonly called **twin lead,** cut to $\lambda/2$ at midband and shorting together the two at each end. Folded dipoles are also invariably used as the driven element in Yagi–Uda antennas. This helps to maintain a reasonably high input impedance because the addition of each director lowers this array's input impedance. It also gives the antenna a broader band of operation.

Twin Lead
standard 300-Ω parallel wire transmission line

In applications where a folded dipole with other than a 288-Ω impedance is desired, a larger-diameter wire for one length of the antenna is used. Impedances up to about 600 Ω are possible in this manner.

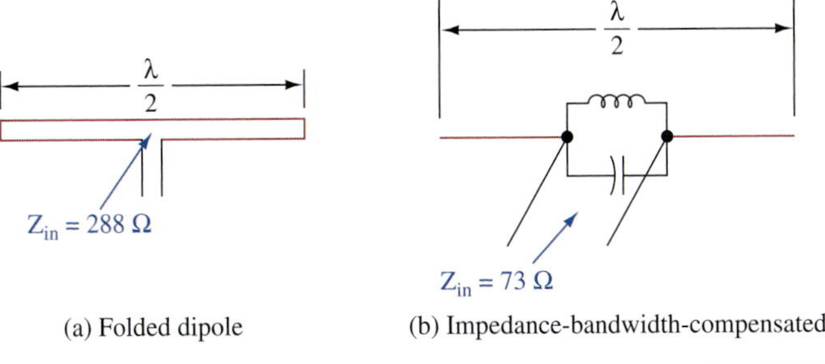

FIGURE 14-23 Dipoles.

Slot Antenna

Coupling RF energy into a slot in a large metallic plane can result in radiated energy with a pattern similar to a dipole antenna mounted over a reflecting surface. The length of the slot is typically one half-wavelength. These antennas function at UHF and microwave frequencies with energy coupled into the slot by waveguides or coaxial line feed connected directly across the short dimension of a rectangular slot. These antennas are commonly used in modern aircraft in an array module as shown

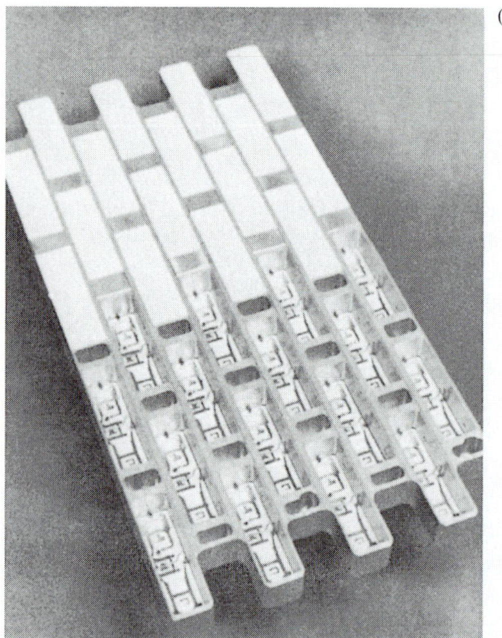
(a)

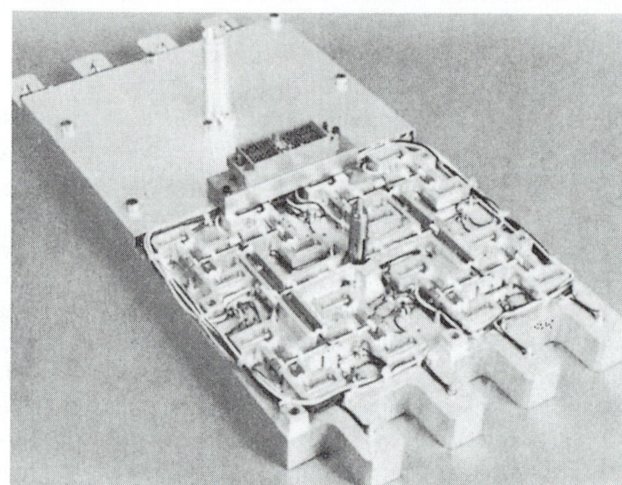
(b)

FIGURE 14-24 Slot antenna array.

in Fig. 14-24(a). This 32-element (slot) array shows half the slots filled with dielectric material (to provide the required smooth airplane surface) and the others open to show the phase-shifting circuitry used to drive the slots. The rear view in Fig. 14-24(b) shows the coaxial feed connectors used for this antenna array.

The individual drive to each slot is controlled by phase-shifting networks. Proper phasing allows production of a directive radiation pattern that can be swept through a wide angle without physically moving the antenna. This allows a convenient mobile scanning radar system without mechanical complexities. These *phased array* antennas are typically built right into the wings of aircraft, with the dielectric window filling eliminating aerodynamic drag.

14-8 TROUBLESHOOTING

Antenna installation is often part of the technician's job. For the antenna to function at peak performance, technicians must follow proper installation techniques. Equipment manufacturers publish guidelines for antenna installation. This section discusses general installation procedures and looks at general troubleshooting techniques.

Often antennas are mounted high and are not easily accessible, making them difficult to inspect. The condition of the antenna and the transmission line can be determined, however, by measuring antenna emission and standing waves on the transmission line. In this section you will discover how some simple-to-use basic communications test equipment can check and troubleshoot the antenna systems.

After completing this section you should be able to:

- Identify safety precautions to observe for an antenna installation
- Describe proper antenna grounding

- Describe correct transmission line installation
- Troubleshoot typical antenna problems
- Explain the use of the SWR meter
- Explain how a grid-dip meter is used
- Describe how the SWR meter can help find antenna faults

Installing the Antenna

TV and FM antenna installation practices will be referred to throughout this discussion. Installing TV and FM antennas portrays fairly standard practices that should be followed when doing any kind of communications antenna installation.

Never neglect safety. Standard operating procedure should always make safety first on the list of things to do. Locate power lines, telephone lines, and obstacles that can interfere with the installation or present a hazard to the installers. A tower structure requires a concrete base as the supporting structure. Guy wires need room for proper mounting. Anchor ladders securely. Use safety belts or harnesses whenever climbing towers or other structures. Be aware of building codes and follow installation procedures supplied by the equipment manufacturer. Heed the equipment manufacturer's warnings.

Grounding and Lightning Protection Antennas on high exposed metal masts are subject to being struck by lightning. Ground the mast by connecting a wire to it and to a ground rod. When the mast is struck by lightning, the surge of electricity is shorted to ground. Some antenna installations require several such ground connections. A typical ground wire size for a TV or FM antenna is 10 AWG. Local ordinances should specify wire size and type for grounding applications. A good grounding system also protects against static charge buildup.

Lightning follows a second path if a strike occurs. The second path is down the lead-in wire and into the equipment that is connected to it. To protect against this lightning, surge protectors and static discharge devices are placed between the antenna and the receiving equipment. Check the installation manual for recommended lightning surge protectors and static discharge devices.

Planning the Installation An antenna installation site must be safe, as previously mentioned. Keep a proper distance from power lines and telephone lines. For good reception, no obstructions should exist between the antenna and the receiving direction. Generally, the antenna requires a base, a mast, and a good supporting structure. The base may be concrete with anchors for footing, as in a tower structure. The mast may be telescoping poles that are secured to a building or other structure.

If guy wires are used, there must be enough room around the mast to install them. Guy wires usually extend in three equally spaced directions. The wires should intercept the mast at about 45° angles for proper support. The guy wires must be well anchored at the ground points where they attach to eyelets. Anchor the eyelets in concrete. This makes a very secure guy wire support system. Tighten the guy wires using turnbuckles.

Lead-in Wires Proper care should be exercised when running the lead-in line. Improper installation can result in troubles later. Lead-in wire should be kept as short as possible. Twin-lead 300-Ω antenna wire, often used with TV and FM antennas, must be installed using standoff insulators. This wire should never touch metal. The

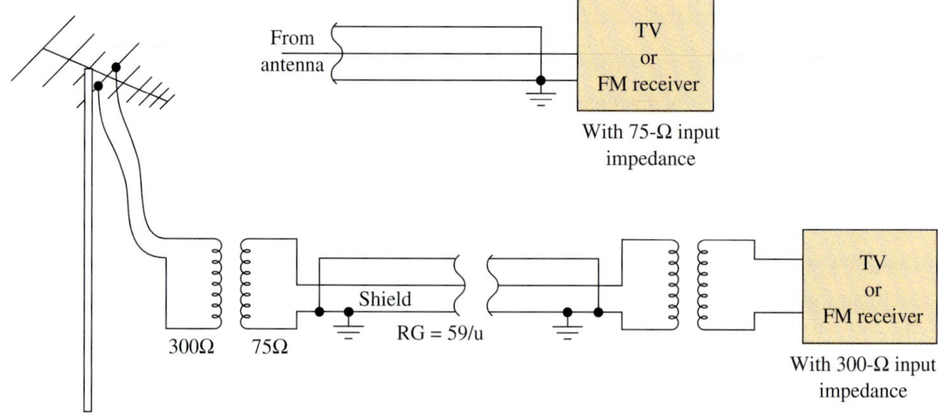

FIGURE 14-25 Matching antenna to receiver.

metal will influence the transmission line's characteristic impedance and cause attenuation and reflections. For areas of high interference, substitute coaxial cable for the twin-lead wire. Coaxial cable does not require insulated standoffs. Taping it to the mast and running it along rain gutters makes for a good installation.

When using coaxial cable instead of twin lead, remember that the antenna impedance must be matched to the transmission line. If the receiver does not permit direct connection of the coaxial cable, an impedance match must be made there also. Coaxial cable normally used with TV and FM antenna installation is RG-59/U. Its characteristic impedance is 75 Ω. Balun transformers match the antenna impedance and receiver input impedance, as illustrated in Fig. 14-25. If the TV or FM receiver has a 75-Ω connection, then an antenna balun is all that is necessary.

Avoid running lead-in wires through windows. Run lines through a special tube that feeds the line into the building via the wall. Some installations may specify using conduit to protect the lead-in line from weathering. Once the line is in the building, distribution outlets can serve to distribute the signal to more than one receiver. If outlets are not used, cut off excess lead-in. Do not curl it up behind the receiving equipment.

Typical Troubleshooting Techniques

1. Is the VSWR as low as it should be? Most antennas are designed to operate with a specific type of transmission line; 50-Ω coax is common. A directional wattmeter can be used to measure VSWR, as shown in Fig. 14-26.

   Sometimes the directional wattmeter takes the form of a directional coupler and a power meter, but the principle is the same. Measure the incident and reflected power and calculate the VSWR with the following formula:

$$\text{VSWR} = \frac{1 + \sqrt{\dfrac{P_r}{P_i}}}{1 - \sqrt{\dfrac{P_r}{P_i}}} \tag{14-7}$$

   where P_i = incident power
   P_r = reflected power

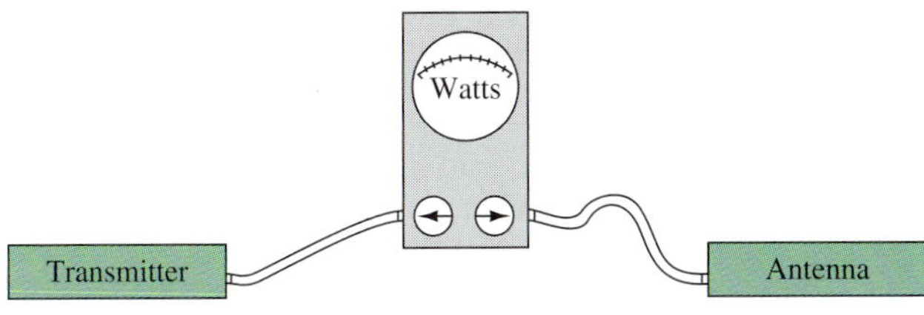

FIGURE 14-26 VSWR test.

Ideally, there is no reflected power and the VSWR is 1.

2. Assume you find a very high VSWR. A common ohmmeter is a very good piece of test equipment. Look for bad connections and open or shorted tuning elements such as capacitors and inductors.

   Another trick is to sweep the frequency to see if there is a frequency where the VSWR is low. This may lead you to broken elements or show that the design is improper.

   If this antenna is associated with a transmitter, visually inspect all insulators for tracks made by arcs. These arcs may not show up in a low-power test. Attempts to repair such insulators are seldom successful. They should be replaced.
3. Has this antenna been subject to severe weather conditions such as ice, wind, and rain? Parts may be broken or full of water.
4. Is the antenna able to handle power applied to it? Antennas such as the Yagi–Uda and log-periodic have extremely high voltages at the ends of the elements. Corona will form and sometimes actually melt the ends of the antenna. If this occurs at an AM station, you can hear the audio on the arc. The problem is cured by adding rings or balls to the elements.
5. Are there installation problems? It is quite often a good idea to question physical dimensions if the antenna is new. People make mistakes during assembly that you can find with a tape measure.

   Consider a parabolic reflector used in a TV satellite system. With a tape measure and some string, you can find the focal point if you know the following formula:

$$\text{Focal length} = \frac{\text{diameter}^2}{16 \times \text{depth}} \qquad \textbf{(14-8)}$$

   Simply stretch the string across the dish and find the depth, measure the diameter, and solve for the focal length. Usually the face of the feed is at the focal point or a very small distance closer to the parabola. Refer to Fig. 14-27. Additional information on parabolic antennas is provided in Chapter 16.

Antenna Measurements

Antenna performance predictions are based on free-space operation. Structures located around the antenna can severely affect the performance of the antenna. Building construction, tree growth, towers, and wires can alter the original antenna installation performance. In addition to changes in the area surrounding the antenna,

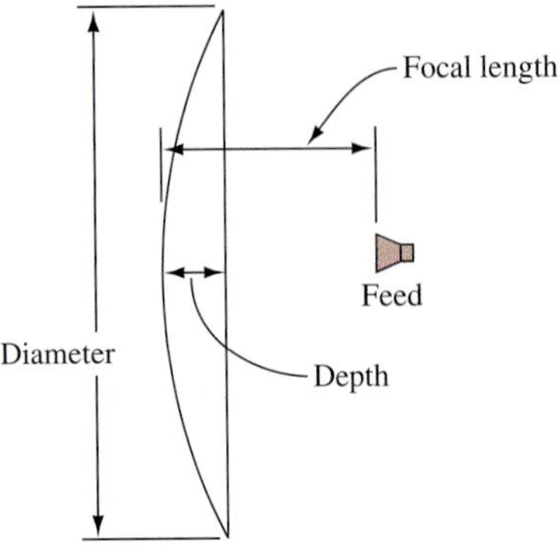

FIGURE 14-27 Parabolic reflector.

characteristics of the antenna or the transmission line are subject to change over time. Remember, antenna efficiency relies on several factors. Antenna performance relies on electrical length, physical length, and matching the antenna impedance to that of the transmission line. All of these characteristics are subject to change due to storms, weathering, and mishaps.

Grid-Dip Meter
device that measures the resonant frequency of tuned circuits and antennas without power being applied to them

Grid-Dip Meter The **grid-dip meter** has been used for years to measure radio frequencies. It is a handheld device that can measure the resonant frequency of tuned circuits and antennas without power being applied to them. Battery-powered and equipped with a tunable oscillator and a scale calibrated in frequency, the meter can measure very accurately the frequency of tuned circuits. It is equipped with plug-in loop coils that serve as probes. The loop couples energy into or out of the grid-dip oscillator circuit. Tuned circuits are checked by bringing the loop coil near them and adjusting the grid-dip oscillator until a dip is seen on the meter (see Fig. 14-28). The scale indicates the tuned circuit's resonant frequency when the dip occurs. A specific frequency is measured by choosing a loop coil for that desired frequency. The grid-dip meter comes with several such loop coils. Sometimes technicians construct their own coil to measure a specific frequency.

The grid-dip meter can measure the resonant frequency of the antenna by connecting it directly to the antenna terminals. The grid-dip can also act as an absorption meter that measures radiated antenna energy. By bringing the meter's loop coil

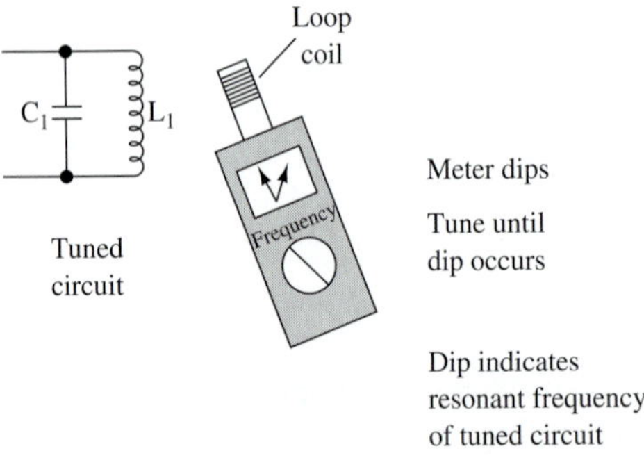

FIGURE 14-28 Grid-dip meter test for a tuned circuit.

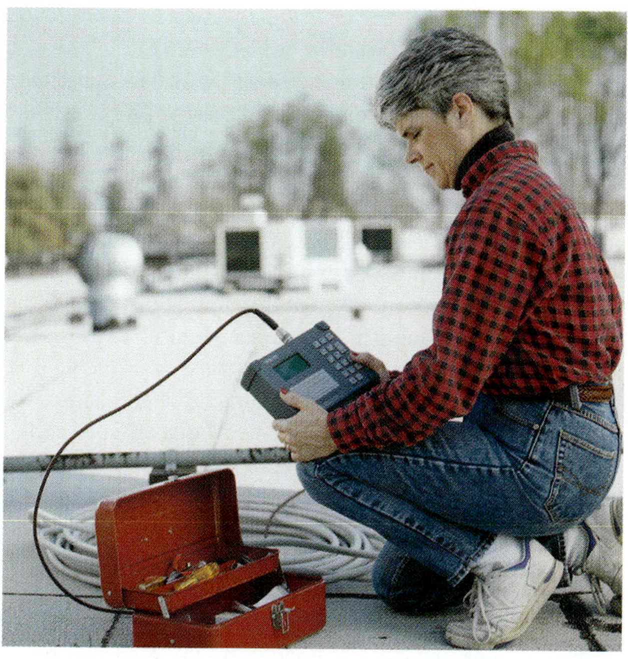

A technician tests antenna/transmission lines at a remote field location. (Courtesy of Anritsu Co.)

close to a radiating antenna and tuning the grid-dip meter for a peak meter indication, the radiating frequency can be determined. Field strength measurements can aid in determining the radiating pattern of the antenna. To do a field strength reading, take a position near the antenna and read the meter after carefully tuning for a peak reading on the scale. After several readings, the antenna radiating pattern can be plotted from the findings.

SWR Meter Another very useful piece of test equipment is the SWR (standing wave ratio) meter. Figure 14-29 shows that the SWR meter is inserted between the transmitter and the antenna. A test cable is attached to the transmitter and to the SWR meter. On the antenna side of the SWR meter, connect the transmission line that normally runs from the transmitter to the antenna. An impedance mismatch at the antenna and the

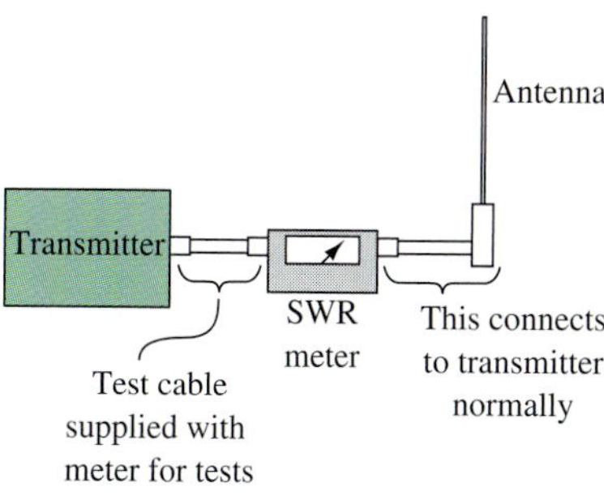

FIGURE 14-29 SWR meter in line between the antenna and transmitter.

transmission line results in the existence of a voltage standing wave ratio (VSWR) on the transmission line. This test is similar to the one shown in Fig. 14-26.

To measure VSWR with the SWR meter, first calibrate the meter. Follow the instructions supplied with the SWR meter regarding the calibration process. Once the meter is calibrated, switch the meter to the SWR setting to make a reading. A good SWR reading will indicate 1.5 or less. Impedance adjusting of the transmission line or the antenna can be done with the SWR meter in line between them. Changes can be made to the transmission line or the antenna until a minimum SWR reading is obtained. SWR meters can be equipped with an antenna probe to make field strength measurements. These readings are taken at different points around the radiating antenna, and a plot is made to establish the antenna pattern.

Troubleshooting with the SWR Meter A high VSWR reading, greater than 1.5, on a transmission line indicates a problem. The problem may be from a crimped cable, a crushed cable, an impedance mismatch at the transmitter or antenna, or moisture in the cable. The antenna should be inspected if it is suspected of being faulty. If a problem exists in the transmission line, the coaxial cable should be tested. Use an ohmmeter to check the cable. Figure 14-30(a) illustrates how the cable should be tested. First, measure the resistance end-to-end from the metal outside portions of each connector. Then measure resistance from the connector's center pin to the other center pin. A small resistance reading indicates a good cable. A high resistance reading indicates an open cable. A cable continuity test can also determine the quality of the cable [see Fig. 14-30(b)]. Short-circuit one end of the coaxial cable using a jumper wire while measuring from the other end. A high-resistance reading or open reading indicates an open cable.

Anechoic Chamber When precise measurement of an antenna's gain and directivity is necessary, an **anechoic chamber** is used. This is basically a large enclosed room that prevents any reflected electromagnetic waves and shields out any interfering waves from the outside world. If you look carefully at the anechoic chamber shown in Fig. 14-31, you will see that the walls are lined with foam material. The foam has multiple pyramid shapes that are impregnated with carbon. This absorbs electromagnetic waves and prevents test data from being adversely affected by reflections. The foam and a grounded metallic enclosure for the chamber prevent any outside radiation from interfering with the testing. The size of the chamber is dictated by the antenna size and the wavelength. The test measurement "receiver" needs to be in the far field [see Eq. (14-1)] so that the induction field (sometimes referred to as the near field) does not affect the desired radiation field (far field) readings.

Anechoic Chamber
a large enclosed room that prevents reflected electromagnetic waves and shields out interfering waves from the outside world; used for radiation measurements

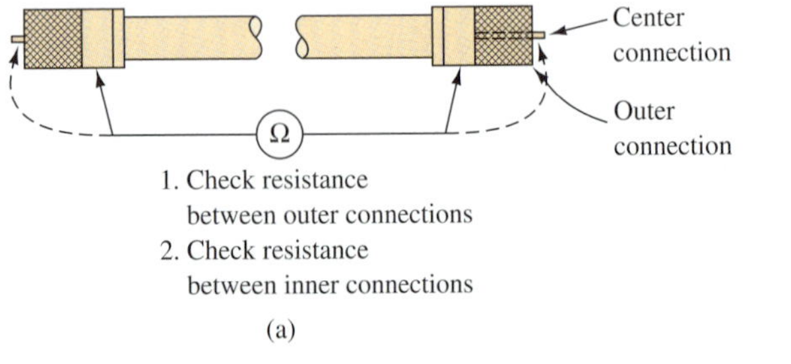

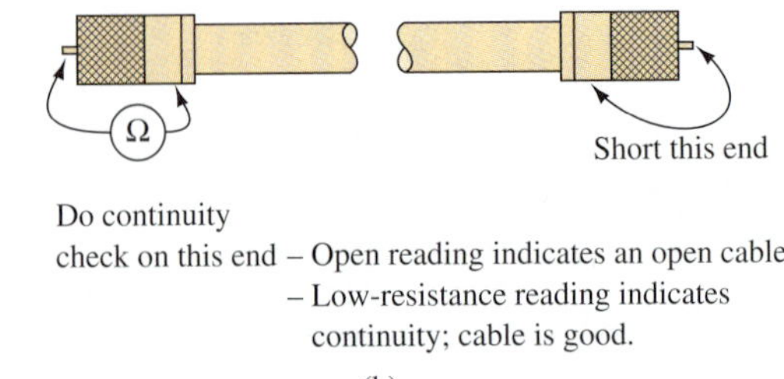

FIGURE 14-30 Testing coaxial cable.

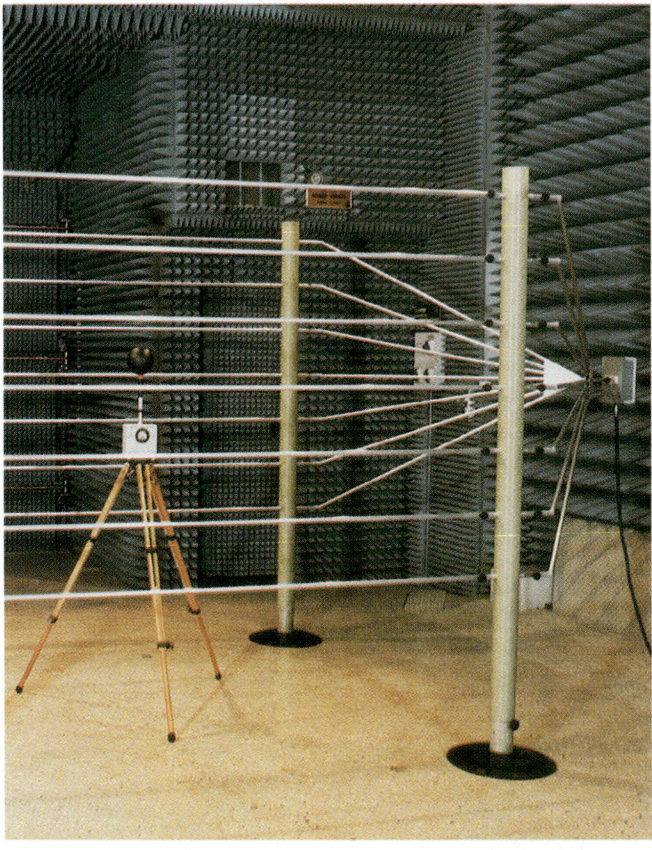

FIGURE 14-31 Anechoic chamber for antenna testing. (Reprinted from EMC Test & Design Measurement, April 1995, Copyright. Courtesy of Intertec Publishing Corp., Overland Park, KS 55212. All rights reserved.)

14-9 TROUBLESHOOTING WITH ELECTRONICS WORKBENCH™ MULTISIM

The basic concepts of antennas were introduced in this chapter. A fundamental antenna that you should understand is the dipole. This exercise further investigates the half-wave dipole using the Multisim tools by incorporating the use of the network analyzer. Begin the exercise by opening **Fig14-32.ms7 (.msm)** on your EWB Multisim CD. This circuit is shown in Fig. 14-32.

This circuit contains a model of a 100-MHz half-wave dipole. This 100-MHz frequency is in the middle of the FM radio band. The operational characteristics of a half-wave dipole were discussed in Sec. 14-2. The 73-Ω resistor is used to model the radiation resistance of the dipole, whereas the 1-μH inductor and 2.5-pF capacitor were selected so that the resonant frequency of the antenna model is approximately 100 MHz. A network analyzer is connected to the model of the dipole for analyzing the antenna's characteristics.

Before you start the simulation, double-click on the network analyzer and click on the **Setup** button next to **Measurement.** You need to set the start and stop

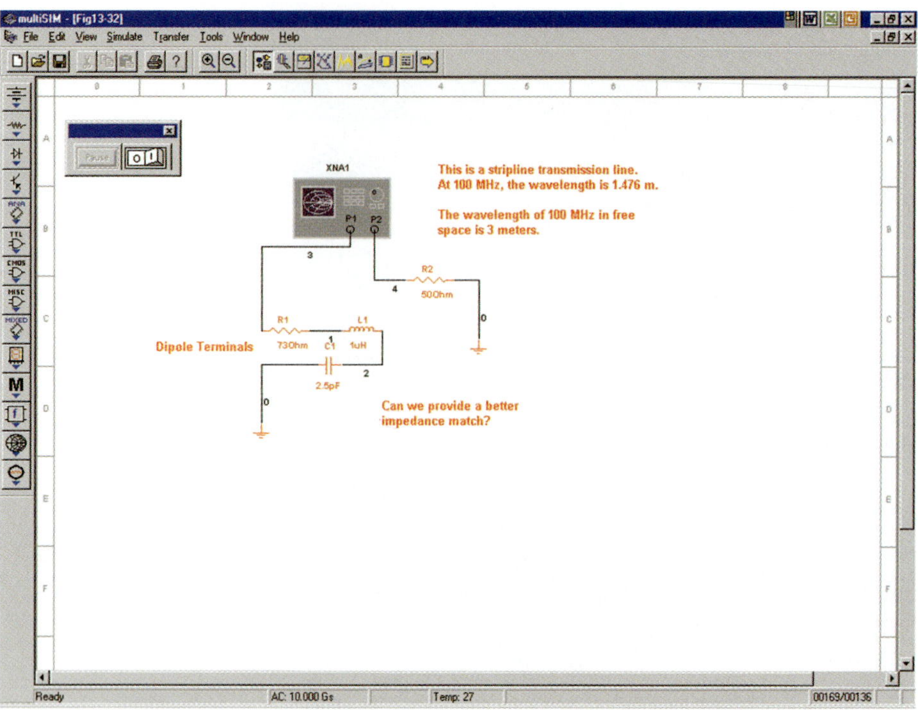

FIGURE 14-32 The Multisim circuit for modeling a 100-MHz half-wave dipole.

frequencies to 90 MHz and 110 MHz, respectively. You also need to change the number of points per decade to 500. Set the characteristic impedance of the network analyzer to 50 Ω. Click **OK** to close the **Measurement setup** window. What do you expect to see on the Smith chart when the simulation is performed? At the resonant frequency of 100 MHz, you should expect the plot of the antenna on the chart to show a real or resistive component of 73 Ω, whereas frequencies above and below 100 MHz will show that the dipole is reactive. Start the simulation and view the results. You should see a display similar to that shown in Fig. 14-33. Use the slider to adjust the frequency to 100 MHz. In the example shown, the normalized input impedance is $Z_{11} = 1.46 \times j0.0354$ at 100.5177 MHz. Multiplying gives $1.46 \times 50 = 73\ \Omega$, which is the characteristic impedance of the half-wave dipole. The reactive term $j0.0354$ is nearly zero, as expected.

What would need to be done if a better impedance match were needed? The next example demonstrates how a single stub tuner can be used to provide a better impedance match to an antenna. Open the file **Fig14-34.ms7 (.msm)** on your EWB CD. The dipole circuit previously analyzed has been modified to include a single stub tuner. The model of the transmission line is provided by the Multisim stripline transmission line element. The dielectric constant of the stripline is $\epsilon_r = 4.13$, which results in a wavelength of 1.476 meters at 100 MHz. The wavelength of 100 MHz in free space is 3 meters. The circuit is shown in Fig. 14-34. You need to double-click on the network analyzer, change the frequency range to (90 to 110) MHz, and change the number of points per decade to 500 points to get a smooth plot.

In real applications, adjusting the stub tuner is a mechanical adjustment; however, in this exercise the adjustment is provided through varying the model characteristics of the transmission line. Double-click on the ground leg of the stub tuner.

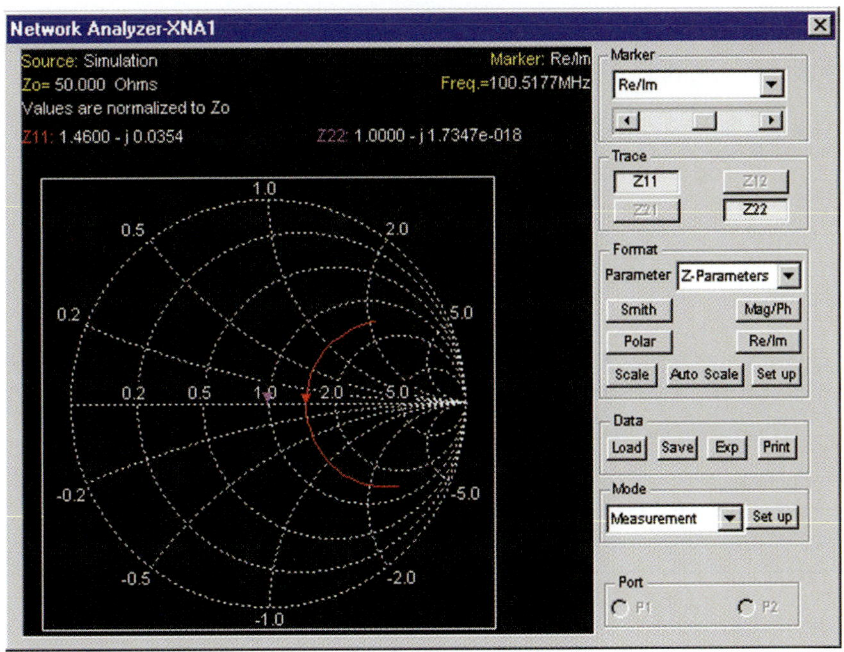

FIGURE 14-33 The network analyzer view of the simulation of a 100-MHz half-wave dipole.

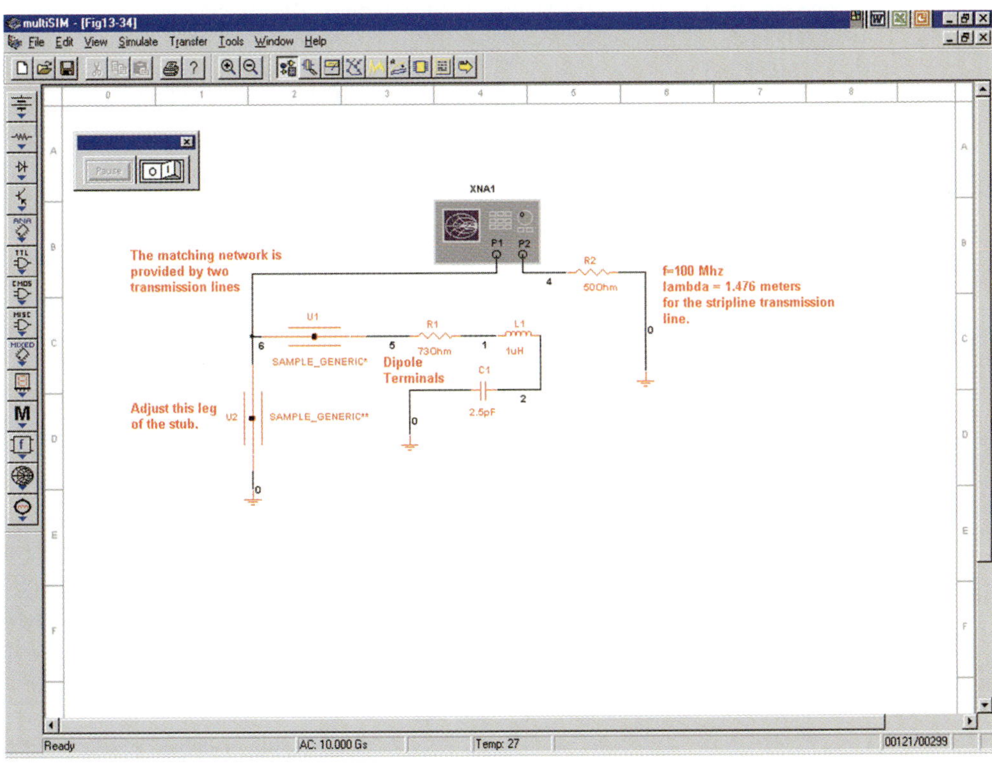

FIGURE 14-34 The model of a single stub tuner using the Multisim stripline transmission-line elements.

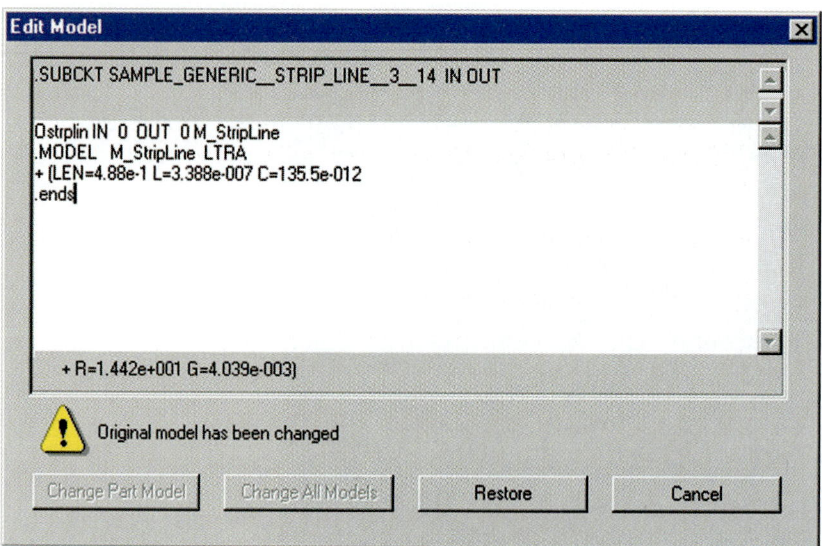
Edit Model
.SUBCKT SAMPLE_GENERIC__STRIP_LINE__3__14 IN OUT
Ostrplin IN 0 OUT 0 M_StripLine
.MODEL M_StripLine LTRA
+ (LEN=4.88e-1 L=3.388e-007 C=135.5e-012
.ends
+ R=1.442e+001 G=4.039e-003)
Original model has been changed
Change Part Model
Change All Models
Restore
Cancel

FIGURE 14-35 The model screen for the stripline element.

You should see a menu for the **strip_line.** Click on the **Value** tab and then click on **Edit Model.** You will see a screen image like the one in Fig. 14-35. These are the model parameters for the stripline.

To begin the tuning, make a small change to the LEN value in the model. This change effectively changes the length of the stub tuner. Remember, the wavelength of 100 MHz in the stripline is 1.476 meters. Change the value to 4.0e-1; click on **Change Part Model** and then **OK** to save your changes. Restart the simulation and view the changes in the simulation results on the network analyzer. The result now shows a very inductive circuit. You need to try the adjustment in the other direction. Once again, double-click on the ground leg of the stub tuner to obtain the menu for the **strip_line.** Click on the **Value** tab and then click on **Edit Model** to view the model parameters for the stripline. Change the LEN value to 5.5e-1. Restart the simulation and view the results on the network analyzer. The results show that we are getting closer to matching the antenna to a 50-Ω load. The tuning requires that this adjustment process be repeated several times until a satisfactory result is obtained. A LEN setting of 5.3e-1 for the ground leg provides a good match for the antenna. The exercises provide additional opportunities for you to experiment with dipole antennas.

Electronics Workbench™ Exercises

1. Design a half-wave 92-MHz dipole antenna. Use the file **Fig14-32.ms7 (.msm)** as a sample. This file is for a 100-MHz half-wave dipole. Modify the values for L1 and C1 so that the resonant frequency is 92 MHz. (L1 = 1.198 μH, C1 = 2.5 pF)
2. Design a half-wave 450-MHz dipole antenna. Use the file **Fig14-32.ms7 (.msm)** as a sample. This file is for a 100-MHz half-wave dipole. You must modify the values for L1 and C1 in this file so that the resonant frequency is 450 MHz. (L1 = 50 nH, C1 = 2.5 pF)
3. Open the Multisim simulation of a single stub tuner found in the **FigE14-1.ms7 (.msm)** file found in your EWB CD. Use the technique presented in the text to provide a match for this 100-MHz dipole antenna.

Summary

In Chapter 14 we examined antenna operation and many of the possible configurations. Be sure to keep in mind that antenna properties apply identically when both transmitting and receiving—the principle of reciprocity. The major topics you should now understand include:

- the analysis of the half-wave dipole antenna, including its impedance, radiation field, radiation pattern, and gain
- the definition of radiation resistance and related calculations
- the description of antenna feed lines, including resonant and nonresonant
- the operation of impedance matching devices, including the delta match and quarter-wave matching transformer
- the analysis of monopole antennas, including ground effects, counterpoise effects, radiation pattern, and loading effects
- the effects of parasitic elements, including discussion of the Yagi–Uda antenna, driven collinear array, broadside array, and vertical array
- the description and operation of various special-purpose antennas, including the log-periodic, loop, ferrite loop, folding dipole, and slot antennas

Questions and Problems

Section 14-1

*1. How should a transmitting antenna be designed if a vertically polarized wave is to be radiated, and how should the receiving antenna be designed for best performance in receiving the ground wave from this transmitting antenna?

*2. If a field intensity of 25 mV/m develops 2.7 V in a certain antenna, what is its effective height? (108 m)

*3. If the power of a 500-kHz transmitter is increased from 150 W to 300 W, what would be the percentage change in field intensity at a given distance from the transmitter? What would be the decibel change in field intensity? (141%, 3 dB)

*4. If a 500-kHz transmitter of constant power produces a field strength of 100 μV/m at a distance of 100 mi from the transmitter, what would be the theoretical field strength at a distance of 200 mi from the transmitter? (50 μV/m)

*5. If the antenna current at a 500-kHz transmitter is reduced 50 percent, what would be the percentage change in the field intensity at the receiving point? (50%)

*6. Define *field intensity*. Explain how it is measured.

*7. Define *polarization* as it refers to broadcast antennas.

8. Explain how antenna reciprocity occurs.

Section 14-2

9. Explain the development of a half-wave dipole antenna from a quarter-wavelength, open-circuited transmission line.

*10. Draw a diagram showing how current varies along a half-wavelength dipole antenna.

*An asterisk preceding a number indicates a question that has been provided by the FCC as a study aid for licensing examinations.

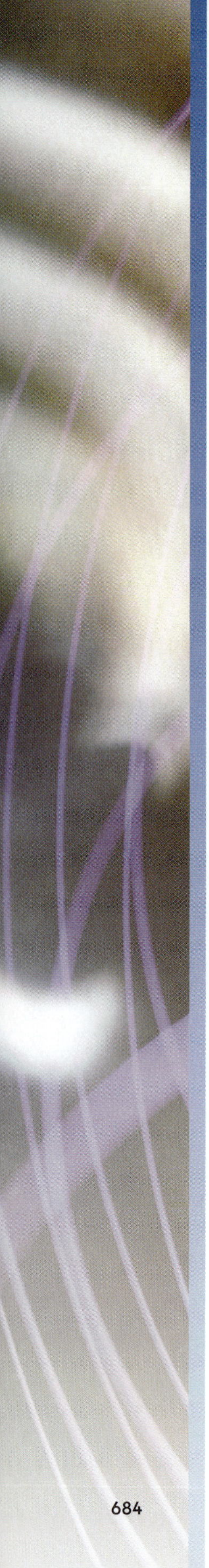

*11. Explain the voltage and current relationships in a one-wavelength antenna, one half-wavelength (dipole) antenna, and one quarter-wavelength *grounded* antenna.

*12. What effect does the magnitude of the voltage and current at a point on a half-wave antenna in *free space* (a dipole) have on the impedance at that point?

*13. Can either of the two fields that emanate from an antenna produce an EMF in a receiving antenna? If so, how?

14. Draw the three-dimensional radiation pattern for the half-wave dipole antenna, and explain how it is developed.

15. Define antenna *beamwidth.*

*16. What is the effective radiated power of a television broadcast station if the output of the transmitter is 1000 W, antenna transmission line loss is 50 W, and the antenna power gain is 3? (2850 W)

17. A $\lambda/2$ dipole is driven with a 5-W signal at 225 MHz. A receiving dipole 100 km away is aligned so that its gain is cut in half. Calculate the received power and voltage into a 73-Ω receiver. (7.57 pW, 23.5 μV)

18. An antenna with a gain of 4.7 dBi is being compared with one having a gain of 2.6 dBd. Which has the greater gain?

19. Explain why a monopole antenna is used below 2 MHz.

20. Explain what is meant by half-wave dipole. Calculate the length of a 100-MHz $\frac{2}{3}\lambda$ antenna.

21. Determine the distance from a $\lambda/2$ dipole to the boundary of the far-field region if the $\lambda/2$ dipole is being used in the transmission of a 90.7-MHz FM broadcast band signal. (R = 1.653 m)

22. Determine the distance from a parabolic reflector of diameter D = 10 m to the boundary of the far-field region. The antenna is being used to transmit a 4.1-GHz signal. (R = 2733.3 m)

23. Define *near* and *far fields.*

Section 14-3

24. Define *radiation resistance* and explain its significance.

*25. The ammeter connected at the base of a vertical antenna has a certain reading. If this reading is increased 2.77 times, what is the increase in output power? (7.67)

26. How is the operating power of an AM transmitter determined using antenna resistance and antenna current?

27. Explain what happens to an antenna's radiation resistance as its length is continuously increased.

28. Explain the effect that ground has on an antenna.

29. Calculate the efficiency of an antenna that has a radiation resistance of 73 Ω and an effective dissipation resistance of 5 Ω. What factors could enter into the dissipation resistance? (93.6%)

*30. Explain the following terms with respect to antennas (transmission or reception):
   (a) Physical length.
   (b) Electrical length.
   (c) Polarization.
   (d) Diversity reception.
   (e) Corona discharge.

31. What is the relationship between the electrical and physical length of a half-wave dipole antenna?

*32. What factors determine the resonant frequency of any particular antenna?

*33. If a vertical antenna is 405 ft high and is operated at 1250 kHz, what is its physical height expressed in wavelengths? (0.54λ)

*34. What must be the height of a vertical radiator one half-wavelength high if the operating frequency is 1100 kHz? (136 m)

Section 14-4

35. What is an antenna feed line? Explain the use of resonant antenna feed lines, including advantages and disadvantages.

36. What is a nonresonant antenna feed line? Explain its advantages and disadvantages.

37. Explain the operation of a delta match. Under what conditions is it a convenient matching system?

*38. Draw a simple schematic diagram of a push-pull, neutralized radio-frequency amplifier stage, coupled to a vertical antenna system.

*39. Show by a diagram how a two-wire radio-frequency transmission line may be connected to feed a half-wave dipole antenna.

*40. Calculate the characteristic impedance of a quarter-wavelength section used to connect a 300-Ω antenna to a 75-Ω line. (150 Ω)

41. Explain how delta matching is accomplished.

Section 14-5

*42. Which type of antenna has a minimum of directional characteristics in the horizontal plane?

*43. If the resistance and the current at the base of a monopole antenna are known, what formula can be used to determine the power in the antenna?

*44. What is the difference between a half-wave dipole and a monopole antenna?

*45. Draw a sketch and discuss the horizontal and vertical radiation patterns of a quarter-wave monopole antenna. Would this also apply to a similar type of receiving antenna?

46. What is an image antenna? Explain its relationship to the monopole antenna.

*47. What would constitute the ground plane if a quarter-wave grounded (whip) antenna, 1 m in length, were mounted on the metal roof of an automobile? Mounted near the rear bumper of an automobile?

*48. What is the importance of the ground radials associated with standard broadcast antennas? What is likely to be the result of a large number of such radials becoming broken or seriously corroded?

*49. What is the effect on the resonant frequency of connecting an inductor in series with an antenna?

*50. What is the effect on the resonant frequency of adding a capacitor in series with an antenna?

*51. If you desire to operate on a frequency lower than the resonant frequency of an available monopole antenna, how may this be accomplished?

*52. What is the effect on the resonant frequency if the physical length of a $\lambda/2$ dipole antenna is reduced?

*53. Why do some standard broadcast stations use top-loaded antennas?

*54. Explain why a *loading coil* is sometimes associated with an antenna. Under this condition, would absence of the coil mean a capacitive antenna impedance?

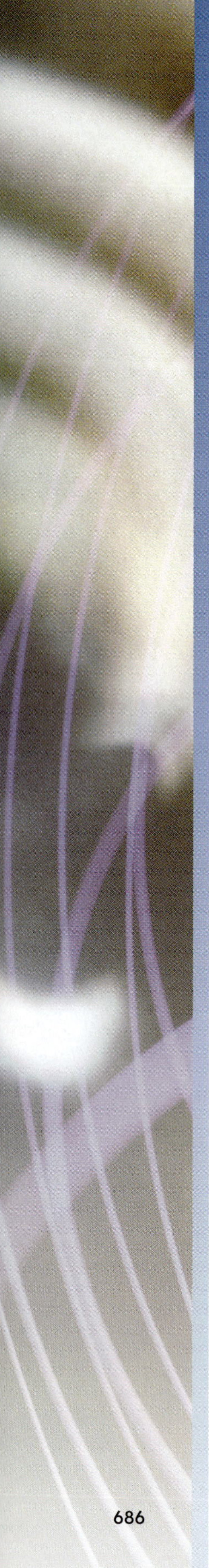

Section 14-6

55. Explain how the directional capabilities of the elementary antenna array shown in Fig. 14-16 are developed.
56. Define the following terms:
    (a) Driven elements.
    (b) Parasitic elements.
    (c) Reflector.
    (d) Director.
57. Calculate the ERP from a Yagi–Uda antenna (illustrated in Fig. 14-17) driven with 500 W. (2500 W)
58. Calculate the *F*/*B* ratio for an antenna with
    (a) Forward gain of 7 dB and reverse gain of −3 dB.
    (b) Forward gain of 18 dB and reverse gain of 5 dB.
59. Sketch a Yagi–Uda configuration.
60. Describe the physical configuration of a collinear array. What is the effect of adding more elements to this antenna?
61. Describe the physical configuration of a broadside array. Explain the major advantage they have compared to collinear arrays.

*62. What is the direction of maximum radiation from two vertical antennas spaced $\lambda/2$ and having equal currents in phase?

*63. How does a directional antenna array at an AM broadcast station reduce radiation in some directions and increase it in other directions?

*64. What factors can cause the directional pattern of an AM station to change?

65. Define *phased array.*
66. Explain how a parasitic array can be developed.

Section 14-7

67. Describe the major characteristics of a log-periodic antenna. What explains its widespread use? Explain the significance of its shortest and longest elements.

*68. Describe the directional characteristics of the following types of antennas:
    (a) Horizontal half-wave dipole antenna.
    (b) Vertical half-wave dipole antenna.
    (c) Vertical loop antenna.
    (d) Horizontal loop antenna.
    (e) Monopole antenna.

*69. What is the directional reception pattern of a loop antenna?

70. What is a ferrite loop antenna? Explain its application and advantages.
71. What is the radiation resistance of a standard folded dipole? What are its advantages over a standard dipole? Why is it usually used as the driven element for Yagi–Uda antennas instead of the half-wave dipole antenna?
72. Describe the operation of a slot antenna and its application with aircraft in a driven array format.
73. An antenna has a maximum forward gain of 14 dB at its 108-MHz center frequency. Its reverse gain is −8 dB. Its beamwidth is 36° and the bandwidth extends from 55 to 185 MHz. Calculate:
    (a) Gain at 18° from maximum forward gain. (11 dB)
    (b) Bandwidth. (130 MHz)

(c) *F/B* ratio. (22 dB)

(d) Maximum gain at 185 MHz. (11 dB)

74. Explain the difference between antenna beamwidth and bandwidth.

Section 14-8

75. Explain VSWR and tell why it is important in troubleshooting antenna problems.
76. Explain how to proceed in troubleshooting the antenna to determine causes of transmission problems.
77. Describe how to use a grid-dip meter, and give examples of where it is most commonly used.
78. What is an SWR meter and how is it used in troubleshooting?
79. Explain what happens when the VSWR is too high.
80. What is an anechoic chamber? Explain what factors to consider with respect to its size requirements.

Questions for Critical Thinking

*81. A ship radio-telephone transmitter operates on 2738 kHz. At a distant point from the transmitter, the 2738-kHz signal has a measured field of 147 mV/m. The second harmonic field at the same point is measured as 405 μV/m. How much has the harmonic emission been attenuated below the 2738-kHz fundamental? (51.2 dB)

*82. You are asked to calculate effective radiated power. What data do you need to collect and how do you perform the calculation?

83. Design a log-periodic antenna to cover the complete VHF TV band. (See Chapter 7 for the frequencies involved.) Use a design factor (τ) of 0.7, and provide a scaled sketch of the antenna with all dimensions indicated.

84. A loop antenna used for DF purposes detects a null from a signal with the loop rotated 35° CCW from a line of latitude. When the antenna is moved 3 mi west along the same line of latitude, it detects a null from the same signal source when rotated 45° CW from the line of latitude. You have been asked to identify the exact location of the signal source with respect to the two points when readings were taken. Provide this information. (You may use a sketch.)

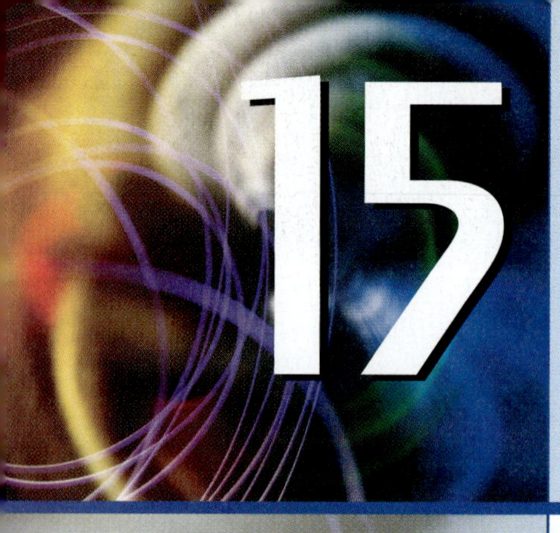

15 Waveguides and Radar

Chapter Outline

- 15-1 Comparison of Transmission Systems
- 15-2 Types of Waveguides
- 15-3 Physical Picture of Waveguide Propagation
- 15-4 Other Types of Waveguides
- 15-5 Other Waveguide Considerations
- 15-6 Termination and Attenuation
- 15-7 Directional Coupler
- 15-8 Coupling Waveguide Energy and Cavity Resonators
- 15-9 Radar
- 15-10 Microintegrated Circuit Waveguiding
- 15-11 Troubleshooting
- 15-12 Troubleshooting with Electronics Workbench™ Multisim

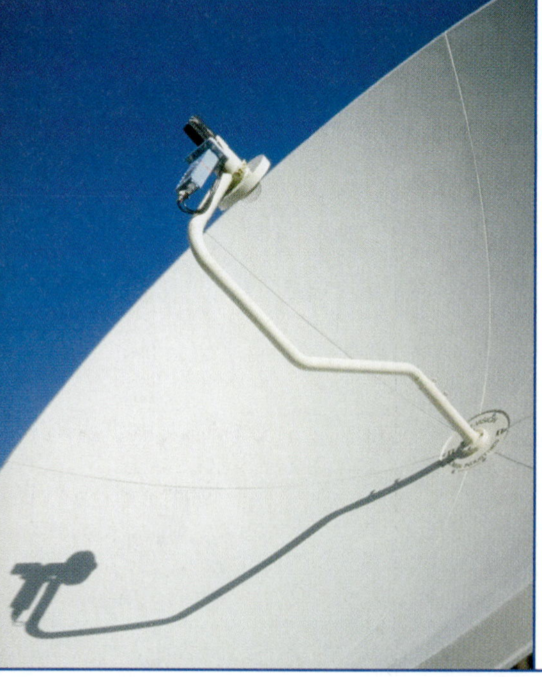

A dish antenna with a waveguide.
(Courtesy of Photo Researchers, Inc. ©Alex Bartel, photographer.)

Objectives

- Differentiate among sending signals on transmission lines, antennas, and waveguides based on power and distance
- Describe basic modes of operation for rectangular waveguides
- Calculate the cutoff wavelength for the dominant mode of operation
- Provide a physical picture of waveguide propagation, including the concepts of guide wavelength and velocity
- Describe other types of waveguides including circular, ridged, flexible, bends, twists, tees, tuners, terminations, attenuators, and directional couples
- Explain three methods for coupling energy into or out of a waveguide and the uses for cavity resonators
- Draw a block diagram for a radar system and explain its operation, including the concepts of range, echoes, and duty cycle
- Calculate an object's velocity when using a Doppler radar system
- Calculate the characteristic impedance for microstrip and stripline

Key Terms

waveguide
characteristic wave impedance
vane
radar
echo signal
pulse repetition frequency (PRF)
pulse repetition rate (PRR)
pulse repetition time (PRT)
rest time
receiver time
radar mile
second return echoes
maximum usable range
double range echoes
peak power
duty cycle
Doppler effect
stripline
microstrip
dielectric waveguide

15-1 Comparison of Transmission Systems

The mode of energy transmission chosen for a given application normally depends on the following factors: (1) initial cost and long-term maintenance, (2) frequency band to be used and its information-carrying capacity, (3) selectivity or privacy offered, (4) reliability and noise characteristics, and (5) power level and efficiency. Naturally, any one mode of energy transmission has only some of the desirable features. It therefore becomes a matter of sound technical judgment when choosing the mode of energy transmission best suited for a particular application.

Transmission lines, antennas, and fiber optics are the more commonly known means of high-frequency transmission, but waveguides also play an important role. The following examples show that each method of transmission has its proper place. Fiber optics has been left out of this comparison, but this mode of transmission is discussed fully in Chapter 18.

It is desired to transmit a 1-GHz signal between two points 30 mi apart. If the received energy in each case were chosen to be 1 nW (10^{-9} W), then for comparison it would be found that for reasonably typical installations, the required *transmitted* power would be on the order of

1. *Transmission lines:* 10^{1500} nW (15,000-dB loss)
2. *Waveguides:* 10^{150} nW (1500-dB loss)
3. *Antennas:* 100 mW (80-dB loss)

Clearly, the transmission of energy without any electrical conductors (antennas) exceeds the efficiency of waveguides and transmission lines by many orders of magnitude.

If the transmission path length of the preceding example were shortened by a factor of 100:1, to a distance of 1500 ft, the comparison becomes

1. *Transmission lines:* 1 MW (150-dB loss)
2. *Waveguides:* 30 nW (15-dB loss)
3. *Antennas:* 10 μW (40-dB loss)

Quite clearly the waveguide now surpasses either the transmission line or antenna for efficiency of energy transfer.

A comparison of the energy input required versus distance to obtain a received power of 1 nW for these three modes of energy transmission is shown in Fig. 15-1. The frequency is 1 GHz, and the results are expressed on a decibel scale with a 0-dB reference at the required receiver power level of 1 nW. The dashed section of the antenna curve, somewhat beyond 30 mi, indicates that the attenuation becomes severe beyond the line-of-sight distance, which is typically 50 mi.

One final comparison, and then a specific look at waveguides. Transmission of energy down to zero frequency is practical with transmission lines, but waveguides, antennas, and fiber optics inherently have a practical low-frequency limit. In the case of antennas, this limit is about 100 kHz, and for waveguides, it is about 300 MHz. Fiber-optic transmissions take place at the frequency of light, or greater than 10^{14} Hz! Theoretically the antennas and waveguides could be made to work at arbitrarily low frequencies, but the physical sizes required would become excessively large. However, with the low gravity and lack of atmosphere on the moon, it may be feasible to have an antenna 10 mi high and 100 mi long for

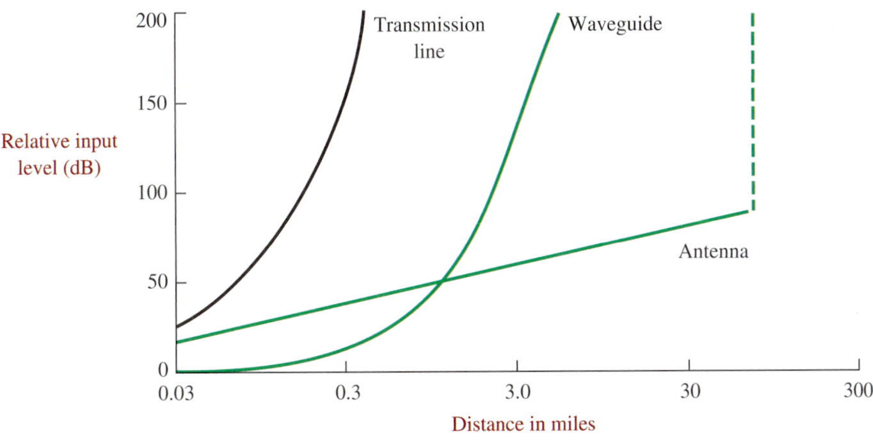

FIGURE 15-1 Input power required versus distance for fixed receiver power.

frequencies as low as a few hundred cycles per second. As an indication of the sizes involved, it may be noted that for either waveguides or antennas, the important dimension is normally a half-wavelength. Thus, a waveguide for a 300-MHz signal would be about the size of a roadway drainage culvert, and an antenna for 300 MHz would be about $1\frac{1}{2}$ ft long.

15-2 Types of Waveguides

Any surface separating two media of distinctly different conductivities or permitivities has a guiding effect on electromagnetic waves. For example, a rod of dielectric material, such as polystyrene, can carry a high-frequency wave, somewhat as a glass fiber conducts a beam of light. These phenomena will be further explored in Sec. 15-10 and Chapter 18. The best guiding surface, however, is that between a good dielectric and a good conductor.

In a broad sense, all kinds of transmission lines, including coaxial cables and parallel wires, are waveguides. In practice, however, the term **waveguide** has come to signify a hollow metal tube or pipe used to conduct electromagnetic waves through its interior. They were first used extensively in radar sets during World War II, operating at wavelengths of between 10 and 3 cm. They are commonly called *plumbing* in the trade.

Waveguide
a microwave transmission line consisting of a hollow metal tube or pipe that conducts electromagnetic waves through its interior

A waveguide can be almost any shape. The most popular shape is rectangular, but circular and even more exotic shapes are used. We shall mainly study the rectangular waveguide operating in the TE_{10} mode. We shall learn more about this terminology shortly.

Like coaxial lines, waveguides are perfectly shielded—hence, no radiation loss. The attenuation of a hollow pipe is less, and the power capacity is greater, than that of a coaxial line of the same size at the same frequency. Most of the copper loss of a coaxial line occurs in the thin inner conductor; hence, its elimination in a waveguide reduces attenuation and increases the power capacity. It also simplifies the construction and makes the line more rugged.

Waveguide Operation

A rigorous mathematical demonstration of waveguide operation is beyond our intentions. A practical explanation is possible by starting out with a normal two-wire transmission line. You may recall from Chapter 12 that a quarter-wavelength shorted stub looks like an open circuit and in fact is often used as an insulating support for transmission lines. If an infinite number of these supports were added both above and below the two-wire transmission line, as shown in Fig. 15-2, you can visualize it turning into a rectangular waveguide. If the shorted stubs were less than a quarter-wavelength, operation would be drastically impaired. The same is true of a rectangular waveguide. The *a* dimension of a waveguide, shown in Fig. 15-3, must be at least a half-wavelength at the operating frequency, and the *b* dimension is normally about one-half the *a* dimension.

The wave that is propagated by a waveguide is electromagnetic and, therefore, has electric (*E* field) and magnetic (*H* field) components. In other words, energy propagates down a waveguide in the form of a radio signal. The configuration of these two fields determines the mode of operation. If no component of the *E* field is in the direction of propagation, we say that it is in the TE mode. TE is the abbreviation for transverse electric. *Transverse* means "at right angles." TM is the mode of waveguide operation whereby the magnetic field has no component in the direction of propagation. Two-number subscripts normally follow the TE or TM designations, and they can be interpreted as follows: For TE modes, the first subscript is the number of one half-wavelength *E*-field patterns along the *a* (longest) dimension, and the second subscript is the number of one half-wavelength *E*-field patterns along the *b* dimension. For TM modes, the number of one half-wavelength *H* fields along

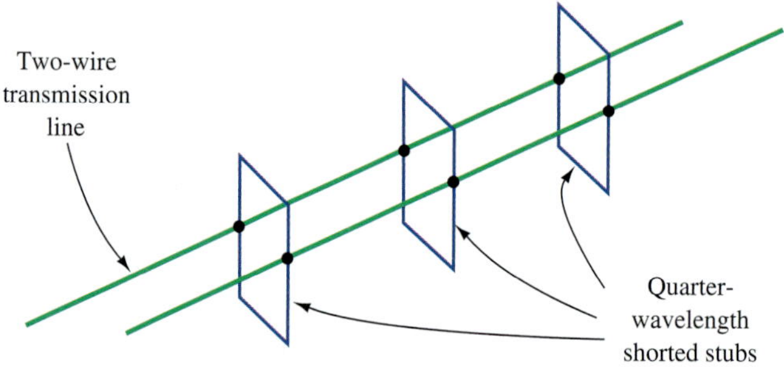

FIGURE 15-2 Transforming a transmission line into a waveguide.

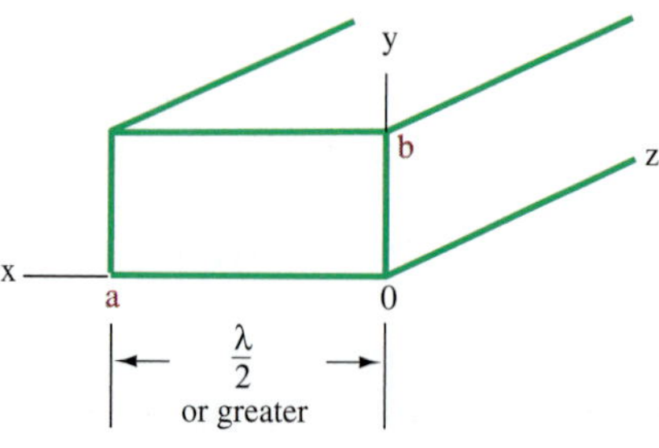

FIGURE 15-3 Waveguide dimension designation.

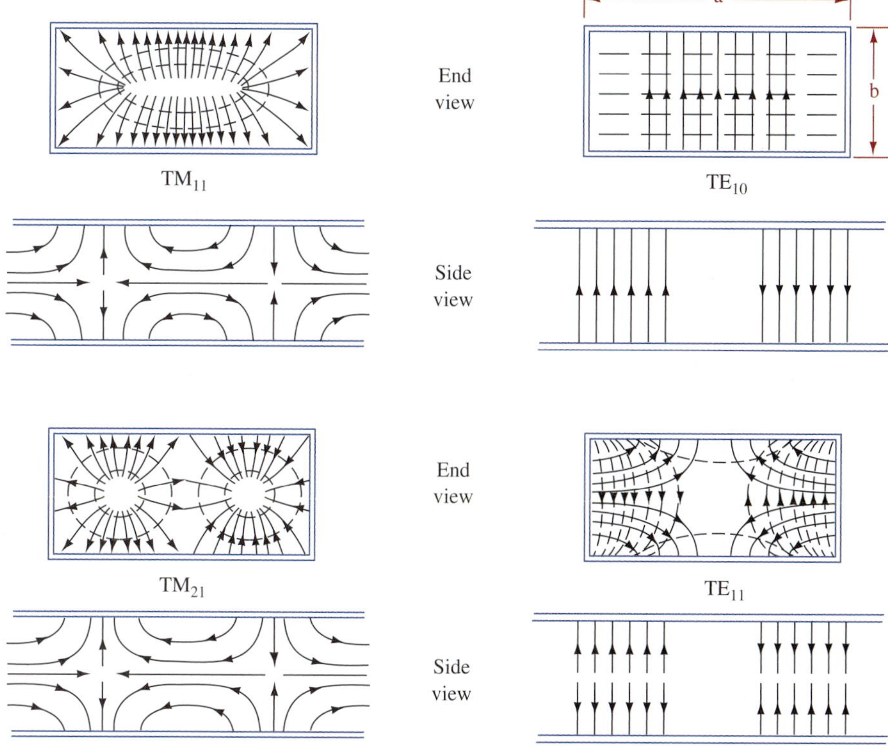

FIGURE 15-4 Examples of modes of operation in rectangular waveguides.

the *a* and *b* dimensions determine the subscripts. Refer to Fig. 15-4 for further illustration of this process.

The electric field is shown with solid lines, and the magnetic field is shown with dashed lines. Notice the end view for the TE_{10} mode in Fig. 15-4. The electric field goes from a minimum at the ends along the *a* dimension to a maximum at the center. This is equivalent to one half-wavelength of *E* field along the *a* dimension, while no component exists along the *b* dimension. This is, therefore, called the TE_{10} mode of operation. In the TM_{21} mode of operation, note that along the *a* dimension the *H* field (dashed lines) goes from zero to maximum to zero to maximum to zero. That's two half-wavelengths. Along the *b* dimension, one half-wavelength of the *H* field occurs—thus the TM_{21} designation. Note that in the side views, the *H* fields (dashed lines) are not shown for the sake of simplicity. In these side views, the TE modes have no *E* field in the direction of propagation (right to left or left to right), while in the TM modes the *E* field (solid lines) does exist in the propagation direction.

Dominant Mode of Operation

The TE_{10} mode of operation is called the dominant mode because it is the most "natural" one for operation. A waveguide is often thought of as a high-pass filter because only very high frequencies can be propagated. The TE_{10} mode has the lowest cutoff frequency of any of the possible modes of propagation, including both TM and

TE types. It is of special interest because there will exist a frequency range between its cutoff frequency (f_c, the lowest frequency that a given waveguide propagates) and that of the next higher-order mode in which this is the only possible mode of transmission. Thus, if a waveguide is excited within this frequency range, energy propagation must take place in the dominant mode, regardless of the way in which the guide is excited. Control of the mode of operation is important in any practical transmission system, and thus the TE_{10} mode has a distinct advantage over the other possible modes in a rectangular waveguide. Even more important, however, is the fact that TE_{10} operation allows use of the physically smallest waveguide for a given frequency of operation.

The dimensions for an RG-52/U waveguide are 0.9×0.4 in. This is one of the standard sizes used in the X-band frequency range and is usually called the X-band waveguide. The recommended frequency range for this waveguide size is 8.2 to 12.4 GHz. As a result of this limited range of usefulness, standard sizes of waveguides have been established, each having a specified frequency range. Table 15-1 provides the frequency range and size of the various waveguide bands.

The formula for cutoff wavelength is

$$\lambda_{co} = 2a \tag{15-1}$$

for the TE_{10} mode, where a is the long dimension of the waveguide rectangle. Thus, for an RG-52/U waveguide, λ_{co} is 2(0.9) or 1.8 in., or 4.56 cm. Therefore,

$$f_{co} = \frac{c}{\lambda_{co}} = \frac{3 \times 10^{10}\ \text{cm/s}}{4.56\ \text{cm}} = 6.56\ \text{GHz} \tag{15-2}$$

The lowest frequency of propagation (without considerable attenuation) is 6.56 GHz, but the recommended range is 8.2 to 12.4 GHz. The next higher-order mode is the TE_{20}, which has a cutoff frequency of 13.1 GHz. Thus, within the frequency range from 6.56 to 13.1 GHz, only the TE_{10} mode can propagate within the X-band waveguide, in the ordinary sense of the word.

Table 15-1 Waveguide Bands/Sizes

Band	Frequency Range in GHz	Type	Size of Waveguide in.	Size of Waveguide cm
L	1.12–1.7	WR650	6.5 × 3.25	16.5 × 8.26
S	1.7–2.6	WR430	4.3 × 2.15	10.9 × 8.6
S	2.6–3.95	WR284	2.84 × 1.34	7.21 × 3.40
G	3.95–5.85	WR187	1.87 × 0.87	4.75 × 2.21
C	4.9–7.05	WR159	1.59 × 0.795	4.04 × 2.02
J	5.85–8.2	WR137	1.37 × 0.62	3.48 × 1.57
H	7.05–10.0	WR112	1.12 × 0.497	2.84 × 1.26
X	8.2–12.4	WR90	0.9 × 0.4	2.29 × 1.02
M	10.0–15.0	WR75	0.75 × 0.375	1.91 × 0.95
P	12.4–18.0	WR62	0.62 × 0.31	1.57 × 0.79
N	15.0–22.0	WR51	0.51 × 0.255	1.30 × 0.65
K	18.0–26.5	WR42	0.42 × 0.17	1.07 × 0.43
R	26.5–40.0	WR28	0.28 × 0.14	0.71 × 0.36

Note: WR = waveguide rectangular

15-3 Physical Picture of Waveguide Propagation

For a wave to exist in a waveguide, it must satisfy Maxwell's equations throughout the waveguide. These mathematically complex equations are beyond the scope of this book, but one boundary condition of these equations can be put into plain language: There can be no tangential component of electric field at the walls of the waveguide. This makes sense because the conductor would then *short* out the E field. An exact solution for the field existing within a waveguide is a relatively complicated mathematical expression. It is possible, however, to obtain an understanding of many of the properties of waveguide propagation from a simple physical picture of the mechanisms involved. The fields in a typical TE_{10} waveguide can be considered as the resultant fields produced by an ordinary plane electromagnetic wave that travels back and forth between the sides of the guide, as illustrated in Fig. 15-5.

The electric and magnetic component fields of this plane wave are in time phase but are geometrically at right angles to each other and to the direction of propagation. Such a wave travels with the velocity of light and, upon encountering the conducting walls of the guide, is reflected with a phase reversal of the electric field and with an angle of reflection equal to the angle of incidence. A picture of the wavefronts involved with such propagation for a rectangular waveguide is shown in Fig. 15-6.

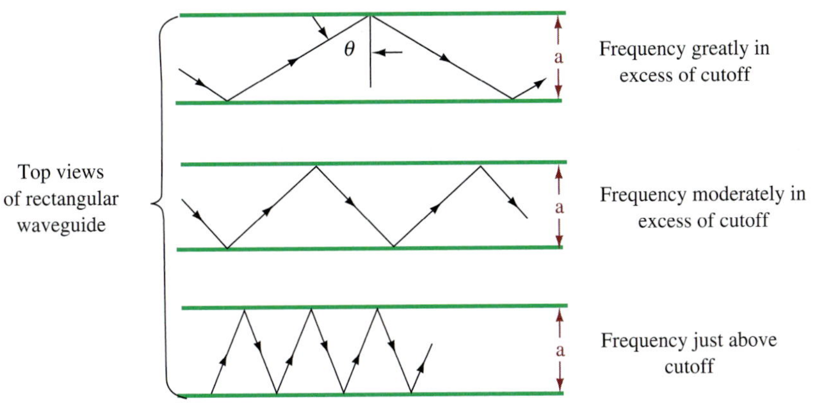

FIGURE 15-5 Paths followed by waves traveling back and forth between the walls of a waveguide.

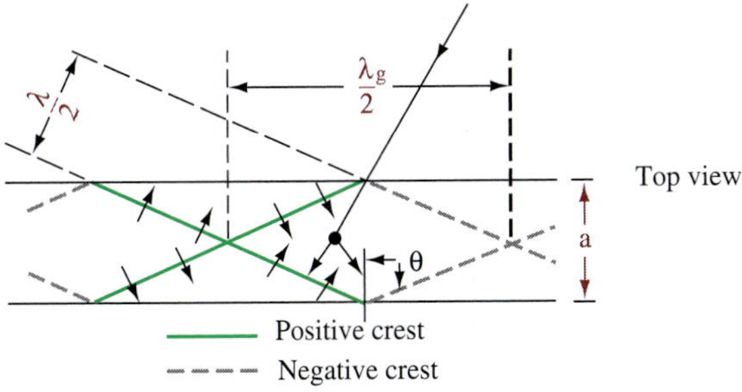

FIGURE 15-6 Wavefront reflection in a waveguide.

When the angle θ (see Fig. 15-5) is such that the successive positive and negative crests traveling in the same direction just fail to overlap inside the guide, it can be shown that the summation of the various waves and their reflections leads to the field distribution of the TE_{10} mode, which travels down the waveguide and represents propagation of energy. The angle that the component waves must have with respect to the waveguide to satisfy the conditions for waveguide propagation in a rectangular guide is given by the relation

$$\cos\theta = \frac{\lambda}{2a} \tag{15-3}$$

where a is the width of the waveguide and λ is the wavelength of the wave on the basis of the velocity of light.

Because the component waves that can be considered as building up the actual field in the waveguide all travel at an angle with respect to the axis of the guide, the rate at which energy propagates down the guide is less than the velocity of light. This velocity with which energy propagates is termed group velocity (V_g) and in the case of Fig. 15-6 is given by the relation

$$\frac{V_g}{c} = \sin\theta = \sqrt{1 - \left(\frac{\lambda}{2a}\right)^2} \tag{15-4}$$

The guide wavelength (λ_g) is greater than the free-space wavelength (λ). A study of the $\lambda_g/2$ and $\lambda/2$ shown in Fig. 15-6 should help in visualizing this situation. Thus,

$$\frac{\text{wavelength in guide}}{\text{wavelength in free space}} = \frac{\lambda_g}{\lambda} = \frac{1}{\sin\theta} \tag{15-5}$$

and therefore

$$\frac{c}{V_g} = \frac{\lambda_g}{\lambda} = \frac{1}{\sqrt{1-(\lambda/2a)^2}} \tag{15-6}$$

In Smith chart solutions of waveguide problems, λ_g should be used for making moves, not the free-space wavelength λ. The velocity with which the wave appears to move past the guide's side wall is termed the phase velocity, V_p. It has a value greater than the speed of light. It is only an "apparent" velocity; however, it is the velocity with which the wave is changing phase at the side wall. The phase and group velocities, V_p and V_g, respectively, are related by the fact that

$$\sqrt{V_p V_g} = \text{velocity of light} \tag{15-7}$$

As the wavelength is increased, the component waves must travel more nearly at right angles to the axis of the waveguide, as shown in the bottom portion of Fig. 15-5. This causes the group velocity to be lowered and the phase velocity to be still greater than the velocity of light, until finally one has $\theta = 0°$. The component waves then bounce back and forth across the waveguide at right angles to its axis and do not travel down the guide at all. Under these conditions, the group velocity is zero, the phase velocity becomes infinite, and propagation of energy ceases. This occurs at the frequency that was previously defined as the cutoff frequency, f_{co}. The cutoff frequency for the TE_{10} mode of operation can be determined from the relationship given in Eq. (15-1). Note that the waveguide acts as a high-pass filter, with the cutoff frequency determined by the waveguide dimensions. To obtain propagation, the waveguide must have dimensions comparable to a half-wavelength, and that limits its practical use to frequencies above 300 MHz.

At frequencies very much greater than cutoff frequency, it is possible for the higher-order modes of transmission to exist in a waveguide. Thus, if the frequency is high enough, propagation of energy can take place down the guide when the system of component waves that are reflected back and forth has the form of the TE_{20} mode. It has a field distribution that is equivalent to two distributions of the dominant TE_{10} mode placed side by side, but each with reversed polarity. This conceptual presentation of waveguide propagation, involving a wave suffering successive reflections between the sides of the guide, can be applied to all types of waves and to other than rectangular guides. The way in which the concept works out in these other cases is not so simple, however, as for the TE_{10} mode.

15-4 Other Types of Waveguides

Circular

The dominant mode (TE_{10}) for rectangular waveguides is by far the most widely used. The use of other modes or of other shapes is extremely limited. However, the use of a circular waveguide is found in radar applications where it is necessary to have a continuously rotating section like that in Fig. 15-7. Modes in circular waveguides can be rotationally symmetrical, which means that a radar antenna can physically rotate with no electrical disturbance. While a circular waveguide is actually simpler to manufacture than a rectangular one, for a given frequency of operation, its cross-sectional area must be more than double that of a rectangular guide. It is, therefore, more expensive and takes up more space than a rectangular guide. Typical radar systems, therefore, consist of a main run with a rectangular waveguide and a circular rotating joint. The transition between rectangular and circular waveguides is accomplished with the circular-to-rectangular taper shown in Fig. 15-8. The transition is accomplished as gradually as possible to minimize reflections.

One of the limitations of circular waveguides relates to the range of frequencies it can propagate. From Table 15-1, we can see that rectangular waveguide has a useful bandwidth of about 50 percent of the frequency range that can be propagated.

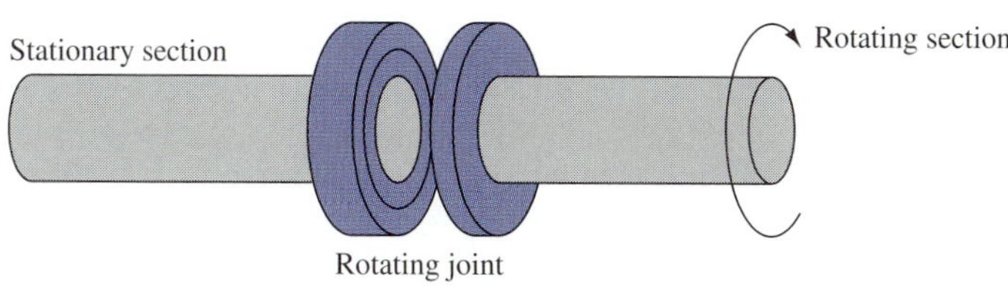

FIGURE 15-7 Circular waveguide rotating joint.

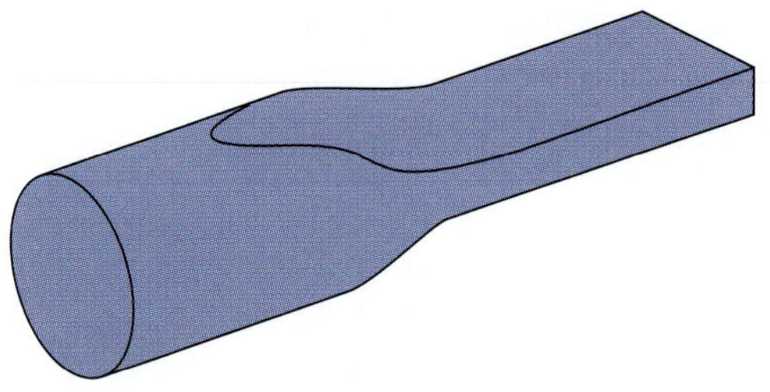

FIGURE 15-8 Circular-to-rectangular taper.

For example, at L-band frequencies, the range is 1.12 to 1.7 GHz, or 0.58 GHz, which is slightly less than half of the center frequency (1.41 GHz) of the waveguide frequency range.

$$\text{center frequency} = 1.12 \text{ GHz} + \frac{(1.7 - 1.12) \text{ GHz}}{2} = 1.41 \text{ GHz}$$

In this case, 0.58 GHz is 41 percent of the center frequency:

$$\frac{0.58 \text{ GHz}}{1.41 \text{ GHz}} \times 100\% = 41\%$$

A study of the modes in a circular waveguide shows the bandwidth to be about 15 percent, a much reduced amount compared to a rectangular waveguide.

Ridged

Two types of ridged waveguides are shown in Fig. 15-9. While it is obviously more expensive to manufacture than a standard rectangular waveguide, it does provide one unique advantage. It allows operation at lower frequencies for a given set of outside dimensions, which means that smaller overall external dimensions are made possible. This property is advantageous in applications where space is at a premium, such as space probes and the like. A ridged waveguide has greater attenuation, and this, combined with its higher cost, limits it to special applications. This means that a ridged waveguide has the greater bandwidth as a percentage of center frequency.

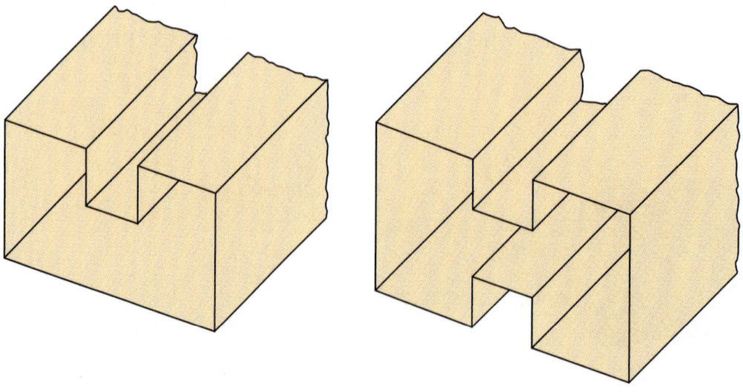

FIGURE 15-9 Ridged waveguides.

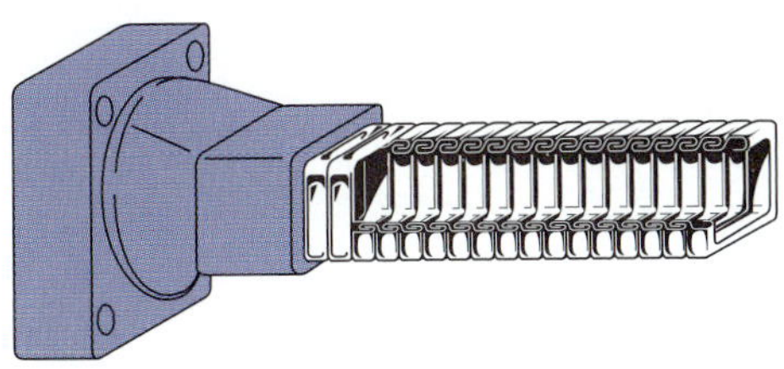

FIGURE 15-10 Flexible waveguide.

Flexible

It is sometimes desirable to have a section of waveguide that is flexible, as shown in Fig. 15-10. This configuration is often useful in the laboratory or in applications where continuous flexing occurs. Flexible waveguides consist of spiral-wound ribbons of brass or copper. The outside section is covered with a soft dielectric such as rubber to maintain air- and watertight conditions and prevent dust contamination, which can encourage arcs to form from one side to the other in high-power situations such as radar. A waveguide is often pressurized with nitrogen to prevent such contamination. In the case of a leak, gas rushes out, preventing anything from entering.

Corrosion would cause an increase in attenuation through surface current losses and increased reflections. In critical applications, waveguides are filled with inert gas and/or their inside walls are coated with noncorrosive (but expensive) metals such as gold or silver.

15-5 Other Waveguide Considerations

Waveguide Attenuation

Waveguides are capable of propagating huge amounts of power. For example, typical X-band (0.9 × 0.4 in.) waveguides can handle 1 million W if operated at 1.5 times f_{co} and an air dielectric strength of 3×10^6 V/m is assumed. At frequencies below cutoff, the attenuation in any waveguide is very large, as previously explained. At frequencies above cutoff, the guide supports traveling waves, and they are slightly attenuated because of losses in the conducting walls and in the dielectric that fills the guide. For air-filled guides, the dielectric loss is normally negligible, but if dielectric other than air is used, these losses are often greater than the conductor losses. As the frequency increases, attenuation drops to a broad minimum and then increases slowly with increasing frequency.

The conductor losses are governed in part by the skin effect described in Chapter 12. (At high frequencies the current tends to flow only at the surface of a conductor.) Current that flows in the guide walls is concentrated near the inner surface.

Bends and Twists

It is often necessary to change the physical direction of propagation or the wave's polarization in waveguides.

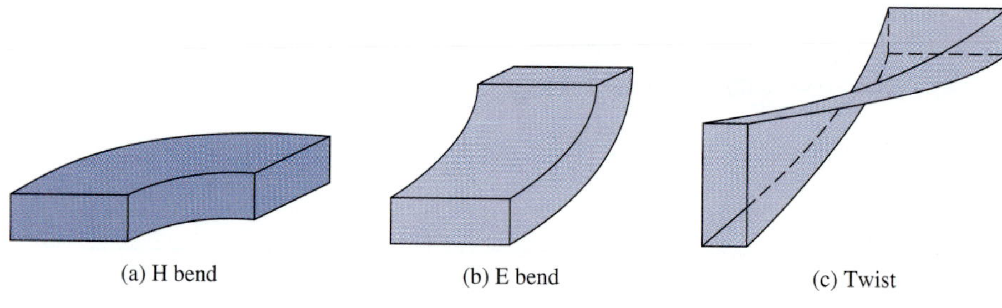

FIGURE 15-11 Waveguide bends and twists.

1. *H bend* [Fig. 15-11(a)]: It is used to change the physical direction of propagation. It derives its name from the fact that the *H* lines are bent in this transition, while the *E* lines remain vertical for the dominant mode.
2. *E bend* [Fig. 15-11(b)]: This section is also used to change the physical direction of propagation. The choice between an *E* or *H* bend is normally governed by mechanical considerations (plumbing considerations, if you will) because neither produces large discontinuities if the bends are gradual enough.
3. *Twist* [Fig. 15-11(c)]: A twist section is used to change the plane of polarization of the wave.

You can see that any desired angular orientation of the wave may be obtained with an appropriate combination of the three types of sections just discussed.

Tees

1. *Shunt tee* [Fig. 15-12(a)]: A shunt tee is so named because of the side arm shunting of the *E* field for TE modes, which is analogous to voltage in a transmission line. You can see that if two input waves at arms *A* and *B* are in phase,

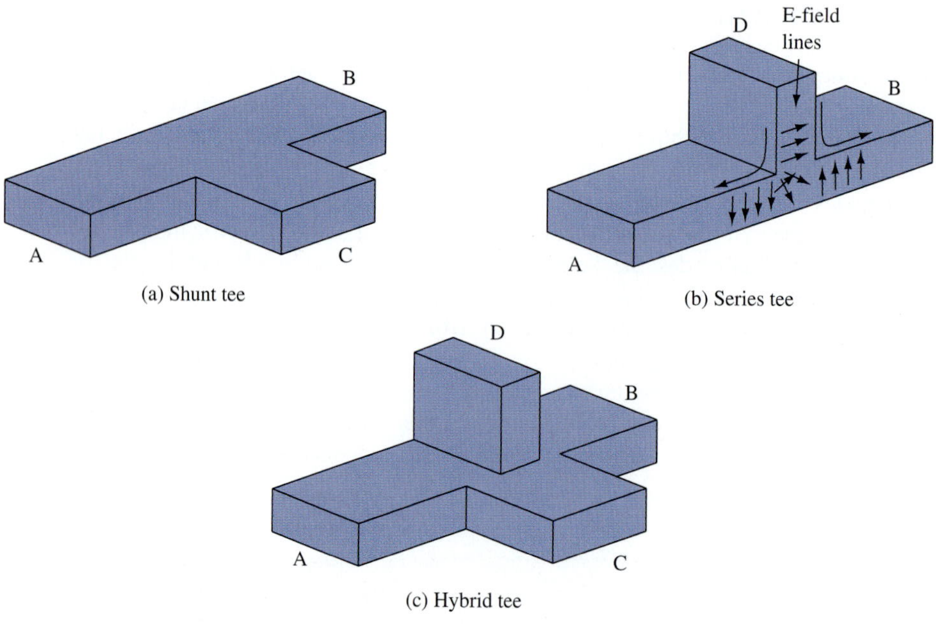

FIGURE 15-12 Shunt, series, and hybrid tees.

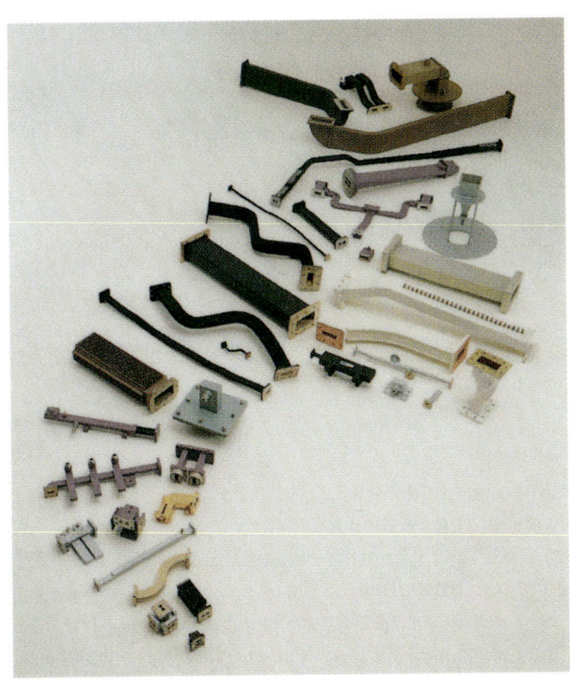

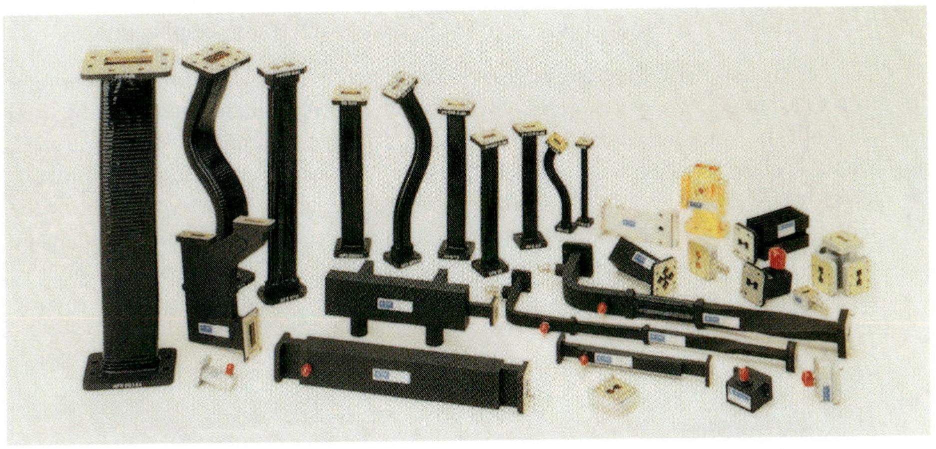

Rectangular and double-ridge seamless flexible waveguide assemblies, microwave components, waveguide subsystems, and special-purpose waveguide components designed and manufactured to customer specifications. (Courtesy of Continental Microwave & Tool Co.)

the portions transmitted into arm *C* will be in phase and thus will be additive. On the other hand, an input at *C* results in two equal, in-phase outputs at *A* and *B*. Of course, the *A* and *B* outputs have half the power (neglecting losses) of the *C* input.

2. *Series tee* [Fig. 15-12(b)]: If you consider the *E* field of an input at *D*, you should be able to visualize that the outputs at *A* and *B* are equal and 180° out of phase, as shown. Once again, the two outputs are equal but are now 180° out of phase. The series tee is often used for impedance matching just as the single-stub

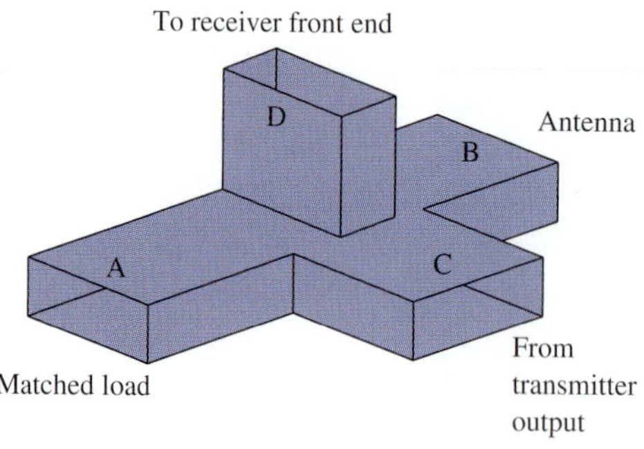

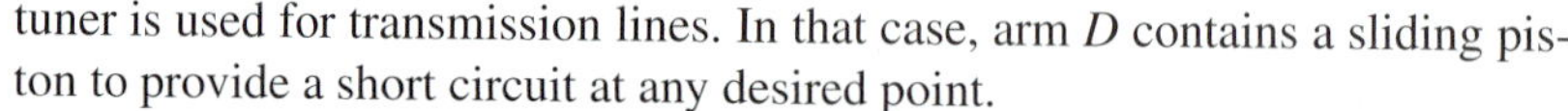

FIGURE 15-13 Hybrid-tee TR switch.

tuner is used for transmission lines. In that case, arm D contains a sliding piston to provide a short circuit at any desired point.

3. *Hybrid or magic tee* [Fig. 15-12(c)]: This is a combination of the first two tees mentioned and exhibits properties of each. From previous consideration of the shunt and series tees, you can see that if two equal signals are fed into arms A and B in phase, there is cancellation in arm D and reinforcement in arm C. Thus, all the energy will be transmitted to C and none to D. Similarly, if energy is fed into C, it will divide evenly between A and B, and none will be transmitted to D. The hybrid tee has many interesting applications.

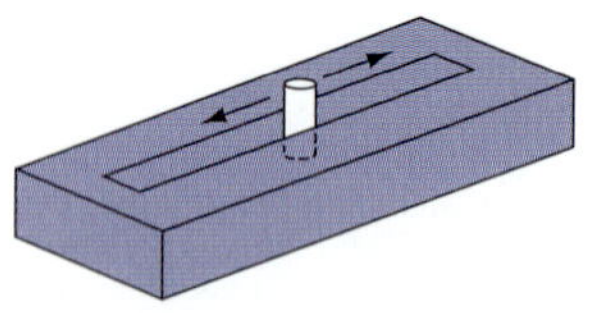

(a) Slide-screw tuner

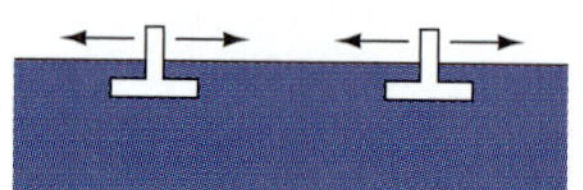

(b) Double-slug tuner

FIGURE 15-14 Tuners.

A typical hybrid tee application is illustrated in Fig. 15-13. It is functioning as a transmit/receive switch (TR switch), which allows a single antenna to be used for both transmission and reception. The transmitter's output is fed into arm C, where it splits between the A and B outputs, with almost no power going to the sensitive receiver at D. When the antenna receives a signal at B, energy is sent to the receiver at D as well as to arms A and C. The low received power does no damage to the powerful transmitter output. The matched load at A is necessary to prevent reflections. Problem 60 at the end of the chapter introduces you to another hybrid tee application.

Tuners

A metallic post inserted in the broad wall of a waveguide provides a lumped reactance at that point. The action is similar to the addition of a shorted stub along a transmission line. When the post extends less than a quarter-wavelength, it appears capacitive, while exceeding a quarter-wavelength makes it appear inductive. Quarter-wavelength insertion causes a series resonance effect whose sharpness (Q) is inversely proportional to the diameter of the post.

The primary usage of posts is in matching a load to a guide to minimize the VSWR. The most often used configurations are shown in Fig. 15-14.

1. *Slide-screw tuner* [Fig. 15-14(a)]: The slide-screw tuner consists of a screw or metallic object of some sort protruding vertically into the guide and adjustable both longitudinally and in depth. The effect of the protruding object is to produce shunting reactance across the guide—thus, it is analogous to a single-stub tuner in transmission line theory.

2. *Double-slug tuner* [Fig. 15-14(b)]: This type of tuner involves placing two metallic objects, called slugs, in the waveguide. The necessary two degrees of freedom to effect a match are obtained by making adjustable both the longitudinal position of the slugs and the spacing between them. Thus, it is somewhat analogous to the transmission line double-stub tuner but differs because the position of the slugs and not the effective shunting reactance is variable.

15-6 Termination and Attenuation

Because a waveguide is a single conductor, it is not as easy to define its characteristic impedance (Z_0) as it is for a coaxial line. Nevertheless, you can think of the characteristic impedance of a waveguide as being approximately equal to the ratio of the strength of the electric field to the strength of the magnetic field for energy traveling in one direction. This ratio is equivalent to the voltage-to-current ratio in coaxial lines on which there are no standing waves. For an air-filled rectangular wave-guide operating in the dominant mode, its characteristic impedance is given by

$$Z_0 = \frac{\mathscr{Z}}{\sqrt{1 - (\lambda/2a)^2}} \tag{15-8}$$

where $\mathscr{Z}$ is the characteristic impedance of free space $= 120\pi = 377\ \Omega$. The guide's characteristic impedance is affected by the frequency of the energy in it because $\lambda = c/f$. Therefore, the guide's impedance is variable and more correctly termed **characteristic wave impedance** rather than just characteristic impedance.

Characteristic Wave Impedance
characteristic impedance of a waveguide; affected by the frequency of operation

On a waveguide there is no place to connect a fixed resistor to terminate it in its characteristic (wave) impedance as there is on a coaxial cable. But several special arrangements accomplish the same result. One consists of filling the end of the waveguide with graphited sand, as shown in Fig. 15-15(a). As the fields enter the sand, currents flow in it. These currents create heat, which dissipates the energy.

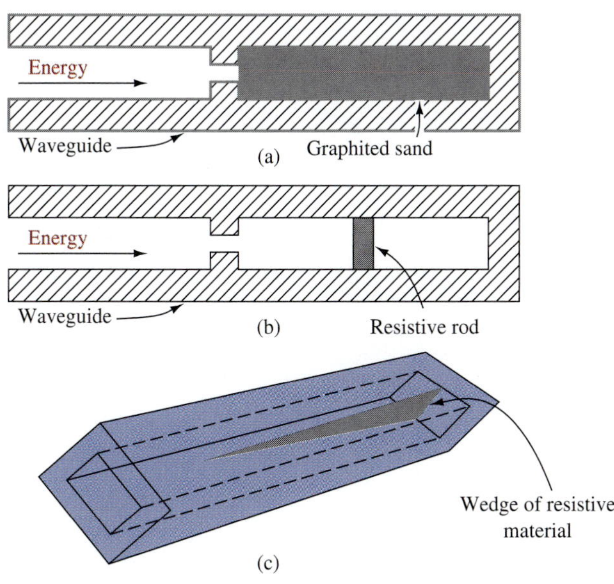

FIGURE 15-15 Termination for minimum reflections.

None of the energy dissipated as heat is reflected back into the guide. Another arrangement [Fig. 15-15(b)] uses a high-resistance rod, which is placed at the center of the E field. The E field (voltage) causes current to flow through the rod. The high resistance of the rod dissipates the energy as an I^2R loss.

Still another method for terminating a waveguide is to use a wedge of resistive material [Fig. 15-15(c)]. The plane of the wedge is placed perpendicular to the magnetic lines of force. When the H lines cut the wedge, a voltage is induced in it. The current produced by the induced voltage flowing through the high resistance of the wedge creates an I^2R loss. This loss is dissipated in the form of heat. This permits very little energy reaching the closed end to be reflected.

Each of the preceding terminations is designed to match the impedance of the guide to ensure a minimum of reflection. On the other hand, there are many instances where it is desirable for all the energy to be reflected from the end of the waveguide. The best way to accomplish this is to simply attach or weld a metal plate at the end of the waveguide.

Variable Attenuators

Variable attenuators find many uses at microwave frequencies. They are used to (1) isolate a source from reflections at its load to preclude frequency pulling; (2) adjust the signal level, as in one arm of a microwave bridge circuit; and (3) measure signal levels, as with a calibrated attenuator.

There are two versions of variable attenuators:

Vane
a thin card of resistive material used as a variable attenuator in a waveguide

1. *Flap attenuator* [Fig. 15-16(a)]: Attenuation is accomplished by insertion of a thin card of resistive material (often referred to as a **vane**) through a slot in

Card

(a) Flap attenuator

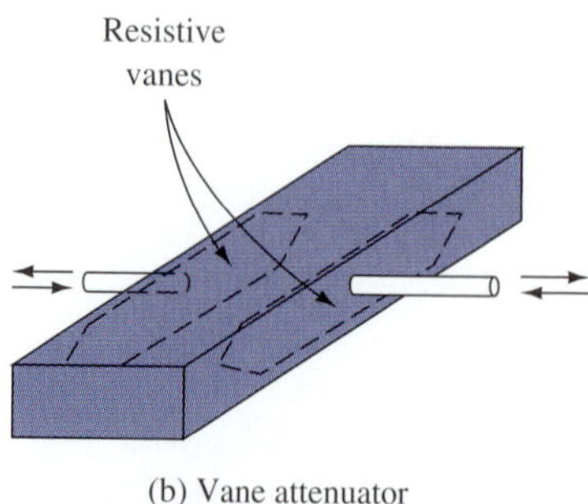

(b) Vane attenuator

FIGURE 15-16 Attenuators.

the top of the guide. The amount of insertion is variable, and the attenuation can be made approximately linear with insertion by proper shaping of the resistance card. Notice the tapered edges, which minimize unwanted reflections.

2. *Vane attenuator* [Fig. 15-16(b)]: In this type of attenuator the resistance card or vanes move in from the sides, as shown in the figure. You can see that the losses (and thus attenuation) are at a minimum when the vanes are close to the side walls where E is small, and maximum when the vanes are in the center.

15-7 Directional Coupler

The two-hole directional coupler consists of two pieces of waveguide with one side common to both guides and two holes in this common side. Its function is analogous to directional couplers used for transmission lines. The sections may be arranged physically either side by side or one over the other. The directional properties of such a device can be seen by looking at the wave paths labeled A, B, C, and D in Fig. 15-17.

Waves A and B follow equal-length paths and thus combine in phase in the secondary guide. If the spacing between holes is $\lambda_g/4$, waves C and D (which are of equal strength) are $\lambda_g/2$ or 180° out of phase and thus cancel. Therefore, if the field within the main guide consists of a superposition of incident and reflected waves, a certain fraction of the wave moving left to right will be coupled out through the secondary guide, and the same fraction of the right-to-left wave will be dissipated in the vane. The wave traveling from right to left causes this energy to be cancelled in the secondary guide's output due to the 180° phase shift caused by the $\lambda_g/2$ path difference of its two equal components through the two coupling holes. This type of coupler is frequency-sensitive because the spacing between holes must be $\lambda_g/4$ or an odd multiple thereof. The addition of more holes properly spaced can improve both the operable frequency range and directivity. This is known as a multihole coupler.

Thus, we see that a directional coupler transfers energy from a primary to an adjacent—otherwise independent—secondary waveguide for energy traveling in the main guide in one direction only. The energy that flows toward the left in the secondary guide is absorbed by the vane in Fig. 15-17, which is a matched load to prevent reflections.

The ratio of P_{out} and the incident power, P_{in}, is known as the *coupling*:

$$\text{coupling (dB)} = 10 \log \frac{P_{in}}{P_{out}} \tag{15-9}$$

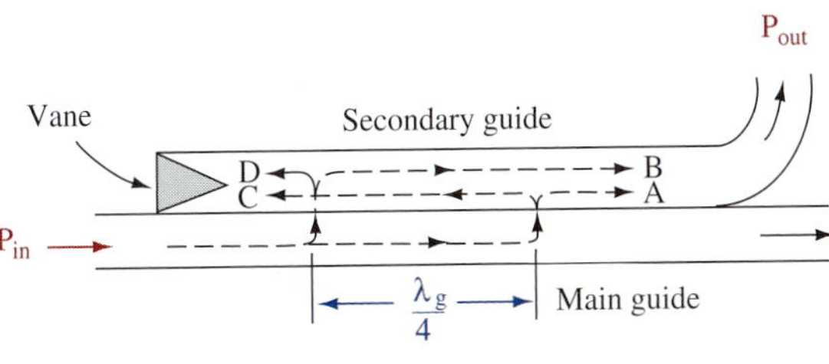

FIGURE 15-17 Two-hole directional coupler.

We now can understand that a directional coupler can distinguish between the waves traveling in opposite directions. It can be arranged to respond to either incident or reflected waves. By connecting a microwave power meter to the output of the secondary guide, a measure of power flow can be made. The coupling is normally less than 1 percent so that the power meter has negligible loading effect on the operation in the main guide. By physically reversing the directional coupler, a power flow in the opposite direction is determined, and the level of reflections and SWR can be determined.

15-8 Coupling Waveguide Energy and Cavity Resonators

We have described waveguide operation in terms of *E* and *M* fields, but how do we form these fields within the guide? In other words, how do we get energy into and out of a waveguide? Fundamentally, there are three methods of coupling energy into or out of a waveguide: *probe, loop,* and *aperture*. Probe, or capacitive, coupling is illustrated in Fig. 15-18. Its action is the same as that of a quarter-wave monopole antenna. When the probe is excited by an RF signal, an electric field is set up [Fig. 15-18(a)]. The probe should be located in the center of the *a* dimension and a quarter-wavelength, or odd multiple of a quarter-wavelength, from the short-circuited end, as illustrated in Fig. 15-18(b). This is a point of maximum *E* field and, therefore, is a point of maximum coupling between the probe and the field. Usually, the probe is fed with a short length of coaxial cable. The outer conductor is connected to the waveguide wall, and the probe extends into the guide but is insulated from it, as shown in Fig. 15-18(c). The degree of coupling may be varied by varying the length of the probe, removing it from the center of the *E* field, or shielding it.

In a pulse-modulated radar system, there are wide sidebands on either side of the carrier frequency. For a probe not to discriminate too sharply among frequencies

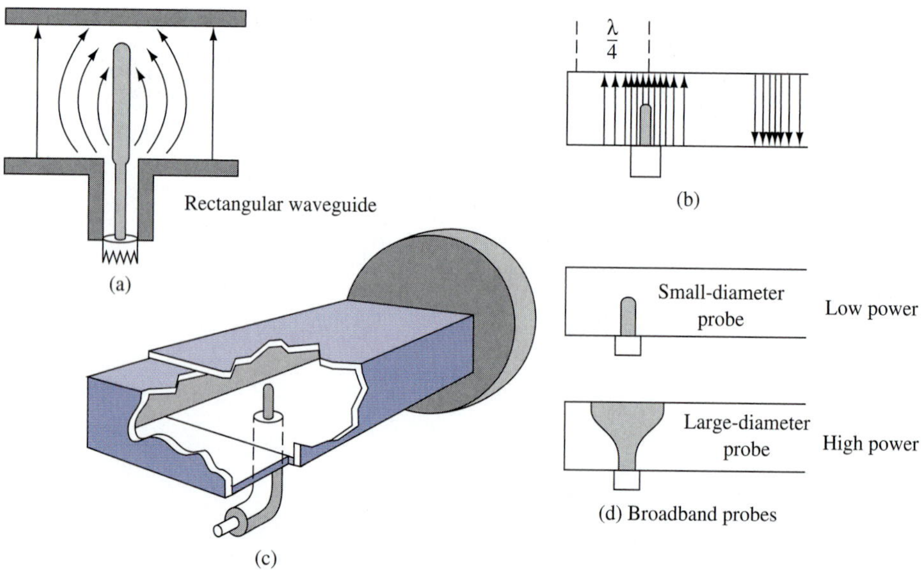

FIGURE 15-18 Probe, or capacitive, coupling.

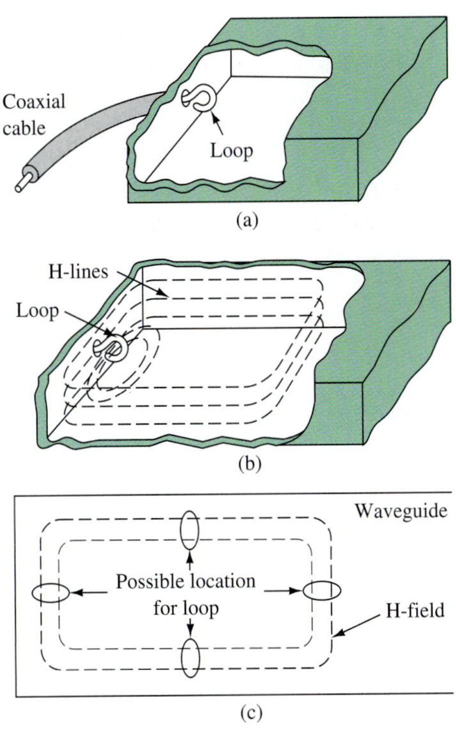

FIGURE 15-19 Loop, or inductive, coupling.

that differ from the carrier frequency, a wideband probe may be used. This type of probe is illustrated in Fig. 15-18(d) for both low- and high-power usage.

Figure 15-19 illustrates loop, or inductive, coupling. The loop is placed at a point of maximum H field in the guide. As shown in Fig. 15-19(a), the outer conductor is connected to the guide, and the inner conductor forms a loop inside the guide. The current flow in the loop sets up a magnetic field in the guide. This action is illustrated in Fig. 15-19(b). As shown in Fig. 15-19(c), the loop may be placed in several locations. The degree of loop coupling may be varied by rotation of the loop.

The third method of coupling is aperture, or slot, coupling. This type of coupling is shown in Fig. 15-20. Slot A is at an area of maximum E field and is a form of electric field coupling. Slot B is at an area of maximum H field and is a form of magnetic field coupling. Slot C is at an area of maximum E and H field and is a form of electromagnetic coupling.

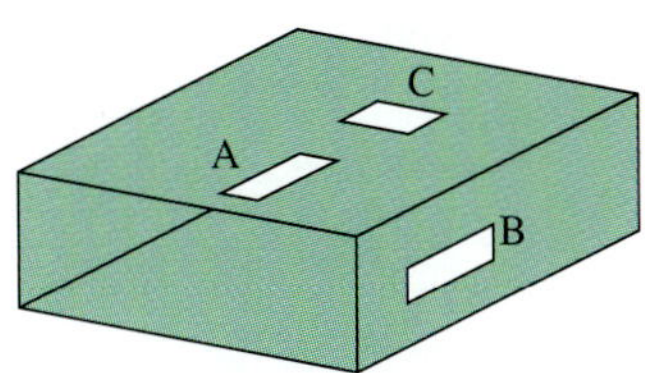

FIGURE 15-20 Aperture, or slot, coupling.

Waveguide to coaxial conversion. (Courtesy of Microwave Development Laboratories.)

Cavity Resonators

Circuits composed of lumped inductance and capacitance elements may be made to resonate at any frequency from less than 1 Hz to many thousand megahertz. At extremely high frequencies, however, the physical size of the inductors and capacitors becomes extremely small. Also, losses in the circuit become extremely great. Resonant devices of different construction are therefore preferred at extremely high frequencies. In the UHF range, sections of parallel wire or coaxial transmission line are commonly employed in place of lumped constant resonant circuits. In the microwave region, cavity resonators are used. Cavity resonators are metal-walled chambers fitted with devices for admitting and extracting electromagnetic energy. The Q of these devices may be much greater than that of conventional LC tank circuits.

Although cavity resonators, built for different frequency ranges and applications, have various physical forms, the basic principles of operation are essentially the same for all. Operating principles of cavity resonators are explained in this chapter. These principles are applied in Chapter 16 to the study of important microwave components employing cavity resonators.

Resonant cavity walls are made of highly conductive material and enclose a good dielectric, usually air. One example of a cavity resonator is the rectangular box shown in Fig. 15-21. It may be thought of as a section of rectangular waveguide closed at both ends by conducting plates. Because the end plates are short circuits for waves traveling in the Z direction, the cavity is analogous to a transmission line section with short circuits at both ends. Resonant modes occur at frequencies for which the distance between end plates is a half-wavelength or multiple of the half-wavelength.

Cavity modes are designated by the same numbering system that is used with waveguides, except that a third subscript is used to indicate the number of half-wave patterns of the transverse field along the axis of the cavity (perpendicular to the transverse field). The rectangular cavity is only one of many cavity devices useful as high-frequency resonators. By appropriate choice of cavity shape, advantages such as compactness, ease of tuning, simple mode spectrum, and high Q may be

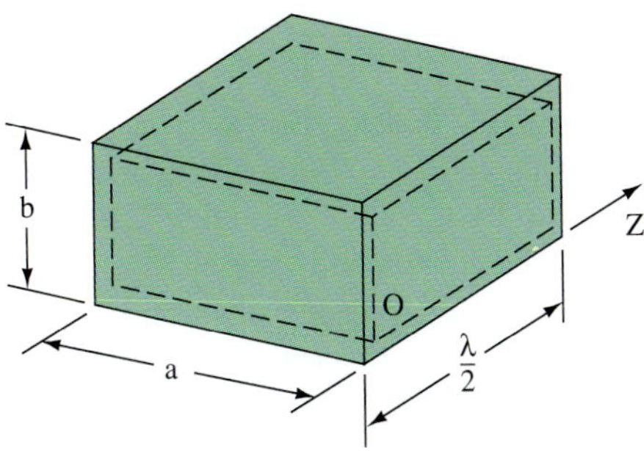

FIGURE 15-21 Rectangular waveguide resonator.

secured as required for special applications. Coupling energy to and from the cavity is accomplished just as for the standard waveguide, as shown in Fig. 15-19.

Cavity Tuning

The resonant frequency of a cavity may be varied by changing any of three parameters: *cavity volume, cavity inductance*, or *cavity capacitance.* Although the mechanical methods for tuning cavity resonators may vary, they all utilize the electrical principles explained below.

Figure 15-22 illustrates a method of tuning a cylindrical-type cavity by varying its volume. Varying the distance *d* results in a new resonant frequency. Increasing distance *d* lowers the resonant frequency, while decreasing *d* causes an increase in resonant frequency. The movement of the disk may be calibrated in terms of frequency. A micrometer scale is usually used to indicate the position of the disk, and a calibration chart is used to determine frequency.

A second method for tuning a cavity resonator is to insert a nonferrous metallic screw (such as brass) at a point of maximum *H* field. This decreases the permeability of the cavity and decreases its effective inductance, which raises its resonant frequency. The farther the screw penetrates into the cavity, the higher is the resonant frequency. A paddle can be used in place of the screw. Turning the paddle to a position more nearly perpendicular to the *H* field increases resonant frequency.

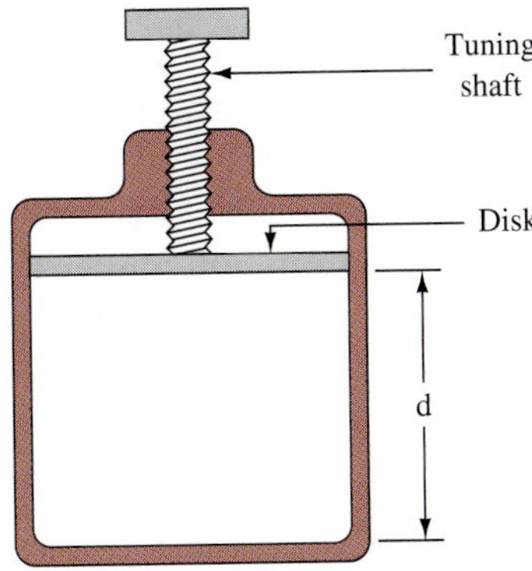

FIGURE 15-22 Cavity tuning by volume.

15-9 Radar

Radar
using radio waves to detect and locate objects by determining the distance and direction from the radar equipment to the object

Echo Signal
part of the returning radar energy collected by the antenna and sent to the receiver

The first practical use of waveguides occurred with the development of radar during World War II. The high powers and high frequencies involved in these systems are much more efficiently carried by waveguides than by transmission lines. The word **radar** is an acronym formed from the words *ra*dio *d*etection *a*nd *r*anging. Radar is a means of employing radio waves to detect and locate objects such as aircraft, ships, and land masses. Location of an object is accomplished by determining the distance and direction from the radar equipment to the object. The process of locating objects requires, in general, the measurement of three coordinates: range, angle of azimuth (horizontal direction), and angle of elevation.

A radar set consists fundamentally of a transmitter and a receiver. When the transmitted signal strikes an object (target), some of the energy is sent back as a reflected signal. The small-beamwidth transmit/receive antenna collects a portion of the returning energy (called the **echo signal**) and sends it to the receiver. The receiver detects and amplifies the echo signal, which is then used to determine object location.

Military use of radar includes surveillance and tracking of air, sea, land, and space targets from air, sea, land, and space platforms. It is also used for navigation, including aircraft terrain avoidance and terrain following. Many techniques and applications of radar developed for the military are now found in civilian equipment. These applications include weather observation, geological search techniques, and air traffic control units, to name just a few. All large ships at sea carry one or more radars for collision avoidance and navigation. Certain frequencies see better through rain; others resolve closely spaced targets better; still others are suited for long-range operation. Generally, the larger a radar antenna, the better the system's resolution. In space, radars are used for spacecraft rendezvous, docking, and landing, as well as for remote sensing of the earth's environment and planetary exploration.

Radar Waveform and Range Determination

Pulse Repetition Frequency (PRF)
the number of radar pulses (waveforms) transmitted per second

Pulse Repetition Rate (PRR)
the pulse repetition rate

Pulse Repetition Time (PRT)
the time from the beginning of one pulse to the beginning of the next

Rest Time
the time between pulses

Receiver Time
rest time

Radar Mile
unit of measurement equal to 2000 yd (6000 ft)

A representative radar pulse (waveform) is shown in Fig. 15-23. The number of these pulses transmitted per second is called the **pulse repetition frequency (PRF)** or **pulse repetition rate (PRR).** The time from the beginning of one pulse to the beginning of the next pulse is called the **pulse repetition time (PRT).** The PRT is the reciprocal of the PRF (PRT = 1/PRF). The duration of the pulse (the time the transmitter is radiating energy) is called the *pulse width* (PW). The time between pulses is called **rest time** or **receiver time.** The pulse width plus the rest time equals the PRT (PW + rest time = PRT). For radar to provide an accurate directional picture, a highly directive antenna is necessary. The desired directivity can be provided only by microwave antennas (see Chapter 16), and thus the RF energy shown in Fig. 15-23 is usually in the GHz (microwave) range.

The distance to the target (range) is determined by the time required for the pulse to travel to the target and return. The velocity of electromagnetic energy is 186,000 statute mi/s, or 162,000 nautical mi/s. (A nautical mile is the accepted unit of distance in radar and is equal to 6076 ft.) In many instances, however, measurement accuracy is secondary to convenience, and as a result a unit known as the **radar mile** is commonly used. A radar mile is equal to 2000 yd, or 6000 ft. The small difference between a radar mile and a nautical mile introduces an error of about 1 percent in range determination.

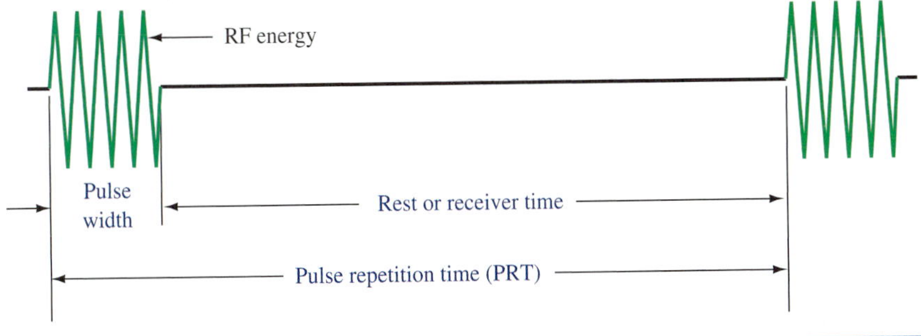

FIGURE 15-23 Radar pulses.

For purposes of calculating range, the two-way travel of the signal must be taken into account. It can be found that it takes approximately 6.18 μs for electromagnetic energy to travel 1 radar mile. Therefore, the time required for a pulse of energy to travel to a target and return is 12.36 μs/radar mile. The range, in miles, to a target may be calculated by the formula

$$\text{range} = \frac{\Delta t}{12.36} \tag{15-10}$$

where Δt is the time between transmission and reception of the signal in microseconds. For shorter ranges and greater accuracy, however, range is measured in meters.

$$\text{range (meters)} = \frac{c\Delta t}{2} \tag{15-11}$$

where c is the speed of light and Δt is in seconds.

Radar System Parameters

Once the pulse of electromagnetic energy is emitted by the radar, a sufficient length of time must elapse to allow any echo signals to return and be detected before the next pulse is transmitted. Therefore, the PRT of the radar is determined by the longest range at which targets are expected. If the PRT were too short (PRF too high), signals from some targets might arrive after the transmission of the next pulse. This could result in ambiguities in measuring range. Echoes that arrive after the transmission of the next pulse are called **second return echoes** (also *second time around* or *multiple time around echoes*). Such an echo would appear to be at a much shorter target range than actually exists and could be misleading if not identified as a second return echo. The range beyond which targets appear as second return echoes is called the *maximum unambiguous range.* Maximum unambiguous range may be calculated by the formula

Second Return Echoes echoes that arrive after the transmission of the next pulse

$$\text{maximum unambiguous range} = \frac{\text{PRT}}{12.2} \tag{15-12}$$

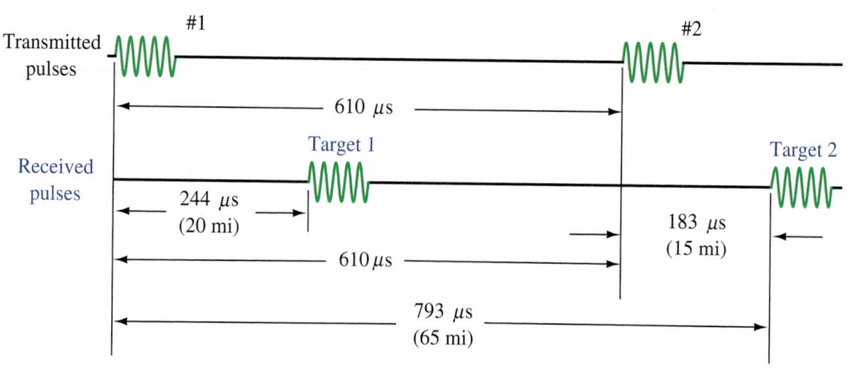

FIGURE 15-24 Second return echo.

where range is in miles and the PRT is in microseconds. Figure 15-24 illustrates the principles of the second return echo.

Figure 15-24 shows a signal with a PRT of 610 μs, which results in a maximum unambiguous range of 50 mi. Target number 1 is at a range of 20 mi. Its echo signal takes 244 μs to return. Target number 2 is actually 65 mi away, and its echo signal takes 793 μs to return. However, this is 183 μs after the next pulse was transmitted; therefore, target number 2 appears to be a weak target 15 mi away. Thus, the maximum unambiguous range is the **maximum usable range** and will be referred to from now on as simply *maximum range.* (It is assumed here that the radar has sufficient power and sensitivity to achieve this range.)

Maximum Usable Range
the maximum distance before second return echoes start occurring

If a target is so close to the transmitter that its echo is returned to the receiver before the transmitter is turned off, the reception of the echo will be masked by the transmitted pulse. In addition, almost all radars utilize an electronic device to block the receiver for the duration of the transmitted pulse. However, **double range echoes** are frequently detected when there is a large target close by. Such echoes are produced when the reflected beam is strong enough to make a second trip, as shown in Fig. 15-25. Double range echoes are weaker than the main echo and appear at twice the range.

Double Range Echoes
echoes produced when the reflected beam makes a second trip

Minimum range is measured in meters and may be calculated by the formula

$$\text{minimum range} = 150\ \text{PW} \tag{15-13}$$

where range is in meters and pulse width (PW) is in microseconds. Typical pulse widths range from fractions of a microsecond for short-range radars to several microseconds for high-power long-range radars.

A radar transmitter generates RF energy in the form of extremely short pulses with comparatively long intervals of rest time. The useful power of the transmitter is that contained in the radiated pulses and is termed the **peak power** of the system. Because the radar transmitter is resting for a time that is long with respect to the pulse time, the average power delivered during one cycle of operation is relatively low compared with the peak power available during the pulse time.

Peak Power
the useful power of the transmitter contained in the radiated pulses

The **duty cycle** of radar is

Duty Cycle
the ratio of pulse width to pulse repetition time

$$\text{duty cycle} = \frac{\text{pulse width}}{\text{pulse repetition time}} \tag{15-14}$$

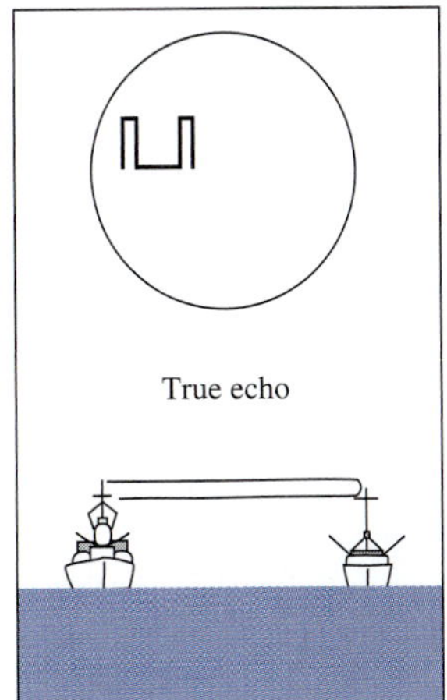

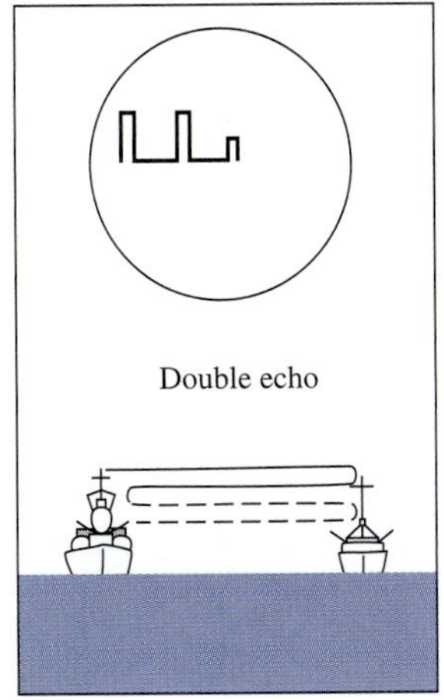

FIGURE 15-25 Double range echo.

For example, the duty cycle of a radar having a pulse width of 2 μs and a pulse repetition time of 2 ms is

$$\frac{2 \times 10^{-6}}{2 \times 10^{-3}} \text{ or } 0.001$$

Similarly, the ratio between the average power and peak power may be expressed in terms of the duty cycle. In a system with peak power of 200 kW, a PW of 2 μs, and a PRT of 2 ms, a peak power of 200 kW is supplied to the antenna for 2 μs, while for the remaining 1998 μs the transmitter output is zero. Because average power equals peak power times duty cycle, the average power equals $(2 \times 10^5) \times (1 \times 10^{-3})$, or 200 W.

High peak power is desirable for producing a strong echo over the maximum range of the equipment. Conversely, low average power enables the transmitter output circuit components to be made smaller and more compact. Thus, it is advantageous to have a low duty cycle. A short pulse width is also advantageous with respect to being able to "see" (resolve) closely spaced objects.

Basic Radar Block Diagram

A block diagram of a basic radar system is shown in Fig. 15-26. The pulse repetition frequency is controlled by the *timer* (also called *trigger generator* or *synchronizer*) in the modulator block. The pulse-forming circuits in the modulator are triggered by the timer and generate high-voltage pulses of rectangular shape and short duration. These pulses are used as the supply voltage for the transmitter and, in effect, turn it on and off. The modulator, therefore, determines the pulse width of the system. The transmitter generates the high-frequency, high-power RF carrier and determines the carrier frequency. The duplexer is an electronic switch that allows the use of a common antenna for both transmitting and receiving. It prevents the strong transmitted signal from being received by the sensitive receiver. The receiver section is basically a conventional superheterodyne receiver. In older radars, no RF amplifier is found because of noise problems with the RF amplifiers of the World War II era.

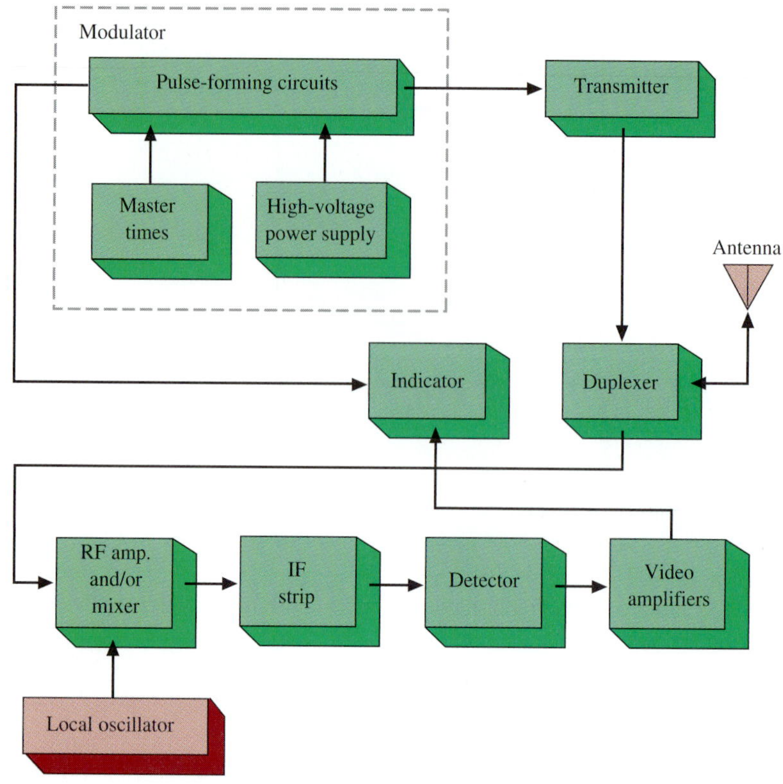

FIGURE 15-26 Radar system block diagram.

Doppler Effect

Doppler Effect phenomenon whereby the frequency of a reflected signal is shifted if there is a relative motion between the source and the reflecting object

The **Doppler effect** is the phenomenon whereby the frequency of a reflected signal is shifted if there is relative motion between the source and reflecting object. This is the same effect whereby the pitch of a train's whistle is shifted as the train moves toward and then away from the listener. Doppler radar, or CW radar, is always on. It is not turned off and on as pulsed radar is, hence the name, continuous wave. Only moving targets are "seen" by CW radar because only moving targets cause a Doppler shift. CW radars use two antennas, one each for transmitting and receiving.

The amount of frequency shift encountered is determined by the relative velocity between transmitter and target. It is predicted by

$$f_d = \frac{2v \cos \theta}{\lambda} \tag{15-15}$$

where f_d = frequency change between transmitted and reflected signal
v = relative velocity between radar and target
λ = wavelength of transmitted wave
θ = angle between target direction and radar system

If you have ever received a speeding ticket, in a radar trap, you now have a better understanding of your downfall.

15-10 MICROINTEGRATED CIRCUIT WAVEGUIDING

The field of communications now makes heavy use of the frequencies from 1 up to 300 GHz—we shall loosely refer to these as *microwave* frequencies. At microwave frequencies, even the shortest circuit connections must be carefully considered due to the extremely small wavelengths involved. In Chapter 16 we shall provide a general study of the microwave field.

The thin-film hybrid and monolithic integrated circuits used at microwave frequencies are called microwave integrated circuits (MICs). Obviously the use of short chunks of coaxial transmission line or waveguides is not practical for the required connections of mass-produced miniature circuits. Instead, either a **stripline** or **microstrip** connection is often used. They are shown in Fig. 15-27. They both lend themselves to mass-produced circuitry and can be thought of as a cross between waveguides and transmission lines with respect to their propagation characteristics.

Stripline transmission line used at microwave frequencies that has two ground planes "sandwiching" a conducting strip

Microstrip transmission line used at microwave frequencies that has one or two conducting strips over a ground plane

The stripline consists of two ground planes (conductors) that "sandwich" a smaller conducting strip with constant separation by a dielectric material (printed circuit board). The two types of microstrip shown in Fig. 15-27 consist of either one or two conducting strips separated from a single ground plane by a dielectric. One conducting strip is analogous to an unbalanced transmission line. The two-conducting-strip version is analogous to a balanced transmission line. While stripline offers somewhat better performance due to lower radiation losses, the simpler and thus more economical microstrip is the prevalent construction technique.

In either case, the losses exceed those of either waveguides or coaxial transmission lines, but the miniaturization and cost savings far outweigh the loss considerations. This is especially true when the very short connection paths are considered.

As with waveguides and transmission lines, the characteristic impedances of stripline and microstrip are determined by physical dimensions and the type of dielectric. The most often used dielectric is alumina, with a relative dielectric constant of 9.6. Proper impedance matching, to minimize standing waves, is still an important consideration.

Figure 15-28 provides end views of the three lines. The formulas for calculating Z_0 for them are provided in the figure. In the formulas, ln is the natural logarithm and ϵ is the dielectric constant of the board.

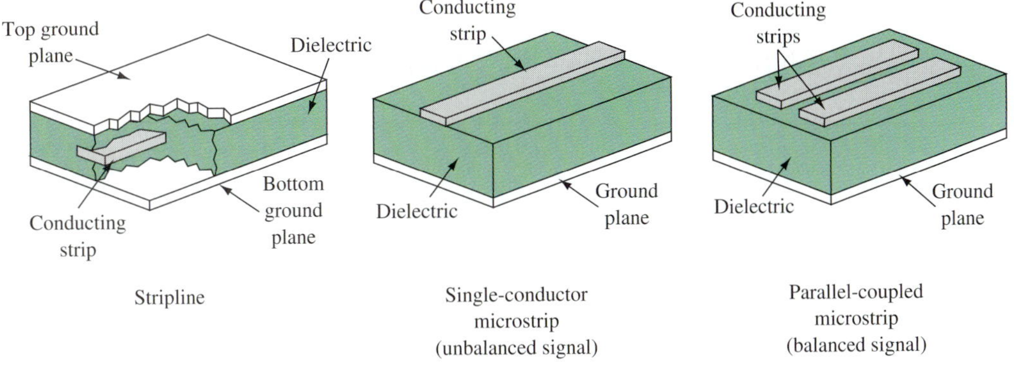

FIGURE 15-27 Stripline and microstrip.

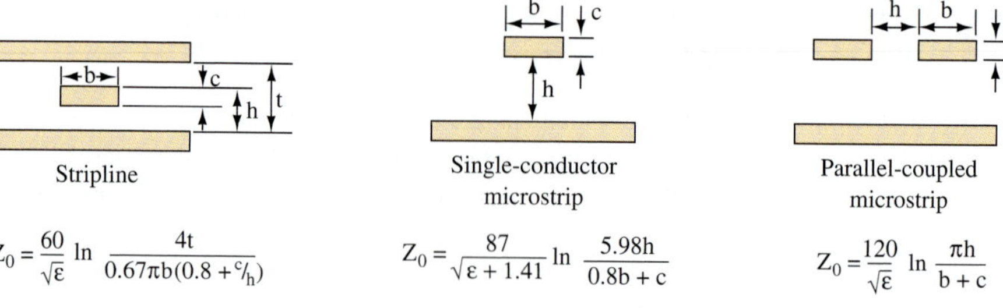

FIGURE 15-28 Characteristic impedance.

Microstrip Circuit Equivalents

Microstrip can be used to simulate circuit elements just as previously discussed for transmission lines and waveguides. The physical layout for some single-conductor microstrip simulations is shown in Fig. 15-29. The series capacitance in Fig. 15-29(c) shows that an actual break in the conductor is used. This concept can be extended to allow coupling between two microstrips by running the two conductors close together. The amount of coupling can be accurately controlled by the length and spacing of the parallel segments, as shown in Fig. 15-29(f). The configuration in Fig. 15-29(e) simulates a series *LC* circuit connected to ground. Almost any type of *LC* circuit can be fabricated with microstrip.

Dielectric Waveguide

Dielectric Waveguide
a waveguide with just a dielectric (no conductors) used to guide electromagnetic waves

A more recent contender for "wiring" of miniature millimeter wavelength circuits is the **dielectric waveguide.** Its operation depends on the principle that two dissimilar dielectrics have a guiding effect on electromagnetic waves.

The dielectric waveguide should not be confused with the dielectric-filled waveguide. Both are shown in Fig. 15-30. A regular metallic waveguide is sometimes

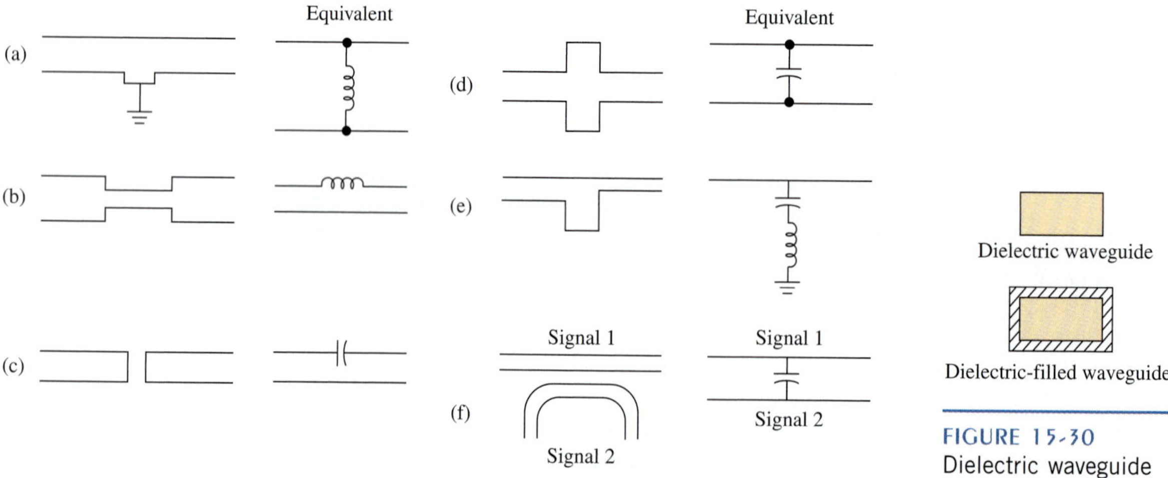

FIGURE 15-29 Microstrip circuit equivalents.

FIGURE 15-30 Dielectric waveguide and dielectric-filled waveguide.

filled with dielectric because it decreases the size necessary to allow propagation of a given frequency.

The dielectric waveguide is obviously easy to mass-produce within integrated circuits and offers an advantage over microstrip. At frequencies above 20 to 30 GHz, the losses with microstrip become excessive for many system applications compared to the dielectric waveguide. For example, at 60 GHz microstrip typically attenuates 0.15 dB/cm, while the dielectric waveguide attenuates only 0.06 dB/cm. The figure for a standard rectangular waveguide at 60 GHz is about 0.02 dB/cm but would be used only in systems where cost is not a factor.

Alumina is commonly used as the dielectric material for dielectric waveguides. However, semiconductors such as silicon and gallanium arsenide (GaAs) will undoubtedly be the dielectrics used in the future. This is dictated by the fact that ultimately semiconductor devices will be fabricated directly into the dielectric waveguide.

15-11 Troubleshooting

After completing this section, you should be able to troubleshoot waveguide systems. Waveguide problems are very similar to ordinary transmission line problems. The test equipment may look different, but it is doing the same things.

A word of caution: Waveguide is commonly used to carry large amounts of microwave power. Microwaves are capable of burning skin and damaging eyesight. Never work on waveguide runs or antennas connected to a transmitter or radar until you are sure the system is off and cannot be turned on by another person.

After completing this section you should be able to

- Identify problems caused by joints and flanges
- Detect arcing problems
- Troubleshoot rotary joint failures
- Detect malfunctions by determining VSWR in the guide

Some Common Problems

1. The joints or flanges between two waveguide sections are the most likely source of a problem. Waveguides are sometimes pressurized to increase the power rating of the guide and keep water out. Improperly fitted joints can let water in and gas out. Water raises the VSWR, which can damage most microwave tubes.

   There are two types of flanges: choke and cover. Choke flanges have a groove cut into the face of the flange to keep microwave energy from escaping. There is a second groove for a gasket. Cover flanges are simply smooth. Both must be clean and flat. The screws must be the correct size because they help align the two pieces and keep them tightly sealed together.
2. Arcs can occur at improperly fitted joints and actually burn holes in the guide. Arcs occur in the waveguide under high power if some component such as the antenna has failed. You might see evidence of an arc on the broad wall right in the center of the guide.
3. Worn-out components must be checked and replaced. Radar antennas generally have one or more rotary joints. Rotary joints have bearings and sometimes moving contacts. Rotary joints sometimes fail only after the transmitter has

had time to heat them sufficiently. By the time the technician has opened the guide and installed test equipment, the joint will cool and test well. It is best to have a directional coupler in line while running the transmitter and watch for an increase in reflected power.

Rotary joints can also be tested on the bench by connecting the joint to a dummy load and measuring VSWR while turning the joint. Still, there is no substitute for the operational test.

Flexible waveguide is subject to cracking and corrosion. Normally, the loss of a two-foot-long piece of rigid waveguide is so low that the loss of the guide is extremely difficult to measure. A bad piece of flexible guide can usually be detected by connecting a dummy load to it and measuring VSWR while flexing the guide.

Test Equipment

Figure 15-31 shows how to connect the test equipment for a VSWR test. First connect the power meter to the forward (incident) power coupler and note the reading. Then connect the power meter to the reflected coupler and note the reading.

Reflected power should be very low, and the VSWR should be nearly 1. VSWR is given by the following equation (from Chapter 14):

$$\text{VSWR} = \frac{1 + \sqrt{\dfrac{P_r}{P_i}}}{1 - \sqrt{\dfrac{P_r}{P_i}}} \tag{15-7}$$

where P_i = incident power
P_r = reflected power

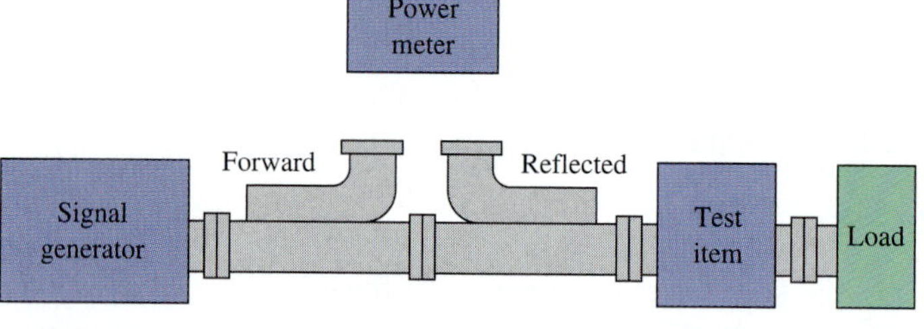

FIGURE 15-31 VSWR test.

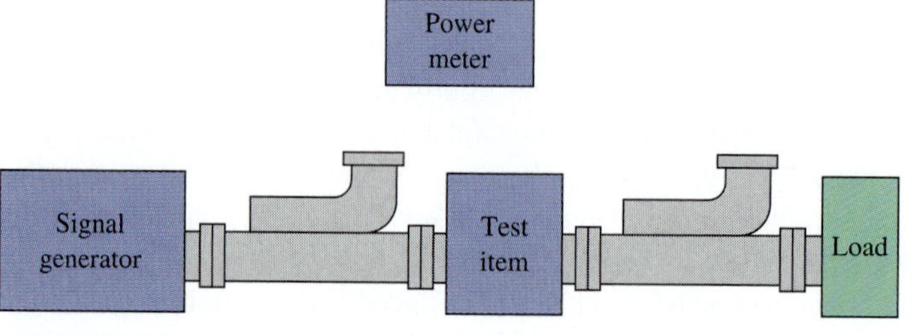

FIGURE 15-32 Loss test.

The same test equipment is used to measure loss by reversing the reflected coupler and putting the test item between the couplers, as shown in Fig. 15-32. No two couplers are exactly alike, so you must first connect them together without the test item and determine the difference. You should be able to make the loss measurement to within 0.2 dB is this manner.

15-12 Troubleshooting with Electronics Workbench™ Multisim

This Multisim exercise explores the properties of a lossy transmission line and a low-loss waveguide. Begin the exercise by opening **Fig15-33.ms7 (.msm)** in your EWB Multisim CD. This circuit contains a sample waveguide attached to the network analyzer. The circuit is shown in Fig. 15-33. Both ends of the waveguide are connected to the ports of the network analyzer. What results do you expect to see from the network analyzer?

Before starting the simulation, click on the network analyzer and change the number of points per decade to 200, the start frequency to 1 GHz, and the stop frequency to 2 GHz. This change provides a smoother plot of the simulation results and a realistic frequency range. Start the simulation and view the results on the network analyzer. You should see a result similar to the one shown in Fig. 15-34.

The plot of the data is along the outside perimeter of the Smith chart, as it is in Fig. 15-34. This indicates that the line is low-loss. Move the frequency marker on the network analyzer to 1 GHz. The input impedance of the waveguide at 1 GHz

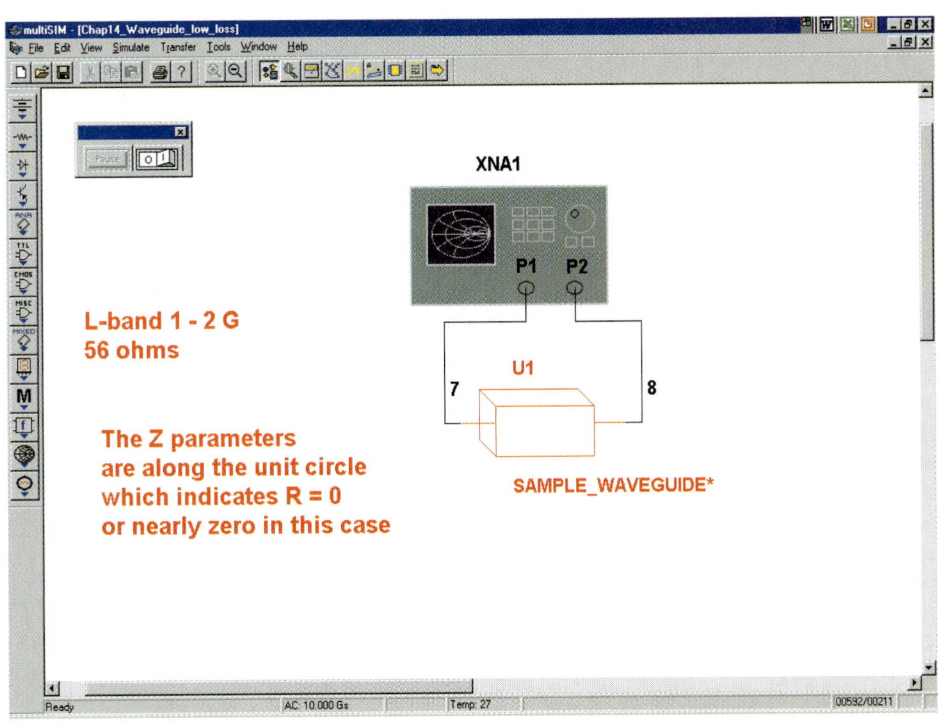

FIGURE 15-33 The circuit example of a low-loss waveguide section connected to a network analyzer.

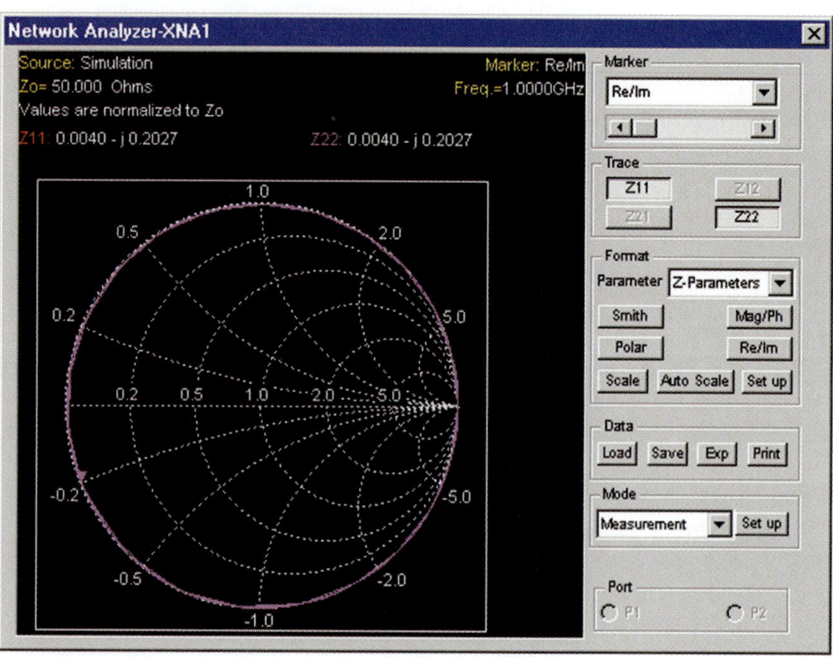

FIGURE 15-34 The simulation of a low-loss waveguide as viewed with the network analyzer.

is $Z_{11} = 0.0040 - j0.2027$ or very little resistive loss. Double-click on the waveguide section and then click on **Edit Model.** You should see an $R = 5.543\text{e-}001$, which is telling us that this waveguide has about 0.55-Ω of resistance per meter. Look at the **LEN** value in the model, which is the specification for the length (in meters) of the section we are analyzing. For this example, the length is 0.012 m **(LEN = 1.200e-002).** Change the value of LEN to 1.200. Click on **Change Part Model** and then click on **OK.** Restart the simulation and compare this result with the previous example. You may need to change the setup on the network analyzer to sweep from 1 to 2 GHz and change the number of data points to 800. The result of the simulation is shown in Fig. 15-35.

There is an obvious difference in Figs. 15-34 and 15-35. Fig. 15-35 is showing a very lossy waveguide. Electronics Workbench™ Multisim also provides a model of a lossy transmission line. Open the file **Fig15-36.ms7 (.msm).** The new circuit with the lossy transmission line model is shown in Fig. 15-36.

Change the setup on the network analyzer to display 200 points per decade and start the simulation. The network analyzer should show a result similar to that shown in Fig. 15-37. This result does not show as lossy a transmission line as the one shown in Fig. 15-35. However, this result is not as good as the one shown in Fig. 15-34.

This material has demonstrated how to identify a lossy or low-loss waveguide or transmission line using the Multisim network analyzer. The Multisim exercises in this chapter provide the opportunity to test your ability to identify these characteristics in waveguide or transmission lines. Refer back to the figures in this example as needed to confirm your observations.

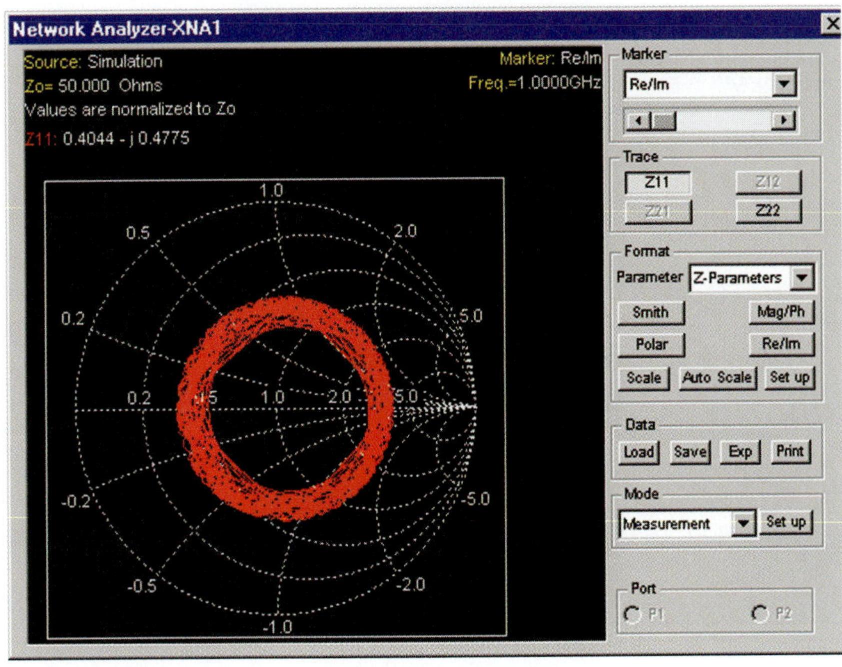

FIGURE 15-35 The simulation of a very lossy waveguide.

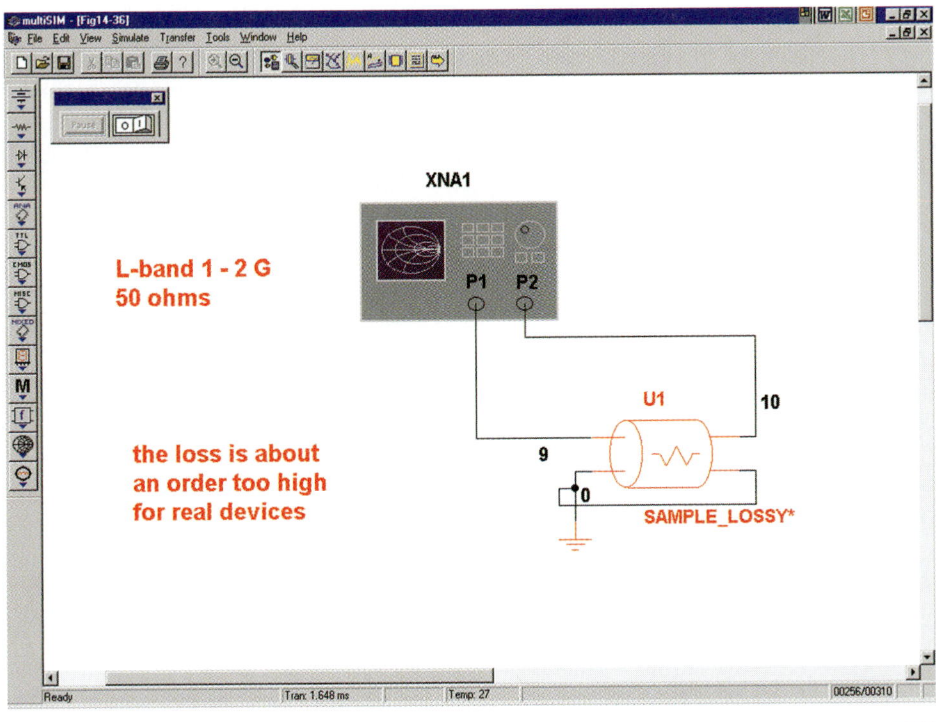

FIGURE 15-36 An example of a test on the Multisim sample lossy transmission line.

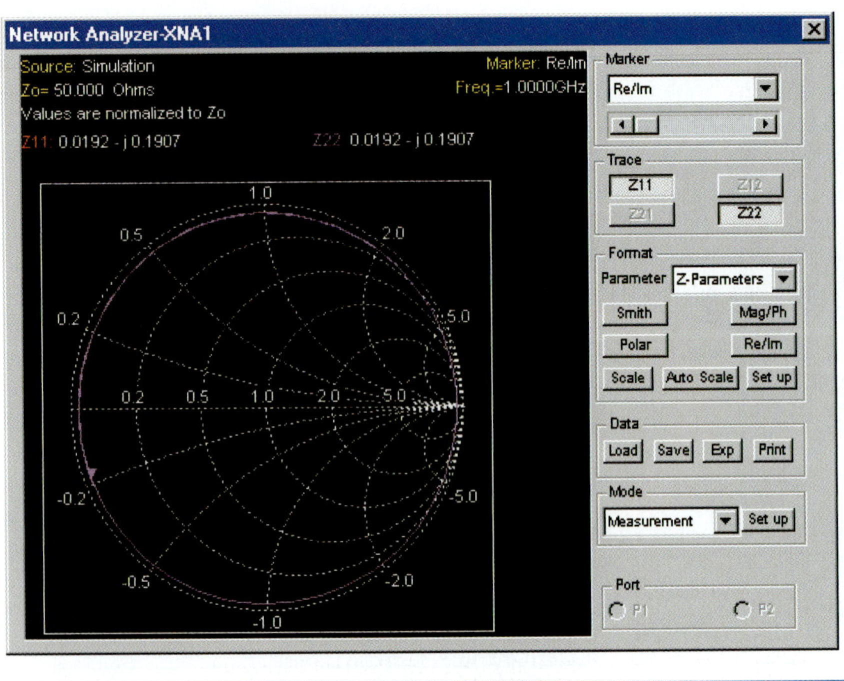

FIGURE 15-37 The simulation results of the lossy transmission line provided by EWB Multisim.

Electronics Workbench™ Exercises

1. Open the file **FigE15-1.ms7 (.msm)** in your EWB CD. Three waveguides are shown. Use the techniques described in the text to determine which waveguide has the least loss. (B)
2. Open the file **FigE15-2.ms7 (.msm)** in your EWB CD. Use the Bode plotter to determine the first resonant frequency of the sample waveguide. *Hint:* The first resonant frequency is at the point where the signal reaches a maximum level. Run the Bode plot and view the results. You will see the first maximum point. (1.48 GHz)
3. Open the file **FigE15-3.ms7 (.msm)** in your EWB CD. Determine the normalized impedance of the waveguide at 1.122 GHz and at 226.4 MHz. What would you conclude about the waveguide at these frequencies? (0.0100 − j0.0081, 0.0728 − j3.4652)

Summary

In Chapter 15 we studied waveguides and radar. We discovered that waveguides can be derived from transmission line analysis and are capable of handling large amounts of power. The major topics you should now understand include:

- the comparison of transmission via waveguides, antennas, and transmission lines
- the analysis of waveguide operation and mode designations

- the description and use of rectangular, circular, ridged, and flexible waveguides
- the application of waveguide bends, twists, tees, and tuners
- the calculation of characteristic impedance for waveguides
- the techniques of waveguide terminations and variable attenuators
- the description and application of directional couplers
- the analysis of coupling waveguide energy using probes, loops, or apertures
- the description and applications for cavity resonators
- the calculation of radar range
- the operational description of radar parameters' maximum range, minimum range, duty cycle, and Doppler effect
- the construction and application of stripline, microstrip, and dielectric waveguides

Questions and Problems

Section 15-1

1. Discuss the relative merits and drawbacks of using antennas, waveguides, and transmission lines as the media for a communications link.

Section 15-2

2. Provide a broad definition of a waveguide. What is normally meant by the term *waveguide*?
3. Explain the basic difference between propagation in a waveguide versus a transmission line.
4. Explain why the different mode configurations are termed *transverse electric* or *transverse magnetic*.
5. What are the *modes* of operation for a waveguide? Explain the subscript notation for TE and TM modes.
6. What is the *dominant mode* in rectangular waveguides? What property does it have that makes it dominant? Show a sketch of the electric field at the mouth of a rectangular waveguide carrying this mode.
7. Describe the significance of the cutoff wavelength.
8. A rectangular waveguide is 1 cm by 2 cm. Calculate its cutoff frequency, f_{co}. (7.5 GHz)
9. How does energy propagate down a waveguide? Explain what determines the angle this energy makes with respect to the guide's sidewalls.
10. For TE_{10}, $a = \lambda/2$. What is a for TE_{20}? (Assume a rectangular waveguide.)

Section 15-3

11. Why is the velocity of energy propagation usually significantly less in a waveguide than in free space? Calculate this velocity (V_g) for an X-band waveguide for a 10-GHz signal. Calculate guide wavelength (λ_g) and phase velocity (V_p) for these conditions. (2.26×10^8 m/s, 3.98 cm, 3.98×10^8 m/s)
12. Why are free-space wavelength (λ) and guide wavelength (λ_g) different? Explain the significance of this difference with respect to Smith chart calculations.

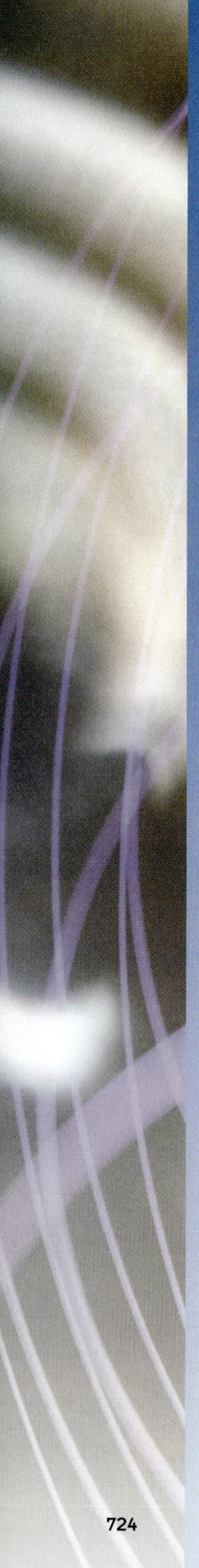

* 13. Why are rectangular cross-sectional waveguides generally used in preference to circular cross-sectional waveguides?

Section 15-4

14. Why are circular waveguides used much less than rectangular ones? Explain the application of a circular rotating joint.
15. Describe the advantages and disadvantages of a ridged waveguide.
16. Describe the physical construction of a flexible waveguide, and list some of its applications.
17. Describe some advantages a ridged waveguide has over a rectangular waveguide.

Section 15-5

18. List some of the causes of waveguide attenuation. Explain their much greater power handling capability as compared to coaxial cable of similar size.

* 19. Describe briefly the construction and purpose of a waveguide. What precautions should be taken in the installation and maintenance of a waveguide to ensure proper operation?

20. Why are waveguide bend and twist sections constructed to alter the direction of propagated energy gradually?
21. Describe the characteristics of shunt and series tee sections. Explain the operation of a hybrid tee when it is used as a TR switch.
22. Discuss several types of waveguide tuners in terms of function and application.

Section 15-6

23. Verify the characteristic wave impedance of 405 Ω for the data given in Problem 59.
24. Calculate the characteristic wave impedance for an X-band waveguide operating at 8, 10, and 12 GHz. (663 Ω, 501 Ω, 450 Ω)
25. Explain various ways of terminating a waveguide to minimize and maximize reflections.
26. Describe the action of flap and vane attenuators.

Section 15-7

27. Describe in detail the operation of a directional coupler. Include a sketch with your description. What are some applications for a directional coupler? Define the *coupling* of a directional coupler.
28. Calculate the coupling of a directional coupler that has 70 mW into the main guide and 0.35 mW out the secondary guide. (23 dB)

---

*An asterisk preceding a number indicates a question that has been provided by the FCC as a study aid for licensing examinations.

SECTION 15-8

29. Explain the basics of capacitively coupling energy into a waveguide.
30. Explain the basics of inductively coupling energy into a waveguide.
31. What is slot coupling? Describe the effect of varying the position of the slot.
* 32. Discuss the following with respect to waveguides:
   (a) Relationship between frequency and size.
   (b) Modes of operation.
   (c) Coupling of energy into the waveguide.
   (d) General principles of operation.
33. What is a cavity resonator? In what ways is it similar to an *LC* tank circuit? How is it dissimilar?
* 34. Explain the operating principles of a cavity resonator.
* 35. What are waveguides?
36. Describe a means whereby a cavity resonator can be used as a waveguide frequency meter.
37. Explain three methods of tuning a cavity resonator.

SECTION 15-9

* 38. Explain briefly the principle of operation for a radar system.
* 39. Why are waveguides used in preference to coaxial lines for the transmission of microwave energy in radar installations?
40. With respect to a radar system, explain the following terms:
   (a) Target.
   (b) Echo.
   (c) Pulse repetition rate.
   (d) Pulse repetition time.
   (e) Pulse width.
   (f) Rest time.
   (g) Range.
41. Calculate the range in miles and meters for a target when Δt is found to be 167 μs. (13.5 mi, 25,050 m)
* 42. What is the distance in nautical miles to a target if it takes 123 μs for a radar pulse to travel from the radar antenna to the target and back to the antenna, and be displayed on the PPI scope? (10 mi)
43. What are double range echoes?
44. Why does a radar system have a minimum range? Calculate the minimum range for a system with a pulse width of 0.5 μs.
45. In detail, discuss the various implications of duty cycle for a radar system.
* 46. What is the peak power of a radar pulse if the pulse width is 1 μs, the pulse repetition rate is 900, and the average power is 18 W? What is the duty cycle? (20 kW, 0.09%)
47. For the radar block diagram in Fig. 15-26, explain the function of each section.
48. A police radar speed trap functions at a frequency of 1.024 GHz in direct line with your car. The reflected energy from your car is shifted 275 Hz in frequency. Calculate your speed in miles per hour. Are you going to get a ticket? (90 mph, *yes!*)
49. What is the Doppler effect? What are some other possible uses for it other than police speed traps?
50. Why is a Doppler radar often called a CW radar?

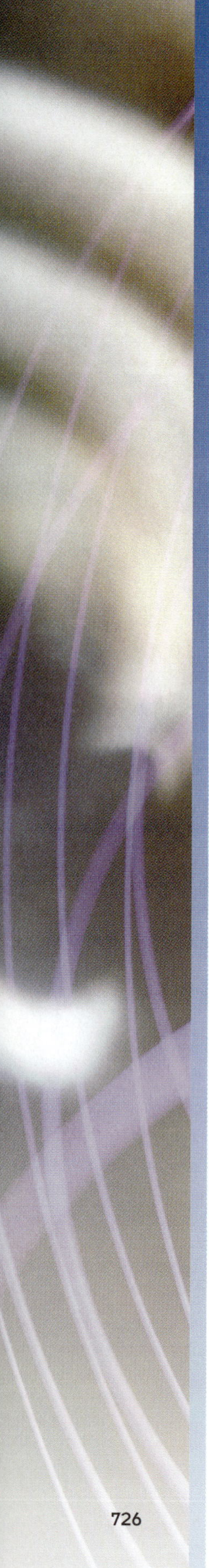

Section 15-10

51. Using sketches, explain the physical construction of stripline, single-conductor microstrip, and parallel-coupled microstrip. Discuss their relative merits, and also compare them to transmission lines and waveguides.
52. What is a dielectric waveguide? Discuss its advantages and disadvantages with respect to regular waveguides.
53. Calculate Z_0 for stripline constructed using a circuit board with a dielectric constant of 2.1, $b = 0.1$ in., $c = 0.006$ in., and $h = 0.08$ in. The conductor is spaced equally from the top and bottom ground planes. (50 Ω)
54. Make a sketch of a single-conductor microstrip that simulates an inductor to ground followed by a series capacitance.

Section 15-11

55. Describe the proper procedure for troubleshooting waveguides.
56. Explain how to prevent arcs.
57. Explain where problems are most likely to occur with waveguides and describe a process to prevent these problems.
58. Describe how to test a waveguide.

Questions for Critical Thinking

59. A 9-GHz signal is operating in the dominant mode in a rectangular waveguide 3 by 4.5 cm. The characteristic (wave) impedance is 405 Ω. Provide a report that includes the λ_g, λ, V_g, and V_p for this system; the SWR that would be caused by a horn antenna load of 350 Ω + j100 Ω; and the impedance in the guide 4 cm from the antenna load. Include a Smith chart analysis with the report.
60. Can a hybrid tee be used to feed the first stage of a microwave receiver (the mixer—no RF stage) with the antenna signal and local oscillator signal without any local oscillator radiation off the receiving antenna? Provide a sketch to illustrate.
61. Analyze the relationship between multiple targets and *maximum range.* Use the calculated maximum unambiguous range for a radar system with PRT equal to 400 μs to illustrate.
62. You must use a directional coupler to measure VSWR and to determine the loss introduced by a device in a waveguide system. Describe the tests you will use and how they differ from each other.

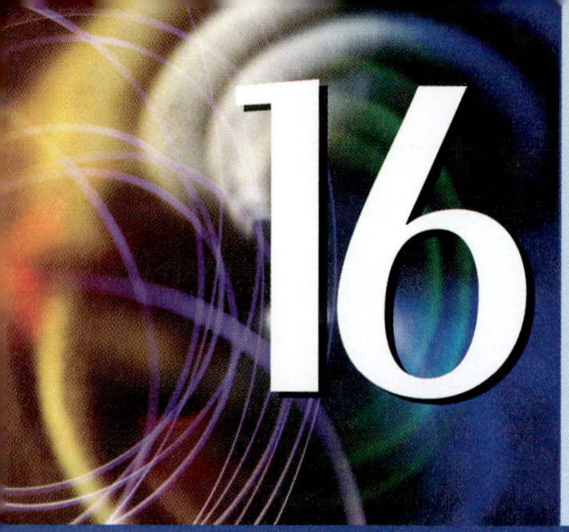

16 Microwaves and Lasers

Chapter Outline

A microwave communications tower.
(Courtesy of Tony Stone. Photo by Peter Poulides.)

Objectives

- Explain the operation of common microwave antennas
- Calculate the gain and beamwidth for a parabolic antenna
- Describe the operation of TWT and magnetron microwave tubes
- Describe the operation of some common solid-state microwave devices, including the Gunn oscillator, IMPATT diode, *p-i-n* diode, and microwave transistors and ICs
- Explain the basic operation of ferrites and some applications
- Describe the operation of the two extremely low-noise microwave amplifiers—the parametric and maser amplifiers
- Explain the operation of a laser and some of its applications

Key Terms

millimeter waves
circular horn
pyramidal horn
sectoral horn
microwave dish
Cassegrain feed
polar pattern
radome
zoning
patch antenna
interaction space
critical value
backward-wave oscillator
velocity modulation
replacement energy
IMPATT diode
p-i-n diodes
precession
Faraday rotation effect
ferrite bead
parametric amplifier
maser
quantum
laser
heterojunction
transphasor

16-1 MICROWAVE ANTENNAS

The antennas studied in Chapter 14 bear little resemblance to those used for microwave frequencies (>1 GHz). Microwave antennas actually use optical theory more than standard antenna theory. These antennas tend to be highly directive and therefore provide high gain compared to the reference half-wavelength dipole. The reasons for this include the following:

1. Because of the short wavelengths involved, the physical sizes required are small enough to allow "peculiar" arrangements not practical at lower frequencies.
2. There is little need for omnidirectional patterns because no broadcasting takes place at these frequencies. Microwave communications are generally of a point-to-point nature. The exception is telemetry applications.
3. Because of increased device noise at microwave frequencies, receivers require the highest possible input signal. Highly directional antennas (and thus high gain) make this possible.
4. Microwave transmitters are limited in their output power due to the cost and/or availability problems of microwave power devices. This low output power is compensated for by a highly directional antenna system.

Microwaves are divided into bands as shown in Table 16-1. The frequencies above 40 GHz are called **millimeter** (mm) **waves** because their wavelength is described in millimeters.

Millimeter Waves microwave frequencies above 40 GHz; wavelength is often expressed in millimeters

Table 16-1 MICROWAVE FREQUENCY DESIGNATIONS

Band	Frequency(GHz)
L	1–2
S	2–4
C	4–8
X	8–12
Ku	12–18
K	18–27
Ka	27–40

HORN ANTENNA

Open-ended sections of waveguides can be used as radiators of electromagnetic energy. The three basic forms of horn antennas are shown in Fig. 16-1. They all provide a gradual flare to the waveguide to allow maximum radiation and thus minimum reflection back into the guide.

Recall that a plain open-circuit waveguide theoretically reflects 100 percent of the incident energy. In practice, however, the open-circuit guide "launches" a fair amount of energy, while the short-circuit guide does provide the theoretical 100 percent reflection. By gradually flaring out the open circuit, the goal of total radiation is nearly attained. The flared end of the horn antenna acts as an impedance transformer between the waveguide and free space. For a proper transformation ratio, the linear dimension of each side must be at least a half-wavelength.

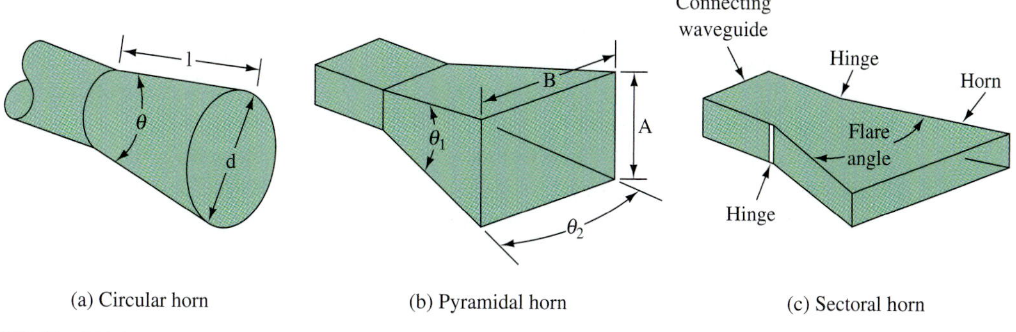

FIGURE 16-1 Horn antennas.

The **circular horn** in Fig. 16-1(a) provides efficient radiation from a circular waveguide. The flare angle θ and length l are important to the amount of gain it can provide. Generally, the greater the diameter d, the greater the gain.

Circular Horn
type of horn antenna that provides radiation from a circular waveguide

For the **pyramidal horn** in Fig. 16-1(b), the radiation pattern depends on the area of the aperture. The effect of horn length is similar to that with the circular horn. Wider horizontal patterns are obtained by increasing θ_2, while wider vertical

Pyramidal Horn
type of horn antenna with two flare angles

Standard Waveguide Horn Antennas
The SWH Series operates from 750 MHz to 40 GHz in contiguous waveguide bands. The VSWR is <1.5:1. All horns have removable waveguide-to-coaxial adapters and can be mounted for both vertical and horizontal polarity.

Horn antennas for 0.75 to 40 GHz. (Courtesy of Antenna Research Associates.)

patterns are possible by increasing θ_1. In Fig. 16-1(b), when the ratio $B/A = 1.35$, a symmetrical radiation pattern is realized.

Sectoral Horn type of horn antenna with top and bottom walls at a 0° flare angle

The **sectoral horn** in Fig. 16-1(c) has the top and bottom walls at a 0° flare angle. The side walls are sometimes hinged (as shown) to provide adjustable flare angles. Maximum radiation occurs for angles between 40° and 60°.

The horns just described provide a maximum gain on the order of 20 dB compared to the half-wavelength dipole reference. While they do not provide the amounts of gain of subsequently described microwave antennas, their simplicity and low cost make them popular for noncritical applications.

Parabolic Antenna

The ability of a paraboloid to focus light rays or sound waves at a point is common knowledge. Some common applications include dentists' lights, flashlights, and automobile headlamps. The same ability is applicable to electromagnetic waves of lower frequency than light as long as the paraboloid's mouth diameter is at least 10 wavelengths. This precludes their use at low radio frequencies but allows use at microwave frequencies.

Microwave Dish paraboloid antenna

Cassegrain Feed a method of feeding a paraboloid antenna by using a secondary reflector

There are various methods of feeding the **microwave dish,** as the paraboloid antenna is commonly called. Figure 16-2(a) shows the dish being fed with a simple dipole/reflector combination at the paraboloid's focus. A horn-fed version is shown in Fig. 16-2(b). The **Cassegrain feed** in Fig. 16-2(c) is used to shorten the length of feed mechanism in highly critical applications. It uses a hyperboloid

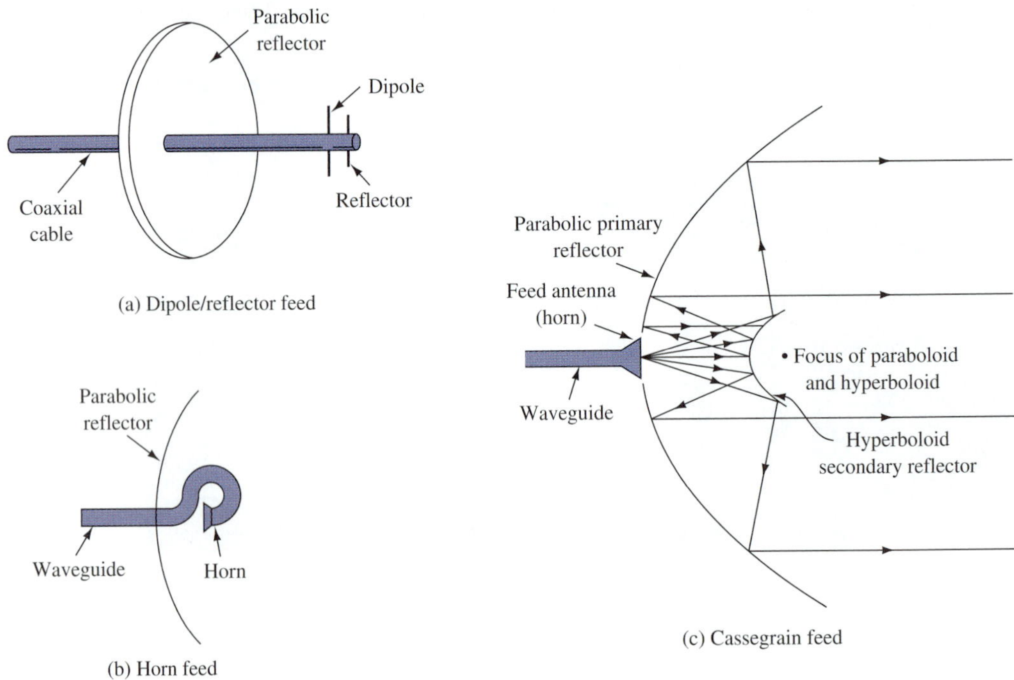

FIGURE 16-2 Microwave dish antennas.

L-Band dish antenna for 1.530 to 1.6605 GHz operation. (Courtesy of Seavey Engineering Associates, Inc.)

secondary reflector whose focus coincides with that of the parabola. Those transmitted rays obstructed by the hyperboloid are generally such a small percentage as to be negligible.

These dish antennas perform equally well transmitting or receiving, as predicted by antenna reciprocity. They provide huge power gains, with a close approximation provided by the following equation:

$$A_p \simeq 6\left(\frac{D}{\lambda}\right)^2 \tag{16-1}$$

where A_p = power gain with respect to a half-wavelength dipole
D = mouth diameter of primary reflector
λ = free-space wavelength of carrier frequency

An accurate approximation of the beamwidth in degrees between half-power points is

$$\text{beamwidth} \simeq \frac{70\lambda}{D} \tag{16-2}$$

EXAMPLE 16-1

Calculate the power gain and beamwidth of a microwave dish antenna with a 3-m mouth diameter when the antenna is used at 10 GHz.

Solution

$$A_p \simeq 6\left(\frac{D}{\lambda}\right)^2 \qquad \textbf{(16-1)}$$

$$\lambda = \frac{c}{f} = \frac{3 \times 10^8 \text{ m/s}}{10 \times 10^9} = 0.03 \text{ m}$$

$$A_p = 6 \times \left(\frac{3 \text{ m}}{0.03 \text{ m}}\right)^2 = 60{,}000 \quad (47.8 \text{ dB})$$

$$\text{beamwidth} \simeq \frac{70\lambda}{D} \qquad \textbf{(16-2)}$$

$$= \frac{70 \times 0.03 \text{ m}}{3 \text{ m}} = 0.7^\circ$$

Polar Pattern
a circular graph that indicates the direction of antenna radiation

Radome
a low-loss dielectric material used as a cover over a microwave antenna

Example 16-1 shows the extremely high gain capabilities of these antennas. In this particular case, the dish with a 1-W output is equivalent to a half-wave dipole with 60,000-W output. This power gain is effective, however, only if the receiver is within the 0.7° beamwidth of the dish. Figure 16-3 shows a **polar pattern** for this antenna. It is typical of parabolic antennas and shows the 47.8-dB gain at the 0° reference. Notice the three side lobes on each side of the main one. As you might expect from the antenna's physical construction, there can be no radiated energy from 90° to 270°.

Microwave dish antennas are widely used in satellite communications because of their high gain; they are also used for satellite tracking and radio astronomy. They are also used at 30- to 50-mi intervals (point-to-point line-of-sight conditions) to carry telephone and broadcast TV and other signals throughout the world. Often these antennas have a "cover" over the dish. This is a low-loss dielectric material known as a **radome.** Its purpose may be maintenance of internal pressure or, more simply, environmental protection. The construction of a bird's nest within the dish is undesirable for the bird as well as the antenna user. More detail on these microwave applications was provided in Sec. 15-6.

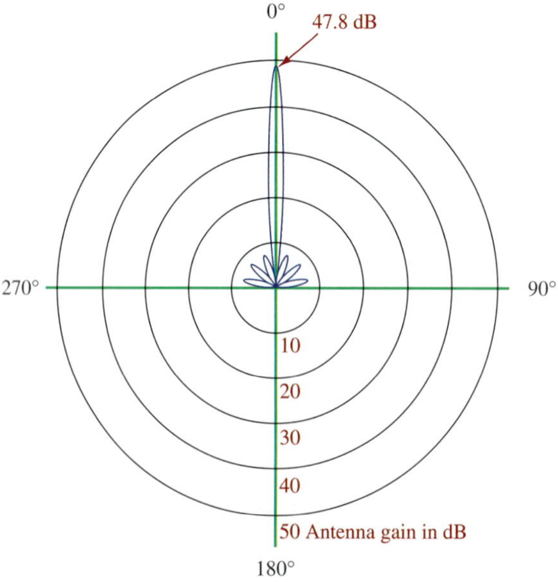

FIGURE 16-3 Polar pattern for parabolic antenna in Ex. 16-1.

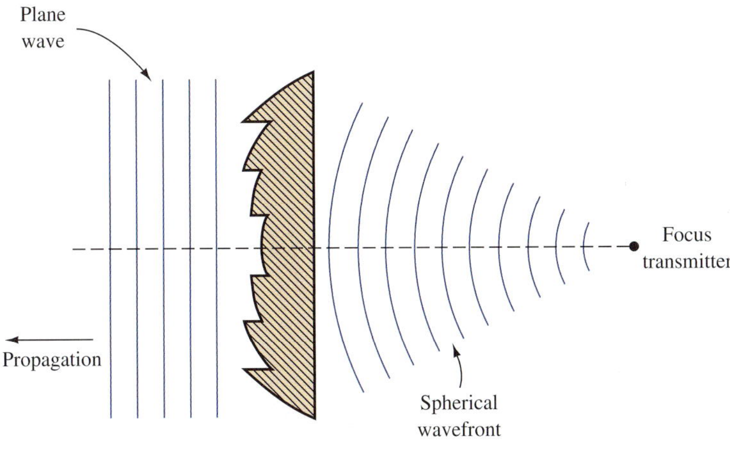

FIGURE 16-4 Zoned lens antenna.

Lens Antenna

We have all witnessed the effect of focusing the sun's rays into a point using a simple magnifying glass. The effect can also be accomplished with microwave energy, but because of the much higher wavelength, the lens must be large and bulky to be effective. The same effect can be obtained with much less bulk using the principle of **zoning,** as shown in Fig. 16-4.

Zoning
a fabrication process that allows a dielectric to change a spherical wavefront into a plane wave

If an antenna launches energy from the focus as shown in Fig. 16-4, its spherical wavefront is converted into a plane (and thus highly directive) wave. The inside section of the lens is made thick at the center and thinner toward the edge to permit the lagging portions of the spherical wave to catch up with the faster portions at the center of the wavefront. The other lens sections have the same effect, but they work on the principle that for all sections of the wavefront to be in phase, it is not necessary for all paths to be the same. A 360° phase difference (or multiple) will provide correct phasing. Thus, the plane wavefronts are made up of parts of two, three, or more of the spherical wavefronts.

Obviously the thickness of the steps is critical because of its relationship to wavelength. This is, therefore, not a broadband antenna, as is a simple magnifying glass antenna. However, the savings in bulk and expense justify the use of the zoned lens. Keep in mind that these antennas need not be glass because microwaves pass through any dielectric material, though at a reduced velocity compared to free space.

Patch Antenna

The **patch antenna** is simply a square or round "island" of conductor on a dielectric substrate backed by a conducting ground plane. A square patch antenna is shown in Fig. 16-5. The square's side is made equal to one half-wavelength and the antenna has a bandwidth less than 10 percent of its resonant frequency. The circular patch antenna is constructed with a diameter equal to about 0.6 wavelength and has about half the bandwidth or less than 5 percent of its resonant frequency.

Patch Antenna
square or round "island" of conductor on a dielectric substrate backed by a conducting ground plane

Notice the antenna feedpoint in Fig. 16-5. The process of matching impedances between a coaxial cable and the antenna can be precisely achieved with proper positioning. The same can be said if the feed is to be with a microstrip line

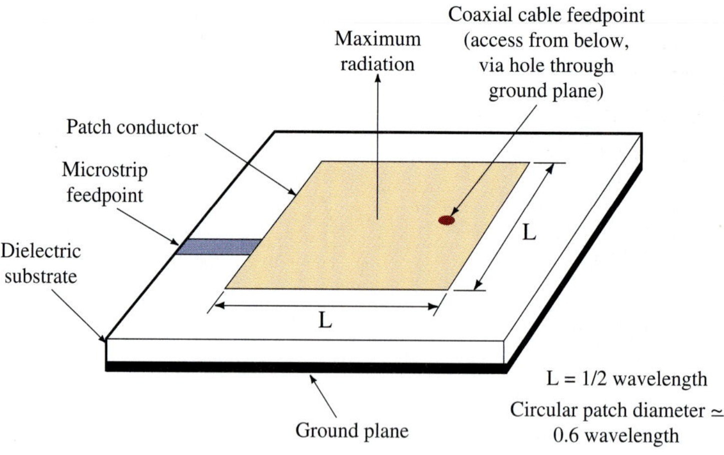

FIGURE 16-5 Patch antenna.

as shown in blue in Fig. 16-5. Keep in mind that the antenna will be fed by one or the other method (coaxial or microstrip) but not both. The radiation pattern is circular and transverse (at right angles) to the antenna away from the ground plane.

The patch antenna is extremely cheap to fabricate using printed circuit boards (PCBs) as the dielectric substrate and using standard microstrip fabrication techniques. In fact, a large number of patch antennas can be fabricated easily on a single PCB so that a phased array antenna can be manufactured inexpensively. As described in Chapter 14, a phased array consists of multiple antennas where each antenna signal can be controlled for power and phase. This allows the transmitted (or received) signal to be electronically "steered."

16-2 Microwave Tubes

Yes, tubes still live. In the case of microwave applications, standard triodes or pentodes are not effective due to the interelectrode capacitances and the associated losses. The special-effect tubes presented here do not suffer in that respect and are still in widespread use.

Magnetron

The magnetron (commonly called "maggies" in the field) is an oscillator unlike any other that has previously been discussed in this text. The magnetron is a self-contained unit. That is, it produces a microwave frequency output within its enclosure without the use of external components such as crystals, inductors, capacitors, etc.

Basically, the magnetron is a diode and has no grid. A magnetic field in the space between the plate (anode) and the cathode serves as a grid. The plate of a magnetron does not have the same physical appearance as the plate of an ordinary electron tube. Because conventional *LC* networks become impractical at microwave frequencies, the plate is fabricated into a cylindrical copper block containing resonant cavities that serve as tuned circuits. The magnetron base differs greatly from the conventional base. It has short, large-diameter leads that are carefully sealed into the tube and shielded, as shown in Fig. 16-6.

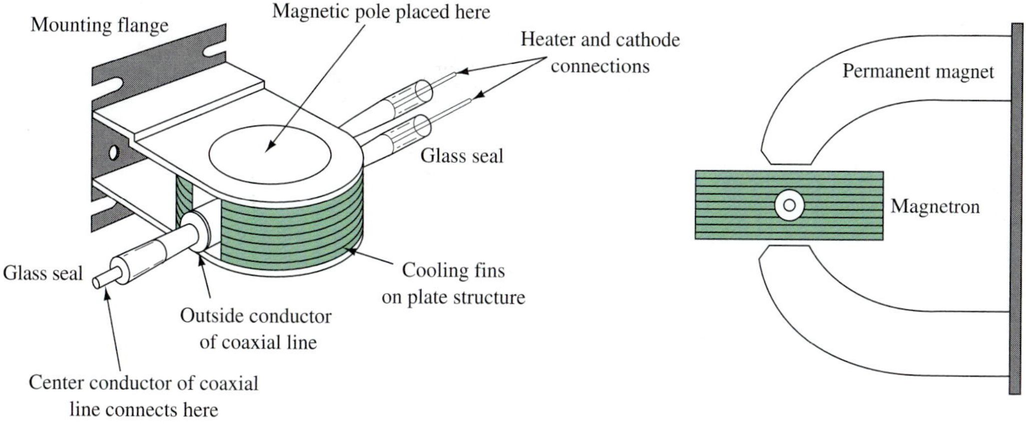

FIGURE 16-6 Magnetron.

The cathode and filament are at the center of the tube. The cathode is supported by the filament leads, which are large and rigid enough to keep the cathode and filament structure fixed in position. The output lead is usually a probe or loop extending into one of the tuned cavities and coupled into a waveguide or coaxial line. The phase structure, as shown in Fig. 16-7, is a solid block of copper. The cylindrical holes around its circumference are resonant cavities. A narrow slot runs from each cavity into the central portion of the tube. Note in the figure how these slots divide the inner structure into as many segments as there are cavities. Alternate segments are strapped together to put the cavities in parallel with regard to the output. The dimensions of the cavities determine their resonant frequency and hence the operating frequency of the magnetron. The straps are circular metal bands that are placed across the top of the block at the entrance slots to the cavities. Because the cathode must operate at high power, it must be fairly large and must be able to withstand high operating temperatures. It must also have good emission characteristics, particularly under back bombardment, because much of the output power is

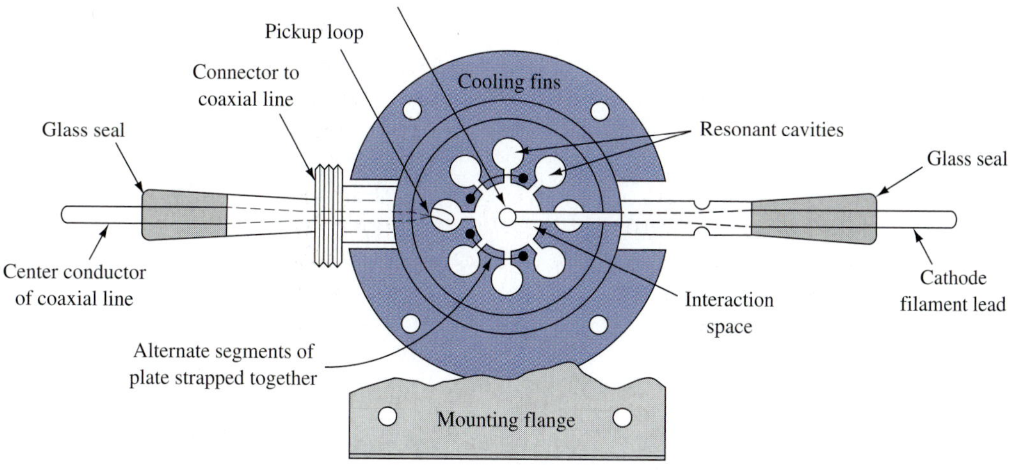

FIGURE 16-7 Cutaway view of a magnetron.

derived from the large number of electrons emitted when high-velocity electrons return to strike the cathode. The cathode is indirectly heated and is constructed of a high-emission material. The open space between the plate and the cathode is called the **interaction space** because in this space, the electric and magnetic fields interact to exert force on the electrons.

Interaction Space
in a magnetron, the open space between the plate and the cathode where the electric and magnetic fields exert force on the electrons

The magnetic field is usually provided by a strong permanent magnet mounted around the magnetron so that the magnetic field is parallel with the axis of the cathode. The cathode is mounted in the center of the interaction space.

Basic Magnetron Operation The theory of operation of the magnetron is based on the motion of electrons under the influence of combined electric and magnetic fields. The direction of an electric field is from the positive electrode to the negative electrode. The law governing the motion of an electron in an electric, or *E,* field states that the force exerted by an electric field on an electron is proportional to the strength of the field. Electrons tend to move from a point of negative potential toward a positive potential. In other words, electrons tend to move against the *E* field. When an electron is being accelerated by an *E* field, energy is taken from the field by the electron. The law of motion of an electron in a magnetic, or *H,* field states that the force exerted on an electron in a magnetic field is at right angles to both the field and the path of the electron.

A schematic diagram of a basic magnetron is shown in Fig. 16-8(a). The tube consists of a cylindrical anode with a cathode placed coaxially with it. The tuned circuit (not shown) in which oscillations take place is a cavity physically located in the anode.

When no magnetic field exists, heating the cathode results in a uniform and direct movement in the field from the cathode to the plate, as illustrated in Fig. 16-8(b). However, as the magnetic field surrounding the tube is increased, a single electron is affected, as shown in Fig. 16-9. In Fig. 16-9(a), the magnetic field has been increased to a point where the electron proceeds to the plate in a curve rather than a direct path.

In Fig. 16-9(b), the magnetic field has reached a value great enough to cause the electron to just miss the plate and return to the filament in a circular orbit. This value is the **critical value** of field strength. In Fig. 16-9(c), the value of the field strength has been increased to a point beyond the critical value, and the electron is made to travel to the cathode in a circular path of smaller diameter.

Critical Value
when the magnetic field reaches a value great enough to cause electrons to just miss the plate and return to the filament in a circular orbit

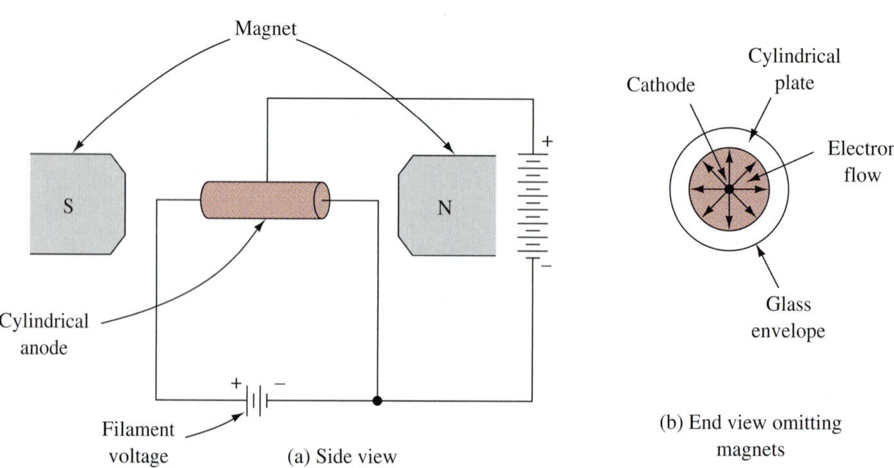

FIGURE 16-8 Basic magnetron.

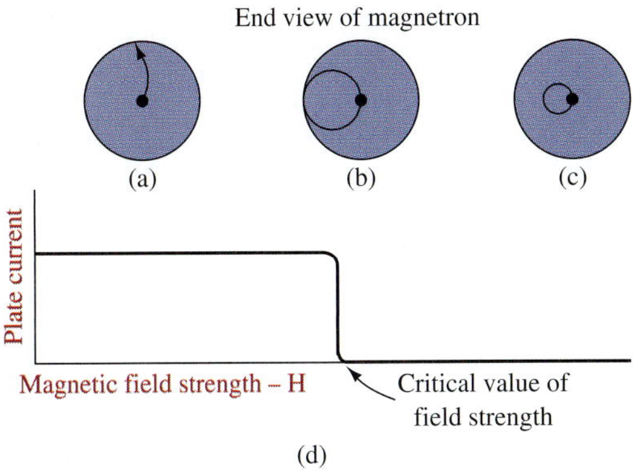

FIGURE 16-9 Effect of magnetic field on single direct path.

Figure 16-9(d) shows how the magnetron plate current varies under the influence of the varying magnetic field. In Fig. 16-9(a), the electron flow reaches the plate so that there is a large amount of plate current flowing. However, when the critical field value is reached, as shown in Fig. 16-9(b), the electrons are deflected away from the plate, and the plate current drops abruptly to a very small value. When the field strength is made still larger [Fig. 16-9(c)] the plate current drops to zero.

When the magnetron is adjusted to the plate current cutoff or critical value and the electrons just fail to reach the plate in their circular motion, the magnetron can produce oscillations at microwave frequency by virtue of the currents induced electrostatically by the moving electrons. This frequency is determined by the size of the cavities. Electrons are accelerated toward the anode by the electric field and bent by the magnetic field so that they travel parallel to the anode. If they pass the anode gap at a time when they are traveling in the same direction as the electric field in the gap, they slow down and thus give up energy (kinetic) to the electric field in the cavity. This reinforces the oscillations and is the basis of operation. A transfer of microwave energy to a load is made possible by connecting an external circuit between the cathode and plate of the magnetron. Magnetrons are widely used as sources of microwave power up to 100 GHz. They can provide up to 25 kW of continuous power at efficiencies up to 80 percent. Pulsed magnetrons are used in radar applications up to 10,000 kW with low duty cycles. They are also widely used for microwave ovens. They operate at 2.45 GHz with continuous outputs of 400 to 1000 W. Their electromagnetic energy is radiated through the food, which is heated (cooked) from the inside out. These tubes have been highly refined because of this volume application, and they can be expected to outlive the average 10- to 15-year life for a home appliance. Because the magnetron is functionally a diode, the only other power supply component required is a transformer for the 2 to 4 kV anode voltage and the filament voltage.

Traveling Wave Tube

The traveling wave tube (TWT) is a high-gain, low-noise, wide-bandwidth microwave amplifier. TWTs are capable of gains of 40 dB or more, with bandwidths of over an octave. (A bandwidth of one octave is one in which the upper frequency is twice the lower frequency.) TWTs have been designed for frequencies as low as 300 MHz and as high as 150 GHz and continuous outputs to 5 kW. Their wide-

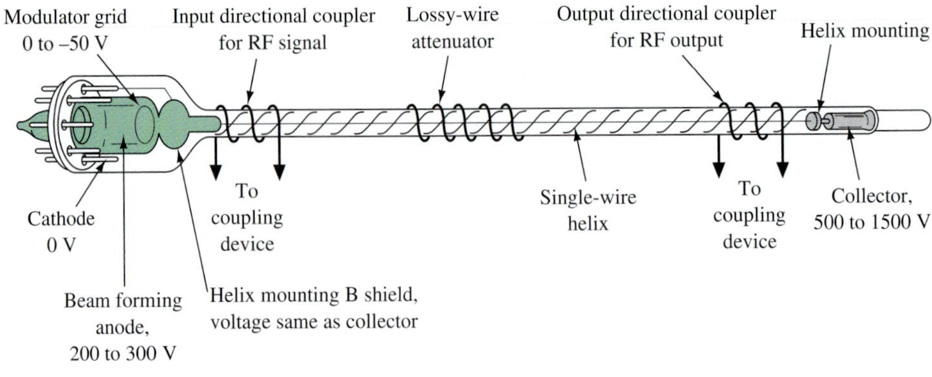

FIGURE 16-10 Pictorial diagram of a traveling wave tube.

bandwidth and low-noise characteristics make them ideal for use as RF and medium-power amplifiers in microwave and electronic countermeasure equipment. They are widely used as the power output stage in orbiting satellites.

Construction Figure 16-10 is a pictorial diagram of a traveling wave tube. The electron gun produces a stream of electrons that are focused into a narrow beam by an axial magnetic field. The field is produced by a permanent magnet or electromagnet (not shown) that surrounds the helix portion of the tube. The narrow beam is accelerated by a high potential on the helix and collector as it passes through the helix.

Operation The beam in a TWT is continually interacting with an RF electric field propagating along an external circuit surrounding the beam. To obtain amplification, the TWT must propagate a wave whose phase velocity is nearly synchronous with the dc velocity of the electron beam. It is difficult to accelerate the beam to greater than about one-fifth the velocity of light. The forward velocity of the RF field propagating along the helix is slowed to nearly that of the beam due to its travel along the helix. Changing the pitch of the helix changes the speed of the RF field.

The electron beam is focused and constrained to flow along the axis of the helix. The longitudinal components of the input signal's RF electric field, along the axis of the helix or slow wave structure, continually interact with the electron beam to provide the gain mechanism of TWTs. This interaction mechanism is pictured in Fig. 16-11. This figure illustrates the RF electric field of the input signal, propagating along the helix, infringing into the region occupied by the electron beam.

Consider first the case where the electron velocity is exactly synchronous with the RF signal passing through the helix. Here, the electrons experience a steady dc electric force that tends to bunch them around position *A* and debunch them around position *B* in Fig. 16-11. This action is due to the accelerating and decelerating electric fields. In this case, as many electrons are accelerated as are decelerated; hence, there is no net energy transfer between the beam and the RF electric field. To achieve amplification, the electron beam is adjusted to travel slightly faster (by increasing the anode voltage) than the RF electric field propagating along the helix. The bunching and debunching mechanisms just discussed are still at work, but the bunches now move slightly ahead of the fields on the helix. Under these conditions more electrons are in the decelerating field to the right of *A* than in the accelerating

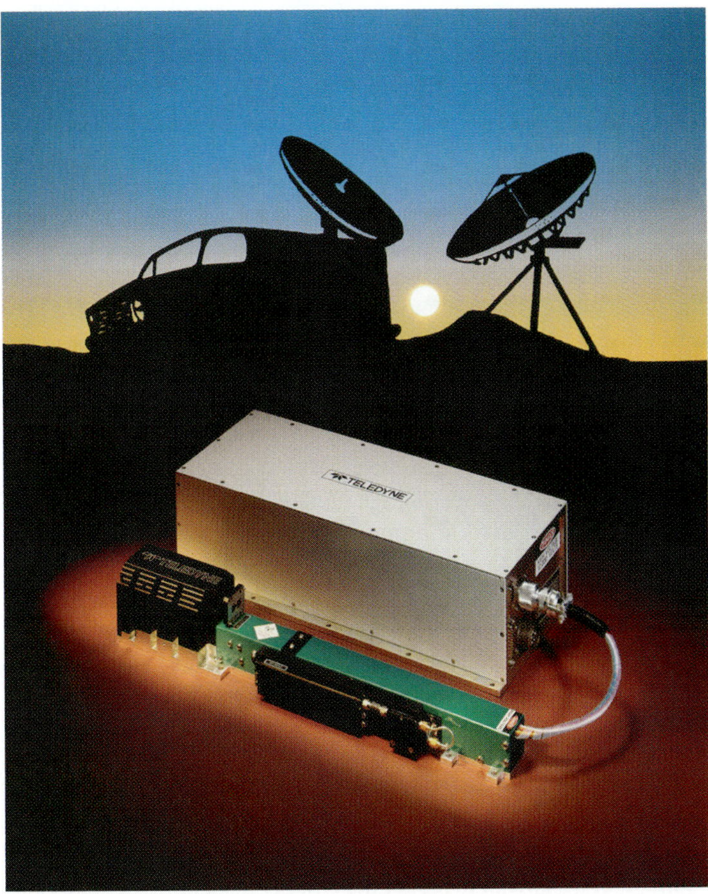

TWT amplifier system provides up to 400 W at 14.5 GHz. (Courtesy of Teledyne Electronic Technologies. Photo by Jon Ho.)

field to the right of B. Because more electrons are decelerated than are accelerated, the energy balance is no longer maintained. Thus, energy transfers from the beam to the RF field, the field grows, and amplification occurs.

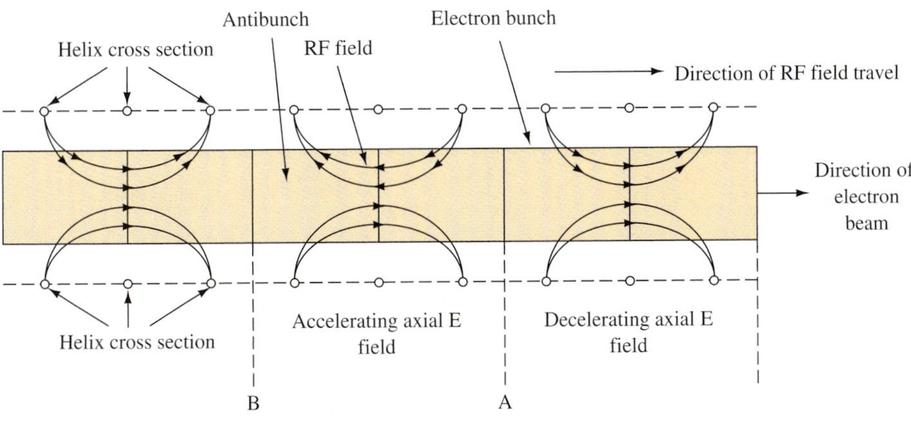

FIGURE 16-11 Helix field interaction.

Fields may propagate in either direction along the helix. This leads to the possibility of oscillation due to reflections back along the helix. This tendency is minimized by placing resistive materials near the input end of the slow wave structure. This resistance may take the form of a lossy-wire attenuator (Fig. 16-10) or a graphite coating placed on insulators adjacent to the helix. Such lossy sections completely absorb any backward-traveling wave. The forward wave is also absorbed to a great extent, but the signal is carried past the attenuator by the bunches of electrons. These bunches are not affected by the attenuator and therefore reinstitute the signal on the helix after they have passed the attenuator.

The traveling wave tube has also found application as a microwave mixer. By virtue of its wide bandwidth, the TWT can accommodate the frequencies generated by the heterodyning process (provided, of course, that the frequencies have been chosen to be within the range of the tube). The desired frequency is selected by the use of a filter on the output of the helix. A TWT mixer has the added advantage of providing gain as well as providing mixer action.

A TWT may be modulated by applying the modulating signal to a modulator grid. The modulator grid may be used to turn the electron beam on and off, as in pulsed microwave applications, or to control the density of the beam and its ability to transfer energy to the traveling wave. Thus, the grid may be used to amplitude-modulate the output. The TWT offers wideband performance with high-power outputs up to 150 GHz. TWTs are widely used in wideband communications repeater links. They offer low-noise performance and high-power gains. Their high reliability dictates their use as power amplifiers in communications satellites, where a lifetime in excess of 10 years can be expected.

TWT Oscillator A forward wave, traveling wave tube may be constructed to serve as a microwave oscillator. Physically, a TWT amplifier and oscillator differ in three major ways. The helix of the oscillator is longer than that of the amplifier, there is no input connection to the oscillator, and the lossy-wire attenuator shown in Fig. 16-10 is eliminated. The tube now allows both forward and backward waves and is usually called a **backward-wave oscillator** (BWO). The operating frequency of a BWO is determined by the pitch of the tube's helix. The oscillator frequency may be fine tuned, within limits, by adjusting the operating potentials of the tube.

Backward-Wave Oscillator
TWT that allows both forward and backward waves and can therefore be used as an oscillator

The electron beam, passing through the helix, induces an electromagnetic field in the helix. Although initially weak, this field, through the action previously described, causes bunching of succeeding portions of the electron beam. With the proper potentials applied, the bunches of electrons reinforce the signal on the helix. This, in turn, increases the bunching of succeeding portions of the electron beam. The signal on the helix is sustained and amplified by this positive feedback resulting from the exchange of energy between electron beam and helix.

Klystron

Another common microwave tube is the klystron. It has been widely used in the past and has certain similarities to the TWT. The klystron gets its name from the greek verb "klyzo," which is a term used to describe the sound that waves make breaking on the shore. Conversion of a low-power RF input signal (e.g., 1 W) to high output power (e.g., 55 kW) in the klystron results in a beam that contains bunches of electrons. The bunching of electrons is created as the electron beam travels along an area called a drift tube. The bunches of waves or electrons resemble ocean waves; the bunching is called **velocity modulation.** The bunching of electrons is maintained by focus magnets. The high-power klystrons are being replaced by either magnetrons or TWTs in new equipment, and solid-state microwave devices are replacing them in low-power applications. The reasons for these replacements are because of the klystron's large size and the complex, costly sources of dc required for operation.

Velocity Modulation an electron beam moving along in bursts of electrons, as in a klystron

16-3 Solid-State Microwave Devices

The advances made over the past fifteen years with microwave solid-state devices have been truly startling. This includes the work with bipolar and field effect transistors as well as several special two-terminal devices.

Gunn Oscillator

The Gunn oscillator is a solid-state bulk-effect source of microwave energy. The discovery that microwaves could be generated by applying a steady voltage across a chip of *n*-type gallium arsenide crystal was made in 1963 by J. B. Gunn. The operation of this device results from the excitation of electrons in the crystal to energy states higher than those they normally occupy. A common application of this device is in the handheld radar "guns" used by the police.

In a gallium arsenide semiconductor are empty electron valence bands, higher than those occupied by electrons. These higher valence bands have the property that electrons occupying them are less mobile under the influence of an electric field than when they are in their normal state at a lower valence band.

To simplify the explanation of this effect, assume that electrons in the higher valence band have essentially no mobility. If an electric field is applied to the gallium arsenide semiconductor, the current that flows increases with an increase in voltage, provided the voltage is low. If the voltage is made high enough, however, it may be possible to excite electrons from their initial band to the higher band, where they become immobile. If the rate at which electrons are removed is high enough, the current decreases even though the electric field is being increased. Thus, the device displays the effect of negative resistance.

If a voltage is applied across an unevenly doped *n*-type gallium arsenide crystal, the crystal breaks up into regions with different intensity electric fields across them. In particular, a small domain forms within which the field is very strong, whereas in the rest of the crystal, outside this domain, the electric field is weak. The domains formed in the gallium arsenide crystal are not stationary because the electric field acting on the electron energy causes the domain to move across the crystal. The domain travels across the crystal from one electrode to the other, and as it disappears at the anode, a new domain forms near the cathode.

The Gunn oscillator has a frequency inversely proportional to the time required for a domain to cross the crystal. This time is proportional to the length of the crystal and, to some degree, to the potential applied. Each domain results in a pulse of current at the output; hence, the output of the Gunn oscillator is a microwave frequency that is determined, for the most part, by the physical length of the chip.

The Gunn oscillator has delivered power outputs of 3 or 4 W at 18 GHz (continuous operation) and up to 1000 W in pulsed operation. The power output capability of this device is limited by the difficulty of removing heat from the small chip. At frequencies above 35 GHz, indium phosphate (InP)-based units are used in place

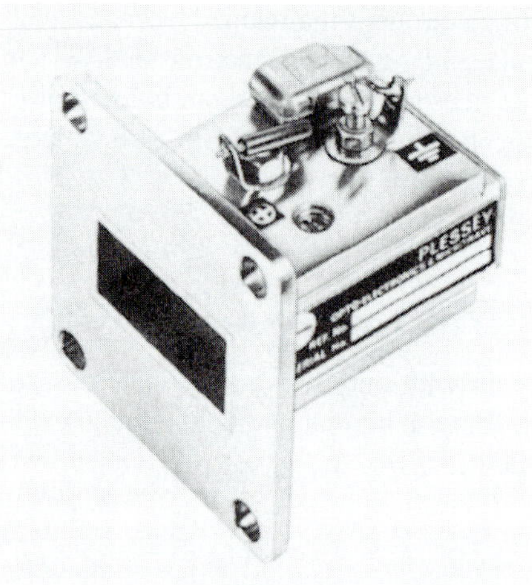

FIGURE 16-12 Gunn oscillator assembly. (Courtesy of Plessey Semiconductor.)

of gallium arsenide (GaAs). These InP devices can deliver continuous powers of 500 mW at 355 GHz and 50 mW at 140 GHz.

The advantages of the Gunn oscillator are its small size, ruggedness, low cost of manufacture, lack of vacuum or filaments, and relatively good efficiency. These advantages open a wide range of application for this device in all phases of microwave operations. This bulk device is the workhorse of the microwave-oscillator field at frequencies above 8 GHz. Below 8 GHz, it competes directly with transistor oscillators.

A commercially available Gunn oscillator assembly is shown in Fig. 16-12. It is included within a waveguide section and can be tuned by an included varactor diode over a 4 percent range at 10 GHz. They are also available in microstrip and coaxial line configurations. Since the Gunn device is a two-terminal solid-state device, it is often termed a Gunn diode. They are also identified as transferred electron devices (TEDs) or limited space-charge accumulation devices (LSAs).

The Gunn device can also function as an amplifier. Its negative resistance characteristic allows it to replace the energy consumed by the positive resistance (loss) of either an *LC* tank circuit, shorted transmission line section, or resonant cavity. This **replacement energy** supplied by the negative resistance can also be used for amplification of applied energy. The major problem encountered with amplification using a two-terminal device is the isolation of input and output energy. This is usually accomplished with a device known as a *circulator*. It is similar to the hybrid tee (described in the previous chapter) in function but differs in physical construction. Its operation is explained in Sec. 16-4.

Replacement Energy
in a Gunn oscillator, energy supplied by the negative resistance to allow amplification

IMPATT Diode

IMPATT is an acronym for *imp*act ionization *a*valanche *t*ransit *t*ime. The theory of this device was presented in 1958, and the first experimental diode was described in 1965. The basic structure of a silicon *pn* junction **IMPATT diode,** from the semiconductor point of view, is identical to that of varactor diodes. The important differences between IMPATT and varactor diodes are in their modes of operation and in thermal design.

IMPATT Diode
*imp*act ionization *a*valanche *t*ransit *t*ime; used in the generation of microwave signals

Figure 16-13 shows a typical dc current versus voltage (*I–V*) characteristic for a *pn* junction diode. In the forward-bias direction, the current increases rapidly for

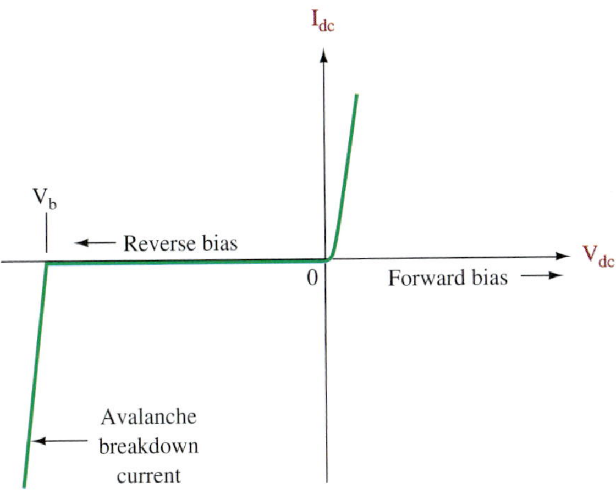

FIGURE 16-13 Terminal *I–V* characteristics of a *pn* junction diode.

voltages above 0.5 V or so. In the reverse direction, a very small current (the *saturation* or *leakage current*) flows until the breakdown voltage, V_b, is reached. Varactor diodes normally operate reverse-biased with a dc operating point well away from V_b. IMPATT diodes, on the other hand, operate in the avalanche breakdown region, that is, with a dc reverse voltage greater than V_b and substantial reverse current flowing.

Figure 16-14 shows a schematic representation of an IMPATT diode reverse-biased into avalanche breakdown. As in any reverse-biased *pn* junction, a *depletion zone* forms in the *n*-type region of the diode; its width depends on the applied reverse voltage. The depletion zone acts as a nonlinear capacitor if V_{dc} is less than V_b.

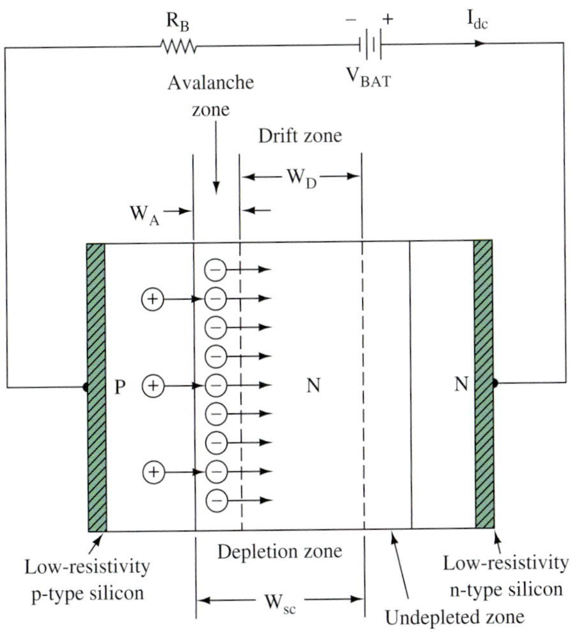

FIGURE 16-14 Schematic representation of reverse-biased *pn* junction diode.

This property is utilized in varactor diodes. The saturation current, which flows while the reverse voltage is less than V_b, is usually on the order of 10 to 100 nA and is depicted in Fig. 16-14 by a small number of electrons flowing to the right from the p^+ region into the *avalanche zone*. When V_{dc} is more negative than V_b, the small number of electrons comprising the saturation current have a very high probability of creating additional electrons and holes in the avalanche zone by the process of avalanche multiplication. The additional electrons are shown in Fig. 16-14 flowing from the avalanche zone into the *drift zone*. In this condition, a large current can flow in the reverse direction with little increase in applied voltage. This is the *avalanche breakdown current,* which is shown in Fig. 16-13. The typical dc operating voltage across a diode is between 70 and 100 V, depending on the diode type, temperature, and the value of the bias current, I_{dc}. Typical avalanche breakdown currents (usually called the *bias current*) range from 20 to 150 mA.

Microwave Properties of the IMPATT Diode Let us assume that somehow an RF voltage, in addition to the dc breakdown voltage, exists across the depletion region of the IMPATT diode. This voltage can be expressed mathematically as

$$V_T(t) = V_b + V_D \sin \omega t \tag{16-3}$$

This form of voltage is illustrated in Fig. 16-15(a) and would exist in practice in the common case where the diode is operated in a single resonant circuit, with Q greater

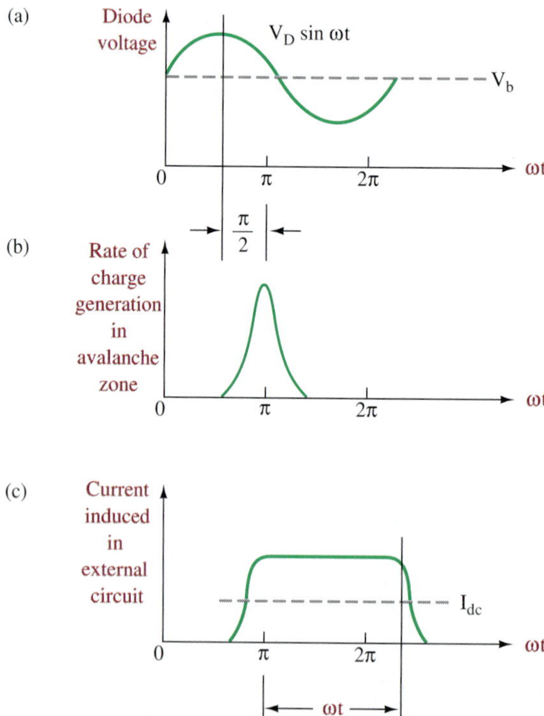

FIGURE 16-15 (a) Voltage across the IMPATT diode depletion layer during oscillation; (b) rate of charge generation in the avalanche zone; (c) current induced in the external circuit by the avalanche-generated charge drifting across the drive zone.

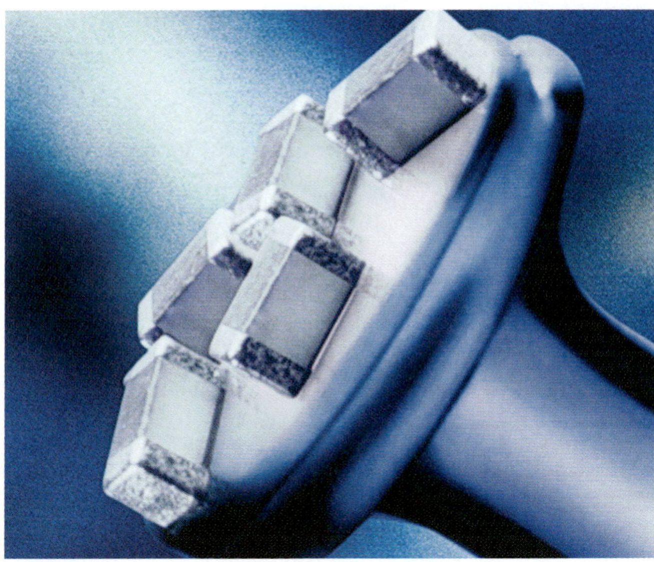

Miniature inductors (1 to 20 nH) for microwave circuits on the head of a pin. (Courtesy of Penstock.)

than 10 or so. Under certain conditions, the RF portion of this voltage induces an RF current that is more than 90° out of phase with the voltage, and therefore the diode has negative resistance. The arguments leading to this conclusion are conveniently divided into two steps:

1. First, as the voltage rises above the dc breakdown voltage during the positive half-cycle of the RF voltage, excess charge builds up in the avalanche region, slowly at first, and reaches a sharply peaked maximum at $\omega t \simeq \pi$, that is, in the middle of the RF voltage cycle when the RF voltage is zero. This is shown in Fig. 16-15(b). Thus, the charge generation waveform, in addition to being sharply peaked, *lags* the RF voltage by 90°. This behavior arises because of the highly nonlinear nature of the avalanche generation process.
2. The second step in the analysis is to consider the behavior of the generated charge subsequent to $\omega t = \pi$. The direction of the field is such that the electrons *drift* to the right (refer to Fig. 16-14). The equal number of generated holes move to the left, back into the p^+ contact, and are not considered further in this simple model. The electrons drift at constant (*saturated*) velocity, v_{sat}, across the drift zone. The time, t, they take to traverse the drift zone is simply the width of the drift zone divided by the constant velocity of the electrons:

$$t = \frac{\text{width}}{v_{sat}} \tag{16-4}$$

While the electrons are drifting through the diode, they induce a current in the external circuit, as shown in Fig. 16-15(c). The current is approximately a square wave. By examining Figs. 16-15(a) and (c), you can see that the combined delay of the avalanche process and the finite transit time across the drift zone has caused *positive* current to flow in the external circuit while the diode's RF voltage is going through its *negative* half-cycle. The diode is thus delivering

RF energy to the external circuit or, in circuit terms, is exhibiting *negative* resistance. Maximum negative resistance is obtained when

$$\omega t \simeq 0.74\pi \tag{16-5}$$

The term ωt is called the *transit angle*; IMPATT diodes are normally designed so that Eq. (16-5) is satisfied at or near the center of the desired operating frequency range.

IMPATT diodes are finding wide application in microwave oscillator and amplification schemes. They can produce 20-W continuous output at 10 GHz and are useful up to 300 GHz. While many two-terminal microwave devices exhibiting negative resistance have been developed in recent years, it appears that the Gunn diode (bulk-effect device) and IMPATT diode (true diode) have the most promising future. Among the other devices that have also been utilized are the following:

1. *TRAPATT diode:* It is similar to the IMPATT diode.
2. *Baritt diode:* It has two junctions separated by a transit time region.
3. *Tunnel diode:* Its use as a microwave power source/amplifier has diminished because Gunn devices and IMPATTs have become readily available.

P-I-N Diode

p-i-n Diodes
diodes used as RF and microwave switches that consist of *p*-type, *i*ntrinsic (lightly doped), and *n*-type material

***P-i-n* diodes** are used as RF and microwave switches whose resistance values are controlled by forward current levels. As shown in Fig. 16-16, a *p-i-n* diode is built from high-resistivity silicon and has an *intrinsic* (very lightly doped) layer sandwiched between a *p* and an *n* layer. When the diode has a forward current, holes and electrons are injected in the *i* region. They do not completely recombine but rather form a stored charge. This stored charge causes the effective resistivity of the *i* region to be much lower than the intrinsic resistivity.

Up to about 100 MHz, this diode acts basically like a conventional rectifier. At higher frequencies, however, it ceases to rectify because of the stored charge in and the transit time across the intrinsic region. To high frequencies, it acts as a variable resistance, easily controlled by the amount of dc forward bias. Resistances of less than an ohm are possible at high-dc forward bias, while a small forward bias may cause it to look like 1 kΩ of resistance to a high-frequency signal.

The *p-i-n* diode is used whenever there is the need to switch microwave energy at frequencies up to 100 GHz, even at high-power levels. A common application is their use as transmit/receive (TR) switches in transceivers operating from 100 MHz and up. As described in Chapter 18, they are also used as photo-detectors in fiber-optic systems.

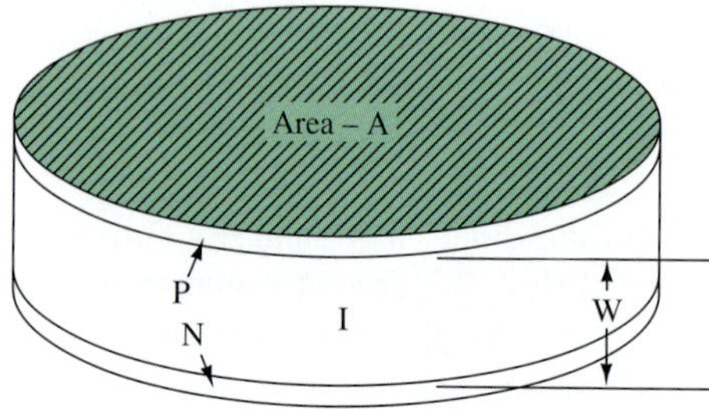

FIGURE 16-16 *P-i-n* diode construction.

Microwave Transistor

The three-terminal devices are used primarily to build amplifiers where their inherent input/output isolation permits simpler designs than two-terminal devices. Bipolar transistors are preferred at frequencies below 5 GHz because they provide higher output power and similar noise performance with respect to FETs. Above 5 GHz, bipolars lose power output capability due to inherent high-frequency limitations, and their noise performance is severely degraded. Since bipolar technology is relatively mature, it is not expected that these conditions will be improved to any great extent in the future.

On the other hand, FET technology is still growing. At high frequencies, gallium arsenide (GaAs) FETs offer superior performance over the standard silicon devices. Above 5 GHz, the GaAs FETs offer superior noise performance and output powers compared to the bipolar transistor. The GaAs FET is the most important amplifying device in the 5- to 20-GHz region. Above 20 GHz, it is necessary to go to a two-terminal device such as an IMPATT diode or a tube such as the TWT. If high-power output is required (>20 W), the tube device is likely to be used for frequencies down to 2 GHz.

Microwave Integrated Circuits

Microwave monolithic integrated circuits (MMICs) are making inroads compared to discrete devices, especially in the lower microwave frequencies of 1–3 GHz. These devices are generally GaAs-based and are popular in high-volume applications such as cellular communications.

Figure 16-17 shows the block diagram for a Philips DCS-1800 MMIC. It is a GaAs MOSFET device used in cellular communications and includes a power amplifier, transmitter up-conversion mixer, LO, and other related components. The LO feeds

Miniature MMICs amplify up to 6 GHz. (Courtesy of Penstock.)

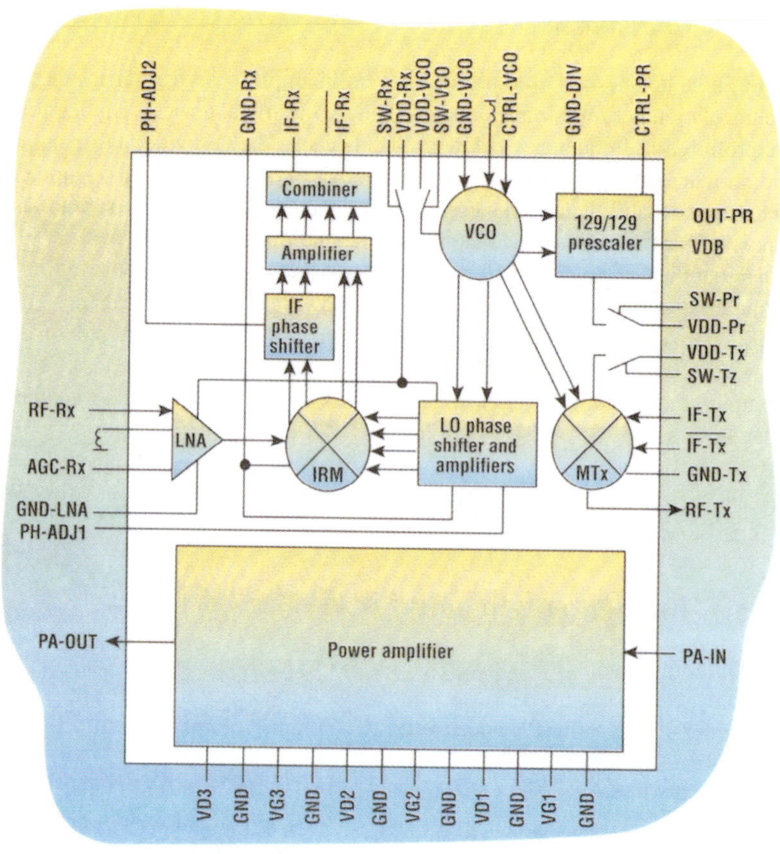

FIGURE 16-17 MMIC used in cellular communications. (Courtesy of Microwaves and RF, from LEP/Phillips Microwave.)

the up-conversion mixer to produce an IF at 400 MHz. The LO is a VCO that is tuned via an external resonator into a frequency ranging from 2110 to 2185 MHz. An external filter is inserted between the transmitter-mixer (MTx) output (RF-Tx) and the PA input to remove signals generated in the image bandwidth of 2510 to 2585 MHz.

The receiver stage consists of a low-noise amplifier (LNA) and an image-reject mixer (IRM). The RF input (RF-Rx) has a bandwidth from 1805 to 1880 MHz, while the IF output (IF-Rx) is set to 300 MHz, thus resulting in an LO range on the order of 2105 to 2180 MHz. Since the same LO is used for both transmit and receive, it must tune from at least 2105 to 2185 MHz.

The MMIC includes a divide-by-128/129 dual-modulus prescaler. The prescaler converts signals from the local oscillator to an output signal of about 17 MHz, which is compatible with most standard frequency synthesizers. The supply voltages on the receiver and transmitter chains are alternately switched on and off by FETs controlled by external signals.

The power amplifier (PA) is designed to provide output of up to +27 dBm (0.5 W) with a 3.3-V supply. It has an efficiency of more than 30 percent at 1700 MHz. The input power requirement is 0 dBm. It includes three stages operating Class AB.

16-4 FERRITES

Ferrites are compounds of iron, zinc, manganese, magnesium, cobalt, aluminum, and nickel oxides. They are manufactured by pressing into shape the required mixture of the finely divided metallic oxide powders and then firing the shaped mixture at about 2000°F. The product is a ceramic with high electrical resistance. Ferrites behave as iron alloys at low frequencies, but at high frequencies their high electrical resistance prevents eddy currents, and resonance takes place within the iron atoms themselves. These unusual effects make it possible to use ferrites for special applications in microwave circuits. The most popular ferrite compounds are manganese ($MnFe_2O_3$), zinc ($ZnFe_2O_3$), and yttrium–iron–garnet [$Y_3Fe_2(FO_4)_3$], which is called *yig*. Ferrites are dielectrics, so they support and propagate the electromagnetic energy of microwave signals. When they are placed in waveguides or coaxial circuits, they can be transparent, reflective, or absorptive, depending on their magnetic characteristics.

Fundamental Theory of Ferrites

A fundamental property of atoms is that both electrons and protons spin on their own axes. In addition, of course, the electron revolves around the nucleus. An analogy is the solar system, where the earth rotates on its axis as it revolves around the sun. As the electron spins, it creates a magnetic moment, or field, along its spin axis. This spinning charge appears as a current flowing around a loop. The atoms having more electrons spinning in one direction than another act as small magnets. The mutual action of all these atoms explains the magnetic properties of magnetic materials.

If a spinning electron is placed in a static magnetic field, the electron's magnetic moment becomes aligned with the static field. The magnetic moment and its alignment with a dc magnetic field are shown in Fig. 16-18.

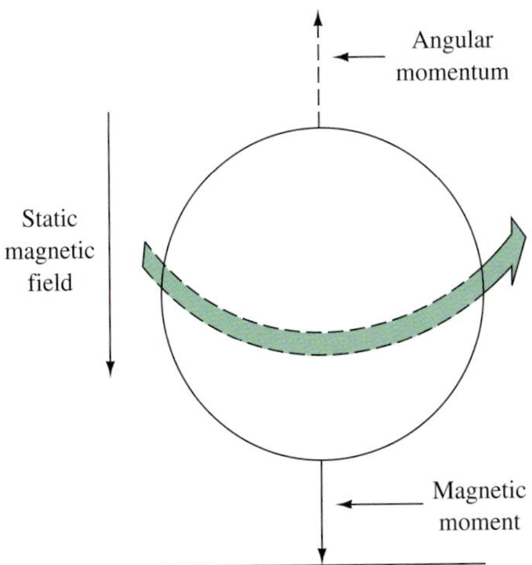

FIGURE 16-18 Electrons in a dc magnetic field.

Gyroscopic Action

Because of their spinning motion, electrons behave like very small gyroscopes. When a force is applied to the spin axis of an electron that causes it to tilt, the electron behaves like any other gyroscope: it precesses, or wobbles. **Precession** is

Precession
movement of the axis of rotation at right angles to its original axis

defined as a movement of the axis of rotation at right angles to its original axis. Figure 16-19 shows a gyroscope mounted to a stick so that the stick can pivot freely. Even with the gyroscope spinning, the stick hangs straight down because of gravity. If you try to move the stick from side to side, however, the gyroscope forces the stick to move around in a circle, or precess.

The direction is determined by the direction of rotation of the gyro rotor, and the frequency is determined by the gravitational force and the momentum of the gyroscope. This is shown in Fig. 16-19. The natural precession frequency could be increased by increasing the force of gravity. A rotational force applied to the stick at the natural precession frequency displaces it from the vertical by a large amount. (The precession path shown in Fig. 16-19 would have a greater diameter.) A rotational force applied at any other frequency produces a much lower displacement. This is similar to the feedback in an oscillator. With feedback at the right frequency, the amplitude of oscillation is much larger than when the feedback is off frequency.

Electrons behave much like gyroscopes, but gravity has little effect on them. Instead, a steady magnetic field is applied to line up the axes of the spinning electrons. This field causes any precession to die out quickly. Now when an alternating field is applied at right angles to the dc field, the electrons precess, or wobble, just the way the gyroscope and stick did when a sideways force was applied.

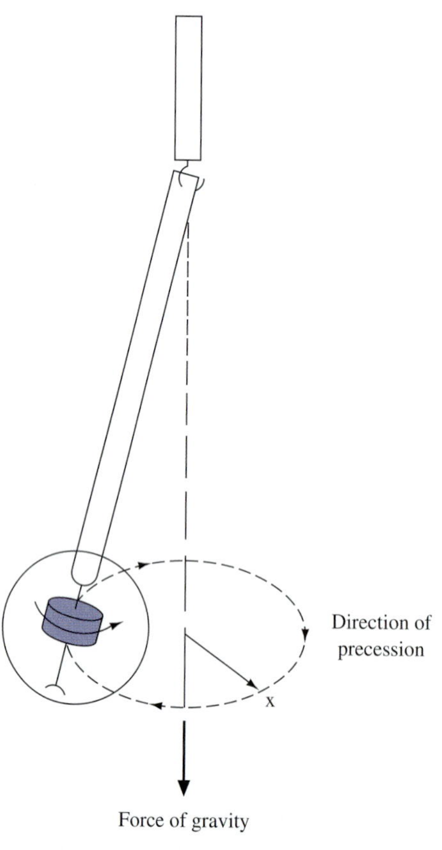

FIGURE 16-19 Precession of a gyroscope.

The natural precession frequency of the electrons in a ferrite depends on the dc magnetic field strength and the type of ferrite material. If an ac field is applied at the natural frequency, the precessional motion builds up. This increases the frictional damping effects because the entire atom is vibrating and the ferrite dissipates as heat energy extracted from the ac field. The range of natural precession frequencies available with presently used ferrites is from about 30 up to over 200 GHz.

Applications

Attenuator One application of ferrites is as an attenuator. Figure 16-20 shows a piece of ferrite placed in the center of a waveguide; a steady magnetic field is applied as shown. This arrangement attenuates frequencies at the resonant frequency of the electrons in the ferrite, whereas other frequencies are attenuated only slightly. Changing the strength of the dc field produces a change in the frequency that is attenuated, although this occurs over a limited range.

The dc field is produced by current flowing through a coil wound around the waveguide. The strength of the field, which depends on the current flowing through the coil, determines the frequency of precession. Usually, the ferrite attenuator is in the form of an adjustable vane of ferrite extending into the waveguide. The farther the vane extends into the waveguide, the greater the attenuation because more of the RF energy must travel through the ferrite.

Isolator Another application of ferrites is that of an isolator. When used as an isolator, the ferrite allows energy to travel in one direction but absorbs energy traveling in the opposite direction. Figure 16-21 depicts an electromagnetic wave traveling from right to left. It illustrates how the wave, at a point off the center line of the guide, appears as a rotating magnetic field. At one instant shown in Fig. 16-21, the magnetic field at point X is pointed up. When the magnetic field at point 2 reaches point X, the magnetic field is directed to the right. When point 3 reaches X, the

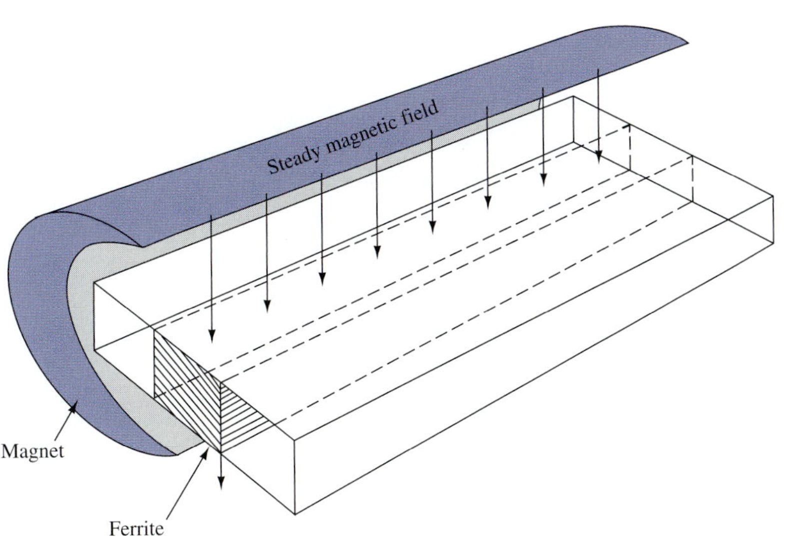

FIGURE 16-20 Ferrite slab mounted in waveguide.

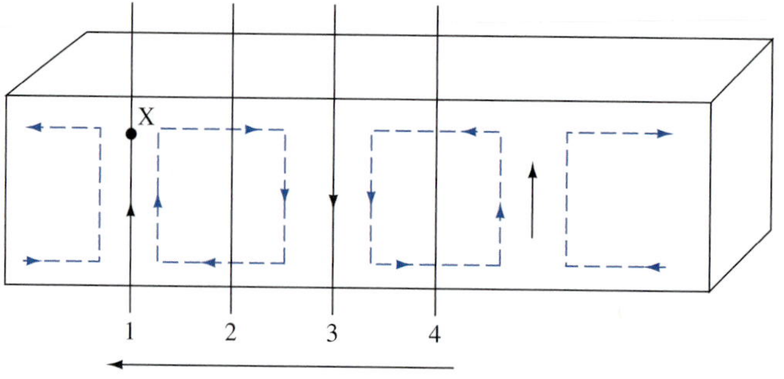

FIGURE 16-21 Effective clockwise rotation of a magnetic field.

magnetic field is downward, and when point 4 on the wave arrives at *X*, the field is directed to the left.

Thus, as the wave passes point *X*, the magnetic field appears to rotate in a clockwise direction. At any point off the center of the waveguide, the magnetic field appears to rotate as the electromagnetic wave passes. This same analysis can be used to show that, with a wave traveling from left to right, the magnetic field appears to rotate counterclockwise at point *X*.

Now let us place a section of ferrite in the waveguide at *X*. See Fig. 16-22, which illustrates a simple isolator consisting of a piece of waveguide, a permanent magnet, and a section of ferrite. The ferrite's electron resonant frequency and the microwave frequency are made the same by changing either the magnetic field strength or the microwave frequency. When the frequencies are the same, a wave traveling from left to right in the waveguide produces a rotating force in the direction of the natural precession of the electrons in the ferrite. The amplitude of precession increases, taking power from the electromagnetic wave. This power is dissipated as heat in the ferrite.

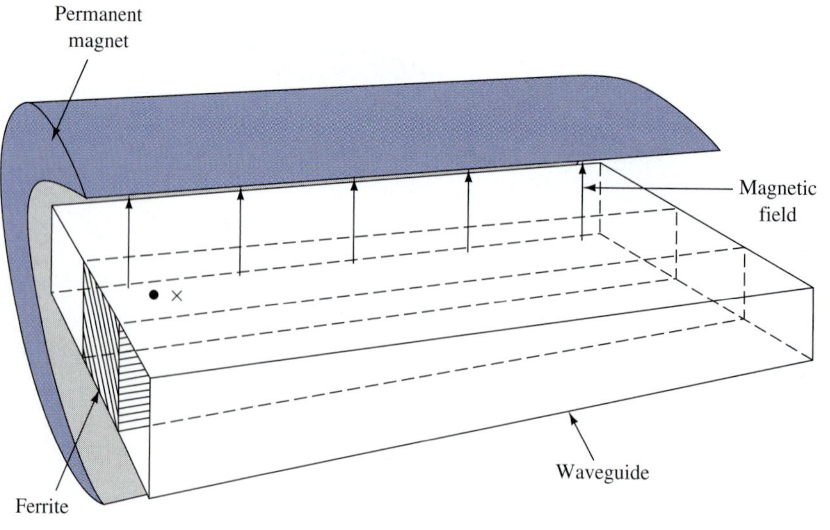

FIGURE 16-22 Simple isolator.

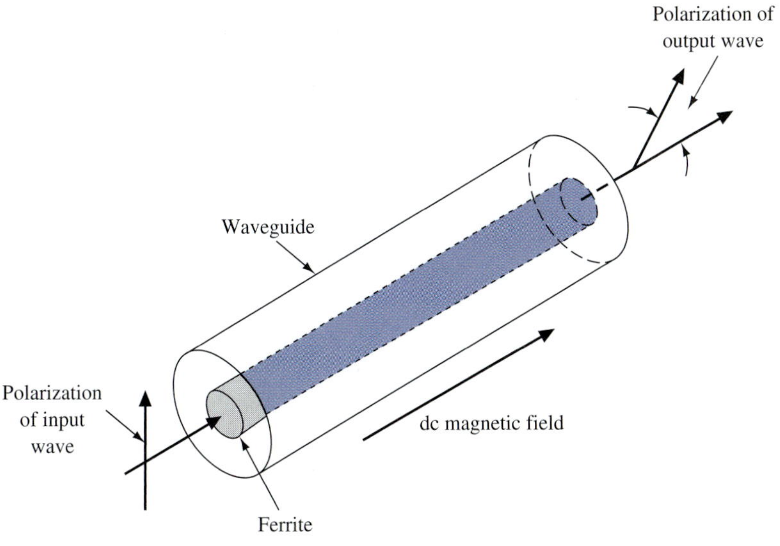

FIGURE 16-23 Faraday rotation.

A wave that is traveling from right to left in this waveguide acts as a rotating force on the electrons to oppose the natural precession. This does not increase the amplitude of the precession, and energy is not absorbed from the electromagnetic field. About 0.4-dB attenuation takes place in a wave traveling from right to left, but as much as 10-dB attenuation occurs in a wave traveling from left to right.

Faraday Rotation Another effect takes place when microwaves are passed through a piece of ferrite in a magnetic field. The plane of polarization of the wave is rotated if the frequency of the microwave is above the resonant frequency of the ferrite electrons. This is known as the **Faraday rotation effect.** When RF energy enters the ferrite material, the magnetic moment of the electron precesses as usual but at a frequency different from the RF. The *H* lines within the ferrite now are the resultant produced by vector addition of the rotating magnetic moment and the RF field. A new RF field, which is rotated from the original RF field, results. The amount of rotation is determined by the dc magnetic field and the length of the ferrite.

Faraday Rotation Effect
when microwaves are passed through a piece of ferrite in a magnetic field and their frequency is above the resonant frequency of the ferrite electrons, the plane of polarization of the wave is rotated

Figure 16-23 shows a ferrite rod that is placed lengthwise in the waveguide. The dc magnetic field is set up by a current-carrying coil. Now assume that a wave that is vertically polarized enters the left end of the waveguide. As it enters the ferrite section, it sets up limited precession motion of the electrons. The interaction between the magnetic fields of the wave and the precessing electrons rotates the polarization of the wave. With the correct dimensions of the ferrite rod, the wave is polarized at a 45° angle from the original. Different dimensions of the rod and magnetic field strengths produce other shifts in polarization.

Circulator As mentioned in Sec. 16-3, two-terminal amplifying devices require a means of isolation between input and output power. A circulator is a ferrite device that is the commonly used solution to that problem. The most popular type of circulator is the Y circulator shown in Fig. 16-24.

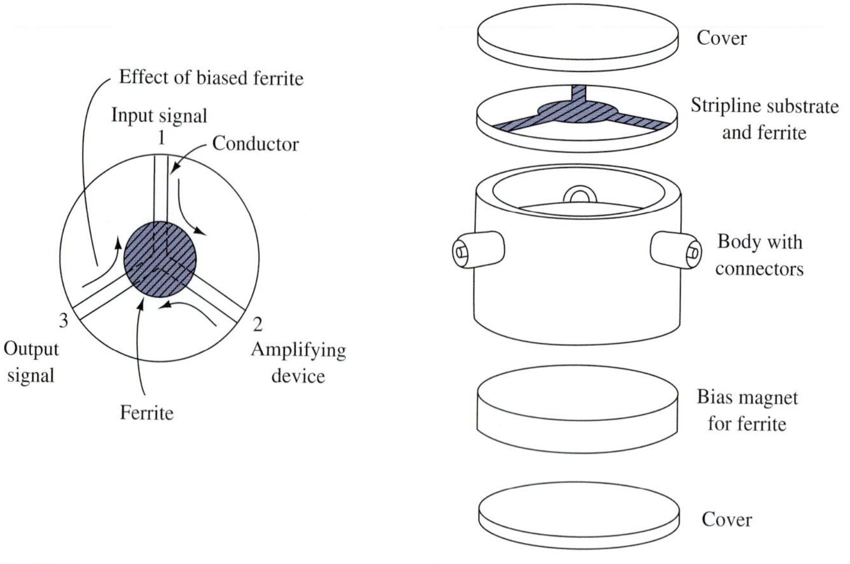

FIGURE 16-24 Y circulator.

Y circulators come in waveguide, coaxial line, or microstrip versions, with the latter shown in Fig. 16-24. With the three lines arranged 120° apart as shown, energy coupled into arm 1 goes only to arm 2, while 2 feeds only 3 and 3 feeds only 1. The ferrite provides the correct rotational shift to provide this operation.

Ferrite Bead
small bead of ferrite material that can be threaded onto a wire to form a device that offers no impedance to dc and low frequencies, but a high impedance at RF

Other Ferrite Applications Ferrites are used in many nonmicrowave applications. They are widely used in portable radio antennas, as the core for winding, IF transformer cores, TV flyback and deflection coil cores, magnetic memory cores in computers, and tape recorder heads. Another application to the communications field is the **ferrite bead.** It is a small donut of ferrite material with a hole through its center so that it can be threaded onto the wires of an electronic circuit. Its effect is to offer almost no impedance to dc and low frequencies but a relatively high impedance to radio frequencies. Ferrite beads are widely used as inexpensive replacements for radio-frequency chokes (RFC) to obtain effective RF decoupling, shielding, and parasitic suppression without an attendant sacrifice in dc or low-frequency power.

A ferrite bead on a conductor and its inductive effect are shown in Fig. 16-25. As the unwanted high frequency flows through the conductor, it creates a magnetic field around the wire. As the field passes into the ferrite bead, the higher (than air)

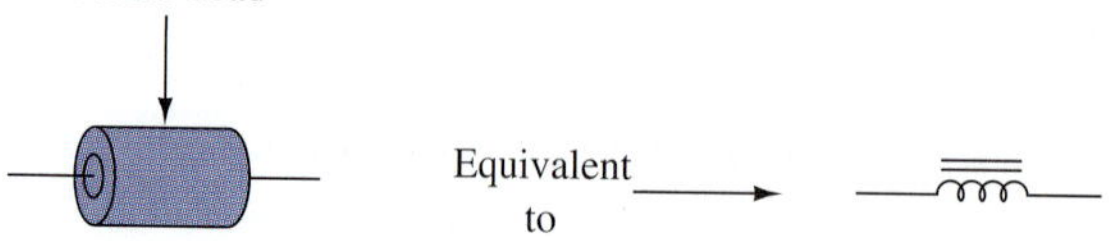

Ferrite bead

Equivalent to

FIGURE 16-25 Ferrite bead.

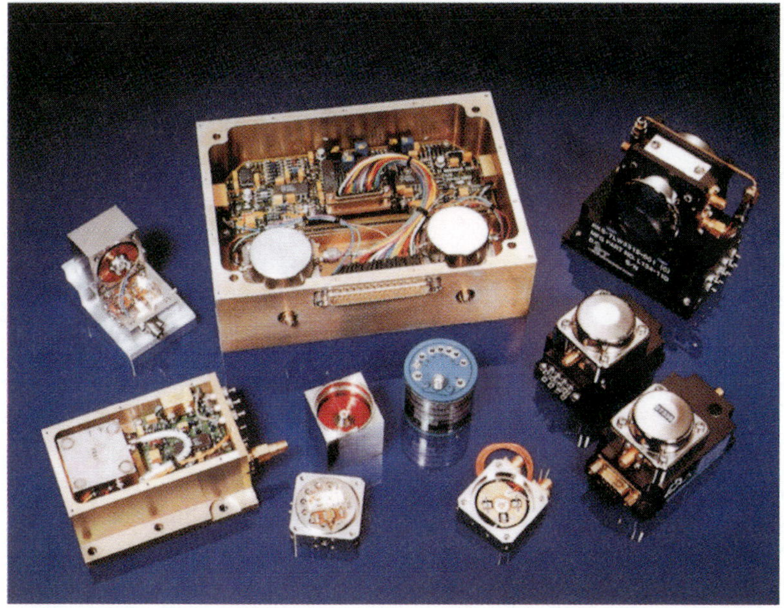

Ferrite applications. (Courtesy of ST Microwave Corp.)

permeability of the bead causes the local impedance to rise and to create the effect of an RFC in that location.

Because ferrite materials can attenuate specific microwave frequencies, they are being used for filter applications in place of resonant cavities. The highest-Q ferrite filters are the yig materials. These filters offer small size and electronic (magnetic) tuning advantages over the resonant cavities but are not as high-Q in response. Yig filters are commonly used with the Gunn or IMPATT devices in electronically controlled solid-state microwave sources, as shown in Fig. 16-26. The electromagnet that controls the frequency has been omitted in the figure but must surround the yig sphere shown. The output energy is taken by the RF coupling loop as shown. The simplicity of this variable-frequency microwave source is apparent from the figure.

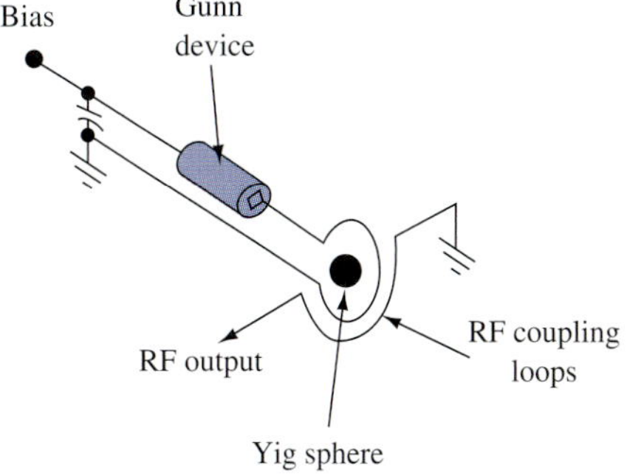

FIGURE 16-26 Variable-frequency microwave signal source.

16-5 Low-Noise Amplification

Microwave receivers usually deal with a very small input signal. The limiting factor on how small the signal may be is primarily determined by the noise figure of the receiver's first amplifier stage. Several new approaches to microwave amplification, including parametric amplifiers and the maser, offer extremely low-noise characteristics.

Parametric Amplifier

Parametric Amplifier provides low-noise amplification at microwave frequencies via the variation of reactance

A **parametric amplifier** provides amplification via the variation of a reactance. This reactance is a *parameter* of a tuned circuit—thus the amplifier's name. Consider an *LC* tank circuit that is oscillating at some microwave frequency. If the capacitor's plates are pulled apart at the instant of time that the voltage across them is maximum positive, work has been accomplished. Pulling the capacitor plates apart decreases capacitance (capacitance is inversely proportional to distance between plates). The charge, q, must remain the same, so the voltage across the capacitor, V, must have been increased because $V = q/C$. This is the first step in the amplification provided by a parametric amplifier.

Now the plates are returned to their original separation as the oscillator's signal causes the voltage across the plates to pass through zero. This effort requires no work because now there is no force exerted between the plates. As the voltage across the plates reaches maximum negative, the plates are once again pushed apart, causing the voltage to increase once again. The process is repeated continuously and amplification has occurred.

The force causing the plates to be pushed apart occurs twice for every cycle of the oscillator's signal and is called the *pump force*. The pump force is thus a signal at twice the oscillator's frequency. It is invariably an ac voltage applied to a varactor diode that is part of the oscillator's tank circuit. The voltage changes the capacitance of the diode at just the right time (as previously described to allow voltage gain). Whereas normally encountered amplifiers provide ac gain with external power obtained from a dc source, the parametric amplifier provides ac gain with external power from an ac source at twice the frequency of the signal being amplified.

A simplified 10-GHz amplifier as just described is shown in Fig. 16-27. Here, it is used as the front end (RF stage) of a 10-GHz receiver. As shown, a four-port circulator is required to keep the various signals from interfering with each other. The low-noise characteristic is the result of amplification via a variable reactance, with the noise from resistance being almost negligible. These amplifiers are capable of noise figures of 0.3 dB, which is an order of magnitude better than that possible with the microwave amplifiers previously discussed. Power gains of 20 dB are realized.

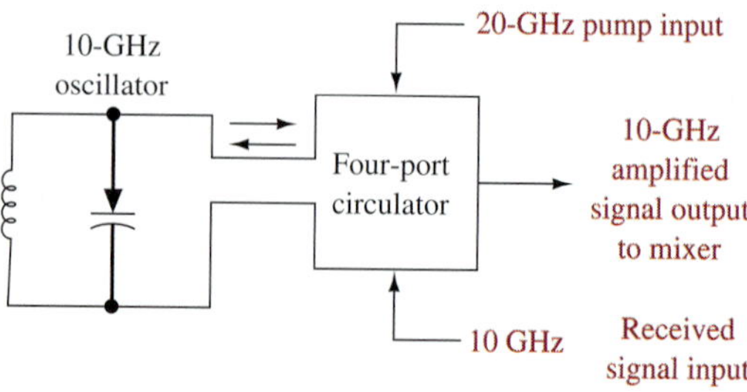

FIGURE 16-27 Parametric amplifier receiver front end.

The noise of microwave systems is often given in terms of *noise temperature*. (See Chapter 1.) This is sometimes a more convenient method of dealing with noise because the noise temperature of two devices is directly additive. Keep in mind, however, that noise temperature, like noise resistance, is simply convenient fiction. Noise temperature is related to noise figure (expressed as a ratio and not in decibels) by $T = T_0$ (NF $-$ 1), where T is noise temperature (K) and T_0 is 290 K. See Sec. 1-4 for a discussion of noise figure.

Parametric amplifiers with pump frequencies double the signal frequency are termed the *degenerate mode*. Parametric amplification is also possible (at reduced gain) with pump frequencies other than 2 times the signal frequency. They are termed the *nondegenerate mode*. In this case, beating between the two frequencies occurs, and a difference signal called the *idler* signal occurs. This mode of operation allows the parametric amplifier to function directly as a mixer, with its output being the first IF frequency.

The Maser

Maser
a low-noise microwave amplifier; similar to a laser that is used with light

An even lower-noise microwave amplifier is the **maser.** It stands for *m*icrowave *a*mplification by *s*timulated *e*mission of *r*adiation. It was developed in 1954 by Professor C. H. Townes, who also advanced the theory for the *laser* in 1958. The laser is the optical version of the maser, where the *l* stands for "light."

Quantum
bundle of energy

Most of the electrons of an atom exist at the lowest energy level when the substance is at a very low temperature (close to absolute zero). If, however, a **quantum,** or bundle of energy, is provided to the atom, the electrons may be raised to a much higher energy level. The applied energy to make this happen is radiation at the frequency of magnetic resonance for the material, as was discussed with ferrites. In this case, however, what is desired is an *emission* of energy, which occurs when the excited electrons return back to the lowest energy level or some intermediate level. The applied frequency is the *pump signal*, while the emitted energy is at some intermediate frequency when the electrons fall back to an intermediate energy level. If an input signal (not the pump signal) is applied at the same frequency as the intermediate frequency, amplification is possible.

A maser amplifier scheme using ruby is shown in Fig. 16-28. Ruby is a crystalline form of silica (Al_2O_3) that has a slight doping of chromium. Its atomic structure has suitably arranged energy levels, and the presence of chromium allows a tuning of usable frequencies of from about 1 up to 6 GHz.

The entire amplifier shown in Fig. 16-28 is enclosed in liquid helium, which maintains it at 4.2 K. This is necessary for the proper electron action within the ruby material, and even the magnetic core is sometimes enclosed to take advantage of *superconductivity*. Because the required magnetic field is extremely high, this setup allows a reduction in power for its maintenance.

The resonant cavity in Fig. 16-28 should be resonant at both the frequency of the pump signal and the signal being amplified. The pump signal, which is a higher frequency than the signal being amplified, is applied to the cavity via a waveguide. The signal being amplified and the output are connected via a circulator and a coupling probe.

These amplifiers are capable of 25-dB gains, with noise figures as low as 0.2 dB. Needless to say, this is an expensive proposition, and their use is reserved for severe applications such as radio astronomy. Other materials may be used in place of ruby, including gases such as ammonia. Ammonia was used in the first practical

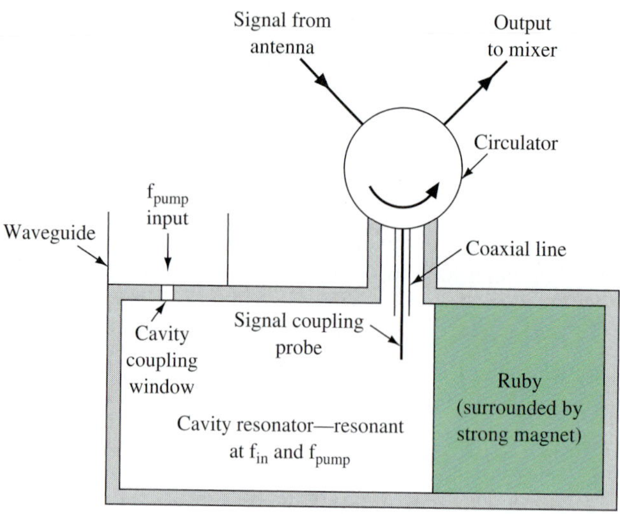

FIGURE 16-28 Ruby maser amplifier.

maser but is capable of amplifying only one frequency—about 24 GHz—because no tuning is possible with an external magnetic field. It does find use, however, in *atomic clock* frequency standards. The accuracy of these clocks is about 1 part per million (1 part per 10^{12}), which means that an error of 1 s every 30,000 years can be expected. Figure 16-29 shows an atomic clock based on atomic transitions of

FIGURE 16-29 Rubidium atomic clock. (Courtesy of Temex Electronics, Inc.)

elemental rubidium. Rubidium is an unstable element that has a relatively long atomic half-life. This compact clock weighs just about a pound and can be used for critical measurement or timing applications.

16-6 LASERS

The **laser** (*l*ight *a*mplification by *s*timulated *e*mission of *r*adiation) is similar to the maser except for the frequencies involved. Visible light is electromagnetic radiation at 430 to 730 $\times$ 10^{12} Hz, as opposed to the microwaves we have been dealing with at up to about 0.3 $\times$ 10^{12} Hz (300 GHz). We are concerned with the communications applications of the laser, but you are undoubtedly aware of other uses, such as:

Laser
low-noise light wave amplifier

1. Distance-measuring equipment
2. Industrial welding
3. Surgical procedures
4. Military applications
5. Production of holograms (three-dimensional photography)
6. Pickup devices in video-disc playback units and compact disc players

As further explained in Chapter 18, the laser is used as the light source for optical-fiber communications systems. The laser can be used to communicate directly by simply transmitting a modulated laser beam through the atmosphere, but too many interferences (fog, dust, rain, and clouds) generally preclude that application. Outer-space communications do not present these limitations. We will look at an application later in this section.

LASER SOURCES

Many different types of materials have been successfully stimulated to exhibit laser action, including solids, liquids, and gases. The brilliant red, green, blue, and yellow beams seen at laser light shows are produced by gaseous materials. In terms of importance to electronic communications using optical fibers, the semiconductor injection laser is the only useful type. These devices are actually members of the light-emitting diode (LED) family. When the current injected into a diode laser is below a critical value, termed the *threshold* (I_{th}), the diode behaves just like an LED and emits a relatively broad spectrum of wavelengths in a wide radiation pattern. As the current is increased to I_{th}, however, the light narrows into a distinct beam and is confined to an extremely narrow spectrum. *Lasing* action has begun and the device is then functioning as a laser.

The early solid-state lasers were short-lived and could be operated only under short-duty-cycle (pulsed) conditions. The key to the development of continuous-duty, long-lifetime devices was the stripe-geometry injection double-heterojunction (DH) laser. **Heterojunction** refers to a junction of two dissimilar semiconductors, such as gallium arsenide (GaAs) and aluminum gallium arsenide (AlGaAs). This allows the light-emitting *pn* junction region to be sandwiched between two or more semiconductor layers that confine the generation and emission of light to the junction region. This allows a low threshold current and high efficiency.

Heterojunction
a junction of two dissimilar semiconductors

The confinement of light between two heterojunctions results when the refractive index of a *pn* junction material (GaAs) is higher than the semiconductor bordering the *pn* junction (AlGaAs). This causes the heterojunctions to appear as

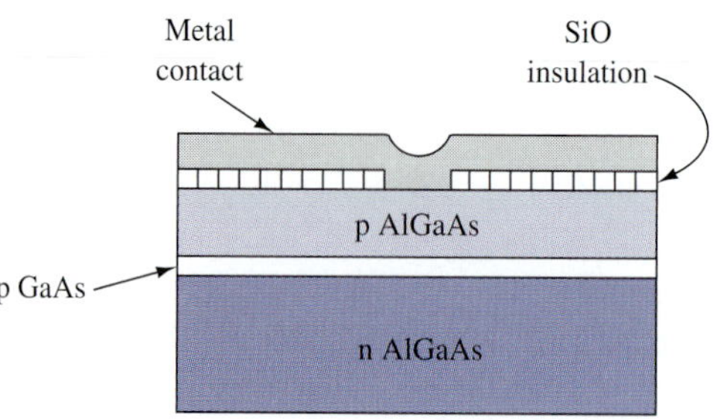

FIGURE 16-30 Stripe-geometry DH laser.

guides to the emitted light just as the clad/core junction does in optical cable (see Chapter 18). The construction of the DH stripe-geometry laser is shown in Fig. 16-30. A *stripe-geometry* device indicates that all but a narrow stripe of one electrode is insulated from the upper surface of one laser. The current flow through the *pn* junction is thereby confined to a thin stripe between the mirror surfaces placed at the two ends. The resulting high-current density in the "stripe" provides very low "lasing" threshold current. These devices can generate several milliwatts of laser light at forward currents of about 100 mA.

The DH lasers are ideally suited for communications through optical fibers (see Sec. 18-5). They are also finding use in laser printing systems and video/audio disc players. Because these devices operate continuously at room temperature, you may think that you need only connect it to a dc source with an adjustable current-limiting resistance and you're in business. But setup is more difficult than that because of the temperature sensitivity of solid-state lasers. A change of just 1° can halt the lasing action or even destroy the device. Because of this, a DH laser is usually operated in a temperature-controlled oven and/or its forward current is temperature compensated.

Laser Communications

As stated earlier, the laser can be used directly as a communication link. The laser can be modulated to contain a large amount of information. As will be described in Chapter 18, optical fibers are usually used to guide the light to its destination. The direct transmission has its problems—it is limited to line of sight, a bird can easily destroy a communication by flying through the laser beam, and severe attenuation by rainfall or snow cannot be circumvented. These conditions are not a problem in outer space, however.

A laser communication system is illustrated in Fig. 16-31. This laser provides an excellent communications link between orbiting communications satellites. The tracking and data acquisition satellite system shown is used by NASA. The need for a cross-link between two satellites at up to 2-Gb/s data rate can be handled by a laser link using laser dishes 1 to 2 ft in diameter. By comparison, a microwave link requires dish antennas 6 to 9 ft in diameter.

Laser Computers

Transphasor
an optical switch using a laser beam

Future computers may be based on optical switches. Such a switch, called a **transphasor,** is represented in Fig. 16-32. A laser beam is applied to a special crystal made of indium antimonide. Most of the laser beam bounces off, but some enters the crystal, where it is trapped, bouncing back and forth. The transphasor is now

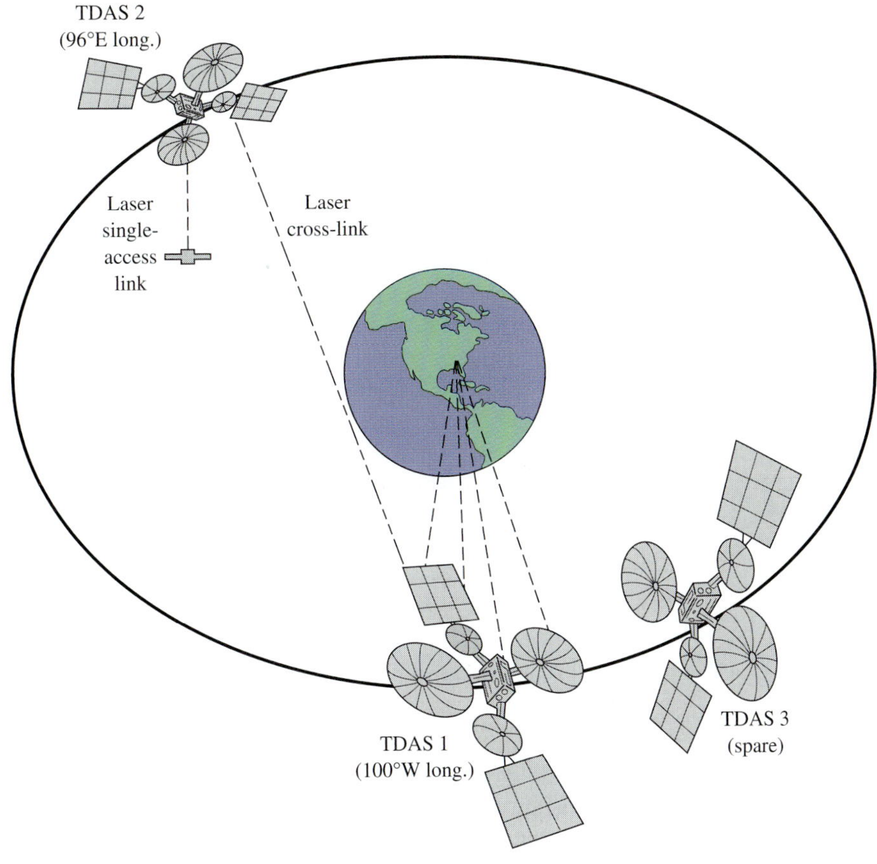

FIGURE 16-31 Laser/satellite communication. (Courtesy of Microwaves and RF, April 1985.)

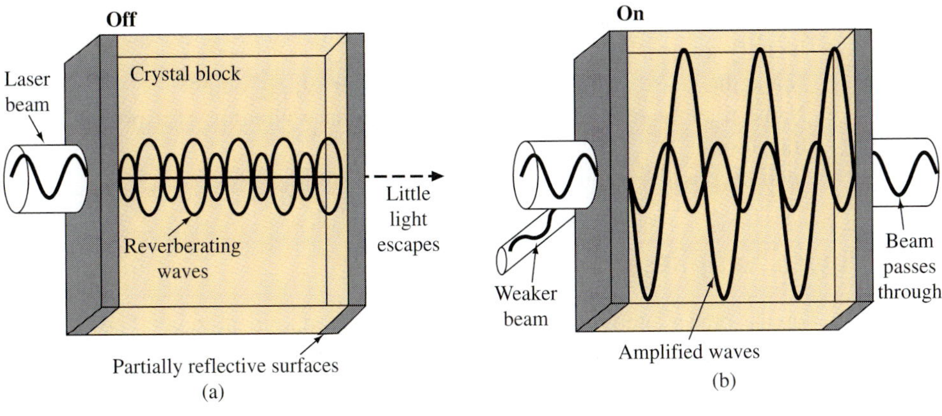

FIGURE 16-32 Transphasor.

"off," as shown in Fig. 16-32(a). In Fig. 16-32(b), a second, weaker laser is directed at the crystal. It increases the light intensity only slightly within the crystal but causes the reverberating light waves to start reinforcing one another, which in turn causes the laser light suddenly to flash out of the crystal's other side. In effect, a weak beam of photons exerts control over a strong one. This is analogous to the weak base or gate current of a transistor having control over a larger current.

The foremost advantage of optical computers is operating speed. It is expected that speeds 1000 times faster than electronic computers will be attained. The second main advantage comes from the fact that photons have no charge or mass. Unlike electrons, photons have little effect on other nearby photons and can even pass right through each other. This means that multiple beams of light in an optical switch could remain separate, whereas several currents in a single transistor become mixed. This characteristic means that optical computers will lend themselves to parallel processing architecture. Instead of solving problems step by step, parallel machines break apart computational puzzles and solve the many parts or steps all at once, much as the human brain can do. This new technology may also prove useful in fiber-optic systems. The expensive translations from electrons to photons and then (after transmission) back to electrons may be eliminated.

16-7 Troubleshooting

Up to this point, we have not considered the power supply's role in electronic communications equipment. The power supply furnishes the voltage and current requirements for electronic circuit operation. Special high-voltage power supplies used in microwave systems and laser systems enable large power handling tube and semiconductor circuits to operate. In this section, we will look at two popular types of power supply circuits and some faults that can occur in them.

In this section, we'll also learn to troubleshoot a traveling wave tube amplifier (TWTA). Improper operation of a TWTA can nearly always be traced to a power supply. There are a few rules you must know to do the job. Violation of the rules results in loss of the tube. Please be aware that extreme caution must be observed when dealing with power supplies. The ac line voltage and any high-voltage outputs can be lethal.

After completing this section you should be able to

- Identify a switching power supply
- Identify a linear power supply
- Name the cause of high ac ripple on a power supply output
- Name a cause for a blown fuse in the primary winding of a power supply transformer
- Identify possible failure modes in a TWT amplifier

Power Supplies

Two popular power supplies are found in electronic equipment today. These are the linear power supply and the switching power supply. Many variations of both kinds of supplies exist. The linear power supply usually has a power transformer that is large and heavy. The linear supply furnishes a constant output voltage to a load. Excessive power in the form of heat is usually wasted in this kind of power supply.

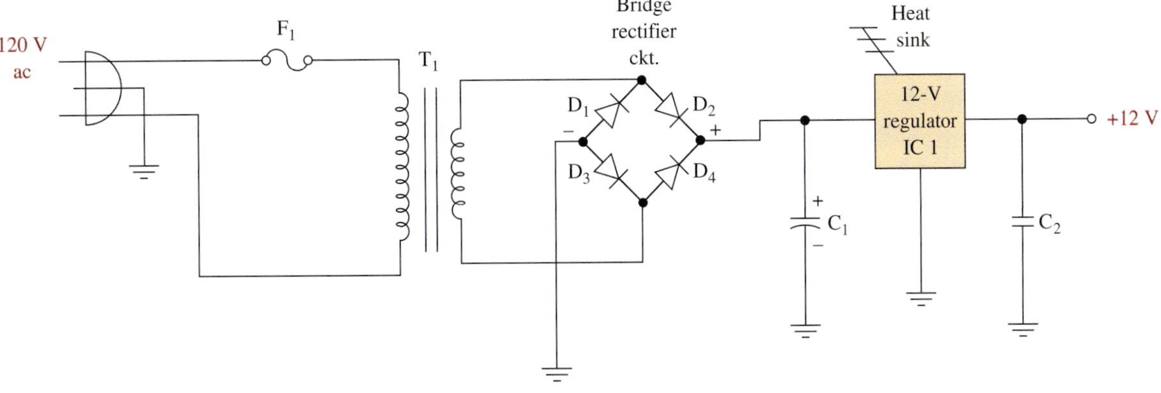

FIGURE 16-33 A linear 12-V regulated power supply using an IC 12-V regulator.

Because of this wasted power, efficiency is low. A typical linear power supply is illustrated in Fig. 16-33.

The switching power supply is light and does not use a large bulky power transformer. Instead, a smaller transformer is used. A diagram of the switching power supply is shown in Fig. 16-34. The input ac voltage is rectified, filtered, and applied to the transformer, T_1. The power transistor, Q_1, in the negative supply line is switched on and off at a high frequency rate, 20 to 40 kHz. The load requirement

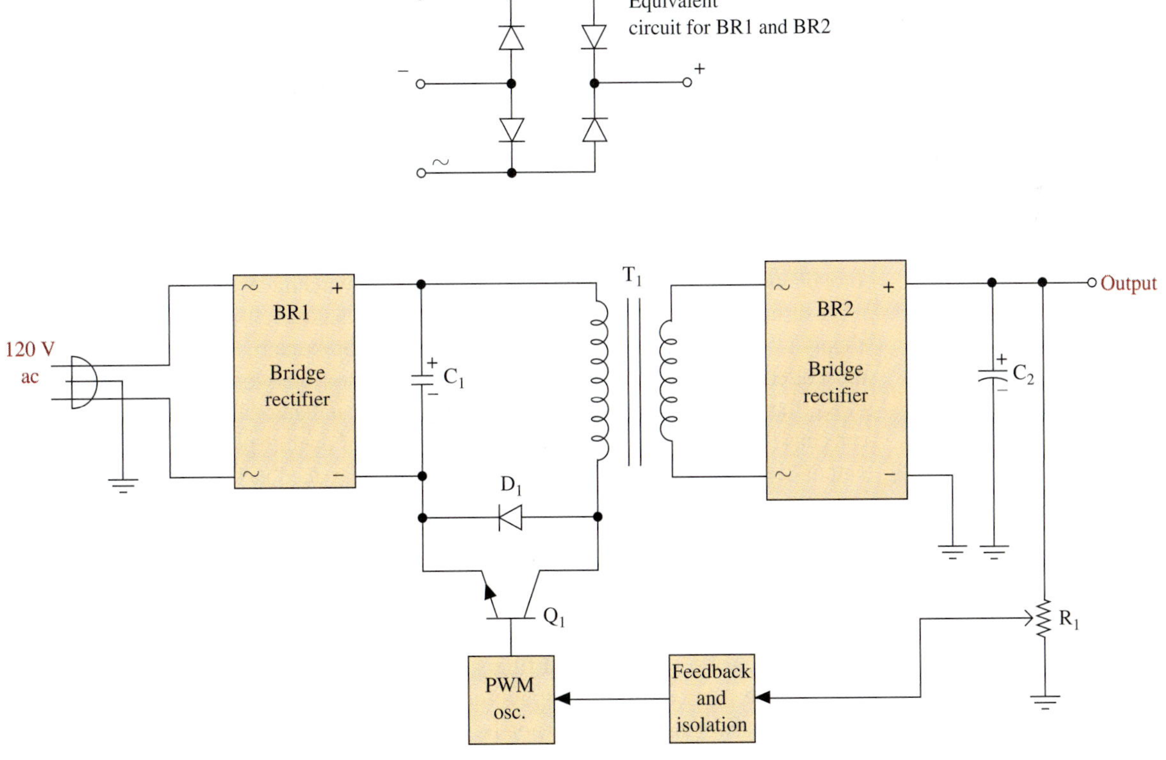

FIGURE 16-34 Simplified diagram of a switching power supply.

determines the output of the supply. If the load requirement goes up, the supply furnishes more power. If the load goes down, less power is supplied. Power is not wasted, as in the linear power supply.

High efficiency rates are achieved in the switching power supply by feeding a portion of the output voltage back to control an oscillator. The oscillator is part of a pulse-width-modulation circuit. A pulse-width-modulated signal is supplied to the base of the power transistor Q_1, which regulates the transistor's duty cycle. An increase of Q_1's duty cycle increases the output voltage to maintain regulation. A decrease of the duty cycle decreases the output voltage.

Troubleshooting the Linear Supply The linear power supply not only provides operating voltage and current to the electronic circuits connected to it but must also provide decoupling for them. The well-designed power supply appears as a low impedance to the decoupled frequencies. If the power supply does not provide decoupling, problems like low-frequency oscillation and distortion can occur in the audio circuits. Hum in the audio output is caused by a bad filter capacitor. Capacitor C_1 in Fig. 16-33 is the input filter capacitor. This capacitor reduces the ac ripple in the dc output of the power supply. For a high-ripple problem, replace C_1. The service manual normally specifies the maximum allowable ripple on the dc output voltages. Voltage is regulated by IC 1 in Fig. 16-33. If IC 1 fails to regulate, a higher than normal voltage appears at the output. Many power supplies have several regulator circuits operating. Check each specified output voltage against actual measurements to ensure that the supply is operating correctly.

The diodes, D_1–D_4, form a full-wave bridge circuit that rectifies the ac from the transformer's secondary winding. If any of the diodes should short, the fuse in the transformer's primary winding will blow. If the power supply blows fuses, suspect a shorted diode. High-wattage resistors in the power supply circuits are subject to changes in value that could change an output voltage. Power transistors are often used in regulator circuits. These are usually mounted on heat sinks and may short out. A shorted power transistor will most certainly blow a fuse and may even burn up any resistor in series with it. When troubleshooting power supplies, look for blown fuses, burned resistors, corroded solder joints, and leaky filter capacitors.

Troubleshooting the Switching Power Supply Switching power supplies are often more difficult to troubleshoot because of the feedback circuit used to regulate the power supply's output. The feedback circuit is a closed loop system. Breaking the loop is the most effective way to isolate a feedback problem. The closed loop in Fig. 16-34 consists of R_1, the feedback and isolation block, the PWM oscillator, and Q_1. Any one of these can cause the switching power supply to shut down or operate poorly. For example, if all outputs are low or all outputs are high, suspect a feedback loop problem. If the protection diode, D_1, continually blows, the feedback loop is the likely candidate. If only one or two output voltages are low, check the filter capacitors associated with those outputs. The switching power supply can emit electromagnetic interference (EMI) radiation into nearby communications gear or the circuits it is supplying if it is not properly filtered. EMI filters protect other circuits from this interference by passing the interference to ground. Bad filters let the power supply generate noise. Use the oscilloscope to monitor the outputs for noise, ripple, and unusual interference. The switching power supply must not be operated without a load connected to it or damage will most likely occur. As stressed in earlier troubleshooting sections, follow a logical troubleshooting plan when tracking down a power supply fault.

Troubleshooting a Traveling Wave Tube Amplifier

Some initial considerations include:

1. To make construction of the tube RF output circuit simple, the collector and helix are grounded. The cathode is above ground and has a negative voltage on it (see Fig. 16-35).
2. The helix is delicate and helix current is small. The collector draws nearly all the current. The helix is protected with an overcurrent relay that kills the power supplies. Never defeat this relay because the tube can be lost in almost zero time.
3. Always operate the TWT amplifier into a good dummy load. High VSWR causes high helix current. The RF input should also be terminated.

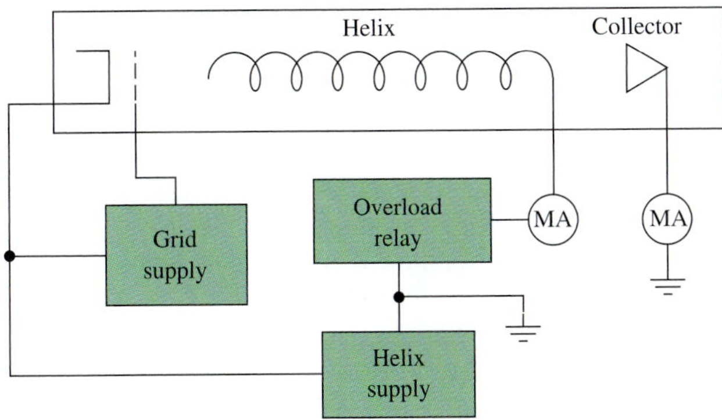

FIGURE 16-35 TWT amplifier dc voltages.

Typical DC Problems

1. Low gain is present. Gain is a function of helix voltage. Make sure it is correct.
2. The amplifier does not stay on and overloads relay trips. Check the overload trip point with an external power supply and milliampere meter, as shown in Fig. 16-36. The relay contacts should open when the desired overload current is reached. The resistor may be a potentiometer. You will find the specification for helix current in the TWT manual.

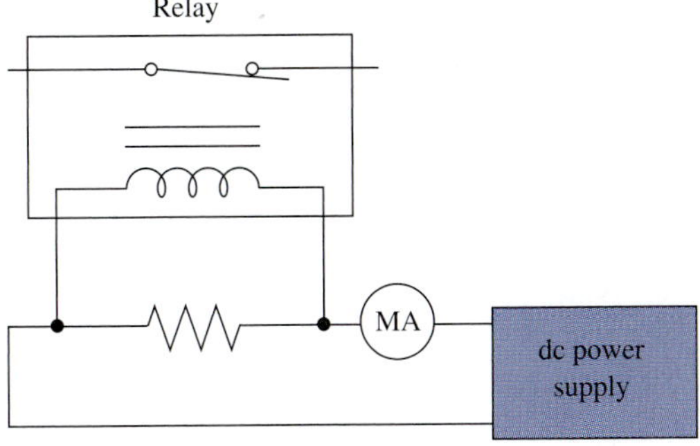

FIGURE 16-36 Overload relay.

Check all power supply components with the power off. If you cannot find the problem, you may have to construct a bank of resistors to use in place of the tube to troubleshoot the power supplies while they are operating.

3. Excessive collector current is present. Make sure the grid supply is okay. Remember, both sides of the grid supply are above ground.
4. You cannot turn the helix supply on. Look for malfunctioning fault relays. TWTs usually have relays that turn the power supplies on in a sequence: heater first, grid next, and helix last.
5. Spurious modulation is present. Hum or ac ripple on the power supplies modulates the RF output. The helix supply is a regulated supply and should have almost no hum. Check this with an oscilloscope with a high-voltage probe.
6. The power output is low. Is too much RF drive being applied? The power output of a TWT increases as drive is increased, until saturation is reached. If drive is increased past that point, power output falls off.

Typical RF Troubles

1. Poor frequency response or low gain. Most TWTs have an RF bandpass filter in the input to correct the frequency response of the tube. Check the filter for excessive loss at the center frequency and for proper bandpass response. Generally, there should be a loss of several dB at the band edges and very little loss at the band center.
2. Low RF output. At microwave frequencies, coaxial cables and connectors cause a lot of trouble. Assembly procedure is critical for proper operation. Look for loose shields at the connectors, and check the cable for loss and VSWR.

   Good TWTs have an isolator in the output circuit. Isolators should show a loss of 1 dB or less in the forward direction and at least a 20-dB loss in the reverse direction.

   These components are best checked with a network analyzer and a sweep oscillator. The test methods shown in Chapters 12, 14, and 15 work too, but they yield less information. Unfortunately, network analyzers are very expensive and at times you will have to do it the hard way.

16-8 Troubleshooting with Electronics Workbench™ Multisim

The concept of microwave devices has been introduced in this chapter. This exercise is used to explore the characteristics of microwave devices, including the RF capacitor, the RF inductor, and the RF transistor. To begin the exercise, open the file **Fig16-37.ms7 (.msm)** in your EWB Multisim CD. This file contains three circuits. Circuit A contains an ideal capacitor with a value of 0.3223 pF, whereas circuit B contains an RF capacitor of the same value. The circuit is shown in Fig. 16-37. The last circuit is a component view of the model of an RF capacitor.

A Bode plotter instrument has been connected to each RC circuit and each is terminated with a 1-kΩ resistor. Start the simulation and compare the results of the two Bode plots. The Bode plots are shown in Fig. 16-38. The top plot is for the ideal capacitor, whereas the bottom plot is for the RF capacitor.

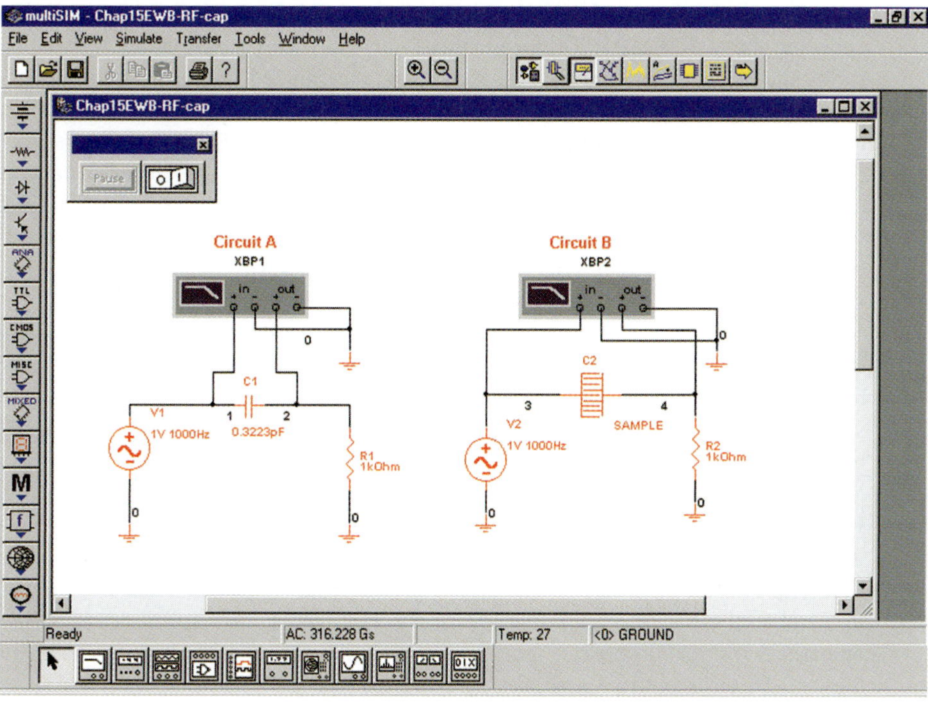

FIGURE 16-37 The simulation circuit for comparing the ideal and RF capacitors.

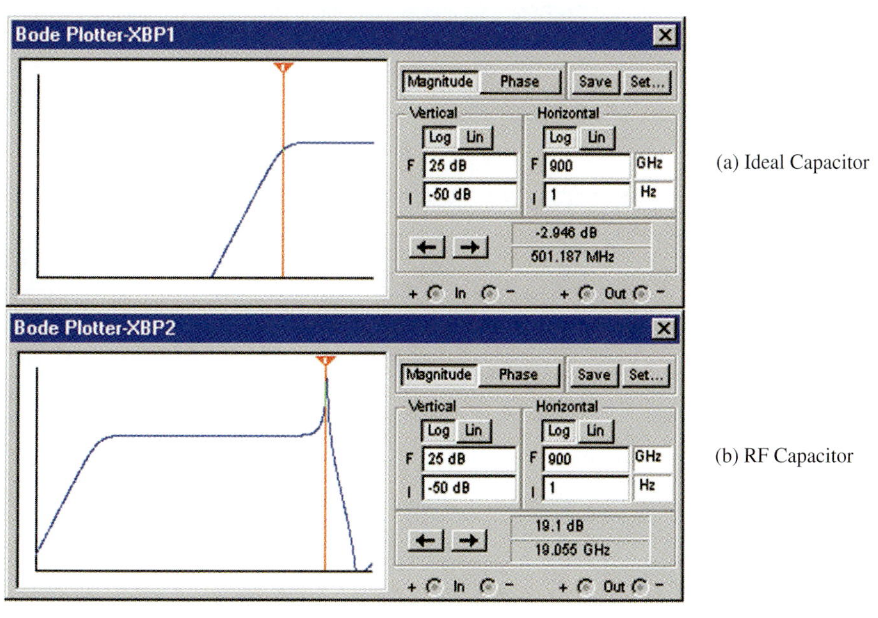

FIGURE 16-38 The Bode plots for the ideal and RF capacitors.

The Bode plot for the ideal capacitor (circuit A) shows that it passes all frequencies above 500 MHz. This is ideal but not realistic. The RF model (circuit B) shows that at very low frequencies, the signal is attenuated by the capacitor, as expected. At high frequencies, the signal has a flat response until about 19 GHz, which is the resonant frequency of the capacitor. To have a resonant frequency implies that the capacitor has an inductance. In fact, all components have resistance, capacitance, and inductance, but most of these characteristics are not significant unless you are operating the circuit at high frequencies. The component view of the RF model for a capacitor is provided in **Fig16-37.ms7 (.msm).** The circuit is also provided in Fig. 16-39.

The RF capacitor is quite complex at high frequencies, as can be seen by the model. The model shows that the RF capacitor is resistive and inductive, in addition to being capacitive. This is an important concept to remember, especially when you are working with high-frequency circuits. Even a simple test lead can alter the tuning of a high-frequency circuit.

The next exercise examines the characteristics of an RF amplifier. Open **Fig16-40.ms7 (.msm)** in your EWB CD. This circuit contains two simple BJT common emitter amplifiers. A 2N2222 BJT transistor is used in circuit A; a BF517 RF transistor is used in circuit B. The circuit is shown in Fig. 16-40. Start the simulation and verify that each amplifier is working. Use the oscilloscope to verify proper operation. You should observe gain and you should see a 180° phase inversion of the signal from input to output.

Next, generate Bode plots for each circuit. You will see that the circuit containing the 2N2222 transistor (circuit A) has a 3-dB upper cutoff frequency of about

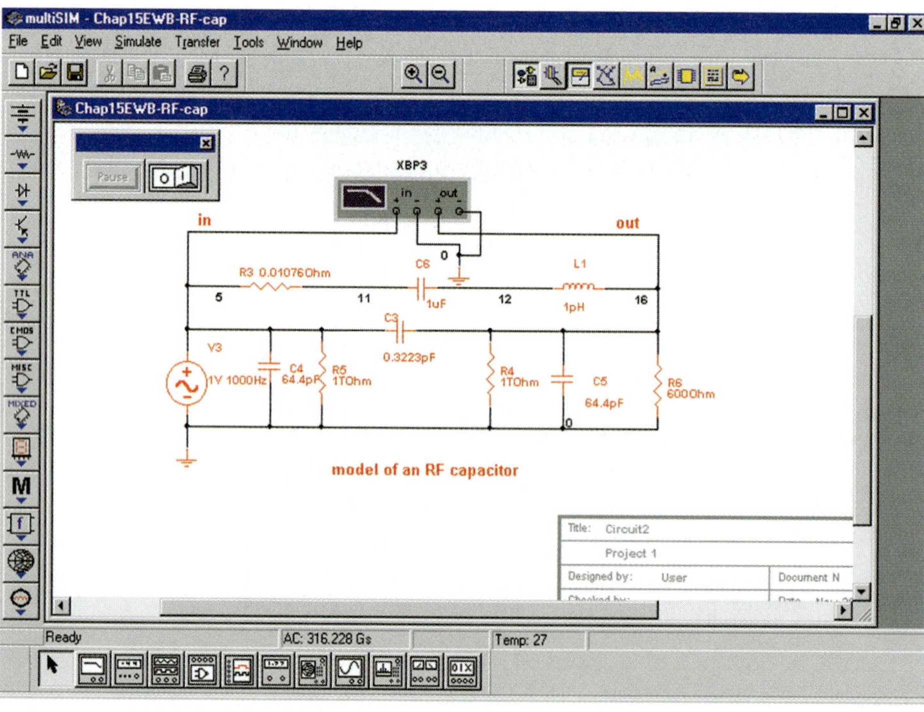

FIGURE 16-39 The component view of the model for an RF capacitor.

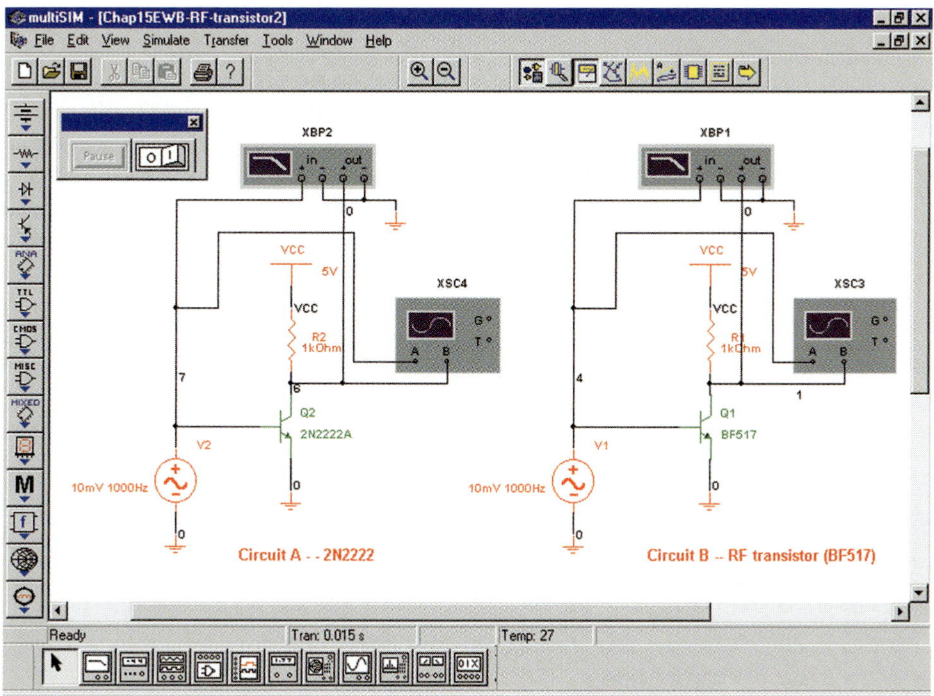

FIGURE 16-40 The example amplifier circuits that incorporate either a low-frequency or a high-frequency RF transistor.

35.5 MHz, whereas circuit B, which is using the BF517 RF transistor, has a 3-dB upper cutoff frequency of about 240 MHz. This demonstrates the vast improvement in the frequency response of an amplifier with the use of an RF circuit.

The following exercises provide you with an opportunity to explore the characteristics of an RF inductor and troubleshoot an RF amplifier.

Electronics Workbench™ Exercises

1. Open the file **FigE16-1.ms7 (.msm)** in your EWB CD. This circuit provides a comparison of an ideal and an RF inductor. Determine the upper 3-dB cutoff frequencies for the inductors. (194 kHz, approx. 1.5 GHz)
2. Open the file **FigE16-2.ms7 (.msm)** in your EWB CD. Determine the resonant frequency of this dipole antenna. (f = 1.071 GHz).
3. Open the file **FigE16-3.ms7 (.msm)** in your EWB CD. Determine if the RF amplifier is working properly. If it isn't, locate and correct the fault and retry the simulation. Report on your findings.

Summary

In Chapter 16 we studied microwaves and lasers. We learned that microwaves share many properties with light waves. The major topics you should now understand include:

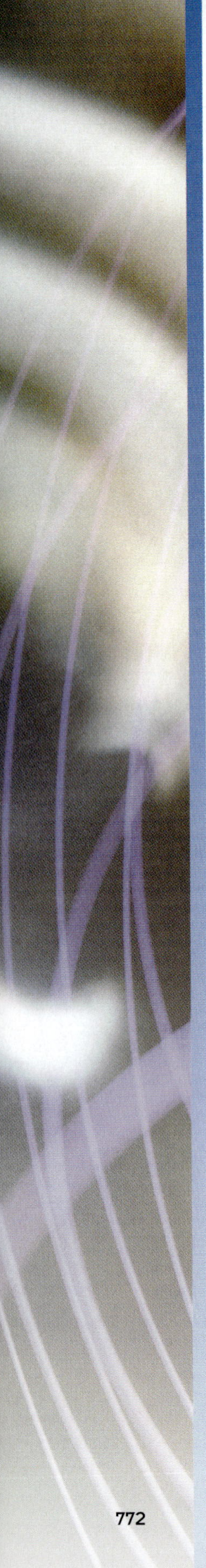

- the description and analysis of microwave antennas, including parabolic, horn, and lens varieties
- the calculation of power gain and beamwidth for parabolic antennas
- the description, operation, and application of magnetrons and traveling wave tubes (TWTs)
- the description of solid-state microwave devices, including the Gunn diode, IMPATT diode, *p-i-n* diode, transistors, and monolithic microwave integrated circuits (MMICs)
- the operation and use of ferrites as attenuators, isolators, and filters
- the applications of circulators and ferrite beads
- the description of low-noise amplification techniques using parametric amplifiers
- the description and application of lasers
- the description of laser sources, laser communications, and laser computers

Questions and Problems

Section 16-1

1. Microwave antennas tend to be highly directive and provide high gain. Discuss the reasons for this.
2. What is a horn antenna? Provide sketches of three basic types, and explain their important characteristics.

* 3. Describe how a radar beam is formed by a paraboloidal reflector.

4. With sketches, explain three different methods of feeding parabolic antennas.
5. A160-ft-diameter parabolic antenna is driven by a 10-W transmitter at 4.3 GHz. Calculate its effective radiated power (ERP) and its beamwidth. (29.3 MW, 0.10°)
6. A parabolic antenna has a 0.5° beamwidth at 18 GHz. Calculate its gain in dB. (50.7)
7. What is a radome? Explain why its use is often desirable in conjunction with parabolic antennas.
8. Explain the principles of a zoned lens antenna, including the transformation of a spherical wave into a plane wave.
9. The antenna in Fig. 16-5 has a resonant frequency of 1.3 GHz. Calculate the bandwidth of this patch antenna. (≃130 MHz)
10. The beamwidth of a 4.0 GHz signal is 1.1°. Determine the dB power gain. (43.8 dB)

Section 16-2

* 11. Explain briefly the principle of operation of the magnetron.

* 12. Draw a simple cross-sectional diagram of a magnetron, showing the anode, the cathode, and the direction of electron movement under the influence of a strong magnetic field.

---

*An asterisk preceding a number indicates a question that has been provided by the FCC as a study aid for licensing examinations.

13. Explain how electric and magnetic fields influence electron travel in a magnetron.
14. List and explain the differences between the two classes of magnetron oscillators.
* 15. Draw a diagram showing the construction and explain the principles of operation of a traveling wave tube (TWT).
16. Describe four methods of coupling for a TWT.
17. Describe how a TWT can be used as an oscillator.
18. What are some practical applications for a TWT? What are some advantages of this device?
19. Describe the operation of a BWO.
20. Describe velocity modulation in a klystron.

Section 16-3

21. In general, describe some advantages and disadvantages of solid-state microwave devices with respect to tube devices.
22. Describe the principle of operation of the Gunn diode. List some of its applications and its important characteristics.
23. What does IMPATT stand for? Describe the basic operation of the IMPATT diode.
24. Explain the two basic arguments leading to the existence of negative resistance in IMPATT diodes.
25. Explain how an IMPATT diode differs from a varactor diode.
26. Draw a sketch showing the construction of a *p-i-n* diode. Describe its operation at low frequencies and at microwave frequencies. List several possible applications for this device.

Section 16-4

27. Describe the composition of a ferrite material. Explain the process of precession as related to ferrite materials.
28. What is an isolator? Describe how a ferrite waveguide isolator works.
29. Describe the Faraday rotation effect.
30. What is a circulator? Draw a sketch of a ferrite Y circulator, and use it to explain the theory of operation.
31. What is a ferrite bead? How is it able to replace the function of a radio frequency choke?

Section 16-5

32. What is a parametric amplifier? In detail, discuss its fundamentals and explain how it differs from a traditional amplifier.
33. Explain the difference between degenerate and nondegenerate mode parametric amplifiers. List some applications for these amplifiers.
34. Sketch a ruby maser amplifier and explain its operation. Why is it necessary to maintain the ruby at extremely low temperatures? What side benefits can be obtained from the cooling?
35. What does the acronym *maser* stand for? Why is the ruby maser more useful in communications than the originally developed ammonia maser?

Section 16-6

36. What is a laser? How does it differ from a maser? List some applications of a laser.
37. Describe the construction of a stripe-geometry DH laser. Explain why temperature stabilization is critical to its operation.
38. Explain why direct laser communications schemes are not widely used.
39. Describe the operation of the transphasor. What are the expected advantages of optical computers?

Section 16-7

40. List the two commonly used types of regulated power supplies and briefly describe their operation.
41. Suppose that each time an active device turns on, the supply voltage drops. Describe possible problems.
42. In the switching power supply in Fig. 16-34, describe what would happen if the diode D_1 were bad.
43. Describe possible problems with the TWT if no output is present.
44. Explain what happens when there is low RF out from the TWT.

Questions for Critical Thinking

45. Write a report on the minimum acceptable parabolic diameter for a signal at 1.5 GHz and the resultant antenna gain and beamwidth. If the antenna diameter were to be increased by a factor of 10, discuss how the gain and beamwidth would be affected.
46. Compare the use of BJT and FET devices at microwave frequencies. Under what frequency-power output conditions would tubes be used instead?
47. How does ferrite material placed in a waveguide provide attenuation? Explain why this attenuation is effective only at one specific frequency range. How can the frequencies attenuated be varied? Describe a method whereby the amount of attenuation can be varied.
48. You are working with a parametric amplifier with a 0.3-dB noise figure. When the signal-to-noise ratio is 7:1, what will the output signal-to-noise ratio be? The noise temperature? Why are these two measurements important? (6.53, 20.7°)

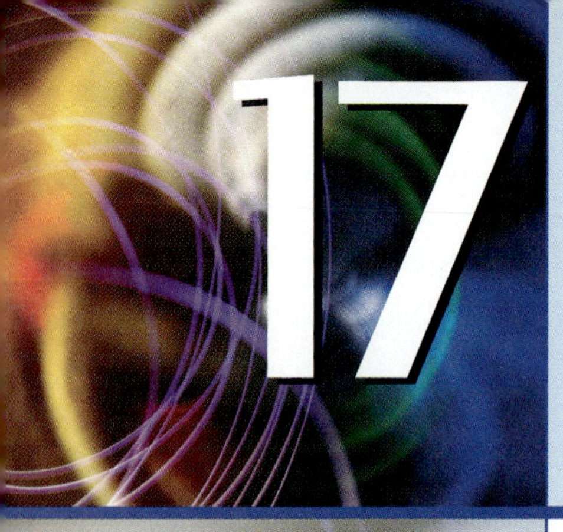

17 Television

Chapter Outline

Harris provides robust, reliable television equipmen ranging from analog to digital technology. (Courtesy of Harris Broadcast Communications. Reprinted with permission.)

Objectives

- Describe the operation of a TV system, including the separation of audio and video functions
- Explain the interlaced scanning process and transmitter/receiver synchronization
- Calculate the effect between resolution and bandwidth
- Describe the operation of a receiver using a detailed block diagram
- Explain the basis of adding color without adding additional bandwidth to the signal
- Describe the color CRT construction and operation
- Analyze the audio system when the stereo sound feature is included
- Develop an understanding of a digital television system, including HDTV, SDTV, MPEG2, AC-3, and 8VSB
- Develop an understanding of digital television transmission and reception
- Explain the steps for troubleshooting a receiver system logically

Key Terms

aural signal
video signal
diplexer
change couple device (CCD)
bucket brigade
pixel
retrace interval
aspect ratio
synchronizing
horizontal retrace
vertical retrace interval
persistence
frame frequency
flicker
interlaced scanning
field
front porch
back porch
color burst
video amplifiers
resolution
vertical resolution
horizontal resolution
vestigial-sideband operation
yoke
flyback transformer
tuner
intercarrier systems
stagger tuning
surface acoustic wave filters
wavetrap
trap
dc restoration
sync separator
clipper
integrator
differentiator
damper
monochrome
interleaving
triads
matrix
luminance
chroma
color killer
confetti
shadow mask
static convergence
dynamic convergence
DTV
HDTV
Advanced Television System Committee (ATSC)
SDTV
4:2:2
MPEG2
AC-3
5.1 Channel Input
8VSB
ATSC pilot carrier
segment sync
frame sync
pixelate
digital on-channel repeater (DOCR)
frequency and phase-locked loop (FPLL)
raster

17-1 Introduction

Television is a field of electronic technology that has more direct effect on the people of our world than any other. It is a very specialized branch of technology that utilizes many of the principles already explained and many new ones.

The concept of television was developed in the 1920s, feasibility was shown in the 1930s, commercial broadcasting started in the 1940s, and the ensuing years have seen the mushrooming growth of an industry so far-reaching that some sociologists make the study of its effects their life's work. The technology, while still undergoing continued improvements, has reached a certain level of maturity. The recent improvements are in the support of digital television. This exciting new area of television is examined in Sec. 17-12.

17-2 Transmitter Principles

Aural Signal
sound or audio portion of a TV signal, transmitted by frequency modulation

A TV transmitter is actually two separate transmitters. The **aural** or sound transmitter is an FM system similar to broadcast FM radio. It is still a high-fidelity system because the same 30-Hz to 15-kHz audio range is transmitted. The major difference between broadcast FM and TV audio systems is that TV uses a ±25-kHz deviation. Recall that broadcast FM uses a ±75-kHz deviation. Thus, the TV aural signal has the same fidelity but is less effective in canceling the indirect noise effects explained in Chapter 5.

Video Signal
picture portion of a TV signal, which is amplitude-modulated onto a carrier

The **video,** or picture, signal is amplitude-modulated onto a carrier. Thus, the composite transmitted signal is a combination of both AM and FM principles. This is done to minimize interference effects between the two at the receiver because an FM receiver is relatively insensitive to amplitude modulation and an AM receiver has rejection capabilities to frequency modulation.

Figure 17-1 shows a simplified block diagram for a TV system. The TV camera converts a visual picture or scene into an electrical signal. The camera is thus a transducer between light energy and electrical energy. At the receiver, the CRT picture tube is the analogous transducer that converts the electrical energy back into light energy.

The microphone and speaker shown in Fig. 17-1 are the similarly related transducers for the sound transmission. There are actually two more transducers shown, the sending and receiving antennas. They convert between electrical energy and the electromagnetic energy required for transmission through the atmosphere.

Diplexer
filter in a TV transmitter that allows both the video AM signal and the audio FM signal to feed the same antenna

The **diplexer** shown in Fig. 17-1 feeding the transmitter antenna feeds both the visual and aural signals to the antenna while not allowing either to be fed back into the other transmitter. Without the diplexer, the low-output impedance of either transmitter's power amplifier would dissipate much of the output power of the other transmitter. The synchronizing signal block will be explained in the next section.

TV Cameras

Charge Couple Device (CCD)
a light-sensitive chip used to convert optical images to an electronic form

The most widely used image pickup device is the **charge couple device (CCD).** CCD cameras are used in many applications such as broadcasting, imaging, scientific

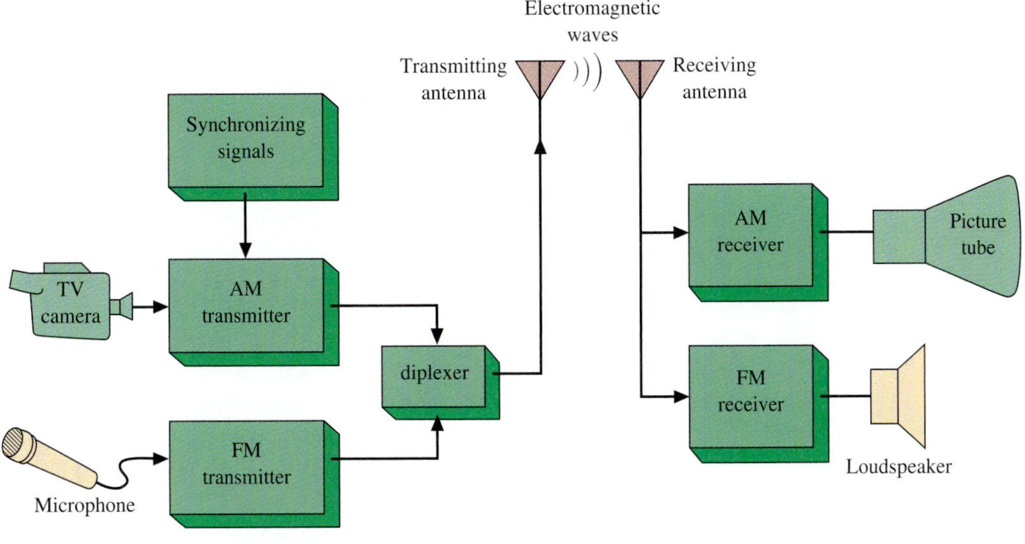

FIGURE 17-1 Simplified TV system.

studies, security, and military applications. The CCD is a solid-state chip consisting of thousands or millions of photosensitive cells arranged in a two-dimensional array. An example of a CCD imaging device is shown in Fig. 17-2. When light (photons) strike the CCD surface, the light information is converted to an electronic analog of the light. The electronic information is then shifted out of the device serially in what is call a **bucket brigade.** The clocking of the bucket brigade is controlled by the timing of the particular system being used. An important limitation of CCD devices is the maximum speed at which the information placed on the device can be serially

Bucket Brigade
the process of serially shifting data out of a CCD

FIGURE 17-2 An example of a CCD imaging device. (Courtesy of Hamaqmatsu Corp.)

shifted out to storage. Undesired characteristics such as smearing can result if the information is not transferred correctly.

Scanning

To understand how these tiny individual outputs can serve to represent an entire scene, refer to Fig. 17-3. In this simplified system, the camera focuses the letter "T" onto the photosensitive cells in the CCD imaging device, but instead of a million cells, this system has just 30, arranged in 6 rows with 5 cells per row. Each separate area is called a **pixel,** which is short for "picture element." The greater the number of pixels, the better the quality (or resolution) of the transmitted picture.

Pixel
picture element; the smallest resolved area in a video scanning technique

The letter "T" is focused on the light-sensitive area so that all of rows 1 and 6 are illuminated [Fig. 17-3(b)], while all of row 2 is dark and the centers of rows 3, 4, and 5 are dark. Now, if we scan each row sequentially and if the *retrace* time is essentially zero, then Fig. 17-3(c) shows the sequential breakup of information. The **retrace interval** is the time it takes to move from the end of one line back to the start of the next lower line. It is usually accomplished very rapidly. The variable light on the photosensitive cells results in a similar variable voltage being developed at the CCD's output, as shown in Fig. 17-3(d). The visual scene has been converted to a video (electrical) signal and can now be suitably amplified and used to amplitude-modulate a carrier for broadcast.

Retrace Interval
the time it takes an electron beam to move from the end of one line to the start of the next line

The picture for broadcast National Television Systems Committee (NTSC) TV has been standardized at a 4:3 ratio of the width to height. This is termed the **aspect ratio** and was selected as the most pleasing picture orientation to the human eye.

Aspect Ratio
in a TV picture, the ratio of frame width distance to frame height distance

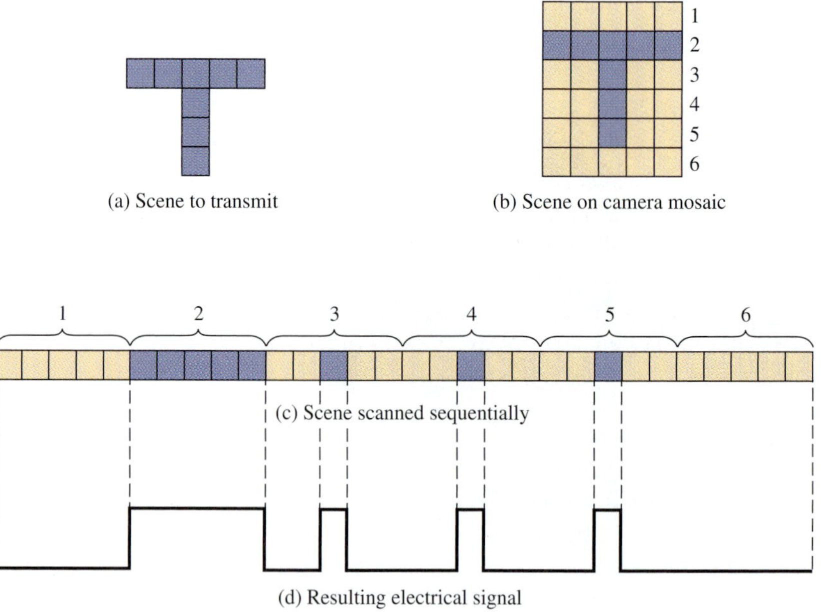

FIGURE 17-3 Simplified scanning representation.

17-3 TRANSMITTER/RECEIVER SYNCHRONIZATION

When the video signal is detected at the receiver, some means of **synchronizing** the transmitter and receiver is necessary:

1. When the TV camera starts scanning line 1, the receiver must also start scanning line 1 on the CRT output display. You do not want the top of a scene appearing at the center of the TV screen.
2. The speed that the transmitter scans each line must be exactly duplicated by the receiver scanning process to avoid distortion in the receiver output.
3. The **horizontal retrace,** or time when the electron beam is returned back to the left-hand side to start tracing a new line, must occur coincidentally at both transmitter and receiver. You do not want the horizontal lines starting at the center of the TV screen.
4. When a complete set of horizontal lines has been scanned, moving the electron beam from the end of the bottom line to the start of the top line (vertical flyback or retrace) must occur simultaneously at both transmitter and receiver.

Synchronizing
in TV, precisely matching the movement of the electron beam horizontally and vertically in the recording camera with the electron beam in the receiver

Horizontal Retrace
in TV, the amount of time it takes to move the electron beam from the right back to the left to start a new line

Visual transmissions are more complex than audio because of these synchronization requirements. At this point, voice transmission seems elementary because it can be sent on a continuous basis without synchronization. Thus, the other major function of the transmitter besides developing the video and audio signals is to generate synchronizing signals that can be used by the receiver so that it stays in step with the transmitter.

In the scanning process for a television, the electron beam starts at the upper left-hand corner and sweeps horizontally to the right side. It then is rapidly returned to the left side, and this interval is termed *horizontal retrace.* An appropriate analogy to this process is the movement of your eye as you read this line and rapidly retrace to the left and drop slightly for the next line. When all the horizontal lines have been traced, the electron beam must move from the lower right-hand corner up to the upper left-hand corner for the next "picture." This **vertical retrace interval** is analogous to the time it takes the eye to move from the bottom of one page to the top of the next.

Vertical Retrace Interval
in TV, the amount of time it takes to move the electron beam from the bottom right corner to the top left corner to start another field

Federal Communications Commission (FCC) regulations stipulate that U.S. NTSC TV broadcasts shall consist of 525 horizontal scanning lines. Of these, about 40 lines are lost as a result of the vertical retrace interval. This leaves 485 visible lines that you can actually see if a TV screen is viewed at close range. The number of visible lines does not depend on the TV screen size. Because this scanning occurs rapidly, persistence of vision and CRT phosphor persistence cause us to perceive these 485 lines as a complete image. **Persistence** is the length of time an image stays on the screen after the electrical signal is removed.

Persistence
length of time an image stays on the screen after the electrical signal is removed

INTERLACED SCANNING

Frame Frequency
number of times per second that a complete set of 485 horizontal lines are traced in a TV receiver

Flicker
motion appears jerky due to insufficient scanning frequency

The **frame frequency** is the number of times per second that a complete set of 485 lines (complete picture) is traced. That rate for broadcast TV is 30 times per second. Stated another way, a complete scene (frame) is traced every $\frac{1}{30}$ s (second). Thirty frames per second is not enough to keep the human eye from perceiving **flicker** as a result of a noncontinuous visual presentation. This flicker effect is observed when watching old-time movies. If the frame frequency were increased to 60 per second,

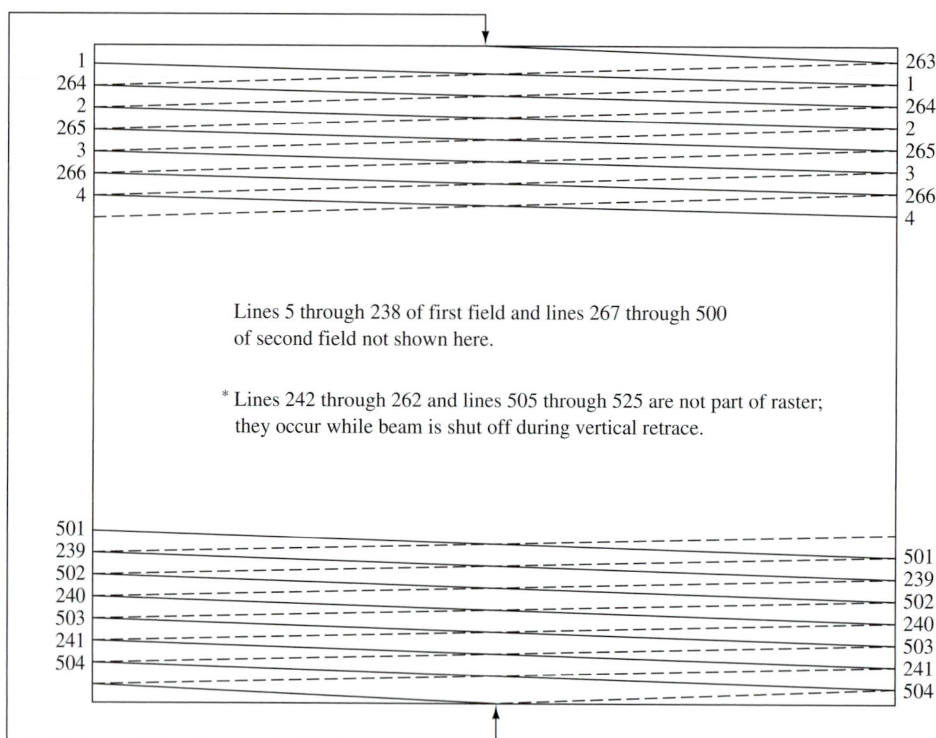

FIGURE 17-4 Interlaced scanning.

Interlaced Scanning
interleaving two fields of 242.5 horizontal lines to form a video image of 485 horizontal lines, so the human eye thinks it is seeing 60 pictures per second

Field
set of lines in a scene

the flicker would no longer be apparent, but the video signal bandwidth would have to be doubled. Instead of that solution, the process of **interlaced scanning** is used to "trick" the human eye into thinking it is seeing 60 pictures per second.

Figure 17-4 illustrates the process of interlaced scanning. The first set of lines (the first **field**) is traced in $\frac{1}{60}$ s, and then the second set of lines (the second *field*) that comprises a full scene (485 lines total) is interleaved between the first lines in the next $\frac{1}{60}$ s. Therefore, lines 2, 4, 6, etc., occur during the first field, with lines 1, 3, 5, etc., interleaved between the even-numbered lines. The field frequency is thus 60 Hz (the actual field rate is 59.94 Hz) with a frame frequency of 30 Hz. This illusion is enough to convince the eye that 60 pictures per second occur when, in fact, there are only 30 full pictures per second.

The process of interlacing in TV is analogous to a trick used in motion picture projection to prevent flicker (noncontinuous motion). In motion pictures, the goal is to conserve film rather than bandwidth, and this is accomplished by flashing each of the 24 frames per second onto the screen twice to create the illusion of 48 pictures per second.

Horizontal Synchronization

To accommodate the 525 lines (485 visible) every $\frac{1}{30}$ s, the transmitter must send a synchronization (sync) pulse between every line of video signal so that perfect transmitter-receiver synchronization is maintained. The detail of these pulses is

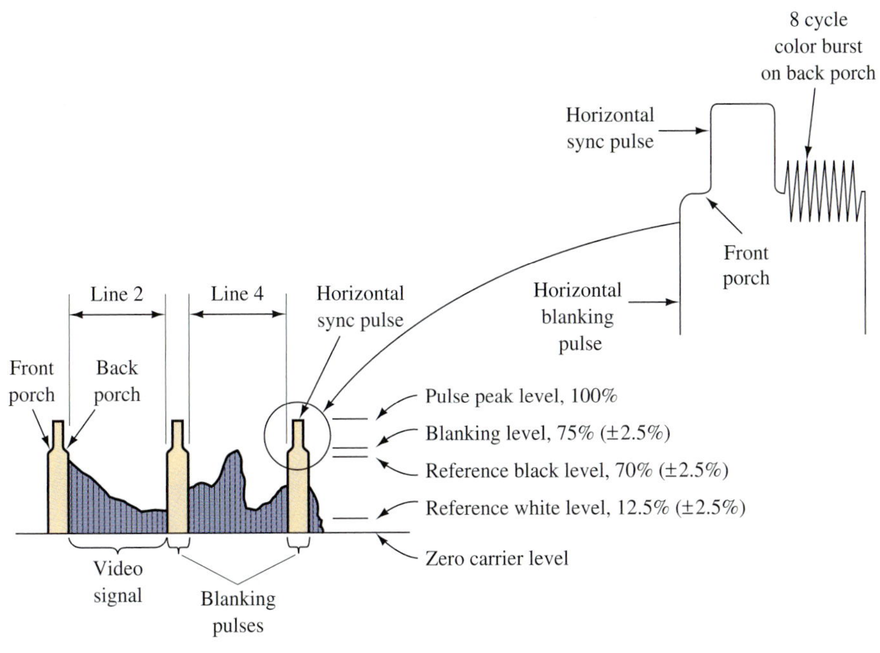

FIGURE 17-5 Horizontal sync pulses.

shown in Fig. 17-5. Three horizontal sync pulses are shown along with the video signal for two lines. The actual horizontal sync pulse rides on top of a so-called blanking pulse, as shown in the figure. The blanking pulse is a strong enough signal so that the electron beam retrace at the receiver is blacked out and thus invisible to the viewer. The interval before the horizontal sync pulse appears on the blanking pulse is termed the **front porch,** while the interval after the end of the sync pulse, but before the end of the blanking pulse, is called the **back porch.** Notice in Fig. 17-5 that the back porch includes an eight-cycle sine-wave burst at 3,579,545 Hz. It is appropriately called the **color burst,** because it is used to calibrate the receiver color subcarrier generator. Further explanation on it will be provided in Sec. 17-6. Naturally enough, a black-and-white broadcast does not include the color burst.

Front Porch
interval before the horizontal sync pulse appears on the blanking pulse in a TV receiver

Back Porch
interval just after the horizontal sync pulse appears on the blanking pulse in a TV receiver

Color Burst
eight-cycle sine-wave burst that occurs on the back porch of the horizontal sync pulse in a color TV broadcast signal

The two lines of video picture signal shown in Fig. 17-5 can be described as follows:

> *Line 2:* It starts out nearly full black at the left-hand side and gradually lightens to full white at the right-hand side.
>
> *Line 4:* It starts out medium gray and stays there until one-third of the way over, when it gradually becomes black at the picture center. It suddenly shifts to white and gradually turns darker gray at the right-hand side.

Since the horizontal sync pulses occur once for each of the 525 lines every $\frac{1}{30}$ s, the frequency of these pulses will be

$$525 \times 30 = 15.75 \text{ kHz}$$

Thus, both transmitter and receiver must contain 15.75-kHz horizontal oscillators to control horizontal electron beam movement.

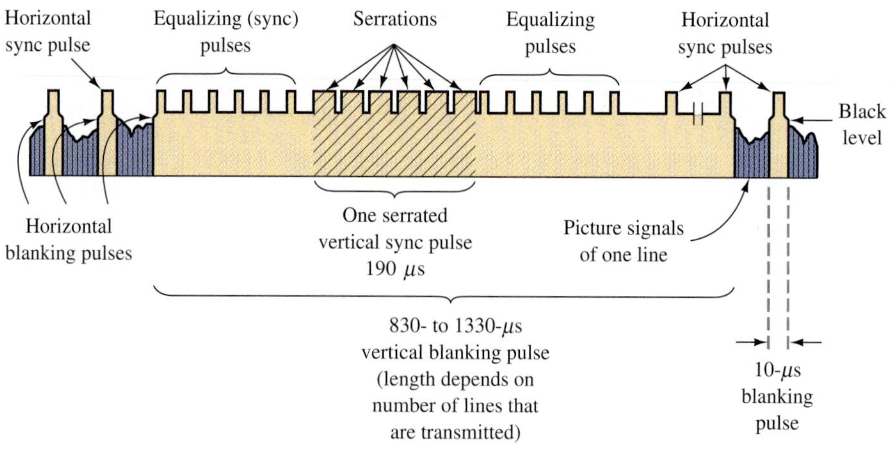

FIGURE 17-6 Vertical retrace interval video signal.

Vertical Synchronization

The vertical retrace and thus vertical sync pulses must occur after each $\frac{1}{60}$ s since the two interlaced fields that make up one frame (picture) occur 60 times per second. The video signal just before, during, and after vertical retrace is shown in Fig. 17-6. Notice that two horizontal sync pulses and the last two lines of video information of a field are initially shown. These are followed in succession by

1. Equalizing pulses at a frequency double the horizontal sweep rate, or 15.75 kHz $\times$ 2 = 31.5 kHz. They each have a duration of about 2.7 μs with a period of 1/31.5 kHz, or 31.75 μs. They are used to keep the receiver horizontal oscillator in sync during the relatively long (830 to 1330 μs) vertical blanking period.
2. One vertical sync pulse with a 190-μs pulse width and five serrations having a duration of 4.4 μs at 27.3-μs intervals. These serrations are used to keep the horizontal oscillator synchronized during the vertical sync pulse interval.
3. More equalizing pulses.
4. Horizontal sync pulses until the entire vertical blanking period has elapsed.

Notice that the vertical blanking period is variable because the number of visible lines transmitted can vary between 482 and 495 at the discretion of the station. All other aspects of the pulses such as number, width, and rise and fall times are tightly specified by the FCC so that all receiver manufacturers know precisely what type of signals their sets have to process.

The vertical sync pulses occur at a frequency of 60 Hz (the exact rate is 59.94), which is the same frequency as the ac line voltage in North America. This allows for good stability of the vertical oscillator in the receiver. In Europe, where 50-Hz line voltage exists, a 50-Hz vertical oscillator system is used.

17-4 Resolution

Video Amplifiers amplifiers with bandpass characteristics from dc up into the MHz region

To provide adequate resolution, the video signal must include modulating frequency components from dc up to 4 MHz. This requires a truly wideband amplifier, and amplifiers that have bandpass characteristics from dc up into the MHz region have come to be known as **video amplifiers.**

Resolution is the ability to resolve detailed picture elements. We already have an idea about resolution in the vertical direction. Since about 485 separate horizontal lines are traced per picture, it might seem that the vertical resolution would be 485 lines. **Vertical resolution** may be defined as the number of horizontal lines that can be resolved. However, the actual resolution turns out to be about 0.7 of the number of horizontal lines, or

Resolution
ability to resolve detailed picture elements in a TV picture

Vertical Resolution
number of horizontal lines that actually make up a TV display

$$0.7 \times 485 = 339$$

Thus, the vertical resolution of broadcast TV is about 339 lines.

Horizontal resolution is defined as the number of vertical lines that can be resolved. A little mathematical analysis will show this capability. The maximum modulating frequency has already been stated as 4 MHz. The more vertical lines to resolve, the higher the frequency of the resulting video signal. The horizontal trace occurs at a 15.75-kHz frequency, and thus each line is 63.5 μs (1/15.75 kHz) in duration. The horizontal blanking time is about 10 μs, leaving 53.5 μs. Since two consecutive lines can be converted into the highest rate video signal, the number of vertical lines resolvable is

Horizontal Resolution
number of vertical lines that can be resolved in a TV display

$$4 \text{ MHz} \times 53.5\ \mu\text{s} \times 2 = 428$$

Thus, the horizontal resolution is about 428 lines. Note that the 428 vertical lines conform nicely to the 339 horizontal lines when one remembers that a TV screen has a 4:3 width to height (aspect) ratio (428/339 $\simeq$ 4/3). Thus, equal resolution exists in both directions, as is desirable. Increased modulating signal rates above 4 MHz allow for increased vertical or horizontal resolutions or some increase for both. This is shown in the following examples.

Example 17-1

Calculate the increase in horizontal resolution possible if the video modulating signal bandwidth were increased to 5 MHz.

Solution

The 53.5 μs allocated for each visible trace could now develop a maximum 5-MHz video signal. Thus, the total number of vertical lines resolvable is

$$53.5\ \mu\text{s} \times 5 \text{ MHz} \times 2 = 535 \text{ lines}$$

Example 17-2

Determine the possible increase in vertical resolution if the video frequency were allowed up to 5 MHz.

Solution

The visible horizontal trace time can now be decreased if the horizontal resolution can stay at 428 lines. That new trace time is

$$\text{trace time} \times 5 \text{ MHz} \times 2 = 428$$
$$\text{trace time} = 42.8\ \mu\text{s}$$

Once again, assuming that 10 μs is used for horizontal blanking, that means 52.8 μs total can be allocated for each horizontal trace. With $\frac{1}{30}$ s available for a full picture, that implies a total number of horizontal traces of

$$\frac{\frac{1}{30}\text{ s}}{52.8\ \mu\text{s}} = 632 \text{ lines}$$

Allowing 32 lines for vertical retrace means a vertical resolution of

$$600 \times 0.7 = 420 \text{ lines}$$

Examples 17-1 and 17-2 are excellent proofs of Hartley's law (see Sec. 1-6). It is plain to see that an increase in bandwidth led to the possibility of greater transmitted information (in the form of increased resolution).

17-5 THE TELEVISION SIGNAL

The maximum modulating rate for the video signal is 4 MHz. Because it is amplitude-modulated onto a carrier, a bandwidth of 8 MHz is implied. However, the FCC allows only a 6-MHz bandwidth per TV station, and that must also include the FM audio signal (*only* is a relative term here because 6 MHz is enough to contain 600 AM radio broadcast stations of 10 kHz each). The TV signal that is transmitted is shown in Fig. 17-7.

The lower visual sideband extends only 1.25 MHz below its carrier with the remainder filtered out, but the upper sideband is transmitted in full. The audio carrier is 4.5 MHz above the picture carrier with FM sidebands as created by its ±25-kHz deviation. The 54- to 60-MHz limit shown in Fig. 17-7 is the allocation

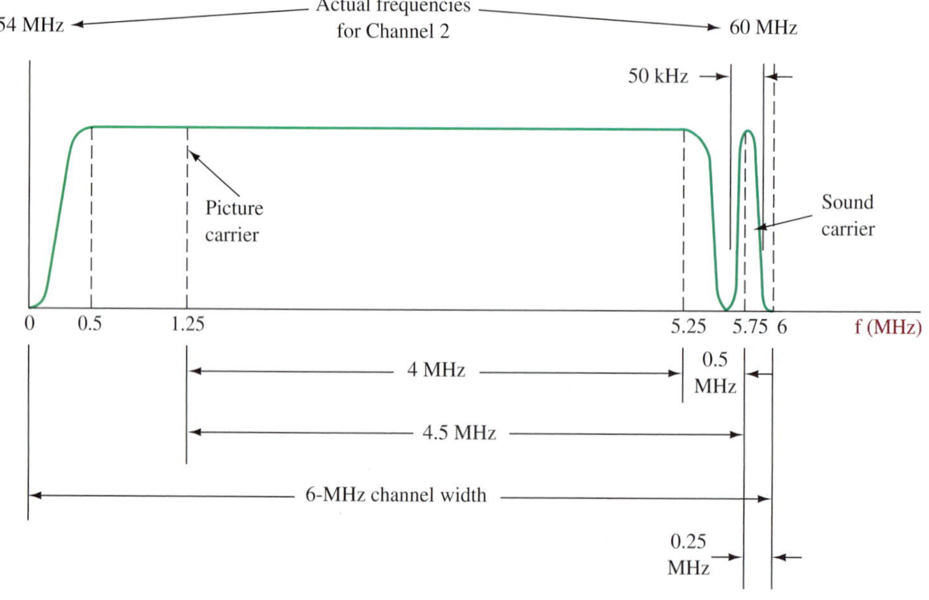

FIGURE 17-7 Transmitted TV signal.

Table 17-1 TV Channel Allocations

Lower VHF Band		Upper VHF Band		UHF Band	
Channel	*Lowest Frequency (MHz)*	*Channel*	*Lowest Frequency (MHz)*	*Channel*	*Lowest Frequency (MHz)*
2	54	7	174	14	470
3	60	8	180	24	530
4	66	9	186	34	590
	(4 MHz skipped)	10	192	44	650
5	76	11	198	54	710
6	82	12	204	64	770
		13	210	69	800

for channel 2. Table 17-1 shows the complete allocation for all the VHF and UHF channels. Notice the VHF channels are broken up into two bands—54 to 88 MHz and 174 to 216 MHz. The UHF band (channels 14 to 69) is continuous and eats up a tremendous chunk of the usable frequency spectrum, as you can see.

The lower sideband is mostly removed by filters that occur near the transmitter output. While only one sideband is necessary, it would be impossible to filter out the entire lower sideband without affecting the amplitude and phase of the lower frequencies of the upper sideband and the carrier. Thus, part of the 6-MHz bandwidth is occupied by a "vestige" of the lower sideband (about 0.75 MHz out of 4 MHz). It is therefore commonly referred to as **vestigial-sideband operation.** It offers the added advantage that carrier reinsertion at the receiver is not necessary as in SSB because the carrier is not attenuated in vestigial-sideband systems.

Vestigial-Sideband Operation
a form of amplitude modulation in which one of the sidebands is partially attenuated

Once the entire TV signal is generated, it is amplified and driven into an antenna that converts the electrical energy into radio (electromagnetic) waves. These waves travel through the atmosphere to be intercepted by a TV receiving antenna and fed into the receiver once again as an electrical signal. That signal consists of the video, audio, and synchronizing signals. The synchronizing signals are contained in the video signal, as previously shown.

17-6 Television Receivers

A TV receiver utilizes the superheterodyne principle, as do almost all other types of receivers. It does become a bit more complex than most others because it must handle video and synchronizing signals as well as the audio that previously studied receivers do. A block diagram for a typical TV receiver is shown in Fig. 17-8.

The incoming signal is selected and amplified by the RF amplifier and stepped down to the IF frequency by the mixer-local oscillator blocks. The IF amplifiers handle the composite TV signal, and then the video detector separates the sound and video signals. The sound signal detected out of the video detector is the FM signal that is sent into the sound channel block, which is a complete FM receiver system in itself. The other video detector output is the video (plus sync) signal. The actual video portion of the video signal is amplified in the video amplifier and subsequently controls the strength of the electron beam that is scanning the phosphor of the CRT. The sync separator separates the horizontal and vertical sync signals,

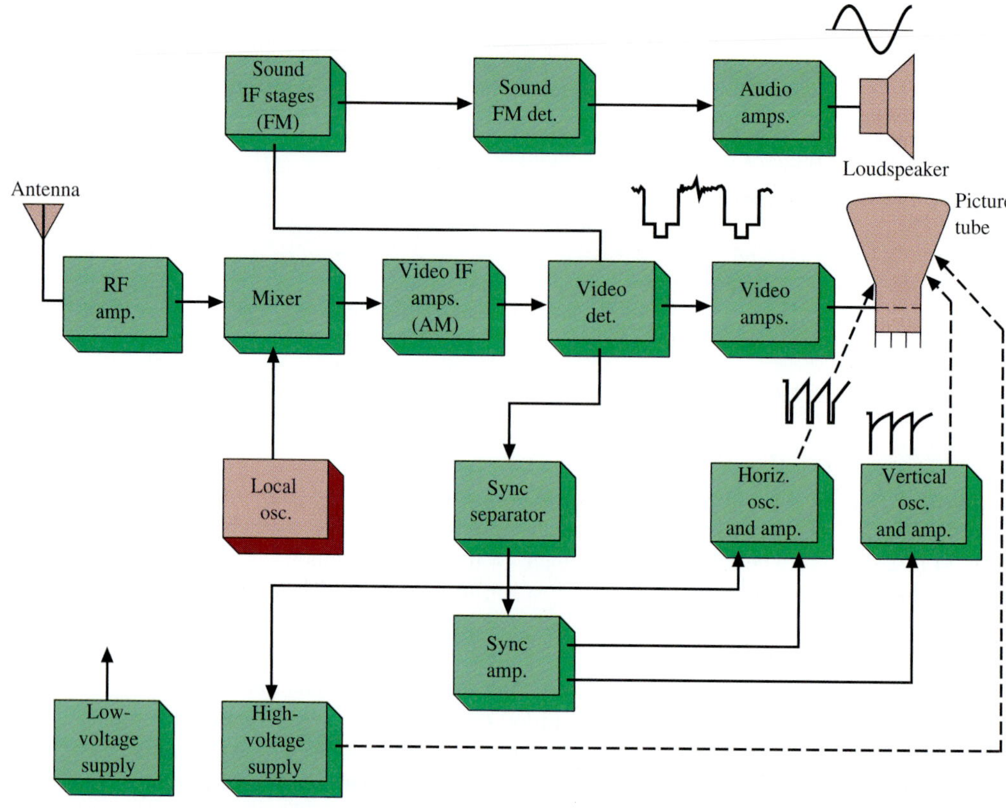

FIGURE 17-8 TV receiver block diagram.

which are then used to calibrate precisely and periodically the horizontal and vertical oscillators. The oscillator outputs are then amplified and used to control precisely the horizontal and vertical movement of the electron beam that is scanning the phosphor of the CRT. They are applied to a coil around the yoke of the CRT tube, whose magnetic fields cause the electron beam to be deflected in the proper fashion. This coil around the tube yoke is commonly referred to as the **yoke.**

Yoke
coil around the CRT tube that deflects the electron beam with its magnetic field

The low-voltage power supply shown in the receiver block diagram is used to power all the electronic circuitry. The high-voltage output is derived by stepping up the horizontal output signal (15.75 kHz) via transformer action. This transformer, usually termed the **flyback transformer,** has an output of 10 kV or more, which is required by the CRT anode to make the electron beam travel from its cathode to the phosphor. Approximately 1 kV is required for each diagonal inch of picture tube. The following sections provide greater detail on the basic operation just presented.

Flyback Transformer
used in TV receivers to produce the high voltage needed for the picture tube anode

 17-7 THE FRONT END AND IF AMPLIFIERS

Tuner
front end of a TV receiver that selects desired station

The front end of a TV receiver is also called the **tuner** and contains the RF amplifier, mixer, and local oscillator. Its output is fed into the first IF amplifier. It is the obvious function of the tuner to select the desired station and to reject all others, but the following important functions are also performed:

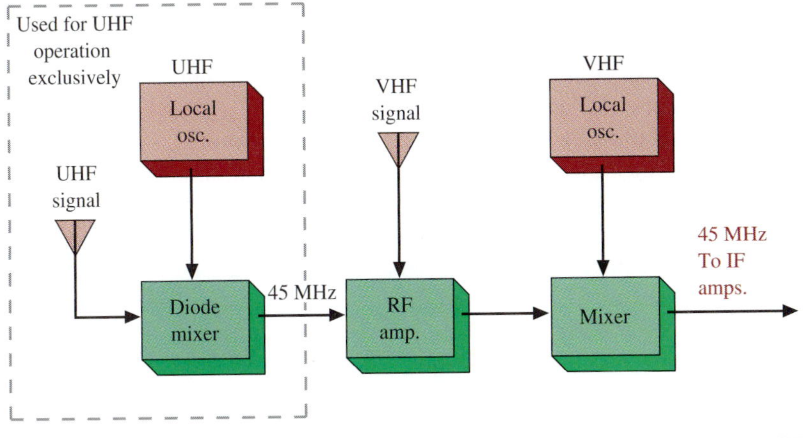

FIGURE 17-9 VHF/UHF tuner block diagram.

1. It provides amplification.
2. It prevents the local oscillator signal from being driven into the antenna and thus radiating unwanted interference.
3. It steps the received RF signal down to the frequency required for the IF stages.
4. It provides proper impedance matching between the antenna–feed line combination into the tuner itself. This allows for the largest possible signal into the tuner and thus the largest possible signal-to-noise ratio.

Figure 17-9 provides a block diagram of a VHF/UHF tuner. The large majority of tuners are synthesized, which allows for the remote control feature that is found on most sets. We will analyze Fig. 17-9 in two steps, starting with the UHF portion inside the dashed lines.

Note that there is no RF amplifier for the UHF front end; the antenna signal goes directly to the mixer. RF amplifiers at these frequencies are expensive and suffer from relatively poor noise performance.

When the tuner is switched to VHF channel 1, power is automatically removed from the VHF local oscillator and applied to the UHF local oscillator. At the same time, the tuned circuits of the RF amplifier and the mixer to its right are switched to 45 MHz, effectively converting them to IF amplifiers for the output of the UHF diode mixer. This compensates for the gain lost because of the missing RF amplifier and brings the UHF output up to equal the level of the VHF output.

Switching the tuner to a VHF channel removes power from the UHF oscillator and sends it to the VHF LO. In this way, only one oscillator operates at a time, thereby preventing the generation of signals that could cause interference. VHF signals are RF amplified, mixed, and sent to the 45 MHz IF.

IF Amplifiers

The IF amplifier section is fed from the mixer output of the tuner. It is often referred to as the video IF, even though it is also processing the sound signal. Sets that process the sound and video in the same IF stages are known as **intercarrier systems.** Very early sets used completely separate IF amps for the sound and video

Intercarrier Systems
TV receivers that process sound and video signals within the same IF amplifier stages

signals. The IF stages of intercarrier sets are often referred to as the video IF, even though they also handle the sound signal, because the sound signal is also processed by another IF stage after it has been extracted from the video signal. From now on the video IF will be referred to simply as the IF.

The major functions of the TV IF stage are the same as in a regular radio receiver: to provide the bulk of the set's selectivity and amplification. The standard IF frequencies are 45.75 MHz for the picture carrier and 41.25 MHz (45.75 MHz minus 4.5 MHz) for the sound carrier. Recall that mixer action causes a reversal in frequency when the IF amplifier accepts the difference between the higher local oscillator frequency and the incoming RF signal. Therefore, the sound carrier that is 4.5 MHz above the picture carrier in the RF signal ends up being 4.5 MHz below it in the mixer output into the first IF stage. The inversion effect of IF frequencies when receiving channel 5 is shown in Table 17-2. The IF frequencies are always equal to the difference between the local oscillator and RF frequencies.

Table 17-2 IF Signal Frequency Inversion

Channel 5 76–82 MHz	Transmitted RF Frequency (MHz)	Local Oscillator Frequency (MHz)	IF Frequency (MHz)
Upper-channel frequency	82	123	41
Sound carrier	81.75	123	41.25
Picture carrier	77.25	123	45.75
Lower-channel frequency	76	123	47

Stagger Tuning

A major difference between radio and TV IF amplifiers is that most radio receivers require relatively high-Q tuned circuits because the desired bandwidth is often less than 10 kHz. A TV IF amp requires a passband of about 6 MHz because of the wide frequency range necessary for video signals. Hence, the problem here is not how to get a very narrow bandwidth with high-Q components, but instead how to get a wide enough bandwidth but still have relatively sharp falloff at the passband edges. Most TV IF amplifiers solve this problem through the use of **stagger tuning.** Stagger tuning is the technique of cascading several tuned circuits with slightly different resonant frequencies, as shown in Fig. 17-10. The response of three separate *LC* tuned circuits is used to obtain the total resultant passband shown with dashed lines. The use of a lower-Q tuned circuit in the middle helps provide a flatter overall response than would otherwise be possible.

Stagger Tuning
cascading a number of tuned bandpass filters, each having a slightly offset bandpass frequency, to form a wider flat bandpass with steep high- and low-frequency roll-off skirts

Another interesting point illustrated in Fig. 17-10 is the attenuation given to the video side frequencies right around the picture carrier. This is done to reverse the vestigial-sideband characteristic generated at the transmitter. Refer back to Fig. 17-7 to refresh your memory about the transmitted characteristic. If the receiver IF response were equal for all the video frequencies, the lower ones (up to 0.75 MHz) would have excessive output because they have both upper- and lower-sideband components.

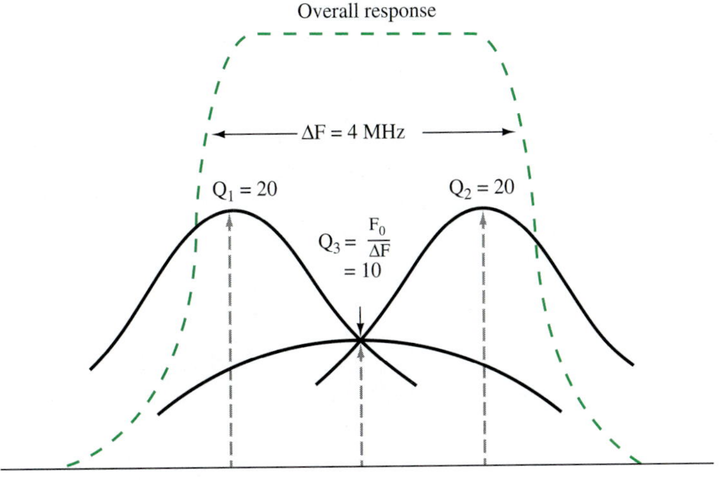

FIGURE 17-10 Stagger tuning.

SAW Filters

Color television receivers require a very complex IF alignment procedure because of the critical nature of their required bandpass characteristic. High-quality sets are now using **surface acoustic wave** (SAW) **filters.** They also find use in modern radar equipment because their characteristics can be matched to the reflected pulse from a target. This relatively new technology is now spreading into many applications.

Surface Acoustic Wave Filters
extremely high-Q filters often used in TV and radar applications

Recall that crystals rely on the effects in an entire solid piezoelectric material to develop a frequency sensitivity. SAW devices instead rely on the surface effects in a piezoelectric material such as quartz or lithium niobate. It is possible to cause mechanical vibrations (i.e., surface acoustic waves) that travel across the solid's surface at about 3000 m/s.

The process for setting up the surface wave is illustrated in Fig. 17-11. A pattern of interdigitated metal electrodes is deposited by the same photolithography process used to produce integrated circuits, and great precision is therefore possible. Because the frequency characteristics of the SAW device are determined by the geometry of the electrodes, an accurate and repeatable response is provided. When an ac signal is applied, a surface wave is set up and travels toward the output electrodes. The surface wave is converted back to an electrical signal by these electrodes. The

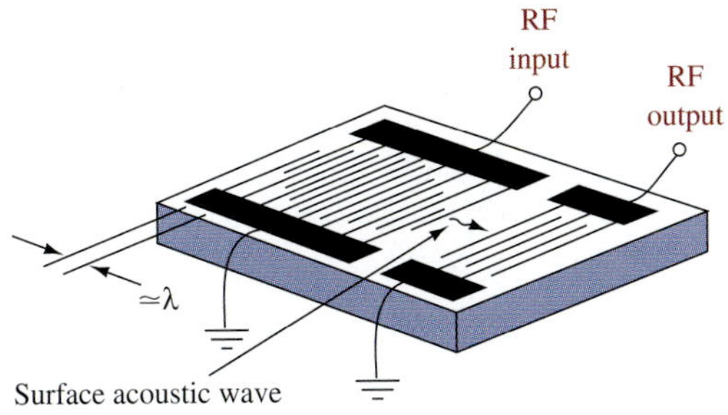

FIGURE 17-11 SAW filter.

Surface-mount SAW IF bandpass filters. (Courtesy of Sawtek Inc.)

length of the input and output electrodes determines the strength of a transmitted signal. The spacing between the electrode "fingers" is approximately one wavelength for the center frequency of interest. The number of fingers and their configuration determines the bandwidth, shape of the response curve, and phase relationships.

IF Amplifier Response

The ideal overall IF response curve in Fig. 17-12 provides some interesting food for thought. The sound IF carrier and its narrow sidebands are amplified at only one-tenth the midband IF gain. This is done to minimize interference effects that the sound

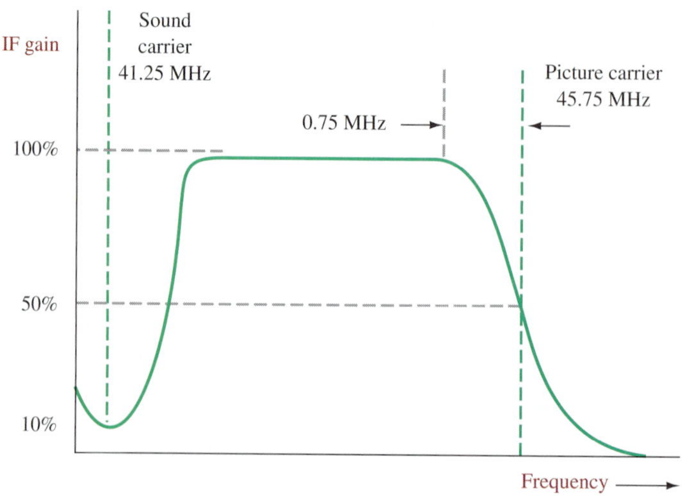

FIGURE 17-12 Ideal IF response curve.

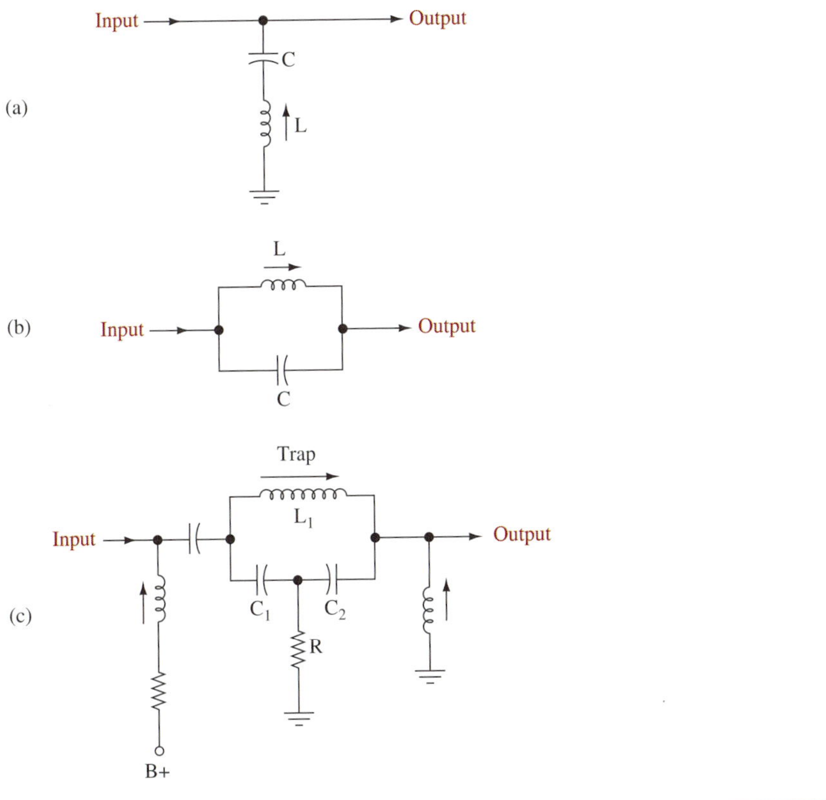

FIGURE 17-13 Wavetraps.

would otherwise have on the picture. You may have noticed a TV with normal picture when no audio is present but visual interference in step with the sound output. This is an indication that the sound signal in the IF is not attenuated enough, and it can often be remedied by adjustment of the set's fine-tuning control.

Wavetraps

To obtain the steep attenuation curve for the sound carrier shown in Fig. 17-12 it is necessary to incorporate a **wavetrap,** more simply termed **trap,** in the IF stage. A trap is a high-Q bandstop circuit that attenuates a narrow band of frequencies. It can be a series resonant circuit that shorts a specific frequency to ground, as in Fig. 17-13(a), or a parallel resonant circuit that blocks a specific frequency, as in Fig. 17-13(b). Even greater attenuation to a specific frequency is obtained with the *bridged-T* trap in Fig. 17-13(c). Traps are also employed in high-quality sets to eliminate carrier signals of adjacent channels. The carrier signal of an upper adjacent channel occurs at 39.75 MHz. While adjacent channels are not assigned in the same city, it is possible for a location midway between two adjacent channel stations to receive severe interference without a 39.75-MHz trap. A similar problem can exist with the sound carrier of a lower adjacent channel, which would occur at 47.25 MHz.

Wavetrap
high-Q bandstop circuit that attenuates a narrow band of frequencies

Trap
another name for wavetrap

17-8 The Video Section

The function of the video section is outlined in the block diagram in Fig. 17-14. It takes the output of the video detector (0 to 4 MHz) and amplifies it to sufficient level to be applied to the picture-tube cathode. Once applied to the cathode, this signal varies or *modulates* the electron beam strength so that white and black spots of a scene are white and black spots on the CRT face. Of course, it causes electron beam strengths of in-between magnitudes to provide various shades of gray.

The contrast control block in Fig. 17-14 is analogous to the volume control of a radio receiver—it simply varies the amplitude of the signal applied to the CRT. The larger the difference in amplitude between maximum and minimum, the greater the picture contrast (difference between black and white).

The sync takeoff block in Fig. 17-14 is the point where the horizontal and vertical sync pulses are extracted from the video signal. Section 17-9 provides further elaboration on this subject. In some receivers the sync takeoff occurs after the final video amplifier stage. The sound takeoff may occur after several stages of video amplification rather than at the video detector, as shown in Fig. 17-14. In color sets, however, the sound takeoff must occur before the video detector.

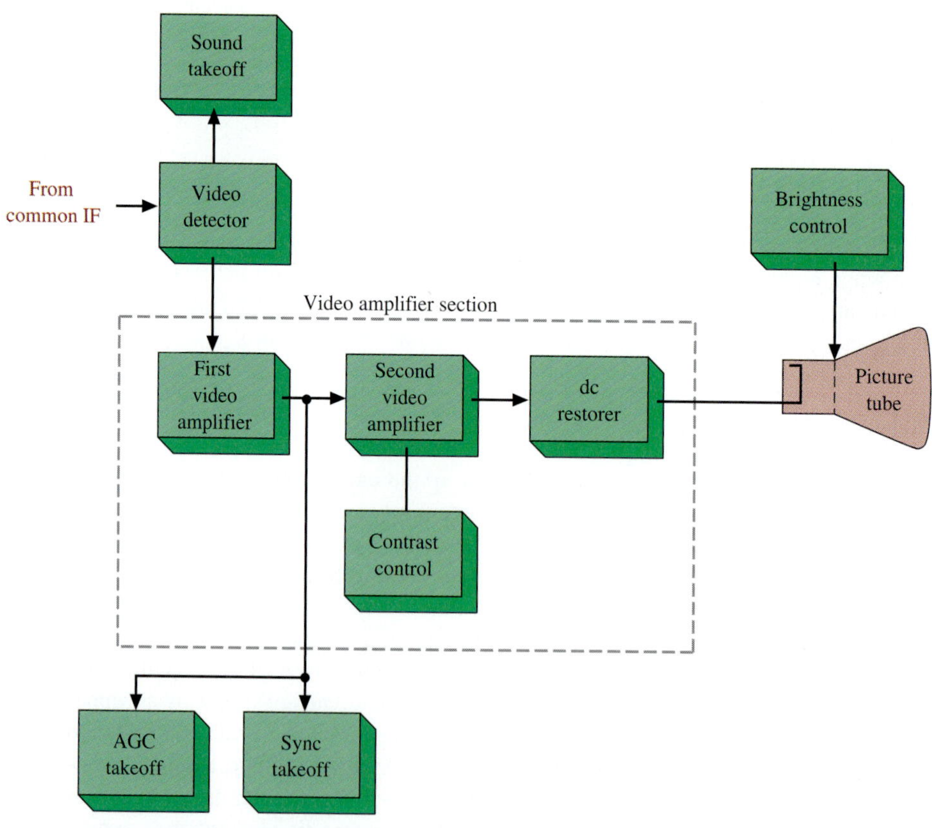

FIGURE 17-14 Video section block diagram.

The **dc restoration** block in Fig. 17-14 is not necessary if the video amplifiers use direct coupling. However, if capacitive coupling is used, as is usually more economical, then the dc portion of the video signal is lost. Without dc restoration, the picture background levels will be erroneous, and their color in color sets will be incorrect.

DC Restoration
process of restoring dc portion of video signal that is often removed by amplifier coupling

The brightness control (Fig. 17-14) is a user adjustment, just as is the contrast control. The brightness control simply varies the dc level applied to the control grid or cathode of the CRT. It is *not* connected to the video signal amplitude in any way. It controls the overall picture brightness and *not* the min–max video signal level, as does the contrast control.

The video section also provides a takeoff point for the automatic gain control (AGC) signal. That signal is used to control the gain of previous amplifying stages such as the RF amp, mixer, and IF stages. This is necessary so that both strong and weak stations end up supplying the CRT cathode with approximately the same signal level and thus providing the same picture illumination. If the received signal is too weak, however, the electrical noise predominates over the desired signal and results in a "snowy" picture.

17-9 Sync and Deflection

The video section provides a takeoff point for the sync signals. Because the set needs both vertical (at 60 Hz or 59.94 Hz) and horizontal (at 15.75 kHz) sync pulses, a means to separate one from the other is necessary. The **sync separator** is the circuit that performs this function. The key factor that enables separation is the fact that the vertical sync pulse is of long duration, while the horizontal sync pulse is of extremely short duration. In addition to separating one from the other, the sync separator *clips* the sync pulse off the video signal. This prevents the sweep instability that could occur because of false synchronization of the sweep oscillators by spurious video signals. Because of this, the sync separator is sometimes referred to as the **clipper.**

Sync Separator
circuit in a TV receiver that separates the horizontal and vertical sync pulses from the video signal

Clipper
another name for sync separator

Once the sync separator has clipped the sync signals from the lower-level video signal, the two types of sync pulses are applied to both low- and high-pass filters. The output of the low-pass filter will be the lower-frequency vertical sync pulse at 60 Hz (59.94 Hz) because it is a wide pulse rich in low-frequency components. The output from the high-pass filter will be the horizontal sync pulse at 15.75 kHz because it is a very narrow pulse that is rich in high-frequency content. A low-pass filter is also termed an **integrator,** while a high-pass filter is classified as a **differentiator.**

Integrator
a low-pass filter

Differentiator
a high-pass filter

This entire process of clipping and separation is shown in Fig. 17-15. The low-pass filter can simply be a shunt capacitance that shorts high frequencies to ground. The high-pass filter includes a series capacitance that blocks low frequencies from reaching its output.

The vertical sync pulse is applied to the vertical oscillator. The vertical oscillator by itself generates a signal at *about* 60 Hz but must be at *precisely* the same frequency as the transmitter's vertical oscillator to prevent the picture from "rolling" in a vertical direction. The vertical adjustment control available to the set user adjusts the vertical oscillator's frequency to enable it to be brought into the range necessary so that it can "lock" onto the frequency of the sync pulse. The foregoing discussion also applies to the horizontal oscillator section except that the frequency is 15.75 kHz, and

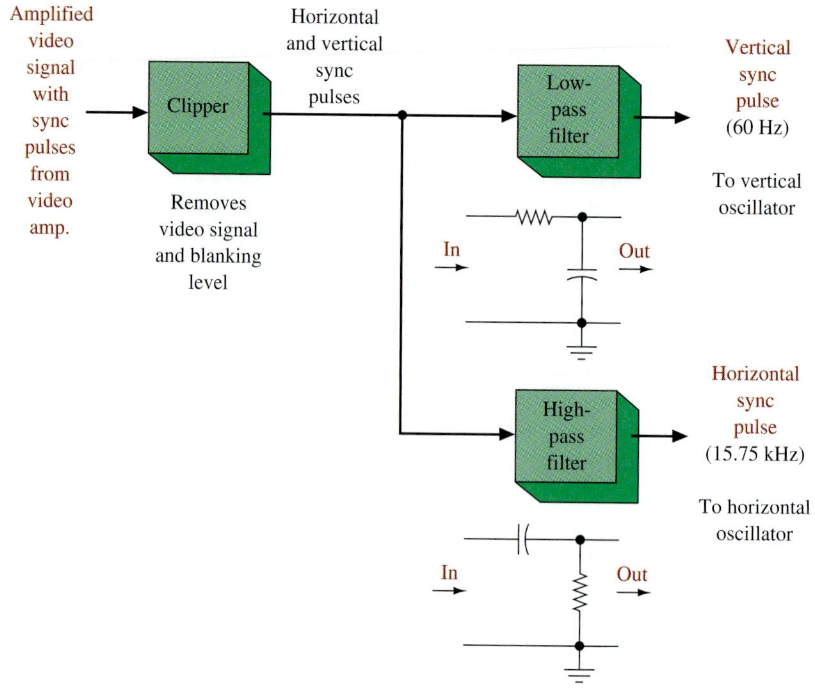

FIGURE 17-15 Sync separator.

loss of horizontal sync results in heavy slanting streaks across the screen. This phenomenon can be witnessed by simply misadjusting the horizontal control on a TV set.

Horizontal Deflection and High Voltage

A typical block diagram for the horizontal deflection and high-voltage systems is shown in Fig. 17-16. The horizontal sync pulses are used to calibrate the horizontal oscillator, which is then amplified to a powerful level by the horizontal output amplifier and then applied to a high-voltage transformer commonly referred to as the *flyback transformer*. Its outputs drive the horizontal yoke windings and provide the high voltages for the CRT after rectification. A **damper** function (which will be subsequently explained) is also provided. The horizontal system is seen to be a complex one, and because of this and the high voltages and powers involved, it is probably the most failure-prone section of a TV receiver.

Damper
a diode in the high-voltage oscillator of a TV receiver that shorts out unwanted damped oscillations during the flyback period

As with vertical scanning, a linear sawtooth (current) waveform is required for linear horizontal deflection. If such a waveform is not provided, distortion in the picture results, as indicated in Fig. 17-17. A horizontal linearity control is sometimes provided at the rear of the set to correct for these conditions.

Horizontal Deflection Circuit

Figure 17-18 provides a schematic for a typical horizontal system. The horizontal oscillator frequency is held in sync by comparing the horizontal sync pulses to a signal fed back from the horizontal output in the phase detector diodes, D_1 and D_2. Any difference is detected as a phase difference and is applied as a dc level to the base of the horizontal oscillator to correct its frequency. A phase-locked loop (PLL) IC is used

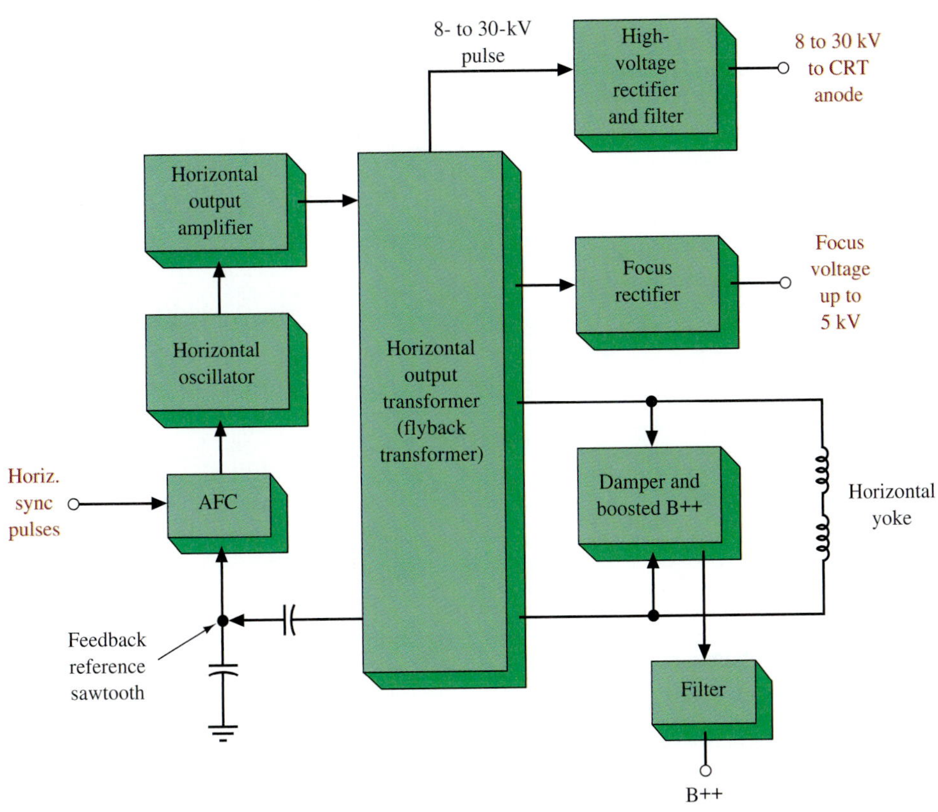

FIGURE 17-16 Horizontal deflection block diagram.

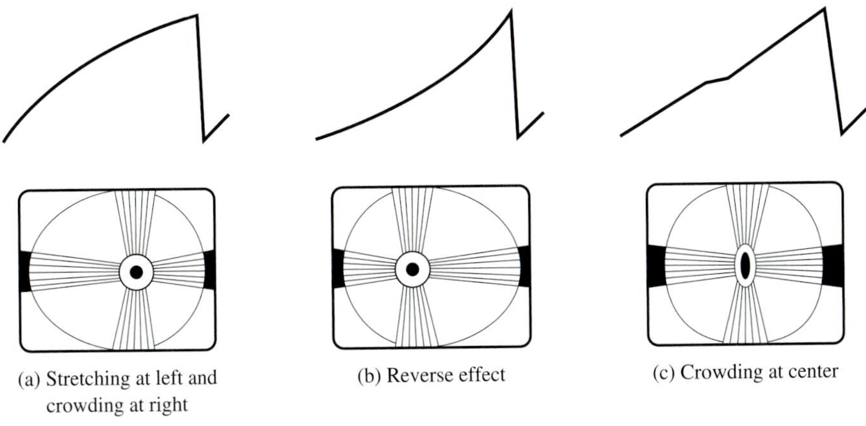

(a) Stretching at left and
crowding at right
(b) Reverse effect
(c) Crowding at center

FIGURE 17-17 Nonlinear horizontal scanning. (From Bernard Grob, Basic Television Principles and Servicing, 4th ed., 1977; Courtesy of McGraw-Hill, Inc., New York.)

in many sets for this application. Note also the user-controlled variable inductor that adjusts the frequency of oscillation into the range that allows the sync pulses to exercise control. The horizontal oscillator signal is transformer-coupled into the base of the horizontal output transistor for amplification so that the signal has sufficient strength to drive the flyback transformer.

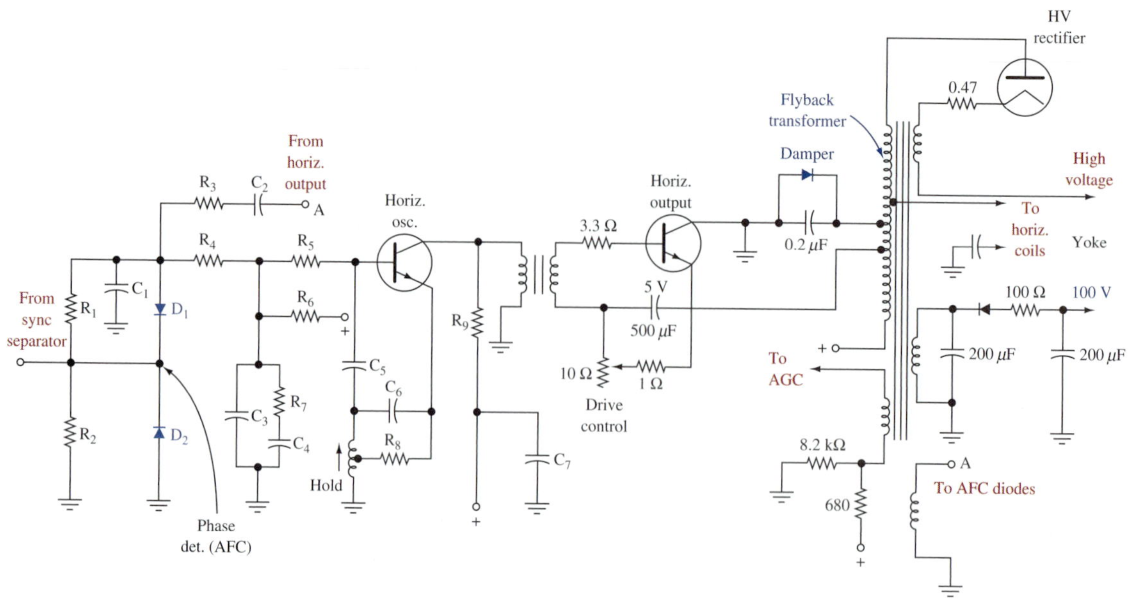

FIGURE 17-18 Horizontal system schematic.

As the sawtooth level builds up on the horizontal amp base, its collector current builds up through the transformer primary and damper diode. When the sawtooth level suddenly changes (during retrace), the collector current drops to zero. The magnetic field around the horizontal yoke coils collapses, rapidly inducing a high-amplitude flyback electromagnetic frequency (EMF) across the transformer secondary. This induces a pulse of current in the secondary of the transformer and a high-induced flyback EMF in the primary. The kilovolts of ac thus induced are rectified by the high-voltage rectifier and applied to the CRT as its required dc anode voltage.

During this flyback period, the energy of the horizontal yoke coils' collapsing magnetic field tends to produce damped oscillations that interfere with the start of the next sawtooth waveform. The *damper* diode serves as a short during this flyback interval so that the unwanted oscillations are quickly damped. An auxiliary secondary winding on the flyback transformer provides a stepped *voltage boost* dc level of about 100 V for all the circuitry requiring more than the 12 to 20 V dc used elsewhere.

17-10 PRINCIPLES OF COLOR TELEVISION

Monochrome black-and-white TV

We have thus far been concerned mainly with black-and-white or **monochrome** television. While color TV presents a much greater degree of sophistication, the student who has mastered monochrome principles reasonably well can advance to the color set by adding a few more basic ideas.

Our system for color TV was instituted in 1953 and is termed *compatible*. That is, a color transmission can be reproduced in black-and-white shades by a monochrome receiver, and a monochrome transmission is reproduced in black and white by a color receiver. To remain compatible, the same total 6-MHz bandwidth must be used, but more information (color) must be transmitted. This problem is

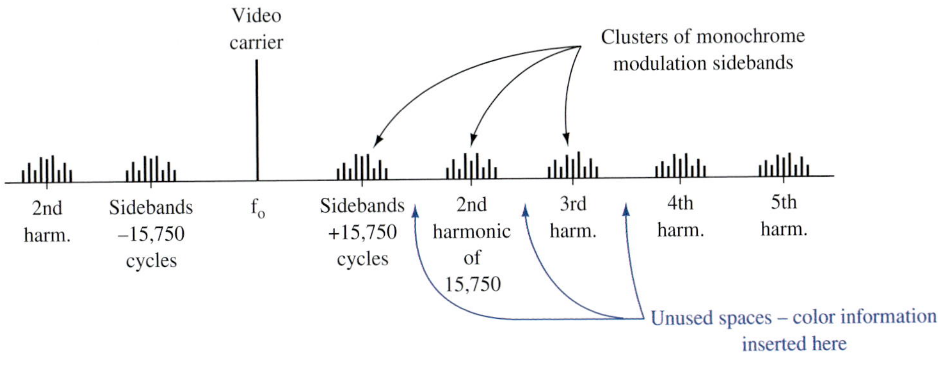

FIGURE 17-19 Interleaving process.

overcome by a form of multiplexing, as when FM stereo was added to FM broadcasting. It turns out that the video signal information is clustered at 15.75-kHz (the horizontal oscillator frequency) intervals throughout its 4-MHz bandwidth. Midway between these 15.75-kHz clusters (harmonics) of information are unused frequencies, as indicated in Fig. 17-19. By generating the color information around just the right color subcarrier frequency (3.579545 MHz), it becomes centered in clusters exactly between the black-and-white signals. This is known as **interleaving.**

Interleaving
generating color information around just the right frequency so that it becomes centered in clusters between the black-and-white signals

At the color TV transmitter, the scene to be televised is actually scanned by three separate pickup sensors in the camera, each camera sensitive to just one of the three primary colors: red, blue, and green. Because various combinations of these three colors can be mixed to form any color to which the human eye is sensitive, an electrical representation of a complete color scene is possible. The three color cameras scan the scene in unison, with the red, green, and blue color content separated into three different signals. This process is accomplished within the color TV camera as shown in Fig. 17-20. The lens focuses the scene onto a beam splitter

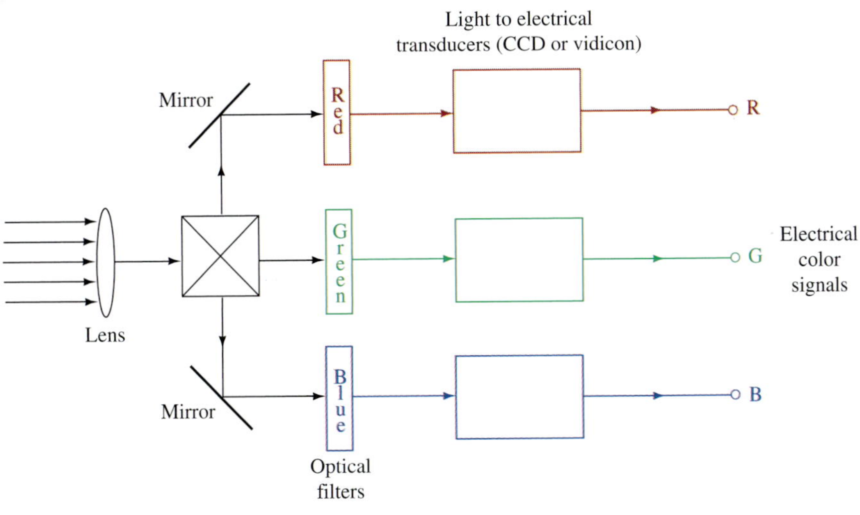

FIGURE 17-20 Generating the electrical color signals (color camera).

that feeds three separate light filters. The red filter passes only the red portion of the scene, resulting in the R (red) electrical signal. A blue and a green filter accomplish the same process to generate the B and G signals. At the receiver, these three separate signals are made to illuminate properly groups of red, green, and blue phosphor dots (called **triads**), and the original scene is reproduced in color.

Triads
individual groups of red, green, and blue phosphor dots on the CRT face

Matrix
transmitter signal processing circuits

Luminance
the Y signal

Chroma
the color signals I and Q

After generation these three separate color signals are fed into the transmitter signal processing circuits (**matrix**) and create the Y, or **luminance,** signal and the **chroma,** or color, signals I and Q. The Y signal contains just the right proportion of red, blue, and green so that it creates a normal black-and-white picture. This proportion is:

$$Y = 0.3R + 0.59G + 0.11B$$

It modulates the video carrier just as does the signal from a single black-and-white camera with a 4-MHz bandwidth. The chroma signals, I and Q, are used to phase-modulate the 3.58-MHz color subcarrier, which then *interleaves* their color information in the gaps left by luminance Y signal's sidebands. The proportions for I and Q are:

$$I = 0.6R + 0.28G + 0.32B$$
$$Q = 0.21R - 0.52G + 0.31B$$

This modulation by the I and Q signals is accomplished in a balanced modulator, thus suppressing the 3.58-MHz subcarrier because it would cause interference at the receiver. The composite transmitted signal is shown in Fig. 17-21.

At the receiver, a monochrome set simply detects the Y signal and thus presents a normal black-and-white rendition of a color picture. The chroma signals (I and Q) cannot be detected in a monochrome set because their 3.58-MHz subcarrier was suppressed and is not present in the received signal. Thus, a color set must have a means to generate and reinject the 3.58-MHz subcarrier to enable detection of the I and Q signals. Notice in Fig. 17-21 that the Q signal is a full DSB signal with sidebands extending ±500 kHz around the color subcarrier, which is 3.58 MHz above the overall carrier frequency. The I signal has a lower sideband 1.5 MHz

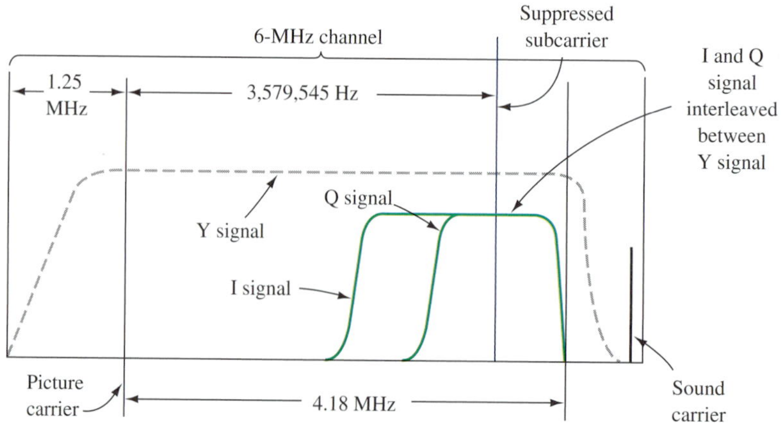

FIGURE 17-21 Composite color TV transmission.

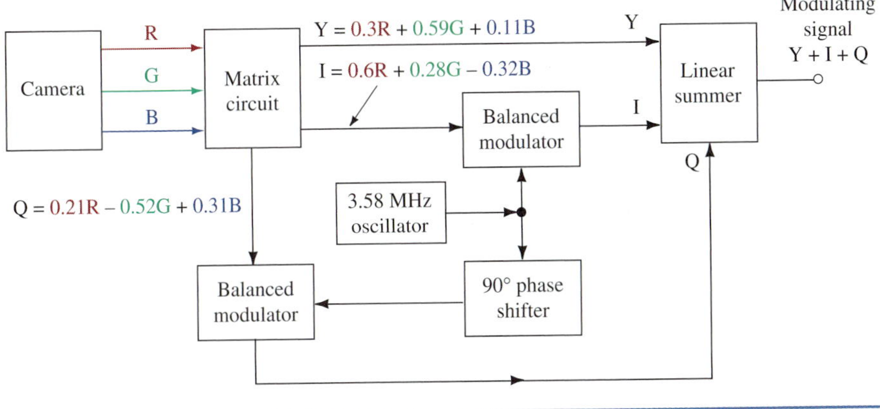

FIGURE 17-22 The composite color modulating signal.

below the color subcarrier. It is a vestigial sideband signal, however, because the upper sideband is attenuated after 500 kHz.

A block diagram showing the generation of the composite color TV modulating signal is shown in Fig. 17-22. It is called the *NTSC* (National Television Systems Committee) *signal* and was approved by the FCC in 1953. Notice that the I and Q signals are summed with the Y signal to modulate the TV carrier frequency. The chrominance signals (I and Q) modulate the 3.58 MHz subcarrier in separate balanced modulators, as shown in Fig. 17-22. These subcarriers are 90° out of phase (in quadrature). The two double-sideband signals created (I and Q) can be separately recovered at the receiver because of this quadrature modulation process.

Color Receiver Block Diagram

A block diagram of a color receiver, from the video detector onward, is shown in Fig. 17-23. After video amplification, the Y signal is immediately available. It is given a delay of about 1 μs, as shown, so that it will arrive at the CRT at the same time as the I and Q signals. This is necessary because the I and Q signals undergo considerably more processing, which takes about 1 μs. The chroma signals are amplified and then sent into a 2- to 4.2-MHz bandpass amplifier and then to the I and Q detectors. These detectors also have inputs from the 3.58-MHz crystal oscillator so that the difference signal in the I detector is the 0- to 1.5-MHz I signal, and in the Q detector, it is the 0- to 0.5-MHz Q signal. Notice in Fig. 17-23 that the 3.58-MHz signal for the Q detector is given a 90° phase shift, which is how the Q signal was generated at the transmitter. This phase shift makes them separable at the receiver.

Once the I and Q signals are detected and passed through their respective low-pass filters, they are given a phase inversion that allows for both + and − chroma signals. This is necessary because

$$\begin{aligned} \text{green} &= -0.64\text{Q} - 0.28\text{I} + \text{Y} \\ \text{blue} &= 1.73\text{Q} - 1.11\text{I} + \text{Y} \\ \text{red} &= 0.62\text{Q} + 0.95\text{I} + \text{Y} \end{aligned}$$

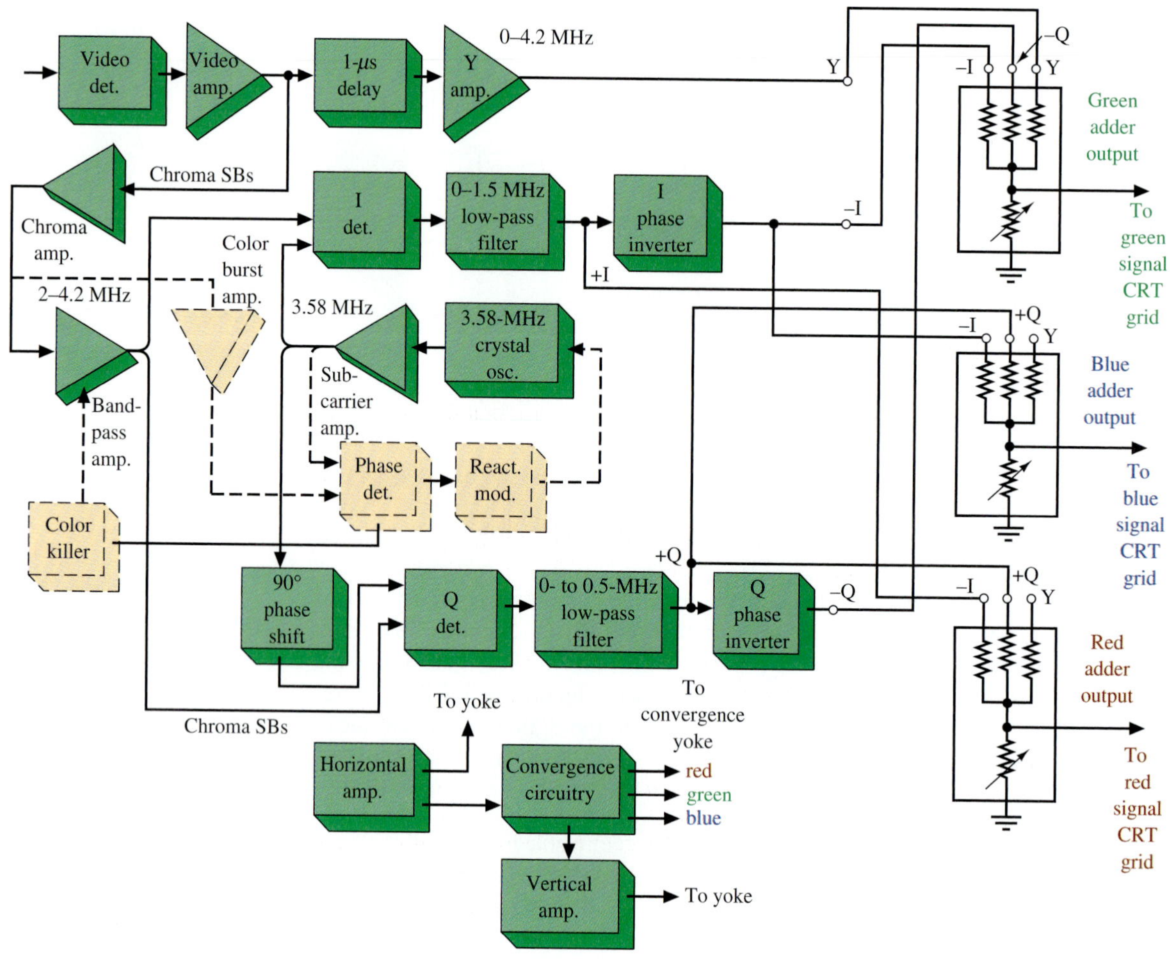

FIGURE 17-23 Color receiver block diagram.

The I, Q, and Y signals are summed in the three-color adder circuits, with the resistor values providing the proper proportion of each signal. The output of each color adder is then applied to the appropriate CRT grid to control beam intensity. Notice the rheostat in each adder circuit. It allows for the intensity of each color signal to be varied in proportion to the other colors.

The color subcarrier crystal oscillator is not precise enough by itself to allow proper chroma signal detection. This is surprising because crystal oscillators are extremely stable and accurate. An accuracy of 1 part of 10^{12} is necessary to obtain the correct chroma signal. Recall that color transmissions eliminate this carrier from the video signal but do include a sample of it on the back porch of the horizontal blanking pulse, as shown in Fig. 17-24. The color burst amp shown in Fig. 17-23 is receptive to that portion of the overall video signal. Its frequency is compared with the 3.58-MHz crystals in the phase detector, and if they are not precisely equal, the phase detector applies a dc level to vary the reactance of the reactance modulator. It, in turn, causes the crystal's frequency to "pull" in the proper direction to bring it back into precise synchronization with the color burst frequency.

Color Killer
circuit that prevents output from the chroma circuits to a monochrome broadcast

Notice that the phase detector in Fig. 17-23 also has an output that is applied to the **color killer.** The name is descriptive because a monochrome broadcast has

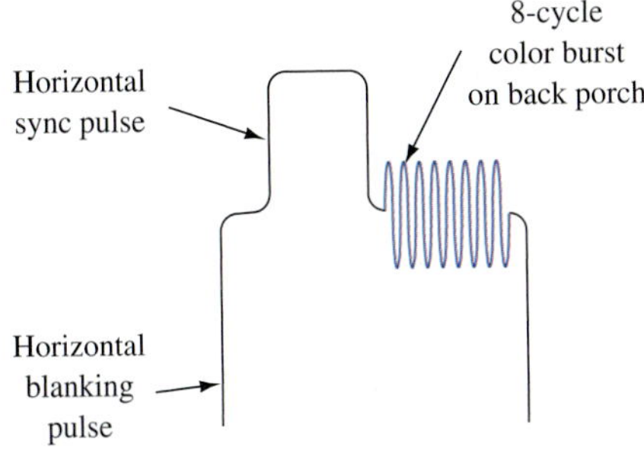

FIGURE 17-24 Color burst.

no color burst, and thus the phase detector has a large dc output that the color killer circuit uses to "kill" the 2- to 4.2-MHz bandpass amplifier. The purpose is to prevent any signals out of the chroma circuits during a monochrome broadcast. A defective color killer results in colored noise, called **confetti,** on the screen of a color receiver during a black-and-white transmission. The confetti looks like snow but with larger spots, in color. The color killer also kills the color if a weak RF signal is received.

Confetti
colored noise on the screen of a color receiver during a black-and-white transmission

The Color CRT and Convergence

Color receiver CRTs are a marvel of engineering precision. As previously mentioned, they are made up of triads of red, blue, and green phosphor dots. The trick is to get the proper electron beam to strike its respective colored phosphor dot. This is accomplished by passing the three beams through a single hole in the **shadow mask,** as shown in Fig. 17-25. The shadow mask prevents the "red" beam from spilling over onto an adjacent blue or green phosphor dot, which would certainly destroy the

Shadow Mask
screen used in a color CRT to prevent an electron beam from striking the wrong color phosphor triad

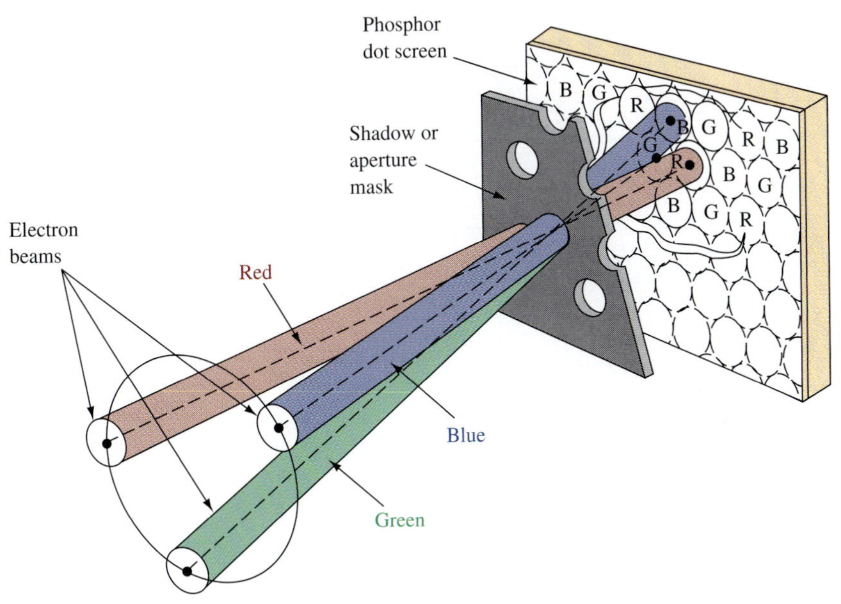

FIGURE 17-25 Color CRT construction.

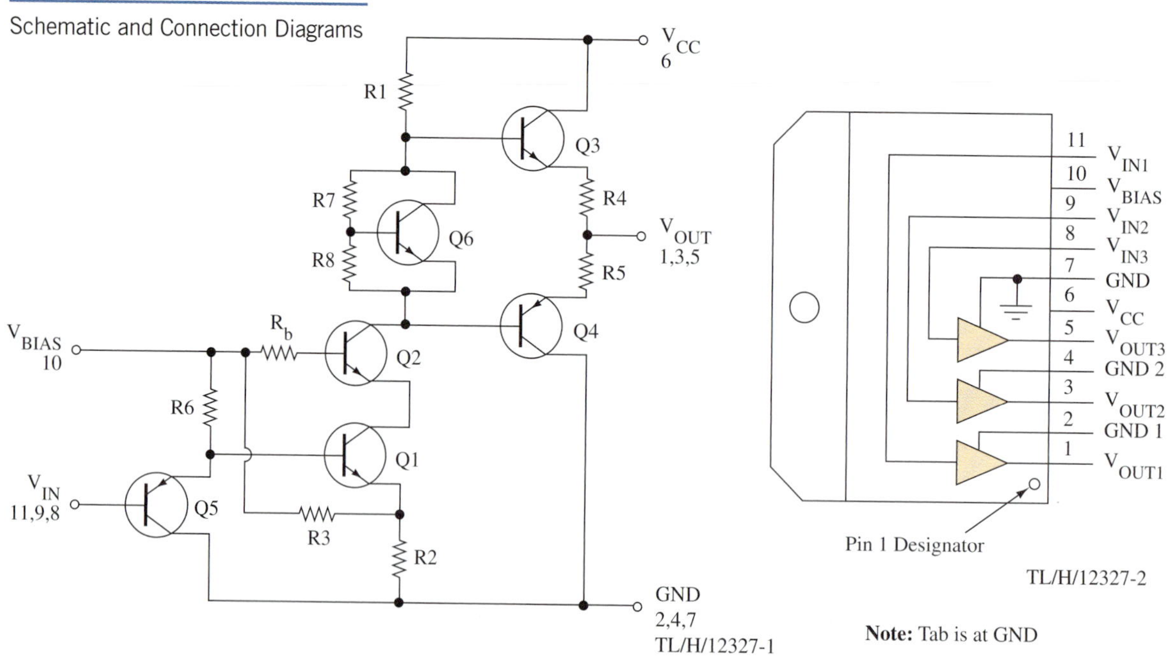

(a)

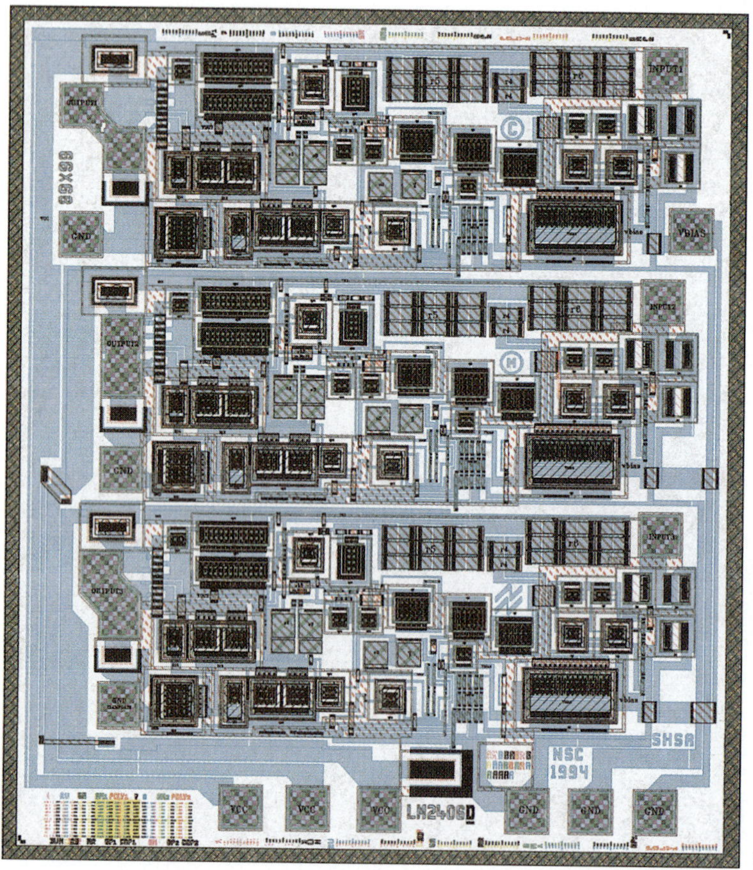

(b)

The LM2406: first single IC RGB CRT driver amplifier schematic (a) and layout (b). (Courtesy of National Semiconductor.)

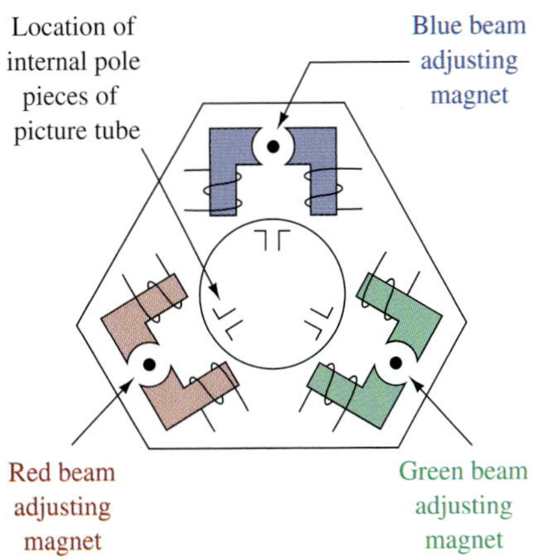

FIGURE 17-26 Color convergence yoke.

color rendition. A typical color CRT has over 200,000 holes in the shadow mask and triads of phosphor dots. To make the three beams converge correctly on their color dot of phosphor throughout the face of the tube requires special modification to the horizontal and vertical deflection systems.

Static convergence refers to proper beam convergence at the center of the CRT's face. This adjustment is made by dc level changes in the horizontal and vertical amplifiers. Convergence away from the center becomes more of a problem and is referred to as **dynamic convergence.** It is necessary because the tube face away from the center is not a perfectly spherical shape (it is more nearly flat), and thus the beams tend to converge in front of the shadow mask away from the tube center. Special dynamic convergence voltages are derived from the horizontal and vertical amplifier signals and are applied to a special color convergence yoke placed around the tube yoke, as shown in Fig. 17-26. The dynamic convergence of a set involves the shown magnet adjustment and several adjustments (usually 12) on the convergence board that have interaction effects. The process is quite involved and time consuming.

Static Convergence proper beam convergence at the center of a CRT

Dynamic Convergence beam convergence away from the center of a CRT

 17-11 Sound and Picture Improvements

Modern television systems have entered a period of rapid change and improvement. Low-power broadcast stations are planning to offer programming tailored to their community. Cable operators have wired cities and suburbs with cable systems that provide more than 70 channels of entertainment programming and data services. Satellite system operators offer direct broadcast systems (DBS) that broadcast television directly to individual subscribers from geostationary satellites. The use of videocassette recorders (VCRs) and digital video disk (DVD) players has become commonplace, providing the user with improved flexibility. With all of these changes there is also a move to improve the quality of TV audio and video. The audio improvement was accomplished when the FCC approved a new system in 1984. The video revolution is currently developing.

Enhanced Audio

The system approved by the FCC in 1984 is called Zenith/dbx. It has similarities to the FM stereo system but it provides a better signal. The multiplexing scheme used is

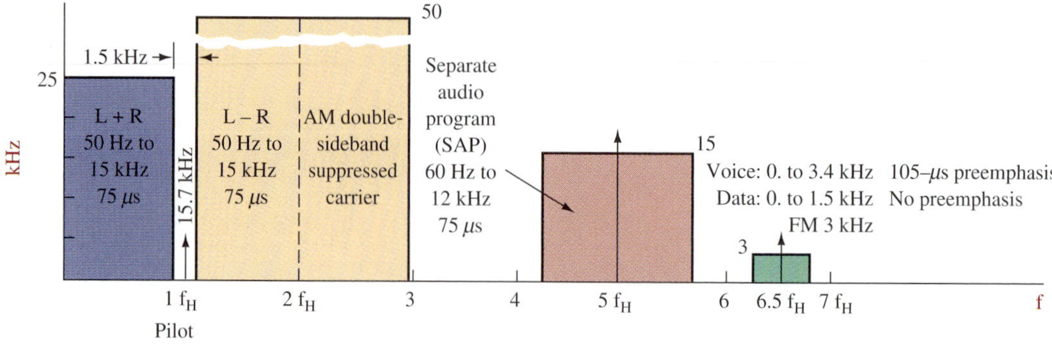

FIGURE 17-27 Zenith/dbx stereo system.

illustrated in Fig. 17-27. Recall that in FM radio, a double-sideband carrier centered around 38 kHz is used. In the Zenith/dbx system, the new component, (L − R), is centered at two times the horizontal scan rate, f_H, about 2 × 15.7 kHz, or ≈31.4 kHz. Notice also the separate audio program (SAP) and the voice/data channels centered at 5 and 6.5 times f_H. With SAP, a station can transmit a simultaneous foreign-language translation or an entirely unrelated service. The subcarrier at $6.5f_H$, known as the professional subchannel, is intended for transmitting voice or data wholly unrelated to video programming. This could include radio reading services, market and financial data, paging and calling, and traffic-control signal switching.

The stereo subcarrier, (L − R), centered at $2f_H$, deviates at ±50 kHz, or twice the regular monophonic signal (L + R). The composite modulating signal creates a bandwidth that exceeds the 200 kHz allowed for FM radio. This is allowable because the sound track is transmitted 4.5 MHz above the picture carrier and 250 kHz below the top of the channel. This increased bandwidth is one reason that TV stereo is a better signal than FM radio, but the main reason is the dbx companding system.

As explained in Chapter 4, "compand" is from "*com*press" and "ex*pand*," in which a variable-gain circuit at the transmitter increases its gain for low-level input signals to provide better noise performance. A complementary circuit at the receiver reverses the process. The monophonic audio channel (L + R) is not companded to maintain compatibility with nonstereo TVs.

The reduced level of the (L − R) signal compared to (L + R) in FM radio has been one of its major drawbacks. In TV, as shown in Fig. 17-27, it is actually increased in amplitude, which was possible due to the increased bandwidth TV has available. As the left–right separation decreases, the difference channel amplitude declines with respect to the sum channel. The increased amplitude of the (L − R) signal in TV is therefore especially beneficial. In the dbx system, the gain control signal is a function of the rms audio signal and compression is 2:1. This means that for every 2-dB decrease in audio level below maximum, the transmitted audio is decreased 1 dB from the maximum. Although the effectiveness of companding has long been known, it has proven difficult to implement. Now that the complex circuitry required can be inexpensively fabricated on ICs, its use in mass-produced equipment is possible (see Sec. 4-4). In areas of good signal reception, a receiver can provide a good signal with or without companding. The companded TV audio signal, however, remains noise-free well beyond the point at which a stereo FM radio signal is noticeably degraded.

17-12 Digital Television

The words *digital television* (**DTV**) have many meanings for the public. Is DTV high-definition television (**HDTV**)? Does this new technology improve the quality of reception? Can our existing televisions, cameras, and VCRs still be used? The answers are yes and no. This is not a very clear answer, but neither is the planned deployment of digital television in the United States. This section will introduce the basics of DTV, including the new screen-size format, data compression, digital transmission and reception, and digital data formatting.

DTV
digital television

HDTV
high definition television

The 30 largest broadcast markets in the United States began digital transmission in 1999, and all stations were scheduled to be on the air with digital television by 2003. Television stations will continue the simultaneous transmission of NTSC (analog) and digital television (DTV) at least through 2006. Most broadcast facilities are for-profit, and for broadcasters to make money, they must have an audience. It is estimated that 70 to 80 percent of U.S. homes receive their local television broadcasts via their local cable television provider. The good news is the majority of cable systems are offering HDTV programs.

Advanced Television System Committee (ATSC)
developed to make recommendations for advanced television in the United States

SDTV
standard definition television, digital television with resolution comparable to NTSC

4:2:2
international standard for digitizing component video

The Basics of Digital Television: Introduction

The DTV standard is based on the standard recommendations by the **Advanced Television System Committee (ATSC).** This standard provides for the transmission of television programs in the HDTV screen format, 16 × 9, as shown in Fig. 17-28. It also provides for the transmission of a **standard definition television (SDTV)** format that provides a digital picture with comparable resolution to analog NTSC formats. The new standard also provides the capability for broadcasters to transmit multiple SDTV programs over a single television channel. The FCC-mandated digital television transmission did not require HDTV. However, most broadcasters commit part of their broadcast day to HDTV transmission.

The Basics of Digital Television: The Video Signal

The format typically used to convert the analog video to a digital format is the ITU-R 601 **4:2:2** format. This is an international standard for digitizing component video. The base sampling frequency for the ITU-R 601 standard is 3.375 MHz. The 4:2:2 represents the sample rate for the following elements of a component video signal.

$$\begin{matrix} 4 & : & 2 & : & 2 \\ \text{luminance}\,[Y] & : & \text{red}-\text{luminance}\,[R-Y] & : & \text{and blue}-\text{luminance}\,[B-Y] \end{matrix}$$

A video signal is composed of green, red, and blue components. In addition to providing green color information, the green channel provides the luminance information. Luminance is the black-and-white detail. The R–Y and B–Y values provide the color-difference values. These components, the Y, R–Y, and B–Y, are then converted to a digital signal using a PCM technique. The base sample rate for the ITU-R 601 standard is 3.375 MHz, and 10-bit sampling is used. This means that the luminance channel is sampled at four times the base rate, and the R–Y and B–Y channels are sampled at two times the base rate. The calculations for the sample frequencies and bit rates are provided in Table 17-3.

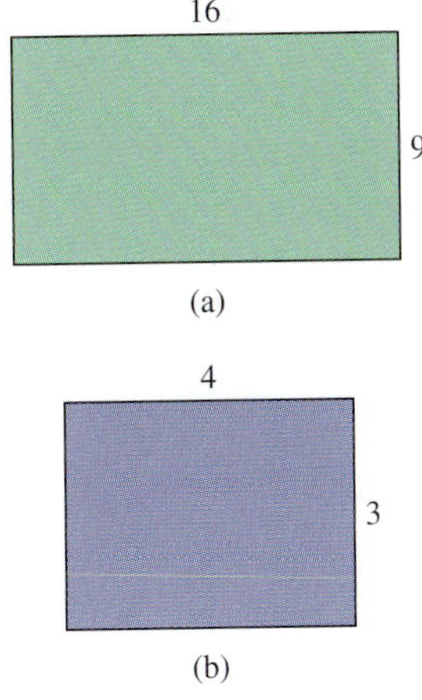

FIGURE 17-28 The (a) HDTV screen ratio and (b) NTSC screen ratio format, shown for comparison.

Table 17-3 Sample Frequencies for the ITU-R 601 4:2:2 Digital Video Signal

Channel	Sample Rate	Bit Rate
Luminance channel	4 × 3.375 MHz = 13.5 MHz	×10 bits/sample = 135 Mbps
R–Y channel	2 × 3.375 MHz = 6.75 MHz	×10 bits/sample = 67.5 Mbps
B–Y channel	2 × 3.375 MHz = 6.75 MHz	×10 bits/sample = 67.5 Mbps

The three digital samples (Y, R–Y, and B–Y) are time-division-multiplexed together with a resulting serial data bit rate of 270 Mbps (135 + 67.5 + 67.5). This data rate must undergo some form of data compression so that the data will fit into the 6-MHz bandwidth available for broadcast television. The video-compression technique selected for DTV transmissions is **MPEG2.** (MPEG is an abbreviation for the Motion Pictures Expert Group.) The compression techniques rely on the redundancies in the video signal. The redundancies in a video signal are summarized as follows:

MPEG2
a video-compression technique used in DTV transmission

Redundancies in a Video Signal

Statistical data redundancy	Only a portion of a video signal is constantly changing; therefore, it is necessary to output only the changing data.
Psychvisual redundancy	The human eye does not perceive all detail in a picture; therefore, the resolution of certain details can be minimized without compromise of video clarity.
Entropy	This is the unpredictable information within a picture. This information must be maintained to provide reconstruction of the image.
Luminance/chrominance contrast	Details are significant only if there is significant contrast. The eye is very sensitive to luminance (black/white) detail and not very sensitive to detail in the chrominance (color).
Spatial redundancies	This defines the areas that are to detect errors—e.g., textured regions, edge errors, fine picture details.
Temporal redundancies	Flicker of the video signal is observable below 50 Hz. Flicker is more noticeable in video signals that are very bright.

Note: This information was adapted from "Compression Concepts, Part 1—Transition to Digital," *Broadcast Engineering* (November 1999).

The Basics of Digital Television: The Audio Signal

The digital compression technique specified for digital television, as defined by ATSC document A/52, details the digital audio compression (**AC-3**) standard developed by Dolby Laboratories. This system provides five full-bandwidth audio channels (3 Hz to 20 kHz). The five channels are for the left, center, right, and left-right surround-sound channels. The standard also provides one low-frequency enhancement channel, which has a reduced bandwidth (3 Hz to 120 Hz). The new audio system is commercially called the **5.1 Channel Input.** The standard provides for various sample rates and input word lengths (up to 24 bits) for compatibility to the

AC-3
the Dolby Laboratories audio-compression technique for digital television

5.1 Channel Input
the commercial name for the AC-3 audio standard

many available digital audio encoding formats. The six audio outputs are multiplexed together, which results in a 5.184-Mbps data stream. This data stream is then compressed to a 384-kbps data stream.

The Basics of Digital Television: Transmission

The output of the MPEG2 video encoder is multiplexed with the AC-3 audio encoded data stream. Additional data (e.g., control, program, and auxiliary data) are also multiplexed with the MPEG2 and AC-3 data streams to form a 19.39-Mbps ATSC data stream. The 19.39-Mbps multiplexed data stream is input into the exciter of an 8VSB transmitter, which fits the data stream into a 6-MHz bandwidth. A block diagram of a digital television transmission system is shown in Fig. 17-29.

8VSB is the ATSC approved method for transmitting digital television. It is an 8-level vestigial-sideband modulator, hence the 8VSB abbreviation (see Sec. 17-5 for a discussion on vestigial sideband). The 8VSB signal constellation somewhat resembles a 64-QAM constellation, except the 8VSB requires the decoding of only the I channel (see Chapter 10, I and Q signals). This helps to minimize the electronics required in the receiver. A picture of the 8VSB constellation is shown in Fig. 17-30.

8VSB
the RF modulation technique for ATSC DTV transmission

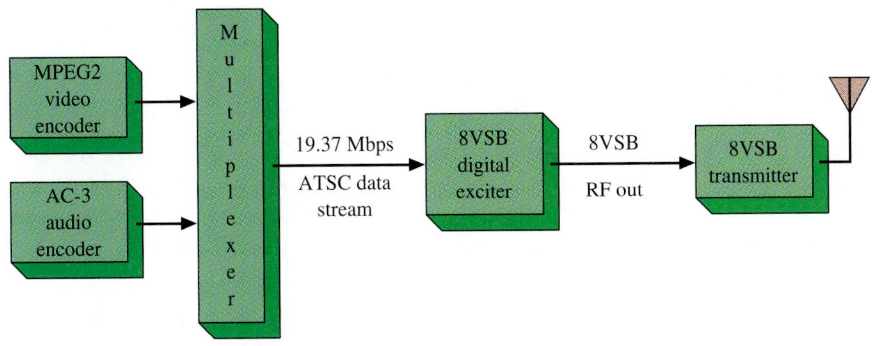

FIGURE 17-29 A block diagram of the ATSC digital transmission system.

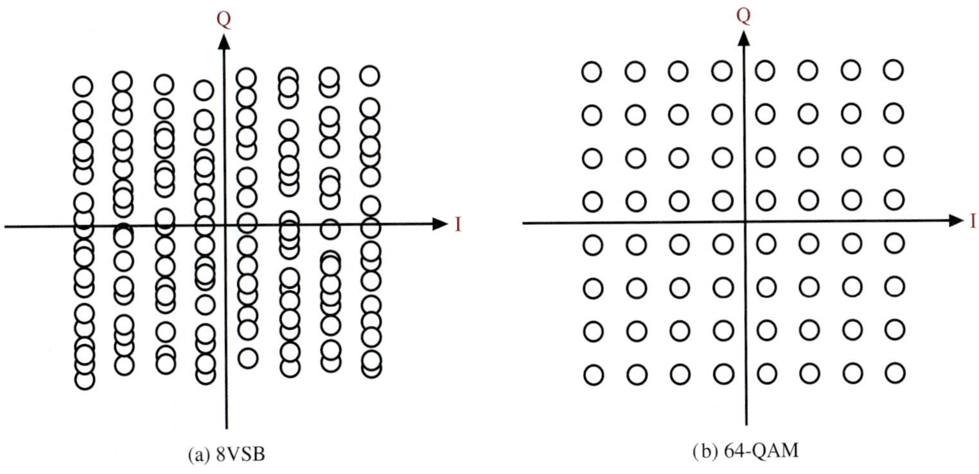

FIGURE 17-30 (a) The 8VSB and (b) the 64-QAM constellations. (Courtesy of Harris Broadcast.)

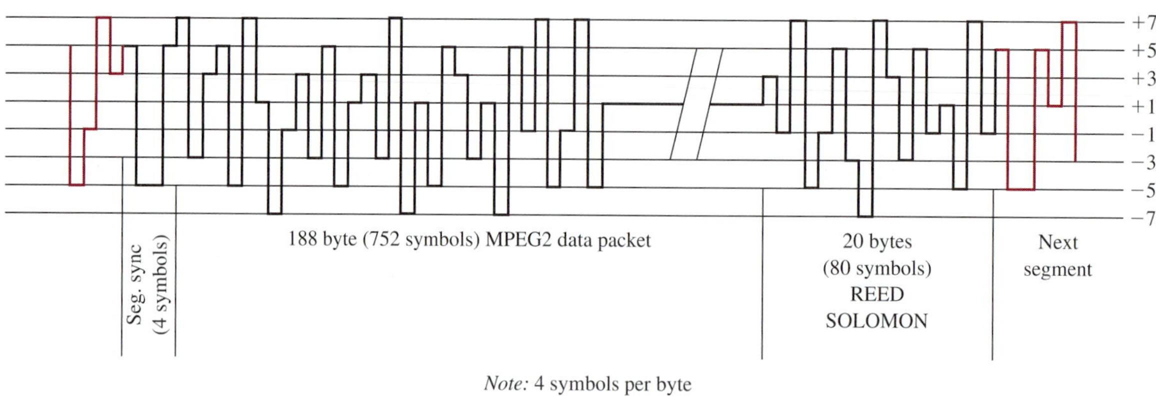

FIGURE 17-31 The 8VSB segment sync. (Courtesy of Harris Broadcast.)

ATSC Pilot Carrier
provides a clock reference for the 8VSB receiver to lock onto

Segment Sync
a repetitive 1-byte pulse that is added to the front of the data segment

Frame Sync
data pulse repeated once every 313 data segments to identify that a frame has been completed

The 8VSB exciter uses the first byte in each MPEG2 packet to synchronize its circuits. The 8VSB exciter then inserts an **ATSC pilot carrier, segment sync,** and **frame sync** to aid the 8VSB receiver in accurately recovering the transmitted data.

ATSC pilot: The pilot provides a clock for the 8VSB receiver to lock onto. The pilot is located at the zero-frequency point of the 8VSB spectrum. For example, if the assigned frequency spectrum is 518–524 MHz, then the pilot is located at 518 MHz. The pilot signal is similar to the color-burst signal in NTSC transmissions, which provides a stable reference for the color signal reconstruction.

Segment sync: This repetitive 1-byte pulse is added to the front of the data segment. The frame sync is used by the 8VSB receiver to generate the receiver clock and ultimately recover the data. The segment sync is similar to the horizontal sync pulse in an NTSC signal. See Fig. 17-31.

Frame sync: The ATSC data frame consists of 313 data segments. The frame sync is repeated once per frame. In an NTSC world, this is comparable to the vertical sync. The frame sync has a "known" data pattern, which can be used to help the receiver remove signal irregularities, such as ghosting, introduced by poor reception. See Fig. 17-32.

The Basics of Digital Television: Reception and Coverage Area

The issue associated with DTV reception is whether or not the licensed coverage area will be the same as it was for analog broadcast. Distortions in the received signal will impair the quality of the received signal. ATSC has recommended a signal-to-noise (SNR) ratio of 27 dB or better at the reception point. For analog transmissions, poor SNR values result in picture roll or a noisy picture. In a digital broadcast, poor SNR ratios result in frozen pictures and repeated pixel values. This is called **pixelate,** where the picture freezes into a series of patterns even when there is motion in the video. Also, multipath or noisy reception can lead to no signal at the receiver, which results in a complete loss of data. When this happens, many receivers insert a blue screen to substitute for the lost data. This screen is humorously called the "blue screen of death" by broadcasters.

Pixelate
when a digital picture freezes even when there is motion in the video, usually due to a poor SNR

Some VHF NTSC broadcasts will be assigned UHF DTV broadcasts. This means that the assigned coverage area of the DTV broadcasts may be affected by the terrain differently than are the coverage areas of the NTSC analog

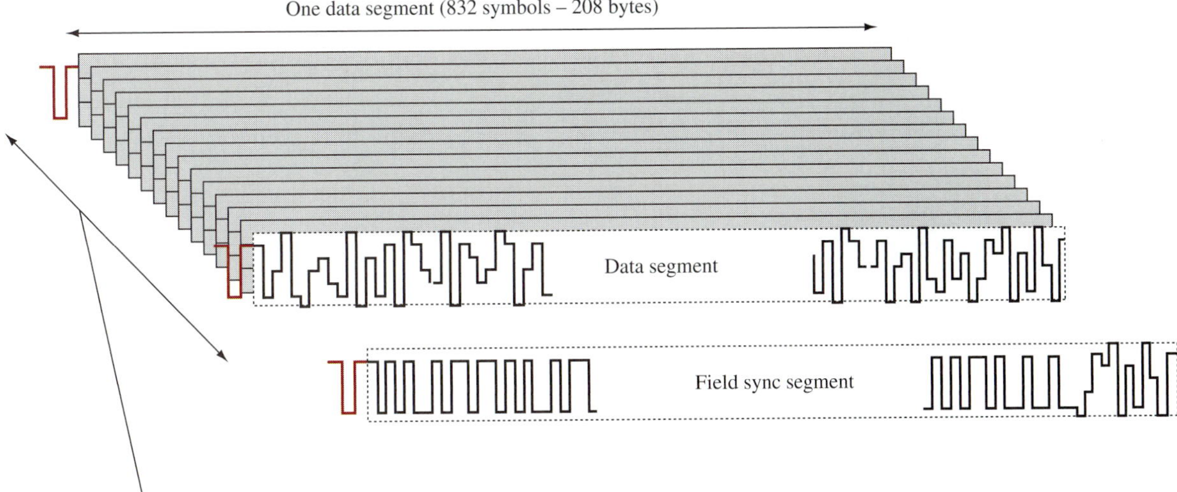

FIGURE 17-32 The 8VSB frame sync. (Courtesy of Harris Broadcast.)

transmissions. Tests have shown that broadcasters might be able to use a technique called **digital on-channel repeater (DOCR)** to reach remote or isolated areas that might not be able to receive DTV transmission. This technique allows for the retransmission of the DTV signal on the same channel while introducing little distortion or ghosting (echoes) of the video image.

Digital On-Channel Repeater (DOCR) allows for the retransmission of the DTV signal on the same channel

Many of the new channel assignments are on adjacent channels. For example, a channel 22 UHF station might receive authorization to originate digital broadcasts on channel 23. Adjacent transmissions (DTV and NTSC) may introduce interference into the NTSC transmission.

DTV receivers reverse the procedure introduced at the DTV transmitter. The receivers must have the capability to display 1920 × 1080, 1280 × 720, or 720 × 480 pixel formats. These formats support the aspect ratio of both HDTV (16 × 9) and SDTV (4 × 3). The ATSC standards support 18 digital video picture formats. These primary video formats are shown in Table 17-4 and are listed by their pixel/line values and associated bit rates. A block diagram of a digital TV receiver is shown in Fig. 17-33.

An example of the chip set for the front-end of a hybrid TV receiver system using the Philips TDA8980 and 8961 is shown in Fig. 17-34. This system allows for the processing of both NTSC and ATSC signals. The TV RF signal is converted to a fixed IF centered at 44 MHz. The signal from the tuner is processed

Table 17-4 Primary Digital Video Format and Bit Rates for the ATSC-Supported Formats

Digital Video Format (Pixel × Lines)	Bit Rate (Mbps)
640 × 480	184
720 × 480	207
1280 × 720	553
1920 × 1080	1244

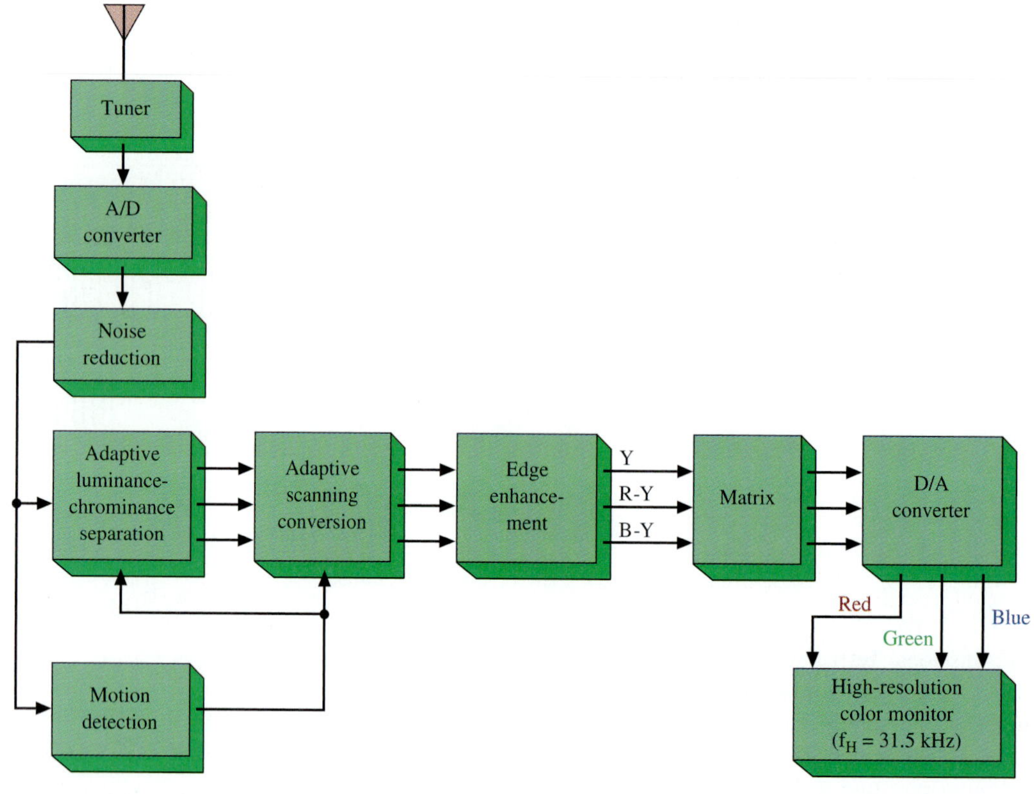

FIGURE 17-33 Digital TV processing.

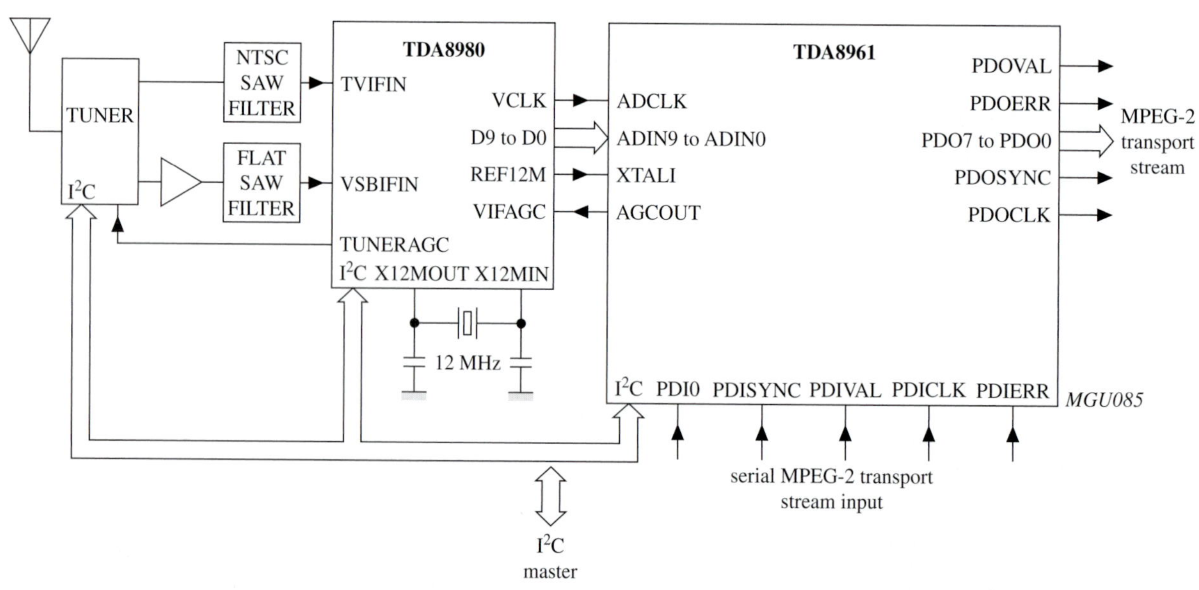

FIGURE 17-34 Front-end design for a hybrid TV system using the TDA8980 and TDA8961.

by two SAW filters and down-converted to a 4 MHz IF by the TDA8980. This chip contains an internal switch that enables it to process both the NTSC IF (analog) and 8VSB IF (digital) signals. The TDA8980 uses on-chip 10-bit digitizing at a 36 MHz sample rate to convert the IF signal to a digital data stream. In the case of an NTSC signal, the digitized signal is bypassed directly to the TDA8961 MPEG-2 output to be displayed by the digital television. In the case of an 8VSB TV signal, carrier recovery is provided by a **frequency and phase-locked loop (FPLL)** circuit within the TDA8961. The FPLL uses the chip's internal clock and the incoming carrier frequencies to stabilize the VCO. This results in a reduction of the VCO phase noise and minimizes or removes NTSC cochannel interference. The 8VSB digital TV signal is processed by the TDA8961, and the chip's MPEG-2 output is a fully compliant ATSC-compliant signal that can be displayed by the digital television.

Frequency and Phase-Locked Loop (FPLL)
a circuit that uses both the internal clock and incoming carrier frequency to stabilize the VCO

17-13 TROUBLESHOOTING

As you recall from Sec. 17-2, a TV is basically an AM receiver for picture information and an FM receiver for sound reception. The basic approach for troubleshooting a TV is to proceed as if it were two radios. In this section, you will learn techniques that will help you identify and isolate faulty sections within the TV receiver. These techniques may be used when troubleshooting any kind of television.

After completing this section you should be able to

- Identify defective stages in a TV receiver
- Describe faulty sections within the TV based on viewed symptoms

Looking into the back of a TV for the first time can be a fearful sight. Inside are many complex circuits and hundreds of components. To be an effective service technician, you must have a thorough knowledge and understanding of how a TV works. It becomes quite apparent that service literature is needed when servicing a TV. Make sure you get the service literature pertaining to the model and chassis that you are repairing. The TV setup should be as close as you can get it to normal viewing. Organize your thinking in a logical pattern *before* making any circuit measurements. Observe and classify any abnormality such as sound, raster, video, or color problems. Study the service literature and identify functional sections in the TV. For example, identify the location of the horizontal output section, vertical section, video section, and others. Try to localize the problem to a specific section from the symptoms being observed.

The television stands out from other communications receivers because circuit defects often show up on the screen. From these visual symptoms, faulty sections can be singled out before you open the back of the TV. For example, symptoms like no video, no sound, and good raster would lead us to check the IF section of the TV because both picture information and sound information are amplified there. The **raster** refers to CRT illumination by the scan lines when no signal is received and/or being displayed.

Raster
illuminated area on the picture of a TV receiver when no signal is being received

Study the problem. Make a decision where to start troubleshooting based on the viewed symptoms. Try to isolate the defective stage from the symptoms on the screen. Is a picture present? Does the picture roll up and down or left to right? Does

Table 17-5

Symptom	Cause	Stage/Area of Trouble
Set is dead, no sound, no video, no raster	No power to circuits	Check main power supply, start-up circuits, main fuses, and line cord
Set blows fuses	A short circuit exists in the main power supply or horizontal output	Check for shorted diodes, shorted regulator transistors, and shorted filter capacitors in the main power supply, or shorted horizontal output transistor and shorted horizontal output transformer
Sound normal, no video, no raster	No high voltage	Check the horizontal output circuit/ high-voltage section
Normal raster, no video, no sound	Video and sound signal missing	Check antenna, tuner, and IF amplifiers
Raster and video normal	Sound signal missing	Check sound IF amplifiers, detector section, audio amplifiers, and speaker
Raster, video, and sound normal, no color	Color signal missing	Check color killer and color processing circuits
Picture has snow, noise heard in sound	Signal-to-noise ratio high	Check RF amplifier in tuner
Vertical roll (up and down)	Vertical sync missing	Check sync separator, vertical oscillator
Horizontal white line across screen	Vertical output signal missing	Check the vertical output circuit, vertical oscillator, or yoke
Horizontal roll (left and right)	Horizontal sync missing	Check sync separator circuit, horizontal AFC, horizontal oscillator
Vertical white line on screen	Horizontal output signal missing	Check horizontal output circuit, horizontal oscillator, yoke

the picture have snow in it? Is the sound present or not? By observing these signposts you can quickly determine the defective section within the TV. Table 17-5 gives symptoms, causes, and the area in which to look. It is not all-inclusive of the problems that might occur in a TV.

Consult the manufacturer's service literature for diagnostic charts or other troubleshooting guidelines. When these troubleshooting aids are available, they often relate to common failures incurred for that model TV receiver or manufacturing defects that may exist in the set.

Pull the back off the set, and with the power off, look for obvious problems. Look for loose wires or connectors, burned components, broken or burned PCB traces, and cold solder joints. With the power on, listen for unusual sounds like hissing (normally associated with horizontal output transformers that have developed high-voltage leaks), arcing, and high-pitched squeals from the horizontal oscillator. Do you smell anything burning? Look for brown areas on the PCB indicating overheating components.

If the preliminary inspection fails to localize a defective component, continue troubleshooting by doing voltage and resistance measurements on the suspected stage in the TV set. Compare the results with specified values from the service literature. Use the oscilloscope to check for proper waveforms in the defective section and associated sections. Schematic diagrams usually give pictures indicating what the correct waveform looks like for all the common signals in the set. Base your troubleshooting on an organized approach and not a disorganized one. Having an organized strategy will save you valuable time and enhance your troubleshooting.

17-14 Troubleshooting with Electronics Workbench™ Multisim

This Electronics Workbench™ Multisim exercise provides an opportunity to use a spectrum analyzer to view the UHF television spectrum. Begin the exercise by opening **Fig17-35.ms7 (.msm)** on your EWB Multisim CD. The circuit is shown in Fig. 17-35.

This circuit is used to demonstrate the frequency spectra for an NTSC television signal. The NTSC frequency spectra includes a visual carrier and an aural carrier. Double-click on the AM and FM sources to view the settings. The amplitude modulated visual carrier frequency has been set to 519.25 MHz. The aural carrier is a frequency-modulated signal, and the carrier frequency is 523.75 MHz. These are the visual and aural carriers for channel 22 in the UHF television spectrum. The two carriers are connected by a summing amplifier. The combined signal, as viewed by the spectrum analyzer, is shown in Fig. 17-36. The visual carrier is shown on the left, and the aural carrier is shown on the right. Use the cursor to verify the center frequency for each carrier.

Next, open the **Fig17-37.ms7 (.msm)** circuit on your EWB CD. This is called a bandstop, or wavetrap, and it is commonly used to attenuate a narrow band of frequencies. This is an example of a series resonant circuit similar to the example shown in Fig. 17-13(a). The EWB Multisim circuit is shown in Fig. 17-37.

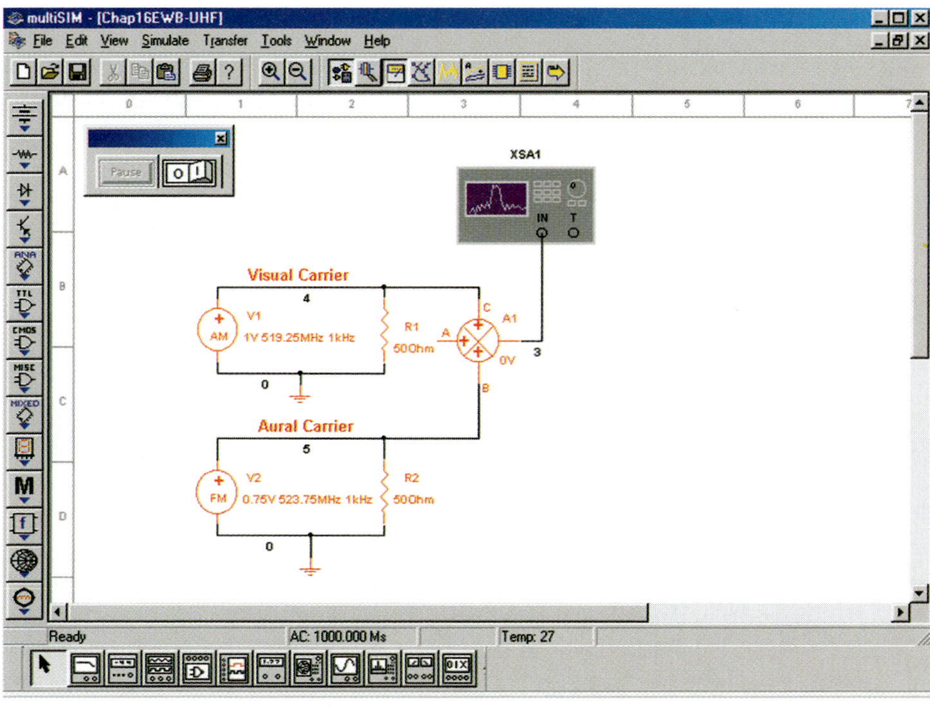

FIGURE 17-35 The Electronics Workbench™ Multisim circuit used to simulate the frequency spectra for a UHF television signal.

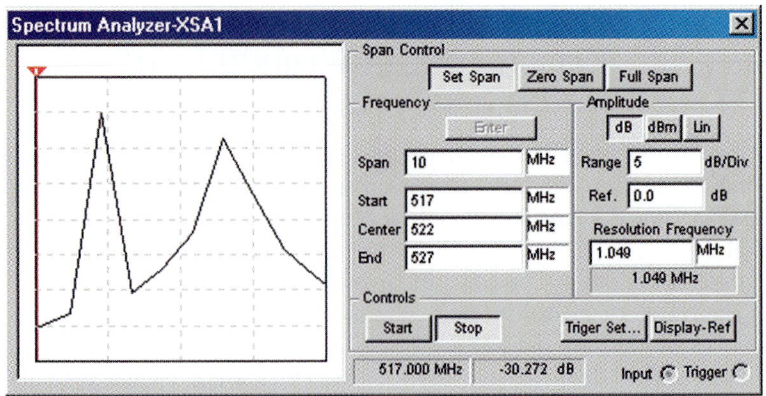

FIGURE 17-36 The Electronics Workbench™ simulation of the frequency spectra for a channel 22 television signal.

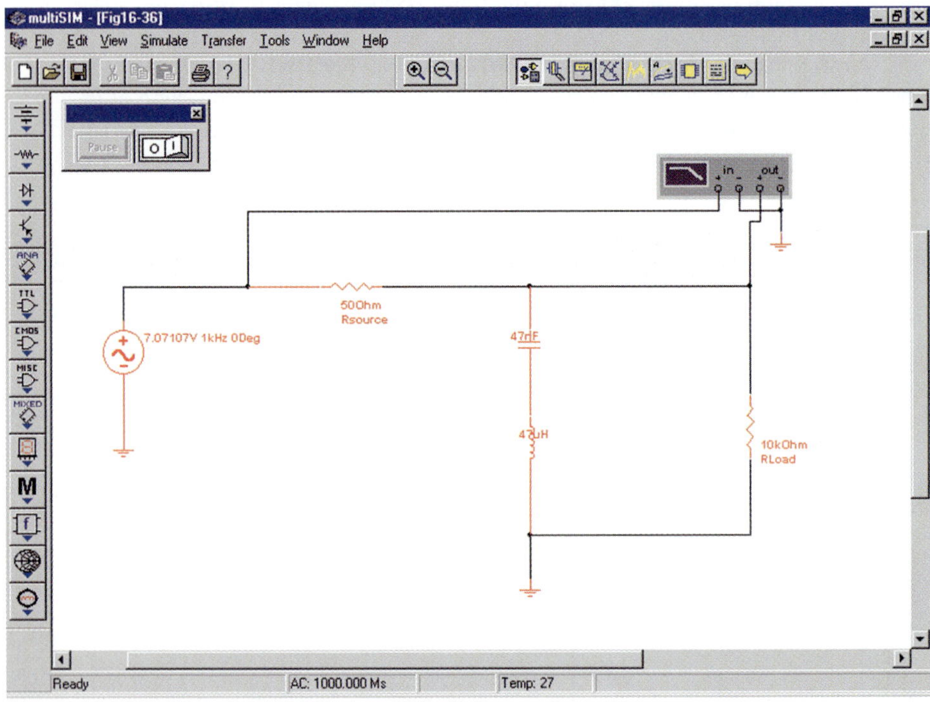

FIGURE 17-37 The Electronics Workbench™ Multisim circuit of a high-Q bandstop circuit, or wavetrap.

Start the simulation and observe the output from the Bode plotter. This circuit provides more than 60 dB of attenuation at 107 kHz. Use the cursor to verify this measurement. You can also use the cursor to obtain the 3-dB corner frequencies, which are 53.7 kHz and 218.7 kHz. The output of the Bode plotter is shown in Fig. 17-38.

How would the degradation of one of the filter components affect the frequency response? Open the file **FigE17-1.ms7 (.msm)** on your EWB CD. Start the simulation

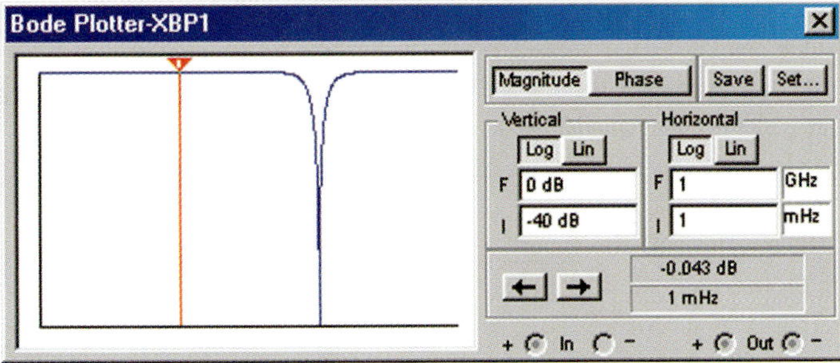

FIGURE 17-38 The Bode plotter output for the wavetrap filter.

and observe the output of the Bode plotter. Is this circuit functioning properly? If not, troubleshoot the circuit to determine the problem. You will find that L_1 is defective. Can you explain why the Bode plot looks like it does? If the inductor, L_1, is leaky or partially shorted, then at high frequencies the inductor appears as a short or a low impedance instead of being reactive. Remember, for a properly functioning inductor, the inductive reactance increases ($X_L = 2\pi fL$) as the frequency increases. This explains why the output is severely attenuated at high frequencies.

Electronics Workbench™ Exercises

1. Open the file **FigE17-2.ms7 (.msm)** in your EWB CD. Determine the center frequencies for the visual and aural carriers. Verify your results with the spectrum analyzer. What frequency band and television channel is this? (67.25 MHz, 71.75 MHz, VHF, channel 4)
2. Open the file **FigE17-3.ms7 (.msm)** in your EWB CD. Determine if the filter circuit is working properly. If the circuit is not working, troubleshoot it and correct the problem.
3. Open the file **FigE17-4.ms7 (.msm)** in your EWB CD. Determine if the filter circuit is working properly. If the circuit is not working, troubleshoot it and correct the problem. Explain why this type of failure might have caused the problem.

Summary

In Chapter 17, the complete NTSC TV and DTV systems were introduced, including NTSC signal generation and transmission, and reception. The video (AM) and the audio (FM) are combined to make up the composite NTSC TV signal. For digital television, the video (MPEG2) and the audio (AC-3) were presented. The signals paths were discussed through compression and multiplexing and into the 8VSB exciter. The characteristics of the received digital television signal were explored. The major topics you should now understand include:

- the description of a complete but simplified NTSC TV system
- the operation of the charge-coupled device (CCD) in a TV camera
- the definition of scanning, pixel, horizontal/vertical retrace interval, and the NTSC TV aspect ratio
- the explanation of interlaced scanning with frame definition and timing

- the calculation and definition of horizontal and vertical resolution
- the explanation of NTSC TV receiver operation using a block diagram
- the changes necessary to the black-and-white signal to allow inclusion of color information in a compatible fashion for NTSC television
- the description of MPEG2 and the AC-3 standards for video and audio in DTV
- drawing a block diagram of the basic operation of an 8VSB transmitter
- describing the techniques used in DTV to provide frame and segment sync
- drawing the picture of the processing section for a DTV receiver

Questions and Problems

Section 17-2

*1. Does the sound transmitter at a television broadcast station employ frequency or amplitude modulation?

2. In what way is TV audio equivalent to broadcast FM radio, and in what way is it inferior?

*3. Does the video transmitter at a television broadcast station employ frequency or amplitude modulation?

4. Explain the major benefit of combining AM and FM techniques in television broadcasting.

5. List and explain the functions of the six transducers used in a complete TV system.

*6. Why is a diplexer a necessary stage of most TV transmitters?

7. Describe the operation of a CCD image pickup device.

*8. What is a mosaic plate in a television camera?

9. Sketch the electrical video signal that would result from scanning the letter "E" in the setup shown in Fig. 17-3.

*10. In television broadcasting, what is the meaning of the term *aspect ratio*?

*11. Numerically, what is the aspect ratio of a picture as transmitted by a television broadcast station?

Section 17-3

*12. What is the purpose of synchronizing pulses in a television broadcast signal?

13. Provide an analogy between horizontal and vertical retrace and reading a book.

*14. If the cathode-ray tube in a television receiver is replaced by a larger tube so that the size of the picture is changed from 6 by 8 in. to 12 by 16 in., what change, if any, is made in the number of scanning lines per frame?

*15. How many frames per second do television broadcast stations transmit?

16. What is *interlacing*?

*17. Why is interlacing used in television broadcasting?

*18. What are synchronizing pulses in a television broadcast and receiving system?

19. Why does flicker occur?

---

* An asterisk preceding a number indicates a question that has been provided by the FCC as a study aid for licensing examinations.

*20. What are blanking pulses in a television broadcasting and receiving system?
21. Calculate the frequency required for the horizontal sync pulses. (15.8 kHz)
*22. What is the field frequency of a television broadcast transmitter?
*23. In television broadcasting, why is the field frequency made equal to the frequency of the commercial (ac) power source?
*24. Besides the camera signal, what other signals and pulses are included in a complete television broadcast signal?

Section 17-4

25. Describe the characteristics of a video amplifier.
26. Define *resolution, vertical resolution,* and *horizontal resolution.*
27. Calculate the horizontal resolution of a broadcast TV picture. (≈428 lines)
28. Calculate the decrease in horizontal resolution if the video signal bandwidth were reduced from 4 to 3.5 MHz. (from 428 to 375 lines)
29. Calculate the vertical resolution if the video signal bandwidth were reduced from 4 to 3.5 MHz, assuming that the horizontal resolution was not to change. (307 lines)

Section 17-5

*30. If a television broadcast station transmits the video signals on channel 6 (82 to 88 MHz), what is the center frequency of the aural transmitter?
*31. What is meant by 100 percent modulation of the aural transmitter at a television broadcast station?
32. What TV channel is most likely to be heard on an FM broadcast receiver? Explain why.
*33. What is *vestigial-sideband transmission* of a television broadcast station?

Section 17-6

34. Draw a TV receiver block diagram, and briefly explain the function of each block.
35. What is the typical output voltage of the flyback transformer for a 14-in. (diagonal) CRT? (14 kV)

Section 17-7

36. State what a TV *front end* consists of and the important functions it performs.
37. Show how the VHF tuner is used in conjunction with VHF reception. Why is the VHF signal stepped down in frequency before it is given any amplification?
38. Explain *stagger tuning* and why it is often used in TV IF amplifiers.
39. Calculate the approximate "finger" spacing for a SAW filter operating at 44 MHz. (0.0682 mm)
40. Why are the sound carrier and its sidebands given only one-tenth the amplification of the video by the IF response curve? Explain why part of the video signal is also given less amplification.
41. Discuss the function of a wavetrap and the need for such traps in TV receivers.

Section 17-8

42. If an amplifier stage of the video section shown in Fig. 17-14 became inoperative, would the receiver's sound be affected?
43. What is the function of dc restoration, and what kind of video sections require it?

Section 17-9

44. What is the function of the sync separator? How is it able to differentiate between the horizontal and vertical sync pulses? What types of circuits are used for each?
45. What is another name for the sync separator?
46. Explain the relationship between the horizontal deflection system and the CRT anode high-voltage supply. Why is this a failure-prone area in a TV receiver?
47. What are the possible effects of a nonlinear deflection waveform?
48. Explain the operation of the horizontal system schematic shown in Fig. 17-18.
49. Explain the function of the damper system and flyback transformer.

Section 17-10

*50. Describe the scanning process employed in connection with color TV broadcast transmission.
51. Describe the important features of the Y, I, and Q signals in a color TV broadcast.
52. Define the meaning of *compatibility* with respect to color and monochrome TV. How does a monochrome set properly display a color transmission?
*53. Describe the composition of the chrominance subcarrier used in the authorized system of color television.
54. Explain how the Y, I, and Q signals are processed by a color TV receiver.
55. Why is extreme accuracy required of the color subcarrier oscillator within a color TV receiver? Explain how this accuracy is obtained in the receiver.
56. Explain the operation of the color killer. Describe the effect of a defective color killer.
57. Describe the important characteristics and construction of the color CRT. Include the need for convergence and how it is accomplished in this discussion.

Section 17-11

58. Describe the stereo audio system used for TV and explain its superior performance compared to broadcast FM stereo.
59. The audio signal for TV stereo reaches a maximum level of 20 dBm at the transmitter. Calculate the companded audio level for 4 dBm and 15 dBm. (12 dBm, 17.5 dBm)
60. What is HDTV? Give a reason why it has an increased aspect ratio over regular TV.

Section 17-12

61. Describe the 4:2:2 format used to digitize component video.
62. Describe the compression techniques used by the MPEG2 format.
63. Why is the AC-3 digital audio compression technique also called 5.1 Channel Input?

64. Draw a block diagram of the ATSC digital transmission system and identify the data rate at the output of the multiplexer.
65. Describe the operation of the 8VSB exciter. Draw a picture of the 8VSB constellation.
66. Describe the data format for the 8VSB data stream. Identify the purpose of the ATSC pilot, the segment sync, and the frame sync.

Section 17-13

67. In several paragraphs, describe a general troubleshooting procedure to be used for repair of a TV receiver.
68. List the probable defective stage(s) for the following symptoms:
    (a) Video and raster normal, sound dead.
    (b) Sound normal, video and raster dim.
    (c) Raster normal, sound and video dead.
    (d) Bent and "contrasty" picture.
    (e) Floating picture, sound and raster normal.
    (f) Loss of vertical sync.
    (g) Loss of horizontal sync.
    (h) Normal sound, no raster.
    (i) No sound or raster.
    (j) No color, black and white normal.
    (k) Loss of one color.
    (l) Loss of color sync.
69. Explain what is meant by the term *raster*. List the probable defective stage(s) if a set's raster is normal but the sound and video are dead.
70. Explain a possible problem if there is no audio output.
71. Describe what would happen if the diode mixer in the VHF/UHF tuner in Fig. 17-9 was not functioning.
72. What would the picture be on the color TV if one color was out?
73. Explain what would happen if the AGC was not operating in Fig. 17-14.
74. What would the output look like if the brightness control in Fig. 17-14 was adjusted to minimum?

Questions for Critical Thinking

75. Explain why vertical resolution is less than the number (about 0.7) of horizontal lines. (*Hint:* Consider what might happen if a pattern of 495 alternate black-and-white horizontal lines were scanned by a TV camera so that each scan saw half of a white-and-black line.)
76. Calculate the sound and picture carrier frequency for channel 10 before and after frequency translation to the IF frequency. What is the required local oscillator frequency? (41 to 47 MHz, 197.75 MHz, 193.25 MHz, 41.25 MHz, 45.75 MHz, 239 MHz)
77. In detail, explain the difference between adjustment of the brightness and contrast controls.
78. Describe the process of interleaving and analyze its importance in enabling the broadcast of color TV on the same bandwidth used in the monochrome system.
79. Describe what problems can be expected with digital television reception.

Fiber Optics

Chapter Outline

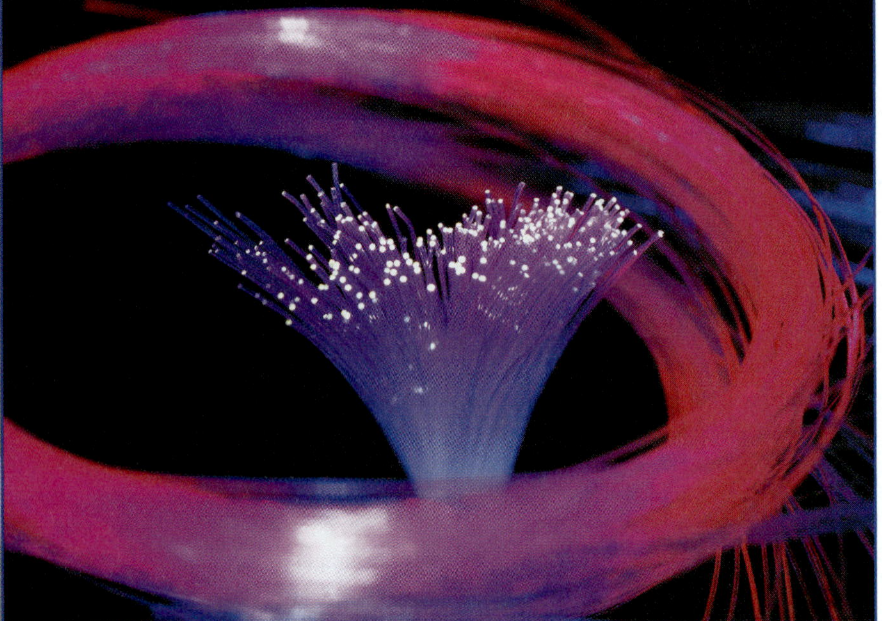

A bundle of fiber-optic cable. (Courtesy of International Stock Photography Ltd. Photo by Johnny Stockshooter.)

Objectives

- Describe a basic fiber-optic communication system and provide nine advantages of glass fiber versus copper conductors
- Provide a physical description of light propagating in an optical fiber, including the concepts of reflection, refraction, critical angle, acceptance cone, attenuation, dispersion, and numerical aperture
- Explain the physical characteristics of the three types of communications grade fiber and provide their relative advantages
- Calculate the power loss for optical fiber
- Describe the physical construction and operation of a semiconductor laser
- Describe the physical construction and operation of a *p-i-n* diode and avalanche photodiode
- Discuss various considerations when making fiber connections
- Calculate a complete power budget analysis for a fiber-optic system
- Explain the incorporation of fiber optics into local area networks
- Understand the safety issues when working with fiber optics

Key Terms

refractive index
infrared light
optical spectrum
O-, E-, S-, C-, L-, and U-bands
core
cladding
numerical aperture
multimode fibers
step-index fibers
pulse dispersion
graded-index fiber
single-mode fibers
long haul
mode field diameter
zero-dispersion wavelength
attenuation
scattering
absorption
macrobending
microbending
dispersion
modal dispersion
chromatic dispersion
polarization mode dispersion
dispersion compensating fiber
fiber Bragg grating
coherent
distributed feedback (DFB) laser
dense wavelength division multiplex (DWDM)
vertical cavity surface emitting lasers (VCSELs)
tunable laser
fiber, light pipe, glass
isolators
received signal level (RSL)
dark current
fusion splicing
mechanical splices
index-matching gel
SC, ST
small-form factor
backhoe fading
OTDR
event
SONET
OC
STS
FTTC
FTTH
air fiber

18-1 Introduction

Recent advances in the development and manufacture of fiber-optic systems have made them the latest frontier in the field of telecommunications. They are being used extensively for both military and commercial data links and have replaced a lot of copper wire. They have also taken over almost all the point-to-point long-distance communications traffic previously handled by microwave and satellite links, particularly transoceanic.

A fiber-optic communications system is surprisingly simple, as shown in Fig. 18-1. It comprises the following elements:

1. A fiber-optic transmission strand can carry the signal (in the form of a light beam modulated by an analog waveform or by digital pulses) a few feet or even hundreds or thousands of miles. A cable may contain three or four hair-like fibers or a bundle of hundreds of such fibers.
2. A source of invisible infrared radiation—usually a light-emitting diode (LED) or a solid-state laser—that can be modulated to impress digital data or an analog signal on the light beam.
3. A photosensitive detector to convert the optical signal back into an electrical signal at the receiver. The most often used detectors are *p-i-n* or avalanche photodiodes.
4. Efficient optical connectors at the light source–cable interface and at the cable–photodetector interface. These connectors are also critical when splicing the optical cable due to excessive loss that can occur at connections.

The advantages of optical communications links compared to waveguides or copper conductors are enormous and include the following:

1. *Extremely wide system bandwidth:* The intelligence is impressed on the light by varying the light's amplitude. Since the best LEDs have a 5-ns response

Optical-fiber system. (Courtesy of AMP.)

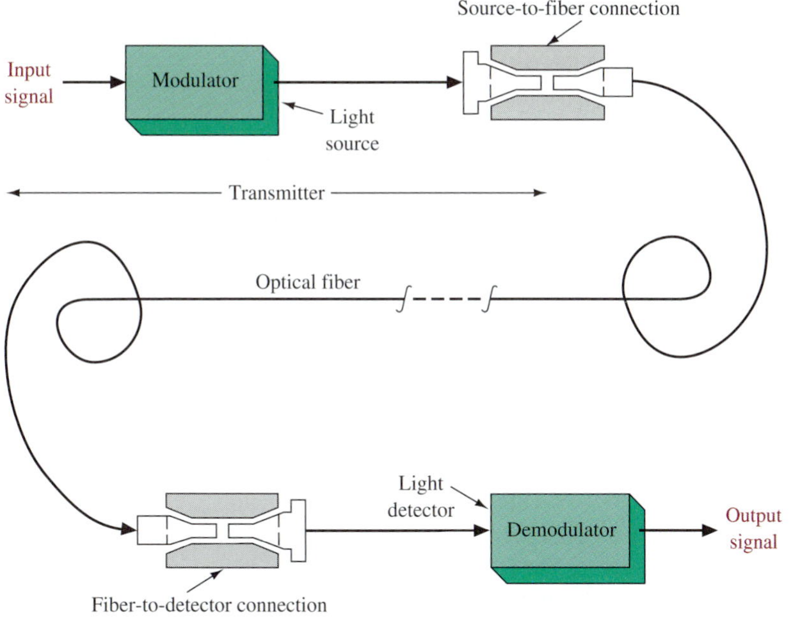

FIGURE 18-1 Fiber-optic communication system.

time, they provide a maximum bandwidth of about 100 MHz. Using laser light sources, however, bandwidths of up to 10 Gbps are possible on a single glass fiber and several lasers can be combined on one fiber. The amount of information multiplexed on such a system, in the tens of Gbps, is indeed staggering.

2. *Immunity to electrostatic interference:* External electrical noise and lightning do not affect energy in a fiber-optic strand. This is true only for the optical strands, however, not the metallic cable components or connecting electronics.
3. *Elimination of crosstalk:* The light in one glass fiber does not interfere with, nor is it susceptible to, the light in an adjacent fiber. Recall that crosstalk results from the electromagnetic coupling between two adjacent copper wires (see Chapter 12).
4. *Lower signal attenuation than other propagation systems:* Typical attenuation of a fiber-optic strand varies from 0.1 to 0.01 dB per 100 feet, depending on the wavelength of operation. By way of contrast, the loss of RG-6 and RG-59 75-ohm coaxial cable at 1 GHz is approximately 11.5 dB per 100 feet. The loss of $\frac{1}{2}$" coaxial cable is approximately 4.2 dB per 100 ft.
5. *Substantially lighter weight and smaller size:* The U.S. Navy replaced conventional wiring on the A-7 airplane with fiber that carries data between a central computer and all its remote sensors and peripheral avionics. In this case, 224 ft of fiber optics weighing 1.52 lb replaced 1900 ft of copper wire weighing 30 lb.
6. *Lower costs:* Optical-fiber costs are continuing to decline. The costs of many systems are declining with the use of fiber, and that trend is accelerating.
7. *Safety:* In many copper wired systems, the potential hazard of short circuits requires precautionary designs, whereas the dielectric nature of optic fibers eliminates the spark hazard.
8. *Corrosion:* Glass is basically inert, so the corrosive effects of certain environments are not a problem.

9. *Security:* Due to its immunity to and from electromagnetic coupling and radiation, optical fiber can be used in most secure environments. Although it can be intercepted or tapped, it is difficult to do so.

18-2 THE NATURE OF LIGHT

Before one can understand the propagation of light in a glass fiber, it is necessary to review some basics of light refraction and reflection. The speed of light in free space is 3×10^8 m/s but is reduced in other media. The reduction as light passes into denser material results in refraction of the light. Refraction causes the light wave to be bent, as shown in Fig. 18-2(a). The speed reduction and subsequent refraction is different for each wavelength, as shown in Fig. 18-2(b). The visible light striking the prism causes refraction at both air/glass interfaces and separates the light into its various frequencies (colors) as shown. This same effect produces a rainbow, with water droplets acting as prisms to split the sunlight into the visible spectrum of colors (the various frequencies).

Refractive Index
ratio of the speed of light in free space to its speed in a given material

The amount of bend provided by refraction depends on the **refractive index** of the two materials involved. The refractive index, n, is the ratio of the speed of light in free space to the speed in a given material. It is slightly variable for different frequencies of light, but for most purposes a single value is accurate enough. The refractive index for free space (a vacuum) is 1.0, while air is 1.0003, and water is 1.33; for the various glasses used in fiber optics, it varies between 1.42 and 1.50.

Snell's law [Eq. (13-6) from Chapter 13] predicts the refraction that takes place when light is transmitted between two different materials:

$$n_1 \sin \theta_1 = n_2 \sin \theta_2 \qquad \textbf{(13-6)}$$

This effect was shown in Fig. 13-4. Figure 18-3 shows the case where an incident ray is at an angle so that the refracted ray goes along the interface and so θ_2 is 90°. When θ_2 is 90°, the angle θ_1 is at the critical angle (θ_c) and defines the angle at which the incident rays no longer pass through the interface. When θ_1 is equal to or greater than θ_c, all the incident light is reflected and the angle of the incidence equals the angle of reflection, as we saw in Fig. 13-4.

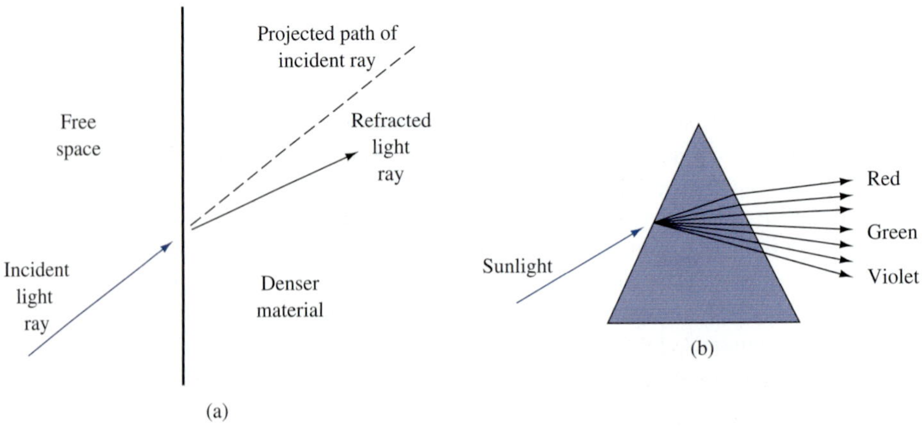

FIGURE 18-2 Refraction of light.

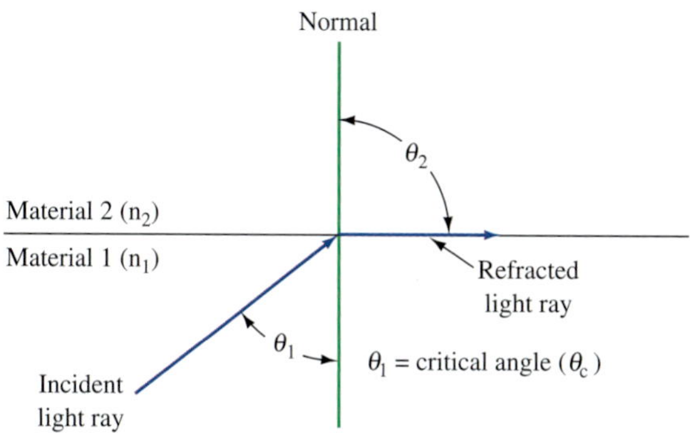

FIGURE 18-3 Critical angle.

The frequency of visible light ranges from about 4.4×10^{14} Hz for red up to 7×10^{14} Hz for violet.

Example 18-1

Calculate the wavelengths of red and violet light.

Solution

For red,

$$\lambda = \frac{c}{f}$$
$$= \frac{3 \times 10^8 \text{ m/s}}{4.4 \times 10^{14} \text{ Hz}} = 6.8 \times 10^{-7} \text{ m}$$
$$= 0.68 \ \mu\text{m or } 0.68 \text{ micron or } 680 \text{ nm}$$

For violet,

$$\lambda = \frac{3 \times 10^8 \text{ m/s}}{7 \times 10^{14} \text{ Hz}} = 0.43 \text{ micron or } 430 \text{ nm}$$

In the fiber-optics industry, spectrum notation is stated in nanometers (nm) rather than in frequency (Hz) simply because it is easier to use, particularly in spectral-width calculations. A convenient point of commonality is that 3×10^{14} Hz, or 300 THz, is equivalent to 1 μm, or 1000 nm. This relationship is shown in Fig. 18-4. The one exception to this naming convention is when discussing dense wavelength division multiplexing (DWDM), which is the transmission of several optical channels, or wavelengths, in the 1550-nm range, all on the same fiber. For DWDM systems, notations, and particularly channel separations, are stated in terahertz (THz). Wave division multiplexing (WDM) systems are discussed in Sec. 18-9.

An electromagnetic wavelength spectrum chart is provided in Fig. 18-4. The electromagnetic light waves just below the frequencies in the visible spectrum are called **infrared light** waves. Whereas visible light has a wavelength from approximately 390 nm up to 770 nm, infrared light extends from 680 nm up to the wavelengths

Infrared Light
extending from 680 nm up to the wavelengths of the microwaves

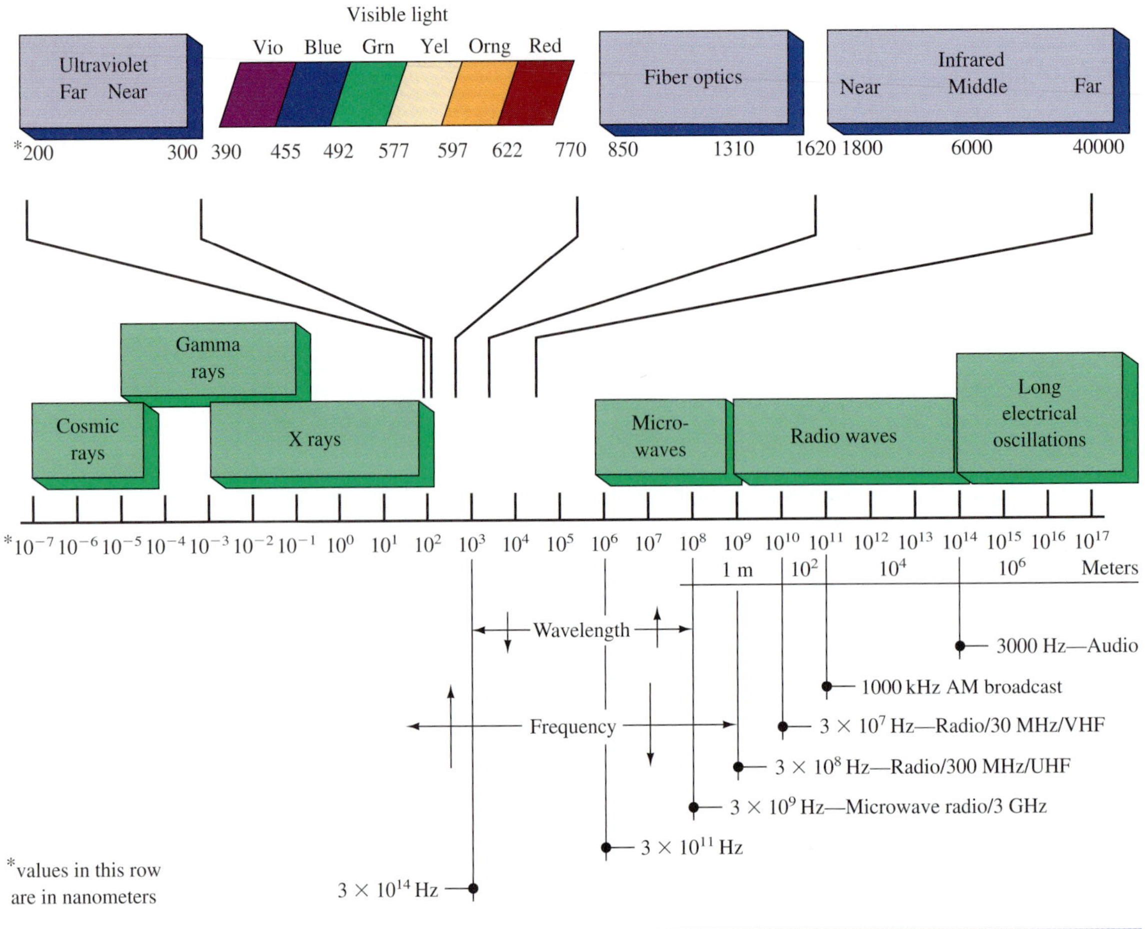

FIGURE 18-4 The electromagnetic wavelength spectrum.

of the microwaves. For the frequencies above visible light, the electromagnetic spectrum includes the ultraviolet (UV) rays and X rays. The frequencies from the infrared on up are termed the **optical spectrum.**

Optical Spectrum
light frequencies from the infrared on up

The most commonly used wavelengths in today's fiber-optic systems are 750 to 850 nm, 1310 nm, and 1530 to 1560 nm. However, industry has categorized the entire spectrum in terms of **O-, E-, S-, C-, L-,** and **U-bands.** Fixed fiber-optic wavelength specifications are simply stated in terms of fixed wavelengths as 850, 1310, or 1550 nm. The band designations shown in Table 18-1 have been proposed to the International Telecommunications Union.

O-, E-, S-, C-, L-, and U-bands
new optical band designations that have been proposed

Construction of the Fiber Strand

Typical construction of an optical fiber is shown in Fig. 18-5. The **core** is the portion of the fiber strand that carries the transmitted light. Its chemical composition is simply a very pure glass: silicon dioxide, doped with small amounts of germanium, boron, and phosphorous. Plastic fiber is used only in short lengths in industrial applications due to its high attenuation. (See Sec. 18-3 for more information on plastic

Core
the portion of the fiber strand that carries the light

Table 18-1 The Optical Bands

Band	Wavelength Range (nm)
O	1260 to 1360
E	1360 to 1460
S	1460 to 1530
C	1530 to 1565
L	1565 to 1625
U	1625 to 1675

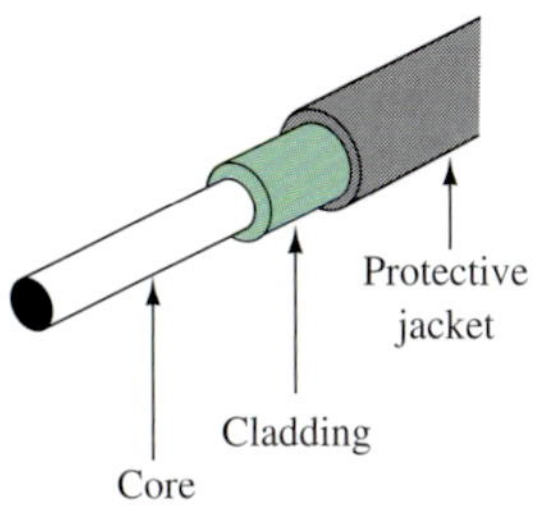

FIGURE 18-5 Single-fiber construction.

fiber.) The **cladding** is the material surrounding the core. It is almost always glass, although plastic cladding of a glass fiber is available but rarely used. In any event, the refractive index for the core and the cladding are different. The cladding must have a lower index of refraction to keep the light in the core. A plastic coating surrounds the cladding to provide protection.

Cladding
the material surrounding the core of an optical waveguide; it must have a lower index of refraction to keep the light in the core

As shown in Fig. 18-6(a), propagation results from the continuous reflection at the core/clad interface so that the ray "bounces" down the fiber length by the process of total internal reflection (TIR). If we consider point *P* in Fig. 18-6(a), the critical angle value for θ_3 is, from Snell's law,

$$\theta_c = \theta_3(\text{min}) = \sin^{-1}\frac{n_2}{n_1}$$

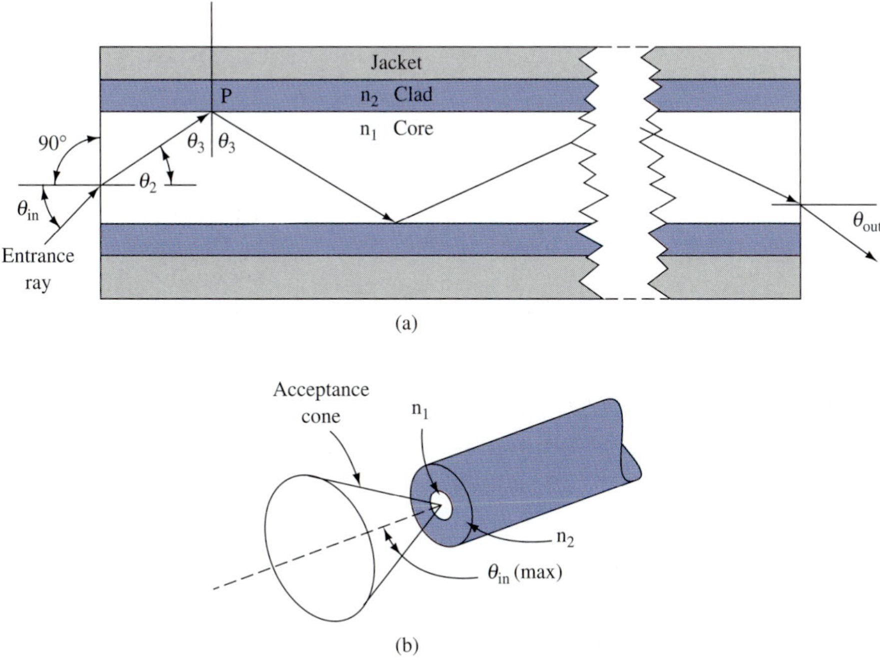

FIGURE 18-6 (a) Development of numerical aperture; (b) acceptance cone.

Because θ_2 is the complement of θ_3,

$$\theta_2(\max) = \sin^{-1}\frac{(n_1^2 - n_2^2)^{1/2}}{n_1}$$

Now applying Snell's law at the entrance surface and because $n_{air} \simeq 1$, we obtain

$$\sin\theta_{in}(\max) = n_1 \sin\theta_2(\max)$$

Combining the two preceding equations yields

$$\sin\theta_{in}(\max) = \sqrt{n_1^2 - n_2^2} \qquad \textbf{(18-1)}$$

Numerical Aperture
a number less than 1 that indicates the range of angles of light that can be introduced to a fiber for transmission

Therefore, $\theta_{in}(\max)$ is the largest angle with the core axis that allows propagation via total internal reflection. Light entering the cable at larger angles is refracted through the core/clad interface and lost. The value $\sin\theta_{in}(\max)$ is called the **numerical aperture** (NA) and defines the half-angle of the cone of acceptance for propagated light in the fiber. This is shown in Fig. 18-6(b). The preceding analysis might lead you to think that crossing over $\theta_{in}(\max)$ causes an abrupt end of light propagation. In practice, however, this is not true; thus, fiber manufacturers usually specify NA as the acceptance angle where the output light is no greater than 10 dB down from the peak value. The NA is a basic specification of a fiber provided by the manufacturer that indicates its ability to accept light and shows how much light can be off-axis and still be propagated.

Example 18-2

An optical fiber and its cladding have refractive indexes of 1.535 and 1.490, respectively. Calculate NA and $\theta_{in}(max)$.

Solution

$$\begin{aligned} NA &= \sin\theta_{in}(\max) = \sqrt{n_1^2 - n_2^2} \qquad \textbf{(18-1)} \\ &= \sqrt{(1.535)^2 - (1.49)^2} = 0.369 \\ \theta_{in}(\max) &= \sin^{-1} 0.369 \\ &= 21.7^\circ \end{aligned}$$

18-3 Optical Fibers

Three types of optical fibers are available, with significant differences in their characteristics. The first communication-grade fibers (early 1970s) had light-carrying core diameters about equal to the wavelength of light. They could carry light in just a single waveguide mode. The difficulty of coupling significant light into such a small fiber led to the development of fibers with cores of about 50 to 100 μm. These fibers support many waveguide modes and are called **multimode fibers.** The first commercial fiber-optic systems used multimode fibers with light at 800 to 900 nm wavelengths. A variation of the multimode fiber, termed *graded-index fiber,* was subsequently developed. This afforded greater bandwidth capability.

Multimode Fibers
fibers with cores of about 50 to 100 μm that support many waveguide modes; light takes many paths

As the technology became more mature, the single-mode fibers were found to provide lower losses and even higher bandwidth. This has led to their use at 1300

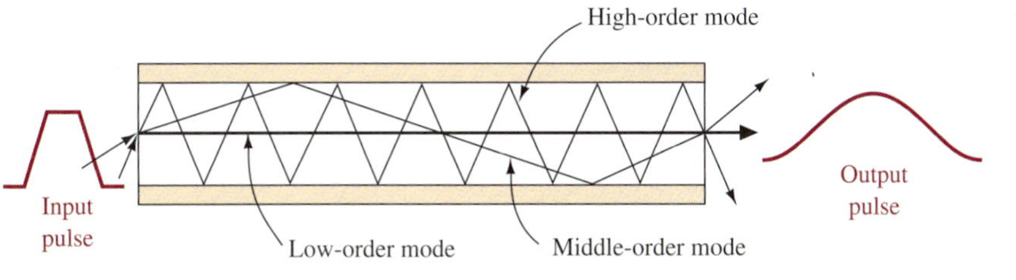

FIGURE 18-7 Modes of propagation for step-index fiber.

and 1500 nm in many telecommunications applications. The new developments have not made old types of fiber obsolete. The application now determines the type used. The following major criteria affect the choice of fiber type:

1. Signal losses, with respect to distance
2. Ease of light coupling and interconnection
3. Bandwidth

Multimode Step-Index Fiber

A fiber showing three different modes (i.e., multimode) of propagation is presented in Fig. 18-7. The lowest-order mode is seen traveling along the axis of the fiber, and the middle-order mode is reflected twice at the interface. The highest-order mode is reflected many times and makes many trips across the fiber. This type of fiber is called **step-index** because of the abrupt change in the refractive index at the core-cladding boundary. As a result of the variable path lengths, the light entering the fiber takes a variable length of time to reach the detector. This results in a pulse-broadening or dispersion characteristic, as shown in Fig. 18-7. This effect is termed **pulse dispersion** and limits the maximum distance and rate at which data (pulses of light) can be practically transmitted. Also note that the output pulse has reduced amplitude as well as increased width. The greater the fiber length, the worse this effect. As a result, manufacturers rate their fiber in bandwidth per length, such as 400 MHz/km. That fiber can successfully transmit pulses at the rate of 400 MHz for 1 km, 200 MHz for 2 km, and so on. In fact, current networking standards limit multimode fiber distances to 2 km. Of course, longer transmission paths are attained by locating regenerators at appropriate locations.

Step-Index Fibers
fibers in which there is an abrupt change in the refractive index from core to clad

Pulse Dispersion
a broadening of received pulse width because of the multiple paths taken by the light

Step-index multimode fibers are rarely used in telecommunications because of their very high amounts of pulse dispersion and minimal bandwidth capability.

Multimode Graded-Index Fiber

In an effort to overcome the pulse dispersion problem, the **graded-index fiber** was developed. In the manufacturing process for this fiber, the index of refraction is tailored to follow the parabolic profile shown in Fig. 18-9 (see page 833). This results in low-order modes traveling through the constant-density material in the center (Fig. 18-8). High-order modes see lower index of refraction material farther from the core axis, and thus the velocity of propagation increases away from the center. Therefore, all modes, even though they take various paths and travel different distances, tend to traverse the fiber length in about the same amount of time. These fibers can therefore handle higher bandwidths and/or provide longer lengths of transmission before pulse dispersion effects destroy intelligibility and introduce bit errors.

Graded-Index Fiber
the index of refraction is gradually varied with a parabolic profile; the highest index occurs at the fiber's center

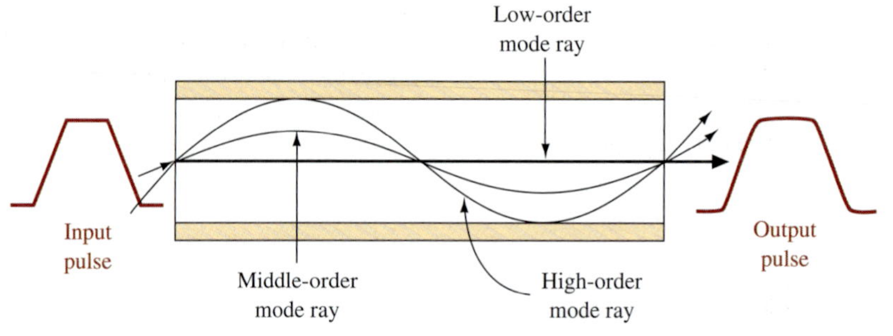

FIGURE 18-8 Modes of propagation for graded-index fiber.

In the telecommunications industry, two core sizes for graded-index fiber are commonly used: 50 and 62.5 μm. Both have 125-μm cladding. The large core diameter and the high NA of these fibers simplifies input cabling and allows the use of relatively inexpensive connectors. Fibers are specified by the diameters of their core and cladding. For example, the fibers just described would be called 50/125 fiber and 62/125 fiber.

The 62.5 μm fiber was standardized for data networks several years ago. Typical bandwidths at 850 nm are up to 180 MHz/km, and at 1300 nm they are up to 600 MHz/km. The 50 μm fiber has become increasingly attractive because of the advent of Gbit and 10 Gbit networks and systems. The smaller core allows greater bandwidth: up to 600 MHz/km at 850 nm and 1000 MHz/km at 1300 nm.

Single-Mode Fibers

A technique used to minimize pulse dispersion effects is to make the core extremely small—on the order of a few micrometers. This type accepts only a low-order mode, thereby allowing operation in high-data-rate, long-distance systems. This fiber is typically used with high-power, highly directional modulated light sources such as a laser. Fibers of this variety are called **single-mode,** or monomode, fibers. Core diameters of only 7 to 10 μm are typical.

Single-Mode Fibers
fiber cables with core diameters of about 7 to 10 μm; light follows a single path through the core

Long Haul
a term used to describe the transmission of data over hundreds or thousands of miles

Mode Field Diameter
the actual guided optical power distribution, which is typically a micron or so larger than the core diameter; single-mode fiber specification sheets typically list the mode field diameter

This type of fiber is also termed a **step-index** fiber. Step index refers to the abrupt change in refractive index from core to clad, as shown in Fig. 18-9. The single-mode fiber, by definition, carries light in a single waveguide mode. A single-mode fiber transmits a single mode for all wavelengths longer than the cutoff wavelength (λ_c). A typical cutoff wavelength is 1260 nm. At wavelengths shorter than the cutoff, the fiber supports two or more modes and becomes multimode in operation.

Single-mode fibers are widely used in **long-haul** telecommunications. They permit transmission of over 1 Gbps and a repeater spacing of over 80 km. These bandwidth and repeater-spacing capabilities are constantly being upgraded by new developments.

When describing the core size of single-mode fibers, the term **mode field diameter** is the more common term used. Mode field diameter is the actual guided optical power distribution diameter. In a typical single-mode fiber—the mode field diameter is 1 μm or so larger than the core diameter—the actual value depends on the wavelength being transmitted. In fiber specification sheets, the core diameter is stated for multimode fibers, but the mode field diameter is typically stated for single-mode fibers.

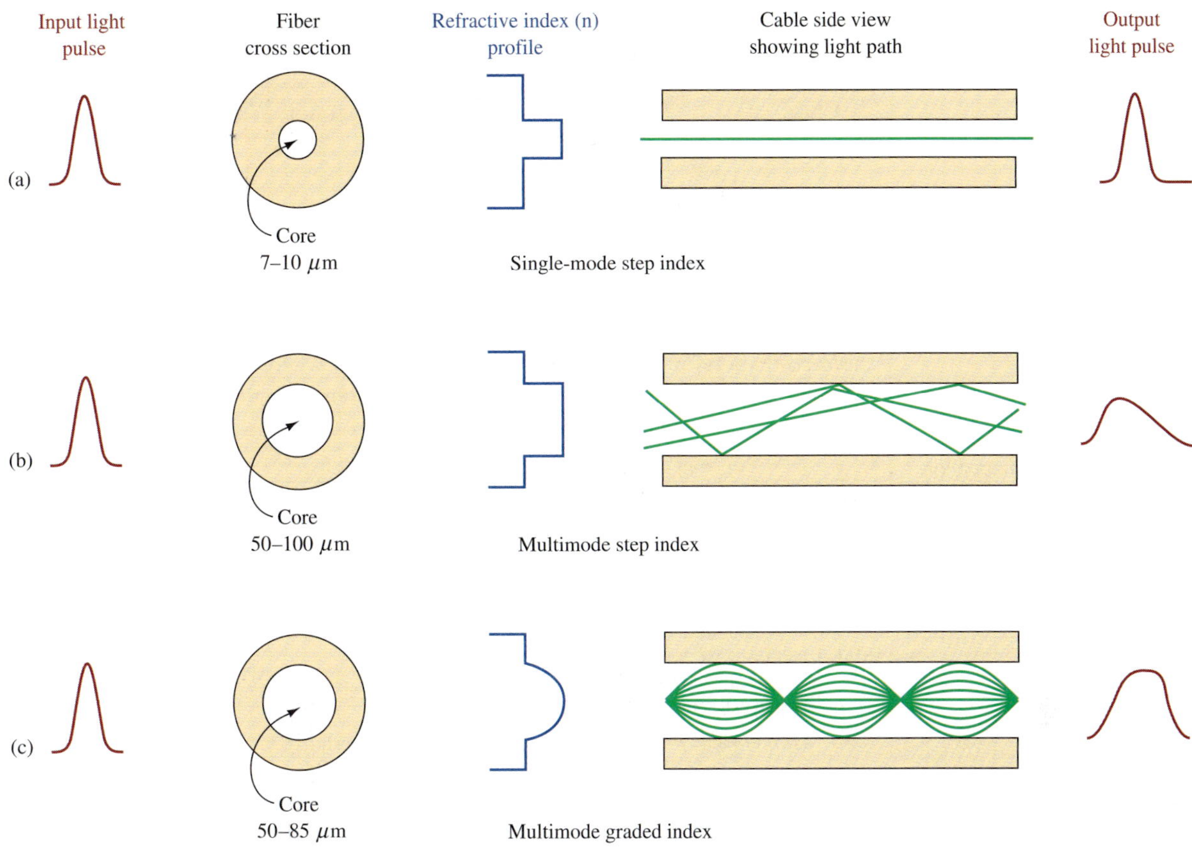

FIGURE 18-9 Types of optical fiber.

Figure 18-9 provides a summary of the three types of fiber discussed, including typical core/clad relationships, refractive index profiles, and pulse-dispersion effects.

Fiber Classification

The various types of fiber strands, both multimode and single mode, are categorized by the Telecommunications Industry Association according to the lists provided in Tables 18-2 and 18-3. Single-mode fiber is classified by dispersion characteristic and

Table 18-2 Multimode Classifications (by the Refractive Index Profile and Composition)

Class	Index	Core	Cladding
Ia	Graded	Glass	Glass
Ib	Quasi-graded	Glass	Glass
Ic	Step	Glass	Glass
IIa	Step	Glass	Plastic cladding retained for connectorization
IIb	Step	Glass	Plastic cladding removed for connectorization
III	Step/graded	Plastic	Plastic

Table 18-3 Single-Mode Classifications

Class	Dispersion Characteristic	Region of Zero-Dispersion Wavelength
IVa	Unshifted	1310 nm
IVb	Shifted	1550 nm
IVc	Flattened	Low values in both 1310 and 1550 nm ranges
IVd	Near zero	Adjacent to but outside the 1530–1560-nm operational range

Zero-Dispersion Wavelength
the wavelength at which the material dispersion and waveguide dispersion cancel one another

the **zero-dispersion wavelength,** which is the point where material and waveguide dispersion cancel one another. A general comparison of single-mode and multimode fiber is provided in Table 18-4.

Plastic Optical Fiber

Plastic fiber is used in short-range markets such as sensors, robotics, displays, automotive applications, and, to a limited extent, in data links under 100 m. It has the same advantages as glass fiber versus copper except for two primary exceptions: high loss and low bandwidth.

Features

Materials—Polymers such as polymethyl acrylate

Core size—Up to 1000 μm

Numerical aperture—0.3 to 0.8 dB

Bandwidth—Up to 3 Gb/s at 100 m but more realistically a few hundred megabits at a few hundred meters

Attenuation—120 to 180 dB/km but optimized at a 650-nm wavelength

Plastic fibers are well supported by connectorization and splicing components, which are not as critical as glass. This results in a less expensive installation.

Table 18-4 Generalized Comparisons of Single-Mode and Multimode Fiber

Feature	Single-Mode	Multimode
Core size	Smaller (7.5 to 10 μm)	Larger (50 to 100 μm)
Numerical aperture	Smaller (0.1 to 0.12)	Larger (0.2 to 0.3)
Index of refraction profile	Step	Graded
Attenuation (dB/km) (a function of wavelength)	Smaller (0.25 to 0.5 dB/km)	Larger (0.5 to 4.0 dB/km)
Information-carrying capacity (a function of distance)	Very large	Small to medium
Usage	Long-haul carriers and CATV, CCTV	Short-haul LAN
Capacity/distance characterization	Expressed in bits/second (bps)	BW in MHz/km
Which to use (this is a judgment call)	Over 2 km	Under 2 km

18-4 FIBER ATTENUATION AND DISPERSION

There are two key distance-limiting parameters in fiber-optic transmissions: attenuation and dispersion.

ATTENUATION

Attenuation is the loss of power introduced by the fiber. This loss accumulates as the light is propagated through the fiber strand. The loss is expressed in dB/km (decibels per kilometer) of length. The loss, or attenuation, of the signal is due to the combination of four factors: scattering, absorption, macrobending, and microbending.

Attenuation
the loss of power as a signal propagates through a fiber strand

Scattering: This is the primary loss factor over the three wavelength ranges used in telecommunications systems. It accounts for 85 percent of the loss and is the basis for the attenuation curves and values, such as that shown in Fig. 18-10 and industry data sheets. The scattering is known as Rayleigh scattering and is caused by refractive index fluctuations. Rayleigh scattering decreases as wavelength increases, as shown in Fig. 18-10.

Scattering
caused by refractive index fluctuations and accounts for 85 percent of the attenuation loss

Absorption: The second loss factor is a composite of light interaction with the atomic structure of the glass. It involves the conversion of optical power to heat. One portion of the absorption loss is due to the presence of OH hydroxol ions dissolved in the glass during manufacture. These ions cause the water attenuation or OH peaks shown in Fig. 18-10 and other attenuation curves. A recent and significant development in the manufacture of optical fiber has been the removal of these hydroxol ions in the 1380-nm region. By eliminating the attenuation peak, the fiber can effectively be utilized continuously

Absorption
light interaction with the atomic structure of the fiber material and also involves the conversion of optical power to heat

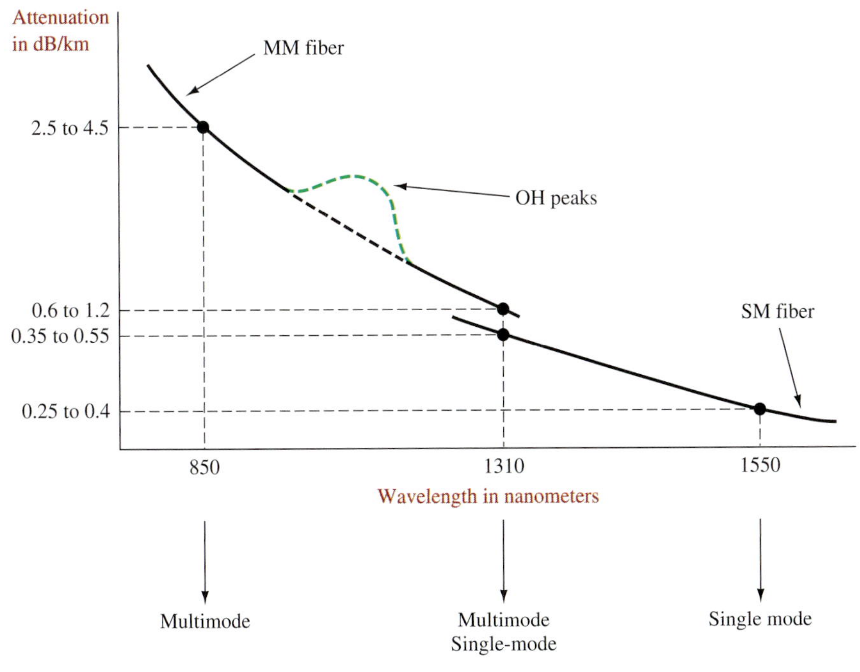

FIGURE 18-10 Typical attenuation of cabled fiber strand.

from 1260 to 1675 nm, a significant increase in the bandwidth capability of the newer fiber (see Table 18-1).

Macrobending
loss due to the light breaking up and escaping into the cladding

Macrobending: The loss caused by the light mode breaking up and escaping into the cladding when the fiber bend becomes too tight. As the wavelength increases, the loss in a bend increases. Although losses are in fractions of dB, the bend radius in small splicing trays and patching enclosures should be kept as large as possible.

Microbending
loss caused by very small mechanical deflections and stress on the fiber

Microbending: A type of loss caused by mechanical stress placed on the fiber strand, usually in terms of deformation resulting from too much pressure being applied to the cable. For example, excessively tight tie wrap or clamps contribute to this loss. This loss is noted in fractions of a dB.

Dispersion

Dispersion
the broadening of a light pulse as it propagates through a fiber strand

Dispersion, or pulse broadening, is the second of the two key distance-limiting parameters in a fiber-optic transmission system. It is a phenomenon in which the light pulse spreads out in time as it propagates along the fiber strand. This results in a broadening of the pulse. If the pulse broadens excessively, it can blend into the adjacent digital time slots and cause bit errors. Pulse dispersion is measured in terms of picoseconds (ps) of pulse broadening per spectral width expressed in nanometers (nm) of the pulse times fiber length (km). The total dispersion is then obtained by multiplying the pulse dispersion value times the length of the fiber (L). Manufacturers's specification sheets are available that provide an estimate of the dispersion for a given wavelength. A summary of values of dispersion for the primary wavelength used in fiber-optic transmission is provided in Table 18-5. The effects of dispersion on a light pulse are shown in Fig. 18-11.

Table 18-5 Dispersion Values for Common Optical Wavelengths for Class IVa Fiber

Wavelength (nm)	Pulse Dispersion [ps/(nm·km)]
850	80 to 100
1310	± 2.5 to 3.5
1550	+ 17

The equation for calculating the total dispersion is

$$\text{pulse dispersion} = \text{ps}/(\text{nm}\cdot\text{km}) \times \Delta\lambda \quad \textbf{(18-2)}$$

The pulse dispersion value is obtained from Table 18-5:

$$\Delta\lambda = \text{spectral width of the light source}$$

$$\text{total dispersion} = \text{pulse dispersion} \times \text{length (km)} \quad \textbf{(18-3)}$$

Example 18-3

Determine the amount of pulse spreading of an 850-nm LED that has a spectral width of 22 nm when run through a fiber 2 km in length. Use a dispersion value of 95 ps/(nm·km).

Solution

The dispersion value is 1 = 95 ps/(nm × km). Use Eq. (18-2), L = 2 km, and $\Delta\lambda$ = 22 nm.

$$\text{pulse dispersion} = \text{ps/(nm·km)} \times \Delta\lambda = (95)(22) = 2090 \text{ ps/km} \quad \textbf{(18-2)}$$

$$\text{total pulse dispersion} = \text{pulse dispersion} \times \text{length } (L) = (2090)(2) = 4.18 \text{ ns/km} \quad \textbf{(18-3)}$$

There are three types of dispersion: modal, chromatic, and polarization.

Modal dispersion: The broadening of a pulse due to different path lengths taken through the fiber by different modes.

Chromatic dispersion: The broadening of a pulse due to different propagation velocities of the spectral components of the light pulse

Polarization mode dispersion: The broadening of a pulse due to the different propagation velocities of the X and Y polarization components of the light pulse.

Modal Dispersion
broadening of a pulse due to different path lengths taken through the fiber by different modes

Chromatic Dispersion
broadening of a pulse due to the different propagation rates of the spectral components of the light

Polarization Mode Dispersion
broadening of a pulse due to the different propagation velocities of the X and Y polarization components

Modal dispersion occurs predominantly in multimode fiber. From a light source, the light modes can take many paths as they propagate along the fiber. Some light rays do travel in a straight line, but most take variable-length routes. As a result, the rays arrive at the detector at different times, and the result is pulse broadening. This is shown in Fig. 18-11. The use of graded-index fiber greatly reduces the effects of modal dispersion and therefore increases the bandwidth to about 1 GHz/km. On the other hand, single-mode fiber does not exhibit modal dispersion because only a single mode is transmitted.

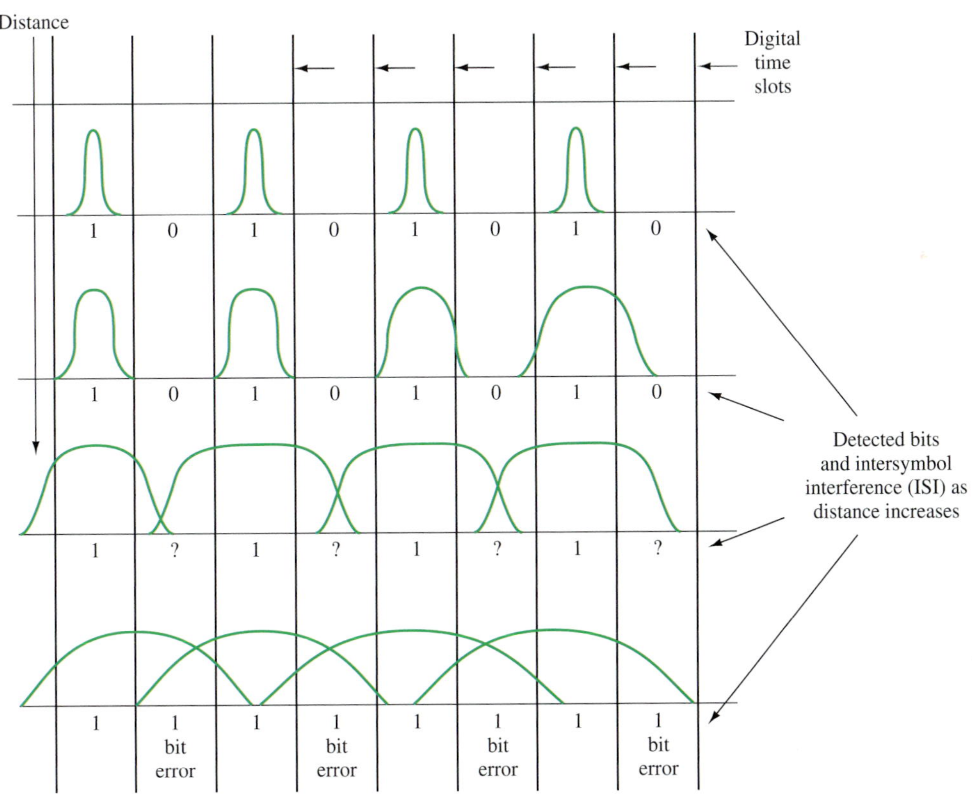

FIGURE 18-11 Pulse broadening or dispersion in optical fibers.

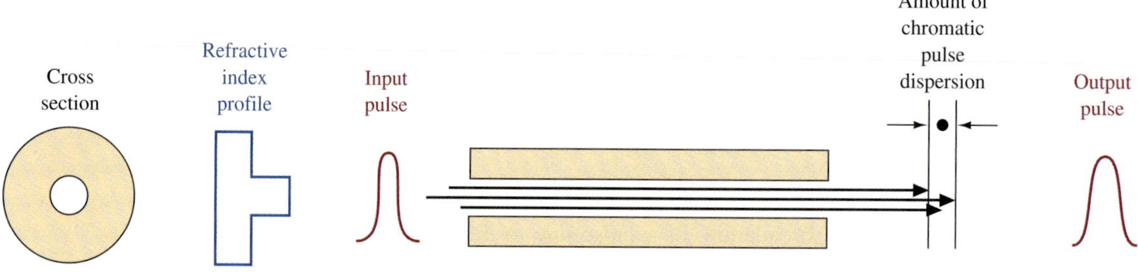

FIGURE 18-12 Spectral component propagation: single-mode, step index.

A second equally important type of dispersion is chromatic. Chromatic dispersion is present in both single-mode and multimode fibers. Basically, the light from both lasers and LEDs produces several different-wavelength light rays. Each light ray travels at a different velocity and, as a result, these rays arrive at the receiver detector at different times, causing the broadening of the pulse (see Figs. 18-11 and 18-12).

There is a point where dispersion is actually at zero, this being determined by the refractive index profile. This happens near 1310 nm and is called the zero-dispersion wavelength. Altering the refractive index profile shifts this zero-dispersion wavelength to the 1550-nm region. Such fibers are called dispersion-shifted. This is significant because the 1550-nm region exhibits a lower attenuation than at 1310 nm. This becomes an operational advantage, particularly to long-haul carriers, because repeater and regenerator spacing can be maximized with minimum attenuation and minimum dispersion in the same wavelength region.

To illustrate chromatic dispersion further, Fig. 18-12 shows a step-index single-mode fiber with different spectral components propagating directly along the core. Because they travel at different velocities, they arrive at the receiver detector at different times, causing a wider pulse to be detected than was transmitted. Again, this broadening is measured in picoseconds per kilometer of length times the spectral width in nanometers, as presented in Eq. (18-3).

Polarization mode is the type of dispersion found in single-mode systems and becomes of particular concern in long-haul, high-data-rate digital and high-bandwidth analog video systems. In a single-mode fiber, the single propagating mode has two polarizations, horizontal and vertical, or X axis and Y axis. The index of refraction can be different for the two components, and this affects their relative velocity. This is shown in Fig. 18-13.

Dispersion Compensation

A considerable amount of fiber in use today is the class IVa variety, installed in the 1980s and early 1990s. These fibers were optimized to operate at the 1310-nm region, which means that their zero-dispersion point was in the 1310-nm wavelength range. As a result of the considerable and continuous network expansion needs in recent years, it is often desired to add transmission capacity to the older fiber cables by using the 1550-nm region, particularly because the attenuation at 1550 nm is less than at 1310 nm. One major problem arises at this point. The dispersion value of the class IVa fiber in the 1550-nm region is approximately +17 ps/nm·km, which severely limits its distance capability.

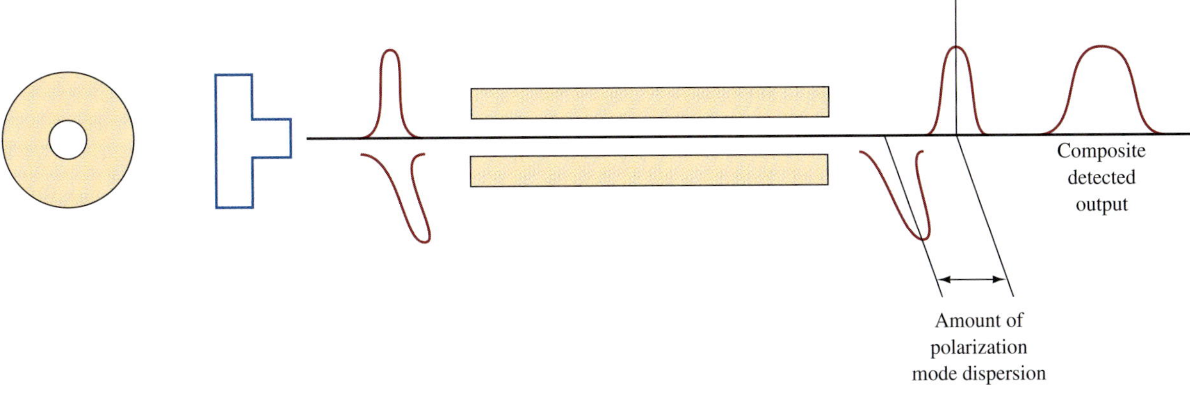

FIGURE 18-13 Polarization mode dispersion in single-mode fiber.

To overcome this problem, a fiber was developed to provide approximately −17 ps of dispersion in the 1550-nm range. Called **dispersion compensating fiber,** it acts like an equalizer, negative dispersion canceling positive dispersion. The result is close to zero dispersion in the 1550-nm region. This fiber consists of a small coil normally placed in the equipment rack just prior to the optical receiver input. This does introduce some insertion loss (3 to 10 dB) and may require the addition of an optical-line amplifier.

Dispersion Compensating Fiber
acts like an equalizer, canceling dispersion effects and yielding close to zero dispersion in the 1550-nm region

Also on the market is a **fiber Bragg grating.** This technology involves etching irregularities onto a short strand of fiber, which changes the index of refraction and, in turn, accelerates slower wavelengths toward the output. This results in a less dispersed, or narrower, light pulse, minimizing intersymbol interference (ISI).

Fiber Bragg Grating
a short strand of modified fiber that changes the index of refraction and minimizes intersymbol interference

18-5 Optical Components

Two kinds of light sources are used in fiber-optic communication systems: the diode laser (DL) and the high-radiance light-emitting diode (LED). In designing the optimum system, the special qualities of each light source should be considered. Diode lasers and LEDs bring to systems different characteristics:

1. Power levels
2. Temperature sensitivities
3. Response times
4. Lifetimes
5. Characteristics of failure

The diode laser is a preferred source for moderate-band to wideband systems. It offers a fast response time (typically less than 1 ns) and can couple high levels of useful optical power (usually several mW) into an optical fiber with a small core and a small numerical aperture. Recent advances in DL fabrication have resulted in predicted lifetimes of 10^5 to 10^6 hours at room temperature. Earlier DLs were of such limited life that it restricted their use. The DL is usually used as the source for single-mode fiber because LEDs have a low input coupling efficiency.

Some systems operate at a slower bit rate and require more modest levels of fiber-coupled optical power (50 to 250 μW). These applications allow the use of high-radiance LEDs. The LED is cheaper, requires less complex driving circuitry than a DL, and needs no thermal or optical stabilizations. In addition, LEDs have longer operating lives (10^6 to 10^7 h) and fail in a more gradual and predictable fashion than do DLs.

Both LEDs and DLs are multilayer devices most frequently fabricated of AlGaAs on GaAs. They both behave electrically as diodes, but their light-emission properties differ substantially. A DL is an optical oscillator; hence it has many typical oscillator characteristics: a threshold of oscillation, a narrow emission bandwidth, a temperature coefficient of threshold and frequency, modulation nonlinearities, and regions of instability.

The light output wavelength spread, or spectrum, of the DL is much narrower than that of LEDs: about 1 nm compared with about 40 nm for an LED. Narrow spectra are advantageous in systems with high bit rates because the dispersion effects of the fiber on pulse width are reduced, and thus pulse degradation over long distances is minimized.

Light is emitted from an LED as a result of the recombining of electrons and holes. Electrically, an LED is a *pn* junction. Under forward bias, minority carriers are injected across the junction. Once across, they recombine with majority carriers and give up their energy. The energy given up is about equal to the material's energy gap. This process is radiative for some materials (such as GaAs) but not so for others, such as silicon. LEDs have a distribution of nonradiative sites—usually crystal lattice defects, impurities, and so on. These sites develop over time and explain the finite life/gradual deterioration of light output.

Figure 18-14 shows the construction of a typical semiconductor laser used in fiber-optic systems. A variation of this laser, the stripe laser, was described in Sec. 16-6. The semiconductor laser uses the properties of the junction between heavily doped layers of *p*- and *n*-type materials. When a large forward bias is applied, a large number of free holes and electrons are created in the immediate vicinity of the junction. When a hole and electron pair collide and recombine, they produce a photon of light. The *pn* junction in Fig. 18-14 is sandwiched between layers of material with different optical and dielectric properties. The material that shields the junction is typically aluminum gallium arsenide, which has a lower index of refraction than gallium arsenide. This difference "traps" the holes and electrons in

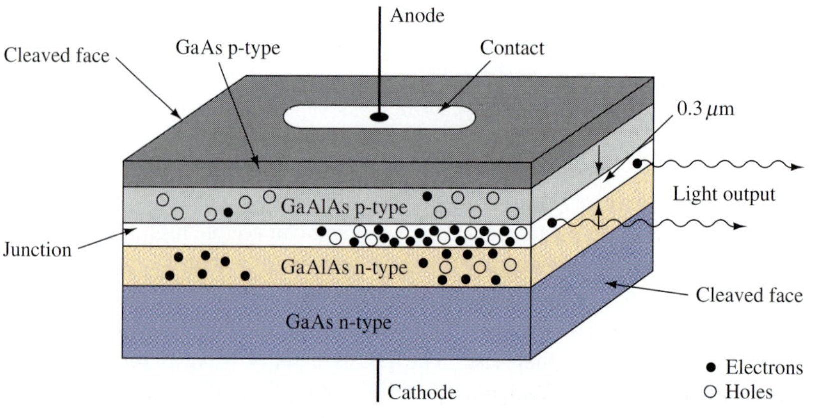

FIGURE 18-14 Semiconductor laser.

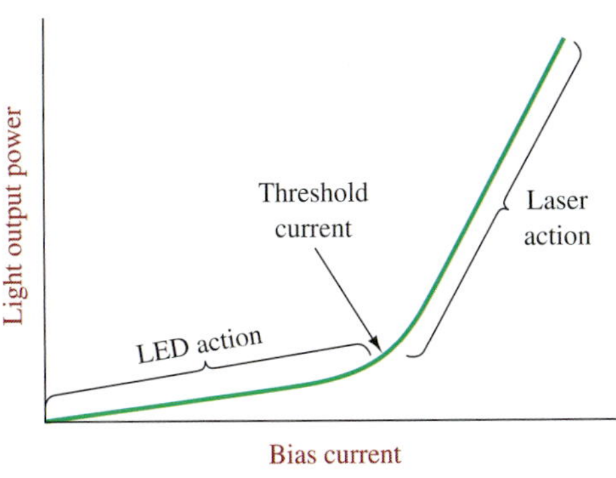

FIGURE 18-15 Light output versus bias current for a laser diode.

the junction region and thereby improves light output. When a certain level of current is reached, the population of minority carriers on either side of the junction increases, and photon density becomes so high that they begin to collide with already excited minority carriers. This causes a slight increase in the ionization energy level, which makes the carrier unstable. It thus recombines with a carrier of the opposite type at a slightly higher level than if no collision had occurred. When it does, two equal-energy photons are released.

The carriers that are "stimulated" (remember, *laser* is an acronym for *l*ight *a*mplification by *s*timulated *e*mission of *r*adiation) as described in the preceding paragraph may reach a density level so that each released photon may trigger several more. This creates an avalanche effect that increases the emission efficiency exponentially with current above the initial emission threshold value. This behavior is usually enhanced by placing mirrored surfaces at each end of the junction zone. These mirrors are parallel, so generated light bounces back and forth several times before escaping. The mirrored surface where light emits is partially transmissive (i.e., partially reflective).

The laser diode functions as an LED until its threshold current is reached. At that point, the light output becomes **coherent** (spectrally pure or only one frequency), and the output power starts increasing rapidly with increases in forward current. This effect is shown in Fig. 18-15. The typical spectral purity of these lasers yields a line width of about 1 nm versus about 40 nm for LED sources. Recall that this is critical for minimizing pulse dispersion. The wavelength of light generated is determined by the materials used. The "short-wavelength" lasers at 780 to 900 nm use gallium arsenide (GaAs) and aluminum gallium arsenide (AlGaAs). "Long-wavelength" (infrared) devices at 1300 to 1600 nm are made of layers of indium gallium arsenide phosphide (InGaAsP) and indium phosphide (InP).

A new device, called a **distributed feedback (DFB) laser** uses techniques that provide optical feedback in the laser cavity. This enhances output stability, which produces a narrow and more stable spectral width. Widths are in the range of 0.01 to 0.1 nm. This allows the use of more channels in **dense wavelength division multiplex (DWDM)** systems.

Another recent development is an entirely new class of laser semiconductors called **vertical cavity surface emitting lasers (VCSELs).** These lasers can support a much faster signal rate than can LEDs, including gigabit networks. They do not have some of the operational and stability problems of conventional lasers; however,

Coherent
light that is spectrally pure

Distributed Feedback (DFB) Laser
a more stable laser suitable for use in DWDM systems

Dense Wavelength Division Multiplex (DWDM)
incorporates the propagation of several wavelengths in the 1550-nm range of a single fiber

Vertical Cavity Surface Emitting Lasers (VCSELs)
lasers with the simplicity of LEDs and the performance of lasers

VCSELs have the simplicity of LEDs with the performance of lasers. Their primary wavelength of operation is in the 750- to 850-nm region, although development work is underway in the 1310-nm region. Reliabilities approaching 10^7 hours are projected. Table 18-6 provides a comparison of laser and LED optical transmitters.

Table 18-6 A Comparison of Laser and LED Optical Transmitters

	Laser	LED
Usage	High bit rate, long haul	Low bit rate, short haul, LAN
Modulation rates	<40 Mbps to gigabits	<400 Mbps
Wavelength	Single-mode at 1310 and 1550 nm	Single-mode and multimode at 850/1310 nm
Rise time	<1 ns	10 to 100 ns
Spectral width	<1 nm up to 4 nm	40 to 100 nm
Spectral content	Discrete lines	Broad spectrum/continuous
Power output	0.3 to 1 mW (−5 to 0 dBm)	10 to 150 μW (−20 to −8 dBm)
Reliability	Lower	Higher
Linearity	40 dB (good)	20 dB (moderate)
Emission angle	Narrow	Wide
Coupling efficiency	Good	Poor
Temperature/humidity	Sensitive	Not sensitive
Durability/life	Medium (10^5 h)	High (>10^6 h)
Circuit complexity	High	Low
Cost	High	Low

Note: The values shown depend, to some extent, on the associated electronic circuitry.

Modulating the Light Source

Most fiber-optic communication occurs using digital pulse (on–off) systems. Pulse-code modulation is most often used, with RZ or Manchester coding. The transmission of analog signals can be accomplished by varying the amplitude of the light output. This is used largely by CATV systems and can be described as an amplitude modulation (AM) system. A very simple AM system using an LED light source is shown in Fig. 18-16. The use of frequency modulation is not possible because the frequency of the light output from a laser or LED cannot be varied in a modulation sense. However, there is a class of lasers called **tunable lasers** in which the fundamental wavelength

Tunable Laser
a laser in which the fundamental wavelength can be shifted a few nanometers; ideal for traffic routing in DWDM systems

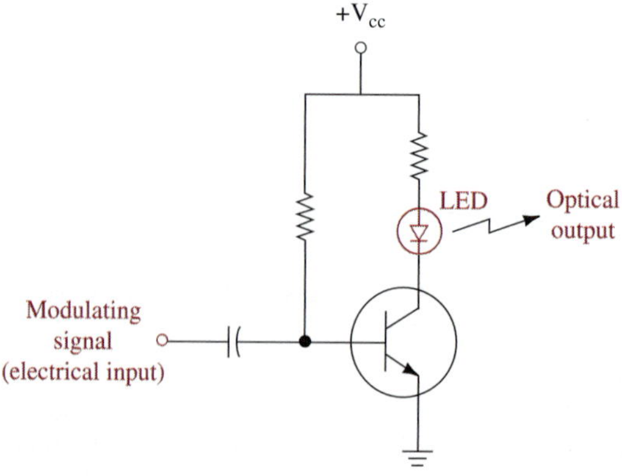

FIGURE 18-16 LED modulator.

can be shifted a few nanometers, but not from a modulation point of view. The primary market for these devices is in a network operations environment where DWDM is involved. Traffic routing is often made by wavelength, and, as such, wavelengths or transmitters must be assigned and reassigned to accommodate dynamic routing or networking, bandwidth on demand, seamless restoration (serviceability), optical packet switching, etc. Tunable lasers are used along with either passive or tunable WDM filters.

Intermediate Components

The typical fiber-optic telecommunications link is—as shown in Fig. 18-1—a light source or transmitter and light detector or receiver, interconnected by a strand of optical **fiber,** or **light pipe,** or **glass.** An increasing number of specialized networks and system applications have various intermediate components along the span between the transmitter and the receiver. A brief review of these devices and their uses is provided.

Fiber, Light Pipe, Glass
common synonymous terms for a fiber-optic strand

Isolators An **isolator** is an in-line passive device that allows optical power to flow in one direction only. Typical forward-direction insertion losses are less than 0.5 dB, with reverse-direction insertion losses of at least 40 to 50 dB. They are polarization-independent and available for all wavelengths. One popular use is preventing reflections caused by optical span irregularities getting back into the laser transmitter. Distributed feedback lasers are particularly sensitive to reflections, which can result in power instability, phase noise, line-width variations, etc.

Isolators
an in-line passive device that allows optical power to flow in one direction only

Attenuators Attenuators are used to reduce the **received signal level (RSL).** They are available in fixed and variable configurations. The fixed attenuators are for permanent use in point-to-point systems to reduce the RSL to a value within the receiver's dynamic range. Typical values of fixed attenuators are 3 dB, 5 dB, 10 dB, 15 dB, and 20 dB. Variable attenuators are typically for temporary use in calibration, testing, and laboratory work but more recently are being used in optical networks, where changes are frequent and programmable.

Received Signal Level (RSL)
the input signal level to an optical receiver

Branching Devices Branching devices are used in simplex systems where a single optical signal is divided and sent to several receivers, such as point-to-multipoint data or a cable TV distribution system. They can also be used in duplex systems to combine or divide several inputs. The units are available in single-mode and multimode units. The primary optical parameters are insertion loss and return loss, but the values for each leg may vary slightly due to differences in the device mixing region.

Splitters Splitters are used to split, or divide, the optical signal for distribution to any number of places. The units are typically simplex and come in various configurations, such as $1 \times 4, 1 \times 8, \ldots, 1 \times 64$.

Couplers Couplers are available in various simplex or duplex configurations, such as 1×2, 2×2, 1×4, and various combinations up to 144×144. There are both passive and active couplers, the latter most often associated with data networks. Couplers can be wavelength dependent or independent.

Wavelength Division Multiplexers Wavelength division multiplexers combine or divide two or more optical signals, each having a different wavelength. They are sometimes called optical beamsplitters. They use dichroic filtering, which passes light selectively by wavelength, or diffraction grating, which refracts light beams at an angle, selectively by wavelength. An additional optical parameter of importance is port-to-port crosstalk coupling, where wavelength number 1 leaks out of or into the port of wavelength number 2. A port is the input or output of the device.

Optical-Line Amplifiers Optical-line amplifiers are not digital regenerators, but analog amplifiers. Placement can be at the optical transmitter output, midspan, or near the optical receiver. They are currently used by high-density long-haul carriers, transoceanic links, and, to some extent, the cable TV industry.

Detectors

The devices used to convert the transmitted light back into an electrical signal are a vital link in a fiber-optic system. This important link in a fiber-optic communications system is often overlooked in favor of the light source and fibers. However, simply changing from one photodetector to another can increase the capacity of a system by an order of magnitude. Because of this, current research is accelerating to allow production of improved detectors. For most applications, the detector used is a *p-i-n* diode. Chapter 16 provided details on *p-i-n* diodes used as microwave switches. The avalanche photodiode is also used in photodetector applications.

Dark Current
the very little current that flows when a *pn* junction is reverse-biased and under dark conditions

Just as a *pn* junction can be used to generate light, it can also be used to detect light. When a *pn* junction is reversed-biased and under dark conditions, very little current flows through it. This is termed the **dark current.** However, when light shines on the device, photon energy is absorbed and hole–electron pairs are created. If the carriers are created in or near the junction depletion region, they are swept across the junction by the electric field. This movement of charge carriers across the junction causes a current flow in the circuitry external to the diode and is proportional to the light power absorbed by the diode.

The important characteristics of light detectors are:

1. *Responsivity:* This is a measure of output current for a given light power launched into the diode. It is given in amperes per watt at a particular wavelength of light.
2. *Dark current:* This is the thermally generated reverse leakage current (under dark conditions) in the diode. In conjunction with the response current as predicted by device responsivity and incident power, it provides an indication of on–off detector output range.
3. *Response speed:* This determines the maximum data-rate capability of the detector.
4. *Spectral response:* This determines the responsitivity that is achieved relative to the wavelength at which responsitivity is specified. Figure 18-17 provides a spectral response versus light wavelength for a typical *p-i-n* photodiode. The curve shows that its relative response at 900 nm (0.9 μm) is about 80 percent of its peak response at 800 nm.

Figure 18-18 shows the construction of a *p-i-n* diode used as a photodetector. As mentioned previously, light falling on a reverse-biased *pn* junction produces hole–electron pairs. The ability of a generated hole–electron pair to contribute to current flow depends on the hole and electron being rapidly separated from each other before they collide and cancel each other out. The reverse-biased diode creates a depletion region at the *pn* junction. The reverse-biased junction can be thought of as a capacitor, with the depletion region acting as the dielectric. The hole

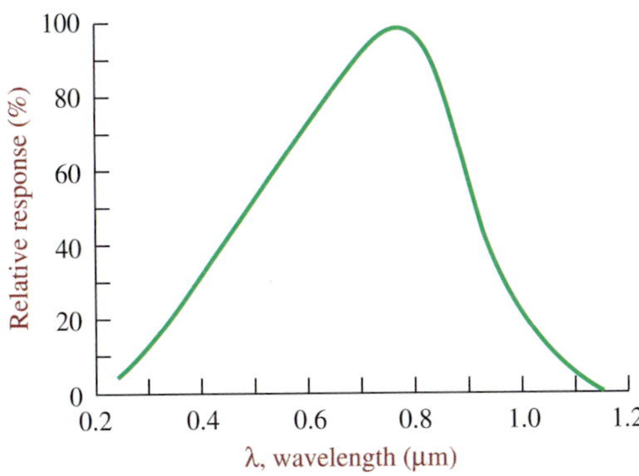

FIGURE 18-17 Spectral response of a *p-i-n* diode.

and electron created in the depletion region are rapidly pulled apart by the *p* and *n* materials that act as the capacitor's plates. Widening the depletion region gives more opportunity for hole–electron pairs to form and thus enhances the photodetector operation. The intrinsic (*i*) layer of the *p-i-n* diode in Fig. 18-18 performs that function. The intrinsic layer is a very lightly doped semiconductor material.

The operation of the avalanche photodiode is illustrated in Fig. 18-19. The diode is operated at a reverse voltage near the breakdown of the junction. At that potential, the electrons can be pulled from the atomic structure. With a small amount of additional energy, electrons are dislodged from their orbits, producing free electrons and resulting holes. As shown in Fig. 18-19, a photon of light incident on the junction generates a hole–electron pair in the depletion region. Because of the large electric field, the electron movement is accelerated and the electrons collide with other bound electrons. This creates additional hole–electron pairs that are also accelerated. This produces still more hole–electron pairs, and an avalanche multiplication process (gain) occurs. One electron may produce up to 100 electrons in the avalanche photodiode. The avalanche photodiode is 5 to 7 dB more sensitive than the *p-i-n* diode. This advantage is maintained except when extremely high data rates exceeding 4 Gbps are experienced. In these cases, the better frequency response of the *p-i-n* diode favors its use.

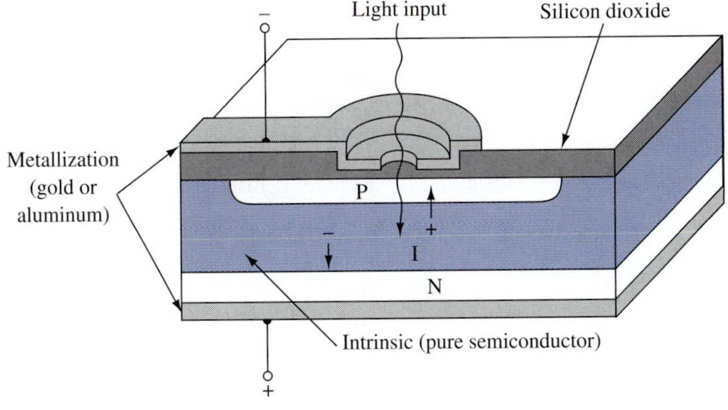

FIGURE 18-18 *p-i-n* diode.

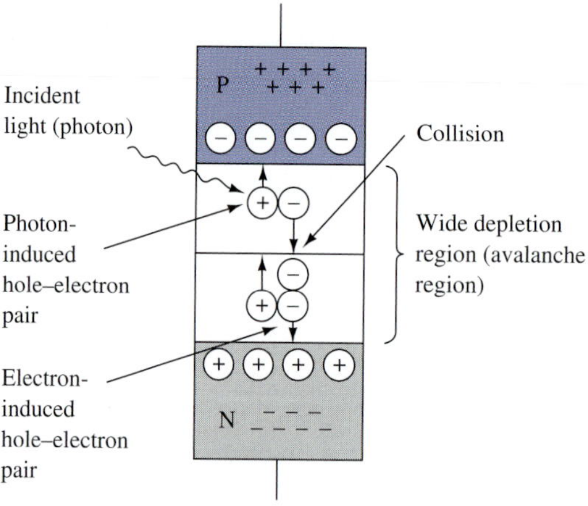

FIGURE 18-19 Avalanche photodiode.

It should be noted that a second role for light detectors in fiber-optic systems exists. Detectors are used to monitor the output of laser diode sources. A detector is placed in proximity to the laser's light output. The generated photocurrent is used in a circuit to maintain the laser's light output constant under varying temperature and bias conditions. This is necessary to keep the laser just above its threshold forward bias current and to enhance its lifetime by not allowing the output to increase to higher levels. Additionally, the receiver does not want to see the varying light levels of a noncompensated laser.

The output current of photodiodes is at a very low level—on the order of 10 nA up to 10 μA. As a result, the noise benefits of fiber optics can be lost at the receiver connection between diode and amplifier. Proper design and shielding can minimize that problem, but an alternative solution is to integrate the first stage of amplification into the same circuit as the photodiode. These integrations are termed *integrated detector preamplifiers* (IDPs) and provide outputs that can drive TTL logic circuits directly.

The complete specifications for the MFOD 3100 photodiode are provided in Fig. 18-20. These devices have a typical responsivity of 0.3 μA/μW and a 2-ns response time. A comparison of *p-i-n* and APD detectors is provided in Table 18-7.

Table 18-7 A Comparison of Detectors

Parameter	*P-I-N*	APD
Bandwidth	Low bit rate <200 Mbps	High bit rate >200 Mbps to Gbps
Wavelength	850 and 1310 nm	1310 and 1550 nm
Sensitivity	Low, −35 dBm to −40 dBm	High, −45 dBm
Dynamic range	Low	High
Dark current	High	Low, less noise
Circuit complexity	Low	Medium
Temperature sensitivity	Low	High
Cost	Low	High
Life	10^9 h	10^6 h
Photon and electron conversion gain	1	3 to 5
Operating voltages	Low	High

Note: The values depend, to some extent, on the associated electronic circuitry.

MOTOROLA
SEMICONDUCTOR
TECHNICAL DATA

MFOD3100

Fiber Optics — MOD Family Photo Detector

Diode Output

MOD FAMILY FIBER OPTICS PHOTO DETECTOR DIODE OUTPUT

. . . designed for low cost infrared radiation detection in high frequency Fiber Optics Systems. Motorola's package fits directly into standard fiber optics connectors. Metal connectors provide excellent RFI immunity. Major applications are: CATV, video systems, M68000 microprocessor systems, industrial controls, computer and peripheral equipment, etc.

- Fast Response — 5 ns Max @ 5 Volts
- Analog Bandwidth (−3 dB) Greater Than 100 MHz
- Performance Matched to Motorola Fiber Optics Emitters
- Plastic Package — Small, Rugged and Inexpensive
- Mates snugly with AMP #228756-1, Amphenol #905-138-5001, OFTI #PCR001 Receptacles
- Low Cost

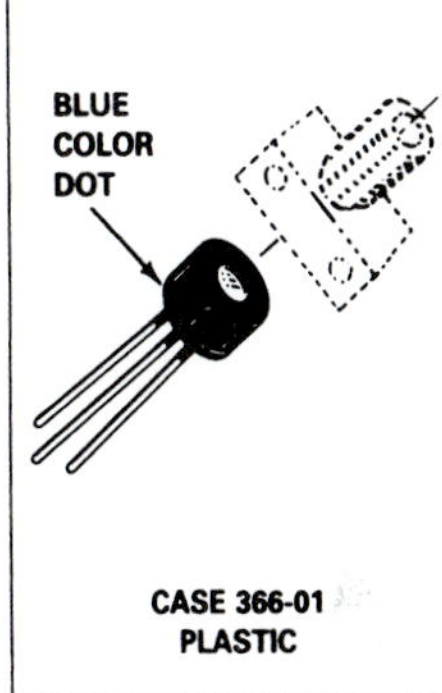

CASE 366-01
PLASTIC

5

MAXIMUM RATINGS (T_A = 25°C unless otherwise noted)

Rating	Symbol	Value	Unit
Reverse Voltage	V_R	50	Volts
Total Device Dissipation @ T_A = 25°C Derate above 25°C	P_D	50 0.67	mW mW/°C
Operating Temperature Range	T_A	−40 to +100	°C
Storage Temperature Range	T_{stg}	−40 to +100	°C

ELECTRICAL CHARACTERISTICS (T_A = 25°C)

Characteristic	Symbol	Min	Typ	Max	Unit
Dark Current (V_R = 5 V, H ≈ 0, Figure 2)	I_D	—	—	1	nA
Reverse Breakdown Voltage (I_R = 10 μA)	$V_{(BR)R}$	50	—	—	Volts
Total Capacitance (V_R = 5 V, f = 1 MHz)	C_T	—	—	5	pF
Noise Equivalent Power	NEP	—	50	—	$fW/\sqrt{Hz}$

OPTICAL CHARACTERISTICS (T_A = 25°C)

Characteristic	Symbol	Min	Typ	Max	Unit
Responsivity @ 850 nm (V_R = 5 V, P = 10 μW, Figure 3)	R	0.2	0.3	—	μA/μW
Response Time @ 850 nm (V_R = 5 V)	t_r, t_f	—	2	5	ns

FIGURE 18-20 MFOD 3100 photodiode specifications. (Courtesy of Motorola, Inc.)

 18-6 Fiber Connections and Splices

Optical fiber is made of ultrapure glass. Optical fiber makes window glass seem opaque by comparison. It is therefore not surprising that the process of making connections from light source to fiber, fiber to fiber, and fiber to detector becomes critical in a system. The low-loss capability of the glass fiber can be severely compromised if these connections are not accomplished in exacting fashion.

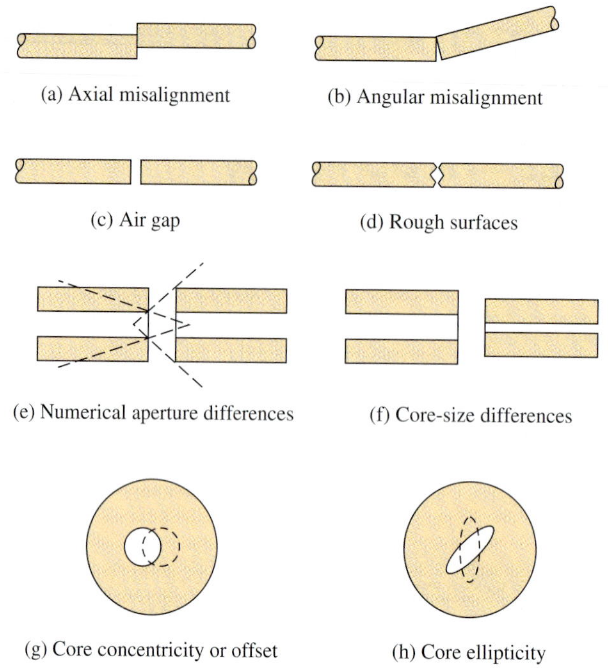

FIGURE 18-21 Sources of connection loss.

Optical fibers are joined either in a permanent fusion splice or with a connector. The connector allows repeated matings and unmatings. Above all, these connections must lose as little light as possible. Low loss depends on correct alignment of the core of one fiber to another, or to a source or detector. Loss occurs when two fibers are not perfectly aligned within a connector. Axial misalignment typically causes the greatest loss—about 0.5 dB for a 10 percent displacement. This condition and other loss sources are illustrated in Fig. 18-21. Most connectors leave an air gap, as shown in Fig. 18-21(c). The amount of gap affects loss because light leaving the transmitting fiber spreads conically. Angular misalignment [Fig. 18-21(b)] can usually be well controlled in a connector.

The losses due to rough end surfaces shown in Fig. 18-21(d) are often caused by a poor cut, or "cleave," but can be minimized by polishing. Polishing typically takes place after a fiber has been placed in a connector.

The source of connection losses shown in Figs. 18-21(a) to (d) can, for the most part, be controlled by a skillful cable splicer. There are four other situations that can cause additional connector or splice loss, although in smaller values. These are shown in Figs. 18-21(e), (f), (g), and (h). These are related to the nature of the fiber strand at the point of connection and are beyond the control of the cable splicer. The effect of these losses can be minimized somewhat by the use of a rotary mechanical splice, which by the joint rotation will get a better core alignment.

Fusion Splicing
a long-term method where two fibers are fused or welded together

In regard to connectorization and splicing, there are two techniques to consider for splicing. **Fusion splicing** is a long-term method in which two fibers are fused or welded together. The two ends are stripped of their coating, cut or cleaved, and inserted into the splicer. The ends of the fiber are aligned and an electric arc is fired across the ends, melting the glass and fusing the two ends together. There are both manual and automatic splicers; the choice usually depends on the number of splices to be done on a given job, technician skill levels available, and, of course, the budget. Typical insertion losses of less than 0.1 dB—frequently in the 0.05-dB range—can be achieved consistently.

Mechanical splices can be permanent and an economical choice for certain fiber-splicing applications. Mechanical splices also join two fibers together, but they differ from fusion splices because an air gap exists between the two fibers. This results in a glass–air–glass interface, causing a severe double change in the index of refraction. This change results in an increase in insertion loss and reflected power. The condition can be minimized by applying an **index-matching gel** to the joint. The gel is a jellylike substance that has an index of refraction much closer to the glass than air. Therefore, the index change is much less severe.

Mechanical Splices
two fibers joined together with an air gap, thereby requiring an index-matching gel to provide a good splice

Index-Matching Gel
a jellylike substance that has an index of refraction much closer to the glass than air

Mechanical splices are universally popular for repair and for temporary or laboratory work. They are quick, cheap, easy, and quite appropriate for small jobs. The best method for splicing depends on the application, including the expected future bandwidth (i.e., gigabit), traffic, the job size, and economics.

Fiber Connectorization

For fiber connectorization, there are several choices on the market. Currently, the most popular are the **SC** and **ST.** A family of smaller connectors is also on the market; it is called **small-form factor.** The connectors are about one-half the size of conventional SC and ST units and are being developed for use in local area networks in the home and office. Three designs, the types LC, MT-RJ, and VF-45 are recognized by the Telecommunications Industry Association but are not yet standardized. Examples of SC, ST, and MT-RJ connectors are provided in Figs. 18-22(a), (b), and (c). Some general requirements for fiber connectors are provided in Table 18-8.

SC, ST
currently the most popular full-size fiber connectors on the market

Small-Form Factor
a family of connectors about half the size of ST and SC connectors

In preparing the fiber for splicing or connectorization, only the coating is removed from the fiber strand. The core and the clad are not separable. The 125-μm clad diameter is the portion that fits into the splice or connector, and therefore most devices can handle both single and multimode fiber.

Sometimes the issue of splicing together fibers of different core sizes arises. The one absolute rule is, do not splice single and multimode fiber together! Similarly, good professional work does not allow different sizes of multimode fiber to be spliced together. However, in an emergency, different sizes of multimode fiber can be spliced together if the following limitations are recognized:

> When transmitting from a small- to a larger-core diameter, there will be minimal, if any, increase in insertion loss. However, when the transmission is from a larger to a smaller core size, there will be added insertion loss, and a considerable increase in reflected power should be expected.

Table 18-8 General Fiber Connector Requirements

Easy and quick to install.
Low insertion loss. A properly installed connector has as little as 0.25 dB insertion loss.
High return loss greater than 50 dB. This is increasingly important in gigabit networks, DWDM systems, high-bandwidth video, etc.
Repeatability.
Economical.

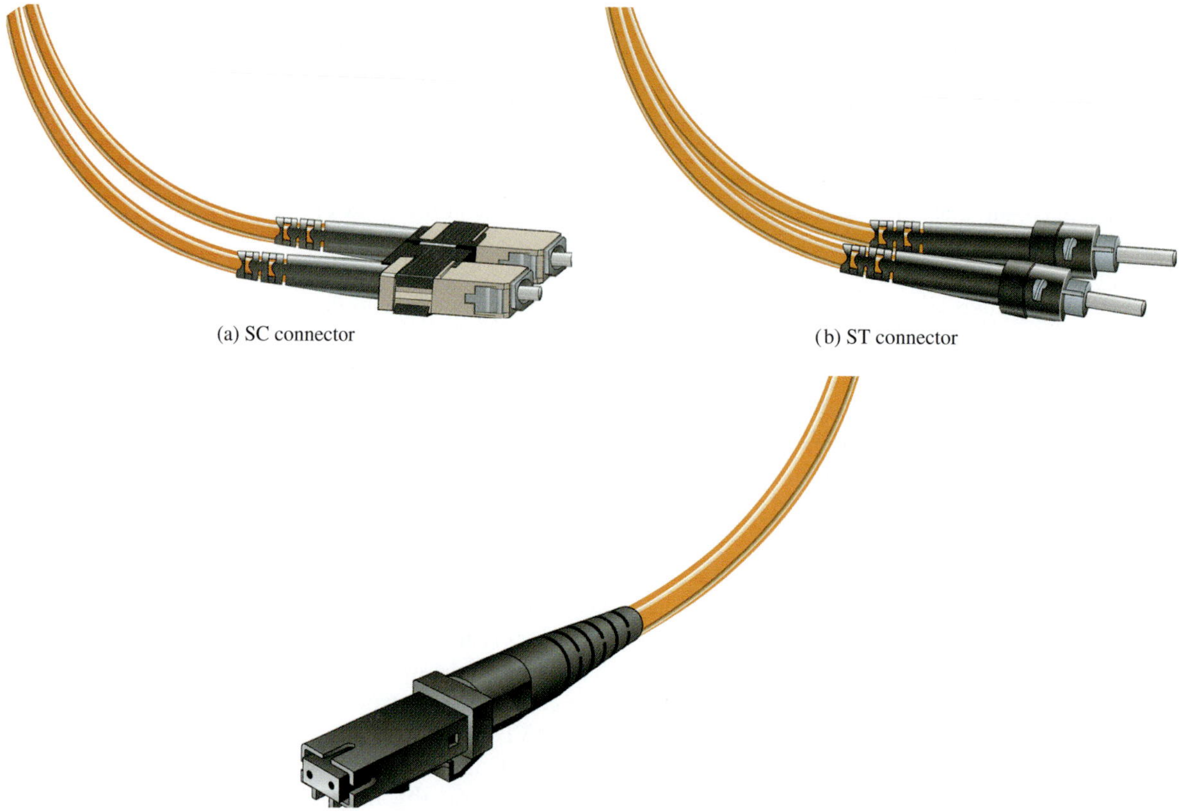

(a) SC connector

(b) ST connector

(c) MT-RJ connector

FIGURE 18-22 Fiber connectors.

Industrial practice has confirmed the acceptability of different-core size interchangability for emergency repairs in the field, mainly as the result of lab tests with 50- and 62.5-μm multimode fiber for a local area net environment.

18-7 SYSTEM DESIGN AND OPERATIONAL ISSUES

When designing a fiber-optic transmission link, the primary performance issue is the bit error rate (BER) for digital systems and the carrier-to-noise ratio (*C/N*) for analog systems. In either case, performance degrades as the link length increases. As stated in Sec. 18-4, attenuation and dispersion are the two distance-limiting factors in optical transmissions. The distance limit is the span length at which the BER or *C/N* degrades below some specified point. From an engineering point of view, there are two different types of environments for fiber links: long haul and local area networks (LANs).

Long Haul
the intercity or interoffice class of system used by telephone companies and long-distance carriers.

A **long-haul** system is the intercity or interoffice class of system used by telephone companies and long-distance carriers. These systems typically have high channel density and high bit rate, are highly reliable, incorporate redundant equipment, and involve extensive engineering studies.

Local area networks (LANs) take a less strict position on the issues stated under long-haul applications. They typically have lower channel capacity and minimal redundance and are restricted to building-to-building or campus environments. Some LANs are becoming very large, including metropolitan area networks (MANs) and wide area networks (WANs); as such, they usually rely on long-haul carriers for their connectivity.

From a design standpoint, those involved in long-haul work actually perform the studies on a per-link basis. LANs typically are prespecified and preengineered as to length, bit-rate capability, performance, etc. The following is an example of a system designed for an installation typical of a LAN or a short distance communication link.

In this example, each of the many factors that make up the link calculation, power budget, or light budget is discussed along with its typical contribution. A minimal received signal level (RSL) must be obtained to ensure that the required BER is satisfied. For example, if the minimum RSL is -40 dBm for a BER of 10^{-9}, then this value is the required received optical power. If, after the initial calculations are

An optical power meter test set. (Courtesy of Wandel & Golterman.)

ATTENUATION OR LINK LOSS
Sample system

[1] Transmitter power output (module, not LD or LED)		−15 dBm
[2] Losses: Cable, 18.6 Mi/30 km @ 0.4 dB/km	12.0	
[3] 8 splices @ 0.2 dB ea	1.6	
[4] 2 connectors @ 0.5 dB ea	1.0	
[5] Extra for two pigtails and inside cable	2.0	
Total losses	16.6	16.6 dB
Received signal power		−31.6 dBm
[6] Operational margin	3.0	
[7] Maintenance margin	3.0	
Total margin	6.0	6.0 dB
[8] Design receive signal power		−37.6 dBm
[9] Minimum receiver sensitivity(RSL) for 10^{-9} BER (module, not APD or PIN)		−40.0 dBm
[10] Extra margin		2.4 dB

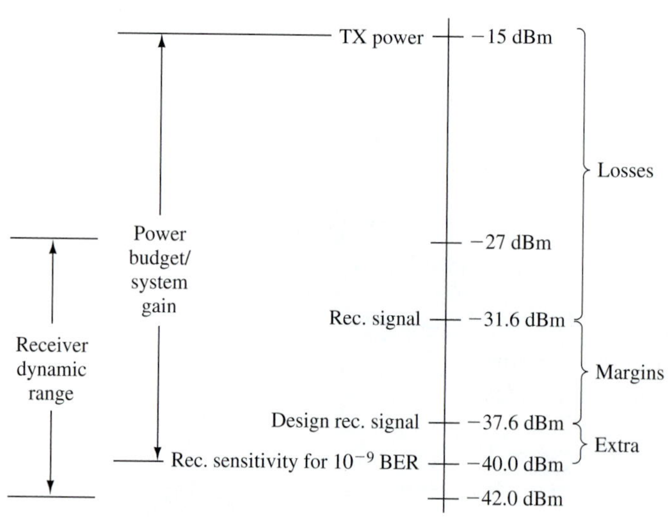

FIGURE 18-23 System design.

completed, the projected performance is not as expected, then go back and adjust any of the parameters, recognizing that there are trade-off issues.

Refer to the system design shown in Fig. 18-23:

1. *Transmitter power output:* A value usually obtained from the manufacturer's specification or marketing sheet. *Caution:* Be sure the value is taken from the output port of the transmit module or rack. This is point 1 in Fig. 18-23. This is the point where the user can access the module for measurement and testing. Otherwise the levels can be off as much as 1 dB due to pigtail or coupling losses between the laser or LED and the actual module output.

2. *Cable losses:* The loss in dB/km obtained from the cable manufacturer's sheet. This value is multiplied by the length of the cable run to obtain the total loss. An example of this calculation is provided in Ex. 18-4.

   Note that the actual fiber length can exceed the cable run length by 0.5 percent to 3 percent due to the construction of the fiber cable (in plastic buffer tubes). Fiber cables are loosely enclosed in buffer tubes to isolate the fiber from construction stress when the cable is pulled.
3. *Splice losses:* Values depend on the method used for splicing as well as the quality of splicing provided by the technician. Losses can vary from 0.2 dB to 0.5 dB per splice.
4. *Connector losses:* A value depending on the type and quality of the connector used as well as the skill level of the installer. Losses can vary from 0.25 dB to 0.5 dB.
5. *Extra losses:* A category used for miscellaneous losses in passive devices such as splitters, couplers, WDM devices, optical patch panels, etc.
6. *Operational margin:* Accounts for system degradation due to equipment aging, temperature extremes, power supply noise and instability, timing errors in regenerators, etc.
7. *Maintenance margin:* Accounts for system degradation due to the addition of link splices, added losses because of wear and misalignment of patch cords and connectors, etc. This includes the loss generated from repairing cables that have been dug up by a backhoe.

   The term **backhoe fading** *is used to indicate that the system has had total loss of data flow because the fiber cable has been dug up by a backhoe.*
8. *Design receive signal power:* A value obtained from summing the gains and losses in items 1–7. This value should exceed the specified received signal level (RSL), as specified in item 10.
9. *Receiver sensitivity for a 10^{-9} BER:* The minimum RSL for the receiver to perform at the specified BER. If the design receive signal power (item 8) does not meet this requirement, then adjustments must be made. For example, transmit power can be increased, splice loss estimates can be reduced, a more optimistic maintenance margin can occur, etc. Also, the receiver may have a maximum RSL, and a system may require attenuators to decrease the RSL so that it falls within a specified operating range (receiver dynamic range).
10. *Extra margin:* The difference between the design receive signal power (item 8) and the receiver sensitivity (item 9). Item 8 should be greater than item 9; for example, −37.6 dBm is larger than −40 dBm. One or two dB is a good figure.
11. *Optional optical attenuator:* A place where an optional attenuator can be installed and later removed as aging losses begin to increase.

Backhoe Fading
total loss of data flow because the fiber cable has been dug up by a backhoe

Example 18-4

Determine the loss in dB for a 30-km fiber cable run that has a loss of 0.4 dB/km.

Solution

$$\text{total cable loss} = 30\text{ km} \times 0.4\text{ dB/km} = 12\text{ dB}$$

A graphical view of the previous system design problem is shown in Fig. 18-24. Figure 18-25 provides another way to describe the system design problem graphically. Notice that the values are placed along the distance covered by the fiber. This provides the maintenance staff and the designer with a clear picture of how the system was put in place.

Another system design consideration is dispersion, the second distance-limiting factor in a fiber-optic system. The concept of dispersion was first examined in Sec. 18-4. The practical significance of dispersion is that it is desirable to have the operating wavelength of a fiber-optic system in the zero dispersion wavelength region (see Table 18-3). The formula for calculating the fiber span length when taking dispersion into account is provided in Eq. (18-4).

$$L = \frac{440{,}000}{BR \times D \times SW} \tag{18-4}$$

where L = span length in kilometers
BR = line bit rate in megabits/second
D = cable dispersion in picoseconds/nanometer/kilometer
SW = spectral width of the transmitter in nanometers
440,000 = an assumed Gaussian constant based on a 3 dB optical bandwidth using a full-width-half-max (FWHM) pulse shape

Example 18-5 demonstrates how to use Eq. (18-4).

Example 18-5

Determine the fiber span length given the following two sets of manufacturers' information. Compare the results for the two span length calculations.
(a) Line bit rate = 565 Mbps
Cable dispersion = 3.5 ps/nm/km
Transmitter spectral width = 4 nm
(b) Line bit rate = 1130 Mbps
Cable dispersion = 3.5 ps/nm/km
Transmitter spectral width = 2 nm

Solution

(a) $L = \dfrac{440{,}000}{565 \times 3.5 \times 4} = 55.6 \text{ km}$

(b) $L = \dfrac{440{,}000}{1130 \times 3.5 \times 2} = 55.6 \text{ km}$

By upgrading the transmitter's spectral width from 4 nanometers to 2 nanometers, the bit rate can be doubled without having to upgrade the cable facility.

Dispersion is typically a single-mode, long-haul, high-bit-rate consideration. The person planning the system should seek advice from cable and optoelectronic equipment manufacturers and experienced system designers.

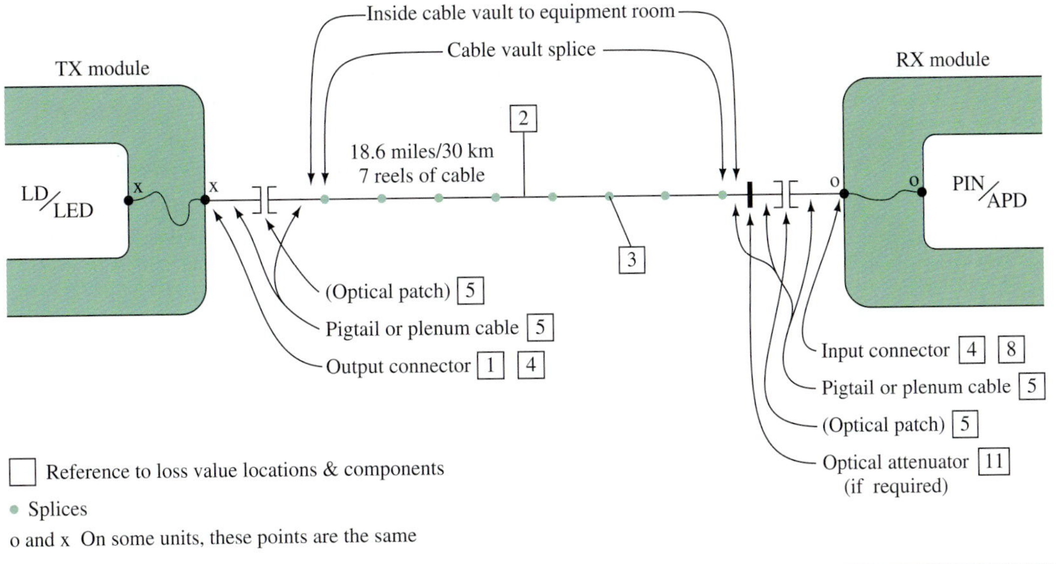

FIGURE 18-24 A graphical view of the system design problem shown in Fig. 18-23.

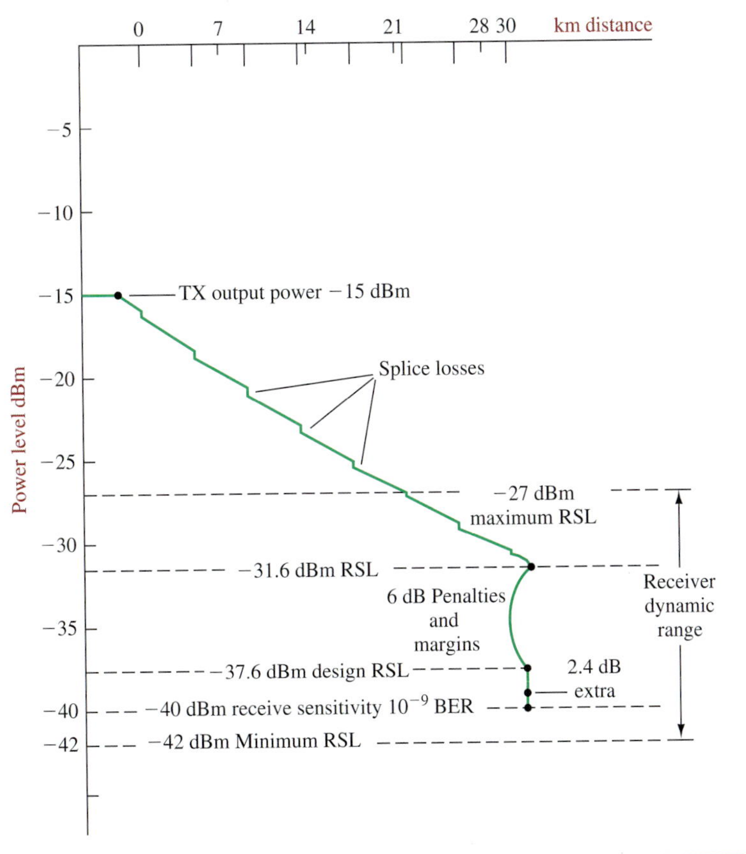

FIGURE 18-25 An alternative view of the system design problem.

 18-8 CABLING AND CONSTRUCTION

This section provides a brief outline of the issues associated with the exterior or interior installation of fiber cable. Even though the installation techniques are well established, new products and tools are being brought to the market to improve the installation. Trade journals and Internet sites provide an excellent and reliable way to keep informed about the changes.

Exterior (Outdoor) Installations

Fiber can be installed on poles or underground in ducts, in utility tunnels, or by direct burial. You must be aware of the exposure factors for each installation. These factors can include temperature, humidity, chemicals, rodents, abrasion, water, ice, wind, mechanical vibration, and lightning, to name a few. Some of the ways to protect an exterior cable are provided by armored cable, a water-resistant sheath, close adherence to pulling tensile and bend radius specs, and frequent grounding of metallic cable components.

Interior (Indoor) Installations

The environment for interior installations is usually well controlled. However, exposure factors include mechanical vibrations, heat, and possibly fire. In all cases, a cable with a fire-retardant sheath and sheaths that generate low or minimal smoke and toxic fumes are required. Plenum cable is available for installation in air ducts, air-handling spaces, and raised floors. Manufacturers' data sheets and local code requirements should be referenced to find the proper cable for an installation.

Testing the Fiber Installation

OTDR
sends a light pulse down the fiber and measures the reflected light, which provides some measure of performance for the fiber

Figures 18-26(a) and (b) shows traces obtained from an **OTDR** (optical time-domain reflectometer) for two different sets of multimode fibers. In field terms, this is called "shooting" the fiber. The OTDR sends a light pulse down the fiber and measures the reflected light. The OTDR enables the installer or maintenance crew to verify the quality of each fiber span and obtain some measure of performance. The *X* axis on the traces indicates the distance, whereas the *Y* axis indicates the measured optical power value in dB. Both OTDR traces are for 850-nm multimode fiber.

Event
a disturbance in the light propagating down a fiber span that results in a disturbance on the OTDR trace

In Fig. 18-26(a), Point A is a "dead" zone or a point too close to the OTDR for a measurement to be made. The measured value begins at about 25 dB and decreases in value as the distance traveled increases. An **event,** or a disturbance in the light propagating down the fiber, occurs at point *B*. This is an example of what a poor-quality splice looks like (in regard to reflection as well as insertion loss). Most likely, this is a mechanical splice. The same type of event occurs at points *C* and *D*. These are also most likely mechanical splices. Points *F* and *G* are most likely the jumpers and patch-panel connections at the fiber end. The steep drop at point *H* is actually the end of the fiber. Point *I* is typical noise that occurs at the end of an

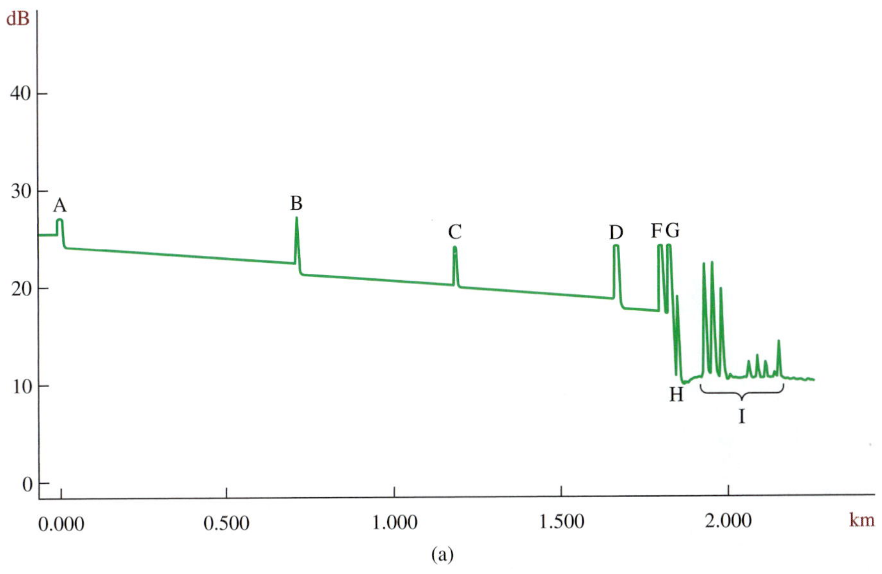

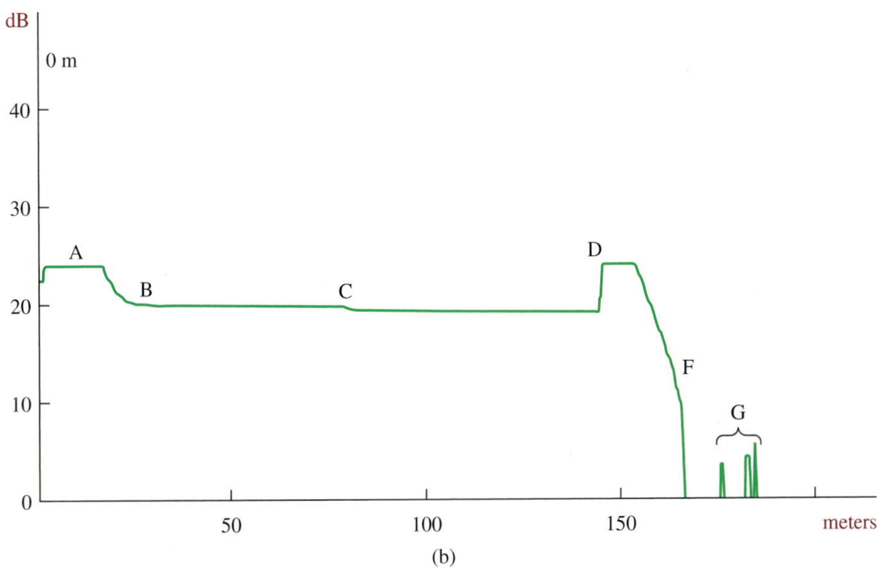

FIGURE 18-26 An OTDR trace of an 850-nm fiber.

"unterminated" fiber. Notice at point *G* that the overall value of the trace has dropped to about 17 dB. There has been about 8 dB of optical power loss in the cable in a 1.7 km run.

An OTDR trace for another multimode fiber is shown in Fig. 18-26(b), the hump at point *A* is basically a "dead" zone. The OTDR cannot typically return accurate measurement values in this region. This is common for most OTDRs, and the dead zone varies for each OTDR. The useful trace information begins at point *B* with a measured value of 20 dB. Point *C* shows a different type of event. This

The FT300 is ideal for inspecting multimode fiber, while the new FT400 offers superior fiber vision for singlemode applications. (Courtesy of Fluke Networks. Reprinted with permission.)

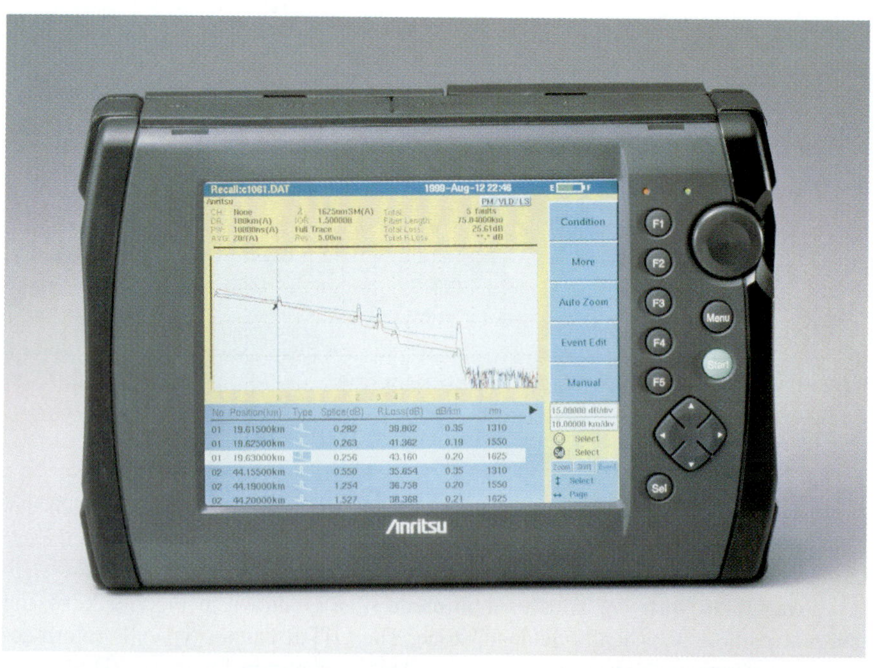

The MW9076 series shows superior performance and supports the construction and maintenance of optical fiber cables. (Courtesy of Anritsu Corporation. Reprinted with permission.)

type of event is typical of coiled fiber, or fiber that has been tightly bound, possibly with a tie-wrap, or that has had some other disturbance affecting the integrity of the fiber. Points *D* and *F* are actually the end of the fiber. At point *D* the trace level is about 19 dB for a loss of about 1 dB over the 150-m run. Point *G* is just the noise that occurs at the end of a "terminated" fiber.

18-9 Optical Networking

The need for increased bandwidth is pushing the fiber-optic community into optical networking solutions that are almost beyond the imagination of even the most advanced networking person. Optical solutions for long-haul, metropolitan, and local area networks are available. Cable companies are already using the high-bandwidth capability of fiber to distribute cable programming as well as data throughout their service areas.

The capital cost differences between a fiber system and a coaxial cable system are diminishing, and the choice of networking technology for new networks is no longer just budgetary. Fiber has the capacity to carry more bandwidth; as the fiber infrastructure cost decreases, fiber will be chosen to carry the data. Of course, the copper infrastructure is already in place, and new developments are providing tremendous increases in data speed over copper. However, optical fiber is smaller and eases the installation in already crowded ducts and conduits. And security is enhanced because it is difficult to tap optical fiber without detection.

Defining Optical Networking

Optical networks are becoming a major part of data delivery in homes, in businesses, and for long-haul carriers. The telecommunications industry has been using fiber for carrying long-haul traffic for many years. Some major carriers are merging with cable companies so that they are poised to provide high-bandwidth capabilities to the home. Developments in optical technologies are reshaping the way we will use fiber in future optical networks.

Yes, fiber provides additional bandwidth, but do we keep using the same approaches to solve networking problems? The answer is no; we need a new set of rules to define optical networking. Sprint Corporation has defined a new foundation for optical networking ["Changing the rules for developing optical solutions," *Lightwave* (October 1999)]. Five of the rules for optical networking are summarized as follows:

1. The next generation of optical networks must be able to carry multiple protocols. For example, optical networks should be able to carry IP Internet traffic and asynchronous transfer mode (ATM).
2. The architecture for the next generation of optical networks must be flexible.
3. The network must be manageable, including diagnostic capabilities for signal quality and faults.
4. The data transport must provide high speed and be invisible to the user. For example, the user should not be concerned how the data are being transported or what protocol is being used.

A fiber network distribution facility. (Courtesy of The Stock Market.)

5. The implementation of optical networks must provide for compatible interfacing with today's data-transport methodologies while providing the flexibility to incorporate future developments.

In addition to these five rules, the issues of chromatic and polarization mode dispersion become an increasing problem because of the need for greater transmission capability coupled with the restricted economic capability to install more fiber.

Basically, these rules for optical networking maintain a level of reliability and flexibility in the transport of data. But there is a new slant with optical networks. Dense wave division multiplexing and tunable lasers have changed the way optical networks can be implemented. It is now possible to transport many wavelengths over a single fiber. Lab tests at AT&T have successfully demonstrated the transmission of 1,022 wavelengths over a single fiber; however, conventional systems are limited to approximately 32 wavelengths.

The transport of multiple wavelengths over a single fiber opens up the possibilities to routing or switching many different data protocols over the same fiber but on different wavelengths. The development of cross connects that allow data to arrive on one wavelength and leave on another opens other possibilities.

SONET
protocol standard for optical transmission in long-haul communication

Synchronous optical network (**SONET**) is currently the standard for the long-haul optical transport of telecommunications data. SONET defines a standard for:

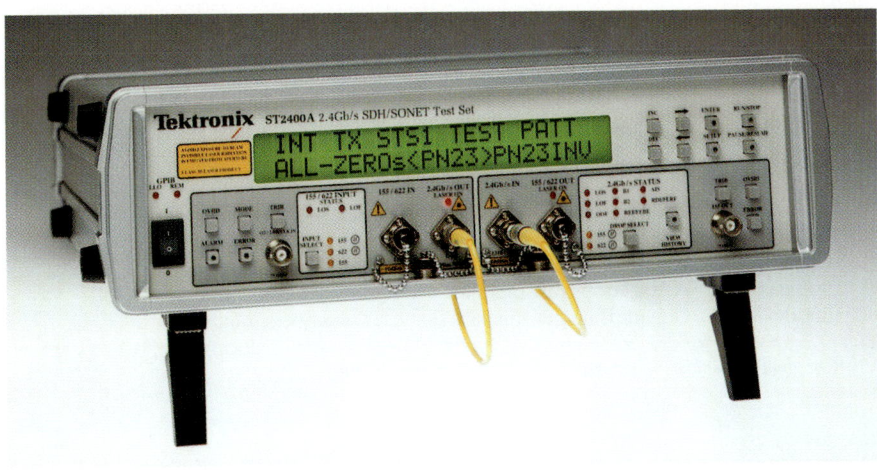

SONET test set. (Courtesy of Tektronix Inc.)

- Increase in network reliability
- Network management
- Defining methods for the synchronous multiplexing of digital signals such as DS-1 (1.544 Mbps) and DS-3 (44.736 Mbps)
- Defining a set of generic operating/equipment standards
- Flexible architecture

SONET specifies the various optical carrier (**OC**) levels and the equivalent electrical synchronous transport signals (**STS**) used for transporting data in a fiber-optic transmission system. Optical network data rates are typically specified in terms of the SONET hierarchy. Table 18-9 lists the more common data rates.

OC
optical carrier

STS
synchronous transport signals

The architectures of fiber networks for the home include providing fiber to the curb (**FTTC**) and fiber to the home (**FTTH**). FTTC is being deployed today. It provides high bandwidth to a location with proximity to the home and provides a high-speed data link, via copper (twisted pair), using very high-data digital subscriber line (VDSL). This is a cost-effective way to provide large-bandwidth capabilities to a home. FTTH will provide unlimited bandwidth to the home; however,

FTTC
fiber to the curb

FTTH
fiber to the home

Table 18-9 SONET HIERARCHY DATA RATES

Signal	Bit Rate	Capacity
OC-1 (STS-1)	51,840 Mbps	28DS-1s or 1 DS-3
OC-3 (STS-3)	155.52 Mbps	84DS-1s or 3 DS-3s
OC-12 (STS-12)	622.080 Mbps	336 DS-1s or 12 DS-3s
OC-48 (STS-48)	2.48832 Gbps	1344 DS-1s or 48 DS-3s
OC-192 (STS-192)	9.95328 Gbps	5376 DS-1s or 192 DS-3s
OC-768 (STS-768)	39.81312 Gbps	768 DS-3s

OC: Optical carrier
STS: Synchronous transport signals

DS-1: 1.544 Mbps
DS-3: 44.736 Mbps

Table 18-10 ETHERNET/FIBER NUMERICS

Numeric	Description
10Base F	10-Mbps Ethernet over fiber—generic specification for fiber
10BaseFB	10-Mbps Ethernet over fiber—part of the IEEE 10BaseF specification. Segments can be up to 2 km.
10BaseFL	10-Mbps Ethernet over fiber—segments can be up to 2 km in length. It replaces the FOIRL specification.
10BaseFP	A passive fiber star network. Segments can be up to 500 m in length.
100BaseFX	A 100-Mbps fast Ethernet standard that uses two fiber strands.
1000BaseLX	Gigabit Ethernet standard that uses two fiber strands.

Note: multimode fiber—2 km length; single-mode fiber—10 km.

the key to its success is the development of a low-cost optical-to-electronic converter in the home and laser transmitters that are tunable to any desired channel.

Conventional high-speed Ethernet local area networks operating over fiber use the numerics listed in Table 18-10 for describing the network configuration.

Fiber helps to eliminate the 100-m distance limit associated with unshielded twisted-pair (UTP) copper cable. This is possible because fiber has a lower attenuation loss. In a star network, the computer and the hub (or switch) are directly connected. If the fiber is used in a star network, a media converter may be required. The media converter converts the electronic signal to an optical signal, and vice versa. A media converter is required at both ends, as shown in Fig. 18-27.

Another example of how fiber is currently used in Ethernet LANs is for the high-speed transport of data, point-to-point, over longer distances. For example, the output of an Ethernet switch might be sent via fiber to a local router, as shown in Fig. 18-28(a). In this example, the inputs to the Ethernet switch are 10BaseT (10-Mbps twisted-pair) lines coming from computers on the LAN. The output of the Ethernet switch leaves via fiber at a 100-Mbps data rate (100BaseFX). The fiber makes it easy to increase the data rate over increased distances. As shown in Fig. 18-28(b), multiple-switch outputs, connected through a router, might be connected to a central router via fiber at a gigabit data rate.

The fiber provides substantially increased bandwidth for the combined traffic of the Ethernet switches and PCs [Figs. 18-28(a) and (b)]. Fiber has greater capacity, which enables faster transfer of data, minimizes congestion problems, and provides tremendous growth potential for each of the fiber runs.

Conventional high-speed Ethernet local area networks operating over fiber use the numerics listed in Table 18-10 for describing the network configuration.

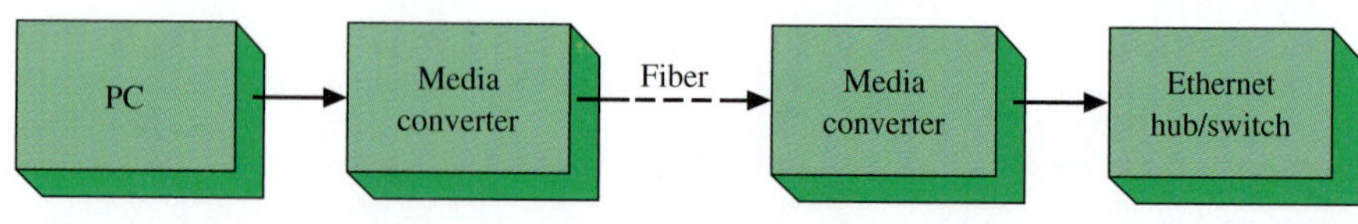

FIGURE 18-27 An example of connecting a PC to an Ethernet hub or switch via fiber.

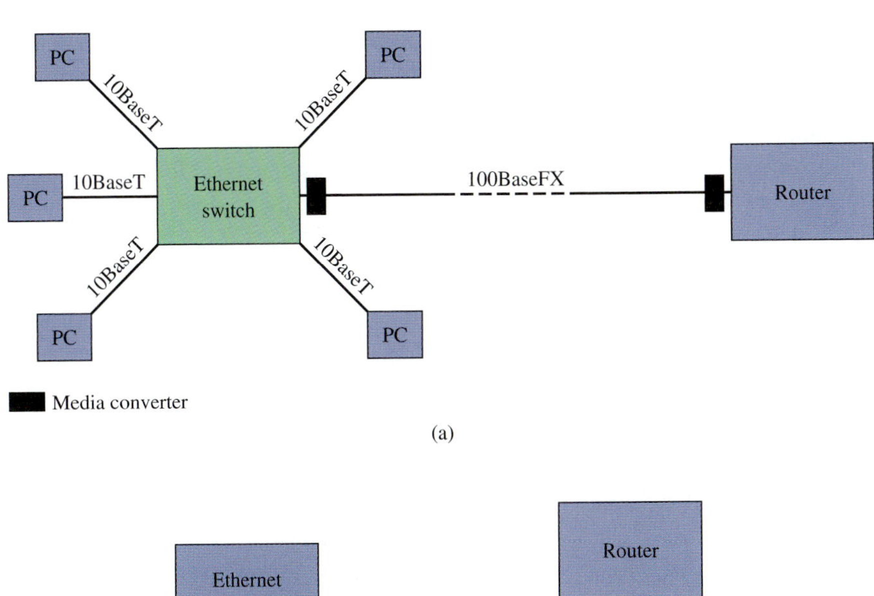

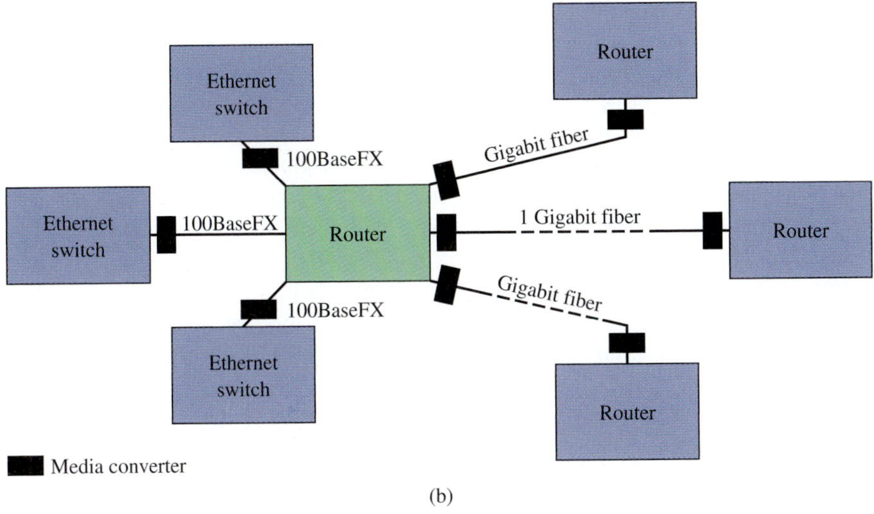

FIGURE 18-28 Examples of point-to-point connections using fiber in local area networks.

Air Fiber

Another form of optical networking involves the propagation of laser energy through the atmosphere, a line-of-sight technique similar to microwave radio. This application (usually called **air fiber,** free space optics, or a similar expression) uses a parabolic lens to focus the laser energy in a narrow beam. The beam is then aimed through the air to a receive parabolic lens a short distance away. This short distance is conservatively about 3 km, or a little farther depending on laser power and detector sensitivity used as well as the reliability/bit error rate desired.

Air Fiber
a form of optical networking that involves the propagation of laser energy through the atmosphere; also called free space optics

The normal long-term 99.999 percent reliability is achievable, but it is difficult. The expected degrading effects of simple rain and fog are often not particularly troublesome, but when a high-moisture-density cell crosses the propagation path, high signal attenuation can be noted. A good weather pattern study is advisable when planning an optical path.

The optical transmission equipment needs a stable mounting platform due to the degrading effects of building movement and vibration. Most optical platforms have auto-tracking options available to optimize antenna (lens) alignment constantly and minimize bit errors and outages.

Wavelengths available are from 800 to 1500 nanometers, all with their own pros and cons. System planners should be aware of laser safety when planning open-air optical spans.

The use of this media for networking is ideal between tall buildings in urban areas, in short metro spans, and on industrial and college compuses. It is particularly well suited for temporary service and disaster recovery operations. An additional advantage is that FCC licensing is not required. These applications are enhanced due to monetary savings in construction cost, physical infrastructure disruptions, and additional fill of cable duct facilities. This type of optical networking equipment is capable of handling a wide variety of data protocols, such as FDDI, DS–3, ATM, and gigabit formats.

FDDI

The American National Standards Institute (ANSI) developed the Fiber Distributed Data Interface (FDDI) standard that is now in widespread use. FDDI utilizes two 100-Mbps token-passing rings. The two independent counter-rotating rings are connected to a certain number of nodes (stations) in the network. The primary ring connects only *class A* stations—those offering a high level of protection because of their ability to transfer operation to the secondary ring should the primary ring fail. The secondary ring reaches all stations and carries data in the opposite direction of the primary ring. The secondary ring can also be used with the primary ring operating to allow increased data throughput. The switchover to the secondary path in the event of failure is accomplished by a pivoting spherical mirror within a *dual-bypass* switch. The changeover takes 5 to 10 μs, and a loss of about 1 dB results from the presence of the dual-bypass switch.

A twisted-pair-to-fiber media converter. (Courtesy of Black Box Corporation.)

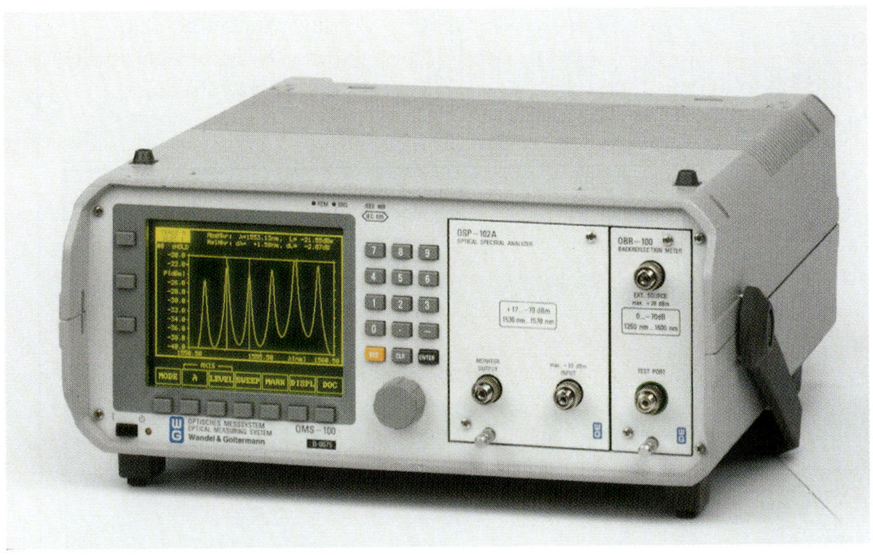

Optical spectral analyzer for measurements on wavelength division multiplexing (WDM) systems. (Courtesy of Wandel & Goltermann.)

Stations on the FDDI ring can be separated by up to 2 km as long as the average distance between nodes is less than 200 m. These limits are imposed to minimize the time it takes a signal to move around the ring. A total of 1000 physical connections and a total fiber path of 200 km are allowed. This allows 500 stations because each represents two physical connections. The type of fiber used is not dictated and is chosen by the user based on the performance required. The fibers used most often are 62.5/125 or 100/140 multimode fiber. LEDs are specified as light sources at 1300 nm.

FDDI cabling. (Courtesy of Black Box Corporation.)

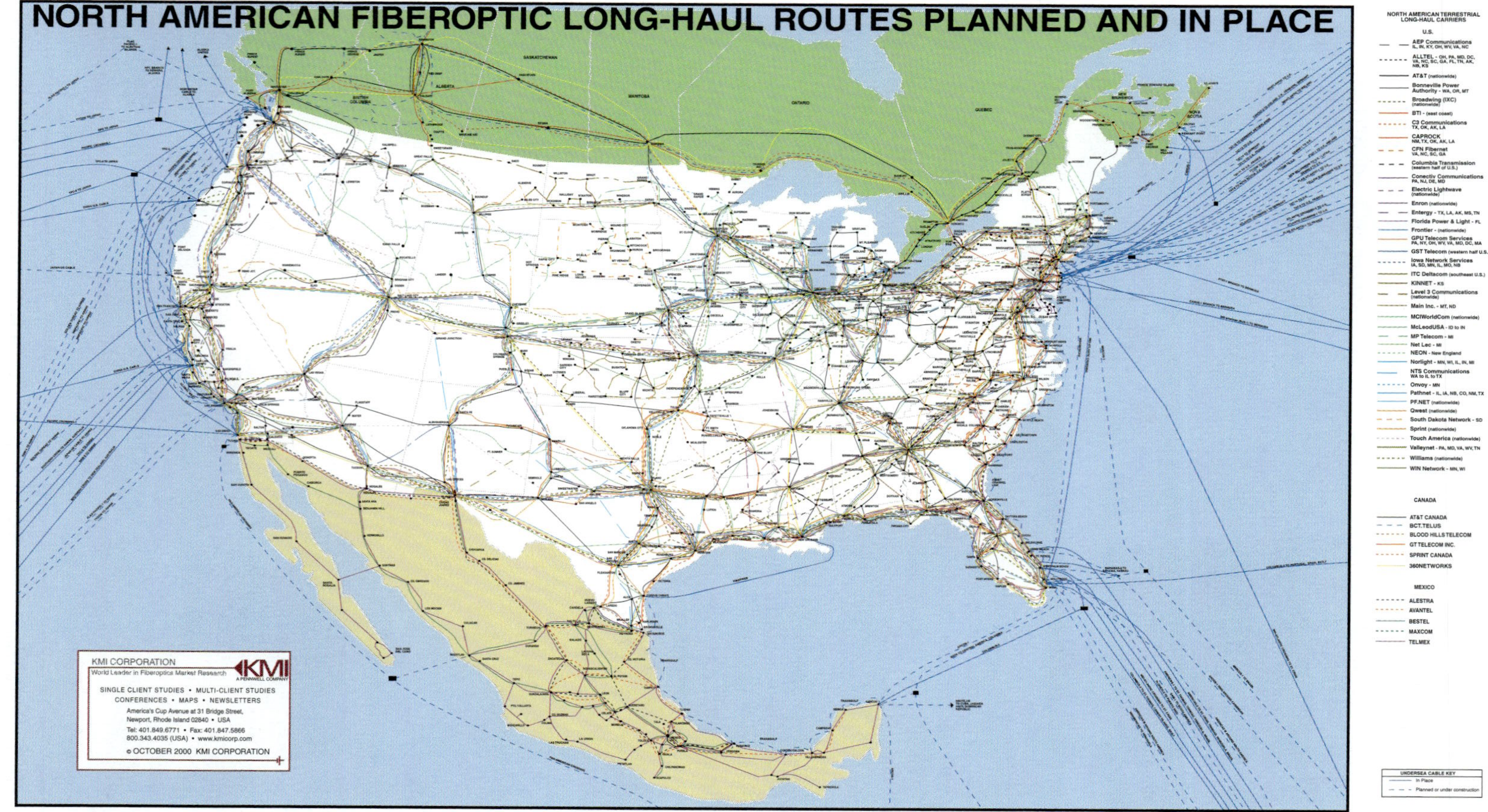

North American fiber-optic routes. (Courtesy of KMI Corporation.)

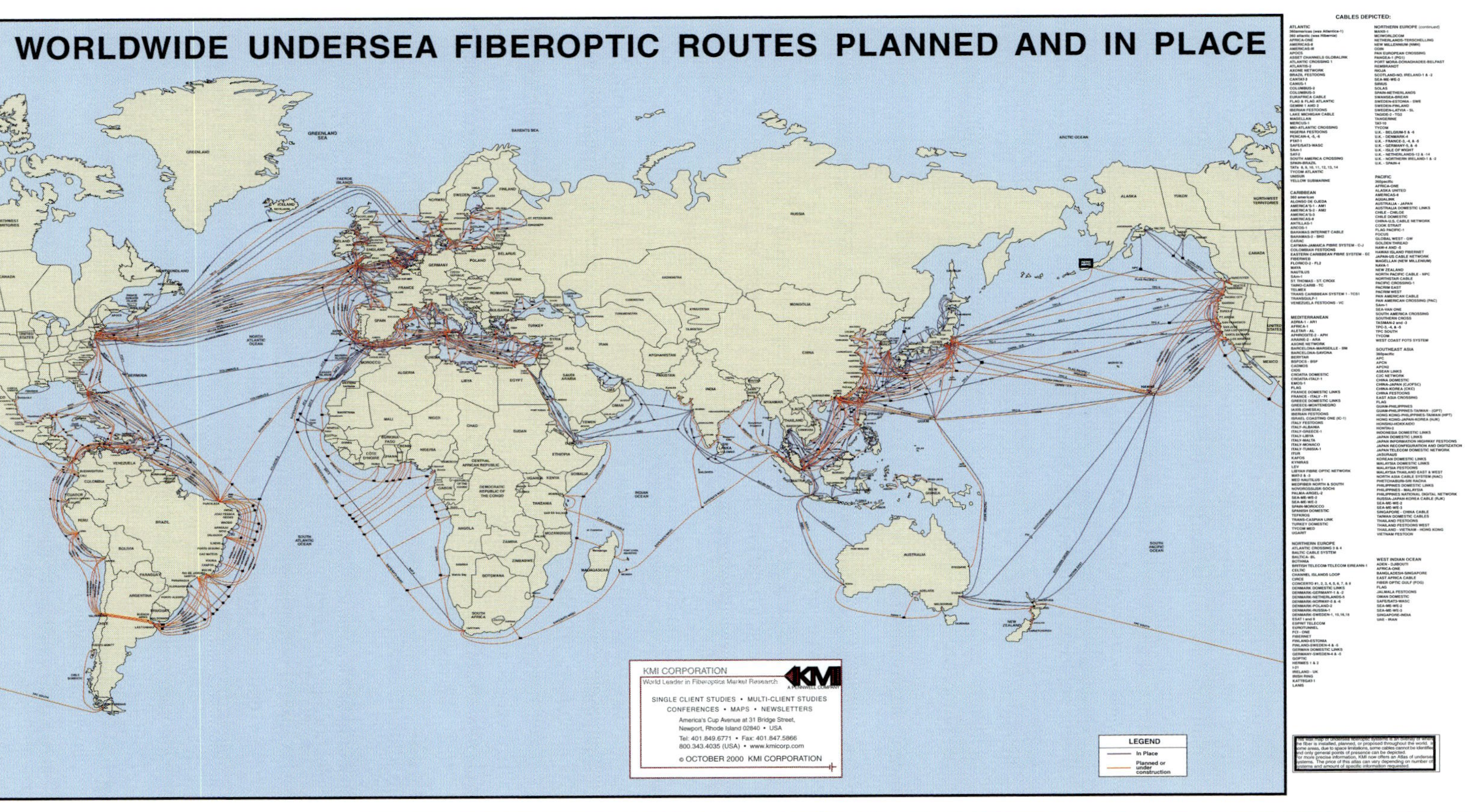

Worldwide fiber-optic routes. (Courtesy of KMI Corporation.)

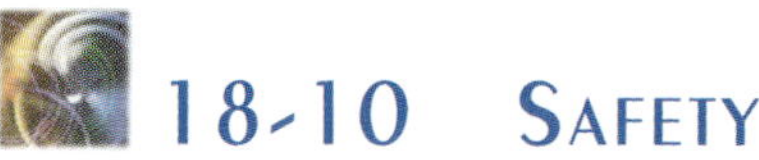

18-10 Safety

Any discussion of fiber optics is not complete unless it addresses safety issues, even if only briefly. As the light propagates through a fiber, two factors will further attenuate the light if there is an open circuit.

1. A light beam will disperse or fan out from an open connector.
2. If a damaged fiber is exposed on a broken cable, the end will likely be shattered, which will considerably disperse the light. In addition, there will be a small amount of attenuation from the strand within the cable, plus any connections or splices along the way.

However, two factors can increase the optical power at an exposed fiber end.

1. There could be a lens in a pigtail that could focus more optical rays down the cable.
2. In the newer DWDM systems, several optical signals are in the same fiber; although separate, they are relatively close together in wavelength. The optical power incident upon the eye is then multiplied.

Be aware of two factors:

1. The eye can't see fiber-optic communications wavelengths, so there is no pain or awareness of exposure. However, the retina can still be exposed and damaged. (Refer to Fig. 18-4, the electromagnetic spectrum.)
2. Eye damage is a function of the optical power, wavelength, source or spot diameter, and the duration of exposure.

For those working on fiber-optic equipment:

1. NEVER look into the output connector of energized test equipment. Such equipment can have higher powers than the communications equipment itself, particularly OTDRs.
2. If you need to view the end of a fiber, *ALWAYS turn off the transmitter,* particularly if you don't know whether the transmitter is a laser or LED, because lasers are higher power sources. If you are using a microscope to inspect a fiber, the optical power will be multiplied.

From a mechanical point of view:

1. Good work practices are detailed in safety, training, and installation manuals. *READ AND HEED.*
2. Be careful with machinery, cutters, chemical solvents, and epoxies.
3. Fiber ends are brittle and break off easily, including the ends cut off from splicing and connectorization. These ends are extremely difficult to see and can become "lost" and/or easily embedded in your finger. You won't know until your finger becomes infected. Always account for all scraps.
4. Use safety glasses specifically designed to protect the eye when working with fiber-optic systems.
5. Obtain and *USE* an optical safety kit.
6. Keep a *CLEAN* and orderly work area.

In all cases, be sure the craft personnel have the proper training for the job!

18-11 Troubleshooting

Today, optical fiber is the infrastructure of many communications hubs. Fiber carries billions of telephone calls a day. Optical fiber makes up the backbone structure of many local area networks currently in use. In this section we will look at planning an optical-fiber installation and maintaining it once it is in place.

Remember, the diode lasers (DLs) can emit radiation with a far higher energy-density than sunlight, and even though you can't see the radiation, it can easily cause blindness by retinal heating. You should always wear eye protection when working on laser systems. A 1-W or more CW output laser, such as that used in medical imaging products, can produce a stunningly high-power density when focused. Even when poorly focused, 1 W across 0.125-in. diameter (like the pupil of your eye) means more than 100 times the power density of sunlight! You need to respect the device and follow the rules for working with it.

After completing this section you should be able to

- Draw a fiber link showing all components
- Explain the use of the optical power meter
- Describe rise-time measurement
- Troubleshoot fiber-optic data links

System Testing

Once a system is installed, it should be tested thoroughly to ensure compliance with the contract specifications and performance in accordance with the manufacturer's manual. The following lists of tests provide a good measure of performance. Also, they are the start of a maintenance database for future reference. Not all these tests apply to all optical networks. In the realm of testing and evaluation, you will find that testing is expensive; from an experience perspective, you will find that not testing is more expensive.

General Guidelines

Tests on the cable plant itself should include:

- Measuring fiber insertion loss, which should be compared to the engineering system design. An optical test set is preferred, but an OTDR can be used to perform simple tests.
- Gathering OTDR traces, noting the loss slope plus return loss.
- Testing of all wavelengths planned or projected for use.

Overall system tests should include:

- Bit error rate (BER) tests
- Central wavelength
- Spectral width tests
- Transmitter output power (average, not peak)
- Receiver sensitivity (to some BER)
- Input voltage tolerance
- Protection and alarms
- System restoration

Note: Specifics on the requirements and guidelines for these tests are available from the manufacturers.

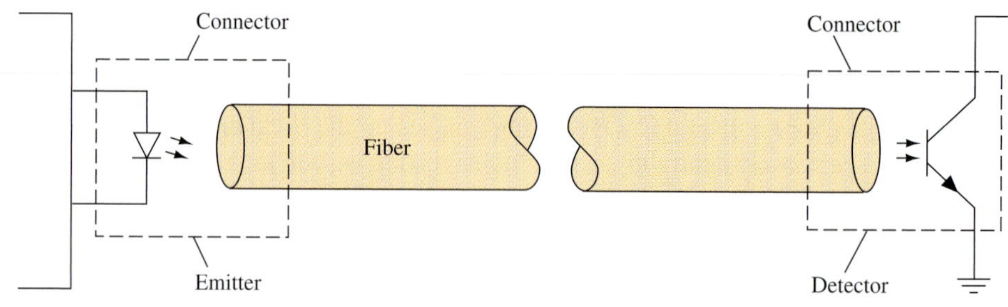

FIGURE 18-29 A fiber link showing emitter, detector, connectors, and fiber cable.

Losses in an Optical-Fiber System

The optical-fiber system in Fig. 18-29 has an emitter, two connectors, the fiber, and the detector. The proper performance of this fiber link depends on the total power losses of the light signal through the link being less than the specified maximum allowable loss. Power is lost in all of the components that make up the system. A connector may have a power loss of 1.5 dB and a splice with 0.5 dB, and the fiber cable itself will also attenuate the light signal. As an example, if a fiber system's maximum allowable losses are 20 dB, and total power losses add up to 17 dB, the system still has a 3-dB working margin. Of course, this is a small working margin and does not take into account emitters and detectors weakening over time.

Calculating Power Requirements

A power budget should be prepared when a fiber-based system is installed. The power budget specifies the maximum losses that can be tolerated in the fiber system. This helps ensure that losses stay within the budgeted power allocation. Once the optical-fiber system is in place, the optical power meter is used to determine the actual power being lost in the system. A calibrated light source injects a known amount of light into the fiber, and the optical power meter connected to the other end of the fiber measures the light power reaching it. Periodic checks should be scheduled as preventive maintenance to keep the fiber system in peak performance. Weakening emitters should be replaced before they degrade the system's performance.

Connector and Cable Problems

Some of the problems associated with fiber-optic links are caused by contact of a foreign substance with the fiber (even the oil from your skin can cause serious trouble). Connectors and splices are potential trouble spots. A back-biased photodiode and an op-amp can be used as a relative signal strength indicator. Looking for the signal while gently flexing cables and connectors can help pinpoint problem areas.

Characteristics of LEDs and DLs

These special-purpose diodes are nonetheless diodes and should exhibit the familiar exponential *I* versus *V* curve. These diodes do not draw current when forward biased until the voltage reaches about 1.4 V. Some ohmmeters do not put out sufficient voltage to turn an LED on; you may have to use a power supply, a current-limiting resistor, and a voltmeter to test the diode.

Time-domain reflectometer for optical systems. (Courtesy of Anritsu Company. Reprinted with permission.)

Reverse voltage ratings are very low compared to ordinary silicon rectifier diodes—as little as 6 V. More voltage may destroy the diode.

A Simple Test Tool

Some systems use visible wavelengths; most use invisible infrared. Another diode of the same type or of similar emission wavelength can be used as a detector to check for output. Use a meter in current mode, not voltage, and compare a good system to the troublesome one.

To increase sensitivity, a simple current-to-voltage converter circuit, made with an op-amp, a feedback resistor, and the detector diode pumping current to the op-amp input, converts the current from the detector diode to a voltage out of the op-amp. The circuit for this is shown in Fig. 18-30. If signal levels are high, just a resistor across the detector diode is appropriate. Remember to keep the bias voltage small enough so that the voltage developed is well below the maximum reverse voltage allowed for the diode.

If you wish to see the signal modulation, try using an oscilloscope in place of a simple multimeter. A less quantitative check for emitted output can be made using a test card of the type used in TV-repair shops to check for output from infrared remote controls. These cards are coated with a special chemical that, in the simultaneous presence of visible and infrared illumination, emits an orange glow.

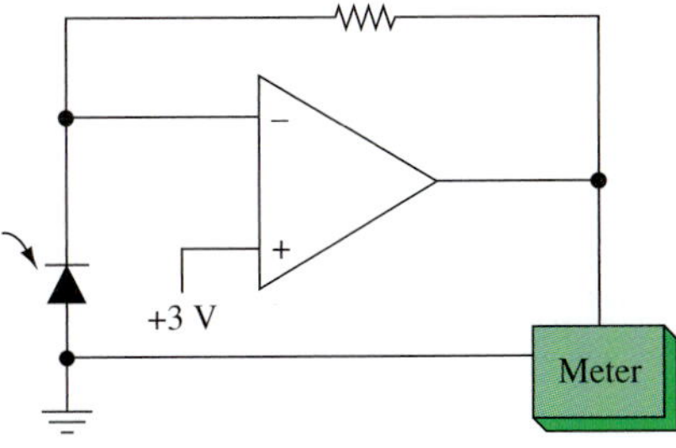

FIGURE 18-30 Light probe.

18-12 Troubleshooting with Electronics Workbench™ Multisim

The concept of preparing a system design for a fiber installation was presented in this chapter. This section presents a simulation exercise of a system design. Open the file **Fig18-31.ms7 (.msm)** on your EWB Multisim CD. This exercise provides you with the opportunity to study a fiber-optic system design in more depth. The circuit for the light-budget simulation is shown in Fig. 18-31.

Electronics Workbench™ Multisim does not contain simulation models or instruments for lightwave communications, but with a little creativity, a system design for a fiber installation can be modeled. This example is patterned after Fig. 18-23. The function generator models the output of a fiber-optic transmitter. The generator is outputting a square wave to model the pulsing of light. The settings for the function generator for three possible operating levels have been provided.

1. The maximum received signal level (RSL): −27 dBm
2. The designed operating level: −31.6 dBm
3. The minimum received signal level (RSL) for a BER of 10^{-9}: −40 dBm

A 16-dB T-type attenuator has been provided to simulate the fiber cable and splice loss. The system is terminated with a 600-Ω resistor for consistency with the analog model, but this resistor does not exist in a real optical system. A voltage-controlled sine-wave oscillator has been provided to simulate the optical receiver. The settings for the voltage-controlled sine-wave oscillator are shown in Fig. 18-32.

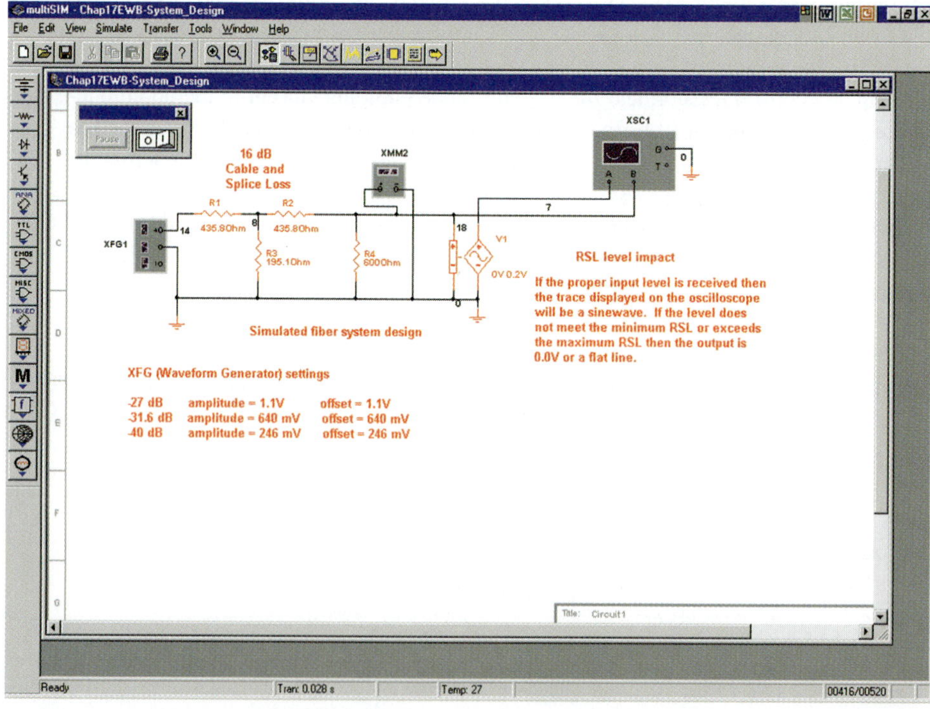

FIGURE 18-31 The Multisim circuit for the light-budget simulation.

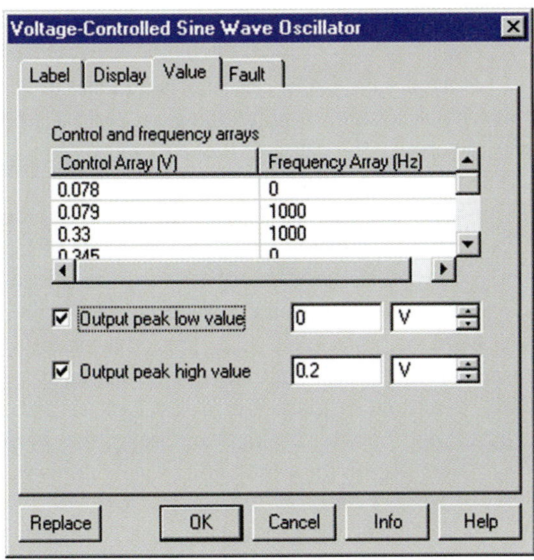

FIGURE 18-32 The settings for the voltage-controlled sine-wave generator that is being used to model an optical receiver with a minimum and maximum RSL.

Double-click on the voltage-controlled sine-wave oscillator to view or change the settings. These settings for the voltage-controlled sine-wave oscillator provide for a sine-wave output as long as the received signal level is within the −40-dB to −27-dB operating range. If the input level falls outside this range, then the oscillator outputs a flat line.

Verify that the function generator is set for an amplitude of 640 mV and an offset voltage of 640 mV. Start the simulation and view the level on the multimeter and the traces on the oscilloscope. The multimeter should show a −31.6-dB level, and the oscilloscope should show a pulse signal on channel B, which is the input to the voltage-controlled sine-wave oscillator. Channel A is connected to the output of the voltage-controlled sine-wave oscillator, and it should show a sine wave. Change the function generator levels to the maximum and minimum receive signal levels (RSLs) of −27 dBm and −40 dBm and view the output of the voltage-controlled sine-wave oscillator.

Settings

maximum RSL = −27 dBm	amplitude = 240 mV	offset = 240 mV
minimum RSL = −40 dBm	amplitude = 1.1 V	offset = 1.1 V

You should see a flat line on the channel A trace for both cases, which indicates that the input signal level does not meet specifications. This example demonstrates that if the input level does not meet the required input signal level specification (RSL), then some information will be lost due to an increase in the bit error rate (BER). The following Electronics Workbench™ Multisim exercises provide you with additional opportunity to troubleshoot the fiber system model when the signal is lost. An optical time domain reflectometer (OTDR) is not available with the Multisim tools, but you can use the multimeter and the oscilloscope to measure signal levels throughout the system and verify the system for proper operation.

Electronics Workbench™ Exercises

1. Open the file **FigE18-1.ms7 (.msm)** in your EWB CD. This is a model of the fiber system design. Determine if the system is functioning properly. If the system is not working properly, determine the cause of the fault and fix it. Rerun the simulation to verify the system is functioning properly.
2. Open the file **FigE18-2.ms7 (.msm)** in your EWB CD. This is a model of the fiber system design. Determine if the system is functioning properly. If the system is not working properly, determine the cause of the fault and fix it. Rerun the simulation to verify the system is functioning properly.
3. Open the file **FigE18-3.ms7 (.msm)** in your EWB CD. This is a model of the fiber system design. Determine if the system is functioning properly. If the system is not working properly, determine if the problem is a fault or if the problem is a system setup error. Report on your findings and return the setup levels back to proper operating points. Rerun the simulation to verify the system is functioning properly.

Summary

In Chapter 18 we introduced the field of fiber optics. We learned that many applications exist in electronic communications for these optical devices. The major topics you should now understand include:

- the advantages offered by fiber-optic communication
- the analysis and properties of light waves
- the physical and optical characteristics of optical fibers, including multimode, graded index, and single-mode fibers
- the attenuation and dispersion effects in fiber
- the description and operation of the diode laser (DL) and high-radiance light-emitting diode (LED) light sources
- the application of *p-i-n* diodes as light detectors
- the description of common techniques used to connect fibers
- the general applications of fiber-optic systems
- the power considerations and calculations in fiber-optic systems
- the usage of fiber optics in local area networks (LANs)
- the description of LAN components, including wavelength-dependent and independent couplers and optical switches

Questions and Problems

Section 18-1

1. List the basic elements of a fiber-optic communications system. Explain its possible advantages compared to a more standard communications system.
2. List five advantages of an optical communications link.

Section 18-2

3. Define *refractive index*. Explain how it is determined for a material.
4. A green LED light source functions at a frequency between red and violet. Calculate its frequency and wavelength. (5.7×10^{14} Hz, 526 nm)
5. A fiber cable has the following index of refraction: core, 1.52, and cladding, 1.31. Calculate the numerical aperture for this cable. (0.77)
6. Determine the critical angle beyond which an underwater light source will not shine into the air. (48.7°)
7. Define *infrared light* and the *optical spectrum*.
8. What are the six fixed wavelengths commonly used today?
9. What are the wavelength ranges for the optical bands O, E, S, C, L, and U?
10. Draw a picture of the construction of a single fiber.

Section 18-3

11. Define *pulse dispersion* and the effect it has on the transmission of data.
12. What is multimode fiber, and what is the range for the core size?
13. Why was graded index fiber developed? What are the two typical core sizes and the cladding size for graded index fiber?
14. What are the applications for single-mode fibers?
15. What are the core/cladding sizes for single-mode fibers?
16. Define *mode field diameter* for fiber-optic cable.
17. Define *zero-dispersion wavelength* for fiber-optic cable.
18. Describe two applications that would be suitable for using plastic optical fiber.

Section 18-4

19. What are the two key distance-limiting parameters in fiber-optic transmission?
20. What are the four factors that contribute to attenuation?
21. Define *dispersion*. What are typical dispersion values for 850-, 1310-, and 1550-nm-wavelength fibers?
22. Determine the amount of pulse spreading of an 850-nm LED that has a spectral width of 18 nm when run through a 1.5-km filter. Use a pulse-dispersion value of 95 ps/(nm·km). (2.565 ns/km)
23. What are the three types of dispersion?
24. What is dispersion-compensating fiber?

Section 18-5

25. Compare the diode laser and the LED for use as light sources in optical communication systems.
26. Explain the process of lasing for a semiconductor diode laser. What is varied to produce light at different wavelengths?
27. Define *dense wavelength division multiplexing*.
28. What are tunable lasers and what is the primary market for them?
29. What are isolators?
30. Attenuators are used to do what?
31. List five intermediate components for fiber-optic systems.
32. What is an optical detector?
33. What are the benefits of a distributed feedback laser?
34. List the advantages of VCSEL.

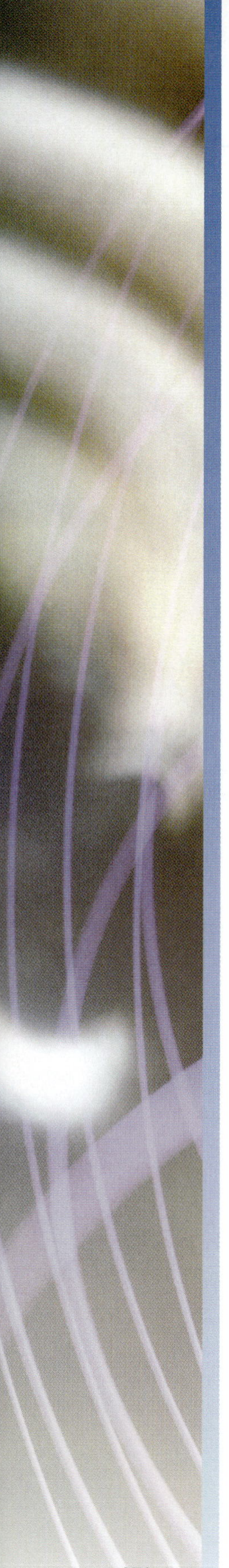

Section 18-6

35. List the eight sources of connection loss in fiber.
36. Compare the advantages and disadvantages of fusion and mechanical splicing. Which would you select if you were splicing many fiber strands? Explain your choice.
37. List the three most popular fiber-end connectors.
38. Describe the procedure for preparing the fiber for splicing or connectorization.
39. What are the general rules for splicing single-mode and multimode fiber together?

Section 18-7

40. What are the primary performance issues when designing a fiber-optic transmission link?
41. Define a *long-haul* and a *local area network*.
42. Define *RSL*.
43. Define *maintenance margin*.
44. When testing a fiber with an OTDR, it was determined that the actual length of fiber used for a 20-km span was 20.34 km. Is this actually possible and why?
45. Determine the cable loss in dB for a 10-km fiber run. The fiber has a loss of 0.4 dB/km. (4 dB)
46. For power budgeting of fiber-optic transmission systems, what are four components that contribute to power loss between a transmitter and a receiver?

Section 18-8

47. List four tips for installing fiber-optic cable.
48. What is an OTDR and how is it used?
49. Examine the OTDR trace provided in Fig. 18-33. Explain the trace behavior at points A, B, C, D, and E.

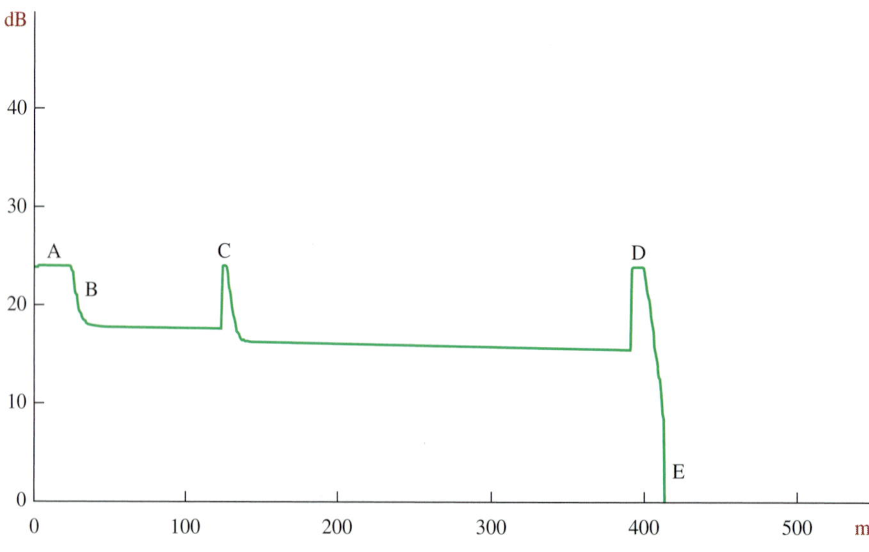

FIGURE 18-33 Figure for Problem 49.

Section 18-9

50. What are the changes in optical solutions that may greatly affect the design of optical networks?
51. Define *FTTC* and *FTTH*.
52. What is OC-192?
53. Describe how fiber can be used in a LAN to increase the data capacity and potentially minimize congestion problems.
54. What is FDDI? Provide a basic description of an FDDI system.
55. A seven-station FDDI system has spacings of 150 m, 17 m, 270 m, 235 m, and 320 m for six of the stations. Determine the maximum spacing for the seventh station. (208 m)

Questions for Critical Thinking

56. Analyze the *NA* and cutoff wavelengths for single-mode fiber with a core of 2.5 μm and refraction indexes of 1.515 and 1.490 for core and cladding, respectively. (0.274, 1.73 μm)
57. A system operating at 1550 nm exhibits a loss of 0.35 dB/km. If 225 μW of light power is fed into the fiber, analyze the received power through a 20-km section. (44.9 μW)
58. A fiber-optic system uses a cable with an attenuation of 3.2 dB/km. It is 1.8 km long and has one splice with an 0.8-dB loss. Due to the source/receiver connection, it has a 2-dB loss at both transmitter and receiver. It requires 3 μW of received optical power at the detector. Report on the level of optical power required from the light source. (34.1 μW)
59. Provide a complete power budget analysis for a system with the following losses and specifications:

Losses:	
Pigtail losses:	6.5 dB
Two connections:	1.0 dB each
Three splices:	0.5 dB each
20 km of fiber:	0.35 dB/km
Specifications:	
Laser power output:	−2 dBm
Minimum RSL:	−33 dBm
Maximum RSL:	−22 dBm
Maintenance margin:	3 dB
Power margin	1 dB
Operational margin	3 dB

Acronyms and Abbreviations

A

AAL	ATM adaptation layer
AC	alternating current
ACA	adaptive channel allocation
ACIL	trade association (formerly the American Council of Independent Laboratories)
ACK	acknowledgment
ACL	advanced CMOS logic
ACR	attenuation and crosstalk measurement
A/D	analog-to-digital
ADC	analog-to-digital converter
ADCCP	advanced digital communications control protocol
ADSL	asymmetric digital subscriber line
AF	audio frequency
AFC	automatic frequency control
AFSK	audio-frequency shift keying
AGC	automatic gain control
AIAA	American Institute of Aeronautics and Astronautics
AlGaAs	aluminum gallium arsenide
ALC	automatic level control
ALU	arithmetic logic unit
AM	amplitude modulation
AMI	alternate mark inversion
AML	automatic-modulation-limiting
AMPS	Advanced Mobile Phone Service
ANSI	American National Standards Institute
APC	angle-polished connectors
APD	avalanche photodiode
AP-S	Antennas and Propagation Society
ARPA	Advanced Research Projects Agency (now DARPA)
ARQ	automatic repeat request
ARRL	American Radio Relay League
ASCII	American Standard Code for Information Interchange
ASIC	application-specific integrated circuit
ASK	amplitude-shift keying
ASSP	application-specific standard products
ATC	adaptive transform coding
ATE	automatic test equipment
ATG	automatic test generation
ATM	asynchronous transfer mode
ATSC	Advanced Television Systems Committee
ATV	advanced television
AWGN	additive white Gaussian noise

B

B	byte
BAW	bulk acoustic wave
BBNS	broadband network services
BCC	block check character
BCCH	broadcast control channel
BCD	binary-coded decimal
B-CDMA	broadband CDMA
BCI	broadcast interference
BeCu	beryllium copper
B8ZS	bipolar 8 zero substitution
BER	bit-error rate
BERT	bit-error-rate tester
BFO	beat-frequency oscillator
BiCMOS	bipolar-CMOS
BIOS	basic input/output system
BIS	buffer information specification
B-ISDN	broadband integrated-services digital network (an ATM protocol model)
BJT	bipolar junction transistor
BPSK	binary phase-shift keying
BRI	basic-rate interface
BS	base station
BSC	base-station controller
BSS	Broadcasting Satellite Service

BW	bandwidth
BWO	backward-wave oscillator

C

CAD	computer-aided design
CAE	computer-aided engineering
CAM	computer-aided manufacturing
CAT	computer-aided test
CAT5	category 5
CAT6/5e	category 6 and category 5e
CATV	community-access (cable) television
CCA	clear-channel assortment
CCD	charge-coupled device
CDM	code-division multiplex
CDMA	code-division multiple access
CDMA2000	a 3G wireless development popular in the United States
CDPD	cellular digital packet data
CE	compliance engineering
CELP	code-excited linear prediction (coding)
CHBT	complementary heterojunction bipolar transistor
C/I	carrier/interference ratio
CMOS	complementary metal-oxide semiconductor
CODEC	coder/decoder
CPU	central processing unit
CRC	cyclic redundancy check
CSMA	carrier-sense multiple access
CSMA/CA	carrier sense multiple access collision avoidance
CSMA/CD	CSMA with collision detection
CSU/DSU	channel service unit/data service unit
CTI	computer telephone integration
CTIA	Cellular Telecommunications Industry Association
CT2	second-generation cordless telephone
CVBS	composite video blanking and synchronization
CVD	chemical-vapor deposition
CW	continuous wave

D

D/A	digital-to-analog
DAC	digital-to-analog converter
DARPA	Defense Advanced Research Projects Agency
DAS	data-acquisition system
dB	decibel
dBc	decibels with respect to carrier
dBi	antenna gain in decibels, with respect to isotropic antenna
dBm	decibels with respect to 1 mW
DBPSK	differential binary phase-shift keying
DBR	distributed Bragg reflector
DBS	direct-broadcast satellite
DC	direct current
DCCH	digital control channel
DCR	direct current receiver
DDC	direct digital control
DDS	direct digital synthesizer (or synthesis) *or* digital-data systems
DECT	Digital European Cordless Telecommunications
DELTIC	delay-line time compression
DFB	distributed feedback
DFD	digital frequency discriminators
DI	dielectric isolation
DIL	dual in-line
DIP	dual in-line package
DLVA	detector log video amplifier
DMA	direct memory access
DMM	digital multimeter
DMT	discrete multitone
DMUX	demultiplexer
DNL	differential nonlinearity
DOCR	digital on-channel repeater
DOD	direct outward dialing
DPDT	double-pole, double-throw
DPSK	differential phase-shift keying
DPST	double pole, single throw
DQPSK	differential quadrature phase-shift keying
DRAM	dynamic random-access memory
DRO	dielectric resonator oscillator
DSL	digital subscriber line
DSO	digital storage oscilloscope
DSP	digital signal processing
DSSS	direct sequence spread spectrum
DTCXO	digital temperature-compensated crystal oscillator
DTH	digital to home
DTMF	dual-tone multifrequency
DTV	digital television
DUT	device under test
DVB	digital video broadcast
DVM	digital voltmeter
DWDM	dense wavelength-division multiplexer

E

EBCDIC	Extended Binary Coded Decimal Interchange Code
ECC	error-correction coding
ECL	emitter-coupled logic
EDA	electronic design automation
EDC	error detection and correction
EDFA	erbium-doped fiber amplifier

EEPROM	electrically-erasable programmable read-only memory
EHF	extremely-high frequency
EIA	Electronic Industries Association
EIRP	effective isotropic radiated power
EISA	extended industry standard architecture
ELF	extremely-low frequency
EM	electromagnetic
EMC	electromagnetic compatibility
EMI	electromagnetic interference
ENOB	effective number of bits
ENR	excess noise ratio
EPROM	erasable programmable read-only memory
ESD	electrostatic discharge
ESF	extended superframe framing
ESI	equivalent series inductance
ESMR	enhanced specialized mobile radio
ESR	electrostatic resistance
ETACS	extended total access communications systems
ETDMA	enhanced time-division multiple access
E-3	industry standard for ATM (34.736 Mb/s)
ETSI	European Telecommunications Standards Institute
eV	electron volts
EVM	error vector magnitude

F

FCC	Federal Communications Commission
FDD	frequency division duplex
FDDI	fiber-distributed data interface
FDM	frequency division multiplex
FDMA	frequency-division multiple access
FEC	forward error correction (*or* control)
FEM	finite-element method
FER	frame error rate
FET	field-effect transistor
FFSK	fast frequency-shift keying
FFT	fast Fourier transform
FHMA	frequency-hopping multiple access
FHSS	frequency hopping spread spectrum
FIFO	first-in, first-out
FITS	failures in 10^9 hours
FLOPs	floating-point operations
FM	frequency modulation or frequency-modulated
4FSK	four-level frequency-shift keying
FPGA	field-programmable gate array
FQPSK	filtered quadrature phase-shift keying
FSF	frequency scaling factor
FSK	frequency-shift keying
FSR	full-scale range
FTTC	fiber to the curb
FTTH	fiber to the home

G

GaAs	gallium arsenide
GEO	geostationary earth orbit
GFSK	Gaussian frequency-shift keying
GHz	gigahertz
GMSK	Gaussian minimum-shift keying
GPS	global positioning system
GSGSG	ground-signal ground-signal ground
GSM	Global System for Mobile Communications
GSSG	ground-signal signal-ground

H

HBT	heterojunction bipolar transistor
HDLC	high-level data link control
HDSL	high-data-rate digital subscriber line
HDTV	high-definition television
HEMT	high-electron mobility transistor
HF	high frequency
HFC	hybrid fiber coaxial
HPA	high-power amplifier
HTS	high-temperature superconductor
Hz	hertz, originally cycles per second

I

IAGC	instantaneous automatic gain control
IANA	Internet Assigned Numbers Authority
IBIS	input/output buffer information specification
IC	integrated circuit
IDSL	ISDN digital subscriber line
IEEE	Institute of Electrical and Electronics Engineers
IESS	Intelsat Earth Station Standards
IF	intermediate frequency
IFM	instantaneous frequency measurement
IM	intermodulation
IMD	intermodulation distortion
IMPATT	impact ionization avalanche transit time diode
IMTS	improved mobile telephone service
IMT-2000	international mobile telecommunications
InGaAs	indium gallium arsenide
INL	integral nonlinearity
InP	indium phosphide
INTELSAT	International Telecommunications Satellite Organization
I/O	input/output
IOC	integrated optical circuit
IP	Internet protocol

I/Q	in-phase/quadrature
IQST	INTELSAT qualified satellite terminals
IR	infrared
IrDA	Infrared Data Association
IS	international standards
ISA	industry-standard architecture
ISDN	integrated-services digital network
IS-54	Interim Standard 54 (dual-mode TDMA/AMPS)
ISHM	International Society for Hybrid Microelectronics
ISI	intersymbol interference
ISL	intersatellite link
ISM	industrial, scientific, and medical
IS-95	Interim Standard 95 (dual-mode CDMA/AMPS)
ISP	Internet service provider
ITFS	instructional television fixed services
ITS	intelligent transportation systems
ITU	International Telecommunications Union
ITV	industrial television

J

JDC	Japanese digital cellular
JFET	junction field-effect transistor

L

LAN	local area network
LC	inductive-capacitive *or* liquid crystal
LCC	leadless ceramic chip carrier
LCD	liquid-crystal display
LDCC	leaded ceramic chip carrier
LDMOS	laterally-diffused metal oxide silicon
LED	light-emitting diode
LF	low frequency
LHCP	left-hand circular polarization
LiIon	lithium ion
LMDS	local multichannel distribution system
LMR	land mobile radio
LNA	low-noise amplifier
LNB	low-noise block down-converter
LNBF	low-noise block feedhorn
LO	local oscillator
LOS	line of sight
LPTV	low-power television
LSB	least-significant bit
LSI	large-scale integration
LVDS	low-voltage differential signaling

M

MAC	medium-access control
MAP	mobile application part
MBE	molecular beam epitaxy
MCA	multichannel amplifier
MCW	modulated continuous wave
MDAC	multiplying digital-to-analog converter
MDS	multipoint distribution systems
MDSL	medium-speed digital subscriber line
MESFET	metal semiconductor field-effect transistor
MFLOPS	million floating-point operations per second
MIC	microwave integrated circuit
MIL	military specification
MIPS	million instructions per second
MMDS	multichannel, multipoint distribution systems
MMIC	monolithic microwave integrated circuit
MOCVD	metal-organic chemical-vapor deposition
modem	modulator/demodulator
MOS	metal-oxide semiconductor
MOSFET	metal-oxide semiconductor field-effect transistor
MPSD	masked-programmable system devices
MPSK	minimal phase-shift keying
MSA	metropolitan statistical area
MSK	minimum shift keying
MSPS	million samples per second
MSS	mobile satellite service
MTA	major trading area
MTBF	mean time between failures
MTTF	mean time to failure
MUX	multiplexer
MVDS	microwave video-distribution system

N

NAB	National Association of Broadcasters
NADC	North American Digital Cellular
NAMPS	narrowband Advanced Mobile Phone Service
NASA	National Aeronautics and Space Administration
NBX	network branch exchange
NCO	numerically controlled oscillator
NEMA	National Electrical Manufacturers Association
NEMI	National Electronics Manufacturing Initiative, Inc.
NEXT	near-end crosstalk
NF	noise figure
NIC	network interface card
NiCd	nickel cadmium
NiMH	nickel metal hydride
NIST	National Institute of Standards & Technology (formerly NBS)
NLSP	network-link services protocol
NMT-900	Nordic Mobile Telephone
NNI	network-node interface
NRZ	nonreturn-to-zero code
NRZI	nonreturn-to-zero inverted code

NRZ-L nonreturn to zero-low
NTSC National Television Systems Committee (U.S. television broadcast standard)

O

OC optical carrier
OC-48 2.4-Gb/s optical-carrier industry standard
OC-192 Optical Carrier 192
OCR optical character recognition
OC-3 155-Mb/s optical-carrier industry standard
OC-12 622-Mb/s optical-carrier industry standard
OCXO oven-controlled crystal oscillator
OEM original-equipment manufacturer
OOK on-off keying (modulation)
OPDAR optical radar
OQPSK offset quadrature phase-shift keying *or* orthogonal quadrature phase-shift keying
OSI-7 open system interconnection
OTA over the air
OTDR optical time-domain reflectometer

P

ϕM phase modulation
PABX private automatic branch exchange
PACS personal advanced communications systems
pACT personal Air Communications Technology
PAE power-added efficiency
PAL phase-alternation-line (a 625-line 50-field color television system)
PAM pulse-amplitude modulation
PBX private branch exchange
PC personal computer
PC convex-polished
PCB printed-circuit board
PCI peripheral component interconnect
PCIA Personal Communications Industry Association
PCM pulse-code modulation
PCMCIA Portable Computer Memory Card International Association
PCN personal communications network
PCS personal communications services *or* plastic-clad silica (fiber)
PCU programmer control unit
PDA personal digital assistant
PDBM pulse-delay binary modulation
PDC personal digital cordless
PDF probability density function
PDH piesiochronous digital hierarchy
PECL positive emitter-coupled logic
PEP peak envelope power
PFM pulse-frequency modulation
PGBM pulse-gated binary modulation
PHEMT pseudomorphic high-electron-mobility transistor
PHP Personal HandyPhone
PHS Personal HandyPhone System
PICD personal information and communication device
PIM passive intermodulation
PIN positive-intrinsic-negative
pixel picture element
PLCC plastic leaded-chip carrier
PLD programmed logic device
PLL phase-locked loop
PLMR public land mobile radio
PLO phase-locked oscillator
PM phase modulation
PMR professional mobile radio
PN pseudorandom noise
PolSK polarization-shift keying
POTS plain old telephone service
p-p peak-to-peak
PPB parts per billion
PPBM pulse-polarization binary modulation
PPM parts per million *or* periodic permanent magnet *or* pulse-position modulation
PPP point-to-point protocol
PQFP plastic quad-leaded flat pack
PRBS pseudorandom-bit sequence
PRF pulse-repetition frequency
PRI pulse-repetition interval
PRK phase-reversal keying
PRL preferred roaming list
PRML partial-response maximum likelihood
PSK phase-shift keying
PSNEXT power sum NEXT test
PSTN public-switched telephone network
PTFE polytetrafluoroethylene
PTM pulse-time modulation
PWM pulse-width modulation

Q

Q quality factor
QAM quadrature amplitude modulation
QFP quad flat pack
QPSK quadrature phase-shift keying
QSOP quarter-sized outline package
QUIL quad in-line

R

RAC reflective array compressor
RADAR radio detecting and ranging
RAM random-access memory
RC resistance-capacitance
RF radio frequency
RFI radio-frequency interference

RFID	radio-frequency identification
RHCP	right-hand circular polarization
RIC	remote intelligent communications
RIS	random interleaved sampling
RISC	reduced-instruction set computer
RMS	root mean square
ROM	read-only memory
RS	Reed-Solomon
RSA	rural statistical area
RS-422, RS-485	balanced-mode serial communications standards that support multidrop applications
RSL	received signal level
RSSI	received signal-strength indicator
RZ	return-to-zero code

S

SAT	signal-audio tone
SAW	surface acoustic wave
SCADA	supervisory control and data-acquisition systems
SCPI	Standard Commands for Programmable Instruments *or* small-computer programmable instrument
SCR	silicon-controlled rectifier
SCSA	Signal Computing Systems Architecture [industry-standard architecture for deploying computer telephone integration (CTI)]
SCSI	small computer system interface
SDH	synchronous digital hierarchy
SDMA	space-division multiple access
SDR	signal-to-distortion ratio
SDTV	standard definition television
SFDR	spurious-free dynamic range
S/H	sample and hold
SHF	super-high frequency
SIMOX	separation by implantation of oxygen
SINAD	signal to noise plus distortion
SLA	sealed lead acid
SLIC	subscriber-line interface circuit
SMART	system monitoring and remote tuning
SMD	surface-mount device
SMP	surface-mount package
SMR	specialized mobile radio
SMSR	side-mode suppression ratio
SMT	surface-mount technology or surface-mount toroidal
S/N	signal to noise
SNR	signal-to-noise ratio
SOE	stripline opposed emitter (package)
SOI	silicon-on-insulator
SOIC	small-outline integrated circuit
SONET	Synchronous Optical Network
SOS	silicon on sapphire
SPDT	single pole, double throw
SPICE	Simulation Program with Integrated Circuit Emphasis
SPST	single pole, single throw
SRAM	static random-access memory
SS	spread spectrum
SSB	single sideband
SS/TDMA	satellite-switched TDMA
SSTV	slow-scan television
STM-1	synchronous transmission module, level one
STS	synchronous transport signal
SVC	switched virtual circuit
S-video	separate luminance and chrominance
SWR	standing wave ratio

T

TACS	Total Access Communication System (U.K. analog)
T&M	test and measurement
TBR	Technical Basis for Regulation (European TETRA Standards)
TC	temperature coefficient
TCR	temperature coefficient of resistance
TCVCXO	temperature-compensated voltage-controlled crystal oscillator
TCXO	temperature-compensated crystal oscillator
TDD	time-division duplex
TDM	time-division multiplex
TDMA	time-division multiple access
TDR	time-domain reflectometer
TEM	transverse electromagnetic
TETRA	trans-European trunked radio system (for public service applications)
THD	total harmonic distortion
3D	three-dimensional
3G	the third generation in wireless connectivity
TIA	Telecommunications Industry Association
TIMS	transmission-impairment measurement set (an interface for PCM)
TQFP	thin-quad flat pack
T/R	transmit/receive
TSS	tangential signal sensitivity
TSSOP	thin-shrink small-outline package
TT&C	telemetry, tracking, and control (*or* command)
TTC&M	telemetry, tracking, control, and monitoring
T-3	ATM industry standard—44.736 Mb/s
TTL	transistor-transistor logic
TVI	television interference
TVRO	television receive only
2D	two-dimensional
TWT	traveling-wave tube
TWTA	traveling-wave-tube amplifier

U

UART universal asynchronous receiver-transmitter
UDLT universal digital-loop transceiver
UHF ultra-high frequency
U-NII unlicensed national information infrastructure
USB universal serial bus
UTOPIA Universal Test and Operations Physical-Layer Interface for ATM
UTP unshielded twisted pairs

V

VA voltampere
VAR value-added resellers
VCC virtual channel connection
VCI virtual channel identifier
VCO voltage-controlled oscillator
VCSEL vertical cavity surface-emitting laser
VCXO voltage-controlled crystal oscillator
VDSL very-high data-rate digital subscriber line
V/F voltage-to-frequency
VGA video graphics array
VHF very-high frequency
V/I voltage/in-current
VLB vesa local bus
VLF very-low frequency
VNA vector network analyzer
VPC virtual path connection
VPI virtual path identifier
VPSK variable phase-shift keying
VSAT very-small-aperture terminal
VSB vestigial sideband (modulation)
VSWR voltage-standing-wave ratio
VVA voltage-variable attenuator

W

WAN wide-area network
WAP wireless application protocol
W-CDMA wideband code division multiple access
WCPE wireless customer premises equipment
WDM wavelength-division multiplex(er)
WLAN wireless local-area network
WLL wireless local loop
WML wireless markup language
WTLS wireless transport layer security
WTP wireless transaction protocol

X

xDSL a generic type of digital subscriber line
X.25 a packet-switched protocol designed for data transmission over analog lines

Y

YAG yttrium-aluminum garnet
YIG yttrium-iron garnet

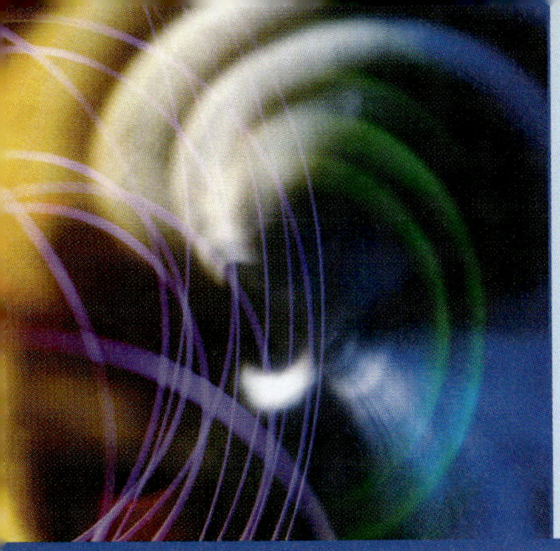

Glossary

acoustic coupler supports a telephone handpiece and uses sound transducers to send and receive audio tones

acquisition time amount of time it takes for the hold circuit to reach its final value

ACR manufacturer combined measurement of attenuation and crosstalk. A large ACR indicates greater bandwidth

AC3 the Dolby Laboratory's audio compression technique for digital television

ADSL provision of up to 1.544 Mbps from the user to the service provider and up to 8 Mbps back to the user from the service provider

advanced mobile phone service (AMPS) cellular mobile radio that uses 12-kHz peak deviation channels, which are spaced 30-kHz apart in the 800–900-MHz band

Advanced Television Systems Committee (ATSC) developed to make recommendations for advanced television in the United States

air interface used by PCS systems to manage the transfer of information

algorithms a plan or set of instructions to achieve a specific goal

alias frequency an undesired frequency produced when the Nyquist sampling rate is not attained

aliasing errors that occur when the input frequency exceeds one-half the sample rate

aliasing distortion the distortion that results if Nyquist criteria are not met in a digital communications system using sampling of the information signal; the resulting alias frequency equals the difference between the input intelligence frequency and the sampling frequency

AMI alternate mark inversion

amplitude companding process of volume compression before transmission and volume expansion after detection

amplitude compandored single sideband (ACSSB) sideband transmission with speech compression in the transmitter and speech expansion in the receiver

amplitude modulation (AM) the process of impressing low-frequency intelligence onto a high-frequency carrier so that the instantaneous changes in the amplitude of the intelligence produce corresponding changes in the amplitude of the high-frequency carrier

anechoic chamber a large enclosed room that prevents reflected electromagnetic waves and shields out interfering waves from the outside world; used for radiation measurements

angle modulation superimposing the intelligence signal on a high-frequency carrier so that its phase angle or frequency is altered as a function of the intelligence amplitude

antenna a device that generates and/or collects electromagnetic energy

antenna array group of antennas or antenna elements arranged to provide the desired directional characteristics

antenna coupler an impedance matching network in the output stage of an RF amplifier or transmitter that ensures maximum power is transferred to the antenna by matching the input impedance of the antenna to the output impedance of the transmitter

antenna gain a measure of how much more power in dB an antenna will radiate in a certain direction with respect to that which would be radiated by a reference antenna, i.e., an isotropic point source or dipole

antialiasing filter a sharp-cutoff low-pass filter used to make sure no frequencies above one-half the sampling rate reach the ADC converter

aperture time the time that the S/H circuit must hold the sampled voltage

apogee farthest distance of a satellite's orbit to earth

Armstrong transmitter FM transmitter that uses a phase modulator to feed the intelligence signal through a low-pass filter integrator network to convert PM to FM

array a group of antennas or antenna elements arranged to provide the desired directional characteristics

ASCII (American Standard Code for Information Interchange) a standardized coding scheme for alphanumeric symbols

aspect ratio in a television picture, the ratio of frame width distance to frame height distance; in the United States, it is standardized at 4/3, HDTV is at 16/9

asymmetric operation a term used to describe the modem connection when the data transfer rates to and from the service provider differ

asynchronous a mode of operation implying that the transmit and receiver clocks are not locked together and the data must provide start and stop information to lock the systems together temporarily

asynchronous system the transmitter and receiver clocks free-run at approximately the same speed

asynchronous transfer mode (ATM) a cell relay network designed for voice, data, and video traffic

ATM asynchronous transfer mode; a cell relay network designed for voice, data, and video traffic

atmospheric noise external noise caused by naturally occurring disturbances in the earth's atmosphere

ATSC pilot carrier provides a clock for the 8VSB receiver to lock onto

attenuation the loss of power as a signal propagates through a medium such as copper, fiber, and free space

attenuation distortion in telephone lines, the difference in gain at some frequency with respect to a reference tone of 1004 Hz

aural signal sound or audio portion of a TV signal, transmitted by frequency modulation

autodyne mixer another name for self-excited mixer

automatic frequency control negative feedback control system in FM receivers used to achieve stability of the local oscillator

automatic gain control (AGC) the function in a receiver that allows weak RF signals to be amplified to a high degree and strong RF signals to be amplified to a lesser degree so that a near constant output level is produced

auxiliary AGC diode reduces receiver gain for very large signals

backhoe fading total loss of data flow because the cable was dug up by a backhoe

back porch interval just after the horizontal sync pulse appears on the blanking pulse in a TV receiver

backward-wave oscillator TWT that allows both forward and backward waves and can therefore be used as an oscillator

balanced line the same current flows in each of two wires but 180° out of phase

balanced mode neither wire in the wire pairs connects to ground

balanced modulator modulator stage that mixes intelligence with the carrier to produce both sidebands with the carrier eliminated

balanced ring modulator balanced modulator design that connects four matched diodes in a ring configuration

balanced transmission line parallel conductors, such as open wire feedline, that carry two equal but opposite phase electrical signals with respect to ground

baluns circuits that convert between balanced and unbalanced operation

Barkhausen criteria two requirements for oscillations: loop gain must be at least unity and loop phase shift must be zero degrees

baseband the signal is transmitted at its base frequency with no modulation to another frequency range

base modulation a modulation system in which the intelligence is injected into the base of a transistor

Baudot code fairly obsolete coding scheme for alphanumeric symbols

baud rate symbol rate

BCC block check code, the code generated when creating the CRC transmit code

beamwidth the angular separation between the half-power points on an antenna's radiation pattern

B8ZS bipolar 8 zero substitution

Bessel functions mathematical functions used for determining the exact bandwidth of an FM signal

binary phase-shift keying (BPSK) a form of phase shift keying in which the binary "1" and "0" states are represented as no phase shift or phase inversion of the carrier signal

biphase code an encoding format for PCM systems that is popular for use with optical systems, satellite telemetry links, and magnetic recording systems

bipolar coding successive 1s are represented by pulses in the opposite voltage direction

bipolar violation the pulse is in the same voltage direction as the previous pulse

bit unit of information required to allow proper selection of one out of two equally probable events

bit error rate similar to error probability; the number of bit errors that occur for a given number of bits transmitted

bit stuffing another name for character insertion

block check character method of error detection involving sending a block of data, then an end of message indicator, then a block check character representing characteristics of the data that was sent

BORSCHT the functions produced on subscriber loop interface circuits in PBX or central office systems; these functions are Battery feeding, Overvoltage protection, Ringing, Supervision, Coding, Hybrid, and Testing

bps bits per second

broadcast address setting the destination MAC address to all 1s broadcasts the message to all computers on the LAN

bucket brigade the process of serially shifting data out of a CCD

buffer amplifier an amplifier designed to prevent any amplitude or frequency loading from occurring; it typically has a very high input impedance and low input capacitance to remove any amplitude reduction or frequency drifting of the desired signal being amplified

bursty a state in which the data rates can momentarily exceed the leased data rate of the service

Butterworth filter a constant-*k* type of *LC* filter

cable modems use of the high bandwidth of the cable television system to deliver high-speed data to and from the service provider

cables optical fibers that have a protective covering

capture effect an FM receiver phenomenon that involves locking onto the stronger of two received signals of the same frequency and suppressing the weaker signal

capture range the range of applied input frequencies to a PLL that will cause it to lock up

capture ratio the necessary difference (in dB) of signal strength to allow suppression of a weaker signal from a stronger one in FM systems

capture state when the phase comparator of a PLL generates a signal that forces the VCO to equal the input frequency

carrier a radio wave of constant amplitude, frequency, and phase at the frequency of operation for the radio, television, or other type of communication system; this radio wave's amplitude, frequency, or phase is altered by an information signal so that it can carry the information to a distant receiver

carrier leakthrough the amount of carrier not suppressed by the balanced modulator

carrier sense, multiple access with collision detection a channel access method where users "listen" for an opening to make a transmission

Carson's rule equation for approximating the bandwidth of an FM signal

Cassegrain feed a method of feeding a paraboloid antenna by using a secondary reflector

cathode ray tube (CRT) a vacuum tube in which the electron can be focused in a small spot on a fluorescent screen at the opposite end of the structure; used in television receivers and oscilloscopes to form the display

CAT6/5e category 6 and category 5e computer networking cable capable of handling a 1000 MHz bandwidth up to a length of 100 m

cavity resonator metal-walled chambers in microwave installations fitted with devices for admitting and extracting electromagnetic energy

CDMA code division multiple access; each station uses a different binary sequence to modulate the same carrier

CDMA2000 a 3G wireless development popular in the United States

cell sites a regular array of transmitter-receiver stations

cell splitting for cellular phones, if all traffic in a given cell increases beyond a reasonable capacity, the cell is split into smaller coverage areas

cellular telephone modern mobile telephone system characterized by a network of cell sites, each serving a hexagon-shaped coverage area, and automatic switching of service from one cell site to another as the mobile customer travels from one coverage area to another

centralized network a basic network configuration

central office a telephone exchange on phone company property that simply switches one telephone line to another so that phone calls can be made

ceramic filter a filter network made from lead zirconate-titanate that exhibits the piezoelectric effect and makes effective filters to convert DSB-SC to SSB in an SSB transmitter

channel a band of frequencies

channel access how a user gets control of a channel to allow transmission

channel guard in a transceiver, causing a specific audio frequency to be encoded onto the carrier together with the audio

character insertion insertion of a bit or character so that a data stream is not mistaken for a control character

characteristic impedance the input impedance of a transmission line either infinitely long or terminated in a pure resistance exactly equal to its characteristic impedance

characteristic wave impedance characteristic impedance of a waveguide; affected by the frequency of operation

character stuffing another name for character insertion

charge couple device (CCD) a light-sensitive device used to convert optical images to an electronic form

chips in the transmission of digital bits, pulses shorter than the message bits

chroma the color signals I and Q

chromatic dispersion broadening of a pulse due to the different propagation rates of the spectral components of the light

circular horn type of horn antenna that provides radiation from a circular waveguide

citizen's band a radio service for personal communication; often used by truck drivers during interstate highway travel

cladding the material surrounding the core of an optical waveguide; the cladding must have a lower index of refraction to keep the light in the core

Clapp oscillator a Colpitts oscillator having a third capacitor in series with its inductor in order to produce a more stable output frequency

clipper another name for sync separator

coaxial cable two conductors, a center conductor and an outer shield, separated by a dielectric at a fixed distance from one another; provides for minimal noise pickup and minimal radiation

codec a single LSI chip containing both the ADC and DAC circuitry

code-division multiple-access (CDMA) communications system in which spread-spectrum techniques are used to multiplex more than one signal within a single channel

coding transforming messages or signals in accordance with a definite set of rules

coefficient of reflection ratio of the reflected electric field intensity divided by the incident intensity

coherent light that is spectrally pure

collinear array any combination of half-wave elements in which all the elements are excited by a connected transmission line

color burst eight-cycle sine-wave burst that occurs on the back porch of the horizontal sync pulse in a color TV broadcast signal

color killer circuit that prevents output from the chroma circuits during a monochrome broadcast

Colpitts oscillator a popular *LC* oscillator, easily recognized by its two capacitors providing the positive feedback path for oscillation

committed burst information rate (CBIR) enables subscribers to exceed the committed information rate (CIR) during times of heavy traffic

committed information rate the guaranteed data rate or bandwidth to be used in the frame relay connection

common mode rejection when signals that are 180° out of phase cancel each other

compandor compress/expand; to provide better noise performance, a variable-gain circuit at the transmitter increases its gain for low-level signals; a complementary circuit in the receiver reverses the process to restore the original signal

confetti colored noise on the screen of a color receiver during a black-and-white transmission

constant-*k* filter filter whose capacitive and inductive reactances are equal to a constant value *k*

constellation pattern display used to monitor QAM data signals on an oscilloscope to provide information on linearity and noise

continuously variable slope delta (CVSD) increasing the step-size in a delta modulation system when three or more consecutive ones or zeros occur

continuous wave a type of transmission where a continuous sinusoidal waveform is interrupted to convey information

convergence in a multibeam cathode-ray tube, a condition in which the beams are adjusted so that they all cross at a specific point; in color television, an alignment procedure used to form the clearest image

conversion frequency another name for the carrier in a balanced modulator

converters another name for mixers

converter stage a stage that serves the purpose of converting one frequency into another

core the portion of the fiber strand that carries the light

corona discharge luminous discharge of energy by an antenna caused by ionization of the air around the surface of the conductor

cosmic noise space noise originating from stars other than the sun

counterpoise reflecting surface of a monopole antenna if the actual earth ground cannot be used; a flat structure of wire or screen placed a short distance above ground with at least a quarter-wavelength radius

critical angle the highest angle with respect to a vertical line at which a radio wave of a specified frequency can be propagated and still be returned to the earth from the ionosphere

critical frequency the highest frequency that will be returned to the earth when transmitted vertically under given ionospheric conditions

critical value in a magnetron, when the magnetic field reaches a value great enough to cause electrons to just miss the plate and return to the filament in a circular orbit

Crosby systems FM systems using direct FM modulation with AFC to control for carrier drift

cross-modulation distortion that results from undesired mixer outputs

crosstalk unwanted coupling caused by overlapping electric and magnetic fields

crystal filter a crystal network commonly used to provide high-*Q* filtering

crystal-lattice filter filter containing at least two but usually four crystals

crystal oscillator an oscillator that uses a piezoelectric crystal in place of the inductor to produce a stable output frequency

CSMA/CA carrier sense multiple access with collision avoidance; the protocol used by wireless LANs

CSMA/CD the Ethernet LAN protocol carrier sense multiple access with collision detection

CSU/DSU channel service unit/data service unit, providing the data interface to the communications carrier providing framing and line management

cyclic prefix the end of a symbol is copied to the beginning of the data stream, thereby increasing its overall length and thus removing any gaps in the data transmission

cyclic redundancy check method of error detection involving performing repetitive binary division on each block of data and checking the remainders

damped the gradual reduction of a repetitive signal due to resistive losses

damper a diode in the high-voltage oscillator of a TV receiver that shorts out unwanted damped oscillations during the flyback period

DAMPS (digital advanced mobile phone service) the digital system for mobile phone service

dark current the very little current that flows when a *pn* junction is reverse-biased and under dark conditions

data bandwidth compression in QPSK transmission, the compression of more data into the same available bandwidth as compared to BPSK

data communications equipment refers to peripheral computer equipment such as a modem, printer, mouse, etc.

data encapsulation properly formatting the data for transport over a serial communications line

data terminal equipment refers to a computer, terminal, personal computer, etc.

dB decibel

dBd antenna gain relative to a dipole antenna

dBi antenna gain relative to an isotropic radiator

dBm a method of rating power or voltage levels with respect to 1 mW of power

dBm(50) a measurement made using a 1-mW reference with respect to a 50-Ω load

dBm(75) a measurement made using a 1-mW reference with respect to a 75-Ω load

dBm(600) decibel measurement using a 1-mW reference with respect to a 600-Ω load

dB(V) dB value measured relative to a 1-V reference

dBW a measurement made using a 1-W reference

DCCH digital control channel

dc restoration process of restoring the dc portion of the video signal that is often removed by amplifier coupling

decade a range of frequencies in which the upper limit is ten times as large as the lower limit

deemphasis process in an FM receiver that reduces the amplitudes of high-frequency audio signals down to their original values to counteract the effect of the preemphasis network in the transmitter

delay distortion when various frequency components of a signal are delayed different amounts during transmission

delayed AGC an AGC that does not provide gain reduction until an arbitrary signal level is attained

delay equalizer an *LC* filter that removes delay distortion from signals on phone lines by providing increased delay to those frequencies least delayed by the line, so that all frequencies arrive at nearly the same time

delay line a length of a transmission line designed to delay a signal from reaching a point by a specific amount of time

delay skew measure of the difference in time for the fastest to the slowest pair in a UTP cable

delta match an impedance matching device that spreads the transmission line as it approaches the antenna

delta modulation digital modulation technique in which the encoder transmits information regarding whether the analog information increases or decreases in amplitude

demodulation process of removing intelligence from the high-frequency carrier in a receiver

demultiplexer (DMUX) a device that recovers the individual groups of data from the TDMA serial data stream

dense wave division multiplexing (DWDM) incorporation of the propagation of several wavelengths in the 1550-nm range of a single fiber

despread return the DSSS signal back to its original modulated format

deviation constant definition of how much the carrier frequency will deviate for a given modulating input voltage level

deviation ratio (DR) the maximum possible frequency deviation divided by the maximum input frequency

device under test an electronic part or system that is being tested

D4 framing the original data framing used in T1 circuits

diagonal clipping distortion that occurs in a diode detector if the time-constant of the low-pass filter is set too high

dibits data sent two bits at a time

dielectric waveguide a waveguide with just a dielectric (no conductors) used to guide electromagnetic waves

differential GPS a technique where GPS satellite clocking corrections are transmitted so that the position error can be minimized

differentiator a high-pass filter

diffraction the phenomenon whereby waves traveling in straight paths bend around an obstacle

digital communication the transfer of information from transmitter to receiver by representing it in a digital format before it is transmitted and then converting it back to its original form after it is detected by the receiver

digital on-channel repeater (DOCR) a device that allows for the retransmission of the DTV signal on the same channel

digital signal processing using programming techniques to process a signal while in digital form

diode detector the simplest method for detecting an AM signal, consisting of a diode in series with a low-pass filter

diplexer filter in a TV transmitter that allows both the video AM signal and the audio FM signal to feed the same antenna

dipole antenna straight radiator one half-wavelength long, usually separated at the center by an insulator and fed by a balanced transmission line

direct digital synthesis frequency synthesizer design that has better repeatability and less drifting but limited maximum output frequencies, greater phase noise, and greater complexity and cost

directional concentrating antenna energy in certain directions at the expense of lower energy in other directions

directional coupler a device that senses how much signal is moving in one direction in a transmission line or waveguide

director the parasitic element that effectively directs energy in the desired direction

discriminator stage in an FM receiver that creates an output signal that varies as a function of its input frequency; recovers the intelligence signal

dispersion the broadening of a light pulse as it propagates through a fiber strand

dispersion compensating fiber acts like an equalizer canceling dispersion effects and yielding close to zero dispersion in the 1550-nm region

dissipation inverse of quality factor

distributed feedback laser (DFB) a more stable laser suitable for use in DWDM systems

distributed network an interconnection of more than one centralized network

dit decimal digit

diversity reception transmitting and/or receiving several signals and either adding them together at the receiver or selecting the best one at any given instant

DMT (discrete multitone) an industry standard data-modulation technique used by ADSL that uses the multiple subchannel frequencies to carry the data

Dolby system advanced noise reduction system in FM systems in which the preemphasis and deemphasis networks work in a dynamic manner

Doppler effect phenomenon whereby the frequency of a reflected signal is shifted if there is a relative motion between the source and the reflecting object

double conversion superheterodyne receiver design with two separate mixers, local oscillators, and intermediate frequencies

double range echoes echoes produced when the reflected beam makes a second trip

double-sideband suppressed carrier output signal of a balanced modulator

double-stub tuner has fixed stub locations, but the position of the short circuits is adjustable to allow a match between line and load

downlink a satellite sending signals to earth

downward modulation the decrease in dc output current in an AM modulator usually caused by low excitation

driven array multi-element antenna in which all the elements are excited through a transmission line

driver amplifier amplifier stage that amplifies a signal prior to reaching the final amplifier stage in a transmitter

DSL a digital subscriber line

DSSS direct sequence spread spectrum

DTV digital television

dual-band a phone operating in two different bands

dummy antenna resistive load used in place of an antenna to test a transmitter without radiating the output signal

DUT device under test

duty cycle the ratio of pulse width to pulse repetition time

dwell time the time each carrier spends at a specific frequency

dynamic convergence beam convergence away from the center of a CRT

dynamic range in a PCM system, the ratio of the maximum input or output voltage level to the smallest voltage level that can be quantized and/or reproduced by the converters; for a receiver, the decibel difference between the largest tolerable receiver input level and its sensitivity (smallest useful input level)

EBCDIC standardized coding scheme for alphanumeric symbols

E_b/N_o the bit energy to noise ratio

echo signal part of the returning radar energy collected by the antenna and sent to the receiver

8VSB the RF modulation technique for ATSC DTV transmission

electrical length the length of a line in wavelengths, not physical length

electrical noise any undesired voltages or currents that end up appearing in a circuit

electromagnetic interference unwanted signals from devices that produce excessive electromagnetic radiation

energy per bit, or bit energy amount of power in a digital bit for a given amount of time for that bit

envelope detector another name for diode detector

equivalent noise resistance used by some manufacturers of microwave devices to represent how noisy a device is by comparing its noise to the resistance value that would produce the same amount of noise

equivalent noise temperature a method of representing how noisy a microwave device actually is

error probability in a digital system, the number of errors per total number of bits received

error voltage output of the phase comparator in a PLL

ESF extended superframe framing

evanescent field the field outside an optical fiber's core–cladding boundary

event a disturbance in the light propagating down a fiber span, which results in a disturbance on the OTDR trace

excess loss a measure of added losses in addition to the desired splitting ratio in an optical coupler

excess noise noise occurring at frequencies below 1 kHz, varying in amplitude and inversely proportional to frequency

exciter stages necessary in a transmitter to create the modulated signal before subsequent amplification

external noise noise in a received radio signal that has been introduced by the transmitting medium

eye patterns using the oscilloscope to display overlayed received data bits that provide information on noise, jitter, and linearity

facsimile system of transmitting images in which the image is scanned at the transmitter, reconstructed at the receiving end, and duplicated on paper

fading variations in signal strength that may occur at the receiver over a period of time

Faraday rotation effect when microwaves are passed through a piece of ferrite in a magnetic field and their frequency is above the resonant frequency of the ferrite electrons, the plane of polarization of the wave is rotated

far field region greater than $2D^2/\lambda$ from the antenna; effect of induction field is negligible

fax abbreviation for facsimile

FDD frequency division duplex

feed line transmission line that transfers energy from the generator to the antenna

ferrite compounds of iron, zinc, manganese, magnesium, cobalt, aluminum, and nickel oxides used in special applications of microwave circuits

ferrite bead small bead of ferrite material that can be threaded onto a wire to form a device that offers no impedance to dc and low frequencies, but a high impedance at RF

FFT (fast Fourier transform) a technique for converting time-varying information to its frequency component

FHSS frequency hopping spread spectrum

fiber Bragg grating a short strand of fiber that changes the index of refraction and minimizes intersymbol interference

fiber, light pipe, glass common synonymous terms for a fiber-optic strand

fiber-optic detector the device in a fiber-optic system that converts the modulated light wave signal back into an electrical information signal

fiber-optic emitter the device in a fiber-optic system that converts the information signal into a modulated light wave signal

fiber-optics the use of light wave radiation to send information from a transmitter site to a receiver via fiber-optic cable

field set of lines in a TV scene

field frequency the number of times per second that a field of 242.5 horizontal lines forms a video image on a television display

filter method the method of creating SSB in a transmitter by first creating DSB-SC and then filtering out the undesired sideband

Firewire A (IEEE 1394a) a high-speed serial connection that supports data transfers up to 400 Mbps

Firewire B (IEEE 1394b) a high-speed serial connection that supports data transfers up to 800 Mbps

first detector the mixer stage in a superheterodyne receiver that mixes the RF signal with a local oscillator signal to form the intermediate frequency signal

5.1 channel input the commercial name for AC-3 audio standard

flash OFDM a spread-spectrum version of OFDM

flat line condition of no reflection; VSWR is 1

flat-top sampling holding the sample signal voltage constant during samples, creating a staircase that tracks the changing input signal

flicker motion appears jerky due to insufficient scanning frequency

flow control protocol used to monitor and control rates at which receiving devices can accept data

flyback transformer used in TV receivers to produce the high voltage needed for the CRT

flywheel effect repetitive exchange of energy in an *LC* circuit between the inductor and the capacitor

forward error-correcting error-checking techniques that permit correction at the receiver, rather than retransmitting the data

Foster-Seely discriminator an outdated FM discriminator design using two tuned *LC* networks and two diode detectors to recover the original intelligence in an FM receiver; requires a separate limiter stage but does provide very low distortion

Fourier analysis method of representing complex repetitive waveforms by sinusoidal components

4:2:2 international standard for digitizing component video

fractional T1 a term used to indicate that only a portion of the data bandwidth of a T1 line is being used

frame frequency number of times per second that a complete set of 485 horizontal lines are traced in a TV receiver

frame relay a packet switching network designed to carry data traffic over the public data network

frame sync a signal repeated once every 313 data segments to identify that a frame has been completed

framing separation of blocks of data into information and control sections

free-running frequency the frequency at which the PLL runs with the input signal removed

free-running state the undesired unstable operating mode of a PLL (when it is not locked up)

frequency deviation amount of carrier frequency increase or decrease around its center reference value

frequency-division multiple-access operating on different frequencies based on which channel is available

frequency-division multiplexing simultaneous transmission of two or more signals on one carrier, each on its own separate frequency range; also called frequency multiplexing

frequency domain record the data points generated by the time to frequency conversion using the FFT

frequency hopping spread spectrum transmitting data by a carrier that is switched in frequency in a pseudorandom fashion

frequency modulation superimposing the intelligence signal on a high-frequency carrier so that the carrier's frequency departs from its reference value by an amount proportional to the intelligence amplitude

frequency multiplexing process of combining signals that are at slightly different frequencies to allow transmission over a single medium

frequency multipliers amplifiers designed so that the output signal's frequency is an integer multiple of the input frequency

frequency reuse in cellular phones, the process of using the same carrier frequency in different cells that are geographically separated

frequency shift keying a form of data transmission in which the modulating wave shifts the output between two predetermined frequencies

frequency synthesizer oscillator that generates a wide range of output frequencies using one reference crystal

Friiss's formula method of determining the total noise produced by amplifier stages in cascade

front end the first amplifier stage of a receiver that receives its input signal from the antenna

front porch interval before the horizontal sync pulse appears on the blanking pulse in a TV receiver

front-to-back ratio the difference in antenna gain in dB from the forward to the reverse direction

FTTC fiber to the curb

FTTH fiber to the home

fusion splicing a long-term splicing method where the two fibers are fused or welded together

generating polynomial defines the feedback paths to be used in the CRC generating circuit

geosynchronous orbit another name for synchronous orbit

ghosting when the same signal arrives at the TV receiver at two different times; the reflected signal has farther to travel and is weaker than the direct signal, resulting in a double image

glass, fiber, light pipe common synonymous terms for fiber-optic strand

GPIB general-purpose interface bus; another name for the IEEE-488 interface standard

graded-index fiber the index of refraction is gradually varied with a parabolic profile; the highest index occurs at the fiber's center

Gray code numeric code for representing decimal values from 0 to 9

grid-dip meter device that measures the resonant frequency of tuned circuits and antennas without power being applied to them

ground wave radio wave that travels along the earth's surface

GSM the global system for mobile communications

guard bands 25-kHz bands at each end of a broadcast FM channel to help minimize interference with adjacent stations

guard times time added to the TDMA frame to allow for the variation in data arrival

Gunn oscillator solid-state bulk-effect source of microwave energy due to the excitation of electrons in the crystal to energy states higher than those they normally occupy

half-wave antenna an antenna whose receive elements are one half-wavelength in length

Hamming code a forward error-checking technique named for R. W. Hamming

Hamming distance the logical distance between defined states; also called minimum distance and Dmin

handoff the process of changing channels to a new cell site

handshaking procedures allowing for orderly exchange of information between a central computer and remote sites

harmonics sinusoidal waves whose frequencies are a multiple of the fundamental frequency

Hartley oscillator a popular *LC* oscillator, easily recognized by its inductor that is tapped to form positive feedback

Hartley's law information that can be transmitted is proportional to the product of the bandwidth times the time of transmission

HDLC high-level data link control; a synchronous proprietary protocol

HDTV high-definition television

heterodyne detector another name for synchronous detector or product detector

heterojunction a junction of two dissimilar semiconductors

high-definition television (HDTV) a new standard being developed that will offer a television picture with the same resolutions as motion picture presentations

high-level modulation in an AM transmitter, intelligence superimposed on the carrier at the last possible point before the antenna

hit when two spread-spectrum transmitters momentarily transmit at the same frequency; they coincide at that instant

hold-in range the range of frequencies in which the PLL will remain locked

hopping sequence the order of frequency changes

horizontal resolution number of vertical lines that can be resolved in a TV display

horizontal retrace in TV, the amount of time it takes to move the electron beam from the right back to the left to start a new line

horn antenna microwave antenna consisting of a waveguide that gradually flares out to allow for maximum radiation and minimum reflection back into the guide

hot-swappable a term used to describe that an external device can be plugged in or unplugged at any time

human-made noise external noise produced by human-made devices that is often due to inherent spark-producing mechanisms

IANA the agency that assigns the computer network IP address

iconoscope an early, now obsolete, TV camera design

idle channel noise small-amplitude signal that exists due to the noise in the system

IEEE-488 Interface a standard used in the transmission of parallel data signals from one device to another

image antenna the simulated $\lambda/4$ antenna resulting from the earth's conductivity with a monopole antenna

image frequency undesired input frequency in a superheterodyne receiver that produces the same intermediate frequency as the desired input signal

image orthicon an early TV camera

IMPATT diode *imp*act *i*onization *a*valanche *t*ransit *t*ime; used in the generation of microwave signals

IMT-2000 international mobile telecommunications; the standard defining 3G wireless

independent sideband transmission another name for twin-sideband suppressed carrier transmission

index-matching gel a jellylike substance that has an index of refraction much closer to the glass than air

induction field radiation that surrounds an antenna and collapses its field back into the antenna

information theory the branch of learning concerned with optimization of transmitted information

infrared light extending from 680 nm up to the wavelengths of the microwaves

input intercept another name for third-order intercept point

insertion loss the attenuation of a signal within its specified bandwidth

integrator a low-pass filter

intelligence low-frequency information that can be modulated onto a high-frequency carrier in a transmitter

intelligence signal low-frequency information modulated onto a high-frequency carrier in a transmitter

interaction space in a magnetron, the open space between the plate and the cathode where the electric and magnetic fields exert force on the electrons

intercarrier systems TV receivers that process sound and video signals within the same IF amplifier stages

interlaced scanning interleaving two fields of 242.5 horizontal lines to form a video image of 485 horizontal lines, so the human eye thinks it is seeing 60 pictures per second

interleaving generating color information around just the right frequency so that it becomes centered in clusters between the black-and-white signals

intermod intermodulation distortion

intermodulation distortion undesired mixing of two signals in a receiver resulting in an output frequency component equal to that of the desired signal

internal noise noise in a radio signal that has been introduced by the receiver

interrupted continuous wave a more accurate name for a continuous wave transmission

intersymbol interference (ISI) the overlapping of data bits that can increase the bit error rate

IP telephony (voice-over IP) the telephone system for computer networks

ISM industrial, scientific, and medical

isolators in-line passive devices that allow power to flow in one direction only

isothermal region the stratosphere, considered to have a constant temperature

isotropic point source a point in space that radiates electromagnetic radiation equally in all directions

jitter the undesired shift or width change in digital bits of data

Johnson noise another name for thermal noise, first studied by J. B. Johnson

keying ensuring that an oscillator starts by turning the dc on and off

klystron an electron tube used at microwave frequencies; it consists of cavity resonators and uses velocity modulation of an electron stream flowing from a heated cathode to a collector anode

laser low-noise light wave generator

latency the time delay from the request for information until a response is obtained

lattice modulator another name for balanced ring modulator

leakage loss of electrical energy between the plates of a capacitor

light pipe, glass, fiber common synonymous terms for a fiber-optic strand

limiter stage in an FM receiver that removes any amplitude variations of the received signal before it reaches the discriminator

limiting knee voltage another term for quieting voltage

linear quantization level another name for uniform quantization level

line control procedure that decides which device has permission to transmit at a given time

line-hybrid transformer permits full-duplex operation by providing isolation between the transmit and receive legs of the phone system

loaded cable cable with added inductance every 6000, 4500, or 3000 feet

loading coil a series inductance used to tune out the capacitive appearance of an antenna or transmission line

lobes small amounts of RF radiation shown on a radiation pattern; generally undesirable

local area network network of users that share computers in a limited area

local loop another name for connection from the central office to the end user

local oscillator (LO) an oscillator used in a superheterodyne receiver to generate a signal to mix with the received RF signal in order to generate a constant intermediate frequency

local oscillator reradiation undesired radiation of the local oscillator signal through a receiver's antenna

locked a PLL in the capture state

lock range range of frequencies over which PLL can track an input signal and remain locked

log periodic antenna a number of dipoles of different lengths and spacing designed to achieve a fairly constant gain and match over a wide range of frequencies

long haul the intercity or interoffice class of system used by telephone companies and long-distance carriers

longitudinal redundancy check extending parity into two dimensions

loop antenna an antenna consisting of a single or multi-turn of wire with dimensions much smaller than a wavelength that provides a sharply bidirectional radiation pattern; often used in direction-finding applications

loopback test configuration for a data link; the receiver takes the data and sends it back to the transmitter, where it is compared with the original data to indicate system performance; also, when data are routed back to the sender

loop gain the total gain of all internal blocks in a PLL

lower sideband band of frequencies produced in a modulator from the creation of difference frequencies between the carrier and information signals

low excitation improper bias or low carrier signal power in an AM modulator

low-level modulation in an AM transmitter, intelligence superimposed on the carrier; then the modulated waveform is amplified before reaching the antenna

low-noise resistor a resistor that exhibits low levels of thermal noise

luminance the Y signal

MAC address a unique 6-byte address assigned by the vendor of the network interface card

macrobending loss due to the light escaping into the cladding

magnetron an electron tube that is surrounded by an electronmagnet that controls the electron flow from cathode to anode; used to generate microwave frequencies in a radio transmitter

Manchester code a popular name for the biphase L-code used on Ethernet systems for local area networks

Marconi antenna another name for vertical antenna; usually a quarter-wave grounded antenna

mark, space analog signal representations of digital high or low states, respectively; usually sine waves of specific frequencies

maser a low-noise microwave amplifier, similar to a laser that is used with light

material dispersion the spreading of light in fiber-optic cable caused by the slight variation of refractive index with wavelength for glass

matrix transmitter signal processing circuits

matrix network adds and/or subtracts and/or inverts electrical signals

maximal length indicates that the PN code has a length of $2^n - 1$

maximum usable frequency the highest frequency that is returned to the earth from the ionosphere between two specific points on earth

maximum usable range the maximum distance before second return echoes start occurring in radar

***m*-derived filter** filter that uses a tuned circuit to provide nearly infinite attenuation at a specific frequency

mechanical filter a mechanically resonant device that is often used as a sharp bandpass filter to convert DSB to SSB in an SSB transmitter

mechanical splices splices joining two fibers together with an air gap, thereby requiring an index matching gel to provide a good splice

metropolitan area network two or more LANs linked together in a limited geographical area

microbending loss caused by very small mechanical deflections and stress on the fiber

microbrowser analogous to a web browser that has been adapted for the wireless environment

microcellular systems another name for personal communication networks

microcontrollers microprocessors that are programmed to do a specific task, such as instrument control

microstrip transmission line used at microwave frequencies that has one or two conducting strips over a ground plane

microwave frequencies above 1 GHz having wavelengths between 1 mm and 30 cm; any radio equipment and antennas associated with these frequencies of operation

microwave dish paraboloid antenna

microwave monolithic integrated circuit (MMIC) integrated circuits that are used at microwave frequencies

Miller code another name for biphase codes in PCM systems

millimeter waves microwave frequencies above 40 GHz; wavelength is often expressed in millimeters

minimum distance (D_{min}) the minimum distance between defined logical states

minimum ones density a pulse is intentionally sent in the data stream even if the data being transmitted is a series of 0s only

modal dispersion the different paths taken by the various propagation modes in fiber-optic cable

mode field diameter the actual guided optical power distribution, which is typically about 1 μm or so larger than the core diameter; single-mode fiber specification sheets typically list the mode field diameter

modem device that converts digital data to an analog signal for transmission and converts the received analog signal to a digital signal

modulated amplifier stage that generates the AM signal

modulation impressing a low-frequency intelligence signal onto a higher-frequency carrier signal

modulation factor another name for modulation index

modulation index measure of the extent to which a carrier is varied by the intelligence

monochrome black-and-white TV

monopole antenna usually a quarter-wave grounded antenna

MPEG2 a video-compression technique used in DTV transmission

multilevel binary codes that have more than two levels representing the data

multimode fibers fibers with cores of about 50 to 100 μm that support many modes; light takes many paths

multiple access any method of multiplexing many signals in one communications channel

multiplexing simultaneous transmission of two or more signals in a single medium

multiplex operation simultaneous transmission of two or more signals in a single medium

multipoint circuits systems with three or more devices

multitone modulation another name for orthogonal frequency division multiplexing

muting the squelch capability of better-quality broadcast FM receivers

NAMPS (narrowband mobile phone service) a system that provides triple the capacity of the AMPS system

narrowband FM FM signals used for voice transmissions such as public service communication systems

natural sampling when the tops of the sampled waveform or analog input signal retain their natural shape

near-end crosstalk (NEXT) a measure of the level of crosstalk or signal coupling within the cable; a high NEXT (dB) value is desirable

near field region less than $2D^2/\lambda$ from the antenna

network an interconnection of users that allows communication among them

network interface card (NIC) the electronic hardware used to interface the computer to the network

neutralization a procedure for tuning up a transmitter in which a negative feedback path is introduced in order to counteract the tendency for an amplifier to self-oscillate due to positive feedback in the semiconductor junctions or tube's interelectrode capacitances

neutralizing capacitor a capacitor that cancels fed-back signals to suppress self-oscillation

(n, k) cyclic code nomenclature used to identify cyclic codes in terms of their transmitted code length (n) and message length (k)

noise figure a figure describing how noisy a device is in decibels

noise floor the baseline on a spectrum analyzer display, representing input or output noise of the system under test

noise limiter a circuit that cuts off all noise pulse peaks that exceed the highest peaks of the desired signal in a receiver

noise ratio a figure describing how noisy a device is as a ratio having no units

nonlinear coding each quantile interval step-size may vary in magnitude

nonlinear device characterized by a nonlinear output versus input signal relationship

nonresonant line one of infinite length or that is terminated with a resistive load equal in ohmic value to its characteristic impedance

nonuniform coding another name for nonlinear coding

normalizing dividing impedances by the characteristic impedance

NRZ code (nonreturn to zero) a popular encoding format for digital systems

null a direction in space with minimal signal level

numerical aperture a number less than 1 that indicates the range of angles of light that can be introduced to a fiber for transmission

Nyquist rate the sampling frequency must be at least twice the highest frequency of the intelligence signal or there will be distortion that cannot be corrected by the receiver

OC optical carrier

OC-1 optical carrier level 1, which operates at 51.84 Mbps

octave range of frequency in which the upper frequency is double the lower frequency

O-, E-, S-, C-, L-, and U-bands new optical band designations that have been proposed

omnidirectional a spherical radiation pattern

100BaseT 100-Mbps baseband data over twisted-pair cable

open systems interconnection reference model to allow different types of networks to be linked together

open-wire cable two conductors spaced a fixed distance apart from each other that connect a transmitter/receiver to an antenna

optical spectrum light frequencies from the infrareds on up

optimum working frequency the frequency that provides for the most consistent communication path via sky waves

orthogonal two signals are orthogonal if the signals can be sent over the same medium without interference

orthogonal frequency division multiplexing (OFDM) a technique used in digital communications to transmit the data on multiple carriers over a single communications channel

oscillator circuit capable of converting electrical energy from dc to ac

OTA over the air

OTDR optical time-domain reflectometer; an instrument that sends a light pulse down the fiber and measures the reflected light, which provides some measure of performance for the fiber

overmodulation when an excessive intelligence signal overdrives an AM modulator producing percentage modulation exceeding 100 percent

packets segments of data

packet switching packets are processed at switching centers and directed to the best network for delivery

padder capacitor small variable capacitor in series with each ganged tuning capacitor in a superheterodyne receiver to provide near-perfect tracking at the low end of the tuning range

parabolic antenna microwave antenna consisting of a paraboloid-shaped sheet of metal that reflects received energy to a single point called the focal point

parametric amplifier provides low-noise amplification at microwave frequencies using the variation of reactance

parasitic array when one or more of the elements in an antenna array is not electrically connected

parasitic element nondriven element of an antenna

parasitic oscillations undesired higher frequency self-oscillations in amplifiers

parity a common method of error detection, adding an extra bit to each code representation to give the word either an even or odd number of 1s

patch antenna square or round "island" of a conductor on a dielectric substrate backed by a conducting ground plane

payload another name for the data being transported

PCS (personal communications services) a 1900-MHz mobile phone service that has enhanced features such as messaging, paging, and data service

peak envelope power method used to rate the output power of an SSB transmitter

peak power the useful power of the transmitter contained in the radiated pulses

peak-to-valley ratio another name for ripple amplitude

percentage modulation measure of the extent to which a carrier voltage is varied by the intelligence for AM systems

perigee closest distance of a satellite's orbit to earth

persistence length of time an image stays on the screen after the electrical signal is removed

personal communications network system that permits communication from small portable radios on microwave frequencies

phase comparator circuit that provides an output proportional to the phase difference of two inputs

phased array combination of antennas in which there is control of the phase and power of the signal applied at each antenna resulting in a wide variety of possible radiation patterns

phase detector another term for phase comparator

phase-locked loop closed-loop control system that uses negative feedback to maintain constant output frequency

phase method a method of creating SSB in a transmitter without the need for high-Q bandpass filters

phase modulation superimposing the intelligence signal on a high-frequency carrier so that the carrier's phase angle departs from its reference value by an amount proportional to the intelligence amplitude

phase noise spurious changes in the phase of a frequency synthesizer's output that produce frequencies other than the desired one

phase shift keying method of data transmission in which data causes the phase of the carrier to shift by a predefined amount

phasing capacitor cancels the effect of another capacitance by a 180° phase difference

piezoelectric effect the property exhibited by crystals that causes a voltage to be generated when they are subject to mechanical stress and, conversely, a mechanical stress to be produced when they are subjected to a voltage

pilot carrier a reference carrier signal

***p-i-n* diodes** diodes used as RF and microwave switches and as AM modulators that consist of *p*-type, *i*ntrinsic (lightly doped), and *n*-type material

pixel picture element; the smallest resolved area in a video scanning technique

pixelate occurs when a digital picture freezes even when there is motion in the video; usually due to poor SNR

PN sequence length the number of times a PN generating circuit must be clocked before repeating the output data sequence

point of presence the point where the user connects data to the communications carrier

polarization the direction of the electric field of an electromagnetic wave

polarization dispersion broadening of the pulse due to the different propagation velocities of the X and Y polarization components

polar pattern a circular graph that indicates the direction of antenna radiation

poles number of RC or LC sections in a filter

power-sum NEXT testing (PSNEXT) a measure of the total crosstalk of all cable pairs to ensure that the cable can carry data traffic on all four pairs at the same time with minimal interference

ppm (parts per million) preferred method for rating the frequency stability of crystals

PPP point-to-point protocol

precession movement of the axis of rotation at right angles to its original axis

preemphasis process in an FM transmitter that amplifies high frequencies more than low-frequency audio signals to reduce the effect of noise

preferred roaming list (PRL) the roaming list of available cell towers

preselector the tuned circuits prior to the mixer in a superheterodyne receiver

private branch exchange (PBX) a telephone exchange on the user's premises that simply switches one telephone line to another so that phone calls can be made

product detector oscillator, mixer, and low-pass filter stage used to obtain the intelligence from an AM or SSB signal

protocols set of rules to allow devices sharing a channel to observe orderly communication procedures

pseudonoise (PN) codes digital codes with pseudorandom output data streams that appear to be noiselike

pseudorandom the number sequence appears random but actually repeats

public data network the local telephone company or a communications carrier

pulse-amplitude modulation sampling short pulses of the intelligence signal; the resulting pulse amplitude is directly proportional to the intelligence signal's amplitude

pulse code modulation (PCM) most common technique for converting an analog signal into a digital word; consists of a sample-hold circuit followed by the actual analog-to-digital converter circuit

pulse dispersion a stretching of received pulse width because of the multiple paths taken by the light

pulse-duration modulation another name for pulse-width modulation

pulse-length modulation another name for pulse-width modulation

pulse modulation the process of using some characteristic of a pulse (amplitude, width, position) to carry an analog signal

pulse-position modulation sampling short pulses of the intelligence signal; the resulting position of the pulses is directly proportional to the intelligence signal's amplitude

pulse repetition frequency (PRF) the number of radar pulses (waveforms) transmitted per second

pulse repetition rate (PRR) the pulse repetition rate

pulse repetition time (PRT) the time from the beginning of one pulse to the beginning of the next

pulse-time modulation modulation schemes that vary the timing (not amplitude) of pulses

pulse-width modulation sampling short pulses of the intelligence signal; the resulting pulse-width is directly proportional to the intelligence signal's amplitude

pump chain the electronic circuitry used to increase the operating frequency up to a specified level

pyramidal horn type of horn antenna with two flare angles

quadrature signals at a 90° angle

quadrature amplitude modulation method of achieving high data rates in limited bandwidth channels, characterized by data signals that are 90° out of phase with each other

quadrature detector a popular integrated circuit FM detector that employs two signals that are 90 degrees out of phase with one another to recover the original intelligence in an FM receiver

quadrature phase-shift keying (QPSK) a form of phase-shift keying that uses four vectors to represent binary data, resulting in reduced bandwidth requirements for the data transmission channel

quality ratio of energy stored to energy lost in an inductor or capacitor

quality of service expected quality of the service

quantile a quantization level step-size

quantile interval another name for quantile

quantization process of segmenting a sampled signal in a PCM system into different voltage levels, each level corresponding to a different binary number

quantization levels another name for quantile

quantizing error an error resulting from the quantization process

quantizing noise another name for quantizing error

quantum bundle of energy

quarter-wavelength matching transformer quarter-wavelength piece of transmission line of specified line impedance used to force a perfect match between a transmission line and its load resistance

quieting the tendency for an FM receiver's audio output signal to turn off as the detector responds to a low input carrier level or no carrier input

quieting voltage the minimum FM receiver input signal that begins the limiting process

quiet zone between the point where the ground wave is completely dissipated and the point where the first sky wave is received

radar using radio waves to detect and locate objects by determining the distance and direction from the radar equipment to the object

radar mile unit of measurement equal to 2000 yd (6000 ft)

radiation the propagation of energy through space or a material

radiation field radiation that surrounds an antenna but does not collapse its field back into the antenna

radiation pattern diagram indicating the intensity of radiation from a transmitting antenna or the response of a receiving antenna as a function of direction

radiation resistance the portion of an antenna's input impedance that results in power radiated into space

radio-frequency interference (RFI) undesired radiation from a radio transmitter

radio horizon a distance about 4/3 greater than line-of-sight; approximate limit for direct space wave propagation

radio telemetry gathering data on some phenomenon without the presence of human monitors and transmitting the data to another site via radio

radome a low-loss dielectric material used as a cover over a microwave antenna

ranging a technique used by a cable modem to determine the time it takes for data to travel to the cable-head end

raster illuminated area on the CRT of a TV receiver when no signal is being received

ratio detector an outdated FM discriminator similar to the Foster-Seely design

Rayleigh fading rapid variation in signal strength received by mobile units in urban environments

Rayleigh scattering the scattering of light waves in a fiber that decreases rapidly with increasing wavelength

reactance modulator amplifier designed so that its input impedance has a reactance that varies as a function of the amplitude of the applied input voltage

received signal level the input signal level

receiver time in radar, rest time

reciprocity an antenna's ability to transfer energy from the atmosphere to its receiver with the same efficiency with which it transfers energy from the transmitter into the atmosphere

reflection abrupt reversal in direction of voltage and current

reflection coefficient the ratio of the reflected voltage to the incident voltage at a termination point in a transmission line

reflector the parasitic element that effectively reflects energy from the driven element

refraction when electromagnetic waves pass from one density to another, the direction of propagation is altered; i.e., the wave bends

refractive index ratio of the speed of light in free space to its speed in a given material

regeneration restoring a noise-corrupted signal to its original condition

rejection notch a narrow range of frequencies attenuated by a filter

relative harmonic distortion expression specifying the fundamental frequency component of a signal with respect to its largest harmonic; in dB

repeater radio installation consisting of a receiver to pick up a signal from one site and a transmitter to amplify and send the same message to another site

replacement energy in a Gunn oscillator, energy supplied by the negative resistance to allow amplification

resistor noise another name for thermal noise, due to the fact that it is produced in resistors, especially carbon resistors

resolution ability to resolve detailed picture elements in a visual display

resonance balanced condition between the inductive and capacitive reactance of a circuit

resonant line a transmission line terminated with an impedance that is not equal to its characteristic impedance

rest time the time between pulses

retrace interval the time it takes an electron beam to move from the end of one line to the start of the next line

return loss a measure of the ratio of power transmitted into a cable to the amount of power returned or reflected

RF shield box isolates the mobile unit under test from any possible interference from nearby towers

ring the nongrounded wire in two-wire phone service

ring modulator another name for balanced ring modulator

ripple amplitude variation in attenuation of a sharp bandpass filter within its 6-dB bandwidths

RJ-45 the four-pair termination commonly used for terminating CAT6/5e cable

roll-off the rate of attenuation in a filter

RS-422, RS-485 balanced-mode serial communications standards that support multidrop applications

RS-232 a standard of voltage levels, timing, and connector pin assignments for serial data transmission

RZ code (return to zero) an encoding format for PCM systems

scattering caused by refractive index fluctuations and accounts for 85 percent of the attenuation loss

Schottky diode specially fabricated majority carrier device formed from a metal-semiconductor interface, with extremely low junction capacitance

SC, ST currently the most popular full-size fiber connectors on the market

SDTV standard definition television

second return echoes echoes that arrive after the transmission of the next pulse

sectoral horn type of horn antenna with top and bottom walls at a 0° flare angle

segment sync a repetitive 1-byte pulse that is added to the front of a data segment in DTV

selectivity the extent to which a receiver can differentiate between the desired signal and other signals

self-excited mixer single stage in a superheterodyne receiver that creates the LO signal and mixes it with the applied RF signal to form the IF signal

sensitivity the minimum input RF signal to a receiver required to produce a specified audio signal at the output

sequence control keeps message blocks from being lost or duplicated and ensures that they are received in the proper sequence

shadow mask screen used in a color CRT to prevent an electron beam from striking the wrong color phosphor triad

shadow zone an area following an obstacle that does not receive a wave by diffraction

shape factor ratio of the 60-dB and 6-dB bandwidths of a high-Q bandpass filter

shorted-stub matching section a shorted transmission line of specified length that can be used to force a perfect match between a transmission line and its load impedance

shot noise noise introduced by carriers in the *pn* junctions of semiconductors

sideband splatter distortion resulting in an overmodulated AM transmission creating excessive bandwidths

signal injection troubleshooting by injecting an input signal and tracing through the circuit to locate the failed component

signal spectrum method of representing a signal by plotting its amplitude versus frequency characteristics

signal-to-noise ratio relative measure of desired signal power to noise power

signature sequence the pseudorandom digital sequence used to spread the signal

single-mode fibers fiber cables with core diameters of about 5 μm; light follows a single path through the core

single sideband (SSB) a form of amplitude modulation whereby the carrier and one sideband are filtered out, leaving the other sideband as the only remaining RF signal

single-stub tuner the stub's distance from the load and the location of its short circuit are adjustable to allow a match between line and load

skin effect the tendency for high-frequency electric current to flow mostly near the surface of the conductive material

skipping the alternate refracting and reflecting of a sky wave signal between the ionosphere and the earth's surface

skip zone another name for quiet zone

sky wave those radio waves radiated from the transmitting antenna in a direction toward the ionosphere

slope detector a simple FM discriminator that detects FM by first converting FM to AM and then detecting the intelligence by a simple diode detector; usually creates too much distortion to be an acceptable design

slope modulation another name for delta modulation

slope overload in delta modulators, when the analog signal has a high rate of amplitude change, the encoder can produce a distorted analog signal

slot antenna array UHF or microwave antenna, often used on aircraft, that couples RF energy into a slot in a large metallic plane

slotted line section of coaxial line with a lengthwise slot cut in the outer conductor to allow measurement of the standing wave pattern

small-form factor a family of connectors about half the size of ST and SC connectors

S meter signal strength meter that responds to the received signal level in a receiver

Smith chart impedance chart developed by P. H. Smith, useful for transmission line analysis

solar noise space noise originating from the sun

SONET protocol standard for optical transmission in long-haul communication

space noise external noise produced outside the earth's atmosphere

space wave a radio wave that travels in straight lines between transmitter and receiver, not necessarily close to the ground; it is typically a line-of-sight transmission. If any obstruction exists, the signal is blocked from reaching the receiver.

spectrum analyzer instrument used to measure the harmonic content of a signal by displaying a plot of amplitude versus frequency

spread the RF signal is spread randomly over a range of frequencies in a noiselike manner

spread spectrum communication systems in which the carrier is periodically shifted about at different nearby frequencies in a random-like manner determined by a hidden code; the receiver must decode the sequence so that it can follow the transmitter's frequency hops to the various values within the specified bandwidth

spurious frequencies extra frequency components that appear in the spectral display of a signal, signifying distortion

spurs undesired frequency components of a signal

square-law device a device that exhibits an output versus input signal response resembling a parabola; often used as mixers and detectors in receivers due to its minimum distortion characteristics

squelch a circuit in a receiver that cuts off the background noise in the absence of a desired signal; often found in FM receivers

stagger tuning cascading a number of tuned bandpass filters, each having a slightly offset bandpass frequency, to form a wider flat bandpass with steep high- and low-frequency roll-off skirts

standing wave waveforms that apparently seem to remain in one position, varying only in amplitude

standing wave ratio another name for voltage standing wave ratio

start bit, stop bit used to precede and follow each transmitted data word

static electrical noise that may occur in the output of a receiver

static convergence proper beam convergence at the center of a CRT

statistical concentration processors at switching centers directing packets so that a network is used most efficiently

step-index referring to the abrupt change in refractive index from core to clad in a fiber

step-index fibers an abrupt change in the refractive index from core to clad in fiber

stereo a radio transmission of two separate signals, left and right, used to create a three-dimensional effect for the receiver's audience

stray capacitance undesired capacitance between two points in a circuit or device

stripline transmission line used at microwave frequencies that has two ground planes sandwiching a conducting strip

STS synchronous transport signals used for transporting data in fiber-optic transmission; has equivalence to OC-number specifications

subsidiary communication authorization an additional channel of multiplexed information authorized by the FCC for stereo FM radio stations to feed services to selected customers

superheterodyne receiver receiver design superior to the simple TRF design due to its mixer and local oscillator stages and its ganged tuning characteristics; provides for easier tuning and near-constant selectivity at all frequencies within its tuning range

surface acoustic wave filters extremely high-Q filters often used in TV and radar applications

surface wave another name for ground wave

surge impedance another name for characteristic impedance

SVC switched virtual circuit

symbol substitution displaying an unused symbol for the character with a parity error

synchronizing in TV, precisely matching the movement of the electron beam horizontally and vertically in the recording camera with the electron beam in the receiver

synchronous a system in which the transmitter and receiver clocks run at exactly the same frequency because the receiver derives its clock signal from the received data stream

synchronous detector a complex method of detecting an AM signal that gives low distortion, fast response, and amplification

synchronous orbit when a satellite's position remains fixed with respect to the earth's rotation

synchronous system the transmitter and receiver clocks run at exactly the same frequency

sync separator circuit in a TV receiver that separates the horizontal and vertical sync pulses from the video signal

syndrome the value left in the CRC dividing circuit after all data have been shifted in

systematic codes the message and block-check character transmitted as separate parts within the same transmitted code

tangential method method of measuring the amplitude of noise on a signal using an oscilloscope display

tank circuit parallel *LC* circuit

TCXO (temperature compensated crystal oscillator) a crystal oscillator that contains circuitry to keep the output frequency constant with respect to changes in temperature

TDD time division duplex

TDMA (time-division multiple-access) a technique used to transport data from multiple users over the same data channel

Telco the local telephone company

telemetry remote metering; gathering data on some phenomenon without the presence of human monitors

thermal noise internal noise caused by thermal interaction between free electrons and vibrating ions in a conductor

third-order intercept point receiver figure of merit describing how well it rejects intermodulation distortion from third-order products resulting at the mixer output

3G the third generation in wireless connectivity

threshold in FM, the point where *S/N* in the output rapidly degrades as *S/N* of received signal is degrading

threshold sensitivity minimum VCO input to allow the PLL to be in the locked mode

threshold voltage another term for quieting voltage

time-division multiple-access a single transmitter can service multiple stations simultaneously on the same frequency based on available bursts of time

time-division multiplexing two or more intelligence signals are sequentially sampled to modulate the carrier in a continuous, repeating fashion

time domain reflectometry technique of sending short pulses of electrical energy down a transmission line to determine its characteristics by observing on an oscilloscope for resulting reflections

time slot a fixed location (relative in time to the start of a data frame) provided for each group of data

tip the grounded wire in two-wire phone service

token passing a channel access method suited to ring network topology where a user waits for "token" possession to make a transmission

topology architecture of a network

total harmonic distortion a measure of distortion that takes all significant harmonics into account

total internal reflection a light wave traveling down a glass fiber by constant reflection off its side walls

tracking ADC an ADC whose output indicates input changes rather than absolute values of input

tracking filter able to provide a fixed bandwidth output even as the center frequency varies

tracking range once a PLL is locked up, this is the range of input frequencies that can be applied without having it lose its lock and return to the free-running state

transceiver transmitter and receiver sharing a single package and some circuits

transducer device that converts energy from one form to another

transit-time noise noise produced in semiconductors when the transit time of the carriers crossing a junction is close to the signal's period and some of the carriers diffuse back to the source or emitter of the semiconductor

transmission line the conductive connections between system elements that carry signal power

transparency control character recognition by using the character insertion process

transphasor an optical switch using a laser beam

transponder electronic system that performs reception, frequency translation, and retransmission of received radio signals

transverse when the oscillations of a wave are perpendicular to the direction of propagation

trap another name for wavetrap

trapezoidal pattern a measurement technique for checking the purity of an AM modulator by use of the oscilloscope in the X:Y mode

traveling waves voltage and current waves moving through a transmission line

traveling wave tube (TWT) a high-gain, low-noise, wide bandwidth microwave amplifier

triads individual groups of red, green, and blue phosphor dots on the CRT face

trimmer small variable capacitance in parallel with each section of a ganged capacitor

tropospheric scatter a phenomenon whereby small amounts of radiation are scattered by the troposphere and picked up by receivers at ground level when the transmitted signal is set at a very high power and selected microwave frequencies are used

trunk the circuit connecting one central office to another

T3 a digital data rate of 44.736 Mbps

tunable laser when a laser's fundamental wavelength can be shifted a few nanometers; ideal for traffic routing in DWDM systems

tuned radio frequency receiver the most elementary receiver design, consisting of RF amplifier stages, a detector, and audio amplifier stages

tuner front end of a TV receiver that selects the desired station

twin lead standard 300-Ω parallel wire transmission line

twin-sideband suppressed carrier the transmission of two independent sidebands, containing different intelligence, with the carrier suppressed to a desired level

Type A connector the UBS upstream connection that connects to the computer

Type B connector the UBS downstream connection to the peripheral

UART device that converts parallel computer data into serial data

unbalanced line the electrical signal in a coaxial line is carried by the center conductor with respect to the grounded outer conductor

uniform quantization level each quantile interval is the same step-size

U-NII unlicensed national information infrastructure

universal asynchronous receiver/transmitter device that converts parallel computer data into serial data

universal serial bus (UBS) a hot-swappable, high-speed serial communications interface

up-conversion mixing the received RF signal with an LO signal to produce an IF signal higher in frequency than the original RF signal

uplink sending signals to a satellite

upper sideband band of frequencies produced in a modulator from the creation of sum-frequencies between the carrier and information signals

vane a thin card of resistive material used as a variable attenuator in a waveguide

varactor diode diode with a small internal capacitance that varies as a function of its reverse bias voltage

variable bandwidth tuning technique to obtain variable selectivity to accommodate reception of variable bandwidth signals

varicap diodes another name for varactor diodes

VCI virtual channel identifier

velocity constant ratio of actual velocity to velocity in free space

velocity factor another name for velocity constant

velocity modulation an electron beam moving along in bursts of electrons as in a klystron

velocity of propagation the speed at which an electrical or optical signal travels

vertical antenna an antenna that consists of a vertical tower, wire, or rod, usually a quarter-wavelength in length, that is fed at the ground and uses the ground as a reflecting surface

vertical cavity surface emitting lasers (VCSELs) a light source exhibiting the simplicity of LEDs combined with the performance of lasers

vertical resolution number of horizontal lines that actually make up a TV display

vertical retrace interval in TV, the amount of time it takes to move the electron beam from the bottom right corner to the top left corner to start another field

vestigial-sideband operation a form of amplitude modulation in which one of the sidebands is partially attenuated

V.44 (V.34) the standard for an all analog modem connection with a maximum data rate of up to 34 Kbps

video amplifiers amplifiers with bandpass characteristics from dc up into the MHz region

video signal picture portion of a TV signal, which is amplitude-modulated onto a carrier

virtual channel connection (VCC) carries the ATM cell from user to user

virtual path connection (VPC) used to connect the end users

V.92 (V.90) the standard for a combination analog and digital modem connection with a maximum data rate up to 56 Kbps

voltage-controlled oscillator designed so that its output voltage varies as a function of the amplitude of the applied input voltage

voltage standing wave ratio ratio of the maximum voltage to minimum on a line

VPI virtual path identifier

VVC diodes another name for varactor or varicap diodes

water mark sticker a sticker used to detect water damage

wavefront a plane joining all points of equal phase in a wave

waveguide a microwave transmission line consisting of a hollow metal tube or pipe that conducts electromagnetic waves through its interior

waveguide dispersion the dispersion of light in fiber-optic cable caused by a portion of the light energy traveling in the cladding

wavelength the distance traveled by a wave during a period of one cycle

wave propagation movement of radio signals through the atmosphere from transmitter to receiver

wavetrap high-Q bandstop circuit that attenuates a narrow band of frequencies

W-CDMA wideband code division multiple access

white noise another name for thermal noise because its frequency content is uniform across the spectrum

wide area network two or more LANs linked together over a wide geographical area

wideband FM FM transmitter/receiver systems that are set up for high-fidelity information, such as music, high-speed data, stereo, etc.

wired digital communications digital communications over a wired connection

wireless the term used today to describe telecommunications technology that uses radio waves, rather than cables, to carry the signal

wireless digital communications the transport of digital data over a wireless medium

wireless markup language (WML) the hypertext language for the wireless environment

WMLScript the WML comparable version of Javascript

xDSL a generic representation of the various DSL technologies available

X.25 a packet-switched protocol designed for data transmission over analog lines

Yagi-Uda antenna a popular type of directional antenna consisting of a half-wave dipole as the driven element, one reflector, and several directors

yoke coil around the CRT tube that deflects the electron beam with its magnetic field

0 dBm 1 mW measured relative to a 1-mW reference

zero-dispersion wavelength the wavelength where material and waveguide dispersion cancel each other

zoning a fabrication process that allows a dielectric to change a spherical wavefront into a plane wave

Index

B

C

D

E

F

G

H

M

O

P

Q

R

S

T

U

X

Y

Z